HANDBOOK OF
TOXIC AND HAZARDOUS CHEMICALS
AND CARCINOGENS

ABOUT THE AUTHOR

Marshall Sittig, a chemical engineer, is President and Managing Director of Sittig & Noyes, International Chemical and Process Industries Consultants, and was formerly with E.I. Du Pont de Nemours & Co., Inc. in chemicals manufacturing, Ethyl Corporation in liaison between research and sales, and Princeton University as Director of Governmental Relations.

HANDBOOK OF
TOXIC AND HAZARDOUS
CHEMICALS AND
CARCINOGENS

Second Edition

by

Marshall Sittig

Princeton University

 NOYES PUBLICATIONS
Park Ridge, New Jersey, U.S.A.

Library of Congress Catalog Card Number: 84-22755
ISBN: 0-8155-1009-8
Printed in the United States

Published in the United States of America by
Noyes Publications
Mill Road, Park Ridge, New Jersey 07656

10 9 8 7 6 5 4 3 2 1

Library of Congress Cataloging in Publication Data

Sittig, Marshall.
    Handbook of toxic and hazardous chemicals and
carcinogens.

    Bibliography: p.
    Includes index.
    1. Poisons--Dictionaries. 2. Hazardous substances--
Dictionaries. 3. Carcinogens--Dictionaries. I. Title.
RA1193.S58 1985   615.9'02   84-22755
ISBN 0-8155-1009-8

Foreword

In April 1980, an explosive fire broke out at an abandoned chemical dump site in Elizabeth, New Jersey. When thousands of barrels containing unidentified toxic waste erupted in flames, an ominous black cloud filled the sky. No one knew for sure whether lethal fumes would jeopardize the surrounding area—one of America's most densely populated. Fortunately, nobody was hurt seriously, perhaps in part because more than 500 pounds of such dangerous agents as nitroglycerine, picric acid and mustard gas were removed prior to the fire by the state environmental protection personnel.

That near-disaster gave the nation fresh evidence of the urgent need to control the disposal of toxic wastes. In New Jersey alone, there are 85 toxic waste sites on the Environmental Protection Agency's list of 546 worst sites identified so far, more than any other state. Of the 15 worst sites in the nation, 6 are in New Jersey.

Toxic wastes are an incredible problem in our country. They threaten our water, our air, and our lives. We must find better ways to handle toxic chemicals and dispose of our toxic wastes. Unfortunately, we have just begun to recognize the dangers posed by years of irresponsible disposal of wastes. And, too frequently our belated response has been to push pollution around from one place to another. When we recognized the danger of air pollution, we devised systems to prevent the discharge of particulates. The systems gathered soot and sludge, so we took that soot and sludge and dumped them in the ocean. When we became concerned about pollution in the ocean, we ordered the wastes buried in the ground. Now we have learned that what is buried in the ground can seep into and poison our drinking water. Our problems have only moved with each new form of disposal.

One reason for the severity of this problem is that it has only recently risen to public consciousness, and years of uncontrolled dumping have taken their toll. But incidents such as the fire at Elizabeth and the contamination of the Love Canal area in New York brought the issue national attention and forced public officials, scientists, and representatives of industry to consider its implications.

The most obvious are the potential effects on national health and the environment. Although the link is poorly understood, there is evidence that many of these chemicals can cause cancer. Just as frightening is the possibility of acute toxic effects such as brain and nerve damage, sterility, and birth defects.

In response to these concerns, Congress has initiated numerous legislative and regulatory measures. I was a strong supporter of the Superfund legislation creating a fund to clean up existing dangerous waste disposal sites and deal with emergency situations in the future. This was a major step forward in providing the funding needed to successfully attack the problem.

The Superfund legislation is finally beginning to produce results. Hazardous waste sites are beginning to be cleaned up. In October, 1983 after much too long a wait, EPA awarded $5 million to New Jersey to pay for most of the cost of relocating Atlantic City's drinking water wells that are threatened by hazardous waste deposited at Price's Pit.

These remedies have often come with agonizing slowness. The intent of legislation such as Superfund is not to unnecessarily restrict the use of chemicals. Modern society has come to depend on them for scientific and technological advancement as well as health, recreation, and many other benefits. But the increase in their production has led to large-scale use of some chemicals known to have adverse health effects and others whose effects especially in the long run, are uncertain. The recent development of a vast number of new chemicals has also added greatly to the difficulties of evaluating which chemicals to regulate and how to establish safeguards for their manufacture, use and disposal. Humans can be exposed to chemicals in many ways with results that can be immediate but years of research are often required to determine whether a particular substance is hazardous.

In addition to environmental concerns, dangerous levels of exposure which can threaten the health of workers must be prevented. Many workers die each year as a result of physical and chemical hazards at work, and the long-term effects of certain occupational conditions are unknown.

Protecting against potential public health hazards requires widespread knowledge about commercial chemicals—their mixtures, by-products and uses. We must know more about their persistence and fate in the environment, what effects they will have, and, most importantly, how we can minimize the risks they pose.

For these reasons, Marshall Sittig's book is an important addition to the literature on toxic chemicals. It provides access to information contained in hundreds of government publications, with particular attention to the identification of carcinogenic materials. References at the end of many entries provide useful bibliographies listing thousands of original publications describing the effects of toxic and hazardous chemicals on man and environment. This comprehensive information makes it a valuable desk reference.

In short, Marshall Sittig has made it easier for us to understand how to control the handling of toxic chemicals and the disposal of toxic wastes. My hope is that this understanding will produce more effective action.

January 1985

Bill Bradley
United States Senator
New Jersey

Preface

This handbook presents concise chemical, health and safety information on nearly 800 toxic and hazardous chemicals (up from nearly 600 in the first edition so that responsible decisions can be made by chemical manufacturers, safety equipment producers, toxicologists, industrial safety engineers, waste disposal operators, health care professionals, and the many others who may have contact with or interest in these chemicals due to their own or third party exposure.

Included in the book are:
   all of the **EPA Priority Toxic Pollutants,**
   all of the substances whose allowable concentrations in workplace air are adopted or proposed by **ACGIH** (1983/1984),
   all of the substances considered to date in the Standards Completion Program of **NIOSH,** and
   most of the chemicals classified as **EPA "hazardous wastes,"**
   most of the chemicals classified as **EPA "hazardous substances,"**
   most of the chemicals reviewed in **EPA "CHIPS"** documents, and
   most of the chemicals reviewed in **NIOSH Information Profiles.**

In addition, this Second Edition includes:
   all of the carcinogens identified by the **U.S. National Toxicology Program,**
   many of the chemicals profiled by the **Dutch Association of Safety Experts,** the **Dutch Chemical Industry Association,** and the **Dutch Safety Institute,**
   most of the chemicals described in the **ILO** *Encyclopedia of Occupational Health and Safety* (1983),
   most of the chemicals in the **United Nations'** *IRPTC Legal File* (1984), and
   most of the chemicals described in the journal publication: *Dangerous Properties of Industrial Materials Report.*

The necessity for informed handling and controlled disposal of hazardous and toxic materials has been spotlighted over and over in recent days as news of fires and explosions at factories and waste sites and groundwater contamination near dump sites has been widely publicized. In late 1980 the EPA imposed long-delayed regulations governing the handling of hazardous wastes—from creation to disposal. Prerequisite to control of hazardous substances, however, is knowledge of the extent of possible danger and toxic effects posed by any particular chemical. This book provides the prerequisites.

The 1984 tragedy at Bhopal, India involving methyl isocyanate (see entry on page 609) may will stimulate the enactment of "right-to-know" legislation requiring that workers be furnished data akin to the entries in this volume on chemicals they encounter in the workplace.

The chemicals are presented alphabetically and each is classified as a "carcino-gen," "hazardous substance," "hazardous waste," and/or a "priority toxic pollu-tant"—as defined by the various federal agencies, and explained in the compre-hensive Introduction to the book.

Particular attention is given in the second edition to delineation of the iden-tity and properties of those chemicals now known to be carcinogens.

Data is furnished, to the extent currently available, on any or all of these im-portant categories:

Chemical Description   Routes of Entry
Code Numbers   Harmful Effects and Symptoms
DOT Designation   Points of Attack
Synonyms   Medical Surveillance
Potential Exposure   First Aid
Incompatibilities   Personal Protective Methods
Permissible Exposure Limits in Air   Respirator Selection
Determination in Air   Disposal Method Suggested
Permissible Concentration in Water   References
Determination in Water

Essentially the book attempts to answer seven questions about each com-pound (to the extent information is available):

(1) What is it?
(2) Where do you encounter it?
(3) How much can one tolerate?
(4) How does one measure it?
(5) What are its harmful effects?
(6) How does one protect against it?
(7) Where can I learn more?

Under category (7), "where can I learn more?" this second edition provides hundreds of new citations to secondary reference sources which in turn provide access to thousands of references on properties and toxicology and safe handling of all the compounds listed.

An outstanding and noteworthy feature of this book is the Index of Carcino-gens. There were 92 listed in the first edition, 178 in the second edition.

This book will thus be a valuable addition to industrial and medical libraries.

Advanced composition and production methods developed by Noyes Publica-tions are employed to bring these durably bound books to you in a minimum of time. Special techniques are used to close the gap between "manuscript" and "completed book." Industrial technology is progressing so rapidly that time-hon-ored, conventional typesetting, binding and shipping methods are no longer suit-able. We have bypassed the delays in the conventional book publishing cycle and provide the user with an effective and convenient means of reviewing up-to-date information in depth.

The alphabetical table of contents serves as a subject index and provides easy access to the information contained in the book.

Contents

• = EPA priority pollutant

Contents

Contents

Contents

Contents

xxvi   Contents

Introduction

The toxic chemicals problem in the United States and indeed in all the world is a frightening problem with news stories about Love Canal, the Valley of the Drums, the Valley of Death in Brazil and the like. All these generate emotional responses, often from people uninformed about science or technology. On the other hand, one encounters some industrialists who tell us that toxic chemicals are present in nature and that industrial contributions are just the price we have to pay for progress. Somewhere in between lies the truth—or at least an area in which we can function. It is the aim of this book to present data on specific industrial chemicals from an unemotional point of view so that decisions can be made by:

> Chemicals manufacturers
> Protective safety equipment producers
> Toxicologists
> Industrial hygienists
> Lawyers
> Doctors
> Industrial safety engineers
> Analytical chemists
> Industrial waste disposal operators
> Legislators
> Enforcement officials
> First aid squad members
> Fire department personnel
> Schoolteachers
> The informed public

This book gives the highlights of available data on nearly 800 important toxic and hazardous chemicals. Importance is defined by inclusion in official and semi-official listings as follows:

- All the substances whose allowable concentrations in workplace air are adopted or proposed by the American Conference of Government Industrial Hygienists as of 1983/84 (over 500 sub-

1

stances). In the interests of conserving space, however, detailed entries are not given for most "nuisance particulates" nor for most "simple asphyxiants." The following are the categories as defined by ACGIH (A-6).

Nuisance Particulates

α-Alumina (Al$_2$O$_3$)
Calcium carbonate
Calcium silicate
Cellulose (paper fiber)
Emery
*Glycerin mist
*Graphite (synthetic)
Gypsum
Kaolin
Limestone
Magnesite
Marble
Mineral wool fiber
*Pentaerythritol

Plaster of Paris
*Portland cement
Rouge
*Silicon
*Silicon carbide
Starch
Sucrose
*Titanium dioxide
Vegetable oil mists
(except castor, cashew nut, or similar irritant oils)
Zinc stearate
Zinc oxide dust

Simple Asphyxiants

*Acetylene
*Argon
Ethane
Ethylene
Helium

Hydrogen
Methane
Neon
*Propane
*Propylene

Some of these compounds (marked by asterisks) have been included because of the availability of other information on these materials. For the general picture on "nuisance particulates," see The entry under "Particulates."

- All the substances considered to date in the Standards Completion Program by the National Institute of Occupational Safety and Health (some 380 substances). The second edition includes updated data sheets to 1981 (A-61).

- All the priority toxic water pollutants defined by the U.S. Environmental Protection Agency as a result of consent decrees in 1976 and 1979 which resulted in draft criteria in 1979 and final criteria in 1980 (for 65 pollutants and classes of pollutants which yielded 129 specific substances).

- Most of the chemicals in the following categories: (1) EPA "Hazardous wastes" as defined under the Resource Conservation and Recovery Act in April 1980 (A-52); (2) EPA "hazardous substances" as defined under the Clean Water Act; (3) chemicals which EPA has made the subject of Chemical Hazards Information Profiles or "CHIPS" review documents; and (4) chemicals which NIOSH has made the subject of "Information Profile" review documents.

- All of the carcinogens identified by the National Toxicology Program of the U.S. Department of Health and Human Services at Research Triangle Park, N.C. (A-62)(A-64).

- Many of the chemicals profiled by the combination of the Dutch Association of Safety Experts, the Dutch Chemical Industry Association and the Dutch Safety Institute (A-60). These provide a source similar to the U.S. Coast Guard CHRIS system of material data sheets.

- Most of the chemicals described in the current and timely periodical entitled "Dangerous Properties of Industrial Materials Report" edited by N. Irving Sax.

- Most of the chemicals described in the 2-volume "Encyclopedia of Occupational Health and Safety" published by the International Labor office in 1983 (A-69).

- Most of the chemicals covered in the Legal File published by International Register of Potentially Toxic Chemicals Program (IRPTC) of the United Nations in 1984 (A-70). The reader who is particularly concerned with legal standards (allowable concentration in air, in water or in foods) is advised to check these most recent references because data may exist in this UN publication which has not been quoted in toto in this volume because of time and space limitations.

All this information from U.S. Government sources has been supplemented by a careful search of, and citations to, publications from the United Kingdom, European and Japanese sources as well as United Nations and World Health Organization publications.

The result, we believe, is a handbook that is more than a handbook. When one looks at most handbooks, one simply expects to get numerical data. Here, we have tried wherever possible to reference the source of the numerical values and to provide a literature reference to some sort of timely review document which opens the door to a much wider field of published materials.

As pointed out in May 1980 by the Toxic Substances Strategy Committee, an interagency group, in a report to President Carter (A-47):

Exposures to toxic substances are linked to a variety of health problems. The immediate effects of high level exposures for a short time include burns, rashes, nausea, loss of eyesight, and fatal poisoning. Prolonged exposure to low doses can cause chronic lung disease (e.g., from coal or cotton dust), heart disease (from exposure to cadmium or carbon monoxide), sterility (from dibromochloropropane—DBCP), and kidney, liver, brain, nerve, and other damage. Exposure to industrial solvents can cause depression, and carbon disulfide workers are associated with a higher suicide rate than the general population. Although most chemicals do not cause cancers, exposure to some has been linked to cancer. Some workers exposed to asbestos, even for a short time, have developed a rare cancer of the chest and stomach linings 30 to 40 years after initial exposure. Vinyl chloride gas is linked to a rare liver cancer, to a brain cancer, and possibly to lung cancer. Diethylstilbestrol (DES), when taken by pregnant women to prevent miscarriages, led to increased risk of vaginal cancer in their daughters and abnormal sexual organs in their sons. Methylmercury, formed by the action of bacteria in sediments on mercury metal and on mercuric ions, can cause acute poisoning, deafness, brain damage, and a range of birth defects.

A single substance can have several kinds of adverse effects, depending on the route and level of exposure. Some effects of exposure to chemicals which are produced in substantial quantities are shown in Table 1. Chemicals listed in the National Occupational Hazard Survey were preselected for all chemicals to which more than 100,000 workers are potentially exposed. The Toxicology Data Bank of the National Library of Medicine was then searched for evidence (human and animal) of chronic or acute effects for all targets, and the 19 substances were chosen to illustrate the multiple target effects. Other data bases could suggest different target specificity.

Table 1: Selected High Volume Substances and Effects of Exposure

Substance	Kidney	Liver	Central Nervous System	Reproductive System	Pulmonary System	Skin
Acetone	x	x			x	x
Acrylonitrile					x	x
Ammonia	x	x			x	x
Asbestos					x	
Cresol	x	x	x		x	x
Dichloromethane	x	x	x		x	x
Diethylene glycol	x	x	x	x	x	x
2-Ethoxyethanol	x	x	x	x	x	x
Ethylene glycol	x	x			x	
Lead	x	x	x	x		
Methyl ethyl ketone	x	x	x		x	x
2-Methyl-2,4-pentanediol	x	x			x	x
Oxalic acid	x		x	x	x	x
Phenol	x	x	x		x	x
Sodium hydroxide					x	x
Sulfuric acid	x	x			x	x
Talc					x	x
1,1,1-Trichloroethane	x	x	x		x	x
Trichloroethylene	x	x	x		x	

Source: National Institute for Occupational Safety and Health, 1979

Each chemical covered in this volume has been discussed using the same outline or checklist of topics, and these have been filled in to the extent information was available.

Under the title of each substance, there are first four "bulleted" designations indicating whether the substance is:

- A carcinogen (the agency making such a determination, the nature of the carcinogenicity—whether human or animal and whether positive or suspected, and a literature reference are given in each case);

- A "hazardous substance" as defined by the U.S. Environmental Protection Agency primarily on the basis of toxicity to aquatic life (A-51).

- A "hazardous waste" or "hazardous waste constituent" as defined by the U.S. Environmental Protection Agency (A-52).

• A "priority toxic pollutant" as defined by the U.S. Environmental Protection Agency (A-53).

Then each substance has data furnished to the extent available, under each of the following categories:

> Description
> Code Numbers
> DOT Designation
> Synonyms
> Potential Exposure
> Incompatibilities
> Permissible Exposure Limits in Air
> Determination in Air
> Permissible Concentration in Water
> Determination in Water
> Routes of Entry
> Harmful Effects and Symptoms
> Points of Attack
> Medical Surveillance
> First Aid
> Personal Protective Methods
> Respirator Selection
> Disposal Method Suggested
> References

In the pages which follow, these categories will be discussed with reference to scope, sources, nomenclature employed, and the like. Omission of a category indicates a lack of available information.

Description: The chemical formula, the color, the odor and the melting or boiling point are given. Structural formulas are given in the cases of complex molecules.

Code Numbers: Three different code numbers are given for each material if they have been assigned:

(1) The Chemical Abstract Service Registry number. It is simply given as CAS XXX-XX-X. It can be used to provide access to the MEDLARS® computerized literature retrieval services of the National Library of Medicine in Washington, DC.

(2) The NIOSH Registry of Toxic Effects of Chemical Substances (RTECS) number is simply given as RTECS ABXXXXXXX (A-49). It is a 9-digit number and can also be used to provide access to updated detailed printouts from the MEDLARS® services cited above.

(3) The United Nations numbers (A-46) for individual chemical commodities. These numbers are now being utilized by the U.S. Department of Transportation (A-50) to assist in the designation of hazardous materials. In some cases an NA number which is a North American identification number (A-50) is cited in parentheses when a UN number has not been assigned.

Use of these identification numbers for hazardous materials will (a) serve to verify descriptions of chemicals; (b) provide for rapid identification of materials when it might be inappropriate or confusing to require the display of lengthy chemical names on vehicles; (c) aid in speeding communication of information on materials from accident scenes and in the receipt of more accurate emergency response information; and (d) provide a means for quick access to immediate emergency response information in the Emergency Response Guidebook (A-56). In this latter volume, the various compounds are simply given "ID Numbers" or identification numbers which correspond closely (but not precisely) to the UN listing (A-46) and to the earlier DOT listing (A-50).

DOT Designation: The U.S. Department of Transportation (A-50) has published listings of chemical substances which give a hazard classification and required labels. The DOT designations are defined as follows:

Hazardous Material — A substance or material which has been determined by the U.S. Secretary of Transportation to be capable of posing an unreasonable risk to health, safety, and property when transported in commerce, and which has been so designated.

Explosive — Any chemical compound, mixture, or device, the primary or common purpose of which is to function by explosion, i.e., with substantially instantaneous release of gas and heat, unless such compound, mixture, or device is otherwise specifically classified.

Class A Explosive — Detonating or otherwise of maximum hazard. The nine types of Class A explosives are defined in Section 173.53 of CFR Title 49-Transportation.

Class B Explosive — In general, function by rapid combustion rather than detonation and include some explosive devices such as special fireworks, flash powders, etc. Flammable hazard.

Class C Explosive — Certain types of manufactured articles containing Class A or Class B explosives, or both, as components but in restricted quantities, and certain types of fireworks. Minimum hazard.

Blasting Agents — A material designed for blasting which has been tested and found to be so insensitive that there is very little probability of accidental initiation to explosion or of transition from deflagration to detonation.

Combustible Liquid — Any liquid having a flash point above 100°F and below 200°F.

Corrosive Material — Any liquid or solid that causes visible destruction of human skin tissue or a liquid that has a severe corrosive rate on steel.

Flammable Liquid — Any liquid having a flash point below 100°F.

Pyrophoric Liquid — Any liquid that ignites spontaneously in dry or moist air at or below 130°F.

Compressed Gas — Any material or mixture having in the container a pressure exceeding 40 psia at 70°F, or a pressure exceeding 104 psia at 130°F; or any liquid flammable material having a vapor pressure exceeding 40 psia at 100°F.

Flammable Gas — Any compressed gas meeting the requirements for lower flammability limit, flammability limit range, flame projection, or flame propagation criteria as specified in Section 173.300(b) of CFR Title 49.

Nonflammable Gas — Any compressed gas other than a flammable compressed gas.

Flammable Solid — Any solid material, other than an explosive, which is liable to cause fires through friction, retained heat from manufacturing or processing, or which can be ignited readily and when ignited burns so vigorously and persistently as to create a serious transportation hazard.

Organic Peroxide — An organic compound containing the bivalent −O−O− structure and which may be considered a derivative of hydrogen peroxide where one or more of the hydrogen atoms have been replaced by organic radicals must be classed as an organic peroxide.

Oxidizer — A substance such as chlorate, permanganate, inorganic peroxide, or a nitrate, that yields oxygen readily to stimulate the combustion of organic matter.

Poison A — Extremely Dangerous Poisons: Poisonous gases or liquids of such nature that a very small amount of the gas, or vapor of the liquid, mixed with air is dangerous to life.

Poison B — Less Dangerous Poisons: Substances, liquids, or solids (including pastes and semisolids), other than Class A or Irritating Materials, which are known to be so toxic to man as to afford a hazard to health during transportation; or which, in the absence of adequate data on human toxicity, are presumed to be toxic to man.

Irritating Material — A liquid or solid substance which upon contact with fire or when exposed to air gives off dangerous or intensely irritating fumes, but not including any poisonous material, Class A.

Etiologic Agent — A viable microorganism, or its toxin which causes or may cause human disease.

Radioactive Material — Any material, or combination of materials, that spontaneously emits ionizing radiation, and having a specific activity greater than 0.002 microcuries per gram.

Other Regulated Materials — Any material that does not meet the definition of a hazardous material, other than a combustible liquid in packagings having a capacity of 110 gallons or less, and is specified as an ORM material or that possesses one or more of the characteristics described in ORM-A through E below. Note: an ORM with a flash point of 100° to 200°F, when transported with more than 110 gallons in one container shall be classed as a combustible liquid.

> ORM-A: A material which has an anesthetic, irritating, noxious, toxic, or other similar property and which can cause extreme annoyance or discomfort to passengers and crew in the event of leakage during transportation.

> ORM-B: A material (including a solid when wet with water) capable of causing significant damage to a transport vehicle or vessel from leakage during transportation. Materials meeting one or both of the following criteria are ORM-B materials: (a) a liquid substance that has a corrosion rate exceeding 0.250 inch per year (IPY) on aluminum (nonclad 7075-T6) at a test temperature of 130°F. An acceptable test is described in NACE Standard TM-01-69.

> ORM-C: A material which has other inherent characteristics not described as an ORM-A or ORM-B but which make it unsuitable

for shipment, unless properly identified and prepared for transportation.

ORM-D: A material such as a consumer commodity which, though otherwise subject to the regulations of this section, presents a limited hazard during transporation due to its form, quantity and packaging.

ORM-E: A material that is not included in any other hazard class. Materials in this class include: (a) hazardous waste; (b) hazardous substances, which means a quantity of a material offered for transportation in one package, or transport vehicle when the material is not packaged, that equals or exceeds the reportable quantity (RQ) specified for the material in EPA regulations at 40 CFR Parts 116 and 117. "Hazardous waste" means any material that is subject to the hazardous waste manifest requirements of the EPA specified in 40 CFR Part 262 or would be subject to these requirements absent an interim authorization to a state under 40 CFR Part 123, Subpart F.

The following are offered to explain additional terms used in preparation of hazardous materials for shipment.

Consumer Commodity — (See ORM-D.) A material that is packaged or distributed in a form intended and suitable for sale through retail sales agencies or instrumentalities for consumption by individuals for purposes of personal care or household use. This term also includes drugs and medicines.

Flash Point — The minimum temperature at which a substance gives off flammable vapors which in contact with spark or flame will ignite.

Forbidden — The hazardous material is one that must not be offered or accepted for transportation.

Limited Quantity — The maximum amount of a hazardous material; as specified in those sections applicable to the particular hazard class, for which there are specific exceptions from the requirements.

Spontaneously Combustible Material (Solid) — A solid substance (including sludges and pastes) which may undergo spontaneous heating or self-ignition under conditions normally incident to transportation or which may upon contact with the atmosphere undergo an increase in temperature and ignite.

Water Reactive Material (Solid) — Any solid substance (including sludges and pastes) which, by interaction with water, is likely to become spontaneously flammable or to give off flammable or toxic gases in dangerous quantities.

Synonyms: Some of the more commonly used synonyms are given for each compound as well as some of the more common registered trade names.

Potential Exposure: A brief indication is given of the nature of exposure to each compound in the industrial environment. Where pertinent, some indications are given of background concentration and occurrence from other than industrial discharges such as water purification plants. Obviously in a volume of this size, this coverage must be very brief. Additional information on numbers of workers exposed and specific occupations which may be exposed are available from the National Institute of Occupational Safety and Health.

It is of course recognized that nonoccupational exposures may be important as well and the reader is referred to an EPA review document on the topic (A-55) for more detail.

Incompatibilities: Important, potentially hazardous incompatibilities of each substance are listed where available. These are primarily drawn from information developed under the NIOSH Standards Completion Program (A-4).

Permissible Exposure Limits in Air: The permissible exposure limit (PEL), as found in 29 CFR 1910.1000 as of January 1977, has been cited as the Federal standard where one exists. Where NIOSH has published a recommended revision to the OSHA regulation, the NIOSH recommended level is also noted. Where the American Conference of Governmental Industrial Hygienists (ACGIH) has recommended revision of the OSHA regulation and NIOSH has not, the ACGIH revised threshold limit value (TLV®) is also noted.

Except where otherwise noted, the PELs are work-shift time-weighted average (TWA) levels. Ceiling levels and TWAs averaged over other than full work-shifts are noted.

Where the Federal standard and the ACGIH (1983/84) value are identical, the ACGIH designation is implied but not stated, in order to conserve space.

The short-term exposure limit (STEL) values are derived from the ACGIH publication (A-6). This value is the maximal concentration to which workers can be exposed for a period up to 15 minutes continuously without suffering from: irritation; chronic or irreversible tissue change; or narcosis of sufficient degree to increase accident proneness, impair self-rescue, or materially reduce work efficiency, provided that no more than four excursions per day are permitted, with at least 60 minutes between exposure periods, and provided that the daily TWA also is not exceeded.

The "Immediately Dangerous to Life or Health" (IDLH) concentration (A-4) is listed in either ppm or mg/m^3. This concentration represents a maximum level from which one could escape within 30 minutes without any escape-impairing symptoms or any irreversible health effects.

Determination in Air: The citations to analytical methods are drawn from various sources, such as:

- The NIOSH "Pocket Guide" (A-4) which summarizes information from the Standards Completion Program. Where references are given to NIOSH Method Set letters or numbers, the complete methods may be found in PB reports available from the National Technical Information Service, Springfield, Virginia, using the following order numbers:

Set Code	NTIS No.	Set Code	NTIS No.
A	PB245850	G	PB265026
B	PB245851	H	PB265027
C	PB245852	I	PB265028
D	PB245935	J	PB263959
E	PB246148	K	PB254227
F	PB246149	L	PB250159
M	PB265029	T	PB262404
N	PB258433	U	PB262405
O	PB262402	V	PB262542
P	PB258434	W	PB262406
Q	PB258435		

Set Code	NTIS No.	Set Code	NTIS No.
R	PB262403	Set 1	PB271712
S	PB263871	Set 2	PB271464

● A publication of the DuPont Company (A-1) which summarizes in tabular form most of the ACGIH threshold limit values, sampling methods, sample sizes, suggested sampling rates and analytical techniques for airborne substances.

Permissible Concentrations in Water: The permissible concentrations in water are drawn from various sources also, including:

— The U.S. Environmental Protection Agency's "red book" published in 1976 (A-3).

— The National Academy of Sciences/National Research Council publication, *Drinking Water and Health* published in 1977 (A-2).

— The priority toxic pollutant criteria published by U.S. EPA in draft form in 1979 and in final form in 1980 (A-53).

— The multimedia environmental goals for environmental assessment study conducted by EPA (A-37). Values are cited from this source when not available from other sources.

Determination in Water: The sources of information in this area have been primarily U.S. EPA publications including the test procedures for priority pollutant analysis (A-54).

Routes of Entry: The toxicologically important routes of entry of each substance are listed. These are primarily taken from the NIOSH Pocket Guide (A-4) but are drawn from other sources as well (A-5).

Harmful Effects and Symptoms: These are primarily drawn from NIOSH publications (A-4) and (A-5) but are supplemented from information from the draft criteria documents for priority toxic pollutants (A-33) and from other sources. The other sources include:

— EPA Chemical Hazard Information Profiles (CHIPS) cited under individual entries.

— NIOSH Information Profiles cited under individual entries.

— EPA Health and Environmental Effect Profiles cited under individual entries.

Particular attention has been paid in this second revised edition to cancer as a "harmful effect" and special effort has been expended to include the latest data on carcinogenicity (A-62)(A-63)(A-64).

Points of Attack: This category is based in part on the "Target Organs" in the NIOSH Pocket Guide (A-4) but the title has been changed as many of the points of attack are not organs (blood, for example).

Medical Surveillance: This information is drawn primarily from a NIOSH publication (A-5). Where additional information is desired in areas of diagnosis, treatment and medical control, the reader is referred to a private publication (A-35) which is adapted from the products of the NIOSH Standards Completion Program.

First Aid: Simple first aid procedures are listed for response to eye contact, skin contact, inhalation, and ingestion of the toxic substance as drawn to a large extent from the NIOSH Pocket Guide (A-4) but supplemented by information from recent commercially available volumes in the U.S. (A-39), in the U.K. (A-44)(A-48) and in Japan (A-38).

Personal Protective Methods: This information is drawn heavily from NIOSH publications (A-4)(A-5) and supplemented by information from the U.S. (A-39), the U.K. (A-48) and Japan (A-38).

There are indeed other "personal protective methods" which space limitations prohibit describing here in full. One of these involves limiting the quantities of carcinogens to which a worker is exposed in the laboratory. Such numerical limits have been discussed in an N.I.H. publication (A-67).

Respirator Selection: Whereas in all the other categories, it has been attempted to provide guidance in brief readable English, it has been necessary in the interest of brevity to use hieroglyphics to designate respirator selection. A condensed table of allowable respirator use is provided from a NIOSH tabulation (A-4) when available. Each line of this item lists a maximum use concentration (in ppm, mg/m^3, $\mu g/m^3$, or mppcf) or condition (e.g., escape) followed by a series of codes representing classes of respirators. Individual respirator codes are separated by slanted (/) lines.

The recommendations for respirator use are based upon the OSHA permissible exposure level. Any approved respirator of a given category can be utilized at any concentration equal to, or less than, the category's listed maximum use concentration for the toxic substance of interest. Codes employed for the various categories of respirator are as follows:

CCR	Chemical cartridge respirator
CCRAG	Chemical cartridge respirator with acid gas cartridge(s)
CCRAGF	CCRAG with a full facepiece
CCRAGFHiE	CCRAG with a full facepiece and with a high-efficiency filter
CCROV	Chemical cartridge respirator with organic vapor cartridge(s)
CCROVAG	CCROV with acid gas cartridge(s)
CCROVAGF	CCROV with acid gas cartridge(s) and full facepiece
CCROVD	CCROV with dust filter
CCROVDM	CCROV with dust and mist filter
CCROVDMF	CCROV with dust and mist filter and full facepiece
CCROVDMFuPest	CCROV with dust, mist, and fume filter, including pesticide respirators meeting these requirements
CCROVDMPest	CCROV with dust and mist filter, including pesticide respirators meeting these requirements
CCROVDPest	CCROV with dust filter, including pesticide respirators
CCROVF	CCROV with full facepiece
CCROVFuHiE	CCROV with fume or high-efficiency filter
CCROVFD	CCROV with full facepiece and dust filter
CCROVFDM	CCROV with full facepiece and dust and mist filter
CCROVFDMFuPest	CCROV with full facepiece and dust, mist and fume filter, including pesticide respirators meeting these requirements
CCROVFDMPest	CCROV with full facepiece and dust and mist filter, including pesticide respirators meeting these requirements
CCROVFDPest	CCROV with dust filter and full facepiece, including pesticide respirators with full facepiece
CCROVFFuHiE	CCROV with full facepiece and fume or high-efficiency filter
CCROVFHiEP	CCROV with high-efficiency particulate filter and full facepiece

CCROVFS	CCROV with full facepiece providing protection against the specific compound of concern
CCROVHiEP	CCROV with high-efficiency particulate filter
CCROVHiEPest	CCROV with high-efficiency filter, including pesticide respirators meeting these requirements
CCROVS	CCROV providing protection against the specific compound of concern
D	Dust mask
DM	Dust and mist respirator
DMFu	Dust, mist, and fume respirator
DMXS	DM, except single-use respirators
DMXSPest	DM, except single-use respirators, including pesticide respirators
DMXSQ	DM, except single-use and quarter-mask respirators
DMXSQPest	DM, except single-use and quarter-mask respirators including pesticide respirators
DXS	Dust mask, except single-use
DXSPest	Dust mask, except single-use, including pesticide respirators
DXSQ	Dust mask, except single-use and quarter-mask respirators
FuHiEP	Fume or high-efficiency particulate respirator
FuHiEPS	FuHiEP providing protection against the specific compound of concern
GMAG	Gas mask with an acid gas canister (chin-style or front- or back-mounted canister)
GMAGHiE	GMAG with high-efficiency filter
GMAGP	GMAG with particulate filter
GMAGS	GMAG providing protection against the specific compound of concern
GMOV	Gas mask with organic vapor canister (chin-style or front- or back-mounted canister)
GMOVc	Chin-style GMOV
GMOVfb	Front- or back-mounted GMOV
GMOVAG	GMOV providing protection against acid gases
GMOVAGF	GMOV, with full facepiece, providing protection against acid gases
GMOVAGHiE	GMOV with high-efficiency filter and acid gas canister
GMOVAGP	GMOV with acid gas canister and particulate filter
GMOVD	GMOV with dust filter
GMOVDFuMPest	GMOV with dust, fume, and mist filter, including pesticide respirators meeting these requirements
GMOVDM	GMOV with dust and mist filter
GMOVDMPest	GMOV with dust and mist filter, including pesticide respirators
GMOVF	GMOV with full facepiece
GMOVFFuHiE	GMOV with full facepiece and fume or high-efficiency filter
GMOVFHiE	GMOV with full facepiece and high-efficiency filter
GMOVFP	GMOV with full facepiece and particulate filter
GMOVHiEP	GMOV with high-efficiency particulate filter
GMOVP	GMOV with particulate filter
GMOVPPest	GMOV with particulate filter, including pesticide respirators
GMPest	Gas mask with pesticide canister (chin-style or front- or back-mounted canister)
GMS	Gas mask with canister providing protection against the compound of concern (chin-style or front- or back-mounted canister)
GMSc	GMS with chin-mounted canister

GMSfb	GMS with front- or back-mounted canister
GMSF	GMS with full facepiece
GMSHiE	GMS with high-efficiency filter
GMSOVPPest	GMS with organic vapor canister and particulate filter, including pesticide respirators meeting these requirements
GMSP	GMS with particulate filter
HiEP	High-efficiency particulate respirator
HiEPAG	HiEP with acid gas cartridge
HiEPF	HiEP with full facepiece
HiEPFu	HiEP or a fume filter respirator
HiEPFPest	HiEP with full facepiece, including pesticide respirators meeting these requirements
HiEPPest	HiEP, including pesticide respirators meeting these requirements
MXS	Mist respirator, except single-use
MXSQ	Mist respirator, except single-use and quarter-mask respirator
PAPCCROVHiEP	Powered air-purifying respirator with organic vapor cartridge and high-efficiency particulate filter
PAPCCROVFHiEP	PAPCCROVHiEP with full facepiece
PAPCCROVFHiEPPest	PAPCCROVHiEP with full facepiece, including pesticide respirators meeting these requirements
PAPHiE	Powered air-purifying respirator with high-efficiency filter
PAPHiEF	PAPHiE with full facepiece
PAPHiEOV	PAPHiE with organic vapor cartridge
PAPHiEOVF	PAPHiE with organic vapor cartridge and full facepiece
PAPHiEPest	PAPHiE, including pesticide respirators meeting these requirements
SA	Supplied-air respirator
SA:PD,PP,CF	Type C SA operated in pressure-demand or other positive pressure or continuous-flow mode
SAF	SA with full facepiece, helmet, or hood
SAF:PD,PP,CF	Type C SA with full facepiece operated in pressure-demand or other positive pressure mode or with full facepiece, helmet, or hood operated in continuous-flow mode
SCBA	Self-contained breathing apparatus
SCBAF	SCBA with full facepiece
SCBAF:PD,PP	SCBA with full facepiece operated in pressure-demand or other positive pressure mode

Disposal Method Suggested: The disposal methods for various chemical substances have been drawn from earlier works by this author on disposal, incineration and landfill disposal (A-31) as well as a more recent volume (A-32) which treats pesticide disposal methods more specifically. Another source is a recent book on the degradation of chemical carcinogens (A-43). A still more recent source on disposal of chemicals from laboratories has been published by the National Research Council (A-66).

References: The general bibliography for this volume follows immediately. It includes general reference sources and references dealing with analytical methods.

The references at the end of individual product entries are generally restricted to:

— references dealing only with that particular compound;

— references which in turn contain bibliographies giving references to the original literature on toxicological and other behavior of the substance in question.

Bibliography

(A-1)   E.I. Du Pont de Nemours and Co., *Industrial Hygiene Sampling and Analytical Guide for Airborne Health Hazards,* Wilmington, DE, Applied Technology Division (1979).

(A-2)   National Research Council, *Drinking Water and Health*, Washington, DC, National Academy of Sciences (1977).

(A-3)   U.S. Environmental Protection Agency, *Quality Criteria for Water,* Washington, DC, Office of Water and Hazardous Materials (July 1976).

(A-4)   National Institute for Occupational Safety and Health, *NIOSH/OSHA Pocket Guide to Chemical Hazards,* DHEW (NIOSH) Publication No. 78-210, Washington, DC (September 1978).

(A-5)   National Institute for Occupational Safety and Health, *Occupational Diseases: A Guide to Their Recognition,* DHEW (NIOSH) Publication No. 77-181, Washington, DC (June 1977).

(A-6)   American Conference of Governmental Industrial Hygienists, *Threshold Limit Values for Chemical Substances and Physical Agents in the Workroom Environment with Intended Changes for 1983/84,* Cincinnati, Ohio, ACGIH (1983).

(A-7)   Worthing, C.R., Editor, *The Pesticide Manual,* Croydon, England, The British Crop Protection Council (1979).

(A-8)   Altshuller, A.P. and Cohen, I.R., "Spectrophotometric Determination of Crotonaldehyde with 4-Hexylresorcinol," *Analytical Chemistry,* 33, 1180 (August 1961).

(A-9)   Diggle, W.M. and Gage, J.C., "Determination of Ketone and Acetic Anhydride in the Atmosphere," *The Analyst,* 78, 473 (August 1953).

(A-10) National Institute for Occupational Safety and Health, *NIOSH Manual of Analytical Methods,* 2nd Edition (4 volumes), DHEW (NIOSH) Publication No. 77-157A (April 1977).

(A-11)   Leichnitz, K.R., "Detector Tubes and Prolonged Air Sampling," *National Safety News*, 115, 59 (April 1977).

(A-12)   Leithe, W., *The Analysis of Air Pollutants,* Ann Arbor, MI, Ann Arbor-Humphrey Science Publishers (1970).

(A-13)   White, L.D. et al, "A Convenient Optimized Method for the Analysis of Selected Solvent Vapors in the Industrial Atmosphere," *American Industrial Hygiene Association Journal*, 31, 225 (March/April 1970).

(A-14)   American Public Health Association, *Standard Methods for the Examination of Water and Wastewater*, 14th Edition, Washington, DC (1975).

(A-15)   Orion Research, Inc., *Orion Research: Analytical Methods Guide,* Cambridge, MA (1973).;

(A-16)   Feinsilver, L. and Oberst, F.W., "Microdetermination of Chloropicrin Vapor in Air," *Analytical Chemistry*, 25, 820 (May 1953).

(A-17)   Ruch, W.E., *Quantitative Analysis of Gaseous Pollutants,* Ann Arbor, MI, Ann Arbor-Humphrey Science Publishers (1970).

(A-18)   Scherberger, R.F.,et al, "The Determination of n-Butylamine in Air," *American Industrial Hygiene Association Journal,* 21, 471 (December 1960).

(A-19)   Dahlgren, G., "Spectrophotometric Determination of Ethyl-, Diethyl- and Triethylamine in Aqueous Solution," *Analytical Chemistry*, 36, 596 (March 1964).

(A-20)   Food and Drug Administration, *Pesticide Analytical Manual,* Washington, DC [undated; quoted in Reference (A-1)].

(A-21)   Perkin-Elmer Corp., *Analytical Methods for Atomic Absorption Spectrophotometry*, Norwalk, CT [undated, quoted in Reference (A-1)].

(A-22)   Warner, B.R. and Raptis, L.Z., "Determination of Formic Acid in the Presence of Acetic Acid," *Analytical Chemistry,* 27, 1783 (November 1955).

(A-23)   Houghton, J.A. and Lee, G., "Ultraviolet Spectrophotometric and Fluorescence Data," *American Industrial Hygiene Association Journal,* 22, 296 (August 1961).

(A-24)   Johannesson, J.K., "Determination of Microgram Quantities of Free Iodine Using o-Tolidine Reagent," *Analytical Chemistry* 28, 1475 (September 1956).

(A-25)   Brief, R.S. et al, "Iron Pentacarbonyl: Its Toxicity Detection and Potential for Formation," *American Industrial Hygiene Association Journal,* 28, 21 (January/February 1967).

(A-26)   Stewart, J.T. et al, "Spectrophotometic Determination of Primary Aromatic Amines with 9-Chloroacridine," *Analytical Chemistry,* 41, 360 (February 1969).

(A-27)   National Institute for Occupational Safety and Health, *NIOSH Manual of Sampling Data Sheets 1977 Edition*, DHEW (NIOSH) Publication No. 77-159, Cincinnati, OH (1977).

(A-28)   Horwitz, W., Editor, *Official Methods of Analysis of the Association of Official Analytical Chemists,* 11th Ed., Washington, DC, AOAC (1970).

(A-29) Moss, R. and Browett, E.V., "Determination of Tetraalkyl Lead Vapour and Inorganic Lead Dust in Air," *The Analyst*, 91, 428 (July 1966).

(A-30) American Industrial Hygiene Association, *Hygienic Guide Series,* Akron, OH.

(A-31) Powers, P.W., *How to Dispose of Toxic Substances and Industrial Wastes,* Park Ridge, NJ, Noyes Data Corp. (1976) as well as Sittig, M., *Incineration of Industrial Hazardous Wastes and Sludges* and *Landfill Disposal of Hazardous Wastes and Sludges,* Park Ridge, NJ, Noyes Data Corp. (1979).

(A-32) Sittig, M., Editor, *Pesticide Manufacturing and Toxic Materials Control Encyclopedia,* Park Ridge, NJ, Noyes Data Corp. (1980).

(A-33) Sittig, M., Editor, *Priority Toxic Pollutants: Health Impacts and Allowable Limits,* Park Ridge, NJ, Noyes Data Corp. (1980).

(A-34) American Council of Governmental Industrial Hygienists, *Documentation of the Threshold Limit Values,* Cincinnati, OH (Third Edition-1971 with supplements through 1979).

(A-35) Proctor, N.H. and Hughes, J.P., *Chemical Hazards of the Workplace,* Philadelphia, PA, J.B. Lippincott and Co. (1978).

(A-36) Verscheuren, K., *Handbook of Environmental Data on Organic Chemicals,* NY, Van Nostrand Reinhold Co. (1977).

(A-37) U.S. Environmental Protection Agency, *Multimedia Environmental Goals for Environmental Assessment,* Report EPA-600/7-77-136, Research Triangle Park, NC (November 1977).

(A-38) The International Technical Information Institute, *Toxic and Hazardous Industrial Chemicals Safety Manual for Handling and Disposal with Toxicity and Hazard Data,* Tokyo, ITII (1980).

(A-39) Plunkett, E.R., *Handbook of Industrial Toxicology,* New York, NY, Chemical Publishing Co., Inc. (1976).

(A-40) U.S. Environmental Protection Agency, *The Carcinogen Assessment Group's List of Carcinogens,* Washington, DC, USEPA (July 14, 1980).

(A-41) Sittig, M., *Pharmaceutical Manufacturing Encyclopedia*, Park Ridge, NJ, Noyes Data Corp. (1979).

(A-42) Banks, R.E., Editor, *Organofluorine Chemicals and Their Industrial Application,* Chichester, West Sussex, Ellis Horwood, Ltd. (1979).

(A-43) Slein, M.W. and Sansome, E.B., *Degradation of Chemical Carcinogens— An Annotated Bibliography,* NY, Van Nostrand Reinhold Co. (1980).

(A-44) Muir, G.D., Editor, *Hazards in the Chemical Laboratory,* 2nd Ed., London, The Chemical Society (1976).

(A-45) Schroeder, W.A. et al, "Ultraviolet and Visible Absorption Spectra in Ethyl Alcohol," *Analytical Chemistry,* 23, 1740 (December 1951).

(A-46) United Nations *Transport of Dangerous Goods,* NY, U.N. (1977).

(A-47) Toxic Substances Strategy Committee, *Toxic Chemicals and Public Protection: A Report to the President,* Washington, DC, Council on Environmental Quality (May 1980).

(A-48) Chemical Industries Association *CEFIC Tremcards: Reference Edition,* London, C.I.A. (1979).

(A-49) National Institute for Occupational Safety and Health, *Registry of Toxic Effects of Chemical Substances,* DHEW (NIOSH) Publication No. 81-116 (February 1982).

(A-50) U.S. Department of Transportation, "Transport of Hazardous Wastes and Chemical Substances," *Federal Register,* 45, No. 101, 34560-34705 (May 22, 1980).

(A-51) U.S. Environmental Protection Agency, "Water Programs: Hazardous Substances," *Federal Register,* 43, No. 49, 10474-10508 (March 13, 1978).

(A-52) U.S. Environmental Protection Agency, "Identification and Testing of Hazardous Waste," *Federal Register,* 45, No. 98, 33084-33133 (May 19, 1980).

(A-53) U.S. Environmental Protection Agency, *Federal Register,* 43, 4109 (January 31, 1978). See also *Federal Register,* 44, 44501 (July 30, 1979) and also *Federal Register,* 45, 79318-79379 (November 28, 1980).

(A-54) U.S. Environmental Protection Agency, "Guidelines Establishing Test Procedures for the Analysis of Pollutants; Proposed Regulations," *Federal Register,* 44, No. 233, 69464-69575 (December 3, 1979) and also a corrected version in *Federal Register,* 44, No. 244, 75028-75052 (December 18, 1979).

(A-55) Holleman, J.W., Ryon, M.G. and Hammons, A.S., (Oak Ridge National Laboratory), *Chemical Contaminants in Nonoccupationally Exposed U.S. Residents,* Report EPA-600/1-80-001, Research Triangle Park, NC (May 1980).

(A-56) Materials Transportation Bureau, U.S. Department of Transportation, *Hazardous Materials: Fmergency Response Guidebook,* Publication No. DOT P 5800.2, Washington DC (1980).

(A-57) Sittig, M., *Metal and Inorganic Waste Reclaiming Encyclopedia,* Park Ridge, NJ, Noyes Data Corp. (1980).

(A-58) Sittig, M., *Organic and Polymer Waste Reclaiming Encyclopedia,* Park Ridge, N.J, Noyes Data Corp. (1981).

(A-59) National Research Council, *Prudent Practices for Handling Hazardous Chemicals in Laboratories,* Committee on Hazardous Substances in the Laboratory, Washington, DC (1981).

(A-60) Dutch Association of Safety Experts, Dutch Chemical Industry Association, and Dutch Safety Institute, *Handling Chemicals Safely,* Amsterdam, The Netherlands (1980).

(A-61) Mackison, F.W. et al, Eds., *NIOSH/OSHA Occupational Health Guidelines for Chemical Hazards,* Publication No. 81-123, Washington, DC, National Institute for Occupational Safety and Health (January 1981).

(A-62) U.S. Dept. of Health and Human Services, *Second Annual Report on Carcinogens,* Research Triangle Park, N.C. (December 1981).

(A-63)   Sax, N.I., *Cancer Causing Chemicals,* New York, Van Nostrand Reinhold (1981).

(A-64)   U.S. Dept. of Health and Human Services, *Third Annual Report on Carcinogens,* Research Triangle Park, N.C. (December 1982).

(A-65)   National Research Council, *Quality Criteria for Water Reuse,* Washington, DC, National Academy Press (1982).

(A-66)   National Research Council, *Prudent Practices for Disposal of Chemicals from Laboratories,* Washington, DC, National Academy Press (1983).

(A-67)   Division of Safety, National Institutes of Health, *NIH Guidelines for the Laboratory Use of Chemical Carcinogens,* Washington, DC, NIH Publication No. 81–2385, U.S. Department of Health and Human Services (May 1981).

(A-68)   National Institute for Occupational Safety and Health, *Occupational Health and Safety Guidelines for Chemical Hazards: Ketones,* Rockville, Maryland, NIOSH (August 1983).

(A-69)   Parmeggiani, L., Ed., *Encyclopedia of Occupational Health and Safety,* Third Edition, Geneva, International Labor Office (ILO) (1983).

(A-70)   United Nations Environment Programme, *IRPTC Legal File 1983,* Geneva, Switzerland, International Register of Potentially Toxic Chemicals (1984).

A

ABATE®

Abate® is a registered trademark name for the compound with the generic name of Temephos (which see). ACGIH lists this compound under the trademarked name rather than the generic name; hence it is cross-referenced here.

ABIETIC ACID

Description: $C_{20}H_{30}O_2$ with the structural formula

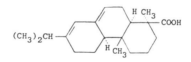

forms crystals melting at 172° to 175°C.

Code Numbers: CAS 514-10-3   RTECS TP8580000

DOT Designation: —

Synonyms: Sylvic acid.

Potential Exposure: In manufacture of esters for use in lacquers and varnishes. Used in making paper sizes, soaps and detergents.

Permissible Exposure Limits in Air: No standards set.

Permissible Concentration in Water: No criteria set.

Routes of Entry: Skin contact, dust inhalation.

Harmful Effects and Symptoms: Irritating to skin and mucous membranes.

Points of Attack: Skin and mucous membranes.

Disposal Method Suggested: Incineration.

References
(1)   Sax, N.I., Ed., *Dangerous Properties of Industrial Materials Report,* 3, No. 3, 31-32, New York, Van Nostrand Reinhold Co. (1983).

ACENAPHTHENE

● Priority toxic pollutant (EPA)

Description: Acenaphthene, $C_{12}H_{10}$, is a white crystalline solid melting at 95° to 97°C.

Code Numbers: CAS 83-32-9

DOT Designation: —

Synonyms: 1,8-Ethylenenaphthalene, 1,2-dehydroacenaphthalene.

Potential Exposure: Acenaphthene occurs in coal tar produced during the high-temperature carbonization or coking of coal. It is used as a dye intermediate, in the manufacture of some plastics, as an insecticide and fungicide, and has been detected in cigarette smoke and gasoline exhaust condensates.

Permissible Exposure Limits in Air: No standards exist.

Permissible Concentration in Water (1): To protect freshwater aquatic life—1,700 $\mu g/\ell$. To protect saltwater aquatic life—on an acute basis 970 $\mu g/\ell$ and on a chronic basis 520 $\mu g/\ell$. To protect human health—20.0 $\mu g/\ell$ (based on organoleptic data).

Determination in Water: Gas chromatography or high performance liquid chromatography (EPA Method 610) or gas chromatography and mass spectrometry (EPA Method 625).

Routes of Entry: Ingestion from water or foods, inhalation.

Harmful Effects and Symptoms: Acenaphthene is irritating to skin and mucous membranes and may cause vomiting if swallowed in large quantities.
(The most thoroughly investigated effect of acenaphthene is its ability to produce nuclear and cytological changes in microbial and plant species. Most of these changes, such as an increase in cell size and DNA content, are associated with disruption of the spindle mechanism during mitosis and the resulting induction of polyploidy. While there is no known correlation between these effects and the biological impact of acenaphthene on mammalian cells, these effects are reported here because they are the only substantially investigated effects of acenaphthene.)

Points of Attack: Liver, kidneys, skin.

Medical Surveillance: Preplacement and regular physical examinations are indicated for workers having contact with acenaphthene in the workplace.

First Aid: If this chemical gets into the eyes, irrigate immediately. If this chemical contacts the skin, wash with soap immediately. When this chemical has been swallowed, get medical attention. Give large quantities of water and induce vomiting. Do not make an unconscious person vomit.

Personal Protective Methods: Prevent repeated or prolonged skin contact. Wear goggles when eye exposure is reasonably probable. Wash skin and change clothing upon contamination.

Respirator Selection: Chemical cartridge respirator, gas mask or supplied air respirator indicator—precise recommendation awaits definition of allowable limits in air.

Disposal Method Suggested: Incineration.

References

(1) U.S. Environmental Protection Agency, *Acenaphthene: Ambient Water Criteria,* Report PB 296-782, Washington, DC (1980).

(2) Sax, N.I., Ed., *Dangerous Properties of Industrial Materials Report, 4,* No. 1, 38-41, New York, Van Nostrand Reinhold Co. (1984).

ACENAPHTHYLENE

See "Polynuclear Aromatic Hydrocarbons."

ACETALDEHYDE

● Hazardous substance (EPA)
● Hazardous waste (EPA)

Description: CH_3CHO, acetaldehyde, is a flammable, volatile colorless liquid with a characteristic penetrating, fruit odor. It boils at $20°$ to $21°C$.

Code Numbers: CAS 75-07-0 RTECS AB1925000 UN 1089

DOT Designation: Flammable liquid.

Synonyms: Acetic aldehyde, aldehyde, ethanal, ethyl aldehyde.

Potential Exposure: Acetaldehyde can be reduced or oxidized to form acetic acid, acetic anhydride, acrolein, aldol, butanol, chloral, paraldehyde, and pentaerythritol. It is also used in the manufacture of disinfectants, drugs (A-41), dyes, explosives, flavorings, lacquers, mirrors (silvering), perfume, photographic chemicals, phenolic and urea resins, rubber accelerators and antioxidants, varnishes, vinegar, and yeast. It is also a pesticide intermediate (A-32).

NIOSH estimates that 2,430 workers are exposed to acetaldehyde. Acetaldehyde is the product of most hydrocarbon oxidations; it is a normal intermediate product in the respiration of higher plants; it occurs in traces in all ripe fruits and may form in wine and other alcoholic beverages after exposure to air. Acetaldehyde is an intermediate product in the metabolism of sugars in the body and hence occurs in traces in blood. It has been reported in fresh leaf tobacco as well as in tobacco smoke and in automobile and diesel exhaust (A-5). It has been found in 5 of 10 water supplies surveyed by EPA with the highest concentrations in Philadelphia and Seattle at 0.1 µg/ℓ (A-2).

Incompatibilities: Strong oxidizers, acids, bases, alcohol, ammonia, amines, phenols, ketones, HCN, H_2S.

Permissible Exposure Limits in Air: The Federal standard (TWA) is 200 ppm (360 mg/m^3); however, the ACGIH 1983/84 recommended TLV is 100 ppm (180 mg/m^3). The tentative STEL is 150 ppm (270 mg/m^3) and the IDLH level is 10,000 ppm.

Determination in Air: Acetaldehyde may be collected by impinger or fritted bubbler and then determined colorimetrically (A-8).

Permissible Concentration in Water: Human exposure to acetaldehyde probably antedates recorded history, inasmuch as acetaldehyde is the major metabolite of ethyl alcohol. An additional source of widespread human exposure is tobacco smoke. The pharmacology and toxicology of acetaldehyde have been studied most extensively in its relationship to alcohol toxicity and human metabolism. Because of this background of human and laboratory experience, there appears to be no need to establish limits for acetaldehyde in drinking water (A-2). However, EPA (A-37) has set an ambient environmental goal of 2,480 μg/ℓ for acetaldehyde on a health basis.

Routes of Entry: Inhalation of vapor, ingestion.

Harmful Effects and Symptoms: *Local* — The liquid and the fairly low levels of the vapor are irritating to the eyes, skin, upper respiratory passages, and bronchi. Repeated exposure may result in dermatitis, rarely, and skin sensitization.

Systemic — Acute involuntary exposure to high levels of acetaldehyde vapors may result in pulmonary edema, preceded by excitement, followed by narcosis. It has been postulated that these symptoms may have been similar to those of alcohol, which is converted to acetaldehyde and acetic acid. Chronic effects have not been documented, and seem unlikely, since voluntary inhalation of toxicologically significant levels of acetaldehyde are precluded by its irritant properties at levels as low as 200 ppm (360 mg/m^3 of air).

Points of Attack: Respiratory system, lungs, skin, kidneys.

Medical Surveillance: Consideration should be given to skin, eyes, and respiratory tract in any preplacement or periodic examinations.

First Aid: If this chemical gets into the eyes, irrigate immediately. If this chemical contacts the skin, flush with water promptly. If a person breathes in large amounts of this chemical, move the exposed person to fresh air at once and perform artificial respiration. When this chemical has been swallowed, get medical attention. Give large quantities of water and induce vomiting. Do not make an unconscious person vomit.

Personal Protective Methods: Wear appropriate clothing to prevent repeated or prolonged skin contact. Wear eye protection to prevent any possibility of eye contact. Employees should wash promptly when skin is wet. Remove clothing immediately if wet or contaminated to avoid flammability hazard. Provide eyewash.

Respirator Selection (A-4):
    1,000 ppm: CCROVF
    5,000 ppm: GMOVc
  10,000 ppm: GMOVfb/SAF/SCBAF
      Escape: GMOV/SCBA

Disposal Method Suggested: Incineration (A-31).

References
(1) U.S. Environmental Protection Agency, *Chemical Hazard Information Profile: Acetaldehyde,* (Preliminary), Washington, DC (1979).
(2) U.S. Environmental Protection Agency, *Acetaldehyde,* Health and Environmental Effects Profile No. 1, Washington, DC, Office of Solid Waste (April 30, 1980).
(3) Sax, N.I., Ed., *Dangerous Properties of Industrial Materials Report, 1,* No. 1, 25-6, New York, Van Nostrand Reinhold Co. (1980).
(4) See Reference (A-61).
(5) See Reference (A-60).
(6) U.S. Environmental Protection Agency, *Chemical Hazard Information Profile Draft Report: Acetaldehyde,* Washington, DC (April 29, 1983).

(7) Sax, N.I., Ed., *Dangerous Properties of Industrial Materials Report, 3,* No. 6, 23–27, New York, Van Nostrand Reinhold Co. (Nov./Dec. 1983).

(8) Parmeggiani, L., Ed., *Encyclopedia of Occupational Health & Safety,* Third Edition, Vol. 1, pp 35–37, Geneva, International Labour Office (1983).

ACETAMIDE

● Carcinogen (Animal Positive, IARC)(2)

Description: CH_3CONH_2, colorless crystals with a mousy odor, melting at 81°C.

Code Numbers: CAS 60-35-5   RTECS AB402500

DOT Designation: —

Synonyms: Acetic Acid Amide, Ethanamide.

Potential Exposure: Used as a solvent in molten state for many chemicals in plastics and chemical manufacturing.

Permissible Exposure Limits in Air: No standards set.

Permissible Concentration in Water: No criteria set; not very toxic to fish (A-36) but increases B.O.D. (1).

Harmful Effects and Symptoms: Is a mild irritant of low toxicity (1). However, IARC has determined it to be an animal positive carcinogen.

Personal Protective Methods: Self-contained breathing apparatus required (1). Skin protection required.

Respirator Selection: See above.

Disposal Method Suggested: Add to alcohol or benzene as a flammable solvent and incinerate; oxides of nitrogen produced may be scrubbed out with alkaline solution.

References
(1) Sax, N.I., Ed., *Dangerous Properties of Industrial Materials Report, 1,* No. 4, 20-21, New York, Van Nostrand Reinhold Co. (1981).

(2) International Agency for Research on Cancer, *IARC Monographs on the Carcinogenic Risks of Chemicals to Humans, 7,* 197, Lyon, France (1974).

(3) See Reference (A-63).

(4) Sax, N.I., Ed., *Dangerous Properties of Industrial Materials Report, 3,* No. 6, 29-31 New York, Van Nostrand Reinhold Co. (Nov./Dec. 1983).

ACETANILIDE

Description: $C_6H_5NHCOCH_3$, white crystals melting at 113.5°C.

Code Numbers: CAS 103-84-4   RTECS AD735000   UN

DOT Designation: —

Synonyms: Acetamidobenzene, acetic acid anilide, acetanil, acetylaniline; antifebrin.

Potential Exposure: In rubber industry as accelerator, in plastics industry as cellulose ester stabilizer, in pharmaceutical manufacture.

Incompatibilities: Alkyl nitrates, alkalies (liberate aniline), chloral hydrate, phenols, ferric salts (1).

Permissible Exposure Limits in Air: No standards set.

Permissible Concentration in Water: No criteria set.

Harmful Effects and Symptoms: Causes contact dermatitis; inhalation or ingestion can cause eczema and cyanosis and methemoglobinemia. Animals tolerate doses of 200 to 400 mg/kg for many weeks (A-36).

Target Organs: Skin and blood stream.

Personal Protective Methods: Wear skin protection, avoid dust inhalation (see respirator selection below).

Respirator Selection: Wear filter mask unless high vapor concentrations are encountered; then use canister or self-contained breathing apparatus.

Disposal Method Suggested: Add to flammable solvents (alcohol or benzene) and incinerate. Oxides of nitrogen may be scrubbed from combustion gases with alkali.

References

(1) Sax, N.I., Ed., *Dangerous Properties of Industrial Materials Report, 1,* No. 4, 21-23, New York, Van Nostrand Reinhold Co. (1981).

(2) Sax, N.I., Ed., *Dangerous Properties of Industrial Materials Report, 3,* No. 6, 27–29 New York, Van Nostrand Reinhold Co. (Nov./Dec. 1983).

ACETATES

See separate entries under amyl acetate, n-butyl acetate, ethyl acetate, sec-hexyl acetate, isopropyl acetate, methyl acetate and n-propyl acetate. The following are a few generalizations about acetates as a class of compounds:

Potential Exposures: The acetates are a group of solvents for cellulose nitrate, cellulose acetate, ethylcellulose, resins, rosin, Cumar, elemi, phenolics, oils, fats, and celluloid. They are also used in the manufacture of lacquers, paints, varnishes, enamel, perfumes, dyes, dopes, plastic and synthetic finishes (e.g., artificial leather), smokeless powder, photographic film, footwear, pharmaceuticals, food preservatives, artificial glass, artificial silk, furniture polish, odorants, and other organic syntheses.

Harmful Effects and Symptoms: *Local —* In higher concentrations, acetates are irritants to the mucous membranes. All irritate eyes and nasal passages in varying degrees. Prolonged exposure can cause irritation of the intact skin. These local effects are the primary risk in industry.

References

(1) National Institute for Occupational Safety and Health, *Profiles on Occupational Hazards for Criteria Document Priorities:Acetates,* pp 62-67, Report PB-274,073, Cincinnati, Ohio (1977).

ACETIC ACID

● Hazardous substance (EPA)

Description: CH_3COOH, acetic acid, is a colorless liquid with a pungent vinegarlike odor. Glacial acetic acid contains 99% acid. It boils at 117° to 118°C.

Code Numbers: CAS 64-19-7   RTECS AF1225000   UN 2789

DOT Designation: Corrosive material.

Synonyms: Ethanoic acid, ethylic acid, methane carboxylic acid, pyroligneous acid, vinegar acid.

Potential Exposure: Acetic acid is widely used as a chemical feedstock for the production of vinyl plastics, acetic anhydride, acetone, acetanilide, acetyl chloride, ethyl alcohol, ketene, methyl ethyl ketone, acetate esters, and cellulose acetates. It is also used alone in the dye, rubber, pharmaceutical (A-41), food preserving, textile, and laundry industries. It is utilized, too, in the manufacture of Paris green, white lead, tint rinse, photographic chemicals, stain removers, insecticides (A-32) and plastics.

Incompatibilities: Strong oxidizers, chromic acid, sodium peroxide, nitric acid, strong caustics.

Permissible Exposure Limits in Air: The Federal standard (TWA) and ACGIH 1983/84 value is 10 ppm (25 mg/m^3). The tentative STEL value is 15 ppm (37 mg/m^3). The IDLH value is 1,000 ppm.

Determination in Air: Acetic acid may be collected by impinger or fritted bubbler and then determined by titration (A-1). See also reference (A-10).

Permissible Concentration in Water: No limit has been established. However, EPA (A-37) has proposed an ambient environmental goal of 345 µg/ℓ based on health effects.

Determination in Water: Acetic acid in water may be determined by titration.

Route of Entry: Inhalation of vapor.

Harmful Effects and Symptoms: *Local* — Acetic acid vapor may produce irritation of the eyes, nose, throat, and lungs. Inhalation of concentrated vapors may cause serious damage to the lining membranes of the nose, throat, and lungs. Contact with concentrated acetic acid may cause severe damage to the skin and severe eye damage, which may result in loss of sight. Repeated or prolonged exposure to acetic acid may cause darkening, irritation of the skin, erosion of the exposed front teeth, and chronic inflammation of the nose, throat, and bronchi (A-5). See also (A-35).

Systemic — Bronchopneumonia and pulmonary edema may develop following acute overexposure. Chronic exposure may result in pharyngitis and catarrhal bronchitis. Ingestion, though not likely to occur in industry, may result in penetration of the esophagus, bloody vomiting, diarrhea, shock, hemolysis, and hemoglobinuria which is followed by anuria.

Points of Attack: Respiratory system, skin, eyes, teeth.

Medical Surveillance: Consideration should be given to the skin, eyes, teeth, and respiratory tract in placement or periodic examinations.

First Aid: If this chemical gets into the eyes, irrigate immediately. If this

chemical contacts the skin, flush with water immediately. If a person breathes in large amounts of this chemical, move the exposed person to fresh air at once and perform artificial respiration. When this chemical has been swallowed, get medical attention. Give large quantities of water and do not induce vomiting.

Personal Protective Methods: When working with glacial acetic acid, personal protective equipment, protective clothing, gloves, and goggles should be worn. Eye fountains and showers should be available in areas of potential exposure. Wear appropriate clothing to prevent any possibility of skin contact with liquids of >50% content or repeated or prolonged contact with liquids of 10 to 49% content. Wear eye protection to prevent any possibility of eye contact. Employees should wash immediately with soap when skin is wet or contaminated with liquids of >50% content and promptly if liquids of 10 to 49% acetic acid are involved. Remove clothing immediately if wet or contaminated with liquids containing 50% and promptly remove if liquid contains 10 to 49% acetic acid. Provide emergency eyewash if liquids containing >5% acetic acid are involved, drench if >50% acetic acid is involved.

Respirator Selection:
>   500 ppm: CGROVF/GMOV/SAF/SCBAF
>   1,000 ppm: SAF:PD,PP,CF
>   Escape: GMOV/SCBA

Disposal Method Suggested: Incineration (A-31).

References

(1)   Sax, N.I., Ed., *Dangerous Properties of Industrial Materials Report, 1,* No. 4, 23-25, New York, Van Nostrand Reinhold Co. (1981).
(2)   See Reference (A-61).
(3)   See Reference (A-60).
(4)   Sax, N.I., Ed., *Dangerous Properties of Industrial Materials Report, 3,* No. 6, 31–35, New York, Van Nostrand Reinhold Co. (Nov./Dec. 1983)
(5)   Parmeggiani, Ed., *Encyclopedia of Occupational Health & Safety,* Third Edition, Vol. 1, pp 37–38, Geneva, International Labour Office (1983).

ACETIC ANHYDRIDE

● Hazardous substance (EPA)

Description: $(CH_3CO)_2O$, acetic anhydride, is a colorless, strongly refractive liquid which has a strongly irritating odor. It boils at $140°C$.

Code Numbers: CAS 108-24-7   RTECS AK1925000   UN 1715

DOT Designation: Corrosive material.

Synonyms: Acetic oxide, acetyl oxide, ethanoic anhydride.

Potential Exposures: Acetic anhydride is used as an acetylating agent or as a solvent in the manufacture of cellulose acetate, acetanilide, synthetic fibers, plastics, explosives, resins, perfumes, and flavorings; and it is used in the textile dyeing industry. It is widely used as a pharmaceutical intermediate (A-41) and as a pesticide intermediate (A-32).

Incompatibilities: Water, alcohols, strong oxidizers, chromic acid, amines, strong caustics.

Permissible Exposure Limits in Air: The Federal standard (TWA) and ACGIH 1983/84 value is 5 ppm (20 mg/m^3) as a ceiling value. There is no proposed STEL value. The IDLH level is 1,000 ppm.

Determination in Air: The sample is bubbled through hydroxylamine, worked up with FeCl$_3$ and analyzed colorimetrically. See NIOSH Method, Set L. See also references (A-9) and (A-10).

Permissible Concentration in Water: No criteria set.

Routes of Entry: Inhalation of vapor, ingestion and eye and skin contact.

Harmful Effects and Symptoms: *Local* — In high concentrations, vapor may cause conjunctivitis, photophobia, lacrimation, and severe irritation of the nose and throat. Liquid acetic anhydride does not cause a severe burning sensation when it comes in contact with the skin. If it is not removed, the skin may become white and wrinked, and delayed severe burns may occur. Both liquid and vapor may cause conjunctival edema and corneal burns, which may develop into temporary or permanent interstitial keratitis with corneal opacity due to progression of the infiltration. Contact and, occasionally, hypersensitivity dermatitis may develop (A-5).

Systemic — Immediate complaints following concentrated vapor exposure include conjunctival and nasopharyngeal irritation, cough, and dyspnea. Necrotic areas of mucous membranes may be present following acute exposure.

Points of Attack: Respiratory system, lungs, eyes, skin.

Medical Surveillance: Consideration should be given to the skin, eyes, and respiratory tract in any placement or periodic examinations.

First Aid: If this chemical gets into the eyes, irrigate immediately. If this chemical contacts the skin, flush with water immediately. If a person breathes in large amounts of this chemical, move the exposed person to fresh air at once and perform artificial respiration. When this chemical has been swallowed, get medical attention. Give large quantities of water and do not induce vomiting.

Personal Protective Methods: Wear appropriate clothing to prevent any reasonable probability of skin contact. Wear eye protection to prevent any possibility of eye contact. Employees should wash immediately with soap when skin is wet or contaminated. Remove nonimpervious clothing immediately if wet or contaminated. Provide emergency showers and eyewash.

Respirator Selection:
    250 ppm: CCROVF/GMOV/SAF/SCBAF
  1,000 ppm: SAF:PD,PP,CF
    Escape: GMOV/SCBA

Disposal Method Suggested: Incineration.

References

(1) See Reference (A-61).
(2) Sax, N.I., Ed., *Dangerous Properties of Industrial Materials Report, 3,* No. 3, 32-35, New York, Van Nostrand Reinhold Co. (1983).
(3) See Reference (A-60).

ACETOL

See "Acetylsalicylic Acid."

ACETONE

- Hazardous waste (EPA)

Description: Acetone, CH_3COCH_3, is a colorless liquid with a sweetish odor. It boils at 56.5°C.

Code Numbers: CAS 67-64-1   RTECS AL3150000   UN 1090

DOT Designation: Flammable liquid.

Synonyms: Dimethylketone, β-ketopropane, 2-propanone, pyroacetic ether.

Potential Exposure: NIOSH has estimated worker exposure to acetone at 2,816,000. It is used as a solvent; it is used in the production of lubricating oils and as an intermediate in the manufacture of chloroform and of various pharmaceuticals (A-41) and pesticides (A-32).

Incompatibilities: Oxidizing material, acids.

Permissible Limits in Air: The Federal standard is 1,000 ppm (2,400 mg/m^3). The ACGIH, as of 1983/84 (A-6), has proposed a TWA of 750 ppm (1,780 mg/m^3) and a STEL of 1,000 ppm (2,375 mg/m^3). The IDLH level is 20,000 ppm.

Determination in Air: Charcoal adsorption followed by CS_2 treatment and gas chromatographic analysis. See NIOSH Method, Set A. See also reference (A-10).

Permissible Concentration in Water: No criteria set.

Routes of Entry: Inhalation, ingestion and skin or eye contact.

Harmful Effects and Symptoms: Irritation of eyes, nose and throat, headaches, dizziness and dermatitis.

Points of Attack: Respiratory system and skin.

Medical Surveillance Suggested: Preplacement examinations should evaluate skin and respiratory conditions. Acetone can be detected in the blood, urine and expired air and has been used as an index of exposure.

First Aid: If this chemical gets into the eyes, irrigate immediately. If this chemical contacts the skin, wash with soap immediately. If a person breathes in large amounts of this chemical, move the exposed person to fresh air at once and perform artificial respiration. When this chemical has been swallowed, get medical attention. Give large quantities of water and induce vomiting. Do not make an unconscious person vomit.

Personal Protective Methods: Wear appropriate clothing to prevent repeated or prolonged skin contact. Wear eye protection to prevent any reasonable probability of eye contact. Employees should wash promptly when skin is wet. Remove clothing promptly if wet or contaminated to avoid flammability hazard. Provide emergency showers.

Respirator Selection:
    5,000 ppm: GMOVc
  20,000 ppm: GMOVfb/SAF/SCBAF
      Escape: GMOV/SCBA

Disposal Method Suggested: Incineration.

References

(1) Nat. Inst. for Occup. Safety and Health, *Criteria for a Recommended Standard: Occupational Exposure to Ketones,* NIOSH Pub. No. 78-173, Washington, DC (1978).

(2) Sax, N.I., Ed., *Dangerous Properties of Industrial Materials Report, 1,* No. 4, 25-27, New York, Van Nostrand Reinhold Co. (1981).

(3) See Reference (A-60).

(4) See Reference (A-68).

(5) Parmeggiani, L., Ed., *Encyclopedia of Occupational Health & Safety,* Third Edition, Vol. 1, pp 38–39, Geneva, International Labour Office (1983).

ACETONITRILE

● Hazardous waste (EPA)

Description: CH_3CN, acetonitrile, as a colorless liquid with an etherlike odor.

Code Numbers: CAS 75-05-8   RTECS AL7700000   UN 1648

DOT Designation: Flammable liquid.

Synonyms: Methyl cyanide, ethanenitrile, cyanomethane.

Potential Exposures: Acetonitrile is used as an extractant for animal and vegetable oils, as a solvent, particularly in the pharmaceutical industry, and as a chemical intermediate in pesticide manufacture, e.g., (A-32). It is present in cigarette smoke.

Incompatibilities: Strong oxidizers.

Permissible Exposure Limits in Air: The Federal standard is 40 ppm (70 mg/m^3). This is the 1983/84 ACGIH TWA value with the notation that skin absorption may be significant. The STEL value proposed is 60 ppm (105 mg/m^3). The IDLH level is 4,000 ppm.

Determination in Air: Charcoal adsorption followed by benzene workup and gas chromatographic analysis. See NIOSH Methods, Set L. See also reference (A-10).

Permissible Concentration in Water: Acetonitrile is infinitely soluble and stable in water. No criteria have been set, but EPA has proposed (A-37) an ambient environmental goal of 970 $\mu g/\ell$ based on health effects.

Routes of Entry: Inhalation, percutaneous absorption, ingestion and skin and eye contact.

Harmful Effects and Symptoms: *Local* — At high concentrations, nose and throat irritation have been reported. Splashes of the liquid in the eyes may cause irritation. Acetonitrile may cause slight flushing of the face and a feeling of chest tightness.

Systemic — Acetonitrile has a relatively low acute toxicity, but there have been reports of severe and fatal poisonings in man after inhalation of high concentrations. Acetonitrile is metabolized to HCN which can be found in high levels in the brain, heart, kidney and spleen. Signs and symptoms may include nausea, vomiting, respiratory depression, weakness, chest or abdominal pain, hematemesis, convulsions, shock, unconsciousness, and death. In most cases there is a latent period of several hours between exposure and onset of symptoms. It has been thought that acetonitrile itself has relatively little toxic effect and that the delayed response is due to the slow release of cyanide. No chronic disease has been reported.

Points of Attack: Kidneys, liver, lungs, skin, eyes, central nervous system, cardiovascular system.

Medical Surveillance: Consider the skin, respiratory tract, heart, central nervous system, renal and liver function in placement and periodic examinations. A history of fainting spells or convulsive disorders might present an added risk to persons working with toxic nitriles.

First Aid: If this chemical gets into the eyes, irrigate immediately. If this chemical contacts the skin, flush with water immediately. If a person breathes in large amounts of this chemical, move the exposed person to fresh air at once and perform artificial respiration. When this chemical has been swallowed, get medical attention. Give large quantities of water and induce vomiting. Do not make an unconscious person vomit.

Personal Protective Methods: Wear appropriate clothing to prevent repeated or prolonged skin contact. Wear eye protection to prevent any reasonable probability of eye contact. Employees should wash immediately when skin is wet or contaminated. Remove clothing immediately if wet or contaminated to avoid flammability hazard. Provide emergency showers.

Respirator Selection:

400 ppm:	CCROV/SA/SCBA
1,000 ppm:	CCROVF
2,000 ppm:	GMOV/SAF/SCBAF
4,000 ppm:	SA: PD,PP,CF
Escape:	GMOV/SCBA

Disposal Method Suggested: Incineration with nitrogen oxide removal from effluent gases by scrubbers or incinerators.

References

(1) U.S. Environmental Protection Agency, *Chemical Hazard Information Profile: Acetonitrile,* Washington, DC (March 9, 1979).

(2) U.S. Environmental Protection Agency, *Acetonitrile,* Health and Environmental Effects Profile No. 2, Washington, DC, Office of Solid Waste (April 30, 1980).

(3) See Reference (A-61).

(4) See Reference (A-60).

(5) Sax, N.I., Ed., *Dangerous Properties of Industrial Materials Report, 4,* No. 1, 44–46, New York, Van Nostrand Reinhold Co. (Jan./Feb. 1984).

(6) United Nations Environment Programme, *IRPTC Legal File 1983,* Vol. I, pp VII/21–22, Geneva, Switzerland, International Register of Potentially Toxic Chemicals (1984).

ACETOPHENETIDIN

- Carcinogen (Animal Positive, IARC) (Human Suspected) (1-3)
- Hazardous Waste Constituent (EPA)

Description: $C_{10}H_{13}NO_2$ with the formula $C_2H_5OC_6H_4NHCOCH_3$ is a crystalline solid melting at 134° to 135°C.

Code Numbers: CAS 62-44-2   RTECS AM4375000

DOT Designation: —

Synonyms: Phenacetin; p-ethoxyacetanilide; acetophenetide.

Potential Exposure: Phenacetin is used as an analgesic and antipyretic drug. It is used alone or in combination with aspirin and caffeine for mild to moderate muscle pain relief. Phenacetin has also been used as a stabilizer for hydrogen peroxide in hair bleaching preparations. In veterinary medicine, it is used as an analgesic and antipyretic.

Permissible Exposure Limit in Air: No standards set.

Permissible Concentration in Water: No criteria set.

Routes of Entry: Ingestion.

Harmful Effects and Symptoms: Rats fed a diet containing phenacetin had an excess of nasal and urinary tract tumours. N-Hydroxyphenacetin (a possible metabolite of phenacetin) produced liver carcinomas in rats following oral administration. Phenacetin alone enhanced the urinary bladder carcinogenesis of N-nitrosobutyl-N-(4-hydroxybutyl)amine in rats (4).

Several studies indicate that the chronic abuse of analgesic mixtures containing phenacetin is associated with papillary necrosis of the kidney, and suggest a relationship between papillary necrosis and the subsequent development of transitional-cell carcinoma of the renal pelvis (1). These compounds contain phenacetin with other antiinflammatory drugs (often salicylates or antipyrine (phenazone) and caffeine (2,3).

References

(1) IARC Monographs 13:141-55, 1977.
(2) IARC Monographs 24:135-61, 1980.
(3) IARC Monographs on the Evaluation of the Carcinogenic Risk of Chemicals to Humans, Supplement I. IARC, Lyon, France, p 39, 1979.
(4) See Reference (A-62).
(5) Sax, N.I., Ed., *Dangerous Properties of Industrial Materials Report, 1,* No. 1, 26-27, New York, Van Nostrand Reinhold Co. (1980).

ACETOPHENONE

● Hazardous waste (EPA)

Description: $CH_3COC_6H_5$ is a colorless liquid with a pleasant odor, melting at 19°C and boiling at 202°C.

Code Numbers: CAS 98-86-2   RTECS AM5250000

DOT Designation: −

Synonyms: Acetylbenzene, methyl phenyl ketone and hypnone.

Potential Exposures: Acetophenone is used in perfume manufacture to impart a pleasant jasmine or orange-blossom odor. It is used as a catalyst in olefin polymerization and as a flavorant in tobacco. It is used in the synthesis of pharmaceuticals.

Permissible Exposure Limits in Air: No U.S. standards set. A TLV has been set in the U.S.S.R. at 1 ppm (5 mg/m^3) (A-36).

Permissible Concentration in Water: No criteria set.

Routes of Entry: Inhalation, skin and eye contact.

Harmful Effects and Symptoms: Skin and eye irritation have been observed in test rabbits (1). Acetophenone is a hypnotic in high concentrations and was used as an anesthetic in the 19th century before less toxic substances were found. Acetophenone is highly toxic to aquatic life. It can cause dermatitis in humans.

Points of Attack: Skin, eyes, central nervous system.

Medical Surveillance: Hippuric acid in the urine may be monitored (A-39).

First Aid: Flush eyes, wash contaminated areas of body with soap and water.

Personal Protective Methods: Wear rubber protective clothing.

Respirator Selection: Use chemical cartridge respirator (A-39).

Disposal Method Suggested: Incineration.

References

(1) U.S. Environmental Protection Agency, *Acetophenone,* Health and Environmental Effects Profile No. 3, Washington, D.C., Office of Solid Waste (April 30, 1980).
(2) See Reference (A-60).

ACETYLACETONE

Description: $CH_3COCH_2COCH_3$, colorless liquid boiling at 140.5°C, with pleasant ketone odor.

Code Numbers: CAS 123-54-6   RTECS SA1925000

DOT Designation: —

Synonyms: 2,4-Pentanedione; acetyl-2-propanone.

Potential Exposure: Those engaged in manufacture and handling of gasoline and lubricant additives, pesticides and varnish and ink driers.

Permissible Exposure Limits in Air: No standard set.

Permissible Concentration in Water: No criteria set; BOD effects summarized (A-36).

Routes of Entry: Inhalation and ingestion.

Harmful Effects and Symptoms: Irritant to skin and mucous membranes. Vapor inhalation may cause dizziness, headache, nausea, vomiting and then loss of consciousness.

Target Organs: Skin, mucous membrane, CNS.

First Aid: Remove contaminated clothing, shower, rinse eyes and mouth, give access to fresh air (A-60).

Personal Protective Methods: Use face shield and protective gloves (A-60).

Disposal Method Suggested: Incineration (1) (A-38).

References

(1) Sax, N.I., Ed., *Dangerous Properties of Industrial Materials Report, 1,* No. 7, 24-26, New York, Van Nostrand Reinhold Co. (1981).
(2) See Reference (A-60).

ACETYLAMINOFLUORENE

See "N-2-Fluorenyl Acetamide."

ACETYL BROMIDE

Description: CH_3COBr, colorless fuming liquid boiling at 76.7°C.

Code Numbers: CAS 506-96-7 RTECS AO5955000 UN 1716

DOT Designation: Corrosive liquid (IATA).

Synonyms: Ethanoyl bromide, acetic acid bromide.

Potential Exposure: Those involved in using this chemical in dye manufacture and other organic syntheses.

Incompatibilities: Water, alcohols.

Permissible Exposure Limits in Air: No standard set.

Permissible Concentration in Water: No criteria set.

Routes of Entry: Eye and skin contact, inhalation and ingestion.

Harmful Effects and Symptoms: Inhalation and swallowing are very toxic and acutely irritating to eyes, skin and mucous membranes (A-38).

Target Organs: Skin, respiratory system.

First Aid: Remove contaminated clothing; flush with water (A-60).

Personal Protective Methods: Wear eye protection and bromine-resistant protective clothing (1). Use rubber gloves (A-38).

Respirator Selection: Self-contained breathing apparatus is recommended (1).

Disposal Method Suggested: Slow addition to sodium bicarbonate solution (1) (A-38).

References

(1) Sax, N.I., Ed., *Dangerous Properties of Industrial Materials Report, 1,* No. 8, 29-30, New York, Van Nostrand Reinhold Co. (1981).
(2) See Reference (A-60).

ACETYL CHLORIDE

- Hazardous substance (EPA)
- Hazardous waste (EPA)

Description: CH_3COCl is a colorless fuming liquid with a pungent odor boiling at 51° to 52°C.

Code Numbers: CAS 75-36-5 RTECS AO6390000 UN 1717

DOT Designation: Flammable liquid.

Synonym: Ethanoyl chloride.

Potential Exposures: Acetyl chloride is used in organic synthesis as an acetylating agent, in the pharmaceutical industry (A-41), and in pesticide manufacture (A-32) for example.

Incompatibilities: Water, alcohols, bases.

Permissible Exposure Limits in Air: No standards for acetyl chloride have been reported. However, a ceiling limit of 5 ppm has been reported for hydrogen chloride (the most irritating hydrolysis product of acetyl chloride) in industrial exposures (1).

Permissible Concentration in Water: No criteria set. However, acetyl chloride reacts violently with water. Thus, its half-life in ambient water should be short and exposure from water should be nil. The degradation products should likewise pose no exposure problems if the pH of the water remains stable (1).

Routes of Entry: Inhalation and ingestion.

Harmful Effects and Symptoms: Acetyl chloride is an irritant and a corrosive. Cutaneous exposure results in skin burns, while vapor exposure causes extreme irritation of the eyes and mucous membranes. Inhalation of 2 ppm acetyl chloride has been found irritating to humans. Death or permanent injury may result after short exposures to small quantities of acetyl chloride (1). An aquatic toxicity rating has been estimated to range from 10 to 100 ppm.

Personal Protective Methods: Wear rubber gloves and coveralls (A-39).

Respirator Selection: Use of self-contained breathing apparatus is recommended (A-39).

Disposal Method Suggested: May be mixed slowly with sodium bicarbonate solution and then flushed to sewer with large volumes of water (A-38). May also be incinerated (A-31).

References

(1) U.S. Environmental Protection Agency, *Acetyl Chloride,* Health and Environmental Effects Profile No. 4, Washington, D.C., Office of Solid Waste (April 30, 1980).
(2) Sax, N.I., Ed., *Dangerous Properties of Industrial Materials Report, 1,* No. 8, 30-32, New York, Van Nostrand Reinhold Co. (1981).
(3) Sax, N.I., Ed., *Dangerous Properties of Industrial Materials Report, 3,* No. 3, 35-36, New York, Van Nostrand Reinhold Co. (1983).
(4) See Reference (A-60).

ACETYLENE

Description: HC≡CH, acetylene, is a colorless gas with a faint ethereal odor.

Code Numbers: CAS 74-86-2   RTECS AO9600000   UN 1001

DOT Designation: Flammable gas.

Synonyms: Ethine, ethyne, narcylene.

Potential Exposures: Acetylene can be burned in air or oxygen and is used for brazing, welding, cutting, metallizing, hardening, flame scarfing, and local heating in metallurgy. The flame is also used in the glass industry. Chemically, acetylene is used in the manufacture of vinyl chloride, acrylonitrile, synthetic rubber, vinyl acetate, trichloroethylene, acrylate, butyrolactone, 1,4-butanediol, vinyl alkyl ethers, pyrrolidone, and other substances.

Incompatibilities: Heat and pressure, copper and silver which form explosive acetylides.

Permissible Exposure Limits in Air: No Federal standard has been established. NIOSH has recommended a ceiling limit of 2,500 ppm. ACGIH classifies acetylene as a simple asphyxiant with no TLV value.

Permissible Concentration in Water: No criteria set but EPA (A-37) suggests an ambient water limit of 73,000 μg/ℓ based on health effects.

Route of Entry: Inhalation of gas.

Harmful Effects and Symptoms: *Local* — Acetylene is nonirritating to skin or mucous membranes.

Systemic — At high concentrations pure acetylene may act as a mild narcotic and asphyxiant. Most accounted cases of illness or death can be attributed to acetylene containing impurities of arsine, hydrogen sulfide, phosphine, carbon disulfide, or carbon monoxide.

Initial signs and symptoms of exposure to harmful concentrations of impure acetylene are rapid respiration, air hunger, followed by impaired mental alertness and muscular incoordination. Other manifestations include cyanosis, weak and irregular pulse, nausea, vomiting, prostration, impairment of judgment and sensation, loss of consciousness, convulsions, and death. Low order sensitization of myocardium to epinephrine resulting in ventricular fibrillation may be possible.

Medical Surveillance: No specific considerations are needed.

First Aid: See Reference (1).

Personal Protective Methods: Acetylene poisoning can quite easily be prevented if (1) there is adequate ventilation and (2) impurities are removed when acetylene is used in poorly ventilated areas. General industrial hygiene practices for welding, brazing, and other metallurgical processes should also be observed.

Respirator Selection: See Reference (1).

Disposal Method Suggested: Incineration.

References

(1) National Institute for Occupational Safety and Health, *Criteria for a Recommended Standard: Occupational Exposure to Acetylene,* NIOSH Doc. No. 76-195, Wash., DC (1976).

(2) Sax, N.I., Ed., *Dangerous Properties of Industrial Materials Report, 1,* No. 2, 23-25, New York, Van Nostrand Reinhold Co. (1980).

(3) See Reference (A-60).

(4) Parmeggiani, L., Ed., *Encyclopedia of Occupational Health & Safety,* Third Edition, Vol. 1, pp 40-41, Geneva, International Labour Office (1983).

ACETYLENE TETRABROMIDE

Description: Acetylene tetrabromide, $CHBr_2CHBr_2$, is a colorless to yellow liquid. It melts at -1°C and boils at 239°C.

Code Numbers: CAS 79-27-6   RTECS KI8225000   UN 2504

DOT Designation: ORM-A.

Synonyms: Tetrabromoethane, 1,1,2,2-tetrabromoethane, symmetrical tetrabromoethane.

Potential Exposure: Acetylene tetrabromide is used as a gauge fluid, as a solvent, and as a refractive index liquid in microscopy.

Incompatibilities: Chemically active metals, strong caustics, hot iron, aluminum, zinc in the presence of steam.

Permissible Exposure Limits in Air: The Federal standard and ACGIH 1983/84 value is 1 ppm (15 mg/m^3). STEL proposed is 1.5 ppm (20 mg/m^3). The IDLH value is 10 ppm.

Determination in Air: Silica adsorption followed by THF treatment and gas chromatographic analysis. See NIOSH Methods, Set I. See also reference (A-10).

Permissible Concentration in Water: No criteria set.

Routes of Entry: Inhalation, ingestion, skin or eye contact.

Harmful Effects and Symptoms: Irritation of eyes and nose, anorexia, nausea, severe headaches, abdominal pains, jaundice, monocytosis. See also (A-35).

Points of Attack: Eyes, upper respiratory system, liver.

Medical Surveillance: Consider the points of attack in preplacement and periodic physical examinations.

First Aid: If this chemical gets into the eyes, irrigate immediately. If this chemical contacts the skin, flush with water promptly. If a person breathes in large amounts of this chemical, move the exposed person to fresh air at once and perform artificial respiration. When this chemical has been swallowed, get medical attention. Give large quantities of saltwater and induce vomiting. Do not make an unconscious person vomit.

Personal Protective Methods: Wear appropriate clothing to prevent repeated or prolonged skin contact. Wear eye protection to prevent any reasonable probability of eye contact. Employees should wash promptly when skin is wet or contaminated. Remove nonimpervious clothing promptly if wet or contaminated.

Respirator Selection:
10 ppm: SA/SCBA
Escape: GMOV/SCBA

Disposal Method Suggested: Incineration in admixture with combustible fuel and with scrubber to remove halo acids produced.

References
(1)   See Reference (A-61).
(2)   U.S. Environmental Protection Agency, *Chemical Hazard Information Profile Draft Report: Tetrabromoethane,* Washington, DC (June 14, 1983).

ACETYL PEROXIDE

Description: $CH_3CO-O-O-COCH_3$ is a low melting solid which melts at 30°C.

Code Numbers: CAS 110-22-5   RTECS AP8500000   UN 2084

DOT Designation: —

Synonyms: Diacetylperoxide.

Potential Exposure: Those using this material as a polymerization catalyst for acrylic esters.

Incompatibilities: Organic materials (A-59).

Permissible Exposure Limit in Air: No standard set.

Permissible Concentration in Water: No criteria set.

Routes of Entry: Inhalation, ingestion, skin and eye contact.

Harmful Effects and Symptoms: Very irritating to eye, skin and mucous membranes. Severe corneal damage can result from eye contact (A-59). Some unreported data on mouse tests were suggestive of carcinogenicity (A-63).

Points of Attack: Eyes, skin, respiratory system.

Personal Protective Methods: Use face shield and rubber gloves (A-59). Also use safety shield or hood door in front of apparatus containing acetyl peroxide.

References

(1) See Reference (A-59).

ACETYLSALICYLIC ACID

Description: $HOOCC_6H_4OCOCH_3$ is a white crystalline solid melting at 135°C. It is odorless but hydrolyzes in moist air to give an acetic acid odor.

Code Numbers: CAS 50-78-2 RTECS VO0700000

DOT Designation: —

Synonyms: 2-(Acetyloxybenzoic) acid, aspirin, acetol.

Potential Exposure: Those engaged in manufacture of aspirin or, more likely, in its consumption in widespread use as an analgesic, antipyretic and anti-inflammatory agent.

Permissible Exposure Limits in Air: There are no Federal standards. ACGIH for the first time in 1980 set a TWA value of 5 mg/m^3 but no tentative STEL value.

Permissible Concentration in Water: No criteria set.

Routes of Entry: Primarily oral in medicinal use.

Harmful Effects and Symptoms: Adverse effects from the usual doses of aspirin are infrequent; most common are gastrointestinal disturbances. Prolonged administration of large doses results in occult bleeding and may result in anemia. See (A-34).

References

(1) Sax, N.I., Ed., *Dangerous Properties of Industrial Materials Report, 1,* No. 3, 20-22, New York, Van Nostrand Reinhold Co. (1981). (As Acetol).

ACONITINE

Description: $C_{34}H_{47}NO_{11}$ with the structural formula

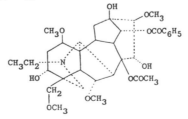

forms hexagonal plates melting at 204°C.

Code Numbers: CAS 302-27-2   RTECS AR5960000

DOT Designation: —

Potential Exposure: Used as an antipyretic and experimentally in producing heart arrythmia.

Permissible Exposure Limits in Air: No standards set.

Permissible Concentration in Water: No criteria set.

Harmful Effects and Symptoms: Very toxic. Can be absorbed through the skin to cause death (1). Ingestion of small quantities can cause blindness.

Personal Protective Methods: Avoid skin contact. Take every precaution to avoid ingestion.

References

(1) Sax, N.I., Ed., *Dangerous Properties of Industrial Materials Report, 1,* No. 3, 22, New York, Van Nostrand Reinhold Co. (1981).

ACRIDINE

Description: $C_{13}H_9N$, acridine, is a colorless or light yellow crystal, very soluble in boiling water. It melts at 110° to 111°C.

Code Numbers: CAS 260-94-6   RTECS AR-7175000   UN 2713

DOT Designation: —

Synonyms: Dibenzopyridine, 10-azaanthracene, 2,3-benzoquinoline.

Potential Exposures: Acridine and its derivatives are widely used in the production of dyestuffs such as acriflavine, benzoflavine, and chrysaniline, and in the synthesis of pharmaceuticals such as aurinacrine, proflavine, and rivanol.

Permissible Exposure Limits in Air: There is no Federal standard for acridine, nor are there ACGIH values. However, EPA (A-37) suggests an ambient air limit of 162 $\mu g/m^3$ based on health effects.

Determination in Air: By fluorometry.

Permissible Concentration in Water: No criteria set but EPA (A-37) suggests an ambient level goal of 800 $\mu g/\ell$ based on health effects.

Route of Entry: Inhalation of vapor.

Harmful Effects and Symptoms: *Local* — Acridine is a severe irritant to the conjunctiva of the eyes, the mucous membranes of the respiratory tract, and the skin. It is a powerful photosensitizer of the skin. Acridine causes sneezing on inhalation.

Systemic — Yellowish discoloration of sclera and conjunctiva may occur. Mutational properties have been ascribed to acridine, but its effect on humans is not known.

Points of Attack: Eyes, skin and respiratory tract.

Medical Surveillance: Evaluate the skin, eyes, and respiratory tract in the course of any placement or periodic examinations.

Personal Protective Methods: Prevent skin, eye, or respiratory contact with protective clothing, gloves, goggles, and appropriate dust respirators. In case of spills or splashes, the skin area should be thoroughly washed and the contaminated clothing changed. Clean work clothing should be supplied on a daily basis, and the worker should shower prior to changing to street clothes.

Respirator Selection: Wear filter mask (2). Fire or heat may dictate use of self-contained breathing apparatus.

Disposal Method Suggested: Incineration with nitrogen oxide removal from the effluent gas by scrubber, catalytic or thermal device.

References

(1) Sawicki, E., and Engel, C.R. 1969. Fluorimetric estimation of acridine in airborne and other particulates. *Mikrochim. Acta.* 1:91.

(2) Sax, N.I., Ed., *Dangerous Properties of Industrial Materials Report, 1,* No. 8, 32-33, New York, Van Nostrand Reinhold Co. (1981).

(3) Parmeggiani, L., Ed., *Encyclopedia of Occupational Health & Safety,* Third Edition, Vol. 1, pp 49, Geneva, International Labour Office (1983).

ACROLEIN

- Hazardous substance (EPA)
- Hazardous waste (EPA)
- Priority toxic pollutant (EPA)

Description: CH_2CHCHO, acrolein, is a clear, yellowish liquid. It has a piercing, disagreeable odor and causes tears. It boils at $53°C$.

Code Numbers: CAS 107-02-8   RTECS AS1050000   UN 1092

Synonyms: Acrylaldehyde, acrylic aldehyde, allyl aldehyde, propenal.

Potential Exposure: Acrolein is primarily used as an intermediate in the production of glycerine and in the production of methionine analogs (poultry feed protein supplements). It is also used in chemical synthesis (1,3,6-hexanetriol and glutaraldehyde), as a liquid fuel, antimicrobial agent, in algae and aquatic weed control, and as a slimicide in paper manufacture. Worker exposure to acrolein has been estimated at 7,550 per year by NIOSH.

Incompatibilities: Oxidizers, acids, alkalies, ammonia.

Permissible Exposure Limits in Air: The Federal standard for exposure to acrolein is 0.1 ppm (0.25 mg/m^3). This is the TWA value as of 1983/84. The STEL value is 0.3 ppm (0.8 mg/m^3). The IDLH value is 5.0 ppm.

Determination in Air: Impingement in sodium bisulfite, workup with TCA and colorimetric analysis based on reaction with 4-hexylresorcinol in the presence of $HgCl_2$ to give a blue color; NIOSH Method, Set C. See also reference (A-10).

Permissible Concentration in Water: To protect freshwater aquatic life— on an acute basis 68 μg/ℓ and on a chronic basis 21 μg/ℓ. To protect saltwater aquatic life—55 μg/ℓ on an acute toxicity basis. To protect human health— 320 μg/ℓ.

Determination in Water: Gas chromatography (EPA Method 603) or gas chromatography and mass spectrometry (EPA Method 624).

Routes of Entry: Inhalation of vapor and percutaneous absorption, ingestion and skin or eye contact.

Harmful Effects and Symptoms: *Local* — In the liquid or pungent vapor form, acrolein produces intense irritation to the eye and mucous membranes of the respiratory tract. Skin burns and dermatitis may result from prolonged or repeated exposure. Sensitization in a few individuals may also occur.

Systemic — Because of acrolein's pungent, offensive odor and the intense irritation of the conjunctiva and upper respiratory tract, severe toxic effects from acute exposure are rare, as workmen will not tolerate the vapor even in minimal concentration. Acute exposure to acrolein may cause bronchial inflammation, resulting in bronchitis or pulmonary edema. Subchronic and chronic exposures have resulted in the development of metaplastic and hyperplastic changes in the trachea and nasal cavities of dogs, monkeys, and hamsters, but acrolein has not given any indication of carcinogenic activity. See (A-59) for further discussion of toxicity and hazards.

Points of Attack: Heart, lungs, eyes, skin, respiratory system.

Medical Surveillance: Preplacement and periodic medical examinations should consider respiratory, skin, and eye disease. Pulmonary function tests may be helpful.

First Aid: If this chemical gets into the eyes, irrigate immediately. If this chemical contacts the skin, flush with water immediately. If a person breathes in large amounts of this chemical, move the exposed person to fresh air at once and perform artificial respiration. When this chemical has been swallowed, get medical attention. Give large quantities of saltwater and induce vomiting. Do not make an unconscious person vomit.

Personal Protective Methods: Wear appropriate clothing to prevent any possible skin contact. Wear eye protection to prevent any possible eye contact. Employees should wash immediately when skin is wet or contaminated. Remove clothing immediately if wet or contaminated to avoid flammability hazard. Provide emergency showers and eyewash.

Respirator Selection:
5 ppm: CCROVF/GMOV/SAF/SCBAF
Escape: GMOVF/SCBAF

Disposal Method Suggested: Incineration. Conditions are 1500°F, 0.5 second minimum for primary combustion; 2000°F, 1.0 second for secondary combustion.

References

(1)   U.S. Environmental Protection Agency, *Chemical Hazard Information Profile: Acrolein,* Washington, DC (March 10, 1978).
(2)   U.S. Environmental Protection Agency, *Acrolein: Ambient Water Quality Criteria,* Washington, DC (1980).
(3)   Nat. Inst. for Occup. Safety and Health, *Information Profiles on Potential Occupational Hazards—Single Chemicals: Acrolein,* Report TR 79-607, Rockville, MD, pp 1-18 (December 1979).
(4)   U.S. Environmental Protection Agency, *Acrolein,* Health and Environmental Effects Profile No. 3, Washington, DC, Office of Solid Waste (April 30, 1980).
(5)   Sax, N.I., Ed., *Dangerous Properties of Industrial Materials Report, 1,* No. 4, 28-31, New York, Van Nostrand Reinhold Co. (1981).
(6)   See Reference (A-61).

(7)   Sax, N.I., Ed., *Dangerous Properties of Industrial Materials Report, 3,* No. 3, 36-41, New York, Van Nostrand Reinhold Co. (1983).

(8)   See Reference (A-60).

(9)   Parmeggiani, L., Ed., *Encyclopedia of Occupational Health & Safety,* Third Edition, Vol. 1, pp 49–51, Geneva, International Labour Office (1983).

(10) United Nations Environment Programme, *IRPTC Legal File 1983,* Vol. I, pp VII/23– 25, Geneva, Switzerland, International Register of Potentially Toxic Chemicals (1984).

ACRYLAMIDE

● Hazardous waste (EPA)

Description: $CH_2CHCONH_2$, acrylamide, in monomeric form consists of flakelike crystals which melt at 84.5°C. It may be stored in a cool, dark place. The monomer readily polymerizes at the melting point or under UV light.

Code Numbers: CAS 79-06-1   RTECS AS 3325000   UN 2074

DOT Designation: —

Synonyms: Propenamide, acrylic amide, acrylamide monomer.

Potential Exposure: The major application for monomeric acrylamide is in the production of polymers as polyacrylamides. Polyacrylamides are used for soil stabilization, gel chromatography, electrophoresis, papermaking strengtheners, clarification and treatment of potable water, and foods. Approximately 70 million pounds of acrylamide were produced in 1974 in the United States. NIOSH estimates that approximately 20,000 workers in the United States are potentially exposed to acrylamide.

Incompatibilities: Strong oxidizers.

Permissible Exposure Limits in Air: NIOSH recommends adherence to the Federal standard of 0.3 mg/m^3 as a time-weighted average concentration for up to a 10-hour workday, 40-hour workweek. ACGIH adds the notation "skin" indicating possibility of cutaneous absorption and sets a proposed STEL of 0.6 mg/m^3. There is no IDLH level set.

Determination in Air: Collection by impinger or fritted bubbler or by charcoal tube followed by gas liquid chromatography (A-1).

Permissible Concentration in Water: No criteria set.

Routes of Entry: Acrylamide can be absorbed through unbroken skin. Inhalation, ingestion and eye and skin contact are other routes of entry.

Harmful Effects and Symptoms: *Local* — Localized effects include peeling and redness of the skin of the hands and less often of the feet, numbness of the lower limbs, and excessive sweating of the feet and hands.

Systemic — The systemic effects due to acrylamide intoxication involve central and peripheral nervous system damage manifested primarily as ataxia, weak or absent reflexes, positive Romberg's sign and loss of vibration and position senses.

Based on laboratory data, EPA has concluded that acrylamide is a potent neurotoxicant at very low levels (2). In humans, the predominant signs of neurotoxicity are related to peripheral nerve involvement and, to a lesser extent, central nervous system involvement. A variety of other signs and symptoms also are generally reported, the most common ones occurring in the skin, hands and feet. The onset of effects is delayed following initial exposure, and the effects may be reversible, although this is not always the case.

The environmental effects of acrylamide have not been evaluated completely. Its high water-solubility, known toxicity to mammals, and possible slow degradation rate under certain environmental conditions (e.g., low temperatures or low oxygen levels) indicate that the compound may pose a hazard upon its release to the environment.

Points of Attack: Central nervous system, peripheral nervous system, skin and eyes.

Medical Surveillance: Since skin contact with the substance may result in localized or systemic effects, NIOSH recommends that medical surveillance be made available to all employees working in an area where acrylamide is stored, produced, processed, or otherwise used, except as an unintentional contaminant in other materials at a concentration of less than 1% by weight.

First Aid: If this chemical gets into the eyes, irrigate immediately. If this chemical contacts the skin, flush with water immediately. If a person breathes in large amounts of this chemical, move the exposed person to fresh air at once and perform artificial respiration. When this chemical has been swallowed, get medical attention. Give large quantities of water and induce vomiting. Do not make an unconscious person vomit.

Personal Protective Methods: Engineering controls should be used wherever feasible to maintain airborne acrylamide concentrations below the prescribed limit, and respirators should be used only in nonroutine or emergency situations which may result in exposure concentrations in excess of the TWA environmental limit. Wear appropriate clothing to prevent repeated or prolonged skin contact. Wear eye protection to prevent any reasonable probability of eye contact. Employees should wash immediately when skin is wet or contaminated. Work clothing should be changed daily if clothing is contaminated. Remove nonimpervious clothing immediately if wet or contaminated. Provide emergency showers.

Respirator Selection:
$15 \, mg/m^3$: SAF/SCBAF
$600 \, mg/m^3$: SAF: PD,PP,CF
Escape: GMOV/SCBA

Disposal Method Suggested: Incineration with provision for scrubbing of nitrogen oxides from flue gases.

References

(1) National Institute for Occupational Safety and Health, *Criteria for a Recommended Standard: Occupational Exposure to Acrylamide,* NIOSH Doc. No. 77-112, Washington, DC (1977).

(2) U.S. Environmental Protection Agency, *Assessment of Testing Needs: Acrylamide,* Report No. EPA-560/11-80-016, Washington, DC, Office of Toxic Substances (July 1980).

(3) Sax, N.I., Ed., *Dangerous Properties of Industrial Materials Report, 2,* No. 4, 24-27, New York, Van Nostrand Reinhold Co. (1982).

(4) See Reference (A-61).

(5) See Reference (A-60).

(6) United Nations Environment Programme, *IRPTC Legal File 1983,* Vol. I, pp VII/26–27, Geneva, Switzerland, International Register of Potentially Toxic Chemicals (1984).

ACRYLIC ACID

- Hazardous waste (EPA)

Description: $CH_2CHCOOH$ is a colorless liquid melting at $12°$ to $14°C$ and boiling at $141°C$. It has an acrid odor.

Code Numbers: CAS 79-10-7   RTECS AS4375000   UN 2218

DOT Designation: Corrosive material.

Synonyms: Propenoic acid, ethylene carboxylic acid, vinylformic acid.

Potential Exposure: Acrylic acid is chiefly used as a monomer in the manufacture of acrylic resins. It is also used as a tackifier and flocculant.

Permissible Exposure Limits in Air: There is no Federal limit but ACGIH as of 1983/84 has proposed a TWA of 10 ppm (30 mg/m^3) but no STEL value.

Permissible Concentration in Water: No U.S. criteria set, but the U.S.S.R. has reportedly set 0.5 mg/ℓ as a limit in drinking water (A-36).

Routes of Entry: Inhalation, skin and eye contact, ingestion.

Harmful Effects and Symptoms: In animal experiments, vapors have caused nasal and eye irritation; direct contact with the liquid has caused skin burns and blindness (A-34). Medical reports of acute human exposures include moderate and severe skin burns, moderate eye burns and mild inhalation effects.

Points of Attack: Skin, eyes, respiratory system.

Medical Surveillance: Consider the points of attack in preplacement and regular physical examinations.

First Aid: Irrigate eyes with water. Wash contaminated areas of body with soap and water.

Personal Protective Methods: Wear chemical goggles, rubber protective clothing and a face shield (A-38).

Respirator Selection: Use of a self-contained breathing apparatus is recommended (A-38).

Disposal Method Suggested: Incineration.

References

(1) Sax, N.I., Ed., *Dangerous Properties of Industrial Materials Report, 1,* No. 7, 26-28, New York, Van Nostrand Reinhold Co. (1981).
(2) See Reference (A-60).
(3) Parmeggiani, L., Ed., *Encyclopedia of Occupational Health & Safety,* Third Edition, Vol. 1, pp 52-54, Geneva, International Labour Office (1983).

ACRYLONITRILE

- Carcinogen (EPA-CAG), (Probable, IARC) (9) (A-62) (A-64)
- Hazardous substance (EPA)
- Hazardous waste (EPA)
- Priority toxic pollutant (EPA)

Description: CH_2CHCN, acrylonitrile, is a colorless liquid with a faint acrid odor. It is both flammable and explosive.

Code Numbers: CAS 107-13-1   RTECS AT5250000   UN 1093

DOT Designation: Flammable liquid and poison.

Synonyms: Vinyl cyanide, cyanoethylene, propene nitrile, AN.

Potential Exposure: Acrylonitrile is used in the manufacture of synthetic fibers, acrylostyrene plastics, acrylonitrile-butadiene-styrene plastics, nitrile rubbers, chemicals, and adhesives. It is also used as a pesticide. NIOSH estimates that approximately 125,000 persons are potentially exposed to acrylonitrile in the workplace.

Incompatibilities: Strong oxidizers (especially bromine), strong bases, copper, copper alloys, ammonia, amines (A-59).

Permissible Exposure Limits in Air: An emergency temporary standard (1) set the TWA at 2 ppm, down from 20 ppm in a previous determination (2). The economic impact of these standards has been assessed (3) and the ETS was later made permanent (4). The ACGIH as of 1983/84 (A-6) has set a TLV of 2 ppm (4.5 mg/m^3) with the notation that acrylonitrile is a human carcinogen, but the IDLH level is only 4 ppm. The notation "skin" is added by ACGIH indicating the possibility of cutaneous absorption.

Determination in Air: Charcoal adsorption followed by methanol extraction and gas chromatographic analysis. See NIOSH Methods, Set K. See also references (A-10) and (2).

Permissible Concentration in Water: To protect freshwater aquatic life— on an acute basis, 7,550 μg/ℓ and on a chronic basis, 2,600 μg/ℓ over 30 days. To protect saltwater aquatic life—insufficient data to yield a value. To protect human health—preferably zero. Water concentration should be below 0.58 μg/ℓ to keep lifetime cancer risk below 10^{-5}.

Determination in Water: By gas chromatography (EPA Method 603) or gas chromatography plus mass spectrometry (EPA Method 624).

Routes of Entry: Inhalation and percutaneous absorption. It may be absorbed from contaminated rubber or leather. Routes include ingestion and eye and skin contact.

Harmful Effects and Symptoms: *Local* — Acrylonitrile may cause irritation of the eyes. Repeated and prolonged exposure may produce skin irritation. When acrylonitrile is held in contact with the skin (e.g., after being absorbed into shoe leather or clothing), it may produce blistering after several hours of no apparent effect. Unless the contaminated clothing is removed promptly and the area washed off, blistering will occur.

Systemic — Acrylonitrile exposure may produce nausea, vomiting, headache, sneezing, weakness, and light-headedness. Exposure to high concentrations may produce profound weakness, asphyxia, and death.

Points of Attack: Cardiovascular system, liver, kidneys, central nervous system, skin.

Medical Surveillance: Consider the skin, respiratory tract, heart, central nervous system, renal and liver function in placement and periodic examinations. A history of fainting spells or convulsive disorders might present an added risk to persons working with toxic nitriles. Diagnostic tests suggested (A-39) include determination of cyanide in blood and cyanomethemoglobin in blood and increased excretion of thiocyanate in urine.

First Aid: If this chemical gets into the eyes, irrigate immediately. If this chemical contacts the skin, flush with water immediately. If a person breathes in large amounts of this chemical, move the exposed person to fresh air at once and perform artificial respiration. When this chemical has been swallowed, get medical attention. Give large quantities of water and induce vomiting. Do not make an unconscious person vomit.

Personal Protective Methods: Wear appropriate clothing to prevent repeated or prolonged skin contact. Leather should not be used in protective clothing since it is readily penetrated by acrylonitrile. Rubber clothing should be frequently washed and inspected because it will soften and swell. Wear eye protection to prevent any reasonable probability of eye contact. Employees should wash immediately when skin is wet or contaminated. Remove clothing immediately if wet or contaminated to avoid flammability hazard. Provide emergency showers.

Respirator Selection: Above 4 ppm only SCBAF: PD,PP,CF or SAF: PD, PP,CF with auxiliary SCBA: PD,PP should be used.

Disposal Method Suggested: Incineration with provision for NO_x removal from effluent gases by scrubbers or afterburners (A-31). A chemical disposal method has also been suggested (A-38) involving treatment with alcoholic NaOH; the alcohol is evaporated and calcium hypochlorite added; after 24 hours the product is flushed to the sewer with large volumes of water. Recovery of acrylonitrile from acrylonitrile process effluents is an alternative to disposal (A-58).

References

(1) *Federal Register,* 43, No. 11, 2586-2621 (January 17, 1978).
(2) National Institute for Occupational Safety and Health, *Criteria for a Recommended Standard: Occupational Exposure to Acrylonitrile,* NIOSH Doc. No. 78-116, Washington, DC (1978).
(3) Department of Labor, *Economic Impact Assessment for Acrylonitrile,* Washington, DC, Occupational Safety and Health Administration (February 21, 1978).
(4) *Federal Register,* 43, No. 192, 45762-45819 (October 3, 1978).
(5) U.S. Environmental Protection Agency, *Status Assessment of Toxic Chemicals: Acrylonitrile,* Report EPA-600/2-79-210A, Washington, DC (December 1979).
(6) U.S. Environmental Protection Agency, *Acrylonitrile: Ambient Water Quality Criteria,* Washington, DC (1980).
(7) U.S. Environmental Protection Agency, *Investigation of Selected Potential Environmental Contaminants: Acrylonitrile,* Report EPA-560/2-78-003, Washington, DC (May 1978).
(8) U.S. Environmental Protection Agency, *Acrylonitrile,* Health and Environmental Effects Profile No. 7, Washington, DC, Office of Solid Water (April 30, 1980).
(9) International Agency for Research on Cancer, *IARC Monographs on the Carcinogenic Risks of Chemicals to Humans,* 19, 73, Lyon, France (1979).
(10) Sax, N.I., Ed., *Dangerous Properties of Industrial Materials Report, 1,* No. 2, 25-27, New York, Van Nostrand Reinhold Co. (1980).
(11) See Reference (A-61).
(12) Sax, N.I., Ed., *Dangerous Properties of Industrial Materials Report, 3,* No. 3, 41-46, New York, Van Nostrand Reinhold Co. (1983).
(13) See Reference (A-63).
(14) See Reference (A-60).
(15) Parmeggiani, L., Ed., *Encyclopedia of Occupational Health & Safety,* Third Edition, Vol. 1, pp 55–56, Geneva, International Labour Office (1983).
(16) United Nations Environment Programme, *IRPTC Legal File 1983,* Vol. I, pp VII/30–33, Geneva, Switzerland, International Register of Potentially Toxic Chemicals (1984).

ACTINOMYCIN D

● Carcinogen (Animal Positive, IARC) (3)

Description: $C_{62}H_{86}N_{12}O_{16}$ has the structural formula

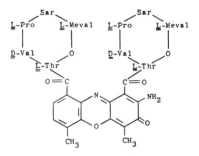

The trihydrate forms bright red crystals melting at 241.5° to 243°C.

Code Numbers: CAS 50-76-0   RTECS AU1575000

DOT Designation: –

Synonyms: Dactinomycin, Meractinomycin.

Potential Exposure: Used as anticancer drug.

Permissible Exposure Limits in Air: No standards set.

Permissible Concentration in Water: No criteria set.

Harmful Effects and Symptoms: Highly toxic. Is experimental carcinogen and teratogen in animals (1).

References
(1)   Sax, N.I., Ed., *Dangerous Properties of Industrial Materials Report, 1,* No. 3, 23-24, New York, Van Nostrand Reinhold Co. (1981).
(2)   See Reference (A-63).
(3)   International Agency for Research on Cancer, *IARC Monographs on the Carcinogenic Risks of Chemicals to Humans, 10,* 29, Lyon, France (1976).

ADIPATE ESTER PLASTICIZERS

Description: Adipate ester plasticizers have the general formula

$$ROOC-(CH_2)_4-COOR$$

They are nonvolatile oily liquids or low melting solids. The most widely used is di-(2-ethylhexyl) adipate which accounts for about 67% of the market. Next in importance is n-octyl, n-decyl adipate which accounts for 13% of the market.

Code Numbers: Di-2(ethylhexyl) adipate   CAS 103-23-1
RTECS AU9700000

DOT Designation: –

Synonyms: Dialkyl adipates.

Potential Exposures: Plasticizer production and compounding of vinyl resins are the primary areas. Compounding of cellulose ester plastics and synthetic elastomers are areas of lesser importance. Contact with fabricated polymer products containing these materials.

Permissible Exposure Limits in Air: No standard exists.

Permissible Concentration in Water: No criteria set.

Routes of Entry: Inhalation, eye and skin contact.

Harmful Effects and Symptoms: Low to moderate acute toxicity. Deleterious embryotoxic and teratogenic effects but significantly less than with similar doses of phthalate ester analogs.

Target Organs: Respiratory system.

Disposal Method Suggested: Incineration.

References
(1) U.S. Environmental Protection Agency, *Chemical Hazard Information Profile: Adipate Ester Plasticizers,* Washington, DC (January 5, 1978).

ADIPIC ACID

Description: $HOOC(CH_2)_4COOH$, white crystals melting at 152°C.

Code Numbers: CAS 124-04-9 RTECS AU8400000

DOT Designation: —

Synonyms: Hexanedioic acid, 1,4 butane dicarboxylic acid.

Potential Exposures: Workers in manufacture of nylon, plasticizers, urethanes, adhesives and food additives.

Permissible Exposure Limits in Air: No standards set.

Permissible Concentration in Water: No criteria set. Effects summarized (A-36).

Harmful Effects and Symptoms: Inhalation can cause burns to respiratory tract (1).

Personal Protective Methods: Wear safety goggles, rubber gloves (A-60) as well as coveralls (A-38).

Respirator Selection: Use self-contained breathing apparatus (A-38).

Disposal Method Suggested: Incineration (A-38).

References
(1) Sax, N.I., Ed., *Dangerous Properties of Industrial Materials Report, 1,* No. 7, 28-29, New York, Van Nostrand Reinhold Co. (1981).
(2) Sax, N.I., Ed., *Dangerous Properties of Industrial Materials Report, 3,* No. 3, 46-49, New York, Van Nostrand Reinhold Co. (1983).

ADRIAMYCIN

● Carcinogen (Animal Positive) (A-63)

Description: $C_{27}H_{29}O_{11}N$ has the structural formula

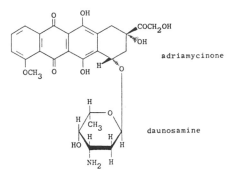

It forms red crystals which melt at 205°C.

Code Numbers: CAS 23214-92-8   RTECS AV9800000

DOT Designation: —

Synonyms: Doxorubicin, 14-hydroxydaunomycin, adriablastine.

Potential Exposure: Used as an anticancer drug.

Permissible Exposure Limits in Air: No standards set.

Permissible Concentration in Water: No criteria set.

Harmful Effects and Symptoms: Highly toxic. Causes baldness, stomatitis and bone marrow aplasia in humans (1). Fatal human cardiac disturbances have been reported.

Target Organs: Cardiovascular system.

References

(1)   Sax, N.I., Ed., *Dangerous Properties of Industrial Materials Report, 1,* No. 3, 24-25, New York, Van Nostrand Reinhold Co. (1981).

AFLATOXINS

- Carcinogens (Animal, Positive, IARC) (A-62, A-64)
- Hazardous Waste (EPA)

Description: $C_{17}H_{12}O_6$, $C_{17}H_{14}O_6$ and $C_{17}H_{12}O_7$ with a typical structural formula of

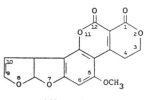

Aflatoxin G_1

form crystals melting at 268° to 269°C (B-1) and 237° to 240°C (G-2).

Code Numbers: CAS 1162-65-9 (B-1) and 7241-98-7 (G-2)
RTECS GY1925000 (B-1) and LV1700000 (G-2).

DOT Designation: —

Potential Exposure: Aflatoxins are a group of toxic metabolites produced by certain types of fungi. Aflatoxins are not commercially manufactured; they are naturally occurring contaminants that are formed by fungi on food during conditions of high temperatures and high humidity.

Human exposure to aflatoxins occurs through ingestion of contaminated food. The estimated amount of aflatoxins that Americans consume daily is estimated to be from 0.15 to 0.50 μg. Grains, peanuts, tree nuts, and cottonseed meal are among the more common foods on which these fungi grow. Meat, eggs, milk, and other edible products from animals that consume aflatoxin-contaminated feed may also contain aflatoxins.

Permissible Exposure Limits in Air: No standards set.

Permissible Concentration in Water: No criteria set.

Routes of Entry: As unavoidable contaminant in foods. The FDA limits the levels of aflatoxin contamination that are permitted in food. The complete elimination of aflatoxin contamination of food is probably not technically feasible. The FDA has lowered the maximum amount allowed in food products as methods of detection and methods of control have improved. Upper limits of 20 ppb (total B_1, B_2, G_1, and G_2) in foods and feeds and 0.5 ppb (M_1) in milk are now in effect.

Harmful Effects and Symptoms: Aflatoxins are carcinogenic in mice, rats, fish, ducks, marmosets, tree shrews and monkeys by several routes of administration (including oral), producing mainly cancers of the liver, colon and kidney.

Epidemiological studies have shown a positive correlation between the average dietary concentrations of aflatoxins in populations and the incidence of primary liver cancer. These studies were undertaken to test this specific hypothesis; however, no studies have been carried out which could link an increased risk of liver cancer to actual aflatoxin intake in individuals (1).

References

(1) See Reference (A-62), also Reference (A-64).
(2) Sax, N.I., Ed., *Dangerous Properties of Industrial Materials Report, 1,* No. 4, 31-33, New York, Van Nostrand Reinhold Co. (1981).
(3) United Nations Environment Programme, *IRPTC Legal File 1983,* Vol. 1, pp VII/34, 243-8; Vol. II, pp VII/381-2, Geneva, Switzerland, International Register of Potentially Toxic Chemicals (1984).

ALACHLOR

Description: Alachlor

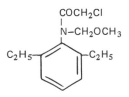

is a crystalline solid melting at 40° to 41°C.

Code Numbers: CAS 15972-60-8   RTECS AE1225000   UN 2588

DOT Designation: —

Synonyms: Lasso®, 2-chloro-2',6'-diethyl-N-(methoxymethyl)acetanilide.

Potential Exposure: In manufacture (A-32), formulation and application of this preemergence herbicide, personnel may be exposed.

Permissible Exposure Limits in Air: No standards set.

Permissible Concentration in Water: No adverse effect level in drinking water has been calculated by NAS/NRC (A-2) as 0.7 mg/ℓ. Allowable daily intake (ADI) has been calculated at 0.1 mg/kg/day.

Harmful Effects and Symptoms: Alachlor is generally well tolerated in animals according to NAS/NRC in a discussion of acute effects (A-2).

Points of Attack: Unknown.

Disposal Method Suggested: This compound is hydrolyzed under strongly acid or alkaline conditions, to chloroacetic acid, methanol, formaldehyde and 2,6-diethylaniline. Incineration is recommended as a disposal procedure (A-32).

References

(1)   United Nations Environment Programme, *IRPTC Legal File 1983*, Vol. I, pp VII/3, Geneva, Switzerland, International Register of Potentially Toxic Chemicals (1984).

ALDICARB

Description: Aldicarb

$$CH_3S-\overset{\overset{\displaystyle CH_3}{|}}{\underset{\underset{\displaystyle CH_3}{|}}{C}}-CH=NO\overset{\overset{\displaystyle O}{\|}}{C}NHCH_3$$

is a systemic carbamate insecticide which forms colorless crystals melting at 98° to 100°C.

Code Numbers: CAS 116-06-3   RTECS VE-2275000   UN 2588

DOT Designation: —

Synonyms: Temik®, [2-methyl-2(methyl-thio)propionaldehyde O-(methyl carbamoyl)oxime].

Potential Exposures: Personnel involved in manufacture (A-32), formulation or application of this insecticide to crops.

Permissible Exposure Limits in Air: No standards set.

Permissible Concentration in Water: A suggested no-adverse effect level in drinking water has been calculated by NAS/NRC (A-2) as 0.007 mg/ℓ. An acceptable daily intake (ADI) of 0.001 mg/kg/day has been calculated.

Determination in Water: Aldicarb may be determined in water by gas-liquid chromatography with flame photometric detection after oxidation to the sulfone (aldoxycarb) by peracetic acid or 3-chloro-perbenzoic acid. Colorimetric methods have also been used based on hydrolysis to hydroxylamine which is oxidized to nitrous acid, the latter used to diazotize sulfanilic acid which is then coupled to give a dye (A-7).

Harmful Effects and Symptoms: As with most carbamate insecticides, aldicarb is metabolized both by oxidative and hydrolytic processes. Oxidation results in compounds which are also cholinesterase inhibitors which hydrolysis gives non-toxic products. Inhibition of cholinesterase by aldicarb appears to be rapidly reversible in man. The acute toxicity of aldicarb is probably the highest of any widely used pesticide, however, according to NAS/NRC (A-2).

References

(1)   Sax, N.I., Ed., *Dangerous Properties of Industrial Materials Report, 4,* No. 2, 37–41, New York, Van Nostrand Reinhold Co. (1984).

(2)   United Nations Environment Programme, *IRPTC Legal File 1983,* Vol. II, pp VII/685–7, Geneva, Switzerland, International Register of Potentially Toxic Chemicals (1984).

ALDRIN

- Carcinogen (EPA-CAG, NCI) (3)
- Hazardous substance (EPA)
- Hazardous waste (EPA)
- Priority toxic pollutant (EPA)

Description: Aldrin, $C_{12}H_8Cl_6$, has the structural formula

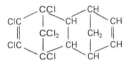

It is a colorless, crystalline solid, melting at 104°C. The technical grade is a tan to dark brown solid, melting at 49° to 60°C.

Code Numbers: CAS 309-00-2   RTECS IO2100000   UN 1542

DOT Designation: Poison B.

Synonyms: HHDN, octalene.

Potential Exposure: Pesticide manufacturers (A-32), formulators and applicators. The people in the U.S. are exposed to aldrin (and dieldrin) in air, water and food because of its persistence in the environment. See dieldrin for more details.

Incompatibilities: None.

Permissible Exposure Limits in Air: The TWA is 0.25 mg/m³. ACGIH as of 1983/84 adds the notation "Skin" indicating possibility of cutaneous absorption and suggests a STEL of 0.75 mg/m³. NIOSH has recommended, as of October 1978, that aldrin be held to the lowest reliably detectable level in air which is 0.15 mg/m³ on a time-weighted average basis. The IDLH level is 100 mg/m³.

Determination in Air: A filter plus bubbler containing isooctane followed by workup with isooctane and analysis by gas chromatography. See NIOSH Methods, Set S. See also reference (A-10).

Permissible Concentration in Water: To protect freshwater aquatic life— not to exceed 3.0 μg/ℓ at any time. To protect saltwater aquatic life—not to exceed 1.3 μg/ℓ at any time. To protect human health—preferably zero. A limit of 0.00074 μg/ℓ is believed to keep lifetime cancer risk below 10⁻⁵.

Determination in Water: Gas chromatography (EPA Method 608) or gas chromatography plus mass spectrometry (EPA Method 625).

Routes of Entry: Inhalation, skin absorption, ingestion and eye and skin contact.

Harmful Effects and Symptoms: Headaches, dizziness, nausea, vomiting, malaise, myoclonic jerks of limbs, clonic, tonic convulsions, coma, hematuria, azotemia. Aldrin and dieldrin have been the subject of litigation bearing upon the contention that these substances cause severe aquatic environmental change and are potential carcinogens. In 1970, the U.S. Department of Agriculture cancelled all registrations of these pesticides based upon a concern to limit dispersal in or on aquatic areas. In 1972, under the authority of the Fungicide, Insecticide, Rodenticide Act as amended by the Federal Pesticide Control Act of 1972, USCS Section 135, et. seq., an EPA order lifted cancellation of all registered aldrin and dieldrin for use in deep ground insertions for termite control, nursery clipping of roots and tops of nonfood plants, and mothproofing of woolen textiles and carpets where there is no effluent discharge.

In 1974, cancellation proceedings disclosed the severe hazard to human health and suspension of registration of aldrin and dieldrin use was ordered; production was restricted for all pesticide products containing aldrin or dieldrin. However, formulated products containing aldrin and dieldrin are imported from Europe each year solely for subsurface soil injection for termite control. Therefore, limits that protect all receiving water uses must be placed on aldrin and dieldrin. The litigation has produced the evidentiary basis for the Administrator's conclusions that aldrin/dieldrin are carcinogenic in mice and rats, approved the EPA's extrapolation to humans of data derived from tests on animals and affirmed the conclusions that aldrin and dieldrin pose a substantial risk of cancer to humans, which constitutes an "imminent hazard" to man (A-33).

Points of Attack: Central nervous system, liver, kidneys, skin.

Medical Surveillance: Consider the points of attack in preplacement and periodic physical examinations.

First Aid: If this chemical gets into the eyes, irrigate immediately. If this chemical contacts the skin, wash with soap immediately. If a person breathes in large amounts of this chemical, move the exposed person to fresh air at once and perform artificial respiration. When this chemical has been swallowed, get medical attention. Give large quantities of water and induce vomiting. Do not make an unconscious person vomit.

Personal Protective Methods: Wear appropriate clothing to prevent any possibility of skin contact. Wear eye protection to prevent any reasonable probability of eye contact. Employees should wash immediately when skin is wet or contaminated. Work clothing should be changed daily if it is possible that clothing is contaminated. Remove nonimpervious clothing immediately if wet or contaminated. Provide emergency showers.

Respirator Selection:
>   2.5 mg/m^3: CCROVDMPest/SA/SCBA
>   12.5 mg/m^3: CCROVFDMPest/GMOVFDMPest/SAF/SCBAF
>   100 mg/m^3: SA: PD,PP,CF/CCROVHiEPest
>   Escape: GMOVPPest/SCBA

Disposal Method Suggested: Aldrin is very stable thermally with no decomposition noted at 250°C. Aldrin (along with the structurally related compounds dieldrin and isodrin) is remarkably stable to alkali (in contrast to chlordane and heptachlor) and refluxing with aqueous or alcoholic caustic has no effect.

Incineration methods for aldrin disposal involving 1500°F, 0.5 second minimum for primary combustion, 3200°F, 1.0 second for secondary combustion, with adequate scrubbing and ash disposal facilities have been recommended (A-31). The combustion of aldrin in polyethylene on a small scale gave more than 99% decompostion. Aldrin can be degraded by active metals such as sodium in alcohol (a reaction which forms the basis of the analytical method for total chlorine), but this method is not suitable for the layman.

A disposal method suggested for materials contaminated with aldrin, dieldrin or endrin consists of burying 8 to 12 feet underground in an isolated area away from water supplies, with a layer of clay, a layer of lye and a second layer of clay beneath the wastes (A-32).

References

(1) U.S. Environmental Protection Agency, *Aldrin/Dieldrin: Ambient Water Quality Criteria,* Washington, DC (1979).

(2) U.S. Environmental Protection Agency, *Aldrin,* Health and Environmental Effects Profile No. 8, Washington, DC, Office of Solid Waste (April 30, 1980).

(3) National Cancer Institute, *Bioassays of Aldrin and Dieldrin for Possible Carcinogenicity,* Technical Report Series No. 21, Bethesda, Maryland (1978).

(4) Sax, N.I., Ed., *Dangerous Properties of Industrial Materials Report, 1,* No. 5, 31-32, New York, Van Nostrand Reinhold Co. (1981).

(5) Sax, N.I., Ed., *Dangerous Properties of Industrial Materials Report, 3,* No. 5, 25-29, New York, Van Nostrand Reinhold Co. (1983).

(6) United Nations Environment Programme, *IRPTC Legal File 1983,* Vol. I, pp VII/279-82, Geneva, Switzerland, International Register of Potentially Toxic Chemicals (1984).

ALKANES (C$_5$-C$_8$)

Description: The alkanes have the formula C$_n$H$_{2n+2}$ and are colorless, flammable liquids, such as pentane, C$_5$H$_{12}$, boiling at 36.1°C and octane, C$_8$H$_{18}$, boiling at 125.6°C.

Synonyms: Paraffins, paraffin hydrocarbons. See also entries under hexane, heptane, octane and pentane.

Harmful Effects and Symptoms: Aliphatic hydrocarbons are asphyxiants and central nervous system depressants. Lower members of the series, methane and ethane, are pharmacologically less active than higher members of the series, their main hazards resulting from the simple displacement of oxygen and from fire and from explosion. Higher members of the series cause narcosis. At least one member (hexane) has neurotoxic properties. Another common effect is irritation of the skin and mucous membranes of the upper respiratory tract. Repeated and prolonged skin contact may result in dermatitis, due to the defatting of skin. Due to its low viscosity, aspiration of liquid may result in diffuse chemical pneumonitis, pulmonary edema, and hemorrhage. Contamination of aliphatic hydrocarbons by benzene significantly increases the hazard. Therefore, it is important that benzene content, if suspected, be determined (A-5).

Reference

(1) National Institute for Occupational Safety and Health. *Criteria for a Recommended Standard: Occupational Exposure to Alkanes.* NIOSH Doc. No. 77-151, Wash., DC (1977).

ALKYL PHTHALATES

See individual entries under:
"Dibutyl Phthalate," "Dimethyl Phthalate," "Di-sec-Octyl Phthalate."

References

(1) U.S. Environmental Protection Agency, *A Study of Industrial Data on Candidate Chemicals for Testing,* (Alkyl Phthalates and Cresols), Report No. EPA-560/5-78-002, Washington, DC (June 1978).

(2) National Institute for Occupational Safety and Health, *Profiles on Occupational Hazards for Criteria Document Priorities: Phthalates,* pp 97-103, Report PB-274,073, Cincinnati, OH (1977).

ALLOXAN

Description: $C_4H_2N_2O_4$, with the structural formula

forms crystals which melt at 230°C and decompose at 256°C.

Code Numbers: CAS 50-71-5   RTECS BA4200000

DOT Designation: —

Synonyms: Mesoxalylurea, mesoxalylcarbamide, 2,4,5,6-tetraoxohexahydropyrimidine.

Potential Exposure: Used as an anticancer drug; used in organic synthesis; used to reduce diabetes in experimental animals.

Permissible Exposure Limits in Air: No standards set.

Permissible Concentration in Water: No criteria set.

Harmful Effects and Symptoms: Highly toxic to experimental rats, mice, rabbits and birds (1).

References

(1) Sax, N.I., Ed., *Dangerous Properties of Industrial Materials Report, 1,* No. 4, 33-34, New York, Van Nostrand Reinhold Co. (1981).

ALLYL ALCOHOL

● Hazardous substance (EPA)
● Hazardous waste (EPA)

Description: CH_2CHCH_2OH, allyl alcohol, is a colorless liquid with a pungent odor. The odor and irritant properties of allyl alcohol should be sufficient warning to prevent serious injury. It boils at 96° to 97°C.

Code Numbers: CAS 107-18-6   RTECS BA5075000   UN 1098

DOT Designation: Flammable liquid and poison.

Synonyms: Vinyl carbinol, propenyl alcohol, 2-propenol-1, propenol-3.

Potential Exposure: Allyl alcohol is used in the production of allyl esters. These compounds are used as monomers and prepolymers in the manufacture of resins and plastics. Allyl alcohol is also used in the preparation of pharmaceuticals, in organic syntheses of glycerol and acrolein and as a fungicide and herbicide.

Incompatibilities: Strong oxidizers.

Permissible Exposure Limits in Air: The Federal standard (TWA) is 2 ppm (5 mg/m^3). ACGIH adds the notation "skin" indicating possibility of cutaneous absorption. The tentative STEL is 4 ppm (10 mg/m^3). The IDLH value is 150 ppm.

Determination in Air: Adsorption on charcoal, workup with CS_2 and gas chromatographic analysis. See NIOSH Methods, Set E. See also reference (A-10).

Permissible Concentration in Water: No criteria set.

Routes of Entry: Inhalation of vapor, percutaneous absorption of liquid and ingestion.

Harmful Effects and Symptoms: *Local* — Liquid and vapor are highly irritating to eyes and upper respiratory tract. Skin irritation and burns have occurred from contact with liquid but are usually delayed in onset and may be prolonged. *Systemic* — Local muscle spasms occur at sites of percutaneous absorption. Pulmonary edema, liver and kidney damage, diarrhea, delirium, convulsions, and death have been observed in laboratory animals, but have not been reported in man.

Points of Attack: Eyes, skin, respiratory system, lungs.

Medical Surveillance: Preplacement and periodic examinations should include the eyes, skin, respiratory tract, and liver and kidney function.

First Aid: If this chemical gets into the eyes, irrigate immediately. If this chemical contacts the skin, flush with water immediately. If a person breathes in large amounts of this chemical, move the exposed person to fresh air at once and perform artificial respiration. When this chemical has been swallowed, get medical attention. Give large quantities of saltwater and induce vomiting. Do not make an unconscious person vomit.

Personal Protective Methods: Wear appropriate clothing (neoprene) to prevent any possible skin contact with liquid or repeated or prolonged contact with vapor >25 ppm. Wear eye protection to prevent any reasonable probability of eye contact. Employees should wash immediately when skin is wet or contaminated. Remove clothing immediately if wet or contaminated to avoid flammability hazard. Provide emergency showers.

Respirator Selection:
100 ppm: CCROVF/GMOV/SAF/SCBAF
150 ppm: SAF: PD,PP,CF
Escape: GMOV/SCBA

Disposal Method Suggested: Incineration after dilution with a flammable solvent.

References

(1)   U.S. Environmental Protection Agency, *Allyl Alcohol*, Health and Environmental Effects Profile No. 9, Washington, DC, Office of Solid Waste (April 30, 1980).
(2)   Sax, N.I., Ed., *Dangerous Properties of Industrial Materials Report, 1*, No. 7, 29-31, New York, Van Nostrand Reinhold Co. (1981).
(3)   See Reference (A-61).
(4)   See Reference (A-60).
(5)   United Nations Environment Programme, *IRPTC Legal File 1983*, Vol. I, pp VII/37– 38, Geneva, Switzerland, International Register of Potentially Toxic Chemicals (1984).

ALLYL AMINE

Description: $CH_2=CHCH_2NH_2$, a liquid with a strong ammonia odor boiling at $55°$ to $58°C$.

Code Numbers: CAS 107-11-9   RTECS BA5425000   UN 2334

DOT Designation: Flammable liquid, poison.

Synonyms: 3-aminopropylene, 2-propen-1-amine, 2-propenylamine.

Potential Exposure: Used in manufacture of pharmaceuticals (mercurial diuretics, e.g.) and in organic synthesis. Used to improve dyeability of acrylic fibers.

Incompatibilities: Oxidizing materials, acids.

Permissible Exposure Limits in Air: No standards set.

Permissible Concentration in Water: No criteria set. Biological effects reviewed (A-36).

Routes of Entry: Inhalation and ingestion.

Harmful Effects and Symptoms: Strong irritant to eyes, mucous membranes and skin. Highly toxic via oral inhalation and dermal routes. Extended exposure produces irregular respiration, cyanosis, excitement, convulsions and death (1).

Points of Attack: Skin, pulmonary system, cardiovascular system.

First Aid: Get to fresh air, remove contaminated clothing, flush with water (A-60).

Personal Protective Methods: Butyl rubber gloves, coveralls and face shield.

Respirator Selection: Multipurpose gas mask required.

Disposal Method Suggested: High temperature incineration; encapsulation by resin or silicate fixation (1).

References

(1)   Sax, N.I., Ed., *Dangerous Properties of Industrial Materials Report, 2*, No. 6, 28-30, New York, Van Nostrand Reinhold Co. (1982).
(2)   See Reference (A-60).

ALLYL BROMIDE

Description: $CH_2=CHCH_2Br$ is a colorless liquid boiling at $71°C$.

Code Numbers: CAS 106-95-6   RTECS UC7090000   UN1099

DOT Designation: —

Synonyms: 3-Bromopropene, bromallylene.

Potential Exposure: Used in synthetic perfume and synthetic resin manufacture.

Incompatibilities: Oxidants.

Permissible Exposure Limits in Air: No standards set.

Permissible Concentration in Water: No criteria set.

Routes of Entry: Inhalation, ingestion, skin contact.

Harmful Effects and Symptoms: Corrosive to eyes, skin and respiratory tract (A-44). Liver and kidney injury may occur and serious cases may be fatal (A-60).

Points of Attack: Eyes, skin, respiratory system, liver, kidneys.

First Aid: Flush eyes and body areas with water.

Personal Protective Methods: Wear protective clothing, gloves and face shield or goggles (A-60).

Respirator Selection: Use chemical cartridge respirator.

Disposal Method Suggested: Incineration (A-38).

References
(1)   See Reference (A-60).

ALLYL CHLORIDE

● Hazardous substance (EPA)

Description: Allyl chloride, CH_2CHCH_2Cl, is a highly-reactive liquid halogenated hydrocarbon (boiling at 44° to 45°C) with an unpleasant, pungent odor.

Code Numbers: CAS 107-05-1   RTECS UC7350000   UN 1100

DOT Designation: Flammable liquid.

Synonyms: 1-Chloro-2-propene, 3-chloropropylene, chlorallylene, α-chloropropylene.

Potential Exposure: Allyl chloride is utilized primarily in the manufacture of epichlorohydrin and glycerol. Potential exposure extends to no more than 5,000 workers.

Incompatibilities: Strong oxidizers, acids, aluminum, zinc, amines, peroxides, chlorides of iron and aluminum.

Permissible Exposure Limits in Air: NIOSH recommends adherence to the present Federal standard of 1 ppm (3 mg/m³) of allyl chloride as a time-weighted average for up to a 10-hour workday, 40-hour workweek and proposes the addition of a 3 ppm ceiling concentration for any 15-minute period. The tentative STEL value is 2 ppm (6 mg/m³). The IDLH level is 300 ppm.

Determination in Air: Charcoal adsorption, working up with C_2H_6 and gas chromatography. See NIOSH Methods, Set I. See also reference (A-10).

Permissible Concentration in Water: No criteria set.

Routes of Entry: Employees may be exposed by dermal or eye contact, inhalation, or ingestion, or by absorption through the skin.

Harmful Effects and Symptoms: The potential for explosions and for damage to the respiratory tract, liver, and kidneys from inhalation was recognized through experimental evidence during the early commercial development of the industry involving this compound and is reflected in the precautions taken during its manufacture and use. NIOSH has found no reports of known cases of either acute or chronic exposure in the United States leading to impairment of the abovementioned organs. Industrial experience in the United States has pointed to such problems as orbital pain and deep-seated aches after eye or skin contact. Both of these phenomena are believed to be transient when they occur and have been minimized through improved work practices.

Points of Attack: Respiratory system, lungs, skin, eyes, liver and kidneys.

Medical Surveillance: Preplacement and periodic physical examinations have been detailed by NIOSH (1). They give special attention to the respiratory system, liver, kidneys, skin and eyes.

First Aid: If this chemical gets into the eyes, irrigate immediately. If this chemical contacts the skin, wash with soap immediately. If a person breathes in large amounts of this chemical, move the exposed person to fresh air at once and perform artificial respiration. When this chemical has been swallowed, get medical attention. Give large quantities of saltwater and induce vomiting. Do not make an unconscious person vomit.

Personal Protective Methods: Appropriate protective apparel, including gloves, aprons, suits, boots and face shields (8-inch minimum) shall be provided and worn where needed to prevent skin contact with liquid allyl chloride. Protective apparel shall be made of materials which most effectively prevent skin contact under the conditions for which it is deemed necessary. Since leather articles cannot be effectively decontaminated, they shall be prohibited for use as protective apparel. Rubber articles may be used provided care is taken to ensure that permeation does not occur during usage. Protective apparel should be discarded at the first sign of deterioration.

Full-facepiece respirators or chemical safety goggles shall be provided and worn for operations in which allyl chloride may splash into the eyes. Face shields may be used to augment chemical safety goggles where full facial protection is needed, but face shields, used alone are not adequate for eye protection.

Employees should wash immediately when skin is wet or contaminated. Remove clothing immediately if wet or contaminated to avoid flammability hazard. Emergency showers should be provided.

Respirator Selection: Engineering controls shall be used to maintain allyl chloride vapor concentrations below the permissible exposure limits. Compliance with the permissible exposure limits may be achieved by the use of respirators only in the following situations:

(1) During the time necessary to install or test the required engineering controls.

(2) For nonroutine operations, such as maintenance or repair activities, in which concentrations in excess of the permissible exposure limits may occur.

(3) During emergencies when air concentrations of allyl chloride may exceed the permissible limits.

When a respirator is permitted, it shall be selected as follows:
       50 ppm: SAF/SCBAF
      300 ppm: SAF/PD,PP,CF
      Escape: GMOV/SCBA

Disposal Method Suggested: Incineration at a temperature of 1800°F for 2 seconds minimum (A-31).

References

(1) National Institute for Occupational Safety and Health, *Criteria for a Recommended Standard: Occupational Exposure to Allyl Chloride,* NIOSH Doc. No. 76-204, Washington, DC (1976).

(2) U.S. Environmental Protection Agency, *Chemical Hazard Information Profile: Allyl Chloride,* Washington, DC (July 1979).

(3) Sax, N.I., Ed., *Dangerous Properties of Industrial Materials Report, 1,* No. 7, 32-34, New York, Van Nostrand Reinhold Co. (1981).

(4) See Reference (A-61).

(5) See Reference (A-60).

(6) United Nations Environment Programme, *IRPTC Legal File 1983,* Vol. II, pp VII/681–82, Geneva, Switzerland, International Register of Potentially Toxic Chemicals (1984).

ALLYL GLYCIDYL ETHER

Description: Allyl glycidyl ether, which is 1-allyloxy-2,3-epoxypropane is a colorless liquid with a strong odor, boiling at 154°C.

Code Numbers: CAS 106-92-3   RTECS RR0875000   UN 2219

DOT Designation: Flammable liquid.

Synonyms: (2-Propenyloxy)methyloxirane, AGE.

Potential Exposure: Workers engaged in the production of allyl glycidyl ether and in the manufacture of epoxy resins are those exposed. People exposed to AGE have been estimated at 6,450 by NIOSH.

Incompatibilities: Strong oxidizers.

Permissible Exposure Limits in Air: The Federal limit is a 10 ppm (45 mg/m³) ceiling. The ACGIH-TWA as of 1983/84 is 5 ppm (22 mg/m³) and the tentative STEL is 10 ppm (44 mg/m³). ACGIH adds the notation "skin" indicating the possibility of cutaneous absorption. The IDLH level is 270 ppm.

Determination in Air: Adsorption in a Tenax-filled tube, workup with ethyl ether, analysis by gas chromatography. See NIOSH Methods, Set 2. See also reference (A-10).

Permissible Concentration in Water: No criteria set.

Routes of Entry: Inhalation, ingestion, eye and skin contact and skin absorption.

Harmful Effects and Symptoms: Dermatitis, eye and nose irritation, pulmonary irritation, edema, narcosis.

Points of Attack: Lungs, respiratory system, skin.

Medical Surveillance: Consider the points of attack in preplacement and periodic physical examinations.

First Aid: If this chemical gets into the eyes, irrigate immediately. If this chemical contacts the skin, flush with water. If a person breathes in large amounts of this chemical, move the exposed person to fresh air at once and perform artificial respiration. When this chemical has been swallowed, get medical attention. Give large quantities of water and induce vomiting. Do not make an unconscious person vomit.

Personal Protective Methods: Wear appropriate clothing to prevent any reasonable probability of skin contact. Wear eye protection to prevent any possibility of eye contact. Employees should wash promptly when skin is wet or contaminated. Remove nonpervious clothing promptly if wet or contaminated. Provide eyewash.

Respirator Selection:
100 ppm: CCROV/SA/SCBA
270 ppm: CCROVF/GMOV/SAF/SCBAF
Escape: GMOV/SCBA

Disposal Method Suggested: Incineration.

References
(1)   Nat. Inst. for Occup. Safety and Health, *Information Profiles on Potential Occupational Hazards: Glycidyl Ethers,* Report No. PB 276-678, Rockville, MD, pp 116-123 (October 1977).
(2)   See Reference (A-60).

ALLYL ISOTHIOCYANATE

Description: $CH_2=CHCH_2NCS$, a colorless to yellow liquid with a pungent, irritating odor and acrid taste boiling at 148° to 154°C.

Code Numbers: CAS 57-06-7   RTECS NX8225000

DOT Designation: —

Synonyms: Allyl mustard oil; allyl isosulfocyanate; 3-isothiocyanato-1-propane; isothiocyanic acid allyl ester.

Potential Exposure: Used as fumigant, in the production of counterirritants in medicine.

Permissible Exposure Limits in Air: No standards set.

Permissible Concentration in Water: No criteria set.

Harmful Effects and Symptoms: Mild allergies; inhalation causes eye watering, sneezing and asthma. There is some indication it is a tumorigenic agent (1).

Points of Attack: Eyes, respiratory system.

References
(1)   Sax, N.I., Ed., *Dangerous Properties of Industrial Materials Report, 1,* No. 1, 28-29, New York, Van Nostrand Reinhold Co. (1980).

ALLYL PROPYL DISULFIDE

Description: $CH_2=CHCH_2SSC_3H_7$ is a pale yellow liquid with a pungent odor.

Code Numbers: CAS 2179-59-1   RTECS JO0350000

DOT Designation: —

Synonyms: Onion oil.

Potential Exposure: Workers in onion processing (slicing and dehydration) operations.

Permissible Exposure Limits in Air: There is no Federal standard but ACGIH as of 1983/84 has set a TWA value of 2 ppm (12 mg/m^3) and set a tentative STEL value of 3 ppm (18 mg/m^3).

Determination in Air: Sample collection by charcoal tube, analysis by gas liquid chromatography (A-13).

Permissible Concentration in Water: No criteria set.

Routes of Entry: Inhalation, ingestion, skin and eye contact.

Harmful Effects and Symptoms: Irritation of eyes, nose and throat.

Points of Attack: Eyes, nose and throat.

Disposal Method Suggested: Incineration.

References
(1) Sax, N.I., Ed., *Dangerous Properties of Industrial Materials Report, 1,* No. 5, 32-33, New York, Van Nostrand Reinhold Co. (1981).

ALUMINUM AND COMPOUNDS

Description: Al, aluminum, is a light, silvery-white, soft, ductile, malleable amphoteric metal, soluble in acids or alkali, insoluble in water. The primary sources are the ores cryolite and bauxite; aluminum is never found in the elemental state.

Code Numbers: For aluminum metal powder—CAS 7429-90-5, RTECS-BD0330000, UN 1396.

DOT Designation: Al powder: flammable solid.

Potential Exposure: Most hazardous exposures to aluminum occur in smelting and refining processes. Aluminum is mostly produced by electrolysis of Al_2O_3 dissolved in molten cryolite (Na_3AlF_6). Aluminum is alloyed with copper, zinc, silicon, magnesium, manganese, and nickel; special additives may include chromium, lead, bismuth, titanium, zirconium, and vanadium. Aluminum and its alloys can be extruded or processed in rolling mills, wireworks, forges, or foundries, and are used in the shipbuilding, electrical, building, aircraft, automobile, light engineering, and jewelry industries. Aluminum foil is widely used in packaging. Powdered aluminum is used in the paints and pyrotechnic industries. Alumina has been used for abrasives, refractories, and catalysts, and in the past in the first firing of china and pottery. Aluminum chloride is used in petroleum processing and in the rubber industry. Alkyl aluminum compounds find use as catalysts in the production of polyethylene.

Permissible Exposure Limits in Air: There is no Federal standard specifically for metallic aluminum. It may be considered as a nuisance dust, the applicable standards being respirable fraction, 15 mppcf or 5 mg/m^3; total dust, 50 mppcf or 15 mg/m^3. A TWA value of 10 mg/m^3 and a proposed STEL value of 20 mg/m^3 were set by ACGIH as of 1983/84 for aluminum metal and oxide. A TWA value of 5 mg/m^3 was set for aluminum pyro powders and for aluminum welding fumes. A TWA value of 2 mg/m^3 was adopted by ACGIH as of 1983/84 for soluble aluminum salts and for aluminum alkyls.

Determination in Air: Filter collection and atomic absorption analysis (A-10).

Permissible Concentration in Water: No criteria set but an ambient water limit of 73 μg/ℓ for aluminum and aluminum compounds has been suggested by EPA (A-37) based on health effects. A value of 138 μg/ℓ has been suggested (A-37) for aluminum oxide.

Routes of Entry: Inhalation of dust or fume.

Harmful Effects and Symptoms: *Local* — Particles of aluminum deposited in the eye may cause necrosis of the cornea. Salts of aluminum may cause dermatoses, eczema, conjunctivitis, and irritation of the mucous membranes of the upper respiratory system by the acid liberated by hydrolysis.

Systemic — The effects on the human body caused by the inhalation of aluminum dust and fumes are not known with certainty at this time. Present data suggest that pneumoconiosis might be a possible outcome. In the majority of causes investigated, however, it was found that exposure was not to aluminum dust alone, but to a mixture of aluminum, silica fume, iron dusts, and other materials.

Medical Surveillance: Preemployment and periodic physical examinations should give special consideration to the skin, eyes, and lungs. Lung function should be followed.

Personal Protective Methods: Workers in electrolysis manufacturing plants should be provided with respirators for protection from fluoride fumes. Dust masks are recommended in areas exceeding the nuisance levels. Aluminum workers generally should receive training in the proper use of personal protective equipment. Workers involved with salts of aluminum may require protective clothing, barrier creams, and where heavy concentrations exist, full-face air supplied respirators may be indicated.

Disposal Method Suggested (A-31):

Aluminum Sulfate—Pretreatment involves hydrolysis followed by neutralization with NaOH. The insoluble aluminum hydroxide formed is removed by filtration and can be heated to decomposition to yield alumina which has valuable industrial applications. The neutral solution of sodium sulfate can be discharged into sewers and waterways as long as its concentration is below the recommended provisional limit of 250 mg/ℓ.

Aluminum Fluoride—Disposal in a chemical waste landfill.

Aluminum Oxide—Disposal in a sanitary landfill. Mixing of industrial process wastes and municipal wastes at such sites is not encouraged however.

Aluminum powder may be recovered and sold as scrap. Recycle and recovery is a viable option to disposal for aluminum metal and aluminum fluoride (A-57).

References

(1) U.S. Environmental Protection Agency, *Chemical Hazard Information Profile: Aluminum and Aluminum Compounds,* Washington, DC (September 1976).

(2) National Institute for Occupational Safety and Health, *Profiles on Occupational Hazards for Criteria Document Priorities: Aluminum and Its Compounds,* Report PB 274-073, Washington, DC, pp 80-84 (1977).

(3) U.S. Environmental Protection Agency, *Toxicology of Metals, Vol. II: Aluminum*, Report EPA 600/1-77-022, Research Triangle Park, NC, pp 4-14 (May 1977).

(4) Sax, N.I., Ed., *Dangerous Properties of Industrial Materials Report, 1*, No. 4, 34, New York, Van Nostrand Reinhold Co. (1981) (Aluminum).

(5) Sax, N.I., Ed., *Dangerous Properties of Industrial Materials Report, 1*, No. 5, 33-34, New York, Van Nostrand Reinhold Co. (1981) (Aluminum Oxide and Aluminum Silicate).

(6) See Reference (A-60) for citations to aluminum chloride, aluminum nitrate, aluminum oxide, aluminum phosphide, aluminum potassium sulfate, aluminum powder and aluminum sulfate.

(7) Parmeggiani, L., Ed., *Encyclopedia of Occupational Health & Safety*, Third Edition, Vol. 1, pp 131–134, Geneva, International Labour Office (1983) (Aluminum, Alloys & Compounds).

2-AMINOANTHRAQUINONE

● Carcinogen (Animal Positive) (A-64)

Description: $C_{14}H_9NO_2$ with the formula $C_6H_4(CO)_2C_6H_3NH_2$ forms red needle-like crystals and melts at $302°C$.

Code Numbers: CAS 117-79-3   RTECS CB5120000

DOT Designation: —

Synonyms: 2-Amino-9,10-anthracenedione, AAQ.

Potential Exposures: AAQ is used as an intermediate in the industrial synthesis of anthraquinone dyes. It is the precursor of five dyes and one pigment, including Colour Index Vat Blues 4, 6, 12, and 24; Vat Yellow 1; and Pigment Blue 22.

Because AAQ is used on a commercial scale solely by the dye industry, the potential for exposure to the compound is greatest for workers at dye manufacturing facilities. However, no additional data are available on the number of facilities using AAQ.

The National Occupational Hazard Survey in 1974 made no estimate of the potential occupational exposure to AAQ (NIOSH). However, the Consumer Product Safety Commission staff believes that trace amounts of unreacted AAQ may possibly be present in some dyes based on this chemical and in the final consumer product. Exposure even to trace amounts may be a cause for concern. This concern is based on experience with other dyes derived from aromatic amines.

Permissible Exposure Limits in Air: No standards set.

Permissible Concentration in Water: No criteria set.

Routes of Entry: Inhalation and skin contact.

Harmful Effects and Symptoms: Technical grade 2-aminoanthraquinone (impurities unspecified), administered in the feed, was carcinogenic in male Fischer 344 rats, causing a combination of hepatocellular carcinomas and neoplastic nodules of the liver. The compound was also carcinogenic in B6C3F1 mice, causing hepatocellular carcinomas in both sexes and malignant hematopoietic lymphomas in females (1). An IARC working group considered that the evidence for the carcinogenicity in experimental animals of the material tested was limited (2). In view of another evaluation of NCI bioassay results, the evidence can be considered as sufficient (3).

References

(1) National Cancer Institute, *Bioassay of 2-Aminoanthraquinone for Possible Carcinogenicity,* Technical Report Series No. 144, DHEW Publication No. (NIH) 78-1399, Bethesda, Maryland (1978).

(2) International Agency for Research on Cancer, *IARC Monographs on the Evaluation of the Carcinogenic Risk of Chemicals to Humans,* Vol. 27, Lyon, France, IARC, pp 191-198 (1982).

(3) Griesemer, R.A., and C. Cueto, Toward a Classification Scheme for Degrees of Experimental Evidence for the Carcinogenicity of Chemicals for Animals, In: R. Montesano, H. Bartsch, and L. Tomatis, eds., *Molecular and Cellular Aspects of Carcinogen Screening Tests,* IARC Scientific Publications, No. 27, Lyon, France, International Agency for Research on Cancer, pp 259-281 (1980).

4-AMINOBIPHENYL

- Carcinogen (Human Positive, IARC) (1) (A-62) (A-64)
- Hazardous waste constituent (EPA)

Description: $C_6H_5C_6H_4NH_2$, 4-aminobiphenyl, as a yellowish brown crystalline solid. It melts at 50° to 52°C.

Code Numbers: CAS 92-67-1   RTECS DU8925000

DOT Designation: —

Synonyms: Biphenyline, p-phenylaniline, xenylamine, 4-aminodiphenyl, 4-biphenylamine, p-aminobiphenyl, p-aminodiphenyl, p-biphenylamine, aminodiphenyl.

Potential Exposure: It is no longer manufactured commercially and is only used for research purposes. 4-Aminobiphenyl was formerly used as a rubber antioxidant and as a dye intermediate. Is a contaminant in 2-aminobiphenyl.

Permissible Exposure Limits in Air: 4-Aminobiphenyl is included in the Federal standards for carcinogens; all contact with it should be avoided. The compound also has the notation "skin" indicating the possibility of cutaneous absorption.

Determination in Air: Collection on a filter and colorimetric analysis (A-8). See also reference (A-10) which describes collection by filter, workup with 2-propanol and analysis by gas chromatography.

Permissible Concentration in Water: No criteria set, but EPA (A-37) has suggested an ambient water limit of 200 $\mu g/\ell$ based on health effects.

Routes of Entry: Inhalation and percutaneous absorption.

Harmful Effects and Symptoms: *Local* — None reported.
Systemic — 4-Aminobiphenyl is a known human bladder carcinogen. An exposure of only 133 days has been reported to have ultimately resulted in a bladder tumor. The latent period is generally from 15 to 35 years. Acute exposure produces headaches, lethargy, cyanosis, urinary burning, and hematuria. Cystoscopy reveals diffuse hyperemia, edema, and frank slough.

Medical Surveillance: Placement and periodic examinations should include an evaluation of exposure to other carcinogens; use of alcohol, smoking, and medications; and family history. Special attention should be given on a regular basis to urine sediment and cytology. If red cells or positive smears are seen, cystoscopy should be done at once. The general health of exposed persons should also be evaluated in periodic examinations.

Personal Protective Methods: These are designed to supplement engineering controls (such as the prohibition of open-vessel operations) and to prevent all skin or respiratory contact. Full body protective clothing and gloves should be used by those employed in handling operations. Fullface, supplied air respirators of continuous flow or pressure demand type should also be used. On exit from a regulated area, employees should shower and change into street clothes, leaving their clothing and equipment at the point of exit to be placed in impervious containers at the end of the work shift for decontamination or disposal. Effective methods should be used to clean and decontaminate gloves and clothing.

Disposal Method Suggested: Controlled incineration whereby oxides of nitrogen are removed from the effluent gas by scrubber, catalytic or thermal devices.

References

(1) International Agency for Research on Cancer, *IARC Monographs on the Carcinogenic Risks of Chemicals to Humans,* 1, 74, Lyon, France (1972).

3-AMINO-2,5-DICHLOROBENZOIC ACID

See "Chloramben."

3-AMINO-9-ETHYLCARBAZOLE

● Carcinogen (Animal Positive, NCI) (2)

Description: 3-Amino-9-ethylcarbazole, $C_{14}H_{14}N_2$ is a crystalline compound melting at $127°C$.

Code Numbers: CAS 132-32-1   RTECS FE3590000

DOT Designation: —

Synonyms: 3-Amino-N-ethylcarbazole.

Potential Exposures: Plant workers engaged in the manufacture of this compound and its use in pigment manufacture. Laboratory workers using this material in colorimetric enzyme assays and as a biological stain.

Permissible Exposure Limits in Air: No standards set.

Permissible Concentration in Water: No criteria set.

Harmful Effects and Symptoms: The NCI carcinogenic assay (2) proved positive.

Disposal Method Suggested: Incinerator is equipped with a scrubber or thermal unit to reduce NO_x emissions.

References

(1) U.S. Environmental Protection Agency, *Chemical Hazard Information Profile: 3-Amino-9-Ethylcarbazole,* Washington, DC (1979).
(2) National Cancer Institute, *Bioassay of 3-Amino-9-ethylcarbazole Hydrochloride for Possible Carcinogenicity,* Technical Report Series No. 93, Bethesda, MD (1978).

2-AMINO ETHYL ETHANOL AMINE

Description: $HO(CH_2)_2NH(CH_2)_2NH_2$, a colorless liquid boiling at 244°C.

Code Numbers: CAS 111-41-1 RTECS KJ6300000

DOT Designation: U.S. Coast Guard: Grade E Combustible Liquid.

Synonyms: Monoethanol ethylene diamine; β-hydroxyethyl ethylene diamine; 2-[(2-aminoethyl)amino] ethanol.

Potential Exposure: In manufacture of dyestuffs, textile compounds, surfactants, polymers, pesticides and pharmaceuticals.

Permissible Exposure Limits in Air: No standards set.

Permissible Concentration in Water: No criteria set.

Harmful Effects and Symptoms: Mild skin and eye irritant.

Points of Attack: Eyes, skin, pulmonary system.

Personal Protective Methods: Wear safety goggles or face shield, butyl rubber gloves (A-38).

Respirator Selection: Use self-contained breathing apparatus.

Disposal Method Suggested: Add to sodium bisulfate, spray with water, neutralize and sewer or incinerate.

References
(1) Sax, N.I., Ed., *Dangerous Properties of Industrial Materials Report, 2,* No. 3, 29-30, New York, Van Nostrand Reinhold Co. (1982).

1-AMINO-2-METHYLANTHRAQUINONE

● Carcinogen (Animal Positive) (A-64)

Description: $C_{15}H_{11}NO_2$ is a crystalline substance melting at 205° to 206°C.

Code Numbers: CAS 82-28-0 RTECS CB5740000

DOT Designation: —

Synonyms: 2-Methyl-1-anthraquinonylamine; Acetate Fast Orange R.

Potential Exposure: 1-Amino-2-methylanthraquinone is used almost exclusively as a dye intermediate for the production of a variety of anthraquinone dyes. The Society of Dyers and Colourists reported that it can be used as a dye for a variety of synthetic fibers, expecially acetates, as well as wool, sheepskins, furs, and surface dyeing of thermoplastics. None of the dyes that can be prepared from it are presently produced in commercial quantities.

1-Amino-2-methylanthraquinone had been produced commercially in the United States since 1948, but production was last reported by one company in 1970.

The potential for exposure is greatest among workers engaged in the dyeing of textiles. The National Occupational Hazard Survey of 1974 contains no information on 1-amino-2-methylanthraquinone, but does estimate that 6,400 workers may have been exposed to anthraquinone dyes. 1-Amino-2-methylanthraquinone is not presently used in consumer products according to the CPSC.

Permissible Exposure Limits in Air: No standards set.

Permissible Concentration in Water: No criteria set.

Routes of Entry: Inhalation of vapors.

Harmful Effects and Symptoms: Technical-grade 1-amino-2-methylanthraquinone (impurities unspecified), administered in the feed, was carcinogenic in Fischer 344 rats, inducing hepatocellular carcinomas in rats of both sexes, and kidney tumors (such as tubular-cell adenomas and adenocarcinomas) in males. The compound was carcinogenic in female B6C3F1 mice, producing an increased combined incidence of hepatocellular carcinomas and neoplastic nodules (1). An IARC working group considered that the evidence for the carcinogenicity in experimental animals of the material tested was limited (2), but in view of another evaluation of NCI bioassay results, the evidence can be considered as sufficient (3).

References

(1) National Cancer Institute, *Bioassay of 1-Amino-2-methyl-anthraquinone for Possible Carcinogenicity,* Technical Report Series No. 111, DHEW Publication No. (NIH) 78-1366, Washington, D.C. (1978).
(2) International Agency for Research on Cancer, *IARC Monographs on the Evaluation of the Carcinogenic Risk of Chemicals to Humans,* Vol. 27, Lyon, France, IARC, pp 199-204, (1982).
(3) Griesemer, R.A., and C. Cueto, Toward a Classification Scheme for Degrees of Experimental Evidence for the Carcinogenicity of Chemicals for Animals. In: R. Montesano, H. Bartsch, and L. Tomatis, eds., *Molecular and Cellular Aspects of Carcinogen Screening Tests,* IARC Scientific Publications, No. 27, Lyon, France, International Agency for Research on Cancer, pp 259-281 (1980).

4-AMINO-2-NITROPHENOL

● Carcinogen (Rat Positive, Mouse Negative, NCI) (A-63)

Description: $C_6H_6N_2O_3$, $C_6H_3NO_2NH_2OH$, as dark red crystals melting at 131°C.

Code Numbers: CAS 119-34-6   RTECS SJ6125000

DOT Designation: —

Synonyms: o-Nitro-p-Aminophenol, 4-hydroxy-3-nitroaniline.

Potential Exposure: In dye formulation for furs and hair.

Permissible Exposure Limits in Air: No standards set.

Permissible Concentration in Water: No criteria set.

Harmful Effects and Symptoms: Severe eye irritant in rabbits. Rat carcinogenicity positive but mouse carcinogenicity negative according to NCI (1).

References

(1) Sax, N.I., Ed., *Dangerous Properties of Industrial Materials Report, 1,* No. 7, 34-35, New York, Van Nostrand Reinhold Co. (1981).

AMINOPHENOLS

Description: C_6H_7NO [$C_6H_4(OH)(NH_2)$]; the amine group and the hydroxyl group may be in the ortho or para positions. o-Aminophenol is a white crystalline substance melting at 170°-174°C; p-Aminophenol is a white or reddish yellow crystalline substance, melting at 184°C with decomposition.

Code Numbers:

ortho: CAS 95-55-6   RTECS SJ4725000   UN 2512
para: CAS 123-30-8   RTECS SJ5075000   UN 2512

DOT Designation: —

Synonyms:

ortho: 2-amino-1-hydroxybenzene; o-hydroxyaniline.
para: 4-amino-1-hydroxybenzene; p-hydroxyaniline.

Potential Exposure: Workers may be exposed to o-aminophenol during its use as a chemical intermediate, in the manufacture of azo and sulfur dyes, and in the photographic industry.

There is potential for consumer exposure to o-aminophenol because of its use in dyeing hair, fur, and leather. The compound is a constituent of 75 registered cosmetic products suggesting the potential for widespread consumer exposure.

p-Aminophenol is used mainly as a dye and dye intermediate and as a photographic developer, and in small quanitities in analgesic drug preparation.

Consumer exposure to p-aminophenol may occur from use as a hair-dye or as a component in cosmetic preparations.

Incompatibilities: Strong oxidants.

Permissible Exposure Limits in Air: No limits set.

Permissible Concentration in Water: No criteria set. Organoleptic limit is 0.05 mg/ℓ in the USSR (A-36).

Routes of Entry: Inhalation, ingestion and skin absorption (A-60).

Harmful Effects and Symptoms: o-Aminophenol has an oral LD_{50} for rats of 1,300 mg/kg and p-aminophenol has produced LD_{50} values in rats of 375, 671 and 1,270 mg/kg which are all of low acute toxicity.

May produce dermatitis, methemoglobinemia, bronchial asthma and restlessness (A-39). Produces lung irritation (A-60).

Points of Attack: Blood, liver, kidneys, brain (A-60), skin and lungs.

First Aid: Remove contaminated clothing, flush with water (A-60).

Personal Protective Methods: Wear protective clothing and goggles (A-60).

Respirator Selection: Chemical cartridge respirator recommended (A-39).

Disposal Method Suggested: Incineration.

References

(1) U.S. Environmental Protection Agency, *Chemical Hazard Information Profile Draft Report: o-Aminophenol, sulfate & hydrochloride,* Wash., D.C. (Mar. 29, 1984).

(2) U.S. Environmental Protection Agency, *Chemical Hazard Information Profile Draft Report: p-Aminophenol, hydrochloride & sulfate,* Wash. D.C. (Mar. 29, 1984).

2-AMINOPYRIDINE

Description: 2-Aminopyridine, $H_2NC_5H_4N$, is a colorless solid with a characteristic odor. It melts at 56°C.

Code Numbers: CAS 504-29-0 RTECS US1575000 UN 2671

DOT Designation: —

Synonyms: α-Aminopyridine.

Potential Exposures: 2-Aminopyridine is used in the manufacture of pharmaceuticals, especially antihistamines (A-41).

Incompatibilities: Strong oxidizers.

Permissible Exposure Limits in Air: The Federal standard and 1983/84 ACGIH value is 0.5 ppm (2 mg/m^3). The proposed STEL is 2.0 ppm (4 mg/m^3). The IDLH level is 5.0 ppm.

Determination in Air: Adsorption on Tenax GC, thermal desorption and gas chromatographic analysis (A-10).

Permissible Concentration in Water: No criteria set.

Routes of Entry: Inhalation, ingestion, eye and skin contact and absorption through the skin.

Harmful Effects and Symptoms: Headaches, dizziness, excited state, nausea, flushed appearance, high blood pressure, respiratory distress, weakness, convulsions, stupor.

Points of Attack: Central nervous system, respiratory system.

Medical Surveillance: Consider the points of attack in preplacement and periodic physical examinations.

First Aid: If this chemical gets into the eyes, irrigate immediately. If this chemical contacts the skin, flush with water immediately. If a person breathes in large amounts of this chemical, move the exposed person to fresh air at once and perform artificial respiration. When this chemical has been swallowed, get medical attention. Give large quantities of water and induce vomiting. Do not make an unconscious person vomit.

Personal Protective Methods: Wear appropriate clothing to prevent any reasonable probability of skin contact. Wear eye protection to prevent any reasonable probability of eye contact. Employees should wash immediately when skin is wet or contaminated. Work clothing should be changed daily if it may be that clothing is contaminated. Remove nonimpervious clothing immediately if wet or contaminated. Provide emergency showers.

Respirator Selection:
    5 ppm: SAF/SCBAF
    Escape: GMOVP/SCBA

Disposal Method Suggested: Incineration with NO$_x$ removal from effluent gas.

References
(1)   See Reference (A-61).

3-AMINO-1,2,4-TRIAZOLE

● Carcinogen (Animal Positive, IARC) (1) (3) (A-62) (A-64)

Description: $C_2H_4N_4$ with the structural formula

This is a colorless crystalline solid melting at 158° to 159°C.

Code Numbers: CAS 61-82-5   RTECS XZ3850000   UN 2588

DOT Designation: —

Synonyms: Amitrole, Weedazol®, Cytrol®, Azolan®, Amitrol-T®, Amizol®.

Potential Exposures: Those involved in the manufacture (A-32), formulation and application of this herbicide, which is now limited to noncrop applications.

Permissible Exposure Limits in Air: ACGIH has classified amitrole as an industrial substance suspect of carcinogenic potential for man with no assigned TLV.

Permissible Concentration in Water: No criteria set.

Harmful Effects and Symptoms: Carcinogenicity is the primary observed effect.

Amitrole is carcinogenic in mice and rats, producing thyroid and liver tumours following oral or subcutaneous administration.

Railroad workers who were exposed to amitrole and other herbicides showed a slight (but statistically significant) excess of cancer when all sites were considered together. Because the workers were exposed to several different herbicides however, no conclusions could be made regarding the carcinogenicity of amitrole alone.

Disposal Method Suggested: Amitrole is resistant to hydrolysis and the action of oxidizing agents. Burning the compound with polyethylene is reported to result in >99% decomposition (A-32).

References

(1)   International Agency for Research on Cancer, *IARC Monographs on the Carcinogenic Risks of Chemicals to Humans, 7*, 31, Lyon, France (1974).
(2)   Sax, N.I., Ed., *Dangerous Properties of Industrial Materials Report, 1*, No. 4, 34-35, New York, Van Nostrand Reinhold Co. (1981).
(3)   Parmeggiani, L., Ed., *Encyclopedia of Occupational Health & Safety*, Third Edition, Vol. 1, pp 147–148, Geneva, International Labour Office (1983).
(4)   Sax, N.I., Ed., *Dangerous Properties of Industrial Materials Report, 4*, No. 2, 41–43, New York, Van Nostrand Reinhold Co. (1984).

AMITRAZ

Description: $C_{19}H_{23}N_3$, N,N'-(Methyliminodimethylidyne)bis-2,4-xylidine, with the structural formula

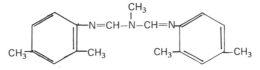

forms colorless needle-like crystals melting at 86° to 87°C.

Code Numbers: CAS 33089-61-1   RTECS ZF0480000   UN 2588

DOT Designation: —

Synonyms: Baam, ENT 27967, Mitac®, Taktic®, Triatox®.

Potential Exposure: Those engaged in the manufacture (A-32), formulation and application of this insecticide and acaricide.

A rebuttable presumption against registration for amitraz has been issued on April 6, 1977 by US EPA on the basis of oncogenicity (1).

Permissible Exposure Limits in Air: No standards set.

Permissible Concentration in Water: No criteria set.

Harmful Effects and Symptoms: Amitraz metabolizes to 2,4-dimethyl-aniline which is a potential human carcinogen. A mouse oncogenic bioassay was conducted by Boots Chemical Company and reported by EPA (1); the results of that study have been disputed.

References

(1)   U.S. Environmental Protection Agency, *Rebuttable Presumption Against Registration (RPAR) of Pesticide Products Containing Amitraz,* Washington, DC (April 6, 1977).

AMITROLE

See "3-Amino-1,2,4-Triazole."

AMMONIA

● Hazardous substance (EPA)

Description: NH_3, ammonia is a colorless, strongly alkaline, and extremely soluble gas with a characteristic pungent odor.

Code Numbers: CAS 7664-41-7   RTECS BO0875000   UN 1005

DOT Designation: Nonflammable gas.

Synonyms: Anhydrous ammonia.

Potential Exposures: Ammonia is used as a nitrogen source for many nitrogen-containing compounds. It is used in the production of ammonium sulfate and ammonium nitrate for fertilizers and in the manufacture of nitric acid, soda, synthetic urea, synthetic fibers, dyes, and plastics. It is also utilized as a refrigerant and in the petroleum refining and chemical industries. It is used in the production of many drugs (A-41) and pesticides (A-32).

Other sources of occupational exposure include the silvering of mirrors, glue-making, tanning of leather, and around nitriding furnaces. Ammonia is produced as a by-product in coal distillation and by the action of steam on calcium cyanamide, and from the decomposition of nitrogenous materials.

Incompatibilities: Strong oxidizers, calcium, hypochlorite bleaches, gold, mercury, silver, halogens.

Permissible Exposure Limits in Air: The Federal standard for ammonia is an 8-hour time-weighted average of 50 ppm (35 mg/m³). NIOSH has recommended 50 ppm expressed as a ceiling and determined by a 5-minute sampling period. ACGIH as of 1983/84 has set TWA values of 25 ppm (18 mg/m³). The tentative STEL value is 35 ppm (27 mg/m³). The IDLH level is 500 ppm.

Determination in Air: Collection by midget impinger and colorimetric analysis using Nessler's reagent (A-10). Ammonia may also be determined using long-duration detector tubes (A-11).

Permissible Concentration in Water: EPA in 1976 (A-3) proposed a limit of 0.02 mg/ℓ (as unionized ammonia) for the protection of freshwater aquatic life. As of 1980, EPA (2) first proposed adding ammonia to the list of priority toxic pollutants and developing criteria for it, but then withdrew the proposal. NAS/NRC proposed (A-2) a limit of 0.5 mg/ℓ for drinking water.

Routes of Entry: Inhalation of gas, ingestion, skin and eye contact.

Harmful Effects and Symptoms: *Local* — Contact with anhydrous liquid ammonia or with aqueous solutions is intensely irritating to the mucous membranes, eyes, and skin. Eye symptoms range from lacrimation, blepharospasm, and palpebral edema to a rise of intraocular pressure, and other signs resembling acute-angle closure glaucoma, corneal ulceration, and blindness. There may be corrosive burns of skin or blister formation. Ammonia gas is also irritating to the eyes and to moist skin.

Systemic — Mild to moderate exposure to the gas can produce headache, salivation, burning of throat, anosmia, perspiration, nausea, vomiting, and substernal pain. Irritation of ammonia gas in eyes and nose may be sufficiently intense to compel workers to leave the area. If escape is not possible, there may be severe irritation of the respiratory tract with the production of cough, glottal edema, bronchospasm, pulmonary edema, or respiratory arrest. Bronchitis or pneumonia may follow a severe exposure if patient survives. Urticaria is a rare allergic manifestation from inhalation of the gas.

Points of Attack: Lungs, respiratory system, eyes.

Medical Surveillance: Preemployment physical examinations for workers in ammonia exposure areas should be directed toward significant changes in the skin, eyes, and respiratory system. Persons with corneal disease, and glaucoma, or chronic respiratory diseases may suffer increased risk. Periodic examinations should include evaluation of skin, eyes, and respiratory system, and pulmonary function tests to compare with baselines established at preemployment examination.

First Aid: If this chemical gets into the eyes, irrigate immediately. If this chemical contacts the skin, flush with water immediately. If a person breathes in large amounts of this chemical, move the exposed person to fresh air at once and perform artificial respiration. When this chemical has been swallowed, get medical attention. Do not induce vomiting.

Personal Protective Methods: Where ammonia hazards exist in concentrations above the standard, respiratory, eye, and skin protection should be provided. Fullface gas masks with ammonia canister or supplied air respirators, both with full facepieces, afford good protection. In areas where exposure to liquid ammonia occurs, goggles or face shields, as well as protective clothing impervious to ammonia and including gloves, aprons, and boots should be required. Where ammonia gas or concentrated ammonia solution is splashed in eyes, immediate flooding of the eyes with large quantities of water for 15 minutes or longer is advised, followed at once by medical examination.

In heavy concentrations of ammonia gas, workers should be outfitted with complete self-contained protective suits impervious to ammonia, with supplied air source, and full headpiece and facepiece. Appropriate clothing should be worn to prevent any possible skin contact with liquids of >10% content or reasonable probability of contact with liquids of <10% content. Wear eye protection to prevent any possibility of eye contact with liquids of >10% NH_3 content. Employees should wash immediately when skin is wet or contaminated with liquids of >10% content. Remove nonimpervious clothing immediately if wet or contaminated with liquids containing >10% and promptly remove if liquid contains <10% NH_3. Provide emergency showers and eyewash if liquids containing >10% NH_3 are involved.

Respirator Selection:

    100 ppm: CCRS/SA/SCBA
    300 ppm: CCRSF
    500 ppm: GMS/SAF/SCBAF
     Escape: GMS/SCBA

Disposal Method Suggested: Dilute with water, neutralize with HCl and discharge to sewer (A-38). Recovery is an option to disposal which should be considered for paper manufacture, textile treating, fertilizer manufacture and chemical process wastes (A-57).

References

(1) National Institute for Occupational Safety and Health, *Criteria for a Recommended Standard: Occupational Exposure to Ammonia,* NIOSH Doc. No. 74-136, Washington, DC (1974).

(2) U.S. Environmental Protection Agency, "Toxic Pollutant List: Proposal to Add Ammonia," *Federal Register,* 45, No. 2, 803-806 (January 3, 1980) Rescinded by *Federal Register,* 45, No. 232, 79692-79693 (December 1, 1980).

(3) National Research Council, Committee on Medical and Biologic Effects of Environmental Pollutants, *Ammonia,* Baltimore, MD, University Park Press (1979).

(4) Sax, N.I., Ed., *Dangerous Properties of Industrial Materials Report, 2,* No. 1, 65-68, New York, Van Nostrand Reinhold Co. (1982).

(5) See Reference (A-61).

(6) Sax, N.I., Ed., *Dangerous Properties of Industrial Materials Report, 3,* No. 3, 49-53, New York, Van Nostrand Reinhold Co. (1983).

(7) See Reference (A-60).

(8) Parmeggiani, L., Ed., *Encyclopedia of Occupational Health & Safety,* Third Edition, Vol. 1, pp 148-150, Geneva, International Labour Office (1983).

AMMONIUM CHLORIDE

- Hazardous substance (EPA)

Description: NH_4Cl is a white crystalline solid which melts at 338°C with decomposition.

Code Numbers: CAS 12125-02-9   RTECS BP4550000   UN (NA 9085)

DOT Designation: ORM-E.

Synonyms: Sal ammoniac.

Potential Exposure: Ammonium chloride is used in galvanizing and soldering as a flux.

Permissible Exposure Limits in Air: There is no Federal standard but ACGIH as of 1983/84 has set a TWA of 10 mg/m^3 for ammonium chloride fume. The tentative STEL value is 20 mg/m^3.

Determination in Air: Collection on a filter and colorimetric analysis (A-1).

Permissible Concentration in Water: No criteria set.

Routes of Entry: Inhalation of fume, ingestion, skin and eye contact.

Harmful Effects and Symptoms: Ammonium chloride is a mild skin and respiratory system irritant with a low grade systemic toxicity by ingestion (A-34).

Points of Attack: Skin, respiratory system.

Disposal Method Suggested: Pretreatment involves addition of sodium hydroxide to liberate ammonia and form the soluble sodium salt. The liberated ammonia can be recovered and sold. After dilution to the permitted provisional limit, the sodium salt can be discharged into a stream or sewer (A-31).

References

(1)   Sax, N.I., Ed., *Dangerous Properties of Industrial Materials Report, 2,* No. 3, 34-36, New York, Van Nostrand Reinhold Co. (1982).

(2)   See Reference (A-60).

AMMONIUM NITRATE

Description: Ammonium nitrate, NH_4NO_3, is a white crystalline solid melting at about 155°C.

Code Numbers: CAS 6484-52-2   RTECS BR9050000   UN 2067

DOT Designation: Oxidizer.

Synonyms: Nitric acid, ammonium salt.

Potential Exposure: Workers engaged in the manufacture of liquid and solid fertilizer compositions and industrial explosives and blasting agents from ammonium nitrate are exposed. About 45,000 workers are potentially exposed, primarily in th manufacture of fertilizers.

Incompatibilities: Acetic acid, heat, powdered metals, organic matter.

Permissible Exposure Limits in Air: No standards set.

Permissible Concentration in Water: No criteria set.

Harmful Effects and Symptoms: Human and animal toxicity data on ammonium nitrate are virtually nonexistent (1) but it is a practically nontoxic substance. The possibility of ammonia poisoning in the course of NH_4NO_3 and fertilizer manufacture is the chief toxic effect associated with ammonium nitrate.

Points of Attack: Respiratory system for ammonia as discussed above.

Medical Surveillance: Consider the points of attack in preplacement and periodic physical examinations.

Personal Protective Methods: Wear rubber gloves and chemical goggles (A-38).

Disposal Method Suggested: Pretreatment involves addition of sodium hydroxide to liberate ammonia and form the soluble sodium salt. The liberated ammonia can be recovered and sold. After dilution to the permitted provisional limit, the sodium salt can be discharged into a stream or sewer (A-31).

References

(1) National Institute for Occupational Safety and Health, *Profiles on Occupational Hazards for Criteria Document Priorities: Ammonium Nitrate,* pp 281-285, Report PB-274,073, Washington, DC (1977).

(2) Sax, N.I., Ed., *Dangerous Properties of Industrial Materials Report, 2,* No. 3, 44-46, New York, Van Nostrand Reinhold Co. (1982).

AMMONIUM SALTS

A variety of ammonium salts have been reviewed in the periodical "Dangerous Properties of Industrial Materials Report" and, in the interest of economy of space, are simply referenced here:

	Volume	Number	Pages
Ammonium acetate	2	3	30–31
Ammonium bicarbonate	4	2	43–45
Ammonium bichromate	3	5	29–32
Ammonium carbamate	2	3	31–33
Ammonium carbonate	2	3	33–34
Ammonium chromate	2	3	36–38
Ammonium dichromate	2	3	38–40
Ammonium ferricyanide	2	3	40–41
Ammonium fluoride	3	5	32–34
Ammonium hydrogen fluoride	3	5	34–37
Ammonium hydroxide	2	3	41–44
Ammonium perchlorate	2	3	46–48
Ammonium peroxysulfate	2	3	48–49
Ammonium picrate	2	3	49–51
Ammonium silicofluoride	4	3	36–38
Ammonium stearate	2	3	51–52
Ammonium sulfide	2	4	27–29
Ammonium thiocyanate	2	3	54–55

Other citations may be given to reference (A-60) for ammonium bicarbonate, ammonium fluoride, ammonium hexafluorosilicate, ammonium hydrogen sulfide, ammonium oxalate, and ammonium sulfate.

AMMONIUM SULFAMATE

● Hazardous substance (EPA)

Description: Ammonium sulfamate, $NH_2SO_3NH_4$, is a colorless, odorless solid melting at 131°C (with decomposition).

Code Numbers: CAS 7773-06-0   RTECS WO6125000   UN (NA 9089)

DOT Designation: ORM-E.

Synonyms: Ammate® and Amicide® herbicides, monoammonium sulfamate, sulfamic acid, monoammonium salt.

Potential Exposure: Ammonium sulfamate is used as a weed killer and in fire retardant compositions.

Incompatibilities: Strong oxidizers, hot water.

Permissible Exposure Limits in Air: The Federal TWA limit is 15 mg/m³. The ACGIH-TWA value as of 1983/84 is 10 mg/m³. The tentative STEL is 20 mg/m³. The IDLH level is 5,000 mg/m³.

Determination in Air: Collection on a filter followed by gravimetric analysis (A-1).

Permissible Concentration in Water: No criteria set.

Routes of Entry: Inhalation, ingestion, skin and eye contact.

Harmful Effects and Symptoms: This material is moderately toxic by ingestion and may cause gastrointestinal disease (A-38).

Medical Surveillance: Nothing special indicated.

First Aid: If this chemical gets into the eyes, irrigate immediately. If this chemical contacts the skin, wash with soap promptly. If a person breathes in large amounts of this chemical, move the exposed person to fresh air at once and perform artificial respiration. When this chemical has been swallowed, get medical attention. Give large quantities of water and induce vomiting. Do not make an unconscious person vomit.

Personal Protective Methods: Not applicable, according to NIOSH (A-4).

Respirator Selection:
```
    75 mg/m³:  DM
   150 mg/m³:  DMXSQ/HiEP/SA/SCBA
   750 mg/m³:  HiEPF/SAF/SCBAF
 5,000 mg/m³:  PAPHiE/SA:PD,PP,CF
    Escape:    DMXS/SCBA
```

Disposal Method Suggested: Dilute with water, make neutral with acid or base and flush into sewer with more water (A-38).

References
(1)   See Reference (A-61).
(2)   Sax, N.I., Ed., *Dangerous Properties of Industrial Materials Report, 2,* No. 3, 52-54, New York, Van Nostrand Reinhold Co. (1982).

AMYL ACETATES

● Hazardous substances (EPA)

Description: The amyl acetates, $CH_3COOC_5H_{11}$, are clear, colorless liquids with a fruity, bananalike odor. There are three amyl acetates:

n-Amyl	$CH_3COOCH_2CH_2CH_2CH_2CH_3$
sec-Amyl	$CH_3COOCH(CH_3)CH_2CH_2CH_3$
Isoamyl	$CH_3COOCH_2CH_2CH(CH_3)CH_3$

Code Numbers:

n-Amyl	CAS 628-63-7	RTECS AJ1925000	
sec-Amyl	CAS 626-38-0	RTECS AJ2100000	UN 1104
Isoamyl	CAS 123-92-2	RTECS NS9800000	

DOT Designation: Flammable liquid.

Synonyms: n-Amyl—1-pentanol acetate; acetic acid, pentyl ester. sec-Amyl—2-pentanol acetate; acetic acid, 2-pentyl ester. Isoamyl—3-methyl-1-butanol acetate; isopentyl alcohol acetate; 2-methylbutyl ethanoate; banana oil.

Potential Exposure: Amyl acetates are used as solvents. Isoamyl acetate is used as a flavor in water and syrups.

Incompatibilites: Nitrates, strong oxidizers, strong alkalies, strong acids.

Permissible Exposure Limits in Air:

	TWA	Tentative STEL	IDLH
n-Amyl	100 ppm (530 mg/m³)	150 ppm (800 mg/m³)	4,000 ppm
sec-Amyl	125 ppm (670 mg/m³)	150 ppm (800 mg/m³)	9,000 ppm
Isoamyl	100 ppm (525 mg/m³)	125 ppm (655 mg/m³)	3,000 ppm

Determination in Air: Charcoal adsorption, workup with CS_2 and gas chromatographic analysis. See NIOSH Methods, Set D (sec-amyl and isoamyl) and Set E (n-amyl). See also reference (A-10).

Permissible Concentration in Water: No criteria set.

Routes of Entry: Inhalation, ingestion, eye and skin contact.

Harmful Effects and Symptoms: Irritation of eyes and nose, narcosis, dermatitis. Throat irritation also in the case of isoamyl acetate.

Points of Attack: Eyes, skin and respiratory system.

Medical Surveillance: Consider the points of attack in preplacement and periodic physical examinations.

First Aid: If this chemical gets into the eyes, irrigate immediately. If this chemical contacts the skin, flush with water promptly. If a person breathes in large amounts of this chemical, move the exposed person to fresh air at once and perform artificial respiration. When this chemical has been swallowed, get medical attention. Give large quantities of saltwater and induce vomiting. Do not make an unconscious person vomit.

Personal Protective Methods: Wear appropriate clothing to prevent repeated or prolonged skin contact. Wear eye protection to prevent any reasonable probability of eye contact. Employees should wash promptly when skin is wet or contaminated. Remove clothing immediately if wet or contaminated to avoid flammability hazard.

Respirator Selection:

n-Amyl	1,000 ppm:	DDROVF
	4,000 ppm:	GMOV/SAF/SCBAF
	Escape:	GMOV/SCBA
sec-Amyl	1,000 ppm:	CCROVF
	5,000 ppm:	GMOV/SAF/SCBAF
	9,000 ppm:	SAF:PD,PP,CF
	Escape:	GMOV/SCBA
Isoamyl	1,000 ppm:	CCROVF/SA/SCBA
	3,000 ppm:	GMOV/SAF/SCBAF
	Escape:	GMOV/SCBA

Disposal Method Suggested: Incineration.

References

(1) See Reference (A-61).
(2) Sax, N.I., Ed., *Dangerous Properties of Industrial Materials Report, 2,* No. 2, 39-41, New York, Van Nostrand Reinhold Co. (1982) (Isoamyl Acetate).
(3) See Reference (A-60).
(4) Sax, N.I., Ed., *Dangerous Properties of Industrial Materials Report, 3,* No. 6, 37-40, New York, Van Nostrand Reinhold Co. (Nov./Dec. 1983) (t-Amyl Acetate).

AMYL ALCOHOLS

Description: $C_5H_{11}OH$, amyl alcohol, has eight isomers. All are colorless liquids, except the isomer 2,2-dimethyl-1-propanol, which is a crystalline solid. Amyl alcohols are obtained from fusel oil which forms during the fermentation of grain, potatoes, or beets for ethyl alcohol. The fusel oil is a mixture of amyl alcohol isomers, and the composition is determined somewhat by the sugar source.

See "Isoamyl Alcohol" for an important isomer and the only one for which a Federal exposure standard exists.

References

(1) Sax, N.I., Ed., *Dangerous Properties of Industrial Materials Report, 2,* No. 3, 55-56, New York, Van Nostrand Reinhold Co. (1982) (n-Amyl Alcohol).
(2) Parmeggiani, L., Ed., *Encyclopedia of Occupational Health & Safety,* Third Edition, Vol. 2, pp 1602-3, Geneva, International Labour Office (1983) (Pentyl Alcohols).

ANILINE

- Hazardous substance (EPA)
- Hazardous waste (EPA)

Description: $C_6H_5NH_2$, aniline, is a clear, colorless, oily liquid with a characteristic odor. It boils at 184°C.

Code Numbers: CAS 62-53-3   RTECS BW6650000   UN 1547

Type of Compound/Label Designation: Poison.

Synonyms: Aminobenzene, phenylamine, aniline oil, aminophen, arylamine.

Potential Exposure: Aniline is widely used as an intermediate in the synthesis of dyestuffs. It is also used in the manufacture of rubber accelerators and antioxidants, pharmaceuticals, marking inks, tetryl, optical whitening agents, photographic developers, resins, varnishes, perfumes, shoe polishes, and many organic chemicals.

Incompatibilities: Strong acids, strong oxidizers.

Permissible Exposure Limits in Air: The Federal standard is 5 ppm (19 mg/m³). ACGIH (1983/84) has adopted TWA values of 2 ppm (10 mg/m³) and an STEL value of 5 ppm (20 mg/m³). ACGIH adds the notation "skin" indicating the possibility of cutaneous absorption. The IDLH level is 100 ppm.

Determination in Air: Silica adsorption, workup in n-propanol, followed by gas chromatography. See NIOSH Methods, Set K. See also reference (A-10).

Permissible Concentration in Water: No criteria set but EPA (A-37) has suggested an ambient limit in water of 262 $\mu g/\ell$ based on health effects.

Routes of Entry: Inhalation of vapors, percutaneous absorption of liquid and vapor, ingestion, skin and eye contact.

Harmful Effects and Symptoms: *Local* — Liquid aniline is mildly irritating to the eyes and may cause corneal damage.

Systemic — Absorption of aniline, whether from inhalation of the vapor or from skin absorption of the liquid, causes anoxia due to the formation of methemoglobin. Moderate exposure may cause only cyanosis. As oxygen deficiency increases, the cyanosis may be associated with headache, weakness, irritability, drowsiness, dyspnea, and unconsciousness. If treatment is not given promptly, death can occur. The development of intravascular hemolysis and anemia due to aniline-induced methemoglobinemia has been postulated, but neither is observed often in industrial practice, despite careful study of numerous cases.

Points of Attack: Blood, cardiovascular system, liver, kidneys.

Medical Surveillance: Preplacement and periodic physical examinations should be performed on all employees working in aniline exposure areas. These should include a work history to elicit information on all past exposures to aniline, other aromatic amines, and nitro compounds known to cause chemical cyanosis, and the clinical history of any occurrence of chemical cyanosis; a personal history to elicit alcohol drinking habits; and general physical examination with particular reference to the cardiovascular system. Persons with impaired cardiovascular status may be at greater risk from the consequences of chemical cyanosis. A preplacement complete blood count and methemoglobin estimation should be performed as baseline levels, also follow-up studies including periodic blood counts and hematocrits.

First Aid: If this chemical gets into the eyes, irrigate immediately. If this chemical contacts the skin, wash with soap promptly. If a person breathes in large amounts of this chemical, move the exposed person to fresh air at once and perform artificial respiration. When this chemical has been swallowed, get medical attention. Give large quantities of water and induce vomiting. Do not make an unconscious person vomit.

Personal Protective Methods: In areas of vapor concentration, the use of respirators alone is not sufficient; skin protection by protective clothing should be provided even though there is no skin contact with liquid aniline. Butyl rubber protective clothing is reportedly superior to other materials. In severe exposure situations, complete body protection has been employed, consisting of air-conditioned suit with air supplied helmet and cape. Personal hygiene practices including prompt removal of clothing which has absorbed aniline, thorough showering after work and before changing to street clothes and clean working clothes daily are essential.

Wear appropriate clothing to prevent any reasonable probability of skin contact with liquid aniline. Wear eye protection to prevent any reasonable probability of eye contact. Provide emergency showers.

Respirator Selection:
    100 ppm: CCROVF/GMOV/SAF/SCBAF
      Escape: GMOV/SCBA

Disposal Method Suggested: Incineration with provision for NO_x removal from flue gases by scrubber, catalytic or thermal device (A-31).

References

(1)   U.S. Environmental Protection Agency, *Chemical Hazard Information Profile: Aniline,* Washington, DC (January 20, 1978).
(2)   See Reference (A-61).
(3)   Sax, N.I., Ed., *Dangerous Properties of Industrial Materials Report, 1,* No. 3, 29-31, New York, Van Nostrand Reinhold Co. (1981).
(4)   See Reference (A-60).
(5)   Sax, N.I., Ed., *Dangerous Properties of Industrial Materials Report, 3,* No. 5, 37-40, New York, Van Nostrand Reinhold Co. (1983).
(6)   Parmeggiani, L., Ed., *Encyclopedia of Occupational Health & Safety,* Third Edition, Vol. 1, pp 153–155, Geneva, International Labour Office (1983).
(7)   United Nations Environment Programme, *IRPTC Legal File 1983,* Vol. 1, pp VII/42–44, Geneva, Switzerland, International Register of Potentially Toxic Chemicals (1984).

ANISIDINES

● Carcinogen (Positive for o-anisidine.HCl, NCI) (1) (A-62) (A-64).

Description: Anisidine, $H_2NC_6H_4OCH_3$, exists as ortho- and para-isomers which are respectively a colorless to pink liquid, boiling at 225°C and a reddish brown solid, melting at 57°C. They both have characteristic amine odors (fishy).

Code Numbers:

ortho-Isomer	CAS 92-15-9	RTECS BZ6500000 (HCl)	UN 2431
para-Isomer	CAS 104-94-9	RTECS BZ5450000	–
mixed isomers	CAS 29191-52-4	–	–

DOT Designation: St. Andrew's Cross required on label (X).

Synonyms: o-Methoxyaniline, p-methoxyaniline.

Potential Exposure: Anisidines are used in the manufacture of azo dyes.

Incompatibilities: Strong oxidizers.

Permissible Exposure Limits in Air: The Federal standard and ACGIH 1983/84 value of 0.1 ppm (0.5 mg/m³) has been set for the mixed isomers. No STEL value has been suggested. The IDLH level is 50 mg/m³. The TWA value bears the notation "skin" indicating the possibility of cutaneous absorption.

Determination in Air: Adsorption on silica gel, elution by ethanol and gas chromatographic analysis (A-10).

Permissible Concentration in Water: No criteria set.

Routes of Entry: Inhalation, ingestion, skin absorption, skin or eye contact.

Harmful Effects and Symptoms: Headaches, dizziness, cyanosis, red blood cell Heinz bodies.

Points of Attack: Blood, kidneys, liver, cardiovascular system.

Medical Surveillance: Consider the points of attack in preplacement and periodic physical examinations.

First Aid: If this chemical gets into the eyes, irrigate immediately. If this

chemical contacts the skin, wash with soap immediately. If a person breathes in large amounts of this chemical, move the exposed person to fresh air at once and perform artificial respiration. When this chemical has been swallowed, get medical attention. Give large quantities of water and induce vomiting. Do not make an unconscious person vomit.

Personal Protective Methods: Wear appropriate clothing to prevent any reasonable probability of skin contact. Wear eye protection to prevent any reasonable probability of eye contact. Employees should wash immediately when skin is wet or contaminated. Work clothing should be changed daily if clothing may be contaminated. Remove nonimpervious clothing immediately if wet or contaminated. Provide emergency showers.

Respirator Selection:
    2.5 mg/m^3: DMXS
     5 mg/m^3: DMXSQ/SA/SCBA
    25 mg/m^3: HiEPF/SAF/SCBAF
    50 mg/m^3: PAPHiE/SA:PD,PP,CF

Disposal Method Suggested: Incineration with provision for NO_x removal from flue gases by scrubbing or incineration.

References

(1) National Cancer Institute, *Bioassay of o-Anisidine Hydrochloride for Possible Carcinogenicity,* Technical Report Series No. 89, Bethesda, MD (1978).

(2) National Cancer Institute, *Bioassay of p-Anisidine Hydrochloride for Possible Carcinogenicity,* Technical Report Series No. 116, Bethesda, MD (1978).

(3) Sax, N.I., Ed., *Dangerous Properties of Industrial Materials Report, 1,* No. 5, 34-36, New York, Van Nostrand Reinhold Co. (1981).

(4) See Reference (A-61).

ANTIMONY AND COMPOUNDS

● Carcinogen (Antimony trioxide production only, ACGIH)
● Hazardous substances (Some compounds, EPA)
    Antimony compounds classified by EPA as hazardous substances are: antimony pentachloride, antimony potassium tartrate, antimony tribromide, antimony trichloride, antimony trifluoride, and antimony trioxide.
● Hazardous waste constituents (EPA)
● Priority toxic pollutant (EPA)

Description: Sb, antimony, is a silvery-white, soft metal insoluble in water and organic solvents.

Code Numbers: (Sb metal) CAS 7440-36-0 RTECS CC4025000 UN 1549 (for Sb compounds, n.o.s.) (Antimony trioxide) CAS 1309-64-4.

DOT Designation: (Antimony compounds n.o.s.) Poison. Use St Andrew's Cross (X) on label.

Synonyms: None.

Potential Exposure: Exposure to antimony may occur during mining, smelting or refining, alloy and abrasive manufacture, and typesetting in printing. Antimony is widely used in the production of alloys, imparting increased hardness, mechanical strength, corrosion resistance, and a low coefficient of friction.

Some of the important alloys are babbitt, pewter, white metal, Britannia metal and bearing metal (which are used in bearing shells), printing-type metal, storage battery plates, cable sheathing, solder, ornamental castings, and ammunition. Pure antimony compounds are used as abrasives, pigments, flameproofing compounds, plasticizers, and catalysts in organic synthesis; they are also used in the manufacture of tartar emetic, paints, lacquers, glass, pottery, enamels, glazes, pharmaceuticals, pyrotechnics, matches, and explosives.

In addition, they are used in dyeing, for blueing steel, and in coloring aluminum, pewter, and zinc. A highly toxic gas, stibine, may be released from the metal under certain conditions.

Incompatibilities: Oxidizers, acids, halogenated acids.

Permissible Exposure Limits in Air: The Federal standard for antimony and its compounds is 0.5 mg/m³, expressed as Sb (see also Stibine). No STEL values have been set by ACGIH. Further, antimony trioxide production has been categorized as "Industrial Substances Suspect of Carcinogenic Potential for Man" by ACGIH as of 1983/84. The IDLH value is 80 mg/m³.

Determination in Air: Collection by a particulate filter, workup with acid, analysis by atomic absorption. See NIOSH Methods, Set A. See also reference (A-10).

Permissible Concentration in Water: To protect freshwater aquatic life— on an acute basis, 9,000 µg/ℓ and on a chronic basis, 1,600 µg/ℓ. To protect saltwater aquatic life—no criterion due to insufficient data. To protect human health—146 µg/ℓ (A-33). EPA has also suggested an ambient limit of 7 µg/ℓ (A-37) based on health effects.

Determination in Water: Digestion followed by atomic absorption. See EPA Methods 204.1 and 204.2.

Routes of Entry: Inhalation of dust or fume, skin and eye contact.

Harmful Effects and Symptoms: *Local* — Antimony and its compounds are generally regarded as primary skin irritants. Lesions generally appear on exposed, moist areas of the body, but rarely on the face. The dust and fumes are also irritants to the eyes, nose, and throat, and may be associated with gingivitis, anemia, and ulceration of the nasal septum and larynx. Antimony trioxide causes a dermatitis known as "antimony spots." This form of dermatitis results in intense itching followed by skin eruptions. A diffuse erythema may occur, but usually the early lesions are small erythematous papules. They may enlarge, however, and become pustular. Lesions occur in hot weather and are due to dust accumulating on exposed areas that are moist due to sweating. No evidence of eczematous reaction is present, nor an allergic mechanism.

Systemic — Systemic intoxication is uncommon from occupational exposure. However, miners of antimony may encounter dust containing free silica; cases of pneumoconiosis in miners have been termed "silico-antimoniosis." Antimony pneumoconiosis, per se, appears to be a benign process.

Antimony metal dust and fumes are absorbed from the lungs into the blood stream. Principal organs attacked include certain enzyme systems (protein and carbohydrate metabolism), heart, lungs, and the mucous membrane of the respiratory tract. Symptoms of acute oral poisoning include violent irritation of the nose, mouth, stomach, and intestines, vomiting, bloody stools, slow shallow respiration, pulmonary congestion, coma, and sometimes death due to circulatory or respiratory failure. Chronic oral poisoning presents symptoms of dry throat, nausea, headache, sleeplessness, loss of appetite, and dizziness. Liver and kidney degenerative changes are late manifestations.

Antimony compounds are generally less toxic than antimony. Antimony trisulfide, however, has been reported to cause myocardial changes in man and experimental animals. Antimony trichloride and pentachloride are highly toxic and can irritate and corrode the skin. Antimony fluoride is extremely toxic, particularly to pulmonary tissue and skin.

Points of Attack: Respiratory system, cardiovascular system, skin, eyes and lungs.

Medical Surveillance: Preemployment and periodic examinations should give special attention to lung disease, skin disease, diseases of the nervous system, heart and gastrointestinal tract. Lung function, EKGs, blood, and urine should be evaluated periodically. See also reference (6).

First Aid: If this chemical contacts the skin, wash with soap immediately. When this chemical has been swallowed, get medical attention. Give large quantities of water and induce vomiting. Do not make an unconscious person vomit.

Personal Protective Methods: A combination of protective clothing, barrier creams, gloves, and personal hygiene will protect the skin. Washing and showering facilities should be available and eating should not be permitted in exposed areas. Dust masks and supplied air respirators should be available in all areas where the Federal standard is exceeded.

Wear appropriate clothing to prevent any reasonable possibility of skin contact. Wear eye protection to prevent any reasonable probability of eye contact. Employees should wash promptly when skin is contaminated. Work clothing should be changed daily if it may be contaminated. Remove nonimpervious clothing promptly if contaminated.

Respirator Selection:

DM, 2.5 mg/m^3:	DMXS
5 mg/m^3:	DMXSQ
DMFu, 5 mg/m^3:	FuHiEP/SA/SCBA
25 mg/m^3:	HiEPF/SAF/SCBAF
80 mg/m^3:	PAPHiEF/SAF.PD,PP,CF
Escape:	DMXS/SCBA

Disposal Method Suggested: Antimony pentasulfide, antimony sulfate, antimony trisulfide—Disposal in a chemical waste landfill.

Antimony trifluoride, antimony pentafluoride—The compound is dissolved in dilute HCl and saturated with H_2S. The precipitate (antimony sulfide) is filtered, washed and dried. The filtrate is air stripped of dissolved H_2S and passed into an incineration device equipped with a lime scrubber. The stripped filtrate is reacted with excess lime, the precipitate ($CaF_2 \cdot CaCl_2$) mixture is disposed of by land burial.

Antimony pentachloride—When dissolved in water and neutralized, the slightly soluble oxide is formed. Removal of the oxide is followed by sulfide precipitation to ensure the removal of the metal ion from solution. The antimony oxides can be sent to a refiner or placed in long-term storage.

Antimony potassium tartrate—Dissolve wastes in water, acidify and precipitate the sulfide using hydrogen sulfide as the reactant. The antimony sulfide precipitate should be returned to suppliers or manufacturers for reprocessing or be placed into long-term storage.

Antimony trichloride—Same as antimony pentachloride above.

Recovery and recycle is an option to disposal which should be considered for scrap antimony and spent catalysts containing antimony (A-57).

References

(1) U.S. Environmental Protection Agency, *Antimony: Ambient Water Quality Criteria,* Washington, DC (1980).

(2) Environmental Protection Agency, *Literature Study of Selected Potential Environmental Contaminants: Antimony and Its Compounds,* Report No. EPA-560/2-76-002, Washington, DC, Office of Toxic Substances (February 1976).

(3) Nat. Inst. for Occup. Safety and Health, *Environmental Exposure to Airborne Contaminants in the Antimony Industry,* NIOSH Publ. No. 79-140, Cincinnati, OH (August 1979).

(4) U.S. Environmental Protection Agency, *Toxicology of Metals, Vol II: Antimony,* Report EPA-600-/1-77-022, Research Triangle Park, NC, pp 15-29 (May 1977).

(5) U.S. Environmental Protection Agency, *Antimony,* Health and Environmental Effects Profile No. 10, Washington, DC, Office of Solid Waste (April 30, 1980).

(6) Nat. Inst. for Occup. Safety and Health, *A Guide to the Work Relatedness of Disease,* Revised Edition, DHEW (NIOSH) Publication No. 79-116, Cincinnati, OH, pp. 28-39 (January 1979).

(7) Sax, N.I., Ed., *Dangerous Properties of Industrial Materials Report, 2,* No. 1, 68-76, New York, Van Nostrand Reinhold Co. (1982). (Antimony, Antimony Trichloride, Antimony Trioxide).

(8) Sax, N.I., Ed., *Dangerous Properties of Industrial Materials Report, 1,* No. 8, 33-36, New York, Van Nostrand Reinhold Co. (1981). (Antimony Tartrate, Antimony Trifluoride).

(9) See Reference (A-60).

(10) Sax., N.I., Ed., *Dangerous Properties of Industrial Materials Report, 3,* No. 5, 40-42, New York, Van Nostrand Reinhold Co. (1983). (Antimony Trifluoride).

(11) Sax, N.I., Ed., *Dangerous Properties of Industrial Materials Report, 3,* No. 5, 42-43, New York, Van Nostrand Reinhold Co. (1983). (Antimony Tribromide).

(12) Parmeggiani, L., Ed., *Encyclopedia of Occupational Health & Safety,* Third Edition, Vol. 1, pp 176–179, Geneva, International Labour Office (1983).

(13) United Nations Environment Programme, *IRPTC Legal File 1983,* Vol. I, pp VII/47–49, Geneva, Switzerland, International Register of Potentially Toxic Chemicals (1984).

ANTU

Description: ANTU,

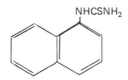

α-naphthylthiourea, is a colorless, odorless crystalline compound melting at 198°C.

Code Numbers: CAS 86-88-4   RTECS YT9275000   UN 1651

DOT Designation: —

Synonyms: α-naphthylthiocarbamide, Krysid®.

Potential Exposure: In production of ANTU (A-32) or its formulations and use as a rodenticide.

Incompatibilities: Strong oxidizers.

Permissible Exposure Limits in Air: The TWA value is 0.3 mg/m³ as of

1983/84 (ACGIH). The STEL value proposed is 0.9 mg/m^3. The IDLH level is 100 mg/m^3.

Determination in Air: Collection on a filter and analysis by gas-liquid chromatography (A-1).

Permissible Concentration in Water: No criteria set.

Routes of Entry: Inhalation and ingestion.

Harmful Effects and Symptoms: Vomiting, dyspnea, cyanosis, coarse pulmonary rales after ingestion of large doses.

Points of Attack: Respiratory system.

Medical Surveillance: Consider the points of attack in preplacement and periodic physical examinations.

First Aid: If this chemical gets into the eyes, irrigate immediately. If this chemical contacts the skin, dust off solid, flush with water, wash with soap promptly. If a person breathes in large amounts of this chemical, move the exposed person to fresh air at once and perform artificial respiration. When this chemical has been swallowed, get medical attention. Give large quantities of water and induce vomiting. Do not make an unconscious person vomit.

Personal Protective Methods: Change clothing daily after work if it may be contaminated.

Respirator Selection:
```
     3 mg/m³:  CCROVDMPest/SA/SCBA
    15 mg/m³:  CCROVFDMPest/GMOVDMPest/SAF/SCBAF
   100 mg/m³:  SA:PD,PP,CF/CCROVHiEPest
     Escape:  GMOVPPest/SCBA
```

Disposal Method Suggested: Incinerate in a furnace equipped with an alkaline scrubber.

References

(1) See Reference (A-61).
(2) Sax, N.I., Ed., *Dangerous Properties of Industrial Materials Report, 4,* No. 2, 83–86, New York, Van Nostrand Reinhold Co. (1984).
(3) United Nations Environment Programme, *IRPTC Legal File 1983,* Vol. II, pp VII/787–88, Geneva, Switzerland, International Register of Potentially Toxic Chemicals (1984).

ARAMITE®

See "Sulfurous Acid, 2-(p-t-Butylphenoxy)-1-methylethyl-2-chloroethyl ester."

ARGON

Description: With the symbol A, argon is one of the elements in the inert gas category. It is colorless.

Code Numbers: CAS 7440-37-1   RTECS CF2300000   UN 1951 (liquid) UN 1006 (compressed gas).

DOT Designation: Nonflammable gas.

Potential Exposures: Argon is used as an inert gas shield in arc welding; it is used to fill electric lamps. It is used as a blanketing agent in metals refining (especially titanium and zirconium).

Permissible Exposure Limits in Air: There is no Federal standard. ACGIH lists argon as a simple asphyxiant with no specified TLV.

Permissible Concentration in Water: No criteria set.

Routes of Entry: Inhalation and possibly skin contact with liquid argon.

Harmful Effects and Symptoms: The gas is a simple asphyxiant as noted above. The liquid can cause frostbite.

Disposal Method Suggested: Vent to atmosphere.

References

(1) Sax, N.I., Ed., *Dangerous Properties of Industrial Materials Report, 1,* No. 5, 36-37, New York, Van Nostrand Reinhold Co. (1981).

ARSENIC AND ARSENIC COMPOUNDS

- Carcinogen (IARC) (11)
- Hazardous substances (Some compounds, EPA)
    Arsenic compounds classified by EPA as hazardous substances are: arsenic disulfide, arsenic pentoxide, arsenic trichloride, arsenic trioxide and arsenic trisulfide. Also the EPA has issued rebuttable presumptions against registration (RPAR's) for several arsenic-containing pesticides as follows: arsenic acid, cacodylic acid, calcium arsenate, DSMA, lead arsenate, MSMA and sodium arsenite.
- Hazardous waste constituents (EPA)
- Priority toxic pollutant (EPA)

Description: As, elemental arsenic, occurs to a limited extent in nature as a steel-gray metal that is insoluble in water. Arsenic in this discussion includes the element and any of its inorganic compounds excluding arsine. Arsenic trioxide (As_2O_3), the principal form in which the element is used, is frequently designated as arsenic, white arsenic, or arsenous oxide. Arsenic is present as an impurity in many other metal ores and is generally produced as arsenic trioxide as a by-product in the smelting of these ores, particularly copper. Most other arsenic compounds are produced from the trioxide.

Code Numbers: (Element) CAS 7440-38-2   RTECS CG0525000   UN 1558

Type of Compound/Label Designation: Poison.

Synonyms: None.

Potential Exposure: Arsenic compounds have a variety of uses. Arsenates and arsenites are used in agriculture as insecticides, herbicides, larvicides, and pesticides. Arsenic trichloride is used primarily in the manufacture of pharmaceuticals. Other arsenic compounds are used in pigment production, the manufacture of glass as a bronzing or decolorizing agent, the manufacture of opal glass and enamels, textile printing, tanning, taxidermy, and antifouling paints. They are also used to control sludge formation in lubricating oils. Metallic arsenic is used as an alloying agent to harden lead shot and in lead-base bearing materials. It is also alloyed with copper to improve its toughness and corrosion resistance.

EPA estimates that more than 6 million people living within 12 miles of major sources—copper, zinc, and lead smelters—may be exposed to 10 times the average U.S. atmospheric levels of arsenic. The agency says that 40,000 people living near some copper smelters may be exposed to 100 times the national atmospheric average.

Permissible Exposure Limits in Air: The Federal standard for arsenic and its compounds was previously 0.5 mg/m^3 of air as As. In 1973, NIOSH proposed (1) the lower recommended standard of 0.05 mg As/m^3 of air determined as a time-weighted average (TWA) exposure for up to a 10-hour workday, 40-hour workweek. Then, in November 1975, OSHA proposed a workplace exposure limit for inorganic arsenic at 4 μg/m^3 (8-hour, TWA). The economic impact of such a standard has been assessed (2). The previous standard of 500 μg/m^3 for all forms of arsenic would remain in effect only for organic forms.

A 1975 NIOSH document (3) proposed that inorganic arsenic be controlled so that no worker is exposed to a concentration of arsenic in excess of 0.002 mg/m^3 (2.0 μg) as determined by a 15-minute sampling period. Finally in 1978 a standard was promulgated (4) which limits occupational exposure to inorganic arsenic to 10 μg/m^3 (μg/m^3 of air) based on an 8-hour time-weighted average.

The ACGIH (1983/84) TWA value for arsenic and soluble compounds (as As) is 0.2 mg/m^3. Arsenic trioxide production is categorized as "suspect of carcinogenic potential for man." As a first step toward regulating industrial emissions of inorganic arsenic, EPA has listed the substance as a hazardous air pollutant, as defined under the Clean Air Act and the agency's proposed airborne carcinogen policy.

Determination in Air: Collection on a filter and analysis by atomic absorption spectrometry (A-1). See also (A-10).

Permissible Concentration in Water: To protect freshwater aquatic life—total recoverable trivalent inorganic arsenic never to exceed 440 μg/ℓ. To protect saltwater aquatic life—508 μg/ℓ on an acute basis. To protect human health—preferably zero. A value of 0.02 μg/ℓ corresponds to a human health risk of 1 in 100,000. EPA has established a maximum arsenic level of 0.05 mg/ℓ. This does not address carcinogenicity and is under review.

Allowable arsenic levels in drinking water have also been set as follows (A-65):

South African Bureau of Standards	0.05 mg/ℓ
World Health Organization	0.05 mg/ℓ
Federal Republic of Germany (1975)	0.04 mg/ℓ

Determination in Water: Total arsenic may be determined by digestion followed by silver diethyldithiocarbamate; an alternative is atomic absorption; another is inductively coupled plasma (ICP) optical emission spectrometry.

Routes of Entry: Inhalation and ingestion of dust and fumes.

Harmful Effects and Symptoms: *Local* — Trivalent arsenic compounds are corrosive to the skin. Brief contact has no effect, but prolonged contact results in a local hyperemia and later vesicular or pustular eruption. The moist mucous membranes are most sensitive to the irritant action. Conjunctiva, moist and macerated areas of the skin, eyelids, the angles of the ears, nose, mouth, and respiratory mucosa are also vulnerable to the irritant effects. The wrists are common sites of dermatitis, as are the genitalia if personal hygiene is poor. Perforations of the nasal septum may occur. Arsenic trioxide and pentoxide are capable of producing skin sensitization and contact dermatitis. Arsenic is also capable of producing keratoses, especially of the palms and soles. Arsenic has been cited as a cause of skin cancer, but the incidence is low.

Systemic — The acute toxic effects of arsenic are generally seen following ingestion of inorganic arsenical compounds. This rarely occurs in an industrial setting. Symptoms develop within ½ to 4 hours following ingestion and are usually characterized by constriction of the throat followed by dysphagia, epigastric pain, vomiting, and watery diarrhea. Blood may appear in vomitus and stools. If the amount ingested is sufficiently high, shock may develop due to severe fluid loss, and death may ensue in 24 hours. If the acute effects are survived, exfoliative dermatitis and peripheral neuritis may develop.

Cases of acute arsenical poisoning due to inhalation are exceedingly rare in industry. When it does occur, respiratory tract symptoms—cough, chest pain, dyspnea—giddiness, headache, and extreme general weakness precede gastrointestinal symptoms. The acute toxic symptoms of trivalent arsenical poisoning are due to severe inflammation of the mucous membranes and greatly increased permeability of the blood capillaries.

Chronic arsenical poisoning due to ingestion is rare and generally confined to patients taking prescribed medications. However, it can be a concomitant of inhaled inorganic arsenic from swallowed sputum and improper eating habits. Symptoms are weight loss, nausea and diarrhea alternating with constipation, pigmentation and eruption of the skin, loss of hair, and peripheral neuritis. Chronic hepatitis and cirrhosis have been described. Polyneuritis may be the salient feature, but more frequently there are numbness and paresthesias of "glove and stocking" distribution. The skin lesions are usually melanotic and keratotic and may occasionally take the form of an intradermal cancer of the squamous cell type, but without infiltrative properties. Horizontal white lines (striations) on the fingernails and toenails are commonly seen in chronic arsenical poisoning and are considered to be a diagnostic accompaniment of arsenical polyneuritis.

Inhalation of inorganic arsenic compounds is the most common cause of chronic poisoning in the industrial situation. This condition is divided into three phases based on signs and symptoms.

First Phase: The worker complains of weakness, loss of appetite, some nausea, occasional vomiting, a sense of heaviness in the stomach, and some diarrhea.

Second Phase: The worker complains of conjunctivitis, and a catarrhal state of the mucous membranes of the nose, larynx, and respiratory passages. Coryza, hoarseness, and mild tracheobronchitis may occur. Perforation of the nasal septum is common, and is probably the most typical lesion of the upper respiratory tract in occupational exposure to arsenical dust. Skin lesions, eczematoid and allergic in type, are common.

Third Phase: The worker complains of symptoms of peripheral neuritis, initially of hands and feet, which is essentially sensory. In more severe cases, motor paralyses occur; the first muscles affected are usually the toe extensors and the peronei. In only the most severe cases will paralysis of flexor muscles of the feet or of the extensor muscles of hands occur.

Liver damage from chronic arsenical poisoning is still debated, and as yet the question is unanswered. In cases of chronic and acute arsenical poisoning, toxic effects to the myocardium have been reported based on EKG changes. These findings, however, are now largely discounted and the EKG changes are ascribed to electrolyte disturbances concomitant with arsenicalism. Inhalation of arsenic trioxide and other inorganic arsenical dusts does not give rise to radiological evidence of pneumoconiosis. Arsenic does have a depressant effect upon the bone marrow, with disturbances of both erythropoiesis and myelopoiesis. Evidence is now available incriminating arsenic compounds as a cause of lung cancer as well as skin cancer.

Skin cancer in humans is causally associated with exposure to inorganic arsenic compounds in drugs, drinking water and the occupational environment. The risk of lung cancer was increased 4 to 12 times in certain smelter workers who inhaled high levels of arsenic trioxide. However, the influence of other constituents of the working environment cannot be excluded in these studies. Case reports have suggested an association between exposure to arsenic compounds and blood dyscrasias and liver tumours (14).

Points of Attack: Skin, eyes, respiratory system.

Medical Surveillance: In preemployment physical examinations, particular attention should be given to allergic and chronic skin lesions, eye disease, psoriasis, chronic eczematous dermatitis, hyperpigmentation of skin, keratosis and warts, baseline weight, baseline blood and hemoglobin count, and baseline urinary arsenic determinations. In annual examinations, the worker's general health, weight, and skin condition should be checked, and the worker observed for any evidence of excessive exposure or absorption of arsenic. Chest x-rays and lung function should be evaluated; analysis of urine, hair, or nails for arsenic should be made every 60 days as long as exposure continues. See also reference (10).

First Aid: Irrigate eyes with water. Wash contaminated areas of body with soap and water.

Personal Protective Methods: Workers should be trained in personal hygiene and sanitation, the use of personal protective equipment, and early recognition of symptoms of absorption, skin contact irritation, and sensitivity. With the exception of arsine and arsenic trichloride, the compounds of arsenic do not have odor or warning qualities. In case of emergency or areas of high dust or spray mist, workers should wear respirators that are supplied-air or self-contained positive-pressure type with fullface mask. Where concentrations are less than 100 x standard, workers may be able to use halfmask respirators with replaceable dust or fume filters. Protective clothing, gloves, goggles and a hood for head and neck should be provided. When liquids are processed, impervious clothing should be supplied. Clean work clothes should be supplied daily and the workers should shower prior to changing to street clothes.

Respirator Selection: See reference (1).

Disposal Method Suggested (A-31): Arsenic—elemental arsenic wastes should be placed in long-term storage or returned to suppliers or manufacturers for reprocessing. Arsenic pentaselenide—wastes should be placed in long-term storage or returned to suppliers or manufacturers for reprocessing. Arsenic trichloride—hydrolyze to arsenic trioxide utilizing scrubbers for hydrogen chloride abatement. The trioxide may then be placed in long-term storage. Arsenic trioxide—long-term storage in large siftproof and weatherproof silos. This compound may also be dissolved, precipitated as the sulfide and returned to the suppliers (A-38).

References

(1) National Institute for Occupational Safety and Health, *Criteria for a Recommended Standard: Occupational Exposure to Inorganic Arsenic,* NIOSH Doc. No. 74-110, Washington, DC (1973).

(2) U.S. Department of Labor, *Inflationary Impact Statement: Inorganic Arsenic,* Washington, DC, Occupational Safety and Health Administration (undated, presumed 1975).

(3) National Institute for Occupational Safety and Health, *Criteria for a Recommended Standard: Occupational Exposure to Inorganic Arsenic (Revised),* NIOSH Doc. No. 75-149, Washington, DC (1975).

(4) *Federal Register, 43,* No. 88, 19584-19631 (May 5, 1978).

(5) U.S. Environmental Protection Agency, *Arsenic: Ambient Water Quality Criteria,* Washington, DC (1979).

(6) U.S. Environmental Protection Agency, *Status Assessment of Toxic Chemicals: Arsenic,* Report No. EPA-600/2-79-210B, Washington, DC (December 1979).

(7) U.S. Environmental Protection Agency, *Toxicology of Metals, Vol II: Arsenic,* Report EPA-600/1-77-022, Research Triangle Park, NC, pp 30-70 (May 1977).

(8) National Academy of Sciences, *Medical and Biological Effects of Environmental Pollutants: Arsenic,* Washington, DC (1977).

(9) U.S. Environmental Protection Agency, *Arsenic,* Health and Environmental Effects Profile No. 11, Office of Solid Waste, Washington, DC (April 30, 1980).

(10) National Inst. for Occup. Safety and Health, *A Guide to the Work Relatedness of Disease,* Revised Edition, DHEW (NIOSH) Publication No. 79-116, Cincinnati, OH, pp 40-53, (January 1979).

(11) International Agency for Research on Cancer, *IARC Monographs on the Carcinogenic Risks of Chemicals to Humans,* Lyon, France, 2, 48 (1973).

(12) Sax, N.I., Ed., *Dangerous Properties of Industrial Materials Report, 1,* No. 3, 32-34, New York, Van Nostrand Reinhold Co. (1981).

(13) Sax, N.I., Ed., *Dangerous Properties of Industrial Materials Report, 2,* No. 3, 56-63, New York, Van Nostrand Reinhold Co. (1982). (Arsenic Acid, Arsenic Pentoxide, Arsenic Tribromide).

(14) See Reference (A-62). Also Reference (A-64).

(15) See Reference (A-60) for Arsenic Trichloride and Arsenic Trioxide.

(16) Lederer, W.H. and Fensterheim, R.J., *Arsenic: Industrial, Biomedical and Environmental Perspectives,* New York, Van Nostrand Reinhold Co. (1983).

(17) Sax, N.I., Ed., *Dangerous Properties of Industrial Materials Report, 3,* No. 5, 44-50, New York, Van Nostrand Reinhold Co. (1983). (Arsenic Sulfide).

(18) Sax, N.I., Ed., *Dangerous Properties of Industrial Materials Report, 3,* No. 5, 50-58, New York, Van Nostrand Reinhold Co. (1983). (Arsenic Trioxide).

(19) Parmeggiani, L., Ed., *Encyclopedia of Occupational Health & Safety,* Third Edition, Vol. 1, pp 179–182, Geneva, International Labour Office (1983).

(20) United Nations Environment Programme, *IRPTC Legal File 1983,* Vol. I, pp VII/50–55, Geneva, Switzerland, International Register of Potentially Toxic Chemicals (1984).

ARSINE

Description: AsH_3, arsine, is a colorless gas with a slight garliclike odor which cannot be considered a suitable warning property in concentrations below 1 ppm.

Code Numbers: CAS 7784-42-1   RTECS CG6475000   UN 2188

DOT Designation: Poison gas and flammable gas.

Synonyms: Hydrogen arsenide, arseniuretted hydrogen, arsenic trihydride.

Potential Exposure: Arsine is not widely used in any industry and may be generated by side reactions or unexpectedly; e.g., it may be generated in metal pickling operations, metal drossing operations, or when inorganic arsenic compounds contact sources of nascent hydrogen. It has been known to occur as an impurity in acetylene. Most occupational exposure occurs in chemical, smelting, and refining industries. Cases of exposure have come from workers dealing with zinc, tin, cadmium, galvanized coated aluminum, and silicon and steel metals.

Incompatibilities: Strong oxidizers, chlorine, nitric acid.

Permissible Exposure Limits in Air: The Federal standard and ACGIH 1983/84 value for arsine is 0.05 ppm (0.2 mg/m^3). NIOSH has recommended that arsine be controlled to the same concentration as other forms of inorganic arsenic (0.002 mg/m^3). The IDLH level is 6 ppm.

Determination in Air: Charcoal adsorption followed by elution with nitric acid and flawless atomic absorption in a high temperature graphite analyzer. See NIOSH Methods, Set O. See also reference (A-10).

Permissible Concentration in Water: No criteria set, but EPA (A-37) cites the same limits as earlier proposed for arsenic by EPA (A-3) of 50 μg/ℓ based on health effects and 10 μg/ℓ based on ecological effects.

Route of Entry: Inhalation of gas.

Harmful Effects and Symptoms: *Local* — High concentrations of arsine gas will cause damage to the eyes. Most experts agree, however, that before this occurs systemic effects can be expected.

Systemic — Arsine is an extremely toxic gas that can be fatal if inhaled in sufficient quantities. Acute poisoning is marked by a triad of main effects caused by massive intravascular hemolysis of the circulating red cells. Early effects may occur within an hour or two and are commonly characterized by general malaise, apprehension, giddiness, headache, shivering, thirst, and abdominal pain with vomiting. In severe acute cases the vomitus may be blood stained and diarrhea ensues as with inorganic arsenical poisoning. Pulmonary edema has occurred in severe acute poisoning.

Invariably, the first sign observed in arsine poisoning is hemoglobinuria, appearing with discoloration of the urine up to port wine hue (first of the triad). Jaundice (second of triad) sets in on the second or third day and may be intense, coloring the entire body surface a deep bronze hue. Coincident with these effects is a severe hemolytic-type anemia. Severe renal damage may occur with oliguria or complete suppression of urinary function (third of triad), leading to uremia and death. Severe hepatic damage may also occur, along with cardiac damage and EKG changes. Where death does not occur, recovery is prolonged.

In cases where the amount of inhaled arsine is insufficient to produce acute effects, or where small quantities are inhaled over prolonged periods, the hemoglobin liberated by the destruction of red cells may be degraded by the reticuloendothelial system and the iron moiety taken up by the liver, without producing permanent damage. Some hemoglobin may be excreted unchanged by the kidneys. The only symptoms noted may be general tiredness, pallor, breathlessness on exertion, and palpitations as would be expected with severe secondary anemia.

Points of Attack: Blood, kidneys, liver.

Medical Surveillance: In preemployment physical examinations, special attention should be given to past or present kidney disease, liver disease, and anemia. Periodic physical examinations should include tests to determine arsenic levels in the blood and urine. The general condition of the blood and the renal and liver functions should also be evaluated. Since arsine gas is a by-product of certain production processes, workers should be trained to recognize the symptoms of exposure and to use appropriate personal protective equipment.

First Aid: If a person breathes in large amounts of this chemical, move the exposed person to fresh air at once and perform artificial respiration. Dimercaprol treatment is indicated and blood transfusions may be necessary (A-39).

Personal Protective Methods: In most cases, arsine poisoning cannot be anticipated except through knowledge of the production processes. Where arsine is suspected in concentrations above the acceptable standard, the worker should be supplied with a supplied air fullface respirator or a self-contained positive pressure respirator with full facepiece.

Respirator Selection:

0.5 ppm:	SA/SCBA
2.5 ppm:	SAF/SCBAF
6 ppm:	SA:PD,PP,CF
Escape:	GMS/SCBA

Disposal Method Suggested: Arsine may be disposed of by controlled burning. When possible, cylinders should be sealed and returned to suppliers.

References

(1)   See Reference (A-60).

(2)   Parmeggiani, L., Ed., *Encyclopedia of Occupational Health & Safety,* Third Edition, Vol. 1, pp 183–184, Geneva, International Labour Office (1983).

ASBESTOS

- Carcinogen (Human positive, IARC) (10) (A-62) (A-64)
- Hazardous waste (EPA)
- Priority toxic pollutant (EPA)

Description: Asbestos is a generic term that applies to a number of naturally occurring, hydrated mineral silicates incombustible in air and separable into filaments. The most widely used in industry in the United States is chrysotile, a fibrous form of serpentine. Other types include amosite, crocidolite, tremolite, anthophyllite, and actinolite.

Code Numbers:

CAS 1332-21-4	RTECS CI6475000	UN 2212 (Blue)
		2590 (White)

DOT Designation: ORM-C.

Synonyms: None.

Potential Exposure: Most asbestos is used in the construction industry. Approximately 92% of the half million tons used in the U.S. construction industry is firmly bonded, i.e., the asbestos is "locked in" in such products as floor tiles, asbestos cements, and roofing felts and shingles; while the remaining 8% is friable or in powder form present in insulation materials, asbestos cement powders, and acoustical products.

As expected, these latter materials generate more airborne fibers than the firmly bonded products. The 187,400 tons of asbestos used in nonconstruction industries is utilized in such products as textiles, friction material including brake linings, and clutch facings, paper, paints, plastics, roof coatings, floor tiles, and miscellaneous other products (1).

Significant quantities of asbestos fibers appear in rivers and streams draining from areas where asbestos-rock outcroppings are found. Some of these out-croppings are being mined. Asbestos fibers have been found in a number of drinking water supplies, but the health implications of ingesting asbestos are not fully documented. Emissions of asbestos fibers into water and air are known to result from mining and processing of some minerals. Asbestiform fibers in the drinking water of Duluth and nearby communities at levels of 12 million fibers per liter have been attributed to the discharge of 67,000 tons of taconite tailings per day into Lake Michigan.

Exposure to asbestos fibers may occur throughout urban environments. A recent study of street dust in Washington, DC, showed approximately 50,000 fibers per gram, much of which appeared to come from brake linings. Autop-sies of New York City residents with no known occupational exposure showed 24 of 28 lung samples to contain asbestos fibers, perhaps resulting from asbestos from brake linings and the flaking of sprayed asbestos insulation material (2).

Incompatibilities: None.

Permissible Exposure Limits in Air: When an asbestos criteria document was first published by NIOSH in 1972 (1), that agency recommended a standard of 2.0 asbestos fibers/cc of air based on a count of fibers greater than 5 μm in length. This standard was recommended with the stated belief that it would "prevent" asbestosis and with the open recognition that it would not "prevent" asbestos-induced neoplasms.

Furthermore, data were presented which supported the fact that technology was available to achieve that standard and that the criteria would be subject to review and revision as necessary. Since the time that the asbestos criteria were published in 1972, sufficient additional data regarding asbestos-related disease have been developed to warrant reevaluation.

On June 7, 1972, the Occupational Safety and Health Administration (OSHA) promulgated a standard for occupational exposure to asbestos con-taining an 8-hour time-weighted average (TWA) concentration exposure limit of 5 fibers longer than 5 μm/cc of air, with a ceiling limitation against any exposure in excess of 10 such fibers/cc. The standard further provided that the 8-hour TWA was to be reduced to 2 fibers/cc on July 1, 1976.

As the result of a court case, OSHA decided that to achieve the most feasible occupational health protection, a reexamination of the standard's general premises and general structure was necessary. To this end, on October 9, 1975, OSHA announced a proposed rule-making to lower the exposure limit to an 8-hour TWA concentration of 0.5 asbestos fibers longer than 5 μm/cc of air with a ceiling concentration of 5 fibers/cc of air determined by a sampling of up to 15 minutes. On December 2, 1975, OSHA requested NIOSH to reevalu-ate the information available on the health effects of occupational exposure to asbestos fibers and to advise OSHA on the results of this study. In November of 1983 an emergency standard was promulgated by OSHA limiting exposure over an 8-hour period to 0.5 fibers/cc of air.

Indeed NIOSH stated in 1976 (3), that the standard should be set at the lowest level detectable by available analytical techniques, an approach consis-tent with NIOSH's most recent recommendation for other carcinogens (i.e., arsenic and vinyl chloride). Such a standard should also prevent the development of asbestosis.

Since phase contrast microscopy is the only generally available and practical analytical technique at the present time, this level is defined as 100,000 fibers >5 μm in length/m^3 (0.1 fiber/cc), on an 8-hour-TWA basis with peak concentrations not exceeding 500,000 fibers >5 μm in length/m^3 (0.5 fiber/cc) based on a 15-minute sample period.

This recommended standard of 100,000 fibers >5 μm in length/m^3 is intended to (A) protect against the noncarcinogenic effects of asbestos, (B) materially reduce the risk of asbestos-induced cancer (only a ban can assure protection against carcinogenic effects of asbestos) and (C) be measured by techniques that are valid, reproducible, and available to industry and official agencies.

The ACGIH (1983/84) has categorized asbestos (all forms) as a human carcinogen but has specified TLV values as follows:

Amosite 0.5 fiber >5 μm/cc
Chrysotile 2.0 fibers >5 μm/cc
Crocidolite 0.2 fiber >5 μm/cc
Other forms 2.0 fibers >5 μm/cc

The economic impact of the proposed OSHA standard for asbestos has been assayed (4).

Determination in Air: Sampling and analytical techniques should be performed as specified by NIOSH publication USPHS/NIOSH Membrane Filter Method for Evaluating Airborne Asbestos Fibers—T.R. 84 (1976). The actual determination involves a microscopic fiber count. See also reference (A-10).

Permissible Concentration in Water: To protect freshwater and saltwater aquatic life—no criteria have been established due to insufficient data. To protect human health—preferably zero. A lifetime cancer risk of 1 in 100,000 corresponds to a concentration of 300,000 fibers/ℓ (5).

Routes of Entry: Inhalation and ingestion.

Harmful Effects and Symptoms: Available studies provided conclusive evidence (3)(13) that exposure to asbestos fibers causes cancer and asbestosis in man. Lung cancers and asbestosis have occurred following exposure to chrysotile, crocidolite, amosite, and anthophyllite.

Mesotheliomas, lung and gastrointestinal cancers have been shown to be excessive in occupationally exposed persons, while mesotheliomas have developed also in individuals living in the neighborhood of asbestos factories and near crocidolite deposits, and in persons living with asbestos workers. Asbestosis has been identified among persons living near anthophyllite deposits.

Likewise, all commercial forms of asbestos are carcinogenic in rats, producing lung carcinomas and mesotheliomas following their inhalation, and mesotheliomas after intrapleural or i.p. injection. Mesotheliomas and lung cancers were induced following even 1 day's exposure by inhalation.

The size and shape of the fibers are important factors; fibers less than 0.5 μm in diameter are most active in producing tumors. Other fibers of a similar size including glass fibers, can also produce mesotheliomas following intrapleural or i.p. injection.

There are data that show that the lower the exposure, the lower the risk of developing cancer. Excessive cancer risks have been demonstrated at all fiber concentrations studied to date. Evaluation of all available human data provides no evidence for a threshold or for a "safe" level of asbestos exposure.

Point of Attack: Lungs.

Medical Surveillance: Medical surveillance is required (1), except where a variance from the medical requirements of this proposed standard have been granted, for all workers who are exposed to asbestos as part of their work environment. For the purposes of this requirement the term "exposed to asbestos" will be interpreted as referring to time-weighted average exposures above 1 fiber/cc or peak exposures above 5 fibers/cc.

The major objective of such surveillance will be to ensure proper medical management of individuals who show evidence of reaction to past dust exposures, either due to excessive exposures or unusual susceptibility. Medical management may range from recommendations as to job placement, improved work practices, cessation of smoking, to specific therapy for asbestos-related disease or its complications.

Medical surveillance cannot be a guide to adequacy of current controls when environmental data and medical examinations only cover recent work experience because of the prolonged latent period required for the development of asbestosis and neoplasms.

Required components of a medical surveillance program include periodic measurements of pulmonary function [forced vital capacity (FVC) and forced expiratory volume for one second (FEV_1)], and periodic chest roentgenograms (postero-anterior 14 x 17 inches). Additional medical requirement components include a history to describe smoking habits and details on past exposures to asbestos and other dusts and to determine presence or absence of pulmonary, cardiovascular, and gastrointestinal symptoms, and a physical examination, with special attention to pulmonary rales, clubbing of fingers and other signs related to cardiopulmonary systems.

Chest roentgenograms and pulmonary function tests should be performed at least every 2 years on all employees exposed to asbestos (1). Such tests should be made annually on individuals:

     (a) who have a history of 10 or more years of employment involving exposure to asbestos or

     (b) who show roentgenographic findings (such as small opacities, pleural plaques, pleural thickening or pleural calcification) which suggest or indicate pneumoconiosis or other reactions to asbestos or

     (c) who have changes in pulmonary function which indicate restrictive or obstructive lung disease.

Preplacement medical examinations and medical examinations on the termination of employment of asbestos-exposed workers are also required. See also reference (11).

Personal Protective Methods: Use of respirators can be decided on the basis of time-weighted average or peak concentration. When the limits of exposure to asbestos dust cannot be met by limiting the concentration in the workplace, the employer must utilize a program of respiratory protection and furnishing of protective clothing to protect every worker exposed (1).

Protective Clothing — (1) The employer shall provide each employee subject to exposure in a variance area with coveralls or similar full body protective clothing and hat, which shall be worn during the working hours in areas where there is exposure to asbestos dust.

(2) The employer shall provide for maintenance and laundering of the soiled protective clothing, which shall be stored, transported and disposed of in sealed nonreusable containers marked "Asbestos-Contaminated Clothing" in easy-to-read letters.

(3) Protective clothing shall be vacuumed before removal. Clothes shall not be cleaned by blowing dust from the clothing or shaking.

(4) If laundering is to be done by a private contractor, the employer shall inform the contractor of the potentially harmful effects of exposure to asbestos dust and of safe practices required in the laundering of the asbestos-soiled work clothes.

(5) Resin-impregnated paper or similar protective clothing can be substituted for fabric type of clothing.

(6) It is recommended that in highly contaminated operations (such as insulation and textiles) provisions be made for separate change rooms.

Respirator Selection: For the purpose of determining the class of respirator to be used, the employer shall measure the atmospheric concentration of airborne asbestos in the workplace when the initial application for variance is made and thereafter whenever process, worksite, climate or control changes occur which are likely to affect the asbestos concentration. The employer shall test for respirator fit and/or make asbestos measurements within the respiratory inlet covering to insure that no worker is being exposed to asbestos in excess of the standard either because of improper respirator selection or fit. Details of respirator selection are contained in a NIOSH publication (1).

Disposal Method Suggested: Asbestos may be recovered from waste asbestos slurries (A-57) as an alternative to disposal.

References

(1)   National Institute for Occupational Safety and Health, *Criteria for a Recommended Standard: Occupational Exposure to Asbestos,* NIOSH Doc. No. HSM 72-10267, Washington, DC (1972).

(2)   Environmental Protection Agency, *Summary Characterizations of Selected Chemicals of Near-Term Interest,* PB 255-817, Washington, DC (April 1976).

(3)   National Institute for Occupational Safety and Health, *Revised Recommended Asbestos Standard,* NIOSH Doc. No. 77-169, Washington, DC (December 1976).

(4)   Arthur D. Little, Inc., *Impact of Proposed OSHA Standard for Asbestos, First Report to the U.S. Dept. of Labor,* Report PB 283-478, Springfield, VA, Nat. Tech. Info. Service (1972).

(5)   U.S. Environmental Protection Agency, *Asbestos: Ambient Water Quality Criteria,* Washington, DC (1980).

(6)   U.S. Environmental Protection Agency, *Status Assessment of Toxic Chemicals: Asbestos,* Report No. EPA-600/2-79-210C, Washington, DC (December 1979).

(7)   National Academy of Sciences, *Medical and Biologic Effects of Environmental Pollutants: Asbestos,* Washington, DC (1971).

(8)   National Cancer Institute, *Asbestos: An Information Resource,* DHEW Publication No. (NIH)-79-1681, Bethesda, MD (May 1978).

(9)   U.S. Environmental Protection Agency, *Asbestos,* Health and Environmental Effects Profile No. 12, Washington, DC, Office of Solid Waste (April 30, 1980).

(10)   International Agency for Research on Cancer, *IARC Monograph on the Carcinogenic Risks of Chemicals to Humans,* 2, 17; 7, 319; 15, 341; and 17, 351, Lyon, France (1973, 1974, 1977 and 1978).

(11)   National Inst. for Occup. Safety and Health, *A Guide to the Work Relatedness of Disease,* Revised Edition, DHEW (NIOSH) Publ. No. 79-116, Cincinnati, OH, pp 55-62 (February 1979).

(12)   Sax, N.I., Ed., *Dangerous Properties of Industrial Materials Report, 1,* No. 1, 29-31, New York, Van Nostrand Reinhold Co. (1981).

(13)   Parmeggiani, L., Ed., *Encyclopedia of Occupational Health & Safety,* Third Edition, Vol. 1, pp 185–187, Geneva, International Labour Office (1983).

(14)   United Nations Environment Programme, *IRPTC Legal File 1983,* Vol. I, pp VII/60–64, Geneva, Switzerland, International Register of Potentially Toxic Chemicals (1984).

ASPHALT FUMES

Description: Asphalt fumes have been defined by NIOSH (1) as the nimbose effusion of small, solid particles created by condensation from the vapor state after volatilization of asphalt. In addition to particles, a cloud of fume may contain materials still in the vapor state.

The major constituent groups of asphalt are asphaltenes, resins, and oils made up of saturated and unsaturated hydrocarbons. The asphaltenes have molecular weights in the range of 1,000 to 2,600, those of the resins fall in the range of 370 to 500, and those of the oils in the range of 290 to 630.

Asphalt has often been confused with tar because the two are similar in appearance and have sometimes been used interchangeably as construction materials. Tars are, however, produced by destructive distillation of coal, oil or wood whereas asphalt is a residue from fractional distillation of crude oil.

The amounts of benzo(a)pyrene found in fumes collected from two different plants that prepared hot mix asphalt ranged from 3 to 22 ng/m^3; this is approximately 0.03% of the amount in coke oven emissions and 0.01% of that emitted from coal-burning home furnaces.

Code Numbers: (Petroleum asphalt fumes) CA 8052-42-4

DOT Designation: –

Synonyms: None.

Potential Exposure: Occupational exposure to asphalt fumes can occur during the transport, storage, production, handling, or use of asphalt. The composition of the asphalt that is produced is dependent on the refining process applied to the crude oil, the source of the crude oil, and the penetration grade (viscosity) and other physical characteristics of the asphalt required by the consumer.

The process for production of asphalt is essentially a closed-system distillation. Refinery workers are therefore potentially exposed to the fumes during loading of the asphalt for transport from the refinery during routine maintenance, such as cleaning of the asphalt storage tanks, or during accidental spills. Most asphalt is used out of doors, in paving and roofing, and the workers' exposure to the fumes is dependent on environmental conditions, work practices, and other factors. These exposures are stated to be generally intermittent and at low concentrations. Workers are potentially exposed also to skin and eye contacts with hot, cut-back, or emulsified asphalts. Spray application of cut-back, or emulsified asphalts may involve respiratory exposure also.

Because of the nature of the major uses of asphalt and asphalt products, it is not possible to determine accurately the number of workers potentially exposed to asphalt fumes in the United States, but an estimate of 500,000 can be derived from estimates of the number of workers in various occupations involved.

Permissible Exposure Limits in Air: Occupational exposure to asphalt fumes shall be controlled so that employees are not exposed to the airborne particulates at a concentration greater than 5 mg/m^3 of air, determined during any 15-minute period. ACGIH gives a tentative STEL of 10 mg/m^3 as of 1983/84.

Occupational exposure to asphalt fumes is defined as exposure in the workplace at a concentration of one-half or more of the recommended occupational exposure limit. If exposure to other chemicals also occurs, as is the case when asphalt is mixed with a solvent, emulsified, or used concurrently with other materials such as tar or pitch, provisions of any applicable standard for the other chemicals shall also be followed.

Determination in Air: A gravimetric method is recommended for estimation of the air concentration of asphalt fumes (A-1). When large amounts of dust are present in the same atmosphere in which the asphalt fume is present, which may occur in road-building operations, the gravimetric method may lead to errone-ously high estimates for asphalt fumes, and to possibly undeserved sanctions and citations for ostensibly exceeding the environmental limit for asphalt fumes or nuisance particulates.

NIOSH recommends (1) that where the resolution of such problems becomes necessary, a more specific procedure which involves solvent extraction and gravimetric analysis, be employed for the determination of asphalt fumes. The best procedure now available seems to be ultrasonic agitation of the filter in benzene and weighing of the dried residue from an aliquot on the clear ben-zene extract. NIOSH is attempting to devise an even more specific method for asphalt fumes for use under such conditions.

Permissible Concentration in Water: No criteria set.

Routes of Entry: Inhalation of dusts and fumes. Skin exposure can cause thermal burns from hot asphalt.

Harmful Effects and Symptoms: The principal adverse effects on health from exposure to asphalt fumes are irritation of the serous membranes of the conjunctivae and the mucous membranes of the respiratory tract. Hot asphalt can cause burns of the skin. In animals, there is evidence that asphalt left on the skin for long periods of time may result in local carcinomas, but there have been no reports of such effects on human skin that can be attributed to as-phalt alone. No reliable reports of malignant tumors of parenchymatous organs due to exposure to asphalt fumes have been found, but there has been no ex-tensive study of this possible consequence of occupational exposure in the asphalt industry.

Points of Attack: Skin, respiratory system.

Medical Surveillance: Details of recommended preplacement and periodic physical examinations and record-keeping have been set forth by NIOSH (1).

Personal Protective Methods: Employees shall wear appropriate protective clothing, including gloves, suits, boots, face shields (8-inch minimum), or other clothing as needed, to prevent eye and skin contact with asphalt.

Respirator Selection: (1) Engineering controls shall be used when needed to keep concentrations of asphalt fumes below the recommended exposure limit. The only conditions under which compliance with the recommended exposure limit may be achieved by the use of respirators are:

(a) During the time required to install or test the necessary engineer-ing controls.

(b) For operations such as nonroutine maintenance or repair activities causing brief exposure at concentrations above the environmental limit.

(c) During emergencies when concentrations of asphalt fumes may exceed the environmental limit.

(2) When a respirator is permitted by (1) above, it shall be selected from a list of respirators approved by NIOSH.

Disposal Method Suggested: Incineration.

References

(1) National Institute for Occupational Safety and Health, *Criteria for a Recommended Standard: Occupational Exposure to Asphalt Fumes,* NIOSH Doc. No. 78-106, Washington, DC (September 1977).

(2) Sax, N.I., Ed., *Dangerous Properties of Industrial Materials Report, 2,* No. 1, 76-77, New York, Van Nostrand Reinhold Co. (1982).

(3) Parmeggiani, L., Ed., *Encyclopedia of Occupational Health & Safety,* Third Edition, Vol. 1, pp 198–200, Geneva, International Labour Office (1983).

ASPIRIN

See "Acetylsalicylic Acid."

ATRAZINE

Description: Atrazine, $C_8H_{14}ClN_5$, has the structural formula

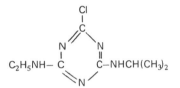

It is a herbicide which forms colorless crystals melting at 175° to 177°C.

Code Numbers: CAS 1912-24-9   RTECS XY5600000   UN 2763

DOT Designation: —

Synonyms: 6-Chloro-N-ethyl-N'-(1-methylethyl)-1,3,5-triazine-2,4-diamine, Gesaprim®, Primatol®, 2-chloro-4-ethylamino-6-isopropylamino-s-triazine.

Potential Exposure: Personnel involved in manufacture (A-32), formulation or application of this herbicide.

Permissible Exposure Limits in Air: ACGIH as of 1983/84 has set a TWA value of 5 mg/m^3 but no STEL value.

Permissible Concentration in Water: A suggested no-adverse effect level in drinking water has been calculated by NAS/NRC (A-2) as 0.15 mg/ℓ. An acceptable daily intake (ADI) of 0.0215 mg/kg/day has been calculated for atrazine (A-2).

Harmful Effects and Symptoms: Dermatitis. The acute oral toxicity to rats is very low.

Disposal Method Suggested: Atrazine is hydrolyzed by either acid or base as shown in the following formula:

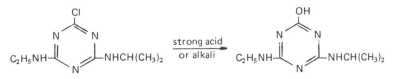

The hydroxy compounds are generally herbicidally inactive, but their complete environmental effects are uncertain. However, the method appears suitable for limited use and quantities of triazine (A-32).

Atrazine underwent >99% decomposition when burned in a polyethylene bag, and combustion with a hydrocarbon fuel would appear to be a generally suitable method for small quantities. Combustion of larger quantities would probably require the use of a caustic wet scrubber to remove nitrogen oxides and HCl from the product gases (A-32).

AURAMINE

- Carcinogen (Animal Proven, Human Suspected) (2)
- Hazardous Waste Constituent (EPA)

Description: $C_{17}H_{21}N_3$, $(CH_3)_2N-C_6H_4-CNH-C_6H_4-N(CH_3)_2$ is a yellow crystalline substance melting at 136°C.

Code Numbers: CAS 2465-27-2   RTECS BY3675000

DOT Designation: Imidocarbonyl

Synonyms: 4,4'-(Imidocarbonyl)bis(N,N-dimethylaniline); tetramethyldiaminodiphenylketoimine; Basic Yellow 2.

Potential Exposure: Auramine is used industrially as a dye or dye intermediate for coloring textiles, paper, and leather. According to 1979 U.S. International Trade Commission data, auramine was domestically produced, and 165,278 lb were imported.

Human exposure to auramine occurs principally through skin absorption or inhalation of vapors. Approximately 3,000 workers in the textile, paper, and leather industries may be exposed to auramine. Low-level dermal exposure to the consumer may occur but would be limited to any migration of auramine from fabric, leather, or paper goods.

Auramine is also used (1) as a powerful antiseptic in ear and nose surgery and in gonorrhea treatment.

Permissible Exposure Limits in Air: No standards set.

Permissible Concentration in Water: No criteria set.

Routes of Entry: Inhalation, ingestion, skin absorption.

Harmful Effects and Symptoms: Skin absorption may result in dermatitis and burns, nausea and vomiting (A-44).

Commercial auramine is carcinogenic in mice and rats after oral administration, producing liver tumours, and after subcutaneous injection in rats, producing local sarcomas.

The manufacture of auramine (which also involves exposure to other chemicals) has been shown in one study to be causally associated with an increase in bladder cancer. The actual carcinogenic compound(s) has not been specified precisely (2).

Points of Attack: Liver, bladder.

Medical Surveillance: Monthly urinalysis (A-38). Physical exam every 6 months focussed on bladder.

First Aid: Wash with soap and water (A-38).

Personal Protective Methods: Wear face shield or goggles (A-44), rubber gloves and protective clothing (A-38).

Respirator Selection: Use self-contained breathing apparatus.

Disposal Method Suggested: Incinerate (A-38).

References

(1) Sax, N.I., Ed., *Dangerous Properties of Industrial Materials Report, 1,* No. 5, 37-38, New York, Van Nostrand Reinhold Co. (1981).

(2) See Reference (A-62).

AZINPHOS-METHYL

Description: Azinphos-methyl, $C_{10}H_{12}N_3O_3PS_2$, has the structural formula

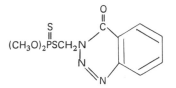

It is a colorless crystalline compound melting at 73° to 74°C. Its technical form is a brown, waxy solid.

Code Numbers: CAS 86-50-0   RTECS TE1925000   UN (NA 2783)

DOT Designation: Poison B.

Synonyms: O,O-dimethyl S-[(4-oxo-1,2,3-benzotriazin-3(4H)-yl)methyl] phosphorodithioate, Guthion®, Gusathion-M®, metiltriazotion (in U.S.S.R.).

Potential Exposure: Personnel engaged in the manufacture (A-32), formulation and application of this insecticide and acaricide.

Incompatibilities: Strong oxidizers.

Permissible Exposure Limits in Air: The 1983/84 ACGIH TWA value is 0.2 mg/m^3 with the notation "skin" indicating the possibility of cutaneous absorption. The STEL value proposed is 0.6 mg/m^3. The IDLH level is 5 mg/m^3.

Determination in Air: Collection by impinger or fritter bubbler, analysis by gas liquid chromatography (A-1).

Permissible Concentration in Water: For the protection of freshwater and marine aquatic life, a criterion of 0.01 μg/ℓ has been suggested by EPA (A-3). For the protection of human health, a no-adverse effect level in drinking water has been calculated by NAS/NRC (A-2) as 0.088 mg/ℓ. An allowable daily intake of 0.0125 mg/kg/day was calculated.

Determination in Water: Pesticide residue methods (A-7) which should be applicable involve hydrolysis with KOH in isopropanol to give anthranilic acid which is diazotized and coupled to give a measurable color.

Routes of Entry: Inhalation, skin absorption, ingestion, skin and eye contact.

Harmful Effects and Symptoms: Miosis, aches of the eyes, rhinorrhea of the front of the head, chest wheezing, laryngeal spasms, salivation, cyanosis, anorexia, nausea, vomiting, diarrhea, sweating, twitching, paralysis, convulsions, low blood pressure, cardiac problems.

Points of Attack: Respiratory system, lungs, central nervous system, cardiovascular system, blood cholinesterase.

Medical Surveillance: Consider the points of attack in preplacement and periodic physical examinations.

First Aid: If this chemical gets into the eyes, irrigate immediately. If this chemical contacts the skin, wash with soap immediately. If a person breathes in large amounts of this chemical, move the exposed person to fresh air at once and perform artificial respiration. When this chemical has been swallowed, get medical attention. Give large quantities of water and induce vomiting. Do not make an unconscious person vomit.

Personal Protective Methods: Wear appropriate clothing to prevent any reasonable probability of skin contact. Wear eye protection to prevent any reasonable probability of eye contact. Employees should wash immediately when skin is wet or contaminated. Work clothing should be changed daily if it is at all possible that clothing is contaminated. Remove nonimpervious clothing immediately if wet or contaminated. Provide emergency showers.

Respirator Selection:

2 mg/m^3: CCROVDMPest/SA/SCBA
5 mg/m^3: CCROVFDMPest/GMOVDMPest/SAF/SA:PD,PP,CF/SCBAF
Escape: GMOVPPest/SCBA

Disposal Method Suggested: Although this compound is chemically stable in storage, it is decomposed at elevated temperatures with evolution of gas, and rapidly decomposed in cold alkali to form anthranilic acid and other decomposition products. 50% hydrolysis at pH 9 and 70°C requires 0.6 hour; 8.9 hours at pH 5 and 70°C, 240 days at pH 5 and 20°C (A-32).

References
(1)   See Reference (A-61).
(2)   Sax, N.I., Ed., *Dangerous Properties of Industrial Materials Report, 3,* No. 4, 60-65, New York, Van Nostrand Reinhold Co. (1983). (As Guthion).
(3)   United Nations Environment Programme, *IRPTC Legal File 1983,* Vol. II, pp VII/578–80, Geneva, Switzerland, International Register of Potentially Toxic Chemicals (1984).

AZIRIDINE

See "Ethylene Imine."

AZOBENZENE

● Carcinogen (Animal Positive, IARC) (A-63)

Description: $C_6H_5N{=}NC_6H_5$ is an orange-red crystalline substance melting at 68°C.

Code Numbers: CAS 103-33-3   RTECS CN1400000

DOT Designation: —

Synonyms: Diphenyldiazene

Potential Exposure: Those engaged in azobenzene use in dye, rubber, chemical and pesticide manufacturing.

Permissible Exposure Limits in Air: No standards set.

Permissible Concentration in Water: No criteria set.

Routes of Entry: Inhalation, ingestion, skin absorption.

Harmful Effects and Symptoms: Carcinogenesis bioassays by NCI were positive for rats and negative for mice. Azobenzene irritates the eyes, skin and respiratory tract (A-60). Can cause blood disorders.

First Aid: Remove contaminated clothes; wash with soap and water; induce vomiting if ingested (A-60).

Personal Protective Methods: Wear safety goggles and protective gloves (A-60).

References

(1) Sax, N.I., Ed., *Dangerous Properties of Industrial Materials Report, 1,* No. 3, 35-36, New York, Van Nostrand Reinhold Co. (1981).

(2) See Reference (A-60).

B

BARIUM AND COMPOUNDS

- Hazardous substance (Barium cyanide only, EPA)
- Hazardous waste constituents (EPA)

Description: Ba, barium, a silver white metal, is produced by reduction of barium oxide. The primary sources are the minerals barite ($BaSO_4$) and witherite ($BaCO_3$). Barium may ignite spontaneously in air in the presence of moisture, evolving hydrogen. Barium is insoluble in water but soluble in alcohol. Most of the barium compounds are soluble in water. The peroxide, nitrate, and chlorate are reactive and may present fire hazards in storage and use.

Code Numbers: (for the most common compound, barium carbonate)
CAS 513-77-9   RTECS CQ8600000   UN 1564

DOT Designation: (Barium compounds n.o.s.) Poison. [St. Andrew's Cross (X) on label.]

Synonyms: Vary depending on specific compound.

Potential Exposure: Metallic barium is used for removal of residual gas in vacuum tubes and in alloys with nickel, lead, calcium, magnesium, sodium, and lithium.

Barium compounds are used in the manufacture of lithopone (a white pigment in paints), chlorine, sodium hydroxide, valves, and green flares; in synthetic rubber vulcanization, x-ray diagnostic work, glassmaking, papermaking, beet-sugar purification, animal and vegetable oil refining. They are used in the brick and tile, pyrotechnics, and electronics industries. They are found in lubricants, pesticides, glazes, textile dyes and finishes, pharmaceuticals, and in cements which will be exposed to saltwater; and barium is used as a rodenticide, a flux for magnesium alloys, a stabilizer and mold lubricant in the rubber and plastics industries, an extender in paints, a loader for paper, soap, rubber, and linoleum, and as a fire extinguisher for uranium or plutonium fires.

Potential occupational exposure is as follows:

Barium sulfonate	2,800,000
Barium phenate	2,300,000
Barium sulfate	900,000
Barium carbonate	200,000
Barium alkyl phenolate	115,000
Barium phosphite	35,000
Barium oxide	30,000

Incompatibilities: Vary with specific compounds.

Permissible Exposure Limits in Air: The Federal standard and the ACGIH 1983/84 value for soluble barium compounds is 0.5 mg/m^3. There is no STEL value. The IDLH level is 250 mg/m^3.

Determination in Air: Collection on a cellulose membrane filter, workup with hot water, analysis by atomic absorption. See NIOSH Methods, Set N. See also reference (A-10).

Allowable Concentration in Water: A criterion of 1.0 mg/ℓ was set by EPA for domestic water supplies on a health basis. A drinking-water guideline was derived from the 8-hour weighted maximum allowable concentration (TLV) in industrial air of 0.5 mg/m^3 set by the American Conference of Governmental Industrial Hygienists. It was assumed that, with an 8-hour inhalation of 10 m^3 of air, the daily intake would be 5 mg of barium, of which 75% was absorbed in the bloodstream and 90% transferred across the gastrointestinal tract. Based on the above assumptions, it was reasoned that a concentration of about 2 mg/ℓ of water would be safe for adults. To provide added safety for more susceptible members of the population, such as children, a level of 1 mg/ℓ was recommended, according to NAS/NRC (A-2).

Determination in Water: Conventional flame atomization does not have sufficient sensitivity to determine barium in most water samples; however, a barium detection limit of 10 μg/ℓ can be achieved, if a nitrous oxide flame is used. A concentration procedure for barium uses thenoyltrifluoroacetone-methylisobutylketone extraction at a pH of 6.8.

With a tantalum liner insert, the barium detection limit of the flameless atomic absorption procedure can be improved to 0.1 μg/ℓ according to NAS/NRC (A-2).

Routes of Entry: Ingestion or inhalation of dust or fume, skin or eye contact.

Harmful Effects and Symptoms: *Local* – Alkaline barium compounds, such as the hydroxide and carbonate, may cause local irritation to the eyes, nose, throat, and skin.

Systemic – Barium poisoning is virtually unknown in industry, although the potential exists when the soluble forms are used. When ingested or given orally, the soluble, ionized barium compounds exert a profound effect on all muscles and especially smooth muscles, markedly increasing their contractility. The heart rate is slowed and may stop in systole. Other effects are increased intestinal peristalsis, vascular constriction, bladder contraction, and increased voluntary muscle tension.

The inhalation of the dust of barium sulfate may lead to deposition in the lungs in sufficient quantities to produce "baritosis"—a benign pneumoconiosis. This produces a radiologic picture in the absence of symptoms and abnormal physical signs. X-rays, however, will show disseminated nodular opacities throughout the lung fields, which are discrete, but sometimes overlap.

Points of Attack: Heart, lungs, central nervous system, skin, respiratory system, eyes.

Medical Surveillance: Consideration should be given to the skin, eye, heart, and lung in any placement or periodic examination.

First Aid: If a soluble barium compound gets into the eyes, irrigate immediately. If a soluble barium compound contacts the skin, flush with water immediately. If a person breathes in large amounts of a soluble barium compound, move the exposed person to fresh air at once and perform artificial respiration. When a soluble barium compound has been swallowed, get medical attention. Give large quantities of water and induce vomiting. Do not make an unconscious person vomit.

Personal Protective Methods: Employees should receive instruction in personal hygiene and the importance of not eating in work areas. Good housekeeping and adequate ventilation are essential. Dust masks, respirators, or goggles may be needed where amounts of significant soluble or alkaline forms are encountered, as well as protective clothing.

Respirator Selection:
    2.5 mg/m^3: DMXS
      5 mg/m^3: DMXSQ/FuHiEP/SA/SCBA
     25 mg/m^3: HiEPF/SAF/SCBAF
    250 mg/m^3: Sa:PD,PP,SAF:CF/PAPHiE
        Escape: HiEP/SCBA

Disposal Method Suggested: Barium fluoride—Precipitation with soda ash or slaked lime. The resulting sludge should be sent to a chemical waste landfill. Barium nitrate, barium sulfide—Chemical reaction with water, caustic soda, and slaked lime, resulting in precipitation of the metal sludge, which may be landfilled. Barite (barium sulfate) may be recovered from drilling muds for reuse (A-57) as an alternative to disposal.

References

(1) U.S. Environmental Protection Agency, *Toxicology of Metals, Vol. 2: Barium,* pp 71-84, Report EPA-600/1-77-022, Research Triangle Park, NC (May 1977).
(2) U.S. Environmental Protection Agency, *Barium,* Health and Environmental Effects Profile No. 13, Wash., DC, Office of Solid Waste (April 30, 1980).
(3) Sax, N.I., Ed., *Dangerous Properties of Industrial Materials Report, 1,* No. 7, 35-40, New York, Van Nostrand Reinhold Co. (1981).
(4) Sax, N.I., Ed., *Dangerous Properties of Industrial Materials Report, 1,* No. 1, 31, New York, Van Nostrand Reinhold Co. (1980). (Barium Sulfate).
(5) See Reference (A-61).
(6) Sax, N.I., Ed., *Dangerous Properties of Industrial Materials Report, 3,* No. 4, 29–32, New York, Van Nostrand Reinhold Co. (1983). (Barium and Barium Cyanide).
(7) See Reference (A-60) for entries under:

Barium	Barium Hydroxide
Barium Acetate	Barium Nitrate
Barium Azide	Barium Perchlorate
Barium Carbonate	Barium Peroxide
Barium Chlorate	Barium Sulfide
Barium Chloride	Barium Sulfite
Barium Diphenylamine Sulfonate	

(8) Parmeggiani, L., Ed., *Encyclopedia of Occupational Health & Safety,* Third Edition, Vol. 1, pp 242-44, Geneva, International Labour Office (1983).

BAYGON®

Baygon® is a registered trademark name for the compound with the generic name of Propoxur (which see). ACGIH has listed the compound under the trademarked name rather than the generic name; hence it is cross-referenced here.

BENOMYL

Description: Methyl 1-[(butylamino)carbonyl] 1H-benzimidazol-2-ylcarbamate, $C_{14}H_{18}N_4O_3$, with the structural formula

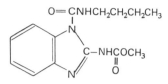

Benomyl is a colorless crystalline solid with a faint acrid odor which decomposes before melting.

Code Numbers: CAS 17804-35-2   RTECS DD6475000   UN 2588

DOT Designation: –

Synonyms: Benlate®.

Potential Exposure: Those involved in the manufacture (A-32), formulation and application of this fungicide.

Permissible Exposure Limits in Air: There is no Federal standard but ACGIH (1983/84) has set a TWA of 0.8 ppm (10 mg/m³) and a tentative STEL of 1.3 ppm (15 mg/m³).

Permissible Concentration in Water: No criteria set.

Harmful Effects and Symptoms: Benomyl is generally felt to have a low order of acute and chronic toxicity (A-34). However, a rebuttable presumption against registration for benomyl was issued on December 6, 1978 by U.S. EPA on the basis of reduction in nontarget species, mutagenicity, teratogenicity, reproductive effects, and hazard to wildlife.

References

(1) United Nations Environment Programme, *IRPTC Legal File 1983,* Vol. I, pp VII/110–11, Geneva, Switzerland, International Register of Potentially Toxic Chemicals (1984).

BENSULIDE

Description: $C_{14}H_{24}NOPS_3$ with the structural formula

$$(CH_3)_2CHO \quad (CH_3)_2CHO \quad \overset{S}{\underset{\|}{P}}SCH_2CH_2NH\overset{O}{\underset{\underset{O}{\|}}{\overset{\|}{S}}}-\bigcirc$$

forms crystals melting at 34.4°C.

Code Numbers: CAS 741-58-2   RTECS TE0250000

DOT Designation: —

Synonyms: Betasan®, Prefar®, bensulfide, Presan®, Exposan®.

Potential Exposure: Those involved in the manufacture and application of this preemergence herbicide.

Permissible Exposure Limits in Air: No standards set.

Permissible Concentration in Water: 0.1 ppm in drinking water (1).

Harmful Effects and Symptoms: Bensulide is slightly toxic (LD_{50} for rats is 770 mg/kg). Exhibits general effects of organophosphate poisoning including nausea, vomiting, diarrhea, excessive salivation, bronchoconstriction, muscle twitching, convulsions, coma and death at high doses.

Points of Attack: Gastrointestinal system, central nervous system.

Disposal Method Suggested: Acid or alkaline hydrolyses or incineration.

References

(1)   Sax, N.I., Ed., *Dangerous Properties of Industrial Materials Report, 2,* No. 4, 29-31, New York, Van Nostrand Reinhold Co. (1982).

BENZAL CHLORIDE

● Hazardous waste (EPA)

Description: $C_6H_5CHCl_2$ is a fuming colorless liquid which boils at 207°C.

Code Numbers: CAS 98-87-3   RTECS CZ5075000

DOT Designation: —

Synonyms: (Dichloromethyl)benzene; Benzylidene chloride

Potential Exposure: Benzal chloride is used almost exclusively for the manufacture of benzaldehyde. It can also be used to prepare cinnamic acid and benzoyl chloride.

Permissible Exposure Limits in Air: No standard set. (Inhaled benzal chloride will hydrolyze and probably produce effects similar to those of inhaled hydrogen chloride.)

Permissible Concentration in Water: No criteria set. (Benzal chloride hydrolyzes to benzaldehyde and HCl on contact with water.)

Harmful Effects and Symptoms: Benzal chloride is irritating to the skin. Benzal chloride was found to induce carcinomas, leukemia, and papillomas in mice. Benzal chloride was shown to possess a longer latency period than benzotrichloride before the onset of harmful effects (1).

Disposal Method Suggested: 1500°F, 0.5 second minimum for primary combustion; 2200°F, 1.0 second for secondary combustion; elemental chlorine formation may be alleviated through injection of steam or methane into the combustion process.

References

(1) U.S. Environmental Protection Agency, *Benzal Chloride*, Health and Environmental Effects Profile No. 14, Wash., DC, Office of Solid Waste (April 30, 1980).

BENZALDEHYDE

Description: C_6H_5CHO is a clear to yellowish liquid with an almond odor boiling at 179°C.

Code Numbers: CAS 100-52-7   RTECS CU4375000   UN 1990

DOT Designation: —

Synonyms: Artificial essential oil of almond; benzoic aldehyde, benzene carboxaldehyde.

Potential Exposure: In manufacture of perfumes, dyes and cinnamic acid; as a solvent; in flavors.

Permissible Exposure Limits in Air: No standards set.

Permissible Concentration in Water: No criteria set.

Routes of Entry: Inhalation, ingestion and skin absorption.

Harmful Effects and Symptoms: May cause contact dermatitis. Acts as a narcotic in high concentrations.

Target Organs: Skin, central nervous system.

First Aid: Remove contaminated clothing, flush with water, get to fresh air (A-60).

Personal Protective Methods: Safety goggles and protective clothing.

Respirator Selection: Self-contained breathing apparatus.

Disposal Method Suggested: Incineration.

References
(1) Sax, N.I., Ed., *Dangerous Properties of Industrial Materials Report, 1,* No. 8, 36-38, New York, Van Nostrand Reinhold Co. (1981).
(2) See Reference (A-60).

BENZ[a] ANTHRACENE

● Carcinogen (EPA-CAG) (A-62, A-64)

Description: $C_{18}H_{12}$, a condensed ring aromatic with the structure

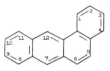

forms crystals which sublime at 155° to 157°C.

DOT Designation: —

Synonyms: benzphenanthrene, tetraphene, benzanthrene, naphthanthracene.

Potential Exposure: BA is a contaminant and does not have any reported commercial use or application, although one producer did report the substance for the Toxic Substances Control Act Inventory.

BA has been reported present in cigarette smoke condensate, automobile exhaust gas, soot, and the emissions from coal and gas works and electric plants. BA also occurs in the aromatic fraction of mineral oil, commercial solvents, waxes, petrolatum, creosote, coal tar, petroleum asphalt, and coal tar pitch. Microgram quantities of BA can be found in various foods such as charcoal broiled, barbecued, or smoked meats and fish; certain vegetables and vegetable oils; and roasted coffee and coffee powders.

Human subjects are exposed to BA through either inhalation or ingestion. Workers at facilities with likely exposure to fumes from burning or heating of organic materials have a potential for exposure to BA. Consumers can be exposed to this chemical through ingestion of various food, with concentrations of 100 μg/kg in some instances. Cigarette smoke condensate has quantities of BA that range from 0.03 to 4.6 μg/g.

BA is found in the atmosphere at levels that vary with geography and climatology. These values can range from up to 136 μg/1,000 m^3 in summer to 361 μg/1,000 m^3 in winter. Drinking water samples may contain up to 23 ng/ℓ BA, and surface waters have been found to contain 4 to 185 ng/ℓ. The soil near industrial centers has been shown to contain as much as 390 μg/kg of BA, whereas soil near highways can have levels of up to 1,500 μg/kg, and areas polluted with coal tar pitch can reach levels of 2,500 mg/kg.

Permissible Exposure Limits in Air: NIOSH has recommended a workplace standard for coal tar products of 0.1 mg/m^3 as a time-weighted average for a 10-hour workshift in a 40-hour workweek. OSHA indirectly limits exposure to BA by requiring that occupational exposure to coal tar pitch volatiles not exceed, as a time-weighted average, 0.2 mg/m^3 over an 8-hour period.

Permissible Concentration in Water: Water quality criteria document for polynuclear aromatic hydrocarbons published in final 11/28/80. Total PAH addressed. A concentration of 2.8 ng PAH/ℓ is estimated to limit cancer risk to one in a million (EPA).

Harmful Effects and Symptoms: BA given by several routes of administration has proved to be carcinogenic in the mouse. It produced hepatomas and lung adenomas following repeated oral administration to young mice (1). In a parallel experiment with methylcholanthrene, the carcinogenic effect upon the liver and lung was similar for the two compounds at the same dose level. In the same experiment, BA did not produce tumours of the gastrointestinal tract, whereas methylcholanthrene induced them consistently.

BA is a complete carcinogen for the mouse skin. The fact that the tumour yield was higher when using a dodecane solution than with toluene is related to the co-carcinogenic effect of dodecane. Benzo(a)pyrene given at a lower dose level produced more skin tumours with a shorter latency period than did BA. BA is also an initiator of skin carcinogenesis in mice.

BA produced tumours in mice following subcutaneous injections. Fifty μg BA was the lowest dose tested, and it was effective in newborn and in adult animals. It produced bladder tumours in mice following implantation.

References
(1)   See Reference (A-62). Also see Reference (A-64).

BENZENE

- Carcinogen (IARC) (5) (12)
- Hazardous substance (EPA)
- Hazardous waste (EPA)
- Priority toxic pollutant (EPA)

Description: C_6H_6, benzene, is a clear, volatile, colorless, highly flammable liquid with a characteristic odor. The most common commercial grade contains 50 to 100% benzene, the remainder consisting of toluene, xylene, and other constituents which distill below 120°C.

Code Numbers: CAS 71-43-2   RTECS CY1400000   UN 1114

DOT Designation: Flammable liquid

Synonyms: Benzol, phenyl hydride, coal naphtha, phene, benxole, cyclohexatriene.

Potential Exposure: Benzene is used as a constituent in motor fuels, as a solvent for fats, inks, oils, paints, plastics, and rubber, in the extraction of oils from seeds and nuts, and in photogravure printing (1). It is also used as a chemical intermediate. By alkylation, chlorination, nitration, and sulfonation, chemicals such as styrene, phenols, and maleic anhydride are produced. Benzene is also used in the manufacture of detergents, explosives, pharmaceuticals, and dyestuffs.

NIOSH estimates that approximately 2 million workers in the U.S. are potentially exposed to benzene. Increased concern for benzene as a significant environmental pollutant arises from public exposure to the presence of benzene in gasoline and the increased content in gasoline due to requirements for unleaded fuels for automobiles equipped with catalytic exhaust converters.

Incompatibilities: Strong oxidizers; chlorine, bromine with iron.

Permissible Exposure Limits in Air: In 1974 NIOSH published a criteria document (2) for occupational exposure to benzene which recommended adherence to an existing Federal standard of 10 ppm (30 mg/m^3) as a time-weighted average with a ceiling of 25 ppm.

The Federal emergency standard for benzene effective May 21, 1977, is 1 ppm for an 8-hour TWA, with 5 ppm as a maximum peak above the acceptable ceiling for a maximum duration of 15 minutes.

OSHA again has amended its permanent standard as of mid-1978 limiting worker exposure to benzene to 1 ppm to exempt workplaces where benzene levels in liquids do not exceed 0.5%. After three years the minimum allowable level would be 0.1%. When OSHA first issued the standard no exemption for benzene mixtures was allowed. Later it amended the standard to exempt workplaces where the level of benzene in mixtures was less than 0.1%. In July 1980, the U.S. Supreme Court struck down the OSHA standard because it failed to show a cost/benefit analysis and the OSHA standard has reverted to a 10 ppm 8-hour TWA limit. In addition, ACGIH designated benzene as an "Industrial Substance Suspect of Carcinogenic Potential for Man," with a TLV of 10 ppm (30 mg/m^3) and an STEL of 25 ppm (75 mg/m^3) as of 1983/84. The IDLH level is 2,000 ppm.

Determination in Air: Adsorption on charcoal, workups with CS_2, analysis by gas chromatography. See NIOSH Methods, Set U. See also reference (A-10). Benzene may also be determined with long duration detector tubes (A-11).

Permissible Concentration in Water : To protect freshwater aquatic life: 5,300 $\mu g/\ell$ on an acute basis. To protect saltwater aquatic life: 5,100 $\mu g/\ell$ on an acute basis. To protect human health: preferably zero. An additional lifetime cancer risk of 1 in 100,000 results from a concentration of 6.6 $\mu g/\ell$.

Determination in Water: Gas chromatography (EPA Method 602) or gas chromatography plus mass spectrometry (EPA Method 624).

Routes of Entry: Inhalation of vapor which may be supplemented by percutaneous absorption although benzene is poorly absorbed through intact skin; ingestion; skin and eye contact.

Harmful Effects and Symptoms: *Local* — Exposure to liquid and vapor may produce primary irritation to skin, eyes, and upper respiratory tract. If the liquid is aspirated into the lung, it may cause pulmonary edema and hemorrhage. Erythema, vesiculation, and dry, scaly dermatitis may also develop from defatting of the skin.

Systemic — Acute exposure to benzene results in central nervous system depression. Headache, dizziness, nausea, convulsions, coma, and death may result. Death has occurred from large acute exposure or as a result of ventricular fibrillation, probably caused by myocardial sensitization to endogenous epinephrine.

Early reported autopsies revealed hemorrhages (nonpathognomonic) in the brain, pericardium, urinary tract, mucous membranes, and skin. Chronic exposure to benzene is well documented to cause blood changes. Benzene is basically a myelotoxic agent. Erythrocyte, leukocyte, and thrombocyte counts may first increase, and then aplastic anemia may develop with anemia, leukopenia, and thrombocytopenia. The bone marrow may become hypo- or hyperactive and may not always correlate with peripheral blood.

Recent epidemiologic studies along with case reports of benzene related blood dyscrasias and chromosomal aberrations have led NIOSH to conclude that benzene is leukemogenic (11). The evidence is most convincing for acute myelogenous leukemia and for acute erythroleukemia, but a connection with chronic leukemia has been noted by a few investigators.

Recent work has shown increases in the rate of chromosomal aberrations associated with benzene myelotoxicity. These changes in the bone marrow are stable and may occur several years after exposure has ceased. "Stable" changes may give rise to leukemic clones and seem to involve chromosomes of the G group.

Points of Attack: Blood, central nervous system, skin, bone marrow, eyes, respiratory system.

Medical Surveillance: Preplacement and periodic examinations should be concerned especially with effects on the blood and bone marrow and with a history of exposure to other myelotoxic agents or drugs or of other diseases of the blood. Preplacement laboratory exams should include: (a) complete blood count (hematocrit, hemoglobin, mean corpuscular volume, white blood count, differential count, and platelet estimation); (b) reticulocyte count; (c) serum bilirubin; and (d) urinary phenol.

The type and frequency of periodic hematologic studies should be related to the data obtained from biologic monitoring and industrial hygiene studies, as well as any symptoms or signs of hematologic effects. Recommendations for proposed examinations have been made in the criteria for a recommended standard. Examinations should also be concerned with other possible effects such as those on the skin, central nervous system, and liver and kidney functions.

Biologic monitoring should be provided to all workers subject to benzene exposure. It consists of sampling and analysis of urine for total phenol content. The objective of such monitoring is to be certain that no worker absorbs an unacceptable amount of benzene. Unacceptable absorption of benzene, posing a risk of benzene poisoning, is considered to occur at levels of 75 mg phenol per liter of urine (with urine specific gravity corrected to 1.024), when determined by methods specified in the NIOSH "Criteria for Recommended Standard—Benzene." Alternative methods shown to be equivalent in accuracy and precision may also be useful. Biological monitoring should be done at quarterly intervals. If environmental sampling and analysis are equal to or exceed accepted safe limits, the urinary phenol analysis should be conducted every two weeks. This increased monitoring frequency should continue for at least 2 months after the high environmental level has been demonstrated.

Two follow-up urines should be obtained within one week after receipt of the original results, one at the beginning and the other at the end of the work week. If original elevated findings are confirmed, immediate steps should be taken to reduce the worker's absorption of benzene by improvement in environment control, personal protection, personal hygiene, and administrative control. See also reference (7).

First Aid: If this chemical gets into the eyes, irrigate immediately. If this chemical contacts the skin, wash with soap promptly. If a person breathes in large amounts of this chemical, move the exposed person to fresh air at once and perform artificial respiration. When this chemical has been swallowed, get medical attention. Do not induce vomiting.

Personal Protective Methods: Wear appropriate clothing to prevent repeated or prolonged skin contact. Wear eye protection to prevent any reasonable probability of eye contact. Employees should wash promptly with soap when skin is wet or contaminated. Remove clothing immediately if wet or contaminated to avoid flammability hazard.

Respirator Selection:
              10 ppm: SA/SCBA
              50 ppm: SAF/SCBAF
           1,000 ppm: SA:PD,PP,CF
           2,000 ppm: SAF:PD,PP,CF
             Escape: GMS/SCBA

Disposal Method Suggested: Incineration.

References

(1) National Academy of Sciences, *A Review of Health Effects of Benzene*, National Academy of Sciences, Washington, DC (June 1976).

(2) National Institute for Occupational Safety and Health, *Criteria for a Recommended Standard: Occupational Exposure to Benzene*, NIOSH Doc. No. 74-137, Washington, DC (1974).

(3) U.S. Environmental Protection Agency, *Benzene: Ambient Water Quality Criteria*, Wash., DC (1980).

(4) U.S. Environmental Protection Agency, *Status Assessment of Toxic Chemicals: Benzene*, Report EPA-600/2-79-210D, Wash., DC (Dec. 1979).

(5) International Agency for Research on Cancer, *IARC Monographs on the Carcinogenic Risk of Chemicals to Humans*, 7, 203, Lyon, France (1974) and 11, 295, Lyon, France (1976).

(6) U.S. Environmental Protection Agency, *Benzene*, Health and Environmental Effects Profile No. 15, Wash., DC, Office of Solid Waste (April 30, 1980).

(7) Nat. Inst. for Occup. Safety and Health, *A Guide to the Work Relatedness of Disease*,

Revised Edition, DHEW (NIOSH) Publication No. 79-116, pp 63-72, Cincinnati, OH (Jan. 1979).

(8) Sax, N.I., Ed., *Dangerous Properties of Industrial Materials Report, 1,* No. 4, 38-41, New York, Van Nostrand Reinhold Co. (1981).

(9) Sax, N.I., Ed., *Dangerous Properties of Industrial Materials Report, 2,* No. 4, 33-38, New York, Van Nostrand Reinhold Co. (1982).

(10) Sax., N.I., Ed., *Dangerous Properties of Industrial Materials Report, 3,* No. 3, 53-59, New York, Van Nostrand Reinhold Co. (1983).

(11) See Reference (A-62). Also see Reference (A-64).

(12) Mehlman, M.A., Ed., *Carcinogenicity and Toxicity of Benzene,* Princeton, N.J., Princeton Scientific Publishers (1983).

(13) Parmeggiani, L., Ed., *Encyclopedia of Occupational Health & Safety,* Third Edition, Vol. 1, pp 257-61, Geneva, International Labour Office (1983).

(14) United Nations Environment Programme, *IRPTC Legal File 1983,* Vol. I, pp VII/69-73, Geneva, Switzerland, International Register of Potentially Toxic Chemicals (1984).

BENZETHONIUM CHLORIDE

Description: $C_{27}H_{42}ClNO_2$ with the following structural formula

forms crystals which melt at 164° to 166°C.

Code Numbers: CAS 121-54-0   RTECS BO7175000

DOT Designation: —

Synonyms: Phemerol chloride®, Phemeride®, Phemerol®.

Potential Exposure: Used as topical antiinfective.

Incompatibilities: Soap, anionic detergents.

Permissible Exposure Limits in Air: No standards set.

Permissible Concentration in Water: No criteria set.

Harmful Effects and Symptoms: Ingestion may cause vomiting, convulsions, collapse and coma.

Personal Protective Methods: Wear skin protection.

Respirator Selection: Hot conditions may necessitate use of self-contained breathing apparatus.

Disposal Method Suggested: Incineration.

References
(1) Sax, N.I., Ed., *Dangerous Properties of Industrial Materials Report, 1,* No. 1, 32-33, New York, Van Nostrand Reinhold Co. (1980).

BENZIDINE

- Carcinogen (Human positive, IARC) (3)
- Hazardous waste (EPA)
- Priority toxic pollutant (EPA)

Description: $NH_2C_6H_4C_6H_4NH_2$, benzidine, is a crystalline solid with a significant vapor pressure. The salts are less volatile, but tend to be dusty.

Code Numbers: CAS 92-87-5   RTECS DC9625000   UN 1885

DOT Designation: Poison B

Synonyms: 4,4'-Biphenyldiamine, p-diaminodiphenyl, 4,4'-diaminobiphenyl, 4,4'-diphenylenediamine, benzidine base.

Potential Exposures: Benzidine is used primarily in the manufacture of azo dyestuffs; there are over 250 of these produced. Other uses, including some which may have been discontinued, are in the rubber industry as a hardener, in the manufacture of plastic films, for detection of occult blood in feces, urine, and body fluids, in the detection of H_2O_2 in milk, in the production of security paper, and as a laboratory reagent in determining HCN, sulfate, nicotine, and certain sugars. No substitute has been found for its use in dyes.

Free benzidine is present in the benzidine-derived azo dyes. According to industry, quality control specifications require that the level not exceed 20 ppm, and in practice the level is usually below 10 ppm. Industry has estimated a total environmental discharge at the 300 user facility sites of 450 pounds per year or about 1.5 pounds per year per facility, assuming all of the free benzidine is discharged in the liquid effluent.

No measurements for benzidine in ambient air, surface water, or drinking water have been reported. Further, no measurements for free benzidine in finished products containing azo dyes have been reported.

Estimates of the number of people exposed to benzidine and 3,3'-dichlorobenzidine are difficult to obtain. It has been suggested that 62 people in the U.S. are exposed to the former and between 250 and 2,500 to the latter.

Permissible Exposure Limits in Air: Benzidine and its salts are included in a Federal standard for carcinogens; all contact with them should be avoided. Cutaneous absorption is possible. ACGIH (1983/84) lists benzidine as a human carcinogen with the notation that skin absorption may be significant.

Determination in Air: Collection on a filter followed by colorimetric analysis (A-12). See also reference (A-10).

Permissible Concentration in Water: To protect freshwater aquatic life—2,500 $\mu g/\ell$ on an acute basis; insufficient data to yield a value for saltwater aquatic life. To protect human health—preferably zero. An additional lifetime cancer risk of 1 in 100,000 results from a concentration of 0.0012 $\mu g/\ell$.

Determination in Water: High performance liquid chromatography (EPA Method 605) or an oxidation/colorimetric method using Chloramine T (available from EPA) or a gas chromatography/mass spectrometric method (EPA Method 625).

Routes of Entry: Inhalation, percutaneous absorption, and ingestion of dust.

Harmful Effects and Symptoms: *Local* — Contact dermatitis to primary irritation or sensitization has been reported.

Systemic — Benzidine is a known human urinary tract carcinogen (11) with an average latent period of 16 years. The first symptoms of bladder cancer usually are hematuria, frequency of urination, or pain.

Points of Attack: Skin, bladder, inhalation.

Medical Surveillance: Placement and periodic examinations should include an evaluation of exposure to other carcinogens; use of alcohol, smoking, and medications; and family history. Special attention should be given on a regular basis to urine sediment and cytology. If red cells or positive smears are seen, cystoscopy should be done at once. The general health of exposed persons should also be evaluated in periodic examinations.

First Aid: Flush eyes with water. Wash contaminated areas of body with soap and water. Use gastric lavage if ingested followed by saline catharsis.

Personal Protective Methods: These are designed to supplement engineering controls (such as a prohibition on open-vessel operation) and to prevent all skin or respiratory contact. Full body protective clothing and gloves should also be used. On exit from a regulated area employees should shower and change into street clothes, leaving their protective clothing and equipment at the point of exit to be placed in impervious containers at the end of the work shift for decontamination or disposal. Effective methods should be used to clean and decontaminate gloves and clothing.

Respirator Selection: Self-contained breathing apparatus or mechanical filter respirators are recommended (A-38, A-39).

Disposal Method Suggested: Incineration; oxides of nitrogen are removed from the effluent gas by scrubber, catalytic or thermal device.

References

(1) U.S. Environmental Protection Agency, *Benzidine: Ambient Water Quality Criteria*, Wash., DC (1980).

(2) U.S. Environmental Protection Agency, *Status Assessment of Toxic Chemicals: Benzidine*, Report EPA-600/2-79-210E, Wash., DC (Dec. 1979).

(3) International Agency for Research on Cancer, *IARC Monographs on the Carcinogenic Risks of Chemicals to Humans*, 1, 80, Lyon, France (1972).

(4) U.S. Environmental Protection Agency, *Reviews of the Environmental Effects of Pollutants: II, Benzidine*, Cincinnati, Ohio, Health Effects Research Laboratory, Report No. EPA-600/1-78-024 (1978).

(5) U.S. Environmental Protection Agency, *Benzidine*, Health and Environmental Effects Profile No. 16, Wash., DC, Office of Solid Waste (April 30, 1980).

(6)   U.S. Environmental Protection Agency, *Benzidine, Its Congeners and Their Derivative Dyes and Pigments*, TSCA Chemical Assessment Series; Preliminary Risk Assessment Phase I, Report EPA-560/11-80-019, Wash., DC (June 1980).

(7)   See Reference (A-61).

(8)   Sax, N.I., Ed., *Dangerous Properties of Industrial Materials Report, 1*, No. 5, 38-39, New York, Van Nostrand Reinhold Co. (1981).

(9)   Sax, N.I., Ed., *Dangerous Properties of Industrial Materials Report, 2*, No. 4, 38-43, New York, Van Nostrand Reinhold Co. (1982).

(10)   Sax, N.I., Ed., *Dangerous Properties of Industrial Materials Report, 3*, No. 4, 32-37, New York, Van Nostrand Reinhold Co. (1983).

(11)   See Reference (A-62). Also see Reference (A-64).

(12)   United Nations Environment Programme, *IRPTC Legal File 1983*, Vol. I, pp VII/101–3, Geneva, Switzerland, International Register of Potentially Toxic Chemicals (1984).

BENZO(b)FLUORANTHENE AND BENZO(j)FLUORANTHENE

See "Polynuclear Aromatic Hydrocarbons."

References
(1) See References (A-62) and (A-64).

BENZOIC ACID

Description: C_6H_5COOH is a crystalline solid melting at 122.4°C.

Code Numbers: CAS 65-85-0   RTECS DG0875000   UN 9094

DOT Designation: —

Synonyms: Benzene carboxylic acid; phenylformic acid; dracylic acid.

Potential Exposure: Those involved in the manufacture of benzoates; in the manufacture of food preservatives.

Permissible Exposure Limits in Air: No standards set.

Permissible Concentration in Water: No criteria set. Biological effects reviewed (A-36).

Routes of Entry: Inhalation and ingestion.

Harmful Effects and Symptoms: Mildly irritating to skin, eyes and mucous membranes. Ingestion causes nausea and G.I. troubles (A-38).

Points of Attack: Skin, eyes and mucous membranes.

First Aid: Get to fresh air, remove contaminated clothes and flush with water (A-60).

Personal Protective Methods: Safety goggles and protective outerwear are recommended.

Respirator Selection: Cannister mask recommended.

Disposal Method Suggested: Incineration.

References
(1) Sax, N.I., Ed., *Dangerous Properties of Industrial Materials Report, 1,* No. 8, 38-40, New York, Van Nostrand Reinhold Co. (1981).
(2) Sax, N.I., Ed., *Dangerous Properties of Industrial Materials Report, 3,* No. 4, 37-40, New York, Van Nostrand Reinhold Co. (1983).
(3) See Reference (A-60).

BENZONITRILE

Description: C_6H_5CN is a colorless, almond-smelling oil which boils at 191°C.

Code Numbers: CAS 100-47-0   RTECS DI2450000   UN2224

DOT Designation: —

Synonyms: Phenyl cyanide; cyanobenzene; benzoic acid nitrile.

Potential Exposure: Workers in organic synthesis of pharmaceuticals, dyestuffs, rubber chemicals. Used as a solvent and chemical intermediate.

Permissible Exposure Limits in Air: No standards set.

Permissible Concentration in Water: No criteria set. Biological effects reviewed (A-36).

Routes of Entry: Inhalation, ingestion, skin absorption.

Harmful Effects and Symptoms: Toxic when ingested or inhaled.

Personal Protective Methods: Wear long rubber gloves, overalls and apron (A-38).

Respirator Selection: Use self-contained breathing apparatus.

Disposal Method Suggested: (1) Mix with calcium hypochlorite and flush to sewer with water or (2) incinerate.

References

(1)   Sax., N.I., Ed., *Dangerous Properties of Industrial Materials Report, 1,* No. 8, 40-42, New York, Van Nostrand Reinhold Co. (1981).
(2)   See Reference (A-60).
(3)   Sax, N.I., Ed., *Dangerous Properties of Industrial Materials Report, 3,* No. 4, 40-42, New York, Van Nostrand Reinhold Co. (1983).

BENZO[a]PYRENE

See "Polynuclear Aromatic Hydrocarbons" also.

● Carcinogen (EPA-CAG) (IARC) (1)
● Hazardous Waste Constituent (EPA)

Description: $C_{20}H_{12}$ with the structure

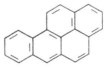

forms yellowish crystals melting at 179°C.

Code Numbers: CAS 50-32-8

DOT Designation: −

Synonyms: 3,4-Benzpyrene.

Potential Exposure: Benzo(a)pyrene [B(a)P] is a polycyclic aromatic hydrocarbon (PAH) that has no commercial-scale production.

B(a)P is produced in the United States by one chemical company and distributed by several specialty chemical companies in quantities from 100 mg to 5 g for research purposes.

Although not manufactured in great quantity, B(a)P is a by-product of combustion. It is estimated that 1.8 million pounds per year are released from stationary sources, with 96% coming from: (1) coal refuse piles, outcrops, and abandoned coal mines; (2) residential external combustion of bituminous coal; (3) coke manufacture; and (4) residential external combustion of anthracite coal.

Human exposure to B(a)P can occur from its presence as a by-product of chemical production. The number of persons exposed is not known. Persons working at airports in tarring operations, refuse incinerator operations, power plants, and coke manufacturers may be exposed to higher B(a)P levels than the general population. Scientists involved in cancer research or in sampling toxic materials may also be occupationally exposed. The general population may be exposed to B(a)P from air pollution, cigarette smoke, and food sources. B(a)P has been detected in cigarette smoke at levels ranging from 0.2 to 12.2 μg per 100 cigarettes. B(a)P has been detected at low levels in foods ranging from 0.1 to 50 ppb.

Permissible Exposure Limits in Air: 0.2 mg/m^3 8-hr TWA (coal tar pitch volatiles) (OSHA). ACGIH (1983/84) designates Benzo(a)pyrene as an industrial substance suspect of carcinogenic potential for man with no TWA value set.

Permissible Concentration in Water: Water quality criteria document for PAH published in final 11/2/80. Total PAH addressed. A concentration of PAH 2.8 ng/ℓ is estimated to limit a cancer risk to one in a million (EPA).

Harmful Effects and Symptoms: B(a)P has produced tumours in all of the nine species for which data are reported following different administrations including oral, skin and intratracheal routes. It has both a local and a systemic carcinogenic effect. In sub-human primates, there is convincing evidence of the ability of B(a)P to produce local sarcomas following repeated subcutaneous injections and lung carcinomas following intratracheal instillation. It is also an initiator of skin carcinogenesis in mice, and it is carcinogenic in single-dose experiments and following prenatal exposure.

In skin carcinogenesis studies in mice, B(a)P was consistently found to produce more tumours in a shorter period of time than did other polycyclic aromatic hydrocarbons, with the possible exception of DB(a,h)A. In a dose-response study involving subcutaneous injection in mice, the minimal dose at which carcinogenicity was detected was higher for B(a)P than for DB(a,h)A and for MC. However, the latent periods were shorter for B(a)P than for DB(a,h)A. In studies using intratracheal administration, B(a)P appeared to be less effective than 7H-dibenzo(c,g)carbazole in the hamster (1).

No epidemiological studies on the significance of B(a)P exposure to man are available, and studies are insufficient to prove that B(a)P is carcinogenic for man. However, coal-tar and other materials which are known to be carcinogenic to man may contain B(a)P. The substance has also been detected in other environmental situations.

References

(1) IARC Monographs on the *Evaluation of Carcinogenic Risk of Chemicals to Man,* vol. 3, IARC, Lyon, France, pp 91-136 (1973).
(2) See Reference (A-62). Also see reference (A-64).
(3) United Nations Environment Programme, *International Register of Potentially Toxic Chemicals,* Geneva, Switzerland (1979).
(4) United Nations Environment Programme, *IRPTC Legal File 1983,* Vol. I, pp VII/121–22, Geneva, Switzerland, International Register of Potentially Toxic Chemicals (1984).

BENZOTRIAZOLE

Description: $C_6H_4NHN_2$ with a structural formula

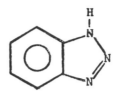

is a white to tan odorless crystalline compound melting at 98° to 99°C.

Code Numbers: CAS 95-14-7   RTECS DM1225000

DOT Designation: —

Synonyms: 1,2-Aminoazophenylene; azimidobenzene; benzisotriazole; benzene azimide; 1-H-benzotriazole.

Potential Exposures: The major uses for benzotriazole are as anticorrosion agents, plastics stabilizers, and photographic antifoggants.

Permissible Exposure Limits in Air: No standards set.

Permissible Concentration in Water: No criteria set.

Determination in Water: Benzotriazole may be determined by polarographic methods (sensitive to 0.4 ppm), spectrophotometric methods (sensitive to 0.05 ppm) or gas chromatography plus mass spectrometry (sensitive to 1 ppb) (1).

Harmful Effects and Symptoms: Benzotriazole and its derivatives have not been extensively studied for biological activity, especially in recent years. They are generally considered to be relatively hazard-free materials. Toxicity testing in humans has been limited mainly to patch tests for allergic sensitization, with negative results. Manufacturers have indicated no experiences with occupational poisoning.

Benzotriazoles may present a severe respiratory hazard if inhaled. Studies of rats indicate that severe local respiratory damage, rather than absorption and systemic poisoning, is responsible for the high toxicity of benzotriazole dusts. The dust is also a severe explosion hazard. In situations outside the manufacturing or direct usage environments, it is not likely that exposure to benzotriazole dusts would occur.

Personal Protective Methods: Spills of the solids should be swept up promptly. The personnel involved should use approved respirators. Local high concentrations of dust should be dampened with water and the area well-ventilated.

Spills of solutions involve accident procedures based mainly on the nature of the particular solvent. The benzotriazoles themselves appear to present no particular hazard in solutions.

Disposal Method Suggested: Recommended methods for disposal of benzotriazole waste solids are: dispose with normal plant waste; bury in a landfill or incinerate (but not in a closed container); or dissolve in a flammable solvent and spray into an incinerator with an afterburner and scrubber. Because the dust is considered an explosion hazard, disposal should be handled in a way which does not encourage the production of dust. Incineration or ordinary combustion of benzotriazole produces carbon dioxide, nitrogen, and nitrogen oxides. In limited air, combustion products may include carbon monoxide and possibly hydrogen cyanide.

References

(1)   U.S. Environmental Protection Agency, *Investigation of Selected Environmental Contaminants: Benzotriazoles,* Report EPA-560/2-77-001, Washington, DC (February 1977).

BENZOTRICHLORIDE

● Hazardous waste (EPA)

Description: $C_6H_5CCl_3$ is a colorless oily fuming liquid which boils at 221°C.

Code Numbers: CAS 98-07-7   RTECS XT9275000   UN 2226

DOT Designation: Corrosive material.

Synonyms: α,α,α-Trichlorotoluene, phenyl chloroform, trichloromethylbenzene.

Potential Exposure: Benzotrichloride is used extensively in the dye industry for the production of Malachite green, Rosamine, Quinoline red, and Alizarine yellow A. It can also be used to produce ethyl benzoate.

Permissible Exposure Limits in Air: No standards set.

Permissible Concentration in Water: No criteria set. (Benzotrichloride decomposes in the presence of water to benzoic and hydrochloric acids.)

Harmful Effects and Symptoms: Benzotrichloride was severely irritating to the skin of rabbits that received dermal applications of 10 mg for 24 hours and to the eyes of rabbits that received instillations of 50 μg to the eye (1).

In a 1975 Japanese study, benzotrichloride was found to induce carcinomas, leukemia, and papillomas in mice. The details of the study were not available for assessment (1). Hence a definitive judgment on carcinogenicity has not been made.

Personal Protective Methods: Wear rubber gloves and coveralls.

Respirator Selection: Use of self-contained breathing apparatus is recommended (A-38).

Disposal Method Suggested: Incineration with flammable solvent added in incinerator with afterburner and alkaline scrubber.

References

(1)   U.S. Environmental Protection Agency, *Benzotrichloride,* Health and Environmental Effects Profile No. 20, Wash., DC, Office of Solid Waste (April 30, 1980).
(2)   See Reference (A-60).

BENZOYL CHLORIDE

Description: C_6H_5COCl is a liquid with a penetrating odor which boils at 197.2°C at 760 mm.

Code Numbers: CAS 98-88-4   RTECS DM6600000   UN 1736

DOT Designation: Corrosive material.

Synonyms: Benzenecarbonyl chloride; α-chlorobenzaldehyde.

Potential Exposure: To workers in organic synthesis. Used in producing dye intermediates.

Incompatibilities: Water, strong oxidants, alcohols (A-60).

Permissible Exposure Limits in Air: The TLV set in the USSR is 0.9 ppm (5 mg/m^3) (A-36).

Determination in Air: By photometry (A-36).

Permissible Concentration in Water: No criteria set.

Routes of Entry: Ingestion and inhalation.

Harmful Effects and Symptoms: Very irritating to skin, eyes and mucous membranes. Inhalation can cause lung edema and serious cases may be fatal (A-60).

Points of Attack: Eyes, skin and mucous membranes.

Personal Protective Methods: Safety goggles and protective outerwear recommended.

Respirator Selection: Self-contained breathing apparatus recommended.

Disposal Method Suggested: Pour into sodium bicarbonate solution and flush to sewer.

References

(1) Sax, N.I., Ed., *Dangerous Properties of Industrial Materials Report, 2*, No. 1, 78-80, New York, Van Nostrand Reinhold Co. (1982).
(2) See Reference (A-60).

BENZOYL PEROXIDE

Description: Benzoyl peroxide, $C_6H_5CO-O-O-COC_6H_5$, is a crystalline solid, melting at 103° to 106°C which may explode when heated.

Code Numbers: CAS 94-36-0   RTECS DM8575000   UN 2085, 2087, 2088, 2089, 2090.

DOT Designation: Organic peroxide

Synonyms: Dibenzoyl peroxide, benzoyl superoxide.

Potential Exposures: Benzoyl peroxide is used in the bleaching of fats, waxes, flour and edible oils; it is also used as a polymerization catalyst.

NIOSH estimates that 25,000 workers in the United States are potentially exposed to benzoyl peroxide or its formulations.

Incompatibilities: Combustible substances, wood, paper, lithium aluminum hydride. Explosive at high temperatures.

Permissible Exposure Limits in Air: Exposure to benzoyl peroxide shall be controlled so that employees are not exposed at a concentration greater than 5 mg/m^3 of air, ceiling determined as a time-weighted average (TWA) concentration for up to a 10-hour shift in a 40-hour workweek (1). ACGIH (1983/84) has also adopted this value.

Determination in Air: Collection on a filter workup with ethyl ether, analysis by high performance liquid chromatography (HPLC). See NIOSH Methods, Set 2. See also reference (A-10). A colorimetric method may also be used according to Du Pont (A-1).

Permissible Concentration in Water: No criteria set.

Routes of Entry: The major concerns from occupational exposure to benzoyl peroxide are the hazards arising from its instability, flammability, and explosive properties. In addition, benzoyl peroxide may cause local irritation of the eyes and skin. Inhalation and skin contact are the main routes of entry. Ingestion is a possibility.

Harmful Effects and Symptoms: Irritation of skin and mucous membranes; dermatitis; eye irritation; decreased pulse and temperature; dyspnea; stupor.

Points of Attack: Skin, respiratory system and eyes.

Medical Surveillance: Preplacement and periodic medical examinations should be conducted with particular attention to skin conditions (1).

First Aid: If this chemical gets into the eyes, irrigate immediately. If this chemical contacts the skin, wash with soap promptly. When this chemical has been swallowed, get medical attention. Give large quantities of water and induce vomiting. Do not make an unconscious person vomit.

Personal Protective Methods: Protective clothing and safety glasses with side shields or safety goggles should be worn by employees to reduce the possibility of skin contact and eye irritation. Such protection is especially important where benzoyl peroxide and other powder or granular benzoyl peroxide formulations may become airborne or where liquid or paste formulations of benzoyl peroxide might be spattered or spilled.

Protective clothing should be fire resistant. Any fabric that generates static electricity is not recommended. To prevent the buildup of static electricity, appropriate conductive footwear should be worn. Gloves made of rubber, leather, or other appropriate material should be worn by employees for protection when they are opening shipping boxes of pure benzoyl peroxide or otherwise handling pure benzoyl peroxide. Aprons made of rubber or another appropriate material are recommended for added protection when handling benzoyl peroxide and its formulations. Plastic aprons that may generate static electricity should not be used.

Employees should wash promptly when skin is contaminated. Work clothing should be changed daily if it may be contaminated. Remove nonimpervious clothing promptly if contaminated.

Respirator Selection: Respiratory protection as follows must be used whenever airborne concentrations of benzoyl peroxide cannot be controlled to the recommended workplace environmental limit by either engineering or administrative controls.

$25 \ mg/m^3$: DMXS
$50 \ mg/m^3$: DMXSQ/HiEPFu/SA/SCBA
$250 \ mg/m^3$: HiEPF/SAF/SCBA
$1,000 \ mg/m^3$: PAPHiEF
Escape: DMXS/SCBA

Do not use oxidizable sorbents.

Disposal Method Suggested: Pretreatment involves decomposition with sodium hydroxide. The final solution of sodium benzoate, which is very biodegradable, may be flushed into the drain. Disposal of large quantities of solution may require pH adjustment before release into the sewer or controlled incineration after mixing with a noncombustible material (A-31).

References

(1)   National Institute for Occupational Safety and Health, *Criteria for a Recommended Standard: Occupational Exposure to Benzoyl Peroxide,* NIOSH Doc. No. 77-166, Wash., DC (1977).
(2)   See Reference (A-61).
(3)   Sax, N.I., Ed., *Dangerous Properties of Industrial Materials Report, 2,* No. 1, 80-82, New York, Van Nostrand Reinhold Co. (1982).
(4)   See Reference (A-60).
(5)   Parmeggiani, L., Ed., *Encyclopedia of Occupational Health & Safety,* Third Edition, Vol. 1, pp 260–262, Geneva, International Labour Office (1983).

BENZYL CHLORIDE

- Carcinogen (Animal positive, IARC) (4)
- Hazardous substance (EPA)
- Hazardous waste (EPA)

Description: $C_6H_5CH_2Cl$, benzyl chloride is a colorless liquid with an unpleasant, irritating odor. It boils at 179.4°C.

Code Numbers: CAS 100-44-7   RTECS XS8925000   UN 1738

DOT Designation: Corrosive material.

Synonyms: α-chlorotoluene.

Potential Exposures: In contrast to phenyl halides, benzyl halides are very reactive. Benzyl chloride is used in production of benzal chloride, benzyl alcohol, and benzaldehyde. Industrial usage includes the manufacture of plastics, dyes, synthetic tannins, perfumes and resins. It is used in the manufacture of many pharmaceuticals (A-41).

Suggested uses of benzyl chloride include: the vulcanization of fluororubbers and the benzylation of phenol and its derivatives for the production of possible disinfectants.

Incompatibilities: Active metals: copper, aluminum, magnesium, iron, zinc, tin; strong oxidizers.

Permissible Exposure Limits in Air: The Federal standard and ACGIH 1983/84 value is 1 ppm (5 mg/m^3). There is no STEL value proposed. The IDLH level is 10 ppm.

Determination in Air: Adsorption on charcoal, workup with CS_2, analysis by gas chromatography. See NIOSH Methods, Set H. See also reference (A-10).

Permissible Concentration in Water: No criteria set but EPA (A-37) has suggested an ambient limit of 69 µg/ℓ based on health effects.

Routes of Entry: Inhalation of vapor, ingestion, eye and skin contact.

Harmful Effects and Symptoms: *Local* — Benzyl chloride is a severe irritant to the eyes and respiratory tract. At 160 mg/m^3 it is unbearably irritating to the eyes and nose. Liquid contact with the eyes produces severe irritation and may cause corneal injury. Skin contact may cause dermatitis.

Systemic — Benzyl chloride is regarded as a potential cause of pulmonary edema. One author has reported disturbances of liver functions and mild leukopenia in some workers, but this has not been confirmed. Benzyl chloride has been shown to induce local sarcomas in rats treated by subcutaneous injection. Benzyl chloride was reported to be weakly mutagenic in *S. typhimurium* TA 100 strain.

Points of Attack: Eyes, skin and respiratory system.

Medical Surveillance: Preplacement and periodic examinations should include the skin, eyes, and an evaluation of the liver, kidney, respiratory tract, and blood.

First Aid: If this chemical gets into the eyes, irrigate immediately. If this chemical contacts the skin, wash with soap immediately. If a person breathes in large amounts of this chemical, move the exposed person to fresh air at once and perform artificial respiration. When this chemical has been swallowed, get medical attention. Give large quantities of salt water and induce vomiting. Do not make an unconscious person vomit.

Personal Protective Methods: Wear appropriate clothing to prevent any reasonable probability of skin contact. Wear eye protection to prevent any possibility of eye contact. Employees should wash immediately when skin is wet or contaminated. Remove nonimpervious clothing immediately if wet or contaminated. Provide emergency showers and eyewash.

Respirator Selection:
>10 ppm: CCROVAGF/GMOVAG/SAF/SCBAF
>Escape: GMOVAGF/SCBAF

Disposal Method Suggested: Incineration at $1500°F$ for 0.5 second minimum for primary combustion and $2200°F$ for 1.0 second for secondary combustion. Elemental chlorine formation may be alleviated by injection of steam or methane into the combustion process.

References

(1) U.S. Environmental Protection Agency, *Chemical Hazard Information Profile: Benzyl Chloride*, Wash., DC (Dec. 9, 1977).
(2) Nat. Inst. for Occup. Safety and Health, *Information Profiles on Potential Occupational Hazards, Benzyl Chloride*, Report PB 276,678, Rockville, MD, pp 2-7 (Oct. 1977).
(3) U.S. Environmental Protection Agency, *Benzyl Chloride*, Health and Environmental Effects Profile No. 21, Wash., DC, Office of Solid Waste (April 30, 1980).
(4) International Agency for Research on Cancer, *IARC Monographs on the Carcinogenic Risks of Chemicals to Humans,* Lyon, France, 11, 217 (1976).
(5) Sax, N.I., Ed., *Dangerous Properties of Industrial Materials Report, 2,* No. 2, 9-11, New York, Van Nostrand Reinhold Co. (1982).
(6) See Reference (A-60).
(7) Parmeggiani, L., Ed., *Encyclopedia of Occupational Health & Safety,* Third Edition, Vol. 1, pp 262, Geneva, International Labour Office (1983).

BERYLLIUM AND COMPOUNDS

● Carcinogen (Animal positive, IARC) (7)

- Hazardous substances (Some compounds, EPA)
  Beryllium compounds designated as hazardous substances by
  EPA include beryllium chloride, beryllium fluoride and beryl-
  lium nitrate.
- Hazardous waste constituents (EPA) and Hazardous waste (Be dust) (EPA)
- Priority toxic pollutant (EPA)

Description: Be, beryllium, is a grey metal which combines the properties of
light weight and high tensile strength. Beryllium is slightly soluble in hot water
and in dilute acids and alkalis. All beryllium compounds are soluble to some degree in water. Beryl ore is the primary source of beryllium, although there are
numerous other sources.

Code Numbers: (for beryllium metal) CAS 7440-41-7   RTECS DS1750000
UN 1567 (for compounds n.o.s.)

DOT Designation: (Compounds n.o.s.) Poison B.

Synonyms: None.

Potential Exposures: Beryllium is used extensively in manufacturing electrical components, chemicals, ceramics, and X-ray tubes. A number of alloys are
produced in which beryllium is added to yield greater tensile strength, electrical
conductivity, and resistance to corrosion and fatigue. The metal is used as a neutron reflector in high-flux test reactors.

Human exposure occurs mainly through inhalation of beryllium dust or
fumes. OSHA estimates that approximately 25,000 workers are exposed to
beryllium. Among these are beryllium ore miners, beryllium alloy makers and
fabricators, phosphor manufacturers, ceramic workers, missile technicians, nuclear reactor workers, electric and electronic equipment workers, and jewelers.

The major source of beryllium exposure of the general population is
through the burning of coal. Approximately 250,000 pounds of beryllium is
released from coal- and oil-fired burners. EPA estimates the total release of
beryllium to the atmosphere from point sources is approximately 5,500 pounds
per year. The principal emissions are from beryllium-copper alloy production.
Approximately 721,000 persons living within 12.5 miles (20 km) of point
sources are exposed to small amounts of beryllium (median concentration
0.005 $\mu g/m^3$). Levels of beryllium have been reported in drinking water supplies
and in small amounts in food (11).

Incompatibilities: Beryllium metal: acids, alkalies, chlorinated hydrocarbons,
oxidizable materials (A-38).

Permissible Exposure Limits in Air: The present Federal standard for beryllium and beryllium compounds is 2 $\mu g/m^3$ as an 8-hour TWA with an acceptable
ceiling concentration of 5 $\mu g/m^3$. The acceptable maximum peak is 25 $\mu g/m^3$
for a maximum duration of 30 minutes. The standard recommended in the
NIOSH Criteria Document (1) is 2 μg Be/m^3 as an 8-hour TWA with a peak
value of 25 μg Be/m^3 as determined by minimum sampling time of 30 minutes. Recently it was recommended that OSHA treat beryllium as a carcinogen;
ACGIH (1983/84) lists beryllium as "an industrial substance suspect of carcinogenic potential for man."

Determination in Air: Collection on a filter, analysis by atomic absorption
spectroscopy (A-1). See also reference (A-10).

Permissible Concentration in Water: The criteria set by EPA (A-3) in 1976 were: 11 μg/ℓ for the protection of aquatic life in soft freshwater; 1,100 μg/ℓ for the protection of aquatic life in hard freshwater; 100 μg/ℓ for continuous irrigation on all soils, except 500 μg/ℓ for irrigation on neutral to alkaline fine-textured soils.

In 1980 the EPA set the following criteria: To protect freshwater aquatic life: 130 μg/ℓ on an acute basis; 5.3 μg/ℓ on a chronic basis. To protect saltwater aquatic life: insufficient data to set criteria. To protect human health: preferably zero. An additional lifetime cancer risk of 1 in 100,000 results from a concentration of 0.037 μg/ℓ.

Determination in Water: Total beryllium may be determined according to EPA, by digestion followed by atomic absorption or by a colorimetric method or by inductively coupled plasma (ZCP) optical emission spectrometry. Dissolved beryllium can be determined by 0.45 micron filtration prior to the above method for total beryllium.

Routes of Entry: Inhalation of fume or dust.

Harmful Effects and Symptoms: *Local* — The soluble beryllium salts are cutaneous sensitizers as well as primary irritants. Contact dermatitis of exposed parts of the body are caused by acid salts of beryllium. Onset is generally delayed about two weeks from the time of first exposure. Complete recovery occurs following cessation of exposure. Eye irritation and conjunctivitis can occur. Accidental implantation of beryllium metal or crystals of soluble beryllium compound in areas of broken or abraded skin may cause granulomatous lesions. These are hard lesions with a central nonhealing area. Surgical excision of the lesion is necessary. Exposure to soluble beryllium compounds may cause nasopharyngitis, a condition characterized by swollen and edematous mucous membranes, bleeding points, and ulceration. These symptoms are reversible when exposure is terminated.

Systemic — Beryllium and its compounds are highly toxic substances. Entrance to the body is almost entirely by inhalation. The acute systemic effects of exposure to beryllium primarily involve the respiratory tract and are manifest by a nonproductive cough, substernal pain, moderate shortness of breath, and some weight loss. The character and speed of onset of these symptoms, as well as their severity, are dependent on the type and extent of exposure. An intense exposure, although brief, may result in severe chemical pneumonitis with pulmonary edema.

Chronic beryllium disease is an intoxication arising from inhalation of beryllium compounds, but it is not associated with inhalation of the mineral beryl (2). The chronic form of this disease is manifest primarily by respiratory symptoms, weakness, fatigue, and weight loss (without cough or dyspnea at the onset), followed by nonproductive cough and shortness of breath. Frequently, these symptoms and detection of the disease are delayed from five to ten years following the last beryllium exposure, but they can develop during the time of exposure. The symptoms are persistent and frequently are precipitated by an illness, surgery, or pregnancy. Chronic beryllium disease usually is of long duration with exacerbations and remissions.

Chronic beryllium disease can be classified by its clinical variants according to the disability the disease process produces.

(1) Asymptomatic nondisabling disease is usually diagnosed only by routine chest x-ray changes and supported by urinary or tissue assay.

(2) In its mildly disabling form, the disease results in some nonproductive cough and dyspnea following unusual levels of exertion. Joint pain and weak-

ness are common complaints. Diagnosis is by x-ray changes. Renal calculi containing beryllium may be a complication. Usually, the patient remains stable for years, but eventually shows evidence of pulmonary or myocardial failure.

(3) In its moderately severe disabling form, the disease produces symptoms of distressing cough and shortness of breath, with marked x-ray changes. The liver and spleen are frequently affected, and spontaneous pneumothorax may occur. There is generally weight loss, bone and joint pain, oxygen desaturation, increase in hematocrit, disturbed liver function, hypercalciuria, and spontaneous skin lesions similar to those of Boeck's sarcoid. Lung function studies show measurable decreases in diffusing capacity. Many people in this group survive for years with proper therapy. Bouts of chills and fever carry a bad prognosis.

(4) The severely disabling disease will show all of the abovementioned signs and symptoms in addition to severe physical wasting and negative nitrogen balance. Right heart failure may appear causing a severe nonproductive cough which leads to vomiting after meals. Severe lack of oxygen is the predominant problem, and spontaneous pneumothorax can be a serious complication. Death is usually due to pulmonary insufficiency or right heart failure.

Points of Attack: Skin, eyes, respiratory system, lungs, liver, spleen, heart.

Medical Surveillance: Preemployment history and physical examinations for worker applicants should include chest x-rays, baseline pulmonary function tests (FVC and FEV_1), and measurement of body weight. Beryllium workers should receive a periodic health evaluation that includes: spirometry (FVC and FEV_1), medical history questionnaire directed toward respiratory symptoms, and a chest x-ray. General health, liver and kidney functions, and possible effects on the skin should be evaluated.

First Aid: Irrigate eyes with water. Wash contaminated areas of body with soap and water.

Personal Protective Methods: Work areas should be monitored to limit and control levels of exposure. Personnel samplers are recommended. Good housekeeping, proper maintenance, and engineering control of processing equipment and technology are essential. The importance of safe work practices and personal hygiene should be stressed. When beryllium levels exceed the accepted standards, the workers should be provided with respiratory protective devices of the appropriate class, as determined on the basis of the actual or projected atmospheric concentration of airborne beryllium at the worksite. Protective clothing should be provided all workers who are subject to exposure in excess of the standard. This should include shoes or protective shoe covers as well as other clothing. The clothing should be reissued clean on a daily basis. Workers should shower following each shift prior to change to street clothes.

Respirator Selection: See Reference (1).

Disposal Method Suggested: For the following: beryllium carbonate, beryllium chloride, beryllium oxide, beryllium (powder) and beryllium selenate, waste should be converted into chemically inert oxides using incineration and particulate collection techniques. These oxides may be landfilled (A-31), but should be returned to suppliers if possible. Recovery and recycle is an alternative to disposal for beryllium scrap and pickle liquors containing beryllium (A-57).

References

(1) National Institute for Occupational Safety and Health, *Criteria for a Recommended Standard: Occupational Exposure to Beryllium*, NIOSH Doc. No. 72-10268, Wash., DC (1972).

(2) Tepper, L.B., Hardy, H.L. and Chamberlin, R.I., *Toxicity of Beryllium Compounds*, Elsevier, New York (1961)

(3) U.S. Environmental Protection Agency, *Beryllium: Ambient Water Quality Criteria*, Wash., DC (1979).

(4) U.S. Environmental Protection Agency, *Toxicology of Metals, Vol. 2: Beryllium*, Report No. EPA-600/1-77-022, Research Triangle Park, pp 85-109 (May 1977).

(5) U.S. Environmental Protection Agency, *Reviews of the Environmental Effects of Pollutants: VI, Beryllium*, Report EPA-600/1-78-028, Cincinnati, Ohio, Health Effects Research Laboratory (1978).

(6) U.S. Environmental Protection Agency, *Beryllium*, Health and Environmental Effects Profile No. 22, Wash., DC, Office of Solid Waste (April 30, 1980).

(7) International Agency for Research on Cancer, *IARC Monographs on the Carcinogenic Risks of Chemicals to Humans,* 1, 17 (1972).

(8) Sax, N.I., Ed., *Dangerous Properties of Industrial Materials Report, 1,* No. 1, 33-36, New York, Van Nostrand Reinhold Co. (1980). (Beryllium Fluoride, Oxide and Sulfate).

(9) Sax, N.I., Ed., *Dangerous Properties of Industrial Materials Report, 2,* No. 1, 84-88, New York, Van Nostrand Reinhold Co. (1982). (Beryllium Nitrate and Sulfate).

(10) Sax, N.I., Ed., *Dangerous Properties of Industrial Materials Report, 2,* No. 2, 13-14, New York, Van Nostrand Reinhold Co. (1982).

(11) See Reference (A-62). Also see Reference (A-64).

(12) See Reference (A-60), citations under Beryllium and Beryllium Oxide.

(13) Sax, N.I., Ed., *Dangerous Properties of Industrial Materials Report, 3,* No. 5, 59-61, New York, Van Nostrand Reinhold Co. (1983). (Beryllium Chloride).

(14) Sax, N.I., Ed., *Dangerous Properties of Industrial Materials Report, 3,* No. 5, 61-64, New York, Van Nostrand Reinhold Co. (1983). (Beryllium Fluoride).

(15) Parmeggiani, L., Ed., *Encyclopedia of Occupational Health & Safety,* Third Edition, Vol. 1, pp 263-68, Geneva, International Labour Office (1983).

(16) United Nations Environment Programme, *IRPTC Legal File 1983,* Vol. I, pp VII/125-28, Geneva, Switzerland, International Register of Potentially Toxic Chemicals (1984).

BIPHENYL

Description: $C_6H_5C_6H_5$ is a colorless to light yellow, leaflet solid with a potent characteristic odor. It melts at 69°C.

Code Numbers: CAS 92-52-4   RTECS DU8050000

DOT Designation: —

Synonyms: Diphenyl, phenylbenzene.

Potential Exposures: Biphenyl is a fungistat for oranges which is applied to the inside of shipping containers and wrappers. It is also used as a heat transfer agent and as an intermediate in organic synthesis. NIOSH estimates that 5,000 to 10,000 workers are potentially exposed to biphenyl each year.

Incompatibilities: Oxidizers

Permissible Exposure Limits in Air: The Federal standard and ACGIH (1983/84) value is 0.2 ppm (1.5 mg/m³). The tentative STEL value is 0.6 ppm (4.0 mg/m³). The IDLH level is 300 mg/m³.

Determination in Air: Sampling by charcoal tube, analysis by gas liquid chromatography (A-13). See also reference (A-10).

Permissible Concentration in Water: No criteria set but EPA has suggested (A-37) an ambient limit of 13.8 μg/ℓ based on health effects.

Routes of Entry: Inhalation of vapor or dust; percutaneous absorption, ingestion, eye and skin contact.

Harmful Effects and Symptoms: *Local* — Repeated exposure to dust may result in irritation of skin and respiratory tract. The vapor may cause moderate eye irritation. Repeated skin contact may produce a sensitization dermatitis.

Systemic — In acute exposure, biphenyl exerts a toxic action on the central nervous system, on the peripheral nervous system, and on the liver. Symptoms of poisoning are headache, diffuse gastrointestinal pain, nausea, indigestion, numbness and aching of limbs, and general fatigue. Liver function tests may show abnormalities. Chronic exposure is characterized mostly by central nervous system symptoms, fatigue, headache, tremor, insomnia, sensory impairment, and mood changes. Such symptoms are rare, however.

Points of Attack: Liver, skin, central nervous system, upper respiratory system, eyes.

Medical Surveillance: Consider skin, eye, liver function and respiratory tract irritation in any preplacement or periodic examination.

First Aid: If this chemical gets into the eyes, irrigate immediately. If this chemical contacts the skin, flush with water immediately. If a person breathes in large amounts of this chemical, move the exposed person to fresh air at once and perform artificial respiration. When this chemical has been swallowed, get medical attention. Do not induce vomiting.

Personal Protective Methods: Because of its low vapor pressure and low order of toxicity, it does not usually present a major problem in industry. Protective creams, gloves, and masks with organic vapor canisters for use in areas of elevated vapor concentrations should suffice. Elevated temperature may increase the requirement for protective methods or ventilation. Wear appropriate clothing to prevent repeated or prolonged skin contact. Wear eye protection to prevent any possibility of eye contact with molten biphenyl. Employees should wash promptly when skin is contaminated. Work clothing should be changed daily if it may be contaminated. Remove nonimpervious clothing immediately if wet or contaminated.

Respirator Selection:
        10 mg/m^3: CCROVDM/SA/SCBA
        50 mg/m^3: CCROVFDM/GMOVDM/SAF/SCBAF
    300 mg/m^3: PAPCCROVFHiEP/SAF:PD,PP,CF
        Escape: GMOVP/SCBA

Disposal Method Suggested: Incineration.

References

(1) Nat. Inst. Of Occup. Safety and Health, *Profiles on Occupational Hazards for Criteria Document Priorities: Diphenyl,* Report PB 274,073, Cincinnati, OH, pp 274-276 (1977).
(2) Sax, N.I., Ed., *Dangerous Properties of Industrial Materials Report, 1,* No. 5, 42-43, New York, Van Nostrand Reinhold Co. (1981).
(3) See Reference (A-60).
(4) Parmeggiani, L., Ed., *Encyclopedia of Occupational Health & Safety,* Third Edition, Vol. 1, pp 642-44, Geneva, International Labour Office (1983) (Diphenyls & Terphenyls).

BIS(2-CHLOROETHOXY)METHANE

- Hazardous waste (EPA)
- Priority toxic pollutant (EPA)

Description: $ClCH_2CH_2OCH_2OCH_2CH_2Cl$ is a colorless liquid with a boiling point of 218°C.

Code Numbers: CAS 112-26-5   RTECS KH4900000

DOT Designation: —

Synonyms: BCEXM, dichloroethyl formal.

Potential Exposures: The chloroalkyl ethers have a wide variety of industrial uses in organic synthesis, treatment of textiles, the manufacture of polymers and insecticides, as degreasing agents and solvents, and in the preparation of ion exchange resins.

Permissible Exposure Limits in Air: No standards set.

Permissible Concentration in Water: No criteria set because of inadequate data, according to EPA (2).

Harmful Effects and Symptoms: Specific data on BCEXM are very sparse. The reader is referred to the sections on other chloroalkyl ethers:

    —Chloromethyl methyl ether, CMME
    —Bis(chloromethyl) ether, BCME
    —Bis(2-chloroethyl) ether, BCEE
    —Bis(2-chloroisopropyl) ether, BCIE

References:

(1) U.S. Environmental Protection Agency, *Chloroalkyl Ethers: Ambient Water Quality Criteria*, Wash., DC (1980).
(2) U.S. Environmental Protection Agency, *Haloethers: Ambient Water Quality Criteria*, Wash., DC (1980).
(3) U.S. Environmental Protection Agency, *Bis(2-Chloroethoxy)Methane*, Health and Environmental Effects Profile No. 23, Wash., DC, Office of Solid Waste (April 30, 1980).

BIS(2-CHLOROETHYL) ETHER

See "Dichloroethyl Ether."

N,N-BIS(2-CHLOROETHYL)-2-NAPHTHYLAMINE

See "Chlornaphazine."

BIS(2-CHLOROISOPROPYL) ETHER

- Carcinogen (Potential, EPA) (1)
- Hazardous waste (EPA)
- Priority toxic pollutant (EPA)

Description:

$$CH_3$$
$$|$$
$(ClCH_2CH)_2O$, boiling at 177° to 178°C is a colorless liquid.

Code Numbers: CAS 108-60-1   RTECS KN1750000   UN 2490

DOT Designation: Corrosive material.

Synonyms: BCIE; dichloroisopropyl ether; bis(2-chloro-1-methylethyl)-ether; BCMEE.

Potential Exposure: BCIE finds use as a solvent.

Permissible Exposure Limits in Air: No standard set. A provisional occupational limit has been suggested of 15 ppm.

Permissible Concentration in Water: To protect freshwater aquatic life— 238,000 $\mu g/\ell$ on an acute basis for chloroalkyl ether in general. No criteria developed for saltwater aquatic life due to lack of data. For protection of human health, the ambient water criterion is 34.7 $\mu g/\ell$.

Determination in Water: Gas chromatography (EPA Method 611) or gas chromatography plus mass spectrometry (EPA Method 625).

Harmful Effects and Symptoms: There is no empirical evidence that BCIE is carcinogenic; however, some chronic toxic effects of the compound have been noted. One approach to estimating a safe level of BCIE in drinking water utilizes the following general equation:

$$NOAEL \times SF \times BW = W \times Z + R \times F \times Z + AD - (R \times F \times Z),$$

where NOAEL is no apparent adverse effect level in mammals; SF is safety factor; BW is body weight of average human (assume 70 kg); W is daily consumption of water (assume 2 liters); Z is safe level for water; R is bioconcentration factor (in ℓ/kg); F is daily consumption of fish (assume 0.0187 kg); A is daily amount absorbed from air; and D is daily amount from total diet (including fish).

Since valid estimates on current exposure from air and total diet cannot be made, the equation can be simplified to $NOAEL \times SF \times BW = (W + R \times F) \times Z$. The lowest dose tested which caused minimum adverse effects was 10 mg/kg/day for the mice. However, even at this dose, there was an increased incidence of centrilobular necrosis of the liver which was not seen in the high dose group. To be conservative, a safety factor of 1/1,000 will be applied. Assuming an average human body weight of 70 kg, acceptable daily intake calculated is 700 μg per day. Using the estimated bioconcentration factor of 106 for BCIE and assuming daily consumption of 0.0187 kg fish and 2 liters of water, the safe level calculated from these data is 175.8 $\mu g/\ell$. Since this safe level is calculated on the basis of several assumptions that cannot be defended, it should be regarded as a very crude estimate. Another approach to deriving a criterion has been suggested by the Carcinogens Assessment Group, EPA.

As previously stated, BCIE has not been empirically proven to be a carcinogen; nevertheless, it is mutagenic and is in a class of compounds that are known as carcinogens. Based on these facts, credence can be lent to deriving a suggested criterion based upon unpublished NCI preliminary data as of 1978 as applied to the linear, nonthreshold model.

Therefore, a lower bound water concentration of 11.5 $\mu g/\ell$ has been calculated such that there is a 95% confidence that this level is lower than the actual level which would produce a 10^{-5} lifetime risk due to exposure to BCIE.

Although both approaches to calculating a criterion are somewhat tenuous, the weight of evidence for the carcinogenic potential of BCIE is sufficient to be qualitatively suggestive and must not be ignored from a public health point of view. Until further conclusive data become available, the EPA feels it is prudent to consider BCIE as a potential carcinogen (A-33).

References

(1) U.S. Environmental Protection Agency, *Chloroalkyl Ethers: Ambient Water Quality Criteria,* Wash., DC (1980).

(2) U.S. Environmental Protection Agency, *Bis(2-Chloroisopropyl) Ether,* Health and Environmental Effects Profile No. 25, Wash., DC, Office of Solid Waste (April 30, 1980).

(3) U.S. Environmental Protection Agency, *Chemical Hazard Information Profile Draft Report: Bis(2-Chloro-1-Methylethyl) Ether (BCMEE);* Washington, DC (July 29, 1983).

BIS(CHLOROMETHYL) ETHER

- Carcinogen (Human positive, IARC) (5)
- Hazardous waste (EPA)
- Priority toxic pollutant (EPA) [See (3)]

Description: $ClCH_2OCH_2Cl$, bis(chloromethyl) ether, is a colorless, volatile liquid with a suffocating odor. This substance may form spontaneously in warm moist air by the combination of formaldehyde and hydrogen chloride.

Code Numbers: CAS 542-88-1   RTECS KN1575000   UN 2249

DOT Designation: Poison.

Synonyms: BCME, sym-dichloromethyl ether, dichloromethyl ether.

Potential Exposures: Exposure to bis(chloromethyl) ether may occur in industry and in the laboratory. This compound is used as an alkylating agent in the manufacture of polymers, as a solvent for polymerization reactions, in the preparation of ion exchange resins, and as an intermediate for organic synthesis.

Haloethers, primarily α-chloromethyl ethers, represent a category of alkylating agents of increasing concern due to the establishment of a causal relationship between occupational exposure to two agents of this class and lung cancer in the United States and abroad. The cancers are mainly oat cell carcinomas (6).

Potential sources of human exposure to BCME appear to exist primarily in areas including (a) its use in chloromethylating (crosslinking) reaction mixtures in anion-exchange resin production; (b) segments of the textile industry using formaldehyde-containing reactants and resins in the finishing of fabric and as adhesives in the laminating and flocking of fabrics; and (c) the nonwoven industry which uses as binders, thermosetting acrylic emulsion polymers comprising methylol acrylamide, since a finite amount of formaldehyde is liberated on the drying and curing of these bonding agents.

NIOSH has confirmed the spontaneous formation of BCME from the reaction of formaldehyde and hydrochloric acid in some textile plants and is now investigating the extent of possible worker exposure to the carcinogen. However, this finding has recently been disputed by industrial tests in which BCME was not formed in air by the reaction of textile systems employing hydrochloric acid and formaldehyde.

Permissible Exposure Limits in Air: Bis(chloromethyl) ether is included in the Federal standard for carcinogens; all contact with it should be avoided. The ACGIH as of 1983/84 has set a TWA limit of 0.001 ppm (0.005 mg/m^3) with the notation that the material is a human carcinogen.

Determination in Air: Collection by charcoal tube, analysis by gas liquid chromatography (A-1). See also reference (A-10).

Permissible Concentration in Water: For maximum protection of human health from potential carcinogenic effects of exposure to BCME through ingestion of water and contaminated aquatic organisms, the ambient water concentration is zero. Concentrations of BCME estimated to result in additional lifetime cancer risks of 1 in 100,000 are presented by a concentration of 0.038 ng/ℓ (3.8 x 10^{-5} μg/ℓ).

Determination in Water: Gas chromatography (EPA Method 611) or gas chromatography plus mass spectrometry (EPA Method 625).

Routes of Entry: Inhalation of vapor and perhaps, but to a lesser extent, percutaneous absorption.

Harmful Effects and Symptoms: *Local* — Vapor is severely irritating to the skin and mucous membranes and may cause corneal damage which may heal slowly.

Systemic — Bis(chloromethyl) ether has an extremely suffocating odor even in minimal concentration so that experience with acute poisoning is not available. It is not considered a respiratory irritant at concentrations of 10 ppm. Bis(chloromethyl) ether is a known human carcinogen. Animal experiments have shown increases in lung adenoma incidence; olfactory esthesioneuroepitheliomas which invaded the sinuses, cranial vault, and brain; skin papillomas and carcinomas; and subcutaneous fibrosarcomas. There have been several reports of increased incidence of human lung carcinomas (primarily small cell undifferentiated) among ether workers exposed to bis(chloromethyl) ether as an impurity. The latency period is relatively short—10 to 15 years. Smokers as well as nonsmokers may be affected.

Points of Attack: Skin, respiratory tract, eyes, lungs.

Medical Surveillance: Preplacement and periodic medical examinations should include an examination of the skin and respiratory tract, including chest x-ray. Sputum cytology has been suggested as helpful in detecting early malignant changes, and in this connection a smoking history is of importance. Possible effects on the fetus should be considered.

Personal Protective Methods: These are designed to supplement engineering controls and should be appropriate for protection of all skin or respiratory contact. Full body protective clothing and gloves should be used on entering areas of potential exposure. Those employed in handling operations should be provided with full face, supplied air respirators of continuous flow or pressure demand type. On exit from a regulated area, employees should remove and leave protective clothing and equipment at the point of exit, to be placed in impervious containers at the end of the work shift for decontamination or disposal. Showers should be taken before dressing in street clothes.

Respirator Selection: Use chemical cartridge respirator.

Disposal Method Suggested: Incineration, preferably after mixing with another combustible fuel. Care must be exercised to assure complete combustion to prevent the formation of phosgene. An acid scrubber is necessary to remove the halo acids produced (A-31).

References

(1) U.S. Environmental Protection Agency, *Chloroalkyl Ethers: Ambient Water Quality Criteria*, Wash., DC (1980).

(2) U.S. Environmental Protection Agency, *Haloethers: Ambient Water Quality Criteria*, Wash., DC (1980).

(3) EPA has proposed removing bis(chloromethyl) ether from the priority toxic pollutants listed under the Clean Water Act. Although the compound is a known carcinogen, the agency says that any possibility of aquatic life or humans being measurably exposed to the substance through water is extremely remote since it decomposes rapidly in water. The compound is neither produced nor used in significant quantities in the U.S. See *Federal Register*, 45, No. 180, 60942-45 (Sept. 15, 1980).

(4) U.S. Environmental Protection Agency, *Bis(Chloromethyl) Ether*, Health and Environmental Effects Profile No. 26, Wash., DC, Office of Solid Waste (April 30, 1980).

(5) International Agency for Research on Cancer, *IARC Monographs on the Carcinogenic Risks of Chemicals to Humans*, 4, 231, Lyon, France (1974) and 13, 243, Lyon, France (1977).

(6) See Reference (A-62). Also see Reference (A-64).

(7) United Nations Environment Programme, *IRPTC Legal File 1983,* Vol. I, pp VII/341-43, Geneva, Switzerland, International Register of Potentially Toxic Chemicals (1984).

BIS(DIETHYLTHIOCARBAMOYL)DISULFIDE

See "Disulfiram."

BIS(DIMETHYLTHIOCARBAMOYL)DISULFIDE

See "Thiram."

BISMUTH AND COMPOUNDS

Description: Bi, bismuth, is a pinkish-silver, hard, brittle metal melting at 271°C. It is found as the free metal in ores such as bismutite and bismuthinite and in lead ores. Bismuth is soluble in some mineral acids and insoluble in water. Most bismuth compounds are soluble in water.

DOT Designation: —

Synonyms: None.

Potential Exposures: Bismuth is used as a constituent of tempering baths for steel alloys, in low melting point alloys which expand on cooling, in aluminum and steel alloys to increase machinability, and in printing type metal. Bismuth compounds are found primarily in pharmaceuticals as antiseptics, antacids, antiluetics, and as a medicament in the treatment of acute angina. They are also used as a contrast medium in roentgenoscopy and in cosmetics.

For the general population the total intake from food is 5 to 20 μg with much smaller amounts contributed by air and water (1).

Permissible Exposure Limits in Air: There is no Federal standard for bismuth or its compounds. ACGIH has set TWA values of 10 mg/m^3 for bismuth telluride and 5 mg/m³ for Se-doped bismuth telluride and STEL values of 20 mg/m³ for bismuth telluride and 10 mg/m³ for Se-doped bismuth telluride as of 1983/84.

Determination in Air: Bismuth telluride may be collected on a filter and analyzed by atomic absorption analysis (A-10).

Permissible Concentration in Water: No criteria set but EPA (A-37) has suggested an ambient limit of 3.5 μg/ℓ based on health effects.

Determination in Water: Atomic absorption spectrophotometry may be used (1). Spark source mass spectrometry may also be used (1).

Routes of Entry: Ingestion of powder or inhalation of dust.

Harmful Effects and Symptoms: *Local* — Bismuth and bismuth compounds have little or no effect on intact skin and mucous membrane. Absorption occurs only minimally through broken skin.

Systemic — There is no evidence connecting bismuth and bismuth compounds with cases of industrial poisoning. All accounts of bismuth poisoning are from the soluble compounds used previously in therapeutics. Fatalities and near fatalities have been recorded chiefly as a result of intravenous or intramuscular injection of soluble salts. Principal organs affected by poisoning are the kidneys and liver. Chronic intoxication from repeated oral or parenteral doses causes "bismuth line." This is a gum condition with black spots of buccal and colonic mucosa, superficial stomatitis, foul breath, and salivation.

Medical Surveillance: No special considerations are necessary other than following good general health practices. Liver and kidney function should be followed if large amounts of soluble salts are ingested.

First Aid: Dimercaptol (BAL) brings good results in the treatment of bismuth poisoning if given early (1). Other measures include atropine and meperidine to relieve gastrointestinal discomfort.

Personal Protective Methods: Personal hygiene should be stressed, and eating should not be permitted in work areas. Dust masks should be worn in dusty areas to prevent inadvertent ingestion of the soluble bismuth compounds.

References

(1)   U.S. Environmental Protection Agency, *Toxicology of Metals, Vol. II: Bismuth,* Report EPA-600/1-77-022, Research Triangle Park, NC, pp 110-123 (May 1977).

(2)   Sax, N.I., Ed., *Dangerous Properties of Industrial Materials Report, 1,* No. 5, 43-45, New York, Van Nostrand Reinhold Co. (1981).

(3)   Sax, N.I., Ed., *Dangerous Properties of Industrial Materials Report, 3,* No. 5, 64-65, New York, Van Nostrand Reinhold Co. (1983).

(4)   Parmeggiani, L., Ed., *Encyclopedia of Occupational Health & Safety,* Third Edition, Vol. 1, pp 291–292, Geneva, International Labour Office (1983).

BISPHENOL-A

Description: Bisphenol-A has the following formula:

$$HOC_6H_4-\overset{\displaystyle CH_3}{\underset{\displaystyle CH_3}{\overset{|}{\underset{|}{C}}}}-C_6H_4OH$$

It has the white or tan crystals or flakes with a mild phenolic odor, melting at 153°C.

Code Numbers: CAS 80-05-7   RTECS SL6300000

DOT Designation: —

Synonyms: 4,4'-isopropylidenediphenol.

Potential Exposure: Workers engaged in the manufacture of epoxy or polycarbonate resins. NIOSH has estimated (1) that 203,000 workers are potentially exposed per year.

Permissible Exposure Limits in Air: A standard of 5 mg/m^3 has been proposed in the U.S.S.R. Rumania (1)

Determination in Air: Collection by charcoal tube, analysis by gas liquid chromatography (A-1).

Permissible Concentration in Water: No criteria set.

Routes of Entry: Skin contact.

Harmful Effects and Symptoms: Bisphenol-A and its resins produce a typical contact dermatitis: redness and edema with weeping, followed by crusting and scaling, usually confined to the area of contact. Since the face is frequently affected, this may indicate that vapors are the cause, although contact with contaminated clothing can also be a factor. Seldom are areas other than the face and neck, back of hands, and forearms involved (1).

Points of Attack: Skin.

Medical Surveillance: Consider the points of attack in preplacement and periodic physical examinations.

References

(1)   National Institute for Occupational Safety and Health, *Information Profiles on Potential Occupational Hazards (Bisphenol-A),* Rockville, MD (March 29, 1978).

BORON, BORIC ACID AND BORATES

Description: Boron, B, is a brownish-black powder and may be either crystalline or amorphous. It does not occur free in nature and is found in the minerals borax, colemanite, boronatrocalcite, and boracite. Boron is slightly soluble in water under certain conditions.

Boric acid, H_3BO_3, is a white, amorphous powder. Saturated solutions at 0°C contain 2.6% acid; at 100°C, 28% acid. Boric acid is soluble, 1 g/18 ml in cold water.

Borax, $Na_2B_4O_7 \cdot 5H_2O$, is a colorless, odorless crystalline solid. Borax is slightly soluble in water.

Code Numbers:

Elemental boron	CAS 7440-42-8	RTECS ED7350000
Boric acid	CAS 10043-35-3	RTECS ED4550000
Borax	CAS 1303-96-4	RTECS VZ2275000

DOT Designation: —

Synonyms:

Elemental boron: None
Boric acid: Boracic acid, orthoboric acid
Borax: Sodium borate, sodium tetraborate

Potential Exposures: Boron is used in metallurgy as a degasifying agent and is alloyed with aluminum, iron, and steel to increase hardness. It is also a neutron absorber in nuclear reactors. Occupational exposures to elemental boron are estimated at 23,000 by NIOSH.

Boric acid is a fireproofing agent for wood, a preservative, and an antiseptic. It is used in the manufacture of glass, pottery, enamels, glazes, cosmetics, cements, porcelain, borates, leather, carpets, hats, soaps, and artificial gems, and in tanning, printing, dyeing, painting, and photography. It is a constituent in powders, ointments, nickeling baths, electric condensers and is used for impregnating wicks and hardening steel.

Borax is used as a soldering flux, preservative against wood fungus, and as an antiseptic. It is used in the manufacture of enamels and glazes and in tanning, cleaning compounds, for fireproofing fabrics and wood, and in artificial aging of wood. Occupational exposures to borax have been estimated by NIOSH at 2,490,000.

Permissible Exposure Limits in Air: The TWA set by ACGIH for anhydrous sodium tetraborate is 1.0 mg/m^3; for sodium tetraborate decahydrate it is 5.0 mg/m^3; for the pentahydrate it is 1.0 mg/m^3 as of 1983/84.

Permissible Concentration in Water: EPA in July 1976 established a criterion for boron of 750 μg/ℓ for long-term irrigation on sensitive crops. More recently (A-37), EPA has suggested an ambient water limit of 43 μg/ℓ based on health effects.

Routes of Entry: Inhalation of dust, fumes, and aerosols; ingestion.

Harmful Effects and Symptoms: *Local* — These boron compounds may produce irritation of the nasal mucous membranes, the respiratory tract, and eyes.

Systemic — These effects vary greatly with the type of compound. Acute poisoning in man from boric acid or borax is usually the result of application of dressings, powders, or ointment to large areas of burned or abraded skin, or accidental ingestion. The signs are: nausea, abdominal pain, diarrhea and violent vomiting, sometimes bloody, which may be accompanied by headache and weakness. There is a characteristic erythematous rash followed by peeling. In severe cases, shock with fall in arterial pressure, tachycardia, and cyanosis occur. Marked CNS irritation, oliguria, and anuria may be present. The oral lethal dose in adults is over 30 g. Little information is available on chronic oral poisoning, although it is reported to be characterized by mild GI irritation, loss of appetite, disturbed digestion, nausea, possibly vomiting, and erythematous rash. The rash may be "hard" with a tendency to become purpuric. Dryness of skin and mucous membranes, reddening of tongue, cracking of lips, loss of hair, conjunctivitis, palpebral edema, gastro-intestinal disturbances, and kidney injury have also been observed.

Although no occupational poisonings have been reported, it was noted that workers manufacturing boric acid had some atrophic changes in respiratory mucous membranes, weakness, joint pains, and other vague symptoms. The biochemical mechanism of boron toxicity is not clear but seems to involve action on the nervous system, enzyme activity, carbohydrate metabolism, hormone function, and oxidation processes; coupled with allergic effects. Borates are excreted principally by the kidneys. No toxic effects have been attributed to elemental boron.

Medical Surveillance: No specific considerations are needed for boric acid or borates except for general health and liver and kidney function. In the case of boron trifluoride, the skin, eyes, and respiratory tract should receive special attention. In the case of the boranes, central nervous system and lung function will also be of special concern.

Personal Protective Methods: Exposed workers should be educated in the proper use of protective equipment and there should be strict adherence to ventilating provisions in work areas. Workers involved with the manufacture of boric acid should be provided with masks to prevent inhalation of dust and fumes.

Disposal Method Suggested: Borax, dehydrated: The material is diluted to the recommended provisional limit (0.10 mg/ℓ) in water. The pH is adjusted to between 6.5 and 9.1 and then the material can be discharged into sewers or natural streams.

Boric acid may be recovered from organic process wastes (A-57) as an alternative to disposal.

References

(1) Environmental Protection Agency, *Preliminary Investigation of Effects on the Environment of Boron, Indium, Nickel, Selenium, Tin, Vanadium and Their Compounds, Volume 1: Boron*, Report EPA-560/2-75-005A, Wash., DC, Office of Toxic Substances (August 1975).

(2) Nat. Inst. for Occup. Safety and Health, *Information Profiles on Potential Occupational Hazards: Boron and Its Compounds*, Report PB 276,678, Rockville, MD, pp 63-75 (Oct. 1977).

(3) Sax, N.I., Ed., *Dangerous Properties of Industrial Materials Report, 1*, No. 8, 44-45, New York, Van Nostrand Reinhold Co. (1981).

(4) Sax, N.I., Ed., *Dangerous Properties of Industrial Materials Report, 3*, No. 5, 65-67, New York, Van Nostrand Reinhold Co. (1983).

(5) Parmeggiani, L., Ed., *Encyclopedia of Occupational Health & Safety*, Third Edition, Vol. 1, pp 319–22, Geneva, International Labour Office (1983).

BORON OXIDE

Description: B_2O_3. Odorless, colorless glassy, hygroscopic granules or flakes.

Code Numbers: CAS 1303-86-2   RTECS ED7900000

DOT Designation: −

Synonyms: Anhydrous boric acid, boric oxide, boric anhydride.

Potential Exposure: Boron oxide is used in glass manufacture and the production of other boron compounds. It is used in fluxes, enamels, drying agents and as a catalyst. NIOSH has estimated occupational exposures at 21,000.

Incompatibilities: None hazardous.

Permissible Exposure Limits in Air: The Federal standard is 15 mg/m^3, ACGIH has set a TWA of 10 mg/m^3 and a tentative STEL of 20 mg/m^3 as of 1983/84.

Determination in Air: Collection on a filter and gravimetric analysis. See NIOSH Methods, Set S. See also reference (A-10).

Permissible Concentration in Water: No criteria set but EPA has suggested (A-37) an ambient water limit of 138 $\mu g/\ell$ based on health effects.

Routes of Entry: Ingestion, inhalation, skin and eye contact.

Harmful Effects and Symptoms: Nasal irritation, conjunctivitis, erythema, low toxicity.

Points of Attack: Skin, eyes.

Medical Surveillance: Consider the points of attack in preplacement and periodic physical examinations.

First Aid: If this chemical gets into the eyes, irrigate immediately. If this chemical contacts the skin, flush with water promptly. If a person breathes in large amounts of this chemical, move the exposed person to fresh air at once and perform artificial respiration. When this chemical has been swallowed, get medical attention. Give large quantities of water and induce vomiting. Do not make an unconscious person vomit.

Personal Protective Methods: Wear appropriate clothing to prevent repeated or prolonged skin contact. Wear eye protection to prevent any reasonable probability of eye contact. Employees should wash promptly when skin is contaminated. Remove nonimpervious clothing promptly if contaminated.

Respirator Selection:
>   D: 75 mg/m^3:   DXS
>   150 mg/m^3:   DXSQ
> DFu: 150 mg/m^3:   FuHiEP/SA/SCBA
>   750 mg/m^3:   HiEPF/SAF/SCBAF
>   7,500 mg/m^3:   PAPHiEF/SAF:PD,PP,CF

References

(1)   See Reference (A-61).
(2)   Sax, N.I., Ed., *Dangerous Properties of Industrial Materials Report, 1,* No. 8, 42-43, New York, Van Nostrand Reinhold Co. (1981). (Boric Acid).

BORON TRIBROMIDE

Description: BBr$_3$ is a colorless fuming liquid boiling at 90°C.

Code Numbers: CAS 10294-33-4   RTECS ED7400000   UN 2692

DOT Designation: Corrosive material.

Synonyms: None.

Potential Exposures: Boron tribromide is used as a catalyst in organic synthesis.

Permissible Exposure Limits in Air: There are no Federal standards but ACGIH as of 1983/84 has set a TLV of 1 ppm (10 mg/m^3) and an STEL of 3 ppm (30 mg/m^3).

Permissible Concentration in Water: No criteria set.

Harmful Effects and Symptoms: Acute local effects are considered to be high for irritation, ingestion and inhalation (A-34). Reference should be made to hydrogen bromides for toxicity since hydrolysis produces 3 mols of HBr.

Points of Attack: See Hydrogen Bromide.

Medical Surveillance: See Hydrogen Bromide.

First Aid: See Hydrogen Bromide.

Personal Protective Methods: See Hydrogen Bromide.

Respirator Selection: See Hydrogen Bromide.

BORON TRIFLUORIDE

Description: Boron trifluoride, BF_3, is a colorless gas with a pungent, suffocating odor. It is corrosive to the skin and fumes in moist air.

Code Numbers: CAS 7637-07-2   RTECS ED2275000   UN 1008

DOT Designation: Nonflammable gas and poison.

Synonyms: None.

Potential Exposures: Boron trifluoride is a highly reactive chemical used primarily as a catalyst in chemical synthesis. It is stored and transported as a gas but can be reacted with a variety of materials to form both liquid and solid compounds. The magnesium industry utilizes the fire-retardant and antioxidant properties of boron trifluoride in casting and heat treating. Nuclear applications of boron trifluoride include neutron detector instruments, boron-10 enrichment and the production of neutron-absorbing salts for molten-salt breeder reactors. Boron trifluoride is known to be produced at the present time as a specialty chemical by only one United States company whose production figures are not available. NIOSH estimates that 50,000 employees are potentially exposed to boron trifluoride in the United States.

Incompatibilities: Reacts with alkalies, fumes in moist air, particulates reduce visibility.

Permissible Exposure Limits in Air: Sufficient technology exists to prevent adverse effects in workers, but techniques to measure airborne levels of boron trifluoride for compliance with an environmental limit are not adequate. Therefore, an environmental limit is not recommended by NIOSH (1), in part because of the unavailability of adequate monitoring methods. ACGIH as of 1983/84 has set a ceiling value of 1 ppm (3 mg/m^3) but no STEL value.

Determination in Air: Collection by an impinger preceded by a filter followed by colorimetric analysis (1).

Permissible Concentration in Water: No criteria set.

Routes of Entry: Inhalation of vapors on skin and eye contact.

Harmful Effects and Symptoms: Boron trifluoride gas, upon contact with air, immediately reacts with water vapor to form a mist which, if at a high enough concentration provides a visible warning of its presence. The gas or mist is irritating to the skin, eyes, and respiratory system.

The toxic action of the halogenated borons (boron trifluoride and trichloride) is considerably influenced by their halogenated decomposition products. They are primary irritants of the nasal passages, respiratory tract, and eyes in man. Animal experiments showed a fall in inorganic phosphorus level in blood and on autopsy, pneumonia, and degenerative changes in renal tubules. Long-term exposure leads to irritation of the respiratory tract, dysproteinemia, reduction in cholinesterase activity, increased nervous system liability. High concentrations showed a reduction of acetyl carbonic acid and inorganic phosphorus in blood, and dental fluorosis.

Points of Attack: Respiratory system, kidneys, eyes, skin.

Medical Surveillance: In the absence of a suitable monitoring method, NIOSH recommends that medical surveillance, including comprehensive pre-placement and annual periodic examinations be made available to all workers employed in areas where boron trifluoride is manufactured, used, handled, or is evolved as a result of chemical processes.

First Aid: If this chemical gets into the eyes, irrigate immediately. If this chemical contacts the skin, flush with water immediately. If a person breathes in large amounts of this chemical, move the exposed person to fresh air at once and perform artificial respiration.

Personal Protective Methods: Engineering controls should be used to maintain boron trifluoride concentrations at the lowest feasible level. Air supplied respirators should only be used in certain nonroutine situations or in emergency situations when air concentrations of boron trifluoride are sufficient to form a visible mist.

Proper impervious protective clothing, including gloves, aprons, suits, boots, goggles, and face shields, shall be worn as needed to prevent skin and eye contact with boron trifluoride.

Respirator Selection:
            10 ppm: SA/SCBA
            50 ppm: SAF/SCBAF
        100 ppm: SAF:PD,PP,CF
        Escape: GMS/SCBA

Disposal Method Suggested: Chemical reaction with water to form boric acid, and fluoroboric acid. The fluoroboric acid is reacted with limestone forming boric acid and calcium fluoride. The boric acid may be discharged into a sanitary sewer system while the calcium fluoride may be recovered or landfilled.

References
(1) National Institute for Occupational Safety and Health, *Criteria for a Recommended Standard: Occupational Exposure to Boron Trifluoride,* NIOSH Doc. No. 77-122, Wash., DC (1977).
(2) See Reference (A-61).
(3) See Reference (A-60).

BRASS

Description: Brass is a term used for alloys of copper and zinc. The ratio of the two compounds is generally 2 to 1, although different types of brass may have different proportions. Brass may contain significant quantities of lead. Bronze is also a copper alloy, usually with tin; however, the term bronze is applied to many other copper alloys, some of which contain large amounts of zinc.

DOT Designation: –

Synonyms: None.

Potential Exposures: Brass may be cast into bearings and other wearing surfaces, steam and water valves and fittings, electrical fittings, hardware, ornamental castings, and other equipment where special corrosion-resistance properties, pressure tightness, and good machinability are required. Wrought forms of brass such as sheets, plates, bars, shapes, wire, and tubing are also widely used.

Permissible Exposure Limits in Air: There is no Federal standard for brass; however, there are standards for its constituents: lead (inorganic) (0.2 mg/m³) (see the entry under Lead for further details); zinc oxide fume (5 mg/m³); copper fume (0.1 mg/m³).

Permissible Concentration in Water: No criteria set.

Route of Entry: Inhalation of fume.

Harmful Effects and Symptoms: *Local* — Brass dust and slivers may cause dermatitis by mechanical irritation.

Systemic — Since zinc boils at a lower temperature than copper, the fusing of brass is attended by liberation of considerable quantities of zinc oxide. Inhalation of zinc oxide fumes may result in production of signs and symptoms of metal fume fever (see Zinc Oxide). Brass founder's ague is the name often given to metal fume fever occurring in the brass-founding industry.

Brass foundings may also release sufficient amounts of lead fume to produce lead intoxication (See Lead—Inorganic).

Medical Surveillance: See Zinc Oxide and/or Lead—Inorganic.

Blood lead values may be useful if lead fume or dust exposure is suspected (see Lead).

Personal Protective Methods: See Zinc Oxide and/or Lead—Inorganic.

Disposal Method Suggested: Landfill.

BROMACIL

Description: This herbicide, $C_9H_{13}BrN_2O_2$, has the structural formula

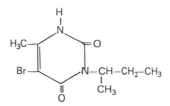

and is a colorless, crystalline solid melting at 158° to 159°C.

Code Numbers: CAS 314-40-9 RTECS YQ9100000 UN 2588

DOT Designation: —

Synonyms: 5-bromo-3-sec-butyl-6-methyluracil, Hyvar X®.

Potential Exposure: Bromacil is used primarily for the control of annual and perennial grasses and broadleaf weeds, both nonselectively on noncrop lands and selectively for weed control in a few crops (citrus and pineapple). Those involved in manufacture (A-32), formulation and application may be exposed to bromacil.

Incompatibilities: Decomposes slowly in strong acids.

Permissible Exposure Limits in Air: A TWA value of 1 ppm (10 mg/m^3) has been adopted by ACGIH as of 1983/84 and an STEL value set at 2 ppm (20 mg/m^3).

Permissible Concentration in Water: A no-adverse effect level in drinking water has been calculated by NAS/NRC (A-2) as 0.086 mg/ℓ.

Harmful Effects and Symptoms: Bromacil is low in both acute and chronic toxicity.

References

(1)   United Nations Environment Programme, *IRPTC Legal File 1983,* Vol. II, pp VII/776, Geneva, Switzerland, International Register of Potentially Toxic Chemicals (1984).

BROMINE

Description: Br, bromine, is a dark reddish-brown, fuming, volatile liquid with a suffocating odor. Bromine is soluble in water and alcohol.

Code Numbers: CAS 7726-95-6   RTECS EF9100000   UN 1744

DOT Designation: Corrosive material.

Synonyms: None.

Potential Exposures: Bromine is primarily used in the manufacture of gasoline antiknock compounds (1,2-dibromoethane). Other uses are for gold extraction, in brominating hydrocarbons, in bleaching fibers and silk, in the manufacture of military gas, dyestuffs, and as an oxidizing agent. It is used in the manufacture of many pharmaceuticals (A-41) and pesticides (A-32).

Incompatibilities: Combustible organics, oxidizable material, aqueous ammonia; anhydrous Br$_2$ reacts with aluminum, titanium, mercury, potassium; wet Br$_2$ with other metals.

Permissible Exposure Limits in Air: The Federal standard for bromine is 0.1 0.1 ppm (0.7 mg/m^3). The STEL value adopted by ACGIH (1983/84) is 0.3 ppm (2.0 mg/m^3). The IDLH level is 10 ppm.

Determination in Air: Sample collection by impinger or fritted bubbler, colorimetric analysis (A-1).

Allowable Concentration in Water: No criteria set.

Routes of Entry: Inhalation, ingestion, eye and skin contact. Bromine may be absorbed through the skin also.

Harmful Effects and Symptoms: *Local* – Bromine is extremely irritating to eyes, skin, and mucous membranes of the upper respiratory tract. Severe burns of the eye may result from liquid or concentrated vapor exposure. Liquid bromine splashed on skin may cause vesicles, blisters, and slow healing ulcers. Continued exposure to low concentrations may result in acne-like skin lesions.

Systemic – Inhalation of bromine is corrosive to the mucous membranes of the nasopharynx and upper respiratory tract, producing brownish discoloration of tongue and buccal mucosa, a characteristic odor of the breath, edema and spasm of the glottis, asthmatic bronchitis, and possibly pulmonary edema which may be delayed until several hours following exposure. A measles-like rash may occur. Exposure to high concentrations of bromine can lead to rapid death due to choking caused by edema of the glottis and pulmonary edema.

Bromine has cumulative properties and is deposited in tissues as bromides, displacing other halogens. Exposures to low concentrations result in cough, copious mucous secretions, nose bleeds, respiratory difficulty, vertigo, and headache. Usually these symptoms are followed by nausea, diarrhea, abdominal distress, hoarseness, and asthmatic type respiratory difficulty.

Points of Attack: Respiratory system, eyes, lungs, central nervous system.

Medical Surveillance: The skin, eyes, and respiratory tract should be given special emphasis during preplacement and periodic examinations. Chest x-rays as well as general health, blood, liver, and kidney function should be considered. Exposure to other irritants or bromine compounds in medications may be important.

First Aid: If this chemical gets into the eyes, irrigate immediately. If this chemical contacts the skin, wash with soap immediately. If a person breathes in large amounts of this chemical, move the exposed person to fresh air at once and perform artificial respiration. When this chemical has been swallowed, get medical attention. Give large quantities of milk or if not available, water, and do not induce vomiting.

Personal Protective Methods: Wear appropriate clothing to prevent any possibility of skin contact. Wear eye protection to prevent any possibility of eye contact. Employees should wash immediately when skin is wet or contaminated. Remove clothing immediately if wet or contaminated. Provide emergency showers and eyewash.

Respirator Selection:
   5 ppm: CCRSF/GMS/SAF/SCBAF
   10 ppm: SAF:PD,PP,CF
   Escape: GMS/SCBA
      Note: Do not use oxidizable sorbents

References

(1) U.S. Environmental Protection Agency, *Chemical Hazard Information Profile: Bromine and Bromine Compounds,* Wash., DC (Nov. 1, 1976).
(2) See Reference (A-61).
(3) Sax, N.I., Ed., *Dangerous Properties of Industrial Materials Report, 1,* No. 4, 41-43, New York, Van Nostrand Reinhold Co. (1981).
(4) See Reference (A-60).
(5) Sax, N.I., Ed., *Dangerous Properties of Industrial Materials Report, 3,* No. 5, 67-69, New York, Van Nostrand Reinhold Co. (1983).
(6) Parmeggiani, L., Ed., *Encyclopedia of Occupational Health & Safety,* Third Edition, Vol. 1, pp 326–329, Geneva, International Labour Office (1983).

BROMINE PENTAFLUORIDE

Description: BrF_5 is a dense colorless liquid which boils at $40°$ to $41°C$.

Code Numbers: CAS 7789-30-2   RTECS EF9350000   UN 1745

DOT Designation: Corrosive, poison, oxidizer.

Synonyms: None

Potential Exposures: BrF_5 is used as an oxidizer in liquid rocket propellant combinations; it may also be used in chemical synthesis.

Incompatibilities: BrF_5 reacts with every known element except inert gases, nitrogen and oxygen. It reacts violently with water and may ignite combustible or organic material.

Permissible Exposure Limits in Air: There are no Federal standards but ACGIH as of 1983/84 has set a TWA value of 0.1 ppm (0.7 mg/m³) and an STEL of 0.3 ppm (2.0 mg/m³).

Determination in Air: Sample collection by impinger or fritted bubbler, analysis by ion-specific electrode (A-10).

Permissible Concentration in Water: No criteria set.

Routes of Entry: Inhalation, eye and skin contact.

Harmful Effects and Symptoms: Respiratory irritation is noted (A-34). See also fluorine and chlorine trifluoride.

Points of Attack: Eyes, skin and mucous membranes.

Respirator Selection: Use of self-contained breathing apparatus is recommended (A-38).

Disposal Method Suggested: Allow gas to flow into mixed caustic soda and slaked lime solution (A-38). Return unwanted cylinders to supplier if possible.

BROMOACETONE

Description: CH_3COCH_2Br is a liquid which boils at 137°C. It turns violet.

Code Numbers: CAS 598-31-2   RTECS UC0525000   UN 1569

DOT Designation: Poison A.

Synonyms: Acetonyl bromide; acetyl methyl bromide; 1-bromo-2-propanone.

Potential Exposure: Used as war gas and riot control agent; also used in organic synthesis.

Incompatibilities: Oxidants.

Permissible Exposure Limits in Air: No standards set.

Permissible Concentration in Water: No criteria set.

Routes of Entry: Inhalation, skin and eye contact, and ingestion.

Harmful Effects and Symptoms: Extreme eye irritant. Very toxic. Corrosive to skin and respiratory tract. Inhalation can cause lung edema (A-60).

Points of Attack: Eyes, respiratory system.

First Aid: Remove contaminated clothes, get to fresh air and flush affected areas with water (A-60).

Personal Protective Methods: Gas-tight eye and skin protection should be provided.

Respirator Selection: Self-contained breathing apparatus approved for chemical warfare agents.

Disposal Method Suggested: Send to U.S. Army Chemical Corps. (1).

References

(1) Sax, N.I., Ed., *Dangerous Properties of Industrial Materials Report, 2,* No. 2, 14-15, New York, Van Nostrand Reinhold Co. (1982).

(2) See Reference (A-60).

BROMOBENZENE

Description: C_6H_5Br, a clear, colorless mobile liquid boiling at 156°C.

Code Numbers: CAS 108-86-1   RTECS CY9000000   UN 2514

DOT Designation: Combustible liquid.

Synonyms: Phenyl bromide; monobromobenzene.

Potential Exposure: Bromobenzene is used as an intermediate in organic synthesis, and as additive in motor oil and fuels. During chlorination water treatment, bromobenzene can be formed in small quantities. NIOSH has estimated occupational exposure to bromobenzene at 18,540.

Incompatibilities: Strong oxidants.

Permissible Exposure Limits in Air: No U.S. standard set, but TLV in U.S.S.R. is reportedly 0.5 ppm (3 mg/m^3) (A-36).

Permissible Concentration in Water: No criteria set.

Routes of Entry: Inhalation, ingestion, skin absorption.

Harmful Effects and Symptoms: *Observations in Man* — Bromobenzene irritates the skin and is a central nervous system depressant in humans. Nothing is known about its chronic effects. In view of the relative paucity of data on the carcinogenicity, teratogenicity, and long-term oral toxicity of bromobenzene, estimates of the effects of chronic oral exposure at low levels cannot be made with any confidence. It is recommended that studies to produce such information be conducted before limits in drinking water can be established. Since bromobenzene was negative on the *Salmonella*/microsome mutagenicity test, there should be less concern than with those substances that are positive (A-2).

Target Organs: Skin, lungs and liver.

Personal Protective Methods: Wear protective gloves and goggles or face shield and overalls.

Respirator Selection: Use self-contained breathing apparatus.

Disposal Method Suggested: Incineration (A-38).

References

(1) National Institute for Occupational Safety and Health, *Information Profiles on Potential Occupational Hazards: Brominated Aromatic Compounds,* pp 76-85 incl., Report PB-276,678, Rockville, MD (Oct. 1977).

BROMOBENZYL CYANIDE

Description: $C_6H_5CHBrCN$ is a crystalline solid with the odor of sour fruit which melts at 29°C and boils with decomposition at 242°C.

Code Numbers: CAS 5798-79-8   RTECS AL8050000   UN 1694

DOT Designation: —

Synonyms: Bromophenyl acetonitrile; α-bromo-α-tolunitrile; bromobenzyl-nitrile; BBC; BBN; CA; Camite.

Potential Exposure: Used as a war gas and riot control agent.

Permissible Exposure Limits in Air: No standards set.

Permissible Concentration in Water: No criteria set.

Routes of Entry: Inhalation.

Harmful Effects and Symptoms: Strong lachrymator and highly toxic (0.90 mg/ℓ in air for 30 minutes is lethal to humans).

Points of Attack: Eyes.

Personal Protective Methods: Goggles, rubber suit.

Respirator Selection: Full military gas mask.

References
(1) Sax, N.I., Ed., *Dangerous Properties of Industrial Materials Report, 2,* No. 3, 68, New York, Van Nostrand Reinhold Co. (1982).

BROMODICHLOROMETHANE

- Carcinogen (Potential, EPA) (1)
- Priority toxic pollutant (EPA)

Description: $BrCHCl_2$ is a liquid which boils at $90°C$.

Code Numbers: CAS 75-27-4   RTECS PA5260000

DOT Designation: —

Synonyms: Dichlorobromomethane

Potential Exposure: This compound may find application in organic synthesis.

Permissible Exposure Limits in Air: No standard set.

Permissible Concentration in Water: Bromodichloromethane, along with chlorodibromomethane, trichloromethane (chloroform) and tribromomethane form the group of halomethanes termed total trihalomethanes (TTHM), which are to be regulated in finished drinking water in the U.S. The maximum permissible concentration set for TTHM in the proposed regulations is 0.100 mg/ℓ.

For the protection of human health, the ambient water quality level suggested by EPA (1) is zero. An additional lifetime cancer risk of 1 in 100,000 is posed by a concentration of 1.9 $\mu g/ℓ$.

Measurement in Water: Gas chromatography (EPA Method 601) or gas chromatography plus mass spectrometry (EPA Method 624).

Harmful Effects and Symptoms: Bromodichloromethane is acutely toxic to mice. It was mutagenic in the *Salmonella typhimurium* TA 100 bacterial test system and carcinogenic in mice with the same qualification for result significance as for dichloromethane noted. Positive correlations between cancer mortality rates and levels of brominated trihalomethanes in drinking water in epidemiological studies have been reported (A-33).

References

(1) U.S. Environmental Protection Agency, *Halomethanes: Ambient Water Quality Criteria*, Wash., DC (1980).

BROMOFORM

- Carcinogen (Potential, EPA) (1)
- Hazardous waste (EPA)
- Priority toxic pollutant (EPA)

Description: $CHBr_3$, a colorless to yellow liquid with a chloroformlike odor boiling at 149.5°C and freezing at 6° to 7°C to hexagonal crystals.

Code Numbers: CAS 75-25-2   RTECS PB5600000   UN 2515

DOT Designation: Use St. Andrew's Cross (X) on label.

Synonyms: Tribromomethane, methyl tribromide.

Potential Exposure: Bromoform is used in pharmaceutical manufacturing, as an ingredient in fire-resistant chemicals and gauge fluid, and as a solvent for waxes, greases, and oils.

Incompatibilities: Chemically active metals: sodium, potassium, calcium, powdered aluminum, zinc, magnesium, strong caustics.

Permissible Exposure Limits in Air: The Federal limit is 0.5 ppm (5.0 mg/m^3). ACGIH seconds this value as of 1983/84 but adds the notation "skin" indicating the possibility of cutaneous absorption. No STEL value has been set, nor has an IDLH level.

Determination in Air: Adsorption on charcoal, workup with CS_2, analysis by gas chromatography. See NIOSH Methods, Set H. See also reference (A-10).

Permissible Concentration in Water: Bromoform is one of 4 trihalomethanes for which the U.S. EPA has proposed a maximum contaminant level in drinking water of 0.100 mg/ℓ.
For the protection of human health—preferably zero. An additional lifetime cancer risk of 1 in 100,000 results at a level of 1.9 µg/ℓ (1).

Determination in Water: Gas chromatography (EPA Method 601) or gas chromatography plus mass spectrometry (EPA Method 624).

Routes of Entry: Inhalation, ingestion, skin absorption, eye and skin contact.

Harmful Effects and Symptoms: Irritation of eyes and respiratory system, central nervous system depression, liver damage.

Points of Attack: Skin, liver, kidneys, respiratory system, lungs, central nervous system.

Medical Surveillance: Consider the points of attack in preplacement and periodic physical examinations.

First Aid: If this chemical gets into the eyes, irrigate immediately. If this chemical contacts the skin, wash with soap promptly. If a person breathes in

large amounts of this chemical, move the exposed person to fresh air at once and perform artificial respiration. When this chemical has been swallowed, get medical attention. Give large quantities of saltwater and induce vomiting. Do not make an unconscious person vomit.

Personal Protective Methods: Wear appropriate clothing to prevent repeated or prolonged skin contact. Wear eye protection to prevent any reasonable probability of eye contact. Employees should wash promptly when skin is wet or contaminated. Remove nonimpervious clothing promptly if wet or contaminated.

Respirator Selection:

25 ppm:	CCROVF/GMOV/SAF/SCBAF
1,000 ppm:	SAF:PD,PP,CF
Escape:	GMOC/SCBA

Disposal Method Suggested: Purify by distillation and return to suppliers (A-38).

References

(1) U.S. Environmental Protection Agency, *Halomethanes: Ambient Water Quality Criteria,* Wash., DC (1980).
(2) U.S. Environmental Protection Agency, *Bromoform,* Health and Environmental Effects Profile No. 28, Wash., DC, Office of Solid Waste (April 30, 1980).
(3) Sax, N.I., Ed., *Dangerous Properties of Industrial Materials Report, 2,* No. 6, 30-35, New York, Van Nostrand Reinhold Co. (1982).
(4) See Reference (A-61).
(5) United Nations Environment Programme, *IRPTC Legal File 1983,* Vol. II, pp VII/441–42, Geneva, Switzerland, International Register of Potentially Toxic Chemicals (1984).

4-BROMOPHENYL PHENYL ETHER

- Hazardous waste (EPA)
- Priority toxic pollutant (EPA)

Description: $BrC_6H_4OC_6H_5$ is a liquid boiling at $310°C$.

Code Number: CAS 101-55-3

DOT Designation: –

Synonyms: None.

Potential Exposures: Very little information on 4-bromophenyl phenyl ether exists. 4-Bromophenyl phenyl ether has been identified in raw water, in drinking water and in river water.

Permissible Exposure Limits in Air: No standards set.

Permissible Concentration in Water: *Freshwater Aquatic Life* – For 4-bromophenyl phenyl ether the criterion to protect freshwater aquatic life as derived using the guidelines is 6.2 $\mu g/\ell$ as a 24 hour average and the concentration should not exceed 14 $\mu g/\ell$ at any time.

Saltwater Aquatic Life – For saltwater aquatic life, no criterion for 4-bromophenyl phenyl ether can be derived using the guidelines, and there are insufficient data to estimate a criterion using other procedures.

Human Health – Because of a lack of adequate toxicological data on nonhuman mammals and humans, protective criteria cannot be derived at this time for this compound (1).

Determination in Water: Methylene chloride extraction followed by gas chromatography with halogen-specific detector (EPA Method 611) or gas chromatography plus mass spectrometry (EPA Method 625).

Harmful Effects and Symptoms: 4-Bromophenyl phenyl ether has been tested in the pulmonary adenoma assay, a short-termed carcinogenicity assay. Although the results were negative, several known carcinogens also gave negative results. No other health effects were available (2).

References

(1) U.S. Environmental Protection Agency, *Haloethers: Ambient Water Quality Criteria*, Wash., DC (1980).

(2) U.S. Environmental Protection Agency, *4-Bromophenyl Phenyl Ether*, Health and Environmental Effects Profile No. 30, Wash., DC, Office of Solid Waste (April 30, 1980).

BROMOXYNIL

Description: $C_6H_2(OH)(CN)Br_2$ with the structural formula

is a solid melting at 194° to 195°C.

Code Numbers: CAS 1689-84-5   RTECS DI3150000   UN NA2810

DOT Designation: —

Synonyms: 3,5-dibromo-4-hydroxybenzonitrile; 3,5-dibromo-4-hydroxyphenyl cyanide; Buctril®, Brominil®.

Potential Exposure: Those engaged in manufacture and application of this herbicide.

Permissible Exposure Limits in Air: No standards set.

Permissible Concentration in Water: No criteria set.

Harmful Effects and Symptoms: Moderately toxic (LD-50 for rats is 250 mg/kg).

Disposal Method Suggested: Small amounts may be disposed of on the ground. Larger amounts should be incinerated.

References

(1) Sax, N.I., Ed., *Dangerous Properties of Industrial Materials Report, 2*, No. 4, 45-48, New York, Van Nostrand Reinhold Co. (1982).

BRUCINE

Description: $C_{23}H_{26}N_2O_4$ with the structural formula

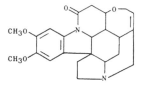

is a crystalline solid melting at 178°C.

Code Numbers: CAS 357-57-3   RTECS EH8925000   UN 1570

DOT Designation: Poison B.

Synonyms: Dimethoxy strychnine; 2,3-dimethoxy-strychnidin-10-one.

Potential Exposure: Used in denaturing alcohol, as a lubricant additive and as a rodent poison.

Permissible Exposure Limits in Air: No standards set.

Permissible Concentration in Water: No criteria set.

Routes of Entry: Inhalation and ingestion.

Harmful Effects and Symptoms: A deadly poison.

Points of Attack: Muscular system (A-60).

First Aid: Remove contaminated clothing and wash with soap and water.

Personal Protective Methods: Goggles, butyl rubber gloves and full protective clothing required (A-38).

Respirator Selection: Self-contained breathing apparatus.

Disposal Method Suggested: Incineration.

References
(1) Sax, N.I., Ed., *Dangerous Properties of Industrial Materials Report, 1,* No. 8, 45-47, New York, Van Nostrand Reinhold Co. (1981).
(2) See Reference (A-60).
(3) Sax, N.I., Ed., *Dangerous Properties of Industrial Materials Report, 3,* No. 5, 70-71, New York, Van Nostrand Reinhold Co. (1983).

BUTACHLOR

Description: $C_{17}H_{26}ClNO_2$ with the structural formula

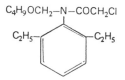

is a light yellow nonvolatile oil boiling at 196°C at 0.5 mm Hg.

Code Numbers: CAS 23184-66-9   RTECS AE1200000

DOT Designation: —

Synonyms: 2-chloro-2',6'-diethyl-N-(butoxymethyl)-acetanilide, Machete®.

Potential Exposure: Butachlor is used as a preemergence herbicide for the control of annual grasses and broad-leaved weeds. Thus, manufacturers, formulators and applicators may be exposed.

Permissible Exposure Limits in Air: No standards set.

Permissible Concentration in Water: A no-adverse effect level in drinking water has been calculated as 0.07 mg/ℓ by NAS/NRC (A-2). This is one order of magnitude lower than the companion compounds, alachlor and propachlor. An allowable daily intake (ADI) of 0.01 mg/kg/day has been calculated.

Disposal Method Suggested: Hydrolyzed by strong acids or alkalis but incineration is the recommended procedure (A-32).

1,3-BUTADIENE

• Carcinogen (Human, Suspected) (ACGIH) (A-6) (3)

Description: $H_2C=CH-CH=CH_2$, 1,3-butadiene, is a colorless, flammable gas with a pungent, aromatic odor. It boils at -4° to -5°C. Because of low flash point, 1,3-butadiene's fire and explosion hazard may be more serious than its health hazard.

Code Numbers: CAS 106-99-0   RTECS EI9275000   UN 1010

DOT Designation: Flammable gas.

Synonyms: Biethylene, bivinyl, butadiene monomer, divinyl, erythrene, methylallene, pyrrolylene, vinylethylene.

Potential Exposures: 1,3-Butadiene is used chiefly as the principal monomer in the manufacture of many types of synthetic rubber. Presently, butadiene is finding increasing usage in the formation of rocket fuels, plastics, and resins.

Incompatibilities: Strong oxidizers, copper, copper alloys.

Permissible Exposure Limits in Air: The Federal standard for 1,3-butadiene is 1,000 ppm (2,200 mg/m^3). The STEL value is 1,250 ppm (2,750 mg/m^3). The IDLH level is 20,000 ppm. The ACGIH (1983/84), however, has indicated its intention to replace TWA values with the notation that butadiene is an "industrial substance suspect of carcinogenic potential for man."

Determination in Air: Adsorption by charcoal, workup with CS_2, analysis by gas chromatography. See NIOSH Methods, Set G. See also reference (A-10).

Permissible Concentration in Water: No criteria set.

Routes of Entry: Inhalation of gas or vapor, eye and skin contact.

Harmful Effects and Symptoms: *Local* — Butadiene gas is slightly irritating to the eyes, nose, and throat. Dermatitis and frostbite may result from exposure to liquid and evaporating gas.

Systemic — In high concentrations, 1,3-butadiene gas can act as an irritant, producing cough, and as a narcotic, producing fatigue, drowsiness, headache, vertigo, loss of consciousness, respiratory paralysis, and death. One report states that chronic exposure may result in central nervous system disorders, diseases of the liver and biliary system, and tendencies toward hypotension, leukopenia, increase in ESR, and decreased hemoglobin content in the blood. These changes have not been seen by most observers in humans.

Points of Attack: Eyes, respiratory system, central nervous system.

Medical Surveillance: No specific considerations are needed, according to NIOSH.

First Aid: If this chemical gets into the eyes, irrigate immediately. If this chemical contacts the skin, flush with water immediately. If a person breathes in large amounts of this chemical, move the exposed person to fresh air at once and perform artificial respiration.

Personal Protective Methods: Wear appropriate clothing to prevent skin freezing. Wear eye protection to prevent any reasonable probability of eye contact. Employees should wash immediately when skin is wet or contaminated. Remove clothing immediately if wet or contaminated to avoid flammability hazard.

Respirator Selection:
5,000 ppm: GMSc
8,000 ppm: SA/SCBA
20,000 ppm: GMSfb/SAF/SCBAF
Escape: GMS/SCBA

Disposal Method Suggested: Incineration.

References

(1)   See Reference (A-61).
(2)   See Reference (A-60).
(3)   National Institute of Occupational Safety and Health, *1,3-Butadiene,* Current Intelligence Bulletin 41, DHHS (NIOSH) Publication No. 84–105, Cincinnati, Ohio (February 9, 1984).
(4)   Parmeggiani, L., Ed., *Encyclopedia of Occupational Health and Safety,* Third Edition, Vol. 1, pp 347–48, Geneva, International Labour Office (1983).

n-BUTANE

Description: With the formula C_4H_{10}, normal butane is $CH_3CH_2CH_2CH_3$.

Code Numbers: CAS 106-97-8   RTECS EJ4200000   UN 1011

DOT Designation: Flammable gas.

Synonyms: Liquified petroleum gas.

Potential Exposures: It is used as a raw material for butadiene, as a fuel for household or industrial purposes (alone or in admixture with propane). It is also used as an extractant, solvent, and aerosol propellant. It is used in plastic foam production as a replacement for fluorocarbons.

Permissible Exposure Limits in Air: There is no Federal standard but ACGIH as of 1983/84 has proposed a TWA value of 800 ppm (1,900 mg/m³) but no STEL value.

Permissible Concentration in Water: No criteria set but EPA (A-37) has suggested an ambient limit of 19,000 $\mu g/\ell$ based on health effects.

Routes of Entry: Inhalation.

Harmful Effects and Symptoms: Butane is not characterized by its toxicity but rather by its narcosis-producing potential at high exposure levels (A-34).

First Aid: Oxygen inhalation and artificial respiration.

Personal Protective Methods: Wear rubber gloves, safety glasses and protective clothing (A-38).

Respirator Selection: Use chemical cartridge mask.

Disposal Method Suggested: Controlled incineration.

2-BUTANONE PEROXIDE

See "Methyl Ethyl Ketone Peroxide."

n-BUTOXYETHANOL

Description: $C_4H_9OCH_2CH_2OH$, a colorless liquid with a mild odor, boiling at 171°C.

Code Numbers: CAS 111-76-2   RTECS KJ8575000   UN 2369

DOT Designation: —

Synonyms: Butyl cellosolve; ethyleneglycol monobutylether; Dowanol EB; butyl oxitol; Jeffersol EB; Ektasolve EB.

Potential Exposures: This material is used as a solvent for resins in lacquers, varnishes and enamels. It is also used in varnish removers and in dry cleaning compounds.

Incompatibilities: Strong oxidizers, strong caustics.

Permissible Exposure Limits in Air: The Federal standard is 50 ppm (240 mg/m³). The ACGIH (as of 1983/84) has proposed a reduction in TWA to 25 ppm (120 mg/m³). ACGIH adds the notation "skin" indicating the possibility of cutaneous absorption. ACGIH has adopted an STEL of 75 ppm (360 mg/m³). The IDLH level is 700 ppm.

Determination in Air: Adsorption on charcoal, workup with methanol and methylene chloride, analysis by gas chromatography. See NIOSH Methods, Set F. See also reference (A-10).

Permissible Concentration in Water: No criteria set.

Routes of Entry: Inhalation, ingestion, skin absorption, eye and skin contact.

Harmful Effects and Symptoms: Irritation of eyes, nose and throat; hemolysis, hemoglobinuria.

Points of Attack: Liver, kidneys, lymphoid system, skin, blood, eyes, respiratory system.

Medical Surveillance: Consider the points of attack in preplacement and periodic physical examinations.

First Aid: If this chemical gets into the eyes, irrigate immediately. If this chemical contacts the skin, wash with soap promptly. If a person breathes in

large amounts of this chemical, move the exposed person to fresh air at once and perform artificial respiration. When this chemical has been swallowed, get medical attention. Give large quantities of saltwater and induce vomiting. Do not make an unconscious person vomit.

Personal Protective Methods: Wear appropriate clothing to prevent repeated or prolonged skin contact. Wear eye protection to prevent any reasonable probability of eye contact. Employees should wash immediately when skin is wet or contaminated. Remove nonimpervious clothing immediately if wet or contaminated. Provide emergency showers.

Respirator Selection:
700 ppm: CCROVF/GMOV/SAF/SCBAF
Escape: GMOV/SCBA

Disposal Method Suggested: Incineration.

References
(1) See Reference (A-61).
(2) See Reference (A-60).

BUTYL ACETATES

● Hazardous substances (EPA)

Description: $C_4H_9OCOCH_3$, these are colorless liquids with pleasant, fruity odors. There are 4 isomers:

		BP, $^\circ$C
n-Butyl	$CH_3CH_2CH_2CH_2OCOCH_3$	126–127
sec-Butyl	$CH_3CH_2CH(CH_3)OCOCH_3$	112–113
Isobutyl	$(CH_3)_2CHCH_2OCOCH_3$	117–118
tert-Butyl	$(CH_3)_3COCOCH_3$	97–98

Code Numbers:

n-Butyl	CAS 123-86-4	RTECS AF7350000	UN 1123
sec-Butyl	CAS 105-46-4	RTECS AF7380000	UN 1124
Isobutyl	CAS 110-19-0	RTECS AI4025000	UN 1213
tert-Butyl	CAS 540-88-5	RTECS AF7400000	

DOT Designation: Flammable liquid.

Synonyms:

n-Butyl	Butyl acetate, butyl ethanoate, acetic acid butyl ester
sec-Butyl	1-Methylpropyl acetate
Isobutyl	2-Methylpropyl acetate, β-methylpropyl ethanoate, acetic acid isobutyl ester
tert-Butyl	Acetic acid tert-butyl ester

Potential Exposures: n-Butyl acetate is an important solvent in the production of lacquers, leather and airplane dopes, and perfumes. sec-Butyl acetate is used as a solvent for nitrocellulose and nail enamels. tert-Butyl acetate is also a solvent and has been suggested as an antiknock agent for motor fuels. Isobutyl acetate is used as a solvent for nitrocellulose, and in perfumes and flavoring materials. According to NIOSH, 1,836,210 workers are exposed to butyl acetate annually and 1,081,689 workers are exposed to sec-butyl acetate annually.

Incompatibilities: Nitrates, strong oxidizers, strong alkalies, strong acids.

Permissible Exposure Limits in Air: n-Butyl and isobutyl acetates have a

Federal limit of 150 ppm (710 mg/m^3). sec-Butyl and tert-butyl have a Federal limit of 200 ppm (950 mg/m^3). The STEL tentative values are:

n-Butyl	200 ppm	(950 mg/m^3)
sec-Butyl	250 ppm	(1,190 mg/m^3)
Isobutyl	187 ppm	(875 mg/m^3)
tert-Butyl	250 ppm	(1,190 mg/m^3)

The IDLH levels are:

n-Butyl	10,000 ppm
sec-Butyl	10,000 ppm
Isobutyl	7,500 ppm
tert-Butyl	8,000 ppm

Determination in Air: Adsorption on charcoal, workup with CS_2, analysis by gas chromatography. See NIOSH Methods, Set D. See also reference (A-10) for all 4 isomers.

Allowable Concentration in Water: No criteria set.

Routes of Entry: Inhalation, ingestion, skin and eye contact.

Harmful Effects and Symptoms: Headaches, drowsiness, eye irritation, irritation of skin and upper respiratory system. Humans and animals that inhale comparatively low doses of n-butyl acetate experience irritation of the nasal and respiratory passages and of the eyes. At higher concentrations narcosis takes place, and repeated exposures have resulted in renal and blood changes in experimental animals.

Points of Attack: Skin, eyes, respiratory system, central nervous system.

Medical Surveillance: Consider initial effects on skin and respiratory tract in any preplacement or periodical examinations, as well as liver and kidney function.

First Aid: If this chemical gets into the eyes, irrigate immediately. If this chemical contacts the skin, flush with water promptly. If a person breathes in large amounts of this chemical, move the exposed person to fresh air at once and perform artificial respiration. When this chemical has been swallowed, get medical attention. Give large quantities of saltwater and induce vomiting. Do not make an unconscious person vomit.

Personal Protective Methods: Wear appropriate clothing to prevent repeated or prolonged skin contact. Wear eye protection to prevent any reasonable probability of eye contact. Employees should wash promptly when skin is wet or contaminated. Remove clothing immediately if wet or contaminated to avoid flammability hazard.

Respirator Selection:

n-Butyl	1,000 ppm:	CCROVF
	5,000 ppm:	GMOVc
	7,500 ppm:	GMOVfb/SAF/SCBAF
	10,000 ppm:	SAF:PD,PP,CF
	Escape:	GMOV/SCBA
sec-Butyl	1,000 ppm:	CCROVF
	5,000 ppm:	CMOVc
	10,000 ppm:	GMOVfb/SAF/SCBAF
	Escape:	GMOV/SCBA
Isobutyl	1,000 ppm:	CCROVF
	5,000 ppm:	GMOVc
	7,500 ppm:	GMOVfb/SAF/SCBAF
	Escape:	GMOV/SCBA

tert-Butyl   1,000 ppm: CCROVF
              5,000 ppm: GMOVc
              8,000 ppm: GMOVfb/SAF/SCBAF
              Escape: GMOV/SCBA

Disposal Methods Suggested: Incineration.

References

(1) Nat. Inst. for Occup. Safety and Health, *Information Profiles on Potential Occupational Hazards—Single Chemicals: n-Butyl Acetate,* Report TR 79-607, Rockville, MD, pp 19-27 (Dec. 1979).
(2) See Reference (A-61).
(3) Sax, N.I., Ed., *Dangerous Properties of Industrial Materials Report, 2,* No. 2, 41-43, New York, Van Nostrand Reinhold Co. (1982).
(4) See Reference (A-60).
(5) Sax, N.I., Ed., *Dangerous Properties of Industrial Materials Report, 3,* No. 6, 35–37, New York, Van Nostrand Reinhold Co. (1983).
(6) Sax, N.I., Ed., *Dangerous Properties of Industrial Materials Report, 4,* No. 3, 38–41, New York, Van Nostrand Reinhold Co. (1984).

n-BUTYL ACRYLATE

Description: $CH_2=CHCOOC_4H_9$ is a colorless liquid boiling at 146° to 148°C.

Code Numbers: CAS 141-32-2   RTECS UD3150000   UN 2348

DOT Designation: Flammable liquid.

Synonyms: 2-Propenoic acid, butyl ester; butyl 2-propenoate.

Potential Exposures: This material is used as a monomer in the production of polymers and copolymers for solvent coatings, adhesives, paints and binders.

Permissible Exposure Limits in Air: There is no Federal standard but ACGIH as of 1983/84 has adopted a TWA value of 10 ppm (55 mg/m^3). There is no tentative STEL value.

Permissible Concentration in Water: No criteria set.

Routes of Entry: Ingestion, skin and eye contact.

Harmful Effects and Symptoms: n-Butyl acrylate was found to be but moderately irritating to the skin. As an eye irritant it produced corneal necrosis in an unwashed rabbit eye, similar to that produced by ethyl alcohol (A-34). Exposure of rats at 1,000 ppm for 4 hours proved lethal to 5 of 6 rats exposed; however, rats survived a 30-minute exposure to 7,000 ppm. There is a close similarity in toxic response by inhalation, skin and eye to methyl acrylate.

Points of Attack: Skin, eyes.

Disposal Method Suggested: Incineration.

References

(1) See Reference (A-60).

BUTYL ALCOHOLS

- Hazardous waste (n-Butanol) (EPA)

Description: C_4H_9OH, these are colorless liquids with strong odors. There are 4 isomers:

n-Butyl	$CH_3CH_2CH_2CH_2OH$	117°–118°C BP
sec-Butyl	$CH_3-CH_2CH(CH_3)OH$	99°–100°C BP
Isobutyl	$(CH_3)_2CHCH_2OH$	108°C BP
tert-Butyl	$(CH_3)_3COH$	82°–83°C BP

Code Numbers:

n-Butyl	CAS 71-36-3	RTECS EO1400000	UN 1120
sec-Butyl	CAS 78-92-2	RTECS EO1750000	UN 1121
Isobutyl	CAS 78-83-1	RTECS NP9625000	UN 1212
tert-Butyl	CAS 75-65-0	RTECS EO1925000	UN 1122

DOT Designations: Flammable liquid.

Synonyms:

n-Butyl	1-Butanol, propylcarbinol, NBA
sec-Butyl	2-Butanol, methyl ethyl carbinol, butylene hydrate, 2-hydroxybutane
Isobutyl	Isobutanol, IBA, 2-methyl-1-propanol, isobutyl carbinol
tert-Butyl	2-Methyl-2-propanol, TBA, trimethyl carbinol

Potential Exposures: Butyl alcohols are used as solvents for paints, lacquers, varnishes, natural and synthetic resins, gums, vegetable oils, dyes, camphor, and alkaloids. They are also used as an intermediate in the manufacture of pharmaceuticals and chemicals and in the manufacture of artificial leather, safety glass, rubber and plastic cements, shellac, raincoats, photographic films, perfumes, and in plastic fabrication.

Incompatibilities: Strong oxidizers in all cases. In the case of tert-butanol, strong mineral acids as well.

Permissible Exposure Limits in Air:

	Federal Limit ppm	. . . . . . ACGIH . . . . . . ppm	mg/m³
n-Butyl	100	50 (Ceiling)	150*
sec-Butyl	150	100 (TWA)	305
Isobutyl	100	50 (TWA)	150
tert-Butyl	100	100 (TWA)	300

*With the notation "skin" indicating that skin absorption may be significant.

The STEL values are:

n-Butyl alcohol	No value set
sec-Butyl alcohol	150 ppm
Isobutyl alcohol	75 ppm
tert-Butyl alcohol	150 ppm

The IDLH levels are:

n-Butyl	8,000 ppm
sec-Butyl	10,000 ppm
Isobutyl	8,000 ppm
tert-Butyl	8,000 ppm

Determination in Air: Adsorption on charcoal, workup with 2-propanol in CS_2, for all 4 isomers except 2-butanone used instead of 2-propanol in tert-butanol measurement, analysis by gas chromatography. See NIOSH Methods, Set E. See also reference (A-10).

Permissible Concentration in Water: No criteria set, but EPA has suggested (A-37) ambient limits as follows, based on health effects:

n-Butanol	2,070 μg/ℓ
sec-Butanol	6,200 μg/ℓ
Isobutanol	2,070 μg/ℓ
tert-Butanol	4,140 μg/ℓ

Routes of Entry: Ingestion, inhalation, skin and eye contact.

Harmful Effects and Symptoms: Irritation of eyes and throat; headaches, drowsiness; skin irritation and cracking.

Points of Attack: Eyes, skin and respiratory system.

Medical Surveillance: Consider irritant effects on eyes, respiratory tract, and skin in any preplacement or periodic examinations.

First Aid: If this chemical gets into the eyes, irrigate immediately. If this chemical contacts the skin, flush with water promptly. If a person breathes in large amounts of this chemical, move the exposed person to fresh air at once and perform artificial respiration. When this chemical has been swallowed, get medical attention. Give large quantities of saltwater and induce vomiting. Do not make an unconscious person vomit.

Personal Protective Methods: Wear appropriate clothing to prevent repeated or prolonged skin contact. Wear eye protection to prevent any reasonable probability of eye contact. Employees should wash promptly when skin is wet or contaminated. Remove clothing immediately if wet or contaminated to avoid flammability hazard.

Respirator Selection:

n-, iso- and tert-butanol	1,000 ppm:	CCROVF
	5,000 ppm:	GMOV/SAF/SCBAF
	8,000 ppm:	SAF:PD,PP,CF
	Escape:	GMOV/SCBA
sec-butanol	1,000 ppm:	CCROVF
	5,000 ppm:	GMOVc
	7,500 ppm:	GMOVfb/SAF/SCBAF
	10,000 ppm:	SAD:PD,PP,CF
	Escape:	GMOV/SCBA

Disposal Method Suggested: Incineration.

References

(1)   See Reference (A-61).
(2)   See Reference (A-60).

n-BUTYLAMINE

● Hazardous substance (EPA)

Description: $CH_3CH_2CH_2CH_2NH_2$, n-butylamine, is a flammable colorless liquid with an ammoniacal odor. The odor and irritation of the mucous membranes cannot be relied upon for exposure control, however.

Code Numbers: CAS 109-73-9   RTECS EO2975000   UN 1125

DOT Designation: Flammable liquid.

Synonyms: 1-Aminobutane.

Potential Exposures: n-Butylamine is used in pharmaceuticals (A-41), dye-stuffs, rubber, chemicals, emulsifying agents, photography, desizing agents for textiles, pesticides, and synthetic agents.

Incompatibilities: Strong oxidizers, strong acids.

Permissible Exposure Limits in Air: The Federal standard and ACGIH (1983/84) value is 5 ppm (15 mg/m^3) as a ceiling value. The notation "skin" is added to indicate the possibility of cutaneous absorption. There is no tentative STEL value. The IDLH level is 2,000 ppm.

Determination in Air: Adsorption on H_2SO_4-treated silica gel, desorption with 50% methanol, analysis by gas chromatography (A-10). There is also a colorimetric method which may be used (A-1).

Permissible Concentration in Water: No criteria set, but EPA (A-37) has suggested ambient water limits for butylamines (n-, iso- or tert-) as 207 µg/ℓ based on health effects.

Routes of Entry: Inhalation and percutaneous absorption, as well as ingestion and eye and skin contact.

Harmful Effects and Symptoms: *Local* – Butylamine vapor is irritating to the nose, throat, and eyes. Contact with the liquid may produce severe eye damage and skin burns.
Systemic – Inhalation of concentrations at or above the threshold limit may produce mild headaches and flushing of the skin and face.
Butylamine vapor has produced pulmonary edema in animal experiments.

Points of Attack: Respiratory system, skin and eyes.

Medical Surveillance: Evaluate risks of eye or skin injury and respiratory irritation in periodic or placement examinations.

First Aid: If this chemical gets into the eyes, irrigate immediately. If this chemical contacts the skin, flush with water immediately. If a person breathes in large amounts of this chemical, move the exposed person to fresh air at once and perform artificial respiration. When this chemical has been swallowed, get medical attention. Give large quantities of water and induce vomiting. Do not make an unconscious person vomit.

Personal Protective Methods: Wear appropriate clothing to prevent any possibility of skin contact . Wear eye protection to prevent any possibility of eye contact. Employees should wash immediately when skin is wet or contaminated. Remove clothing immediately if wet or contaminated to avoid flammability hazard. Provide emergency showers and eyewash.

Respirator Selection:
>   250 ppm: CCRSF/GMS/SAF/SCBAF
>   2,000 ppm: SAF:PD,PP,CF
>   Escape: GMS/SCBA

Disposal Method Suggested: Incineration; incinerator is equipped with a scrubber or thermal unit to reduce NO_x emissions (A-31). Alternatively it may be poured into sodium bisulfate, neutralized and flushed to the sewer with water (A-38).

References

(1) Sax, N.I., Ed., *Dangerous Properties of Industrial Materials Report, 2,* No. 3, 68-70, New York, Van Nostrand Reinhold Co. (1982).
(2) See Reference (A-61).
(3) See Reference (A-60).

tert-BUTYLAMINE

● Hazardous substance (EPA)

Description: $(CH_3)_3CNH_2$ is a liquid which boils at $44°$ to $46°C$.

Code Numbers: CAS 75-64-9   RTECS EO3330000

DOT Designation: −

Synonyms: 2-Aminoisobutane; 2-amino-2-methylpropane; dimethylethylamine; 2-propanamine, 2-methyl; trimethylaminomethane.

Potential Exposures: tert-Butylamine is used as a chemical intermediate in the production of tert-butylaminoethyl methacrylate (a lube oil additive), as an intermediate in the production of rubber and in rust preventatives and emulsion deterrents in petroleum products. It is used in the manufacture of several drugs (A-41).

Permissible Exposure Limits in Air: The Federal standard is 5 ppm (15 mg/m^3). There is no STEL value proposed by ACGIH.

Determination in Air: See n-Butylamine.

Permissible Concentration in Water: No criteria set but see n-Butylamine.

Routes of Entry: Inhalation, ingestion, skin and eye contact.

Harmful Effects and Symptoms: Very little information concerning the biological effects of exposure to tert-butylamine was encountered. Straight chain butylamines cause irritation of the eyes and respiratory tract, excitability, followed by depression, and pulmonary edema in laboratory animals. Danger to humans results from skin and respiratory tract irritation with lung edema being the maximum injury.

Points of Attack: Skin, eyes, respiratory system.

Medical Surveillance: Consider the points of attack in preplacement and periodic physical examinations.

First Aid: See n-Butylamine.

Personal Protective Methods: See n-Butylamine.

Respirator Selection: See n-Butylamine.

Disposal Method Suggested: Incineration; incinerator is equipped with a scrubber or thermal unit to reduce NO_x emissions.

References

(1)   Nat. Inst. for Occup. Safety and Health, *Information Profiles on Potential Occupational Exposures—Single Chemicals: tert-Butylamine,* Report TR79-607, Rockville, MD, pp 28-33 (Dec. 1979).
(2)   See Reference (A-60).

tert-BUTYL CHROMATE

● Carcinogen (Inferred, NIOSH) (1)

Description: $[(CH_3)_3CO]_2CrO_2$, is a liquid.

Code Numbers: CAS 1189-85-1   RTECS GB2900000

DOT Designation −

Synonyms: Bis(tert-butyl)chromate; chromic acid, di-tert-butyl ester.

Potential Exposures: Butyl chromate is used in specialty reactions as an organic source of chromium.

Incompatibilities: Reducing agents, moisture.

Permissible Exposure Limits in Air: The Federal standard and ACGIH 1983/84 value is 0.1 mg/m³ as a ceiling value. ACGIH adds the notation "skin" indicating the possibility of cutaneous absorption. The applicable NIOSH Criteria Document on occupational exposure to Cr(+6) indicates 1.0 $\mu g/m^3$.

Permissible Concentration in Water: No criteria set.

Routes of Entry: Inhalation, skin absorption, ingestion and skin and eye contact.

Harmful Effects and Symptoms: Lung and sinus cancer.

Points of Attack: Respiratory system, lungs, skin, eyes, central nervous system.

Medical Surveillance: Consider the points of attack in preplacement and periodic physical examinations.

First Aid: If this chemical gets into the eyes, irrigate immediately. If this chemical contacts the skin, wash with soap immediately. If a person breathes in large amounts of this chemical, move the exposed person to fresh air at once and perform artificial respiration. When this chemical has been swallowed, get medical attention. Give large quantities of water and induce vomiting. Do not make an unconscious person vomit.

Personal Protective Methods: Wear appropriate clothing to prevent any possibility of skin contact. Wear eye protection to prevent any possibility of eye contact. Employees should wash immediately when skin is wet or contaminated and daily at the end of each work shift. Remove nonimpervious clothing immediately if wet or contaminated. Provide emergency showers and eyewash.

Respirator Selection:
   5 mg/m³: SAF/SCBAF
   200 mg/m³: SAF:PD,PP,CF
   Escape: GMOVP/SCBA

References

(1) Nat. Inst. for Occup. Safety and Health, *Criteria for a Recommended Standard: Occupational Exposure to Chromium,* NIOSH Doc. No. 76-129, Wash., DC (1976).
(2) See Reference (A-61).

1,3-BUTYLENE GLYCOL

Description: $CH_3CHOHCH_2CH_2OH$ is a colorless viscous liquid which boils at 207.5°C.

Code Numbers: CAS 107-88-0 RTECS EK0440000

DOT Designation: (U.S. Coast Guard: Grade E combustible liquid).

Synonyms: 1,3-butanediol; 1,3-dihydroxy-butane; methyl trimethylene glycol.

Potential Exposure: Used as a humectant for cellophane and tobacco. Intermediate in the manufacture of plasticizers.

Incompatibilities: Oxidants.

Permissible Exposure Limits in Air: No standards set.

Permissible Concentration in Water: No criteria set.

Routes of Entry: Inhalation and ingestion.

Harmful Effects and Symptoms: Skin and eye irritant.

Points of Attack: Skin, eyes.

First Aid: Get to fresh air, remove contaminated clothes and rinse with water (A-60).

Personal Protective Methods: Wear goggles.

Respirator Selection: Canister-type masks.

Disposal Method Suggested: Incineration.

References

(1)   Sax, N.I., Ed., *Dangerous Properties of Industrial Materials Report, 3,* No. 2, 35-36, New York, Van Nostrand Reinhold Co. (1983).
(2)   Sax, N.I., Ed., *Dangerous Properties of Industrial Materials Report, 2,* No. 1, 88-91, New York, Van Nostrand Reinhold Co. (1982).

n-BUTYL GLYCIDYL ETHER

Description: $C_4H_9OCH_2CH-CH_2$ is a colorless liquid with a slight irritant odor boiling at 164°C.

Code Numbers: CAS 2426-08-6   RTECS TX4200000

DOT Designation: –

Synonyms: BGE, 1,2-epoxy-3-butoxypropane.

Potential Exposures: NIOSH has estimated human exposures at 17,580. It is used as a viscosity-reducing agent, as an acid acceptor for solvents and as a chemical intermediate.

Incompatibilities: Strong oxidizers, strong caustics.

Permissible Exposure Limits in Air: The Federal standard is 50 ppm (270 mg/m^3). The ACGIH has, as of 1983/84, lowered the TWA value to 25 ppm (135 mg/m^3). There are no STEL values proposed. The IDLH level is 3,500 ppm.

Determination in Air: Adsorption on charcoal, workup with CS_2, analysis by gas chromatography. See NIOSH Methods, Set F. See also reference (A-10).

Permissible Concentration in Water: No criteria set.

Routes of Entry: Inhalation, ingestion, skin and eye contact.

Harmful Effects and Symptoms: Irritation of eyes and nose, skin irritation. sensitizations; narcosis. See also (A-35).

Points of Attack: Eyes, skin, respiratory system, central nervous system.

Medical Surveillance: Consider the points of attack in preplacement and periodic physical examinations.

First Aid: If this chemical gets into the eyes, irrigate immediately. If this chemical contacts the skin, wash with soap immediately. If a person breathes in large amounts of this chemical, move the exposed person to fresh air at once and perform artificial respiration. When this chemical has been swallowed, get medical attention. Give large quantities of saltwater and induce vomiting. Do not make an unconscious person vomit.

Personal Protective Methods: Wear appropriate clothing to prevent any possibility of skin contact. Wear eye protection to prevent any reasonable probability of eye contact. Employees should wash promptly when skin is wet or contaminated. Remove nonimpervious clothing promptly if wet or contaminated.

Respirator Selection:
>2,500 ppm: CCROV/GMOV/SAF/SCBAF
>3,500 ppm: PAPOVF/SAF:PD,PP,CF
>Escape: GMOV/SCBA

Disposal Method Suggested: Incineration.

References

(1) Nat. Inst. for Occup. Safety and Health, *Criteria for a Recommended Standard: Occupational Exposure to Glycidyl Ethers,* NIOSH Doc. No. 78-166, Wash., DC (1978).

(2) Nat. Inst. for Occup. Safety and Health, *Information Profiles on Potential Occupational Hazards: Glycidyl Ethers,* Report PB 276-678, Rockville, MD, pp 116-123 (Oct. 1977).

(3) See Reference (A-60).

BUTYL LACTATE

Description: $CH_3CHOHCOO(CH_2)_3CH_3$ is a liquid boiling at 188° to 190°C.

Code Numbers: CAS 138-22-7 RTECS OD4025000

DOT Designation: –

Synonyms: Lactic acid, butyl ester.

Potential Exposures: Butyl lactate is used as a resin solvent in lacquers, varnishes and inks.

Permissible Exposure Limits in Air: There is no Federal standard but ACGIH as of 1983/84 has adopted a TWA value of 5 ppm (25 mg/m^3). There is no tentative STEL value.

Determination in Air: Collection by filter and analysis by gas liquid chromatography (A-1).

Permissible Concentration in Water: No criteria set.

Harmful Effects and Symptoms: At concentrations of 7 ppm with short peaks of 11 ppm, workers experienced headaches, upper respiratory system irritation and coughing. Some complained of sleepiness and headache in the evening after work and occasional nausea and vomiting was experienced. When exposures were below 1.4 ppm, however, no symptoms were manifested (A-34).

Disposal Method Suggested: Incineration.

BUTYL MERCAPTAN

Description: $CH_3CH_2CH_2CH_2SH$ is a colorless liquid with a strong skunklike odor. It boils at 97° to 98°C.

Code Numbers: CAS 109-79-5   RTECS EK6300000   UN 2347 (formerly 2702).

DOT Designation: Flammable liquid.

Synonyms: Butanethiol, 1-mercaptobutane.

Potential Exposure: The major use is in the production of organophosphorus compounds and thiolcarbamates; more specifically insecticides, herbicides, acaricides and defoliants. Occupational exposures to various butyl mercaptans have been estimated by NIOSH to be 19,410.

Incompatibilities: Strong oxidizers such as dry bleaches.

Permissible Exposure Limits in Air: The Federal standard is 10 ppm (35 mg/m³). The ACGIH has adopted 0.5 ppm (1.5 mg/m³) as a TWA value as of 1983/84. The IDLH level is 2,500 pppm.

Determination in Air: Adsorption on Chemisorb 104, desorption with acetone, analysis by gas chromatography (A-10).

Permissible Concentration in Water: No criteria set, but EPA (A-37) has suggested an ambient water limit of 21 μg/ℓ based on health effects.

Routes of Entry: Inhalation, ingestion, skin and eye contact.

Harmful Effects and Symptoms: In animals–narcosis, incoordination, weakness; cyanosis, pulmonary irritation, eye irritation, paralysis.

Points of Attack: Respiratory system, lungs. In animals–central nervous system, liver, kidneys.

Medical Surveillance: Consider the points of attack in preplacement and periodic physical examinations.

First Aid: If this chemical gets into the eyes, irrigate immediately. If this chemical contacts the skin, wash with soap promptly. If a person breathes in large amounts of this chemical, move the exposed person to fresh air at once and perform artificial respiration. When this chemical has been swallowed, get medical attention. Give large quantities of water and induce vomiting. Do not make an unconscious person vomit.

Personal Protective Methods: Wear appropriate clothing to prevent repeated or prolonged skin contact. Wear eye protection to prevent any reasonable probability of eye contact. Employees should wash promptly when skin is wet or contaminated. Remove clothing immediately if wet or contaminated to avoid flammability hazard.

Respirator Selection:
      100 ppm: SA/SCBA
      500 ppm: SAF/SCBAF
    2,500 ppm: SA:PD,PP,CF
      Escape: GMOV/SCBA

Disposal Method Suggested: Incineration (2000°F) followed by scrubbing with a caustic solution (A-31).

References

(1) Nat. Inst. for Occup. Safety and Health, *Information Profiles on Potential Occupational Hazards: Mercaptans*, Report PB 276,678, Rockville, MD, pp 169-176 (October 1977).

o-sec-BUTYLPHENOL

Description: $C_2H_5(CH_3)CHC_6H_4OH$ is a liquid boiling at $224°$ to $237°C$.

Code Numbers: CAS 89-72-5 RTECS SJ8920000 UN 2228

DOT Designation: Label to bear St. Andrew's cross (X).

Synonyms: None.

Potential Exposures: This material is a chemical intermediate in the preparation of resins, plasticizers and surface active agents.

Permissible Exposure Limits in Air: There is no Federal standard but ACGIH as of 1983/84 has adopted a TWA value of 5 ppm (30 mg/m^3) with the notation "skin" indicating the possibility of cutaneous absorption. There is no tentative STEL.

Permissible Concentration in Water: No criteria set.

Harmful Effects and Symptoms: Skin burns, mild respiratory irritation.

Points of Attack: Skin, respiratory system.

Disposal Method Suggested: Incineration.

p-tert-BUTYLTOLUENE

Description: $p-CH_3C_6H_4C_4H_9$, a colorless liquid which boils at $193°$ to $194°C$. It has an aromatic gasolinelike odor.

Code Numbers: CAS 98-51-1 RTECS XS8400000 UN 2667

DOT Designation: –

Synonyms: 1-Methyl-4-tert-butylbenzene.

Potential Exposures: This material is used as a solvent for resins and as an intermediate in organic synthesis.

Incompatibilities: Oxidizers.

Permissible Exposure Limits in Air: The Federal limit and the ACGIH 1983/84 value is 10 ppm (60 mg/m^3). The STEL is 20 ppm (120 mg/m^3). The IDLH level is 1,000 ppm.

Determination in Air: Adsorption on charcoal, workup with CS_2, analysis by gas chromatography. See NIOSH Methods, Set C. See also reference (A-10).

Permissible Concentration in Water: No criteria set.

Routes of Entry: Inhalation, ingestion, eye and skin contact. See also (A-35).

Harmful Effects and Symptoms: Irritation of eyes and skin; dry nose and throat and headaches; low blood pressure; tachycardia; abnormal cardiovascular system behavior; central nervous system depression; hematopoietic depression.

Points of Attack: Cardiovascular system, central nervous system, skin, bone marrow, eyes, upper respiratory tract.

Medical Surveillance: Consider the points of attack in preplacement and periodic physical examinations.

First Aid: If this chemical gets into the eyes, irrigate immediately. If this chemical contacts the skin, flush with water promptly. If a person breathes in large amounts of this chemical, move the exposed person to fresh air at once and perform artificial respiration. When this chemical has been swallowed, get medical attention. Do not induce vomiting.

Personal Protective Methods: Wear appropriate clothing to prevent repeated or prolonged skin contact. Wear eye protection to prevent any reasonable probability of eye contact. Employees should wash promptly when skin is wet or contaminated. Remove nonimpervious clothing promptly if wet or contaminated.

Respirator Selection:
    500 ppm: CCROVF/GMOV/SAF/SCBAF
    1,000 ppm: SAF,PF,PP,CF
    Escape: GMOV/SCBA

Disposal Method Suggested: Incineration.

References
(1)   See Reference (A-60).

BUTYRIC ACID

Description: $CH_3(CH_2)_2COOH$ is an oily liquid with an unpleasant odor which boils at 163.5°C.

Code Numbers: CAS 107-92-6   RTECS ES5425000   UN 2820

DOT Designation: Corrosive material (U.S. Coast Guard: Grade E combustible liquid).

Synonyms: Butanoic acid; ethylacetic acid.

Potential Exposure: In manufacture of butyrate esters, some of which go into artificial flavoring.

Permissible Exposure Limits in Air: The U.S.S.R. TLV value is 2.5 ppm (10 mg/m^3) (A-36).

Permissible Concentration in Water: No criteria set. Biological effects reviewed (A-36)

Routes of Entry: Inhalation and ingestion.

Harmful Effects and Symptoms: Is a mild irritant for the eyes, skin and respiratory tract.

Points of Attack: Skin, eyes and respiratory system.

Personal Protective Methods: Wear eye and skin protection.

Respirator Selection: Wear self-contained breathing apparatus.

Disposal Method Suggested: Incineration.

References
(1)   Sax, N.I., Ed., *Dangerous Properties of Industrial Materials Report, 2,* No. 3, 71-73, New York, Van Nostrand Reinhold Co. (1982).
(2)   See Reference (A-60).

C

CADMIUM AND COMPOUNDS

- Carcinogen (EPA–CAG) (7)
- Hazardous substances (acetate, bromide and chloride) (EPA)
- Hazardous waste constituents (EPA)
- Priority toxic pollutant (EPA)

Description: Cd, cadmium, is a bluish-white metal. The only cadmium mineral, greenockite, is rare; however, small amounts of cadmium are found in zinc, copper, and lead ores. It is generally produced as a by-product of these metals, particularly zinc. Cadmium is insoluble in water but is soluble in acids. "Cadmium dust" includes dust of various cadmium compounds such as $CdCl_2$. "Cadmium fume" has the composition Cd/CdO.

Code Numbers: (for cadmium metal) CAS 7440-43-9 RTECS EU9800000
(for cadmium compounds) UN 2570

DOT Designation: Label should bear St. Andrews cross (X).

Synonyms: None.

Potential Exposures: Cadmium is highly corrosion resistant and is used as a protective coating for iron, steel, and copper; it is generally applied by electroplating, but hot dipping and spraying are possible. Cadmium may be alloyed with copper, nickel, gold, silver, bismuth, and aluminum to form easily fusible compounds. These alloys may be used as coatings for other materials, welding electrodes, solders, etc. It is also utilized in electrodes of alkaline storage batteries, as a neutron absorber in nuclear reactors, a stabilizer for polyvinyl chloride plastics, a deoxidizer in nickel plating, an amalgam in dentistry, in the manufacture of fluorescent lamps, semiconductors, photocells, and jewelry, in process engraving, in the automobile and aircraft industries, and to charge Jones reductors.

Various cadmium compounds find use as fungicides, insecticides, nematocides, polymerization catalysts, pigments, paints, and glass; they are used in the photographic industry and in glazes. Cadmium is also a contaminant of superphosphate fertilizers.

Human exposure to cadmium and certain cadmium compounds occurs through inhalation and ingestion. OSHA estimates that 360,000 workers are potentially exposed to cadmium and its compounds. The entire population is exposed to low levels of cadmium in the diet because of the entry of cadmium into the food chain as a result of its natural occurrence. Tobacco smokers are exposed to an estimated 1.7 µg/cigarette. Cadmium is present in relatively low amounts in the earth's crust; as a component of zinc ores, cadmium may be released into the environment around smelters.

Incompatibilities: Strong oxiders, elemental sulfur, selenium and tellerium are incompatible with cadmium dust.

Permissible Exposure Limits in Air: The Federal standard for cadmium fume is 0.1 mg (100 μg)/m^3 (as Cd) as an 8-hour TWA with an acceptable ceiling of 0.3 mg/m^3. For cadmium dust, the standard is 0.2 mg/m^3 (Cd) as an 8-hour TWA with an acceptable maximum ceiling of 0.6 mg/m^3. NIOSH has recommended (1) a TWA limit of 40 μg/m^3 with a ceiling limit of 200 μg in a 5-minute sampling period. ACGIH (1983/84) has set a TWA value of 0.05 mg/m^3 for Cd dust, fume and salts and cadmium oxide production. An STEL value of 0.2 mg/m^3 has been set for cadmium dust and salts.

Determination in Air: Collection of particles on a filter, workup with acid and measurement by atomic absorption has been specified by NIOSH. See NIOSH Methods Set W for cadmium dust and Set 1 for cadmium fume. See also reference (A-10).

Permissible Concentration in Water: To protect freshwater aquatic life:

$$e^{[1.05\ \ln(\text{hardness})\ -\ 8.53]}\ \mu g/\ell$$

as a 24-hour average, never to exceed:

$$e^{[1.05\ \ln(\text{hardness})\ -3.73]}\ \mu g/\ell$$

at any time. To protect saltwater aquatic life, 4.5 μg/ℓ as a 24-hour average, never to exceed 59.0 μg/ℓ at any time. To protect human health, 10 μg/ℓ (USEPA). WHO has also set a 10 μg/ℓ standard (A-65). The Federal Republic of Germany has set 6 μg/ℓ and at the other end of the scale, the South African Bureau of Standards has set 50 μg/ℓ as a drinking water standard.

Determination in Water: Total cadmium may be determined by digestion followed by atomic absorption or colorimetric (Dithizone) analysis or by Inductively Coupled Plasma (ICP) Optical Emission Spectrometry. Dissolved cadmium is determined by 0.45 μ filtration followed by the previously cited methods.

Routes of Entry: Inhalation or ingestion of fumes or dust.

Harmful Effects and Symptoms: *Local* — Cadmium is an irritant to the respiratory tract. Prolonged exposure can cause anosmia and a yellow stain that gradually appears on the necks of the teeth. Cd compounds are poorly absorbed from the intestinal tract, but relatively well absorbed by inhalation. Skin absorption appears negligible. Once absorbed Cd has a very long half-life and is retained in the kidneys and liver.

Systemic — Acute toxicity is almost always caused by inhalation of cadmium fumes or dust which are produced when cadmium is heated. There is generally a latent period of a few hours after exposure before symptoms develop. During the ensuing period, symptoms may appear progressively. The earliest symptom is slight irritation of the upper respiratory tract. This may be followed over the next few hours by cough, pain in the chest, sweating, and chills which resemble the symptoms of nonspecific upper respiratory infection. Eight to 24 hours following acute exposure severe pulmonary irritation may develop, with pain in the chest, dyspnea, cough, and generalized weakness. Dyspnea may become more pronounced as pulmonary edema develops. The mortality rate in acute cases is about 15%. Patients who survive may develop emphysema and cor pulmonale; recovery can be prolonged.

Chronic cadmium poisoning has been reported after prolonged exposure to cadmium oxide fumes, cadmium oxide dust, cadmium sulfides, and cadmium stearates. Heavy smoking has been reported to considerably increase tissue Cd levels. In some cases, only the respiratory tract is affected. In others the effects may be systemic due to absorption of the cadmium. Lung damage often results in a characteristic form of emphysema which in some instances is not preceded by a history of chronic bronchitis or coughing. This type of emphysema can be extremely disabling. Some studies have not shown these effects.

Systemic changes due to cadmium absorption include damage to the kidneys with proteinuria, anemia, and elevated sedimentation rate. Of these, proteinuria (low molecular weight) is the most typical. In advanced stages of the disease, there may be increased urinary excretion of amino acids, glucose, calcium, and phosphates. These changes may lead to the formation of renal calculi. If the exposure is discontinued, there is usually no progression of the kidney damage. Mild hypochromic anemia is another systemic condition sometimes found in chronic exposure to cadmium.

In studies with experimental animals, cadmium has produced damage to the liver and central nervous system, testicular atrophy, teratogenic effects in rodents after intravenous injection of Cd, decrease in total red cells, sarcomata, and testicular neoplasms. Hypertensive effects have also been produced. None of these conditions, however, has been found in man resulting from occupational exposure to cadmium. Heavy smoking would appear to increase the risk of cumulative toxic effects.

Cadmium chloride, oxide, sulfate, and sulfide are carcinogenic in rats causing local sarcomas after subcutaneous injection. Cadmium powder and cadmium sulfide produce local sarcomas in rats following intramuscular administration. Cadmium chloride and cadmium sulfate produces testicular tumours in mice and rats following subcutaneous administration.

Early studies suggested that occupational exposure to cadmium in some form (possibly the oxide) increases the risk of prostate cancer in humans. In addition, one of these studies suggested an increased risk of respiratory tract cancer. A later study showed a slight but not statistically significant increase in prostrate cancer in battery plant workers (2 observed vs. 1.2 expected) and cadmium alloy workers (4 observed vs. 2.69 expected).

A case-control study of renal cancer patients showed a 2.5-fold increased risk associated with occupational cadmium exposure. This relative risk doubled when cigarette smoking was included (12).

Points of Attack: Respiratory system, lungs, kidneys, prostate, blood.

Medical Surveillance: In preemployment physical examinations emphasis should be given to a history of, or actual presence of, significant kidney disease, smoking history, and respiratory disease. A chest x-ray and baseline pulmonary function study is recommended. Periodic examinations should emphasize the respiratory system, including pulmonary function tests, kidneys, and blood.

A low molecular weight proteinuria may be the earliest indication of renal toxicity. The trichloroacetic acid test may pick this up, but more specific quantitative studies would be preferable. If renal disease due to cadmium is present, there may also be increased excretion of calcium, amino acids, glucose, and phosphates.

First Aid: *For Cadmium Dust* — If this chemical gets into the eyes, irrigate immediately. If this chemical contacts the skin, wash with soap. If a person breathes in large amounts of this chemical, move the exposed person to fresh air at once and perform artificial respiration. When this chemical has been swallowed, get medical attention. Give large quantities of water and do induce vomiting. Do not make an unconscious person vomit.

For Cadmium Fume — If a person breathes in large amounts of this chemical, move the exposed person to fresh air at once and perform artificial respiration.

Personal Protective Methods: *For Cadmium Dust* — Wear eye protection to prevent any possibility of eye contact with $CdCl_2$. Employees should wash daily at the end of each work shift. Work clothing should be changed daily if it is possible that clothing is contaminated. Provide eyewash if $CdCl_2$ is involved.

Respirator Selection:

For Cadmium Dust —
1 mg/m^3: DXS
2 mg/m^3: DXSQ/HiEP/SA/SCBA
10 mg/m^3: HiEPF/SAF/SCBAF
40 mg/m^3: PAPHiE/SA:PD,PP,CF
Escape: DXS/SCBA

For Cadmium Fume —
1 mg/m^3: FuHiEP/SA/SCBA
5 mg/m^3: HiEPF/SAF/SCBAF
40 mg/m^3: PAPHiE/SA:PD,PP,CF
Escape: HiEPF/SCBA

Disposal Method Suggested: For cadmium fluoride: precipitation with soda ash or slaked lime. The resulting sludge should be sent to a chemical waste landfill (A-31). With cadmium compounds in general, precipitation from solution as sulfides, drying and return of the material to suppliers for recovery is recommended (A-38). Cadmium may be recovered from battery scrap (A-57) as an alternative to disposal.

References

(1) National Institute for Occupational Safety and Health, *Criteria for a Recommended Standard: Occupational Exposure to Cadmium,* NIOSH Doc. No. 76-192, Washington, DC (1976).

(2) U.S. Environmental Protection Agency, *Cadmium: Ambient Water Quality Criteria,* Washington, DC (1980).

(3) U.S. Environmental Protection Agency, *Status Assessment of Toxic Chemicals: Cadmium,* Report EPA-600/2-79-210F, Washington, DC (December 1979).

(4) U.S. Environmental Protection Agency, *Toxicology of Metals, Vol II: Cadmium,* Report EPA-600/1-77-022, Research Triangle Park, NC, pp 124-163 (May 1977).

(5) U.S. Environmental Protection Agency, *Reviews of the Environmental Effects of Pollutants, IV: Cadmium,* Report EPA-600/1-78-026, Health Effects Research Laboratory, Cincinnati, OH (1978).

(6) U.S. Environmental Protection Agency, *Health Assessment Document for Cadmium,* Report EPA-600/8-79-003, Research Triangle Park, NC (1979).

(7) U.S. Environmental Protection Agency, *Cadmium,* Health and Environmental Effects Profile No. 31, Office of Solid Wastes, Washington, DC (April 30, 1980).

(8) Sax, N.I., Ed., *Dangerous Properties of Industrial Materials Report, 1,* No. 1, 36-38, New York, Van Nostrand Reinhold Co. (1980). (Cadmium).

(9) Sax, N.I., Ed., *Dangerous Properties of Industrial Materials Report, 2,* No. 3, 73-78, New York, Van Nostrand Reinhold Co. (1982). (Cadmium Chloride, Fluoborate).

(10) Sax, N.I., Ed., *Dangerous Properties of Industrial Materials Report, 2,* No. 4, 48-53, New York, Van Nostrand Reinhold Co. (1982). (Cadmium Nitrate, Sulfate).

(11) See Reference (A-61). (Cadmium Dust and Fume).

(12) See Reference (A-62). Also see Reference (A-64).

(13) See Reference (A-60) for entries on Cadmium, Cadmium Acetate, Cadmium Hydroxide, Cadmium Oxide and Cadmium Sulfide.

(14) Sax, N.I., Ed., *Dangerous Properties of Industrial Materials Report, 3,* No. 5, 72-76, New York, Van Nostrand Reinhold Co. (1983). (Cadmium).

(15) Sax, N.I., Ed., *Dangerous Properties of Industrial Materials Report, 3,* No. 5, 76-79, New York, Van Nostrand Reinhold Co. (1983). (Cadmium Bromide).

(16) Parmeggiani, L., Ed., *Encyclopedia of Occupational Health & Safety,* Third Edition, Vol. 1, pp 356–58, Geneva, International Labour Office (1983).

(17) United Nations Environment Programme, *IRPTC Legal File 1983,* Vol. I, pp VII/144–52, Geneva, Switzerland, International Register of Potentially Toxic Chemicals (1984).

CALCIUM ARSENATE

- Carcinogen (See "Arsenic.")
- Hazardous substance (EPA)

Description: $Ca_3(AsO_4)_2$, a white flocculent powder.

Code Numbers: CAS 7778-44-1 RTECS CG0830000 UN 1573

DOT Designation: Poison B.

Synonyms: Tricalcium arsenate, tricalcium orthoarsenate; cucumber dust.

Potential Exposure: Workers engaged in manufacture (A-32), formulation and application of pesticides containing calcium arsenate.

Incompatibilities: None hazardous, according to NIOSH.

Permissible Exposure Limits in Air: The Federal limit is 1.0 mg/m^3. NIOSH has recommended a limit of 2 μg/m^3 with a 15-minute ceiling.

Determination in Air: Filter collection followed by atomic absorption analysis (A-10).

Permissible Concentration in Water: No criteria set for calcium arsenate. See "Arsenic and Arsenic Compounds."

Routes of Entry: Inhalation, skin absorption, ingestion, eye and skin contact.

Harmful Effects and Symptoms: Weakness; gastrointestinal distress; peripheral neuropathy; hyperpigmentation; palmar-plantar hyperkeratoses; dermatitis, skin cancer; cancer of the lungs, larynx and lymph.

A rebuttable presumption against registration was issued on October 18, 1978 by U.S. EPA on the basis of oncogenicity, teratogenicity and mutagenicity.

Points of Attack: Eyes, respiratory system, liver, skin, lymphatics, lungs, central nervous system.

Medical Surveillance: Consider the points of attack in preplacement and periodic physical examinations.

First Aid: If this chemical gets into the eyes, irrigate immediately. If this chemical contacts the skin, wash with soap promptly. If a person breathes in large amounts of this chemical, move the exposed person to fresh air at once and perform artificial respiration. When this chemical has been swallowed, get medical attention. Give large quantities of water and induce vomiting. Do not make an unconscious person vomit.

Personal Protective Methods: Wear appropriate clothing to prevent repeated or prolonged skin contact. Wear eye protection to prevent any possibility of eye contact. Employees should wash promptly when skin is wet or contaminated and daily at the end of each work shift. Work clothing should be changed daily if it is possible that clothing is contaminated. Remove nonimpervious clothing promptly if contaminated.

Respirator Selection:
50 mg/m³: HiEPF/SAF/SCBAF
100 mg/m³: PAPHiEF/SAF:PD,PP,CF
Escape: HiEPF/SCBA

Disposal Method Suggested: Long-term storage in large, weatherproof, and sift-proof storage bins or silos; small amounts may be disposed in a chemical waste landfill (A-31). Alternatively, dissolve in HCl, precipitate as sulfide, with H_2S, dry and return to supplier (A-38).

References

(1)   National Institute for Occupational Safety and Health, *Criteria for a Recommended Standard: Occupational Exposure to Inorganic Arsenic,* NIOSH Doc. No. 74-110, Washington, DC (1974).

(2)   Sax, N.I., Ed., *Dangerous Properties of Industrial Materials Report, 2,* No. 1, 89-91, New York, Van Nostrand Reinhold Co. (1982).

(3)   United Nations Environment Programme, *IRPTC Legal File 1983,* Vol. I, pp VII/56–57, Geneva, Switzerland, International Register of Potentially Toxic Chemicals (1984).

CALCIUM CARBIDE

Description: CaC_2 is a grayish-black solid melting at about 2300°C.

Code Numbers: CAS 75-20-7   RTECS EV9400000   UN 1402

DOT Designation: Flammable solid.

Synonyms: Calcium acetylide, acetylenogen.

Potential Exposure: Those involved in the manufacture and handling of carbide and the generation of acetylene.

Incompatibilities: Water.

Permissible Exposure Limits in Air: No standards set.

Permissible Concentration in Water: No criteria set.

Routes of Entry: Inhalation and ingestion.

Harmful Effects and Symptoms: Irritation of skin, eyes and respiratory tract. Inhalation of dust may cause lung edema (A-60).

Points of Attack: Eyes, skin and respiratory tract.

Personal Protective Methods: Goggles and long sleeved shirts are recommended.

Respirator Selection: Dust respirator suggested.

Disposal Method Suggested: Mixing with large quantity of water using pilot flame to ignite evolved acetylene. Lime residue sent to landfill.

References

(1) Sax, N.I., Ed., *Dangerous Properties of Industrial Materials Report, 2,* No. 1, 91-93, New York, Van Nostrand Reinhold Co. (1982).

(2) See Reference (A-60).

(3) Parmeggiani, L., Ed., *Encyclopedia of Occupational Health & Safety,* Third Edition, Vol. 1, pp 359-60, Geneva, International Labour Office (1983).

CALCIUM CYANAMIDE

Description: $CaCN_2$, calcium cyanamide, is a blackish-grey, shiny powder.

Code Numbers: CAS 156-62-7   RTECS GS6000000   UN 1403

DOT Designation: ORM-C

Synonyms: Nitrolime, calcium carbimide, cyanamide.

Potential Exposures: Calcium cyanamide is used in agriculture as a fertilizer, herbicide, defoliant for cotton plants, and pesticide. It is also used in the manufacture of dicyandiamide and calcium cyanide as a desulfurizer in the iron and steel industry, and in steel hardening.

Permissible Exposure Limits in Air: There is no Federal standard for calcium cyanamide. The 1983/84 ACGIH TLV is 0.5 mg/m^3, and the STEL is 1.0 mg/m^3.

Permissible Concentration in Water: No criteria set.

Routes of Entry: Inhalation of dust, ingestion.

Harmful Effects and Symptoms: *Local* — Calcium cyanamide is a primary irritant of the mucous membranes of the respiratory tract, eyes, and skin. Inhalation may result in rhinitis, pharyngitis, laryngitis, and bronchitis. Conjunctivitis, keratitis, and corneal ulceration may occur. An itchy erythematous dermatitis has been reported and continued skin contact leads to the formation of slowly healing ulcerations on the palms and between the fingers. Sensitization occasionally develops. Chronic rhinitis and perforation of the nasal septum have been reported after long exposures. All local effects appear to be due to the caustic nature of cyanamide.

Systemic — Calcium cyanamide causes a characteristic vasomotor reaction. There is erythema of the upper portions of the body, face and arms accompanied by nausea, fatigue, headache, dyspnea, vomiting, oppression in the chest and shivering. Circulatory collapse may follow in the more serious cases. The vasomotor response may be triggered or intensified by alcohol ingestion. Pneumonia or lung edema may develop. Cyanide ion is not released in the body, and the mechanism of toxic action is unknown.

Points of Attack: Skin, eyes, respiratory tract.

Medical Surveillance: Evaluate skin, respiratory tract and history of alcohol intake in placement or periodic examinations.

First Aid: Irrigate eyes with water. Wash contaminated areas of body with soap and water. If ingested, use gastric lavage followed by saline catharsis (A-39).

Personal Protective Methods: In addition to personal protective equipment, waterproof barrier creams may be used to provide additional face and skin protection. Personal hygiene measures are to be encouraged, such as showering after work and a complete change of clothing.

Respirator Selection: In areas of heavy dust concentrations, full-face dust masks are recommended.

References

(1) Sax, N.I., Ed., *Dangerous Properties of Industrial Materials Report, 2,* No. 6, 38-41, New York, Van Nostrand Reinhold Co. (1982).
(2) Parmeggiani, L., Ed., *Encyclopedia of Occupational Health & Safety,* Third Edition, Vol. 1, pp 360-61, Geneva, International Labour Office (1983).

CALCIUM HYDROXIDE

Description: $Ca(OH)_2$ is a soft white crystalline powder with an alkaline, bitter taste. It dehydrates to calcium oxide at $580°C$.

Code Numbers: CAS 1305-62-0   RTECS EW2800000

DOT Designation: —

Synonyms: Hydrated lime, slaked lime, calcium hydrate.

Potential Exposures: Calcium hydroxide is used in agriculture and in fertilizer manufacture; it is used in the formulation of mortar, plasters and cements; it is used as a scrubbing and neutralizing agent in the chemical industry.

Permissible Exposure Limits in Air: There is no Federal standard but ACGIH as of 1983/84 has adopted a TWA value of 5 mg/m^3. They have set no tentative STEL.

Permissible Concentration in Water: No criteria set.

Harmful Effects and Symptoms: It is a moderately caustic irritant to all exposed surfaces of the body including the eyes and respiratory tract (A-34). No published toxicity information, either acute or chronic, has been found (A-34), but on the whole industrial experience has been rather favorable. Calcium hydroxide was originally on the EPA list of hazardous substances but was subsequently removed from the list by EPA.

Personal Protective Method: Wear rubber gloves, large size face mask and protective clothing (A-38).

Disposal Method Suggested: Landfill or admixture with acid industrial wastes prior to lagooning.

References

(1) Sax, N.I., Ed., *Dangerous Properties of Industrial Materials Report, 1,* No. 8, 48-50, New York, Van Nostrand Reinhold Co. (1981).

CALCIUM HYPOCHLORITE

See "Chlorinated Lime."

CALCIUM OXIDE

Description: CaO, calcium oxide, occurs as white or greyish-white lumps or granular powder. The presence of iron gives it a yellowish or brownish tint. It is soluble in water and acids.

Code Numbers: CAS 1305-78-8   RTECS EW3100000   UN 1910

DOT Designation: ORM-B

Synonyms: Lime, burnt lime, quicklime, calx, fluxing lime.

Potential Exposures: Calcium oxide is used as a refractory material, a binding agent in bricks, plaster, mortar, stucco and other building materials, a dehydrating agent, a flux in steel manufacturing, and a laboratory agent to absorb CO_2; in the manufacture of aluminum, magnesium, glass, pulp and paper, sodium carbonate, calcium hydroxide, chlorinated lime, calcium salts, and other chemicals; in the flotation of nonferrous ores, water and sewage treatment, soil treatment in agriculture, dehairing hides, the clarification of cane and beet sugar juice, and in fungicides, insecticides, drilling fluids and lubricants.

Incompatibilities: Water.

Permissible Exposure Limits in Air: The Federal standard for calcium oxide is 5 mg/m^3. ACGIH (1983/84) has set a TWA value of 2 mg/m^3. There is no set STEL value. IDLH level is 250 mg/m^3.

Determination in Air: Filtration, workup with acid and analysis by atomic absorption are specified in NIOSH Methods, Set N. See also reference (A-10).

Permissible Concentration in Water: No criteria set.

Route of Entry: Inhalation of dust.

Harmful Effects and Symptoms: *Local* — The irritant action of calcium oxide is due primarily to its alkalinity and exothermic reaction with water. It is irritating and may be caustic to the skin, conjunctiva, cornea and mucous membranes of upper respiratory tract, may produce burns or dermatitis with desquamation and vesicular rash, lacrimation, spasmodic blinking, ulceration, and ocular perforation, ulceration and inflammation of the respiratory passages, ulceration of nasal and buccal mucosa, and perforation of nasal septum.

Systemic — Bronchitis and pneumonia have been reported from inhalation of dust. The lower respiratory tract is generally not affected because irritation of upper respiratory passages is so severe that workers are forced to leave the area.

Points of Attack: Respiratory system, skin and eyes.

Medical Surveillance: Preemployment physical examinations should be directed to significant problems of the eyes, skin and the upper respiratory tract. Periodic examinations should evaluate the skin, changes in the eyes, especially the cornea and conjunctiva, mucosal ulcerations of the nose, mouth and nasal septum, and any pulmonary symptoms. Smoking history should be known.

First Aid: If this chemical gets into the eyes, irrigate immediately. If this chemical contacts the skin, flush with water immediately. If a person breathes in large amounts of this chemical, move the exposed person to fresh air at once and perform artificial respiration. When this chemical has been swallowed, get medical attention. Give large quantities of water and do NOT induce vomiting.

Personal Protective Methods: Wear appropriate clothing to prevent any reasonable probability of skin contact. Wear eye protection to prevent any possibility of eye contact. Employees should wash promptly when skin is contaminated and daily at the end of each work shift. Work clothing should be changed daily if it is possible that clothing is contaminated. Remove nonimpervious clothing if contaminated. Provide emergency showers and eyewash.

Respirator Selection:

25 mg/m^3: DM
50 mg/m^3: DMXSQ/FuHiEP/SA/SCBA
250 mg/m^3: HiEPF/SAF/SCBAF
Escape: HiEPF/SCBA

Disposal Method Suggested: Pretreatment involves neutralization with hydrochloric acid to yield calcium chloride. The calcium chloride formed is treated with soda ash to yield the insoluble calcium carbonate. The remaining brine solution, when its sodium chloride concentration is below 250 mg/ℓ, may be discharged into sewers and waterways (A-31).

Note: Calcium oxide was originally on the EPA list of hazardous substances but was subsequently removed from that list by EPA.

References

(1)   Sax, N.I., Ed., *Dangerous Properties of Industrial Materials Report, 2,* No. 1, 98-99, New York, Van Nostrand Reinhold Co. (1982).
(2)   See Reference (A-61).

CALCIUM SALTS

A variety of calcium salts have been reviewed in the periodical *Dangerous Properties of Industrial Materials Report* and, in the interests of economy of space, they are simply referenced here.

Salt	Volume	Number	Pages
Calcium chloride	2	1	93–94
Calcium cyanide	2	1	95–96
Calcium dodecyl benzene sulfonate	2	4	53–55
Calcium fluoride	1	8	47–48
Calcium nitrate	2	1	96–98
Calcium phosphate	2	1	99–102
Calcium phosphide	2	1	102–103

See also Reference (A-60) for listings on Calcium Bromate, Calcium Chloate, Calcium Hydride, Calcium Nitrite and Calcium Sulfide.

CAMPHOR

Description: $C_{10}H_{16}O$, camphor, is a colorless glassy solid with a penetrating, characteristic odor. It melts at $180°C$.

Code Numbers: CAS 76-22-2   RTECS EX1225000   UN 2717

DOT Designation: —

Synonyms: 2-camphonone, synthetic camphor, gum camphor, laurel camphor.

Potential Exposure: Camphor is used as a plasticizer for cellulose esters and ethers; it is used in lacquers and varnishes and in explosives and pyrotechnics formulations. It is used as a moth repellent and as a medicinal.

Incompatibilities: Oxidizers, chromic anhydride.

Permissible Exposure Limits in Air: The Federal limit and the ACGIH 1983/84 value is 2 ppm (12 mg/m^3). The tentative STEL value is 3 ppm (18 mg/m^3). The IDLH level is 200 mg/m^3.

Determination in Air: Charcoal adsorption is followed by CS_2 workup and analysis by gas chromatography. See NIOSH Methods, Set B. See also reference (A-10).

Permissible Concentration in Water: No criteria set.

Routes of Entry: Inhalation, ingestion, skin and eye contact.

Harmful Effects and Symptoms: Irritation of eyes, skin and mucous membranes; nausea, vomiting, diarrhea, headaches, dizziness, excitement, irrational behavior, epileptiform convulsions.

Points of Attack: Central nervous system, eyes, skin, respiratory system.

Medical Surveillance: Consider the points of attack in preplacement and periodic physical examinations.

First Aid: If this chemical gets into the eyes, irrigate immediately. If this chemical contacts the skin, wash with soap immediately. If a person breathes in large amounts of this chemical, move the exposed person to fresh air at once and perform artificial respiration. When this chemical has been swallowed, get medical attention. Give large quantities of salt water and induce vomiting. Do not make an unconscious person vomit.

Personal Protective Methods: Wear appropriate clothing to prevent repeated or prolonged skin contact. Wear eye protection to prevent any reasonable probability of eye contact. Employees should wash promptly when skin is wet or contaminated. Work clothing should be changed daily if it is possible that clothing is contaminated. Remove nonimpervious clothing promptly if wet or contaminated.

Respirator Selection:
200 mg/m^3: CCROVFDM/GMOVDM/SAF/SCBAF
    Escape: GMOV/SCBA

Disposal Method Suggested: Incineration of a solution in a flammable solvent.

References

(1) Sax, N.I., Ed., *Dangerous Properties of Industrial Materials Report, 1*, No. 8, 52–54, New York, Van Nostrand Reinhold Co. (1981).
(2) See Reference (A-61).
(3) See Reference (A-60).
(4) Parmeggiani, L., Ed., *Encyclopedia of Occupational Health & Safety,* Third Edition, Vol. 1, pp 365–66, Geneva, International Labour Office (1983).

CAPROLACTAM

Description: $C_6H_{11}NO$, forms white crystals melting at 69°C.

Code Numbers: CAS 105-60-2 RTECS CM3675000

DOT Designation: —

Synonyms: Hexahydro-2H-azepin-2-one; aminocaproic lactam, epsilon-caprolactam; hexamethyleneimine.

Potential Exposure: Caprolactam is used in the manufacture of nylon, plastics, bristles, film, coatings, synthetic leather, plasticizers, and paint vehicles; as a crosslinking agent for curing polyurethanes; and in the synthesis of lysine.

Permissible Exposure Limits in Air: ACGIH has set a TWA value for the vapor of 5 ppm (20 mg/m^3) and for the dust of 1.0 mg/m^3. The STEL values are 10 ppm (40 mg/m^3) for the vapor and 3 mg/m^3 for the dust as of 1983/84.

Determination in Air: Collection on a filter (for the dust) or by impinger (for the vapor) and analysis by gas liquid chromatography (A-1).

Permissible Concentration in Water: In view of the relative paucity of data on the mutagenicity, carcinogenicity, teratogenicity, and long-term oral toxicity of caprolactam, estimates of the effects of chronic oral exposure at low levels cannot be made with any confidence. It is recommended by NAS/NRC (A-2) that studies to produce such information be conducted before limits in drinking water are established.

Routes of Entry: Inhalation, ingestion, skin and eye contact.

Harmful Effects and Symptoms: Skin contact with caprolactam can cause serious burns if contact is prolonged and confined. Exposure in airborne dust at 5 mg/m^3 causes skin irritation in some people but not at 1 mg/m^3. Sensitivity has not been related to race, skin pigmentation, or other common indices of sensitivity. The prevalence of dermatoses among workers in a caprolactam manufacturing plant showed that contact dermatitis and eczema of the hands were most prevalent. Dry erythematous squamous foci on smooth skin was a typical manifestation (A-2).

Light sensitivity of the eyes was produced by inhalation of caprolactam at 0.11 mg/m^3 and higher. The olfactory threshold was 0.30 mg/m^3. An oral dose of 3 to 6 g was given daily for 3 to 5 yr for the treatment of obesity in 90 subjects. No toxic effects were observed. There was no effect on appetite, and only one person developed an allergy to caprolactam.

Disposal Method Suggested: Controlled incineration (oxides of nitrogen are removed from the effluent gas by scrubbers and/or thermal devices). Also, caprolactam may be recovered (A-58) from caprolactam still bottoms or nylon waste.

References

(1) See Reference (A-60).
(2) United Nations Environment Programme, *IRPTC Legal File 1983,* Vol. I, pp VII/65, Geneva, Switzerland, International Register of Potentially Toxic Chemicals (1984).

CAPTAFOL

Description: 3a,4,7,7a-tetrahydro-2-(1,1,2,2-tetrachloroethyl)thio-1H-isoindole-1,3(2H)-dione, $C_{10}H_9Cl_4NO_2S$ with the structural formula:

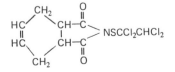

Captafol is a white crystalline solid melting at 160° to 161°C.

Code Numbers: CAS 2425-06-1   RTECS GW4900000   UN 2588

DOT Designation: —

Synonyms: Difolatan (Japan), Difolatan®, N-[(1,1,2,2-tetrachloroethyl)-thio] 4-cyclohexene-1,2-dicarboximide.

Potential Exposure: Those engaged in the manufacture (A-32), formulation and application of this fungicide.

Permissible Exposure Limits in Air: There is no Federal standard but ACGIH as of 1983/84 has adopted a TWA value of 0.1 mg/m^3 with the notation "skin" indicating the possibility of cutaneous absorption. There is no tentative STEL.

Permissible Concentration in Water: No criteria set.

Harmful Effects and Symptoms: Skin and respiratory sensitization and irritation (A-34).

Points of Attack: Skin and respiratory system.

Medical Surveillance: Consider the points of attack in preplacement and periodic physical examinations.

CAPTAN

- Carcinogen (Positive, NCI)(2)
- Hazardous substance (EPA)

Description: N-trichloromethylthio-4-cyclohexene-1,2-dicarboximide, which has the structural formula:

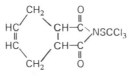

forms colorless crystals melting at 178°C.

Code Numbers: CAS 133-06-2   RTECS GW5075000   UN (NA 9099)

DOT Designation: —

Synonyms: ENT 26538; Orthocide®.

Potential Exposure: Those involved in the manufacture, formulation or application of this contact fungicide. Captan is rapidly degraded in natural soil by chemical as well as biologic means (estimated half-life, days to weeks). It has not been reported to be present in water, air, or nontarget plants. However, captan was reported to be the thirty-ninth most frequently cited pesticide (11 cases in 1973) in the EPA Pesticide Accident Surveillance System (PASS), which lists accidents involving humans, animals, and plants.

Permissible Exposure Limits in Air: The ACGIH as of 1983/84 has set a TWA for captan of 5 mg/m^3. The STEL value is 15 mg/m^3.

Permissible Concentration in Water: A no-adverse-effect level in drinking water has been calculated by NAS/NRC as 0.35 mg/ℓ.

Harmful Effects and Symptoms: The acute oral LD_{50} value for rats is 9,000 mg/kg (insignificantly toxic). Most of the chronic-oral-toxicity data on captan suggest that the no-adverse-effect or toxicologically safe dosage is about 1,000 ppm (50 mg/kg/day). However, on the basis of fetal mortality observed in monkeys exposed to captan (12.5 mg/kg/day), the acceptable daily intake of captan has been established at 0.1 mg/kg of body weight by the FAO/WHO. Based on long-term feeding studies' results in rats and dogs, ADIs were calculated at 0.05 mg/kg/day for captan.

A rebuttable presumption against registration for captan was issued on August 19, 1980 by EPA on the basis of possible oncogenicity, mutagenicity and teratogenicity.

Disposal Method Suggested: Captan decomposes fairly readily in alkaline media (pH >8). It is hydrolytically stable at neutral or acid pH but decomposes when heated alone at its melting point (A-32).

References

(1) U.S. Environmental Protection Agency, "Rebuttable Presumption Against Registration (RPAR) and Continued Registration of Pesticide Products Containing Captan," *Federal Register,* 45, No. 161, 54938-54986 (August 18, 1980).

(2) National Cancer Institute, *Bioassay of Captan for Possible Carcinogenicity,* Technical Report Series No. 15, Bethesda, MD (1977).

(3) Sax, N.I., Ed., *Dangerous Properties of Industrial Materials Report, 1,* No. 4, 93–94, New York, Van Nostrand Reinhold Co. (1981).

(4) Sax, N.I., Ed., *Dangerous Properties of Industrial Materials Report, 3,* No. 5, 80–84, New York, Van Nostrand Reinhold Co. (1983).

(5) United Nations Environment Programme, *IRPTC Legal File 1983,* Vol. I, pp VII/237–38, Geneva, Switzerland, International Register of Potentially Toxic Chemicals (1984).

CARBARYL

● Carcinogen (animal suspected, IARC)(2)
● Hazardous substance (EPA)

Description: $C_{10}H_7OOCNHCH_3$,

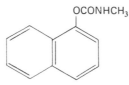

is a crystalline substance melting at 145°C.

Code Numbers: CAS 63-25-2   RTECS FC5950000   UN 2757

DOT Designation: ORM-A

Synonyms: 1-naphthyl-N-methylcarbamate, Sevin®.

Potential Exposure: Workers engaged in production (A-32), formulation and application of carbaryl as a contact insecticide for fruits, vegetables, cotton and other crops. NIOSH estimates that approximately 100,000 workers in the United States are potentially exposed to carbaryl.

Incompatibilities: Strong oxidizers.

Permissible Exposure Limits in Air: NIOSH recommends adherence to the present Federal standard and ACGIH 1983/84 value of 5 milligrams of carbaryl per cubic meter of air as a time-weighted average for up to a 10-hour workday, 40-hour workweek. The STEL is 10 mg/m^3 and the IDLH level is 625 mg/m^3.

Determination in Air: Filtration from air, workup with acid and colorimetric analysis per NIOSH Methods, Set S. See also reference (A-10).

Permissible Concentration in Water: A no-adverse-effect level in drinking water has been calculated as 0.574 mg/ℓ by NAS/NRC (A-2).

Routes of Entry: Inhalation, skin contact or eye contact, skin absorption.

Harmful Effects and Symptoms: The major health problem associated with occupational exposure to carbaryl is related to its inhibition of the enzyme cholinesterase in the central, autonomic and peripheral nervous systems. The inhibition of cholinesterase allows acetylcholine to accumulate at these sites and thereby leads to overstimulation of innervated organs. The signs and symptoms observed as a consequence of exposure to carbaryl in the workplace environment are manifestations of excessive cholinergic stimulation, e.g., nausea, vomiting, mild abdominal cramping, dimness of vision, dizziness, headache, difficulty in breathing, and weakness.

Issuance of a rebuttable presumption against registration was being considered by EPA as of September 1979 on the basis of possible teratogenicity and fetotoxicity. In January 1981, EPA stated that the rebuttable presumption against registration was being withdrawn.

Points of Attack: Respiratory system, skin, central nervous system, cardiovascular system.

Medical Surveillance: NIOSH (1) recommends that workers subject to carbaryl exposure have comprehensive preplacement medical examinations, with subsequent annual medical surveillance.

First Aid: If this chemical gets into the eyes, irrigate immediately. If this chemical contacts the skin, wash with soap promptly. If a person breathes in large amounts of this chemical, move the exposed person to fresh air at once and perform artificial respiration. When this chemical has been swallowed, get medical attention. Give large quantities of water and induce vomiting. Do not make an unconscious person vomit.

Personal Protective Methods: Wear appropriate clothing to prevent repeated or prolonged skin contact. Wear eye protection to prevent any reasonable probability of eye contact. Employees should wash promptly when skin is contaminated. Work clothing should be changed daily if it is possible that clothing is contaminated. Remove nonimpervious clothing promptly if contaminated.

Any employee whose work involves likely exposure of the skin to carbaryl or carbaryl formulations, e.g., mixing or formulating, shall wear full-body coveralls or the equivalent, impervious gloves, i.e., highly resistant to the penetration of carbaryl, impervious footwear, and, when there is danger of carbaryl coming in contact with the eyes, goggles or a face shield.

Any employee engaged in field application of carbaryl shall be provided with, and required to wear, the following protective clothing and equipment: goggles, full-body coveralls, impervious footwear, and a protective head covering.

Employees working as flaggers in the aerial application of carbaryl shall be provided with, and required to wear, full-body coveralls or waterproof rain-suits, protective head coverings, impervious gloves and impervious footwear.

Respirator Selection: Engineering controls should be used wherever feasible to maintain carbaryl concentrations below the prescribed limits, and respirators should only be used in certain nonroutine or emergency situations. During certain agricultural applications, however, respirators must be used.

NIOSH recommendations are as follows:

        50 mg/m³: SA/SCBA
       250 mg/m³: SAF/SCBAF
       625 mg/m³: SA:PD,PP,CF
        Escape: GMOVPPest/SCBA

Disposal Method Suggested: Incineration.

References

(1) National Institute for Occupational Safety and Health, *Criteria for a Recommended Standard: Occupational Exposure to Carbaryl,* NIOSH Document No. 77-107 (1977).

(2) International Agency for Research on Cancer, *IARC Monographs on the Carcinogenic Risks of Chemicals to Humans,* 12, 37, Lyon, France (1976).

(3) See Reference (A-61).

(4) Sax, N.I., Ed., *Dangerous Properties of Industrial Materials Report, 1,* No. 5, 45–46, New York, Van Nostrand Reinhold Co. (1981).

(5) Sax, N.I., Ed., *Dangerous Properties of Industrial Materials Report, 3,* No. 6, 42–48, New York, Van Nostrand Reinhold Co. (Nov./Dec. 1983) (Methyl Carbamic Acid-1-Naphthylester).

(6) United Nations Environment Programme, *IRPTC Legal File 1983,* Vol. I, pp VII/172–74, Geneva, Switzerland, International Register of Potentially Toxic Chemicals (1984).

CARBOFURAN

● Hazardous substance (EPA)

Description: 2,3-dihydro-2,2-dimethyl-7-benzofuranyl methylcarbamate, $C_{12}H_{15}NO_3$, with the structural formula

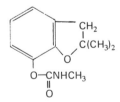

is a colorless crystalline solid melting at 150° to 152°C.

Code Numbers: CAS 1563-66-2 RTECS FB9450000 UN 2757

DOT Designation: Poison-B

Synonyms: ENT 27164, Furadan®, Curaterr®.

Potential Exposures: Those involved in the manufacture (A-32), formulation and application of this insecticide, acaricide and nematocide.

Permissible Exposure Limits in Air: There is no Federal standard but ACGIH as of 1983/84 lists a TWA value of 0.1 mg/m^3. There is no STEL proposed.

Permissible Concentration in Water: No criteria set.

Harmful Effects and Symptoms: The oral and inhalation toxicity to rats is high but the dermal toxicity is low. In rat feeding studies, the highest level having no effect was 25 ppm; at a 50 ppm level, inhibition of plasma, erythrocyte and brain cholinesterase level was observed (A-34).

References

(1)   United Nations Environment Programme, *IRPTC Legal File 1983,* Vol. I, pp VII/165–67, Geneva, Switzerland, International Register of Potentially Toxic Chemicals (1984).

CARBON BLACK

Description: Substantially elemental carbon, C, it is a black, odorless solid.

Code Numbers: CAS 7440-44-0   RTECS FF5250000   UN 1361

DOT Designation: —

Synonyms: Channel black, lamp black, furnace black, thermal black, acetylene black.

Potential Exposure: Workers in carbon black production or in its use in rubber compounding, ink manufacture or paint manufacture, plastics compounding, dry-cell battery manufacture.

Incompatibilities: Strong oxidizers such as chlorates, bromates, nitrates.

Permissible Exposure Limits in Air: The Federal TWA limit and ACGIH 1983/84 value is 3.5 mg/m^3. The STEL value is 7 mg/m^3. No IDLH level has been set.

Determination in Air: Filtration from air is followed by gravimetric analysis as described in NIOSH Methods, Set R. See also reference (A-10).

Permissible Concentration in Water: No standard set.

Routes of Entry: Inhalation, skin and eye contact.

Harmful Effects and Symptoms: None expected, according to NIOSH (A-4) but conjunctivitis, corneal hypoplasia, eczema, bronchitis and pneumatotumors are symptoms reported by a Japanese source (A-38).

Points of Attack: None known.

First Aid: Irrigate eyes promptly.

Personal Protective Methods: Wear goggles to avoid repeated exposure. Employees should wash daily.

Respirator Selection:
    17.5 mg/m^3:   DM
      35 mg/m^3:   DMXSQ/HiEPFu/SA/SCBA
    175 mg/m^3:   HiEPF/SAF/SCBAF
  2,000 mg/m^3:   PAPHiE/SA:PD,PP,CF/SAF:PD,PP,CF

Disposal Method Suggested: Dump into a landfill.

References

(1) National Institute for Occupational Safety and Health, *Information Profiles on Potential Occupational Hazards: Carbon Black*, Report PB-276,678, pp 8–11, Rockville, MD (October 1977).

(2) U.S. Environmental Protection Agency, *Chemical Hazard Information Profile: Carbon Black*, Washington, DC (August 1, 1976).

(3) National Institute for Occupational Safety and Health, *Criteria for a Recommended Standard: Occupational Exposure to Carbon Black*, NIOSH Doc. No. 78-204, Washington, DC (1978).

(4) Parmeggiani, L., Ed., *Encyclopedia of Occupational Health & Safety*, Third Edition, Vol. 1, pp 390–392, Geneva, International Labour Office (1983).

CARBON DIOXIDE

Description: CO_2, carbon dioxide, is a colorless, odorless, noncombustible gas, soluble in water. It is commonly sold in the compressed liquid form, and the solid form (dry ice).

Code Numbers: CAS 124-38-9   RTECS FF6400000   UN 1013 (gas), 2187 (liquid), 1845 (solid).

DOT Designation: Nonflammable gas.

Synonyms: Carbonic acid gas, carbonic anhydride.

Potential Exposure: Gaseous carbon dioxide is used to carbonate beverages, as a weak acid in the textile, leather and chemical industries, in water treatment, and in the manufacture of aspirin and white lead, for hardening molds in foundries, in food preservation, in purging tanks and pipelines, as a fire extinguisher, in foams, and in welding. Because it is relatively inert, it is utilized as a pressure medium. It is also used as a propellant in aerosols, to promote plant growth in green houses; it is used medically as a respiratory stimulant, in the manufacture of carbonates, and to produce an inert atmosphere when an explosive or flammable hazard exists. The liquid is used in fire extinguishing equipment, in cylinders for inflating life rafts, in the manufacturing of dry ice, and as a refrigerant. Dry ice is used primarily as a refrigerant.

Occupational exposure to carbon dioxide may also occur in any place where fermentation processes may deplete oxygen with the formation of carbon dioxide, e.g., in mines, silos, wells, vats, ships' holds, etc.

Incompatibilities: Chemically active metals such as sodium, potassium, hot titanium.

Permissible Exposure Limits in Air: The Federal standard and ACGIH 1983/84 value is 5,000 ppm (9,000 mg/m³). The STEL is 15,000 ppm (18,000 mg/m³). The IDLH level is 50,000 ppm.

Determination in Air: Collection in a bag followed by gas chromatography per NIOSH Methods, Set 1. See also reference (A-10).

Permissible Concentration in Water: No criteria set.

Route of Entry: Inhalation of gas.

Harmful Effects and Symptoms: *Local* — Frostbite may result from contact with dry ice or gas at low temperature.

Systemic — Carbon dioxide is a simple asphyxiant. Concentrations of 10%

(100,000 ppm) can produce unconsciousness and death from oxygen deficiency. A concentration of 5% may produce shortness of breath and headache. Continuous exposure to 1.5% CO_2 may cause changes in some physiological processes. The concentration of carbon dioxide in the blood affects the rate of breathing.

Points of Attack: Lungs, skin, cardiovascular system.

Medical Surveillance: No special considerations are necessary although persons with cardiovascular or pulmonary disease may be at increased risk.

First Aid: If dry ice gets into the eyes, get medical attention. If this chemical contacts the skin, get medical attention for frostbite. If a person breathes in large amounts of this chemical, move the exposed person to fresh air at once and perform artificial respiration.

Personal Protective Methods: Carbon dioxide is a heavy gas and accumulates at low levels in depressions and along the floor. Generally, adequate ventilation will provide sufficient protection for the worker. Where concentrations are of a high order, supplied air respirators are recommended. Wear protective clothing to prevent skin freezing.

Respirator Selection:
50,000 ppm: SA/SCBA
Escape: SCBA

Disposal Method Suggested: Vent to atmosphere.

References

(1) National Institute for Occupational Safety and Health, *Criteria for a Recommended Standard: Occupational Exposure to Carbon Dioxide,* NIOSH Doc. No. 76-194 (1976).
(2) See Reference (A-61).
(3) See Reference (A-60).
(4) Parmeggiani, L., Ed., *Encyclopedia of Occupational Health & Safety,* Third Edition, Vol. 1, pp 392–93, Geneva, International Labour Office (1983).

CARBON DISULFIDE

- Hazardous substance (EPA)
- Hazardous waste (EPA)

Description: CS_2, carbon disulfide, is a highly refractive, flammable liquid boiling at 46°C which in pure form has a sweet odor and in commercial and reagent grades has a foul smell. It can be detected by odor at about 1 ppm but the sense of smell fatigues rapidly and, therefore, odor does not serve as a good warning property. It is slightly soluble in water, but more soluble in organic solvents.

Code Numbers: CAS 75-15-0   RTECS FF6650000   UN 1131

DOT Designation: Flammable liquid, poison.

Synonyms: Carbon bisulfide, dithiocarbonic anhydride.

Potential Exposures: Carbon disulfide is used in the manufacture of viscose rayon, ammonium salts, carbon tetrachloride, carbanilide, xanthogenates, flotation agents, soil disinfectants, dyes, electronic vacuum tubes, optical glass, paints, enamels, paint removers, varnishes, varnish removers, tallow, textiles, explosives, rocket fuel, putty, preservatives, and rubber cement; as a solvent for phosphorus, sulfur, selenium, bromine, iodine, alkali cellulose, fats, waxes, lacquers, camphor, resins, and cold vulcanized rubber. It is also used in degreasing, chemical analysis, electroplating, grain fumigation, oil extraction, and drycleaning. It is widely used as a pesticide intermediate (A-32).

NIOSH estimates that 20,000 workers in the United States are exposed or potentially exposed to carbon disulfide.

Incompatibilities: Strong oxidizers, chemically active metals (such as sodium, potassium, zinc), azides, organic amines.

Permissible Exposure Limits in Air: The Federal standard is 20 ppm (60 mg/m^3) determined as an 8-hour TWA. The acceptable ceiling concentration is 30 ppm (90 mg/m^3) with a maximum peak above this for an 8-hour workshift of 100 ppm (300 mg/m^3) for a maximum duration of 30 minutes.

NIOSH recommends that exposure be limited to a TWA concentration of 3 mg/m^3 (1 ppm). In addition, a ceiling concentration of 30 mg/m^3 (10 ppm), determined on the basis of a 15-minute sampling period, is recommended. Occupational exposure to CS$_2$ is defined as exposure above an action level of 1.5 mg/m^3. ACGIH (1983/84) has recommended a TWA value of 10 ppm (30 mg/m^3). ACGIH adds the notation "skin" indicating the possibility of cutaneous absorption. The IDLH level is 500 ppm.

Determination in Air: Adsorption on charcoal, workup with benzene, gas chromatographic analysis per NIOSH Methods, Set R. See also reference (A-10).

Permissible Concentration in Water: In view of the relative paucity of data on the mutagenicity, carcinogenicity, and long-term oral toxicity of carbon disulfide, it was stated (A-2) that estimates of the effects of chronic oral exposure at low levels cannot be made with any confidence. It was recommended by NAS/NRC (A-2) that studies to produce such information be conducted before limits in drinking water are established. Now, however, EPA (A-37) has suggested a permissible ambient goal of 830 μg/ℓ.

Routes of Entry: Inhalation of vapor which may be compounded by percutaneous absorption of liquid or vapor, ingestion and skin and eye contact.

Harmful Effects and Symptoms: *Local* — Carbon disulfide vapor in sufficient quantities is severely irritating to eyes, skin, and mucous membranes. Contact with liquid may cause blistering with second and third degree burns. Skin sensitization may occur. Skin absorption may result in localized degeneration of peripheral nerves which is most often noted in the hands. Respiratory irritation may result in bronchitis and emphysema, though these effects may be overshadowed by systemic effects.

Systemic — Intoxication from carbon disulfide is primarily manifested by psychological, neurological, and cardiovascular disorders. Recent evidence indicates that once biochemical alterations are initiated they may remain latent; clinical signs and symptoms then occur following subsequent exposure.

Following repeated carbon disulfide exposure, subjective psychological as well as behavioral disorders have been observed. Acute exposures may result in extreme irritability, uncontrollable anger, suicidal tendencies, and a toxic manic depressive psychosis. Chronic exposures have resulted in insomnia, nightmares, defective memory, and impotency. Less dramatic changes include headache, dizziness, and diminished mental and motor ability, with staggering gait and loss of coordination.

Neurological changes result in polyneuritis. Animal experimentation has revealed pyramidal and extrapyramidal tract lesions and generalized degeneration of the myelin sheaths of peripheral nerves. Chronic exposure signs and symptoms include retrobulbar and optic neuritis, loss of sense of smell, tremors, paresthesias, weakness, and, most typically, loss of lower extremity reflexes.

Atherosclerosis and coronary heart disease have been significantly linked to exposure to carbon disulfide. Atherosclerosis develops most notably in the blood vessels of the brain, glomeruli, and myocardium. Abnormal electroencephalograms and retinal hypertension typically occur before renal involvement is noted. Any of the above three areas may be affected by chronic exposure, but most often only one aspect can be observed. A significant increase in coronary heart disease mortality has been observed in carbon disulfide workers. Studies also reveal higher frequency of angina pectoris and hypertension. Abnormal electrocardiograms may also occur and are also suggestive of carbon disulfide's role in the etiology of coronary disease.

Other specific effects include chronic gastritis with the possible development of gastric and duodenal ulcers; impairment of endocrine activity, specifically adrenal and testicular; abnormal erythrocytic development with hypochromic anemia; and possible liver dysfunction with abnormal serum cholesterol. Also in women, chronic menstrual disorders may occur. These effects usually occur following chronic exposure and are subordinate to the other symptoms.

Recently human experience and animal experimentation have indicated several possible biochemical changes. Carbon disulfide and its metabolites (i.e., dithiocarbamic acids and isothiocyanates) show amino acid interference, cerebral monoamine oxidase inhibition, endocrine disorders, lipoprotein metabolism interference, blood protein, and zinc level abnormalities, and inorganic metabolism interference due to chelating of polyvalent ions. The direct relationship between these biochemical changes and clinical manifestations is only suggestive.

Points of Attack: Central nervous system, peripheral nervous system, cardiovascular system, eyes, kidneys, liver, skin.

Medical Surveillance: Preplacement and periodic medical examinations should be concerned especially with skin, eyes, central and peripheral nervous system, cardiovascular disease, as well as liver and kidney function. Electrocardiograms should be taken. CS_2 can be determined in expired air, blood, and urine. The iodine-azide test is most useful although nonspecific, and it may indicate other sulfur compounds.

First Aid: If this chemical gets into the eyes, irrigate immediately. If this chemical contacts the skin, wash with soap immediately. If a person breathes in large amounts of this chemical, move the exposed person to fresh air at once and perform artificial respiration. When this chemical has been swallowed, get medical attention. Give large quantities of water and induce vomiting. Do not make an unconscious person vomit.

Personal Protective Methods: Wear appropriate clothing to prevent any reasonable probability of skin contact. Wear eye protection to prevent any reasonable probability of eye contact. Employees should wash promptly when skin is wet or contaminated. Remove clothing immediately if wet or contaminated to avoid flammability hazard.

Respirator Selection:
        200 ppm: CCROV/SA/SCBA
        500 ppm: CCROVF/GMOV/SAF/SCBAF/SA:PD,PP,CF
        Escape: GMOV/SCBA

Disposal Method Suggested: This compound is a very flammable liquid which evaporates rapidly. It burns with a blue flame to carbon dioxide (harmless) and sulfur dioxide. Sulfur dioxide has a strong suffocating odor; 1,000 ppm in air is lethal to rats. The pure liquid presents an acute fire and explosion hazard. The following disposal procedure is suggested:

All equipment or contact surfaces should be grounded to avoid ignition by static charges. Absorb on vermiculite, sand, or ashes and cover with water. Transfer underwater in buckets to an open area. Ignite from a distance with an excelsior train. If quantity is large, carbon disulfide may be recovered by distillation and repackaged for use.

References

(1) National Institute for Occupational Safety and Health, *Criteria for a Recommended Standard: Occupational Exposure to Carbon Disulfide,* NIOSH Doc. No. 77-156 (1977).

(2) World Health Organization, *Carbon Disulfide,* Environmental Health Criteria No. 10, Geneva (1979).

(3) U.S. Environmental Protection Agency, *Carbon Disulfide,* Health and Environmental Effects Profile No. 32, Office of Solid Waste, Washington, DC (April 30, 1980).

(4) Sax, N.I., Ed., *Dangerous Properties of Industrial Materials Report, 1,* No. 2, 28-30, New York, Van Nostrand Reinhold Co. (1980).

(5) See Reference (A-61).

(6) See Reference (A-60).

(7) Sax, N.I., Ed., *Dangerous Properties of Industrial Materials Report, 3,* No. 5, 84-87, New York, Van Nostrand Reinhold Co. (1983).

(8) Parmeggiani, L., Ed., *Encyclopedia of Occupational Health & Safety,* Third Edition, Vol. 1, pp 393-95, Geneva, International Labour Office (1983).

(9) United Nations Environment Programme, *IRPTC Legal File 1983,* Vol. I, pp VII/176-78, Geneva, Switzerland, International Register of Potentially Toxic Chemicals (1984).

CARBON MONOXIDE

Description: CO, carbon monoxide, is a colorless, odorless, tasteless gas, partially soluble in water.

Code Numbers: CAS 630-08-0   RTECS FG3500000   UN 1016

DOT Designation: Flammable gas.

Synonyms: None.

Potential Exposures: Carbon monoxide is used in metallurgy as a reducing agent, particularly in the Mond process for nickel; in organic synthesis, especially in the Fischer-Tropsch process for petroleum products and in the oxo reaction; and in the manufacture of metal carbonyls.

It is usually encountered in industry as a waste product of incomplete combustion of carbonaceous material (complete combustion produces CO_2). The major source of CO emission in the atmosphere is the gasoline-powered internal combustion engine.

Specific industrial processes which contribute significantly to CO emission are iron foundries, particularly the cupola; fluid catalytic crackers, fluid coking, and moving-bed catalytic crackers in petroleum refining; lime kilns and kraft recovery furnaces in kraft paper mills; furnace, channel, and thermal operations in carbon black plants; beehive coke ovens, basic oxygen furnaces, sintering of blast furnace feed in steel mills; and formaldehyde manufacture. There are numerous other operations in which a flame touches a surface that is cooler than the ignition temperature of the gaseous part of the flame where exposure to CO may occur, e.g., arc welding, automobile repair, traffic control, tunnel construction, fire fighting, mines, use of explosives, etc.

Incompatibilities: Strong oxidizers.

Permissible Exposure Limits in Air: The present Federal standard is 50 ppm (55 mg/m^3). The standard recommended by NIOSH is 35 ppm with a ceiling value of 200 ppm. This latter value is to limit carboxyhemoglobin formation to 5% in a nonsmoker engaged in sedentary activity at normal altitude. The EPA has set an ambient air quality standard of 9 ppm, averaged over an 8-hour period and 35 ppm for 1 hour, not to be exceeded more than once a year. As of 1980, it proposed lowering the 1-hour standard to 25 ppm (4). The STEL value adopted by ACGIH (1983/84) is 400 ppm (440 mg/m^3). The IDLH level is 1,500 ppm.

Determination in Air: Collection in gas sampling bag and electrochemical analysis (A-10). Long duration detector tubes may also be used for CO determination (A-11).

Permissible Concentration in Water: No criteria set, but EPA (A-37) has suggested a permissible ambient level of 552 µg/ℓ based on health effects.

Route of Entry: Inhalation of gas.

Harmful Effects and Symptoms: *Local* — None.

Systemic — Carbon monoxide combines with hemoglobin to form carboxyhemoglobin which interferes with the oxygen-carrying capacity of blood, resulting in a state of tissue hypoxia. The typical signs and symptoms of acute CO poisoning are headache, dizziness, drowsiness, nausea, vomiting, collapse, coma, and death. Initially the victim is pale; later the skin and mucous membranes may be cherry-red in color. Loss of consciousness occurs at about the 50% carboxyhemoglobin level. The amount of carboxyhemoglobin formed is dependent on concentration and duration of CO exposure, ambient temperature, health, and metabolism of the individual. The formation of carboxyhemoglobin is a reversible process. Recovery from acute poisoning usually occurs without sequelae unless tissue hypoxia was severe enough to result in brain cell degeneration.

Carbon monoxide at low levels may initiate or enhance deleterious myocardial alterations in individuals with restricted coronary artery blood flow and decreased myocardial lactate production.

Severe carbon monoxide poisoning has been reported to permanently damage the extrapyramidal system, including the basal ganglia.

Points of Attack: Central nervous system, lungs, blood, cardiovascular system.

Medical Surveillance: Preplacement and periodic medical examinations should give special attention to significant cardiovascular disease and any medical conditions which could be exacerbated by exposure to CO. Heavy smokers may be at greater risk. Methylene chloride exposure may also cause an increase of carboxyhemoglobin. Smokers usually have higher levels of carboxyhemoglobin than nonsmokers (often 5 to 10% or more). Carboxyhemoglobin levels are reliable indicators of exposure and hazard. See reference (5).

First Aid: If a person breathes in large amounts of this chemical, move the exposed person to fresh air at once and perform artificial respiration.

Personal Protective Methods: Under certain circumstances where carbon monoxide levels are not exceedingly high, gas masks with proper canisters can be used for short periods, but are not recommended. In areas with high concentrations, self-contained air apparatus is recommended.

Respirator Selection:
　　　500 ppm:　SA/SCBA
　　1,500 ppm:　SAF/SCBAF/SA:PD,PP,CF
　　　Escape:　GMS/SCBA

Disposal Method Suggested: Incineration. Carbon monoxide can also be recovered from gas mixtures (A-58) as an alternative to disposal.

References

(1) National Institute for Occupational Safety and Health, *Criteria for a Recommended Standard: Occupational Exposure to Carbon Monoxide,* NIOSH Doc. No. 73-11,000 (1973).
(2) National Academy of Sciences, *Medical and Biologic Effects of Environmental Pollutants: Carbon Monoxide,* Washington, DC (1977).
(3) U.S. Environmental Protection Agency, *Air Quality Criteria for Carbon Monoxide,* Report EPA 600/8-79-022, Environmental Criteria and Assessment Office, Research Triangle Park, NC (1979).
(4) U.S. Environmental Protection Agency, "Carbon Monoxide: Proposed Revisions to the National Ambient Air Quality Standards," *Federal Register,* 45, No. 161, 55066-84 (August 18, 1980).
(5) National Institute for Occupational Safety and Health, *A Guide to the Work Relatedness of Disease,* Revised Edition, DHEW (NIOSH) Publication No. 79-116, Cincinnati, OH, pp 73-79 (January 1979).
(6) World Health Organization, *Carbon Monoxide,* Environmental Health Criteria No. 13, Geneva, Switzerland (1979).
(7) Sax, N.I., Ed., *Dangerous Properties of Industrial Materials Report, 1,* No. 7, 43-45, New York, Van Nostrand Reinhold Co. (1981).
(8) See Reference (A-61).
(9) See Reference (A-60).
(10) Sax, N.I., Ed., *Dangerous Properties of Industrial Materials Report, 3,* No. 5, 87-89, New York, Van Nostrand Reinhold Co. (1983).
(11) Parmeggiani, L., Ed., *Encyclopedia of Occupational Health & Safety,* Third Edition, Vol. 1, pp 395-99, Geneva, International Labour Office (1983).

CARBON OXYSULFIDE

Description: COS is a gas or liquid which boils at 50°C.

Code Numbers: CAS 463-58-1   RTECS FG6475000   UN 2204

DOT Designation: —

Synonyms: Carbonyl sulfide; carbon oxide sulfide.

Potential Exposures: Carbon oxysulfide is an excellent source of useable atomic sulfur; therefore, it can be used in various chemical syntheses, such as the production of episulfides, alkenylthiols, and vinylicthiols.

It is probable that the largest source of carbon oxysulfide is as a by-product from various organic syntheses and petrochemical processes.

Carbon oxysulfide is always formed when carbon, oxygen, and sulfur, or their compounds, such as carbon monoxide, carbon disulfide, and sulfur dioxide, are brought together at high temperatures. Hence, carbon oxysulfide is formed as an impurity in various types of manufactured gases and as a by-product in the manufacture of carbon disulfide. Carbon oxysulfide is also often present in refinery gases.

Permissible Exposure Limits in Air: No standards set.

Permissible Concentration in Water: No criteria set.

Harmful Effects and Symptoms: A search of the available literature revealed an almost total lack of information regarding the toxicity of carbon oxysulfide. The NIOSH Registry of Toxic Effects of Chemical Substances (1976) listed the LC_{Lo} for inhalation by mice as 2,900 ppm; this chemical is no longer found in the current (1978) edition of the registry.

The acute toxicity of carbon oxysulfide was examined by Japanese workers. Exposure of laboratory animals to this contaminant of coal gas and petroleum gas was associated with pathological changes in the brain, medulla oblongata, liver, kidney, and lung. When rats were placed in chambers containing 0.05 and 0.2 percent carbon oxysulfide, death occurred in 10 and 0.5 to 1.0 hours, respectively.

Personal Protective Methods: Wear rubber gloves and coveralls (A-38).

Disposal Method Suggested: Incineration (A-38).

References

(1) National Institute for Occupational Safety and Health, *Information Profiles on Potential Occupational Hazards—Single Chemicals: Carbon Oxysulfide,* pp 34-38, Publ. No. TR79-607, Rockville, MD (Dec. 1979).

CARBON TETRABROMIDE

Description: CBr_4 forms colorless crystals melting at 90°C.

Code Numbers: CAS 558-13-4   RTECS FG4725000   UN 2516

DOT Designation: Label requires St. Andrews cross (X).

Synonym: Tetrabromomethane.

Potential Exposure: CBr_4 is used in organic synthesis.

Permissible Exposure Limits in Air: There is no Federal standard but ACGIH (1983/84) has adopted a TWA value of 0.1 ppm (1.4 mg/m^3) and set an STEL of 0.3 ppm (4.0 mg/m^3).

Permissible Concentration in Water: No criteria set.

Harmful Effects and Symptoms: This is a highly toxic material. Acute exposures to high concentrations cause upper respiratory irritation and injury to the lungs, liver and kidneys. Low level chronic exposure produces only liver injury. The material is a potent lachrymator even at low concentrations (A-34).

Personal Protective Methods: Wear rubber gloves and work clothing.

Respirator Selection: Wear self-contained breathing apparatus.

Disposal Method Suggested: Purify by distillation and return to suppliers (A-38).

References

(1) See Reference (A-60).

CARBON TETRACHLORIDE

- Carcinogen (Animal Positive, IARC)(5)
- Hazardous substance (EPA)
- Hazardous waste (EPA)
- Priority toxic pollutant (EPA)

Description: CCl_4, carbon tetrachloride, is a colorless, nonflammable liquid with a characteristic odor. Oxidative decomposition by flame causes phosgene and hydrogen chloride to form.

Code Numbers: CAS 56-23-5   RTECS FG4900000   UN 1846

DOT Designation: ORM-A

Synonyms: Tetrachloromethane, perchloromethane.

Potential Exposures: Carbon tetrachloride is used as a solvent for oils, fats, lacquers, varnishes, rubber, waxes, and resins. Fluorocarbons are chemically synthesized from it. It is also used as an azeotropic drying agent for spark plugs, a dry-cleaning agent, a fire extinguishing agent, a fumigant, and an anthelmintic agent. The use of this solvent is widespread, and substitution of less toxic solvents when technically possible is recommended.

Approximately 4,500 workers are exposed during production processes, and 52,000 more workers are exposed during industrial use and consumption of the chemical. OSHA estimates that 3.4 million workers could be exposed CCl_4 directly or indirectly.

An estimated 5 million lb/yr of CCl_4 are given off as emissions during manufacture and processing, and approximately 60 million lb/yr are released as solvent emissions. The amount of CCl_4 emitted to the environment indicates that a large proportion of the general population has been exposed. EPA estimates that 8 million people living within 12.5 miles of manufacturing sites are exposed to average levels of 0.5 $\mu g/m^3$, with peaks of 1,580 $\mu g/m^3$. CCl_4 has been found in 10% of 113 surveyed public water systems, at mean concentrations ranging from 2.4 to 6.4 $\mu g/\ell$, in 45% of surface water supplies, and in 25% of groundwater supplies, at concentrations of 0.001 to 0.40 mg/ℓ. Estimates indicate that 19 million people are exposed to CCl_4 through the ambient air, 20 million are exposed through drinking contaminated water, and 2 million are exposed to contaminated soil or landfills.

Incompatibilities: Chemically active metals such as sodium, potassium, magnesium.

Permissible Exposure Limits in Air: The Federal standard is 10 ppm (65 mg/m^3) as an 8-hour TWA with an acceptable ceiling concentration of 25 ppm; acceptable maximum peaks above the ceiling of 200 ppm are allowed for one 5-minute duration in any 4-hour period. NIOSH has recommended a ceiling limit of 2 ppm based on a 1-hour sampling period at a rate of 750 ml/min. ACGIH (1983/84) has set a TWA value of 5 ppm (30 mg/m^3) and an STEL of 20 ppm (125 mg/m^3) with the notation that CCl_4 is a substance suspect of carcinogenic potential for man and with the further notation "skin" indicating the possibility of cutaneous absorption. The IDLH level is 300 ppm with the notation "occupational carcinogen."

Determination in Air: Charcoal adsorption followed by workup with CS_2 and analysis by gas chromatography; see NIOSH Methods Set J. See also reference (A-10).

Permissible Concentration in Water: To protect freshwater aquatic life: 35,200 $\mu g/\ell$ on an acute toxicity basis. To protect saltwater aquatic life: 50,000 $\mu g/\ell$ on an acute toxicity basis. To protect human health: preferably zero. An additional lifetime cancer risk of 1 in 100,000 is presented by a concentration of 4.0 $\mu g/\ell$.

Determination in Water: Gas chromatography (EPA Method 601) or gas chromatography plus mass spectrometry (EPA Method 624).

Routes of Entry: Inhalation of vapor, percutaneous absorption, ingestion and skin and eye contact.

Harmful Effects and Symptoms: *Local* — Carbon tetrachloride solvent removes the natural lipid cover of the skin. Repeated contact may lead to a dry, scaly, fissured dermatitis. Eye contact is slightly irritating, but this condition is transient.

Systemic — Excessive exposure may result in central nervous system depression, and gastrointestinal symptoms may also occur. Following acute exposure, signs and symptoms of liver and kidney damage may develop. Nausea, vomiting, abdominal pain, diarrhea, enlarged and tender liver, and jaundice result from toxic hepatitis. Diminished urinary volume, red and white blood cells in the urine, albuminuria, coma, and death may be consequences of acute renal failure. The hazard of systemic effects is increased when carbon tetrachloride is used in conjunction with ingested alcohol.

A rebuttable presumption against registration for CCl_4 in pesticide uses was issued on October 15, 1980 by EPA on the basis of possible oncogenicity, nephrotoxicity, and hepatotoxicity.

Carbon tetrachloride is carcinogenic in mice and rats, producing liver tumours after administration by various routes. It also produced liver tumours in trout and hamsters following oral administration.

Three case reports describe liver tumours associated with cirrhosis in humans exposed to carbon tetrachloride (8).

Points of Attack: Central nervous system, eyes, lungs, liver, kidneys, skin.

Medical Surveillance: Preplacement and periodic examinations should include an evaluation of alcohol intake and appropriate tests for liver and kidney functions. Special attention should be given to the central and peripheral nervous system, the skin, and blood. Expired air and blood levels may be useful as indicators of exposure.

First Aid: If this chemical gets into the eyes, irrigate immediately. If this chemical contacts the skin, wash with soap immediately. If a person breathes in large amounts of this chemical, move the exposed person to fresh air at once and perform artificial respiration. When this chemical has been swallowed, get medical attention. Give large quantities of salt water and induce vomiting. Do not make an unconscious person vomit.

Personal Protective Methods: Wear appropriate clothing to prevent repeated or prolonged skin contact. Wear eye protection to prevent any reasonable probability of eye contact. Employees should wash promptly when skin is wet or contaminated. Remove nonimpervious clothing promptly if wet or contaminated.

Respirator Selection:
        100 ppm: SA/SCBA
        300 ppm: SAF/SCBAF
        Escape: GMOV/SCBA

Disposal Method Suggested: Incineration, preferably after mixing with another combustible fuel; care must be exercised to assure complete combustion to prevent the formation of phosgene; an acid scrubber is necessary to remove the halo acids produced (A-31). Recover and purify by distillation where possible.

References
(1) National Institute for Occupational Safety and Health, *Criteria for a Recommended Standard: Occupational Exposure to Carbon Tetrachloride,* NIOSH Doc. No. 76-133 (1976).

(2) U.S. Environmental Protection Agency, *Carbon Tetrachloride: Ambient Water Quality Criteria,* Washington, DC (1980).

(3) National Academy of Sciences, *Chloroform, Carbon Tetrachloride and Other Halomethanes: An Environmental Assessment,* Washington, DC (1978).

(4) U.S. Environmental Protection Agency, *Carbon Tetrachloride,* Health and Environmental Effects Profile No. 33, Office of Solid Waste, Wash., DC (April 30, 1980).

(5) International Agency for Research on Cancer, *IARC Monographs on the Carcinogenic Risks of Chemicals to Humans,* Lyon, France, 1, 53 (1972); and 20, 371 (1979).

(6) Sax, N.I., Ed., *Dangerous Properties of Industrial Materials Report, 1,* No. 2, 30-32, New York, Van Nostrand Reinhold Co. (1980).

(7) See Reference (A-61).

(8) See Reference (A-62). Also see Reference (A-64).

(9) Sax, N.I., Ed., *Dangerous Properties of Industrial Materials Report, 3,* No. 5, 89-94, New York, Van Nostrand Reinhold Co. (1983).

(10) Parmeggiani, L., Ed., *Encyclopedia of Occupational Health & Safety,* Third Edition, Vol. 1, pp 399–401, Geneva, International Labour Office (1983).

(11) United Nations Environment Programme, *IRPTC Legal File 1983,* Vol. I, pp VII/181–84, Geneva, Switzerland, International Register of Potentially Toxic Chemicals (1984).

CARBONYL FLUORIDE

● Hazardous waste (EPA)

Description: COF_2 is a colorless, nearly odorless gas.

Code Numbers: CAS 353-50-4   RTECS FG6125000   UN 2417

DOT Designation: Poison gas.

Synonyms: Carbon oxyfluoride, fluorophosgene.

Potential Exposures: The major source of exposure to COF_2 results from the thermal decomposition of fluorocarbon plastics such as PTFE in air. Carbonyl fluoride may also find use in organic synthesis. It has been suggested for use as a military poison gas.

Permissible Exposure Limits in Air: There is no Federal standard but as of 1983/84 ACGIH has adopted a TWA of 2ppm (5 mg/m^3) and an STEL of 5 ppm (15 mg/m^3).

Permissible Concentration in Water: No criteria set.

Harmful Effects and Symptoms: On an acute basis, COF_2 is about as toxic as HF as a respiratory irritant gas. The long-term effects are due to the fluoride ion generated by hydrolysis; this inhibits succinic dehydrogenase activity since this is a fluoride-sensitive enzyme (A-34).

Personal Protective Methods: Wear long rubber gloves and full protective clothing.

Respirator Selection: Use self-contained breathing apparatus.

CARBOPHENOTHION

Description: $C_{11}H_{16}ClO_2PS_3$ with the structural formula

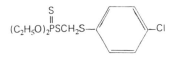

is a light amber liquid with a boiling point of $82°C$ at 0.01 mm Hg.

Code Numbers: CAS 786-19-6   RTECS ID5250000   UN 1615

Dot Designation: —

Synonyms: Trithion®, Garrathion®, Acarithion®.

Potential Exposure: Those engaged in the manufacture or application of this insecticide.

Permissible Exposure Limits in Air: No standards set.

Permissible Concentration in Water: No criteria set.

Harmful Effects and Symptoms: Highly toxic (oral LD-50 for rats is 32 mg/kg). Produces headaches, nausea, weakness and dizziness.

Disposal Method Suggested: Hydrolysis by hypochlorites may be used as may incineration.

References

(1) Sax, N.I., Ed., *Dangerous Properties of Industrial Materials Report, 2,* No. 4, 55-59, New York, Van Nostrand Reinhold Co. (1982).

(2) United Nations Environment Programme, *IRPTC Legal File 1983,* Vol. II, pp VII/563-4, Geneva, Switzerland, International Register of Potentially Toxic Chemicals (1984).

CATECHOL

Description: $C_6H_4(OH)_2$ forms colorless-to-brownish crystals that melt at $104°C$. It sublimes readily.

Code Numbers: CAS 120-80-9   RTECS UX1050000

DOT Designation: —

Synonyms: Pyrocatechol, o-dihydroxybenzene.

Potential Exposures: It is used as an antiseptic, in photography, in dyestuff manufacture and application. It is also used in electroplating, in the formulation of specialty inks and in antioxidants and light stabilizers.

Permissible Exposure Limits in Air: There are no Federal standards but ACGIH has adopted TWA values of 5 ppm (20 mg/m³) as of 1983/84. They have not proposed a STEL value, however.

Permissible Concentration in Water: No criteria set, but EPA (A-37) has suggested a permissible ambient goal of 280 μg/ℓ on a health basis.

Routes of Entry: Skin absorption, skin and eye contact, inhalation of vapors, ingestion.

Harmful Effects and Symptoms: Skin contact causes eczematous dermatitis. Absorption through the skin results in illness akin to that which phenol produces except convulsions are more pronounced. Catechol increases blood pressure, apparently from peripheral vasoconstriction. Catechol can cause death, apparently initiated by respiratory failure (A-34).

References

(1)   See Reference (A-60).

(2)   United Nations Environment Programme, *IRPTC Legal File 1983,* Vol. II, pp VII/702, Geneva, Switzerland, International Register of Potentially Toxic Chemicals (1984).

CDEC

See "Sulfallate."

CEMENT

See "Portland Cement."

CERIUM AND COMPOUNDS

Description: Ce, cerium, a soft, steel-gray metal, is found in the minerals monazite, cerite, and orthite. It may form either tri- or tetravalent compounds. The cerous salts are usually white and the ceric salts are yellow to orange-red. Cerium decomposes in water and is soluble in dilute mineral acids.

Code Numbers: Cerium chloride, $CeCl_3$ CAS 7790-86-5 RTECS FK5075000

DOT Designation: —

Synonyms: None.

Potential Exposures: Cerium and its compounds are used as a catalyst in ammonia synthesis, a deoxidizer to improve the mechanical quality and refine grain size of steel, an opacifier in certain enamels, an arc-stabilizer in carbon arc lamps, an abrasive for polishing mirrors and lenses, a sedative and as a medicinal agent for vomiting during pregnancy. It is used in the manufacture of topaz yellow glass, spheroidal cast iron, incandescent gas mantles and in decolorizing glass, to prevent mildew in textiles, and to produce a vacuum in neon lamps and electronic tubes. Alloyed with aluminum, magnesium, and manganese, it increases resistance to creep and fatigue. Ferrocerium is the pyrophoric alloy in gas cigarette lighters, and an alloy of magnesium, cerium, and zirconium is utilized for jet engine parts.

Permissible Exposure Limits in Air: There is no Federal standard for cerium or its compounds, nor are any values proposed by ACGIH.

Permissible Concentration in Water: No criteria set.

Routes of Entry: Inhalation of dust.

Harmful Effects and Symptoms: *Local* — No local effects have been reported due to cerium and its compounds.

Systemic — There are no records of injury to human beings from either the industrial or medicinal use of cerium. The main risk to workers is from dust in mining and production areas. Recent reports in the literature describe "Cerpneumoconiosis," a condition found in a group of graphic arts workers who use carbon arc lights in their work. Chest x-rays reveal small, miliary, homogeneously distributed infiltrates. Cer-pneumoconiosis cannot be considered a dust disease of the lung similar to silicosis. In the later stages of the reaction to the dust of carbon arc lamps, perifocal emphysema, and slight fibrosis of lungs are noted. It has been speculated that these changes may have been due to inhalation of substances containing radioactive elements of the thorium chain. To date, these views have not been confirmed by animal experimentation, autopsy, or human biopsy. Animal experimentation has demonstrated increased coagulation time from organic preparations of cerium, disturbance of lipid metabolism from cerium and its nitrates, and profound effects on metabolism and intestinal muscle causing loss of motility from cerium chloride.

Medical Surveillance: Chest x-rays should be taken as a part of preemployment and periodic physical examinations.

Personal Protective Methods: In areas of carbon arc lights, workers should wear effective dust filters or respirators. In mining and production areas, workers should wear effective dust filters or respirators suitable for the particulate size of air borne dust.

References

(1) Sax, N.I., Ed., *Dangerous Properties of Industrial Materials Report, 1,* No. 8, 55-56, New York, Van Nostrand Reinhold Co. (1981).

CESIUM HYDROXIDE

Description: CsOH is a colorless-to-yellowish crystalline compound which melts at 272°C.

Code Numbers: CAS 21351-79-1   RTECS FK9800000   UN 2682

DOT Designation: —

Synonyms: Cesium hydrate, caesium hydroxide.

Potential Exposures: CsOH may be used as a raw material for other cesium salts such as the chloride which in turn may be used to produce cesium metal. Cesium metal is used in electronic devices.

Incompatibilities: CsOH is the strongest base known and must be stored in silver or platinum out of contact with air because of its reactivity with glass or CO_2.

Permissible Exposure Limits in Air: There is no Federal standard but ACGIH (1983/84) has adopted a TWA value of 2 mg/m^3.

Permissible Concentration in Water: No criteria set.

Harmful Effects and Symptoms: While CsOH is not a skin sensitizer, it is extremely irritant and corrosive to the eyes in 5% concentration. It is threefold less toxic by ingestion than KOH. Further, 5% CsOH is considered safe for human skin contact whereas 5% KOH is a mild skin irritant (A-34).

CETYL PYRIDINIUM CHLORIDE

Description: $C_{21}H_{38}ClN$ with the following structural formula

is a white powder melting at 77° to 83°C.

Code Numbers: CAS 123-03-5   RTECS UU4900000

DOT Designation —

Synonyms: 1-Hexadecylpyridinium chloride, Cepacol chloride.

Potential Exposure: Those involved in manufacture and use of this topical antiseptic and disinfectant.

Permissible Exposure Limits in Air: No standards set.

Permissible Concentration in Water: No criteria set.

Harmful Effects and Symptoms: Mild skin irritation. Large quantities may cause nausea, vomiting, collapse, convulsions and coma.

Disposal Method Suggested: Incineration.

References

(1)   Sax, N.I., Ed., *Dangerous Properties of Industrial Materials Report, 2,* No. 4, 59-61, New York, Van Nostrand Reinhold Co. (1982).

CHLORAL

● Hazardous waste (EPA)

Description: CCl_3CHO is an oily liquid with a pungent irritating odor which boils at 97° to 98°C.

Code Numbers: (for chloral hydrate) CAS 302-17-0   RTECS FM8750000 (for anhydrous) UN 2075.

DOT Designation: Poison (anhydrous chloral).

Synonym: Trichloroacetaldehyde.

Potential Exposure: Chloral is used as an intermediate in the manufacture of such pesticides as DDT, methoxychlor, DDVP, naled, trichlorfon, and TCA

(A-32). Chloral is also used in the production of chloral hydrate, a therapeutic agent with hypnotic and sedative effects used prior to the introduction of barbiturates.

Permissible Exposure Limits in Air: There are no U.S. standards. The U.S.S.R. has recommended a maximum concentration in workroom air of 220 mg/m^3 (1).

Permissible Concentration in Water: There are no U.S. criteria but the U.S.S.R. reportedly has set 0.2 mg/ℓ (1).

Harmful Effects and Symptoms: Specific information on the pharmacokinetic behavior, carcinogenicity, mutagenicity, teratogenicity, and other reproductive effects of chloral was not found in the available literature. However, the pharmacokinetic behavior of chloral may be similar to chloral hydrate where metabolism to trichloroethanol and trichloroacetic acid and excretion via the urine (and possibly bile) have been observed. Chloral hydrate produced skin tumors in 4 of 20 mice dermally exposed.

Information on the chronic or acute effects of chloral in humans was not found in the available literature. Chronic effects from respiratory exposure to chloral as indicated in laboratory animals include reduction of kidney function and serum transaminase activity, change in central nervous system function (unspecified), decrease in antitoxic and enzyme-synthesizing function of the liver, and alteration of morphological characteristics of peripheral blood. Slowed growth rate, leukocytosis and changes in the arterial blood pressure were also observed.

Alcohol synergistically increases the depressant effect of the compound, creating a potent depressant commonly referred to as "Mickey Finn" or "knockout drops." Addiction to chloral hydrate through intentional abuse of the compound has been reported.

Personal Protective Measures: Wear rubber gloves and protective clothing.

Respirator Selection: Use self-contained breathing apparatus.

Disposal Method Suggested: Incineration after mixing with another combustible fuel; care must be taken to assure complete combustion to prevent phosgene formation; an acid scrubber is necessary to remove the halo acids produced.

References

(1) U.S. Environmental Protection Agency, *Chloral,* Health and Environmental Effects Profile No. 34, Office of Solid Waste, Washington, DC (April 30, 1980).

(2) See Reference (A-60).

CHLORAMBEN

● Carcinogen (Positive, NCI) (1)

Description: 3-Amino-2,5-dichlorobenzoic acid is a colorless crystalline solid melting at 200° to 201°C.

Code Numbers: CAS 133-90-4 RTECS DG1925000

DOT Designation: –

Synonyms: Amben; Amiben®; Amoben®; Vegiben®.

Potential Exposure: Workers involved in the manufacture (A-32), formulation or application of this preemergence herbicide.

Permissible Exposure Limits in Air: No standards set.

Permissible Concentration in Water: The no-adverse effect level in drinking water has been calculated to be 1.75 mg/ℓ by NAS/NRC (A-2).

Harmful Effects and Symptoms: The available data on Chloramben are very sparse. Much additional information is needed regarding its chronic toxicity, teratogenicity, and carcinogenicity before limits can be confidently set (A-2).

No-observed-adverse-effect doses for chloramben were at 250 mg/kg/day and 500 mg/kg/day in dogs and rats, respectively, in feeding studies. Based on these data an ADI was calculated at 0.25 mg/kg/day.

Disposal Method Suggested: Chloramben is stable to heat, oxidation, and hydrolysis in acidic or basic media. The stability is comparable to that of benzoic acid. Chloramben is decomposed by sodium hypochlorite solution (A-32).

References

(1)   National Cancer Institute, *Bioassay of Chloramben for Possible Carcinogenicity,* Technical Report Series No. 25, Bethesda, MD (1977).

(2)   Sax, N.I., Ed., *Dangerous Properties of Industrial Materials Report, 1,* No. 3, 28-29, New York, Van Nostrand Reinhold Co. (1981). (As 3-Amino-2,5-Dichlorobenzoic Acid).

(3)   United Nations Environment Programme, *IRPTC Legal File 1983,* Vol. I, pp VII/117, Geneva, Switzerland, International Register of Potentially Toxic Chemicals (1984).

CHLORAMBUCIL

- Carcinogen (Human and Animal Positive) (A-62) (A-64)
- Hazardous Waste Constituent (EPA)

Description: $(ClCH_2CH_2)_2N-C_6H_4-(CH_2)_3COOH$ is a crystalline solid melting at 64° to 66°C.

Code Numbers: CAS 305-03-3   RTECS ES7525000

DOT Designation: —

Synonyms: Chloraminophene; Leukeran®; 4-[bis(2-chloroethyl)amino]-benzenebutanoic acid.

Potential Exposure: Chlorambucil, a drug against cancer, is a derivative of nitrogen mustard. This drug is primarily used as an antineoplastic agent for the treatment of lymphocytic leukemia, malignant lymphomas, follicular lymphoma, and Hodgkin's disease. The treatments are not curative but do produce some marked remissions. Chlorambucil has also been tested for treatment of chronic hepatitis, rheumatoid arthritis, and as an insect chemosterilant.

All of the chemical used in this country is imported from the United Kingdom. No data are available on the quantity of this importation, but the FDA reported that 6.4 million prescriptions were dispensed by retail pharmacies in 1980. Work exposure in the United States would be limited to workers formulating the tablets, or to patients receiving the drug.

Permissible Exposure Limits in Air: No standards set.

Permissible Concentration in Water: No criteria set.

Harmful Effects and Symptoms: Chlorambucil is carcinogenic in rats and mice following intraperitoneal injection, producing lymphomas in rats, and lymphosarcomas, ovarian tumours, and lung tumours in mice.

Excesses of acute leukemia were reported in a number of epidemiological studies of people treated with chlorambucil, either alone or in combination with other therapies, for both nonmalignant and malignant diseases. Other cancers have also been associated with the use of chlorambucil and other agents. An excess of acute leukemia in association with chlorambucil was seen in a further study in which 431 previously untreated patients with polycythemia vera were given phlebotomy alone or chlorambucil with phlebotomy, and followed for a mean of 6.5 years. Of the 26 cases of acute leukemia that occurred, 16 were in the group receiving chlorambucil. The risk increased with increasing dose and time of treatment.

References

(1) Sax, N.I., Ed., *Dangerous Properties of Industrial Materials Report, 1,* No. 4, 43-44, New York, Van Nostrand Reinhold Co. (1981).

(2) See Reference (A-62). Also see Reference (A-64).

CHLORDANE

- Carcinogen (Positive, NCI)(3)
- Hazardous substance (EPA)
- Hazardous waste (EPA)
- Priority toxic pollutant (EPA)

Description: $C_{10}H_6Cl_8$, a pale-yellow nonvolatile liquid with a structural formula:

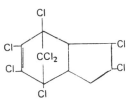

Code Numbers: CAS 57-74-9   RTECS PB9800000   UN 2762

DOT Designation: Flammable liquid.

Synonyms: ENT 9932;   1,2,4,5,6,7,8,8-octachloro-2,3,3a,4,7,7a-hexahydro-4,7-methane-1H-indene; Octachlor®.

Potential Exposures: Chlordane is a broad spectrum insecticide of the group of polycyclic chlorinated hydrocarbons called cyclodiene insecticides. Chlordane has been used extensively over the past 30 years for termite control, as an insecticide for homes and gardens, and as a control for soil insects during the production of crops such as corn. Both the uses and the production volume of chlordane have decreased extensively since the issuance of a registration suspension notice for all food crops and home and garden uses of chlordane by the

U.S. Environmental Protection Agency. However, significant commercial use of chlordane for termite control continues. In addition, under the terms of a settlement which terminated chlordane registration cancellation proceedings, chlordane will be permitted for limited usage through 1980 as an agricultural insecticide (43 *FR* 12372; March 24, 1978).

Special groups at risk include children as a result of milk consumed; fishermen and their families because of the high consumption of fish and shellfish, especially freshwater fish; persons living downwind from treated fields; and persons living in houses treated with chlordane pesticide control agents.

Incompatibilities: Strong oxidizers.

Permissible Exposure Limits in Air: The Federal limit and ACGIH 1983/84 value is 0.5 mg/m^3. The STEL value is 2.0 mg/m^3. ACGIH adds the notation "skin" indicating the possibility of cutaneous absorption. The IDLH level is 500 mg/m^3.

Determination in Air: Collection by impinger or fritted bubbler, analysis by gas liquid chromatography (A-1).

Permissible Concentration in Water: To protect freshwater aquatic life: 0.0043 $\mu g/\ell$ as a 24-hour average, not to exceed 2.4 $\mu g/\ell$ at any time. To protect saltwater aquatic life: 0.0040 $\mu g/\ell$ as a 24-hour average, never to exceed 0.09 $\mu g/\ell$. To protect human health: preferably zero. An additional lifetime cancer risk of 1 in 100,000 is presented by a concentration of 0.0046 $\mu g/\ell$.

Determination in Water: Gas chromatography (EPA Method 608) or gas chromatography plus mass spectrometry (EPA Method 625).

Routes of Entry: Inhalation, skin absorption, ingestion and skin and eye contact.

Harmful Effects and Symptoms: Blurred vision; confusion; ataxia, delirium; coughing; abdominal pain, nausea, vomiting, diarrhea; irritability, tremors, convulsions; anuria.

Points of Attack: Central nervous system, eyes, lungs, liver, kidneys, skin.

Medical Surveillance: Consider the points of attack in preplacement and periodic physical examinations.

First Aid: If this chemical gets into the eyes, irrigate immediately. If this chemical contacts the skin wash with soap immediately. If a person breathes in large amounts of this chemical, move the exposed person to fresh air at once and perform artificial respiration. When this chemical has been swallowed, get medical attention. Give large quantities of water and induce vomiting. Do not make an unconscious person vomit.

Personal Protective Methods: Wear appropriate clothing to prevent any possibility of skin contact. Wear eye protection to prevent any reasonable probability of eye contact. Employees should wash immediately when skin is wet or contaminated. Work clothing should be changed daily if it is possible that clothing is contaminated. Remove nonimpervious clothing immediately if wet or contaminated. Provide emergency showers.

Respirator Selection:
     5 mg/m^3: CCROVDMPest/SA/SCBA
    25 mg/m^3: CCROVFDMPest/GMPest/SAF/SCBAF
   500 mg/m^3: SA:PD,PP,CF/CCROVHiEPest
      Escape: GMOVPPest/SCBA

Disposal Method Suggested: Chlordane is readily dehydrochlorinated in alkali to form "nontoxic" products, a reaction catalyzed by traces of iron. The environmental hazards of the products are uncertain. Chlordane is completely dechlorinated by sodium in isopropyl alcohol. The MCA recommends incineration methods for disposal of chlordane.

References

(1) U.S. Environmental Protection Agency, *Chlordane: Ambient Water Quality Criteria,* Washington, DC (1980).

(2) U.S. Environmental Protection Agency, *Chlordane,* Health and Environmental Effects Profile No. 35, Office of Solid Waste, Washington, DC (April 30, 1980).

(3) National Cancer Institute, *Bioassay of Chlordane for Possible Carcinogenicity,* Tech. Report Series No. 8, Bethesda, MD (1977).

(4) Sax, N.I., Ed., *Dangerous Properties of Industrial Materials Report, 1,* No. 2, 33-35, New York, Van Nostrand Reinhold Co. (1980).

(5) See Reference (A-61).

(6) Sax, N.I., Ed., *Dangerous Properties of Industrial Materials Report, 3,* No. 5, 94-99, New York, Van Nostrand Reinhold Co. (1983).

(7) United Nations Environment Programme, *IRPTC Legal File 1983,* Vol. II, pp VII/447–50, Geneva, Switzerland, International Register of Potentially Toxic Chemicals (1984).

CHLORDECONE

- Carcinogen (Potential Human, NIOSH)(A-5)(Animal Positive, IARC)(5)(7)
- Hazardous substance (EPA)
- Hazardous waste (EPA)

Description: $C_{10}Cl_{10}O$, is a crystalline solid with the structural formula:

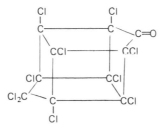

which decomposes at 350°C.

Code Numbers: CAS 143-50-0   RTECS PC8575000   UN 2588

DOT Designation: —

Synonyms: Kepone®; ENT 16,391; 1,1a,3,3a,4,5,5,5a,5b,6-decachloro-octa-hydro-1,3,4-metheno-2H-cyclobuta[c,d]pentalen-2-one.

Potential Exposures: NIOSH has identified fewer than 50 establishments processing or formulating pesticides using chlordecone and has estimated that 600 workers are potentially exposed to chlordecone. (NIOSH is unaware of any plant in the United States which is currently manufacturing chlordecone; the only known plant manufacturing it was closed in July 1975.)

Kepone was registered for the control of rootborers on bananas with a residue tolerance of 0.01 ppm. This constituted the only food or feed use of

Kepone. Nonfood uses included wireworm control in tobacco fields and bait to control ants and other insects in indoor and outdoor areas.

A rebuttable presumption against registration of chloredecone was issued by the U.S. EPA on March 25, 1976 on the basis of oncogenicity. The trademarked Kepone and products of six formulations were the subject of voluntary cancellation according to a U.S. EPA notice dated July 27, 1977. In a series of decisions, the first of which was issued on June 17, 1976, the EPA effectively cancelled all registered products containing Kepone as of May 1, 1978.

Permissible Exposure Limits in Air: NIOSH recommends that the workplace environmental level be limited to 1 $\mu g/m^3$ as a time-weighted average concentration for up to a 10-hour workday, 40-hour workweek, as an emergency standard.

Determination in Air: Collection by membrane filter and backup impinger containing NaOH solution, workup with benzene, analysis by gas chromatography with electron capture detector (A-10).

Permissible Concentration in Water: No criteria set.

Routes of Entry: Inhalation of dust, ingestion, skin absorption.

Harmful Effects and Symptoms: In July 1975, a private physician submitted a blood sample to the Center for Disease Control (CDC) to be analyzed for Kepone, a chlorinated hydrocarbon pesticide. The sample had been obtained from a Kepone production worker who suffered from weight loss, nystagmus, and tremors. CDC notified the State epidemiologist that high levels of Kepone were present in the blood sample, and he initiated an epidemiologic investigation which revealed other employees suffering with similar symptoms. It was evident to the State official after visiting the plant that the employees had been exposed to Kepone at extremely high concentrations through inhalation, ingestion, and skin absorption. He recommended that the plant be closed, and company management complied.

Of the 113 current and former employees of this Kepone-manufacturing plant examined, more than half exhibited clinical symptoms of Kepone poisoning. Medical histories of tremors (called "Kepone shakes" by employees), visual disturbances, loss of weight, nervousness, insomnia, pain in the chest and abdomen and, in some cases, infertility and loss of libido were reported. The employees also complained of vertigo and lack of muscular coordination. The intervals between exposure and onset of the signs and symptoms varied between patients but appeared to be dose related.

NIOSH has received a report on a carcinogenesis bioassay of technical grade Kepone which was conducted by the National Cancer Institute using Osborne-Mendel rats and B6C3F1 mice. Kepone was administered in the diet at two tolerated dosages. In addition to the clinical signs of toxicity, which were seen in both species, a significant increase ($P<0.05$) of hepatocellular carcinoma in rats given large dosages of Kepone and in mice at both dosages was found. Rats and mice also had extensive hyperplasia of the liver.

In view of these findings, NIOSH must assume that Kepone is a potential human carcinogen (A-5).

Medical Surveillance: Employers shall make medical surveillance available to all workers occupationally exposed to Kepone, including personnel periodically exposed during routine maintenance or emergency operations. Periodic examinations shall be made available at least on an annual basis.

Personal Protective Methods: *Protective Clothing* — (a) Coveralls or other full-body protective clothing shall be worn in areas where there is occupational exposure to Kepone. Protective clothing shall be changed at least daily at the end of the shift and more frequently if it should become grossly contaminated.

(b) Impervious gloves, aprons and footwear shall be worn at operations where solutions of Kepone may contact the skin. Protective gloves shall be worn at operations where dry Kepone or materials containing Kepone are handled and may contact the skin.

(c) Eye protective devices shall be provided by the employer and used by the employees where contact of Kepone with eyes is likely. Selection, use, and maintenance of eye protective equipment shall be in accordance with the provisions of the American National Standard Practice for Occupational and Educational Eye and Face Protection, ANSI Z87.1-1968. Unless eye protection is afforded by a respirator hood or facepiece, protective goggles or a face shield shall be worn at operations where there is danger of contact of the eyes with dry or wet materials containing Kepone because of spills, splashes, or excessive dust or mists in the air.

(d) The employer shall ensure that all personal protective devices are inspected regularly and maintained in clean and satisfactory working condition.

(e) Work clothing may not be taken home by employees. The employer shall provide for maintenance and laundering of protective clothing.

(f) The employer shall ensure that precautions necessary to protect laundry personnel are taken while soiled protective clothing is being laundered.

(g) The employer shall ensure that Kepone is not discharged into municipal waste treatment systems or the community air.

Respiratory Protection from Kepone — Engineering controls shall be used wherever feasible to maintain airborne Kepone concentrations at or below that recommended. Compliance with the environmental exposure limit by the use of respirators is allowed only when airborne Kepone concentrations are in excess of the workplace environmental limit because required engineering controls are being installed or tested when nonroutine maintenance or repair is being accomplished, or during emergencies. When a respirator is thus permitted, it shall be selected and used in accordance with NIOSH requirements.

Respirator Selection: See reference (1).

Disposal Method Suggested: A process has been developed (2) which effects chlordecone degradation by treatment of aqueous wastes with UV radiation in the presence of hydrogen in aqueous sodium hydroxide solution. Up to 95% decomposition was effected by this process.

Chlordecone previously presented serious disposal problems because of its great resistance to bio- and photodegradation in the environment. It is highly toxic to normally-occurring degrading microorganisms. Although it can undergo some photodecomposition when exposed to sunlight to the dihydro compound (leaving a compound with 8 chloro substituents) that degradation product does not significantly reduce toxicity.

References

(1) National Institute for Occupational Safety and Health, *Recommended Standard for Occupational Exposure to Kepone,* Washington, DC (January 27, 1976).
(2) Kitchens, J.A.F., U.S. Patent 4,144,152, March 13, 1979, assigned to Atlantic Research Corporation.
(3) U.S. Environmental Protection Agency, *Reviews of the Environmental Effects of Pollutants: I. Mirex and Kepone,* Report EPA-600/1-78-013, Cincinnati, OH (1978).
(4) National Academy of Sciences, *Kepone, Mirex, Hexachlorocyclopentadiene: An Environmental Assessment.* Washington, DC (1978).
(5) International Agency for Research on Cancer, *IARC Monographs on the Carcinogenic Risks of Chemicals to Humans,* 20, Lyon, France (1979).

(6)   Sax, N.I., Ed., *Dangerous Properties of Industrial Materials Report, 1,* No. 4, 77-79, New York, Van Nostrand Reinhold Co. (1981).

(7)   See Reference (A-62). Also see Reference (A-64).

(8)   Parmeggiani, L., Ed., *Encyclopedia of Occupational Health & Safety,* Third Edition, Vol. 1, pp 1169-70, Geneva, International Labour Office (1983).

(9)   United Nations Environment Programme, *IRPTC Legal File 1983,* Vol. II, pp VII/460-61, Geneva, Switzerland, International Register of Potentially Toxic Chemicals (1984).

CHLORFENVINPHOS

Description: $C_{12}H_{14}Cl_3O_4P$ with the following structural formula

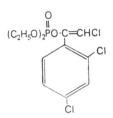

is a yellow liquid boiling at 110°C at 0.001 mm Hg and 168° to 170°C at 0.5 mm Hg.

Code Numbers: CAS 470-90-6   RTECS TB8750000   UN 2783

DOT Designation: —

Synonyms: 2-chloro-1-(2,4-dichlorophenyl)ethenyl diethyl phosphate; Birlane®; Sapecron®; Supona®.

Potential Exposure: Those engaged in the production, formulation and application of this insecticide.

Permissible Exposure Limits in Air: No standards set.

Permissible Concentration in Water: 0.1 mg/ℓ in drinking water (1).

Harmful Effects and Symptoms: Highly toxic (LD-50 for rats is 25 mg/kg).

Disposal Method Suggested: Destruction by alkali hydrolysis or incineration (1).

References

(1)   Sax, N.I., Ed., *Dangerous Properties of Industrial Materials Report, 2,* No. 4, 63-67, New York, Van Nostrand Reinhold Co. (1982).

(2)   United Nations Environment Programme, *IRPTC Legal File 1983,* Vol. II, pp VII/545-7, Geneva, Switzerland, International Register of Potentially Toxic Chemicals (1984).

CHLORINATED BENZENES

See also entries under:

"Chlorobenzene," "Dichlorobenzenes," "1,2,4-Trichlorobenzene," "Pentachlorobenzene," "Hexachlorobenzene."

References

(1) West, W.L. and Ware, S.A., *Preliminary Report, Investigation of Selected Potential Environmental Contaminants: Halogenated Benzenes,* Environmental Protection Agency, Washington, DC (1977).

(2) U.S. Environmental Protection Agency, *Chlorinated Benzenes: Ambient Water Quality Criteria,* Washington, DC (1980).

(3) U.S. Environmental Protection Agency, *Chlorinated Benzenes,* Health and Environmental Effects Profile No. 36, Office of Solid Waste, Washington, DC (April 30, 1980).

(4) Sax, N.I., Ed., *Dangerous Properties of Industrial Materials Report, 4,* No. 3, 89–91, New York, Van Nostrand Reinhold Co. (1984) (1,2,3,4-Tetrachlorobenzene).

(5) Sax, N.I., Ed., *Dangerous Properties of Industrial Materials Report, 4,* No. 3, 91–93, New York, Van Nostrand Reinhold Co. (1984) (1,2,4,5-Tetrachlorobenzene).

CHLORINATED CAMPHENE

See "Toxaphene."

CHLORINATED DIPHENYL OXIDE

● Priority toxic pollutant (EPA)

Description: $C_{12}H_4Cl_6O$ is a waxy solid or liquid.

Code Numbers: CAS 55720-99-5   RTECS KO0875000

DOT Designation: —

Synonyms: Hexachlorodiphenyl oxide, hexachlorophenyl ether.

Potential Exposures: These materials are used as dielectric fluids in the electrical industry; they may be used as organic intermediates.

Incompatibilities: Strong oxidizers.

Permissible Exposure Limits in Air: The Federal limit and the ACGIH 1983/84 TWA value is 0.5 mg/m³. The STEL is 2.0 mg/m³. IDLH level is 5.0 mg/m³.

Determination in Air: Collect on filter, work up with isooctane, analyze by gas chromatography. See NIOSH Methods, Set I. See also reference (A-10).

Permissible Concentration in Water: The TLV for hexachlorophenyl ether is 500 μg/m³. By a process analogous to that used by Stokinger and Woodward, this standard could be used to calculate a water criterion. However, since the TLV for these compounds is based on preventing chloracne, rather than chronic toxicity, such a calculation would not be appropriate. Because of the lack of data on both toxicologic effects and environmental contamination, the hazard posed by these compounds cannot be estimated according to the Environmental Protection Agency.

Routes of Entry: Inhalation, ingestion, eye and skin contact.

Harmful Effects and Symptoms: Acne-form dermatitis and liver damage.

Points of Attack: Skin, liver.

Medical Surveillance: Consider the points of attack in preplacement and periodic physical examinations.

First Aid: If this chemical gets into the eyes, irrigate immediately. If this chemical contacts the skin, wash with soap promptly. If a person breathes in large amounts of this chemical, move the exposed person to fresh air at once and perform artificial respiration. When this chemical has been swallowed, get medical attention. Give large quantities of salt water and induce vomiting. Do not make an unconscious person vomit.

Personal Protective Methods: Wear appropriate clothing to prevent repeated or prolonged skin contact. Wear eye protection to prevent any reasonable probability of eye contact. Employees should wash promptly when skin is wet or contaminated. Work clothing should be changed daily if it is possible that clothing is contaminated. Remove nonimpervious clothing promptly if wet or contaminated.

Respirator Selection:
    5 mg/m³: SA/SCBA
    Escape: GMOVAGP/SCBA

References
(1) U.S. Environmental Protection Agency, *Haloethers: Ambient Water Quality Criteria,* Washington, DC (1980).
(2) See Reference (A-61).

CHLORINATED ETHANES

See also entries under: "Ethyl Chloride," "1,1-Dichloroethane," "Ethylene Dichloride," "1,1,1-Trichloroethane," "1,1,2-Trichloroethane," "1,1,1,2-Tetrachloroethane," "1,1,2,2-Tetrachloroethane," "Pentachloroethane," "Hexachloroethane."

References
(1) National Institute for Occupational Safety and Health, *Chloroethanes: Review of Toxicity,* Current Intelligence Bulletin No. 27, NIOSH Publication No. 78-181, Washington, DC (1978).
(2) National Cancer Institute, *Bioassay of 1,2-Dichloroethane for Possible Carcinogenicity,* DHEW Publication No. (NIH) 78-1305, Washington, DC (1978).
(3) National Cancer Institute, *Bioassay of 1,1,2-Trichloroethane for Possible Carcinogenicity,* DHEW Publication No. (NIH) 78-1324, Washington, DC (1978).
(4) National Cancer Institute, *Bioassay of 1,1,2,2-Tetrachloroethane for Possible Carcinogenicity,* DHEW Publication No. (NIH) 78-827, Washington, DC (1978).
(5) National Cancer Institute, *Bioassay of Hexachloroethane for Possible Carcinogenicity,* DHEW Publication No. (NIH) 78-1318, Washington, DC (1978).
(6) National Institute for Occupational Safety and Health, *Criteria Document: Recommendations for an Occupational Exposure Standard for Ethylene Dichloride,* DHEW Publication No. (NIOSH) 76-139, Washington, DC (1976).
(7) National Institute for Occupational Safety and Health, *Ethylene Dichloride (1,2-Dichloroethane),* Current Intelligence Bulletin No. 25, DHEW (NIOSH) Publication No. 78-149, Washington, DC (1978).
(8) National Institute for Occupational Safety and Health, *Criteria for a Recommended Standard for Exposure to 1,1,1-Trichloroethane (Methyl Chloroform),* DHEW Publication No. (NIOSH) 76-184, Washington, DC (1976).
(9) National Institute for Occupational Safety and Health, *Criteria for a Recommended Standard: Occupational Exposure to 1,1,2,2-Tetrachloroethane,* DHEW (NIOSH) Publication No. 77-121, Washington, DC (1976).

(10) National Academy of Sciences, *Drinking Water and Health,* Washington, DC (1977).
(11) National Cancer Institute, *Bioassay of 1,1,1-Trichloroethane for Possible Carcinogenicity,* Carcinog. Tech. Rept. Ser. NCI-CG-TR-3, Washington, DC (1977).
(12) U.S. Environmental Protection Agency, *Chlorinated Ethanes: Ambient Water Quality Criteria,* Washington, DC (1980).
(13) U.S. Environmental Protection Agency, *Chlorinated Ethanes,* Health and Environmental Effects Profile No. 37, Office of Solid Waste, Washington, DC (April 30, 1980).

CHLORINATED LIME

Description: Chlorinated lime is a white or grayish-white hygroscopic powder with a chlorine odor. It is a relatively unstable chlorine carrier in solid form and is a complex compound of indefinite composition. Chemically, it consists of varying proportions of calcium hypochlorite, calcium chlorite, calcium oxychloride, calcium chloride, free calcium hydroxides, and water. The commercial product generally contains 24-37% available chlorine. On exposure to moisture, chlorine is released.

Code Numbers: 7778-54-3   RTECS NH3485000   UN 2208

DOT Designation: ORM-C

Synonyms: Bleaching powder; HTH®; chloride of lime.

Potential Exposures: Chlorinated lime is a bleaching agent, i.e., it has the ability to chemically remove dyes or pigments from materials. It is used in the bleaching of wood pulp, linen, cotton, straw, oils, and soaps, and in laundering, as an oxidizer in calico printing, a chlorinating agent, a disinfectant, particularly for drinking water and sewage, a decontaminant for mustard gas, and as a pesticide for caterpillars (A-5).

Permissible Exposure Limits in Air: There is no Federal standard for chlorinated lime. (See Chlorine.)

Permissible Concentration in Water: No criteria set.

Routes of Entry: Inhalation of dust. Inhalation of vapor and ingestion.

Harmful Effects and Symptoms: *Local* — The toxic effects of chlorinated lime are due to its chlorine content. The powder and its solutions have corrosive action on skin, eyes, and mucous membranes, can produce conjunctivitis, blepharitis, corneal ulceration, gingivitis, contact dermatitis, and may damage the teeth.
Systemic — The dust is irritating to the respiratory tract and can produce laryngitis and pulmonary edema. Chlorinated lime is extremeley hygroscopic and with the addition of water evolves free chlorine. Inhalation of the vapor is extremely irritating and toxic. (See Chlorine.) Ingestion of chlorinated lime causes severe oral, esophageal, and gastric irritation.

Medical Surveillance: Consider possible effects of skin, teeth, eyes, or respiratory tract. There are no specific diagnostic tests.

Personal Protective Methods: In dusty areas, the worker should be protected by appropriate respirators. Simple dust masks should not be used since the moisture present in expired air will release the chlorine. Skin effects can be minimized with protective clothing. Most important is the fact that free chlorine is liberated when chlorinated lime comes in contact with water. All precautions should be followed to protect the worker under these circumstances. (See Chlorine.)

Disposal Method Suggested: Dissolve the material in water and add to a large volume of concentrated reducing agent solution, then acidify the mixture with H_2SO_4. When reduction is complete, soda ash is added to make the solution alkaline. The alkaline liquid is decanted from any sludge produced, neutralized, and diluted before discharge to a sewer or stream. The sludge is landfilled (A-31).

References

(1) Sax, N.I., Ed., *Dangerous Properties of Industrial Materials Report, 1,* No. 8, 50-52, New York, Van Nostrand Reinhold Co. (1981).

(2) Sax, N.I., Ed., *Dangerous Properties of Industrial Materials Report, 4,* No. 3, 76-79, New York, Van Nostrand Reinhold Co. (1984) (Hypochlorous Acid, Calcium Salt).

CHLORINATED NAPHTHALENES

- Hazardous waste constituents (EPA) (Hazardous waste: 2-chloronaphthalene) (EPA)
- Priority toxic pollutants (EPA)

Description: $C_{10}H_{8-x}Cl_x$, the chlorinated naphthalenes, are naphthalenes in which one or more hydrogen atoms have been replaced by chlorine to form waxlike substances, beginning with monochloronaphthalene and going on to the octochlor derivatives. Their physical states vary from mobile liquids to waxysolids depending on the degree of chlorination. Melting points of the pure compounds range from $17°C$ for 1-chloronaphthalene to $198°C$ for 1,2,3,4-tetrachloronaphthalene.

Code Numbers:

1-Chloro	CAS 90-13-1	RTECS QJ2100000
2-Chloro	CAS 91-58-7	RTECS QJ2275000
Trichloro	CAS 1321-65-9	RTECS QK4025000
Tetrachloro	CAS 1335-88-2	RTECS QK3700000
Pentachloro	CAS 1321-64-8	RTECS QK0300000
Hexachloro	CAS 1335-87-1	RTECS QJ7350000
Octachloro	CAS 2234-13-1	RTECS QK0250000

DOT Designations: —

Synonyms:

1-Chloro	None
2-Chloro	None
Trichloro	Halowax, Seekay Wax, Nibren Wax
Tetrachloro	Halowax, Seekay Wax, Nibren Wax
Pentachloro	Halowax 1013
Hexachloro	Halowax 1014
Octachloro	Halowax 1051

Potential Exposures: Industrial exposure from individual chlorinated naphthalenes is rarely encountered; rather it usually occurs from mixtures of two or more chlorinated naphthalenes. Due to their stability, thermoplasticity, and nonflammability, these compounds enjoy wide industrial application. These compounds are used in the production of electric condensers, in the insulation of electric cables and wires, as additives to extreme pressure lubricants, as supports for storage batteries, and as a coating in foundry use.

Because of the possible potentiation of the toxicity of higher chlorinated naphthalenes by ethanol and carbon tetrachloride, individuals who ingest enough alcohol to result in liver dysfunction would be a special group at risk. Individuals, e.g., analytical and synthetic chemists, mechanics and cleaners, who are routinely exposed to carbon tetrachloride or other hepatotoxic chemicals would also be at a greater risk than a population without such exposure. Individuals involved in the manufacture, utilization, or disposal of polychlorinated naphthalenes would be expected to have higher levels of exposure than the general population.

Incompatibilities: Strong oxidizers.

Permissible Exposure Limits in Air:

	Federal Standards	ACGIH STEL Values	IDLH Levels
	. . . .	(mg/m^3).	
Trichloronaphthalene	5.0	10	50
Tetrachloronaphthalene	2.0	4	20
Pentachloronaphthalene	0.5	2.0	no value set
Hexachloronaphthalene*	0.2	0.6	2
Octachloronaphthalene*	0.1	0.3	200

*Carry the notation "skin" indicating the possibility of cutaneous absorption.

Determination in Air: A filter plus a bubbler followed by workup with isooctane (for tri-, tetra- and pentachloronaphthalenes) or hexane (for hexa- and octachloronaphthalenes) followed by analysis by gas chromatography. See NIOSH Methods, Set I for tri- and tetra-; Set G for penta- and octa-; and Set H for hexachloronaphthalene. See also reference (A-10).

Permissible Concentrations in Water: To protect freshwater aquatic life: 1,600 $\mu g/\ell$ on an acute toxicity basis. To protect saltwater aquatic life: 7.5 $\mu g/\ell$ on an acute toxicity basis.

For the protection of human health from the toxic properties of chlorinated naphthalenes ingested through water and through contaminated aquatic organisms, there are insufficient data to permit establishment of criteria.

Determination in Water: 2-Chloronaphthalene may be determined by gas chromatography (EPA Method 612) or by gas chromatography plus mass spectrometry (EPA Method 625).

Routes of Entry: Inhalation of fumes and percutaneous absorption of liquid, ingestion, and eye and skin contact.

Harmful Effects and Symptoms: *Local* — Chronic exposure to chlorinated naphthalenes can cause chloracne, which consists of simple erythematous eruptions with pustules, papules, and comedones. Cysts may develop due to plugging of the sebaceous gland orifices.

Systemic — Cases of systemic poisoning are few in number and they may occur without the development of chloracne.

It is believed that chloracne develops from skin contact and inhalation of fumes, while systemic effects result primarily from inhalation of fumes. Symptoms of poisoning may include headaches, fatigue, vertigo, and anorexia. Jaundice may occur from liver damage. Highly chlorinated naphthalenes seem to be more toxic than those chlorinated naphthalenes with a lower degree of substitution.

Points of Attack: Skin, liver.

Medical Surveillance: Preplacement and periodic examinations should be concerned particularly with skin lesions such as chloracne and with liver function.

First Aid: If this chemical gets into the eyes, irrigate immediately. If this chemical contacts the skin, wash with soap immediately. If a person breathes in large amounts of this chemical, move the exposed person to fresh air at once and perform artificial respiration. When this chemical has been swallowed, get medical attention. Give large quantities of salt water and induce vomiting. Do not make an unconscious person vomit.

Personal Protective Methods: Wear appropriate clothing to prevent any possibility of skin contact with molten material or reasonable probability of contact with solutions. Wear eye protection to prevent any possibility of eye contact with molten material or any reasonable probability of contact with solutions. Employees should wash promptly when skin is wet or contaminated and daily at the end of each work shift. Work clothing should be changed daily if it is possible that clothing is contaminated. Remove nonimpervious clothing immediately if wet or contaminated with molten material and promptly if liquid solution contamination occurs.

Respirator Selection:

Trichloronaphthalene
    50 mg/m^3: SAF/SCBAF
      Escape: GMOVP/SCBA
Tetrachloronaphthalene
    20 mg/m^3: SAF/SCBAF
      Escape: GMOVP/SCBA
Pentachloronaphthalene
    5 mg/m^3: SA/SCBA
    25 mg/m^3: SAF/SCBAF
    500 mg/m^3: SA:PD,PP,CF
  1,000 mg/m^3: SAF:PD,PP,CF
      Escape: GMOVP/SCBA
Hexachloronaphthalene
    2 mg/m^3: SAF/SCBAF
      Escape: GMPest/SCBA
Octachloronaphthalene
    1 mg/m^3: SA/SCBA
    5 mg/m^3: SAF/SCBAF
    100 mg/m^3: SA:PD,PP,CF
    200 mg/m^3: SAF:PD,PP,CF
      Escape: GMOVP/SCBA

Disposal Method Suggested: Incineration, preferably after mixing with another combustible fuel. Care must be exercised to assure complete combustion to prevent the formation of phosgene. An acid scrubber is necessary to remove the halo acids produced (A-31).

References

(1) U.S. Environmental Protection Agency, *Chlorinated Naphthalenes: Ambient Water Quality Criteria,* Washington, DC (1980).
(2) U.S. Environmental Protection Agency, *Chlorinated Naphthalenes,* Health and Environmental Effects Profile No. 38, Office of Solid Waste, Washington, DC (April 30, 1980).
(3) U.S. Environmental Protection Agency, *2-Chloronaphthalene,* Health and Environmental Effects Profile No. 49, Office of Solid Waste, Washington, DC (April 30, 1980).

(4) Sax, N.I., Ed., *Dangerous Properties of Industrial Materials Report, 3,* No. 2, 77-78, New York, Van Nostrand Reinhold Co. (1983). (1-Chloronaphthalene).

(5) Parmeggiani, L. Ed., *Encyclopedia of Occupational Health and Safety,* Third Edition, Vol. 1, pp 465–66, Geneva, International Labour Office (1983).

(6) Sax, N.I., Ed., *Dangerous Properties of Industrial Materials Report, 4,* No. 3, 53–54, New York, Van Nostrand Reinhold Co. (1984). (1,2-Dichloronaphthalene).

(7) Sax, N.I., Ed., *Dangerous Properties of Industrial Materials Report, 4,* No. 3, 54–55, New York, Van Nostrand Reinhold Co. (1984). (1,3-Dichloronaphthalene).

(8) Sax, N.I., Ed., *Dangerous Properties of Industrial Materials Report, 4,* No. 3, 55–56, New York, Van Nostrand Reinhold Co. (1984). (1,4-Dichloronaphthalene).

(9) United Nations Environment Programme, *IRPTC Legal File 1983,* Vol. II, pp VII/463, Geneva, Switzerland, International Register of Potentially Toxic Chemicals (1984).

CHLORINATED PARAFFINS

Description: These are derivatives of branched and unbranched long-chain paraffin hydrocarbons containing 20 to 30 carbon atoms and perhaps 40 to 50% chlorine by weight. A typical formula might be $C_{22}H_{38}Cl_8$.

Code Numbers: For Chlorowax® 40 ($C_{20}H_{37}Cl_5$): CAS 51990-12-6 RTECS FY2280000

DOT Designation: —

Synonyms: Paroil®; Clorafin®; Cereclor®; Chlorowax®; Chlorez®.

Potential Exposures: Major applications of chlorinated paraffins include uses as lubricating oil additives (45% of total production), secondary vinyl plasticizers (24%), flame retardants in rubber, plastics, and paints (27%), and traffic paint additives (4%). Chlorinated paraffins are (1) produced in larger quantities than PCB's, (2) are likely to be released to the environment, (3) are less mobile and persistent than PCB's, and (4) are less acutely toxic.

Release of chlorinated paraffins used as oil additives to water resources and landfills is probably very sizable, since waste oil is frequently not recovered and this application is a major market for chlorinated paraffins. Chlorinated paraffins may also reach the environment as plasticizers in plastics (discarded in solid waste), by leaching from traffic and other paints, and as components of materials that have chlorinated paraffins incorporated in them for flame retardancy (also discarded in solid waste).

Permissible Exposure Limits in Air: No standards set.

Permissible Concentration in Water: No criteria set.

Determination in Water: Gas chromatography using electron capture detection or combined with mass spectrometry has been evaluated and considered unacceptable for chlorinated paraffins. With the shorter chained (C_9–C_{17}) chlorinated paraffins, where the compounds are volatile enough to pass through a gas chromatographic column, a relatively specific microcoulometric-gas chromatographic technique has been devised that is capable of measuring 0.5 ppm in fish flesh. With the higher chained (C_{20}–C_{30}) chlorinated paraffins, only the very nonspecific direct injection microcoulometric detection method has been used at 1 ppm concentrations. Neither of these methods is capable of measuring chlorinated paraffins at background environmental levels. It has been suggested that a combination of liquid chromatography with microcoulometric detection would be a very appropriate system for analyzing trace amounts of chlorinated paraffins.

Harmful Effects and Symptoms: Dermal application of chlorinated paraffins to human skin apparently does not produce local irritation or allergic sensitization. Furthermore, acute studies in nonhuman mammals have demonstrated that chlorinated paraffins possess extremely low toxicity when administered by oral, topical, and inhalation routes. Clearly lacking in the literature, however, are long term studies and investigations aimed at the determination of toxic reactions to chlorinated paraffin impurities and degradation products. Similarly, the question of biotransformation and metabolic activation of chlorinated paraffins into potentially harmful substances has not been answered as yet.

First Aid: Skin contact—wash with soapy water; Eye contact—flush with warm water; Ingestion—consult a physician.

Personal Protective Method: Most of the commercial formulations have no effects on the skin, on repeated or prolonged contact, but protective gloves are recommended. It is also suggested that safety glasses and body length clothing be worn and that respiratory protection equipment be used when working in aerosol mists of liquid chlorinated paraffins.

Disposal Method Suggested: Because chlorinated paraffins decompose at relatively low temperatures ($300°-400°C$) compared to PCB's ($>800°C$), they can be disposed of with conventional incinerators without the need for special precautions (e.g., afterburners). It is suspected that the chlorinated paraffins would decompose in an incinerator before significant amounts are volatilized. Some control of the hydrogen chloride generated would probably be required.

References

(1) U.S. Environmental Protection Agency, *Investigation of Selected Potential Environmental Contaminants: Chlorinated Paraffins,* Report EPA-560/2-75-007, Wash. DC (Nov. 1975).
(2) See Reference (A-60).

CHLORINATED PHENOLS

See also "2-Chlorophenol," "2,4-Dichlorophenol," "2,6-Dichlorophenol," "2,4,5-Trichlorophenol" and "Pentachlorophenol."

It is well known that the highly toxic polychlorinated dibenzo-p-dioxins may be formed during the chemical synthesis of some chlorophenols and that the amount of contaminant formed is dependent upon the temperature and pressure control of the reaction. The toxicity of the dioxins varies with the position and number of substituted chlorine atoms and those containing chlorine in the 2, 3 and 7 positions are particularly toxic. The 2,3,7,8-tetrachlorodibenzo-p-dioxin (TCDD) is considered the most toxic of all the dioxins.

Evidence has accumulated that the various chlorophenols are formed as intermediate metabolites during the microbiological degradation of the herbicides 2,4-D and 2,4,5-T and pesticides silvex, ronnel, lindane and benzene hexachloride. In view of this, it is clear that chlorinated phenols represent important compounds with regard to potential point source and nonpoint source water contamination.

Chlorophenols may be produced inadvertently by chlorination reactions which take place during the disinfection of wastewater effluents or drinking water sources. Phenol has been reported to be highly reactive to chlorine in dilute aqueous solutions over a considerable pH range.

References

(1) U.S. Environmental Protection Agency, *Chemical Hazard Information Profile: Mono/ Dichlorophenols,* Washington, DC (1979).
(2) U.S. Environmental Protection Agency, *Chlorinated Phenols: Ambient Water Quality Criteria,* Washington, DC (1980).
(3) U.S. Environmental Protection Agency, *Reviews of the Environmental Effects of Pollutants, XI: Chlorophenols,* Report EPA-600/1-79-012, Cincinnati, OH (1979).
(4) U.S. Environmental Protection Agency, *Chlorinated Phenols,* Health and Environmental Effects Profile No. 39, Office of Solid Waste, Washington, DC (April 30, 1980).
(5) Sax, N.I., Ed., *Dangerous Properties of Industrial Materials Report, 2,* No. 6, 46-55, New York, Van Nostrand Reinhold Co. (1982).
(6) United Nations Environment Programme, *IRPTC Legal File 1983,* Vol. II, pp VII/505–9, Geneva, Switzerland, International Register of Potentially Toxic Chemicals (1984).

CHLORINE

● Hazardous substance (EPA)

Description: Cl_2, chlorine, is a greenish-yellow gas with a pungent odor. It is slightly soluble in water and is soluble in alkalis. It is the commonest of the four halogens which are among the most chemically reactive of all the elements.

Code Numbers: CAS 7782-50-5   RTECS FO2100000   UN 1017

DOT Designation: Nonflammable gas, poison and oxidizer.

Synonyms: None.

Potential Exposures: Gaseous chlorine is a bleaching agent in the paper and pulp and textile industries for bleaching cellulose for artificial fibers. It is used in the manufacture of chlorinated lime, inorganic and organic compounds such as metallic chlorides, chlorinated solvents, refrigerants, pesticides, and polymers, e.g., synthetic rubber and plastics; it is used as a disinfectant, particularly for water and refuse, and in detinning and dezincing iron. NIOSH estimated in 1973 that 15,000 workers had potential occupational exposure to chlorine.

Incompatibilities: Combustible substances and finely divided metals.

Permissible Exposure Limits in Air: Federal standard and ACGIH 1983/84 TWA value is 1 ppm (3 mg/m³). NIOSH has recommended a ceiling limit of 0.5 ppm for a 15-minute sampling period. The basis for the NIOSH-recommended environmental limit is the prevention of irritation of the skin, eyes, and respiratory tract. The STEL is 3 ppm (9 mg/m³). The IDLH level is 25 ppm.

Determination in Air: Collection by fritted bubbler, colorimetric analysis using methyl orange which is bleached by free chlorine (A-11).

Permissible Concentration in Water: EPA (A-3) has suggested the following limits: Total residual chlorine: 2.0 μg/ℓ for salmonid fish; 10.0 μg/ℓ for other freshwater and marine organisms.

Routes of Entry: Inhalation, eye and skin contact.

Harmful Effects and Symptoms: *Local* — Chlorine reacts with body moisture to form acids. It is itself extremely irritating to skin, eyes, and mucous membranes, and it may cause corrosion of teeth. Prolonged exposure to low concentrations may produce chloracne.

Systemic — Chlorine in high concentrations acts as an asphyxiant by causing cramps in the muscles of the larynx (choking), swelling of the mucous membranes, nausea, vomiting, anxiety, and syncope. Acute respiratory distress including cough, hemoptysis, chest pain, dyspnea, and cyanosis develop, and later tracheobronchitis, pulmonary edema, and pneumonia may supervene.

Points of Attack: Lungs, respiratory system.

Medical Surveillance: Special emphasis should be given to the skin, eyes, teeth, cardiovascular status in placement and periodic examinations. Chest x-rays should be taken and pulmonary function followed.

First Aid: If this chemical gets into the eyes, irrigate immediately. If this chemical contacts the skin, flush with water immediately. If a person breathes in large amounts of this chemical, move the exposed person to fresh air at once and perform artificial respiration.

Personal Protective Methods: Whenever there is likelihood of excessive gas levels, workers should use respiratory protection in the form of fullface gas masks with proper canisters or supplied air respirators. The skin effects of chlorine can generally be controlled by good personal hygiene practices. Where very high gas concentrations or liquid chlorine may be present, full protective clothing, gloves, and eye protection should be used. Changing work clothes daily and showering following each shift where exposures exist are recommended.

Respirator Selection:
> 25 ppm: CCRSF/GMS/SAF/SCBAF
> Escape: GMSF/SCBA
>
> Note: Do not use oxidizable sorbents.

Disposal Method Suggested: Introduce into large volume and solution of reducing agent (bisulfite, ferrous salts or hypo), neutralize and flush to sewer with water (A-38). Recovery is an option to disposal for chlorine (A-57) in the case of gases from aluminum chloride electrolysis and chlorine in wastewaters.

References

(1) National Institute for Occupational Safety and Health, *Criteria for a Recommended Standard: Occupational Exposure to Chlorine,* NIOSH Document No. 76-170 (1976).

(2) National Academy of Sciences, *Medical and Biological Effects of Environmental Pollutants: Chlorine and Hydrogen Chloride,* Washington, DC (1976).

(3) See Reference (A-61).

(4) Sax, N.I., Ed., *Dangerous Properties of Industrial Materials Report, 1,* No. 3, 41-43, New York, Van Nostrand Reinhold Co. (1981).

(5) Sax, N.I., Ed., *Dangerous Properties of Industrial Materials Report, 2,* No. 4, 67-70, New York, Van Nostrand Reinhold Co. (1982). (Chlorine-36).

(6) See Reference (A-60).

(7) Parmeggiani, L., Ed., *Encyclopedia of Occupational Health & Safety,* Third Edition, Vol. 1, pp 454–459, Geneva, International Labour Office (1983).

CHLORINE DIOXIDE

Description: ClO_2 is a yellow-green to orange gas or liquid with a pungent sharp odor. It boils at $10°C$.

Code Numbers: CAS 10049-04-4   RTECS FO3000000   UN (NA 9091) (for frozen hydrate).

DOT Designation: Forbidden

Synonyms: None

Potential Exposure: Chlorine dioxide is used in bleaching cellulose pulp, bleaching flour, water purification. It is used as an oxidizing agent.

Incompatibilities: Combustible substances, dust, organic matter, sulfur.

Permissible Exposure Limits in Air: Federal limit is 0.1 ppm (0.3 mg/m^3). ACGIH has set an STEL of 0.3 ppm (0.9 mg/m^3) as of 1983/84. The IDLH level is 10 ppm.

Determination in Air: Collection by impinger or fritted bubbler and colorimetric analysis (A-14).

Permissible Concentration in Water: No criteria set.

Routes of Entry: Inhalation, ingestion, eye and skin contact.

Harmful Effects and Symptoms: Irritation of eyes, nose, throat; cough, wheezing, bronchitis, pulmonary edema; chronic bronchitis.

Points of Attack: Respiratory system, lungs, eyes.

Medical Surveillance: Consider the points of attack in preplacement and periodic physical examinations.

First Aid: If this chemical gets into the eyes, irrigate immediately. If this chemical contacts the skin, wash with soap immediately. If a person breathes in large amounts of this chemical, move the exposed person to fresh air at once and perform artificial respiration. When this chemical has been swallowed, get medical attention. Give large quantities of water but do NOT induce vomiting.

Personal Protective Methods: Wear appropriate clothing to prevent any possibility of skin contact. Wear eye protection to prevent any possibility of eye contact. Wash immediately when skin is wet or contaminated. Remove clothing immediately if wet or contaminated. Provide emergency showers and eyewash.

Respirator Selection:
```
 5 ppm:  CCROVFS/GMS/SAF/SCBAF
10 ppm:  SAF:PD,PP,CF
Escape:  GMS/SCBA
Note: Do not use oxidizable sorbents.
```

Disposal Method Suggested: Same as "Chlorine" (which see).

References
(1) See Reference (A-61).
(2) See Reference (A-60).

CHLORINE TRIFLUORIDE

Description: ClF$_3$ is a greenish-yellow, almost colorless, liquid or gas with a sweet, irritating odor. It boils at 11°C.

Code Numbers: CAS 7790-91-2 RTECS FO2800000 UN 1749

DOT Designation: Oxidizer and poison gas, corrosive.

Synonyms: None

Potential Exposure: ClF_3 is used as a fluorinating agent. It may be used as an igniter and propellant in rockets. It is used in nuclear fuel processing.

Incompatibilities: Combustible substances, water, sand, glass, asbestos, silicon-containing compounds.

Permissible Exposure Limits in Air: The Federal limit and the 1983/84 ACGIH TWA value is 0.1 ppm (0.4 mg/m^3) as a ceiling value. There is no tentative STEL value. The IDLH level is 20 ppm.

Determination in Air: Collection by impinger or fritted bubbler, analysis by ion-specific electrode (A-16).

Permissible Concentration in Water: No criteria set.

Routes of Entry: Inhalation, ingestion, eye and skin contact.

Harmful Effects and Symptoms: In animals: irritation of eyes and respiratory tract; burns of eyes and skin.

Points of Attack: Skin, eyes, respiratory tract.

Medical Surveillance: Consider the points of attack in preplacement and periodic physical examinations.

First Aid: If this chemical gets into the eyes, irrigate immediately. If this chemical contacts the skin, flush with water immediately. If a person breathes in large amounts of this chemical, move the exposed person to fresh air at once and perform artificial respiration. When this chemical has been swallowed, get medical attention. Give large quantities of water and do NOT induce vomiting.

Personal Protective Methods: Wear appropriate clothing to prevent any possibility of skin contact. Wear eye protection to prevent any possibility of eye contact. Employees should wash immediately when skin is wet or contaminated. Remove clothing immediately if wet or contaminated. Provide emergency showers and eyewash.

Respirator Selection:
    5 ppm: SAF/SCBAF
    20 ppm: SAF:PD,PP,CF
    Escape: GMS/SCBA

References
(1)   See Reference (A-61).

CHLORNAPHAZINE

- Carcinogen (Human Positive, IARC) (1,2)
- Hazardous Waste Constituent (EPA)

Description: $C_{14}H_{15}Cl_2N$ with the structural formula

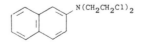

forms crystals melting at 54° to 56°C.

Code Numbers: CAS 494-03-1   RTECS QM2450000

DOT Designation: —

Synonyms: N,N-Bis(2-chloroethyl)-2-naphthylamine; dichloroethyl-β-naphthylamine.

Potential Exposures: Not produced or used commercially in the United States, chlornaphazine has been used in other countries in the treatment of leukemia and related cancers (3). Currently, this drug does not have wide therapeutic usage.

Permissible Exposure Limits in Air: No standards set.

Permissible Concentration in Water: No criteria set.

Harmful Effects and Symptoms: N,N-Bis(2-Chloroethyl)-2-naphthylamine (chlornaphazine) produces lung tumours in mice following intraperitoneal injection, and local sarcomas in rats after subcutaneous administration (1).

The administration of chlornaphazine together with radioactive phosphorus (^{32}P-sodium phosphate) caused bladder cancer in 10 of 61 patients treated for polycythemia vera. In 46 patients treated with ^{32}P-sodium phosphate alone, no cases of bladder cancer were found (1)(2).

References

(1) IARC Monographs 4:119-124 (1974).
(2) IARC Monographs on the Evaluation of the Carcinogenic Risk of Chemicals to Humans, Supplement I, IARC, Lyon, France, p 26, (1979).
(3) See Reference (A-62). Also see Reference (A-64).

CHLOROACETALDEHYDE

● Hazardous waste (EPA)

Description: ClCH$_2$CHO is a colorless liquid with a very sharp, irritating odor. It boils at 85° to 86°C.

Code Numbers: CAS 107-20-0   RTECS AB2450000   UN 2232

DOT Designation: Poison.

Synonyms: 2-Chloroethanal.

Potential Exposure: Chloroacetaldehyde is used as a fungicide, as an intermediate in 2-aminothiazole manufacture and in bark removal from tree trunks (1).

Incompatibilities: Oxidizers, acids, water.

Permissible Exposure Limits in Air: The Federal limit and the ACGIH 1983/84 TWA value is 1 ppm (3 mg/m^3) as a ceiling value. There is no STEL value. The IDLH level is 250 ppm.

Determination in Air: Collection by charcoal tube, analysis by gas liquid chromatography (A-13).

Permissible Concentration in Water: No criteria set.

Routes of Entry: Inhalation, ingestion, eye and skin contact.

Harmful Effects and Symptoms: Irritation of skin, eyes and mucous membrane; skin burns; eye damage; pulmonary edema; sensitization of skin and respiratory system.

Points of Attack: Eyes, skin, respiratory system, lungs.

Medical Surveillance: Consider the points of attack in preplacement and periodic physical examinations.

First Aid: If this chemical gets into the eyes, irrigate immediately. If this chemical contacts the skin, flush with water immediately. If a person breathes in large amounts of this chemical, move the exposed person to fresh air at once and perform artificial respiration. When this chemical has been swallowed, get medical attention. Give large quantities of salt water and induce vomiting. Do not make an unconscious person vomit.

Personal Protective Methods: Wear appropriate clothing to prevent any possibility of skin contact with liquids of >0.1% content or repeated or prolonged contact with liquids of <0.1% content. Wear eye protection to prevent any possibility of eye contact. Employees should wash immediately when skin is wet or contaminated. Remove nonimpervious clothing immediately if wet or contaminated. Provide emergency showers and eyewash.

Respirator Selection:

         50 ppm: CCROVF/GMOV/SAF/SCBAF
        250 ppm: SAF:PD,PP,CF
         Escape: GMOV/SCBA

Disposal Method Suggested: Incineration, preferably after mixing with another combustible fuel; care must be exercised to assure complete combustion to prevent the formation of phosgene; an acid scrubber is necessary to remove the halo acids produced (A-31).

References

(1)   U.S. Environmental Protection Agency, *Chloroacetaldehyde,* Health and Environmental Effects Profile No. 40, Office of Solid Waste, Washington, DC (April 30, 1980).
(2)   See Reference (A-61).
(3)   Sax, N.I., Ed., *Dangerous Properties of Industrial Materials Report, 2,* No. 4, 70-72, New York, Van Nostrand Reinhold Co. (1982).

CHLOROACETIC ACID

Description: $ClCH_2COOH$ is a colorless to white crystalline solid which melts at 63°C and boils at 188°C.

Code Numbers: CAS 79-11-8   RTECS AF8575000   UN 1750 (liquid) UN 1751 (solid)

DOT Designation: Corrosive material.

Synonyms: Monochloroacetic acid; chloroethanoic acid.

Potential Exposure: Monochloroacetic acid is used primarily as a chemical intermediate in the synthesis of sodium carboxymethyl cellulose, and such other diverse substances as ethyl chloroacetate, glycine, synthetic caffeine, sarcosine, thioglycolic acid, and various dyes. Hence, workers in these areas are affected. It is also used as an herbicide. Therefore, formulators and applicators of such herbicides are affected. NIOSH estimates (1) that 100,000-200,000 workers are potentially exposed each year.

Incompatibilities: Alkalies.

Permissible Exposure Limits in Air: No standards set.

Permissible Concentration in Water: No criteria set.

Routes of Entry: Inhalation and ingestion.

Harmful Effects and Symptoms: Monochloroacetic acid is primarily an ir-ritant in humans and can produce severe local skin, eye, and respiratory tract reactions. It is absorbed through the skin and is moderately toxic by inhalation of dust and vapor. It is also a disaster hazard, as it emits phosgene and chloride fumes. Inhalation can cause lung emema (A-60).

Animal toxicity data (1) reported in various species include apathy, weight loss, narcosis, and subsequent death or complete recovery.

Personal Protective Methods: Wear gloves, protective clothing and face shield (A-60).

Disposal Method Suggested: Incineration, preferably after mixing with an-other combustible fuel; care must be exercised to assure complete combustion to prevent the formation of phosgene; an acid scrubber is necessary to remove the halo acids produced.

References

(1) National Institute for Occupational Safety and Health, *Profiles on Occupational Haz-ards for Criteria Document Priorites: Monochloroacetic Acid,* pp 309-11, Report PB-274,073, Rockville, MD (1977).
(2) See Reference (A-60).
(3) Sax, N.I., Ed., *Dangerous Properties of Industrial Materials Report, 3,* No. 5, 99-101, New York, Van Nostrand Reinhold Co. (1983).

2-CHLOROACETOPHENONE

Description: $C_6H_5COCH_2Cl$ is a colorless-to-gray solid with a sharp, irri-tating odor.

Code Numbers: CAS 532-27-4   RTECS AM6300000   UN 1697

DOT Designation: Irritating material and poison.

Synonyms: Phenacyl chloride; omega-chloroacetophenone; chloromethyl phenyl ketone; phenyl chloromethyl ketone; tear gas; CN; Chemical Mace®.

Potential Exposures: Chloroacetophenone is used as a chemical warfare agent (Agent CN) and as a principal ingredient in the riot control agent Mace. It is also used as a pharmaceutical intermediate (A-41).

Incompatibilities: Water or steam.

Permissible Exposure Limits in Air: The Federal limit and the 1983/84 ACGIH TWA value is 0.05 ppm (0.3 mg/m³). There is no STEL value set. The IDLH level is 100 mg/m³.

Determination in Air: Collection by charcoal tube, analysis by gas liquid chromatography (A-13).

Permissible Concentration in Water: No criteria set.

Routes of Entry: Inhalation, skin absorption, ingestion, skin and eye con-tact.

Harmful Effects and Symptons: Skin and eye irritation, respiratory system irritation, pulmonary edema.

Points of Attack: Eyes, skin, respiratory system, lungs.

Medical Surveillance: Consider the points of attack in preplacement and periodic physical examinations.

First Aid: If this chemical gets into the eyes, irrigate immediately. If this chemical contacts the skin, wash with soap immediately. If a person breathes in large amounts of this chemical, move the exposed person to fresh air at once and perform artificial respiration. When this chemical has been swallowed, get medical attention. Give large quantities of salt water and induce vomiting. Do not make an unconscious person vomit.

Personal Protective Methods: Wear appropriate clothing to prevent any possibility of skin contact. Wear eye protection to prevent any possibility of eye contact. Employees should wash immediately when skin is wet or contaminated. Work clothing should be changed daily if it is possible that clothing is contaminated. Remove nonimpervious clothing immediately if wet or contaminated. Provide eyewash.

Respirator Selection:
    15 mg/m³: CCROVHiEF/GMOVHiE/SAF/SCBAF
   100 mg/m³: SAF:PD,PP,CF
    Escape: GMOVFP/SCBA

Disposal Method Suggested: Tear gas-containing waste is dissolved in an organic solvent and sprayed into an incinerator equipped with an afterburner and alkaline scrubber utilizing reaction with sodium sulfide in an alcohol-water solution. Hydrogen sulfide is liberated and collected by an alkaline scrubber (A-31).

References

(1) See Reference (A-61).
(2) Sax, N.I., Ed., *Dangerous Properties of Industrial Materials Report, 4,* No. 1, 48–49, New York, Van Nostrand Reinhold Co. (Jan./Feb. 1984).

CHLOROACETYL CHLORIDE

Description: ClCH$_2$COCl is a colorless-to-yellowish liquid with a pungent odor which boils at 105° to 110°C.

Code Numbers: CAS 79-04-9 RTECS AO6475000 UN 1752

DOT Designation: Corrosive material.

Synonyms: None.

Potential Exposures: Chloroacetyl chloride is used in the manufacture of acetophenone. It is used in the manufacture of a number of pesticides (A-32) including: alachlor, allidochlor, butachlor, dimethachlor, formothion, mecarbam, metolachlor, propachlor. It is also used in the manufacture of pharmaceuticals (A-41) such as chlordiazepoxide hydrochloride, diazepam, lidocaine, mianserin.

Permissible Exposure Limits in Air: There is no Federal standard. ACGIH, as of 1983/84, has adopted a TWA value of 0.05 ppm (0.2 mg/m³), but has set no STEL value.

Permissible Concentration in Water: No criteria set. (Chloroacetyl chloride decomposes in water.)

Routes of Entry: Skin absorption, skin and eye contact, inhalation, ingestion.

Harmful Effects and Symptoms: Medical reports of the effects of acute exposures include: mild-to-moderate skin burns and erythema; lachrymation and mild eye burns; mild-to-moderate respiratory effects with cough, dyspnea and cyanosis; and mild gastrointestinal effects (A-34).

Points of Attack: Skin, eyes, respiratory system.

Medical Surveillance: Should include attention to skin, eyes and respiratory system in preplacement and regular physical examinations.

Personal Protective Methods: Wear rubber gloves and coveralls.

Respirator Selection: Use of self-contained breathing apparatus is recommended.

Disposal Method Suggested: It may be discharged into sodium bicarbonate solution, then flushed to the sewer with water (A-38).

References
(1) See Reference (A-60).

CHLOROALKYL ETHERS

See separate entries under: "Bis(2-Chloroethyl) Ether," "Bis(2-Chloroisopropyl) Ether," "Bis(Chloromethyl) Ether," "Chloromethyl Methyl Ether," "2-Chloroethyl Vinyl Ether," "Bis(2-Chloroethoxy)Methane."

References
(1) U.S. Environmental Protection Agency, *Chloroalkyl Ethers: Ambient Water Quality Criteria,* Washington, DC (1980).
(2) U.S. Environmental Protection Agency, *Chloroalkyl Ethers,* Health and Environmental Effects Profile No. 41, Office of Solid Waste, Washington, DC (April 30, 1980).

CHLOROBENZENE

- Hazardous substance (EPA)
- Hazardous waste (EPA)
- Priority toxic pollutant (EPA)

Description: C_6H_5Cl, a colorless liquid boiling at 131° to 132°C with a mild aromatic odor.

Code Numbers: CAS 108-90-7 RTECS CZ0175000 UN 1134

DOT Designation: Flammable liquid.

Synonyms: Monochlorobenzene, chlorobenzol, phenyl chloride, MCB.

Potential Exposure: Chlorobenzene is used in the manufacture of aniline, phenol, and chloronitrobenzene and as an intermediate in the manufacture of dyestuffs and many pesticides (A-32).

Incompatibilities: Strong oxidizers.

Permissible Exposure Limits in Air: The Federal limit and the 1983/84 ACGIH TWA value is 75 ppm (350 mg/m^3). There is no STEL value set. The IDLH level is 2,400 ppm.

Determination in Air: Charcoal adsorption followed by workup with CS$_2$ and analysis by gas chromatography. See NIOSH Methods, Set I. See also reference (A-10).

Permissible Concentration in Water: To protect freshwater aquatic life: 250 µg/ℓ on an acute basis for chlorobenzenes as a class. To protect saltwater aquatic life: 160 µg/ℓ on an acute basis and 129 µg/ℓ on a chronic basis for chlorinated benzenes as a class. To protect human health: for the prevention of adverse toxicological effects, 488 µg/ℓ; but to prevent adverse organoleptic effects, 20 µg/ℓ.

Determination in Water: Gas chromatography (EPA Methods 601 and 602) or gas chromatography plus mass spectrometry (EPA Method 624).

Routes of Entry: Inhalation, ingestion, eye and skin contact.

Harmful Effects and Symptoms: Irritation of the eyes and nose; drowsiness, incoherence; skin irritation; liver damage.

Points of Attack: Respiratory system, eyes, skin, central nervous system, liver.

Medical Surveillance: Consider the points of attack in preplacement and periodic physical examinations.

First Aid: If this chemical gets into the eyes, irrigate immediately. If this chemical contacts the skin, wash with soap promptly. If a person breathes in large amounts of this chemical, move the exposed person to fresh air at once and perform artificial respiration. When this chemical has been swallowed, get medical attention. Do NOT induce vomiting.

Personal Protective Methods: Wear appropriate clothing to prevent repeated or prolonged skin contact. Wear eye protection to prevent any reasonable probability of eye contact. Employees should wash promptly when skin is wet or contaminated. Remove clothing immediately if wet or contaminated to avoid flammability hazard.

Respirator Selection:
　　　1,000 ppm: CCROVF
　　　2,400 ppm: GMOV/SAF/SCBAF
　　　　Escape: GMOV/SCBA

Disposal Method Suggested: Incineration, preferably after mixing with another combustible fuel; care must be exercised to assure complete combustion to prevent the formation of phosgene; an acid scrubber is necessary to remove the halo acids produced.

References

(1) U.S. Environmental Protection Agency, *Chlorinated Benzenes: Ambient Water Quality Criteria,* Washington, DC (1980).
(2) U.S. Environmental Protection Agency, *Chlorobenzene,* Health and Environmental Effects Profile No. 42, Office of Solid Waste, Washington, DC (April 30, 1980).
(3) See Reference (A-61).
(4) Sax, N.I., Ed., *Dangerous Properties of Industrial Materials Report, 2,* No. 4, 72-75, New York, Van Nostrand Reinhold Co. (1982).

(5) Parmeggiani, L., Ed., *Encyclopedia of Occupational Health & Safety,* Third Edition, Vol. 1, pp 459–61, Geneva, International Labour Office (1983).

(6) United Nations Environment Programme, *IRPTC Legal File 1983,* Vol. I, pp VII/75–78, Geneva, Switzerland, International Register of Potentially Toxic Chemicals (1984).

p-CHLOROBENZOTRICHLORIDE

● Carcinogen (Animal Positive) (1)

Description: $Cl_3C-C_6H_4-Cl$ is a water-white liquid which boils at 245° to 257°.

Code Numbers: CAS 5216-25-1

DOT Designation: —

Synonyms: 1-Chloro-4-(trichloromethyl)benzene; p, alpha-, alpha-, alpha-tetrachlorotoluene; p-chlorophenyltrichloromethane.

Potential Exposure: Used in pesticide manufacture as an intermediate; reaction with HF yields chlorobenzotrifluoride as a major intermediate for several pesticides.

Permissible Exposure Limits in Air: No limits set.

Permissible Concentration in Water: No criteria set.

Routes of Entry: Inhalation, ingestion.

Harmful Effects and Symptoms: If released into the environment, pCBTC could constitute a health hazard for humans since pCBTC is carcinogenic in mice when administered either orally or dermally.

References

(1) U.S. Environmental Protection Agency, *Chemical Hazard Information Profile Draft Report: p-Chlorobenzotrichloride,* Washington, D.C., (February 24, 1983).

o-CHLOROBENZYLIDENE MALONITRILE

Description: $ClC_6H_4CH=C(CN)_2$, o-chlorobenzylidene malonitrile, is a white crystalline solid.

Code Numbers: CAS 2698-41-1   RTECS OO3675000

DOT Designation: —

Synonyms: OCBM, CS, o-chlorobenzalmalononitrile.

Potential Exposures: OCBM is used as a riot control agent.

Incompatibilities: Strong oxidizers.

Permissible Exposure Limits in Air: The Federal standard and ACGIH 1983/84 TWA is 0.05 ppm (0.4 mg/m^3) as a ceiling value. The notation "skin" is added to indicate the possibility of cutaneous absorption. There is no STEL value set. The IDLH level is 2.0 mg/m^3.

Determination in Air: Collection by charcoal tube, analysis by gas liquid chromatography (A-1).

Permissible Concentration in Water: No criteria set.

Routes of Entry: Inhalation, ingestion, eye and skin contact.

Harmful Effects and Symptoms: *Local* — OCBM is extremely irritating and acts on exposed sensory nerve endings (primarily in the eyes and upper respiratory tract). The signs and symptoms from exposure to the vapor are conjunctivitis and pain in the eyes, lacrimation, erythema of the eyelids, blepharospasms, irritation and running of the nose, burning in the throat, coughing and constricted feeling in the chest, and excessive salivation. Vomiting may occur if saliva is swallowed. Most of the symptoms subside after exposure ceases. Burning on the exposed skin is increased by moisture. With heavy exposure, vesiculation and erythema occur. Photophobia has been reported (A-5).

Systemic — Animal experiments indicate that OCBM has a relatively low toxicity. The systemic changes observed in human experiments are nonspecific reactions to stress. OCBM is capable of sensitizing guinea pigs; there also appears to be a cross-reaction in guinea pigs previously sensitized to 1-chloroacetophenone.

Points of Attack: Respiratory system, skin and eyes.

Medical Surveillance: Consideration should be given to the eyes, skin, and respiratory tract in any placement or periodic evaluations.

First Aid: If this chemical gets into the eyes, irrigate immediately. If this chemical contacts the skin, wash with soap immediately. If a person breathes in large amounts of this chemical, move the exposed person to fresh air at once and perform artificial respiration. When this chemical has been swallowed, get medical attention. Give large quantities of water and induce vomiting. Do not make an unconscious person vomit.

Personal Protective Methods: Wear appropriate clothing to prevent repeated or prolonged skin contact. Wear eye protection to prevent any reasonable probability of eye contact. Employees should wash promptly when skin is wet or contaminated and daily at the end of each work shift. Work clothing should be changed daily if it is possible that clothing is contaminated. Remove nonimpervious clothing promptly if wet or contaminated.

Respirator Selection:
    2 mg/m^3: CCROVHiEP/GMOVHiEP/SAF/SCBAF
       Escape: GMOVF/SCBAF

References
(1) See Reference (A-61).

CHLOROBROMOMETHANE

Description: CH_2BrCl is a colorless-to-pale-yellow liquid with a characteristic sweet odor.

Code Numbers: CAS 74-97-5 RTECS PA5250000 UN 1887

DOT Designation: ORM-A

Synonyms: Bromochloromethane, methylene chlorobromide, CB, CBM, Halon 1011.

Potential Exposure: This compound is used as a fire-fighting agent and in organic synthesis.

Incompatibilities: Chemically active metals: calcium, powdered aluminum, zinc, magnesium.

Permissible Exposure Limits in Air: The Federal limit and ACGIH 1983/84 TWA value is 200 ppm (1,050 mg/m^3). The STEL value is 250 ppm (1,300 mg/m^3). The IDLH level is 5,000 ppm.

Determination in Air: Charcoal adsorption, workup with CS_2, followed by gas chromatography. See NIOSH Methods, Set H. See also reference (A-10).

Permissible Concentration in Water: No criteria set.

Routes of Entry: Inhalation, ingestion, eye and skin contact.

Harmful Effects and Symptoms: Disorientation, dizziness; irritation of eyes, throat and skin; pulmonary edema, headaches, anorexia, nausea, vomiting, abdominal pain, weight loss, memory impairment, paralysis, weakness, tremors and convulsions, narcosis.

Points of Attack: Skin, liver, kidneys, respiratory system, lungs, central nervous system.

Medical Surveillance: Consider the points of attack in preplacement and periodic physical examinations.

First Aid: If this chemical gets into the eyes, irrigate immediately. If this chemical contacts the skin, wash with soap promptly. If a person breathes in large amounts of this chemical, move the exposed person to fresh air at once and perform artificial respiration. When this chemical has been swallowed, get medical attention. Give large quantities of salt water and induce vomiting. Do not make an unconscious person vomit.

Personal Protective Methods: Wear appropriate clothing to prevent repeated or prolonged skin contact. Wear eye protection to prevent any reasonable probability of eye contact. Employees should wash promptly when skin is wet or contaminated. Remove nonimpervious clothing promptly if wet or contaminated.

Respirator Selection:
 1,000 ppm: CCROV
 2,000 ppm: SA/SCBA
 5,000 ppm: GMOV
 Escape: GMOV/SCBA

Disposal Method Suggested: Incinerate together with flammable solvent in furnace equipped with afterburner and alkali scrubber.

References
(1) See Reference (A-61).

p-CHLORO-m-CRESOL

● Hazardous waste (EPA)

Description: $ClC_6H_3(CH_3)OH$ is a crystalline solid with a slight phenolic odor, melting at 55° to 56°C.

Code Numbers: CAS 59-50-7   RTECS GO7100000   UN 2669

DOT Designation: —

Synonym: 4-Chloro-m-cresol.

Potential Exposures: p-Chloro-m-cresol is used as an external germicide and as a preservative for glues, gums, paints, inks, textiles and leather goods. It is also used as a preservative in cosmetics. It has been found to be formed by the chlorination of waters receiving effluents from electric power-generating plants and by the chlorination of the effluent from a domestic sewage-treatment facility.

Permissible Exposure Limits in Air: No standards set.

Permissible Concentration in Water: No criteria set.

Harmful Effects and Symptoms: Very little toxicological data for p-chloro-m-cresol are available (1). One source has rated p-chloro-m-cresol as very toxic, with a probable lethal dose to humans of 50 to 500 mg/kg. p-Chloro-m-cresol was also reported as nonirritating to skin in concentrations of 0.5 to 1.0% in alcohol.

References

(1)   U.S. Environmental Protection Agency, *p-Chloro-m-Cresol*, Health and Environmental Effects Profile No. 43, Office of Solid Waste, Washington, DC (April 30, 1980).
(2)   See Reference (A-60).

CHLORODIFLUOROMETHANE

Description: $CHClF_2$ is a colorless, nearly odorless gas.

Code Numbers: CAS 75-45-6   RTECS PA6390000   UN 1018

DOT Designation: Nonflammable gas.

Synonyms: Difluorochloromethane, difluoromonochloromethane, Propellant 22, Refrigerant 22, Fluorocarbon 22, F-22.

Potential Exposures: $CHClF_2$ is used as an aerosol propellant, refrigerant and low-temperature solvent. It is used in the synthesis of polytetrafluoroethylene (PTFE).

Permissible Exposure Limits in Air: There is no Federal standard, however ACGIH has as of 1983/84 adopted TWA values of 1,000 ppm ($3,500$ mg/m^3) and set an STEL of 1,250 ppm ($4,375$ mg/m^3).

Permissible Concentration in Water: No criteria set.

Harmful Effects and Symptoms: In animal studies, stimulation and then depression were produced by concentrations of 100,000 ppm; 200,000 ppm produced narcosis; and death resulted at 300,000 to 400,000 ppm. At 14,000 ppm, pathological changes were noted in the lungs, central nervous system, heart, liver, kidneys and spleen. At 2,000 ppm no effects were noted (A-34).

Personal Protective Methods: Wear rubber gloves.

Respirator Selection: Use air mask (A-38).

Disposal Method Suggested: Vent to atmosphere.

CHLORODIPHENYL

See "Polychlorinated Biphenyls (PCB's)."

2-CHLOROETHYL VINYL ETHER

- Hazardous waste (EPA)
- Priority toxic pollutant (EPA)

Description: $ClCH_2CH_2OCH=CH_2$ is a colorless liquid boiling at 109°C.

Code Numbers: CAS 110-75-8 RTECS KN6300000

DOT Designation: —

Synonym: Vinyl 2-chloroethyl ether.

Potential Exposures: The compound finds use in the manufacture of anesthetics, sedatives, and cellulose ethers. NIOSH estimates annual exposure at 23,500 workers. The number of potentially exposed individuals is greatest for the following areas: fabricated metal products; wholesale trade; leather, rubber and plastic, and chemical products (1).

Permissible Exposure Limits in Air: No standards set.

Permissible Concentration in Water: For the protection of freshwater aquatic life: 50,000 $\mu g/\ell$ (2). No criteria were developed for saltwater aquatic life or for the protection of human health.

Determination in Water: Inert gas purge followed by gas chromatography with halide specific detection (EPA Method 601) or gas chromatography plus mass spectrometry (EPA Method 624).

Harmful Effects and Symptoms: Very little toxicological data for 2-chloroethyl vinyl ether is available. The oral LD_{50} for 2-chloroethyl vinyl ether in rats is 250 mg/kg (moderately toxic). Primary skin irritation and eye irritation studies have also been conducted for 2-chloroethyl vinyl ether. Dermal exposure to undiluted 2-chloroethyl vinyl ether did not cause even slight erythema. Application of undiluted 2-chloroethyl vinyl ether to the eyes of rabbits resulted in severe eye injury.

References

(1) U.S. Environmental Protection Agency, *2-Chloroethyl Vinyl Ether,* Health and Environmental Effects Profile No. 46, Office of Solid Waste, Washington, DC (April 30, 1980).

(2) U.S. Environmental Protection Agency, *Chloroalkyl Ethers: Ambient Water Quality Criteria,* Washington, DC (1980).

CHLOROFORM

- Carcinogen (Animal Suspected, IARC) (6)
- Hazardous substance (EPA)
- Hazardous waste (EPA)
- Priority toxic pollutant (EPA)

Description: $CHCl_3$, chloroform, is a clear, colorless liquid with a characteristic odor. Though nonflammable, chloroform decomposes to form hydrochloric acid, phosgene, and chlorine upon contact with a flame.

Code Numbers: CAS 67-66-3   RTECS FS9100000   UN 1888

DOT Designation: ORM-A, poison.

Synonym: Trichloromethane.

Potential Exposures: Chloroform was one of the earliest general anesthetics, but its use for this purpose has been abandoned because of toxic effects. Chloroform is widely used as a solvent (especially in the lacquer industry); in the extraction and purification of penicillin and other pharmaceuticals; in the manufacture of artificial silk, plastics, floor polishes, and fluorocarbons; and in sterilization of catgut. The wide industrial usage of chloroform potentially exposes 360,000 workers (OSHA). Chemists and support workers as well as hospital workers are believed to be at a higher risk than the general population.

Chloroform is widely distributed in the atmosphere and water (including municipal drinking water primarily as a consequence of chlorination). A survey of 80 American cities by EPA found chloroform in every water system in levels ranging from <0.3 to 311 ppb.

Incompatibilities: Strong caustics, chemically active metals such as aluminum, magnesium powder, sodium, potassium.

Permissible Exposure Limits in Air: The Federal standard is 50 ppm (240 mg/m^3). The ACGIH recommended 1976 TLV was 25 ppm. NIOSH's recommended limit is a ceiling of 2 ppm based on a one-hour sample collected at 750 ℓ/min. ACGIH (1983/84) has set 10 ppm (50 mg/m^3) as a TWA with the notation that chloroform is an "Industrial Substance Suspect of Carcinogenic Potential for Man." The STEL value is 50 ppm (225 mg/m^3). The IDLH value is 1,000 ppm.

Determination in Air: Charcoal adsorption, workup with CS_2, analysis by gas chromatography. See NIOSH Methods, Set J. See also reference (A-10).

Permissible Concentration in Water: To protect freshwater aquatic life: 28,900 μg/ℓ on an acute basis and 1,240 μg/ℓ on a chronic basis. To protect saltwater aquatic life: no value set due to insufficient data. To protect human health: preferably zero. An additional lifetime cancer risk of 1 in 100,000 results at a level of 1.9 μg/ℓ.

Determination in Water: Gas chromatography (EPA Method 601) or gas chromatography plus mass spectrometry (EPA Method 624).

Routes of Entry: Inhalation of vapors, ingestion, skin and eye contact.

Harmful Effects and Symptoms: *Local* — Chloroform may produce burns if left in contact with the skin.

Systemic — Chloroform is a relatively potent anesthetic at high concentrations. Death from its use as an anesthetic has resulted from liver damage and from cardiac arrest. Exposure may cause lassitude, digestive disturbance, dizziness, mental dullness, and coma. Chronic overexposure has been shown to cause enlargement of the liver and kidney damage. Alcoholics seem to be affected sooner and more severely from chloroform exposure. Disturbance of the liver is more characteristic of exposure than central nervous system depression or renal injury. There is animal experimental evidence that indicates chloroform is a carcinogen.

Chloroform was tested in three experiments in mice and in one in rats by oral administration. It produced hepatomas and hepatocellular carcinomas in mice, malignant kidney tumours in male rats and tumours of the thyroid in female rats. In another series of experiments in male mice chloroform administered orally produced benign and malignant kidney tumours (9).

Points of Attack: Liver, kidneys, heart, eyes, skin.

Medical Surveillance: Preplacement and periodic examinations should include appropriate tests for liver and kidney functions, and special attention should be given to the nervous system, the skin, and to any history of alcoholism. Expired air and blood levels may be useful in estimating levels of acute exposure.

First Aid: If this chemical gets into the eyes, irrigate immediately. If this chemical contacts the skin, wash with soap promptly. If a person breathes in large amounts of this chemical, move the exposed person to fresh air at once and perform artificial respiration. When this chemical has been swallowed, get medical attention. Give large quantities of salt water and induce vomiting. Do not make an unconscious person vomit.

Personal Protective Methods: Wear appropriate clothing to prevent any reasonable probability of skin contact. Wear eye protection to prevent any reasonable probability of eye contact. Employees should wash promptly when skin is wet or contaminated. Remove nonimpervious clothing promptly if wet or contaminated.

Respirator Selection:
500 ppm: SA/SCBA
1,000 ppm: SAF/SCBAF
Escape: GMOV/SCBA

Disposal Method Suggested: Incineration, preferably after mixing with another combustible fuel. Care must be exercised to assure complete combustion to prevent the formation of phosgene. An acid scrubber is necessary to remove the halo acids produced.

Where possible it should be recovered, purified by distillation, and returned to the supplier.

References

(1) National Institute for Occupational Safety and Health, *Criteria for a Recommended Standard: Occupational Exposure to Chloroform,* NIOSH Document No. 75-114, Washington, DC (1975).

(2) National Institute for Occupational Safety and Health, *Current Intelligence Bulletin No. 9—Chloroform,* Washington, DC (1976).

(3) U.S. Environmental Protection Agency, *Chloroform: Ambient Water Quality Criteria,* Washington, DC (1980).

(4) National Academy of Sciences, *Chloroform, Carbon Tetrachloride and Other Halomethanes: An Environmental Assessment,* Washington, DC (1978).

(5) U.S. Environmental Protection Agency, *Chloroform,* Health and Environmental Effects Profile No. 47, Office of Solid Waste, Washington, DC (April 30, 1980).

(6) International Agency for Research on Cancer, *IARC Monographs on the Carcinogenic Risks of Chemicals to Humans,* 20, 401, Lyon, France (1979).

(7) See Reference (A-61).

(8) Sax, N.I., Ed., *Dangerous Properties of Industrial Materials Report, 1,* No. 4, 44-47, New York, Van Nostrand Reinhold Co. (1981).

(9) See Reference (A-62). Also see Reference (A-64).

(10) Sax, N.I., Ed., *Dangerous Properties of Industrial Materials Report, 3,* No. 5, 101-106, New York, Van Nostrand Reinhold Co. (1983).

(11) Parmeggiani, L., Ed., *Encyclopedia of Occupational Health & Safety,* Third Edition, Vol. 1, pp 463–464, Geneva, International Labour Office (1983).
(12) United Nations Environment Programme, *IRPTC Legal File 1983,* Vol. I, pp VII/185–88, Geneva, Switzerland, International Register of Potentially Toxic Chemicals (1984).

CHLOROMETHYL METHYL ETHER

- Carcinogen (Suspected Human, IARC)(3)
- Hazardous waste (EPA)
- Priority toxic pollutant (EPA)

Description: $ClCH_2OCH_3$, chloromethyl methyl ether, is a volatile, corrosive liquid boiling at $59°C$. Commercial chloromethyl methyl ether contains from 1 to 7% bis(chloromethyl) ether, a known carcinogen.

Code Numbers: CAS 107-30-2   RTECS KN6650000   UN 1239

DOT Designation: Flammable liquid and poison.

Synonyms: CMME, methyl chloromethyl ether, monochloromethyl ether, chloromethoxymethane.

Potential Exposures: Chloromethyl methyl ether is a highly reactive methylating agent and is used in the chemical industry for synthesis of organic chemicals. Most industrial operations are carried out in closed process vessels so that exposure is minimized.

Permissible Exposure Limits in Air: Chloromethyl methyl ether is included in the Federal standard for carcinogens; all contact with it should be avoided. ACGIH (1983/84) has designated it an "Industrial Substance Suspect of Carcinogenic Potential for Man."

Determination in Air: Collection by impinger, analysis by gas chromatography with electron capture detector (A-10).

Permissible Concentration in Water: No criteria have been set for the protection of freshwater or saltwater aquatic life due to lack of data. For the protection of human health: preferably zero.

Routes of Entry: Inhalation of vapor and possibly percutaneous absorption.

Harmful Effects and Symptoms: *Local* — Vapor exposure results in severe irritation of the skin, eyes and nose. Rabbit skin tests using undiluted material resulted in skin necrosis.

Systemic — Chloromethyl methyl ether is only moderately toxic given orally. Acute exposure to chloromethyl methyl ether vapor may result in pulmonary edema and pneumonia.

Several studies of workers with CMME manufacturing exposure have shown an excess of bronchiogenic cancer predominately of the small cell-undifferentiated type with relatively short latency period (typically 10 to 15 years). Therefore, commercial grade chloromethyl methyl ether must be considered a carcinogen (4). It is not known whether or not chloromethyl methyl ether's carcinogenic activity is due to bis(chloromethyl) ether (BCME) contamination, but this may be a moot question inasmuch as two of the hydrolysis products of CMME can combine to form BCME.

Animal experiments to determine chloromethyl methyl ether's ability to produce skin cancer indicated marginal carcinogenic activity; highly pure CMME was used. Inhalation studies, using technical grade CMME showed only one bronchiogenic cancer and one esthesioneuroepithelioma out of 79 animals exposed.

Medical Surveillance: Preplacement and periodic medical examinations should include an examination of the skin and respiratory tract, including a chest x-ray. Sputum cytology has been suggested as helpful in detecting early malignant changes, and in this connection a detailed smoking history is of importance. Possible effects on the fetus should be considered.

Personal Protective Methods: These are designed to supplement engineering controls and to prevent all skin or respiratory contact. Full body protective clothing and gloves should be used on entering areas of partial exposure. Those employed in handling operations should be provided with fullface, supplied-air respirators of continuous-flow or pressure-demand type. On exit from a regulated area, employees should be required to remove and leave protective clothing and equipment at the point of exit, to be placed in impervious containers at the end of the workshift for decontamination or disposal. Showers should be taken prior to dressing in street clothes.

Disposal Method Suggested: Incineration, preferably after mixing with another combustible fuel. Care must be exercised to assure complete combustion to prevent the formation of phosgene. An acid scrubber is necessary to remove the halo acids produced.

References

(1) U.S. Environmental Protection Agency, *Chloroalkyl Ethers: Ambient Water Quality Criteria,* Washington, DC (1980).
(2) U.S. Environmental Protection Agency, *Chloroalkyl Ethers,* Health and Environmental Effects Profile No. 41, Office of Solid Waste, Washington, DC (April 30, 1980).
(3) International Agency for Research on Cancer, *IARC Monographs on the Carcinogenic Risks of Chemicals to Humans,* 4, 239, Lyon, France (1974).
(4) See Reference (A-62). Also see Reference (A-64).
(5) See Reference (A-60).
(6) United Nations Environment Programme, *IRPTC Legal File 1983,* Vol. I, pp VII/345–46, Geneva, Switzerland, International Register of Potentially Toxic Chemicals (1984).

2-CHLORONAPHTHALENE

See "Chlorinated Naphthalenes."

CHLORONITROBENZENES

See o-Nitrochlorobenzene and p-Nitrochlorobenzene entries in this volume.

1-CHLORO-1-NITROPROPANE

Description: $C_2H_5CHClNO_2$ is a colorless liquid with an unpleasant odor that causes tears (lachrymator). It boils at 171°C.

Code Numbers: CAS 600-25-9   RTECS TX5075000

DOT Designation: —

Synonym: Korax®.

Potential Exposure: This compound is used in the synthetic rubber industry and as a component in rubber cements.

Incompatibilities: Strong oxidizers.

Permissible Exposure Limits in Air: Federal limit is 20 ppm (100 mg/m^3). The TWA set by ACGIH as of 1983/84 is 2 ppm (10 mg/m^3). There is no STEL value set. The IDLH level is 2,000 ppm.

Determination in Air: Collection by charcoal tube, analysis by gas liquid chromatography (A-13).

Permissible Concentration in Water: No criteria set.

Routes of Entry: Inhalation, ingestion, skin and eye contact.

Harmful Effects and Symptoms: In animals: irritation of lungs and eyes; liver, kidney, heart, blood vessel damage.

Points of Attack: In animals: respiratory system, lungs, liver, kidneys, cardiovascular system.

Medical Surveillance: Consider the points of attack in preplacement and periodic physical examinations.

First Aid: If this chemical gets into the eyes, irrigate immediately. If this chemical contacts the skin, wash with soap. If a person breathes in large amounts of this chemical, move the exposed person to fresh air at once and perform artificial respiration. When this chemical has been swallowed, get medical attention. Give large quantities of water and induce vomiting. Do not make an unconscious person vomit.

Personal Protective Methods: Wear appropriate clothing to prevent repeated or prolonged skin contact. Wear eye protection to prevent any reasonable probability of eye contact. Employees should wash promptly when skin is wet or contaminated. Remove nonimpervious clothing promptly if wet or contaminated.

Respirator Selection:
```
1,000 ppm:  CCROVF/GMOV/SAF/SCBAF
2,000 ppm:  SAF:PD,PP,CF
  Escape:   GMOV/SCBA
```

Disposal Method Suggested: Incineration (1500°F, 0.5-second minimum for primary combustion; 2200°F, 1.0 second for secondary combustion) after mixing with other fuel. The formation of elemental chlorine may be prevented by injection of steam or using methane as a fuel in the process. Alternatively (A-38) it may be poured over soda ash, neutralized and flushed into the sewer with large volumes of water.

References

(1)   See Reference (A-61).

CHLOROPENTAFLUOROETHANE

Description: ClF_2CCF_3 is a colorless, odorless gas.

Code Numbers: CAS 76-15-3 RTECS KH7877500 UN 1020

DOT Designation: Nonflammable gas.

Synonyms: Fluocarbon 115, Propellant 115, Refrigerant 115, FC-115.

Potential Exposures: This material is used as a refrigerant, as a dielectric gas, and as a propellant in aerosol food preparations.

Permissible Exposure Limits in Air: There are no Federal standards, but ACGIH, as of 1983/84, has proposed a TWA of 1,000 ppm (6,320 mg/m^3), but no STEL value.

Permissible Concentration in Water: No criteria set.

Harmful Effects and Symptoms: There is little evidence of systemic injury or cardiac sensitization by this compound, although very high concentrations have produced some such phenomena in animals (A-34).

2-CHLOROPHENOL

- Hazardous waste (EPA)
- Priority toxic pollutant (EPA)

Description: HOC_6H_4Cl is a liquid boiling at 175° to 176°C. (It melts at 9°C).

Code Numbers: CAS 95-97-8 RTECS SK2625000 UN 2020

DOT Designation: Label should bear St. Andrews Cross (X).

Synonyms: o-Chlorophenol, 2-chloro-1-hydroxybenzene.

Potential Exposure: Industrial workers involved in the manufacture or handling of 2-chlorophenol constitute the primary group at risk. The generation of waste sources from the commercial production of 2-chlorophenol, its chemically derived products and the inadvertent synthesis of 2-chlorophenol due to chlorination of phenol in effluents and drinking water sources, may clearly indicate its importance in potential point source and nonpoint source water contamination.

Permissible Exposure Limits in Air: No limits set.

Permissible Concentration in Water: To protect freshwater aquatic life: 4,380 µg/ℓ on an acute toxicity basis. To protect saltwater aquatic life: no criteria set due to insufficient data. To protect human health: 0.1 µg/ℓ for the prevention of adverse effects due to organoleptic properties.

Determination in Water: Gas chromatography (EPA Method 604) or gas chromatography plus mass spectrometry (EPA Method 625).

Routes of Entry: Skin absorption, inhalation, ingestion.

Harmful Effects and Symptoms: Although 2-chlorophenol has been reported to be less toxic than the higher chlorophenols, its low odor threshold in water and its tainting properties are considered a potential threat to certain beneficial uses of water and the utilization of aquatic life as a food source.

First Aid: Wash contaminated areas of skin with soap and water immediately.

Personal Protective Methods: Wear butyl gloves, protective clothing and safety shoes (A-38). Remove contaminated clothing immediately.

Respirator Selection: Use self-contained breathing apparatus.

Disposal Method Suggested: Incinerate in admixture with flammable solvent in furnace equipped with afterburner and scrubber.

References

(1) U.S. Environmental Protection Agency, *2-Chlorophenol: Ambient Water Quality Criteria,* Washington, DC (1980).
(2) U.S. Environmental Protection Agency, *2-Chlorophenol,* Health and Environmental Effects Profile No. 50, Office of Solid Waste, Washington, DC (April 30, 1980).

CHLOROPHENOLS

See "Chlorinated Phenols."

CHLOROPICRIN

Description: CCl_3NO_2 is a colorless, oily liquid with a sharp, penetrating odor that causes tears. It boils at 112°C.

Code Numbers: CAS 76-06-2   RTECS PB6300000   UN 1580

DOT Designation: Poison B.

Synonyms: Nitrotrichloromethane, trichloronitromethane, nitrochloroform.

Potential Exposure: Chloropicrin is used in the manufacture of the dye-stuff methyl violet and in other organic syntheses. It has been used as a military poison gas. It is used as a fumigant insecticide (A-32).

Incompatibilities: Strong oxidizers.

Permissible Exposure Limits in Air: The Federal limit and the 1983/84 ACGIH TWA value is 0.1 ppm (0.7 mg/m^3). The STEL value is 0.3 ppm (2.0 mg/m^3). The IDLH level is 4.0 ppm.

Determination in Air: Collection by impinger or fritted bubbler, analysis colorimetrically (A-16).

Permissible Concentration in Water: No criteria set.

Routes of Entry: Inhalation, ingestion, eye and skin contact.

Harmful Effects and Symptoms: Eye irritation, lachrymation; coughing, pulmonary edema; nausea, vomiting; skin irritation.

Points of Attack: Respiratory system, lungs, skin, eyes.

Medical Surveillance: Consider the points of attack in preplacement and periodic physical examinations.

First Aid: If this chemical gets into the eyes, irrigate immediately. If this chemical contacts the skin, wash with soap immediately. If a person breathes in large amounts of this chemical, move the exposed person to fresh air at once and perform artificial respiration. When this chemical has been swallowed, get medical attention. Give large quantities of salt water and induce vomiting. Do not make an unconscious person vomit.

Personal Protective Methods: Wear appropriate clothing to prevent any possibility of skin contact. Wear eye protection to prevent any possibility of eye contact. Employees should wash immediately when skin is wet or contaminated. Remove nonimpervious clothing immediately if wet or contaminated. Provide emergency showers and eyewash.

Respirator Selection:
    4 ppm: CCROVF/GMOV/SAF/SCBAF
    Escape: GMOVF/SCBAF

Disposal Method Suggested: Incineration ($1500°F$, 0.5-second minimum for primary combustion; $2200°F$, 1.0 second for secondary combustion) after mixing with other fuel. The formation of elemental chlorine may be prevented by injection of steam or using methane as a fuel in the process.

Chloropicrin reacts readily with alcoholic sodium sulfite solutions to produce methanetrisulfonic acid (which is relatively nonvolatile and less harmful). This reaction has been recommended for treating spills and cleaning equipment. Although not specifically suggested as a decontamination procedure, the rapid reaction of chloropicrin with ammonia to produce guanidine (LD_{50} = 500 mg/kg) could be used for detoxification.

The Manufacturing Chemists Association has suggested two procedures for disposal of chloropicrin: (a) Pour or sift over soda ash. Mix and wash slowly into large tank. Neutralize and pass to sewer with excess water. (b) Absorb on vermiculite. Mix and shovel into paper boxes. Drop into incinerator with afterburner and scrubber.

References

(1) See Reference (A-61).
(2) Sax, N.I., Ed., *Dangerous Properties of Industrial Materials Report, 2,* No. 2, 17-19, New York, Van Nostrand Reinhold Co. (1982).
(3) Parmeggiani, L., Ed., *Encyclopedia of Occupational Health & Safety,* Third Edition, Vol. 1, pp 466–67, Geneva, International Labour Office (1983).
(4) United Nations Environment Programme, *IRPTC Legal File 1983,* Vol. II, pp VII/443–4, Geneva, Switzerland, International Register of Potentially Toxic Chemicals (1984).

CHLOROPRENE

- Carcinogen (NIOSH)(A-4)

Description: H_2C=CCl–CH=CH_2, chloroprene, is a colorless, flammable liquid possessing a pungent odor.

Code Numbers: CAS 126-99-8   RTECS EI9625000   UN 1991

DOT Designation: Flammable liquid.

Synonyms: 2-Chloro-1,3-butadiene, chlorobutadiene, β-chloroprene.

Potential Exposure: The only major use of chloroprene is in the production of artificial rubber (neoprene, duprene). Chloroprene is extremely reactive, e.g., it can polymerize spontaneously at room temperatures, the process being catalyzed by light, peroxides and other free radical initiators. It can also react with oxygen to form polymeric peroxides and because of its instability, flammability and toxicity, chloroprene has no end product uses as such. An estimated 2,500 workers are exposed to chloroprene in the United States.

Incompatibilities: Peroxides, other oxidizers.

Permissible Exposure Limits in Air: The Federal standard is 25 ppm (90 mg/m^3). NIOSH, in 1977, specified that the employer shall control exposure to chloroprene so that no employee is ever exposed at a concentration greater than 3.6 mg/m^3 of air (1 ppm), determined as a ceiling concentration for any 15-minute sampling period during a 40-hour workweek. ACGIH in 1983/84 has proposed TWA values of 10 ppm (36 mg/m^3), with the notation "skin" indicating the possibility of cutaneous absorption. No STEL values have been set. The IDLH level is 400 ppm.

Determination in Air: Charcoal adsorption, workup with CS$_2$ and analysis by gas chromatography. See NIOSH Methods, Set 1. See also reference (A-10).

Permissible Concentration in Water: No criteria set.

Routes of Entry: Inhalation of vapor and skin absorption, ingestion and skin and eye contact.

Harmful Effects and Symptoms: *Local* — Chloroprene acts as a primary irritant on contact with skin, conjunctiva, and mucous membranes and may result in dermatitis, conjunctivitis, and circumscribed necrosis of the cornea. Temporary hair loss has been reported during the manufacture of polymers.

Systemic — Inhalation of high concentrations may result in anesthesia and respiratory paralysis. Chronic exposure may produce damage to the lungs, nervous system, liver, kidneys, spleen, and myocardium. Chloroprene is designated by NIOSH (A-4) as a carcinogen.

Points of Attack: Respiratory system, skin, eyes and lungs.

Medical Surveillance: Preplacement and periodic examinations should include an evaluation of the skin, eyes, respiratory tract, and central nervous system. Liver and kidney function should be evaluated.

First Aid: If this chemical gets into the eyes, irrigate immediately. If this chemical contacts the skin, wash with soap immediately. If a person breathes in large amounts of this chemical, move the exposed person to fresh air at once and perform artificial respiration. When this chemical has been swallowed, get medical attention. Give large quantities of salt water and induce vomiting. Do not make an unconscious person vomit.

Personal Protective Methods: Wear appropriate clothing to prevent any possibility of skin contact. Wear eye protection to prevent any reasonable probability of eye contact. Employees should wash promptly when skin is wet or contaminated. Remove clothing immediately if wet or contaminated to avoid flammability hazard.

Respirator Selection:
    400 ppm: SAF/SCBAF
     Escape: GMOV/SCBA

Disposal Method Suggested: Incineration, preferably after mixing with another combustible fuel. Care must be exercised to assure complete combustion to prevent the formation of phosgene. An acid scrubber is necessary to remove the halo acids produced.

References

(1)   National Institute for Occupational Safety and Health, *Criteria for a Recommended Standard: Occupational Exposure to Chloroprene,* NIOSH Document No. 77-210 (1977).

(2)   Sax, N.I., Ed., *Dangerous Properties of Industrial Materials Report, 1,* No. 4, 47-49, New York, Van Nostrand Reinhold Co. (1981).

(3) See Reference (A-61).

(4) Parmeggiani, L., Ed., *Encyclopedia of Occupational Health and Safety,* Third Edition, Vol. 1, pp 467–68, Geneva, International Labour Office (1983).

(5) United Nations Environment Programme, *IRPTC Legal File 1983,* Vol. 1, pp. VII/134–35, Geneva, Switzerland, International Register of Potentially Toxic Chemicals (1984).

o-CHLOROSTYRENE

Description: $ClC_6H_4CH=CH_2$ is a liquid.

Code Numbers: CAS 1331-28-8   RTECS WL4150000

DOT Designation: —

Synonyms: None.

Potential Exposures: In organic synthesis; in the preparation of specialty polymers.

Permissible Exposure Limits in Air: There is no Federal standard, however, ACGIH as of 1983/84 has adopted a TWA of 50 ppm (285 mg/m^3) and set an STEL of 75 ppm (430 mg/m^3).

Permissible Concentration in Water: No criteria set.

Harmful Effects and Symptoms: Experiments involving rabbits, guinea pigs, rats and dogs indicated no clear-cut effects due to exposure but there appeared to be a slightly higher incidence of pathological changes in liver and kidneys (A-34).

CHLOROTHION

Description: $C_8H_9ClNO_5PS$; $C_6H_3(NO_2)(Cl)[OPS(OCH_3)_2]$ is a yellow oil which boils at 125°C.

Code Numbers: CAS 500-28-7   RTECS TE8050000

DOT Designation: —

Synonyms: O-(3-chloro-4-nitrophenyl)-O,O-dimethyl phosphorothioate; Compound Bay 22190.

Potential Exposure: Those engaged in the manufacture, formulation and application of this insecticide.

Permissible Exposure Limits in Air: No standards set.

Permissible Concentration in Water: No criteria set.

Routes of Entry: Skin absorption.

Harmful Effects and Symptoms: Quite toxic (fatal dose for man estimated at 850 mg/kg) (1).

Points of Attack: Skin.

Personal Protective Methods: Wear rubber gloves and protective clothing.

Respirator Selection: Self-contained breathing apparatus must be worn.

Disposal Method Suggested: Hydrolysis in alkali.

References

(1)   Sax, N.I., Ed., *Dangerous Properties of Industrial Materials Report, 2,* No. 2, 19-20, New York, Van Nostrand Reinhold Co. (1982).

o-CHLOROTOLUENE

Description: $CH_3C_6H_4Cl$ is a colorless liquid boiling at $159°C$.

Code Numbers: CAS 95-49-8   UN 2238

DOT Designation: Flammable liquid.

Synonyms: 2-Chloro-1-methylbenzene, 2-chlorotoluene.

Potential Exposures: o-Chlorotoluene is widely used as a solvent and intermediate in the synthesis of dyes, pharmaceuticals and other organic chemicals.

Permissible Exposure Limits in Air: There is no Federal standard, however as of 1983/84, ACGIH has adopted a TWA value of 50 ppm (250 mg/m³) with the notation "skin" indicating the possibility of cutaneous absorption. The STEL is 75 ppm (375 mg/m³).

Permissible Concentration in Water: No criteria set, but EPA (A-37) has suggested a permissible ambient goal of 3,450 µg/ℓ based on health effects.

Harmful Effects and Symptoms: Skin and eye irritation, loss of coordination, vasodilation, labored respiration and narcosis have been observed in test animals (A-34) but no adverse effects in human exposure have been reported.

CHLORPYRIFOS

● Hazardous substance (EPA)

Description: O,O-diethyl O-3,5,6-trichloro-2-pyridyl phosphorothioate, with the structural formula

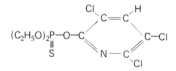

Chlorpyrifos is a colorless crystalline compound with a mild mercaptan odor, melting at 42° to 43°C.

Code Numbers: CAS 2921-88-2   RTECS TF6300000   UN (NA 2783)

DOT Designation: ORM-A.

Synonyms: Chlorpyriphos-ethyl (in France), chlorpyriphos (in Japan), ENT 27311, Dursban®, Lorsban®.

Potential Exposures: Those involved in the manufacture (A-32), formulation and application of this insecticide.

Permissible Exposure Limits in Air: There is no Federal standard, however, ACGIH has set a threshold limit value for chlorpyrifos in air of 0.2 mg/m^3 as of 1983/84. The short-term exposure limit is 0.6 mg/m^3. Both limits bear the notation "skin" indicating that cutaneous absorption should be prevented so the threshold limit value is not invalidated.

Permissible Concentration in Water: No criteria set.

Routes of Entry: Skin absorption, inhalation of dusts, ingestion.

Harmful Effects and Symptoms: This compound is a plasma cholinesterase inhibitor. However, it has only moderate capacity to reduce red blood cell cholinesterase or cause cholinergic symptoms and little capacity to cause systemic injury (A-34).

Disposal Method Suggested: This compound is 50% hydrolyzed in aqueous MeOH solution at pH 6 in 1,930 days, and in 7.2 days at pH 9.96. Spray mixtures of <1% concentration are destroyed with an excess of 5.25% sodium hypochlorite in < 30 minutes at 100°C, and in 24 hours at 30°C. Concentrated (61.5%) mixtures are essentially destroyed by treatment with 100:1 volumes of the above sodium hypochlorite solution and steam in 10 minutes (A-32).

References

(1) United Nations Environment Programme, *IRPTC Legal File 1983,* Vol. II, pp VII/607–8, Geneva, Switzerland, International Register of Potentially Toxic Chemicals (1984).

CHROMIUM AND COMPOUNDS

- Carcinogen (varies with compound, see text)
- Hazardous substance (some compounds)
    Chromium compounds designated by EPA as hazardous substances include: chromic acetate, chromic acid, chromic sulfate and chromous chloride.
- Hazardous waste constituents (EPA)
- Priority toxic pollutant (EPA)

Description: This group includes chromium trioxide (CrO$_3$), and its aqueous solutions (1). Chromium may exist in one of three valence states in compounds, +2, +3, and +6. Chromic acid, along with chromates, is in the hexavalent form. Chromium trioxide is produced from chromite ore by roasting with alkali or lime (calcium oxide), leaching, crystallization of the soluble chromate or dichromate followed by reaction with sulfuric acid.

Code Numbers:

Elemental Cr	CAS 7440-47-3	RTECS GB4200000	
Chromic acid	CAS 7738-94-5	RTECS GB2450000	UN 1463

DOT Designations: Chromic acid: oxidizer.

Synonyms: Chromium trioxide is also known as chromic anhydride, chromic acid and chromium(VI) oxide. Synonyms vary depending upon specific compound.

Potential Exposures: Chromium trioxide is used in chrome plating, copper stripping, aluminum anodizing, as a catalyst, refractories, in organic synthesis, and photography. An estimated 15,000 industrial workers are potentially exposed to chromic acid (1). NIOSH estimates that 175,000 workers are potentially exposed to chromium(VI) in various forms (2). Another NIOSH estimate in 1977 reported exposures to inorganic chromium compounds as follows (3):

	Exposures
Chromic oxide	212,730
Chromic chloride	7,710
Chromic fluoride	15,330
Chromium potassium sulfate	91,380
Chromium sulfate	7,920
Chromic phosphate	20,070
Chromic nitrate	9,960
Chromic sulfate, basic	7,920
Chromous chloride	7,530

Incompatibilities:

> Chromic acid and chromates: Combustible, organic, or other readily oxidizable materials such as paper, wood, sulfur, aluminum, and plastics, etc.
> Chromium metal and insoluble salts: Strong oxidizers.
> Soluble chromic and chromous salts: Water.

Permissible Exposure Limits in Air: The IDLH levels are as follows:

Chromic acid and chromates:	30 mg/m^3
Chromium metal and insoluble salts:	500 mg/m^3
Soluble chromic and chromous salts:	250 mg/m^3

Current TWA standards are:

	Federal Standard	ACGIH (1983/84)
	. . . . (mg/m^3)	
Chromium metal	1.0	0.5
Chromium(II) compounds	–	0.5
Chromium(III) compounds	0.5	0.5
Chromium(VI) compounds		
Water soluble	0.5	0.05
Water insoluble	1.0	0.05 (Carcinogens)
Chromyl chloride	–	0.15

The NIOSH Criteria for a Recommended Standard (1) set work-place limits for chromic acid of 0.05 mg/m^3 as chromium trioxide as a TWA with a ceiling concentration of 0.1 mg/m^3 as chromium trioxide determined by a sampling time of 15 minutes, as of 1973. More recently, in 1976, however (2), certain forms of chromium(VI) have been found to cause increased respiratory cancer mortality among workers.

A table of differentiation between noncarcinogenic and carcinogenic chromium(VI) compounds has been presented by NIOSH (2) as follows:

. . . . Noncarcinogens	
Evident	**Inferred**
Sodium bichromate	Lithium bichromate
Sodium chromate	Lithium chromate
Chromium(VI) oxide	Potassium bichromate
	Potassium chromate
	Rubidium bichromate
	Rubidium chromate
	Cesium bichromate
	Cesium chromate
	Ammonium bichromate
	Ammonium chromate

Evident	Inferred
Calcium chromate	Alkaline earth chromates and bichromates
Sintered calcium chromate	Chromyl chloride
Alkaline lime roasting process residue	tert-Butyl chromate
Zinc potassium chromate	Other chromium(VI) materials not listed
Lead chromate	in this table

NIOSH has not conducted an in-depth study of the toxicity of chromium metal or compounds containing chromium in an oxidation state other than chromium(VI). NIOSH recommends that the permissible exposure limit for carcinogenic chromium(VI) compounds be reduced to 0.001 mg/m^3 and that these compounds be regulated as occupational carcinogens. NIOSH also recommends that the permissible exposure limit for noncarcinogenic chromium(VI) be reduced to 0.025 mg/m^3 averaged over a workshift of up to 10 hours per day, 40 hours per week, with a ceiling level of 0.05 mg/m^3 averaged over a 15-minute period. It is recommended further that chromium(VI) in the workplace be considered carcinogenic, unless it has been demonstrated that only the noncarcinogenic chromium(VI) compounds mentioned above are present. The NIOSH Criteria Documents for Chromic Acid (1) and Chromium(VI)(2) should be consulted for more detailed information.

Determination in Air: For chromic acid and chromates: collection on a filter, followed by workup with H_2SO_4 and diphenylcarbazide, followed by colorimetric analysis; see NIOSH Methods, Set O. For chromium metal and both insoluble and soluble salts: collection on a filter followed by acid workup and analysis by atomic absorption; see NIOSH Methods, Set O. See also reference (A-10).

Permissible Concentration in Water: For the protection of freshwater aquatic life: Trivalent chromium: not to exceed e [1.08 ln (hardness) + 3.48] μg/ℓ. Hexavalent chromium: 0.29 μg/ℓ as a 24-hour average, never to exceed 21.0 μg/ℓ. For the protection of saltwater aquatic life: Trivalent chromium: 10,300 μg/ℓ on an acute toxicity basis. Hexavalent chromium: 18 μg/ℓ as a 24-hour average, never to exceed 1,260 μg/ℓ. To protect human health: Trivalent chromium: 170 μg/ℓ. hexavalent chromium 50 μg/ℓ. The 50 μg/ℓ limit is also established (A-65) as a drinking water standard in South Africa and in Germany.

Determination in Water: Total chromium may be determined by digestion followed by atomic absorption or by colorimetry (diphenylcarbazide) or by inductively coupled plasma (ICP) optical emission spectrometry. Chromium(VI) may be determined by extraction and atomic absorption or colorimetry (using diphenylhydrazide). Dissolved total Cr or Cr(VI) may be determined by 0.45 μ filtration followed by the above-cited methods.

Routes of Entry: Inhalation, ingestion, and eye and skin contact.

Harmful Effects and Symptoms: *Local* — In some workers, chromium compounds act as allergens which cause dermatitis to exposed skin. They may also produce pulmonary sensitization. Chromic acid has a direct corrosive effect on the skin and the mucous membranes of the upper respiratory tract; and, although rare, the possibility of skin and pulmonary sensitization should be considered.

Systemic — Chromium compounds in the +3 state are of a low order of toxicity. In the +6 state, chromium compounds are irritants and corrosive, and can enter the body by ingestion, inhalation, and through the skin. Typical industrial hazards are: inhalation of the dust and fumes released during the manufacture of dichromate from chromite ore; inhalation of chromic acid mist during the electroplating and surface treatment of metals; and skin contact in various manufacturing processes.

Acute exposures to dust or mist may cause coughing and wheezing, headache, dyspnea, pain on deep inspiration, fever, and loss of weight. Tracheobronchial irritation and edema persist after other symptoms subside. In electroplating operations, workers may experience a variety of symptoms including lacrimation, inflammation of the conjunctiva, nasal itch and soreness, epistaxis, ulceration and perforation of the nasal septum, congested nasal mucosa and turbinates, chronic asthmatic bronchitis, dermatitis and ulceration of the skin, inflammation of laryngeal mucosa, cutaneous discoloration, and dental erosion. Hepatic injury has been reported from exposure to chromic acid used in plating baths, but appears to be rare. Working in the chromate-producing industry increases the risk of lung cancer.

Calcium chromate is carcinogenic in rats after administration by several routes, including intrabronchial implantation. Chromium chromate, strontium chromate, and zinc chromate produce local sarcomas in rats at the sites of application. The evidence for the carcinogenicity in mice and rats of barium chromate, lead chromate, chromic acetate, sodium dichromate and chromium carbonyl is inadequate.

There is an increased incidence of lung cancer among workers in the chromate-producing industry and possibly also among chromium platers and chromium alloy workers (10). There is also a suggestion of increased incidence of cancers at other sites. The chromium compound(s) responsible has not been specified precisely (17).

Points of Attack:

Chromic acid and chromates	blood, lungs, respiratory system, liver, kidneys, eyes, skin
Chromium metal and insoluble salts	respiratory system and lungs
Soluble chromic and chromous salts	skin

Medical Surveillance: Preemployment physical examinations should include: a work history to determine past exposure to chromic acid and hexavalent chromium compounds, exposure to other carcinogens, smoking history, history of skin or pulmonary sensitization to chromium, history or presence of dermatitis, skin ulcers, or lesions of the nasal mucosa and/or perforation of the septum, and a chest x-ray. On periodic examinations an evaluation should be made of skin and respiratory complaints, especially in workers who demonstrate allergic reactions. Chest x-rays should be taken yearly for workers over age 40, and every five years for younger workers. Blood, liver, and kidney function should be evaluated periodically.

First Aid: If this chemical gets into the eyes, irrigate immediately. If this chemical contacts the skin, flush with water and wash with soap promptly. If a person breathes in large amounts of this chemical, move the exposed person to fresh air at once and perform artificial respiration. When this chemical has been swallowed, get medical attention. Give large quantities of water and induce vomiting. Do not make an unconscious person vomit.

Personal Protective Methods: Full body protective clothing should be worn in areas of chromic acid exposure, and impervious gloves, aprons, and footwear should be worn in areas where spills or splashes may contact the skin. Where chromic acid may contact the eyes by spills or splashes, impervious protective goggles or face shield should be worn. All clothing should be changed at the end of the shift and showering encouraged prior to change to street clothes. Clean clothes should be reissued at the start of the shift.

Respirators should be used in areas where dust, fumes, or mist exposure exceeds Federal standards or where brief concentrations exceed the TWA, and for emergencies. Dust, fumes and mist filter-type respirators or supplied air respirators should be supplied all workers exposed, depending on concentration of exposure.

Respirator Selection:

Chromic acid and chromates
     50 mg/m³: HiEPF/SAF/SCBAF
     30 mg/m³: PAPHiEPF/SAF:PD,PP,CF
     Escape: HiEP/SCBA
Chromium metal and insoluble salts
     5 mg/m³: DM
     10 mg/m³: DMXSQ/FuHiEP/SA/SCBA
     50 mg/m³: HiEPF/SAF/SCBAF
     500 mg/m³: PAPHiE/SA:PD,PP,CF
Soluble chromic and chromous salts
     2.5 mg/m³: DMXS
     5 mg/m³: DMXSQ/FuHiEP/SA/SCBA
     25 mg/m³: HiEPF/SAF/SCBAF
     250 mg/m³: PAPHiEP/SA:PD,PP,CF

Disposal Method Suggested: For chromic acid (liquids, chromium trioxide): Chemical reduction of concentrated materials to chromium(III) and precipitation by pH adjustment. Precipitates are normally disposed in a chemical waste landfill. For chromic fluoride and chromic sulfate: Alkaline precipitation of the heavy metal gel followed by effluent neutralization and discharge into a sanitary sewer system. The heavy metal may be disposed in a chemical waste landfill.

Recovery and recycle is a viable alternative to disposal (A-57) for chromium in plating wastes, tannery wastes, cooling tower blowdown water and chemical plant wastes.

References

(1) National Institute for Occupational Safety and Health, *Criteria for a Recommended Standard: Occupational Exposure to Chromic Acid,* NIOSH Publication No. 73-11021 (1973).
(2) National Institute for Occupational Safety and Health, *Criteria for a Recommended Standard: Occupational Exposure to Chromium(VI),* NIOSH Document No. 76-129 (1976).
(3) National Institute for Occupational Safety and Health, *Information Profiles on Potential Occupational Hazards: Inorganic Chromium Compounds,* Report PB-276,678, pp 136-142, Rockville, MD (October 1977).
(4) U.S. Environmental Protection Agency, *Chromium: Ambient Water Quality Criteria,* Washington, DC (1980).
(5) U.S. Environmental Protection Agency, *Toxicology of Metals, Vol. II: Chromium,* Report EPA-600/1-77-022, pp 164-187, Research Triangle Park, NC (May 1977).
(6) National Academy of Sciences, *Medical and Biological Effects of Environmental Pollutants: Chromium,* Washington, DC (1974).
(7) U.S. Environmental Protection Agency, *Reviews of the Environmental Effects of Pollutants: III, Chromium,* Report EPA-600/1-78-023, Cincinnati, OH (1978).
(8) U.S. Environmental Protection Agency, *Chromium,* Health and Environmental Effects Profile No. 51, Office of Solid Waste, Washington, DC (April 30, 1980).
(9) See Reference (A-61).
(10) Sax, N.I., Ed., *Dangerous Properties of Industrial Materials Report, 1,* No. 1, 40-41, New York, Van Nostrand Reinhold Co. (1980). (Chromium).

(11) Sax, N.I., Ed., *Dangerous Properties of Industrial Materials Report, 1,* No. 3, 43-45, New York, Van Nostrand Reinhold Co. (1981). (Chromic Acetate).
(12) Sax, N.I., Ed., *Dangerous Properties of Industrial Materials Report, 1,* No. 7, 47-49, New York, Van Nostrand Reinhold Co. (1981). (Chromic Oxide).
(13) Sax, N.I., Ed., *Dangerous Properties of Industrial Materials Report, 2,* No. 2, 21-22, New York, Van Nostrand Reinhold Co. (1982). (Chromic Acid).
(14) Sax, N.I., Ed., *Dangerous Properties of Industrial Materials Report, 3,* No. 3, 59-62, New York, Van Nostrand Reinhold Co. (1983). (Chromic Acid).
(15) Sax, N.I., Ed., *Dangerous Properties of Industrial Materials Report, 3,* No. 3, 62-65, New York, Van Nostrand Reinhold Co. (1983). (Chromic Sulfate).
(16) Sax, N.I., Ed., *Dangerous Properties of Industrial Materials Report, 3,* No. 3, 65-68, New York, Van Nostrand Reinhold Co. (1983). (Chromium).
(17) See Reference (A-62). Also see Reference (A-64).
(18) Sax, N.I., Ed., *Dangerous Properties of Industrial Materials Report, 3,* No. 6, 64-67, New York, Van Nostrand Reinhold Co. (Nov./Dec. 1983) (Sodium Dichromate).
(19) Parmeggiani, L., Ed., *Encyclopedia of Occupational Health & Safety,* Third Edition, Vol. 1, pp 468-73, Geneva, International Labour Office (1983).
(20) United Nations Environment Programme, *IRPTC Legal File 1983,* Vol. I, pp VII/189-94, Geneva, Switzerland, International Register of Potentially Toxic Chemicals (1984).

CHRYSENE

● Carcinogen (Industrial Substance Suspect of Carcinogenic Potential for Man) (A-6).

See "Polynuclear Aromatics."

CINNAMYL ANTHRANILATE

● Carcinogen (Animal Suspected, IARC) (2)

Description: $C_6H_5CH{=}CHCH_2OCOC_6H_4NH_2$ is a brownish powder which melts above 60°C.

Code Numbers: CAS 87-29-6   RTECS CB2725000

DOT Designation: —

Synonyms: Cinnamyl-2-aminobenzoate.

Potential Exposure: This is a flavor and fragrance chemical widely used in foods, beverages, soaps, detergents, perfumes.

Permissible Exposure Limits in Air: No standards set.

Permissible Concentration in Water: No criteria set.

Harmful Effects and Symptoms: Open to debate. The Council of Europe lists it as safe at levels of 25 mg/kg, but there is evidence that it causes lung cancer. More data are needed. Animal carcinogenicity is suspected (A-63).

References
(1) Sax, N.I., Ed., *Dangerous Properties of Industrial Materials Report, 1,* No. 5, 47, New York, Van Nostrand Reinhold Co. (1981).
(2) International Agency for Research on Cancer, *IARC Monographs on the Carcinogenic Risks of Chemicals to Humans, 16,* 27, Lyon, France (1978).

CITRUS RED 2

- Carcinogen (Animal Positive, IARC)
- Hazardous Waste Constituent (EPA)

Description: $(CH_3O)_2C_6H_3-N=N-C_{10}H_6OH$ with the structural formula

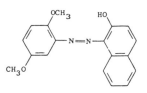

is a solid melting at 156°C.

Code Numbers: CAS 6538-53-8   RTECS QL 3675000

DOT Designation: −

Synonyms: C.I. Solvent Red 80; C.I. 12156; 1-(2,5-dimethoxyphenylazo)-2-naphthol.

Potential Exposure: To those encountering this material as a colorant in foods or cosmetics.

Permissible Exposure Limits in Air: No standards set.

Permissible Concentration in Water: No criteria set.

Routes of Entry: Ingestion in foods.

Harmful Effects and Symptoms: Has produced lung carcinomas, lymphomas and bladder carcinomas in animals.

References
(1) Sax, N.I., Ed., *Dangerous Properties of Industrial Materials Report, 1,* No. 3, 46-47, New York, Van Nostrand Reinhold Co. (1981).
(2) International Agency for Research on Cancer, *IARC Monographs on the Carcinogenic Risks of Chemicals to Humans, 8,* 101, Lyon, France (1975).

CLOPIDOL

Description: 3,5-Dichloro-2,6-dimethyl-4-pyridinol with the structural formula:

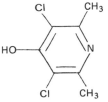

Clopidol is a crystalline solid melting above 320°C.

Code Numbers: CAS 2971-90-6

DOT Designation: −

Synonyms: Meticlorpindol, clopindol, Coyden®.

Potential Exposures: Those engaged in formulation, application or manufacture [see U.S. Patent 3,206,358 (1965)] of this veterinary coccidiostat.

Permissible Exposure Limits in Air: There is no Federal standard, however ACGIH has adopted a TWA value of 10 mg/m^3 and set an STEL of 20 mg/m^3 as of 1983/84.

Permissible Concentration in Water: No criteria set.

Harmful Effects and Symptoms: Clopidol has a low order of toxicity. Rats fed 15 mg/kg/day for 2 years showed no ill effects (A-34).

COAL TAR PITCH VOLATILES

● Carcinogen (Human, ACGIH) (A-6)

Description: The term "coal tar products," as used by NIOSH (1), includes coal tar and two of the fractionation products of coal tar, creosote and coal tar pitch, derived from the carbonization of bituminous coal. Coal tar, coal tar pitch, and creosote derived from bituminous coal often contain identifiable components which by themselves are carcinogenic, such as benzo[a]pyrene, benzanthracene, chrysene, and phenanthrene. Other chemicals from coal tar products such as anthracene, carbazole, fluoranthene, and pyrene may also cause cancer, but these causal relationships have not been adequately documented.

Code Numbers:

Coal tar	CAS 8007-45-2	RTECS GF8600000	UN 1999
Coal tar pitch	CAS - None	RTECS GF8655000	
Coal tar distillate	CAS - None	RTECS GF8617500	UN 1136

DOT Designation: Coal tar distillate: Combustible liquid.

Synonyms: None.

Potential Exposures: The coke-oven plant is the principal source of coal tar. The hot gases and vapors produced during the conversion of coal to coke are collected by means of a scrubber, which condenses the effluent into ammonia, water, crude tar, and other by-products. Crude tar is separated from the remainder of the condensate for refining and may undergo further processing.

Employees may be exposed to pitch and creosote in metal and foundry operations, when installing electrical equipment, and in construction, railway, utility, and briquette manufacturing. A list of primary employment in which the various types of pitch and creosote are encountered is as follows:

Product	User Industry	Percent of Tar Processed	Number of Jobs Affected
Electrode binder pitch	aluminum	43.2	28,000
	steel	3.0	50,000
	graphite	9.2	10,000
Core pitch	foundry	2.2	2,000
Refractory pitch	steel	2.4	50,000
Fiber pitch	electrical	3.5	—
Miscellaneous pitch	various	3.4	—
Roofing pitch	construction	8.8	—
Other tars and fuel residue	fuel	24.3	—
Creosote	railway, utility, construction	—	5,000

Incompatibilities: Strong oxidizers.

Permissible Exposure Limits in Air: In 1967, the American Conference of Governmental Industrial Hygienists (ACGIH) adopted a threshold limit value (TLV) of 0.2 mg/m^3 for coal tar pitch volatiles (CTPV), described as a "benzene-soluble" fraction, and listed certain carcinogenic components of CTPV. The TLV was established to minimize exposure to the listed substances believed to be carcinogens, viz, anthracene, benzo[a]pyrene, phenanthrene, acridine, crysene, and pyrene. This TLV was promulgated as a Federal standard under the Occupational Safety and Health Act of 1970 (29 CFR 1910.1000). No foreign standards were found for exposure to coal tar pitch or creosote.

In 1973, NIOSH published the *Criteria for a Recommended Standard—Occupational Exposure to Coke Oven Emissions,* recommending work practices to minimize the harmful effects of exposure to coke-oven emissions and inhalation of coal tar pitch volatiles. In 1974, OSHA established a Standards Advisory Committee on Coke-Oven Emissions to study the problem of the exposure of coke-oven workers to CTPV and to prepare recommendations for an effective standard in the assigned area. In 1975, the Committee recommended a limit of 0.2 μg/m^3 for benzo[a]pyrene (*Federal Register,* 41:46742-46787, October 22, 1976).

According to a NIOSH publication (1), occupational exposure to coal tar products shall be controlled so that employees are not exposed to coal tar, coal tar pitch, creosote or mixtures of these substances at a concentration greater than 0.1 mg/m^3 of the cyclohexane-extractable fraction of the sample, determined as a time-weighted average (TWA) concentration for up to a 10-hour workshift in a 40-hour workweek.

The OSHA standard for coke oven emissions, which was promulgated in 1977, established a permissible exposure limit of 150 μg/m^3 averaged over any 8-hour period. Under this standard, specific engineering and work practice control requirements became effective in January 1980.

Determination in Air: Collection on a filter, extraction, column chromatography, spectrophotometric measurement (A-10). Benzene solubles may be determined by collection of particulates on a filter, ultrasonic extraction with benzene, evaporation and gravimetric determination (A-10).

Permissible Concentrations in Water: See "Polynuclear Aromatic Hydrocarbons."

Determination in Water: See "Polynuclear Aromatic Hydrocarbons."

Routes of Entry: Inhalation and skin and eye contact.

Harmful Effects and Symptoms: Based on a review of the toxicologic and epidemiologic evidence presented, it has been concluded (1) that some materials contained in coal tar pitch, and, therefore, in coal tar, can cause lung and skin cancer, and perhaps cancer at other sites. Based on a review of experimental toxicologic evidence, it is also concluded that creosote can cause skin and lung cancer. While the evidence on creosote is not so strong as that on pitch (in part because of difficulties in chemical characterization of such mixtures), the conclusion on the carcinogenic potential of creosote is supported by information on the presence of polynuclear aromatic hydrocarbons and imputations and evidence of the carcinogenicity of such hydrocarbons.

The overwhelming scientific evidence in the record supports the finding that coke oven emissions are carcinogenic. This finding rests on epidemiological surveys as well as animal studies and chemical analyses of coke oven emissions.

Coke oven workers have an increased risk of developing cancer of the lung and urinary tract. In addition, observations of animals and of human populations have shown that skin tumors can be induced by the products of coal combustion and distillation. Chemical analyses of coke oven emissions reveal the presence of a large number of scientifically recognized carcinogens as well as several agents known to enhance the effect of chemical carcinogens especially on the respiratory tract (4).

Points of Attack: Respiratory system, lungs, bladder, kidneys, skin.

Medical Surveillance: Medical surveillance shall be made available, as specified below, to all employees occupationally exposed to coal tar products. See also reference (2).

Preplacement Medical Examinations — These examinations shall include:

- Comprehensive initial medical and work histories, with special emphasis directed toward identifying preexisting disorders of the skin, respiratory tract, liver, and kidneys.
- A physical examination giving particular attention to the oral cavity, skin, and respiratory system. This shall include posteroanterior and lateral chest x-rays (35 x 42 cm). Pulmonary function tests, including forced vital capacity (FVC) and forced expiratory volume at 1 second (FEV 1.0), and a sputum cytology examination shall be offered as part of the medical examination of exposed employees. Other tests, such as liver function and urinalysis, should be performed as considered appropriate by the responsible physician. In addition, the mucous membranes of the oral cavity should be examined.
- A judgement of the employee's ability to use positive pressure respirators.

Periodic Examinations — These examinations shall be made available at least annually and shall include:

- Interim medical and work histories.
- A physical examination as outlined above.

Initial Medical Examinations — These examinations shall be made available to all workers as soon as practicable after the promulgation of a standard based on these recommendations.

Pertinent Medical Records — These records shall be maintained for at least 30 years after termination of employment. They shall be made available to medical representatives of the government, the employer or the employee.

First Aid: If this chemical gets into the eyes, irrigate immediately. If this chemical contacts the skin, wash with soap. If a person breathes in large amounts of this chemical, move the exposed person to fresh air at once and perform artificial respiration.

Personal Protective Methods: Employers shall use engineering controls when needed to keep the concentration of airborne coal tar products at, or below, the specified limit. Employers shall provide protective clothing and equipment impervious to coal tar products to employees whenever liquid coal tar products may contact the skin or eyes. Emergency equipment shall be located at well-marked and clearly identified stations and shall be adequate to permit all personnel to escape from the area or to cope safely with the emergency on reentry. Protective equipment shall include: eye and face protection; protective clothing; and respiratory protection as spelled out in detail by NIOSH (1).

Respirator Selection:

2 mg/m^3:   CCROVFuHiE/SA/SCBA
10 mg/m^3:   CCROVFFuHiE/GMOVFFuHiE/SAF/SCBAF
200 mg/m^3:   SA:PD,PP,CF/PAPCCROVHiEP
400 mg/m^3:   SAF:PD,PP,CF
Escape: GMOVPest/SCBA

Disposal Method Suggested: Incineration.

References

(1) National Institute for Occupational Safety and Health, *Criteria for a Recommended Standard: Occupational Exposure to Coal Tar Products,* NIOSH Document No. 78-107, Washington, DC (September 1977).

(2) National Institute for Occupational Safety and Health, *A Guide to the Work Relatedness of Disease,* Revised Edition, pp 80-87, DHEW (NIOSH) Publication No. 79-116, Cincinnati, OH (January 1979).

(3) See Reference (A-61).

(4) See Reference (A-62). Also see Reference (A-64).

(5) Parmeggiani, L., Ed., *Encyclopedia of Occupational Health & Safety,* Third Edition, Vol. 2, pp 2147–49, Geneva, International Labour Office (1983) (Tar & Pitch).

COBALT AND COMPOUNDS

- Hazardous substances (EPA) (some compounds)
  Cobalt compounds designated as hazardous substances by EPA include: cobaltous bromide, cobaltous formate and cobaltous sulfamate.

Description: Co, cobalt, is a silver-grey, hard, brittle, magnetic metal. It is relatively rare; the important mineral sources are the arsenides, sulfides, and oxidized forms. It is generally obtained as a by-product of other metals, particularly copper. Cobalt is insoluble in water, but soluble in acids. Cobalt fume and dust have the composition $Co/CoO/Co_2O_2/Co_2O_4$.

Code Numbers:

Cobalt metal	CAS 7440-48-4	RTECS GF8750000
Cobaltous oxide	CAS 1307-96-6	RTECS GG2800000
Cobaltic oxide	CAS 1308-04-9	RTECS GG2900000

DOT Designation: —

Synonyms: None.

Potential Exposures: Nickel-aluminum-cobalt alloys are used for permanent magnets. Alloys with nickel, aluminum, copper, beryllium, chromium, and molybdenum are used in the electrical, automobile, and aircraft industries. Cobalt is added to tool steels to improve their cutting qualities and is used as a binder in the manufacture of tungsten carbide tools.

Various cobalt compounds are used as pigments in enamels, glazes, and paints, as catalysts in afterburners, and in the glass, pottery, photographic, electroplating industries. Radioactive cobalt (^{60}Co) is used in the treatment of cancer.

Cobalt has been added to beer to promote formation of foam but cobalt acts with alcohol to produce severe cardiac effects at concentrations as low as 1.2 to 1.5 mg/ℓ of beer.

Cobalt is part of the vitamin B_{12} molecule and as such is an essential nutrient.

The requirement of humans for cobalt in the form of vitamin B_{12} is about 0.13 μg/day.

Incompatibilities: Strong oxidizers.

Permissible Exposure Limits in Air: The Federal standard for cobalt, metal fume and dust, is 0.1 mg/m^3. ACGIH has proposed a TWA value of 0.05 mg/m^3 and an STEL value of 0.1 mg/m^3 as of 1983/84.

Determination in Air: Cobalt metal, dust and fume may be determined by filter collection, acid dissolution, digestion and measurement by atomic absorption spectrophotometry (A-10).

Permissible Concentration in Water: The U.S.S.R. has set a limit of 1.0 mg/ℓ of water, according to NAS/NRC (A-2). The Interim Primary Drinking Water Regulations do not limit cobalt, nor has the WHO recommended a limit on its International or European Standards.

Because the maximum no-adverse-health-effect concentration is more than an order of magnitude greater than that found in any natural-water or drinking-water supply, there appears to be no reason at present to regulate the concentration of cobalt in drinking water (A-2).

The EPA (A-37) has suggested a permissible ambient goal of 0.7 μg/ℓ based on health effects.

Determination in Water: Atomic absorption spectroscopy gives a detection limit of 0.05 mg/ℓ in water. Neutron activation can detect cobalt in urine below 0.5 μg/ℓ (1).

Routes of Entry: Inhalation of dust or fume, ingestion, skin or eye contact.

Harmful Effects and Symptoms: *Local* — Cobalt dust is mildly irritating to the eyes and to a lesser extent to the skin. It is an allergen and has caused allergic sensitivity type dermatitis in some industries where only minute quantities of cobalt are used. The eruptions appear in the flexure creases of the elbow, knee, ankle, and neck. Cross sensitization occurs between cobalt and nickel, and to chromium when cobalt and chromium are combined.

Systemic — Inhalation of cobalt dust may cause an asthmalike disease with cough and dyspnea. This situation may progress to interstitial pneumonia with marked fibrosis. Pneumoconiosis may develop which is believed to be reversible. Since cobalt dust is usually combined with other dusts, the role cobalt plays in causing the pneumoconiosis is not entirely clear.

Ingestion of cobalt or cobalt compounds is rare in industry. Vomiting, diarrhea, and a sensation of hotness may occur after ingestion or after the inhalation of excessive amounts of cobalt dust. Cardiomyopathy has also been reported, but the role of cobalt remains unclear in this situation.

Points of Attack: Respiratory system, skin and lungs.

Medical Surveillance: In preemployment examinations, special attention should be given to a history of skin diseases, allergic dermatitis, baseline allergic respiratory diseases, and smoking history. A baseline chest x-ray should be taken. Periodic examinations should be directed toward skin and respiratory symptoms and lung function.

First Aid: If this chemical gets into the eyes, irrigate immediately. If this chemical contacts the skin, wash with soap If a person breathes in large amounts of this chemical, move the exposed person to fresh air at once and perform artificial respiration. When this chemical has been swallowed, get medical attention. Give large quantities of water and induce vomiting. Do not make an unconscious person vomit.

Personal Protective Methods: Where dust levels are excessive, dust respirators should be used by all workers. Protective clothing should be issued to all workers and changed on a daily basis. Showering after each shift is encouraged prior to change to street clothes. Gloves and barrier creams may be helpful in preventing dermatitis.

Respirator Selection: For cobalt metal dust and fume:

    0.5 mg/m^3: DMXS
    1.0 mg/m^3: DMXSQ/FuHiEP
      5 mg/m^3: HiEPF/SAF/SCBAF
   20 mg/m^3: PAPHiEF/SAF:PD,PP,CF
    Escape: HiEPF/SCBA

Disposal Method Suggested: For cobalt chloride: Chemical reaction with water, caustic soda, and slaked lime, resulting in precipitation of the metal sludge, which may be landfilled. Cobalt metal may be recovered from scrap and cobalt compounds from spent catalysts (A-57) as alternatives to disposal.

References

(1) U.S. Environmental Protection Agency, *Toxicology of Metals, Vol. II: Cobalt,* pp 188-205, Report EPA-600/1-77-022, Research Triangle Park, NC (May 1977).

(2) See Reference (A-61). (Cobalt Metal Fume and Dust).

(3) Sax, N.I., Ed., *Dangerous Properties of Industrial Materials Report, 1,* No. 3, 47-48, New York, Van Nostrand Reinhold Co. (1981). (Cobalt).

(4) Sax, N.I., Ed., *Dangerous Properties of Industrial Materials Report, 2,* No. 5, 26-35, New York, Van Nostrand Reinhold Co. (1982). (Cobalt 60, Cobaltous Nitrate, Cobaltous Chloride).

(5) See Reference (A-60) for entries on Cobalt (II) Acetate, Cobaltous Chloride, Cobaltous Nitrate and Cobalt Sulfate.

(6) Sax, N.I., Ed., *Dangerous Properties of Industrial Materials Report, 4,* No. 1, 49-51, New York, Van Nostrand Reinhold Co. (Jan./Feb. 1984) (Cobaltous Formate).

(7) Sax, N.I., Ed., *Dangerous Properties of Industrial Materials Report, 4,* No. 1, 51-53, New York, Van Nostrand Reinhold Co. (Jan./Feb. 1984) (Cobaltous Sulfamate).

(8) Parmeggiani, L., Ed., *Encyclopedia of Occupational Health & Safety,* Third Edition, Vol. 1, pp 493-95, Geneva, International Labour Office (1983).

COBALT CARBONYL

Description: $Co_2(CO)_4$ is a solid which melts at $50°C$ with decomposition.

Code Numbers: RTECS GF 9625000

DOT Designation: —

Permissible Exposure Limits in Air: ACGIH (1983/84) has adopted 0.1 mg/m^3 as a TWA but has set no STEL value.

Permissible Concentration in Water: No criteria set.

Harmful Effects and Symptoms: This compound is highly toxic due to the release of carbon monoxide.

References

(1) American Conference of Governmental Industrial Hygienists, Inc., *Documentation of the Threshold Limit Values: Supplemental Documentation: 1981,* Cincinnati, OH (1981).

COBALT HYDROCARBONYL

Description: $HCO(CO)_4$ is a flammable and toxic gas which decomposes rapidly at room temperature.

Code Numbers: CAS 16842-03-8   RTECS GG0900000

DOT Designation: —

Potential Exposure: Those involved in manufacture and use of this material as a catalyst.

Permissible Exposure Limits in Air: ACGIH (1983/84) has adopted 0.1 mg/m^3 as a TWA value but has set no STEL value.

Permissible Concentration in Water: No criteria set.

Harmful Effects and Symptoms: The 30-minute LD-50 in rats is 165 mg/m^3. The clinical effects are similar to nickel carbonyl and iron pentacarbonyl but it has about one-half the toxicity of nickel carbonyl.

References
(1) American Council of Governmental Industrial Hygienists, Inc., *Documentation of the Threshold Limit Values: Supplemental Documentation: 1981,* Cincinnati, OH (1981).

COKE OVEN EMISSIONS

See "Coal Tar Pitch Volatiles."

COPPER AND COMPOUNDS

● Hazardous substances (EPA) (some compounds)
  Copper compounds listed by EPA as hazardous substances include: Cupric acetate, cupric acetoarsenite, cupric chloride, cupric nitrate, cupric oxalate, cupric sulfate, cupric sulfate ammoniated, cupric tartrate.
● Hazardous waste (copper cyanide) (EPA)
● Priority toxic pollutant (EPA)

Description: Cu, copper, is a reddish-brown metal which occurs free or in ores such as malachite, cuprite, and chalcopyrite. It may form both mono- and divalent compounds. Copper is insoluble in water, but soluble in nitric acid and hot sulfuric acid.

Copper dusts and mists have been assigned the formula $CuSO_4 \cdot 5H_2O/CuCl$ by NIOSH. Copper fume has been designated as $Cu/CuO/Cu_2O$ by NIOSH.

Code Numbers: Copper metal: CAS 7440-50-8   RTECS GL5325000

DOT Designation: —

Synonyms: None.

Potential Exposures: Metallic copper is an excellent conductor of electricity and is widely used in the electrical industry in all gauges of wire for circuitry, coil, and armature windings, high conductivity tubes, commutator bars, etc.

It is made into castings, sheets, rods, tubing, and wire, and is used in water and gas piping, roofing materials, cooking utensils, chemical and pharmaceutical equipment, and coinage. Copper forms many important alloys: Be-Cu alloy, brass, bronze, gun metal, bell metal, German silver, aluminum bronze, silicon bronze, phosphor bronze, and manganese bronze.

Copper compounds are used as insecticides, algicides, molluscicides, plant fungicides, mordants, pigments, catalysts, and as a copper supplement for pastures, and in the manufacture of powdered bronze paint and percussion caps. They are also utilized in analytical reagents, in paints for ships' bottoms, in electroplating, and in the solvent for cellulose in rayon manufacture.

Incompatibilities: Acetylene gas; magnesium metal is incompatible with copper dusts and mists.

Permissible Exposure Limits in Air: The Federal standard for copper fume is 0.1 mg/m^3, and for copper dusts and mists, 1 mg/m^3. ACGIH, as of 1983/84, has set a TWA of 0.2 mg/m^3 for copper fume. There are no STEL values or IDLH levels set.

Determination in Air: Copper dusts and mists are collected on a filter, worked up with acid, measured by atomic absorption. See NIOSH Methods, Set M. See also reference (A-10). For copper fume: filter collection, acid digestion, measurement by atomic absorption (A-10).

Permissible Concentration in Water: To protect freshwater aquatic life: 5.6 μg/ℓ as a 24-hour average, never to exceed e $^{[0.94\ \ln\ (hardness)\ -\ 1.23]}$ μg/ℓ. To protect saltwater aquatic life: 4.0 μg/ℓ as a 24-hour average, not to exceed 23.0 μg/ℓ. To protect human health: 1,000 μg/ℓ.

Determination in Water: Total copper may be determined by digestion followed by atomic absorption or by colorimetry (using neocuproine) or by Inductively Coupled Plasma (ICP) Optical Emission Spectrometry. Dissolved copper may be determined by 0.45 μ filtration followed by the preceding methods.

Routes of Entry: Inhalation of dust or fume, ingestion, or skin or eye contact.

Harmful Effects and Symptoms: *Local* — Copper salts act as irritants to the intact skin causing itching, erythema, and dermatitis. In the eyes, copper salts may cause conjunctivitis and even ulceration and turbidity of the cornea. Metallic copper may cause keratinization of the hands and soles of the feet, but it is not commonly associated with industrial dermatitis.

Systemic — Industrial exposure to copper occurs chiefly from fumes generated in welding copper-containing metals. The fumes and dust cause irritation of the upper respiratory tract, metallic taste in the mouth, nausea, metal fume fever, and in some instances discoloration of the skin and hair. Inhalation of dusts, fumes, and mists of copper salts may cause congestion of the nasal mucous membranes, sometimes of the pharynx, and on occasions ulceration with perforation of the nasal septum. If the salts reach the gastrointestinal tract, they act as irritants producing salivation, nausea, vomiting, gastric pain, hemorrhagic gastritis, and diarrhea. It is unlikely that poisoning by ingestion in industry would progress to a serious point as small amounts induce vomiting, emptying the stomach of copper salts.

Chronic human intoxication occurs rarely and then only in individuals with Wilson's disease (hepatolenticular degeneration). This is a genetic condition caused by the pairing of abnormal autosomal recessive genes in which there is abnormally high absorption, retention, and storage of copper by the body. The disease is progressive and fatal if untreated.

Points of Attack: For copper dusts and mists: respiratory system, lungs, skin, liver, including risk with Wilson's disease, kidneys. For copper fume: respiratory system, skin, eyes, and risk with Wilson's disease.

Medical Surveillance: Consider the skin, eyes, and respiratory system in any placement or periodic examinations.

First Aid: For copper dusts and mists: If this chemical gets into the eyes, irrigate immediately. If this chemical contacts the skin, wash with soap promptly. If a person breathes in large amounts of this chemical, move the exposed person to fresh air at once and perform artificial respiration. When this chemical has been swallowed, get medical attention. Give large quantities of water and induce vomiting. Do not make an unconscious person vomit.

For copper fume: If a person breathes in large amounts of this chemical, move the exposed person to fresh air at once and perform artificial respiration.

Personal Protective Methods: For copper dusts or mists: Wear appropriate clothing to prevent repeated or prolonged skin contact. Wear eye protection to prevent any reasonable probability of eye contact. Employees should wash promptly when skin is contaminated. Work clothing should be changed daily if clothing is contaminated. Remove nonimpervious clothing promptly if contaminated.

Respirator Selection:
Copper dusts and mists
      50 mg/m³: HiEPF/SAF/SCBAF
   2,000 mg/m³: SAF:PD,PP,CF
Copper fume
       1 mg/m³: FuHiEP/SA/SCBA
       5 mg/m³: HiEPF/SAF/SCBAF
   100 mg/m³: PAPHiE/SA:PD,PP,CF
   200 mg/m³: SAF:PD,PP,CF

Disposal Methods Suggested: Copper-containing wastes can be concentrated through the use of ion exchange, reverse osmosis, or evaporators to the point where copper can be electrolytically removed and sent to a reclaiming firm. Details of copper recovery from a variety of industrial wastes have been published (A-57). If recovery is not feasible, the copper can be precipitated through the use of caustics and the sludge deposited in a chemical waste landfill.

References

(1) U.S. Environmental Protection Agency, *Toxicology of Metals, Vol. II: Copper,* Report EPA-600/1-77-022, pp 206-221, Research Triangle Park, NC (May 1977).

(2) U.S. Environmental Protection Agency, *Copper: Ambient Water Quality Criteria,* Washington, DC (1980).

(3) National Academy of Sciences, *Medical and Biologic Effects of Environmental Pollutants: Copper,* Washington, DC (1977).

(4) See Reference (A-61). (Copper Dusts and Mists and Fume).

(5) Sax, N.I., Ed., *Dangerous Properties of Industrial Materials Report, 1,* No. 5, 48-49, New York, Van Nostrand Reinhold Co. (1981). (Copper).

(6) Sax, N.I., Ed., *Dangerous Properties of Industrial Materials Report, 1,* No. 8, 58-60, New York, Van Nostrand Reinhold Co. (1981). (Copper Chloride).

(7) Sax, N.I., Ed., *Dangerous Properties of Industrial Materials Report, 2,* No. 5, 35-39, New York, Van Nostrand Reinhold Co. (1982). (Copper Nitrate).

(8) Sax, N.I., Ed., *Dangerous Properties of Industrial Materials Report, 3,* No. 1, 45-47, New York, Van Nostrand Reinhold Co. (1983). (Copper Naphthenate).

(9)   See Reference (A-60) for entries on Copper Acetate, Copper Chloride, Copper Sulfate, Cupric Carbonate, Cupric Hydroxide, Cupric Nitrate and Cuprous Chloride.

(10) Parmeggiani, L., Ed., *Encyclopedia of Occupational Health & Safety,* Third Edition, Vol. 1, pp 546–48, Geneva, International Labour Office (1983).

(11) United Nations Environment Programme, *IRPTC Legal File 1983,* Vol. I, pp VII/195–99, Geneva, Switzerland, International Register of Potentially Toxic Chemicals (1984).

COTTON DUST

Description: $(C_6H_{10}O_5)_n$, "Cotton dust," is defined as dust generated into the atmosphere as a result of the processing of cotton fibers combined with any naturally occurring materials such as stems, leaves, bracts, and inorganic matter which may have accumulated on the cotton fibers during the growing or harvesting period. Any dust generated from processing of cotton through the weaving of fabric in textile mills and dust generated in other operations or manufacturing processes using new or waste cotton fibers or cotton fiber by-products from textile mills is considered cotton dust.

Code Numbers: CAS: None   RTECS GN2275000

DOT Designation: ORM-C.

Synonyms: None.

Potential Exposures: The Occupational Safety and Health Administration has estimated that 800,000 workers are involved in work with cotton fibers and thus are potentially exposed to cotton dust in the work place.

Incompatibilities: Strong oxidizers.

Permissible Exposure Limit in Air: The current Federal standard for cotton dust was promulgated by OSHA June 23, 1978 and is 200 $\mu g/m^3$ (mean concentration) in textile yarn manufacturing; 750 $\mu g/m^3$ (mean concentration) in textile slashing and weaving operations; and 500 $\mu g/m^3$ (mean concentration) in all other operations except that this standard does not apply to the harvesting of cotton, cotton ginning, the handling and processing of woven or knitted materials, or washed cotton. These concentrations are for lint-free samples averaged over an eight-hour period collected by the vertical elutriator or equivalent method (A-6). This standard is currently in litigation in the courts. The STEL is 0.6 mg/m^3 and the IDLH level is 500 mg/m^3.

Determination in Air: Collection on a filter and gravimetric analysis (A-1).

Permissible Concentration in Water: No criteria set.

Routes of Entry: Inhalation of dust, ingestion, eye and skin contact.

Harmful Effects and Symptoms: The processing of raw cotton is associated with byssinosis (or "brown lung"), a disease characterized by pulmonary dysfunction. There is evidence, but not proof, that cotton dust itself is not the cause of byssinosis but may carry materials into the respiratory tract, possibly foreign protein or proteolytic enzymes.

Points of Attack: Respiratory system, lungs, cardiovascular system.

Medical Surveillance: [See also reference (2).]

(a) Preplacement: A comprehensive physical examination shall be made available to include as a minimum: medical history, baseline forced vital capacity (FVC), and forced expiratory volume at 1 second (FEV_1). The history shall include administration of a questionnaire designed to elicit information regarding symptoms of chronic bronchitis, byssinosis, and dyspnea. If a positive personal history of respiratory allergy, chronic obstructive lung disease, or other diseases of the cardiopulmonary system are elicited, or where there is a positive history of smoking, the applicant shall be counseled on his increased risk from occupational exposure to cotton dust. At the time of this examination, the advisability of the workers using negative or positive pressure respirators shall be evaluated.

(b) Each newly employed person shall be retested for ventilatory capacity (FVC and FEV_1) within 6 weeks of employment. This retest shall be performed on the first day at work after at least 40 hours' absence from exposure to cotton dust and shall be performed before and after at least 6 hours of exposure on the same day.

(c) Each current employee exposed to cotton dust shall be offered a medical examination at least yearly that shall include administration of a questionnaire designed to elicit information regarding symptoms of chronic bronchitis, byssinosis, and dyspnea.

(d) Each current employee exposed to cotton dust shall have measurement of forced vital capacity (FVC) and of forced expiratory volume at 1 sec (FEV_1). These tests of ventilatory function should be performed on the first day of work following at least 40 hours of absence from exposure to cotton dust, and shall be performed before and after at least 6 hours of exposure on the same day.

(e) Ideally, the judgment of the employee's pulmonary function should be based on preplacement values (values taken before exposure to cotton dust). When preplacement values are not available, then reference to standard pulmonary function value tables may be necessary. Note that these tables may not reflect normal values for different ethnic groups. For example, the average healthy black male may have an approximately 15% lower FVC than a Caucasian male of the same body build. A physician shall consider, in cases of significantly decreased pulmonary function, the impact of further exposure to cotton dust and evaluate the relative merits of a transfer to areas of less exposure or protective measures. A suggested plan for the management of cotton workers was proposed as a result of a conference on cotton workers' health.

(f) Medical records, including information on all required medical examinations, shall be maintained for persons employed in work involving exposure to cotton dust. Medical records with pertinent supporting documents shall be maintained at least 20 years after the individual's termination of employment. These records shall be available to the medical representatives of the Secretary of Health, Education, and Welfare; of the Secretary of Labor; of the employee or former employee; and of the employer.

First Aid: If this chemical gets into the eyes, irrigate immediately. If a person breathes in large amounts of this chemical, move the exposed person to fresh air at once and perform artificial respiration.

Personal Protective Methods: Engineering control shall be used wherever feasible to maintain cotton dust concentrations below the prescribed limit. Administrative controls can also be used to reduce exposure.

Respirators shall also be provided and used for nonroutine operations (occasional brief exposures above the environmental limit and for emergencies) and shall be considered for use by employees who have symptoms even when exposed to concentrations below the established environmental limit.

Respirator Selection:
     5 mg/m³: D
     10 mg/m³: DXSQ/SA/SCBA
     50 mg/m³: HiEPF/SAF/SCBAF
     500 mg/m³: PAPHiEP/SAF:PD,PP,CF

References

(1) National Institute for Occupational Safety and Health, *Criteria for a Recommended Standard: Occupational Exposure to Cotton Dust,* NIOSH Document No. 75-118 (1975).

(2) National Institute for Occupational Safety and Health, *A Guide to the Work-Relatedness of Disease,* Revised Edition, DHEW (NIOSH) Publication No. 79-116, pp 88-98, Cincinnati, OH (January 1979).

CREOSOTE

- Carcinogen (EPA-CAG) (A-40)
- Hazardous waste (EPA)

Description: Creosote is a flammable, heavy, oily liquid with a characteristic sharp, smoky smell, and caustic burning taste. In pure form it is colorless, but the industrial product is usually brownish. It is produced by the destructive distillation of wood or coal tar at temperatures above 200°C. The chemical composition is determined by the source and may contain guaiacol, creosols, phenol, cresols, pyridine, and numerous other aromatic compounds.

Code Numbers: CAS 8001-58-9 RTECS GF8615000 UN (NA 1993)

DOT Designation: Combustible liquid.

Synonyms: Creosotum, creosote oil, brick oil, coal tar creosote.

Potential Exposures: Creosote is used primarily as a wood preservative, and those working with the treated wood may be exposed. It is also used as a waterproofing agent, an animal dip, a constituent in fuel oil, a lubricant for die molds, as pitch for roofing, and in the manufacture of chemicals and lampblack. In the pharmaceutical industry, it is used as an antiseptic, disinfectant, antipyretic, astringent, styptic, germicide, and expectorant.

Incompatibilities: Strong oxidizers.

Permissible Exposure Limits in Air: There is no Federal standard. NIOSH recommends a TWA of 0.1 mg/m³.

Permissible Concentration in Water: No criteria set.

Route of Entry: Skin absorption.

Harmful Effects and Symptoms: *Local* — The liquid and vapors are strong irritants producing local erythema, burning, itching, pigmentation (greyish-yellow to bronze), vesiculation, ulceration, and gangrene. Eye injuries include keratitis, conjunctivitis, and permanent corneal scars. Contact dermatitis is reported in industry. Photosensitization has been reported. Skin cancer may occur.

Systemic — Symptoms of systemic illness include salivation, vomiting, vertigo, headache, loss of pupillary reflexes, hypothermia, cyanosis, convulsions, thready pulse, respiratory difficulties, and death.

Medical Surveillance: Consider the skin, eyes, respiratory tract, and central nervous system in placement and periodic examination.

Personal Protective Methods: Protective clothing should be worn where employees are exposed to the liquid or high vapor concentration. Masks with fullface protection and organic vapor canisters should be worn. Gloves and goggles are advisable in any area where spill or splash might occur.

Disposal Method Suggested: Incineration.

References

(1) National Institute for Occupational Safety and Health, *Criteria for a Recommended Standard: Occupational Exposure to Coal Tar Products,* NIOSH Document No. 78-107 (1978).

(2) U.S. Environmental Protection Agency, *Creosote,* Health and Environmental Effects Profile No. 53, Office of Solid Waste, Washington, DC (April 30, 1980).

CRESIDINE

● Carcinogen (Animal Positive) (NCI) (1).

Description: $C_6H_3(NH_2)(OCH_3)(CH_3)$ is a white crystalline solid melting at 51.5°C.

Code Numbers: CAS 120-71-8   RTECS BZ6825000

DOT Designation: —

Synonyms: 5-methyl-o-anisidine; 3-amino-p-cresol methyl ether; 2-methoxy-5-methyl aniline; 4-methoxy-m-toluidine.

Potential Exposure: p-Cresidine appears to be used solely as an intermediate in the production of various azo dyes and pigments, including 11 dyes that are produced commercially in the United States.

The average annual production is not known, but one manufacturer has reported production during the period 1971 to 1976 to be about 590,000 lb. The EPA (1977) reported 14 manufacturers and/or importers.

Human exposure to p-cresidine occurs primarily through inhalation of vapors or skin absorption of the liquid. Exposure to p-cresidine is believed to be limited to workers in dye-production facilities. The Consumer Product Safety Commission staff believes it is possible that residual levels or trace impurities of p-cresidine may be present in some dyes based on this chemical, and it may be present in the final consumer product. The presence of p-cresidine, even as a trace contaminant, may be cause for concern. However, data describing the actual levels of impurities in the final product and the potential for consumer exposure and uptake are currently lacking (2).

Permissible Exposure Limits in Air: No standards set.

Permissible Concentration in Water: No criteria set.

Harmful Effects and Symptoms: When administered in the diet, p-cresidine was carcinogenic to rats, causing increased incidences of carcinomas and papillomas of the urinary bladder in both sexes, increased incidences of olfactory neuroblastomas in both sexes, and of liver tumors in males. p-Cresidine was also carcinogenic in mice, causing carcinomas of the urinary bladders in both sexes and hepatocellular carcinomas in females (1).

References

(1) National Cancer Institute, *Bioassay of p-Cresidine for Possible Carcinogenicity*, DHHS Publication No. (NIH) 78-1394, National Technical Information Service, Springfield, VA (1979).
(2) See Reference (A-62). Also see Reference (A-64).
(3) See Reference (A-63). (As 5-Methyl-o-Anisidine).

CRESOLS

- Hazardous substance (EPA)
- Hazardous waste (EPA)

Description: $CH_3C_6H_4OH$, cresol, is a mixture of the three isomeric cresols, o-, m-, and p-cresol. It is a colorless, yellowish, brownish-yellow, or pinkish liquid with a phenolic odor. Cresols are soluble in alcohol, glycol, and dilute alkalis. Also they may be combustible.

Code Numbers:

Cresol	CAS 1319-77-3	RTECS GO5950000	UN 2076
o-Cresol	CAS 95-48-7	RTECS GO6300000	UN 2076
m-Cresol	CAS 108-39-4	RTECS GO6125000	UN 2076
p-Cresol	CAS 106-44-5	RTECS GO6475000	UN 2076

DOT Designation: Corrosive material.

Synonyms: Cresylic acid, hydroxytoluene, methyl phenol, oxytoluene.

Potential Exposures: Cresol is used as a disinfectant, as an ore flotation agent, and as an intermediate in the manufacture of chemicals, dyes, plastics, and antioxidants. A mixture of isomers is generally used; the concentrations of the components are determined by the source of the cresol.

Incompatibilities: Strong oxidizers.

Permissible Exposure Limits in Air: Federal standard is 5 ppm (22 mg/m^3). There is no STEL value. ACGIH adds the notation "skin" indicating the possibility of cutaneous absorption as of 1983/84.

Determination in Air: Silica adsorption, workup with acetone and analysis by gas chromatography. See NIOSH Methods, Set L. See also reference (A-10).

Permissible Concentration in Water: No criteria set, but EPA (A-37) has suggested a permissible ambient concentration, based on health effects, of 304 $\mu g/\ell$.

Routes of Entry: Inhalation or percutaneous absorption of liquid or vapor, ingestion, eye and skin contact.

Harmful Effects and Symptoms: *Local* — Cresol is very corrosive to all tissues. It may cause burns if it is not removed promptly and completely and in case of extensive exposure, if it is not removed completely from contaminated areas of the body very quickly, death may result. When it contacts the skin, it may not produce any sensation immediately. After a few moments, prickling and intensive burning occur. This is followed by loss of feeling. The affected skin shows wrinkling, white discoloration, and softening. Later gangrene may occur. If the chemical contacts the eyes, it may cause extensive damage and blindness. A skin rash may result from repeated or prolonged exposure of the skin to low concentrations of cresol. Discoloration of the skin may also occur from this type of exposure.

Systemic — When cresol is absorbed into the body either through the lungs, through the skin, mucous membranes, or by swallowing, it may cause systemic poisoning. The signs and symptoms of systemic poisoning may develop in 20 or 30 minutes. These toxic effects include weakness of the muscles, headache, dizziness, dimness of vision, ringing of the ears, rapid breathing, mental confusion, loss of consciousness, and sometimes death.

Prolonged or repeated absorption of low concentrations of cresol through the skin, mucous membranes, or respiratory tract may cause chronic systemic poisoning. Symptoms and signs of chronic poisoning include vomiting, difficulty in swallowing, salivation, diarrhea, loss of appetite, headache, fainting, dizziness, mental disturbances, and skin rash. Death may result if there has been severe damage to the liver and kidneys.

Points of Attack: Central nervous system, respiratory system, liver, kidneys, skin, eyes.

Medical Surveillance: Consider the skin, eyes, respiratory system, and liver and kidney functions in placement or periodic examinations.

First Aid: If this chemical gets into the eyes, irrigate immediately. If this chemical contacts the skin, wash with soap immediately. If a person breathes in large amounts of this chemical, move the exposed person to fresh air at once and perform artificial respiration. When this chemical has been swallowed, get medical attention. Give large quantities of water and induce vomiting. Do not make an unconscious person vomit.

Personal Protective Methods: Wear appropriate clothing to prevent any possibility of skin contact. Wear eye protection to prevent any possibility of eye contact. Employees should wash immediately when skin is wet or contaminated. Work clothing should be changed daily if it is possible that clothing is contaminated. Remove nonimpervious clothing immediately if wet or contaminated. Provide emergency showers and eyewash.

Respirator Selection:
      50 ppm: CCROVDM/SA/SCBA
    250 ppm: CCROVFDM/GMOVDM/SAF/SCBAF
    Escape: GMOVP/SCBA

Disposal Method Suggested: Incineration.

References

(1) National Institute for Occupational Safety and Health, *Criteria for a Recommended Standard: Occupational Exposure to Cresol,* NIOSH Document No. 78-133, Washington, DC (1978).

(2) U.S. Environmental Protection Agency, *A Study of Industrial Data on Candidate Chemicals for Testing (Alkyl Phthalates and Cresols),* Report EPA-560/5-78-002, Washington, DC (June 1978).

(3) U.S. Environmental Protection Agency, *Cresols and Cresylic Acid,* Health and Environmental Effects Profile No. 54, Office of Solid Waste, Washington, DC (April 30, (1980).

(4) See Reference (A-61).

(5) Parmeggiani, L., Ed., *Encyclopedia of Occupational Health & Safety,* Third Edition, Vol. 1, pp 569–70, Geneva, International Labour Office (1983).

(6) United Nations Environment Programme, *IRPTC Legal File 1983,* Vol. I, pp VII/207–15, Geneva, Switzerland, International Register of Potentially Toxic Chemicals (1984).

CROTONALDEHYDE

- Hazardous substance (EPA)
- Hazardous waste (EPA)

Description: $CH_3CH=CHCHO$ is a colorless-to-straw-colored liquid with an irritating, pungent, suffocating odor.

Code Numbers: CAS 123-73-9   RTECS GP9625000   UN 1143

DOT Designation: Flammable liquid and poison.

Synonyms: β-Methylacrolein, propylene aldehyde, crotonic aldehyde, 2-butenal.

Potential Exposure: Crotonaldehyde is used as an intermediate in the manufacture of n-butanol and crotonic and sorbic acids and in resin and rubber antioxidant manufacture; it is also used as a solvent in mineral oil purification, as a warning agent in fuel gas and as an alcohol denaturant (1).

Incompatibilities: Caustics, ammonia, organic amines, mineral acids, strong oxidizers.

Permissible Exposure Limits in Air: The Federal limit and 1983/84 ACGIH TWA value is 2.0 ppm (6 mg/m³). The STEL value is 6.0 ppm (18 mg/m³). The IDLH level is 400 mg/m³.

Determination in Air: Collection by charcoal tube, analysis by gas liquid chromatography (A-13).

Permissible Concentration in Water: No criteria set.

Routes of Entry: Inhalation, ingestion, eye and skin contact.

Harmful Effects and Symptoms: Irritation of eyes, skin and respiratory system.

Points of Attack: Respiratory system, eyes, skin.

Medical Surveillance: Consider the skin, eyes and respiratory system in placement or periodic examinations.

First Aid: If this chemical gets into the eyes, irrigate immediately. If this chemical contacts the skin, flush with water immediately. If a person breathes in large amounts of this chemical, move the exposed person to fresh air at once and perform artificial respiration. When this chemical has been swallowed, get medical attention. Give large quantities of water and induce vomiting. Do not make an unconscious person vomit.

Personal Protective Methods: Wear appropriate clothing to prevent any reasonable probability of skin contact. Wear eye protection to prevent any possibility of eye contact. Employees should wash immediately when skin is wet or contaminated. Remove clothing immediately if wet or contaminated to avoid flammability hazard. Provide emergency showers and eyewash.

Respirator Selection:
```
100 ppm:  CCROVF/GMOV/SAF/SCBAF
400 ppm:  SAF:PD,PP,CF
  Escape: GMOV/SCBA
```

Disposal Method Suggested: Incineration.

References

(1) U.S. Environmental Protection Agency, *Crotonaldehyde,* Health and Environmental Effects Profile No. 55, Office of Solid Waste, Washington, DC (April 30, 1980).
(2) See Reference (A-61).
(3) See Reference (A-60).
(4) Sax, N.I., Ed., *Dangerous Properties of Industrial Materials Report, 4,* No. 1, 56–59, New York, Van Nostrand Reinhold Co. (Jan./Feb. 1984).

CROTOXYPHOS

Description: $C_{14}H_{19}O_6P$ with the following structural formula

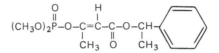

is a straw-colored liquid which boils at 135°C under 0.03 mm Hg pressure.

Code Numbers: CAS 7700-17-6   RTECS GQ5075000

DOT Designation: —

Synonyms: 1-Phenylethyl(E)-[(dimethoxyphosphinyl)oxy] -2-butenoate; Ciodrin®; Ciovap®.

Potential Exposure: Those involved in the manufacture, formulation and application of this insecticide.

Incompatibilities: Dichlorvos and mineral carriers, water, alkalies.

Permissible Exposure Limits in Air: No standards set.

Permissible Concentration in Water: No criteria set.

Harmful Effects and Symptoms: Acute effects include general weakness, headache, tightness in chest, blurred vision, nonreactive pupils, salivation, sweating, nausea, vomiting, diarrhea, abdominal cramps and/or convulsions.

References

(1) Sax, N.I., Ed., *Dangerous Properties of Industrial Materials Report, 2,* No. 5, 39-41, New York, Van Nostrand Reinhold Co. (1982).

CRUFOMATE

Description: 2-Chloro-4-(1,1-dimethylethyl)phenyl methyl methylphosphoramidate, $C_{12}H_{19}ClNO_3P$, with the structural formula:

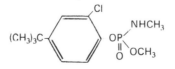

is a colorless crystalline compound melting at 60°C.

Code Numbers: CAS 299-86-5   RTECS TB3850000   UN 2765

DOT Designation: —

Synonyms: ENT 25602-X, Ruelene®.

Potential Exposures: Those involved in the manufacture (A-32), formulation and application of this insecticide and anthelmintic for cattle.

Permissible Exposure Limits in Air: There is no Federal standard, however, as of 1983/84, ACGIH has adopted a TWA of 5 mg/m^3 and set an STEL of 20 mg/m^3.

Permissible Concentration in Water: No criteria set.

Routes of Entry: Skin absorption, inhalation of dust, ingestion.

Harmful Effects and Symptoms: This compound is an active plasma and erythrocyte cholinesterase inhibitor but has only moderate capacity to cause cholinergic symptoms. Human test subjects ingesting 200 mg/day (3 mg/kg/day) did not exhibit any cholinesterase depression or other symptoms (A-34).

Disposal Method Suggested: Crufomate decomposes above pH 7.0 in alkaline media (A-32).

CUMENE

- Hazardous waste (EPA)

Description: $C_6H_5CH(CH_3)_2$ is a colorless liquid with a sharp, penetrating odor.

Code Numbers: CAS 98-82-8   RTECS GR8575000   UN 1918

DOT Designation: Flammable liquid.

Synonyms: Isopropyl benzene, 2-phenylpropane.

Potential Exposure: Cumene is used as a high octane gasoline component; it is used as a thinner for paints and lacquers; it is an important intermediate in phenol manufacture.

Incompatibilities: Oxidizers.

Permissible Exposure Limits in Air: The Federal limit is 50 ppm (245 mg/m^3). ACGIH has set an STEL of 75 ppm (365 mg/m^3) as of 1983/84. ACGIH adds the notation "skin" indicating the possibility of cutaneous absorption. The IDLH level is 8,000 ppm.

Determination in Air: Charcoal adsorption, workup with CS_2, analysis by gas chromatography. See NIOSH Methods, Set C. See also reference (A-10).

Permissible Concentration in Water: No criteria set.

Routes of Entry: Inhalation, ingestion, skin and eye contact.

Harmful Effects and Symptoms: Irritation of eyes and mucous membranes, headaches, dermatitis, narcosis, coma.

Points of Attack: Eyes, upper respiratory system, skin, central nervous system.

Medical Surveillance: Consider the points of attack in preplacement and periodic physical examinations.

First Aid: If this chemical gets into the eyes, irrigate immediately. If this chemical contacts the skin, flush with water promptly. If a person breathes in large amounts of this chemical, move the exposed person to fresh air at once and perform artificial respiration. When this chemical has been swallowed, get medical attention. Do NOT induce vomiting.

Personal Protective Methods: Wear appropriate clothing to prevent repeated or prolonged skin contact. Wear eye protection to prevent any reasonable probability of eye contact. Employees should wash promptly when skin is wet or contaminated. Remove nonimpervious clothing promptly if contaminated or wet.

Respirator Selection:

500 ppm:	CCROV/SA/SCBA
1,000 ppm:	CCROVF
2,500 ppm:	GMOV/SAF/SCBAF
8,000 ppm:	SAF:PD,PP,CF
Escape:	GMOV/SCBA

Disposal Method Suggested: Incineration.

References

(1)   See Reference (A-61).
(2)   Sax, N.I., Ed., *Dangerous Properties of Industrial Materials Report, 4,* No. 1, 59–62, New York, Van Nostrand Reinhold Co. (Jan./Feb. 1984).
(3)   Parmeggiani, L., Ed., *Encyclopedia of Occupational Health & Safety,* Third Edition, Vol. 1, pp 572–73, Geneva, International Labour Office (1983).

CUPFERRON

● Carcinogen (Animal Positive) (A-64)

Description: $C_6H_9N_3O_2$, also $C_6H_5N(NO)ONH_4$, is a crystalline compound melting at 163° to 164°C.

Code Numbers: CAS 135-20-6   RTECS NC4725000

DOT Designation: —

Synonyms: Ammonium N-Nitrosophenylhydroxylamine.

Potential Exposure: Cupferron is used to separate tin from zinc, and copper and iron from other metals in the laboratory. Cupferron also finds application as a quantitative reagent for vanadates and titanium and for the colorimetric determination of aluminum.

EPA has indicated that two companies currently produce cupferron in two regions. Domestic production has been estimated as 37,000 lb.

The potential for exposure appears to be greatest for those engaged in analytical or research studies involving use of the chemical. Workers may also be exposed to the compound during manufacturing processes.

The National Occupational Hazard Survey in 1974 estimated that 4,000 workers were potentially exposed to cupferron.

Permissible Exposure Limits in Air: No standards set.

Permissible Concentration in Water: No criteria set.

Routes of Entry: Human exposure to cupferron occurs mainly through ingestion or inhalation of the dust from the dry salt. Skin absorption is a secondary route of exposure.

Harmful Effects and Symptoms: Cupferron, given in the diet, was carcinogenic to Fischer 344 rats, causing hemangiosarcomas, hepatocellular carcinomas, and squamous-cell carcinomas of the forestomach in males and females, as well as carcinomas of the auditory sebaceous gland in females. The chemical was also carcinogenic to B6C3F1 mice, causing hemangiosarcomas in males; and hepatocellular carcinomas, carcinomas of the auditory sebaceous gland, a combination of hemangiosarcomas and hemangiomas, and adenomas of the Harderian gland in females (1).

References

(1) National Cancer Institute, *Bioassay of Cupferron for Possible Carcinogenicity,* Technical Report Series No. 100, DHEW Publication No. (NIH) 78-1350, Bethesda, MD (1978).

CUTTING FLUIDS

Description: Cutting fluids are liquids applied to a metal cutting tool to assist in the machining operation by washing away metal chips or serving as a coolant or lubricant. Many materials find common usage as cutting fluids: water solutions or emulsions of detergents and oils; mineral oils; fatty oils; chlorinated mineral oils; sulfurized mineral oils; and mixtures of the above.

The term generally applies to substances used in drilling, gear cutting, grinding, lathing, milling, and other machining operations, for the purpose of cooling, lubricating, and removing metal or plastic chips, filings, and cuttings from the contact area.

Commercial cutting fluids can be divided into four categories:

- Cutting Oils or Straight Oils—Contain mineral oil, fat, and additives. These oils are water-insoluble.

- Soluble Cutting Oils—Contain mineral oil, fat, emulsifiers (may include amines), additives (rarely nitrite), and water.

- Semisynthetic Cutting Oils—Contain mineral oil, water, fat, a soluble base (usually including amines), emulsifiers (may include amines), and additives (usually including nitrite).

- Synthetic Cutting Fluids—A soluble base (usually including amines), additives (usually including nitrite) and water.

DOT Designation: —

Synonyms: These substances are variously referred to as cutting, cooling, grinding, industrial, lubricating, and synthetic oils or fluids.

Potential Exposures: Various proprietary cutting fluids are produced by over one thousand companies in the United States. NIOSH estimates that 780,000 persons are occupationally exposed in the manufacture and use of cutting fluids.

Permissible Exposure Limits in Air: No standards set.

Permissible Concentration in Water: No criteria set.

Routes of Entry: Eye and skin contact, ingestion.

Harmful Effects and Symptoms: The problems associated with occupational exposure are many: eye irritation, pneumonitis, allergic skin sensitization, and acne and folliculitis which can lead to keratosis and hyperkeratosis, ultimately resulting in malignant dyskeratosis and squamous cell carcinoma if exposure continues.

Synthetic cutting fluids, semisynthetic cutting oils, and soluble cutting oils may contain nitrosamines (which are potent animal carcinogens) either as contaminants in amines, or as products from the reaction of amines (e.g., triethanolamine) with nitrite. Straight oils do not contain nitrites or amines but may contain polynuclear aromatic compounds (recognized as having carcinogenic potential).

Personal Protective Methods: The following are suggested good industrial hygiene practices that can help in minimizing exposure to cutting fluids. The recent detection of nitrosamines in certain cutting fluids has compounded the recognized problem of cutting oil control.

- Engineering Control—The most effective control of any contaminant is control at the source of generation. Effective engineering measures include the use of local exhaust ventilation, with a suitable collector, or the use of electrostatic precipitator.

- Substitution—The substitution of a cutting fluid that does not contain either nitrosamine contaminated amines, or the necessary ingredients (amines and nitrites) for nitrosamine formation, is another possible control measure. Since many of the proprietary ingredients of cutting fluids have not undergone complete toxicological evaluation, caution should be used when contemplating any change from one cutting fluid formulation to another, giving full consideration to the potential hazards of the substitute.

- Respirators—Personal respiratory protective devices should only be used as an interim measure while engineering controls are being installed, for nonroutine use and during emergencies. Considering the carcinogenic potential and the lack of a standard for nitrosamines as a group, the only available personal respiratory protective measure is the use of a positive pressure supplied air respirator or a positive pressure self-contained breathing apparatus.

- Protective Clothing—Impervious clothing should be provided and should be replaced or repaired as necessary. Nonimpervious clothing is not suggested, but if used, it should be removed and laundered frequently to remove all traces of cutting fluids before being reworn. (Laundry personnel should be made aware of the potential hazard from handling contaminated clothing.)

- Personal Cleanliness—All exposed areas of the body and any area that becomes wet with cutting fluids should be washed with soap or mild detergent. Frequent showering is recommended.

- Isolation—Where possible, any operations involved with cutting fluids should be placed in an isolated area to reduce exposure to employees not directly concerned with the operations.

- Barrier Creams—Barrier creams may provide protection against dermal irritation and skin absorption, however, the barrier cream should not contain secondary or tertiary amines (which may react to form nitrosamines in the presence of nitrites).

Respirator Selection: See the paragraph on Respirators under "Personal Protective Methods" above.

Disposal Method Suggested: Many synthetic cutting fluids can be successfully biodegraded prior to disposal. This is based upon the relative ease with which long-chain fatty acids can be broken down into shorter chained fatty acids, thus reducing the "oily" character of the spent cutting fluid. Oil-base cutting fluids present a special problem because of the low allowable concentration of oil in wastewater. Biological degradation of hydrocarbons proceeds slowly, and therefore it is a common practice to subject spent cutting oils to physicochemical separation procedures. A process such as this attempts to separate the oil and water phases such that each can be dealt with individually. The aqueous phase often retains some oil, however, and must undergo additional treatment prior to disposal. The final disposal method is incineration of either the separated oil or the untreated cutting fluid. In the latter case, water is driven off as steam and the recovered oil is used to fuel the unit. This method solves the water pollution problem but is somewhat costly and may create an air problem.

References

(1) U.S. Environmental Protection Agency, *Chemical Hazard Information Profile: Cutting Fluids,* Washington, DC (May 1, 1977).

(2) National Institute for Occupational Safety and Health, *Nitrosamines in Cutting Fluids,* Current Intelligence Bulletin No. 15, Rockville, MD (Oct. 6, 1976).

CYANAMIDE

Description: H_2NCN is a crystalline solid melting at 42°C.

Code Numbers: CAS 420-04-2   RTECS GS5950000

DOT Designation: —

Synonyms: Amidocyanogen, carbamonitrile, carbimide, cyanogen nitride, cyanogenamide, carbodiimide.

Potential Exposures: Cyanamide may be melted to give a dimer, dicyandiamide or cyanoguanidine. At higher temperatures it gives the trimer, melamine, a raw material for melamine-formaldehyde resins.

Permissible Exposure Limits in Air: There is no Federal standard, however, ACGIH has adopted a TWA of 2 mg/m^3 as of 1983/84. There is no proposed STEL value.

Permissible Concentration in Water: No criteria set.

Harmful Effects and Symptoms: Cyanamide is very irritating and caustic to the skin. Inhalation may cause irritation of the mucous membranes. Ingestion or inhalation may cause transitory intense redness of the face, headache, vertigo, increased respiration, tachycardia and hypotension.

Personal Protective Methods: Wear long rubber gloves and overalls or apron.

Respirator Selection: Use self-contained breathing apparatus (A-38).

Disposal Method Suggested: Add excess alkaline calcium hypochlorite with agitation. Flush to sewer after 24 hours (A-38).

CYANAZINE

Description: $C_9H_{13}ClN_6$ with the following structural formula

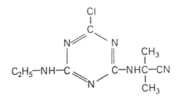

is a white crystalline solid melting at 167° to 169°C.

Code Numbers: CAS 21725-46-2   RTECS UG1490000

DOT Designation: —

Synonyms: Bladex®, Payze®, Fortrol®, 2-{[4-chloro-6-(ethylamino)-1,3,5-triazin-2-yl] amino}-2-methylpropanenitrile.

Potential Exposure: Those involved in the manufacture, formulation and application of this herbicide.

Incompatibilities: Metals.

Permissible Exposure Limits in Air: No standards set.

Permissible Concentration in Water: No criteria set.

Harmful Effects and Symptoms: Moderately toxic (the LD-50 for rats is 182 mg/kg). A major metabolite is diisopropyl atrazine.

Disposal Method Suggested: Incineration.

References
(1)   Sax, N.I., Ed., *Dangerous Properties of Industrial Materials Report, 3,* No. 1, 47-50, New York, Van Nostrand Reinhold Co. (1983).

CYANIDES

● Hazardous substances (EPA)
● Hazardous waste constituents (EPA)
● Priority toxic pollutants (EPA)

Description: KCN and NaCN are white solids with a faint almond odor.

Code Numbers:

KCN	CAS 151-50-8	RTECS TS8750000	UN 1680
NaCN	CAS 143-33-9	RTECS VZ7525000	UN 1689

DOT Designations: KCN and NaCN: Poison B.

Synonyms: KCN = potassium cyanide; NaCN = sodium cyanide.

Potential Exposures: Sodium and potassium cyanides are used primarily in the extraction of ores, electroplating, metal treatment, and various manufacturing processes. The number of workers with potential exposure to NaCN has been estimated by NIOSH to be 20,000.

Incompatibilities: Strong oxidizers such as nitrates, chlorates, acids, acid salts.

Permissible Exposure Limits in Air: The Federal limits are 5 mg/m^3. NIOSH has recommended 5 mg/m^3 as a 10-minute ceiling value. There is no STEL value. The IDLH level is 50 mg/m^3. ACGIH adds the notation "skin" indicating the possibility of cutaneous absorption as of 1983/84.

Determination in Air: Collection with a filter and impinger, workup with NaOH, measurement by ion-specific electrode. See NIOSH Methods, Set R. See also reference (A-10). Also, analysis may be carried out with long-duration detector tubes (A-13).

Permissible Concentration in Water: In 1976 the EPA criterion was 5.0 μg/ℓ for freshwater and marine aquatic life and wildlife. As of 1980, the criteria are: To protect freshwater aquatic life: 3.5 μg/ℓ as a 24-hour average, never to exceed 52.0 μg/ℓ. To protect saltwater aquatic life: 30.0 μg/ℓ on an acute toxicity basis; 2.0 μg/ℓ on a chronic toxicity basis. To protect human health: 200 μg/ℓ. The allowable daily intake for man is 8.4 mg/day.

On the international scene (A-65), the South African Bureau of Standards has set 10 μg/ℓ, the World Health Organization (WHO) 50 μg/ℓ and the Federal Republic of Germany 50 μg/ℓ as drinking water standards.

Determination in Water: Distillation followed by silver nitrate titration or colorimetric analysis using pyridine pyrazolone (or barbituric acid).

Routes of Entry: Inhalation; skin absorption; ingestion; eye, skin contact.

Harmful Effects and Symptoms: Weakness, headaches; confusion; nausea, vomiting; eye and skin irritation; slow gasping respiration.

Points of Attack: Liver, kidneys, skin, cardiovascular system, central nervous system.

Medical Surveillance: Consider the points of attack in preplacement and periodic physical examinations.

First Aid: If this chemical gets into the eyes, irrigate immediately. If this chemical contacts the skin, wash with soap immediately. If a person breathes in large amounts of this chemical, move the exposed person to fresh air at once and perform artificial respiration. When this chemical has been swallowed, get medical attention. Give large quantities of water and induce vomiting. Do not make an unconscious person vomit. Use Amyl Nitrite Pearls.

Personal Protective Methods: Wear appropriate clothing to prevent any possibility of skin contact. Wear eye protection to prevent any possibility of eye contact. Employees should wash immediately when skin is wet or contaminated. Work clothing should be changed daily if it is possible that clothing is contaminated. Remove nonimpervious clothing immediately if wet or contaminated. Provide emergency showers and eyewash.

Respirator Selection:
50 mg/m³: SA/SCBA
Escape: GMSP/SCBA

Disposal Method Suggested: Add strong alkaline hypochlorite and react for 24 hours. Then flush to sewer with large volumes of water.

References

(1) U.S. Environmental Protection Agency, *Cyanides: Ambient Water Quality Criteria,* Washington, DC (1980).

(2) National Institute for Occupational Safety and Health, *Criteria for a Recommended Standard: Occupational Exposure to Hydrogen Cyanide and Cyanide Salts,* NIOSH Document No. 77-108, Washington, DC (1977).

(3) U.S. Environmental Protection Agency, *Reviews of the Environmental Effects of Pollutants; V: Cyanide,* Report No. EPA-600/1-78-027, Washington, DC (1978).

(4) U.S. Environmental Protection Agency, *Cyanides,* Health and Environmental Effects Profile No. 56, Office of Solid Waste, Washington, DC (April 30, 1980).

(5) See Reference (A-61).

(6) Sax, N.I., Ed., *Dangerous Properties of Industrial Materials Report, 3,* No. 6, 56–60, New York, Van Nostrand Reinhold Co. (Nov./Dec. 1983) (Potassium Cyanide).

(7) Sax, N.I., Ed., *Dangerous Properties of Industrial Materials Report, 3,* No. 6, 60–63, New York, Van Nostrand Reinhold Co. (Nov./Dec. 1983) (Sodium Cyanide).

(8) United Nations Environment Programme, *IRPTC Legal File 1983,* Vol. I, pp VII/226–29, Geneva, Switzerland, International Register of Potentially Toxic Chemicals (1984).

CYANOGEN

● Hazardous waste (EPA)

Description: $(CN)_2$ is a gas at room temperature.

Code Numbers: CAS 460-19-5   RTECS GT1925000   UN 1026

DOT Designation: Flammable gas and poison gas.

Synonyms: Dicyanogen, dicyan, ethanedinitrile, oxalonitrile.

Potential Exposure: Cyanogen is currently used as an intermediate in organic syntheses; at one time, it was used in poison gas warfare. NIOSH estimates that 7,500 workers are exposed to cyanogen annually.

Permissible Exposure Limits in Air: There is no Federal standard. ACGIH as of 1983/84 set a TWA of 10 ppm (20 mg/m³) but no STEL value. The Threshold Limit Value is recommended at a level to prevent irritation as well as systemic effects, and is based on cyanogen's similarity to hydrogen cyanide. Some foreign standards are as follows: Rumania (average), 3 mg/m³, (ceiling) 5 mg/m³; Australia, Belgium, Finland, the Netherlands, Switzerland, 2 mg/m³.

Determination in Air: Sample collection by impinger or fritted bubbler, colorimetric analysis (A-17).

Permissible Concentration in Water: No criteria set.

Routes of Entry: Inhalation, skin and eye contact.

Harmful Effects and Symptoms: Cyanogen hydrolyzes to yield one molecule of hydrogen cyanide and one of cyanate; based on this, the toxic effects of $(CN)_2$ are thought to be comparable to HCN. The cyanide ion when released in the body causes a form of asphyxia by inhibiting many enzymes—especially those concerned with cellular respiration. Although the blood is saturated with oxygen, the tissues are not able to use it. Symptoms appear within a few seconds or minutes of ingesting or breathing of vapors. The victims experience constriction of the chest, giddiness, confusion, headache, hypernea, palpitation, unconsciousness, convulsion, feeble and rapid respiration, and an extremely weak pulse. Death occurs within a few minutes after a large dose.

The irritant properties of cyanogen have been tested using both human male and female subjects, 21 to 65 years of age. The distinctive bitter almond smell of cyanogen could not be detected at concentrations of 50, 100 and 250 ppm. When exposed to 8 ppm for 6 minutes or 16 ppm for 6 to 8 minutes, immediate eye and nose irritation occurred.

Disposal Method Suggested: Incineration; oxides of nitrogen are removed from the effluent gas by scrubbers and/or thermal devices.

References

(1) National Institute for Occupational Safety and Health, *Information Profiles on Potential Occupational Hazards—Single Chemicals: Cyanogen,* Report TR 79-607, pp 39-44, Rockville, MD (December 1979).

(2) Sax, N.I., Ed., *Dangerous Properties of Industrial Materials Report, 2,* No. 1, 103-105, New York, Van Nostrand Reinhold Co. (1982).

(3) Parmeggiani, L., Ed., *Encyclopedia of Occupational Health & Safety,* Third Edition, Vol. 1, pp 574-77, Geneva, International Labour Office (1983) (Cyanogen, HCN and Cyanides).

CYANOGEN BROMIDE

● Hazardous Waste Constituent (EPA)

Description: BrCN is a volatile solid which melts at 52°C and boils at 61° to 62°C.

Code Numbers: CAS 506-68-3   RTECS GT2100000   UN 1889

DOT Designation: Poison B.

Synonyms: Bromine cyanide, bromocyanogen.

Potential Exposure: Those manufacturing this compound or using it in organic synthesis or as a fumigant, in textile treatment, in gold cyaniding or as a military poison gas.

Incompatibilities: Water, acids.

Permissible Exposure Limits in Air: 5 mg/m^3 (as CN) (A-60).

Permissible Concentration in Water: No criteria set.

Routes of Entry: Inhalation, ingestion, skin absorption.

Harmful Effects and Symptoms: Very toxic by inhalation. Vapors are very irritating to eyes and respiratory tract. Similar to HCN (which see).

Points of Attack: Eyes and respiratory system.

Personal Protective Methods: Wear goggles and protective skin covering.

Respirator Selection: Wear self-contained breathing apparatus.

References

(1) Sax, N.I., Ed., *Dangerous Properties of Industrial Materials Report, 1,* No. 8, 60-62, New York, Van Nostrand Reinhold Co. (1981).

(2) See Reference (A-60).

CYANOGEN CHLORIDE

- Hazardous substance (EPA)
- Hazardous waste (EPA)

Description: CNCl is a colorless gas at room temperature with a pungent odor; the boiling point is 14°C.

Code Numbers: CAS 506-77-4   RTECS GT2275000   UN 1589

DOT Designation: Nonflammable gas and poison gas (Poison A).

Synonyms: Chlorine cyanide, chlorocyanogen.

Potential Exposures: Cyanogen chloride is used as a fumigant, metal cleaner, in ore refining, production of synthetic rubber and in chemical synthesis. Cyanogen chloride can be used in the military as a poison gas.

Permissible Exposure Limits in Air: There is no Federal standard but ACGIH as of 1983/84 has adopted a ceiling value of 0.3 ppm (0.6 mg/m^3).

Permissible Concentration in Water: No criteria set.

Harmful Effects and Symptoms: The toxicity of cyanogen chloride resides very largely on its pharmacokinetic property of yielding readily to hydrocyanic acid in vivo. Inhaling small amounts of cyanogen chloride causes dizziness, weakness, congestion of the lungs, hoarseness, conjunctivitis, loss of appetite, weight loss, and mental deterioration. These effects are similar to those found from inhalation of cyanide. Cyanide chloride is a severe irritant to both eyes and throat.

Ingestion or inhalation of a lethal dose of cyanogen chloride (LD$_{50}$ = 13 mg/kg), as for cyanide or other cyanogenic compounds, causes dizziness, rapid respiration, vomiting, flushing, headache, drowsiness, drop in blood pressure, rapid pulse, unconsciousness, convulsions, with death occurring within 4 hours.

Personal Protective Measures: Wear long rubber gloves and apron (A-38).

Respirator Selection: Use of a self-contained breathing apparatus is recommended.

Disposal Method Suggested: React with strong calcium hypochlorite solution for 24 hours, then flush to sewer with large volumes of water (A-38).

References

(1) U.S. Environmental Protection Agency, *Cyanogen Chloride,* Health and Environmental Effects Profile No. 57, Office of Solid Waste, Washington, DC (April 30, 1980).

(2) Sax, N.I., Ed., *Dangerous Properties of Industrial Materials Report, 1,* No. 8, 62-63, New York, Van Nostrand Reinhold Co. (1981).

(3) See Reference (A-60).

CYCASIN

- Carcinogen (Animal Positive, IARC) (1)
- Hazardous Waste Constituent (EPA)

Description: $C_8H_{16}N_2O_7$ with the following structural formula

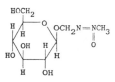

is a crystalline solid melting with decomposition at 154°C.

Code Numbers: CAS 14901-08-7   RTECS LZ5950000

DOT Designation: —

Synonyms: Methylazoxymethanol beta-d-glucoside.

Potential Exposure: Cycasin occurs naturally in the seeds, roots, and leaves of cycad plants which are found in the tropical and subtropical regions of the world. Nuts from the cycads are used to make chips, flour, and starch.

Cycasin is not produced or used commercially. The major potential exposure is the ingestion of the foods containing cycasin. It is estimated that about 50 to 55 percent of the inhabitants of Guam are potentially exposed (50,000 to 60,000 persons) to cycasin. Wastewater from the preparation of the cycad nuts contain large amounts of cycasin and represents a potential secondary exposure source (3).

Permissible Exposure Limits in Air: No standards set.

Permissible Concentration in Water: No criteria set.

Harmful Effects and Symptoms: Cycasin is carcinogenic in 5 animal species, inducing tumors in various organs (3). Following oral exposure, it is carcinogenic in the rat, hamster, guinea pig and fish. By this route, the data in the mouse is of borderline significance and the negative experiment in chickens only lasted 68 weeks. It is active in single-dose experiments and following prenatal exposure. The carcinogenicity of its metabolite, methylazoxymethanol, has been demonstrated in the rat and the hamster and that of a closely related synthetic substance, methylazoxymethanol acetate, in the rat (1).

References

(1) *IARC Monographs on the Evaluation of Carcinogenic Risk of Chemicals to Man,* vol. 1, IARC, Lyon, France, pp 157-163 (1972).
(2) Sax, N.I., Ed., *Dangerous Properties of Industrial Materials Report, 1,* No. 3, 48-49, New York, Van Nostrand Reinhold Co. (1981).
(3) See Reference (A-62). Also see Reference (A-64).

CYCLOHEXANE

- Hazardous substance (EPA)
- Hazardous waste (EPA)

Description: C_6H_{12} is a colorless liquid with a mild, sweet odor. It boils at 80° to 81°C.

Code Numbers: CAS 110-82-7   RTECS GU6300000   UN 1145

DOT Designation: Flammable liquid.

Synonyms: Hexahydrobenzene, hexamethylene, benzene hexahydride.

Potential Exposure: Cyclohexane is used as a chemical intermediate, as a solvent for fats, oils, waxes, resins, and certain synthetic rubbers, and as an extractant of essential oils in the perfume industry.

Incompatibilities: Oxidizers.

Permissible Exposure Limits in Air: The Federal limit and the 1983/84 ACGIH TWA value is 300 ppm (1,050 mg/m^3). The STEL value is 375 ppm (1,300 mg/m^3). The IDLH level is 10,000 ppm.

Determination in Air: Adsorption on charcoal. Workup with CS_2, analysis by gas chromatography. See NIOSH Methods, Set C. See also reference (A-10).

Permissible Concentration in Water: No criteria set.

Routes of Entry: Inhalation, ingestion, skin and eye contact.

Harmful Effects and Symptoms: *Local* — Repeated and prolonged contact with liquid may cause defatting of the skin and a dry, scaly, fissured dermatitis. Mild conjunctivitis may result from acute vapor exposure.

Systemic — Alicyclic hydrocarbons are central nervous system depressants, although their acute toxicity is low. Symptoms of acute exposure are excitement, loss of equilibrium, stupor, coma, and, rarely, death as a result of respiratory failure.

The danger of chronic poisoning is relatively slight because these compounds are almost completely eliminated from the body. Metabolism of cyclohexane, for example, results in cyclohexanone and cyclohexanol entering the bloodstream and does not include the metabolites of phenol, as with benzene. Damage to the hematopoietic system does not occur except when exposure is compounded with benzene, which may be a contaminant. Alicyclic hydrocarbons are excreted in the urine as sulfates or glucuronides, the particular content of each varying. Small quantities of these compounds are not metabolized and may be found in blood, urine, and expired breath.

Points of Attack: Eyes, respiratory system, central nervous system, skin.

Medical Surveillance: Consider possible irritant effects to the skin and respiratory tract in any preplacement or periodic examination, as well as any renal or liver complications.

First Aid: If this chemical gets into the eyes, irrigate immediately. If this chemical contacts the skin, flush with water promptly. If a person breathes in large amounts of this chemical, move the exposed person to fresh air at once and perform artificial respiration. When this chemical has been swallowed, get medical attention. Do NOT induce vomiting.

Personal Protective Methods: Wear appropriate clothing to prevent repeated or prolonged skin contact. Wear eye protection to prevent any reasonable probability of eye contact. Employees should wash promptly when skin is wet or contaminated. Remove clothing immediately if wet or contaminated to avoid flammability hazard.

Respirator Selection:
        3,000 ppm: SA/SCBA
       10,000 ppm: SAF/SCBAF
          Escape: GMOV/SCBA

Disposal Method Suggested: Incineration.

References

(1) See Reference (A-61).
(2) See Reference (A-60).

CYCLOHEXANOL

Description: $C_6H_{11}OH$ is a colorless, viscous liquid or sticky solid with a faint camphor odor. It boils at 161°C.

Code Numbers: CAS 108-93-0   RTECS GV7875000

DOT Designation: —

Synonyms: Hexalin, Hydralin, hydroxycyclohexane, Anol, hexahydrophenol, cyclohexyl alcohol.

Potential Exposure: Cyclohexanol is used as a solvent for ethyl cellulose and other resins; it is used in soap manufacture; it is used as a raw material for adipic acid manufacture as a nylon intermediate.

Incompatibilities: Strong oxidizers.

Permissible Exposure Limits in Air: The Federal limit and the ACGIH 1983/84 TWA value is 50 ppm (200 mg/m³). There is no proposed STEL value. The IDLH level is 3,500 ppm.

Determination in Air: Adsorption on charcoal, workup with 2-propanol in CS_2, analysis by gas chromatography. See NIOSH Methods, Set E. See also reference (A-10).

Permissible Concentration in Water: No criteria set.

Routes of Entry: Inhalation, ingestion, skin and eye contact.

Harmful Effects and Symptoms: Irritation of eyes, nose and throat; skin irritations; narcosis.

Points of Attack: Eyes, skin, respiratory system.

Medical Surveillance: Consider the points of attack in preplacement and periodic physical examinations.

First Aid: If this chemical gets into the eyes, irrigate immediately. If this chemical contacts the skin, flush with water promptly. If a person breathes in large amounts of this chemical, move the exposed person to fresh air at once and perform artificial respiration. When this chemical has been swallowed, get medical attention. Give large quantities of salt water and induce vomiting. Do not make an unconscious person vomit.

Personal Protective Methods: Wear appropriate clothing to prevent repeated or prolonged skin contact. Wear eye protection to prevent any reasonable probability of eye contact. Employees should wash promptly when skin is wet or contaminated. Remove nonimpervious clothing promptly if contaminated or wet.

Respirator Selection:
　　　1,000 ppm: CCROVF
　　　2,500 ppm: GMOV/SAF/SCBAF
　　　3,500 ppm: SAF,PD,PP,CF
　　　　Escape: GMOV/SCBA

Disposal Method Suggested: Incineration.

References

(1)   See Reference (A-61).
(2)   See Reference (A-60).

CYCLOHEXANONE

● Hazardous waste (EPA)

Description: $C_6H_{10}O$ is a water-white-to-slightly-yellow liquid with a pepper-mintlike odor. It boils at 157°C.

Code Numbers: CAS 108-94-1   RTECS GW1050000   UN 1915

DOT Designation: Flammable liquid.

Synonyms: Pimelic ketone, cyclohexyl ketone, ketohexamethylene.

Potential Exposures: It is used in metal degreasing and as a solvent for lacquers, resins and insecticides. It is an intermediate in adipic acid manufacture. NIOSH has estimated annual worker exposures at 1,190,000.

Incompatibilities: Oxidizing agents, nitric acid.

Permissible Exposure Limits in Air: Federal limit is 50 ppm (200 mg/m^3). The ACGIH, as of 1983/84, has proposed a TWA of 25 ppm (100 mg/m^3) and an STEL value of 100 ppm (400 mg/m^3). The IDLH level is 5,000 ppm.

Determination in Air: Charcoal adsorption, workup with CS_2, measurement by gas chromatography. See NIOSH Methods, Set A. See also reference (A-10).

Permissible Concentration in Water: No criteria set.

Routes of Entry: Inhalation, ingestion, eye and skin contact.

Harmful Effects and Symptoms: Irritation of eyes and mucous membranes; irritation of central nervous system and skin; narcosis.

Points of Attack: Respiratory system, eyes, skin, central nervous system.

Medical Surveillance: Consider the points of attack in preplacement and periodic physical examinations.

First Aid: If this chemical gets into the eyes, irrigate immediately. If this chemical contacts the skin, flush with water promptly. If a person breathes in large amounts of this chemical, move the exposed person to fresh air at once and perform artificial respiration. When this chemical has been swallowed, get medical attention. Give large quantities of salt water and induce vomiting. Do not make an unconscious person vomit.

Personal Protective Methods: Wear appropriate clothing to prevent repeated or prolonged skin contact. Wear eye protection to prevent any reasonable probability of eye contact. Employees should wash promptly when skin is wet or contaminated. Remove clothing immediately if wet or contaminated to avoid flammability hazard.

Respirator Selection:

1,000 ppm:	CCROVF
2,500 ppm:	GMOV/SAF/SCBAF
5,000 ppm:	SAF:PD,PP,CF
Escape:	GMOV/SCBA

Disposal Method Suggested: Incineration.

References

(1) National Institute for Occupational Safety and Health, *Criteria for a Recommended Standard: Occupational Exposure to Ketones,* NIOSH Document No. 78-173 (1978).

(2) See Reference (A-60).

(3) See Reference (A-68).

(4) United Nations Environment Programme, *IRPTC Legal File 1983,* Vol. I, pp VII/235–36, Geneva, Switzerland, International Register of Potentially Toxic Chemicals (1984).

CYCLOHEXENE

Description: C_6H_{10} is a colorless liquid with a sweetish odor. It boils at 82° to 83°C.

Code Numbers: CAS 110-83-8 RTECS GW2500000 UN 2256

DOT Designation: Flammable liquid.

Synonyms: Tetrahydrobenzene, benzene tetrahydride.

Potential Exposure: May be used as an intermediate in adipic acid, maleic acid and hexahydrobenzoic acid manufacture.

Incompatibilities: Strong oxidizers.

Permissible Exposure Limits in Air: Federal limit is 300 ppm (1,015 mg/m^3). ACGIH has assigned no STEL values as of 1983/84. The IDLH level is 10,000 ppm.

Determination in Air: Charcoal adsorption followed by workup with CS_2, and analysis by gas chromatography. See NIOSH Methods, Set F. See also reference (A-10).

Permissible Concentration in Water: No criteria set.

Routes of Entry: Inhalation, ingestion, eyes and skin contact.

Harmful Effects and Symptoms: Irritation of eyes and respiratory system; skin irritation; drowsiness.

Points of Attack: Skin, eyes, respiratory system.

Medical Surveillance: Consider the points of attack in preplacement and periodic physical examinations.

First Aid: If this chemical gets into the eyes, irrigate immediately. If this chemical contacts the skin, wash with soap promptly. If a person breathes in large amounts of this chemical, move the exposed person to fresh air at once and perform artificial respiration. When this chemical has been swallowed, get medical attention. Do NOT induce vomiting.

Personal Protective Methods: Wear appropriate clothing to prevent repeated or prolonged skin contact. Wear eye protection to prevent any reasonable probability of eye contact. Employees should wash promptly when skin is wet or contaminated. Remove clothing immediately if wet or contaminated to avoid flammability hazard.

Respirator Selection:

        1,000 ppm: CCROVF
        5,000 ppm: GMOVc
       10,000 ppm: GMOVfb/SAF/SCBAF
          Escape: GMOV/SCBA

Disposal Method Suggested: Incineration.

References

(1)   See Reference (A-61).
(2)   See Reference (A-60).

CYCLOHEXIMIDE

Description: $C_{15}H_{23}NO_4$ with the following structural formula

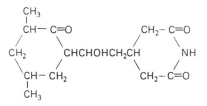

is a crystalline substance melting at 119.5° to 121°C.

Code Numbers: CAS 66-81-9   RTECS MA4375000

DOT Designation: —

Synonyms: 4-[2-(3,5-Dimethyl-2-oxocyclohexyl)-2-hydroxyethyl] -2,6-pi-peridinedione; naramycin A, Actidione®.

Potential Exposure: Those involved in the manufacture, formulation or application of this fungicide.

Incompatibilities: Alkalis.

Permissible Exposure Limits in Air: No standards set.

Permissible Concentration in Water: No criteria set.

Harmful Effects and Symptoms: Extremely toxic (LD-50 value for rats is only 2 mg/kg).

References

(1)   Sax, N.I., Ed., *Dangerous Properties of Industrial Materials Repor*, *2*, No. 5, 41-43, New York, Van Nostrand Reinhold Co. (1982).

CYCLOHEXYLAMINE

Description: $C_6H_{11}NH_2$ is a colorless liquid boiling at 134.5°C with an unpleasant fishy odor.

Code Numbers: CAS 108-91-8   RTECS GX0700000   UN 2357

DOT Designation: Flammable liquid, corrosive, poison.

Synonyms: CHA, aminocyclohexane, hexahydroaniline, cyclohexanamine.

Potential Exposure: CHA is used as a chemical intermediate in the production of cyclamate sweeteners and rubber chemicals. It is also used as a boiler feedwater additive. Annual worker exposure to CHA has been estimated at 6,000 by NIOSH.

Incompatibilities: Strong oxidizers.

Permissible Exposure Limits in Air: A TLV of 10 ppm (40 mg/m^3) (skin) has been adopted by ACGIH as of 1983/84. The notation "skin" is added indicating the possibility of cutaneous absorption. No STEL value has been set.

Determination in Air: Collection by impinger or fritted bubbler followed by colorimetric analysis (A-18). An alternative is silica adsorption followed by acid elution and gas chromatographic analysis (A-10).

Permissible Concentration in Water: No criteria set but EPA (A-37) has suggested a permissible concentration based on health effects of 550 μg/ℓ.

Routes of Entry: Inhalation, ingestion, skin absorption, eye and skin contact.

Harmful Effects and Symptoms: Cyclohexylamine is caustic to skin and mucous membranes, and its systemic effects in man include nausea, vomiting, anxiety, restlessness, and drowsiness. Cyclohexylamine may also be a skin sensitizer (1).

The Food and Drug Administration banned the use of cyclamates as artificial sweeteners in 1969 because of their metabolic conversion to cyclohexylamine, which was thought to be carcinogenic in rats. There is no evidence that CHA is a human carcinogen (1), however.

CHA does not appear to be carcinogenic or teratogenic following oral exposure in rats, mice, hamsters, or rabbits, but is embryotoxic and may affect chromosome and testicular structure (2). Eye and skin irritation from CHA may be severe, and weight loss and corneal opacity are effects of chronic exposure.

First Aid: Flush eyes with water. Wash contaminated areas of body with soap and water.

Personal Protective Methods: Wear butyl rubber gloves, face shield and work clothing.

Respirator Selection: An all-purpose canister mask is recommended (A-38).

Disposal Method Suggested: Incineration; incinerator is equipped with a scrubber or thermal unit to reduce NO$_x$ emissions.

References:

(1) U.S. Environmental Protection Agency, *Chemical Hazard Information Profile: Cyclohexylamine,* Washington, DC (October 21, 1977).

(2) National Institute for Occupational Safety and Health, *Information Profiles on Potential Occupational Hazards—Single Chemicals: Cyclohexylamine,* Report TR 79-607, pp 45-55, Rockville, MD (December 1979).

(3) See Reference (A-60).

CYCLONITE

Description: $\overline{N(NO_2)CH_2N(NO_2)CH_2N(NO_2)CH_2}$, hexahydro-1,3,5-trinitro-s-triazine is a white crystalline compound melting at 203° to 204°C.

Code Numbers: CAS 121-82-4 RTECS XY9450000 UN 0072

DOT Designation: Class A explosive.

Synonyms: Cyclotrimethylenetrinitramine, RDX, trinitrotrimethylenetri-amine, Hexogen, HMX.

Potential Exposure: Those involved in the manufacture of this material and its handling in munitions and solid-propellant manufacture.

Incompatibilities: Heat, shock and detonators, oxidizing materials and com-bustibles.

Permissible Exposure Limits in Air: There is no Federal standard but ACGIH as of 1983/84 has set a TWA of 1.5 mg/m^3 (with the notation "skin" indicating the possibility of cutaneous absorption) and an STEL of 3.0 mg/m^3.

Determination in Air: Collection on a filter, analysis by ultraviolet (A-45).

Permissible Concentration in Water: No criteria set.

Routes of Entry: Inhalation, ingestion, skin and eye contact.

Harmful Effects and Symptoms: Irritation of eyes and respiratory tract; ulceration of mucous membranes; dermatitis; headaches; irritability; salivation and anorexia; asthenia; insomnia; unconsciousness; convulsions (A-39).

First Aid: Irrigate eyes with water; soap wash contaminated areas; administer oxygen (A-39).

Personal Protective Methods: Wear chemical goggles, provide adequate ventilation (A-39). Neoprene gloves and plastic clothing are recommended (A-38).

Respirator Selection: Use chemical cartridge respirator (A-39).

Disposal Method Suggested: Pour over soda ash, neutralize and flush to sewer with water (A-38). Also HMX may be recovered from solid propellant waste (A-38).

CYCLOPENTADIENE

Description: C_5H_6 is a colorless liquid with a sweet odor like turpentine; dimer crystalline, camphorlike odor.

Code Numbers: CAS 542-92-7   RTECS GY1000000

DOT Designation: —

Synonym: 1,3-Cyclopentadiene.

Potential Exposure: Cyclopentadiene is used as an intermediate in the manufacture of resins and of many pesticides (A-32).

Incompatibilities: Strong oxidizers.

Permissible Exposure Limits in Air: The Federal standard and the 1983/84 ACGIH TWA value is 75 ppm (200 mg/m^3). The STEL value is 150 ppm (400 mg/m^3). The IDLH level is 2,000 ppm.

Determination in Air: Sample collection by charcoal tube, analysis by gas liquid chromatography (A-13).

Permissible Concentration in Water: No criteria set.

Routes of Entry: Inhalation, ingestion, eye and skin contact.

Harmful Effects and Symptoms: Irritation of eyes and nose.

Points of Attack: Eyes, respiratory system.

Medical Surveillance: Consider the points of attack in preplacement and periodic physical examinations.

First Aid: If this chemical gets into the eyes, irrigate immediately. If this chemical contacts the skin, wash with soap promptly. If a person breathes in large amounts of this chemical, move the exposed person to fresh air at once and perform artificial respiration. When this chemical has been swallowed, get medical attention. Do NOT induce vomiting.

Personal Protective Methods: Wear appropriate clothing to prevent repeated or prolonged skin contact. Wear eye protection to prevent any reasonable probability of eye contact. Employees should wash promptly when skin is wet or contaminated. Remove clothing immediately if wet or contaminated to avoid flammability hazard.

Respirator Selection:
    750 ppm: CCROV/SA/SCBA
    2,000 ppm: GMOV/SAF/SCBAF
    Escape: GMOV/SCBA

Disposal Method Suggested: Incineration.

References
(1) See Reference (A-61).
(2) See Reference (A-60).

CYCLOPENTANE

Description: $CH_2CH_2CH_2CH_2CH_2$, C_5H_{10}, is a colorless liquid which boils at 49°C.

Code Numbers: CAS 287-92-3   RTECS GY2390000   UN 1146

DOT Designation: Flammable liquid.

Synonym: Pentamethylene.

Potential Exposures: Cyclopentane is used as a solvent.

Permissible Exposure Limits in Air: There is no Federal standard, but ACGIH as of 1983/84 has proposed a TWA of 600 ppm (1,720 mg/m^3) and an STEL of 900 ppm (2,580 mg/m^3).

Permissible Concentration in Water: No criteria set.

Harmful Effects and Symptoms: Alicyclic hydrocarbons are central nervous system depressants, although their acute toxicity is low (A-5). Symptoms of acute exposure are excitement, loss of equilibrium, stupor, coma and, rarely, death as a result of respiratory failure. Severe diarrhea and vascular collapse resulting in heart, lung and liver and brain degeneration have been reported in oral administration of alicyclic hydrocarbons to animals (A-5).

Medical Surveillance: Consider possible irritant effects to the skin and respiratory tract in preplacement and periodic examinations as well as any renal or liver complications.

Personal Protective Methods: Protect the skin with barrier creams or gloves.

Respirator Selection: Workers exposed to high concentrations of gas or vapor may need masks.

Disposal Method Suggested: Incineration.

References

(1)   See Reference (A-60).

CYCLOPHOSPHAMIDE

- Carcinogen (Animal Positive, Human Suspected) (A-62) (A-64)
- Hazardous Waste Constituent (EPA)

Description: $C_7H_{15}Cl_2N_2O_2P$ has the following structural formula

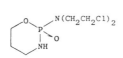

The monohydrate forms crystals melting at 41° to 45°C.

Code Numbers: CAS 50-18-0   RTECS RP5950000

DOT Designation: —

Synonyms: N,N-Bis(2-chloroethyl)tetrahydro-2H-1,3,2-oxazaphosphorin-2-amine-2-oxide; Cytoxan®; Endoxan®; Procytox®; Sendoxan®.

Potential Exposure: Cyclophosphamide is used in the treatment of malignant lymphoma, multiple myeloma, leukemias, and other malignant diseases. Cyclophosphamide has been tested as an insect chemosterilant and for use in the chemical shearing of sheep.

Cyclophosphamide is not produced in the United States. It is produced in the Federal Republic of Germany and exported to the United States where one company has formulated and marketed the drug since 1959. United States sales are approximately 1,300 lb annually.

The FDA estimates that 200,000 to 300,000 patients per year are treated with cyclophosphamide. It is administered orally and through injection. The adult dosage is usually 1 to 5 mg/kg of body weight daily or 10 to 15 mg/kg administered intravenous every 7 to 10 days.

Permissible Exposure Limits in Air: No standards set.

Permissible Concentration in Water: No criteria set.

Harmful Effects and Symptoms: There is sufficient evidence for the carcinogenicity of cyclophosphamide both in humans and in experimental animals. Cyclophosphamide was carcinogenic in rats following administration in drinking water and intravenous injection, and in mice following subcutaneous injection. Dosages were comparable to those used in clinical practice. The chemical produced benign and malignant tumors at various sites, including bladder tumors in the rats.

Five epidemiological studies are available in which persons treated with cyclophosphamide for a variety of medical conditions were compared with similarly afflicted controls. These studies consistently demonstrate an excess of various neoplasms and leukemias in the treated groups, although the number in all five studies was small (2).

References

(1) Sax, N.I., Ed., *Dangerous Properties of Industrial Materials Report, 1,* No. 3, 62-64, New York, Van Nostrand Reinhold Co. (1981). (As Endoxan).

(2) See Reference (A-62). Also see Reference (A-64).

CYHEXATIN

Description: $(C_6H_{11}O)_3SnOH$ is a colorless crystalline powder which melts at 195° to 198°C.

Code Numbers: CAS 13121-70-5   RTECS WH8750000   UN 2788

DOT Designation: —

Synonyms: Tricyclohexyltin hydroxide, tricyclohexylhydroxystannane and ENT 27395-X, Plictran®.

Potential Exposures: Those involved in the manufacture (A-32), formulation and application of this acaricide (miticide).

Permissible Exposure Limits in Air: There are no Federal standards, but ACGIH, as of 1983/84, has adopted a TWA value of 5 mg/m³ and set an STEL value of 10 mg/m³.

Permissible Concentration in Water: No criteria set.

Harmful Effects and Symptoms: Cyhexatin is moderate in acute oral toxicity to animals. This is in contrast to alkyl tin compounds with smaller (methyl and ethyl) radicals which are highly toxic. A diet including 6 mg/kg of body weight of cyhexatin for two years showed no effect in rats (A-34).

References

(1) National Institute for Occupational Safety and Health, *Criteria for a Recommended Standard: Occupational Exposure to Organotin Compounds,* NIOSH Document No. 77-115 (1977).

D

2,4-D

- Hazardous substance (EPA)
- Hazardous waste (EPA)

Description: $Cl_2 C_6 H_3 OCH_2 COOH$ is a colorless powder, melting at 140.5°C with a slight phenolic odor.

Code Numbers: CAS 94-75-7   RTECS AG6825000   UN (NA 2765)

DOT Designation: ORM-A

Synonyms: 2,4-Dichlorophenoxyacetic acid, 2,4-PA (in Japan).

Potential Exposure: 2,4-D, or 2,4-dichlorophenoxyacetic acid, was introduced as a plant growth-regulator in 1942. It is registered in the United States as an herbicide for control of broadleaf plants and as a plant growth-regulator. Thus, workers engaged in manufacture (A-32), formulation or application are affected as may be citizens in areas of application.

Incompatibilities: Strong oxidizers.

Permissible Exposure Limits in Air: The Federal limit and the 1983/84 ACGIH TWA value is 10 mg/m³. The STEL value is 20 mg/m³. The IDLH level is 500 mg/m³.

Determination in Air: Collection by impinger or fritted bubbler, analysis by gas liquid chromatography (A-1).

Permissible Concentration in Water: A level of 100 μg/ℓ for domestic water supplies has been set by EPA on a health basis. NAS/NRC has calculated a no-adverse effect level in water of 0.09 mg/ℓ. The acceptable daily intake of 2,4-D has been set at 0.03 mg/kg by FAO/WHO.

Determination in Water: On the basis of electron-capture gas chromatography, the detection limit for 2,4-D in water is 1 ppb according to NAS/NRC (A-3).

Routes of Entry: Inhalation, skin absorption, ingestion, skin and eye contact.

Harmful Effects and Symptoms: Weakness; stupor; hyporeflexia; muscle twitching; convulsions; dermatitis.

Points of Attack: Skin, central nervous system.

Medical Surveillance: Consider the points of attack in preplacement and periodic physical examinations.

First Aid: If this chemical gets into the eyes, irrigate immediately. If this

288

chemical contacts the skin, wash with soap promptly. If a person breathes in large amounts of this chemical, move the exposed person to fresh air at once and perform artificial respiration. When this chemical has been swallowed, get medical attention. Give large quantities of water and induce vomiting. Do not make an unconscious person vomit.

Personal Protective Methods: Wear appropriate clothing to prevent repeated or prolonged skin contact. Wear eye protection to prevent any reasonable probability of eye contact. Employees should wash promptly when skin is wet or contaminated. Work clothing should be changed daily if clothing is contaminated. Remove nonimpervious clothing promptly if wet or contaminated.

Respirator Selection:
100 mg/m³: CCROVDPest/SA/SCBA
500 mg/m³: CCROVFDPest/DMOVDMPest/SAF/SA:PD,PP,CF/SCBAF
Escape: GMOVPPest/SCBA

Disposal Method Suggested: Incineration of phenoxys is effective in one second at 1800°F using a straight combustion process or at 900°F using catalytic combustion. Over 99% decomposition was reported when small amounts of 2,4-D were burned in a polyethylene bag (A-32).

References

(1) U.S. Environmental Protection Agency, *2,4-Dichlorophenoxy Acetic Acid,* Health and Environmental Effects Profile No. 77, Washington, DC, Office of Solid Waste (April 30, 1980).
(2) See Reference (A-61).
(3) United Nations Environment Programme, *IRPTC Legal File 1983,* Vol. I, pp VII/8–11, Geneva, Switzerland, International Register of Potentially Toxic Chemicals (1984).

DDT

- Carcinogen (Animal suspected, IARC) (4) (Potential, EPA) (A-40). (Negative, NCI) (5)
- Hazardous substance (EPA)
- Hazardous waste (EPA)
- Priority toxic pollutant (EPA)

Description: CCl_3
$|$
$ClC_6H_4CHC_6H_4Cl$ is a waxy solid of indefinite melting point with a weak, chemical odor.

Code Numbers: CAS 50-29-3   RTECS KJ3325000 UN (NA 2761)

DOT Designation: ORM-A

Synonyms: p,p'-DDT, 2,2-bis(p-chlorophenyl)-1,1,1-trichloroethane, dichlorodiphenyltrichloroethane, ENT 1506, dicophane, chlorophenothane, Gesarol®, Guesarol® and Neocid®.

Potential Exposure: DDT is a low-cost broad-spectrum insecticide. However, following an extensive review of health and environmental hazards of the use of DDT, U.S. EPA decided to ban further use of DDT in December 1972. This decision was based on several properties of DDT that had been well evidenced: (1) DDT and its metabolites are toxicants with long-term persistence in soil and water; (2) it is widely dispersed by erosion, runoff and volatilization; and (3) the low-water solubility and high lipophilicity of DDT result in concentrated accumulation of DDT in the fat of wildlife and humans which may be hazardous.

Human exposure to DDT is primarily by ingestion of contaminated food. Air and water intake is negligible and amounts to probably less than 0.01 mg/yr. Therefore, by EPA estimate total intake of DDT/yr for the average U.S. resident will be less than 3 mg/yr.

The entire population of the U.S. thus has some low level exposure to dietary contaminants. Minimal exposure from air and water sources, however, may be more important in previously heavily sprayed agricultural areas, where large amounts of residues may still be present. Groups at special risk are workmen in manufacturing plants and formulating plants and applicators, handlers and sprayers.

During such times as when exceptions are granted by the U.S. EPA for crop usage or during use for public health measures, those involved in handling or applying DDT may have considerable exposure.

Estimating the number of individuals at high risk due to occupational exposure is difficult. It is estimated that 8,700 workers are involved in formulating or manufacturing all pesticides. Since DDT constitutes much less than 10% of the total, the maximal number of exposed workers would be ~500. Since usage of DDT is severely limited, persons exposed by application would probably be fewer.

Incompatibilities: Strong oxidizers.

Permissible Exposure Limits in Air: The Federal limit and the 1983/84 ACGIH TWA value is 1 mg/m^3. The STEL value is 3 mg/m^3. The IDLH level has not been set.

Determination in Air: Collection on a filter, workup with isooctane, analysis by gas chromatography. See NIOSH Methods, Set S. See also reference (A-10).

Permissible Concentration in Water: To protect freshwater aquatic life— 0.0010 μg/ℓ as a 24 hr average; never to exceed 1.1 μg/ℓ. To protect saltwater aquatic life—0.0010 μg/ℓ as a 24 hr average; never to exceed 0.13 μg/ℓ. To protect human health—preferably zero. An additional lifetime cancer risk of 1 in 100,000 is imposed by a level of 0.24 ng/ℓ (0.00024 μg/ℓ).

Determination in Water: Gas chromatography (EPA Method 608) or gas chromatography plus mass spectrometry (EPA Method 625).

Routes of Entry: Inhalation, skin absorption, ingestion, eye and skin contact.

Harmful Effects and Symptoms: DDT is of moderate acute toxicity to man and most other organisms. However, its extremely low solubility in water (0.0012 ppm) and high solubility in fat (100,000 ppm) result in great bioconcentration. Its principal breakdown product, DDE (3), has very similar properties. Both compounds are also highly persistent in living organisms, so the major concern about DDT toxicity is related to its chronic effects (A-3).

Symptoms include paresthesia of tongue, lips and face; tremors; apprehension, dizziness, confusion, malaise, headaches; convulsions; paresis of the hands; vomiting; irritation of eyes and skin.

Points of Attack: Central nervous system, liver, kidneys, skin, peripheral nervous system.

Medical Surveillance: Consider the points of attack in preplacement and periodic physical examinations.

First Aid: If this chemical gets into the eyes, irrigate immediately. If this chemical contacts the skin, wash with soap promptly. If a person breathes in large amounts of this chemical, move the exposed person to fresh air at once and perform artificial respiration. When this chemical has been swallowed, get medical attention. Give large quantities of water and induce vomiting. Do not make an unconscious person vomit.

Personal Protective Methods: Wear appropriate clothing to prevent repeated or prolonged skin contact. Wear eye protection to prevent any reasonable probability of eye contact. Employees should wash promptly when skin is wet or contaminated. Work clothing should be changed daily if it is possible that clothing is contaminated. Remove nonimpervious clothing promptly if wet or contaminated.

Respirator Selection:
10 mg/m^3: CCROVDMPest/SA/SCBA
50 mg/m^3: CCROVFDMPest/GMOVDMPest/SAF/SCBAF
500 mg/m^3: CCROVHiEPest/SA:PD,PP,CF

Disposal Method Suggested: Incineration has been successfully used on a large scale for several years and huge incinerator equipment with scrubbers to catch HCl, a combustion product, are in use at several facilities such as Hooker Chemical, Dow Chemical and other producers of chlorinated hydrocarbon products. One incinerator operates at 900° to 1400°C with air and steam added which precludes formation of Cl$_2$. A few companies also construct incinerator-scrubber combinations of smaller size, e.g., a system built by Garver-Davis, Inc., of Cleveland, Ohio, for the Canadian government can handle 200 to 500 lb DDT/day as a kerosene solution (A-32).

References
(1) U.S. Environmental Protection Agency, *DDT: Ambient Water Quality Criteria,* Washington, DC (1980).
(2) U.S. Environmental Protection Agency, *DDT* Health and Environmental Effects Profile No. 60, Washington, DC, Office of Solid Waste (April 30, 1980).
(3) U.S. Environmental Protection Agency, *DDE* Health and Environmental Effects Profile No. 59, Washington, DC, Office of Solid Waste (April 30, 1980).
(4) International Agency for Research on Cancer, *IARC Monographs on the Carcinogenic Risks of Chemicals to Humans,* Lyon, France, 5, 83 (1974).
(5) National Cancer Institute *Bioassay of DDT, TDE and DDE for Possible Carcinogenicity,* Technical Report Series No. 131, Bethesda, MD (1978).
(6) World Health Organization, *DDT and Derivatives,* Environmental Health Criteria No. 9, Geneva, Switzerland (1979).
(7) Sax, N.I., Ed., *Dangerous Properties of Industrial Materials Report, 1,* No. 3, 51-54, New York, Van Nostrand Reinhold Co. (1981).
(8) Parmeggiani, L., Ed., *Encyclopedia of Occupational Health & Safety,* Third Edition, Vol. 1, pp 592-93, Geneva, International Labour Office (1983).
(9) United Nations Environment Programme, *IRPTC Legal File 1983,* Vol. I, pp VII/328-31, Geneva, Switzerland, International Register of Potentially Toxic Chemicals (1984).

DECABORANE

Description: B$_{10}$H$_{14}$ is a colorless solid with a pungent odor. It melts at 100°C.
Code Numbers: CAS 17702-41-9   RTECS HD1400000   UN 1868
OSHA Designation: Flammable Solid and Poison

Synonyms: None

Potential Exposure: Decaborane is used as a catalyst in olefin polymerization, in rocket propellants, in gasoline additives and as a vulcanizing agent for rubber.

Incompatibilities: Oxidizers, water, halogenated hydrocarbons.

Permissible Exposure Limits in Air: The Federal limit is 0.05 ppm (0.3 mg/m^3). The STEL value from ACGIH as of 1983/84 is 0.15 ppm (0.9 mg/m^3). ACGIH adds the notation "skin" indicating the possibility of cutaneous absorption. The IDLH level is 20 ppm.

Permissible Concentration in Water: No criteria set. (Decaborane hydrolyzes slowly in water.)

Routes of Entry: Inhalation, skin absorption, ingestion, eye and skin contact.

Harmful Effects and Symptoms: Decaborane's toxic effects are similar to pentaborane. Symptoms of central nervous system damage predominate; however, they are not as marked as the pentaborane. See Pentaborane.

Points of Attack: Central nervous system.

Medical Surveillance: Consider the points of attack in preplacement and periodic physical examinations.

First Aid: If this chemical gets into the eyes, irrigate immediately. If this chemical contacts the skin, wash with soap immediately. If a person breathes in large amounts of this chemical, move the exposed person to fresh air at once and perform artificial respiration. When this chemical has been swallowed, get medical attention. Give large quantities of water and induce vomiting. Do not make an unconscious person vomit.

Personal Protective Methods: Wear appropriate clothing to prevent any reasonable probability of skin contact. Wear eye protection to prevent any reasonable probability of eye contact. Employees should wash immediately when skin is wet or contaminated and daily at the end of each work shift. Work clothing should be changed daily if it is possible that clothing is contaminated. Remove nonimpervious clothing immediately if wet or contaminated. Provide emergency showers and eyewash.

Respirator Selection:
> 0.5 ppm: SA/SCBA
> 2.5 ppm: SAF/SCBAF
> 20 ppm: SAF:PD,PP,CF
> Escape: GMOVP/SCBA

Disposal Method Suggested: Incineration with aqueous scrubbing of exhaust gases to remove B_2O_3 particulates.

References
(1)   See Reference (A-61).
(2)   Sax, N.I., Ed., *Dangerous Properties of Industrial Materials Report, 1*, No. 8, 64-65, New York, Van Nostrand Reinhold Co. (1981).

1-DECENE

Description: $CH_3(CH_2)_7CH{=}CH_2$ is a liquid boiling at 171°C.

Code Numbers: CAS 872-05-9

DOT Designation: —

Synonyms: n-Decylene.

Potential Exposure: To those involved in manufacture, separation or utilization of this compound in detergent manufacture.

Permissible Exposure Limits in Air: No standards set.

Permissible Concentration in Water: No criteria set.

Routes of Entry: Skin contact, inhalation.

Harmful Effects and Symptoms: Irritates the skin upon contact and has narcotic properties upon inhalation.

Personal Protective Methods: Wear protective clothing.

Respirator Selection: Wear self-contained breathing apparatus.

Disposal Method Suggested: Incineration.

References

(1) Sax, N.I., Ed., *Dangerous Properties of Industrial Materials Report, 1,* No. 7, 49-50, New York, Van Nostrand Reinhold Co. (1981).
(2) Sax, N.I., Ed., *Dangerous Properties of Industrial Materials Report, 3,* No. 2, 73-74, New York, Van Nostrand Reinhold Co. (1983).

DEMETON

Description:

$$(C_2H_5O)_2 \overset{\overset{\text{S}}{\|}}{P}OCH_2CH_2SC_2H_5 \text{ (Demeton-O) plus}$$

$$(C_2H_5O)_2 \overset{\overset{\text{O}}{\|}}{P}SCH_2CH_2SC_2H_5 \text{ (Demeton-S)}$$

is a light brown liquid with an odor of sulfur compounds.

Code Numbers:

Mixture	CAS 8065-48-3	RTECS TF3150000	
Demeton-O	CAS 298-03-3	RTECS TF3125000	UN 2783
Demeton-S	CAS 126-75-0	RTECS TF3130000	UN 2783

DOT Designation: —

Synonyms: O,O-diethyl O-2(ethylthio)ethyl phosphorothioate (Demeton-O), O,O-diethyl S-2(ethylthio)ethyl phosphorothioate (Demeton-S), mercaptofos (in U.S.S.R.), ENT 17,295, Systox®.

Potential Exposure: Those involved in the manufacture (A-32), formulation and application of this systemic insecticide and acaricide.

Incompatibilities: Strong oxidizers.

Permissible Exposure Limits in Air: The Federal limit is 0.1 mg/m³. The ACGIH TWA value is 0.01 ppm (0.1 mg/m³) and the STEL value is 0.03 ppm (0.3 mg/m³) as of 1983/84. ACGIH adds the notation "skin" indicating the possibility of cutaneous absorption. The IDLH level is 20 mg/m³.

Permissible Concentration in Water: A criterion of 0.1 µg/ℓ demeton for freshwater and marine aquatic life is recommended by U.S. EPA as of 1977 since that concentration will not be expected to significantly inhibit acetylcholinesterase (AChE) over a prolonged period of time. In addition, the criterion recommendation is in close agreement with the criteria for the other organophosphates.

Routes of Entry: Inhalation, skin absorption, ingestion, eye and skin contact.

Harmful Effects and Symptomes: Miosis, aching eyes, rhinorrhea; frontal headaches; wheezing, laryngeal spasms, salivation; cyanosis; anorexia, nausea, vomiting, abdominal cramps, diarrhea; local sweating; muscle fasciculation; paralysis; dizziness, ataxia; convulsions; low blood pressure.

Points of Attack: Respiratory system, lungs, central nervous system, cardiovascular system, skin, eyes, blood cholinesterase.

Medical Surveillance: Consider the points of attack in preplacement and periodic physical examinations.

First Aid: If this chemical gets into the eyes, irrigate immediately. If this chemical contacts the skin, wash with soap immediately. If a person breathes in large amounts of this chemical, move the exposed person to fresh air at once and perform artificial respiration. When this chemical has been swallowed, get medical attention. Give large quantities of water and induce vomiting. Do not make an unconscious person vomit.

Personal Protective Methods: Wear appropriate clothing to prevent any possibility of skin contact. Wear eye protection to prevent any possibility of eye contact. Employees should wash immediately when skin is wet or contaminated. Remove nonimpervious clothing immediately if wet or contaminated. Provide emergency showers and eyewash.

Respirator Selection:
1 mg/m^3: SA/SCBA
5 mg/m^3: SAF/SCBAF
20 mg/m^3: SA:PD,PP,CF
Escape: GMOVPest/SCBA

Disposal Method Suggested: The thiono and thiolo isomers of this mixture are 50% hydrolyzed in 75 minutes and 0.85 minute, respectively at $20°C$ and pH 13. At pH 9 and $70°C$, the half life of demeton is 1.25 hr, but at pH 1 to 5 it is over 11 hr (A-32). Sand and crushed limestone may be added together with a flammable solvent; the resultant mixture may be burned in a furnace with afterburner and alkaline scrubber (A-38).

References
(1)   See Reference (A-61).
(2)   United Nations Environment Programme, *IRPTC Legal File 1983,* Vol. II, pp VII/596–7, Geneva, Switzerland, International Register of Potentially Toxic Chemicals (1984).

DEMETON-METHYL

Description:

$$\underset{(CH_3O)_2 \overset{\overset{O}{\|}}{P} SCH_2CH_2SCH_2CH_3}{} \quad \text{plus} \quad \underset{(CH_3O)_2 \overset{\overset{S}{\|}}{P} OCH_2CH_2SCH_2CH_3}{}$$

is a pale yellow oil.

Code Numbers: CAS 8022-00-2   RTECS TG1760000   UN 2783

DOT Designation: —

Synonyms: Metasystox®.

Potential Exposure: Those engaged in the manufacture (A-32), formulation and application of the insecticide and acaricide on agricultural and horticultural crops.

Incompatibilities: Strong oxidizers.

Permissible Exposure Limits in Air: There is no Federal standard but ACGIH as of 1983/84 has adopted a TWA value of 0.5 mg/m³ with the notation "skin" indicating the possibility of cutaneous absorption. They have also set an STEL value of 1.5 mg/m³.

Permissible Concentrations in Water: See Demeton.

Routes of Entry: Inhalation, skin absorption, ingestion, skin and eye contact.

Harmful Effects and Symptoms: Demeton methyl is basically a less toxic compound than demeton (A-34). With this in mind, refer to the data under Demeton for guidance.

Points of Attack: See Demeton for guidance.

Medical Surveillance: See Demeton for guidance.

First Aid: See Demeton for guidance.

Personal Protective Methods: See Demeton for guidance.

Respirator Selection: See Demeton for guidance.

Disposal Method Suggested: See Demeton for guidance.

References

(1)   Sax, N.I., Ed., *Dangerous Properties of Industrial Materials Report, 1,* No. 5, 68-69, New York, Van Nostrand Reinhold Co. (1981). (As Meta-Systox).

(2)   United Nations Environment Programme, *IRPTC Legal File 1983,* Vol. II, pp VII/621–23, Geneva, Switzerland, International Register of Potentially Toxic Chemicals (1984).

2,4-DES-SODIUM

Description: $Cl_2C_6H_3O(CH_2)_2OSO_3Na$, colorless odorless crystals, MP 170°C.

Code Numbers: CAS 136-78-7   RTECS KK4900000   UN 2765

DOT Designation: —

Synonyms: Sesone, Crag Herbicide No. 1®, sodium-2-(2,4-dichlorophenoxy)-ethyl sulfate, disul sodium.

Potential Exposure: Those involved in manufacture (A-32), formulation and application of this herbicide as well as citizens in the area of application.

Incompatibilities: Strong oxidizers.

Permissible Exposure Limits in Air: The Federal limit is 15 mg/m³. ACGIH as of 1983/84 has set a TWA of 10 mg/m³ and an STEL of 20 mg/m³. The IDLH level is 5,000 mg/m³.

Determination in Air: Collection on a filter and gravimetric analysis (A-1).

Permissible Concentration in Water: No criteria set.

Routes of Entry: Inhalation, ingestion, skin and eye contact.

Harmful Effects and Symptoms: None known in humans.

Point of Attack: None known.

Medical Surveillance: Nothing particularly indicated.

First Aid: If this chemical gets into the eyes, irrigate immediately. If it contacts the skin, flush with water promptly. If a person breathes in large amounts of this chemical, move the exposed person to fresh air at once and perform artificial respiration. When swallowed, get medical attention. Give large quantities of water and induce vomiting. Do not make an unconscious person vomit.

Personal Protective Methods: Wear appropriate clothing to prevent repeated or prolonged skin contact. Wear eye protection to prevent any reasonable probability of eye contact. Employees should wash promptly when skin is wet or contaminated. If it is possible that work clothing is contaminated, it should be changed daily. Remove nonimpervious clothing promptly if wet or contaminated.

Respirator Selection:
75 mg/m^3: DMXS
150 mg/m^3: DMXSQ/FuHiEP/SA/SCBA
750 mg/m^3: HiEPF/SAF/SCBAF
$5,000$ mg/m^3: PAPHiE/SA:PD,PP,CF
Escape: DMXS/SCBA

Disposal Method Suggested: $2,4\text{-}Cl_2C_6H_3OC_2H_4OSO_3Na$, is hydrolyzed by alkali to $NaHSO_4$ and apparently the dichlorophenoxyethanol (A-32).

DIACETONE ALCOHOL

Description: $(CH_3)_2C(OH)CH_2COCH_3$ is a colorless liquid with a mild odor.

Code Numbers: CAS 123-42-2   RTECS SA9100000   UN 1148

DOT Designation: Flammable liquid.

Synonymes: Diacetone; 4-hydroxy-4-methyl-2-pentanone; 2-methyl-2-pentanol-4-one.

Potential Exposure: NIOSH has estimated worker exposures at 1,350,000. It is used as a solvent for pigments, cellulose esters, oils and fats. It is used in hydraulic brake fluids and in antifreeze formulations.

Incompatibilities: Strong oxidizers, strong alkalies.

Permissible Exposure Limits in Air: The Federal limit and the 1983/84 ACGIH TWA value is 50 ppm (240 mg/m^3). The STEL value is 75 ppm (360 mg/m^3). The IDLH level is 2,100 ppm.

Determination in Air: Adsorption on charcoal workup with 2-propanol in CS$_2$, analysis by gas chromatography. See NIOSH Methods, Set E. See also reference (A-10).

Permissible Concentration in Water: No criteria set.

Routes of Entry: Inhalation, ingestion, skin and eye contact.

Harmful Effects and Symptoms: Irritation of eyes, nose, throat, corneal tissue damage, narcosis, skin irritation.

Points of Attack: Eyes, skin, respiratory system.

Medical Surveillance: Consider the points of attack in preplacement and periodic physical examinations.

First Aid: If this chemical gets into the eyes, irrigate immediately. If this chemical contacts the skin, flush with water promptly. If a person breathes in large amounts of this chemical, move the exposed person to fresh air at once and perform artificial respiration. When this chemical has been swallowed, get medical attention. Give large quantities of salt water and induce vomiting. Do not make an unconscious person vomit.

Personal Protective Methods: Wear appropriate clothing to prevent repeated or prolonged skin contact. Wear eye protection to prevent any reasonable probability of eye contact. Employees should wash promptly when skin is wet or contaminated. Remove nonimpervious clothing promptly if wet or contaminated.

Respirator Selection:
    1,000 ppm: CCROVF
    2,100 ppm: GMOV/SAF/SCBAF
    Escape: GMOV/SCBA

Disposal Method Suggested: Incineration.

References

(1) National Institute for Occupational Safety and Health, *Criteria for a Recommended Standard: Occupational Exposure to Ketones,* NIOSH Doc. No. 78-173 (1978).
(2) See Reference (A-60).
(3) See Reference (A-68).

DIALIFOR

Description: $C_{14}H_{17}ClNO_4PS_2$ has the structural formula

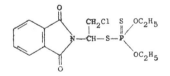

This is a white crystalline solid melting at 67° to 69°C.

Code Numbers: CAS 10311-84-9   RTECS TD5165000

DOT Designation: —

Synonyms: Phosphorodithioic acid 5-[2-chloro-1-(1,3-dihydro-1,3-dioxo-2H-isoindol-2-yl)ethyl] O,O-diethyl ester; Torak®; Dialifos.

Potential Exposure: Those involved in the manufacture, formulation and application of this insecticide.

Permissible Exposure Limits in Air: No standards set.

Permissible Concentration in Water: No criteria set.

Harmful Effects and Symptoms: This material is highly toxic (the LD-50 for rats is 5 mg/kg).

Disposal Method Suggested: Alkaline hydrolysis or incineration.

References

(1) Sax, N.I., Ed., *Dangerous Properties of Industrial Materials Report, 2,* No. 5, 43-45, New York, Van Nostrand Reinhold Co. (1982).

DIALLATE

● Carcinogen (Animal Positive, IARC) (EPA-CAG)
● Hazardous Waste Constituent (EPA)

Description: $C_{10}H_{17}Cl_2NOS$ has the following structural formula

$$[(CH_3)_2CH]_2NCOSCH_2CCl{=}CHCl$$

This is a brown liquid boiling at 150°C under 9 mm Hg pressure.

Code Numbers: CAS 2303-16-4   RTECS EZ8225000

DOT Designation: −

Synonyms: Bis(1-methylethyl)carbamothioic acid S-(2,3-dichloro-2-propenyl)ester; Avadex®.

Potential Exposure: Those involved in the manufacture, formulation and application of this preemergence herbicide.

Incompatibilities: Alkalis.

Permissible Exposure Limits in Air: 2.7 to 4.6 mg/m³ (1).

Permissible Concentration in Water: Below 0.1 ppm (1).

Harmful Effects and Symptoms: This is a moderately toxic compound (LD-50 value for rats is 395 mg/kg). Concentrates may cause irritation to skin, eyes and mucous membranes.

Points of Attack: Skin, eyes, nervous system.

Personal Protective Method: Wear goggles and protective clothing.

Respirator Selection: Wear canister masks.

References

(1) Sax, N.I., Ed., *Dangerous Properties of Industrial Materials Report, 3,* No. 1, 50-53, New York, Van Nostrand Reinhold Co. (1983).

2,4-DIAMINOANISOLE

● Carcinogen (Animal Positive) (A-63) (A-64)

Description: 2,4-Diaminoanisole has the structural formula

Code Numbers: CAS 615-05-4   RTECS ST2690000

DOT Designation: —

Synonyms: 4-methoxy-m-phenylenediamine.

Potential Exposures: The principal use of 2,4-diaminoanisole is as a component of oxidation (permanent) hair and fur dye formulations. Approximately three-quarters of the current oxidation hair dye formulations contain 2,4-diaminoanisole in concentrations ranging from approximately 0.05% to approximately 2%. The concentration is determined by the shade of the dye. Oxidation hair dyes are very common among professional as well as over-the-counter products and account for approximately $200 million of the $280 million annual retail expenditure for hair dyes. NIOSH is unaware of any current domestic production of 2,4-diaminoanisole. Imports of 2,4-diaminoanisole are on the order of 25,000 pounds per year.

NIOSH estimates that approximately 400,000 workers have potential occupational exposure to 2,4-diaminoanisole. Hairdressers and cosmetologists comprise the largest portion of workers with potential exposure. (Gloves are usually worn by hairdressers when applying hair dyes). A relatively small number of fur dyers are probably exposed to higher levels of 2,4-diaminoanisole.

Permissible Exposure Limits in Air: While the carcinogenicity of 2,4-diaminoanisole is being further evaluated, the National Institute for Occupational Safety and Health recommends, as an interim and prudent measure, that occupational exposure to 2,4-diaminoanisole and its salts be minimized. Exposures should be limited to as few employees as possible, while minimizing workplace exposure levels with engineering and work practice controls. In particular, skin exposures should be avoided.

Permissible Concentration in Water: No criteria set.

Routes of Entry: Skin contact.

Harmful Effects and Symptoms: NIOSH has conducted two epidemiologic studies which suggest excess cancer among cosmetologists.

A report (1) based on data from a case-control study of 25,416 hospital admissions between 1956 and 1965 at Roswell Park Memorial Institute suggests an excess of cancer of specific genital sites (corpus uteri, ovaries) among hairdressers and cosmetologists.

Another study currently being conducted by NIOSH is also suggestive of excess cancer among cosmetologists. This study involves a sample of 53,183 records which are representative of the 417,795 Social Security disability awards made to female workers between 1969 and 1972. Age and race adjusted proportional morbidity ratios (PMbR's) have been constructed for 24 selected occupational groups. Among cosmetologists, elevated PMbR's were observed for cancer of the digestive organs, respiratory system, trachea, bronchus and lung, breast, and genital organs. Cosmetologists had a greater number of elevated PMbR's for specific primary malignant neoplasms than any other tabulated occupational group.

Thus, the preliminary analysis of the Social Security Administration disability data is consistent with the hypothesis that persons employed in occupations classified within the broad category of cosmetology may be at elevated risks of developing a neoplasm from exposures of occupational origin. Other relevant epidemiologic studies with conflicting results have been reported. These studies do not clearly demonstrate an association between hair dyes and cancer. NIOSH believes that its studies do suggest an association between cancer and employment as cosmetologists and hairdressers. However, it is recognized that cosmetologists and hairdressers are exposed to a large variety of substances, and it is difficult at this time to attribute any excess incidence of cancer to either hair dyes in general or 2,4-diaminoanisole in particular.

Personal Protective Methods: Gloves are usually worn by hairdressers when applying hair dyes. Beyond that, NIOSH recommends minimization of exposure.

References

(1) *A Retrospective Survey of Cancer in Relation to Occupation,* DHEW (NIOSH) Publication No. 77-178, U.S. Department of Health, Education, and Welfare, Public Health Service, Center for Disease Control, National Institute for Occupational Safety and Health, Cincinnati, OH (1977).

(2) National Institute for Occupational Safety and Health, *2,4-Diaminoanisole in Hair and Fur Dyes,* Current Intelligence Bulletin No. 19, Washington, DC (Jan. 13, 1978).

(3) United Nations Environment Programme, *IRPTC Legal File 1983,* Vol. II, pp VII/531, Geneva, Switzerland, International Register of Potentially Toxic Chemicals (1984).

2,4-DIAMINOAZOBENZENE

● Carcinogen (Animal Positive, IARC) (1)

Description: This material with the formula $C_6H_5N=NC_6H_3(NH_2)_2$ is a yellow crystalline solid melting at 117.5°C. Its hydrochloride forms blue crystals or a red powder melting at 235°C.

Code Numbers:

Base	CAS 495-54-5	RTECS ST3325000
Hydrochloride	CAS 532-82-1	RTECS ST3380000

DOT Designation: —

Synonyms: 4-Phenylazo-m-phenylenediamine; chrysoidine; azobenzene-2,4-diamine.

Potential Exposure: In manufacture or use of this substance in dye manufacture.

Permissible Exposure Limits in Air: No standards set.

Permissible Concentration in Water: No criteria set.

Harmful Effects and Symptoms: This compound is carcinogenic in mice and more recently has been implicated in the induction of human bladder tumors. It is toxic to rat livers and irritating to rabbits eyes.

Points of Attack: Bladder and liver.

Medical Surveillance: Attention should be paid to points of attack.

Disposal Method Suggested: Incineration at high temperature with nitrogen oxide absorption from fluegases.

References

(1) U.S. Environmental Protection Agency, *Chemical Hazard Information Profile Draft Report: 2,4-Diaminoazobenzene,* Washington, DC (June 29, 1983).

4,4'-DIAMINODIPHENYLMETHANE

Description: 4,4'-Diaminodiphenylmethane, $H_2NC_6H_4CH_2C_6H_4NH_2$, is a crystalline solid melting at 91.5° to 92.0°C.

Code Numbers: CAS 101-77-9   RTECS BY5425000   UN 2651

Synonyms: DDM, p,p'-methylenedianiline, MDA, bis(4-aminophenyl)methane, DAPM.

Potential Exposure: Approximately 99% of the DDM produced is consumed in its crude form (occasionally containing not more than 50% DDM and poly-DDM) at its production site by reaction with phosgene in the preparation of isocyanates and polyisocyanates. These isocyanates and polyisocyanates are employed in the manufacture of rigid polyurethane foams which find application as thermal insulation. Polyisocyanates are also used in the preparation of the semiflexible polyurethane foams used for automotive safety cushioning.

DDM is also used as: an epoxy hardening agent; a raw material in the production of polyurethane elastomers; in the rubber industry as a curative for neoprene and as an antifrosting agent (antioxidant) in footwear; a raw material in the production of Qiana® nylon; and a raw material in the preparation of poly(amide-imide) resins (used in magnet wire enamels).

It is estimated that 2,500 workers are exposed to DDM. Many of these exposures to DDM are in the preparation of isocyanates and polyisocyanates and, on construction sites, in the application of epoxy coatings (A-5).

Permissible Exposure Limits in Air: ACGIH (1983/84) has adopted TWA values of 0.1 ppm (0.8 mg/m^3) and an STEL value of 0.5 ppm (4.0 mg/m^3).

Permissible Concentration in Water: No criteria set.

Routes of Entry: Inhalation, skin absorption.

Harmful Effects and Symptoms: In 1965, the hepatotoxic effects of DDM in humans were first seen in the so-called "Epping Jaundice" outbreak in Great Britain. In this incident 84 people who had eaten DDM-contaminated bread, experienced hepatocellular damage. DDM has also been shown to produce liver lesions in a group of intragastrically fed rats and has caused liver degeneration and spleen lesions in another group of DDM fed rats.

DDM in the occupational environment has been implicated in a number of cases of toxic hepatitis. During an 18 month period beginning April 1972, six cases of hepatitis developed among about 300 men who used epoxy resins in the construction of a nuclear power plant in Alabama. Two chemicals, DDM and 2-nitropropane, were held suspect in this study. Although experiments with rats have induced cancer, there have been no reported human cancers associated with DDM.

If, as hypothesized in the Center for Disease Control study of nuclear power plant construction workers, not all workers are susceptible to liver injury after exposure to DDM, and if the 1 to 2% incidence of liver disease seen in this study were applied to all workers with possible exposure to DDM, one would expect to see 25 to 50 cases of DDM-associated toxic hepatitis a year.

Medical Surveillance: See 4,4-methylenebis(2-chloroaniline) for a related material.

Personal Protective Methods: Worker exposure by all routes should be carefully controlled to levels consistent with animal and human experience, according to ACGIH.

Disposal Method Suggested: Controlled incineration (oxides of nitrogen are removed from the effluent gas by scrubbers and/or thermal devices).

2-4-DIAMINOTOLUENE

See "Toluene-2,4-Diamine."

DIANISIDINE

See "3,3'-Dimethoxybenzidine."

DIAZINON

● Hazardous substance (EPA)

Description: O,O-diethyl O-[6-methyl-2(1-methylethyl)-4-pyrimidinyl] phosphorothioate with the structural formula

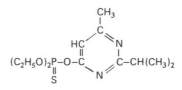

is a colorless liquid.

Code Numbers: CAS 333-41-5   RTECS TF3325000   UN (NA2783)

Dot Designation: ORM-A

Synonyms: Dimpylate, ENT 19507, Basudin®, Diazitol®, Neocidol®, Nucidol®.

Potential Exposures: To producers (A-32), formulators and applicators of this nonsystemic pesticide and acaricide. Diazinon is used in the United States on a wide variety of agricultural crops, ornamentals, domestic animals, lawns and gardens, and household pests.

Incompatibilities: Water.

Permissible Exposure Limits in Air: A TLV of 0.1 mg/m³ and a STEL value of 0.3 mg/m³ have been set by ACGIH as of 1983/84. ACGIH adds the notation "skin" indicating the possibility of cutaneous absorption.

Permissible Concentration in Water: A no-adverse effect level in drinking water has been calculated by NAS/NRC at 0.014 mg/ℓ. The acceptable daily intake has been established by WHO/FAO at 0.002 mg/kg.

Determination in Water: Gas-chromatographic and thin-layer chromatographic methods are available for the analysis of diazinon in water. These methods require extraction of the diazinon from the water before analysis. Sensitivity thus depends on the size of the sample used. The analytical sensitivity of the methods, however, is adequate to detect concentrations lower than the recommended no-adverse-effect concentration in samples of reasonable size, according to NAS/NRC (A-3).

Harmful Effects and Symptoms: The WHO/FAO reported in 1967 that diazinon causes no toxicologic effects in man at 0.02 mg/kg/day. This determination resulted from treatment of human subjects at 0.02, 0.025, and 0.05 mg per kilogram per day for 37 days. Plasma cholinesterase decrease was measured.

The mode of action of diazinon, as with other organophosphorus insecticides, is inhibition of the enzyme cholinesterase. Its acute toxicity, however, in comparison with other organophosphates, is only moderate. Its metabolism is straightforward and leads to metabolites that have little toxic potential. Subchronic- and chronic-feeding studies are sufficiently complete and indicate little problem with the use of diazinon.

Disposal Method Suggested: Diazinon is hydrolyzed in acid media about 12 times as rapidly as parathion, and at about the same rate as parathion in alkaline media. In excess water this compound yields diethylthiophosphoric acid and 2-isopropyl-4-methyl-6-hydroxypyrimidine. With insufficient water, highly toxic tetraethyl monothiopyrophosphate is formed (A-32). Therefore, incineration would be a preferable ultimate disposal method with caustic scrubbing of the incinerator effluent.

Reference
(1)   United Nations Environment Programme, *IRPTC Legal File 1983,* Vol. II, pp VII/598–600, Geneva, Switzerland, International Register of Potentially Toxic Chemicals (1984).

DIAZOMETHANE

● Carcinogen (animal positive, IARC) (1)

Description: CH_2N_2 is a yellow gas or a liquid under pressure.

Code Numbers: CAS 334-88-3   RTECS PA7000000

DOT Designation: —

Synonyms: Azimethylene, diazirine.

Potential Exposure: Diazomethane is a powerful methylating agent for acidic compounds such as carboxylic acids, phenols and enols. It is used in pesticide manufacture (A-32) and pharmaceutic manufacture (A-41).

Incompatibilities: Alkali metals, drying agents such as calcium sulfate.

Permissible Exposure Limits in Air: The Federal limit and the 1983/84 ACGIH TWA value is 0.2 ppm (0.4 mg/m^3). The ACGIH has set no STEL value. The IDLH level is 10 ppm.

Determination in Air: Adsorption on resin, workup with CS_2, analysis by gas chromatography. See NIOSH Methods, Set J. See also reference (A-10).

Permissible Concentration in Water: No criteria set, but EPA (A-37) has suggested a permissible ambient goal of 5.5 $\mu g/\ell$, based on health effects.

Routes of Entry: Inhalation, ingestion, skin and eye contact.

Harmful Effects and Symptoms: Cough, shortness of breath; headaches; flushed skin, fever; chest pain, pulmonary edema, pneumonitis; asthma; eye irritation.

Points of Attack: Respiratory system, lungs, eyes, skin.

Medical Surveillance: Consider the points of attack in preplacement and periodic physical examinations.

First Aid: If this chemical gets into the eyes, irrigate immediately. If this chemical contacts the skin, flush with water immediately. If a person breathes in large amounts of this chemical, move the exposed person to fresh air at once and perform artificial respiration. When this chemical has been swallowed, get medical attention. Do not induce vomiting.

Personal Protective Methods: Wear appropriate clothing to prevent skin contact or freezing. Wear eye protection to prevent any possibility of eye contact. Employees should wash promptly when skin is contaminated. Remove clothing immediately if wet or contaminated to avoid flammability hazard. Provide eyewash.

Respirator Selection:
> 2 ppm: SA/SCBA
> 10 ppm: SAF/SCBAF
> Escape: GMOV/SCBA

Disposal Method Suggested: Decompose chemically with ceric ammonium nitrate under constant agitation and cooling (A-38).

References

(1)   International Agency for Research on Cancer, *IARC Monographs on the Carcinogenic Risks of Chemical to Humans,* Lyon, France 7, 223 (1974).

(2)   Sax, N.I., Ed., *Dangerous Properties of Industrial Materials Report, 1,* No. 3, 55, New York, Van Nostrand Reinhold Co. (1981).

(3)   See Reference (A-61).

(4)   Parmeggiani, L., Ed., *Encyclopedia of Occupational Health & Safety,* Third Edition, Vol. 1, pp 620–21, Geneva, International Labour Office (1983).

DIBENZACRIDINES

See "Polynuclear Aromatic Hydrocarbons."

DIBENZOANTHRACENES

See "Polynuclear Aromatic Hydrocarbons."

DIBENZOCARBAZOLES

See "Polynuclear Aromatic Hydrocarbons."

DIBENZOPYRENES

See "Polynuclear Aromatic Hydrocarbons."

DIBORANE

Description: B_2H_6 is a colorless gas with a nauseating odor. It ignites spontaneously in moist air, and on contact with water, hydrolyzes exothermically forming hydrogen and boric acid.

Code Numbers: CAS 19287-45-7   RTECS HQ9275000   UN 1911

DOT Designation: Flammable gas and poison.

Synonyms: Boroethane, diboron hexahydride.

Potential Exposure: Diborane is used as a catalyst for olefin polymerization, a rubber vulcanizer, a reducing agent, a flame-speed accelerator, a chemical intermediate for other boron hydrides, and as a doping agent; and in rocket propellants and in the conversion of olefins to trialkyl boranes and primary alcohols.

Incompatibilities: Air, haolgenated compounds, aluminum, lithium, active metals, oxidized surfaces.

Permissible Exposure Limits in Air: The Federal limit is 0.1 ppm (0.1 mg/m^3). There is no STEL value set as of 1983/84. The IDLH level is 40 ppm.

Permissible Concentration in Water: No criteria set. (Diborane reacts on contact with water as noted above.)

Routes of Entry: Inhalation, skin and eye contact.

Harmful Effects and Symptoms: Diborane is the least toxic of the boron hydrides. In acute poisoning, the symptoms are similar to metal fume fever: tightness, heaviness and burning in chest, coughing, shortness of breath, chills, fever, pericardial pain, nausea, shivering, and drowsiness. Signs appear soon after exposure or after a latent period of up to 24 hr and persist for 1 to 3 days or more. Pneumonia may develop later. Reversible liver and kidney changes were seen in rats exposed to very high gas levels. This has not been noted in man.

Subacute poisoning is characterized by pulmonary irritation symptoms, and if this is prolonged, central nervous system symptoms such as headaches, dizziness, vertigo, chills, fatigue, muscular weakness, and only infrequent transient tremors, appear. Convulsions do not occur. Chronic exposure leads to wheezing, dyspnea, tightness, dry cough, rales, and hyperventilation which persist for several years.

Points of Attack: Respiratory system, lungs, central nervous system.

Medical Surveillance: Consider the points of attack in preplacement and periodic physical examinations.

First Aid: If this chemical gets into the eyes, irrigate immediately. If a person breathes in large amounts of this chemical, move the exposed person to fresh air at once and perform artificial respiration.

Personal Protective Methods: None applicable according to NIOSH (A-4), except respirator protection. A Japanese source (A-38), however, recommends rubber gloves, gum or teflon-coated coveralls and a full body shield.

Respirator Selection:
```
    1 ppm: SA/SCBA
    5 ppm: SAF/SCBAF
   40 ppm: SA:PD,PP,CF
  Escape: GMS/SCBA
```

Disposal Methods Suggested: Incineration with aqueous scrubbing of exhaust gases to remove B_2O_3 particulates (A-31).

References

(1) Sax, N.I., Ed., *Dangerous Properties of Industrial Materials Report, 2,* No. 1, 105-107, New York, Van Nostrand Reinhold Co. (1982).
(2) See Reference (A-61).
(3) See Reference (A-60).

DIBROMOCHLOROMETHANE

● Hazardous waste (EPA)

Description: $CHBr_2Cl$ is a clear colorless liquid boiling at $119°$ to $120°C$.

Code Numbers: CAS 124-48-1   RTECS PA6360000

DOT Designation: —

Synonyms: Chlorodibromomethane.

Potential Exposures: Dibromochloromethane is used as a chemical intermediate in the manufacture of fire extinguishing agents, aerosol propellants, refrigerants, and pesticides. Dibromochloromethane has been detected in drinking water in the United States. It is believed to be formed by the haloform reaction that may occur during water chlorination. Dibromochloromethane can be removed from drinking water via treatment with activated carbon. There is a potential for dibromochloromethane to accumulate in the aquatic environment because of its resistance to degradation. Volatilization is likely to be an important means of environmental transport.

Permissible Exposure Limits in Air: No standards set.

Permissible Concentration in Water: The Maximum Contaminant Level (MCL) for total trihalomethanes (including dibromochloromethane) in drinking water has been set by the U.S. EPA at 0.10 mg/ℓ (44 FR 68624).

Harmful Effects and Symptoms: Very little toxicity information is available. Dibromochloromethane gave positive results in mutagenicity tests with *Salmonella typhimurium* TA100. It is currently under test by the National Cancer Institute.

References

(1) U.S. Environmental Protection Agency, *Dibromochloromethane,* Health and Environmental Effects Profile No. 61, Washington DC, Office of Solid Waste (Apr. 30, 1980).

DIBROMOCHLOROPROPANE

● Carcinogen (Positive, NCI) (2) (animal positive, IARC) (3);
● Hazardous waste (EPA)

Description: $CH_2BrCHBrCH_2Cl$ is an amber to brown liquid with a pungent odor. It boils at $199°C$ under a pressure of 760 mm of Hg.

Code Numbers: CAS 96-12-8   RTECS TX8750000   UN 2872

DOT Designation: —

Synonyms: DBCP, Fumazone®, Nemagon®.

Potential Exposure: DBCP has been used in agriculture as a nematocide since 1955, being supplied for such use in the forms of liquid concentrate, emulsifiable concentrate, powder, granules, and solid material. Estimates of worker exposure to DBCP are not available but the number of workplaces affected are indicated by the fact that in August 1977, NIOSH alerted approximately 80 manufacturers and formulators to the potential hazard of worker exposure to DBCP.

Permissible Exposure Limits in Air: The National Institute for Occupational Safety and Health (NIOSH) recommended as of 1977 that employee exposure to dibromochloropropane shall be controlled in the workplace so that no employee is exposed to airborne dibromochloropropane at a concentration greater than 10 parts per billion ($\sim$0.1 mg/m^3) determined as a time-weighted average (TWA) concentration for up to a 10 hr work shift, 40 hr workweek. This was decreased to 1.0 ppb in 1978.

Permissible Concentration in Water: No criteria set.

Route of Entry: Inhalation of vapors.

Harmful Effects and Symptoms: The possible effects on the health of employees chronically exposed to DBCP may include sterility, diminished renal function, and degeneration and cirrhosis of the liver. In addition, ingestion of daily doses of DBCP by mice and rats has been found to result in the appearance of gastric cancers in both sexes of both species and in mammary cancers in female rats (5). Although an increased risk for cancer has not been seen with inhalation exposures, these results are not definitive, therefore the risk of cancer due to occupational exposure to DBCP remains a continuing concern. There are indications from in vitro experiments that mutagenic effects may occur also, but there has been no study yet of this possibility with mammalian subjects. Employees should be told of these possible effects and informed that some 20 to 25 years of experience in the manufacture and formulation of DBCP has not yet called such effects in employees of the pesticide industry to the notice of physicians and epidemiologists.

A rebuttable presumption against registration for pesticide uses was issued by U.S. EPA on September 22, 1977, on the basis of oncogenicity and reproductive effects. Then, as of November 3, 1977, EPA in a further action suspended all registrations of end use products, subject to various specific restrictions (A-32). Further, on November 9, 1979, the EPA suspended, unconditionally, all DBCP-containing products for end uses except for use on pineapples in Hawaii. It further cited an intent to cancel unconditionally all remaining end uses of all registered DBCP-containing pesticide products. The FDA established maximum residue levels of 1.5 mg/ℓ in raw milk and 0.05 mg/kg in all other raw agricultural commodities.

Medical Surveillance: Medical surveillance shall be made available to employees as outlined below:

(a) Comprehensive preplacement or initial medical and work histories with emphasis on reproductive experience and menstrual history.

(b) Comprehensive physical examination with emphasis on the genitourinary tract including testicle size and consistency in males.

(1) Semen analysis to include sperm count, motility and morphology.

(2) Other tests, such as serum testosterone, serum follicle stimu-
lating hormone (FSH), and serum lutenizing hormone (LH)
may be carried out if, in the opinion of the responsible physician, they are indicated. In addition, screening tests of the
renal and hepatic systems may be considered.

(3) A judgment of the worker's ability to use positive pressure
respirators.

(c) Employees shall be counseled by the physician to ensure that each
employee is aware that DBCP has been implicated in the production
of effects on the reproductive system including sterility in male
workers. In addition, they should be made aware that cancer was produced in some animals. While the relevancy of these findings is not yet
clearly defined, they do indicate that both employees and employers
should do everything possible to minimize exposure to DBCP.

(d) Periodic examinations containing the elements of the preplacement or
initial examination shall be made available on at least an annual basis.

(e) Examinations of current employees shall be made available as soon as
practicable after the promulgation of a standard for DBCP.

(f) Medical surveillance shall be made available to any worker suspected
of having been exposed to DBCP.

(g) Pertinent medical records shall be maintained for all employees subject to exposure to DBCP in the workplace. Such records shall be
maintained for 30 yr and shall be available to medical representatives
of the U.S. Government, the employer and the employee.

Personal Protective Methods: *Respiratory Protection* — Engineering controls
shall be used wherever needed to keep airborne dibromochloropropane concentrations below the recommended occupational exposure limit. Compliance with
this limit may be achieved by the use of respirators under the following conditions only:

(1) During the time necessary to install or test the required engineering
controls;

(2) For nonroutine operations, such as emergency maintenance or repair
activities; and

(3) During emergencies when air concentrations of dibromochloropropane may exceed the recommended occupational exposure limit.

When a respirator is permitted, it shall be selected and used pursuant to detailed requirements set forth by NIOSH (1).

Eye Protection — Eye protection shall be provided by the employer and used
by the employees where eye contact with liquid dibromochloropropane is likely.
Selection, use, and maintenance of eye protective equipment shall be in accordance with the provisions of the American National Standard Practice for Occupational and Educational Eye and Face Protection, ANSI Z87.1-1968. Unless eye
protection is afforded by a respirator hood or facepiece, protective goggles
[splash-proof safety goggles (cup-cover type dust and splash safety goggles)
which comply with 29 CFR 1910.133(a) (2)-(a) (6)] or a face shield (8 inch
minimum) shall be worn at operations where there is danger of contact of the
eyes with liquid dibromochloropropane because of spills or splashes. If there is
danger of liquid dibromochloropropane striking the eyes from underneath or

around the sides of the face shield, safety goggles shall be worn as added protection.

Protective Clothing – Protective clothing shall be resistant to the penetration and to the chemical action of dibromochloropropane. Additional protection, including gloves, bib-type aprons, boots, and overshoes, shall be provided for, and worn by, each employee during any operation that may cause direct contact with liquid dibromochloropropane. Supplied-air hoods or suits resistant to penetration by dibromochloropropane shall be worn when entering confined spaces, such as pits or storage tanks. In situations where heat stress is likely to occur, supplied-air suits, preferably cooled, are recommended. The employer shall ensure that all personal protective clothing is inspected regularly for defects and is maintained in a clean and satisfactory condition by the employee.

Respirator Selection: See reference (1) for details.

Disposal Method Suggested: Dibromochloropropane is reported to be stable to neutral and acid media. It is hydrolyzed by alkali to 2-bromoallyl alcohol. For recommended disposal procedure see ethylene dibromide (A-32).

References

(1) National Institute for Occupational Safety and Health, *Criteria for a Recommended Standard: Occupational Exposure to Dibromochloropropane,* NIOSH Doc. No. 78-115 (1978).

(2) National Cancer Institute, *Bioassay of Dibromochloropropane for Possible Carcinogenicity,* Technical Report Series No. 28, Bethesda, MD (1978).

(3) International Agency for Research on Cancer, *IARC Monographs on the Carcinogenic Risks of Chemicals to Humans,* Lyon, France 15, 139 (1977).

(4) Sax, N.I., Ed., *Dangerous Properties of Industrial Materials Report, 1,* No. 3, 55-57, New York, Van Nostrand Reinhold Co. (1981).

(5) See Reference (A-62). Also see Reference (A-64).

(6) See Reference (A-60).

(7) Parmeggiani, L., Ed., *Encyclopedia of Occupational Health & Safety,* Third Edition, Vol. 1, pp 621-623, Geneva, International Labour Office (1983).

(8) United Nations Environment Programme, *IRPTC Legal File 1983,* Vol. II, pp VII/669-71, Geneva, Switzerland, International Register of Potentially Toxic Chemicals (1984).

1,2-DIBROMOETHANE

See "Ethylene Dibromide."

DIBUTYLAMINOETHANOL

Description: $(C_4H_9)_2NCH_2CH_2OH$ is a colorless liquid with a faint amine-like odor which boils at 224° to 232°C.

Code Numbers: CAS 102-81-8 RTECS KK3850000

DOT Designation: –

Synonyms: N,N-Dibutylethanolamine; N,N-dibutyl-N-(2-hydroxyethyl)amine.

Potential Exposure: This material is used in organic synthesis.

Permissible Exposure Limits in Air: There is no Federal standard but ACGIH

as of 1983/84 has adopted a TWA value of 2 ppm (14 mg/m^3) with the notation "skin" indicating the possibility of cutaneous absorption. The STEL is 4 ppm (28 mg/m^3).

Permissible Concentration in Water: No criteria set.

Harmful Effects and Symptoms: In animal (rat) experiments, exposure to 22 ppm for 27 weeks resulted in no adverse health effects. Exposure to 33 ppm for 1 week resulted in 3% body weight loss, no significant changes in liver and kidney weight. Exposure to 70 ppm for 6 hr daily for 5 days resulted in the death of one rat, a 57% average body weight loss and a two-fold increase in liver and kidney to body weight ratio (A-34).

DI-tert-BUTYL-p-CRESOL

Description: $[(CH_3)_3C]_2C_6H_2(CH_3)OH$ is a white crystalline solid melting at 70°C.

Code Numbers: CAS 128-37-0   RTECS GO7875000

Synonyms: Butylated hydroxytoluene; DBPC; 2,6-di-tert-butyl methyl-phenol.

Potential Exposures: DBPC is used as an antioxidant to stabilize petroleum fuels, rubber and vinyl plastics. It is also used as an antioxidant in human foods and animal feeds.

Permissible Exposure Limits in Air: There are no Federal limits but ACGIH as of 1983/84 has adopted a TWA value of 10 mg/m^3 and set an STEL of 20 mg/m^3.

Permissible Concentration in Water: No criteria set.

Harmful Effects and Symptoms: This compound is not toxic to goldfish in saturated solution (A-36) and has an acute oral LD$_{50}$ for rats of 1,800 mg/kg which is classified as slightly toxic.

DIBUTYL PHOSPHATE

Description: $(n\text{-}C_4H_9O)_2PO(OH)$ is a colorless to brown, odorless liquid.

Code Numbers: CAS None   RTECS HS6300000

DOT Designation: —

Synonyms: Dibutyl acid o-phosphate; di-n-butyl hydrogen phosphate; dibutyl phosphoric acid.

Potential Exposure: This material is used as a catalyst in organic synthesis.

Incompatibilities: Strong oxidizers.

Permissible Exposure Limits in Air: The Federal value and the ACGIH TWA value (1983/84) is 1.0 ppm (5 mg/m^3). The STEL value is 2.0 ppm (10 mg/m^3). The IDLH level is 125 ppm.

Permissible Concentration in Water: No criteria set.

Routes of Entry: Inhalation, ingestion, eye and skin contact.

Harmful Effects and Symptoms: Headaches and irritation of skin, eyes, nose and respiratory system.

Points of Attack: Respiratory system, skin.

Medical Surveillance: Consider the points of attack in preplacement and periodic physical examinations.

First Aid: If this chemical gets into the eyes, irrigate immediately. If this chemical contacts the skin, wash with soap promptly. If a person breathes in large amounts of this chemical, move the exposed person to fresh air at once and perform artificial respiration. When this chemical has been swallowed, get medical attention. Give large quantities of water and induce vomiting. Do not make an unconscious person vomit.

Personal Protective Methods: Wear appropriate clothing to prevent any reasonable probability of skin contact. Wear eye protection to prevent any reasonable probability of eye contact. Employees should wash promptly when skin is wet or contaminated. Remove nonimpervious clothing promptly if wet or contaminated. Provide emergency showers.

Respirator Selection:

10 ppm:	SA/SCBA
50 ppm:	SAF/SCBAF
125 ppm:	SA:PD,PP,CF
Escape:	GMOVP/SCBA

References

(1) See Reference (A-61).

DIBUTYL PHTHALATE

- Hazardous substance (EPA)
- Hazardous waste (EPA)
- Priority toxic pollutant (EPA)

Description: $C_6H_4(COOC_4H_9)_2$ is a colorless oily liquid with a very weak aromatic odor. It boils at 340°C.

Code Numbers: CAS 84-74-2 RTECS TI0875000 UN (NA9095)

DOT Designation: ORM-E.

Synonyms: DBP; dibutyl 1,2-benzenedicarboxylate; di-n-butyl phthalate.

Potential Exposure: Use in plasticizing vinyl acetate emulsion systems and in plasticizing cellulose esters. Also used as insect repellent. Occupational exposure estimated by NIOSH at 1,000,000 workers.

Incompatibilities: Nitrates, strong oxidizers, strong alkalies, strong acids.

Permissible Exposure Limits in Air: The Federal limit is 5 mg/m^3 which is also the ACGIH TWA value as of 1983/84. The STEL value is 10 mg/m^3. The IDLH level is 9,300 mg/m^3.

Determination in Air: Collection on a filter, workup with CS$_2$, analysis by gas chromatography. See NIOSH Methods, Set D. See also reference (A-10).

Permissible Concentration in Water: To protect freshwater aquatic life—940

$\mu g/\ell$ on an acute basis and 3 $\mu g/\ell$ on a chronic basis for all phthalate esters. On a chronic basis, as low as 2,944 $\mu g/\ell$. To protect human health—34,000 $\mu g/\ell$.

Determination in Water: Gas chromatography (EPA Method 606) or gas chromatography plus mass spectrometry (EPA Method 625).

Routes of Entry: Inhalation, ingestion, eye and skin contact.

Harmful Effects and Symptoms: Irritation of nasal passages and upper respiratory system; stomach irritation; light sensitivity.

Points of Attack: Respiratory system, gastrointestinal system.

Medical Surveillance: Consider the points of attack in preplacement and periodic physical examinations.

First Aid: If this chemical gets into the eyes, irrigate promptly. If this chemical contacts the skin, wash regularly. If a person breathes in large amounts of the chemical, move the exposed person to fresh air at once and perform artificial respiration. When swallowed, get medical attention. Give large quantities of salt water to induce vomiting. Do not make an unconscious person vomit.

Personal Protective Methods: Wear eye protection to prevent any reasonable probability of eye contact.

Respirator Selection:
    250 mg/m³: HiEPF/SAF/SCBAF
    9,300 mg/m³: SAF:PD,PP,CF

Disposal Method Suggested: Incineration.

References
(1) U.S. Environmental Protection Agency, *Phthalate Esters: Ambient Water Quality Criteria,* Washington, DC (1980).
(2) National Institute for Occupational Safety and Health, *Profiles on Occupational Hazards for Criteria Document Priorities: Phthalates,* 97-103, Report PB-274,073, Cincinnati, OH (1977)
(3) U.S. Environmental Protection Agency, *Di-n-butyl Phthalate,* Health and Environmental Effects Profile No. 62, Washington, DC, Office of Solid Waste (April 30, 1980).
(4) See Reference (A-61).
(5) United Nations Environment Programme, *IRPTC Legal File 1983,* Vol. II, pp VII/632–4, Geneva, Switzerland, International Register of Potentially Toxic Chemicals (1984).

DICAMBA

● Hazardous Substance (EPA)

Description: $Cl_2C_6H_2(OCH_3)COOH$ is a colorless solid melting at 114° to 116°C.

Code Numbers: CAS 1918-00-9   RTECS DG7525000   UN: − (NA 2769)

DOT Designation: ORM-E

Synonyms: Dianat (in U.S.S.R.); MDBA (in Japan); 3,6-dichloro-o-anisic acid; 3,6-dichloro-2-methoxybenzoic acid; Banvel®; Mediben®.

Potential Exposure: Those involved in manufacture, formulation and application of this postemergence herbicide.

Permissible Exposure Limits in Air: No standards set.

Permissible Concentration in Water: A no-adverse effect level in drinking water has been claculated by NAS/NRC (A-2) at 0.009 mg/ℓ.

Determination in Water: A detection limit of 1 ppb for Dicamba by electron-capture gas chromatography has been reported by NAS/NRC (A-2).

Harmful Effects and Symptoms: The acute toxicity of Dicamba is relatively low. Dicamba produced no adverse effect when fed to rats at up to 19.3 mg/kg/day and 25 mg/kg/day in subchronic and chronic studies. The no-adverse-effect dose in dogs was 1.25 mg/kg/day in a 2-year feeding study. Based on these data, an ADI was calculated at 0.00125 mg/kg/day. No data were available on human exposure.

References

(1) United Nations Environment Programme, *IRPTC Legal File 1983,* Vol. I, pp VII/118–19, Geneva, Switzerland, International Register of Potentially Toxic Chemicals (1984).

DICHLOROACETYLENE

Description: $ClC{\equiv}CCl$ is a liquid with an isocyanide-like odor boiling at $32°C$.

Code Numbers: CAS 7572-29-4   RTECS AP1080000

DOT Designation: Forbidden.

Synonyms: None.

Potential Exposure: Dichloroacetylene may be produced by incineration of trichloroethylene below optimal furnace temperatures. Also in closed circuit anesthesia with trichloroethylene, heat and moisture produced by soda-lime absorption of CO_2 may produce dichloroacetylene along with phosgene and CO (A-34).

Incompatibilities: Shock, heat, acid, oxidizing materials.

Permissible Exposure Limits in Air: There are no Federal standards but ACGIH (1983/84) has set a ceiling value of 0.1 ppm (0.4 mg/m^3). There is no STEL value given.

Permissible Concentration in Water: No criteria set.

Harmful Effects and Symptoms: Headache, loss of appetite, extreme nausea, vomiting, involvement of the trigeminal nerve and facial muscles and the development of facial herpes. A number of fatalities attributed to dichloroacetylene have been reported (A-34).

DICHLOROBENZENES

- Hazardous substances (EPA)
- Hazardous wastes (EPA)
- Priority toxic pollutants (EPA)

Description: $C_6H_4Cl_2$, there are three isomeric forms: 1,2-dichlorobenzene is a colorless to pale yellow liquid with a pleasant, aromatic odor boiling at $179°C$; 1,3-dichlorobenzene is a liquid boiling at $172°C$; and 1,4-dichlorobenzene is a colorless solid with a mothball-like odor, boiling at $174°C$ but melting at $53°C$.

Code Numbers:

1,2-DCB	CAS 95-50-1	RTECS CZ45	UN 1591
1,3-DCB	CAS none	RTECS none	UN —
1,4-DCB	CAS 106-46-7	RTECS CZ455	UN 1592

DOT Designations: ORM-A.

Synonyms: 1,2-Dichloro—o-DCB, 1,2-DCB, ODB, ODCB, Dowtherm E®; 1,3-dichloro—m-DCB, 1,3-DCB; and 1,4-dichloro—p-DCB, 1,4-DCB, PDB, PDCB.

Potential Exposure: The major uses of 1,2-DCB are as a process solvent in the manufacturing of toluene diisocyanate and as an intermediate in the synthesis of dyestuffs, herbicides, and degreasers. 1,4-Dichlorobenzene is used primarily as an air deodorant and an insecticide, which account for 90% of the total production of this isomer. Information is not available concerning the production and use of 1,3-DCB. However, it may occur as a contaminant of 1,2- or 1,4-DCB formulations. Both 1,2-dichloro- and 1,4-dichlorobenzene are produced almost entirely as by-products during the production of monochlorobenzene. Combined annual production of these two isomers in the United States approaches 50,000 metric tons.

Incompatibilities: For o- or 1,2-dichlorobenzene, strong oxidizers, hot aluminum or aluminum alloys. For p- or 1,4-dichlorobenzene no hazardous incompatibilities are cited.

Permissible Exposure Limits in Air:

	Federal Standard	1983/84 STEL	IDLH Level
o-DCB	50 ppm (300 mg/m^3)*	none	1,700 ppm
p-DCB	75 ppm (450 mg/m^3)	110 ppm (675 mg/m^3)	1,000 ppm

*As a ceiling value.

Determination in Air: Charcoal adsorption followed by CS_2 workup and gas chromatographic analysis. See NIOSH Method, Set J for o-DCB and Set T for p-DCB. See also reference (A-10).

Permissible Concentration in Water: To protect freshwater aquatic life— 1,120 µg/ℓ on an acute toxicity basis and 763 µg/ℓ on a chronic basis. To protect saltwater aquatic life—1,970 µg/ℓ on an acute toxicity basis. To protect human health—400 µg/ℓ for all isomers.

Determination in Water: Gas chromatography (EPA Methods 601, 602, 612) or gas chromatography plus mass spectrometry (EPA Method 625). Gas-chromatographic methods have been developed for PDB with a sensitivity of 380 pg/cm peak height, and PDB concentrations as low as 1.0 ppb in water have been analyzed according to NAS/NRC.

Routes of Entry: Inhalation, ingestion, eye and skin contact for p-DCB. Also skin absorption for o-DCB.

Harmful Effects and Symptoms: Human exposure to dichlorobenzene is reported to cause hemolytic anemia and liver necrosis, and 1,4-dichlorobenzene has been found in human adipose tissue. In addition, the dichlorobenzenes are toxic to nonhuman mammals, birds, and aquatic organisms and impart an offensive taste and odor to water. The dichlorobenzenes are metabolized by mammals, including humans, to various dichlorophenols, some of which are as toxic as the dichlorobenzenes.

Persons with preexisting pathology (hepatic, renal, central nervous system, blood) or metabolic disorders, who are taking certain drugs (hormones or otherwise metabolically active), or who are otherwise exposed to DCBs or related (chemically or biologically) chemicals by such means as occupation, or domestic use or abuse (e.g., pica or "sniffing") of DCB products, might well be considered at increased risk from exposure to DCBs.

For o-DCB, irritation of eyes and nose; liver and kidney damage; skin blistering. For p-DCB headaches; eye irritation, periorbital swelling; profuse rhinitis; anorexia, nausea, vomiting, weight loss, jaundice, cirrhosis.

Points of Attack: For o-DCB—liver, kidneys, skin, eyes. For p-DCB—liver, respiratory system, eyes, kidneys, skin.

Medical Surveillance: Consider the points of attack in preplacement and periodic physical examinations.

First Aid: *For o-DCB* — If this chemical gets into the eyes, irrigate immediately. If this chemical contacts the skin, wash with soap promptly. If a person breathes in large amounts of this chemical, move the exposed person to fresh air at once and perform artificial respiration. When this chemical has been swallowed, get medical attention. Give large quantities of salt water and induce vomiting. Do not make an unconscious person vomit.

For p-DCB — If this chemical gets into the eyes, irrigate immediately. If this chemical contacts the skin, wash with soap. If a person breathes in large amounts of this chemical, move the exposed person to fresh air at once and perform artificial respiration. When this chemical has been swallowed, get medical attention. Give large quantities of water and induce vomiting. Do not make an unconscious person vomit.

Personal Protective Methods: Wear appropriate clothing to prevent repeated or prolonged skin contact. Wear eye protection to prevent any reasonable probability of eye contact. Employees should wash promptly when skin is wet or contaminated (and daily at the end of each work shift in the case of p-DCB). Remove nonimpervious clothing promptly if wet or contaminated with o-DCB.

Respirator Selection: *For o-DCB* —
    1,000 ppm: CCROVF
    1,700 ppm: GMOV/SAF/SCBAF
      Escape: GMOV/SCBA

For p-DCB —
    1,000 ppm: CCROVFD/GMOVD/SAF/SCBAF
      Escape: GMOVP/SCBA

Disposal Method Suggested: Incineration, preferably after mixing with another combustible fuel. Care must be exercised to assure complete combustion to prevent the formation of phosgene. An acid scrubber is necessary to remove the halo acids produced.

References

(1) U.S. Environmental Protection Agency, *Dichlorobenzene: Ambient Water Quality Criteria,* Washington, DC (1980).

(2) U.S. Environmental Protection Agency, *1,2-Dichlorobenzene,* Health and Environmental Effects Profile No. 64, Washington, DC, Office of Solid Waste (April 30, 1980).

(3) U.S. Environmental Protection Agency, *1,3-Dichlorobenzene,* Health and Environmental Effects Profile No. 65, Washington, DC, Office of Solid Waste (April 30, 1980).

(4) U.S. Environmental Protection Agency, *1,4-Dichlorobenzene,* Health and Environmental Effects Profile No. 66, Washington, DC, Office of Solid Waste (April 30, 1980).

(5) U.S. Environmental Protection Agency, *Dichlorobenzenes,* Health and Environmental Effects Profile No. 67, Washington, DC, Office of Solid Waste (April 30, 1980).

(6)   See Reference (A-61).
(7)   Sax, N.I., Ed., *Dangerous Properties of Industrial Materials Report, 4,* No. 2, 45–48, New York, Van Nostrand Reinhold Co. (1984) (1,3-Dichlorobenzene).
(8)   Sax, N.I., Ed., *Dangerous Properties of Industrial Materials Report, 4,* No. 2, 49–52, New York, Van Nostrand Reinhold Co. (1984) (1,4-Dichlorobenzene).
(9)   United Nations Environment Programme, *IRPTC Legal File 1983,* Vol. I, pp VII/79–81, Geneva, Switzerland, International Register of Potentially Toxic Chemicals (1984).

3,3'-DICHLOROBENZIDINE AND SALTS

● Carcinogen (Animal positive, IARC) (3)
● Hazardous waste (EPA)
● Priority toxic pollutant (EPA)

Description: $C_6H_3ClNH_2C_6H_3ClNH_2$, 3,3'-dichlorobenzidine, is a gray or purple crystalline solid, melting at 132° to 133°C.

Code Numbers: CAS 91-94-1   RTECS DD0525000

DOT Designation: —

Synonyms: 4,4'-Diamino-3,3'-dichlorobiphenyl, 3,3'-dichlorobiphenyl-4,4'-diamine, 3,3'-dichloro-4,4'-biphenyldiamine, DCB.

Potential Exposure: The major uses of dichlorobenzidine are in the manufacture of pigments for printing ink, textiles, plastics, and crayons and as a curing agent for solid urethane plastics. There are no substitutes for many of its uses.

It is estimated that between 250 and 2,500 workers receive exposure to DCB in the U.S. compared to 62 for benzidine. Additional groups that may be at risk include workers in the printing or graphic arts professions handling the 3,3'-DCB-based azo pigments. 3,3'-DCB may be present as an impurity in the pigments, and there is some evidence that 3,3'-DCB may be metabolically liberated from the azo pigment. More information on the level of exposure to the pigments is needed.

Permissible Exposure Limits in Air: 3,3'-Dichlorobenzidine and its salts are included in a Federal standard for carcinogens; all contact with it should be avoided. Skin absorption is possible. ACGIH (1983/84) has categorized DCB as an "Industrial Substance Suspect of Carcinogenic Potential for Man."

Determination in Air: Collection on a filter, elution with triethylamine in methanol, analysis by high performance liquid chromatography (A-10).

Permissible Concentration in Water: To protect freshwater and saltwater aquatic life—no criteria developed due to insufficient data. To protect human health—preferably zero. An additional lifetime cancer risk of 1 in 100,000 results at a level of 0.103 $\mu g/\ell$.

Determination in Water: Chloroform extraction followed by concentration and high performance liquid chromatography (EPA Method 605) or gas chromatography plus mass spectrometry (EPA Method 625).

Routes of Entry: Inhalation and probably percutaneous absorption.

Harmful Effects and Symptoms: *Local* — May cause allergic skin reactions.
Systemic — 3,3'-Dichlorobenzidine was shown to be a potent carcinogen in

rats and mice in feeding and injection experiments, but no bladder tumors were produced. However, no cases of human tumors have been observed in epidemiologic studies of exposure to the pure compound (6).

Medical Surveillance: Preplacement and periodic examinations should include history of exposure to other carcinogens, smoking, alcohol, medication, and family history. The skin, lung, kidney, bladder, and liver should be evaluated; sputum or urinary cytology may be helpful.

Personal Protective Methods: These are designed to supplement engineering controls and to prevent all skin or respiratory contact. Full body protective clothing and gloves should be used by those employed in handling operations. Fullface supplied air respirators of continuous flow or pressure demand type should also be used. On exit from a regulated area, employees should shower and change into street clothes, leaving their protective clothing and equipment at the point of exit to be placed in impervious containers at the end of the work shift for decontamination or disposal. Effective methods should be used to clean and decontaminate gloves and clothing.

Disposal Method Suggested: Incineration (1500°F, 0.5 second for primary combustion; 2200°F, 1.0 second for secondary combustion). The formation of elemental chlorine can be prevented through injection of steam or methane into the combustion process. NO_x may be abated through the use of thermal or catalytic devices.

References

(1) U.S. Environmental Protection Agency, *3,3'-Dichlorobenzidine: Ambient Water Quality Criteria,* Washington, DC (1980).

(2) U.S. Environmental Protection Agency, *3,3'-Dichlorobenzidine,* Health and Environmental Effects Profile No. 68, Washington, DC, Office of Solid Waste (April 30, 1980).

(3) International Agency for Research on Cancer, *IARC Monograph on the Carcinogenic Risks of Chemicals to Humans,* Lyon, France, 4, 49 (1974).

(4) Sax, N.I., Ed., *Dangerous Properties of Industrial Materials Report, 2,* No. 5, 45-49, New York, Van Nostrand Reinhold Co. (1982).

(5) Sax, N.I., Ed., *Dangerous Properties of Industrial Materials Report, 3,* No. 2, 79-82, New York, Van Nostrand Reinhold Co. (1983).

(6) See Reference (A-62). Also see Reference (A-64).

DICHLORODIFLUOROMETHANE

● Hazardous waste (EPA)

Description: CCl_2F_2 is a colorless gas with a characteristic ether-like odor at >20% by volume. It boils at -30°C.

Code Numbers: CAS 75-71-8   RTECS PA8200000   UN 1028

DOT Designation: Nonflammable gas.

Synonyms: Refrigerant 12; Freon 12; F-12; Propellant 12; Halon 122.

Potential Exposure: Dichlorodifluoromethane is used as an aerosol propellant, refrigerant and foaming agent (2).

Incompatibilities: Chemically active metals—sodium, potassium, calcium, powdered aluminum, zinc, magnesium.

Permissible Exposure Limits in Air: The Federal limit and the 1983/84 ACGIH TWA value is 1,000 ppm (4,950 mg/m³). The STEL value is 1,250 ppm (6,200 mg/m³).

Determination in Air: Sample collection by charcoal tube, analysis by gas liquid chromatography (A-10).

Permissible Concentration in Water: Human health protection—preferably 0. Additional lifetime cancer risk of 1 in 100,000 results at a level of 1.9 $\mu g/\ell$. In Jan. 1981 EPA (46FR2266) removed F-12 from the priority toxic pollutant list.

Determination in Water: Inert gas purge followed by gas chromatography with halide specific detector (EPA Method 601).

Routes of Entry: Inhalation, eye and skin contact.

Harmful Effects and Symptoms: Dizziness; tremors; unconsciousness; cardiac arrhythmia; cardiac arrest.

Points of Attack: CVS, PNS.

Medical Surveillance: Consider the points of attack in preplacement and periodic physical examination.

First Aid: If this chemical gets into the eyes, irrigate immediately. If this chemical contacts the skin, flush with water immediately. If a person breathes in large amounts of this chemical, move the exposed person to fresh air at once and perform artificial respiration.

Personal Protective Methods: Wear appropriate clothing to prevent skin contact or freezing. Wear eye protection to prevent any reasonable probability of eye contact. Remove clothing immediately if wet or contaminated.

Respirator Selection:
      10,000 ppm: SA/SCBA
      50,000 ppm: SAF/SCBAF/SAF:PD,PP,CF
         Escape: GMOV/SCBA

Disposal Method Suggested: Incineration, preferably after mixing with another combustible fuel. Care must be exercised to assure complete combustion to prevent the formation of phosgene. An acid scrubber is necessary to remove the halo acids produced.

References

(1) U.S. Environmental Protection Agency, *Halomethanes: Ambient Water Quality Criteria,* Washington, DC (1980).
(2) U.S. Environmental Protection Agency, *Trichlorofluoromethane and Dichlorodifluoromethane,* Health and Environmental Effects Profile No. 167, Washington, DC, Office of Solid Waste (April 30, 1980).
(3) See Reference (A-61).
(4) See Reference (A-60).

1,3-DICHLORO-5,5-DIMETHYLHYDANTOIN

Description: $C_5H_6Cl_2N_2O_2$ is a white solid with a chlorine-like odor and a structural formula ClNCONClCOC(CH₃)₂. It melts at 130°C.

Code Numbers: CAS 118-52-5   RTECS MU0700000

DOT Designation: —

Synonyms: Dactin®, Halane®.

Potential Exposures: It is used as a chlorinating agent, disinfectant and laundry bleach. It is used as a polymerization catalyst and in drug and pesticide synthesis.

Incompatibilities: Water, strong acids, easily oxidized materials, such as ammonia salts, sulfides, etc.

Permissible Exposure Limits in Air: The Federal limit is 0.2 mg/m^3. The 1983/84 STEL is 0.4 mg/m^3. The IDLH level is 5 mg/m^3.

Determination in Air: Sample collection by impinger or fritted bubbler, colorimetric determination (A-1).

Permissible Concentration in Water: No criteria set.

Routes of Entry: Inhalation, ingestion, eye and skin contact.

Harmful Effects and Symptoms: Irritation of eyes, mucous membrane and respiratory system.

Points of Attack: Respiratory system, lungs and eyes.

Medical Surveillance: Consider the points of attack in preplacement and periodic physical examinations.

First Aid: If this chemical gets into the eyes, irrigate immediately. If this chemical contacts the skin, wash with soap promptly. If a person breathes in large amounts of this chemical, move the exposed person to fresh air at once and perform artificial respiration. When this chemical has been swallowed, get medical attention. Give large quantities of salt water and induce vomiting. Do not make an unconscious person vomit.

Personal Protective Methods: Wear appropriate clothing to prevent repeated or prolonged skin contact. Wear eye protection to prevent any possibility of eye contact. Employees should wash promptly when skin is wet or contaminated. Work clothing should be changed daily if it is possible that clothing is contaminated. Remove nonimpervious clothing promptly if wet or contaminated. Provide emergency eyewash.

Respirator Selection:
 5 mg/m^3: CCRSFD/GMSHiE/SAF/SCBAF
 Escape: GMSHiE/SCBA

Disposal Method Suggested: Incineration (1500°F, 0.5 sec for primary combustion; 2200°F, 1.0 sec for secondary combustion). The formation of elemental chlorine can be prevented by injection of steam or methane into the combustion process. NO_x may be abated by the use of thermal or catalytic devices.

References

(1) See Reference (A-61).

2,3-DICHLORO-1,4-DIOXANE

● Carcinogen (Animal Positive) (1)

Description: $C_4H_6Cl_2O_2$ has the following structural formula

This is a colorless liquid boiling at 88° to 89°C at 19 mm Hg pressure.

Code Numbers: CAS 95-59-0. Transisomer: CAS 3883-43-0 RTECS JG9800000

DOT Designation: —

Potential Exposure: Used in the manufacture of the pesticide, dioxathion.

Permissible Exposure Limits in Air: No limits set.

Permissible Concentration in Water: No criteria set.

Routes of Entry: Inhalation and skin absorption.

Harmful Effects and Symptoms: The oral LD-50 for rats is about 2,000 mg/kg which is only in the "slightly toxic" range. However, carcinogenicity has been demonstrated in mice and rats and is suspected in humans by analogy with other chloroethers (1).

References

(1)   U.S. Environmental Protection Agency, *Chemical Hazard Information Profile Draft Report: 2,3-Dichloro-1,4-Dioxane*, Washington, DC (August 29, 1983).

1,1-DICHLOROETHANE

- Hazardous waste (EPA)
- Priority toxic pollutant (EPA)

Description: CH_3CHCl_2 is a colorless liquid with a chloroform-like odor boiling at 57.3°C.

Code Numbers: CAS 75-34-3   RTECS KI0175000   UN 2362

DOT Designation: Flammable liquid.

Synonyms: Asymmetrical dichloroethane; ethylidene chloride; 1,1-ethylidene dichloride.

Potential Exposure: NIOSH has estimated the number of workers exposed at 4,600. It is used as a solvent and cleaning and degreasing agent as well as in organic synthesis as an intermediate.

Incompatibilities: Strong oxidizers, strong caustics.

Permissible Exposure Limits in Air: The Federal limit is 100 ppm (400 mg/m³). The ACGIH TWA value as of 1983/84 is 200 ppm (810 mg/m³) and the STEL value is 250 ppm (1,010 mg/m³). The IDLH level is 4,000 ppm.

Determination in Air: Charcoal adsorption, workup with CS_2, analysis by gas chromatography. See NIOSH Methods, Set I. See also reference (A-10).

Permissible Concentration in Water: No criteria set for aquatic life or human health due to insufficient data.

Determination in Water: Inert gas purge followed by gas chromatography with halide specific detection (EPA Method 601) or gas chromatography plus mass spectrometry (EPA Method 624).

Routes of Entry: Inhalation, ingestion, eye and skin contact.

Harmful Effects and Symptoms: Central nervous system depression; skin irritation; drowsiness; unconsciousness; liver and kidney damage.

Points of Attack: Skin, liver, kidneys.

Medical Surveillance: Consider the points of attack in preplacement and periodic physical examinations.

First Aid: If this chemical gets into the eyes, irrigate immediately. If this chemical contacts the skin, flush with soap promptly. If a person breathes in large amounts of this chemical, move the exposed person to fresh air at once and perform artificial respiration. When this chemical has been swallowed, get medical attention. Give large quantities of salt water to induce vomiting. Do not make an unconscious person vomit.

Personal Protective Methods: Wear appropriate clothing to prevent repeated or prolonged skin contact. Wear eye protection to prevent any reasonable probability of eye contact. Employees should wash immediately when skin is wet or contaminated. Remove clothing immediately if wet or contaminated to avoid flammability hazard.

Respirator Selection:
    1,000 ppm: CCROV/SA/SCBA
    4,000 ppm: GMOV/SAF/SAF:PD,PP,CF/SCBAF
      Escape: GMOV/SCBA

Disposal Method Suggested: Incineration; preferably after mixing with another combustible fuel. Care must be exercised to assure complete combustion to prevent the formation of phosgene. An acid scrubber is necessary to remove the halo acids produced.

References

(1) U.S. Environmental Protection Agency, *Chloroethanes: Ambient Water Quality Criteria,* Washington, DC (1980).

(2) U.S. Environmental Protection Agency, *1,1-Dichloroethane,* Health and Environmental Effects Profile No. 69, Washington, DC, Office of Solid Waste (April 30, 1980).

(3) See Reference (A-61).

(4) Sax, N.I., Ed., *Dangerous Properties of Industrial Materials Report, 4,* No. 3, 44–47, New York, Van Nostrand Reinhold Co. (1984).

1,2-DICHLOROETHANE

See "Ethylene Dichloride."

1,2-DICHLOROETHYLENE

- Hazardous waste (EPA)
- Priority toxic pollutant (EPA)

Description: ClCH=CHCl, 1,2-dichloroethylene, exists in two isomers, cis 60% and trans 40%. There are variations in toxicity between these two forms. At room temperature, it is a liquid with a slight acrid, ethereal odor. Gradual decomposition results in hydrochloric acid formation in the presence of ultraviolet light or upon contact with hot metal.

Code Numbers: CAS 540-59-0   RTECS KV9360000   UN 1150

DOT Designation: Flammable liquid.

Synonyms: Acetylene dichloride, sym-dichloroethylene, 1,2-dichloroethylene.

Potential Exposure: 1,2-Dichloroethylene is used as a solvent for waxes, resins, and acetylcellulose. It is also used in the extraction of rubber, as a refrigerant, in the manufacture of pharmaceuticals and artificial pearls, and in the extraction of oils and fats from fish and meat.

Incompatibilities: Strong oxidizers.

Permissible Exposure Limits in Air: Federal standard is 200 ppm (790 mg/m^3). The ACGIH (1983/1984) STEL value is 250 ppm (1,000 mg/m^3). The IDLH level is 4,000 ppm.

Determination in Air: Charcoal adsorption, workup with CS_2, analysis by gas chromatography. See NIOSH Method, Set H. See also reference (A-10).

Permissible Concentration in Water: To protect freshwater aquatic life— 11,600 $\mu g/\ell$ on an acute toxicity basis for dichloroethylenes in general. To protect saltwater aquatic life: 224,000 $\mu g/\ell$ on an acute toxicity basis for dichloroethylenes as a class. To protect human health—no criteria developed due to insufficient data.

Determination in Water: Trans-1,2-dichloroethylene may be determined by inert gas purge followed by gas chromatography with halide specific detection (EPA Method 601) or gas chromatography plus mass spectrometry (EPA Method 624).

Routes of Entry: Inhalation of the vapor, ingestion, skin and eye contact.

Harmful Effects and Symptoms: *Local* — This liquid can act as a primary irritant producing dermatitis and irritation of mucous membranes.
Systemic — 1,2-Dichloroethylene acts principally as a narcotic, causing central nervous system depression. Symptoms of acute exposure include dizziness, nausea and frequent vomiting, and central nervous system intoxication similar to that caused by alcohol. Renal effects, when they do occur, are transient.

Points of Attack: Respiratory system, eyes, central nervous system.

Medical Surveillance: Consider possible irritant effects on skin or respiratory tract as well as liver and renal function in preplacement or periodic examinations. Expired air analyses may be useful in detecting exposure.

First Aid: If this chemical gets into the eyes, irrigate immediately. If this chemical contacts the skin, wash with soap promptly. If a person breathes in large amounts of this chemical, move the exposed person to fresh air at once and perform artificial respiration. When this chemical has been swallowed, get medical attention. Give large quantities of salt water and induce vomiting. Do not make an unconscious person vomit.

Personal Protective Methods: Wear appropriate clothing to prevent repeated or prolonged skin contact. Wear eye protection to prevent any reasonable probability of eye contact. Employees should wash promptly when skin is wet or contaminated. Remove clothing promptly if wet or contaminated to avoid flammability hazard.

Respirator Selection:
1,000 ppm: CCROVF
4,000 ppm: GMOV/SAF/SCBAF
Escape: GMOV/SCBA

Disposal Method Suggested: Incineration, preferably after mixing with another combustible fuel. Care must be exercised to assure complete combustion to prevent the formation of phosgene. An acid scrubber is necessary to remove the halo acids produced.

References

(1) U.S. Environmental Protection Agency, *Dichloroethylenes: Ambient Water Quality Criteria,* Washington, DC (1980).
(2) U.S. Environmental Protection Agency, *Trans-1,2-Dichloroethylene,* Health and Environmental Effects Profile No. 72, Washington, DC, Office of Solid Waste (April 30, 1980).
(3) U.S. Environmental Protection Agency, *Dichloroethylenes,* Health and Environmental Effects Profile No. 73, Washington, DC, Office of Solid Waste (April 30, 1980).
(4) See Reference (A-61).
(5) Sax, N.I., Ed., *Dangerous Properties of Industrial Materials Report, 4,* No. 3, 48–53, New York, Van Nostrand Reinhold Co. (1984).

DICHLOROETHYL ETHER

- Carcinogen (Animal positive, IARC) (3)
- Hazardous waste (EPA)
- Priority toxic pollutant (EPA)

Description: $ClCH_2CH_2OCH_2CH_2Cl$ is a clear, colorless liquid with a pungent, fruity odor. It is also described as having a chlorinated solvent-like odor. It boils at 176° to 178°C.

Code Numbers: CAS 111-44-4 RTECS KN0875000 UN 1916

DOT Designation: Flammable liquid, poison.

Synonyms: Dichloroether, dichloroethyl oxide, sym-dichloroethyl ether, bis(2-chloroethyl) ether, 2,2'dichloroethyl ether, BCEE, Chlorex®.

Potential Exposure: Dichloroethyl ether is used in the manufacture of paint, varnish, lacquer, soap, and finish remover. It is also used as a solvent for cellulose esters, naphthalenes, oils, fats, greases, pectin, tar, and gum; in dry cleaning; in textile scouring; and in soil fumigation.

Incompatibilities: Strong oxidizers.

Permissible Exposure Limits in Air: The Federal standard for dichloroethyl ether is 15 ppm (90 mg/m³); however, the ACGIH recommended TLV as of 1983/84 is 5 ppm (30 mg/m³). ACGIH adds the notation "skin" indicating the possibility of cutaneous absorption. The STEL value is 10 ppm (60 mg/m³). The IDLH level is 250 ppm.

Determination in Air: Charcoal adsorption, workup with CS_2, analysis by gas chromatography. See NIOSH Methods, Set V. See also reference (A-10).

Permissible Concentration in Water: To protect freshwater aquatic life—238,000 $\mu g/\ell$, for chloroalkyl ethers in general. No criteria developed for protection of saltwater aquatic life due to insufficient data. For the protection of human health—preferably zero. An additional lifetime cancer risk of 1 in 100,000 is posed by a concentration of 0.3 $\mu g/\ell$.

Determination in Water: CH_2Cl_2 extraction followed by gas chromatography with halogen; specific detector (EPA Method 611) or gas chromatography plus mass spectrometry (EPA Method 625).

Routes of Entry: Inhalation of vapor, percutaneous absorption, ingestion, skin and eye contact.

Harmful Effects and Symptoms: *Local* — Irritation of the conjunctiva of the eyes with profuse lacrimation, irritation to mucous membranes of upper respiratory tract, coughing, and nausea may result from exposure to vapor. The liquid when placed in animal eyes has produced damage. Vapors in minimal concentrations (3 ppm) are distinctly irritating and serve as a warning property.

Systemic — In animal experiments dichloroethyl ether has caused severe irritation of the respiratory tract and pulmonary edema. Animal experiments have also shown dichloroethyl ether to be capable of causing drowsiness, dizziness, and unconsciousness at high concentrations. Except for accidental inhalation of high concentrations, the chief hazard in industrial practice is a mild bronchitis which may be caused by repeated exposure to low concentrations.

Points of Attack: Respiratory system, skin and eyes.

Medical Surveillance: Consideration should be given to the skin, eyes, and respiratory tract, and to the central nervous system in placement or periodic examinations.

First Aid: If this chemical gets into the eyes, irrigate immediately. If this chemical contacts the skin, wash with soap. If a person breathes in large amounts of this chemical, move the exposed person to fresh air at once and perform artificial respiration. When this chemical has been swallowed, get medical attention. Give large quantities of water and induce vomiting. Do not make an unconscious person vomit.

Personal Protective Methods: Wear appropriate clothing to prevent reasonable probability of skin contact. Wear eye protection to prevent any reasonable probability of eye contact. Employees should wash immediately when skin is wet or contaminated. Remove nonimpervious clothing immediately if wet or contaminated. Provide emergency showers.

Respirator Selection:
　　150 ppm: CCROV/SA/SCBA
　　250 ppm: CCROVF/GMOV/SAF/SCBAF
　　Escape: GMOV/SCBA

Disposal Method Suggested: Incineration, preferably after mixing with another combustible fuel. Care must be exercised to assure complete combustion to prevent the formation of phosgene. An acid scrubber is necessary to remove the halo acids produced.

References

(1) U.S. Environmental Protection Agency, *Chloroalkyl Ethers: Ambient Water Quality Criteria,* Washington, DC (1980).

(2) U.S. Environmental Protection Agency, *Bis(2-Chloroethyl) Ether,* Health and Environmental Effects Profile No. 24, Washington, DC, Office of Solid Waste (April 30, 1980).

(3) International Agency for Research on Cancer, *IARC Monographs on the Carcinogenic Risks of Chemicals to Humans,* Lyon, France 9, 117 (1975).
(4) See Reference (A-61).
(5) See Reference (A-60).

DICHLOROMONOFLUOROMETHANE

Description: $CHCl_2F$ is a colorless liquid or gas with a slight ether-like odor. It boils at $9°C$.

Code Numbers: CAS 75-43-4 RTECS PA8400000 UN 1029

DOT Designation: Nonflammable gas.

Synonyms: Refrigerant 21; Freon 21; F-21; Halon 112; dichlorofluoromethane.

Potential Exposure: This material is used as a refrigerant and a propellent gas.

Incompatibilities: Chemically active metals—sodium potassium, calcium, powdered aluminum, zinc, magnesium.

Permissible Exposure Limits in Air: The Federal limit is 1,000 ppm (4,200 mg/m³). The ACGIH, as of 1983/84, has proposed a TWA value of 10 ppm (40 mg/m³), but has set no STEL value. The IDLH level is 50,000 ppm.

Determination in Air: Charcoal adsorption, workup with CS_2, analysis by gas chromatography. See NIOSH Methods, Set 1. See also reference (A-10).

Permissible Concentration in Water: No criteria set.

Routes of Entry: Inhalation, ingestion, eye and skin contact.

Harmful Effects and Symptoms: Asphyxia, cardiac arrhythmia; cardiac arrest.

Points of Attack: Respiratory system, lungs, cardiovascular system.

Medical Surveillance: Consider the points of attack in preplacement and periodic physical examinations.

First Aid: If this chemical gets into the eyes, irrigate immediately. If this chemical contacts the skin, flush with water immediately. If a person breathes in large amounts of this chemical, move the exposed person to fresh air at once and perform artificial respiration. When this chemical has been swallowed, get medical attention. Give large quantities of salt water and induce vomiting. Do not make an unconscious person vomit.

Personal Protective Methods: Wear appropriate clothing to prevent any possibility of skin contact. Wear eye protection to prevent any reasonable probability of eye contact. Remove clothing immediately if wet or contaminated.

Respirator Selection:
  10,000 ppm: SA/SCBA
  50,000 ppm: SAF/SCBAF/SA:PD,PP,CF
  Escape: GMOV/SCBA

Disposal Method Suggested: Incineration, preferably after mixing with another combustible fuel. Care must be exercised to assure complete combustion to prevent the formation of phosgene. An acid scrubber is necessary to remove the halo acids produced. Alternatively, it may simply be vented to the atmosphere.

References

(1)   See Reference (A-61).
(2)   See Reference (A-60).

1,1-DICHLORO-1-NITROETHANE

Description: $CH_3CCl_2NO_2$ is a colorless liquid with an unpleasant odor; causes tears. It boils at 125°C.

Code Numbers: CAS 594-72-9   RTECS KI1050000   UN 2650

DOT Designation: —

Synonyms: Ethide®.

Potential Exposure: This material is used as a fumigant insecticide. Therefore, those engaged in the manufacture (A-32), formulation and application of this material may be exposed.

Incompatibilities: Strong oxidizers.

Permissible Exposure Limits in Air: The Federal limit is a 10 ppm (60 mg/m³) ceiling value. The ACGIH has adopted as of 1983/84 a TWA of 2 ppm (10 mg/m³) and a STEL value of 10 ppm (60 mg/m³). The IDLH level is 150 ppm.

Determination in Air: Charcoal adsorption, workup with CS_2, analysis by gas chromatography. See NIOSH Methods, Set P. See also reference (A-10).

Permissible Concentration in Water: No criteria set.

Routes of Entry: Inhalation, ingestion, eye and skin contact.

Harmful Effects and Symptoms: In animals—lung irritation; liver, heart, kidney and blood vessel damage.

Points of Attack: Lungs.

Medical Surveillance: Consider the points of attack in preplacement and periodic physical examinations.

First Aid: If this chemical gets into the eyes, irrigate immediately. If this chemical contacts the skin, wash with soap immediately. If a person breathes in large amounts of this chemical, move the exposed person to fresh air at once and perform artificial respiration. When this chemical has been swallowed, get medical attention. Give large quantities of water and induce vomiting. Do not make an unconscious person vomit.

Personal Protective Methods: Wear appropriate clothing to prevent repeated or prolonged skin contact. Wear eye protection to prevent any reasonable probability of eye contact. Employees should wash promptly when skin is wet or contaminated. Remove nonimpervious clothing promptly if wet or contaminated.

Respirator Selection:
    150 ppm: SAF/SCBAF
        Escape: GMOV/SCBA

Disposal Method Suggested: Incineration (1500°F, 0.5 second for primary combustion; 2200°F, 1.0 second for secondary combustion). The formation of elemental chlorine can be prevented through injection of steam or methane into the combustion process. NO_x may be abated through the use of thermal or catalytic devices.

References
(1) See Reference (A-61).

2,4-DICHLOROPHENOL

- Hazardous waste (EPA)
- Priority toxic pollutant (EPA)

Description: $Cl_2C_6H_3OH$ is a colorless crystalline solid melting at 45°C and boiling at 210°C.

Code Numbers: CAS 120-83-2   RTECS SK8575000   UN 2021

DOT Designation: —

Synonyms: DCP

Potential Exposure: 2,4-Dichlorophenol is a commercially produced substituted phenol used entirely in the manufacture of industrial and agricultural products. As an intermediate in the chemical industry, 2,4-DCP is utilized as the feedstock for the manufacture of 2,4-dichlorophenoxyacetic acid (2,4-D), and 2,4-D derivatives (germicides, soil sterilants, etc.) and certain methyl compounds used in mothproofing, antiseptics and seed disinfectants. 2,4-DCP is also reacted with benzene sulfonyl chloride to produce miticides or further chlorinated to pentachlorophenol, a wood preservative. It is thus a widely used pesticide intermediate (A-32). The only group expected to be at risk for high exposure to 2,4-DCP is industrial workers involved in the manufacturing or handling of 2,4-DCP and 2,4-D.

Permissible Exposure Limits in Air: No standards set.

Permissible Concentration in Water: To protect freshwater aquatic life—2,020 $\mu g/\ell$ on an acute toxicity basis and 365 $\mu g/\ell$ on a chronic toxicity basis. To protect saltwater aquatic life—no criteria because of insufficient data. To protect human health—0.3 $\mu g/\ell$ based on organoleptic effects and 3,090 $\mu g/\ell$ based on toxicity data.

Determination in Water: Methylene chloride extraction followed by gas chromatography with flame ionization or electron capture detection (EPA Method 604) or gas chromatography plus mass spectrometry (EPA Method 625).

Harmful Effects and Symptoms: Although a paucity of aquatic toxicity data exists, 2,4-DCP appears to be less toxic than the higher chlorinated phenols. 2,4-DCP's toxicity to certain microorganisms and plant life has been demonstrated and its tumor promoting potential in mice has been reported. In addition, it has been demonstrated that 2,4-DCP can produce objectionable odors when present in water at extremely low levels. These findings, in conjunction with potential 2,4-DCP pollution by waste sources from commercial processes or the inadvertent production of 2,4-DCP due to chlorination of waters containing phenol, lead to the conclusion that 2,4-DCP represents a potential threat to aquatic and terrestrial life, including man. 2,4-DCP can irritate tissue and mucous membranes.

First Aid: Wash contaminated areas immediately with soap and water.

Personal Protective Measures: Butyl rubber gloves, protective clothing and safety shoes (A-38). Remove contaminated clothing immediately.

Respirator Selection: Use of a self-contained breathing apparatus is recommended.

Disposal Method Suggested: Dissolve in a combustible solvent and incinerate in a furnace equipped with afterburner and scrubber.

References

(1)   U.S. Environmental Protection Agency, *2,4-Dichlorophenol: Ambient Water Quality Criteria,* Washington, DC (1980).

(2)   U.S. Environmental Protection Agency, *Chemical Hazard Information Profile: Mono/Dichlorophenols,* Washington, DC (1979).

(3)   U.S. Environmental Protection Agency, *2,4-Dichlorophenol,* Health and Environmental Effects Profile No. 75, Washington, DC, Office of Solid Waste (April 30, 1980).

(4)   Sax, N.I., Ed., *Dangerous Properties of Industrial Materials Report, 1,* No. 7, 50-52, New York, Van Nostrand Reinhold Co. (1981).

(5)   See Reference (A-60).

2,6-DICHLOROPHENOL

● Hazardous waste (EPA)

Description: $Cl_2C_6H_3OH$ is a white crystalline solid melting at $68°$ to $69°C$ and having a strong odor similar to o-chlorophenol.

Code Numbers: CAS 87-65-0   RTECS SK8750000   UN 2021

DOT Designation: —

Synonyms: None

Potential Exposures: 2,6-Dichlorophenol is produced as a by-product from the direct chlorination of phenol. It is used primarily as a starting material for the manufacture of trichlorophenols, tetrachlorophenols, and pentachlorophenols.

Permissible Exposure Limits in Air: No standards set.

Permissible Concentration in Water: Based on the organoleptic properties of 2,6-dichlorophenol, a water quality criterion of 0.2 $\mu g/\ell$ has been recommended by the U.S. EPA in 1980.

Harmful Effects and Symptoms: There is no available information on the possible carcinogenic, teratogenic, or adverse reproductive effects of 2,6-dichlorophenol. The compound did not show mutagenic activity in the Ames assay. A single report has indicated that 2,6-dichlorophenol produced chromosome aberrations in rat bone marrow cells; details of this study were not available for evaluation. Prolonged administration of 2,6-dichlorophenol may produce hepatotoxic effects. Pertinent data on the toxicity of 2,6-dichlorophenol to aquatic organisms were not found in the available literature.

First Aid: See 2,4-Dichlorophenol.

Personal Protective Methods: See 2,4-Dichlorophenol.

Respirator Selection: See 2,4-Dichlorophenol.

Disposal Method Suggested: See 2,4-Dichlorophenol.

References

(1) U.S. Environmental Protection Agency, *2,6-Dichlorophenol,* Health and Environmental Effects Profile No. 76, Washington, DC, Office of Solid Waste (April 30, 1980).

1,2-DICHLOROPROPANE

- Hazardous substance (EPA)
- Hazardous waste (EPA)
- Priority toxic pollutant (EPA)

Description: $ClCH_2CHClCH_3$ is a colorless stable liquid boiling at 96°C. It has an unpleasant odor similar to chloroform.

Code Numbers: CAS 78-87-5   RTECS TX9625000   UN 1279

DOT Designation: Flammable liquid.

Synonyms: Propylene dichloride; ENT 15406.

Potential Exposure: Dichloropropane is used as a chemical intermediate in perchloroethylene and carbon tetrachloride synthesis and as a lead scavenger for antiknock fluids. It is also used as a solvent for fats, oils, waxes, gums and resins, and in solvent mixtures for cellulose esters and ethers. Other applications include the use of dichloropropane as a fumigant, alone and in combination with dichloropropene (1) (4), as a scouring compound, and a metal degreasing agent. It is also used as an insecticidal fumigant. NIOSH estimates that one million workers may be exposed each year.

Incompatibilities: Strong oxidizers, strong acids.

Permissible Exposure Limits in Air: The Federal standard and the ACGIH TWA value (1983/84) is 75 ppm (350 mg/m^3) The STEL value is 110 ppm (510 mg/m^3). The IDLH level is 2,000 ppm.

Determination in Air: Charcoal adsorption, workup with CS_2, analysis by gas chromatography. See NIOSH Methods, Set G. See also (A-13).

Permissible Concentration in Water: To protect freshwater aquatic life— 23,000 μg/ℓ on an acute toxicity basis and 5,700 μg/ℓ on a chronic basis. To protect saltwater aquatic life—10,300 μg/ℓ on an acute toxicity basis and 3,040 μg/ℓ on a chronic basis. To protect human health—no value set because of insufficient data.

Determination in Water: Inert gas purge followed by gas chromatography with halide specific detection (EPA Method 601) or gas chromatography plus mass spectrometry (EPA Method 624).

Routes of Entry: Inhalation of vapor, ingestion, eye and skin contact.

Harmful Effects and Symptoms: *Local* — Propylene dichloride may cause dermatitis by defatting the skin. More severe irritation may occur if it is confined against the skin by clothing. Undiluted, it is moderately irritating to the eyes, but does not cause permanent injury.

Systemic — In animal experiments, acute exposure to propylene dichloride produced central nervous system narcosis, fatty degeneration of the liver and kidneys.

Points of Attack: Skin, eyes, respiratory system, liver, kidneys.

Medical Surveillance: Evaluate the skin and liver and renal function on a periodic basis, as well as cardiac and respiratory status and general health.

First Aid: If this chemical gets into the eyes, irrigate immediately. If this chemical contacts the skin, wash with soap promptly. If a person breathes in large amounts of this chemical, move the exposed person to fresh air at once and perform artificial respiration. When this chemical has been swallowed, get medical attention. Give large quantities of salt water and induce vomiting. Do not make an unconscious person vomit.

Personal Protective Methods: Wear appropriate clothing to prevent repeated or prolonged skin contact. Wear eye protection to prevent any reasonable probability of eye contact. Employees should wash promptly when skin is wet or contaminated. Remove clothing immediately if wet or contaminated to avoid flammability hazard.

Respirator Selection:

        400 ppm: CCROV/SA/SCBA
      2,000 ppm: GMOV/SAF/SCBAF
        Escape: GMOV/SCBA

Disposal Method Suggested: Incineration, preferably after mixing with another combustible fuel. Care must be exercised to assure complete combustion to prevent the formation of phosgene. An acid scrubber is necessary to remove the halo acids produced.

References

(1) U.S. Environmental Protection Agency, *Dichloropropanes/Dichloropropenes: Ambient Water Quality Criteria,* Washington, DC (1980).

(2) National Institute for Occupational Safety and Health, *Profiles on Occupational Hazards for Criteria Document Priorities: Dichloropropane,* pp 292–294, Report PB-274,073, Cincinnati, OH (1977).

(3) U.S. Environmental Protection Agency, *1,2-Dichloropropane,* Health and Environmental Effects Profile No. 78, Washington, DC, Office of Solid Waste (April 30, 1980).

(4) U.S. Environmental Protection Agency, *Dichloropropanes/Dichloropropenes,* Health and Environmental Effects Profile No. 79, Washington, DC, Office of Solid Waste (April 30, 1980).

(5) See Reference (A-61).

(6) See Reference (A-60).

(7) United Nations Environment Programme, *IRPTC Legal File 1983,* Vol. II, pp VII/672–74, Geneva, Switzerland, International Register of Potentially Toxic Chemicals (1984).

DICHLOROPROPANOLS

● Hazardous waste constituents (EPA)

Description: $C_3H_6OCl_2$ is a colorless viscous liquid with a chloroform-like odor. There are 4 isomers as follows:

	Boiling Point (°C)
2,3-Dichloro-1-propanol	182
1,3-Dichloro-2-propanol	174
3,3-Dichloro-1-propanol	82–83
1,1-Dichloro-2-propanol	146–148

Code Numbers:

2,3-Dichloro-1-propanol	CAS 616-23-9	RTECS UB1225000	UN —
1,3-Dichloro-2-propanol	CAS 96-23-1	RTECS UB1400000	UN 2750

DOT Designation: —

Synonyms: Glycerol dichlorohydrin.

Potential Exposures: It is used as a solvent for hard resins and nitrocellulose, in the manufacture of photographic chemicals and lacquer, as a cement for Celluloid, and as a binder for water colors. It occurs in effluents from glycerol and halohydrin production plants.

Permissible Exposure Limits in Air: There are no U.S. standards but, the maximum allowable concentration of dichloropropanol in the working environment air in the U.S.S.R. is 5 mg/m³ (1).

Permissible Concentration in Water: There are no U.S. standards but the organoleptic limit in water set in the U.S.S.R. (1970) is 1.0 ng/ℓ (1).

Harmful Effects and Symptoms: There was no evidence found in the available literature to indicate that exposure to dichloropropanol produces carcinogenic effects. Conclusive evidence of mutagenic, teratogenic, or chronic effects of dichloropropanol was not found in the available literature. Acute exposure re-results in toxicity similar to that induced by carbon tetrachloride, including hepato- and nephrotoxicity. Data concerning the effects of dichloropropanol to aquatic organisms was not found in the available literature.

References

(1) U.S. Environmental Protection Agency, *Dichloropropanol,* Health and Environmental Effects Profile No. 80, Washington, DC, Office of Solid Waste (April 30, 1980).

1,3-DICHLOROPROPENE

- Hazardous substance (EPA)
- Hazardous waste (EPA)
- Priority toxic pollutant (EPA)

Description: $CHCl=CHCH_2Cl$ is a liquid boiling at 103° to 110°C.

Code Numbers: CAS 542-75-6   RTECS UC8310000   UN 2047

DOT Designation: Flammable liquid.

Synonyms: α-Chlorallyl chloride, Telone®.

Potential Exposure: Workers engaged in manufacture, formulation and application of this soil fumigant and nematocide. It is used in combinations with dichloropropanes as a soil fumigant also (1)(4).

Permissible Exposure Limits in Air: There is no Federal limit but ACGIH has adopted a TWA of 1.0 ppm (5 mg/m³) and a STEL of 10 ppm (50 mg/m³) as of 1983/84. The additional notation "skin" indicates the possibility of cutaneous absorption.

Permissible Concentration in Water: To protect freshwater aquatic life— 6,060 μg/ℓ on an acute toxicity basis and 244 μg/ℓ on a chronic basis. To protect saltwater aquatic life—790 μg/ℓ on an acute toxicity basis. To protect human health—87.0 μg/ℓ.

Determination in Water: Inert gas purge followed by gas chromatography with halide specific detection (EPA Method 601) or gas chromatography plus mass spectrometry (EPA Method 624).

Harmful Effects and Symptoms: The actions of dichloropropenes on living organisms seem to depend upon the isomer (volatility, solubility, etc.) and the individual organisms. Additionally, judging by the rapid excretion of dichloropropenes in rats, it is unlikely that these compounds will remain and accumulate in mammals. Dichloropropenes were shown to be mutagenic. However, they have a low tumor-causing potential if any at all. Data on the toxicology of dichloropropenes have been published by Torkelson and Oyen (2). Dichloropropene is irritating to mucous membranes and has caused liver and kidney injury in animals.

Personal Protective Methods: Wear rubber gloves and protective clothing.

Respirator Selection: Use of self-contained breathing apparatus is recommended (A-38).

Disposal Method Suggested: Incineration, preferably after mixing with another combustible fuel. Care must be exercised to assure complete combustion to prevent the formation of phosgene. An acid scrubber is necessary to remove the halo acids produced.

References

(1)   U.S. Environmental Protection Agency, *Dichloropropanes/Dichloropropenes: Ambient Water Quality Criteria,* Washington, DC (1980).

(2)   Torkelson, R.R. and Oyen, F., "The Toxicity of 1,3-Dichloropropene as Determined by Repeated Exposure of Laboratory Animals," *Jour. Am Ind. Hyg. Assoc.,* 38, 217 (1977).

(3)   U.S. Environmental Protection Agency, *1,3-Dichloropropene,* Health and Environmental Effects Profile No. 81, Washington, DC, Office of Solid Waste (April 30, 1980).

(4)   U.S. Environmental Protection Agency, *Dichloropropanes/Dichloropropenes,* Health and Environmental Effects Profile No. 79, Washington, DC, Office of Solid Waste (April 30, 1980).

(5)   See Reference (A-60).

(6)   United Nations Environment Programme, *IRPTC Legal File 1983,* Vol. II, pp VII/683–84, Geneva, Switzerland, International Register of Potentially Toxic Chemicals (1984).

2,2-DICHLOROPROPIONIC ACID

● Hazardous substance (EPA)

Description: CH_3CCl_2COOH is a colorless liquid boiling at 185° to 190°C.

Code Numbers: CAS 75-99-0   RTECS UF0690000   UN (NA 1759)

DOT Designation: Corrosive material.

Synonyms: Proprop (in South Africa); DPA (in Japan); and sodium salt—Dowpon®, Radapon®, Dalzpon®.

Potential Exposure: Those involved in the manufacture (A-32), formulation and application of the herbicide.

Permissible Exposure Limits in Air: There is no Federal standard but ACGIH (1983/84) has adopted a TWA value of 1 ppm (6 mg/m³) but set no STEL value.

Permissible Concentration in Water: No criteria set.

Harmful Effects and Symptoms: This material is corrosive to the skin and can cause permanent eye injury. Medical reports following acute exposure show mild to moderate skin, eye, respiratory and gastrointestinal responses (A-34).

Personal Protective Methods: Wear rubber gloves, goggles, hats, suits and boots.

Respirator Selection: Use self-contained breathing apparatus.

References

(1) Sax, N.I., Ed., *Dangerous Properties of Industrial Materials Report, 3,* No. 2, 74-77, New York, Van Nostrand Reinhold Co. (1983).
(2) United Nations Environment Programme, *IRPTC Legal File 1983,* Vol. II, pp VII/690, 692, Geneva, Switzerland, International Register of Potentially Toxic Chemicals (1984).

DICHLOROTETRAFLUOROETHANE

Description: $F_2ClCCClF_2$ is a colorless liquid or gas with a very slight, ethereal odor.

Code Numbers: CAS 76-14-2   RTECS KI1100000   UN 1958

DOT Designation: Nonflammable gas.

Synonyms: 1,2-Dichlorotetrafluoroethane; Freon 114; Refrigerant 114; Halon 242, F-114.

Potential Exposure: This material is used as a refrigerant and also as a propellant gas.

Incompatibilities: Chemically active metals—sodium potassium, calcium, powdered aluminum, zinc, and magnesium.

Permissible Exposure Limits in Air: The Federal limit and the 1983/84 ACGIH TWA value is 1,000 ppm (7,000 mg/m^3). The STEL value is 1,250 ppm (8,750 mg/m^3). The IDLH level is 50,000 ppm.

Determination in Air: Charcoal adsorption followed by workup with CH_2Cl_2 and analysis by gas chromatography. See NIOSH Method, Set T. See also reference (A-10).

Permissible Concentration in Water: No criteria set.

Routes of Entry: Inhalation, ingestion, eye and skin contact.

Harmful Effects and Symptoms: Respiratory irritation, asphyxia; cardiac arrhythmia, cardiac arrest.

Points of Attack: Respiratory system, lungs, cardiovascular system.

Medical Surveillance: Consider the points of attack in preplacement and periodic physical examinations.

First Aid: If this chemical gets into the eyes, irrigate immediately. If this chemical contacts the skin, flush with water immediately. If a person breathes in large amounts of this chemical, move the exposed person to fresh air at once and perform artificial respiration. When this chemical has been swallowed, get medical attention. Give large quantities of salt water to induce vomiting. Do not make an unconscious person vomit.

Personal Protective Methods: Wear appropriate clothing to prevent any possibility of skin contact with liquid. Wear eye protection to prevent any reasonable probability of eye contact. Remove clothing immediately if wet or contaminated.

Respirator Selection:
10,000 ppm: SA/SCBA
50,000 ppm: SAF/SCBAF/SAF:PD,PP,CF
Escape: GMOV/SCBA

References
(1) See Reference (A-61).

DICHLORVOS

● Hazardous substance (EPA)

Description: $(CH_3O)_2\overset{\overset{\displaystyle O}{\displaystyle \|}}{P}OCH{=}CCl_2$ is a colorless to amber liquid with a mild aromatic odor.

Code Numbers: CAS 62-73-7   RTECS TC0350000   UN 2783

DOT Designation: Poison B.

Synonyms: DDVP; 2,2-dichlorovinyldimethylphosphate, DDVF (in U.S.S.R.), Vapona®, Nogos®, Nuvan®, Dedevap®.

Potential Exposure: Those involved in manufacture (A-32), formulation and application of this fumigant insecticide in household, public health and agricultural uses.

Incompatibilities: None hazardous.

Permissible Exposure Limits in Air: The Federal limit and the 1983/84 ACGIH TWA value is 0.1 ppm (1 mg/m³). The STEL value is 0.3 mg/m³. The IDLH level is 200 mg/m³.

Determination in Air: Sample collection by impinger or fritted bubbler; analysis by gas liquid chromatography (A-1).

Permissible Concentration in Water: No criteria set.

Routes of Entry: Inhalation, skin absorption, ingestion, eye and skin contact.

Harmful Effects and Symptoms: Miosis, aching eyes, rhinorrhea, frontal headaches; wheezing, laryngeal spasms, salivation; cyanosis; anorexia, nausea, vomiting, diarrhea; sweating; muscular fasciculation, paralysis, ataxia; convulsions; low blood pressure. Issuance of a rebuttable presumption against registration for dichlorvos was being considered by U.S. EPA as of October 1980, on the basis of possible mutagenicity, reproductive and fetotoxic effects, oncogenicity and neurotoxicity.

Points of Attack: Respiratory system, cardiovascular system, central nervous system, eyes, skin, blood cholinesterase.

Medical Surveillance: Consider the points of attack in preplacement and periodic physical examinations.

First Aid: If this chemical gets into the eyes, irrigate immediately. If this

chemical contacts the skin, wash with soap immediately. If a person breathes in large amounts of this chemical, move the exposed person to fresh air at once and perform artificial respiration. When this chemical has been swallowed, get medical attention. Give large quantities of water and induce vomiting. Do not make an unconscious person vomit.

Personal Protective Methods: Wear appropriate clothing to prevent repeated or prolonged skin contact. Wear eye protection to prevent any reasonable probability of eye contact. Employees should wash immediately when skin is wet or contaminated. Remove nonimpervious clothing immediately if wet or contaminated.

Respirator Selection:
> 10 mg/m^3: SA/SCBA
> 50 mg/m^3: SAF/SCBAF
> 200 mg/m^3: SA:PD,PP,CF
> Escape: GMOVPPest/SCBA

Disposal Method Suggested: 50% hydrolysis is obtained in pure water in 25 minutes at 70°C and in 61.5 days at 20°C. A buffered solution yields 50% hydrolysis (37.5°C) in 301 minutes at pH 8, 462 minutes at pH 7, 4,620 minutes at pH 5.4. Hydrolysis yields no toxic residues (A-32). Incineration in a furnace equipped with an afterburner and alkaline scrubber is recommended (A-38).

References

(1) U.S. Environmental Protection Agency, *Investigation of Selected Potential Environmental Contaminants: Haloalkyl Phosphates,* Report EPA-560/2-76-007, Washington, DC (August 1976).

(2) See Reference (A-61).

(3) Sax, N.I., Ed., *Dangerous Properties of Industrial Materials Report, 1,* No. 3, 57-59, New York, Van Nostrand Reinhold Co. (1981).

(4) United Nations Environment Programme, *IRPTC Legal File 1983,* Vol. II, pp VII/551–53, Geneva, Switzerland, International Register of Potentially Toxic Chemicals (1984).

DICROTOPHOS

Description: $(CH_3O)_2POC(CH_3)=CHCN(CH_3)_2$, 3-(dimethylamino)-1-methyl-3-oxo-1-propenyl dimethyl phosphate, is an amber liquid with a mild ester odor, boiling at 400°C.

Code Numbers: CAS 141-66-2   RTECS TC3850000   UN 2783

DOT Designation: —

Synonyms: ENT 24482, Carbicron®, Ektafos®, Bidrin®.

Potential Exposure: Those involved in the manufacture (A-32), formulation and application of this insecticide and acaricide.

Permissible Exposure Limits in Air: There is no Federal standard but ACGIH as of 1983/84 has adopted a TWA value of 0.25 mg/m^3. No STEL value is proposed. The TWA bears the notation "skin" indicating the possibility of cutaneous absorption.

Permissible Concentration in Water: No criteria set.

Routes of Entry: Skin absorption and inhalation.

Harmful Effects and Symptoms: Dicrotophos is a cholinestelase inhibitor which can penetrate the skin but effects do not seem to be cumulative (A-34). Symptoms include the following (A-38): headache, anorexia, nausea, vertigo, weakness, abdominal cramps, diarrhea, salivation, lachrymation, ataxia, tremor of lower extremities, muscle twitching, cyanosis, pulmonary edema, areflexia, sphincter ataxia, convulsion, coma, heartblock, shock and oligopnea.

Medical Surveillance: Determine cholinesterase weekly.

First Aid: Flush eyes with water. Wash contaminated areas of body with soap and water. Give artificial respiration and oxygen. Use gastric lavage (stomach wash) if swallowed followed by saline catharsis.

Personal Protective Methods: Wear rubber gloves, safety glasses and protective clothing.

Disposal Method Suggested: Dicrotophos decomposes after 31 days at $75°C$ or 7 days at $90°C$. Hydrolysis is 50% complete in aqueous solutions at $38°C$ after 50 days at pH 9.1 (100 days are required at pH 1.1). Alkaline hydrolysis (NaOH) yields $(CH_3)_2NH$ (A-32). Incineration is also recommended as a disposal method (A-38).

References

(1)   Sax, N.I., Ed., *Dangerous Properties of Industrial Materials Report, 2,* No. 5, 49-54, New York, Van Nostrand Reinhold Co. (1982).

(2)   United Nations Environment Programme, *IRPTC Legal File 1983,* Vol. II, pp VII/557, Geneva, Switzerland, International Register of Potentially Toxic Chemicals (1984).

DICYCLOHEXYLAMINE

Description: $C_6H_{11}-NH-C_6H_{11}$ is a colorless liquid boiling at $256°C$ with a faint amine odor.

Code Numbers: CAS 101-83-7   RTECS HY4025000   UN 2565

DOT Designation: —

Synonyms: Dodecahydrodiphenylamine; di-CHA.

Potential Exposure: Dicyclohexylamine salts of fatty acids and sulfuric acid have soap and detergent properties useful to the printing and textile industries. Metal complexes of di-CHA are used as catalysts in the paint, varnish, and ink industries. Several vapor-phase corrosion inhibitors are solid di-CHA derivatives. These compounds are slightly volatile at normal temperatures and are used to protect packaged or stored ferrous metals from atmospheric corrosion. Dicyclohexylamine is also used for a number of other purposes: plasticizer; insecticidal formulations; antioxidant in lubricating oils, fuels, and rubber; and extractant.

Permissible Exposure Limits in Air: No standards set.

Permissible Concentration in Water: No criteria set.

Harmful Effects and Symptoms: Dicyclohexylamine is somewhat more toxic than cyclohexylamine. Poisoning symptoms and death appear earlier in rabbits injected with 0.5 g/kg di-CHA (as opposed to CHA). Doses of 0.25 g/kg are just sublethal, causing convulsions and reversible paralysis. Dicyclohexylamine is a skin irritant (1). Di-CHA has been found to be weakly carcinogenic in rats.

Disposal Method Suggested: Incineration; incinerator is equipped with a scrubber or thermal unit to reduce NO_x emissions.

References

(1) U.S. Environmental Protection Agency, *Chemical Hazard Information Profile: Cyclohexylamine,* Washington, DC (October 21, 1977).

DICYCLOHEXYLAMINE NITRITE

Description: $C_6H_{11}-NH-C_6H_{11}HNO_2$ is a solid salt which has some volatility at room temperature and higher.

Code Numbers: CAS 3129-91-7   RTECS HY4200000   UN 2687

DOT Designation: −

Synonyms: Dechan, di-Chan; dicyclohexylamine nitrite; dodecahydro-diphenylamine nitrite.

Potential Exposure: It is used as a vapor-phase corrosion inhibitor whereby it vaporizes either from the solid state or from solution and offers protection against atmospheric rusting. Wrapping paper, plastic wraps, and other materials may be impregnated with di-Chan to protect metal parts during packaging and storage.

Permissible Exposure Limits in Air: A maximum concentration of $0.2\,mg/m^3$ has been recommended in the U.S.S.R.

Permissible Concentration in Water: No criteria set.

Harmful Effects and Symptoms: Prolonged exposure to dicyclohexylamine nitrite vapor is reported to lead to changes in the CNS, erythrocytes, and methemoglobinemia and to disturb the functional state of the liver and kidneys of human workers. Di-Chan is quite possibly a weakly active carcinogen in mice and rats.

Disposal Method Suggested: See dicyclohexylamine.

References

(1) U.S. Environmental Protection Agency, *Chemical Hazard Information Profiles: Cyclohexylamine,* Washington, DC (October 21, 1977).

DICYCLOPENTADIENE

Description: $C_{10}H_{12}$ is a crystalline solid melting at 34°C and boiling at 172°C.

Code Numbers: CAS 77-73-6   RTECS PC1050000   UN 2048

DOT Designation: Flammable liquid.

Synonyms: Bicyclopentadiene; 1,3-cyclopentadiene dimer; 3a,4,7,7a-tetrahydro-4,7-methanoindene.

Potential Exposures: This compound is used in the manufacture of cyclopentadiene as a pesticide intermediate. It is used in the production of ferrocene compounds. It is used in paints and varnishes and resin manufacture.

Permissible Exposure Limits in Air: There is no Federal standard but ACGIH

(1983/84) has adopted a TWA value of 5 ppm (30 mg/m^3). There is no STEL value.

Permissible Concentration in Water: No criteria set.

Harmful Effects and Symptoms: Animal experiments produced leucocytosis and kidney lesions. Human subjects reported an odor threshold at 0.003 ppm, slight eye or throat irritation at 1 to 5 ppm. Some headaches and increased urinary frequency have been reported in exposed workers (A-34).

Personal Protective Methods: Wear rubber gloves, face shield and work clothing.

Respirator Selection: An all purpose cannister mask is recommended (A-38).

Disposal Method Suggested: Incineration.

References
(1) See Reference (A-60).

DICYCLOPENTADIENYL IRON

Description: (C$_5$H$_5$)$_2$Fe is an orange crystalline solid with a camphor-like odor which melts at 173° to 174°C.

Code Numbers: CAS 102-54-5   RTECS LK0700000

DOT Designation: —

Synonyms: Ferrocene, bis(cyclopentadienyl) iron, iron dicyclopentadienyl.

Potential Exposures: It is used as an additive for furnace oils and jet fuels to reduce combustion smoke. It has been proposed as an antiknock additive in gasoline. It is used as a curing agent for rubber and silicone resins. It has been proposed as a high temperature lubricant and as a raw material for high temperature polymers.

Permissible Exposure Limits in Air: There is no Federal standard but ACGIH (1983/84) has adopted a TWA value of 10 mg/m^3 and set a tentative STEL of 20 mg/m^3.

Permissible Concentration in Water: No criteria set, but EPA (A-37) has suggested a permissible ambient goal of 530 μg/ℓ based on health effects.

Harmful Effects and Symptoms: Ferrocene is classified as a slightly toxic material, but the toxicological properties have not been extensively investigated (A-34).

References
(1) Sax, N.I., Ed., *Dangerous Properties of Industrial Materials Report, 1,* No. 4, 67-68, New York, Van Nostrand Reinhold Co. (1981).

DIELDRIN

- Carcinogen (Animal positive, IARC) (3) (EPA-CAG) (A-40)
- Hazardous substance (EPA)
- Hazardous waste (EPA)

● Priority toxic pollutant (EPA)

Description: $C_{12}H_8Cl_6O$ with the structural formula

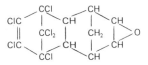

a colorless to light tan solid with a mild chemical odor melting at 175° to 176°C.

Code Numbers: CAS 60-57-1   RTECS IO1750000   UN (NA 2761)

DOT Designation: ORM-A.

Synonyms: 1,2,3,4,10,10-hexachloro-6,7-epoxy-1,4,4a,5,6,7,8,8a-octahydro-1,4-endo-exo-5,8-dimethanonaphthalene; Octalox®; Heod; ENT-16225.

Potential Exposure: Aldrin and dieldrin are manmade compounds belonging to the group of cyclodiene insecticides. They are a subgroup of the chlorinated cyclic hydrocarbon insecticides which include DDT, BHC, etc. They were manufactured in the United States by Shell Chemical Co. until the U.S. EPA prohibited their manufacture in 1974 under the Federal Insecticide, Fungicide and Rodenticide Act. They were then manufactured by Shell Chemical Co. in Holland. Prior to 1974, both insecticides were available in the United States in various formulations for broad-spectrum insect control.

The primary use of the chemicals in the past was for control of corn pests, although they were also used by the citrus industry. Uses are restricted to those where there is no effluent discharge.

Dieldrin's persistence in the environment is due to its extremely low volatility (i.e., a vapor pressure of 1.78 x 10⁻⁷ mm mercury at 20°C), and low solubility in water (186 µg/ℓ at 25° to 29°C). In addition, dieldrin is extremely apolar, resulting in a high affinity for fat which accounts for its retention in animal fats, plant waxes, and other such organic matter in the environment. The fat solubility of dieldrin results in the progressive accumulation in the food chain which may result in a concentration in an organism which would exceed the lethal limit for a consumer.

Waters sampled in the United States contained aldrin or dieldrin in amounts up to 0.05 µg/ℓ. The standard diet in the United States has been calculated to contain about 43 ng/g of dieldrin. Tolerances for dieldrin in cattle meat fat, milk fat, meat and meat by-products have been petitioned for at levels of 0.3, 0.2, and 0.1 ppm respectively.

Children, especially infants, have a high dairy product diet that has been shown to contain dieldrin. It has also been demonstrated that human milk contains dieldrin residues and that some infants may be exposed to high concentrations of dieldrin from that source alone.

Incompatibilities: Strong oxidizers, active metals like sodium, strong acids, phenols.

Permissible Exposure Limits in Air: The Federal limit and the 1983/84 ACGIH TWA value is 0.25 mg/m³. The notation "skin" indicates the possibility of cutaneous absorption. The STEL value is 0.75 mg/m³. The IDLH level is 450 mg/m³.

Determination in Air: Collection on a filter, workup in isooctane, analysis by gas chromatography. See NIOSH Methods, Set T. See also reference (A-10).

Permissible Concentration in Water: To protect freshwater aquatic life— 0.0019 $\mu g/\ell$ as a 24 hr average, never to exceed 2.5 $\mu g/\ell$. To protect saltwater aquatic life—0.0019 $\mu g/\ell$ as a 24 hr average never to exceed 0.71 $\mu g/\ell$. To protect human health—preferably zero. An additional lifetime cancer risk of 1 in 100,000 results at a level of 0.71 ng/ℓ (0.00071 $\mu g/\ell$).

Determination in Water: Methylene chloride extraction followed by gas chromatography with electron capture or halogen specific detection (EPA Method 608) or gas chromatography plus mass spectrometry (EPA Method 625).

Routes of Entry: Inhalation, skin absorption, ingestion, eye and skin contact.

Harmful Effects and Symptoms: During the past decade, considerable information has been generated concerning the toxicity and potential carcinogenicity of the two organochlorine pesticides, aldrin and dieldrin. These two pesticides are usually considered together since aldrin is readily epoxidized to dieldrin in the environment. Both are acutely toxic to most forms of life including arthropods, mollusks, invertebrates, amphibians, reptiles, fish, birds and mammals. Dieldrin is extremely persistent in the environment. By means of bioaccumulation it is concentrated many times as it moves up the food chain.

Symptoms of dieldrin exposure include: headaches, dizziness, nausea, vomiting, malaise and sweating; myoclonic limbjerks; clonic or tonic convulsions; coma.

Points of Attack: Central nervous system, liver, kidneys, skin.

Medical Surveillance: Consider the points of attack in preplacement and periodic physical examinations.

First Aid: If this chemical gets into the eyes, irrigate immediately. If this chemical contacts the skin, wash with soap immediately. If a person breathes in large amounts of this chemical, move the exposed person to fresh air at once and perform artificial respiration. When this chemical has been swallowed, get medical attention. Give large quantities of water and induce vomiting. Do not make an unconscious person vomit.

Personal Protective Methods: Wear appropriate clothing to prevent any possibility of skin contact. Wear eye protection to prevent any possibility of eye contact. Employees should wash immediately when skin is wet or contaminated. Work clothing should be changed daily if it is possible that clothing is contaminated. Remove nonimpervious clothing immediately if wet or contaminated. Provide emergency showers and eyewash.

Respirator Selection:

2.5 mg/m³:	CCROVDMPest/SA/SCBA
12.5 mg/m³:	CCROVFDMPest/GMOVDMPest/SAF/SCBAF
250 mg/m³:	PAPOVHiEPPest/SA:PD,PP,CF
450 mg/m³:	SAF:PD,PP,CF
Escape:	GMOVPPest/SCBA

Disposal Method Suggested: Incineration (1500°F, 0.5 second minimum for primary combustion; 3200°F, 1.0 second for secondary combustion) with adequate scrubbing and ash disposal facilities (A-31).

References

(1)   U.S. Environmental Protection Agency, *Aldrin/Dieldrin: Ambient Water Quality Criteria,* Washington, DC (1980).

(2) U.S. Environmental Protection Agency, *Dieldrin,* Health and Environmental Effects Profile No. 82, Washington, DC, Office of Solid Waste (April 30, 1980).

(3) International Agency for Research on Cancer, *IARC Monographs on the Carcinogenic Risks of Chemicals to Humans,* Lyon, France 5, 125 (1974).

(4) Sax, N.I., Ed., *Dangerous Properties of Industrial Materials Report, 1,* No. 4, 52-55, New York, Van Nostrand Reinhold Co. (1981).

(5) United Nations Environment Programme, *IRPTC Legal File 1983,* Vol. I, pp VII/275–78, Geneva, Switzerland, International Register of Potentially Toxic Chemicals (1984).

DIEPOXYBUTANE

- Carcinogen (Animal Positive) (A-64)
- Hazardous Waste Constituent (EPA)

Description: $C_4H_6O_2$ has the formula

$$O{<}\begin{matrix} CH_2 \\ | \\ CH \\ | \\ CH \\ | \\ CH_2 \end{matrix}{>}O$$

This is a liquid boiling at 138°C.

Code Numbers: CAS 1464-53-5   RTECS EJ8225000

DOT Designation: —

Synonyms: Butadiene dioxide; butadiene diepoxide, DEB.

Potential Exposure: DEB is primarily used in research and experimental work, as a curing agent for polymers, as a crosslinking agent for textile fabrics, and in preventing microbial spoilage in substances. DEB is also used commercially as mixed stereoisomers and as individual isomers in the preparation of erythritol and other pharmaceuticals.

Permissible Exposure Limits in Air: No standards set.

Permissible Concentration in Water: No criteria set.

Routes of Entry: Human exposure to DEB is principally through inhalation and skin absorption.

Harmful Effects and Symptoms: There is sufficient evidence for the carcinogenicity of diepoxybutane in experimental animals (1). Two forms of 1,2:3,4-diepoxybutane (DL and meso) were carcinogenic in mice by skin application. Both compounds produced squamous-cell skin carcinomas. The D,L-racemate also produced local sarcomas in mice and rats by subcutaneous injection (2).

References

(1) International Agency for Research on Cancer, *IARC Monographs on the Evaluation of the Carcinogenic Risk of Chemicals to Humans,* Supplement 4, Lyon, France, IARC (1982).

(2) International Agency for Research on Cancer, *IARC Monographs on the Evaluation of the Carcinogenic Risk of Chemicals to Man,* Vol. 11, Lyon, France, IARC, pp 115-123 (1975).

(3) Sax, N.I., Ed., *Dangerous Properties of Industrial Materials Report, 4,* No. 3, 56–60, New York, Van Nostrand Reinhold Co. (1984).

DIETHYLAMINE

- Hazardous substance (EPA)
- Hazardous waste (EPA)

Description: $(C_2H_5)_2NH$ is a colorless liquid with a fishy ammonia-like odor.

Code Numbers: CAS 109-89-7   RTECS HZ8750000   UN 1154

DOT Designation: Flammable liquid.

Synonyms: DEA.

Potential Exposures: Diethylamine (DEA) is used in the manufacture of the following chemicals: diethyldithiocarbamate and thiurams (rubber processing accelerators), diethylaminoethanol (medicinal intermediate), diethylamino-propylamine (epoxy curing agent), N,N-diethyl-m-toluamide and other pesticides (A-32), and 2-diethylaminoethylmethacrylate. It is used in the manufacture of several drugs (A-41).

Incompatibilities: Strong oxidizers, strong acids.

Permissible Exposure Limits in Air: The Federal limit is 25 ppm (75 mg/m³). The ACGIH as of 1983/84 has adopted a TWA of 10 ppm (30 mg/m³) and an STEL value of 25 ppm (75 mg/m³). The IDLH level is 2,000 ppm.

Determination in Air: Adsorption on silica, workup with H_2SO_4 in CH_3OH, analysis by gas chromatography. See NIOSH Methods, Set J. See also reference (A-10).

Permissible Concentration in Water: No criteria set.

Routes of Entry: Inhalation, ingestion, skin absorption, eye and skin contact.

Harmful Effects and Symptoms: Eye, skin and respiratory irritation.

Points of Attack: Respiratory system, lungs, skin, eyes.

Medical Surveillance: Consider the points of attack in preplacement and periodic physical examinations.

First Aid: If this chemical gets into the eyes, irrigate immediately. If this chemical contacts the skin, flush with water immediately. If a person breathes in large amounts of this chemical, move the exposed person to fresh air at once and perform artificial respiration. When this chemical has been swallowed, get medical attention. Give large quantities of water and induce vomiting. Do not make an unconscious person vomit.

Personal Protective Methods: Wear appropriate clothing to prevent any possibility of skin contact with liquids. Wear eye protection to prevent any possibility/reasonable probability of eye contact with liquids >5% DEA. Employees should wash immediately when skin is wet or contaminated with liquid. Remove clothing immediately if wet or contaminated to avoid flammability hazard. Provide emergency showers and eyewash if liquids containing >0.5% DEA are involved.

Respirator Selection:
    1,500 ppm: CCRSF/GMS/SAF/SCBAF
    2,000 ppm: SAF:PD,PP,CF
        Escape: GMS/SCBA

Disposal Method Suggested: Incineration; incinerator is equipped with a scrubber or thermal unit to reduce NO_x emissions (A-31).

References

(1) National Institute for Occupational Safety and Health, *Profiles on Occupational Hazards for Criteria Document Priorities: Primary Aliphatic Amines,* 154–166, Report PB-274,073, Cincinnati, OH (1977).

(2) U.S. Environmental Protection Agency, *Chemical Hazard Information Profile: Ethylamines,* Washington, DC (April 1, 1978).

(3) See Reference (A-61).

(4) See Reference (A-60).

DIETHYLAMINOETHANOL

Description: $(C_2H_5)_2NCH_2CH_2OH$ is a colorless liquid with a weak ammoniacal odor, boiling at 161°C.

Code Numbers: CAS 100-37-8 RTECS KK5075000 UN 2686

DOT Designation: Flammable liquid.

Synonyms: 2-Diethylaminoethyl alcohol; N,N-diethylethanolamine; diethyl-(2-hydroxyethyl)amine; 2-diethylamino ethanol.

Potential Exposure: This compound is used as a chemical intermediate for the production of emulsifiers, detergents, solubilizers, cosmetics, and textile finishing agents. It is also used in drug manufacture (A-41).

Incompatibilities: Strong oxidizers, strong acids.

Permissible Exposure Limits in Air: The Federal limit and the 1983/84 ACGIH TWA value is 10 ppm (50 mg/m³). The notation "skin" indicates the possibility of cutaneous absorption. There is no tentative STEL value. The IDLH level is 500 ppm.

Determination in Air: Adsorption on silica gel, elution with methanolic acid, derivatization with benzaldehyde and gas chromatographic analysis (A-10).

Permissible Concentration In Water: No criteria set.

Routes of Entry: Inhalation, skin absorption, ingestion, eye and skin contact.

Harmful Effects and Symptoms: Nausea, vomiting; eye, skin and respiratory irritation.

Points of Attack: Respiratory system, skin, eyes.

Medical Surveillance: Consider the points of attack in preplacement and periodic physical examinations.

First Aid: If this chemical gets into the eyes, irrigate immediately. If this chemical contacts the skin, flush with water immediately. If a person breathes in large amounts of this chemical, move the exposed person to fresh air at once and perform artificial respiration. When this chemical has been swallowed, get medical attention. Give large quantities of water and induce vomiting. Do not make an unconscious person vomit.

Personal Protective Methods: Wear appropriate clothing to prevent any reasonable probability of skin contact. Wear eye protection to prevent any possibility of eye contact with solutions of 5% or greater concentration. Employees should wash immediately when skin is wet. Remove nonimpervious clothing immediately if wet or contaminated. Provide emergency showers and eyewash if liquids containing ⩾5% contaminants are involved.

Respirator Selection:
    500 ppm: SAF/SCBAF
    Escape: GMOV/SCBA

Disposal Method Suggested: Controlled incineration; incinerator is equipped with a scrubber or thermal unit to reduce NO_x emissions.

References
(1)   See Reference (A-60).

O,O-DIETHYL DITHIOPHOSPHORIC ACID

Description: $(C_2H_5O)_2\overset{\overset{\displaystyle S}{\|}}{P}SH$ is a liquid.

Code Numbers: CAS 298-06-6

DOT Designation: —

Synonyms: O,O-Diethyl phosphorodithioic acid.

Potential Exposure: This compound is used primarily as an intermediate in the synthesis of several pesticides including: azinphosmethyl, carbophenothion, dialifor, dioxathion, disulfoton, ethion, phorate, phosalone and terbufos (A-32).

Permissible Exposure Limits in Air: No standards set.

Permissible Concentration in Water: No criteria set.

Harmful Effects and Symptoms: There is no available information to indicate that O,O-diethyl dithiophosphoric acid produces carcinogenic, mutagenic, teratogenic, or adverse reproductive effects. The pesticide phorate, which may release O,O-diethyl dithiophosphoric acid as a metabolite, has shown some teratogenic effects in developing chick embryos and adverse reproductive effects in mice.

References
(1)   U.S. Environmental Protection Agency, *O,O-Diethyl Dithiophosphoric Acid,* Health and Environmental Effects Profile No. 83, Washington, DC, Office of Solid Waste (April 30, 1980).

DIETHYLENE TRIAMINE

Description: $(NH_2CH_2CH_2)_2NH$ is a yellow liquid with an ammoniacal odor which boils at 207°C.

Code Numbers: CAS 111-40-0   RTECS IE1225000   UN 2079

DOT Designation: Corrosive material.

Synonyms: Aminoethylethandiamine; 3-azapentane-1,5-diamine; bis(2-amino-ethyl)amine; 2,2'-diaminodiethyl amine; DETA.

Potential Exposure: This material is used as a solvent for sulfur, acid gases, resins and dyes.

Permissible Exposure Limits in Air: There is no Federal standard but ACGIH

as of 1983/84 has adopted a TWA value of 1 ppm ($4mg/m^3$) with the notation "skin" indicating the possibility of cutaneous absorption. There is no STEL value.

Determination in Air: Collection by impinger or fritted bubbler and colorimetric determination (A-1).

Permissible Concentration in Water: A maximum allowable concentration in drinking water has been reported from a German source (A-36) at 0.2 mg/ℓ.

Harmful Effects and Symptoms: Skin sensitization, dermatitis and respiratory tract sensitization and irritation, possibly leading to bronchial asthma (A-34). Also conjunctivitis, keratitis; nausea, vomiting (A-38).

First Aid: Flush eyes with water. Wash contaminated areas of body with soap and water.

Personal Protective Methods: Wear butyl rubber gloves, safety glasses and work clothing (A-38).

Respirator Selection: An all-purpose cartridge mask is recommended.

Disposal Method Suggested: Incinerate in admixture with flammable solvent in furnace equipped with afterburner and scrubber.

References

(1) See Reference (A-60).
(2) United Nations Environment Programme, *IRPTC Legal File 1983,* Vol. I, pp VII/268, Geneva, Switzerland, International Register of Potentially Toxic Chemicals (1984).

DI(2-ETHYLHEXYL) ADIPATE

See "Adipate Ester Plasticizers."

DI(2-ETHYLHEXYL) PHTHALATE

- Carcinogen (NCI) (4)
- Hazardous waste (EPA)
- Priority toxic pollutant (EPA)

Description: $C_6H_4(COOCH_2CHCH_2CH_2CH_2CH_3)_2$ is a colorless oily liquid with almost no odor, boiling at 387°C. [with C_2H_5 group shown above the CH]

Code Numbers: CAS 117-81-7   RTECS TI0350000

DOT Label Description: —

Synonyms: DOP; di-sec-octyl phthalate; DEHP.

Potential Exposure: Di(2-ethylhexyl) phthalate (DEHP) is commercially produced by the reaction of 2-ethylhexyl alcohol and phthalic anhydride. It is used as a plasticizer for resins and in the manufacture of organic pump fluids. Two groups are at risk in regard to phthalic acid esters. These are workers in the industrial environment in which the phthalates are manufactured or used and patients receiving chronic transfusion of blood and blood products stored in PVC blood bags.

Incompatibilities: Nitrates, strong oxidizers, strong acids, strong alkalies.

Permissible Exposure Limits in Air: The TWA value adopted by ACGIH (1983/84) is 5 mg/cm³. The STEL value is 10 mg/m³. The IDLH level has not been defined.

Determination in Air: Collection on a filter, workup with CS_2, analysis by gas chromatography. See NIOSH Methods, Set D. See also (A-13).

Permissible Concentration in Water: For freshwater and saltwater aquatic life, no criteria have been set due to lack of data. For protection of human health, the ambient water criterion is 15 mg/ℓ (15,000 µg/ℓ). A no-adverse effect level in drinking water has been calculated by NAS/NRC to be 4.2 mg/ℓ. An acceptable daily intake (ADI) value of 0.6 µg/kg/day has been calculated by NAS/NRC.

Determination in Water: Gas chromatography (EPA Method 606) or gas chromatography plus mass spectrometry (EPA Method 625).

Routes of Entry: Inhalation, ingestion, skin and eye contact.

Harmful Effects and Symptoms: Irritation of the eyes and mucous membranes; nausea; diarrhea. This compound is also a proven animal carcinogen (7).

Points of Attack: Eyes, upper respiratory system, gastrointestinal system.

Medical Surveillance: Consider the points of attack in preplacement and periodic physical examinations.

First Aid: If this chemical gets into the eyes, irrigate promptly. If this chemical contacts the skin, wash regularly. If a person breathes in large amounts of this chemical, move the exposed person to fresh air at once and perform artificial respiration. When this chemical has been swallowed, get medical attention.

Personal Protective Methods: Wear eye protection to prevent any reasonable probability of eye contact.

Respirator Selection:
50 mg/m³: HiEP/SA/SCBA
250 mg/m³: HiEPF/SAF/SCBAF
500 mg/m³: PAPHiE/SA:PD,PP,CF
10,000 mg/m³: SAF:PD,PP,CF

Disposal Method Suggested: Incineration.

References

(1) U.S. Environmental Protection Agency, *Phthalate Esters: Ambient Water Quality Criteria,* Washington, DC (1980).
(2) National Institute for Occupational Safety and Health, *Profiles on Occupational Hazards for Criteria Document Priorities: Phthalates,* 97–103, Report PB-274,073, Rockville, MD (1977).
(3) U.S. Environmental Protection Agency, *Bis(2-Ethylhexyl) Phthalate,* Health and Environmental Effects Profile No. 27, Washington, DC, Office of Solid Waste (April 30, 1980).
(4) National Cancer Institute, "Bioassay of 2-Ethylhexyl Phthalate for Possible Carcinogenesis" [review completed according to *Chemical Week,* p 14 (Oct. 22, 1980)].
(5) Sax, N.I., Ed., *Dangerous Properties of Industrial Materials Report, 1,* No. 7, 52-54, New York, Van Nostrand Reinhold Co. (1981).
(6) Sax, N.I., Ed., *Dangerous Properties of Industrial Materials Report, 2,* No. 2, 22-25, New York, Van Nostrand Reinhold Co. (1982).
(7) See Reference (A-62). Also see Reference (A-64).
(8) United Nations Environment Programme, *IRPTC Legal File 1983,* Vol. II, pp VII/629–31, Geneva, Switzerland, International Register of Potentially Toxic Chemicals (1984).

DIETHYL KETONE

Description: $C_2H_5COC_2H_5$ is a colorless liquid with an acetone-like odor which boils at 101°C.

Code Numbers: CAS 96-22-0 RTECS SA8050000 UN 1156

DOT Designation: Flammable liquid.

Synonyms: 3-Pentanone, Dimethylacetone, Propione, DEK, Metacetone.

Potential Exposure: This compound is used as a solvent and in organic synthesis.

Permissible Exposure Limits in Air: There is no Federal standard but ACGIH as of 1983/84 has adopted a TWA of 200 ppm (705 mg/m³). There is no STEL proposed.

Permissible Concentration in Water: No criteria set.

Harmful Effects and Symptoms: This material is slightly toxic. A 4 hr exposure to 8,000 ppm reportedly (A-34) was fatal to four of six test rats.

Disposal Method Suggested: Incineration.

References

(1) National Institute for Occupational Safety and Health, *Criteria for a Recommended Standard: Occupational Exposure to Ketones,* NIOSH Doc. No. 78-173 (1978).
(2) See Reference (A-60).

O,O-DIETHYL-S-METHYL PHOSPHORODITHIOATE

- Hazardous waste constituent (EPA)

Description: $(C_2H_5O)_2\overset{\displaystyle S}{\overset{\displaystyle \|}{P}}SCH_3$ is the methyl derivative of O,O-diethyl dithiophosphoric acid (which see).

Code Numbers: CAS 3288-58-2 UN 2783

DOT Designation: —

Synonyms: None.

Potential Exposure: The compound has partly insecticidal, acaricidal and fungicidal activity and is useful as an intermediate for organic synthesis.

Permissible Exposure Limits in Air: No standards set.

Permissible Concentration in Water: No criteria set.

Harmful Effects and Symptoms: There is no available information on the possible carcinogenic, mutagenic, teratogenic or adverse reproductive effects of O,O-diethyl-S-methyl phosphorodithioate. Pesticides containing the O,O-diethyl phosphorodithioate moiety did not show carcinogenic effects in rodents (dioxathion) or teratogenic effects in chick embryos (phorate). The possible metabolite of this compound, O,O-diethyl phosphorothioate, did not show mutagenic activity in *Drosophila, E. coli,* or *Saccharomyces.* O,O-diethyl-S-methyl phosphorodithioate, like other organophosphate compounds, is expected to produce cholinesterase inhibition in humans. There is no available data on the aquatic toxicity of this compound.

Reference

(1) U.S. Environmental Protection Agency, *O,O-Diethyl-S-Methyl Phosphorodithioate,* Health and Environmental Effects Profile No. 84, Washington, DC, Office of Solid Waste (April 30, 1980).

DIETHYL PHTHALATE

- Hazardous waste (EPA)
- Priority toxic pollutant (EPA)

Description: $C_6H_4(OCOC_2H_5)_2$ is a water-white odorless liquid which boils at 296°C.

Code Numbers: CAS 84-66-2   RTECS TI1050000

DOT Designation: —

Synonyms: DEP

Potential Exposure: This material is used as a solvent for cellulose esters, as a vehicle in pesticidal sprays, as a fixative and solvent in perfumery, as an alcohol denaturant and as a plasticizer in solid rocket propellants.

Permissible Exposure Limits in Air: There are no Federal standards but ACGIH (1983/84) has adopted a TWA value of 5 mg/m³ and set an STEL of 10 mg/m³.

Determination in Air: Collection by filter and analysis by gas liquid chromatography (A-27).

Permissible Concentration in Water: Data are insufficient to draft criterion for the protection of either freshwater or marine organisms. The recommended water quality criterion level for protection of human health is 350 mg/ℓ (350,000 µg/ℓ).

Determination in Water: Methylene chloride extraction followed by gas chromatography with flame ionization or electron capture detection (EPA Method 606) or gas chromatography plus mass spectrometry (EPA Method 625).

Harmful Effects and Symptoms: DEP has few acute or chronic toxic properties and seems to be devoid of any major irritating or sensitizing effects on the skin. Exposure to heated vapors may produce transient irritation of the nose and throat (A-34). Conjunctivitis, corneal necrosis, respiratory tract irritation, dizziness, nausea and eczema are symptoms cited by others (A-38). Diethyl phthalate has been shown to produce mutagenic effects in the Ames *Salmonella* assay.

Teratogenic effects were reported following i.p. administration of diethyl phthalate to pregnant rats. This same study has also indicated fetal toxicity and increased resorptions after i.p. administration of DEP. Evidence that diethyl phthalate produces carcinogenic effects has not been found (2).

First Aid: Flush eye with water. Wash contaminated areas of body with soap and water. Gastric lavage if swallowed followed by saline catharsis.

Personal Protective Methods: Wear safety glasses, rubber gloves, face shield and full protective clothing.

Respirator Selection: An all-purpose cannister mask is recommended (A-38).

Disposal Method Suggested: Incineration.

References

(1) U.S. Environmental Protection Agency, *Phthalate Esters: Ambient Water Quality Criteria,* Washington, DC (1980).

(2) U.S. Environmental Protection Agency, *Diethyl Phthalate,* Health and Environmental Effects Profile No. 85, Washington, DC, Office of Solid Waste (April 30, 1980).

(3) United Nations Environment Programme, *IRPTC Legal File 1983,* Vol. II, pp VII/635-7, Geneva, Switzerland, International Register of Potentially Toxic Chemicals (1984).

DIETHYLSTILBESTROL

- Carcinogen (Human Positive, IARC) (EPA-CAG) (A-62) (A-64)
- Hazardous Waste Constituent (EPA)

Description: $C_{18}H_{20}O_2$ has the formula

$$HO-C_6H_4-C=C-C_6H_4-OH$$
$$Et\ \ Et$$

This is a crystalline compound melting at 169° to 172°C.

Code Numbers: CAS 56-53-1   RTECS WJ5600000

DOT Designation: —

Synonyms: (E)-4,4'-(1,2-Diethyl-1,2-ethenediyl)-bisphenol; DES, many trade names.

Potential Exposure: DES had been used extensively as a growth promoter for cattle and sheep and is used in veterinary medicine to treat estrogen-deficiency disorders. It has been used in humans to prevent spontaneous abortions, to treat symptoms associated with menopause, menstrual disorders, postpartum breast engorgement, primary ovarian failure, breast cancer, and prostate cancers in males.

Most exposures to DES occur as a result of its oral administration as a drug. OSHA estimates that 4,000 workers are exposed to it during manufacture or product formulation. An estimated 500,000 patients are treated each year with DES. They are mostly males treated for neoplastic disease. Dosages are in the range of 1-15 mg/day. When it is used as a growth promoter for sheep and cattle, the entire population could be exposed to low levels of less than 10 ppb.

In 1979, the FDA revoked all use of DES in food-producing animals. The Court of Appeals upheld the Commissioner's decision to revoke the use of DES in all food-producing animals on November 24, 1980, the motion to reconsider was denied on December 24, 1980.

Permissible Exposure Limits in Air: No standards set.

Permissible Concentration in Water: No criteria set.

Harmful Effects and Symptoms: Diethylstilbestrol is carcinogenic in mice, rats, hamsters, frogs, and squirrel monkeys, producing tumors principally in oestrogen-responsive tissues.

Diethylstilbestrol causes clear-cell carcinoma of the vagina in females exposed in utero. The evidence for an association with other human cancers is either limited (endometrium) or inadequate (breast, ovary).

References
(1) Sax, N.I., Ed., *Dangerous Properties of Industrial Materials Report, 1,* No. 3, 59-61, New York, Van Nostrand Reinhold Co. (1981).
(2) See Reference (A-62). Also see Reference (A-64).

DIFLUORODIBROMOMETHANE

Description: CBr_2F_2 is a colorless liquid or gas boiling at $24°C$ with a characteristic odor.

Code Numbers: CAS 75-61-6   RTECS PA7525000   UN 1941

DOT Designation: ORM-A

Synonyms: Dibromodifluoromethane; Freon 12B2; Halon 1202.

Potential Exposure: This material is used as a fire-extinguishing agent.

Incompatibilities: Chemically active metals—sodium, potassium, calcium, powdered aluminum, zinc, magnesium.

Permissible Exposure Limits in Air: The Federal limit and the 1983/84 ACGIH TWA value is 100 ppm (860 mg/m^3). The STEL value is 150 ppm ($1,290$ mg/m^3). The IDLH level is 2,500 ppm.

Determination in Air: Charcoal adsorption, workup with isopropanol, analysis by gas chromatography. See NIOSH Methods, Set H. See also reference (A-10).

Permissible Concentration in Water: No criteria set.

Routes of Entry: Inhalation, ingestion, eye and skin contact.

Harmful Effects and Symptoms: Frostbite; irritation of nose and throat; drowsiness; unconsciousness; liver damage.

Points of Attack: Skin, respiratory system.

Medical Surveillance: Consider the points of attack in preplacement and periodic physical examinations.

First Aid: If this chemical gets into the eyes, irrigate immediately. If this chemical contacts the skin, flush with water immediately. If a person breathes in large amounts of this chemical, move the exposed person to fresh air at once and perform artificial respiration. When this chemical has been swallowed, get medical attention. Give large quantities of salt water and induce vomiting. Do not make an unconscious person vomit.

Personal Protective Methods: Wear appropriate clothing to prevent repeated or prolonged skin contact. Wear eye protection to prevent any reasonable probability of eye contact. Remove nonimpervious clothing promptly if wet or contaminated.

Respirator Selection:
    1,000 ppm: SA/SCBA
    2,500 ppm: SAF/SCBAF
        Escape: GMOV/SCBA

Disposal Method Suggested: Vent to atmosphere.

References
(1) See Reference (A-61).

DIGLYCIDYL ETHER

Description: $C_6H_{10}O_3$, with the structural formula:

$$CH_2CH-CH_2-O-CH_2-CH-CH_2$$

is a colorless liquid with a strong, irritant odor.

Code Numbers: CAS 2238-07-5 RTECS-KN2350000

DOT Designation: —

Synonyms: Di(epoxypropyl) ether; bis(2,3-epoxypropyl) ether; 2,3-epoxypropyl ether; diallyl ether dioxide, DGE.

Potential Exposure: This material is used as a diluent for epoxy resins, as a textile treating agent and as a stabilizer for chlorinated organic compounds.

Incompatibilities: Strong oxidizers.

Permissible Exposure Limits in Air: The Federal limit is 0.5 ppm (2.8 mg/m³). The ACGIH as of 1983/84 has proposed a TWA of 0.1 ppm (0.5 mg/m³) but no STEL value. The IDLH level is 85 ppm.

Determination in Air: Adsorption by charcoal tube, analysis by gas liquid chromatography (A-13).

Permissible Concentration in Water: No criteria set.

Routes of Entry: Inhalation, ingestion, eye and skin contact.

Harmful Effects and Symptoms: Irritation of eyes and respiratory system, skin burns.

Points of Attack: Skin, eyes and respiratory system.

Medical Surveillance: Consider the points of attack in preplacement and periodic physical examinations.

First Aid: If this chemical gets into the eyes, irrigate immediately. If this chemical contacts the skin, wash with soap immediately. If a person breathes in large amounts of this chemical, move the exposed person to fresh air at once and perform artificial respiration. When this chemical has been swallowed, get medical attention. Give large quantities of salt water and induce vomiting. Do not make an unconscious person vomit.

Personal Protective Methods: Wear appropriate clothing to prevent any possibility of skin contact. Wear eye protection to prevent any reasonable probability of eye contact. Employees should wash immediately when skin is wet or contaminated and daily at the end of each work shift. Work clothing should be changed daily if it is possible that clothing is contaminated. Remove nonimpervious clothing immediately if wet or contaminated. Provide emergency showers.

Respirator Selection:
           25 ppm: SAF/SCBAF
           85 ppm: SAF:PD,PP,CF
          Escape: GMOV/SCBA

Disposal Method Suggested: Incineration.

References

(1) National Institute for Occupational Safety and Health, *Information Profiles on Potential Occupational Hazards: Glycidyl Ethers,* 116-123, Report PB-276,678, Rockville, MD (October 1977).

(2) National Institute for Occupational Safety and Health, *Criteria for a Recommended Standard: Occupational Exposure to Glycidyl Ethers,* NIOSH Doc. No. 78-166, Washington, DC (1978).

(3) United Nations Environment Programme, *IRPTC Legal File 1983,* Vol. I, pp VII/344, Geneva, Switzerland, International Register of Potentially Toxic Chemicals (1984).

DIHYDROSAFROLE

● Carcinogen (Animal Positive, IARC, NCI)(1)

Description: $C_{10}H_{12}O_2$ has the structural formula

Dihydrosafrole is an oily liquid boiling at 228°C.

Code Numbers: CAS 94-58-6   RTECS DA6125000   UN

DOT Designation: —

Synonyms: 1,2-Methylenedioxy-4-propylbenzene; 5-propyl-1,3-benzodioxole.

Potential Exposure: Workers may be exposed to dihydrosafrole during its use as a chemical intermediate in the production of piperonyl butoxide and related insecticidal synergists, and in the production of fragrances for cosmetic products.

Permissible Exposure Limits in Air: No limits set.

Permissible Concentration in Water: No criteria set.

Determination in Water: By chromatography (1).

Harmful Effects and Symptoms: Dihydrosafrole is recognized as a carcinogen in experimental animals by the International Agency for Research on Cancer (IARC) and the National Cancer Institute Bioassay Program.

An LD_{50} value of 2,260 mg/kg has been reported for rats (slightly toxic).

Points of Attack: Liver, spleen.

Disposal Method Suggested: The U.S.E.P.A. (1) recommends packaging of product residues and sorbent media in epoxy-lined drums and disposal at an EPA-approved site. The compound may be destroyed by permanganate oxidation, high temperature incineration with scrubbing equipment, or microwave plasma treatment.

References

(1) U.S. Environmental Protection Agency, *Chemical Hazard Information Profile Draft Report: Dihydrosafrole,* Wash., D.C. (May 17, 1984).

DIISOBUTYL CARBINOL

Description: $[(CH_3)_2CHCH_2]CHOH$ is a colorless liquid boiling at $176°$ to $177°C$.

Code Numbers: CAS 108-82-7   RTECS MJ3325000

DOT Designation: Grade E Combustible (U.S. Coast Guard).

Synonyms: 2,6-dimethyl-4-heptanol.

Potential Exposure: Used in the manufacture of surfactants, lubricant additives, rubber chemicals, flotation agents; used as an antifoam.

Permissible Exposure Limits in Air: No standards set.

Permissible Concentration in Water: No criteria set.

Routes of Entry: Inhalation and ingestion.

Harmful Effects and Symptoms: Powerful human irritant by inhalation. Medium irritant to animal (rabbit) skin and eyes. Has caused CNS and liver injury in animals.

Personal Protective Methods: Wear protective clothing.

Respirator Selection: Wear canister mask.

Disposal Method Suggested: Incineration.

References

(1) Sax, N.I., Ed., *Dangerous Properties of Industrial Materials Report, 1*, No. 8, 65-67, New York, Van Nostrand Reinhold Co. (1981).
(2) See Reference (A-60).

DIISOBUTYLENE

Description: $(CH_3)_3CCH_2\overset{CH_3}{C}=CH_2$ is a colorless liquid boiling at $102°C$.

Code Numbers: CAS 25167-70-8   RTECS SB2710000   UN 2050

DOT Designation: Grade C Flammable (U.S. Coast Guard).

Synonyms: diisobutene; 2,2,4-trimethylpentene.

Potential Exposure: Used as the raw material for production of isooctane as high octane fuel component.

Permissible Exposure Limits in Air: No standards set.

Permissible Concentration in Water: No criteria set.

Harmful Effects and Symptoms: Moderately irritating by oral in inhalation routes. Irritating and narcotic in high concentrations. Has caused kidney and liver damage in laboratory animals.

Points of Attack: Eyes, skin, respiratory tract, nervous system.

Personal Protective Methods: Full protective clothing required.

Respirator Selection: Self-contained breathing apparatus.

Disposal Method Suggested: Incineration.

References

(1) Sax, N.I., Ed., *Dangerous Properties of Industrial Materials Report, 1,* No. 8, 67-68, New York, Van Nostrand Reinhold Co. (1981).

(2) See Reference (A-60).

DIISOBUTYL KETONE

Description: $[(CH_3)_2CHCH_2]_2CO$ is a colorless liquid with a mild odor.

Code Numbers: CAS 108-83-8   RTECS MJ5775000   UN 1157

DOT Designation: Combustible liquid.

Synonyms: 2,6-Dimethyl-4-heptanone; sym-diisopropyl acetone; isovalerone; valerone.

Potential Exposure: This material is used as a solvent, as a dispersant for resins, and as an intermediate in the synthesis of pharmaceuticals and pesticides.

Incompatibilities: Strong oxidizers.

Permissible Exposure Limits in Air: The Federal limit is 50 ppm (290 mg/m^3). The ACGIH has set a TWA of 25 ppm (150 mg/m^3) but no STEL value as of 1983/84. The IDLH level is 2,000 ppm.

Determination in Air: Charcoal adsorption, workup with CS_2, analysis by gas chromatography. See NIOSH Methods, Set V. See also reference (A-10).

Permissible Concentration in Water: No criteria set.

Routes of Entry: Inhalation, ingestion, eye and skin contact.

Harmful Effects and Symptoms: Irritation of eyes, nose and throat; headaches, dizziness; unconsciousness; dermatitis.

Points of Attack: Respiratory system, skin, eyes.

Medical Surveillance: Consider the points of attack in preplacement and periodic physical examinations.

First Aid: If this chemical gets into the eyes, irrigate immediately. If this chemical contacts the skin, wash with soap promptly. If a person breathes in large amounts of this chemical, move the exposed person to fresh air at once and perform artificial respiration. When this chemical has been swallowed, get medical attention. Give large quantities of water and induce vomiting. Do not make an unconscious person vomit.

Personal Protective Methods: Wear appropriate clothing to prevent repeated or prolonged skin contact. Employees should wash promptly when skin is wet. Remove nonimpervious clothing promptly if wet or contaminated.

Respirator Selection:
    1,000 ppm: CCROVF
    2,000 ppm: GMOV/SAF/SCBAF
      Escape: GMOV/SCBA

Disposal Method Suggested: Incineration.

References

(1) National Institute for Occupational Safety and Health, *Criteria for a Recommended Standard: Occupational Exposure to Ketones,* NIOSH Doc. No. 78-113, Washington, DC (1978).

(2) See Reference (A-68).

DIISOPROPYLAMINE

Description: $(CH_3)_2CHNHCH(CH_3)_2$ is a colorless liquid with an ammoniacal odor. It boils at $83°C$.

Code Numbers: CAS 108-18-9   RTECS IM4025000   UN 1158

DOT Designation: Flammable liquid.

Synonyms: None.

Potential Exposure: This material is used as a chemical intermediate in the synthesis of pharmaceuticals and pesticides (Diallate, Fenamiphos and Triallate, for example) (A-32).

Incompatibilities: Strong oxidizers, strong acids.

Permissible Exposure Limits in Air: The Federal limit and the 1983/84 ACGIH TWA value is 5 ppm (20 mg/m^3). The notation "skin" indicates the possibility of cutaneous absorption. There is no STEL value set. The IDLH level is 1,000 ppm.

Determination in Air: Collection by impinger in sulfuric acid, analysis by gas chromatography (A-10). A colorimetric method is also available (A-19).

Permissible Concentration in Water: No criteria set.

Routes of Entry: Inhalation, ingestion, skin absorption, eye and skin contact.

Harmful Effects and Symptoms: Nausea, vomiting; headaches; eye irritation, visual disturbance; pulmonary irritation.

Points of Attack: Respiratory system, lungs, skin, eyes.

Medical Surveillance: Consider the points of attack in preplacement and periodic physical examinations.

First Aid: If this chemical gets into the eyes, irrigate immediately. If this chemical contacts the skin, flush with water immediately. If a person breathes in large amounts of this chemical, move the exposed person to fresh air at once and perform artificial respiration. When this chemical has been swallowed, get medical attention. Give large quantities of water and induce vomiting. Do not make an unconscious person vomit.

Personal Protective Methods: Wear appropriate clothing to prevent repeated or prolonged skin contact. Wear eye protection to prevent any possibility of eye contact with solutions over 5% concentration. Employees should wash promptly when skin is wet. Remove clothing immediately if wet or contaminated to avoid flammability hazard. Provide emergency eyewash if liquids containing ⩾5% contaminants are involved.

Respirator Selection:

250 ppm: SAF/SCBAF

1,000 ppm: SAF:PD,PP,CF

Escape: GMOV/SCBA

Disposal Method Suggested: Incineration; incinerator is equipped with a scrubber or thermal unit to reduce NO$_x$ emissions.

References

(1)   See Reference (A-61).

DIISOPROPYL ETHER

Description: $[(CH_3)_2CH]_2O$ is a colorless liquid with a sharp sweet ether-like odor. It boils at 68° to 69°C.

Code Numbers: CAS 108-20-3   RTECS TZ5425000   UN 1159

DOT Designation: Flammable liquid.

Synonyms: Isopropyl ether; diisopropyl oxide; 2-isopropoxypropane; 2,2'-oxybispropane.

Potential Exposure: This material is used as a solvent and chemical intermediate.

Incompatibilities: Strong oxidizers.

Permissible Exposure Limits in Air: The Federal limit is 500 ppm (2,100 mg/m³). The ACGIH as of 1983/84 has set a TWA value of 250 ppm (1,050 mg/m³) and an STEL of 310 ppm (1,320 mg/m³). The IDLH level is 10,000 ppm.

Determination in Air: Charcoal adsorption, workup with CS_2, analysis by gas chromatography. See NIOSH Methods, Set V. See also reference (A-10).

Permissible Concentration in Water: No criteria set.

Routes of Entry: Inhalation, ingestion, eye and skin contact.

Harmful Effects and Symptoms: Irritation of eyes and nose; respiratory discomfort; dermatitis. In animals, drowsiness, dizziness and unconsciousness.

Points of Attack: Respiratory system, skin.

Medical Surveillance: Consider the points of attack in preplacement and periodic physical examinations.

First Aid: If this chemical gets into the eyes, irrigate immediately. If this chemical contacts the skin, wash with soap promptly. If a person breathes in large amounts of this chemical, move the exposed person to fresh air at once and perform artificial respiration. When this chemical has been swallowed, get medical attention. Give large quantities of water and induce vomiting. Do not make an unconscious person vomit.

Personal Protective Methods: Wear appropriate clothing to prevent repeated or prolonged skin contact. Wear eye protection to prevent any reasonable probability of eye contact. Employees should wash promptly when skin is wet or contaminated. Remove clothing immediately if wet or contaminated.

Respirator Selection:
  1,000 ppm: CCROVF
  5,000 ppm: GMOVc
  10,000 ppm: GMOVfb/SAF/SCBAF
  Escape: GMOV/SCBA

Disposal Method Suggested: Concentrated waste containing no peroxides—discharge liquid at a controlled rate near a pilot flame. Concentrated waste containing peroxide—perforation of a container of the waste from a safe distance followed by open burning.

References

(1)   See Reference (A-61).

DIMETHOXANE

● Carcinogen (Animal Positive, IARC) (2)

Description: $C_8H_{14}O_4$ with the structural formula

is a clear yellow to light amber liquid which boils at 66° to 68°C.

Code Numbers: CAS 828-00-2   RTECS AH1350000

DOT Designation: —

Synonyms: Acetomethoxane; 2,6-Dimethyl-m-dioxan-4-ol acetate; 4-acetoxy-2,6-dimethyl dioxane; DDOA; Dioxin (note! — given in RTECS).

Potential Exposure: Dimethoxane is an antimicrobial agent that is used for preserving water-based paints, water-based cutting oils, dyes, textile chemicals, inks and cosmetic products (2). Its use in cosmetics was substantially discontinued in 1978 after linkage of dimethoxane to malignant tumors in rats.

Permissible Exposure Limits in Air: No standards set.

Permissible Concentration in Water: No criteria set.

Harmful Effects and Symptoms: An allergic sensitization to commercial dimethoxane has been noted, perhaps caused by acetaldehyde and crotonaldehyde contamination.

The only experimental data available (3) relate to the development of malignant tumors in rats after oral administration of dimethoxane.

Points of Attack: Liver, kidneys, skin.

Medical Surveillance: Consider the points of attack in preplacement and periodic physical examinations.

Disposal Method Suggested: Concentrated waste containing no peroxides: discharge liquid at a controlled rate near a pilot flame. Concentrated waste containing peroxides: perforation of a container of the waste from a safe distance followed by open burning.

References

(1) U.S. Environmental Protection Agency, *Chemical Hazard Information Profile: Dimethoxane,* Washington, DC (1979).

(2) International Agency for Research on Cancer, *IARC Monographs on the Evaluation of Carcinogenic Risk of Chemicals to Man,* Lyon, France, 15, 177-181 (1977).

(3) Hoch-Ligeti, C. et al, *Oncogenic activity of an m-dioxane derivative: 2,6-dimethyl-m-dioxan-4-ol acetate (Dimethoxane),* J. Nat. Cancer Inst. 53, 791 (1974).

3,3'-DIMETHOXYBENZIDINE

● Carcinogen (Animal Positive (A-64)
● Hazardous Waste Constituent (EPA)

Description: $C_{14}H_{16}N_2O_2$ with the following structural formula

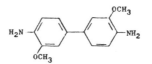

is a violet crystalline compound melting at 137° to 138°C.

Code Numbers: CAS 119-90-4   RTECS DD0875000

DOT Designation: —

Synonyms: Fast Blue B Base; dianisidine.

Potential Exposure: DMOB or its dihydrochloride is used principally as a chemical intermediate for the production of azo dyes. The Society of Dyers and Colorists reported the production of 89 dyes using DMOB as an intermediate in 1971. About 30% of DMOB is used as a chemical intermediate in the production of o-dianisidine diisocyanate (3,3'-dimethoxy-4.4'-diisocyanate-biphenyl) that is used in adhesive systems and also as a component of polyurethane elastomers and resins. DMOB is used as a dye itself for leather, paper, plastics, rubber, and textiles. DMOB has also been used in the detection of metals, thiocyanates, and nitrites.

Workers potentially exposed to DMOB are dye makers and o-dianisidine di-isocyanate production workers. However, current dye production processes for DMOB and dyes made from DMOB are generally closed systems with minimal risk to workers. The National Occupational Hazard Survey in 1974 estimated that 200 workers were potentially exposed to DMOB.

DMOB-based dyes and pigments are metabolized to DMOB. The CPSC staff is concerned that dyes and pigments based on DMOB contain residual levels or trace impurities of DMOB in the parts-per-million range and that traces may be present in the final consumer product. At present, no data exist on the actual quantities present in the final consumer product. A study is now in progress on the dermal penetration of DMOB and DMOB dyes.

Permissible Exposure Limits in Air: No standards set.

Permissible Concentration in Water: No criteria set.

Routes of Entry: Human exposure to DMOB is possible through inhalation of dye particles from equipment vent systems and through skin absorption from the finished dye product, textile processing, mixing operations, or packaging processes.

Harmful Effects and Symptoms: There is sufficient evidence for the carcinogenicity of 3,3'-dimethoxybenzidine (o-dianisidine) in experimental animals [1]. 3,3'-Dimethoxybenzidine administered by stomach tube was carcinogenic in rats, producing tumors at various sites, including intestinal, skin and Zymbal's gland carcinomas. The findings in hamsters exposed to o-dianisidine in the feed also suggest carcinogenicity [2].

The evidence for the carcinogenicity of o-dianisidine in humans is inadequate [1]. Most of the workers exposed to this substance were also exposed to related amines, such as benzidine, which are strongly associated with urinary bladder cancer in humans [2].

References

(1) International Agency for Research on Cancer, *IARC Monographs on the Evaluation of the Carcinogenic Risk of Chemicals to Humans,* Supplement 4, Lyon, France, IARC (1982).

(2) International Agency for Research on Cancer, *IARC Monographs on the Evaluation of the Carcinogenic Risk of Chemicals to Man,* Vol. 4, Lyon, France, IARC, pp 41-47 (1974).

(3) United Nations Environment Programme, *IRPTC Legal File 1983,* Vol. I, pp VII/106, Geneva, Switzerland, International Register of Potentially Toxic Chemicals (1984).

N,N-DIMETHYLACETAMIDE

Description: $CH_3CON(CH_3)_2$, dimethylacetamide, is a colorless, nonvolatile liquid with a faint ammonia-like odor. It boils at 165°C.

Code Numbers: CAS 127-19-5   RTECS AB7700000

DOT Designation: —

Synonyms: Acetic acid dimethylamide, DMA, DMAC, acetyl dimethylamide.

Potential Exposure: Dimethylacetamide is used primarily as a solvent for synthetic and natural resins, especially acrylic fibers and spandex. About 15% of dimethylacetamide production is used to make alkyl (C_{12-14}) dimethylamine oxide (a surfactant) and rubber chemicals. Dimethylacetamide is also used as an extraction solvent for butadiene manufacture. NIOSH estimates that 8,500 workers are exposed annually to DMA.

Incompatibilities: Carbon tetrachloride; other halogenated compounds, when in contact with iron.

Permissible Exposure Limits in Air: The Federal standard and the ACGIH 1983/84 TWA value is 10 ppm (35 mg/m^3). The notation "skin" indicates the possibility of cutaneous absorption. The STEL value is 15 ppm (50 mg/m^3). The IDLH level is 400 ppm.

Determination in Air: Adsorption on silica, workup with methanol, analysis by gas chromatography. See NIOSH Methods, Set R. See also reference (A-10).

Permissible Concentration in Water: No criteria set.

Routes of Entry: Inhalation of vapor and absorption through intact skin, ingestion and eye and skin contact.

Harmful Effects and Symptoms: *Local* — None known.
Systemic — Jaundice has been noted in workers exposed chronically to dimethylacetamide vapor although skin absorption may also have occurred. Liver injury consists of cord-cell degeneration, but recovery is usually rapid. Other symptoms from large oral doses as an anticancer drug include depression, lethargy, and visual and auditory hallucinations (1). Degeneration of the heart, kidney, and brain may also occur (2). Rodent tests indicated no mutagenic or carcinogenic effects; however, DMA was embryotoxic to rats, and had teratogenic effects on rabbits.

Points of Attack: Liver, skin.

Medical Surveillance: Preplacement and periodic medical examinations should give special attention to skin, central nervous system, and liver function or disease.

First Aid: If this chemical gets into the eyes, irrigate immediately. If this chemical contacts the skin, flush with water immediately. When this chemical has been swallowed, get medical attention. Give large quantities of water and induce vomiting. Do not make an unconscious person vomit.

Personal Protective Methods: Organic vapor masks or air supplied respirators may be required in elevated vapor concentrations. Percutaneous absorption should be prevented by gloves and other protective clothing. Goggles should be used to prevent eye splashes. In cases of spills or splashes, the wet clothing should be immediately removed and the involved skin area thoroughly cleaned. Clean clothing should be issued to workers on a daily basis and showers taken before changing to street clothes.

Respirator Selection:
    100 ppm: SA/SCBA
    400 ppm: SAF/SCBAF/SA:PD,PP,CF
    Escape: GMOV/SCBA

Disposal Method Suggested: Controlled incineration (incinerator is equipped with a scrubber or thermal unit to reduce NO_x emissions).

References

(1) Horn, H.J., "Toxicology of Dimethylacetamide," *Toxicol Appl. Pharmacol,* 3:12 (1961).
(2) National Institute for Occupational Safety and Health, *Information Profiles on Potential Occupational Hazards—Single Chemicals: N,N-Dimethyl Acetamide,* 56-64, Report TR79-607, Rockville, MD (December 1979).
(3) Sax, N.I., Ed., *Dangerous Properties of Industrial Materials Report, 1,* No. 5, 50-51, New York, Van Nostrand Reinhold Co. (1981).
(4) See Reference (A-61).
(5) See Reference (A-60).

DIMETHYLAMINE

- Hazardous substance (EPA)
- Hazardous waste (EPA)

Description: $(CH_3)_2NH$ is a colorless liquid or gas with a pungent, fishy, or ammonia-like odor. It boils at 7°C.

Code Numbers: CAS 124-40-3   RTECS IP8750000   UN 1032

DOT Designation: Flammable gas or liquid.

Synonyms: Dimethylamine, anhydrous.

Potential Exposure: This material is used in leather tanning, as an accelerator in rubber vulcanization, in the manufacture of detergents and in drug synthesis (A-41) and pesticide manufacture (A-32).

Incompatibilities: Strong oxidizers, chlorine, mercury.

Permissible Exposure Limits in Air: The Federal limit and the 1983/84 ACGIH TWA value is 10 ppm (18 mg/m³). No STEL value has been set. The IDLH level is 2,000 ppm.

Determination in Air: Adsorption on silica, workup with H_2SO_4 in CH_3OH, analysis by gas chromatography. See NIOSH Methods, Set J. See also reference (A-10).

Permissible Concentration in Water: No criterion set, but EPA (A-37) has suggested a permissible ambient goal of 248 μg/ℓ, based on health effects.

Routes of Entry: Inhalation, ingestion, eye and skin contact.

Harmful Effects and Symptoms: Irritation of eyes and throat, sneezing, coughing and dyspnea; pulmonary edema; conjunctivitis; dermatitis; burns of skin and mucous membranes.

Points of Attack: Respiratory system, lungs, skin, eyes.

Medical Surveillance: Consider the points of attack in preplacement and periodic physical examinations.

First Aid: If this chemical gets into the eyes, irrigate immediately. If this chemical contacts the skin, flush with water immediately. If a person breathes in large amounts of this chemical, move the exposed person to fresh air at once and perform artificial respiration. When this chemical has been swallowed, get medical attention. Give large quantities of water and induce vomiting. Do not make an unconscious person vomit.

Personal Protective Methods: Wear appropriate clothing to prevent any possibility of skin contact with liquid or repeated or prolonged contact with solutions. Wear eye protection to prevent any possibility of eye contact. Employees should wash immediately when skin is wet or contaminated with liquid DMA. Remove clothing immediately if wet or contaminated to avoid flammability hazard. Provide emergency showers and eyewash if liquid DMA is involved.

Respirator Selection:
          500 ppm: CCRSF/GMS/SAF/SCBAF
        2,000 ppm: SAF:PD,PP,CF
          Escape: GMS/SCBA

Disposal Method Suggested: Incineration; incinerator is equipped with a scrubber or thermal unit to reduce NO_x emissions.

References
(1)   See Reference (A-61).
(2)   United Nations Environment Programme, *IRPTC Legal File 1983,* Vol. I, pp VII/283–4, Geneva, Switzerland, International Register of Potentially Toxic Chemicals (1984).

4-DIMETHYLAMINOAZOBENZENE

- Carcinogen (Animal Positive, IARC) (2)
- Hazardous Waste Constituent (EPA)

Description: $C_6H_5NNC_6H_4N(CH_3)_2$, 4-dimethylaminoazobenzene, is a flaky yellow crystal.

Code Numbers: CAS 60-11-7   RTECS BX7350000

DOT Designation: —

Synonyms: Aniline-N,N-dimethyl-p-(phenylazo), benzeneazo dimethylaniline, fat yellow, oil yellow, butter yellow, methyl yellow, dimethyl yellow.

Potential Exposure: DAB is used for coloring polishes and other wax products, polystyrene, soap, and as a chemical indicator.

Human exposure to DAB can occur through either inhalation or skin absorption. OSHA has estimated that 2,500 workers are exposed to DAB.

Permissible Exposure Limits in Air: 4-Dimethylaminoazobenzene is included in the Federal standard for carcinogens; all contact with it should be avoided.

Determination in Air: Collection on a glass fiber filter, elution with 2-propanol, analysis by gas chromatography (A-10).

Permissible Concentration in Water: No criteria set.

Routes of Entry: Probably inhalation and percutaneous absorption.

Harmful Effects and Symptoms: *Local* — Unknown.

Systemic — p-Dimethylaminoazobenzene (DAB) is carcinogenic in rats, producing liver tumors after its administration by several routes, and in dogs, producing bladder tumors following its administration by the oral route.

DAB has also been tested by subcutaneous injection in mice, and the results are suggestive of local and hepatic carcinogenicity. Treatment of newborn animals produced systematic carcinogenic effects in mice. Skin-painting with DAB produced epidermal tumors in rats but not in mice (3).

Target Organs: Liver.

Medical Surveillance: Preplacement and periodic examinations should include a history of exposure to other carcinogens; use of alcohol, smoking, and medications; and family history. Special attention should be given to liver size and liver function tests.

Personal Protective Methods: These are designed to supplement engineering controls and to prevent all contact with skin and the respiratory tract. Protective clothing and gloves should be provided, and also appropriate type dust or supplied air respirators. On exit from a regulated area, employees should shower and change into street clothes, leaving their clothes at the point of exit, to be placed in impervious containers at the end of the work shift for decontamination or disposal.

References

(1) Miller, J.A., and E.C. Miller, The Carcinogenic Aminoazo Dyes, *Adv. Cancer Res.,* 1:339 (1953).
(2) International Agency for Research on Cancer, *IARC Monographs on the Carcinogenic Risks of Chemicals to Humans,* Lyon, France, 8, 125 (1975).
(3) See Reference (A-62). Also see Reference (A-64).
(4) Parmeggiani, L., Ed., *Encyclopedia of Occupational Health & Safety,* Third Edition, Vol. 1 p 633, Geneva, International Labour Office (1983).

N,N-DIMETHYLANILINE

Description: $C_6H_5N(CH_3)_2$ is a straw- to brown-colored liquid with a characteristic amine-like odor. It boils at 194°C.

Code Numbers: CAS 121-69-7   RTECS BX4725000   UN 2253

DOT Designation: Poison

Synonyms: N,N-dimethylphenylamine, DMA.

Potential Exposure: This material is used as an intermediate in the manufacture of many dyes and rubber chemicals. It is also used as an analytical reagent.

Incompatibilities: Strong oxidizers, strong acids.

Permissible Exposure Limits in Air: The Federal limit is 5 ppm (25 mg/m³). The notation "skin" indicates the possibility of cutaneous absorption. The STEL set by ACGIH as of 1983/84 is 10 ppm (50 mg/m³). The IDLH level is 100 ppm.

Determination in Air: Adsorption on charcoal, workup with CS_2, analysis by gas chromatography. See NIOSH Methods, Set L. See also reference (A-10).

Permissible Concentration in Water: No criteria set, but EPA (A-37) has suggested permissible ambient concentrations of 345 µg/ℓ.

Routes of Entry: Inhalation, skin absorption, ingestion, eye and skin contact.

Harmful Effects and Symptoms: Anoxia, cyanosis, weakness, dizziness, ataxia.

Points of Attack: Blood, kidneys, liver, cardiovascular system.

Medical Surveillance: Consider the points of attack in preplacement and periodic physical examinations.

First Aid: If this chemical gets into the eyes, irrigate immediately. If this chemical contacts the skin, wash with soap immediately. If a person breathes in large amounts of this chemical, move the exposed person to fresh air at once and perform artificial respiration. When this chemical has been swallowed, get medical attention. Give large quantities of water and induce vomiting. Do not make an unconscious person vomit.

Personal Protective Methods: Wear appropriate clothing to prevent any reasonable probability of skin contact. Wear eye protection to prevent any reasonable probability of eye contact. Employees should wash immediately when skin is wet or contaminated. Remove nonimpervious clothing immediately if wet or contaminated. Provide emergency shower.

Respirator Selection:
    50 ppm: SA/SCBA
    100 ppm: SAF/SCBAF/SA:PD,PP,CF
    Escape: GMOV/SCBA

Disposal Method Suggested: Incineration in a furnace equipped with afterburner and scrubber.

References
(1)   See Reference (A-61).
(2)   See Reference (A-60).

3,3'-DIMETHYLBENZIDINE

● Carcinogen (A-64)

See "o-Tolidine."

DIMETHYL CARBAMOYL CHLORIDE

● Carcinogen (Animal positive, IARC) (1)
● Hazardous waste (EPA)

Description: $(CH_3)_2NCOCl$ is a liquid which boils at 165° to 167°C.

Code Numbers: CAS 79-44-7   RTECS FD4200000   UN 2262

DOT Designation: —

Synonyms: DMCC, chloroformic acid dimethyl amide, dimethyl carbamyl chloride.

Potential Exposure: DMCC is used as a chemical intermediate in the production of pharmaceuticals, pesticides, rocket fuel, and in dye synthesis. The pharmaceuticals (neostigmine bromide, neostigmine methylsulfate, and pyridostigmine bromide) are used in the treatment of myasthenia gravis. The three U.S. registered pesticides derived from DMCC are known by the trade names: Tandex®, Dimetilan®, and Pirimor®.

Human exposure is limited to but not restricted to chemical workers, pesticide formulators, dye makers, and pharmaceutical workers. OSHA estimated that 200 workers are exposed to DMCC, primarily through inhalation. DMCC has been found at levels up to 6 ppm during the production of phthaloyl chlorides. It is possible that levels of exposure might be higher in facilities in which the chemical is used for further synthesis. When DMCC is used as a dye intermediate, exposure can occur from the amount of residue in the product and its ability to migrate.

Permissible Exposure Limits in Air: There are no Federal standards but ACGIH (1983/84) classifies DMCC as an "industrial substance suspect of carcinogenic potential for man with no assigned TWA.

Permissible Concentration in Water: No criteria set.

Harmful Effects and Symptoms: This is a very toxic material, a proven carcinogen in animals and a suspected carcinogen in man (A-34).

Dimethylcarbamoyl chloride is carcinogenic in mice, producing local carcinomas after application to the skin and local sarcomas after subcutaneous or intraperitoneal injection.

A study of humans exposed to dimethylcarbamoyl chloride was considered inadequate due to the small number of people observed (2).

References
(1)   International Agency for Research on Cancer, *IARC Monographs on the Carcinogenic Risks of Chemicals to Humans,* 12, 77 (1976).
(2)   See Reference (A-62). Also see Reference (A-64).
(3)   Parmeggiani, L., Ed., *Encyclopedia of Occupational Health & Safety,* Third Edition, Vol. 1, pp 633–34, Geneva, International Labour Office (1983).

N,N-DIMETHYLFORMAMIDE

Description: $HCON(CH_3)_2$, dimethylformamide, is a colorless liquid which boils at 153°C. It has a fishy, unpleasant odor at relatively low concentrations.

Code Numbers: CAS 68-12-2   RTECS LQ2100000   UN 2265

DOT Designation: Label with St. Andrews Cross (X).

Synonyms: DMF, formyldimethylamine.

Potential Exposure: Dimethylformamide has powerful solvent properties for a wide range of organic compounds. Because of dimethyl formamide's physical properties, it has been used when solvents with a slow rate of evaporation are required. It finds particular usage in the manufacture of polyacrylic fibers, butadiene, purified acetylene, pharmaceuticals, dyes, petroleum products, and other organic chemicals. NIOSH estimates worker exposure to DMF at 45,660/yr.

Incompatibilities: Carbon tetrachloride; other halogenated compounds, when in contact with iron; strong oxidizers; alkyl aluminums.

Permissible Exposure Limits in Air: The Federal standard is 10 ppm (30 mg/m³). The notation "skin" indicates the possibility of cutaneous absorption. The STEL set by ACGIH as of 1983/84 is 20 ppm (60 mg/m³). The IDLH level is 3,500 ppm.

Determination in Air: Adsorption on silica, workup with methanol, analysis by gas chromatography. See NIOSH Methods, Set R. See also reference (A-10).

Permissible Concentration in Water: No criteria set.

Routes of Entry: Inhalation of vapor, and absorption through intact skin, ingestion and skin and eye contact.

Harmful Effects and Symptoms: *Local* — Dimethylformamide exposure may cause dermatitis.

Systemic — Inhalation of dimethylformamide or skin contact with this chemical may cause colicky abdominal pain, anorexia, nausea, vomiting, constipation, diarrhea, facial flushing (especially after drinking alcohol), elevated blood pressures, hepatomegaly, and other signs of liver damage. This chemical has produced kidney damage in animals.

Points of Attack: Liver, kidneys, skin, cardiovascular system.

Medical Surveillance: Preplacement and periodic examinations should be concerned particularly with liver and kidney function and with possible effects on the skin.

First Aid: If this chemical gets into the eyes, irrigate immediately. If this chemical contacts the skin, flush with water promptly. If a person breathes in large amounts of this chemical, move the exposed person to fresh air at once and perform artificial respiration. When this chemical has been swallowed, get medical attention. Give large quantities of water and induce vomiting. Do not make an unconscious person vomit.

Personal Protective Methods: Organic vapor masks or air supplied respirators may be required in elevated vapor concentrations. Percutaneous absorption should be prevented by gloves and other protective clothing. Goggles should be used to prevent eye splashes. In cases of spills or splashes, the wet clothing should be immediately removed and the skin area thoroughly cleaned. Clean clothing should be issued to workers on a daily basis and showers taken before changing to street clothes.

Respirator Selection:
100 ppm:	SA/SCBA
500 ppm:	SAF/SCBAF
3,500 ppm:	SAF:PD,PP,CF
Escape:	GMOV/SCBA

Disposal Method Suggested: Burn in solution in flammable solvent in furnace equipped with alkali scrubber. Recovery and recycle is an alternative to disposal for DMF from fiber spin baths and from PVC reactor cleaning solvents (A-58).

References

(1) U.S. Environmental Protection Agency, *Chemical Hazard Information Profile: N,N-Dimethylformamide,* Washington, DC (April 13, 1978).

(2)   National Institute for Occupational Safety and Health, *Information Profiles on Potential Occupational Hazards—Single Chemicals: N,N-Dimethyl Formamide,* 65-73, Report TR79-607, Rockville, MD (December 1979).
(3)   Sax, N.I., Ed., *Dangerous Properties of Industrial Materials Report, 1,* No. 3, 61-62, New York, Van Nostrand Reinhold Co. (1981).
(4)   See Reference (A-61).
(5)   See Reference (A-60).

1,1-DIMETHYLHYDRAZINE

- Carcinogen (Animal positive, IARC) (2)
- Hazardous waste (EPA)

Description: $(CH_3)_2NNH_2$ is a fuming colorless liquid with an amine-like odor. It boils at 63°C.

Code Numbers: CAS 57-14-7   RTECS MV2450000   UN 1163

DOT Designation: Flammable liquid and poison.

Synonyms: Unsymmetrical dimethylhydrazine, UDMH, Dimazine®.

Potential Exposure: This material is used as a component in liquid rocket propellant combinations; it is also used as an intermediate in organic synthesis.

Incompatibilities: Oxidizers, halogens, metallic mercury, fuming nitric acid, hydrogen peroxide.

Permissible Exposure Limits in Air: The Federal limit is 0.5 ppm (1 mg/m³). The notation "skin" indicates the possibility of cutaneous absorption. The ACGIH has set an STEL of 1.0 ppm (2 mg/m³) with the notation that UMDH is an "industrial substance suspect of carcinogenic potential for man." NIOSH (1) has proposed a 2-hr ceiling concentration of 0.06 ppm (0.15 mg/m³). The IDLH level is 50 ppm.

Determination in Air: Collection by a bubbler, colorimetric determination with phosphomolybdic acid. See NIOSH Methods, Set K. See also reference (A-10).

Permissible Concentration in Water: No criteria set, but EPA (A-37) has suggested a permissible ambient goal of 13.8 µg/ℓ based on health effects.

Routes of Entry: Inhalation, skin absorption, ingestion, eye and skin contact.

Harmful Effects and Symptoms: Eye irritation; choking, chest pains, dyspnea; lethargy; nausea; skin irritation; anoxia; convulsions; liver injury.

Points of Attack: Central nervous system, liver, gastrointestinal system, blood, respiratory system, eyes, skin.

Medical Surveillance: Consider the points of attack in preplacement and periodic physical examinations.

First Aid: If this chemical gets into the eyes, irrigate immediately. If this chemical contacts the skin, flush with water immediately. If a person breathes in large amounts of this chemical, move the exposed person to fresh air at once and perform artificial respiration. When this chemical has been swallowed, get medical attention. Give large quantities of water and induce vomiting. Do not make an unconscious person vomit.

Personal Protective Methods: Wear appropriate clothing to prevent any possibility of skin contact. Wear eye protection to prevent any possibility of eye contact. Employees should wash immediately when skin is wet or contaminated. Remove clothing immediately if wet or contaminated to avoid flammability hazard. Provide emergency showers and eyewash.

Respirator Selection:
        25 ppm: SAF/SCBAF
        50 ppm: SAF:PD,PP,CF
        Escape: GMSF/SCBAF

Disposal Method Suggested: Controlled incineration (oxides of nitrogen are removed from the effluent gas by scrubbers and/or thermal devices).

References

(1) National Institute for Occupational Safety and Health, *Criteria for a Recommended Standard: Occupational Exposure to Hydrazines,* NIOSH Doc. No. 78-172, Washington, DC (1978).

(2) International Agency for Research on Cancer, *IARC Monographs on the Carcinogenic Risks of Chemicals to Humans,* Lyon, France 4, 137 (1974).

(3) See Reference (A-60).

(4) Sax, N.I., Ed., *Dangerous Properties of Industrial Materials Report, 4,* No. 3, 60–67, New York, Van Nostrand Reinhold Co. (1984).

DIMETHYL METHYLPHOSPHONATE

Description: $(CH_3O)_2\overset{\overset{O}{\|}}{P}CH_3$ is a colorless liquid which boils at $180°C$ at 760 mm pressure.

Code Numbers: CAS 756-79-6   RTECS SZ912000

DOT Designation: —

Synonyms: Phosphonic acid, methyl-, dimethyl ester; DMMP.

Potential Exposure: May be used as a gasoline additive, hydraulic fluid additive, as a heavy metal extractor, as a solvent, as a simulant for nerve gas agents, and as an additive flame retardant in plastics.

Permissible Exposure Limits in Air: No standards set.

Permissible Concentration in Water: No criteria set.

Harmful Effects and Symptoms: DMMP is not very toxic; the oral LD-50 in mice and rats is greater than 5,000 mg/kg. There are some indications of alteration in reproductive functions of rats exposed to doses of 250 mg/kg, however.

References

(1) U.S. Environmental Protection Agency, *Chemical Hazard Information Profile Draft Report: Dimethyl methylphosphonate,* Washington, DC (September 2, 1983).

2,4-DIMETHYLPHENOL

- Hazardous substance (EPA)
- Hazardous waste (EPA)
- Priority toxic pollutant (EPA)

Description: $C_8H_{10}O$, $HOC_6H_3(CH_3)_2$, is a colorless crystalline solid melting at 27° to 28°C. The 2,4-isomer is one of 5 isomers of this formula.

Code Numbers: CAS 105-67-9   RTECS ZE5600000

DOT Designation: —

Synonyms: 2,4-Xylenol; 1-hydroxy-2,4-dimethylbenzene; m-xylenol; 2,4-DMP.

Potential Exposure: 2,4-DMP finds use commercially as an important chemical feedstock or constituent for the manufacture of a wide range of commercial products for industry and agriculture. 2,4-Dimethylphenol is used in the manufacture of phenolic antioxidants, disinfectants, solvents, pharmaceuticals, insecticides, fungicides, plasticizers, rubber chemicals, polyphenylene oxide, wetting agents, and dyestuffs, and is an additive or constituent of lubricants, gasolines, and cresylic acid. 2,4-Dimethylphenol (2,4-DMP) is a naturally occurring, substituted phenol derived from the cresol fraction of petroleum or coal tars by fractional distillation and extraction with aqueous alkaline solutions. It is the cresylic acid or tar acid fraction of coal tar.

Workers involved in the fractionation and distillation of petroleum or coal and coal tar products comprise one group at risk. Workers who are intermittently exposed to certain commercial degreasing agents containing cresol may also be at risk. Cigarette and marijuana smoking groups and those exposed to cigarette smoke inhale μg quantities of 2,4-dimethylphenol. The National Institute for Occupational Safety and Health has estimated that 11,000 people in the United States are occupationally exposed to cresol containing 2,4-dimethylphenol.

Permissible Exposure Limits in Air: No standards set.

Permissible Concentration in Water: To protect freshwater aquatic life—2,120 μg/ℓ on an acute toxicity basis. To protect saltwater aquatic life—no criterion established due to insufficient data. To protect human health—in view of the relative paucity of data on the mutagenicity, carcinogenicity, teratogenicity and long term oral toxicity of 2,4-dimethylphenol, estimates of the effects of chronic oral exposure at low levels cannot be made with any confidence. It is recommended that studies to produce such information be conducted before limits in drinking water are established. A criterion of 400 μg/ℓ is suggested by EPA (1) on an organoleptic basis.

Determination in Water: Methyl chloride extraction followed by gas chromatography with flame ionization or electron capture detection (EPA Method 604) or gas chromatography plus mass spectrometry (EPA Method 625).

Harmful Effects and Symptoms: 2,4-Dimethylphenol appears to be a topical cocarcinogen, but its role as a primary cancer-producing agent is uncertain. Its potential role in cancer production warrants consideration of further testing. An in vitro mutagenicity assay should be carried out to further evaluate its mutagenic potential.

Disposal Method Suggested: Incineration.

Reference

(1) U.S. Environmental Protection Agency, *2,4-Dimethylphenol: Ambient Water Quality Criteria,* Washington, DC (1980).

(2) U.S. Environmental Protection Agency, *2,4-Dimethylphenol,* Health and Environmental Effects Profile No. 87, Washington, DC, Office of Solid Waste (April 30, 1980).

DIMETHYL PHTHALATE

- Hazardous waste (EPA)
- Priority toxic pollutant (EPA)

Description: $C_6H_4(COOCH_3)_2$ is a colorless oily liquid with a slight ester odor. It boils at 285°C.

Code Numbers: CAS 131-11-3   RTECS TI1575000

DOT Designation: —

Synonyms: Phthalic acid, dimethyl ester; dimethyl 1,2-benzenedicarboxylate; DMP; ENT-262.

Potential Exposure: Dimethyl phthalate is used as a plasticizer for cellulose ester plastics and as an insect repellent (3).

Incompatibilities: Nitrates, strong oxidizers, strong alkalies, strong acids.

Permissible Exposure Limits in Air: The Federal limit and the TWA value adopted by ACGIH (1983/84) is 5 mg/m³. The STEL value is 10 mg/m³. The IDLH level is 9,300 mg/m³.

Determination in Air: Collection by charcoal tube, analysis by gas liquid chromatography (A-13).

Permissible Concentration in Water: To protect freshwater and saltwater aquatic life—no criteria developed due to insufficient data. To protect human health—313 mg/ℓ.

Determination in Water: Methylene chloride extraction followed by gas chromatography with flame ionization or electron capture detection (EPA Method 606) or gas chromatography plus mass spectrometry (EPA Method 626).

Routes of Entry: Ingestion, inhalation, eye and skin contact.

Harmful Effects and Symptoms: Irritation of nasal passages and upper respiratory system; stomach irritation; eye pain.

Points of Attack: Respiratory system, gastrointestinal (GI) system.

Medical Surveillance: Consider the points of attack in preplacement and periodic physical examinations.

First Aid: If this chemical gets into the eyes, irrigate promptly. If this chemical contacts the skin, wash regularly. If a person breathes in large amounts of this chemical, move the exposed person to fresh air at once and perform artificial respiration. When this chemical has been swallowed, get medical attention. Give large quantities of salt water and induce vomiting. Do not make an unconscious person vomit.

Personal Protective Methods: Wear eye protection to prevent any reasonable probability of eye contact.

Respirator Selection:
250 mg/m³: HiEPF/SAF/SCBAF
9,300 mg/m³: SAF,PD,PP,CF

Disposal Method Suggested: Incineration.

References

(1) U.S. Environmental Protection Agency, *Phthalate Esters: Ambient Water Quality Criteria,* Washington, DC (1980).

(2)   National Institute for Occupational Safety and Health, *Profiles on Occupational Hazards for Criteria Document Priorities—Phthalates,* 97-103, Report PB-274,073, Cincinnati, OH (1977).

(3)   U.S. Environmental Protection Agency, *Dimethyl Phthalate,* Health and Environmental Effects Profile No. 88, Washington, DC, Office of Solid Waste (April 30, 1980).

(4)   See Reference (A-61).

(5)   Sax, N.I., Ed., *Dangerous Properties of Industrial Materials Report, 2,* No. 4, 80-84, New York, Van Nostrand Reinhold Co. (1982).

(6)   See Reference (A-60).

DIMETHYL SULFATE

- Carcinogen (human suspected, IARC) (2) (EPA-CAG)
- Hazardous waste (EPA)

Description: $(CH_3)_2SO_4$, dimethyl sulfate, is an oily, colorless liquid slightly soluble in water, but more soluble in organic solvents. It boils at 188°C.

Code Numbers: CAS 77-78-1   RTECS WS8225000   UN 1595

DOT Designation: Corrosive material.

Synonyms: Sulfuric acid dimethyl ester, methyl sulfate, DMS.

Potential Exposure: Industrial use of dimethyl sulfate is based upon its methylating properties. It is used in the manufacture of methyl esters, ethers and amines, in dyes, drugs, perfume, phenol derivatives, and other organic chemicals. It is also used as a solvent in the separation of mineral oils. It is used as an intermediate in the manufacture of many pharmaceuticals (A-41) and pesticides (A-32). NIOSH has estimated worker exposure to dimethyl sulfate at 4,200 annually.

Incompatibilities: Strong oxidizers, strong ammonia solutions.

Permissible Exposure Limits in Air: The Federal standard is 1 ppm (5 mg/m^3). The ACGIH as of 1983/84 has adopted a TWA value of 0.1 ppm (0.5 mg/m^3) with the note that dimethyl sulfate is an "industrial substance suspect of carcinogenic potential for man." Skin absorption is possible.

Permissible Concentration in Water: No criteria set.

Routes of Entry: Inhalation of vapor, percutaneous absorption of liquid, ingestion, eye and skin contact.

Harmful Effects and Symptoms: *Local* — Liquid is highly irritating and causes skin vesiculation and analgesia. Lesions are typically slow-healing and may result in scar tissue while analgesia may last several months. Liquid and vapor are irritating to the mucous membranes, and exposure produces lacrimation, rhinitis, edema of the mucosa of the mouth and throat, dysphagia, sore throat, and hoarseness. Irritation of the skin and mucous membranes may be delayed in appearance. Eye irritation may result in conjunctivitis, keratitis, and photophobia. In severe cases corneal opacities, perforation of the nasal septum and permanent or persistent visual disorders have been reported.

Systemic — The toxicity of dimethyl sulfate is based upon its alkylating properties and its hydrolysis to sulfuric acid and methyl alcohol. Acute exposure may cause respiratory dysfunctions such as pulmonary edema, bronchitis, and pneumonitis following a latent period of 6 to 24 hr. Cerebral edema and other central nervous system effects such as drowsiness, temporary blindness, tachycardia or bradycardia may be linked to dimethyl sulfate's effect on nerve endings. Secondary pulmonary effects such as susceptibility to infection, as well as more pronounced effects in those persons with preexisting respiratory disorders, are also noteworthy. Chronic poisoning occurs only rarely and is usually limited to ocular and respiratory disabilities. It has been reported to be carcinogenic in rats, but this has not been verified in man.

Dimethyl sulfate is carcinogenic in rats after inhalation or subcutaneous injection, producing mainly local tumors, and after prenatal exposure, producing tumors of the nervous system.

Four bronchial carcinomas have been reported in men occupationally exposed to dimethyl sulfate. In an epidemiological study, six cancer deaths were found versus 2.4 expected; three of these were cancers of the respiratory tract (1.02 expected). Neither the respiratory tract cancers nor the cancer rate at all sites are statistically significantly increased (5).

Points of Attack: Eyes, respiratory system, liver, kidneys, central nervous system, skin, lungs.

Medical Surveillance: Preplacement and periodic medical examinations should give special consideration to the skin, eyes, central nervous system, lungs. Chest x-rays should be taken and lung, liver, and kidney functions evaluated. Sputum and urinary cytology may be useful in detecting the presence or absence of carcinogenic effects.

First Aid: If this chemical gets into the eyes, irrigate immediately. If this chemical contacts the skin, flush with water immediately. If a person breathes in large amounts of this chemical, move the exposed person to fresh air at once and perform artificial respiration. When this chemical has been swallowed, get medical attention. Give large quantities of water and do not induce vomiting.

Personal Protective Methods: These are designed to supplement engineering controls and to reduce skin, eye, or respiratory contact to a negligible level. The liquid and the vapor of dimethyl sulfate are extremely irritating so that the skin, eyes, as well as the respiratory tract should be protected at all times.

Wear appropriate clothing to prevent any possibility of skin contact. Wear eye protection to prevent any possibility of eye contact. Employees should wash immediately when skin is wet or contaminated. Remove nonimpervious clothing immediately if wet or contaminated. Provide emergency showers and eyewash.

Respirator Selection:
       10 ppm: SAF/SCBAF
       Escape: GMSF/SCBAF

Disposal Method Suggested: Incineration (1800°F, 1.5 seconds minimum) of dilute, neutralized dimethyl sulfate waste is recommended. The incinerator must be equipped with efficient scrubbing devices for oxides of sulfur.

References
(1) National Institute for Occupational Safety and Health, *Information Profiles on Potential Occupational Hazards—Single Chemicals: Dimethyl Sulfate*, 74-84, Report TR79-607, Rockville, MD (December 1979).

(2) International Agency for Research on Cancer, *IARC Monographs on the Carcinogenic Risks of Chemicals to Humans*, Lyon, France, 4, 271 (1974).

(3) Sax, N.I., Ed., *Dangerous Properties of Industrial Materials Report, 1*, No. 5, 51-53, New York, Van Nostrand Reinhold Co. (1981).

(4) See Reference (A-61).

(5) See Reference (A-62). Also see Reference (A-64).
(6) See Reference (A-60).
(7) Parmeggiani, L., Ed., *Encyclopedia of Occupational Health & Safety,* Third Edition, Vol. 1, pp 634–35, Geneva, International Labour Office (1983).
(8) United Nations Environment Programme, *IRPTC Legal File 1983,* Vol. II, pp VII/747–8, Geneva, Switzerland, International Register of Potentially Toxic Chemicals (1984).

DIMETHYL SULFOXIDE

Description: $(CH_3)_2SO$ is a hygroscopic liquid which boils at $189°C$.

Code Numbers: CAS 67-68-5   RTECS PV6210000

DOT Designation: —

Synonyms: Sulfinyl-bis(methane), DMSO.

Potential Exposure: Used as a solvent, as a pharmaceutical, in chemicals production.

Incompatibilities: Strong oxidants.

Permissible Exposure Limits in Air: No standards set.

Permissible Concentration in Water: No criteria set.

Harmful Effects and Symptoms: Irritation results from skin contact. Systemically, it produces anaesthesia, vomiting, chills, cramps and lethargy.

Points of Attack: Skin, eyes.

Personal Protective Methods: Goggles, butyl rubber gloves, protective clothing.

Respirator Selection: Canister-type mask.

References

(1) Sax, N.I., Ed., *Dangerous Properties of Industrial Materials Report, 1,* No. 1, 42-43, New York, Van Nostrand Reinhold Co. (1980).
(2) See Reference (A-60).

DIMETHYL TEREPHTHALATE

Description: $CH_3OCOC_6H_4COOCH_3$ is a solid which melts at $141°$ to $142°C$.

Code Numbers: CAS 120-61-6   RTECS WZ1225000

DOT Designation: —

Synonyms: 1,4-Benzenedicarboxylic acid dimethyl ester; Terephthalic acid dimethyl ester; Dimethyl p-phthalate; DMT.

Potential Exposures: Essentially all DMT is consumed in the production of polyethylene terephthalate, the polymer for polyester fibers and polyester films. Less than 2% of production is used to make polybutylene terephthalate resins and other specialty products.

NIOSH estimates worker exposure to DMT at 4,600 annually.

Permissible Exposure Limits in Air: No standards set.

Permissible Concentration in Water: Although o-phthalic acid esters have been the subject of concern (see "Dimethyl Phthalate"), the p-phthalic acid esters have no criteria set.

Routes of Entry: Inhalation of dust or vapor, ingestion, skin and eye contact.

Harmful Effects and Symptoms: DMT appears to have a very low order of toxicity. Acute animal studies indicate oral, i.p., and dermal LD_{50} values in excess of 3,400 mg/kg, and subchronic oral (10,000 ppm DMT in the diet for 96 days) and inhalation exposures (2–10 ppm, 4 hr/day x 58 days) have not resulted in any hematologic, blood chemical, or pathologic alterations attributable to DMT. Rats ingesting 10,000 ppm DMT in the diet did exhibit, however, a slightly reduced weight gain after 96 days, and offspring of rats fed 5,000 and 10,000 ppm DMT showed lowered weights at weaning, although no effects on parental fertility or reproductive capacity were observed. Results of other experiments with rats and rabbits demonstrated that DMT is rapidly absorbed and excreted (primarily in the urine), and that no significant quantities accumulate in tissues following single or repeated oral, intratracheal, dermal, or ocular administration. DMT does not appear to irritate or sensitize rodent skin. A recently completed NCI bioassay has concluded that DMT is not carcinogenic to rats or mice at dietary levels of 2,500 and 5,000 ppm, and DMT does not appear to be teratogenic in rats. Extremely limited Russian information, available only in undetailed abstract form, suggests that occupational exposure to DMT may influence hematologic parameters and immune responses in workers.

In aggregate, the known biological effects of DMT do not seem to indicate any great toxicological hazard (1).

References

(1) National Institute for Occupational Safety and Health, *Information Profiles on Potential Occupational Hazards-Single Chemicals: Dimethyl Terephthalate,* pp 84–97, Report TR79-607, Rockville, MD (December 1979).

(2) See Reference (A-60).

DINITOLMIDE

Description: $H_3CC_6H_2(NO_2)_2CONH_2$, 3,5-dinitro-o-toluamide is a yellowish crystalline substance melting at 177° to 181°C.

Code Numbers: CAS 148-01-6 RTECS XS4200000

DOT Designation: —

Synonyms: Zoalene®, Zoamix®, 2-methyl-3,5-dinitrobenzamide.

Potential Exposure: Those involved in the manufacture, formulation and application of this veterinary coccidiostat.

Permissible Exposure Limits in Air: There are no Federal standards but ACGIH (1983/84) has adopted a TWA value of 5 mg/m³ and set an STEL of 10 mg/m³.

Permissible Concentration in Water: No criteria set.

Harmful Effects and Symptoms: This material has a moderate acute oral toxicity toward rats. On a chronic basis, rats fed 3 mg/kg/day for 2 yr showed no adverse effects. Rats fed 6 mg/kg/day for 2 yr showed only slight changes in liver weight and fatty content (A-34).

DINITROBENZENES

- Hazardous substances (EPA)
- Hazardous waste constituents (EPA)

Description: $C_6H_4(NO_2)_2$, dinitrobenzene, may exist in three isomers; the meta form is the most widely used. They are solids, melting at: ortho, 117° to 118°C; meta, 90°C; and para, 173° to 174°C.

Code Numbers:

Ortho isomer	CAS 528-29-0	RTECS CZ7450000	UN 1597
Meta isomer	CAS 99-65-0	RTECS CZ7350000	UN 1597
Para isomer	CAS 100-25-4	RTECS CZ7525000	UN 1597

DOT Designations: Poison B

Synonyms: Dinitrobenzol.

Potential Exposure: Dinitrobenzene is used in the synthesis of dyestuffs, dyestuff intermediates, and explosives and in Celluloid production.

Incompatibilities: Strong oxidizers, caustics, metals, such as tin, zinc.

Permissible Exposure Limits in Air: The Federal standard for all isomers of dinitrobenzene is 1 mg/m³ (0.15 ppm). The notation "skin" indicates the possibility of cutaneous absorption. The STEL value adopted by ACGIH (1983/84) is 0.5 ppm (3 mg/m³). The IDLH level is 200 mg/m³.

Determination in Air: Collection by filter in series with an ethylene glycol bubbler, analysis by high performance liquid chromatography (A-10).

Permissible Concentration in Water: No criteria set.

Routes of Entry: Inhalation, percutaneous absorption of liquid, ingestion, eye and skin contact.

Harmful Effects and Symptoms: *Local* – Exposure to dinitrobenzene may produce yellowish coloration of the skin, eyes and hair.
Systemic – Exposure to any isomer of dinitrobenzene may produce methemoglobinemia, symptoms of which are headache, irritability, dizziness, weakness, nausea, vomiting, dyspnea, drowsiness, and unconsciousness. If treatment is not given promptly, death may occur. Consuming alcohol, exposure to sunlight, or hot baths may make symptoms worse. Dinitrobenzene may also cause a bitter almond taste or burning sensation in the mouth, dry throat, and thirst. Reduced vision may occur. In addition liver damage, hearing loss, and ringing of the ears may be produced. Repeated or prolonged exposure may cause anemia.

Points of Attack: Blood, liver, cardiovascular system, eyes, central nervous system.

Medical Surveillance: Preemployment and periodic examinations should be concerned particularly with a history of blood dyscrasias, reactions to medications, alcohol intake, eye disease, and skin and cardiovascular status. Liver and renal functions should be evaluated periodically as well as blood and general health. Methemoglobin levels should be followed until normal in all cases of suspected cyanosis. Dinitrobenzene can be determined in the urine; levels greater than 25 mg/ℓ may indicate significant absorption.

First Aid: If this chemical gets into the eyes, irrigate immediately. If this chemical contacts the skin, wash with soap immediately. If a person breathes in large amounts of this chemical, move the exposed person to fresh air at once

and perform artificial respiration. When this chemical has been swallowed, get medical attention. Give large quantities of water and induce vomiting. Do not make an unconscious person vomit.

Personal Protective Methods: Dinitrobenzene is readily absorbed through intact skin and its vapors are highly toxic. Protective clothing impervious to the liquid should be worn in areas where the likelihood of splash or spill exists. When splash or spill occurs on ordinary work clothes, they should be removed immediately and the area washed thoroughly. In areas of elevated vapor concentrations fullface masks with organic vapor canisters or air supplied respirators with fullface piece should be used. Daily changes of work clothing and mandatory showering at the end of each shift before changing to street clothes should be enforced. Wear eye protection to prevent any reasonable probability of eye contact. Provide emergency showers.

Respirator Selection:

5 mg/m³:	DMXS
10 mg/m³:	DMXSQ/FuHiEP/SA/SCBA
50 mg/m³:	HiEPF/SAF/SCBAF
200 mg/m³:	PAPHiE/SA:PD,PP,CF
Escape:	DMXS/SCBA

Disposal Method Suggested: Incineration (1800°F, 2.0 seconds minimum) followed by removal of the oxides of nitrogen that are formed using scrubbers and/or catalytic or thermal devices. The dilute wastes should be concentrated before incineration.

References

(1) U.S. Environmental Protection Agency, *Dinitrobenzenes,* Health and Environmental Effects Profile No. 89, Washington, DC, Office of Solid Waste (April 30, 1980).
(2) See Reference (A-61).
(3) Sax, N.I., Ed., *Dangerous Properties of Industrial Materials Report, 3,* No. 3, 80-82, New York, Van Nostrand Reinhold Co. (1983). (p-Dinitrobenzene).

DINITRO-o-CRESOL

- Hazardous waste (EPA)
- Priority toxic pollutant (EPA)

Description: $CH_3C_6H_2(NO_2)_2OH$, dinitro-o-cresol, exists in 9 isomeric forms of which 3,5-dinitro-o-cresol is the most important commercially. It is a yellow crystalline solid, melting at 88°C.

Code Numbers: CAS 534-52-1 RTECS GO9625000 UN 1598

DOT Designation: Poison

Synonyms: DNOC; 4,6-dinitro-o-cresol is also known as 3,5-dinitro-o-cresol; 2-methyl-4,6-dinitrophenol; 3,5-dinitro-2-hydroxytoluene.

Potential Exposure: DNOC is widely used in agriculture as a herbicide and pesticide; it is also used in the dyestuff industry. Although 4,6-dinitro-o-cresol (DNOC) is no longer manufactured in the United States, a limited quantity is imported and used as a blossom-thinning agent on fruit trees and as a fungicide, insecticide, and miticide on fruit trees during the dormant season. Hence, individuals formulating or spraying the compound incur the highest risk of exposure to the compound. NIOSH estimates that 3,000 workers in the United States are potentially exposed to DNOC. In view of the small amount of DNOC used in the United States, exposure of the general public is expected to be minimal.

Incompatibilities: Strong oxidizers.

Permissible Exposure Limits in Air: The Federal standard and ACGIH 1983/84 TWA value for all isomers of DNOC is 0.2 mg/m^3. The notation "skin" indicates the possibility of cutaneous absorption. The STEL value is 0.6 mg/m^3. The IDLH level is 5.0 mg/m^3.

Determination in Air: Collection by charcoal tube, analysis by gas liquid chromatography (A-13).

Permissible Concentration in Water: To protect human health, 13.4 μg/ℓ.

Determination in Water: Methylene chloride extraction followed by gas chromatography with flame ionization or electron capture detection (EPA Method 604) or gas chromatography plus mass spectrometry (EPA Method 625).

Routes of Entry: Inhalation, percutaneous absorption, ingestion, eye and skin contact.

Harmful Effects and Symptoms: *Local* — None reported except for staining of skin and hair.

Systemic — DNOC blocks the formation of high energy phosphate compounds, and the energy from oxidative metabolism is liberated as heat. Early symptoms of intoxication by inhalation or skin absorption are elevation of the basal metabolic rate and rise in temperature accompanied by fatigue, excessive sweating, unusual thirst, and loss of weight. The clinical picture resembles in part a thyroid crisis. Weakness, fatigue, increased respiratory rate, tachycardia, and fever may lead to rapid deterioration and death. Bilateral cataracts have been seen following oral ingestion for therapeutic purposes. These have not been seen during industrial or agricultural use.

Points of Attack: Eyes, endocrine system, cardiovascular system.

Medical Surveillance: Consider eyes, thyroid, and cardiovascular system, as well as general health.

First Aid: If this chemical gets into the eyes, irrigate immediately. If this chemical contacts the skin, wash with soap immediately. If a person breathes in large amounts of this chemical, move the exposed person to fresh air at once and perform artificial respiration. When this chemical has been swallowed, get medical attention. Give large quantities of water and induce vomiting. Do not make an unconscious person vomit.

Personal Protective Methods: Wear appropriate clothing to prevent any reasonable probability of skin contact. Wear eye protection to prevent any reasonable probability of eye contact. Employees should wash promptly when skin is wet and daily at the end of each work shift. Work clothing should be changed daily if it is possible that clothing is contaminated. Remove nonimpervious clothing promptly if wet or contaminated.

Respirator Selection:
     5 mg/m^3: HiEPF/SAF/SCBAF

Disposal Method Suggested: Incineration (1100°F minimum) with adequate scrubbing and ash disposal facilities.

References
(1) National Institute for Occupational Safety and Health, *Criteria for a Recommended Standard: Occupational Exposure to Dinitro-ortho-Cresol,* NIOSH Publication No. 78-131, Washington, DC (1978).

(2) U.S. Environmental Protection Agency, *Nitrophenols: Ambient Water Quality Criteria,* Washington, DC (1980).

(3) U.S. Environmental Protection Agency, *4,6-Dinitro-o-Cresol,* Health and Environmental Effects Profile No. 90, Washington, DC, Office of Solid Waste (April 30, 1980).

(4) Sax, N.I., Ed., *Dangerous Properties of Industrial Materials Report, 2,* No. 5, 54-59, New York, Van Nostrand Reinhold Co. (1982).

(5) Sax, N.I., Ed., *Dangerous Properties of Industrial Materials Report, 4,* No. 1, 62-66, New York, Van Nostrand Reinhold Co. (Jan./Feb. 1984).

(6) Parmeggiani, L., Ed., *Encyclopedia of Occupational Health & Safety,* Third Edition, Vol. 1, pp 635-36, Geneva, International Labour Office (1983).

(7) United Nations Environment Programme, *IRPTC Legal File 1983,* Vol. I, pp VII/216-18, Geneva, Switzerland, International Register of Potentially Toxic Chemicals (1984).

2,4-DINITROPHENOL

- Hazardous substance (EPA)
- Hazardous waste (EPA)
- Priority toxic pollutant (EPA)

Description: There are six isomers of dinitrophenol, $C_6H_3(NO_2)_2OH$, of which 2,4-dinitrophenol is the most important industrially. It is an explosive, yellow crystalline solid, melting at $114°$ to $115°C$.

Code Numbers: CAS 51-28-5 RTECS SL2800000 UN 0076 (solution: 1599).

DOT Designation: Dry—explosive, poison; solution—Poison B.

Synonyms: DNP.

Potential Exposure: 2,4-DNP is used in the manufacturing of dyestuff intermediates, wood preservatives, pesticides, herbicides, explosives, chemical indicators, photographic developers, and also in chemical synthesis.

Incompatibilities: Heavy metals and their compounds.

Permissible Exposure Limits in Air: There is no Federal standard for DNP. A useful guideline of 0.2 mg/m³ is based on data for dinitro-o-cresol.

Permissible Concentration in Water: To protect freshwater aquatic life—230 $\mu g/\ell$ on an acute toxicity basis for nitrophenols as a class. To protect saltwater aquatic life—4,850 $\mu g/\ell$ on an acute toxicity basis for nitrophenols as a class. To protect human health—70.0 $\mu g/\ell$. This compares to a limit of 30 $\mu g/\ell$ set in the U.S.S.R.

Determination in Water: Methylene chloride extraction followed by gas chromatography with flame ionization or electron capture detection (EPA Method 604) or gas chromatography plus mass spectrometry (EPA Method 625).

Routes of Entry: Percutaneous absorption and inhalation of dust and vapors.

Harmful Effects and Symptoms: *Local* – DNP causes yellow staining of exposed skin. Dermatitis may be due to either primary irritation or allergic sensitivity.

Systemic – The isomers differ in their toxic effects. In general, DNP disrupts oxidative phosphorylation (as in the case of DNOC) which results in increased metabolism, oxygen consumption, and heat production. Acute intoxication is characterized by sudden onset of fatigue, thirst, sweating, and oppression of the chest. There is rapid respiration, tachycardia, and a rise in body temperature. In less severe poisoning, the symptoms are nausea, vomiting, anorexia, weakness, dizziness, vertigo, headache, and sweating. The liver may be sensitive to pressure, and there may also be jaundice. DNP poisoning is more severe in warm environments. If not fatal, the effects are rapidly and completely reversible. Chronic exposure results in kidney and liver damage and cataract formation. Occasional hypersensitivity reactions, e.g., neutropenia, skin rashes, peripheral neuritis, have been seen after oral use.

Points of Attack: Skin, liver, central nervous system.

Medical Surveillance: Consider skin, eyes, thyroid, blood, central nervous system, liver and kidney function, as well as general health in preplacement and periodic examinations. DNP can be measured in urine as such or as an aminophenol.

First Aid: Flush eyes with water. Wash contaminated areas of body with soap and water. Use gastric lavage if swallowed followed by saline catharsis.

Personal Protective Methods: Because of its wide use in agriculture, lumbering, photography, as well as in the petrochemical industry, worker education on the toxic properties of dinitrophenol is important. Spills and splashes that contaminate clothing require the worker to immediately change clothes and wash the area thoroughly. Workers should have clean work clothes on every shift and should be required to shower prior to changing to street clothing.

Respirator Selection: Fullface masks with organic vapor canisters or air supplied respirators are necessary in areas of high concentration of dust or vapor.

Disposal Method Suggested: Incineration (1800°F, 2.0 seconds minimum) with adequate scrubbing equipment for the removal of NO_x.

References

(1) U.S. Environmental Protection Agency, *Nitrophenols: Ambient Water Quality Criteria,* Washington, DC (1980).
(2) U.S. Environmental Protection Agency, *2,4-Dinitrophenol,* Health and Environmental Effects Profile No. 91, Washington, DC, Office of Solid Waste (April 30, 1980).
(3) Sax, N.I., Ed., *Dangerous Properties of Industrial Materials Report, 2,* No. 2, 25-27, New York, Van Nostrand Reinhold Co. (1982).
(4) Sax, N.I., Ed., *Dangerous Properties of Industrial Materials Report, 3,* No. 2, 38-41, New York, Van Nostrand Reinhold Co. (1983).
(5) See Reference (A-60).
(6) Parmeggiani, L., Ed., *Encyclopedia of Occupational Health & Safety,* Third Edition, Vol. 1, pp 636–37, Geneva, International Labour Office (1983).
(7) United Nations Environment Programme, *IRPTC Legal File 1983,* Vol. II, pp VII/510–12, Geneva, Switzerland, International Register of Potentially Toxic Chemicals (1984).

2,6-DINITROPHENOL

- Carcinogen (Suspected) (NIOSH) (1)
- Hazardous Substance (EPA)

Description: $C_6H_3(NO_2)_2OH$ is a yellow crystalline solid melting at 63°-64°C.

Code Numbers: CAS 573-56-8   RTECS SL2975000

DOT Designation: Poison Class B.

Synonyms: DNP.

Potential Exposure: Those involved in dye manufacture, picric acid manufacture, photographic chemicals.

Permissible Exposure Limits in Air: 0.20 mg/m^3 (1).

Permissible Concentration in Water: 0.001 mg/ℓ (1) on a taste-imparting basis.

Routes of Entry: Skin contact, inhalation of dust.

Harmful Effects and Symptoms: Dermatitis results from skin contact. Cataracts may be produced. A fatal dose in adults is 1 to 3 g by mouth. Symptoms include headache, loss of appetitie, vomiting, abdominal pain, diarrhea, fever, chest pains, dizziness, fatigue, jaundice, leg cramps, cyanosis, anxiety, pulmonary edema, convulsions.

Personal Protective Methods: Safety goggles or face mask, butyl rubber gloves and boots.

Respirator Selection: Self-contained breathing apparatus.

Disposal Method Suggested: Incineration.

References

(1) Sax, N.I., Ed., *Dangerous Properties of Industrial Materials Report, 3,* No. 2, 41-44, New York, Van Nostrand Reinhold Co. (1983).

(2) United Nations Environment Programme, *IRPTC Legal File 1983,* Vol. II, pp VII/510-12, Geneva, Switzerland, International Register of Potentially Toxic Chemicals (1984).

DINITROSOPENTAMETHYLENETETRAMINE

Description: This compound has the formula

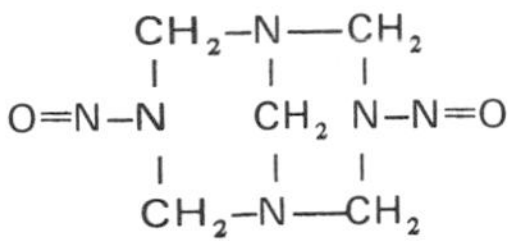

$$\begin{array}{ccc}
CH_2-N-CH_2 & & \\
| \quad | \quad | & & \\
O=N-N \quad CH_2 \; N-N=O & & \\
| \quad | \quad | & & \\
CH_2-N-CH_2 & &
\end{array}$$

It is a light yellow solid which decomposes at 207°C.

Code Numbers: CAS 101-25-7   RTECS XA5250000

DOT Designation: —

Synonyms: 3,7-dinitroso-1,3,5,7-tetrazabicyclononane; DNPT.

Potential Exposure: DNPT is used as a blowing agent in rubbers and plastics. Natural and synthetic unicellular rubber, which is made using DNPT, is used as carpet underlay, weatherstripping, insulation, shoe lining, and cushioning. DNPT is also an effective blowing agent for polyvinyl chloride plastisols and epoxy, polyester, and silicone resins.

Permissible Exposure Limits in Air: No standards set.

Permissible Concentration in Water: No criteria set.

Harmful Effects and Symptoms: Fainting, dizziness, cyanosis and convulsions have been reported by DNPT production workers.

IARC has concluded that DNPT is not a rat carcinogen by oral administration or by intraperitoneal injection.

References

(1)   U.S. Environmental Protection Agency, *Chemical Hazard Information Profile: Dinitrosopentamethylenetetramine,* Washington, DC (June 1, 1978).

DINITROTOLUENES

- Carcinogen (Positive, NCI for 2,4-DNT) (6)
- Hazardous substance (EPA)
- Hazardous wastes (2,4- and 2,6-isomers) (EPA)
- Priority toxic pollutant (EPA)

Description: Six isomers of dinitrotoluene, $C_6H_3(NO_2)_2CH_3$, exist, the most important being 2,4-dinitrotoluene, an orange-yellow solid which has a melting point of 71°C.

Code Numbers: For 2,4-DNT—CAS 121-14-2 RTECS XT1575000 UN 1600 (liquid), UN 2038 (solid).

DOT Designation: ORM-E.

Synonyms: Dinitrotoluol, DNT.

Potential Exposure: DNT is used in the manufacture of dyes and explosives (e.g., trinitrotoluene) and in organic synthesis.

Incompatibilities: Strong oxidizers, caustics, metals, such as tin, zinc.

Permissible Exposure Limits in Air: The Federal standard and the 1983/84 ACGIH TWA value is 1.5 mg/m³. The notation "skin" indicates the possibility of cutaneous absorption. The STEL is 5 mg/m³. The IDLH level is 200 mg/m³.

Determination in Air: Filter and ethylene glycol bubbler collection, analysis by high pressure liquid chromatography (A-10).

Permissible Concentration in Water: To protect freshwater aquatic life— 330 μg/ℓ on an acute toxicity basis and 230 μg/ℓ on a chronic toxicity basis. To protect saltwater aquatic life—590 μg/ℓ on an acute toxicity basis. To protect human health—preferably zero. An additional lifetime cancer risk of 1 in 100,000 results at a concentration of 1.1 μg/ℓ of 2,4-DNT.

Determination in Water: Methylene chloride extraction followed by exchange to toluene and gas chromatography with flame ionization detection (EPA Method 609) or gas chromatography plus mass spectrometry (EPA Method 625).

Routes of Entry: Inhalation of vapor, percutaneous absorption of liquid, ingestion and eye and skin contact.

Harmful Effects and Symptoms: *Local* – None.

Systemic – The effects from exposure to dinitrotoluene are caused by its capacity to produce anoxia due to the formation of methemoglobin. Cyanosis may occur with headache, irritability, dizziness, weakness, nausea, vomiting, dyspnea, drowsiness, and unconsciousness. If treatment is not given promptly, death may occur. The onset of symptoms may be delayed. The ingestion of alcohol may cause increased susceptibility. Repeated or prolonged exposure may cause anemia.

Points of Attack: Blood, liver, cardiovascular system.

Medical Surveillance: Preemployment and periodic examinations should be concerned particularly with a history of blood dyscrasias, reactions to medications, alcohol intake, eye disease, skin and cardiovascular status. Liver and renal functions should be evaluated periodically as well as blood and general health.

First Aid: If this chemical gets into the eyes, irrigate immediately. If this chemical contacts the skin, wash with soap immediately. If a person breathes in large amounts of this chemical, move the exposed person to fresh air at once and perform artificial respiration. When this chemical has been swallowed, get medical attention. Give large quantities of water and induce vomiting. Do not make an unconscious person vomit.

Personal Protective Methods: Wear appropriate clothing to prevent any possibility or probability of skin contact with molten liquids or repeated or prolonged contact with solutions. Wear eye protection to prevent any possibility of eye contact with molten material. Employees should wash immediately when skin is wet or contaminated and daily at the end of each work shift. Work clothing should be changed daily if it is possible that clothing is contaminated. Remove nonimpervious clothing immediately if wet or contaminated. Provide emergency showers.

Respirator Selection:
    15 mg/m^3: SA/SCBA
    75 mg/m^3: SAF/SCBAF
    200 mg/m^3: SAF:PD,PP,CF
    Escape: GMOVP/SCBA

Disposal Method Suggested: Pretreatment involves contact of the dinitrotoluene contaminated waste with $NaHCO_3$ and solid combustibles followed by incineration in an alkaline scrubber equipped incinerator unit.

References

(1) U.S. Environmental Protection Agency, *Dinitrotoluenes: Ambient Water Quality Criteria,* Washington, DC (1980).

(2) U.S. Environmental Protection Agency, *Chemical Hazard Information Profile: 2,4-Dinitrotoluene,* Washington, DC (March 9, 1978).

(3) U.S. Environmental Protection Agency, *Dinitrotoluenes,* Health and Environmental Effects Profile No. 92, Washington, DC, Office of Solid Waste (April 30, 1980).

(4) U.S. Environmental Protection Agency, *2,4-Dinitrotoluene,* Health and Environmental Effects Profile No. 93, Washington, DC, Office of Solid Waste (April 30, 1980).

(5) U.S. Environmental Protection Agency, *2,6-Dinitrotoluene,* Health and Environmental Effects Profile No. 94, Washington, DC, Office of Solid Waste (April 30, 1980).

(6) National Cancer Institute, *Bioassay of 2,4-Dinitrotoluene for Possible Carcinogenicity,* Technical Report Series No. 54, Bethesda, MD (1978).

(7) See Reference (A-61).

(8) Sax, N.I., Ed., *Dangerous Properties of Industrial Materials Report, 3,* No. 2, 70-72, New York, Van Nostrand Reinhold Co. (1983). (2,4-Dinitrotoluene).

DI-n-OCTYL PHTHALATE

- Hazardous waste (EPA)
- Priority toxic pollutant (EPA)

Description: $C_6H_4(COOC_8H_{17})_2$ is a liquid boiling at 220°C.

Code Numbers: CAS 117-84-0 RTECS TI1925000

DOT Designation: —

Synonyms: DOP, benzenedicarboxylic acid di-n-octyl ester.

Potential Exposure: DOP is used as a plasticizer in plastics product manufacture.

Permissible Exposure Limits in Air: No standards set.

Permissible Concentration in Water: Insufficient data are available to permit definition of criteria to prevent damage to aquatic life or harm to humans (1).

Determination in Water: Methylene chloride extraction followed by gas chromatography with flame ionization or electron capture detection (EPA Method 606) or gas chromatography plus mass spectrometry (EPA Method 625).

Harmful Effects and Symptoms: Di-n-octyl phthalate has produced teratogenic effects following i.p. injection of pregnant rats. This same study has also indicated some increased resorptions and fetal toxicity. Evidence is not available indicating mutagenic or carcinogenic effects of di-n-octyl phthalate. Data pertaining to the aquatic toxicity of di-n-octyl phthalate is not available (2).

Disposal Method Suggested: Incineration.

References

(1)   U.S. Environmental Protection Agency, *Phthalate Esters: Ambient Water Quality Criteria,* Washington, DC (1980).
(2)   U.S. Environmental Protection Agency, *Di-n-Octyl Phthalate,* Health and Environmental Effects Profile No. 95, Washington, DC, Office of Solid Waste (April 30, 1980).
(3)   See Reference (A-60).

DIOXANE

● Carcinogen (Positive, NCI) (3) (Animal positive, IARC)
● Hazardous waste (EPA)

Description: $OCH_2CH_2OCH_2CH_2$, dioxane, is a volatile, colorless liquid that may form explosive peroxides during storage. It boils at 101°C.

Code Numbers: CAS 123-91-1   RTECS JG8225000   UN 1165

DOT Designation: Flammable liquid.

Synonyms: 1,4-Diethylenedioxide, diethylene ether, 1,4-dioxane, p-dioxane.

Potential Exposure: Dioxane finds its primary use as a solvent for cellulose acetate, dyes, fats, greases, lacquers, mineral oil, paints, polyvinyl polymers, resins, varnishes, and waxes. It finds particular usage in paint and varnish strippers, as a wetting agent and dispersing agent in textile processing, dye baths, stain and printing compositions, and in the preparation of histological slides.

The National Occupational Hazard Survey estimates that 334,000 workers are potentially exposed to dioxane, 100,000 of whom are exposed as a result of dioxane contamination of 1,1,1-trichloroethane. OSHA estimates that 466,000 workers are potentially exposed.

Incompatibilities: Strong oxidizers.

Permissible Exposure Limits in Air: The Federal standard is 100 ppm (360 mg/m³); however, the ACGIH 1983/84 recommended TLV was 25 ppm (90 mg/m³) of technical grade. The notation "skin" indicates the possibility of cutaneous absorption. NIOSH has promulgated a recommended standard as of 1977 such that occupational exposure to dioxane shall be controlled so that employees are not exposed at airborne concentrations greater than 1 ppm (3.6 mg/m³) based on a 30-minute sampling period. The STEL value adopted by ACGIH is 100 ppm (360 mg/m³) and the IDLH level is 200 ppm.

Determination in Air: Charcoal adsorption, CS_2 workup, gas chromatography analysis. See NIOSH Methods, Set G. See also reference (A-10).

Permissible Concentration in Water: No criteria set, but EPA (A-37) has suggested a permissible ambient goal of 2,480 µg/ℓ based on health effects.

Routes of Entry: Inhalation of vapor, percutaneous absorption, ingestion, eye and skin contact.

Harmful Effects and Symptoms: *Local* — Liquid and vapor may be irritating to eyes, nose and throat.

Systemic — Exposure to dioxane vapor may cause drowsiness, dizziness, loss of appetite, headache, nausea, vomiting, stomach pain, and liver and kidney damage. Prolonged skin exposure to the liquid may cause drying and cracking.

1,4-Dioxane is carcinogenic in rats and guinea-pigs by oral administration: it produced malignant tumors of the nasal cavity and liver in rats and tumors of the liver and gall bladder in guinea-pigs. It was also active as a promoter in a two-stage skin carcinogenesis study in mice (6).

Points of Attack: Liver, kidneys, skin, eyes.

Medical Surveillance: Preplacement and periodic examinations should be directed to symptoms of headache and dizziness, as well as nausea and other gastrointestinal disturbances. The condition of the skin and of renal and liver function should be considered. No specific biomonitoring tests are available. However, the pharmacokinetic and metabolic fate of 1,4-dioxane has shown its limited capacity to be metabolized to β-hydroxyethylacetic acid (HEAA), the major urinary metabolite.

First Aid: If this chemical gets into the eyes, irrigate immediately. If this chemical contacts the skin, flush with water promptly. If a person breathes in large amounts of this chemical, move the exposed person to fresh air at once and perform artificial respiration. When this chemical has been swallowed, get medical attention. Give large quantities of salt water and induce vomiting. Do not make an unconscious person vomit.

Personal Protective Methods: Wear appropriate clothing to prevent any reasonable probability of skin contact. Wear eye protection to prevent any reasonable probability of eye contact. Employees should wash promptly when skin is wet or contaminated. Remove clothing immediately if wet or contaminated to avoid flammability hazard.

Respirator Selection:
   200 ppm: SAF/SCBAF
   Escape: GMOV/SCBA

Disposal Method Suggested: Concentrated waste containing no peroxides—discharge liquid at a controlled rate near a pilot flame. Concentrated waste containing peroxides—perforation of a container of the waste from a safe distance followed by open burning.

References

(1) National Institute for Occupational Safety and Health, *Criteria for a Recommended Standard: Occupational Exposure to Dioxane,* NIOSH Doc. No. 77-226 (1977).
(2) U.S. Environmental Protection Agency, *Chemical Hazard Information Profile: Dioxane,* Washington, DC (1979).
(3) National Cancer Institute, *Bioassay of 1,4-Dioxane for Possible Carcinogenicity,* Technical Report Series No. 80, Bethesda, MD (1978).
(4) International Agency for Research on Cancer, *IARC Monographs on the Carcinogenic Risks of Chemicals to Humans,* Lyon, France 11, 147 (1976).
(5) See Reference (A-61).
(6) See Reference (A-62). Also see Reference (A-64).
(7) See Reference (A-60).
(8) Parmeggiani, L., Ed., *Encyclopedia of Occupational Health & Safety,* Third Edition, Vol. 1, pp 637–38, Geneva, International Labour Office (1983).
(9) United Nations Environment Programme, *IRPTC Legal File 1983,* Vol. I, pp VII/287–88, Geneva, Switzerland, International Register of Potentially Toxic Chemicals (1984).

DIOXATHION

Description: S,S'-1,4-dioxane-2,3-diyl O,O,O',O'-tetraethyl di(phosphorodithioate), $C_{12}H_{26}O_6P_2S_4$ with the structural formula:

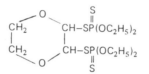

This is a brown liquid.

Code Numbers: CAS 78-34-2   RTECS TE335000   UN 2783

DOT Designation: —

Synonyms: Delnav®, dioxane phosphate (Japan), ENT 22879.

Potential Exposure: Those involved in the manufacture (A-32), formulation or application of this insecticide and acaricide.

Incompatibilities: Dioxathion is unstable to iron or tin surfaces or when mixed with certain carriers (A-7).

Permissible Exposure Limits in Air: There is no Federal Standard but ACGIH (1983/84) has adopted a TWA value of 0.2 mg/m³ with the notation "skin" indicating the possibility of cutaneous absorption. They have proposed no STEL. Other sources (RTECS) have cited 0.02 mg/m³ (20 µg/m³) as a TLV value.

Permissible Concentration in Water: No criteria set.

Harmful Effects and Symptoms: Subacute administration to human volunteers of 0.075 mg/kg/day showed no effect on either plasma or erythrocyte cholinesterase. However, volunteers receiving 0.15 mg/kg/day showed a possibly significant slight depression of plasma cholinesterase activity but no erythrocyte activity. The estimated acceptable daily intake for man set by WHO is 0.0015 mg/kg/day (A-34). The National Cancer Institute found dioxathion not to be a carcinogen (1).

References

(1) National Cancer Institute, *Bioassay of Dioxathion for Possible Carcinogenicity*, Technical Report Series No. 125, Bethesda, MD (1978).

(2) Sax, N.I., Ed., *Dangerous Properties of Industrial Materials Report, 2,* No. 5, 60-63, New York, Van Nostrand Reinhold Co. (1982).

(3) United Nations Environment Programme, *IRPTC Legal File 1983,* Vol. II, pp VII/585-6, Geneva, Switzerland, International Register of Potentially Toxic Chemicals (1984).

DIPHENYLAMINE

Description: $(C_6H_5)_2NH$ is a colorless to gray crystalline solid melting at $53°C$.

Code Numbers: CAS 122-39-4   RTECS JJ7800000

DOT Designation: —

Synonyms: DPA, N-phenylaniline, anilinobenzene.

Potential Exposure: DPA is used in antioxidants and in stabilizers for plastics including solid rocket propellants. It is used in the manufacture of pharmaceuticals, pesticides, explosives and dyes.

Permissible Exposure Limits in Air: There is no Federal standard but ACGIH (1983/84) has adopted a TWA value of 10 mg/m^3 and set an STEL of 20 mg/m^3.

Permissible Concentration in Water: No criteria set.

Routes of Entry: Inhalation of dust, ingestion, skin absorption, eye and skin contact.

Harmful Effects and Symptoms: Industrial poisoning has been encountered and was manifested clinically by bladder symptoms, tachycardia, hypertension and skin trouble (A-34).

First Aid: Flush eyes with water. Wash contaminated areas of body with soap and water.

Personal Protective Methods: Wear safety glasses, rubber gloves and rubber clothing (A-38).

Respirator Selection: Use mechanical filter respirator.

Disposal Method Suggested: Burn in admixture with flammable solvent in furnace equipped with afterburner and scrubber.

References

(1) Sax, N.I., Ed., *Dangerous Properties of Industrial Materials Report, 2,* No. 5, 63-66, New York, Van Nostrand Reinhold Co. (1982).

(2) United Nations Environment Programme, *IRPTC Legal File 1983,* Vol. I, pp VII/289, Geneva, Switzerland, International Register of Potentially Toxic Chemicals (1984).

1,2-DIPHENYLHYDRAZINE

- Carcinogen (Positive, NCI) (3)
- Hazardous waste (EPA)
- Priority toxic pollutant (EPA)

Description: $C_6H_5NHNHC_6H_5$ is a crystalline compound melting at 131°C.

Code Numbers: CAS 122-66-7   RTECS MW2625000

DOT Designation: —

Synonyms: Hydrazobenzene; symmetrical diphenylhydrazine; N,N'-diphenylhydrazine; N,N'-bianiline; 1,1'-hydrazodibenzene; DPH.

Potential Exposure: 1,2-Diphenylhydrazine (DPH) is a precursor in the manufacture of benzidine, an intermediate in the production of dyes. 1,2-Diphenylhydrazine is used in the synthesis of phenylbutazone, a potent antiinflammatory (antiarthritic) drug. Manufacturers of dyes and pharmaceuticals are subject to occupational exposure. Groups working in the laboratory and forensic medicine may also be subject to 1,2-diphenylhydrazine exposure.

Permissible Exposure Limits in Air: No limits set.

Permissible Concentration in Water: To protect freshwater aquatic life—270 µg/ℓ on an acute toxicity basis. To protect saltwater aquatic life—no criterion developed due to insufficient data. To protect human health—preferably zero. An additional lifetime cancer risk of 1 in 100,000 is presented by a concentration of 0.4 µg/ℓ.

Determination in Water: Gas chromatography plus mass spectrometry (EPA Method 625).

Harmful Effects and Symptoms: Diphenylhydrazine is a suspected carcinogen in humans because of its structural relationship to benzidine, which is an established human bladder carcinogen. Recent studies in rats and mice have shown that diphenylhydrazine produces both benign and malignant tumors when administered subcutaneously. Carcinogenicity in both rats of both sexes and female mice has been established (5). In view of the relative paucity of data on the mutagenicity, teratogenicity and long-term oral toxicity of diphenylhydrazine, estimates of the effects of chronic oral exposure at low levels cannot be made with any confidence according to NAS/NRC (A-2).

Disposal Method Suggested: Controlled incineration whereby oxides of nitrogen are removed from the effluent gas by scrubber, catalytic or thermal device.

References

(1) U.S. Environmental Protection Agency, *Diphenylhydrazine: Ambient Water Quality Criteria,* Washington, DC (1980).
(2) U.S. Environmental Protection Agency, *1,2-Diphenylhydrazine,* Health and Environmental Effects Profile No. 96, Washington, DC, Office of Solid Waste (April 30, 1980).
(3) National Cancer Institute, *Bioassay of Hydrazobenzene for Possible Carcinogenicity,* Technical Report Series No. 92, Bethesda, MD (1978).
(4) Sax, N.I., Ed., *Dangerous Properties of Industrial Materials Report, 2,* No. 5, 68-70, New York, Van Nostrand Reinhold Co. (1982).
(5) See Reference (A-62). Also see Reference (A-64).
(6) Sax, N.I., Ed., *Dangerous Properties of Industrial Materials Report, 3,* No. 2, 45-46, New York, Van Nostrand Reinhold Co. (1983).
(7) United Nations Environment Programme, *IRPTC Legal File 1983,* Vol. II, pp VII/390, Geneva, Switzerland, International Register of Potentially Toxic Chemicals (1984).

DIPHENYL OXIDE

Description: $(C_6H_5)_2O$ is a colorless solid or liquid with a somewhat disagreeable odor. It boils at 259°C. It is often used commercially in admixture with diphenyl. The mixture is a straw colored liquid, which darkens on use, with an aromatic, somewhat disagreeable odor. It boils at 257°C.

Code Numbers: CAS 101-84-8   RTECS KN8970000

DOT Designation: —

Synonyms: Phenyl ether, diphenyl ether. The diphenyl/diphenyl oxide mixture is known as Dowtherm A.

Potential Exposure: This material is used as a heat transfer medium; it is used in perfuming soaps and in organic synthesis.

Incompatibilities: Strong oxidizers.

Permissible Exposure Limits in Air: The Federal limit and the 1983/84 ACGIH TWA value is 1.0 ppm (7 mg/m^3). The STEL is 2 ppm (14 mg/m^3).

Determination in Air: Charcoal adsorption, CS_2 workup, gas chromatography. See NIOSH Methods, Set F. For the Dowtherm A mixture, silica adsorption, benzene workup and gas chromatographic analysis are recommended in NIOSH Methods, Set F.

Permissible Concentration in Water: No criteria set.

Routes of Entry: Inhalation, eye and skin contact, ingestion.

Harmful Effects and Symptoms: Nausea; irritation of eyes, nose and skin.

Points of Attack: Eyes, skin, respiratory system.

Medical Surveillance: Consider the points of attack in preplacement and periodic physical examinations.

First Aid: If this chemical gets into the eyes, irrigate immediately. If this chemical contacts the skin, wash with soap promptly. If a person breathes in large amounts of this chemical, move the exposed person to fresh air at once and perform artificial respiration. When this chemical has been swallowed, get medical attention. Give large quantities of salt water and induce vomiting. Do not make an unconscious person vomit.

Personal Protective Methods: Wear appropriate clothing to prevent repeated or prolonged skin contact. Wear eye protection to prevent any reasonable probability of eye contact. Employees should wash promptly when skin is wet or contaminated. Remove nonimpervious clothing promptly if wet or contaminated.

Respirator Selection:
         50 ppm: CCROVDM/GMOVDM/SAF/SCBAF
        100 ppm: SAF:PD,PP,CF/PAPCCROVFHiEP
         Escape: GMOVP/SCBA

Reference

(1)   United Nations Environment Programme, *IRPTC Legal File 1983,* Vol. I, pp VII/348, Geneva, Switzerland, International Register of Potentially Toxic Chemicals (1984).

DIPROPYLENE GLYCOL METHYL ETHER

Description: $CH_3OCH_2CH(CH_3)OCH_2CH(CH_3)OH$ is a colorless liquid with a weak odor boiling at 190°C.

Code Numbers: CAS 34590-94-8   RTECS JM1575000

DOT Designation: —

Synonyms: Dipropylene glycol monomethyl ether, Dowanol DPM®, Dowanol 50B®.

Potential Exposure: This material is used as a solvent for nitrocellulose and other synthetic resins.

Incompatibilities: Strong oxidizers.

Permissible Exposure Limits in Air: The Federal limit and the ACGIH 1983/84 TWA value is 100 ppm (600 mg/m³). The STEL value is 150 ppm (900 mg/m³). The IDLH level has not been defined.

Determination in Air: Charcoal adsorption, workup with CS_2, analysis by gas chromatography. See NIOSH Methods, Set F. See also reference (A-10).

Permissible Concentration in Water: No criteria set.

Routes of Entry: Inhalation, ingestion, skin and eye contact.

Harmful Effects and Symptoms: Irritation of eyes and nose; lightheadedness, headaches.

Points of Attack: Respiratory system, eyes.

Medical Surveillance: Consider the points of attack in preplacement and periodic physical examinations.

First Aid: If this chemical gets into the eyes, irrigate immediately. If this chemical contacts the skin, wash promptly. If a person breathes in large amounts of this chemical, move the exposed person to fresh air at once and perform artificial respiration. When this chemical has been swallowed, get medical attention. Give large quantities of salt water and induce vomiting. Do not make an unconscious person vomit.

Personal Protective Methods: Adequate respirator protection. Wear goggles to avoid eye contact.

Respirator Selection:
    1,000 ppm: SA/SCBA
    5,000 ppm: GMOV/SAF/SCBAF
      Escape: GMOVP/SCBA

Disposal Method Suggested: Concentrated waste containing no peroxides—discharge liquid at a controlled rate near a pilot flame. Concentrated waste containing peroxides—perforation of a container of the waste from a safe distance followed by open burning.

References
(1)   See Reference (A-61).

DIPROPYL KETONE

Description: $C_3H_7COC_3H_7$ is a colorless liquid boiling at 144°C.

Code Numbers: CAS 123-19-3   RTECS MJ5600000   UN 2710

DOT Designation: Flammable liquid.

Synonyms: 4-Heptanone, butyrone, propyl ketone.

Potential Exposures: This compound is used as a solvent for nitrocellulose and many other natural and synthetic resins. It is used in lacquer formulations and in food flavorings.

Permissible Exposure Limits in Air: There is no Federal standard but ACGIH (1983/84) has proposed a TWA of 50 ppm (235 mg/m³). There is no proposed STEL.

Permissible Concentration in Water: No criteria set.

Harmful Effects and Symptoms: Low oral toxicity to rats is indicated. Rats inhaling 2,000 ppm for 4 hr survived but at 4,000 ppm, all died (A-34).

Disposal Method Suggested: Incineration.

DIQUAT

● Hazardous substance (EPA)

Description: 6,7-Dihydrodipyrido[1,2-a:2',1'-c]pyrazidiinium dibromide, $C_{12}H_{12}N_2Br_2$ with the structural formula:

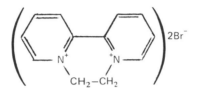

Diquat forms a monohydrate which consists of colorless to yellow crystals decomposing above 300°C.

Code Numbers: CAS 85-00-7   RTECS JM5690000   UN 2781

DOT Designation: ORM-E.

Synonyms: Reglone®, Weedol®, Pathclear®, Aquacide®.

Potential Exposures: Those involved in the manufacture (A-32), formulation and application of this herbicide.

Permissible Exposure Limits in Air: There is no Federal standard but ACGIH (1983/84) has adopted a TWA value of 0.5 mg/m³ and set an STEL of 1.0 mg/m³.

Permissible Concentration in Water: No criteria set.

Harmful Effects and Symptoms: Diquat does not produce in man or animals the lung injury peculiar to paraquat (which see). Some respiratory distress may be present in acute poisoning but it is nonspecific. Diquat produces cataracts in test animals (rats and dogs) after prolonged exposure but this effect has not been noted in man (A-34).

Disposal Method Suggested: Diquat is inactivated by inert clay or by anionic surfactants. Therefore, an effective and environmentally safe disposal method would be to mix the product with ordinary household detergent and bury the mixture in clay soil (A-32).

Reference

(1)   United Nations Environment Programme, *IRPTC Legal File 1983,* Vol. I, pp VII/293–96, Geneva, Switzerland, International Register of Potentially Toxic Chemicals (1984).

DIRECT BLACK 38

● Carcinogen (Animal Positive) (A-64)

Description: $C_{34}H_{25}N_9O_7S_2 \cdot 2Na$ has the structural formula

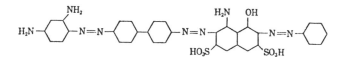

Code Numbers: CAS 1937-37-7   RTECS QJ6160000

DOT Designation: –

Synonyms: 2,7-Naphthalenedisulfonic acid, 4-amino-3-[4'-(2,4-diaminophenylazo)(1,1'-biphenylazo)]-5-hydroxy-6-(phenylazo)-, disodium salt and many trade names including Atlantic Black C, Chloramine Black C, Chrome Leather Black C.

Potential Exposure: Direct Black 38 is possibly being used to dye fabric, leather, cotton, cellulosic materials, and paper; these are then used in consumer products. The chemical may be used by artists (CPSC, EPA). The FDA has indicated that although Direct Black 38 is identified in the literature as a hair-dye component, it is currently not used by the cosmetic industry.

In view of a health hazard alert issued in December 1980 by OSHA, which cautioned workers and employers of the carcinogenic effect of benzidine-derived Direct Black 38, new nonbenzidine Direct Black dyes have been developed and used successfully in commercial applications by the paper and leather industry during the past year. These nonbenzidine dyes were developed with the hope of replacing benzidine-based dyes throughout industry.

In the general population, unspecified exposure levels to Direct Black 38 are thought to occur through the use of textile dyed products and retail packaged dyes for home dyeing and school use. The National Occupational Hazard Survey in 1974 estimated that approximately 16,000 workers were potentially exposed to Direct Black 38. The population exposed consisted of dyers of leather, plastics, cotton, wool, and silks, along with makers of aqueous inks, biological stains, typewriter ribbons, and wood stains.

Permissible Exposure Limits in Air: No standards set.

Permissible Concentration in Water: No criteria set.

Routes of Entry: Human exposure to Black 38 may occur through inhalation, skin absorption, and unintentional ingestion. Consumer exposure to Direct Black 38 depends upon the ability of the dye to migrate out of consumer products and either penetrate the skin or degrade prior to penetrating the skin. No additional data quantifying the rate of migration or degradation of this dye are currently available.

Harmful Effects and Symptoms: There is sufficient evidence that commercial Direct Black 38 is carcinogenic to experimental animals (1). In a thirteen-week feeding study, Direct Black 38 (technical grade) was carcinogenic in male and female Fischer 344 rats, inducing hepatocellular carcinomas and neoplastic nodules in the liver. The test dye was not carcinogenic to B6C3F1 mice under the same test conditions (2).

In a recent occupational hazard review (3), it was concluded that all benzidine-based dyes, including Direct Black 38, regardless of their physical state or proportion in the mixture, should be recognized as potential human carcinogens.

References

(1) International Agency for Research on Cancer, *IARC Monographs on the Evaluation of the Carcinogenic Risk of Chemicals to Humans,* Supplement 4, Lyon, France: IARC (1982).

(2) National Cancer Institute, *Thirteen-Week Subchronic Toxicity Studies of Direct Blue 6, Direct Black 38, and Direct Brown 95 Dyes,* Technical Report Series No. 108, DHEW Publication No. (NIH) 78-1358, Bethesda, MD (1978).

(3) National Institute of Occupational Safety and Health, *Special Occupational Hazard Review for Benzidine-Based Dyes,* Washington, DC: U.S. Government Printing Office (1980).

DIRECT BLUE 6

● Carcinogen (Animal Positive, Human Potential) (A-64)

Description: $C_{32}H_{24}N_6O_{14}S_4 \cdot 4Na$ is a benzidine-based dyestuff of somewhat analagous structure to Direct Black 38 (which see).

Code Numbers: CAS 2602-46-2 RTECS QJ6400000

DOT Designation: —

Synonyms: 2,7-Naphthalenedisulfonic acid, 3,3'-[(4,4'-biphenylene)bis-(azo)] bis(5-amino-4-hydroxy-), tetrasodium salt and many trade names including Blue 2B, Chloramine Blue 2B, Diamine Blue 2B.

Potential Exposure: Direct Blue 6 may be used by artists. It is potentially used to dye fabric, leather, cotton, cellulosic materials, and paper; these are then used in consumer products (CPSC, EPA). The FDA has indicated that although Direct Blue 6 has been identified in the literature as a hair dye component, it is not presently used by the cosmetic industry.

The primary source for exposure to Direct Blue 6 is at the production site. The initial production step is in a closed system. However, other production operations, such as filter press, drying, and blending, may be performed in the open and, therefore, may afford a greater potential for worker exposure. The general population may be exposed to Direct Blue 6 through the use of retail packaged dyes containing this benzidine-based dye.

Consumer exposure to Direct Blue 6 depends upon the ability of the dye to migrate out of the consumer product and either penetrate the skin or break down prior to penetrating the skin. No additional data are currently available. The National Occupational Hazard Survey in 1974 estimated that 1,300 workers were potentially at risk of exposure to Direct Blue 6.

Permissible Exposure Limits in Air: No standards set.

Permissible Concentration in Water: No criteria set.

Routes of Entry: Human exposure to Direct Blue 6 may occur through inhalation, skin absorption, and to a lesser extent, unintentional ingestion, when the dye is in the press cake or dry powder form.

Harmful Effects and Symptoms: There is sufficient evidence for the carcinogenicity of Direct Blue 6 (technical grade) in experimental animals (1). In a thirteen-week feeding study, Direct Blue 6 Dye (technical grade) was carcinogenic to male and female Fischer 344 rats, inducing hepatocellular carcinomas and neoplastic nodules in the liver. The test dye was not carcinogenic to B6C3F1 mice under the same test conditions (2).

In a recent occupational hazard review (3), it was concluded that all benzidine-based dyes, including Direct Blue 6, regardless of their physical state or proportion in the mixture should be recognized as potential human carcinogens.

References

(1) International Agency for Research on Cancer, *IARC Monographs on the Evaluation of the Carcinogenic Risk of Chemicals to Humans,* Supplement 4, Lyon, France: IARC (1982).

(2) National Cancer Institute, *Thirteen-Week Subchronic Tocicity Studies of Direct Blue 6, Direct Black 38, and Direct Brown 95 Dyes,* Technical Report No. 108, DHEW Publication No. (NIH) 78-1358, Bethesda, MD (1979).

(3) National Institute of Occupational Safety and Health, *Special Occupational Hazard Review for Benzidine-Based Dyes,* Washington, DC: U.S. Government Printing Office (1980).

DISULFIRAM

Description: $(C_2H_5)_2\overset{\overset{\displaystyle S}{\|}}{N}C-S-S-\overset{\overset{\displaystyle S}{\|}}{C}N(C_2H_5)_2$, tetraethylthiuram disulfide is a dark brown crystalline solid melting at 70°C.

Code Numbers: CAS 97-77-8   RTECS JO1225000   UN 2588

DOT Designation: —

Synonyms: Bis(diethylthiocarbamoyl)disulfide; Antabuse®.

Potential Exposure: Disulfiram is used as a rubber accelerator and vulcanizer. It is used as a seed disinfectant and fungicide. It is used in therapy as an alcohol deterrent.

Permissible Exposure Limits in Air: There is no Federal standard but ACGIH (1983/84) has adopted a TWA value of 2 mg/m^3 and set an STEL of 5 mg/m^3.

Permissible Concentration in Water: No criteria set.

Harmful Effects and Symptoms: The major characteristics of this compound are those shown upon alcohol ingestion after disulfiram administration. Effects include intense vasodilation of face and neck, tachycardia and tachypnea followed by nausea, vomiting, pallor and hypotension. Occasionally, convulsions, cardiac arrhythmia, and myocardial infarction may occur. Other side effects include a metallic or garlic taste in the mouth, polyneuropathy, optic neuritis and skin eruptions.

Disposal Method Suggested: Disulfiram can be dissolved in alcohol or other flammable solvent and burned in an incinerator equipped with afterburner and scrubber.

References

(1)   Sax, N.I., Ed., *Dangerous Properties of Industrial Materials Report, 1,* No. 5, 40, New York, Van Nostrand Reinhold Co. (1981).

DISULFOTON

- Hazardous substance (EPA)
- Hazardous waste (EPA)

Description: $(C_2H_5O)_2\overset{\overset{\text{S}}{\|}}{P}SCH_2CH_2SC_2H_5$, O,O-diethyl S-2-ethylthioethyl phosphorodithioate is a colorless to yellowish oil with a characteristic odor.

Code Numbers: CAS 298-04-4   RTECS TD9275000   UN 2783

DOT Designation: Poison B.

Synonyms: Ethylthiodemeton (Japan), ENT 23347, Thiodemeton, Di-Syston®, Dithiosystox®, Frumin Al®, Solvirex®.

Potential Exposure: Those involved in the manufacture (A-32), formulation and application of this insecticide and acaricide.

Permissible Exposure Limits in Air: There is no Federal standard but ACGIH (1983/84) has adopted a TWA value of 0.1 mg/m^3 and set an STEL of 0.3 mg/m^3.

Permissible Concentration in Water: No criteria set.

Routes of Entry: Skin absorption, inhalation, ingestion.

Harmful Effects and Symptoms: Disulfoton is a highly toxic organophosphorus insecticide used on many agricultural crops. The human oral LD$_{50}$ is estimated at 5 mg/kg body weight. Exposure results in central nervous system toxicity. The LD$_{50}$ for several animal species ranges from 3.2 to 6 mg/kg. Carcinogenic, mutagenic, and teratogenic studies were not found in the available literature.

According to a Japanese source (A-38), multiple exposures may give a partially cumulative effect and symptoms include: headaches, anorexia, nausea, abdominal cramps, diarrhea, salivation, lachrymation, sweating, shortness of breath, slow pulse, tremor, muscle cramps, fever, cyanosis, pulmonary edema, convulsions, coma, heart block and shock.

Medical Surveillance: Determine cholinesterase weekly.

First Aid: Flush eyes with water, wash contaminated areas of body with soap and water. Use stomach wash if swallowed followed by saline catharsis. If breathed in, use artificial respiration and administer oxygen (A-38).

Personal Protective Methods: Wear goggles, gloves, rubber boots and long sleeved coveralls with tight collars and cuffs.

Respirator Selection: Wear gas mask.

Disposal Method Suggested: Incineration with added flammable solvent in a furnace with alkali scrubber.

Reference

(1)   U.S. Environmental Protection Agency, *Disulfoton,* Health and Environmental Effects Profile No. 97, Washington, DC, Office of Solid Waste (April 30, 1980).

(2)   United Nations Environment Programme, *IRPTC Legal File 1983,* Vol. II, pp VII/568–
      70, Geneva, Switzerland, International Register of Potentially Toxic Chemicals
      (1984).

DIURON

● Hazardous substance (EPA)

Description: $Cl_2C_6H_3NHCON(CH_3)_2$, 3-(3,4-dichlorophenyl)-1,1-dimethylurea is a crystalline solid melting at 158° to 159°C.

Code Numbers: CAS 330-54-1   RTECS YS8925000   UN2767

DOT Designation: ORM-E.

Synonyms: Dichlorfenidim (in U.S.S.R.), DCMU (in Japan), Karmex®.

Potential Exposure: Those involved in the manufacture (A-32), formulation and application of the herbicide.

Permissible Exposure Limits in Air: There is no Federal standard but ACGIH (1983/84) has adopted a TWA value of 10 mg/m³. There is no tentative STEL value.

Permissible Concentration in Water: No criteria set.

Harmful Effects and Symptoms: The concentrated material may cause irritation to the eyes and mucous membranes, but a 50% water paste was not irritating to the intact skin of guinea pigs. In 2 yr feeding trials, the no effect level was 250 mg/kg diet for rats and 125 mg/kg diet for dogs (A-7).

Disposal Method Suggested: Diuron, stable under normal conditions, decomposes on heating to 180° to 190°C giving dimethylamine and 3,4-dichlorophenyl-isocyanate. Treatment at elevated temperatures by acid or base yields dimethylamine and 3,4-dichloroaniline. Hydrolysis is not recommended as a disposal procedure because of the generation of the toxic products, 3,4-dichloroaniline and dimethylamine (A-32)

References

(1)   United Nations Environment Programme, *IRPTC Legal File 1983,* Vol. II, pp VII/781–
      2, Geneva, Switzerland, International Register of Potentially Toxic Chemicals (1984).

1,4-DIVINYL BENZENE

Description: $C_6H_4(CH=CH_2)_2$ is a pale straw colored liquid boiling at 195° to 200°C. It exists as o-, m-, and p-isomers.

Code Numbers: (m-isomer), CAS 108-57-6   RTECS CZ9450000

DOT Designation: —

Synonyms: Vinylstyrene, DVB.

Potential Exposure: This compound is used as a monomer for the preparation of special synthetic rubbers, drying oils, ion-exchange resins and casting resins, and in polyester resin manufacture.

Permissible Exposure Limits in Air: There is no Federal standard but ACGIH

(1983/84) has adopted a TWA of 10 ppm (50 mg/m^3). There is no tentative STEL value.

Permissible Concentration in Water: No criteria set.

Harmful Effects and Symptoms: Skin, eye and respiratory system irritation (A-34).

Disposal Method Suggested: Incineration in a furnace equipped with afterburner and scrubber.

References

(1) See Reference (A-60).

1-DODECENE

Description: $CH_3(CH_2)_9CH=CH_2$ is a colorless liquid boiling at 213°C.

Code Numbers: CAS 112-41-4 RTECS UO1860000

DOT Designation: —

Synonyms: 1-Dodecylene; Adacene®.

Potential Exposure: Those involved in the manufacture of detergents and lubricant additives from dodecene.

Permissible Exposure Limits in Air: No standards set.

Permissible Concentration in Water: No criteria set.

Routes of Entry: Inhalation and ingestion.

Harmful Effects and Symptoms: Irritates the skin on contact; can be a narcotic at high concentrations.

Personal Protective Methods: Skin protection required in high vapor concentrations.

Respirator Selection: Canister mask required for high vapor concentrations.

Disposal Method Suggested: Incineration.

References

(1) Sax, N.I., Ed., *Dangerous Properties of Industrial Materials Report, 3,* No. 2, 37-38, New York, Van Nostrand Reinhold Co. (1983).
(2) See Reference (A-60).

DYFONATE

See "Fonofos."

E

ENDOSULFAN

- Hazardous substance (EPA)
- Hazardous waste (EPA)
- Priority toxic pollutant (EPA)

Description: Endosulfan is a chlorinated cyclodiene insecticide having the molecular formula $C_9Cl_6H_6O_3S$ and the following structural formula:

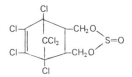

It is a light to dark brown crystalline solid with a terpene-like odor. It melts at 70° to 100°C.

Code Numbers: CAS 115-29-7   RTECS RB9275000   UN (NA 2761)

DOT Designation: Poison B

Synonyms: Thiodan (in USSR), Benzoepin (in Japan), ENT 23979, OMS-570, 6,7,8,9,10,10-hexachloro-1,5,5a,6,9,9a-hexahydro-6,9-methano-2,4,3-benzo-dioxathiepin-3 oxide, Thiodan®, Cyclodan®, Beosit®, Thimul® and Thiofor®.

Potential Exposure: Those engaged in the manufacture (A-32), formulation and application of this material.

Permissible Exposure Limits in Air: The ACGIH (1983/84) has set a TWA of 0.1 mg/m³ and an STEL of 0.3 mg/m³. The notation "skin" is added to indicate the possibility of cutaneous absorption.

Permissible Concentration in Water: To protect freshwater aquatic life—0.056 μg/ℓ as a 24 hr average, never to exceed 0.22 μg/ℓ. To protect saltwater aquatic life—0.0087 μg/ℓ as a 24 hr average, never to exceed 0.034 μg/ℓ. To protect human health—74.0 μg/ℓ. Earlier, EPA (A-3) had established the following criteria—0.003 μg/ℓ for freshwater aquatic life; 0.001 μg/ℓ for marine aquatic life.

Determination in Water: Methylene chloride extraction followed by gas chromatography with electron capture or halogen specific detection (EPA Method 608) or gas chromatography plus mass spectrometry (EPA Method 625).

Routes of Entry: Inhalation, ingestion, eye and skin contact.

Harmful Effects and Symptoms: Workers who failed to use good safety practices (i.e., to cover skin and use respiratory protection) have died from endosulfan exposure. In one incident, three persons exposed showed central nervous system symptoms; two of them died. It therefore appears that the most toxic potential effect to man is that of central nervous system toxicity since the available data indicate a lack of carcinogenic, mutagenic, or teratogenic potential. The absence of reports on toxic effects associated with the proper use of endosulfan (particularly such effects as skin sensitization or other human symptoms) has been noted.

Points of Attack: Central nervous system, lungs, skin.

Medical Surveillance: Consider the points of attack in preplacement and periodic physical examinations.

Disposal Method Suggested: A recommended method for disposal is burial 18 inches deep in noncropland away from water supplies, but bags can be burned (A-32).

References

(1) U.S. Environmental Protection Agency, *Endosulfan: Ambient Water Quality Criteria,* Washington, DC (1980).

(2) U.S. Environmental Protection Agency, *Endosulfan,* Health and Environmental Effects Profile No. 98, Washington, DC, Office of Solid Waste (April 30, 1980).

(3) United Nations Environment Programme, *IRPTC Legal File 1983,* Vol. II, pp VII/479–82, Geneva, Switzerland, International Register of Potentially Toxic Chemicals (1984).

ENDRIN

- Hazardous substance (EPA)
- Hazardous waste (EPA)
- Priority toxic pollutant (EPA)

Description: Endrin is the common name of one member of the cyclodiene group of pesticides. It is a cyclic hydrocarbon having a chlorine-substituted methano bridge structure as follows:

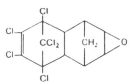

It is a white, crystalline solid melting at 226° to 230°C.

Code Numbers: CAS 72-20-8   RTECS IO1575000   UN (NA 2761)

DOT Designation: Poison B

Synonyms: 1,2,3,4,10,10-Hexachloro-6,7-epoxy-1,4,4a,5,6,7,8,8a-octahydro-1,4-endo-endo-5,8-dimethano-naphthalene, nendrin (in South Africa), ENT 17251.

Potential Exposures: Those involved in manufacture (A-32), formulation and field application of this pesticide.

Incompatibilities: Strong oxidizers, strong acids.

Permissible Exposure Limits in Air: The Federal limit and the ACGIH 1983/84 TWA value is 0.1 mg/m^3. The notation "skin" indicates the possibility of cutaneous absorption. The STEL is 0.3 mg/m^3. The IDLH level is 200 mg/m^3.

Determination in Air: Collection by impinger or fritted bubbler, analysis by gas liquid chromatography (A-1).

Permissible Concentration in Water: To protect freshwater aquatic life—0.0023 µg/ℓ as a 24 hr average, never to exceed 0.18 µg/ℓ. To protect saltwater aquatic life—0.0023 µg/ℓ as a 24 hr average, never to exceed 0.037 µg/ℓ. To protect human health—1.0 µg/ℓ.

Determination in Water: Methylene chloride extraction followed by gas chromatography with electron capture or halogen specific detection (EPA Method 608) or gas chromatography plus mass spectrometry (EPA Method 625).

Routes of Entry: Inhalation, skin absorption, ingestion, eye and skin contact.

Harmful Effects and Symptoms: Quantitative data on endrin toxicity to humans are not available according to EPA (1). However, outbreaks of human poisoning have resulted from accidental contamination of foods and have been traced to doses as low as 0.2 mg/kg body weight. Endrin toxicity seems to result primarily from the effects of endrin and its metabolites on the central nervous system. Symptoms usually observed in victims of endrin poisoning were convulsions, vomiting, abdominal pain, nausea, dizziness and headache. Additional symptoms include insomnia, aggressive confusion, lethargy; weakness and anorexia. Respiratory failure was the most common cause of death. Significantly increased activity of the hepatic microsomal drug-metabolizing enzymes has occurred in individuals employed in the manufacture of endrin. No irreversible adverse effects of occupational exposure to endrin have been reported in the literature so far.

A rebuttable presumption notice against registration was issued on July 27, 1976 by EPA on the basis of oncogenicity, teratogenicity, and reductions in endangered species and nontarget species.

Points of Attack: Liver, central nervous system.

Medical Surveillance: Consider the points of attack in preplacement and periodic physical examinations.

First Aid: If this chemical gets into the eyes, irrigate immediately. If this chemical contacts the skin, wash with soap immediately. If a person breathes in large amounts of this chemical, move the exposed person to fresh air at once and perform artificial respiration. When this chemical has been swallowed, get medical attention. Give large quantities of water and induce vomiting. Do not make an unconscious person vomit.

Personal Protective Methods: Wear appropriate clothing to prevent any possibility of skin contact. Wear eye protection to prevent any possibility of eye contact. Employees should wash immediately when skin is wet or contaminated. Work clothing should be changed daily if it is possible that clothing is contaminated. Remove nonimpervious clothing immediately if wet or contaminated. Provide emergency showers and eyewash.

Respirator Selection:
       1 mg/m^3: CCROVDMPest/SA/SCBA
       5 mg/m^3: CCROVFDMPest/GMOVDMPest/SAF/SCBAF
     100 mg/m^3: CCROVHiEPPest/SA:PD,PP,CF
     200 mg/m^3: SAF:PD,PP,CF
       Escape: GMOVPPest/SCBA

Disposal Method Suggested: A disposal procedure recommended by the manufacturer consists of absorption, if necessary, and burial at least 18 inches deep, preferably in sandy soil in a flat or depressed location away from wells, livestock, children, wildlife, etc. (A-32).

References

(1) U.S. Environmental Protection Agency, *Endrin: Ambient Water Quality Criteria,* Washington, DC (1980).

(2) U.S. Environmental Protection Agency, *Reviews of the Environmental Effects of Pollutants: XIII, Endrin,* Report EPA-600/1-79-005, Cincinnati, OH (1979).

(3) U.S. Environmental Protection Agency, *Endrin,* Health and Environmental Effects Profile No. 99, Washington, DC, Office of Solid Waste (April 30, 1980).

(4) Sax, N.I., Ed., *Dangerous Properties of Industrial Materials Report, 1,* No. 5, 55-57, New York, Van Nostrand Reinhold Co. (1981).

(5) See Reference (A-61).

(6) United Nations Environment Programme, *IRPTC Legal File 1983,* Vol. I, pp VII/269–74, Geneva, Switzerland, International Register of Potentially Toxic Chemicals (1984).

ENFLURANE

Description: CHF_2OCF_2CHFCl, 2-chloro-1,1,2-trifluoroethyl difluoromethyl ether is a liquid boiling at 56° to 57°C.

Code Numbers: CAS 13838-16-9   RTECS KN6800000

DOT Designation: —

Synonyms: Ethrane®, methylflurether.

Potential Exposures: This compound is used as an anesthetic.

Permissible Exposure Limits in Air: There is no Federal standard but ACGIH as of 1983/84 has proposed a TWA of 75 ppm (575 mg/m^3). There is no proposed STEL.

Permissible Concentration in Water: No standards set.

Routes of Entry: Inhalation of vapors.

Harmful Effects and Symptoms: Enflurane is a potent respiratory depressant and has been reported to impair cardiac performance (A-42).

References

(1) Nat. Inst. for Occupational Safety and Health, *Criteria for a Recommended Standard: Occupational Exposure to Waste Anesthetic Gases and Vapors,* NIOSH Doc. No. 77-140 (1977).

EPICHLOROHYDRIN

- Carcinogen (animal positive, IARC) (3)
- Hazardous substance (EPA)
- Hazardous waste (EPA)

Description: CH_2OCHCH_2Cl, epichlorohydrin, is a colorless liquid with a chloroform-like odor. It boils at 115°C.

Code Numbers: CAS 106-89-8   RTECS TX4900000   UN 2023

DOT Designation: Flammable liquid and poison

Synonyms: Epi, chloropropylene oxide, 1-chloro-2,3-epoxypropane, chloromethyloxirane, 2-epichlorohydrin.

Potential Exposures: Epichlorohydrin is used in the manufacture of many glycerol and glycidol derivatives and epoxy resins, as a stabilizer in chlorine-containing materials, as an intermediate in the preparation of cellulose esters and ethers, paints, varnishes, nail enamels, and lacquers, and as a cement for celluloid. It is an intermediate in the manufacture of a number of drugs (A-41).

NIOSH estimates that approximately 50,000 workers in the United States are potentially exposed to epichlorohydrin.

Incompatibilities: Strong oxidizers, strong acids, caustics, zinc, aluminum, chlorides of iron and aluminum.

Permissible Exposure Limits in Air: The Federal standard is 5 ppm (19 mg/m^3). NIOSH has recommended a TWA limit of 2 mg/m^3 with ceiling concentration of 19 mg/m^3 based on a 15-minute sampling period. NIOSH recommended in late 1978 that epichlorohydrin be handled as a human carcinogen.

ACGIH has set a TWA value of 2 ppm (10 mg/m^3) and an STEL of 5 ppm (20 mg/m^3) as of 1983/84. The notation "skin" is added to indicate the possibility of cutaneous absorption. The IDLH level is 100 ppm.

A value of 1 ppm (4 mg/m^3) has been cited in a Dutch Publication (A-60).

Determination in Air: Charcoal adsorption, workup with CS_2, gas chromatography. See NIOSH Methods, Set I. See also reference (A-10).

Permissible Concentration in Water: No criteria set but EPA (A-37) has suggested a permissible ambient goal of 276 $\mu g/\ell$ based on health effects.

Routes of Entry: Inhalation, skin absorption, ingestion, eye and skin contact.

Harmful Effects and Symptoms: *Local* — Epichlorohydrin is highly irritating to eyes, skin, and respiratory tract. Skin contact may result in delayed blistering and deep-seated pain. Allergic eczematous contact dermatitis occurs occasionally.

Systemic — The earliest symptoms of intoxication may be referable to the gastrointestinal tract (nausea, vomiting, abdominal discomfort) or pain in the region of the liver. Labored breathing, cough, and cyanosis may be evident and the onset of chemical pneumonitis may occur several hours after exposure. Animals exposed repeatedly to this chemical have developed lung, kidney, and liver injury.

Points of Attack: Respiratory system, lungs, skin, kidneys.

Medical Surveillance: Consider possible effects on the skin, eyes, lungs, liver, and kidney in preplacement or periodic examinations.

First Aid: If this chemical gets into the eyes, irrigate immediately. If this chemical contacts the skin, wash with soap immediately. If a person breathes in large amounts of this chemical, move the exposed person to fresh air at once and perform artificial respiration. When this chemical has been swallowed, get medical attention. Give large quantities of salt water and induce vomiting. Do not make an unconscious person vomit.

Personal Protective Methods: Goggles and rubber, protective clothing should be worn to prevent any possibility of skin contact or eye contact. Epichlorohydrin slowly penetrates rubber, so all contaminated clothing should be thoroughly washed. Employees should wash immediately when skin is wet or contaminated. Remove nonimpervious clothing immediately if wet or contaminated. Provide emergency showers and eyewash.

Respirator Selection:
50 ppm: SA/SCBA
250 ppm: SAF/SCBAF
Escape: GMOVAG/SCBA

Disposal Method Suggested: Incineration, preferably after mixing with another combustible fuel. Care must be exercised to assure complete combustion to prevent the formation of phosgene. An acid scrubber is necessary to remove the halo acids produced.

References

(1) National Institute for Occupational Safety and Health, *Criteria for a Recommended Standard: Occupational Exposure to Epichlorohydrin,* NIOSH Doc. No. 76-206, Washington, DC (1976).

(2) U.S. Environmental Protection Agency, *Epichlorohydrin,* Health and Environmental Effects Profile No. 100, Washington, DC, Office of Solid Waste (April 30, 1980).

(3) International Agency for Research on Cancer, *IARC Monographs on the Carcinogenic Risks of Chemicals to Humans,* 13, 131 (1976).

(4) Santodonato, J., Lande, S.S., Howard, P.N., Orzel, D. and Bogyo, D. (Syracuse Research Corp. Center for Chemical Hazard Assessment), *Investigation of Selected Potential Environmental Contaminants: Epichlorohydrin and Epibromohydrin,* Report EPA-560/11-80-006, Washington, DC, U.S. Environmental Protection Agency (March 1980).

(5) Sax, N.I., Ed., *Dangerous Properties of Industrial Materials Report, 1,* No. 4, 56-57, New York, Van Nostrand Reinhold Co. (1981).

(6) See Reference (A-61).

(7) Sax, N.I., Ed., *Dangerous Properties of Industrial Materials Report, 3,* No. 3, 68-71, New York, Van Nostrand Reinhold Co. (1983).

(8) United Nations Environment Programme, *IRPTC Legal File 1983,* Vol. II, pp VII/666–68, Geneva, Switzerland, International Register of Potentially Toxic Chemicals (1984).

EPN

Description: $O_2NC_6H_4OP-C_6H_5$ is a light yellow crystalline powder MP 36°C.

(with structure showing $\overset{S}{\underset{OC_2H_5}{\overset{\|}{P}}}$)

Code Numbers: CAS 2104-64-5   RTECS TB1925000   UN 2783

DOT Designation: –

Synonyms: O-Ethyl O-p-nitrophenyl thionobenzene phosphonate, O-ethyl O-p-nitrophenyl benzenephonothioate, ENT 17,298.

Potential Exposure: Those involved in the manufacture (A-32), formulation and field application of this pesticide.

Incompatibilities: Strong oxidizers.

Permissible Exposure Limits in Air: The Federal limit and the ACGIH 1983/84 TWA value is 0.5 mg/m³. The notation "skin" indicates the possibility of cutaneous absorption. The STEL is 2.0 mg/m³. The IDLH level is 50 mg/m³.

Determination in Air: Collection on a filter, workup with iso-octane, determination by gas chromatography. See NIOSH Methods, Set T. See also reference (A-10).

Permissible Concentration in Water: No criteria set.

Routes of Entry: Inhalation, ingestion, skin absorption, eye and skin contact.

Harmful Effects and Symptoms: Miosis, eye irritation, rhinorrhea; headaches; tight chest, wheezing, laryngeal spasm, salivation, cyanosis; anorexia, nausea, abdominal cramps, diarrhea; convulsions, low blood pressure.

A rebuttable presumption against registration was issued on September 19, 1979 by EPA on the basis of neurotoxicity.

Points of Attack: Respiratory system, lungs, cardiovascular system, central nervous system, eyes, skin, blood cholinesterase.

Medical Surveillance: Consider the points of attack in preplacement and periodic physical examinations.

First Aid: If this chemical gets into the eyes, irrigate immediately. If this chemical contacts the skin, wash with soap immediately. If a person breathes in large amounts of this chemical, move the exposed person to fresh air at once and perform artificial respiration. When this chemical has been swallowed, get medical attention. Give large quantities of water and induce vomiting. Do not make an unconscious person vomit.

Personal Protective Methods: Wear appropriate clothing to prevent any possibility of skin contact. Wear eye protection to prevent any possibility of eye contact. Employees should wash immediately when skin is wet or contaminated. Work clothing should be changed daily if it is possible that clothing is contaminated. Remove nonimpervious clothing immediately if wet or contaminated. Provide emergency showers and eyewash.

Respirator Selection:
>   5 mg/m³: SA/SCBA
>   25 mg/m³: SAF/SCBAF
>   50 mg/m³: SA:PD,PP,CF
>   Escape: GMOVPPest/SCBA

Disposal Method Suggested: EPN plant wastes are treated by preaeration, activated sludge treatment, recycle, chlorination and final polishing where additional natural biological stabilization occurs (A-32). EPN is also relatively rapidly hydrolyzed in alkaline solution to benzene thiophosphoric acid, alcohol and p-nitrophenol.

References
(1)   See Reference (A-61).
(2)   United Nations Environment Programme, *IRPTC Legal File 1983,* Vol. II, pp VII/541–2, Geneva, Switzerland, International Register of Potentially Toxic Chemicals (1984).

ERYTHRITOL ANHYDRIDE

See "Diepoxybutane."

ETHANOLAMINES

Description: $H_2NCH_2CH_2OH$—monoethanolamine, $HN(CH_2CH_2OH)_2$—diethanolamine, $N(CH_2CH_2OH)_3$—triethanolamine. All three compounds are water soluble liquids. Monoethanolamine has a low vapor pressure while the vapor pressures of the other ethanolamines are very low. Monoethanolamine and diethanolamine have ammonia odors while triethanolamine has only a faint non-ammonia odor. The acid salts have less odor and are of low volatility. Ethanolamines can be detected by odor as low as 2 to 3 ppm.

Code Numbers:

Mono-	CAS	141-43-5	RTECS	KJ5775000	UN 2491
Di-	CAS	111-42-2	RTECS	KL2975000	UN 2491
Tri-	CAS	102-71-6	RTECS	KL9275000	UN 2491

DOT Designation: Corrosive material.

Synonyms: Monoethanolamine—ethanolamine, 2-aminoethanol, colamine. Diethanolamine—2,2'-iminodiethanol. Triethanolamine—2,2',2"-nitrilotriethanol.

Potential Exposures: Monoethanolamine is widely used in industry to remove carbon dioxide and hydrogen from natural gas, to remove hydrogen sulfide and carbonyl sulfide, as an alkaline conditioning agent, and as an intermediate for soaps, detergents, dyes, and textile agents.

Diethanolamine is an absorbent for gases, a solubilizer for 2,4-dichlorophenoxyacetic acid (2,4-D), and a softener and emulsifier intermediate for detergents. It also finds use in the dye and textile industry.

Triethanolamine is used as a plasticizer, neutralizer for alkaline dispersions, lubricant additive, corrosion inhibitor, and in the manufacture of soaps, detergents, shampoos, shaving preparations, face and hand creams, cements, cutting oils, insecticides, surface active agents, waxes, polishes, and herbicides.

Incompatibilities: Strong oxidizers, strong acids.

Permissible Exposure Limits in Air: The Federal standard for monoethanolamine is 3 ppm ($8mg/m^3$). The STEL for monoethanolamine is 6 ppm ($15 mg/m^3$). The IDLH level is 1,000 ppm. As of 1983/84 the ACGIH TWA value for diethanolamine is 3 ppm ($15 mg/m^3$); there is no STEL value. There are no TWA standards for triethanolamine.

Determination in Air: Adsorption on silica gel, elution with methanolic acid, derivatization with benzaldehyde, analysis by gas chromatography (A-10) for monoethanolamine.

Permissible Concentration in Water: No criteria set, but EPA (A-37) has suggested permissible ambient limits of 83 $\mu g/\ell$.

Routes of Entry: Inhalation of vapor, percutaneous absorption, ingestion and eye and skin contact.

Harmful Effects and Symptoms: *Local* — Ethanolamine has had wide use in industry, yet reports of injury in man are lacking. Ethanolamine in animal experiments was highly irritating to the skin, eyes, and respiratory tract. Diethanolamine and triethanolamine produced much less irritation. In human experiments, ethanolamine produced only redness of the skin.

Systemic — No specific published data on human exposure are available. Animal experiments indicate that it is a central nervous system depressant. Acute high level exposures produced pulmonary damage and nonspecific hepatic and renal lesions in animals.

Points of Attack: Skin, eyes, respiratory system.

Medical Surveillance: Evaluate possible irritant effects on skin and eyes.

First Aid: If this chemical gets into the eyes, irrigate immediately. If this chemical contacts the skin, flush with water promptly. If a person breathes in large amounts of this chemical, move the exposed person to fresh air at once and perform artificial respiration. When this chemical has been swallowed, get medical attention. Give large quantities of water and induce vomiting. Do not make an unconscious person vomit.

Personal Protective Methods: Wear appropriate clothing to prevent repeated or prolonged skin contact. Wear eye protection to prevent any possibility of eye contact. Employees should wash promptly when skin is wet or contaminated. Work clothing should be changed daily if it is possible that clothing is contaminated. Remove nonimpervious clothing promptly if wet or contaminated. Provide emergency eyewash.

Respirator Selection:
　　　30 ppm: CCRS/SA/SCBA
　　150 ppm: CCRSF/GMS/SAF/SCBAF
　1,000 ppm: SAF:PD,PP,CF
　　Escape: GMS/SCBA

Disposal Method Suggested: Controlled incineration; incinerator is equipped with a scrubber or thermal unit to reduce NO_x emissions.

References

(1) U.S. Environmental Protection Agency, *Chemical Hazard Information Profile: Ethanolamines,* Washington, DC (April 14, 1978).
(2) See Reference (A-61).
(3) Sax, N.I., Ed., *Dangerous Properties of Industrial Materials Report, 4,* No. 1, 66–69, New York, Van Nostrand Reinhold Co. (Jan./Feb. 1984).

ETHION

● Hazardous substance (EPA)

Description: $(C_2H_5O)_2\overset{\text{S}}{\overset{\|}{P}}SCH_2S\overset{\text{S}}{\overset{\|}{P}}(OC_2H_5)_2$, O,O,O',O'-tetraethyl-S,S'-methylene di(phosphorodithioate), is a colorless to amber liquid (melting at -12° to -13°C), with a disagreeable odor.

Code Numbers: CAS 563-12-2   RTECS TE4550000   UN 2783

DOT Designation: Poison B

Synonyms: Diethion, ENT 24105, Nialate®

Potential Exposures: Those involved in the manufacture (A-32), formulation and application of this insecticide and acaricide.

Ethion is a preharvest topical insecticide used primarily on citrus fruits, deciduous fruits, nuts and cotton. It is also used as a cattle dip for ticks and as a treatment for buffalo flies.

Permissible Exposure Limits in Air: There are no Federal standards but ACGIH as of 1983/84 has adopted a TWA value of 0.4 mg/m³. There is no tentative STEL value. The notation "skin" is added to the TWA indicating the possibility of cutaneous absorption.

Permissible Concentration in Water: No criteria set, but the World Health Organization has established an ADI level of 0.005 mg/kg for ethion based on cholinesterase inhibition studies.

Harmful Effects and Symptoms: Ethion has not shown mutagenic effects in mice or teratogenic effects in fowl. Subcutaneous injection of the compound into atropinized chickens produced neurotoxic effects. There is no available information on the possible carcinogenic effects of ethion. Ethion will produce anticholinesterase effects in mammals.

Symptoms include headache, anorexia, nausea, weakness, dizziness, blurred vision, salivation, lachrymation, sweating, shortness of breath, muscular cramps, ataxia, fever, cyanosis, pulmonary edema, areflexia, convulsions, heart block, shock and respiratory failure.

Medical Surveillance: Determine cholinesterase weekly.

First Aid: Irrigate eyes with water. Wash contaminated areas of body with soap and water. If inhaled, administer artificial respiration and oxygen. If swallowed use gastric lavage followed by saline catharsis (A-38).

Personal Protective Methods: Wear safety goggles.

Respirator Selection: Use gas mask.

Disposal Method Suggested: Incineration with added solvent in furnace equipped with afterburner and alkali scrubber.

References

(1) U.S. Environmental Protection Agency, *S,S'-Methylene-O,O,O',O'-Tetraethyl Phosphorodithioate*, Health and Environmental Effects Profile No. 127, Washington, DC, Office of Solid Waste (April 30, 1980).

(2) Sax, N.I., Ed., *Dangerous Properties of Industrial Materials Report, 4*, No. 1, 69–74, New York, Van Nostrand Reinhold Co. (Jan./Feb. 1984).

(3) United Nations Environment Programme, *IRPTC Legal File 1983*, Vol. II, pp VII/589–90, Geneva, Switzerland, International Register of Potentially Toxic Chemicals (1984).

ETHOPROP

Description: $C_8H_{19}O_2PS_2$ has the formula

$$(C_3H_7S)_2POCH_2CH_3$$
$$\overset{\|}{O}$$

This is a compound with a melting point of $20°C$ and a boiling point of $86°$ to $91°C$ at 0.2 mm Hg.

Code Numbers: CAS 13194-48-4   RTECS TE4025000

DOT Designation: —

Synonyms: O-Ethyl-S,S,-dipropyl phosphorodithionate.

Potential Exposure: Those involved in the manufacture, formulation and application of this nematocide and soil insecticide.

Permissible Exposure Limits in Air: No standards set.

Permissible Concentration in Water: 0.1 mg/ℓ in drinking water (1).

Harmful Effects and Symptoms: This is a moderately toxic material (LD-50 for rats is 62 mg/kg). Symptoms are typical of cholinesterase inhibitors. They include tightness across the chest, increased salivation, nausea, vomiting, abdominal cramps, diarrhea, abnormal heart rates, involuntary urination, constriction of pupils and arm and leg weakness. Large doses may cause death through asphyxia caused by respiratory failure.

References

(1) Sax, N.I., Ed., *Dangerous Properties of Industrial Materials Report, 2,* No. 4, 85-88, New York, Van Nostrand Reinhold Co. (1982).

2-ETHOXYETHANOL

Description: $C_2H_5OCH_2CH_2OH$ is a colorless liquid with a sweetish odor having a boiling point of 135°C.

Code Numbers: CAS 110-80-5   RTECS KK8050000   UN 1171

DOT Designation: Combustible liquid.

Synonyms: Ethylene glycol monoethyl ether, Cellosolve® solvent.

Potential Exposure: This material is used as a solvent for nitrocellulose and alkyd resins in lacquers. It is used in dyeing leathers and textiles and in the formulation of cleaners and varnish removers.

Incompatibilities: Strong oxidizers.

Permissible Exposure Limits in Air: The Federal limit is 200 ppm (740 mg/m³). The ACGIH has set a TWA value of 50 ppm (185 mg/m³) and a tentative STEL value of 100 ppm (370 mg/m³). A proposed (1983/84) change could lower the TWA to 5 ppm (19 mg/m³). The notation "skin" indicates the possibility of cutaneous absorption. The IDLH level is 6,000 ppm. A value of 100 ppm (370 mg/m³) has been cited in a Dutch Publication (A-60).

Determination in Air: Collection by charcoal tube, analysis by gas liquid chromatography (A-13).

Permissible Concentration in Water: No criteria set.

Routes of Entry: Inhalation, skin absorption, ingestion, eye and skin contact.

Harmful Effects and Symptoms: In animals—hematologic effects; liver, kidneys, lung damage; eye irritation. Has the potential to cause adverse reproductive effects in male and female workers (2).

Points of Attack: In animals—lungs, eyes, blood, kidneys, liver.

Medical Surveillance: Consider the points of attack in preplacement and periodic physical examinations.

First Aid: If this chemical gets into the eyes, irrigate immediately. If this chemical contacts the skin, flush with water promptly. If a person breathes in large amounts of this chemical, move the exposed person to fresh air at once and perform artificial respiration. When this chemical has been swallowed, get medical attention. Give large quantities of water and induce vomiting. Do not make an unconscious person vomit.

Personal Protective Methods: Wear appropriate clothing to prevent repeated or prolonged skin contact. Wear eye protection to prevent any reasonable probability of eye contact. Employees should wash promptly when skin is wet or contaminated. Remove nonimpervious clothing promptly if wet or contaminated.

Respirator Selection:
2,000 ppm: SA/SCBA
6,000 ppm: SAF/SA:PD,PP,CF/SCBAF
Escape: GMOV/SCBA

Disposal Method Suggested: Incineration.

References
(1) See Reference (A-61).
(2) National Institute for Occupational Safety and Health, *Glycol Ethers: 2-Methoxyethanol and 2-Ethoxyethanol,* Current Intelligence Bulletin No. 39, Cincinnati, Ohio (May 2, 1983).
(3) United Nations Environment Programme, *IRPTC Legal File 1983,* Vol. I, pp VII, 337–8, Geneva, Switzerland, International Register of Potentially Toxic Chemicals (1984).

2-ETHOXYETHYL ACETATE

Description: $C_2H_5OCH_2CH_2OCOCH_3$ is a colorless liquid with a mild, nonresidual odor boiling at 156°C.

Code Numbers: CAS 111-15-9 RTECS KK8225000 UN 1172

DOT Designation: Combustible liquid.

Synonyms: Cellosolve® acetate, glycol monoethyl ether acetate, ethylene glycol monoethyl ether acetate.

Potential Exposures: This material is used as a solvent for nitrocellulose and other resins.

Incompatibilities: Nitrates, strong oxidizers, strong alkalies, strong acids.

Permissible Exposure Limits in Air: The Federal limit is 100 ppm (540 mg/m³). The ACGIH has set a TWA of 50 ppm (270 mg/m³) and a tentative STEL value of 100 ppm (540 mg/m³). A proposed (1983/84) change would lower the TWA to 5 ppm (27 mg/m³). The notation "skin" is added indicating the possibility of cutaneous absorption. The IDLH level is 2,500 ppm.

Determination in Air: Charcoal adsorption, workup with CS_2, gas chromatography analysis. See NIOSH Methods, Set D. See also reference (A-10).

Permissible Concentration in Water: No criteria set.

Routes of Entry: Inhalation, ingestion, eyes and skin contact.

Harmful Effects and Symptoms: Irritation of eyes and nose, vomiting; kidney damage; paralysis.

Points of Attack: Respiratory system, eyes, gastrointestinal system.

Medical Surveillance: Consider the points of attack in preplacement and periodic physical examinations.

First Aid: If this chemical gets into the eyes, irrigate immediately. If this chemical contacts the skin, flush with water promptly. If a person breathes in large amounts of this chemical, move the exposed person to fresh air at once and perform artificial respiration. When this chemical has been swallowed, get medical attention. Give large quantities of salt water and induce vomiting. Do not make an unconscious person vomit.

Personal Protective Methods: Wear appropriate clothing to prevent repeated or prolonged skin contact. Wear eye protection to prevent any reasonable probability of eye contact. Employees should wash promptly when skin is wet or contaminated. Remove nonimpervious clothing promptly if wet or contaminated.

Respirator Selection:
- 1,000 ppm: CCROVF
- 2,500 ppm: GMOV/SAF/SCBAF
- Escape: GMOV/SCBA

Disposal Method Suggested: Incineration.

References

(1)   See Reference (A-61).
(2)   See Reference (A-60).

ETHYL ACETATE

- Hazardous waste (EPA)

Description: $CH_3COOC_2H_5$ is a colorless liquid with a pleasant, fruity odor, boiling at 77°C.

Code Numbers: CAS 141-78-6   RTECS AH5425000   UN 1173

DOT Designation: Flammable liquid.

Synonyms: Acetic ester, acetic ether, ethyl ethanoate.

Potential Exposure: This material is used as a lacquer solvent. It is also used in flavoring and perfumery and in smokeless powder manufacture.

Incompatibilities: Nitrates, strong oxidizers, strong alkalies, strong acids.

Permissible Exposure Limits in Air: The Federal limit and the 1983/84 ACGIH TWA value is 400 ppm (1,400 mg/m³). There is no tentative STEL value set. The IDLH level is 10,000 ppm.

Determination in Air: Charcoal adsorption, workup with CS_2, analysis by gas chromatography. See NIOSH Methods, Set E. See also reference (A-10).

Permissible Concentration in Water: No criteria set.

Routes of Entry: Inhalation, ingestion, eye and skin contact.

Harmful Effects and Symptoms: Irritation of eyes, nose and throat; narcosis; dermatitis.

Points of Attack: Eyes, skin, respiratory system.

Medical Surveillance: Consider the points of attack in preplacement and periodic physical examinations.

First Aid: If this chemical gets into the eyes, irrigate immediately. If this chemical contacts the skin, flush with water promptly. If a person breathes in large amounts of this chemical, move the exposed person to fresh air at once and perform artificial respiration. When this chemical has been swallowed, get medical attention. Give large quantities of salt water and induce vomiting. Do not make an unconscious person vomit.

Personal Protective Methods: Wear appropriate clothing to prevent repeated or prolonged skin contact. Wear eye protection to prevent any reasonable probability of eye contact. Employees should wash promptly when skin is wet or contaminated. Remove clothing immediately if wet or contaminated to avoid flammability hazard.

Respirator Selection:
```
 1,000 ppm: CCROVF
 5,000 ppm: GMOVc
10,000 ppm: GMOVfb/SAF/SCBAF
    Escape: GMOV/SCBA
```

Disposal Method Suggested: Incineration.

References

(1) See Reference (A-61).
(2) See Reference (A-60).
(3) Sax, N.I., Ed., *Dangerous Properties of Industrial Materials Report, 4,* No. 1, 75–79, New York, Van Nostrand Reinhold Co. (Jan./Feb. 1984).

ETHYL ACRYLATE

● Hazardous waste (EPA)

Description: $CH_2=CHCOOC_2H_5$ is a colorless liquid with a sharp, acrid odor, boiling at 100°C.

Code Numbers: CAS 140-88-5   RTECS AT0700000   UN 1917

DOT Designation: Flammable liquid.

Synonyms: Ethyl propenoate.

Potential Exposures: This material is used as a monomer in the manufacture of homopolymer and copolymer resins for the production of paints and plastic films.

Incompatibilities: Oxidizers, peroxides, polymerizers, strong alkalies, moisture.

Permissible Exposure Limits in Air: The Federal limit is 25 ppm (100 mg/m^3). The ACGIH as of 1983/84 has set a TWA of 5 ppm (20 mg/m^3) and an STEL of 25 ppm (100 mg/m^3). The IDLH level is 2,000 ppm.

Determination in Air: Charcoal adsorption, workup with CS_2, gas chromatography. See NIOSH Methods, Set D. See also reference (A-10).

Permissible Concentration in Water: No criteria set.

Routes of Entry: Inhalation, ingestion, eye and skin contact.

Harmful Effects and Symptoms: Irritation of eyes, skin and respiratory system.

Points of Attack: Respiratory system, lungs, eyes, skin.

Medical Surveillance: Consider the points of attack in preplacement and periodic physical examinations.

First Aid: If this chemical gets into the eyes, irrigate immediately. If this chemical contacts the skin, flush with water immediately. If a person breathes in large amounts of this chemical, move the exposed person to fresh air at once and perform artificial respiration. When this chemical has been swallowed, get medical attention. Give large quantities of salt water and induce vomiting. Do not make an unconscious person vomit.

Personal Protective Methods: Wear appropriate clothing to prevent repeated or prolonged skin contact. Wear eye protection to prevent any reasonable probability of eye contact. Employees should wash promptly when skin is wet or contaminated. Remove clothing immediately if wet or contaminated to avoid flammability hazard.

Respirator Selection:
>1,000 ppm: CCROVF/GMOV/SAF/SCBAF
>2,000 ppm: SAF:PD,PP,CF
>Escape: GMOV/SCBA

Disposal Method Suggested: Incineration.

References
(1) See Reference (A-61).
(2) Sax, N.I., Ed., *Dangerous Properties of Industrial Materials Report, 1,* No. 2, 35-37, New York, Van Nostrand Reinhold Co. (1980).
(3) See Reference (A-60).
(4) United Nations Environment Programme, *IRPTC Legal File 1983,* Vol. I, pp VII/28– 9, Geneva, Switzerland, International Register of Potentially Toxic Chemicals (1984).

ETHYL ALCOHOL

Description: CH_3CH_2OH, ethyl alcohol, is a colorless, volatile, flammable liquid boiling at 78° to 79°C.

Code Numbers: CAS 64-17-5   RTECS KQ6300000   UN 1170

DOT Designation: Flammable liquid.

Synonyms: Ethanol, grain alcohol, spirit of wine, cologne spirit, ethyl hydroxide, ethyl hydrate.

Potential Exposures: Ethyl alcohol is used in the chemical synthesis of a wide variety of compounds such as acetaldehyde, ethyl ether, ethyl chloride, and butadiene. It is a solvent or processing agent in the manufacture of pharmaceuticals, plastics, lacquers, polishes, plasticizers, perfumes, cosmetics, rubber accelerators, explosives, synthetic resins, nitrocellulose, adhesives, inks, and preservatives. It is also used as an antifreeze and as a fuel. It is an intermediate in the manufacture of many drugs (A-41) and pesticides (A-32).

Permissible Exposure Limits in Air: The Federal standard and the 1983/84 ACGIH TWA value is 1,000 ppm ($1,900 \text{ mg/m}^3$). There is no tentative STEL value.

Determination in Air: Collection by charcoal tube, analysis by gas liquid chromatography (A-10).

Permissible Concentration in Water: No criteria set, but EPA (A-37) has suggested a permissible ambient goal of 26,000 $\mu g/\ell$ based on health effects.

Routes of Entry: Inhalation of vapor and percutaneous absorption, ingestion, eye and skin contact.

Harmful Effects and Symptoms: *Local* — Mild irritation of eye and nose occurs at very high concentrations. The liquid can defat the skin, producing a dermatitis characterized by drying and fissuring.

Systemic — Prolonged inhalation of high concentrations, besides the local effect on the eyes and upper respiratory tract, may produce headache, drowsiness, tremors, and fatigue. Tolerance may be a factor in individual response to a given air concentration.

Bizarre symptoms (other than typical manifestations of intoxication) may result from the denaturants often present in industrial ethyl alcohol. Ethyl alcohol may act as an adjuvant, increasing the toxicity of other inhaled, absorbed, or ingested chemical agents. An exception is methanol where ethyl alcohol counteracts methanol toxicity.

Medical Surveillance: Look for chronic irritation of mucous membranes and signs of chronic alcoholism in regular physical examinations. Ethyl alcohol can readily be determined in blood, urine, and expired air.

First Aid: Irrigate eyes with water.

Personal Protective Methods: Personal protective equipment is recommended where skin contact may occur.

Disposal Method Suggested: Incineration.

References

(1) Sax, N.I., Ed., *Dangerous Properties of Industrial Materials Report, 1,* No. 7, 55-57, New York, Van Nostrand Reinhold Co. (1981).
(2) See Reference (A-60).
(3) Parmeggiani, L., Ed., *Encyclopedia of Occupational Health & Safety,* Third Edition, Vol. 1, pp 790–92, Geneva, International Labour Office (1983).

ETHYLAMINE

● Hazardous substance (EPA)

Description: $C_2H_5NH_2$ is a colorless liquid or gas with a strong ammonia-like odor. It boils at 15° to 16°C.

Code Numbers: CAS 75-04-7   RTECS KH2100000   UN 1036

DOT Designation: Flammable gas.

Synonyms: Ethylamine (anhydrous), aminoethane, monoethylamine.

Potential Exposure: Monoethylamine (MEA) is used as an intermediate in the manufacture of the following chemicals: triazine herbicides (A-32), 1,3-diethyl-thiourea (a corrosion inhibitor), ethylaminoethanol, 4-ethylmorpholine (urethane foam catalyst), ethyl isocyanate, and dimethylolethyltriazone (agent used in wash-and-wear fabrics). The cuprous chloride salts of MEA are used in the refining of petroleum and vegetable oil.

Incompatibilities: Strong acids, strong oxidizers.

Permissible Exposure Limits in Air: The Federal standard and the 1983/84 ACGIH TWA value is 10 ppm (18 mg/m^3). There are no STEL values proposed. The IDLH level is 4,000 ppm.

Determination in Air: Adsorption on silica, workup with H_2SO_4 using ultrasonics, analysis by gas chromatography. See NIOSH Methods, Set K. See also reference (A-10).

Permissible Concentration in Water: No criteria set, but EPA (A-37) has suggested an ambient environmental goal of 248 $\mu g/\ell$ on a health basis.

Routes of Entry: Inhalation, skin absorption, ingestion, eye and skin contact.

Harmful Effects and Symptoms: Eye irritation, skin burns, dermatitis, respiratory irritation.

Points of Attack: Respiratory system, eyes, skin.

Medical Surveillance: Consider the points of attack in preplacement and periodic physical examinations.

First Aid: If this chemical gets into the eyes, irrigate immediately. If this chemical contacts the skin, flush with water immediately. If a person breathes in large amounts of this chemical, move the exposed person to fresh air at once and perform artificial respiration. When this chemical has been swallowed, get medical attention. Give large quantities of water and induce vomiting. Do not make an unconscious person vomit.

Personal Protective Methods: Wear appropriate clothing to prevent any reasonable probability of skin contact. Wear eye protection to prevent any possibility of eye contact. Employees should wash immediately when skin is wet or contaminated. Remove nonimpervious clothing immediately if wet or contaminated. Provide emergency showers and eyewash.

Respirator Selection:

500 ppm:	SAF/SCBAF
4,000 ppm:	SAF:PD,PP,CF
Escape:	GMS/SCBA

Disposal Method Suggested: Controlled incineration; incinerator is equipped with a scrubber or thermal unit to reduce NO_x emissions.

References

(1) U.S. Environmental Protection Agency, *Chemical Hazard Information Profile: Ethylamines,* Washington, DC (April 1, 1978).
(2) See Reference (A-61).
(3) See Reference (A-60).
(4) United Nations Environment Programme, *IRPTC Legal File 1983,* Vol. I, pp VII/297–8, Geneva, Switzerland, International Register of Potentially Toxic Chemicals (1984).

ETHYL AMYL KETONE

See "5-Methyl-3-Heptanone."

ETHYLBENZENE

- Hazardous substance (EPA)
- Priority toxic pollutant (EPA)

Description: $C_6H_5CH_2CH_3$ is a colorless liquid with a pungent aromatic odor. It boils at 136°C.

Code Numbers: CAS 100-41-4   RTECS DA0700000   UN 1175

DOT Designation: Flammable liquid.

Synonyms: Ethylbenzol, phenylethane, EB.

Potential Exposure: Ethyl benzene is used in the manufacture of cellulose acetate, styrene, and synthetic rubber. It is also used as a solvent or diluent and as a component of automotive and aviation gasoline.

Significant quantities of EB are present in mixed xylenes. These are used as diluents in the paint industry, in agricultural sprays for insecticides and in gasoline blends (which may contain as much as 20% EB). In light of the large quantities of EB produced and the diversity of products in which it is found, there exist many environmental sources for ethylbenzene, e.g., vaporization during solvent use, pyrolysis of gasoline and emitted vapors at filling stations.

Groups of individuals who are exposed to EB to the greatest extent and could represent potential pools for the expression of EB toxicity include: (1) individuals in commercial situations where petroleum products or by-products are manufactured (e.g., rubber or plastics industry); (2) individuals residing in areas with high atmospheric smog generated by motor vehicle emissions.

Incompatibilities: Strong oxidizers.

Permissible Exposure Limits in Air: The Federal standard and the 1983/84 ACGIH TWA value is 100 ppm (435 mg/m^3). The STEL is 125 ppm (545 mg/m^3). The IDLH level is 2,000 ppm. The USSR standards are eight times lower than in the United States.

Determination in Air: Charcoal adsorption, workup with CS_2, analysis by gas chromatography. See NIOSH Methods, Set C. See also reference (A-10).

Permissible Concentration in Water: To protect freshwater aquatic life— 32,000 µg/ℓ, on an acute tocicity basis. To protect saltwater aquatic life— 430 µg/ℓ, on an acute toxicity basis. For the protection of human health— 1.4 mg/ℓ.

Determination in Water: Inert gas purge followed by gas chromatography and photoionization detection (EPA Method 602) or gas chromatography plus mass spectrometry (EPA Method 624).

Routes of Entry: Inhalation, ingestion, eye and skin contact.

Harmful Effects and Symptoms: Kidney disease, liver disease, chronic respiratory disease, skin disease—and the facts in brief are as follows: EB is not nephrotoxic. Concern is expressed becuase the kidney is the primary route of excretion of EB and its metabolites. EB is not hepatotoxic. Since EB is metabolized by the liver, concern is expressed for this tissue. Exacerbation of pulmonary pathology might occur following exposure to EB. Individuals with impaired pulmonary function might be at risk. EB is a defatting agent and may cause dermatitis following prolonged exposure. Individuals with preexisting skin problems may be more sensitive to EB.

Irritation of eyes and mucous membranes; headaches, dermatitis; narcosis, coma.

Points of Attack: Eyes, upper respiratory system, skin, central nervous system.

Medical Surveillance: Consider the points of attack in preplacement and periodic physical examinations.

First Aid: If this chemical gets into the eyes, irrigate immediately. If this chemical contacts the skin, flush with water promptly. If a person breathes in large amounts of this chemical, move the exposed person to fresh air at once and perform artificial respiration. When this chemical has been swallowed, get medical attention. Do not induce vomiting.

Personal Protective Methods: Wear appropriate clothing to prevent repeated or prolonged skin contact. Wear eye protection to prevent any reasonable probability of eye contact. Employees should wash promptly when skin is wet or contaminated. Remove clothing immediately if wet or contaminated to avoid flammability hazard.

Respirator Selection:
    1,000 ppm: CCROVF
    2,000 ppm: GMOV/SAF/SCBAF
     Escape: GMOV/SCBA

Disposal Method Suggested: Incineration.

References

(1)   U.S. Environmental Protection Agency, *Ethylbenzene: Ambient Water Quality Criteria*, Washington, DC (1980).
(2)   Sax, N.I., Ed., *Dangerous Properties of Industrial Materials Report, 2,* No. 6, 57-60, New York, Van Nostrand Reinhold Co. (1982).
(3)   See Reference (A-61).
(4)   See Reference (A-60).
(5)   United Nations Environment Programme, *IRPTC Legal File 1983,* Vol. I, pp VII/84–6, Geneva, Switzerland, International Register of Potentially Toxic Chemicals (1984).

ETHYL BROMIDE

Description: C_2H_5Br is a colorless to yellow liquid with an ether-like odor. It boils at 37° to 38°C. It has a burning taste and becomes yellowish on exposure to air.

Code Numbers: CAS 74-96-4   RTECS KH6475000   UN 1891

Dot Designation: None.

Synonyms: Bromoethane, monobromoethane.

Potential Exposure: This chemical is used as an ethylating agent in organic synthesis and gasoline, as a refrigerant, and as an extraction solvent. It has limited use as a local anesthetic.

Incompatibilities: Chemically active metals: sodium, potassium, calcium, powdered aluminum, zinc, magnesium.

Permissible Exposure Limits in Air: The Federal standard and the ACGIH 1983/84 TWA value is 200 ppm (890 mg/m^3). The STEL is 250 ppm (1,110 mg/m^3). The IDLH level is 3,500 ppm.

Determination in Air: Charcoal adsorption, workup with isopropanol, gas chromatography. See NIOSH Methods, Set H. See also reference (A-10).

Permissible Concentration in Water: No criteria set.

Routes of Entry: Inhalation, ingestion, eye and skin contact.

Harmful Effects and Symptoms: Irritation of skin, eyes and respiratory system; central nervous system depression; pulmonary edema; liver, kidney disease; cardiac arrhythmias; cardiac arrest; dizziness.

Points of Attack: Skin, liver, kidneys, respiratory system, lungs, cardiovascular system, central nervous sytem.

Medical Surveillance: No specific considerations needed according to NIOSH.

First Aid: If this chemical gets into the eyes, irrigate immediately. If this chemical contacts the skin, flush with soap promptly. If a person breathes in large amounts of this chemical, move the exposed person to fresh air at once and perform artificial respiration. When this chemical has been swallowed, get medical attention. Give large quantities of salt water and induce vomiting. Do not make an unconscious person vomit.

Personal Protective Methods: Wear appropriate (rubber, not leather) clothing to prevent repeated or prolonged skin contact. Wear eye protection to prevent any reasonable probability of eye contact. Employees should wash promptly when skin is wet. Remove clothing immediately if wet or contaminated to avoid flammability hazard.

Respirator Selection:
 2,000 ppm: SA/SCBA
 3,500 ppm: SAF/SCBAF/SAF:PD,PP,CF
 Escape: GMOV/SCBA

Disposal Method Suggested: Controlled incineration with adequate scrubbing and ash disposal facilities.

References
(1) See Reference (A-61).
(2) See Reference (A-60).

ETHYL BUTYL KETONE

Description: $C_2H_5COC_4H_9$ is a colorless liquid with a mild, fruity odor boiling at 147° to 148°C.

Code Numbers: CAS 106-35-4   RTECS MJ5250000

DOT Designation: —

Synonyms: Butyl ethyl ketone, 3-heptanone.

Potential Exposure: Ethyl butyl ketone is used as a solvent and as an intermediate in organic synthesis. It is a solvent for vinyl and nitrocellulose resins. It is used in food flavoring.

Incompatibilities: Oxidizers.

Permissible Exposure Limits in Air: The Federal standard and the 1983/84 ACGIH TWA value is 50 ppm (230 mg/m^3). The STEL is 75 ppm (345 mg/m^3). The IDLH level is 3,000 ppm.

Determination in Air: Charcoal adsorption, workup with CS_2, analysis by gas chromatography. See NIOSH Methods, Set B. See also reference (A-10).

Permissible Concentration in Water: No criteria set.

Routes of Entry: Inhalation, ingestion, skin and eye contact.

Harmful Effects and Symptoms: Irritation of eyes and mucous membrane; headaches; narcosis, coma; dermatitis.

Points of Attack: Eyes, skin, respiratory system.

Medical Surveillance: Consider the points of attack in preplacement and periodic physical examinations.

First Aid: If this chemical gets into the eyes, irrigate immediately. If this chemical contacts the skin, flush with water. If a person breathes in large amounts of this chemical, move the exposed person to fresh air at once and perform artificial respiration. When this chemical has been swallowed, get medical attention. Give large quantities of salt water and induce vomiting. Do not make an unconscious person vomit.

Personal Protective Methods: Wear appropriate clothing to prevent repeated or prolonged skin contact. Wear eye protection to prevent any reasonable probability of eye contact. Employees should wash promptly when skin is wet or contaminated. Remove nonimpervious clothing promptly if wet or contaminated.

Respirator Selection:

500 ppm:	CCROV
1,000 ppm:	CCROVF
2,500 ppm:	GMOV/SAF/SCBAF
3,000 ppm:	SAF:PD,PP,CF
Escape:	GMOV/SCBA

Disposal Method Suggested: Incineration.

References

(1) Nat. Inst. for Occup. Safety and Health, *Criteria for a Recommended Standard: Occupational Exposure to Ketones,* NIOSH Doc. No. 78-173, Washington, DC (1978).

(2) See Reference (A-61).

ETHYL BUTYRALDEHYDE

Description: $(C_2H_5)_2CHCHO$ is a colorless liquid which boils at $116.8°C$.

Code Numbers: CAS 97-96-1   RTECS ES2625000   UN 1178

DOT Designation: Flammable liquid.

Synonyms: Diethylacetaldehyde.

Potential Exposure: Involved in use in organic synthesis of pharmaceuticals, rubber chemicals.

Permissible Exposure Limits in Air: No standards set.

Permissible Concentration in Water: No criteria set.

Harmful Effects and Symptoms: Mildly toxic by oral and inhalation routes. Skin irritant.

Personal Protective Method: Full protective clothing should be worn.

Respirator Selection: Self-contained breathing apparatus.

Disposal Method Suggested: Incineration.

References

(1) Sax, N.I., Ed., *Dangerous Properties of Industrial Materials Report, 1,* No. 8, 69-71, New York, Van Nostrand Reinhold Co. (1981).

(2) Sax, N.I., Ed., *Dangerous Properties of Industrial Materials Report, 3,* No. 2, 85-87, New York, Van Nostrand Reinhold Co. (1983).

ETHYL CHLORIDE

Description: CH_3CH_2Cl, ethyl chloride, is a flammable gas with an ethereal odor and a burning taste. It is flammable, and the products of combustion include phosgene and hydrogen chloride. It boils at 12° to 13°C.

Code Numbers: CAS 75-00-3   RTECS KH7525000   UN 1037

DOT Designation: Flammable liquid.

Synonyms: Monochloroethane, hydrochloric ether, chloroethane.

Potential Exposures: Ethyl chloride is used as an ethylating agent in the manufacture of tetraethyl lead, dyes, drugs, and ethyl cellulose. It can be used as a refrigerant and as a local anesthetic (freezing).
NIOSH estimates worker exposure at 113,000 per year.

Incompatibilities: Chemically active metals: sodium, potassium, calcium, powdered aluminum, zinc, magnesium.

Permissible Exposure Limits in Air: The Federal standard and the 1983/84 ACGIH TWA value is 1,000 ppm (2,600 mg/m^3). The STEL value is 1,250 ppm (3,250 mg/m^3). The IDLH level is 20,000 ppm.

Determination in Air: Adsorption on charcoal, workup with CS_2, gas chromatograhic analysis (A-10).

Permissible Concentration in Water: No criteria set due to volatility and low specific gravity (2).

Routes of Entry: Inhalation of gas, slight percutaneous absorption, ingestion, eye and skin contact.

Harmful Effects and Symtpoms: *Local* — The liquid form of ethyl chloride is mildly irritating to skin and eyes. Frostbite can occur due to rapid liquid evaporation.
Systemic — Ethyl chloride exposure may produce headache, dizziness, incoordination, stomach cramps, and eventual loss of consciousness. In high concentrations, it is a respiratory tract irritant, and death due to cardiac arrest has been recorded. Renal damage has been reported in animals. Effects from chronic exposure have not been reported.

Points of Attack: Liver, kidneys, respiratory system, cardiovascular system.

Medical Surveillance: Consider possible acute cardiac effects in any preplacement or periodic examination.

First Aid: If this chemical gets into the eyes, irrigate immediately. If this chemical contacts the skin, flush with water promptly. If a person breathes in large amounts of this chemical, move the exposed person to fresh air at once and perform artificial respiration. When this chemical has been swallowed, get medical attention. Give large quantities of salt water and induce vomiting. Do not make an unconscious person vomit.

Personal Protective Methods: Wear appropriate clothing to prevent repeated or prolonged skin contact. Wear eye protection to prevent any reasonable probability of eye contact. Remove clothing immediately if wet or contaminated to avoid flammability hazard.

Respirator Selection:
10,000 ppm: SA/SCBA
20,000 ppm: SAF/SCBAF/SAF:PD,PP,CF
Escape: GMOV/SCBA

Disposal Method Suggested: Incineration, preferably after mixing with another combustible fuel. Care must be exercised to assure complete combustion to prevent the formation of phosgene. An acid scrubber is necessary to remove the halo acids produced.

References

(1) U.S. Environmental Protection Agency, *Chemical Hazard Information Profile: Chloroethane,* Washington, DC (1979).
(2) U.S. Environmental Protection Agency, *Chloroethanes: Ambient Water Quality Criteria,* Washington, DC (1980).
(3) U.S. Environmental Protection Agency, *Chloroethane,* Health and Environmental Effects Profile No. 44, Washington, DC, Office of Solid Waste (April 30, 1980).
(4) Sax, N.I., Ed., *Dangerous Properties of Industrial Materials Report, 1,* No. 4, 64-66, New York, Van Nostrand Reinhold Co. (1981).
(5) See Reference (A-61).
(6) See Reference (A-60).
(7) United Nations Environment Programme, *IRPTC Legal File 1983,* Vol. I, pp VII/299–300, Geneva, Switzerland, International Register of Potentially Toxic Chemicals (1984).

ETHYLENE CHLOROHYDRIN

Description: CH_2ClCH_2OH, ethylene chlorohydrin, is a colorless liquid with an ether-like odor. It boils at $130°C$.

Code Numbers: CAS 107-07-3   RTEC KK0875000   UN 1135

DOT Designation: Flammable liquid.

Synonyms: Glycol chlorohydrin, 2-chloroethanol, β-chloroethyl alcohol.

Potential Exposures: Ethylene chlorohydrin is used in the synthesis of ethylene glycol, ethylene oxide, amines, carbitols, indigo, malonic acid, novocaine, and in other reactions where the hydroxyethyl group is introduced into organic compounds, for the separation of butadiene from hydrocarbon mixtures, in dewaxing and removing cycloalkanes from mineral oil, in the refining of rosin, in the manufacture of certain pesticides (A-32) and in the extraction of pine lignin. In the lacquer industry, it is used as a solvent for cellulose acetate, cellulose esters, resins and waxes, and in the dyeing and cleaning industry, it is used to remove tar spots, as a cleaning agent for machines, and as a solvent in fabric dyeing. It has also found use in agriculture in speeding up sprouting of potatoes and in treating seeds to inhibit biological activity.

Incompatibilities: Strong oxidizers, strong caustics.

Permissible Exposure Limits in Air: The Federal standard is 5 ppm (16 mg/m^3). The ACGIH (1983/84) has set 1 ppm (3 mg/m^3) as a ceiling value with the notation "skin" indicating the possibility of cutaneous absorption. NIOSH has recommended a TWA of 5 ppm (16 mg/m^3) and a 15-minute ceiling value of 15 ppm. ACGIH has no STEL value. The IDLH level is 10 ppm.

Determination in Air: Adsorption on charcoal, workup with isopropanol in CS$_2$, analysis by gas chromatography. See NIOSH Methods, Set H. See also reference (A-10).

Permissible Concentration in Water: No criteria set.

Routes of Entry: Inhalation of vapor, percutaneous absorption of liquid, ingestion, skin and eye contact.

Harmful Effects and Symptoms: *Local* — High vapor concentrations are irritating to the eyes, nose, throat, and skin.
Systemic — Ethylene chlorohydrin is extremely toxic and in addition to local irritation of eyes, respiratory tract, and skin, inhalation of the vapor may produce nausea, vomiting, dizziness, headache, thirst, delirium, low blood pressure, collapse, and unconsciousness. The urine may show red cells, albumin, and casts. Death may occur in high concentrations with damage to the lung and brain. There is little margin of safety between early reversible symptoms and fatal intoxication. The toxic effects may be related to its metabolites, chloroacetaldehdye and chloroacetic acid.

Points of Attack: Respiratory system, liver, kidneys, skin, central nervous system, cardiovascular system.

Medical Surveillance: Preplacement examination, including a complete history and physical should be performed. Examination of the respiratory system, liver, kidneys, and central nervous system should be stressed. The skin should be examined. A chest x-ray should be taken and pulmonary function tests performed (FVC-FEV).
The above procedures should be repeated on an annual basis, except that the x-ray is needed only when indicated by pulmonary function testing.

First Aid: If this chemical gets into the eyes, irrigate immediately. If this chemical contacts the skin, flush with water immediately. If a person breathes in large amounts of this chemical, move the exposed person to fresh air at once and perform artificial respiration. When this chemical has been swallowed, get medical attention. Give large quantities of salt water and induce vomiting. Do not make an unconscious person vomit.

Personal Protective Methods: Wear appropriate clothing to prevent any possibility of skin contact (beware: the liquid penetrates rubber). Wear eye protection to prevent any possibility of eye contact. Employees should wash immediately when skin is wet or contaminated. Remove nonimpervious clothing immediately if wet or contaminated. Provide emergency showers and eyewash.

Respirator Selection:
10 ppm: SAF/SCBAF
Escape: GMOV/SCBA

Disposal Method Suggested: Incineration, preferably after mixing with another combustible fuel. Care must be exercised to assure complete combustion to prevent the formation of phosgene. An acid scrubber is necessary to remove the halo acids produced.

References
(1)   See Reference (A-61).

ETHYLENEDIAMINE

- Hazardous substance (EPA)
- Hazardous waste (EPA)

Description: $H_2NCH_2CH_2NH_2$, ethylenediamine, is a strongly alkaline, colorless, clear, thick liquid with an ammonia odor. It boils at $116°$ to $117°C$.

Code Numbers: CAS 107-15-3   RTECS KH8575000   UN 1604

DOT Designation: Corrosive material.

Synonyms: Ethanediamine, 1,2-diaminoethane.

Potential Exposures: Ethylenediamine is used as a solvent, an emulsifier for casein and shellac solutions, a stabilizer in rubber latex, a chemical intermediate in the manufacture of dyes, corrosion inhibitors, synthetic waxes, fungicides, resins, insecticides, asphalt wetting agents, and pharmaceuticals (A-41).

Ethylenediamine is a degradation product of the agricultural fungicide Maneb.

Incompatibilities: Strong acids, strong oxidizers, chlorinated organic compounds.

Permissible Exposure Limits in Air: The Federal standard and the 1983/84 ACGIH TWA value is 10 ppm (25 mg/m^3). There are no STEL values set. The IDLH level is 2,000 ppm.

Determination in Air: Adsorption on silica gel, elution with methanolic acid, derivatization with benzaldehyde, analysis by gas chromatography (A-10).

Permissible Concentration in Water: No criteria set.

Routes of Entry: Inhalation of vapor, percutaneous absorption, ingestion and skin and eye contact.

Harmful Effects and Symptoms: *Local* — Ethylenediamine vapor may cause irritation of the nose and tingling of the face. Cutaneous sensitivity has been reported.

In animal experiments, the liquid has produced severe irritation of the eyes and corneal damage. It has also produced severe irritation and necrosis.

Systemic — In animal experiments, high concentrations of ethylenediamine vapor have produced damage to liver, lungs, and kidneys.

Points of Attack: Respiratory system, liver, kidneys, skin.

Medical Surveillance: Consider possible irritant effects on skin, eyes and respiratory system. History of allergic redness of skin or asthmatic symptoms may be important in placement and periodic examinations.

First Aid: If this chemical gets into the eyes, irrigate immediately. If this chemical contacts the skin, flush with water immediately. If a person breathes in large amounts of this chemical, move the exposed person to fresh air at once and perform artificial respiration. When this chemical has been swallowed, get medical attention. Give large quantities of water and induce vomiting. Do not make an unconscious person vomit.

Personal Protective Methods: Wear appropriate clothing to prevent any possibility of skin contact. Wear eye protection to prevent any possibility of eye contact. Employees should wash immediately when skin is wet or contaminated and daily at the end of each work shift. Work clothing should be changed daily if it is possible that clothing is contaminated. Remove nonimpervious clothing immediately if wet or contaminated. Provide emergency showers and eyewash if liquids containing >5% contaminants are involved.

Respirator Selection:
     500 ppm: CCRSF/GMS/SAF/SCBAF
  2,000 ppm: SAF:PD,PP,CF
     Escape: GMS/SCBA

Disposal Method Suggested: Controlled incineration (oxides of nitrogen are removed from the effluent gas by scrubbers and/or thermal devices).

References
(1) U.S. Environmental Protection Agency, *Chemical Hazard Information Profile: Ethylenediamine,* Washington, DC (May 9, 1978).
(2) See Reference (A-61).
(3) See Reference (A-60).
(4) Sax, N.I., Ed., *Dangerous Properties of Industrial Materials Report, 4,* No. 2, 54–57, New York, Van Nostrand Reinhold Co. (1984).

ETHYLENE DIAMINE TETRA-
(METHYLENE PHOSPHONIC ACID)

Description: $C_6H_{20}N_2O_{12}P_4$ has the structural formula

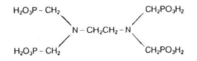

EDTMPA is a white, crystalline solid.

Code Numbers: CAS 1429-50-1   RTECS –

DOT Designation: –

Synonyms: Phosphonic acid, [ethylenebis(nitrilodimethylene)] tetra-

Potential Exposure: About one-third of the EDTMPA sold in the U.S. is used as a stabilizer in hydrogen peroxide, one-third is used by the precious metals industry in plating applications, and one-third is used in miscellaneous applications, primarily in metal chelating applications.

Permissible Exposure Limits in Air: No limits set. Worker exposure during the manufacturing of EDTMPA and its salts should be low as a result of the controls needed to maintain levels of the reactant formaldehyde within the recommended limits of 1 ppm (TLV-TWA) and 2 ppm (TLV-STEL).

Permissible Concentration in Water: No criteria set. Once released into the environment, the compounds are reported to be stable to hydrolysis as are other organophosphonates.

Harmful Effects and Symptoms: Osteosarcomas with metastases were observed in both male and female Sprague-Dawley rats that had died during the first 12 months of EDTMPA treatment at doses ranging from 50–333 mg/kg/day for 30 months.

EDTMPA and its tetrapotassium salt are of low acute toxicity by ingestion and skin absorption, and are considered to be non-irritating to the skin. EDTMPA, as a solid, is rated as moderately irritating to the eye, while the tetrapotassium salt, an alkaline solution, is considered to be slightly irritating.

Points of Attack: Bone, blood, skin, eyes.

References

(1) U.S. Environmental Protection Agency, *Chemical Hazard Information Profile Draft Report: Ethylenediaminetetra(methylenephosphonic acid) and its salts,* Wash., D.C. (Feb. 22, 1984).

ETHYLENE DIBROMIDE

- Carcinogen (animal positive, IARC) (4)
- Hazardous substance (EPA)
- Hazardous waste (EPA)

Description: $BrCH_2CH_2Br$ is a colorless nonflammable liquid with a chloroform-like odor. It boils at $131°C$.

Code Numbers: CAS 106-93-4   RTECS KH9275000   UN 1605

DOT Designation: ORM-A

Synonyms: 1,2-Dibromoethane, ethylene bromide, sym-dibromoethane.

Potential Exposures: 1,2-Dibromoethane is used principally as a fumigant for ground pest control and as a constituent of ethyl gasoline. It is also used in fire extinguishers, gauge fluids, and waterproofing preparations; and it is used as a solvent for celluloid, fats, oils, and waxes.

The concentration of ethylene dibromide in leaded gasoline is variable, but is on the order of 0.025% (wt/vol). (Ethylene dichloride is also used in admixture with DBE.) A major source of ethylene dibromide and dichloride emissions are from automotive sources via evaporation from the fuel tank and carburetor of cars operated on leaded fuel.

It should be noted that the increased use of unleaded gasoline should result in lower ambient air levels of ethylene dibromide from its major sources of emissions (1).

Ethylene dibromide has also been found in concentrations of 96 $\mu g/m^3$, up to a mile away from a U.S. Dept. of Agriculture fumigation center.

Concentrations of ethylene dibromide on the order of 1 ppm have been found in samples from streams of water on industrial sites. Limited information suggests that ethylene dibromide degrades at moderate rates in both water and soil. The use of ethylene dichloride and dibromide in fumigant mixtures of disinfecting fruits, vegetables, food grains, tobacco, seeds, seed beds, mills and warehouses, suggests the possibility that their residues per se or that of their respective hydrolytic products (e.g., ethylene chlorohydrin or bromohydrin) may be present in fumigated materials.

Although materials such as ethylene dichloride and dibromide are volatile, their actual occurrence in processed or cooked foods can possibly be considered negligible. More significant exposure is considered more likely among agricultural workers or those fumigating grain and crops in storage facilities and the field, than among consumers of the food products.

Human exposure is primarily through inhalation, ingestion, or skin absorption. NIOSH estimates that approximately 9,000 employees at production facilities are potentially exposed to ethylene dibromide. OSHA estimates that 218,000 workers are potentially exposed in processes utilizing this compound. The National Occupational Hazard Survey (1974) indicated that 126,000 workers may be exposed. In addition to these potential industrial exposures, NIOSH (1977) estimated that 650,000 gas station attendants may have a potential risk of exposure.

Incompatibilities: Chemically active metals—sodium, potassium, calcium, powdered aluminum, zinc, magnesium, liquid ammonia; strong oxidizers.

Permissible Exposure Limits in Air: The Federal standard is 20 ppm (145 mg/m^3) as an 8-hour TWA with an acceptable ceiling concentration of 30 ppm; acceptable maximum peaks above the ceiling of 50 ppm are allowed for 5 minutes duration. NIOSH has recommended as of 1977 that the employer shall control workplace concentrations of ethylene dibromide so that no employee is exposed in his workplace to concentrations greater than 1.0 mg/m^3 (0.13 ppm) as a ceiling limit, as determined by a sampling period of 15 minutes. ACGIH has cited ethylene dibromide as "an industrial substance suspect of carcinogenic potential for man" with no set TLV value. In November 1978, NCI called EDB "the most potent cancer-causing substance ever found in the animal test program of NCI." Industry sources rebutted this and a decision by EPA on use restriction was pending.

A rebuttable presumption against registration for EDB for pesticide uses was issued on December 14, 1977 by EPA on the basis of oncogenicity, mutagenicity and reproductive effects (A-32).

All soil fumigant use of EDB was suspended by EPA in September 1983. As of September 1983, a new rule limiting EDB to 100 ppb (0.1 ppm) is under consideration by OSHA.

Determination in Air: Charcoal adsorption, workup with CS$_2$, analysis by gas chromatography. See NIOSH Methods, Set H. See also reference (A-10).

Permissible Concentration in Water: No criteria set.

Routes of Entry: Inhalation of the vapor, absorption through the skin, ingestion and eye and skin contact.

Harmful Effects and Symptoms: *Local* — Prolonged contact of the liquid with the skin may cause erythema, blistering, and skin ulcers. These reactions may be delayed 24 to 48 hours. Dermal sensitization to the liquid may develop. The vapor is irritating to the eyes and to the mucous membranes of the respiratory tract.

Systemic — Inhalation of the vapor may result in severe acute respiratory injury, central nervous system depression, and severe vomiting. Animal experiments have produced injury to the liver and kidneys.

When inhaled, 1,2-dibromoethane was carcinogenic for rats, causing increased incidences of tumors in the nasal cavity, circulatory system, and pituitary of males and females; mesotheliomas in the tunica vaginalis of males; and alveolar/bronchiolar carcinomas or adenomas, and fibroadenomas of the mammary gland of females.

1,2-Dibromoethane was also carcinogenic for mice, causing increased incidences of hemangiosarcomas of the circulatory system and alveolar/bronchiolar carcinomas or adenomas in males and females and fibrosarcomas in the subcutaneous tissue, tumors of the nasal cavity, and adenocarcinomas of the mammary gland in females (7) (8).

Points of Attack: Respiratory system, liver, kidneys, skin, eyes.

Medical Surveillance: Preemployment and periodic examinations should evaluate the skin and eyes, respiratory tract, and liver and kidney functions.

First Aid: If this chemical gets into the eyes, irrigate immediately. If this chemical contacts the skin, wash with soap immediately. If a person breathes in large amounts of this chemical, move the exposed person to fresh air at once and perform artificial respiration. When this chemical has been swallowed, get medical attention. Give large quantities of salt water and induce vomiting. Do not make an unconscious person vomit.

Personal Protective Methods: Wear appropriate clothing to prevent repeated or prolonged skin contact. Wear eye protection to prevent any reasonable probability of eye contact. Employees should wash immediately when skin is wet or contaminated. Remove nonimpervious clothing immediately if wet or contaminated. Provide emergency showers.

Respirator Selection:

10 mg/m^3:	SA/SCBA
50 mg/m^3:	SAF/SCBAF
2,000 mg/m^3:	SAF:PD,PP,CF
Escape:	GMOV/SCBA

Disposal Method Suggested: Controlled incineration with adequate scrubbing and ash disposal facilities.

References

(1) Environmental Protection Agency, *Sampling and Analysis of Selected Toxic Substances, Task II—Ethylene Dibromide,* Final Report. Office of Toxic Substances, EPA, Washington, DC (September 1975).

(2) Occupational Health and Safety Administration, *Criteria for a Recommended Standard: Occupational Exposure to Ethylene Dibromide,* NIOSH Doc. No. 77-221 (1977).

(3) Nat. Inst. for Occup. Safety and Health, *Current Intelligence Bulletin No. 3: Ethylene Dibromide,* Rockville, MD (July 7, 1975).

(4) International Agency for Research on Cancer, *IARC Monographs on the Carcinogenic Risks of Chemicals to Humans,* Lyon, France 15, 195 (1977).

(5) Sax, N.I., Ed., *Dangerous Properties of Industrial Materials Report, 1,* No. 5, 58-60, New York, Van Nostrand Reinhold Co. (1981).

(6) See Reference (A-61).

(7) National Cancer Institute, *Bioassay of 1,2-Dibromoethane for Possible Carcinogenicity,* DHHS Publication No. (NIH) 80-1766, National Technical Information Service, Springfield, VA (1980).

(8) See Reference (A-62). Also see Reference (A-64).

(9) Nat. Inst. for Occup. Safety & Health, Current Intelligence Bulletin No. 37, *Ethylene Dibromide,* Cincinnati, Ohio (Oct. 26, 1981).

(10) Sax, N.I., Ed., *Dangerous Properties of Industrial Materials Report,* Special Bulletin, New York, Van Nostrand Reinhold Co. (March 1984).

(11) Parmeggiani, L., Ed., *Encyclopedia of Occupational Health & Safety,* Third Edition, Vol. 1, pp 623–24, Geneva, International Labour Office (1983).

(12) United Nations Environment Programme, *IRPTC Legal File 1983,* Vol. I, pp VII/301–4, Geneva, Switzerland, International Register of Potentially Toxic Chemicals (1984).

ETHYLENE DICHLORIDE

- Carcinogen (positive, NCI) (7) (IARC) (9)
- Hazardous substance (EPA)
- Hazardous waste (EPA)
- Priority toxic pollutant (EPA)

Description: $ClCH_2CH_2Cl$, 1,2-dichloroethane, is a colorless, flammable liquid which has a pleasant odor, sweetish taste. It boils at 84°C.

Code Numbers: CAS 107-06-2 RTECS KI0525000 UN 1184

DOT Designation: Flammable liquid.

Synonyms: 1,2-Dichloroethane, sym-dichloroethane, ethylene chloride, glycol dichloride.

Potential Exposures: In recent years, 1,2-dichloroethane has found wide use in the manufacture of ethylene glycol, diaminoethylene, polyvinyl chloride, nylon, viscose rayon, styrene-butadiene rubber, and various plastics. It is a solvent for resins, asphalt, bitumen, rubber, cellulose acetate, cellulose ester, and paint; a degreaser in the engineering, textile and petroleum industries; and an extracting agent for soybean oil and caffeine. It is also used as an antiknock agent in gasoline, a pickling agent, a fumigant, and a drycleaning agent. It has found use in photography, xerography, water softening, and also in the production of adhesives, cosmetics, pharmaceuticals, and varnishes.

In an early document (1) issued in 1976 NIOSH estimated that 18,000 workers were potentially exposed to ethylene dichloride. In a subsequent document issued in 1978 (2) NIOSH estimates that as many as 2 million workers may have occupational exposure to ethylene dichloride. Of these workers, an estimated 34,000 are exposed to ethylene dichloride 4 hours or more per day.

Incompatibilities: Strong oxidizers, strong caustics, chemically active metals, such as aluminum or magnesium powder, sodium, potassium.

Permissible Exposure Limits in Air: The current Federal standard for ethylene dichloride is 50 parts of ethylene dichloride per million parts of air (ppm) averaged over an eight-hour work shift, with a ceiling level of 100 ppm and a maximum acceptable peak of 200 ppm for 5 minutes in any three-hour period. NIOSH has recommended that the permissible exposure limit be reduced to 5 ppm averaged over a work shift of 10 hours per day, 40 hours per week, with a ceiling level of 15 ppm averaged over a 15-minute period. The NIOSH Criteria Document for Ethylene Dichloride (1) should be consulted for more detailed information.

Neither of these levels may provide adequate protection from potential carcinogenic effects because they were selected to prevent toxic effects other than cancer.

As an interim and prudent measure while the carcinogenicity of ethylene dichloride is being further evaluated, NIOSH recommends (2) that occupational exposure be minimized. Exposures should be limited to as few employees as possible, while minimizing workplace exposure levels with engineering and work practice controls.

The ACGIH has recommended a TWA value of 10 ppm (40 mg/m³) and an STEL value of 15 ppm (60 mg/m³) as of 1983/84. The IDLH level is 1,000 ppm.

Determination in Air: Charcoal adsorption, workup with CS_2, analysis by gas chromatography. See NIOSH Methods, Set J. See also reference (A-10).

Permissible Concentration in Water: To protect freshwater aquatic life— 118,000 μg/ℓ on an acute toxicity basis and 20,000 μg/ℓ on a chronic basis. To protect saltwater aquatic life—113,000 μg/ℓ on an acute toxicity basis. To protect human health—preferably zero. An additional lifetime cancer risk of 1 in 100,000 occurs at a concentration of 9.4 μg/ℓ.

Determination in Water: Inert gas purge followed by gas chromatography with halide specific detection (EPA Method 601) or gas chromatography plus mass spectrometery (EPA Method 624).

Routes of Entry: Inhalation of vapor, skin absorption of liquid, ingestion, eye and skin contact.

Harmful Effects and Symptoms: *Local* — Repeated contact with liquid can produce a dry, scaly, fissured dermatitis. Liquid and vapor may also cause eye damage, including corneal opacity.

Systemic — Inhalation of high concentrations may cause nausea, vomiting, mental confusion, dizziness, and pulmonary edema. Chronic exposure has been associated with liver and kidney damage.

Acute exposures can lead to death from respiratory and circulatory failure. Autopsies in such situations have revealed widespread bleeding and damage in most internal organs. Repeated long-term exposures to ethylene dichloride has resulted in neurologic changes, loss of appetite and other gastrointestinal problems, irritation of the mucous membranes, liver and kidney impairment, and death (1).

1,2-Dichloroethane was tested in one experiment in mice and in one in rats by oral administration (7). In mice, it produced benign and malignant tumors of the lung and malignant lymphomas in animals of both sexes, hepatocellular carcinomas in males and mammary and uterine adenocarcinomas in females. In rats, it produced carcinomas of the forestomach in male animals, benign and malignant mammary tumors in females and haemangiosarcomas in animals of both sexes (9) (10).

Points of Attack: Kidneys, liver, eyes, skin, central nervous sytem.

Medical Surveillance: Preplacement and periodic examinations should include an evaluation of the skin and liver and kidney functions.

First Aid: If this chemical gets into the eyes, irrigate immediately. If this chemical contacts the skin, wash with soap promptly. If a person breathes in large amounts of this chemical, move the exposed person to fresh air at once and perform artificial respiration. When this chemical has been swallowed, get medical attention. Give large quantities of salt water and induce vomiting. Do not make an unconscious person vomit.

Personal Protective Methods: Wear appropriate clothing to prevent repeated or prolonged skin contact. Wear eye protection to prevent any reasonable probability of eye contact. Employees should wash promptly when skin is wet or contaminated. Remove clothing immediately if wet or contaminated to avoid flammability hazard.

Respirator Selection:
      50 ppm: SA/SCBA
    250 ppm: SAF/SCBAF
    Escape: GMOV/SCBA
      Note: For eye irritation, use full facepiece

Disposal Method Suggested: Incinerationn, preferably after mixing with another combustible fuel. Care must be exercised to assure complete combustion to prevent the formation of phosgene. An acid scrubber is necessary to remove the halo acids produced. Alternatively, ethylene dichloride may be recovered from process off gases (A-58).

References

(1) National Institute for Occupational Safety and Health, *Criteria for a Recommended Standard: Occupational Exposure to Ethylene Dichloride,* NIOSH Doc. No. 76-139 (1976).

(2) National Institute for Occupational Safety and Health, *Ethylene Dichloride,* NIOSH Current Intelligence Bulletin No. 25, Washington, DC (April 19, 1978).

(3) National Institute for Occupational Safety and Health, *Chloroethanes: Review of Toxicity,* Current Intelligence Bulletin No. 27, Washington, DC (August 21, 1978).

(4) U.S. Environmental Protection Agency, *Chemical Hazard Information Profile: 1,2-Dichloroethane,* Washignton, DC (September 1, 1977).

(5) U.S. Environmental Protection Agency, *Chlorinated Ethanes: Ambient Water Quality Criteria,* Washington, DC (1980).

(6) U.S. Environmental Protection Agency, *1,2-Dichloroethane,* Health and Environmental Effects Profile No. 70, Washington, DC, Office of Solid Waste (April 30, 1980).

(7) National Cancer Institute, *Bioassay of 1,2-Dichloroethane for Possible Carcinogenicity,* Technical Report Series No. 55, Bethesda, MD (1978).

(8) Sax, N.I., Ed., *Dangerous Properties of Industrial Materials Report, 1,* No. 4, 50-52, New York, Van Nostrand Reinhold Co. (1981).

(9) *IARC Monographs on the Evaluation of Carcinogenic Risk of Chemicals to Man,* Vol. 20, IARC, Lyon, France, pp 429-444 (1979).

(10) See Reference (A-62). Also see Reference (A-64).

(11) See Reference (A-60).

(12) Parmeggiani, L., Ed., *Encyclopedia of Occupational Health & Safety,* Third Edition, Vol. 1, pp 794-96, Geneva, International Labour Office (1983).

(13) United Nations Environment Programme, *IRPTC Legal File 1983,* Vol. I, pp VII/305-8, Geneva, Switzerland, International Register of Potentially Toxic Chemicals (1984).

ETHYLENE GLYCOL

Description: $HOCH_2CH_2OH$, ethylene glycol, is a colorless, odorless, viscous liquid with a sweetish taste.

Code Numbers: CAS 107-21-1 RTECS KW2975000

DOT Designation: —

Synonyms: 1,2-Ethanediol, glycol alcohol, glycol, EG.

Potential Exposures: Because of ethylene glycol's physical properties, it is used in antifreeze, hydraulic fluids, electrolytic condensors, and heat exchangers. It is also used as a solvent and as a chemical intermediate for ethylene glycol dinitrate, glycol esters, resins, and for pharmaceuticals (A-41).

Permissible Exposure Limits in Air: There is no Federal standard; however, ACGIH in 1983/84 has recommended a TLV of 10 mg/m³ for particulate ethylene glycol and an STEL of 20 mg/m³. A ceiling value of 50 ppm (125 mg/m³) was adopted for the vapor. There was no STEL set for the vapor. Consideration is being given to deletion of the particulate TWA value.

Permissible Concentration in Water: No criteria set, but EPA (A-37) has suggested a permissible ambient goal of 140 μg/ℓ based on health effects.

Routes of Entry: Inhalation of particulate or vapor. Percutaneous absorption may also contribute to intoxication.

Harmful Effects and Symptoms: *Local* — None.

Systemic — Ethylene glycol's vapor pressure is such that at room temperature toxic concentrations are unlikely to occur. Poisoning resulting from vapor usually occurs only if ethylene glycol liquid is heated; therefore, occupational exposure is rare. Chronic symptoms and signs include: anorexia, oliguria, nystagmus, lymphocytosis, and loss of consciousness. Inhalation seems to primarily result in central nervous system depression and hematopoietic dysfunction, whereas, ingestion may result in depression followed by respiratory and cardiac failure, renal and brain damage.

Medical Surveillance: Perform functional tests of kidneys and liver in animal examinations.

Urinalysis for oxalic acid, an ethylene glycol metabolite, may be useful in diagnosis of poisoning by oral ingestion.

First Aid: Flush eyes with water. Wash contaminated areas of body with soap and water. If swallowed, gastric lavage followed by saline catharsis.

Personal Protective Methods: Masks should be worn in areas of vapor concentration.

Disposal Method Suggested: Incineration. Alternatively, ethylene glycol can be recovered from polyester plant wastes (A-58).

References

(1) See Reference (A-60).
(2) Parmeggiani, L., Ed., *Encyclopedia of Occupational Health & Safety,* Third Edition, Vol. 1, pp 973–76, Geneva, International Labour Office (1983) (Glycols and Derivatives).
(3) Sax, N.I., Ed., *Dangerous Properties of Industrial Materials Report, 4,* No. 3, 70–74, New York, Van Nostrand Reinhold Co. (1984).

ETHYLENE GLYCOL DINITRATE

Description: $O_2NOCH_2CH_2ONO_2$ is a colorless to yellow oily liquid. It may be detonated by mechanical shock, heat, or spontaneous chemical reaction.

Code Numbers: CAS 628-96-6   RTECS KW5600000

DOT Designation: —

Synonyms: Nitroglycol, glycol dinitrate, ethanediol dinitrate, EGDN, ethylene dinitrate.

Potential Exposures: Although ethylene glycol dinitrate is an explosive in itself, it is primarily used to lower the freezing point of nitroglycerin; together these compounds are the major constituents of commercial dynamite, cordite, and blasting gelatin. Occupational exposure generally involves a mixture of the two compounds. Ethylene glycol dinitrate is 160 times more volatile than nitroglycerin.

Incompatibilities: Acids, heat, mechanical shock.

Permissible Exposure Limits in Air: The Federal standard for ethylene glycol dinitrate and/or nitroglycerin is 0.2 ppm (2 mg/m³) as a ceiling value, and, at concentrations greater than 0.02 ppm, personal protection may be necessary to avoid headache. These levels should be reduced when the substance is also absorbed percutaneously.

The ACGIH has set a TWA value of 0.05 ppm (0.3 mg/m³) and an STEL of 0.1 ppm (0.6 mg/m³). The IDLH level is 80 ppm.

Determination in Air: Adsorption on Tenax, workup with n-propanol, analysis by gas chromatography. See NIOSH Methods, Set P. See also reference (A-10).

Permissible Concentration in Water: No criteria set.

Routes of Entry: Inhalation of dust or vapor; ingestion of dust; percutaneous absorption, eye and skin contact.

Harmful Effects and Symptoms: *Local* — None reported.

Systemic — Exposure to small amounts of ethylene glycol dinitrate and/or nitroglycerin by skin exposure, inhalation, or swallowing may cause severe throbbing headaches. With larger exposure, nausea, vomiting, cyanosis, palpitations of the heart, coma, cessation of breathing, and death may occur. A temporary tolerance to the headache may develop, but this is lost after a few days without exposure. On some occasions a worker may have anginal pains a few days after discontinuing repeated daily exposure.

Points of Attack: Skin, blood, cardiovascular system.

Medical Surveillance: Placement and periodic examinations should be concerned with central nervous system, blood, glaucoma, and especially history of alcoholism.

Urinary and blood ethylene glycol dinitrate may be determined by gas chromatography.

First Aid: If this chemical gets into the eyes, irrigate immediately. If this chemical contacts the skin, wash with soap immediately. If a person breathes in large amounts of this chemical, move the exposed person to fresh air at once and perform artificial respiration. When this chemical has been swallowed, get medical attention. Give large quantities of water and induce vomiting. Do not make an unconscious person vomit.

Personal Protective Methods: Both EGDN and nitroglycerin are readily absorbed through the skin, lungs, and mucous membranes. It is, therefore, essential that adequate skin protection be provided for each worker: impervious clothing where liquids are likely to contaminate and full body clothing where dust creates the problem. All clothing should be discarded at the end of the shift and clean work clothing provided each day. Showers should be taken at the end of each shift and prior to changing to street clothing. In case of spill or splash that contaminates work clothing, the clothes should be changed at once and the skin area washed thoroughly. Masks of the dust type or organic vapor canister type may be necessary in areas of concentration of dust or vapors.

Respirator Selection:
        2 ppm: SA/SCBA
       10 ppm: SAF/SCBAF
       80 ppm: SA:PD,PP,CF
      Escape: GMOVP/SCBA

Disposal Method Suggested: Controlled incineration in the scrubber equipped Deactivation Furnace incinerator (The Chemical Agent Munition Disposal System) (A-31). Also, ethylene glycol dinitrate can be recovered from wastewaters (A-58).

References

(1) Parmeggiani, L., Ed., *Encyclopedia of Occupational Health & Safety,* Third Edition, Vol. 1, pp 796–97, Geneva, International Labour Office (1983).

ETHYLENE GLYCOL MONOMETHYL ETHER AND ACETATE

Description: $CH_3OCH_2CH_2OH$, ethylene glycol monomethyl ether, boils at 124° to 125°C. $CH_3OCH_2CH_2OOCCH_3$, ethylene glycol monomethyl ether acetate, boils at 145°C. These substances are colorless liquids with a slight odor.

Code Numbers:

Ether	CAS 109-86-4	RTECS KL5775000	UN 1188
Ether acetate	CAS 110-49-6	RTECS KL5950000	UN 1189

DOT Designation: Combustible liquid.

Synonyms: Ethylene glycol monomethyl ether—methyl Cellosolve, 2-methoxyethanol. Ethylene glycol monomethyl ether acetate—methyl Cellosolve acetate, 2-methoxyethyl acetate.

Potential Exposures: Ethylene glycol ethers are used as solvents for resins, lacquers, paints, varnishes, gum, perfume, dyes and inks, and as a constituent of painting pastes, cleaning compounds, liquid soaps, cosmetics, nitrocellulose, and hydraulic fluids. Acetate derivatives are used as solvents for oils, greases and ink, in the preparation of lacquers, enamels, and adhesives, and to dissolve resins and plastics.

Incompatibilities: Ethylene glycol monomethyl ether—strong oxidizers, strong caustics. Ethylene glycol monomethyl ether acetate—nitrates, strong oxidizers, strong alkalies, strong acids.

Permissible Exposure Limits in Air: The Federal standards for these compounds are ethylene glycol monomethyl ether, 25 ppm (80 mg/m^3); ethylene glycol monomethyl ether acetate, 25 ppm (120 mg/m^3) (skin absorption possible in both cases). ACGIH (1983/84) has adopted an STEL of 35 ppm (120 mg/m^3) for the ether and 35 ppm (170 mg/m^3) for the acetate. They have proposed lowering the TWA to 5 ppm (16 mg/m^3) for the ether and 5 ppm (24 mg/m^3) for the acetate with no STEL levels. The IDLH levels are 2,000 ppm for the ether and 4,500 ppm for the acetate.

Determination in Air: *Ethylene Glycol Monomethyl Ether* — Adsorption on charcoal, workup with methanol/methylene chloride, analysis by gas chromatography. See NIOSH Methods, Set F.

Ethylene Glycol Monomethyl Ether Acetate — Adsorption on charcoal, workup with CS$_2$, analysis by gas chromatography. See NIOSH Methods, Set D.

Permissible Concentration in Water: No criteria set.

Routes of Entry: Inhalation of vapor, ingestion and eye and skin contact (plus cutaneous absorption of liquid in the case of the ether only).

Harmful Effects and Symptoms: *Local* — Ethylene glycol ethers are only midly irritating to the skin. Vapor may cause conjunctivitis and upper respiratory tract irritation. Temporary corneal clouding may also result and may last several hours. Acetate derivatives cause greater eye irritation than the parent compounds. The butyl and methyl ethers may penetrate skin readily.

Systemic — Acute exposure to these compounds results in narcosis, pulmonary edema, and severe kidney and liver damage. Symptoms from repeated overexposure to vapors are fatigue and lethargy, headache, nausea, anorexia, and tremor. Anemia and encephalopathy have been reported with ethylene glycol monomethyl ether. Acute poisoning by ingestion resembles ethylene glycol toxicity, with death from renal failure.

These compounds have the potential to cause adverse reproductive effects in male and female workers (3).

Points of Attack: For ethylene glycol monomethyl ether—central nervous system, blood, skin, eyes, kidneys. For ethylene glycol monomethyl ether acetate—kidneys, brain, central nervous system, peripheral nervous system.

Medical Surveillance: Preplacement and periodic examinations should evaluate blood, central nervous system, renal and liver functions, as well as the skin and respiratory tract.

First Aid: If this chemical gets into the eyes, irrigate immediately. If this chemical contacts the skin, flush with water promptly. If a person breathes in large amounts of this chemical, move the exposed person to fresh air at once and perform artificial respiration. When this chemical has been swallowed, get medical attention. Give large quantities of salt water and induce vomiting. Do not make an unconscious person vomit.

Personal Protective Methods: Wear appropriate clothing to prevent repeated or prolonged skin contact. Wear eye protection to prevent any reasonable probability of eye contact. Employees should wash immediately when skin is wet or contaminated. Remove nonimpervious clothing promptly if wet or contaminated. Provide emergency showers.

Respirator Selection:
For Ethylene Glycol Monomethyl Ether —
      250 ppm: CCROV/SA/SCBA
    1,250 ppm: SAF/SCBAF
    2,000 ppm: SAF:PD,PP,CF
      Escape: GMOV/SCBA
For Ethylene Glycol Monomethyl Ether Acetate —
      250 ppm: CCROV/SA/SCBA
    1,000 ppm: CCROVF/GMOV/SAF/SBAF
    4,500 ppm: SAF:PD,PP,CF
      Escape: GMOV/SCBA

Disposal Method Suggested: Concentrated waste containing no peroxides: discharge liquid at a controlled rate near a pilot flame. Concentrated waste containing peroxides: perforation of a container of the waste from a safe distance followed by open burning.

References
(1) Nat. Inst. for Occup. Safety and Health, *Profiles on Occupational Hazards for Criteria Document Priorities: Glycol Ethers,* pp 110-115, Report PB 274,073, Cincinnati, OH (1977).
(2) See Reference (A-61).
(3) National Institute for Occupational Safety & Health, *Glycol Ethers: 2-Methoxyethanol and 2-Ethoxyethanol,* Current Intelligence Bulletin No. 39, Cincinnati, Ohio (May 2, 1983).
(4) Sax, N.I., Ed., *Dangerous Properties of Industrial Materials Report, 4,* No. 2, 67–70, New York, Van Nostrand Reinhold Co. (1984).
(5) United Nations Environment Programme, *IRPTC Legal File 1983,* Vol. I, pp VII/339–40, Geneva, Switzerland, International Register of Potentially Toxic Chemicals (1984).

ETHYLENEIMINE

- Carcinogen (animal positive, IARC) (2)
- Hazardous waste (EPA)

Description: H_2CNHCH_2, ethyleneimine, is a colorless volatile liquid with an ammoniacal odor. It boils at 56° to 57°C.

Code Numbers: CAS 151-56-4   RTECS KX5075000   UN 1185

DOT Designation: Flammable liquid and poison.

Synonyms: Azacyclopropane, aziridine, dimethyleneimine, ethylenimine, vinylamine, azirane, dihydroazirine, EI.

Potential Exposures: Ethyleneimine is a highly reactive compound and is used in many organic syntheses. The polymerization products, polyethyleneimines, are used as auxiliaries in the paper industry and as flocculation aids in the clarification of effluents. It is also used in the textile industry for increasing wet strength, flameproofing, shrinkproofing, stiffening, and waterproofing.

Incompatibilities: Acids or acid-forming substances.

Permissible Exposure Limits in Air: Ethyleneimine was included in the Federal standard for carcinogens; all contact with it should be avoided. ACGIH (1983/84) set a TWA of 0.5 ppm (1.0 mg/m^3) with no caution on carcinogenic nature but with the notation "skin" indicating the possibility of cutaneous absorption. No STEL value is proposed.

Permissible Concentration in Water: No criteria set, but EPA (A-37) has suggested a permissible ambient goal of 14 μg/ℓ based on health effects.

Routes of Entry: Inhalation and percutaneous absorption, ingestion, skin and eye contact.

Harmful Effects and Symptoms: *Local* — The vapor is strongly irritating to the conjunctiva and cornea, the mucous membranes of the nose, throat, and upper respiratory tract, and the skin. The liquid is a severe irritant and vesicant in humans, and severe eye burns have followed contact with the cornea. Skin sensitization has occurred.

Systemic — Actue exposures in humans have caused nausea, vomiting, headaches, dizziness, and pulmonary edema. In mice acute lethal exposures to vapor produced pulmonary edema, renal damage, and hematuria.

Points of Attack: Skin, eyes, respiratory tract.

Medical Surveillance: Based partly on animal experimental data, examinations should include history of exposure to other carcinogens, smoking, alcohol, medications, and family history. The skin, eye, lung, liver, and kidney should be evaluated. Sputum or urine cytology may be helpful.

First Aid: Flush eyes with water. Wash contaminated areas of body with soap and water.

Personal Protective Methods: These are designed to supplement engineering control and prevent all skin or respiratory exposure. Full body protective clothing and gloves should be used. Full face supplied air respirators with continuous flow or pressure demand type should also be used. Eyes should be protected at all times. On exit from a regulated area employees should shower and change to street clothes, leaving their protective clothing and equipment at the point of exit, to be placed in impervious containers at the end of the work shift for decontamintion or disposal.

Respirator Selection: Use self-contained breathing apparatus.

Disposal Method Suggested: Controlled incineration; incinerator is equipped with a scrubber or thermal unit to reduce NO$_x$ emissions.

References

(1)   Dermer, O.C. and Ham, G.E., *Ethyleneimine and Other Aziridines,* Academic Press, NY (1969).

(2)   International Agency for Research on Cancer, *IARC Monographs on the Carcinogenic Risks of Chemicals to Humans,* Lyon, France 9, 37 (1975).
(3)   Sax, N.I., Ed., *Dangerous Properties of Industrial Materials Report, 1,* No. 2, 37-38, New York, Van Nostrand Reinhold Co. (1980).
(4)   See Reference (A-60) under Aziridine.

ETHYLENE OXIDE

- Potential Occupational Carcinogen (3)
- Hazardous waste (EPA)

Description: H_2COCH_2, ethylene oxide, is a colorless gas with a sweetish odor.

Code Numbers: CAS 75-21-8   RTECS KX2450000   UN 1040

DOT Designation: Flammable liquid.

Synonyms: 1,2-Epoxyethane, oxirane, dimethylene oxide, anprolene.

Potential Exposures: Ethylene oxide is used as an intermediate in organic synthesis for ethylene glycol, polyglycols, glycol ethers, esters, ethanolamines, acrylonitrile, plastics, and surface-active agents. It is also used as a fumigant for foodstuffs and textiles, an agricultural fungicide, and for sterilization, especially for surgical instruments. It is used in drug synthesis (A-41) and as a pesticide intermediate (A-32).

Incompatibilities: Even small amounts of strong acids, alkalies, oxidizers, catalytic anhydrous chloride of iron, of aluminum, of tin, oxides of iron, aluminum.

Permissible Exposure Limits in Air: In mid-1984, OSHA set a revised TWA value of 1.0 ppm (down from 50 ppm). A short term Federal limit or ceiling value is still under study. NIOSH has recommended a 75 ppm limit as a 15-minute ceiling value. The ACGIH has recommended a TWA value of 1 ppm (2 mg/m³) but no tentative STEL value as of 1983/84 with the notation that ethylene oxide is an "industrial substance suspect of carcinogenic potential for man." The IDLH level is 800 ppm.

Determination in Air: Charcoal adsorption, workup with CS_2, analysis by gas chromatography. See NIOSH Methods, Set T. See also reference (A-10).

Permissible Concentration in Water: No criteria set.

Routes of Entry: Inhalation of gas, ingestion, eye and skin contact.

Harmful Effects and Symptoms: *Local* — Aqueous solutions of ethylene oxide or solutions formed when the anhydrous compound comes in contact with moist skin are irritating and may lead to a severe dermatitis with blisters, blebs, and burns. It is also absorbed by leather and rubber and may produce burns or irritation. Allergic eczematous dermatitis has also been reported. Exposure to the vapor in high concentration leads to irritation of the eyes. Severe eye damage may result if the liquid is splashed in the eyes. Large amounts of ethylene oxide evaporating from the skin may cause frostbite.

Systemic — Breathing high concentrations of ethylene oxide may cause nausea, vomiting, irritation of the nose, throat, and lungs. Pulmonary edema may occur. In addition, drowsiness and unconsciousness may occur. Ethylene oxide has been found to cause cancer in female mice exposed to it for prolonged periods.

Ethylene oxide is a well-known mutagen in commercial use in plants. No mutagenic effect has been demonstrated in man. However, a rebuttable presumption against registration of ethylene oxide for pesticidal applications was issued on January 27, 1978 by EPA on the basis of mutagenicity and testicular effects (A-32).

Points of Attack: Eyes, blood, respiratory system, liver, central nervous system, kidneys, lungs.

Medical Surveillance: Preplacement and periodic examinations should consider the skin and eyes, allergic history, the respiratory tract, blood, liver, and kidney function.

First Aid: If this chemical gets into the eyes, irrigate immediately. If this chemical contacts the skin, flush with water immediately. If a person breathes in large amounts of this chemical, move the exposed person to fresh air at once and perform artificial respiration. When this chemical has been swallowed, get medical attention. Give large quantities of water and induce vomiting. Do not make an unconscious person vomit.

Personal Protective Methods: Wear appropriate clothing to prevent any possibility of skin contact. Wear eye protection to prevent any reasonable probability of eye contact. Employees should wash immediately when skin is wet or contaminated. Remove clothing immediately if wet or contaminated to avoid flammability hazard. Provide emergency eyewash.

Respirator Selection:

    500 ppm: SA/SCBA
    800 ppm: SAF/SCBAF
    Escape: GMOV/SCBA

Disposal Method Suggested: Concentrated waste containing no peroxides—discharge liquid at a controlled rate near a pilot flame. Concentrated waste containing peroxides—perforation of a container of the waste from a safe distance followed by open burning.

References

(1)   Bogyo, D.A., Lande, S.S., Meylan, W.M., Howard, P.H. and Santodonato, J., (Syracuse Research Corp. Center for Chemical Hazard Assessment), *Investigation of Selected Potential Environmental Contaminants: Epoxides,* Report EPA-560/11-80-005, Washington, DC, U.S. Environmental Protection Agency (March 1980).

(2)   See Reference (A-60).

(3)   National Institute for Occupational Safety and Health, *Ethylene Oxide,* Current Intelligence Bulletin No. 35, DHHS (NIOSH) Publication No. 81-130, Cincinnati, Ohio (May 22, 1981).

(4)   Parmeggiani, L., Ed., *Encyclopedia of Occupational Health & Safety,* Third Edition, Vol. 1, pp 797–99, Geneva, International Labour Office (1983).

(5)   Sax, N.I., Ed., *Dangerous Properties of Industrial Materials Report, 4,* No. 2, 70–73, New York, Van Nostrand Reinhold Co. (1984).

(6)   United Nations Environment Programme, *IRPTC Legal File 1983,* Vol. II, pp VII/361–3, Geneva, Switzerland, International Register of Potentially Toxic Chemicals (1984).

ETHYLENE THIOUREA

● Carcinogen (Animal Positive, IARC) (2) (A-62) (A-64)

Description: Ethylene thiourea,

$$\underset{\underset{\displaystyle \mathrm{NH-C-NH-CH_2-CH_2}}{\vert_____\vert}}{\overset{\displaystyle S}{\overset{\Vert}{}}}$$

is a white needle-like crystalline solid which melts at 203° to 204°C. Commercial ethylene thiourea is available as a solid powder, as a dispersion in oil (which retards the formation of fine dust dispersions in workplace air), and "encapsulated" in a matrix of compatible elastomers. In this latter form, ethylene thiourea may be least likely to escape into the workplace air.

Code Numbers: CAS 96-45-7   RTECS NI9625000

DOT Designation: —

Synonyms: 2-Imidazolidinethione, ETU.

Potential Exposure: Ethylene thiourea is used extensively as an accelerator in the curing of polychloroprene (Neoprene) and other elastomers. NIOSH estimates that approximately 3,500 workers in the rubber industry have potential occupational exposure to ethylene thiourea. In addition, exposure to ethylene thiourea also results from the very widely used ethylene bisdithiocarbamate fungicides. Ethylene thiourea may be present as a contaminant in the ethylene bisdithiocarbamate fungicides and can also be formed when food containing the fungicides is cooked.

Permissible Exposure Limits in Air: Pending completion of a NIOSH Special Occupational Hazard Review, NIOSH believes it would be prudent to minimize occupational exposure to ethylene thiourea. There is no current Occupational Safety and Health Administration (OSHA) exposure standard for ethylene thiourea. The ACGIH has set no limits either.

Determination in Air: Filter collection, extraction with water, complexation with pentacyanoamineferrate and spectrophotometric measurement (A-10).

Permissible Concentration in Water: No criteria set, but according to NAS/NRC, a concentration of 10 ppb during a lifetime of exposure to this compound would be expected to produce one excess case of cancer for every 50,000 persons exposed. If the population of the United States is taken to be 220 million people, this translates into 4,400 excess lifetime deaths from cancer or 62.8 per year.

Route of Entry: Inhalation of dust.

Harmful Effects and Symptoms: Ethylene thiourea has been shown to be carcinogenic and teratogenic (causing malformation in offspring) in laboratory animals. In addition, ethylene thiourea can cause myxedema (the drying and thickening of skin, together with a slowing down of physical and mental activity), goiter, and other effects related to decreased output of thyroid hormone.

Medical Surveillance: Initial and routine employee exposure surveys should be made by competent industrial hygiene and engineering personnel. These surveys are necessary to determine the extent of employee exposure and to ensure that controls are effective.

The *NIOSH Occupational Exposure Sampling Strategy Manual,* NIOSH Publication #77-173, may be helpful in developing efficient programs to monitor employee exposures to ethylene thiourea. The manual discusses determination of the need for exposure measurements, selection of appropriate employees for exposure evaluation, and selection of sampling times.

Employee exposure measurements should consist of 8-hour TWA (time-weighted average) exposure estimates calculated from personal or breathing zone samples (air that would most nearly represent that inhaled by the employees). Area and source measurements may be useful to determine problem areas, processes, and operations.

Personal Protective Methods: There are four basic methods of limiting employee exposure to ethylene thiourea. None of these is a simple industrial hygiene or management decision and careful planning and thought should be used prior to implementation of any of these.

Product Substitution—The substitution of an alternative material with a lower potential health and safety risk in one method. However, extreme care must be used when selecting possible substitutes. Alternatives to ethylene thiourea should be fully evaluated with regard to possible human effects. Unless the toxic effects of the alternative have been thoroughly evaluated, a seemingly safe replacement, possibly only after years of use, may be found to induce serious health effects.

Contaminant Controls—The most effective control of ethylene thiourea, where feasible, is at the source of contamination by enclosure of the operation and/or local exhaust ventilation.

If feasible, the process or operation should be enclosed with a slight vacuum so that any leakage will result in the flow of air into the enclosure.

The next most effective means of control would be a well designed local exhaust ventilation system that physically encloses the process as much as possible, with sufficient capture velocity to keep the contaminant from entering the work atmosphere.

To ensure that ventilation equipment is working properly, effectiveness (e.g., air velocity, static pressure, or air volume) should be checked at least every three months. System effectiveness should be checked soon after any change in production, process or control which might result in significant increases in airborne exposures to ethylene thiourea.

Employee Isolation—A third alternative is the isolation of employees. It frequently involves the use of automated equipment operated by personnel observing from a closed control booth or room. The control room is maintained at a greater air pressure than that surrounding the process equipment so that air flow is out of, rather than into, the room. This type of control will not protect those employees that must do process checks, adjustments, maintenance, and related operations.

Personal Protective Equipment—The least preferred method is the use of personal protective equipment. This equipment, which may include respirators, goggles, gloves, and related items, should not be used as the only means to prevent or minimize exposure during routine operations.

Exposure to ethylene thiourea should not be controlled with the use of respirators except:

— During the time necessary to install or implement engineering or work practice controls; or

— In work situation in which engineering and work practice controls are technically not feasible; or

— For maintenance; or

— For operations which require entry into tanks or closed vessels; or

— In emergencies.

Respirator Selection: Only respirators approved by the National Institute for Occupational Safety and Health (NIOSH) should be used. Refer to *NIOSH Certified Equipment, December 15, 1975,* NIOSH Publication #76-145 and *Cumulative Supplement June 1977, NIOSH Certified Equipment,* NIOSH Publication #77-195. The use of faceseal coverlets or socks with any respirator voids NIOSH approvals.

Quantitative faceseal fit test equipment (such as sodium chloride, dioctyl phthalate, or equivalent) should be used. Refer to *A Guide to Industrial Respiratory Protection,* NIOSH Publication #76-189 for guidelines on appropriate respiratory protection programs.

References

(1) National Institute for Occupational Safety and Health,*Ethylene Thiourea,* Current Intelligence Bulletin 22, Washington, DC (April 11, 1978).

(2) International Agency for Research on Cancer, *IARC Monographs on the Carcinogenic Risks of Chemicals to Humans,* Lyon, France 1, 45 (1974).

(3) Sax, N.I., Ed., *Dangerous Properties of Industrial Materials Report, 1,* No. 2, 38-39, New York, Van Nostrand Reinhold Co. (1980).

(4) United Nations Environment Programme, *IRPTC Legal File 1983,* Vol. II, pp VII/399, Geneva, Switzerland, International Register of Potentially Toxic Chemicals (1984).

ETHYL ETHER

● Hazardous waste (EPA)

Description: $CH_3CH_2OCH_2CH_3$, is a colorless, mobile, highly flammable, volatile liquid with a characteristic pungent odor. It boils at 35°C.

Code Numbers: CAS 60-29-7   RTECS KI5775000   UN 1155

DOT Designation: Flammable liquid.

Synonyms: Anesthetic ether, diethyl oxide, ether, ethoxyethane, ethyl oxide, sulfuric ether, 1,1'-oxybisethane, diethyl ether.

Potential Exposures: Ethyl ether is used as a solvent for waxes, fats, oils, perfumes, alkaloids, dyes, gums, resins, nitrocellulose, hydrocarbons, raw rubber, and smokeless powder. It is also used as an inhalation anesthetic, a refrigerant, in diesel fuels, in dry cleaning, as an extractant, and as a chemical reagent for various organic reactions.

Incompatibilities: Strong oxidizers.

Permissible Exposure Limits in Air: The Federal standard and the ACGIH 1983/84 TWA value is 400 ppm (1,200 mg/m³). The STEL value is 500 ppm (1,500 mg/m³). The IDLH level is 19,000 ppm.

Determination in Air: Charcoal adsorption, workup with ethyl acetate, analysis by gas-liquid chromatograhy. See NIOSH Methods, Set F. See also reference (A-10).

Permissible Concentration in Water: No criteria set.

Routes of Entry: Inhalation of vapor, ingestion, eye and skin contact.

Harmful Effects and Symptoms: *Local* — Ethyl ether vapor is midly irritating to the eyes, nose, and throat. Contact with liquid may produce a dry, scaly, fissured dermatitis.

Systemic — Ethyl ether has predominantly narcotic properties. Overexposed individuals may experience drowsiness, vomiting, and unconsciousness. Death may result from severe overexposure. Chronic exposure results in some persons in anorexia, exhaustion, headache, drowsiness, dizziness, excitation, and psychic disturbances. Albuminuria has been reported. Chronic exposure may cause an increased susceptibility to alcohol.

Points of Attack: Central nervous system, skin, respiratory system, eyes.

Medical Surveillance: Preplacement or periodic examinations should evaluate the skin and respiratory tract, liver, and kidney function. Persons with a past history of alcoholism may be at some increased risk due to possibility of ethyl ether addiction (known as "ether habit").

Tests for exposure may include expired breath for unmetabolized ethyl ether and blood for ethyl ether content by oxidation with chromate solution or by gas chromatographic methods.

First Aid: If this chemical gets into the eyes, irrigate immediately. If this chemical contacts the skin, flush with water promptly. If a person breathes in large amounts of this chemical, move the exposed person to fresh air at once and perform artificial respiration. When this chemical has been swallowed, get medical attention. Give large quantities of salt water and induce vomiting. Do not make an unconscious person vomit.

Personal Protective Methods: Wear appropriate clothing to prevent repeated or prolonged skin contact. Wear eye protection to prevent reasonable probability of eye contact. Remove clothing immediately if wet or contaminated to avoid flammability hazard.

Respirator Selection:

1,000 ppm:	CCROV
4,000 ppm:	SA/SCBA
19,000 ppm:	GMOVfb/SAF/SCBAF
Escape:	GMOV/SCBA

Disposal Method Suggested: Concentrated waste containing no peroxides—discharge liquid at a controlled rate near a pilot flame. Concentrated waste containing peroxides—perforation of a container of the waste from a safe distance followed by open burning.

References

(1) See Reference (A-61).
(2) See Reference (A-60).
(3) Sax, N.I., Ed., *Dangerous Properties of Industrial Materials Report, 4,* No. 1, 81–84, New York, Van Nostrand Reinhold Co. (Jan./Feb. 1984).

ETHYL FORMATE

Description: $HCOOC_2H_5$ is a colorless liquid with a fruity odor. It boils at 54° to 55°C.

Code Numbers: CAS 109-94-4   RTECS LQ8400000   UN 1190

DOT Designation: Flammable liquid.

Synonyms: Ethyl methanoate, formic acid ethyl ester.

Potential Exposure: This material is used as a solvent for cellulose nitrate and acetate; it is used as a fumigant and in the production of synthetic flavors. It is also a raw material in pharmaceutical manufacture (A-41).

Incompatibilities: Nitrates, strong oxidizers, strong alkalies, strong acids.

Permissible Exposure Limits in Air: The Federal standard is 100 ppm (300 mg/m^3). The STEL set by ACGIH as of 1983/84 is 150 ppm (450 mg/m^3). The IDLH level is 8,000 ppm.

Determination in Air: Charcoal adsorption, workup with CS_2, analysis by gas chromatography. See NIOSH Methods, Set D. See also reference (A-10).

Permissible Concentration in Water: No criteria set.

Routes of Entry: Inhalation, ingestion, eye and skin contact.

Harmful Effects and Symptoms: Irritation of eyes and respiratory system; narcosis.

Points of Attack: Eyes, respiratory system.

Medical Surveillance: Consider the points of attack in preplacement and periodic physical examinations.

First Aid: If this chemical gets into the eyes, irrigate immediately. If this chemical contacts the skin, flush with water immediately. If a person breathes in large amounts of this chemical, move the exposed person to fresh air at once and perform artificial respiration. When this chemical has been swallowed, get medical attention. Give large quantities of salt water and induce vomiting. Do not make an unconscious person vomit.

Personal Protective Methods: Wear appropriate clothing to prevent repeated or prolonged skin contact. Wear eye protection to prevent any reasonable probability of eye contact. Employees should wash promptly when skin is wet or contaminated. Remove clothing immediately if wet or contaminated to avoid flammability hazard.

Respirator Selection:
```
1,000 ppm:  CCROVF
5,000 ppm:  GMOV/SAF/SCBAF
8,000 ppm:  SAF:PD,PP,CF
  Escape:   GMOV/SCBA
```

Disposal Method Suggested: Incineration.

References

(1) See Reference (A-61).
(2) See Reference (A-60).

2-ETHYL HEXALDEHYDE

Description: $(C_2H_5)(C_4H_9)CHCHO$ is a colorless liquid boiling at 163°C.

Code Numbers: CAS 123-05-7   RTECS MN7525000   UN 1191

DOT Designation: Combustible Liquid.

Synonyms: Butyl ethyl acetaldehyde; 2-ethyl-caproaldehyde; 3-formyl-heptane.

Potential Exposure: Uses include organic synthesis and perfume formulation.

Incompatibilities: Oxidants.

Permissible Exposure Limits in Air: No standards set.

Permissible Concentration in Water: No criteria set.

Routes of Entry: Skin absorption, inhalation, ingestion.

Harmful Effects and Symptoms: Toxic and irritating material.

Points of Attack: Eyes, mucous membranes.

Personal Protective Methods: Full protective clothing should be used.

Respirator Selection: Self-contained breathing apparatus.

References

(1)   Sax, N.I., Ed., *Dangerous Properties of Industrial Materials Report, 1,* No. 8, 71-72, New York, Van Nostrand Reinhold Co. (1981).
(2)   Sax, N.I., Ed., *Dangerous Properties of Industrial Materials Report, 3,* No. 2, 47-48, New York, Van Nostrand Reinhold Co. (1983).
(3)   See Reference (A-60).

2-ETHYLHEXYL ACRYLATE

Description: $CH_2=CHCOOCH_2CH(C_2H_5)(C_4H_9)$ is a colorless liquid with a pleasant odor which boils at 130°C.

Code Numbers: CAS 103-11-7   RTECS AT0855000

DOT Designation: —

Synonyms: Octyl acrylate; 2-ethylhexyl-2-propenoate.

Potential Exposure: Those involved in the use of this monomer in plastics manufacture, protective coatings, paper treatment and water-based paints.

Permissible Limits in Air: A TLV of 50 ppm has been suggested (A-36).

Permissible Concentration in Water: No criteria set.

Routes of Entry: Skin absorption.

Harmful Effects and Symptoms: Skin and eye irritant. General toxicity fairly low.

Points of Attack: Skin, eyes.

Personal Protective Methods: Protective clothing should be worn.

Respirator Selection: Self-contained breathing apparatus should be used.

Disposal Method Suggested: Incineration.

References

(1)   Sax, N.I., Ed., *Dangerous Properties of Industrial Materials Report, 1,* No. 7, 57-59, New York, Van Nostrand Reinhold Co. (1981).
(2)   Sax, N.I., Ed., *Dangerous Properties of Industrial Materials Report, 3,* No. 2, 83-85, New York, Van Nostrand Reinhold Co. (1983).
(3)   See Reference (A-60).

ETHYLIDENE NORBORNENE

Description: This compound has the structural formula:

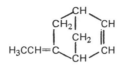

Code Numbers: CAS 16219-75-3   RTECS RB9450000

DOT Designation: —

Synonyms: 5-Ethylidenebicyclo[2,2,1]hept-2-ene

Potential Exposures: Those engaged in the synthesis of pharmaceuticals or pesticides or in the preparation of specialty resins.

Permissible Exposure Limits in Air: There are no Federal standards but ACGIH has set a ceiling value of 5 ppm (25 mg/m^3) as of 1983/84. There is no STEL value.

Permissible Concentration in Water: No criteria set.

Harmful Effects and Symptoms: This is a rather toxic hydrocarbon as shown by animal tests. After 4-hours exposure to 4,000 ppm, 5 of 6 rats died. Repeated 7-hour exposures to 237 ppm, 5 days a week for 88 days were fatal to 21 of 24 rats. At 90 ppm, no deaths resulted but renal lesions were evident. Human volunteers have noted nose and eye irritation (A-34).

References

(1)   See Reference (A-60).

ETHYL MERCAPTAN

Description: C_2H_5SH is a colorless liquid with a strong skunk-like odor. It boils at 35°C.

Code Numbers: CAS 75-08-1   RTECS KI9625000   UN 2363

DOT Designation: Flammable liquid.

Synonyms: Ethanethiol, ethyl sulfhydrate.

Potential Exposures: This material is used as a warning odorant for liquefied petroleum gases. It is used as an intermediate in the manufacture of many pesticides (A-32) and other organic chemicals.

Incompatibilities: Strong oxidizers.

Permissible Exposure Limits in Air: The Federal standard is a 10 ppm (25 mg/m^3) ceiling value. The ACGIH as of 1983/84 has set a 0.5 ppm (1 mg/m^3) TWA value and an STEL of 2 ppm (3 mg/m^3). The IDLH level is 2,500 ppm.

Permissible Concentration in Water: No criteria set but EPA (A-37) has suggested a permissible ambient goal of 13.8 μg/ℓ based on health effects.

Routes of Entry: Inhalation, ingestion, eye and skin contact.

Harmful Effects and Symptoms: Headaches, nausea; irritation of throat and mucous membranes. In animals: incoordination; paralysis; pulmonary irritation; liver and kidney damage.

Points of Attack: Respiratory system, lungs; in animals: liver, kidneys.

Medical Surveillance: Consider the points of attack in preplacement and periodic physical examinations.

First Aid: If this chemical gets into the eyes, irrigate immediately. If this chemical contacts the skin, wash with soap immediately. If a person breathes in large amounts of this chemical, move the exposed person to fresh air at once and perform artificial respiration. When this chemical has been swallowed, get medical attention. Give large quantities of salt water and induce vomiting. Do not make an unconscious person vomit.

Personal Protective Methods: Wear appropriate clothing to prevent repeated or prolonged skin contact. Wear eye protection to prevent any reasonable probability of eye contact. Employees should wash promptly when skin is wet or contaminated. Remove clothing immediately if wet or contaminaed to avoid flammability hazard.

Respirator Selection:

100 ppm:	SA/SCBA
500 ppm:	SAF/SCBAF
2,500 ppm:	SA:PD,PP,CF
Escape:	GMOV/SCBA

Disposal Method Suggested: Incineration (2000°F) followed by scrubbing with a caustic solution.

References

(1)   See Reference (A-60).

ETHYL METHACRYLATE

● Hazardous waste (EPA)

Description: $CH_2=C(CH_3)COOC_2H_5$ is a colorless liquid BP 117° to 119°C.

Code Numbers: CAS 97-63-2   RTECS OZ4550000   UN 2277

DOT Designation: Flammable liquid.

Synonyms: Ethyl α-methyl acrylate, ethyl 2-methyl-2-propenoate.

Potential Exposures: Widely known as "Plexiglass" (in the polymer form), ethyl methacrylate is used to make polymers, which in turn are used for building, automotive, aerospace, and furniture industries. It is also used by dentists as dental plates, artificial teeth, and orthopedic cement.

Permissible Exposure Limits in Air: No standards set.

Permissible Concentration in Water: No criteria set.

Harmful Effects and Symptoms: Ethyl methacrylate in the polymerized form is not toxic; however, chemicals used to produce ethyl methacrylate are extemely toxic.

Lower molecular weight acrylic monomers such as ethyl methacrylate cause systemic toxic effects. Its administration to animals results in an immediate increase in respiration rate, followed by a decrease after 15 to 40 minutes. A prompt fall in blood pressure also occurs, followed by recovery in 4 to 5 minutes. As the animal approaches death, respiration becomes labored and irregular, lacrimation may occur, defecation and urinaton increase, and finally reflex activity ceases, and the animal lapses into a coma and dies.

Acrylic monomers are irritants to the skin and mucous membranes. When placed in the eyes of animals, they elicit a very severe response and, if not washed out, can cause permanent damage.

Information on the carcinogenic and mutagenic effects of ethyl methacrylate was not found in the available literature. Ethyl methacrylate has, however, been shown to cause teratogenic effects in rats.

References

(1) U.S. Environmental Protection Agency, *Ethyl Methacrylate,* Health and Environmental Effects Profile No. 101, Washington, DC, Office of Solid Waste (April 30, 1980).

N-ETHYLMORPHOLINE

Description: $C_6H_{13}ON$, structural formula $\overline{CH_2CH_2OCH_2CH_2NCH_2CH_3}$, is a colorless liquid with an ammonia-like odor boiling at 138°C.

Code Numbers: CAS 100-74-3   RTECS QE4025000

DOT Designation: —

Synonyms: 4-Ethylmorpholine.

Potential Exposure: This material is used as a catalyst in polyurethane foam production. It is a solvent for dyes and resins. It is used as an intermediate in surfactant and rubber chemical manufacture.

Incompatibilities: Strong acids, strong oxidizers.

Permissible Exposure Limits in Air: The Federal standard is 20 ppm (94 mg/m³). The ACGIH as of 1983/84 has proposed a TWA of 5 ppm (23 mg/m³) and an STEL of 20 ppm (95 mg/m³). The notation "skin" is added to indicate the possibility of cutaneous absorption. The IDLH level is 2,000 ppm.

Determination in Air: Adsorption on SiO_2, workup with H_2SO_4 using ultrasonics; analysis by gas chromatography. See NIOSH Methods, Set K. See also reference (A-10).

Permissible Concentration in Water: No criteria set.

Routes of Entry: Inhalation, skin absorption, ingestion, eye and skin contact.

Harmful Effects and Symptoms: Irritation of eyes, nose and throat; visual disturbance; severe eye irritation from splashes.

Points of Attack: Respiratory system, eyes, skin.

Medical Surveillance: Consider the points of attack in prelacement and periodic physical examinations.

First Aid: If this chemical gets into the eyes, irrigate immediately. If this chemical contacts the skin, flush with water promptly. If a person breathes in large amounts of this chemical, move the exposed person to fresh air at once

and perform artificial respiration. When this chemical has been swallowed, get medical attention. Give large quantities of water and induce vomiting. Do not make an unconscious person vomit.

Personal Protective Methods: Wear appropriate clothing to prevent repeated or prolonged skin contact. Wear eye protection to prevent any possibility of eye contact. Employees should wash promptly when skin is wet or contaminated. Remove clothing immediately if wet or contaminated to avoid flammability hazard. Provide emergency showers and eyewash if liquid containing >15% contaminants are involved.

Respirator Selection:
```
1,000 ppm:  SAF/SCBAF
2,000 ppm:  SAF:PD,PP,CF
  Escape:  GMOV/SCBA
```

Disposal Method Selected: Controlled incineration (oxides of nitrogen are removed from the effluent gas by scrubbers and/or thermal devices).

References
(1)   See Reference (A-61).

2-ETHYL-3-PROPYL ACROLEIN

$$\overset{\displaystyle CHO}{|}$$

Description: $C_2H_5C=CHC_3H_7$ is a crystalline solid melting at $100°C$ and boiling at $175°C$ with a powerful odor.

Code Numbers: CAS 645-62-5   RTECS MP6300000

DOT Designation: Grade E Combustible Liquid (U.S. Coast Guard).

Synonyms: 2-ethyl-2-hexenal.

Potential Exposure: Those involved in organic synthesis operations and use of this material as a warning agent.

Permissible Exposure Limits in Air: No standards set.

Permissible Concentration in Water: No criteria set.

Harmful Effects and Symptoms: Is an irritant. Is a moderate inhalative and ingestive toxin. Is slightly toxic (LD-50 for rats is 3,000 mg/kg).

Personal Protective Methods: Full protective clothing recommended.

Respirator Selection: Self-contained breathing apparatus recommended.

Disposal Method Suggested: Incineration.

References
(1)   Sax, N.I., Ed., *Dangerous Properties of Industrial Materials Report, 1,* No. 8, 72-73, New York, Van Nostrand Reinhold Co. (1981).
(2)   Sax, N.I., Ed., *Dangerous Properties of Industrial Materials Report, 3,* No. 2, 48-50, New York, Van Nostrand Reinhold Co. (1983).

ETHYL SILICATE

Description: $(C_2H_5O)_4Si$, ethyl silicate, is a colorless, flammable liquid with a sharp odor detectable at 85 ppm. It boils at $169°C$.

Code Numbers: CAS 78-10-4   RTECS VV9450000   UN 1292

DOT Designation: Combustible liquid.

Synonyms: Tetraethyl orthosilicate, tetraethoxy silane.

Potential Exposures: Ethyl silicate is used in production of cases and molds for investment casting of metals. The next largest application is in corrosion-resistant coatings, primarily as a binder for zinc dust paints. Miscellaneous uses include the protection of white-light bulbs, the preparation of soluble silicas, catalyst preparation and regeneration, and as a crosslinker and intermediate in the production of silicones. NIOSH estimates worker exposure at 461,000 annually.

Incompatibilities: Strong oxidizers, water.

Permissible Exposure Limits in Air: The Federal standard is 100 ppm (approximately 850 mg/m^3) determined as a time-weighted average. This TLV has not been confirmed in human exposure. At 3,000 ppm, ethyl silicate vapors are intolerable. ACGIH (1983/84) has set a TWA of 10 ppm (85 mg/m^3) and an STEL of 30 ppm (255 mg/m^3). The IDLH level is 1,000 ppm.

Determination in Air: Adsorption on resin, workup with CS_2, analysis by gas chromatography. See NIOSH Methods, Set S. See also reference (A-10).

Permissible Concentration in Water: No criteria set.

Routes of Entry: Inhalation of vapor, ingestion, skin and eye contact.

Harmful Effects and Symptoms: *Local* — Ethyl silicate is a primary irritant to the eyes and the nose.

Systemic — Damage to the lungs, liver, and kidneys, and anemia have been observed in animal experiments but have not been reported for human exposure.

Symptoms of exposure in humans consist of irritation of eyes and respiratory passages, fatigue, tremors, and narcosis. If exposure persists, narcosis will eventually lead to death (1).

Points of Attack: Respiratory system, liver, kidneys, blood, skin.

Medical Surveillance: Placement or periodic examinations should include the skin, eyes, respiratory tract, as well as liver and kidney functions.

First Aid: If this chemical gets into the eyes, irrigate immediately. If this chemical contacts the skin, wash with soap promptly. If a person breathes in large amounts of this chemical, move the exposed person to fresh air at once and perform artificial respiration. When this chemical has been swallowed, get medical attention. Give large quantities of water and induce vomiting. Do not make an unconscious person vomit.

Personal Protective Methods: Wear appropriate clothing to prevent repeated or prolonged skin contact. Wear eye protection to prevent any reasonable probability of eye contact. Employees should wash promptly when skin is wet or contaminated. Remove clothing immediately if wet or contaminated to avoid flammability hazard.

Respirator Selection:
    1,000 ppm: CCROVF/GMOV/SAF/SCBAF
       Escape: GMOV/SCBA

Disposal Method Suggested: Incineration in admixture with a more flammable solvent.

References
(1)   Nat. Inst. for Occup. Safety and Health, *Information Profiles on Potential Occupational Hazards—Single Chemicals: Ethyl Silicate,* pp 98-105, Report No. TR79-607, Rockville, MD (December 1979).
(2)   See Reference (A-60).
(3)   See Reference (A-61).

F

FENAMIPHOS

Description: $C_{13}H_{22}NO_3PS$ has the following structural formula

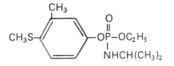

This is a crystalline solid melting at 40°C (technical grade) and 49.2°C (pure compound).

Code Numbers: CAS 22224-92-6   RTECS TB3675000

DOT Designation: —

Synonyms: Ethyl-3-methyl-4-(methylthio)phenyl-1-methylethylphosphoramidate, Nemacur®.

Potential Exposure: Those involved in the manufacture, formulation or application of this nematocide.

Permissible Exposure Limits in Air: 0.1 mg/m³ according to ACGIH (1983/84) with the notation that skin absorption is possible.

Permissible Concentration in Water: 0.1 mg/ℓ maximum (1).

Harmful Effects and Symptoms: This is a highly toxic chemical (the LD_{50} for rats is 15-20 mg/kg). It is a cholinesterase inhibitor with effects typical of such compounds.

Disposal Method Recommended: Incineration.

References
(1) Sax, N.I., Ed., *Dangerous Properties of Industrial Materials Report, 3,* No. 1, 52-56, New York, Van Nostrand Reinhold Co. (1983).

FENITROTHION

Description: $C_9H_{12}NO_5PS$ has the following structural formula

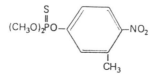

This is a volatile oil boiling at $118°C$ at 0.05 mm Hg pressure.

Code Numbers: CAS 122-14-5   RTECS TG0350000

DOT Designation: —

Synonyms: O,O-dimethyl O-(3-methyl-4-nitrophenyl)phosphorothionate, Folithion®, Accothion® Cytel®, Cyfen®, Sumithion®, and other trade names.

Potential Exposure: Those involved in the manufacture, formulation and application of this insecticide.

Permissible Exposure Limits in Air: No standards set.

Permissible Concentration in Water: No criteria set.

Harmful Effects and Symptoms: This is a moderately toxic compound (LD_{50} value for rats is 250–500 mg/kg). The estimated acceptable daily intake for man is 0.005 mg/kg. This is a cholinesterase inhibitor with the accompanying effects.

Disposal Method Suggested: Incineration.

References

(1)   Sax, N.I., Ed., *Dangerous Properties of Industrial Materials Report, 2,* No. 4, 88-92, New York, Van Nostrand Reinhold Co. (1982).

(2)   United Nations Environment Programme, *IRPTC Legal File 1983,* Vol. II, pp VII/618–19, Geneva, Switzerland, International Register of Potentially Toxic Chemicals (1984).

FENSULFOTHION

Description: $(C_2H_5O)_2\overset{S}{\overset{\|}{P}}OC_6H_4\overset{O}{\overset{\|}{S}}CH_3$, O,O-diethyl O-4-(methylsulfinyl)phenyl phosphorothioate, is a yellow oil.

Code Numbers: CAS 115-90-2   RTECS TF3850000   UN 2765

DOT Designation: —

Synonyms: ENT 24945; Dasanit®; Terracur P®.

Potential Exposure: Those involved in the manufacture (A-32), formulation or application of this insecticide and nematocide.

Permissible Exposure Limits in Air: There is no Federal standard but ACGIH (1983/84) has adopted a TWA value of 0.1 mg/m^3 but no STEL value.

Permissible Concentration in Water: No criteria set.

Harmful Effects and Symptoms: The acute oral LD_{50} value for rats is 5 to 10 mg/kg which is highly to extremely toxic. It is a cholinesterase inhibitor.

References

(1)   United Nations Environment Programme, *IRPTC Legal File 1983,* Vol. II, pp VII/601–2, Geneva, Switzerland, International Register of Potentially Toxic Chemicals (1984).

FENTHION

Description: $(CH_3O)_2\overset{S}{\overset{\|}{P}}OC_6H_3(CH_3)SCH_3$, O,O-dimethyl O-[3-methyl-4-methylthio)phenyl] phosphorothioate, is a colorless liquid.

Code Numbers: CAS 55-38-9 RTECS TF9625000 UN 2765

DOT Designation: –

Synonyms: MPP (in Japan); ENT 25440; Baycid®; Baytex®; Entex®; Lebaycid®; Mercaptophos®; Queletox®; Tiguvon®.

Potential Exposure: Those involved in the manufacture (A-32), formulation or application of this insecticide.

Permissible Exposure Limits in Air: There is no Federal standard but ACGIH (1983/84) has adopted a TWA value of 0.2 mg/m^3 but set no STEL value. They have added the notation "skin" indicating the possibility of cutaneous absorption.

Permissible Concentration in Water: No criteria set.

Harmful Effects and Symptoms: The acute oral LD_{50} value for rats is 215 mg/kg which is moderately toxic.

References

(1) Sax, N.I., Ed., *Dangerous Properties of Industrial Materials Report, 3,* No. 1, 56-61, New York, Van Nostrand Reinhold Co. (1983).

(2) United Nations Environment Programme, *IRPTC Legal File 1983,* Vol. II, pp VII/613-14, Geneva, Switzerland, International Register of Potentially Toxic Chemicals (1984).

FERBAM

Description: $[(CH_3)_2NCSS]_2$ Fe is an odorless black solid.

Code Numbers: CAS 14484-64-1 RTECS NO8750000 UN 2771

DOT Designation: –

Synonyms: Ferric dimethyldithiocarbamate; tris(dimethyldithiocarbamate)-iron; ENT 14689; Fermate®.

Potential Exposure: Those involved in the production, formulation and application of this dithiocarbamate fungicide.

Incompatibilities: Strong oxidizers.

Permissible Exposure Limits in Air: The Federal standard is 15 mg/m^3. The ACGIH (1983/84) has adopted a TWA value of 10 mg/m^3 and an STEL value of 20 mg/m^3. There is no IDLH value available.

Determination in Air: Collection by filter, colorimetric analysis (A-20).

Permissible Concentration in Water: No criteria set but degradation produces ethylene thiourea.

Routes of Entry: Inhalation, ingestion, eye and skin contact.

Harmful Effects and Symptoms: Irritation of eyes and respiratory tract, dermatitis, gastrointestinal disturbances.

Points of Attack: Respiratory system, lungs, skin, gastrointestinal tract.

Medical Surveillance: Consider the points of attack in preplacement and periodic physical examinations.

First Aid: If this chemical gets into the eyes, irrigate immediately. If this chemical contacts the skin, wash with soap promptly. If a person breathes in

large amounts of this chemical, move the exposed person to fresh air at once and perform artificial respiration. When this chemical has been swallowed, get medical attention. Give large quantities of water and induce vomiting. Do not make an unconscious person vomit.

Personal Protective Methods: Wear appropriate clothing to prevent repeated or prolonged skin contact. Wear eye protection to prevent any reasonable probability of eye contact. Employees should wash promptly when skin is wet or contaminated. Work clothing should be changed daily if it is possible that clothing is contaminated. Remove nonimpervious clothing promptly if wet or contaminated.

Respirator Selection:

75 mg/m³:	DXS
150 mg/m³:	DXSQ/FuHIEP/SA/SCBA
750 mg/m³:	HiEPF/SAF/SCBAF
7,500 mg/m³:	PAPHie/SA:PD,PP,CF
Escape:	DXS/SCBA

Disposal Method Suggested: Ferbam is hydrolyzed by alkali and is unstable to moisture, lime and heat. Ferbam can be incinerated (A-32).

References

(1) See Reference (A-61).
(2) United Nations Environment Programme, *IRPTC Legal File 1983,* Vol. II, pp VII/402, Geneva, Switzerland, International Register of Potentially Toxic Chemicals (1984).

FERRIC CYANIDE

● Hazardous waste (EPA)

Description: The compounds are considered here to include ferric ferricyanide, $Fe[Fe(CN)_6]$, and ferric ferrocyanide, $Fe_4[Fe(CN)_6]_3$. The empirical formula of the misnamed ferric cyanide, $Fe(CN)_3$, corresponds actually to one of the ferricyanide compounds, the ferric ferricyanide with the actual formula $Fe[Fe(CN)_6]$.

DOT Designation: —

Synonyms: Berlin green for ferric ferricyanide.

Potential Exposure: These compounds are colored pigments, insoluble in water or weak acids, although they can form colloidal dispersions in aqueous media. These pigments are generally used in paint, printing ink, carbon paper inks, crayons, linoleum, paper pulp, writing inks and laundry blues.

Incompatibilities: These compounds are sensitive to alkaline decomposition.

Permissible Exposure Limits in Air: No standards set.

Permissible Concentration in Water: No criteria set.

Harmful Effects and Symptoms: No adequate toxicity data are available. All ferrocyanide and ferricyanide salts are reported as possibly moderately toxic (from 0.5 to 5.0 mg/kg as a probable lethal dose in humans).

References

(1) U.S. Environmental Protection Agency, *Ferric Cyanide,* Health and Environmental Effects Profile No. 102, Office of Solid Waste, Washington, DC (April 30, 1980).

FERRIC SALTS

See also "Iron and Iron Compounds."

FERROUS SALTS

See also "Iron and Iron Compounds."

FERROVANADIUM DUST

Description: This material consists of dark, odorless solid particles having the composition 35 to 85% vanadium with iron and trace silicon, manganese, chromium, nickel, etc.

Code Numbers: CAS 12604-58-9   RTECS LK2900000   UN none

DOT Designation: —

Potential Exposure: This material is added to steel to produce fineness of grain, toughness, torsion properties, and resistance to high temperatures.

Incompatibilities: Strong oxidizers.

Permissible Exposure Limits in Air: The Federal standard is 1 mg/m^3. The ACGIH (1983/84) has also set a TWA of 1 mg/m^3 and a STEL value of 0.3 mg/m^3 (less than the TWA). There is no IDLH level available.

Determination in Air: Collection by filter, analysis by atomic absorption spectrophotometry (A-21).

Permissible Concentration in Water: No criteria set.

Routes of Entry: Inhalation, eye and skin contact.

Harmful Effects and Symptoms: Irritation of eyes and respiratory system.

Points of Attack: Respiratory system, eyes.

Medical Surveillance: Consider the points of attack in preplacement and periodic physical examinations.

First Aid: If this chemical gets into the eyes, irrigate immediately. If a person breathes in large amounts of this chemical, move the exposed person to fresh air at once and perform artificial respiration.

Personal Protective Methods: No particular precautions recommended by NIOSH. Thick working gloves and safety goggles are recommended by others (A-38).

Respirator Selection:
        5 mg/m^3: DM
        10 mg/m^3: DMXSQ/FuHiEP/SA/SCBA
        50 mg/m^3: HiEPF/SAF/SCBAF
        500 mg/m^3: PAPHiE/SA:PD,PP,CF

Disposal Method Suggested: Disposal in a sanitary landfill.

References

(1)   National Institute for Occupational Safety and Health, *Criteria for a Recommended Standard: Occupational Exposure to Vanadium,* NIOSH Document No. 77-222, Washington, DC (1977).
(2)   See Reference (A-61).

FIBROUS GLASS

Description: Fibrous glass is the name for a manufactured fiber in which the fiber-forming substance is glass. Glasses are a class of materials made from silicon dioxide with oxides of various metals and other elements, that solidify from the molten state without crystallization. A fiber is considered to be a particle with a length-to-diameter ratio of 3 to 1 or greater.

Most fibrous glass that is manufactured consists of fibers with diameters 3.5 μm or larger. The volume of small diameter fiber production has not been determined. Fibers with diameters less than 1 μm are estimated to comprise less than 1% of the fibrous glass market.

Code Numbers: CAS 14808-60-7   RTECS VV7330000   UN none

DOT Designation: —

Synonyms: Glass fiber, glass wool, Fiberglas®.

Potential Exposure: It is estimated that fibrous glass is used in over 30,000 product applications. The major uses of fibrous glass are in thermal, electrical, and acoustical insulation, weatherproofing, plastic reinforcement, filtration media, and in structural and textile materials. NIOSH estimates that 200,000 workers are potentially exposed to fibrous glass.

Permissible Exposure Limits in Air: NIOSH recommends that occupational exposure to fibrous glass be controlled so that no worker is exposed at an airborne concentration greater than 3,000,000 fibers/m^3 of air (3 fibers/cc of air) having a diameter equal to or less than 3.5 μm and a length equal to or greater than 10 μm determined as a time weighted average (TWA) concentration for up to a 10-hour work shift in a 40-hour work week; airborne concentrations determined as total fibrous glass shall be limited to a TWA concentration of 5 mg/m^3 of air. This differs from the present Federal standard which classifies fibrous glass an an inert or nuisance dust with the limits of exposure being 15 million particles per cubic foot or 5 mg/m^3 for the respirable fraction and 50 million particles per cubic foot or 15 mg/m^3 total dust, both as 8-hour TWA concentrations. ACGIH as of 1983/84 gives a TWA of 10 mg/m^3 for glass in fibrous or dust form.

Determination in Air: Collection on a filter and gravimetric analysis (A-1).

Permissible Concentration in Water: No criteria set.

Routes of Entry: Inhalation and skin contact.

Harmful Effects and Symptoms: Different dimensions of fibrous glass will produce different biologic effects. Large diameter (greater than 3.5 μm) glass fibers have been found to cause skin, eye, and upper respiratory tract irritation; a relatively low frequency of fibrotic changes; and a very slight indication of an excess mortality due to nonmalignant respiratory disease. Smaller diameter (less than 3.5 μm) fibrous glass has not been conclusively related to health effects in humans but glass fibers of this dimension have only been regularly produced since the 1960s.

Smaller diameter fibers have the ability to penetrate to the alveoli. This potential is cause for concern and the primary reason that fibers 3.5 μm or smaller are subject to special controls. Experimental studies in animals have demonstrated carcinogenic effects with the long (greater than 10 μm) and thin fibers (usually less than 1 μm in diameter). However, these studies were performed by implanting fibrous glass in the pleural or peritoneal cavities.

The data from studies with these routes of exposure cannot be directly extrapolated to conditions of human exposure. On the basis of available information, NIOSH does not consider fibrous glass to be a substance that produces cancer as a result of occupational exposure. The data on which to base this conclusion are limited. Fibrous glass does not appear to possess the same potential as asbestos for causing health hazard. Glass fibers are not usually of the fine submicron diameters as are asbestos fibrils and the concentrations of glass fibers in workplace air are generally orders of magnitude less than for asbestos. In one study, glass fibers were found to be cleared from the lungs more readily than asbestos.

Medical Surveillance: NIOSH recommends that workers subject to fibrous glass exposure have comprehensive preplacement medical examinations with emphasis on skin susceptibility and prior exposure in dusty trades. Subsequent annual examinations should give attention to the skin and respiratory system with attention to pulmonary function.

Personal Protective Methods: Protective clothing shall be worn to prevent fibrous glass contact with skin especially hands, arms, neck, and underarms. Safety goggles or face shields and goggles shall be worn during tear-out or blowing operations or when applying fibrous glass materials overhead. They should be used in all areas where there is a likelihood that airborne glass fibers may contact the eyes. Engineering controls should be used wherever feasible to maintain fibrous glass concentrations at or below the prescribed limits. Respirators should only be used when engineering controls are not feasible; for example, in certain nonstationary operations where permanent controls are not feasible.

Respirator Selection: See Reference (2).

Disposal Method Suggested: Fibrous glass waste and scrap should be collected and disposed of in a manner which will minimize its dispersal into the atmosphere. Emphasis should be placed on covering waste containers, proper storage of materials, and collection of fibrous glass dust. Cleanup of fibrous glass dust should be performed using vacuum cleaners or wet cleaning methods. Dry sweeping should not be performed.

References

(1) National Institute for Occupational Safety and Health, *Occupational Exposure to Fibrous Glass: A Symposium.* NIOSH Doc. No. 76-151 (1976).
(2) National Institute for Occupational Safety and Health, *Criteria for a Recommended Standard: Occupational Exposure to Fibrous Glass,* NIOSH Doc. No. 77-152 (1977).
(3) National Institute for Occupational Safety and Health, *Criteria for a Recommended Standard: Occupational Exposure to Crystalline Silica,* NIOSH Doc. No. 75-120, Washington, DC (1975).

FLUORANTHENE

- Hazardous waste (EPA)
- Priority toxic pollutant (EPA)

Description: Fluoranthene, $C_{16}H_{10}$ has the structural formula:

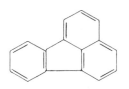

It melts at 111°C.

Code Numbers: CAS 206-44-0   RTECS LL4025000   UN none

DOT Designation: —

Synonyms: Idryl; benzo[jk]fluorene; 1,2-benzacenaphthene; 1,2-(1,8-naphthylene)benzene.

Potential Exposure: Fluoranthene, a polynuclear aromatic hydrocarbon, is produced from the pyrolytic processing of organic raw materials such as coal and petroleum at high temperatures. It is also known to occur naturally as a product of plant biosynthesis. Fluoranthene is ubiquitous in the environment and has been detected in U.S. air, in foreign and domestic drinking waters and in foodstuffs. It is also contained in cigarette smoke.

Individuals living in areas which are heavily industrialized, and in which large amounts of fossil fuels are burned, would be expected to have greatest exposure from ambient sources of fluoranthene. In addition, certain occupations (e.g., coke oven workers, steelworkers, roofers, automobile mechanics) would also be expected to have elevated levels of exposure relative to the general population.

Exposure to fluoranthene will be considerably increased among tobacco smokers or those who are exposed to smokers in closed environments (i.e., indoors).

Permissible Exposure Limits in Air: No standards set.

Determination in Air: Collection on a filter, extraction with benzene, chromatographic separation, spectrophotometric analysis (A-10).

Permissible Concentration in Water: The only existing standard which takes fluoranthene into consideration is a drinking water standard for PAHs. The 1970 World Health Organization European Standards for Drinking Water recommends a concentration of PAHs not exceeding 0.2 $\mu g/\ell$. This recommended standard is based upon the analysis of six PAHs in drinking water as follows: fluoranthene; benzo[a]pyrene; benzo[ghi]perylene; benzo[b]fluoranthene; benzo[k]fluoranthene; and indeno[1,2,3-cd]pyrene.

More recently EPA has established ambient water criteria as follows: To protect freshwater aquatic life: 3,980 $\mu g/\ell$ based on acute toxicity. To protect saltwater aquatic life: 40 $\mu g/\ell$ based on acute toxicity and 16 $\mu g/\ell$ based on chronic toxicity. To protect human health: 42 $\mu g/\ell$.

Determination in Water: Methylene chloride extraction followed by high pressure liquid chromatography with fluorescence as UV detection; or gas chromatography (EPA Method 610), or gas chromatography plus mass spectrometry (EPA Method 625).

Harmful Effects and Symptoms: There is concern about the toxicity of fluoranthene because it is widespread in the human environment and belongs to a class of compounds (polynuclear aromatic hydrocarbons) that contain numerous potent carcinogens. Experimentally, fluoranthene does not exhibit properties of a mutagen or primary carcinogen but it is a potent cocarcinogen. In the laboratory, fluoranthene has also demonstrated toxicity to various freshwater and marine organisms. This finding, coupled with the cocarcinogenic properties of the compound, points out the need to protect humans and aquatic organisms from the potential hazards associated with fluoranthene in water.

Disposal Method Suggested: Incineration.

References

(1) U.S. Environmental Protection Agency, *Fluoranthene: Ambient Water Quality Criteria,* Washington, DC (1980).
(2) U.S. Environmental Protection Agency, *Fluoranthene,* Health and Environmental Effects Profile No. 103, Office of Solid Waste, Washington, DC (April 30, 1980).

N-2-FLUORENYL ACETAMIDE

● Carcinogen (HHS-NTP) (A-62, A-64)

Description: $C_{15}H_{13}NO$ is a crystalline solid melting at 194°C with the following structural formula

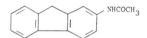

Code Numbers: CAS 56-96-3   RTECS AB9450000

DOT Designation: —

Synonyms: 2-Acetylaminofluorene; AAF.

Potential Exposure: 2-Acetylaminofluorene (AAF) was intended to be used as a pesticide, but it was never marketed because this chemical was found to be carcinogenic.

AAF is used frequently by biochemists and technicians engaged in the study of liver enzymes and the carcinogenicity and mutagenicity of aromatic amines as a positive control. Therefore, these persons may be exposed to AAF.

AAF is distributed by several companies that deal in specialty chemicals. Information obtained from these distributors indicates that AAF is imported from Europe. A typical chemical distributor keeps approximately 9 lb of AAF in stock. The chemical is usually sold in 1-, 5-, or 25-g quantities. Based on this information, it is estimated that the total U.S. usage is less than 20 lb per year.

Human exposure to AAF may occur through inhalation and skin absorption. The occupations at greatest risk to AAF exposure are organic chemists, chemical stockroom workers, and biomedical researchers. Although neither NIOSH nor OSHA has estimated the number of U.S. workers exposed to AAF, perhaps fewer than 1,000 workers in 200 laboratories may come in contact with this animal carcinogen.

Permissible Exposure Limits in Air: No standards set.

Permissible Concentration in Water: No criteria set.

Harmful Effects and Symptoms: Incorporation of this compound in feed caused increased incidences of malignant tumors in a variety of organs in the rat (1). Long-term studies in which mice were given 2-acetylaminofluorene in their diet showed that this compound caused increased incidences of liver and urinary bladder cancers (1).

Personal Protective Methods: Because AAF is a carcinogen, on February 11, 1974, OSHA promulgated a standard for this chemical designating protective clothing, hygiene procedures for workers, and special engineering requirements for the manufacture or processing of AAF. Open vessel operations are prohibited.

Respirator Selection: See NIOSH requirements.

References

(1) See Reference (A-62).
(2) Parmeggiani, L., Ed., *Encyclopedia of Occupational Health & Safety,* Third Edition, Vol. 1, pp 39–40, Geneva, International Labour Office (1983).

FLUORIDES

Description: Of the general formula M_xF_y, appearance, odor and properties vary with specific compounds.

DOT Designation: —

Synonyms: Vary depending on specific compound.

Potential Exposure: Fluorides are used as an electrolyte in aluminum manufacture, a flux in smelting nickel, copper, gold, and silver, as a catalyst for organic reactions, a wood preservative, fluoridation agent for drinking water, a bleaching agent for cane seats, in pesticides, rodenticides, and as a fermentation inhibitor. They are utilized in the manufacture of steel, iron, glass, ceramics, pottery, enamels, in the coagulation of latex, in coatings for welding rods, and in cleaning graphite, metals, windows, and glassware. Exposure to fluorides may also occur during preparation of fertilizer from phosphate rock by addition of sulfuric acid.

Major users of fluoride compounds are the aluminum, chemical, and steel industries, and NIOSH estimates that 350,000 workers are potentially exposed to fluorides.

Air pollution by fluoride dusts and gases has done substantial damage to vegetation and to animals in the vicinity of industrial fluoride sources. However, the contribution of ambient air to human fluoride intake is only a few hundredths of a milligram per day, an amount that is insignificant in comparison with other sources of fluoride.

Operations that introduce fluoride dusts and gases into the atmosphere include: grinding, drying, and calcining of fluoride-containing minerals; acidulation of the minerals; smelting; electrochemical reduction of metals with fluoride fluxes or melts as in the aluminum and steel industry; kiln firing of brick and other clay products and the combustion of coal.

Generally speaking, good progress has been made in reducing fluoride exposure to industrial workers by ventilation and emission control practices.

Incompatibilities: Strong acids.

Permissible Exposure Limits in Air: The Federal standard is 2.5 mg/m^3 which is also the value set by NIOSH and ACGIH (1983/84). There is no tentative STEL. The IDLH level is 500 mg/m^3.

Determination in Air: Gaseous fluorides collected by impinger using caustic; particulates by filter. Analysis is by low specific electrode (A-10).

Permissible Concentration in Water: No criteria set.

Routes of Entry: Inhalation, ingestion, eye and skin contact.

Harmful Effects and Symptoms: *Local* – Fluorine and some of its compounds are primary irritants of skin, eyes, mucous membranes, and lungs. Thermal or chemical burns may result from contact; the chemical burns cause deep tissue destruction and may not become symptomatic until several hours after contact, depending on dilution. Nosebleeds and sinus trouble may develop on chronic exposure to low concentration of fluoride or fluorine in air. Accidental fluoride burns, even when they involve small body areas (less than 3%), can cause systemic effects of fluoride poisoning by absorption of the fluoride through the skin.

Systemic – Inhalation of excessive concentration of elemental fluorine or of hydrogen fluoride can produce bronchospasm, laryngospasm, and pulmonary edema. Gastrointestinal symptoms may be present. A brief exposure to 25 ppm has caused sore throat and chest pain, irreparable damage to the lungs, and death.

Most cases of acute fluoride intoxication result from ingestion of fluoride compounds. The severity of systemic effects is directly proportional to the irritating properties and the amount of the compound that has been ingested. Gastrointestinal symptoms of nausea, vomiting, diffuse abdominal cramps, and diarrhea can be expected. Large doses produce central nervous system involvement with twitching of muscle groups, tonic and clonic convulsions, and coma.

The systemic effects of prolonged absorption of fluorides from either dusts or vapors have long been a source of some uncertainty. Fluorides are retained preferentially in bone, and excessive intake may result in an osteosclerosis that is recognizable by x-ray. The first signs of changes in density appear in the lumbar spine and pelvis. Usually some ossification of ligaments occurs. Recent investigations suggest that rather severe skeletal fluorosis can exist in workers without any untoward physiological effects, detrimental effects on their general health, or physical impairment.

Fluorides occur in nature and enter the human body through inhalation or ingestion (natural dusts and water). In children, mottling of the dental enamel may occur from increased water concentrations. These exposures are usually minimal and occur over extended periods. Residential districts which adjoin manufacturing areas can be subjected to continual exposures at minimal levels, or to heavy exposure in the event of accident or plant failure, as in the case of the Meuse Valley disaster.

Points of Attack: Eyes, respiratory system, central nervous system, skeleton, kidneys, skin.

Medical Surveillance: Preemployment and periodic examinations should consider possible effects on the skin, eyes, teeth, respiratory tract, and kidneys. Chest x-rays and pulmonary function should be followed. Kidney function should be evaluated. If exposures have been heavy and skeletal fluorosis is suspected, pelvic x-rays may be helpful. Intake of fluoride from natural sources in food or water should be known.

In the case of exposure to fluoride dusts, periodic urinary fluoride excretion levels have been very useful in evaluating industrial exposures and environmental dietary sources.

First Aid: If this chemical gets into the eyes, irrigate immediately. If this chemical contacts the skin, wash with soap promptly. If a person breathes in

large amounts of this chemical, move the exposed person to fresh air at once and perform artificial respiration. When this chemical has been swallowed, get medical attention. Give large quantities of water and induce vomiting. Do not make an unconscious person vomit.

Personal Protective Methods: In areas with excessive gas or dust levels for any type of fluorine, worker protection should be provided. Respiratory protection by dust masks or gas masks with an appropriate canister or supplied air respirator should be provided. Goggles or fullface masks should be used. In areas where there is a likelihood of splash or spill, acid resistant clothing including gloves, gauntlets, aprons, boots, and goggles or face shields should be provided to the worker.

Personal hygiene should be encouraged, with showering following each shift and before change to street clothes. Work clothes should be changed following each shift, especially in dusty areas. Attention should be given promptly to any burns from fluorine compounds due to absorption of the fluorine at the burn site and the possibility of developing systemic symptoms from absorption from burn sites.

Respirator Selection:

12.5 mg/m³:	D
25 mg/m³:	DXSQ*/SA/SCBA
125 mg/m³:	HiEPF/SAF/SCBAF*
250 mg/m³:	PAPHiEF/SAF:PD,PP,CF
Escape:	GMOV/SCBA
*Note:	May need acid gas sorbent.

Disposal Method Suggested: Reaction of aqueous waste with an excess of lime, followed by lagooning, and either recovery or land disposal of the separated calcium fluoride.

References

(1) National Institute for Occupational Safety and Health, *Criteria for a Recommended Standard: Occupational Exposure to Inorganic Fluoride,* NIOSH Doc. No. 76-103 (1976).

(2) National Academy of Sciences, *Medical and Biologic Effect of Environmental Pollutants: Fluorides,* Washington, DC (1971).

(3) See Reference (A-61).

(4) Sax, N.I., Ed., *Dangerous Properties of Industrial Materials Report, 3,* No. 6, 67–69, New York, Van Nostrand Reinhold Co. (Nov./Dec. 1983) (Sodium Hydrogen Fluoride).

(5) Parmeggiani, L., Ed., *Encyclopedia of Occupational Health & Safety,* Third Edition, Vol. 1, pp 891–95, Geneva, International Labour Office (1983).

FLUORINE

● Hazardous waste (EPA)

Description: F_2, molecular fluorine, is a yellow gas.

Code Numbers: CAS 7782-41-4   RTECS LM6475000   UN 1045 (gas)
UN 9192 (liquid)

DOT Designation: Nonflammable gas, poison and oxidizer.

Potential Occupational Exposure: Elemental fluorine is used in the conversion of uranium tetrafluoride to uranium hexafluoride, in the synthesis of organic and inorganic fluorine compounds, and as an oxidizer in rocket fuel.

Incompatibilities: Water, nitric acid, most oxidizable materials.

Permissible Exposure Limits in Air: The Federal standard is 0.1 ppm (0.2 mg/m^3). The ACGIH (1983/84) has adopted a TWA of 1.0 ppm (2.0 mg/m^3) and an STEL value of 2.0 ppm (4.0 mg/m^3). The IDLH level is 25 ppm.

Permissible Concentration in Water: No criteria set.

Routes of Entry: Inhalation, eye and skin contact.

Harmful Effects and Symptoms: See also "Fluorides." Symptoms include irritation of eyes, nose and respiratory tract; laryngeal spasms, bronchial spasms; pulmonary edema; skin and eye burns. In animals, liver and kidney damage has been observed.

Points of Attack: Respiratory system, lungs, eyes, skin; in animals: liver, kidneys.

Medical Surveillance: Consider the points of attack in preplacement and periodic physical examinations.

First Aid: If this chemical gets into the eyes, irrigate immediately. If this chemical contacts the skin, flush with water immediately. If a person breathes in large amounts of this chemical, move the exposed person to fresh air at once and perform artificial respiration.

Personal Protective Methods: Wear appropriate clothing to prevent any possibility of skin contact. Wear eye protection to prevent any possibility of eye contact. Employees should wash immediately when skin is wet or contaminated. Remove nonimpervious clothing immediately if wet or contaminated. Provide emergency showers and eyewash.

Respirator Selection:
1 ppm:	SA/SCBA
5 ppm:	SAF/SCBAF
25 ppm:	SA:PD,PP,CF
Escape:	GMS/SCBA
Note:	Do not use oxidizable sorbents.

Disposal Method Suggested: Pretreatment involves reaction with a charcoal bed. The product of the reaction is carbon tetrafluoride which is usually vented. Residual fluorine can be combusted by means of a fluorine-hydrocarbon air burner followed by a caustic scrubber and stack (A-31).

References

(1) Sax, N.I., Ed., *Dangerous Properties of Industrial Materials Report, 1,* No. 4, 68-70, New York, Van Nostrand Reinhold Co. (1981).
(2) See Reference (A-61).
(3) See Reference (A-60).
(4) Sax, N.I., Ed., *Dangerous Properties of Industrial Materials Report, 3,* No. 4, 50-53, New York, Van Nostrand Reinhold Co. (1983).

FLUOROTRICHLOROMETHANE

● Hazardous waste (EPA)

Description: CCl_3F is a colorless liquid or gas with a chlorinated solvent odor which is detectable >20% by volume. It boils at 24°C.

Code Numbers: CAS 75-69-4   RTECS PB6125000   UN none

DOT Designation: —

Synonyms: Refrigerant 11; monofluorotrichloromethane; trichlorofluoromethane; trichloromonofluoromethane; Freon® 11; F-11.

Potential Exposure: This material is used as a refrigerant, aerosol propellant and foaming agent.

Incompatibilities: Chemically active metals: sodium, potassium, calcium, powdered aluminum, zinc, magnesium.

Permissible Exposure Limits in Air: The Federal standard is 1,000 ppm (5,600 mg/m^3). ACGIH (1983/84) confirms this as a ceiling value but sets no STEL value. The IDLH level is 10,000 ppm.

Determination in Air: Charcoal adsorption, workup with CS_2, analysis by gas chromatography. See NIOSH Methods, Set 1. See also reference (A-10).

Permissible Concentration in Water: For the protection of human health: preferably zero. An additional lifetime cancer risk of 1 in 100,000 results at a level of 1.9 $\mu g/\ell$. In January 1981 EPA (46FR2266) removed F-11 from the priority toxic pollutant list.

Determination in Water: Inert gas purge followed by gas chromatography with halide specific detection (EPA Method 601) or gas chromatography plus mass spectrometry (EPA Method 624).

Routes of Entry: Inhalation, ingestion, eye and skin contact.

Harmful Effects and Symptoms: Incoherence, tremors, dermatitis, frostbite, cardiac arrythmias, cardiac arrest.

Points of Attack: Skin, cardiovascular system.

Medical Surveillance: Consider the points of attack in preplacement and periodic physical examinations.

First Aid: If this chemical gets into the eyes, irrigate immediately. If this chemical contacts the skin, flush with water immediately. If a person breathes in large amounts of this chemical, move the exposed person to fresh air at once and perform artificial respiration. When this chemical has been swallowed, get medical attention. Give large quantities of salt water and induce vomiting. Do not make an unconscious person vomit.

Personal Protective Methods: Wear appropriate clothing to prevent repeated or prolonged skin contact. Wear eye protection to prevent any possibility of eye contact. Remove nonimpervious clothing promptly if wet or contaminated. Provide emergency showers and eyewash.

Respirator Selection:
      10,000 ppm: SA/SCBA
        Escape: GMOV/SCBA

Disposal Method Suggested: Incineration, preferably after mixing with another combustible fuel. Care must be exercised to assure complete combustion to prevent the formation of phosgene. An acid scrubber is necessary to remove the halo acids produced.

References

(1) U.S. Environmental Protection Agency, *Halomethanes: Ambient Water Quality Criteria,* Washington, DC (1980).
(2) U.S. Environmental Protection Agency, *Trichlorofluoromethane and Dichlorodifluoromethane,* Health and Environmental Effects Profile No. 167, Office of Solid Waste, Washington, DC (April 30, 1980).
(3) See Reference (A-61).

FOLPET

Description: Folpet is a colorless crystalline solid which melts at 177°C. It has the structure

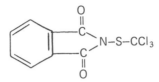

Code Numbers: CAS 133-07-3   RTECS TI5685000   UN 2773

DOT Designation: —

Synonyms: N-[(Trichloromethyl)thio]phthalimide; Folpel (in France); Phaltan®; Folpan®.

Potential Exposure: Those involved in manufacture, formulation and application of this fungicide.

Incompatibilities: Folpet is stable when dry, but hydrolyzes slowly in water at room temperature. It has been reported to undergo photodegradation on plant surfaces to phthalic acid, chloride, and inorganic sulfur compounds. In the presence of sulfhydryl compounds, Folpet degrades rapidly to sulfur, phthalimide and hydrochloric acid.

Permissible Exposure Limits in Air: No standard set.

Permissible Concentration in Water: A no-adverse effect level in drinking water has been calculated by NAS/NRC to be 1.1 mg/ℓ.

Harmful Effects and Symptoms: See Captan. Folpet is insignificantly toxic (LD_{50} for rats is 10,000 mg/kg).

Disposal Method Suggested: Folpet hydrolyzes slowly in water. It hydrolyzed rapidly at elevated temperatures or in alkaline media (A-32).

FONOFOS

Description: O-ethyl S-phenyl ethylphosphonodithioate with the structural formula:

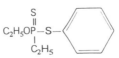

is a pale yellow liquid with a pungent mercaptonlike odor.

Code Numbers: CAS 944-22-9   RTECS TA5950000   UN 2783

DOT Designation: —

Synonyms: Dyfonate®; ENT 25796.

Potential Exposure: Those involved in the manufacture (A-32), formulation and application of this insecticide.

Permissible Exposure Limit in Air: There is no Federal standard but ACGIH (1983/84) has adopted a TWA value of 0.1 mg/m^3 with the notation "skin" indicating the possibility of skin absorption.

Permissible Concentration in Water: No criteria set.

Harmful Effects and Symptoms: There seems to be no data available on human exposure but feeding experiments with both male and female rats for 13 weeks showed no effects at levels as high as 31.6 ppm; only moderate inhibition of serum and red blood cell cholinesterase activity was noted at 100 ppm (A-34).

Disposal Method Suggested: This phosphono compound is reported to be satisfactorily decomposed by hypochlorite (A-32).

FORMALDEHYDE

- Carcinogen (Animal Suspected) (7), (Potential Occupational) (9) (12)
- Hazardous substance (EPA)
- Hazardous waste (EPA)

Description: HCHO, formaldehyde, is a colorless, pungent gas. It is sold in aqueous solution containing 30 to 50% formaldehyde and from 0 to 15% methanol, which is added to prevent polymerization.

Code Numbers: CAS 50-00-0   RTECS LP8925000   UN 2209 (or 1198)

DOT Designation: Combustible liquid.

Synonyms: Oxomethane; oxymethylene; methylene oxide; formic aldehyde; methyl aldehyde. Formaldehyde solution: formalin.

Potential Exposure: Formaldehyde has found wide industrial usage as a fungicide, germicide, and in disinfectants and embalming fluids. It is also used in the manufacture of artificial silk and textiles, latex, phenol, urea, thiourea and melamine resins, dyes, and inks, cellulose esters and other organic molecules, mirrors, and explosives. It is also used in the paper, photographic, and furniture industries. It is an intermediate in drug manufacture (A-41) and is a pesticide intermediate (A-32).

Human exposure to HCHO is principally through inhalation and skin absorption or less frequently by ingestion. Most of the HCHO production in the United States is from methanol in closed automated process systems. Exposure potential during transportation and storage is likely to be minimal. Estimated emission levels from production plants range from 0.0004 to 2,500 μg/m^3 with a median exposure of 0.01 μg/m^3. EPA estimates that 27.7 million people living within 12.5 miles of point sources may be exposed to low levels. NIOSH estimates that 8,000 workers are potentially exposed to HCHO during direct production. In addition, pathologists and histology technicians represent a high exposure group. The National Occupational Hazard Survey estimates that 57,000 full-time and 1.7 million part-time employees may be exposed to HCHO, and OSHA estimates the number of workers exposed as 2.6 million.

Most consumers are exposed to formaldehyde through its use in construction materials, wood products, textiles, home furnishings, paper, cosmetics, and pharmaceuticals. The ambient air levels in the United States, ranging from about 0.001 to 0.030 ppm, expose the entire 220 million population. Two subpopulations have been identified as having particularly high potential for HCHO exposure: 2.2 million residents of mobile homes containing particle board and plywood have an average exposure of 0.4 ppm HCHO, and 1.7 million persons living in conventional homes insulated with urea-formaldehyde foam have a potential average exposure of 0.12 ppm. Inadvertent production of formaldehyde from combustion sources also may contribute to these exposures. An estimated 159 million persons are potentially exposed to ambient air concentrations at maximum 0.25 $\mu g/m^3$ levels of HCHO. Automobiles alone emit 610 million pounds of HCHO each year. HCHO has a short half-life in air, however, because it is degraded by photochemical processes.

Incompatibilities: Strong oxidizers, strong alkalies, acids; phenols; urea.

Permissible Exposure Limits in Air: The Federal standard is 3 ppm determined as a TWA. The acceptable ceiling concentration is 5 ppm with an acceptable maximum peak above this value of 10 ppm for a maximum duration of 30 minutes. ACGIH has adopted a TLV of 2 ppm (3 mg/m^3) as a ceiling value but has set no STEL value. ACGIH has indicated an intended change as of 1983/84 to a TWA of 1.0 ppm (1.5 mg/m^3) and an STEL of 2.0 ppm (3 mg/m^3) with the notation that formaldehyde is "an industrial substance suspect of carcinogenic potential for man." NIOSH has recommended (2) a ceiling of 0.8 ppm (1.2 mg/m^3) for any 30-minute sampling period. The IDLH level is 100 ppm.

Determination in Air: Adsorption on alumina, elution with aqueous methanol, reaction with chromotropic acid in sulfuric acid, spectrophotometric determination (A-10). Polarography may also be employed (A-10).

Permissible Concentration in Water: No criteria set, but EPA (A-37) has suggested a permissible ambient goal of 41.4 $\mu g/\ell$ based on health effects.

Routes of Entry: Inhalation, ingestion, skin and eye contact.

Harmful Effects and Symptoms: *Local* – Formaldehyde gas may cause severe irritation to the mucous membranes of the respiratory tract and eyes. The aqueous solution splashed in the eyes may cause eye burns. Urticaria has been reported following inhalation of gas. Repeated exposure to formaldehyde may cause dermatitis either from irritation or allergy.

Systemic – Systemic intoxication is unlikely to occur since intense irritation of upper respiratory passages compels workers to leave areas of exposure. If workers do inhale high concentrations of formaldehyde, coughing, difficulty in breathing, and pulmonary edema may occur. Ingestion, though usually not occurring in industrial experience, may cause severe irritation of the mouth, throat, and stomach. Formaldehyde has been found to be mutagenic in a variety of tests.

While a full evaluation of the carcinogenicity of formaldehyde vapor must await completion of studies at the Chemical Industry Institute of Toxicology, evidence presented to date demonstrates that inhalation of formaldehyde results in a high incidence of nasal cancers in rats (7).

Points of Attack: Respiratory system, lungs, eyes, skin.

Medical Surveillance: Consider the skin, eyes, and respiratory tract in any preplacement or periodic examination, especially if the patient has a history of allergies.

First Aid: If this chemical gets into the eyes, irrigate immediately. If this chemical contacts the skin, flush with water promptly. If a person breathes in large amounts of this chemical, move the exposed person to fresh air at once and perform artificial respiration. When this chemical has been swallowed, get medical attention. Give large quantities of water and induce vomiting. Do not make an unconscious person vomit.

Personal Protective Methods: Prevention of intoxication may be easily accomplished by supplying adequate ventilation and protective clothing. Barrier creams may also be helpful. In areas of high vapor concentration, full protective face masks with air supply is needed, as well as protective clothing. Wear appropriate clothing to prevent any reasonable probability of skin contact. Wear eye protection to prevent any possibility of eye contact. Employees should wash immediately when skin is wet or contaminated. Remove nonimpervious clothing immediately if wet or contaminated. Provide emergency showers and eyewash.

Respirator Selection:

    50 ppm: CCROVF/GMOV/SAF/SCBAF
  100 ppm: SAF:PD,PP,CF
  Escape: GMOV/SCBA

Disposal Method Suggested: Incineration. Also, formaldehyde may be recovered from wastewaters (A-58).

References

(1) Environmental Protection Agency, *Investigation of Selected Potential Environmental Contaminants—Formaldehyde, Final Report,* Office of Toxic Substances, Environmental Protection Agency, August, 1976.

(2) National Institute for Occupational Safety and Health, *Criteria for a Recommended Standard: Occupational Exposure to Formaldehyde,* NIOSH Doc. No. 77-126 (1977).

(3) U.S. Environmental Protection Agency, *Chemical Hazard Information Profile: Formaldehyde,* Washington, DC (1979).

(4) U.S. Environmental Protection Agency, *Formaldehyde,* Health and Environmental Effects Profile No. 104, Office of Solid Waste, Washington, DC (April 30, 1980).

(5) Sax, N.I., Ed., *Dangerous Properties of Industrial Materials Report, 1,* No. 4, 70-72, New York, Van Nostrand Reinhold Co. (1981).

(6) Sax, N.I., Ed., *Dangerous Properties of Industrial Materials Report, 3,* No. 3, 71-76, New York, Van Nostrand Reinhold Co. (1983).

(7) See Reference (A-62). Also see Reference (A-64).

(8) See Reference (A-60).

(9) National Institute for Occupational Safety and Health, *Formaldehyde: Evidence of Carcinogenicity,* Current Intelligence Bulletin No. 34, DHHS (NIOSH) Publication No. 81-111, Cincinnati, Ohio (April 15, 1981).

(10) Parmeggiani, L., Ed., *Encyclopedia of Occupational Health and Safety,* Third Edition, Vol. 1, pp 914-916, Geneva, International Labour Office (1983).

(11) United Nations Environment Programme, *IRPTC Legal File 1983,* Vol. II, pp VII/374-78, Geneva, Switzerland, International Register of Potentially Toxic Chemicals (1984).

(12) U.S. Environmental Protection Agency, 49 FR 21870 (May 23, 1984).

(13) Clary, J.J., Gibson, J.E. and Waritz, R.S., *Formaldehyde Toxicology, Epidemiology, Mechanisms,* New York, Marcel Dekker, Inc. (1983).

FORMAMIDE

Description: $HCONH_2$ is a clear viscous liquid boiling at $210°C$ with a faint ammonia odor.

Code Numbers: CAS 75-12-7 RTECS LQ0525000

DOT Designation: —

Synonyms: Carbamaldehyde; methanamide

Potential Exposure: Formamide is a powerful solvent. It is also used as an intermediate in pharmaceutical manufacture (A-41). It may be pyrolyzed to give HCN.

Permissible Exposure Limits in Air: There are no Federal standards but ACGIH (1983/84) has adopted a TWA value of 20 ppm ($30 \ mg/m^3$) and set an STEL of 30 ppm ($45 \ mg/m^3$). A reported Soviet limit is $3 \ mg/m^3$.

Permissible Concentration in Water: No criteria set, but EPA (A-37) has suggested a permissible ambient goal of $414 \ \mu g/\ell$ based on health effects.

Harmful Effects and Symptoms: Formamide is mildly irritant to the skin and mucous membranes.

First Aid: Irrigate eyes with water. Wash all contaminated areas of body with soap and water. Treat skin burns as usual (A-38).

Personal Protective Methods: Wear rubber gloves, goggles and overalls (A-38).

Disposal Method Suggested: Dissolve in a combustible solvent and dispose by open burning or preferably incinerate in a furnace equipped with an alkali scrubber for the exit gases.

References
(1) Sax, N.I., Ed., *Dangerous Properties of Industrial Materials Report, 1,* No. 1, 44-45, New York, Van Nostrand Reinhold Co. (1980).
(2) See Reference (A-60).
(3) United Nations Environment Programme, *IRPTC Legal File 1983,* Vol. II, pp VII/379, Geneva, Switzerland, International Register of Potentially Toxic Chemicals (1984).

FORMIC ACID

- Hazardous substance (EPA)
- Hazardous waste (EPA)

Description: HCOOH, formic acid, is a colorless, flammable, fuming liquid, with a pungent odor. It boils at $101°C$.

Code Numbers: CAS 64-18-6 RTECS LQ4900000 UN 1779

DOT Designation: Corrosive material.

Synonyms: Methanoic acid; formylic acid; hydrogen carboxylic acid.

Potential Exposure: Formic acid is a strong reducing agent and is used as a decalcifier. It is used in dyeing color fast wool, electroplating, coagulating latex rubber, regenerating old rubber, and dehairing, plumping, and tanning leather. It is also used in the manufacture of acetic acid, airplane dope, allyl alcohol, cellulose formate, phenolic resins, and oxalate; and it is used in the laundry, textile, insecticide, refrigeration, and paper industries, as well as in drug manufacture (A-41).

Incompatibilities: Strong oxidizers, strong caustics, concentrated sulfuric acid.

Permissible Exposure Limits in Air: The Federal standard and the 1983/84 ACGIH TWA value is 5 ppm (9 mg/m^3). There is no STEL value. The IDLH level is 100 ppm.

Determination in Air: Collection with caustic in an impinger, conversion to ethyl formate, analysis by gas chromatography (A-10). See also (A-22).

Permissible Concentration in Water: No criteria set, but EPA (A-37) has suggested a permissible ambient goal of 124 $\mu g/\ell$ based on health effects.

Routes of Entry: Inhalation of vapor, percutaneous absorption, ingestion, eye and skin contact.

Harmful Effects and Symptoms: *Local* – The primary hazard of formic acid results from severe irritation of the skin, eyes, and mucous membranes. Lacrimation, increased nasal discharge, cough, throat discomfort, erythema, and blistering may occur depending upon solution concentrations.

Systemic – These have not been reported from inhalation exposure and are unlikely due to its good warning properties.

Swallowing formic acid has caused a number of cases of severe poisoning and death. The symptoms found in this type of poisoning include salivation, vomiting, burning sensation in the mouth, bloody vomiting, diarrhea, and pain. In severe poisoning, shock may occur. Later, breathing difficulties may develop. Kidney damage may also be present.

Points of Attack: Respiratory system, lungs, skin, kidneys, liver, eyes.

Medical Surveillance: Consideration should be given to possible irritant effects on the skin, eyes, and lungs in any placement or periodic examinations.

First Aid: If this chemical gets into the eyes, irrigate immediately. If this chemical contacts the skin, flush with water immediately. If a person breathes in large amounts of this chemical, move the exposed person to fresh air at once and perform artificial respiration. When this chemical has been swallowed, get medical attention. Give large quantities of water and do not induce vomiting.

Personal Protective Method: Wear appropriate clothing to prevent any possibility of skin contact. Wear eye protection to prevent any possibility of eye contact. Employees should wash immediately when skin is wet or contaminated. Remove nonimpervious clothing immediately if wet or contaminated. Provide emergency showers and eyewash.

Respirator Selection:
>110 ppm: CCROVFDM/GMOVDM/SAF/SCBAF
>Escape: GMOVP/SCBA

Disposal Method Suggested: Incineration.

References

(1) U.S. Environmental Protection Agency, *Formic Acid,* Health and Environmental Effects Profile No. 105, Office of Solid Waste, Washington, DC (April 30, 1980).
(2) Sax, N.I., Ed., *Dangerous Properties of Industrial Materials Report, 1,* No. 2, 39-41, New York, Van Nostrand Reinhold Co. (1980).
(3) See Reference (A-61).
(4) See Reference (A-60).
(5) Sax, N.I., Ed., *Dangerous Properties of Industrial Materials Report, 3,* No. 4, 53-56, New York, Van Nostrand Reinhold Co. (1983).

FUMARONITRILE

$$NC-CH$$
$$\| $$

Description: HC–CN is a crystalline solid melting at 97°C and boiling at 186°C.

Code Numbers: CAS 17656-09-6

DOT Designation: –

Synonyms: Trans-1,2-dicyanoethylene; 2-butenedinitrile.

Potential Exposure: Fumaronitrile is used as a bactericide and as an antiseptic for metal cutting fluids. It is used to make polymers with styrene and numerous other compounds. This compound is easily isomerized to the cisform, maleonitrile, which is a bactericide and fungicide.

Permissible Exposure Limit in Air: No standards set.

Permissible Concentration in Water: No criteria set.

Harmful Effects and Symptoms: LD_{50} values for injected mice and orally dosed rats were 38 and 50 mg/kg, respectively. Human exposure to fumaronitrile cannot be assessed due to a lack of monitoring data. Data concerning the effects of fumaronitrile to aquatic organisms were not found in the available literature.

References

(1) U.S. Environmental Protection Agency, *Fumaronitrile,* Health and Environmental Effects Profile No. 106, Office of Solid Waste, Washington, DC (April 30, 1980).

FURFURAL

- Hazardous substance (EPA)
- Hazardous waste (EPA)

Description: $C_5H_4O_2$, furfural, is an aromatic heterocyclic aldehyde with an amber color and aromatic odor. It boils at 161° to 162°C.

Code Numbers: CAS 98-01-1   RTECS LT7000000   UN 1199

DOT Designation: Combustible liquid.

Synonyms: Furfurol (a misnomer); furfuraldehyde; artificial ant oil; pyromucic aldehyde; furol; 2-furaldehyde.

Potential Exposure: Furfural is used as a solvent for wood resin, nitrated cotton, cellulose acetate, and gums. It is used in the production of phenolic plastics, thermosetting resins, refined petroleum oils, dyes, and varnishes. It is also utilized in the manufacture of pyromucic acid, vulcanized rubber, insecticides, fungicides, herbicides, germicides, furan derivatives, polymers, and other organic chemicals.

Incompatibilities: Strong acids, oxidizers.

Permissible Exposure Limits in Air: The Federal standard is 5 ppm (20 mg/m³). The ACGIH as of 1983/84 has adopted a TWA value of 2 ppm (8 mg/m³) and an STEL value of 10 ppm (40 mg/m³). The notation "skin" is added to indicate the possibility of cutaneous absorption. The IDLH level is 250 ppm.

Determination in Air: Adsorption by charcoal or alumina, analysis by gas liquid chromatography (A-13).

Permissible Concentration in Water: No criteria set.

Routes of Entry: Inhalation of vapor, percutaneous absorption, ingestion, skin and eye contact.

Harmful Effects and Symptoms: *Local* — Liquid and concentrated vapor are irritating to the eyes, skin, and mucous membranes of the upper respiratory tract. Eczematous dermatitis as well as skin sensitization, resulting in allergic contact dermatitis and photosensitivity, may develop following repeated exposure.

Systemic — Workers chronically exposed to the vapor have had complaints of headache, fatigue, itching of the throat, lacrimation, loss of the sense of taste, numbness of the tongue, and tremor. Occupational overexposure is relatively rare due to the liquid's low vapor pressure, and symptoms usually disappear rapidly after removal from exposure.

Points of Attack: Eyes, respiratory system, skin.

Medical Surveillance: Consider skin irritation and skin allergies (especially to aldehydes) in preplacement or periodic examinations. Also consider possible respiratory irritant effects.

First Aid: If this chemical gets into the eyes, irrigate immediately. If this chemical contacts the skin, flush with water promptly. If a person breathes in large amounts of this chemical, move the exposed person to fresh air at once and perform artificial respiration. When this chemical has been swallowed, get medical attention. Give large quantities of salt water and induce vomiting. Do not make an unconscious person vomit.

Personal Protective Methods: Wear appropriate clothing to prevent repeated or prolonged skin contact. Wear eye protection to prevent any reasonable probability of eye contact. Employees should wash promptly when skin is wet or contaminated. Remove nonimpervious clothing promptly if wet or contaminated.

Respirator Selection:
> 250 ppm: CCROVF/GMOV/SAF/SCBAF
> Escape: GMOV/SCBA

Disposal Method Suggested: Incineration.

References

(1) Sax, N.I., Ed., *Dangerous Properties of Industrial Materials Report, 1,* No. 2, 41-42, New York, Van Nostrand Reinhold Co. (1980).
(2) See Reference (A-61).
(3) See Reference (A-60).
(4) Parmeggiani, L., Ed., *Encyclopedia of Occupational Health & Safety,* Third Edition, Vol. 1, pp 931–32, Geneva, International Labour Office (1983).

FURFURYL ALCOHOL

Description: $C_4H_3OCH_2OH$ with the structural formula:

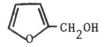

is an amber liquid with a mildly irritating odor. It boils at 170°C.

Code Numbers: CAS 98-00-0   RTECS LU9100000   UN 2874

Synonyms: 2-hydroxymethylfuran; 2-furylmethanol.

Potential Exposure: Furfuryl alcohol is mainly used as a starting monomer in the production of furan resins or furan polymers, and is also used to produce THFA (tetrahydrofurfuryl alcohol).

Incompatibilities: Strong oxidizers, strong acids, organic acids may lead to polymerization.

Permissible Exposure Limit in Air: The Federal standard is 50 ppm (200 mg/m³). The ACGIH as of 1983/84 has adopted a TWA of 10 ppm (40 mg/m³) and an STEL of 15 ppm (60 mg/m³). The IDLH level is 250 ppm. A Dutch publication (A-60) has set 5 ppm (20 mg/m³) as the maximum allowable concentration in air.

Determination in Air: Use of an adsorption tube, workup with acetone, analysis by gas chromatography. See NIOSH Methods, Set 2. See also reference (A-10).

Permissible Concentration in Water: No criteria set.

Routes of Entry: Inhalation, ingestion, skin and eye contact.

Harmful Effects and Symptoms: Dizziness, nausea, diarrhea, vomiting. Respiration and body temperature are affected also.

Points of Attack: Respiratory system, lungs.

Medical Surveillance: Consider the points of attack in preplacement and periodic physical examinations.

First Aid: If this chemical gets into the eyes, irrigate immediately. If this chemical contacts the skin, flush with water immediately. If a person breathes in large amounts of this chemical, move the exposed person to fresh air at once and perform artificial respiration. When this chemical has been swallowed, get medical attention. Give large quantities of water and induce vomiting. Do not make an unconscious person vomit.

Personal Protective Methods: Wear appropriate clothing to prevent any reasonable probability of skin contact. Wear eye protection to prevent any reasonable probability of eye contact. Employees should wash immediately when skin is wet or contaminated. Remove nonimpervious clothing immediately if wet or contaminated. Provide emergency showers.

Respirator Selection:
        250 ppm: CCROVF/GMOV/SAF/SCBAF
        Escape: GMOV/SCBA

Disposal Method Suggested: Incineration in admixture with a more flammable solvent.

References

(1) National Institute for Occupational Safety and Health, *Information Profiles on Potential Occupational Hazards: Furfuryl Alcohol,* Report PB-276,678, pp. 12–15, Rockville, Md. (1977).

(2) National Institute for Occupational Safety and Health, *Criteria for a Recommended Standard: Occupational Exposure to Furfuryl Alcohol,* NIOSH Document No. 79-133 (1979).

G

GALLIC ACID

Description: $C_7H_6O_5$ has the following structural formula

Gallic acid is a crystalline compound which melts with decomposition at 235° to 240°C.

Code Numbers: CAS 149-91-7 RTECS LW7525000

DOT Designation: —

Synonyms: 3,4,5-Trihydroxybenzoic acid.

Potential Exposure: In manufacture of gallic esters as antioxidants, in photographic developing, in tanning.

Incompatibilities: Oxidants.

Permissible Exposure Limits in Air: No standards set.

Permissible Concentration in Water: No criteria set.

Routes of Entry: Inhalation, ingestion.

Harmful Effects and Symptoms: Irritates the eyes, skin and respiratory tract.

First Aid: Remove contaminated clothes, rinse affected areas with water.

Personal Protective Methods: Wear protective gloves and safety goggles.

Disposal Method Suggested: Incineration.

References

(1) Sax, N.I., Ed., *Dangerous Properties of Industrial Materials Report, 3,* No. 4, 56-58, New York, Van Nostrand Reinhold Co. (1983).
(2) See Reference (A-60).

GASOLINE

Description: Gasoline is a highly flammable, mobile liquid with a characteristic odor. It boils over a range from 40° to 200°C.

Code Numbers: CAS 8006-61-9   RTECS LX3300000   UN 1203

DOT Designation: Flammable liquid.

Synonyms: Petrol, motor spirits, benzin.

Potential Exposure: Gasoline is used as a fuel, diluent, and solvent throughout industry.

Permissible Exposure Limits in Air: Presently, the composition of gasoline is so varied that a single Federal standard for all types of gasoline is not applicable. It is recommended, however, that atmospheric concentrations should be limited by the aromatic hydrocarbon content. ACGIH as of 1983/84, however, has adopted a TWA of 300 ppm (900 mg/m^3) and a STEL value of 500 ppm (1,500 mg/m^3).

Permissible Concentration in Water: No criteria set.

Routes of Entry: Most cases of poisoning reported have resulted from inhalation of vapor and ingestion. It is not known whether gasoline poisoning may be compounded by percutaneous absorption.

Harmful Effects and Symptoms: *Local* — Gasoline is irritating to skin, conjunctiva, and mucous membranes. Dermatitis may result from repeated and prolonged contact with the liquid, which may defat the skin. Certain individuals may develop hypersensitivity.

Systemic — Gasoline vapor acts as a central nervous system depressant. Exposure to low concentrations may produce flushing of the face, staggering gait, slurred speech, and mental confusion. In high concentrations, gasoline vapor may cause unconsciousness, coma, and possibly death resulting from respiratory failure.

Other signs also may develop following acute exposure. These signs are early acute hemorrhage of the pancreas, centrilobular cloudy swelling and fatty degeneration of the liver, fatty degeneration of the proximal convoluted tubules and glomeruli of the kidneys, and passive congestion of the spleen.

Ingestion and aspiration of the liquid gasoline usually occurs during siphoning.

Chemical pneumonitis, pulmonary edema, and hemorrhage may follow. Aromatic hydrocarbon content may also cause hematopoietic changes. Absorption of alkyl lead antiknock agents contained in many gasolines poses an additional problem especially where there is prolonged skin contact. The existence of chronic poisoning has not been established.

Medical Surveillance: No special considerations are necessary.

First Aid: Irrigate eyes with running water. Wash contaminated areas of body with soap and water. If swallowed, use gastric lavage (stomach wash) followed by saline catharsis.

Personal Protective Methods: Barrier creams and impervious gloves, protective clothing. Masks in heavy exposure to vapors.

Disposal Method Suggested: Incineration. Alternatively, gasoline vapors may be recovered from fuel transfer operations by various techniques (A-58).

GERMANIUM

Description: Ge, germanium, is a greyish-white, lustrous, brittle metalloid.

It is never found free and occurs most commonly in argyrodite and germanite. It is generally produced from germanium containing minerals or as a by-product in zinc production or coal processing. Germanium is insoluble in water.

Code Numbers: CAS None   RTECS None

DOT Designation: –

Synonyms: None.

Potential Exposure: Because of its semiconductor properties, germanium is widely used in the electronic industry in rectifiers, diodes, and transistors. It is alloyed with aluminum, aluminum-magnesium, antimony, bronze, and tin to increase strength, hardness, or corrosion resistance. In the process of alloying germanium and arsenic, arsine may be released; stibine is released from the alloying of germanium and antimony. Germanium is also used in the manufacture of optical glass for infrared applications, red-fluorescing phosphors, and cathodes for electronic valves, and in electroplating, in the hydrogenation of coal, and as a catalyst, particularly at low temperatures. Certain compounds are used medically.

Industrial exposures to the dust and fumes of the metal or oxide generally occur during separation and purification of germanium, welding, multiple-zone melting operations, or cutting and grinding of crystals. Germanium tetrahydride (germanium hydride, germane, monogermane) and other hydrides are produced by the action of a reducing acid on a germanium alloy.

Permissible Exposure Limits in Air: There is no Federal standard for germanium or its compounds; however, the ACGIH (1983/84) has set a TLV for germanium tetrahydride of 0.2 ppm (0.6 mg/m^3) and an STEL of 0.6 ppm (1.8 mg/m^3).

Permissible Concentration in Water: No criteria set, but EPA (A-37) has suggested a permissible ambient goal of 8 μg/ℓ based on health effects.

Determination in Water: Germanium may be determined by atomic absorption spectroscopy, emission spectography and spectrophotometry with phenylfluorone (1).

Routes of Entry: Inhalation of gas, vapor, fume, or dust.

Harmful Effects and Symptoms: *Local* – The dust of germanium dioxide is irritating to the eyes. Germanium tetrachloride causes irritation of the skin.

Systemic – Germanium tetrachloride is an upper respiratory irritant and may cause bronchitis and pneumonitis. Prolonged exposure to high level concentrations may result in damage to the liver, kidney, and other organs. Germanium tetrahydride is a toxic hemolytic gas capable of producing kidney damage.

Medical Surveillance: Consider respiratory, liver, and kidney disease in any placement or periodic examinations.

Personal Protective Methods: In dust areas, protective clothing and gloves may be necessary to protect the skin, and goggles to protect the eyes. In areas where germanium tetrachloride is in high concentrations, dust-fume masks or supplied air respirators with full facepiece should be supplied to all workers. Personal hygiene is to be encouraged, with change of clothes following each shift and showering prior to change to street clothes.

Disposal Method Suggested: Recovery and return to suppliers for reprocessing is preferable.

References
(1) U.S. Environmental Protection Agency, *Toxicology of Metal, Vol. II: Germanium,* pp 222–223, Report EPA-600/1-77-022, Research Triangle Park, NC (May 1977).
(2) Parmeggiani, L., Ed., *Encyclopedia of Occupational Health & Safety,* Third Edition, Vol. 1, pp 964–65, Geneva, International Labour Office (1983).

GLUTARALDEHYDE

Description: $HCO(CH_2)_3CHO$ is a liquid which readily changes to a glossy polymer.

Code Numbers: CAS 111-30-8   RTECS MA2450000

DOT Designation: —

Synonyms: Glutaric dialdehyde; **1,5-pentanedial.**

Potential Exposure: Glutaraldehyde is used as a cross-linking agent for protein and polyhydroxy materials. It has been used in tanning and as a fixative for tissues. It is also used as an intermediate. Buffered solutions are used as antimicrobial agents in hospitals.

Permissible Exposure Limits in Air: There are no Federal standards but ACGIH (1983/84) has adopted a ceiling value of 0.2 ppm (0.7 mg/m^3) but proposed no STEL.

Permissible Concentration in Water: No criteria set.

Harmful Effects and Symptoms: Irritation of skin, eyes, nasal passages and upper respiratory tract (A-34).

Points of Attack: Skin, eyes and respiratory system.

Disposal Method Suggested: Incineration.

References
(1) See Reference (A-60).

GLYCERIN (MIST)

Description: $HOCH_2CHOHCH_2OH$ is a viscous liquid boiling at 290°C.

Code Numbers: CAS 56-81-5   RTECS MA8050000

DOT Designation: —

Synonyms: Glycerol; Glycyl alcohol; 1,2,3-propanetriol; Trihydroxypropane.

Potential Exposure: Glycerol is used as a humectant in tobacco; it is used in cosmetics, antifreezes and inks. It is used as a fiber lubricant. It is used as a raw material for alkyd resins and in explosives manufacture.

Incompatibilities: Strong oxidizers.

Permissible Exposure Limits in Air: There is no Federal limit but ACGIH (1980) classifies glycerin mist as a nuisance particulate with a TLV of 10 mg/m^3 and a tentative STEL of 10 mg/m^3.

Determination in Air: Collection of the mist on a filter followed by gravimetric analysis (A-1).

Permissible Concentration in Water: No criteria set.

Harmful Effects and Symptoms: Glycerin can be irritating to the eyes and respiratory tract. When swallowed, it can cause insomnia, nausea, vomiting, diarrhea, fever, hemoglobinuria, convulsions and paralysis (A-38).

First Aid: Irrigate eyes. Wash contaminated areas of body with soap and water (A-38).

Disposal Method Suggested: Mixture with a more flammable solvent followed by incineration (A-38).

References

(1)   Sax, N.I., Ed., *Dangerous Properties of Industrial Materials Report, 1,* No. 5, 61-63, New York, Van Nostrand Reinhold Co. (1981).
(2)   See Reference (A-60).
(3)   Sax, N.I., Ed., *Dangerous Properties of Industrial Materials Report, 3,* No. 4, 58-60, New York, Van Nostrand Reinhold Co. (1983).
(4)   Parmeggiani, L., Ed., *Encyclopedia of Occupational Health & Safety,* Third Edition, Vol. 1, pp 971–73, Geneva, International Labour Office (1983).

GLYCIDOL

Description: $C_3H_6O_2$, $HOCH_2\overset{O}{\overset{/\ \ \backslash}{CH-CH_2}}$ is a colorless liquid boiling at 166°C.

Code Numbers: CAS 556-52-5   RTECS UB4375000

DOT Designation: –

Synonyms: 2-Hydroxymethyloxiran; hydroxymethyl ethylene oxide; epoxypropyl alcohol; 3-hydroxypropylene oxide; 2,3-epoxy-1-propanol.

Potential Exposure: Glycidol is used as an intermediate in the synthesis of glycerol, glycidyl ethers, esters and amines.

Incompatibilities: Strong oxidizers, nitrates.

Permissible Exposure Limits in Air: The Federal standard is 50 ppm (150 mg/m³). The ACGIH as of 1983/84 has adopted a TWA of 25 ppm (75 mg/m³) and an STEL of 100 ppm (300 mg/m³). The IDLH is 500 ppm.

Determination in Air: Adsorption on charcoal workup with tetrahydrofuran, analysis by gas chromatography. See NIOSH Method, Set 2. See also reference (A-10).

Permissible Concentration in Water: No criteria set.

Routes of Entry: Inhalation, eye and skin contact, ingestion.

Harmful Effects and Symptoms: Irritation of eyes, nose, throat and skin; narcosis.

Points of Attack: Eyes, skin, respiratory system, central nervous system.

Medical Surveillance: Consider the points of attack in preplacement and periodic physical examinations.

First Aid: If this chemical gets into the eyes, irrigate immediately. If this

chemical contacts the skin, flush with water promptly. If a person breathes in large amounts of this chemical, move the exposed person to fresh air at once and perform artificial respiration. When this chemical has been swallowed, get medical attention. Give large quantities of salt water and induce vomiting. Do not make an unconscious person vomit.

Personal Protective Methods: Wear appropriate clothing to prevent repeated or prolonged skin contact. Wear eye protection to prevent any reasonable probability of eye contact. Wash promptly when skin is wet or contaminated. Remove nonimpervious clothing promptly if wet or contaminated.

Respirator Selection:

> 500 ppm: SAF/SCBAF
> Escape: GMOV/SCBA

Disposal Method Suggested: Concentrated waste containing no peroxides: discharge liquid at a controlled rate near a pilot flame. Concentrated waste containing peroxides: perforation of a container of the waste from a safe distance followed by open burning.

References

(1)   See Reference (A-61).
(2)   See Reference (A-60).

GRAIN DUST

Description: Oats, barley or wheat dust containing microbial flora and fauna.

DOT Designation: —

Potential Exposure: Grain elevator workers, grain harvesters.

Permissible Exposure Limits in Air: 4 mg/m^3 total dust per ACGIH (1983/84).

Permissible Concentration in Water: No criteria set.

Routes of Entry: Inhalation.

Harmful Effects and Symptoms: Impaired lung function (1), chronic bronchitis, both immediate and delayed asthmatic reactions, grain fever upon exposure to dust concentrations in excess of 15 mg/m^3.

Points of Attack: Respiratory system.

Respirator Selection: Dust respirator required.

References

(1)   American Conference of Governmental Industrial Hygienists, *Documentation of the Threshold Limit Values: Supplemental Documentation,* Cincinnati, OH, ACGIH (1982).

GRAPHITE

Description: Graphite is crystallized carbon and usually appears as soft, black scales. There are two types of graphite, natural and artificial.

Code Numbers: CAS None   RTECS VV7780000

DOT Designation: —

Synonyms: Plumbago, black lead, mineral carbon.

Potential Exposure: Natural graphite is used in foundry facings, steelmaking, lubricants, refractories, crucibles, pencil "lead," paints, pigments, and stove polish. Artificial graphite may be substituted for these uses with the exception of clay crucibles; other types of crucibles may be produced from artificial graphite. Additionally, it may be used as a high temperature lubricant or for electrodes. It is utilized in the electrical industry in electrodes, brushes, contacts, and electronic tube rectifier elements; as a constituent in lubricating oils and greases; to treat friction elements, such as brake linings; to prevent molds from sticking together; and in moderators in nuclear reactors.

In addition, concerns have been expressed about synthetic graphite in fibrous form. Those exposed are involved in production of graphite fibers from pitch or acrylonitrile fibers and the manufacture and use of composites of plastics, metals or ceramics reinforced with graphite fibers.

Incompatibilities: Very strong oxidizers, such as fluorine, chlorine trifluoride, potassium peroxide.

Permissible Exposure Limits in Air: The Federal standard for natural graphite is 15 mppcf. Synthetic graphite is designated a "nuisance particulate" by ACGIH with a TLV of 30 mppcf or 10 mg/m^3.

Determination in Air: Collection on a filter, gravimetric analysis (A-1).

Permissible Concentration in Water: No criteria set.

Route of Entry: Inhalation of dust, eye and skin contact.

Harmful Effects and Symptoms: *Local* — None.

Systemic — Exposure to natural graphite may produce a progressive and disabling pneumoconiosis similar to anthracosilicosis. Symptoms include headache, coughing, depression, decreased appetite, dyspnea, and the production of black sputum. Some individuals may be asymptomatic for many years then suddenly become disabled. It has not yet been determined whether the free crystalline silica in graphite is solely responsible for development of the disease. There is evidence that artificial graphite may be capable of producing a pneumoconiosis.

Points of Attack: Respiratory system, lungs, cardiovascular system.

Medical Surveillance: Preemployment and periodic examinations should be directed toward detecting significant respiratory disease, through chest x-rays and pulmonary function tests.

First Aid: If this chemical gets into the eyes, irrigate immediately. If a person breathes in large amounts of this chemical, move the exposed person to fresh air at once and perform artificial respiration.

Personal Protective Methods: Workers in exposed areas should be provided with dust masks with proper cartridges and should be instructed in their maintenance.

Respirator Selection:

```
75 mppcf:  D
150 mppcf: DXSQ/FuHiEP/SA/SCBA
750 mppcf: HiEPF/SAF/SCBAF
7500 mppcf: PAPHiE/SA:PD,PP,CF
```

Disposal Method Suggested: Carbon (graphite) fibers are difficult to dispose of by incineration. Waste fibers should be packaged and disposed of in a landfill authorized for the disposal of special wastes of this nature, or as otherwise may be required by law. Do not incinerate.

References

(1) See Reference (A-61).
(2) Parmeggiani, L., Ed., *Encyclopedia of Occupational Health & Safety,* Third Edition, Vol. 1, pp 978–79, Geneva, International Labour Office (1983).

GUTHION®

See "Azinphos-Methyl."

H

HAFNIUM AND COMPOUNDS

Description: Hf is a refractory metal which occurs in nature in zirconium minerals.

Code Numbers: (Hafnium metal) CAS 7440-58-6 RTECS MG4600000 UN 2545

DOT Designation: Flammable solid.

Potential Exposure: Hafnium metal has been used as a control rod material in nuclear reactors. Thus, those engaged in fabrication and machining of such rods may be exposed.

Incompatibilities: Strong oxidizers and chlorine.

Permissible Exposure Limits in Air: The Federal standard and the ACGIH 1983/84 TWA value is 0.5 mg/m^3. The STEL value is 1.5 mg/m^3. The IDLH level is 250 mg/m^3.

Determination in Air: Collection on a filter, analysis by atomic absorption spectrophotometry (A-21).

Permissible Concentration in Water: No criteria set.

Routes of Entry: Inhalation, ingestion, eye and skin contact.

Harmful Effects and Symptoms: Irritation of eyes, skin and mucous membranes.

Points of Attack: Eyes, skin and mucous membranes.

Medical Surveillance: Consider the points of attack in preplacement and periodic physical examinations.

First Aid: If this chemical gets into the eyes, irrigate immediately. If this chemical contacts the skin, wash with soap promptly. If a person breathes in large amounts of this chemical, move the exposed person to fresh air at once and perform artificial respiration. When this chemical has been swallowed, get medical attention. Give large quantities of water and induce vomiting. Do not make an unconscious person vomit.

Personal Protective Methods: Wear appropriate clothing to prevent any possibility of skin contact. Wear eye protection to prevent any possibility of eye contact. Employees should wash promptly when skin is wet or contaminated and daily at the end of each work shift. Work clothing should be changed daily if it is possible that clothing is contaminated. Remove nonimpervious clothing promptly if wet or contaminated. Provide emergency showers and eyewash.

478

Respirator Selection:

    2.5 mg/m^3 : DMXS
      5 mg/m^3 : DMXSQ/FuHiEP/SA/SCBA
   25 mg/m^3 : HiEPF/SAF/SCBAF
 250 mg/m^3 : SAF:PD,PP,CF
     Escape: HiEP/SCBA

Disposal Method Suggested: Recovery

References

(1) See Reference (A-61).

HALOMETHANES

See also entries under: "Methyl Chloride," "Methylene Chloride," "Chloroform," "Carbon Tetrachloride," "Chlorobromomethane," "Bromodichloromethane," "Chlorodifluoromethane," "Dibromochloromethane," "Fluorotrichloromethane," "Dichlorodifluoromethane," "Dichloromonofluoromethane" and "Trifluorobromomethane."

References

(1) U.S. Environmental Protection Agency, *Halomethanes: Ambient Water Quality Criteria,* Washington, DC (1980).
(2) U.S. Environmental Protection Agency, *Halomethanes,* Health and Environmental Effects Profile No. 107, Office of Solid Waste, Washington, DC (April 30, 1980).

HALOTHANE

Description: $CF_3CHBrCl$, 2-bromo-2-chloro-1,1,1-trifluoroethane is a colorless liquid with a sweetish odor boiling at 50°C.

Code Numbers: CAS 74-96-4 RTECS KH6475000

DOT Designation: —

Synonyms: Fluothane®

Potential Exposure: Halothane is used as an inhalation anesthetic. It has been estimated (A-42) that halothane accounts for two-thirds of all anesthesias.

Permissible Exposure Limits in Air: There is no Federal standard but ACGIH has proposed as of 1983/84 a TWA value of 50 ppm (400 mg/m^3) but no STEL value.

Permissible Concentration in Water: No criteria set.

Harmful Effects and Symptoms: Halothane gives rise to only a very low incidence of postoperative nausea and is generally safe which accounts for its widespread use.

References

(1) Sax, N.I., Ed., *Dangerous Properties of Industrial Materials Report, 1,* No. 5, 63, New York, Van Nostrand Reinhold Co. (1981).

HEMATITE: UNDERGROUND HEMATITE MINING

• Carcinogen (Human, Suspected) (A-62) (A-64)

Description: Hematite is an iron ore composed mainly of ferric oxide, Fe_2O_3.

Code Numbers: CAS 1317-60-8   RTECS MH7875000

DOT Designation: —

Potential Exposure: It, as an iron ore composed mainly of ferric oxide, is a major source of iron and is used as a pigment for rubber, paints, paper, linoleum, ceramics, dental restoratives, and as a polishing agent for glass and precious metals. It is also used in electrical resistors, semiconductors, magnets, and as a catalyst.

The combined U.S. production and imports of hematite approaches 2 billion pounds. Approximately 71 companies produce and import hematite in the United States.

Human exposure to hematite from underground hematite mining is principally through inhalation and/or ingestion of dust. No estimates are available concerning the number of underground miners exposed.

Permissible Exposure Limits in Air: ACGIH (1983/84) has adopted a TWA of 5 mg/m^3 for iron oxide fume with an STEL of 10 mg/m^3.

Permissible Concentration in Water: No criteria set.

Routes of Entry: Dust inhalation.

Harmful Effects and Symptoms: No carcinogenic effects were observed in mice, hamsters, or guinea-pigs given ferric oxide intratracheally.

Underground haematite miners have a high incidence of lung cancer, whereas surface haematite miners do not. It is not known whether this excess risk may be due to haematite; to radon (a known lung carcinogen); to inhalation of ferric oxide or silica; or to a combination of these or other factors. Some studies of metal workers exposed to ferric oxide dusts have shown an increased incidence of lung cancer, while other studies have not. The influence of factors in the workplace other than ferric oxide cannot be eliminated (1).

References
(1)   See Reference (A-62). Also see Reference (A-64).

HEPTACHLOR

- Carcinogen (positive, NCI)(4)
- Hazardous substance (EPA)
- Hazardous waste (EPA)
- Priority toxic pollutant (EPA)

Description: $C_{10}H_5Cl_7$ with the structural formula:

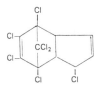

is a light tan waxy solid with a camphorlike odor melting at 95° to 96°C.

Code Numbers: CAS 76-44-8   RTECS PC0700000   UN (NA2761)

DOT Designation: ORM-E.

Synonyms: 1,4,5,6,7,8,8a-Heptachloro-3a,4,7,7a-tetrahydro-4,7-methano-indene (principal ingredient), ENT 15152.

Potential Exposure: Those involved in the manufacture, formulation and application of this insecticide. Infants have been exposed to heptachlor and heptachlor epoxide through mothers' milk, cows' milk, and commercially prepared baby foods. It appears that infants raised on mothers' milk run a greater risk of ingesting heptachlor epoxide than if they were fed cows' milk and/or commercially prepared baby food. Persons living and working in or near heptachlor treated areas have a particularly high inhalation exposure potential.

Incompatabilities: Melted heptachlor with iron and rust.

Permissible Exposure Limits in Air: The Federal standard and the ACGIH TWA value is 0.5 mg/m^3. The STEL value is 2.0 mg/m^3. The notation "skin" indicates the possibility of cutaneous absorption. The IDLH level is 100 mg/m^3.

Determination in Air: Collection by impinger or fritted bubbler, analysis by gas-liquid chromatography (A-1).

Permissible Concentration in Water: To protect freshwater aquatic life: 0.0038 μg/ℓ as a 24 hour average, never to exceed 0.52 μg/ℓ. To protect saltwater aquatic life: 0.0036 μg/ℓ as a 24 hour average, never to exceed 0.053 μg/ℓ. To protect human health: preferably zero. An additional lifetime cancer risk of 1 in 100,000 is imposed by a concentration of 2.78 ng/ℓ (0.00278 μg/ℓ).

Determination in Water: Methylene chloride extraction followed by gas chromatography with electron capture or halogen specific detection (EPA Method 608) or gas chromatography plus mass spectrometry (EPA Method 625).

Routes of Entry: Inhalation, skin absorption, ingestion, eye and skin contact.

Harmful Effects and Symptoms: Heptachlor has been demonstrated to be highly toxic to aquatic life, to persist for prolonged periods in the environment, to bioconcentrate in organisms at various trophic levels, and to exhibit carcinogenic activity in mice. Exposure symptoms in animals include tremors, convulsions and liver damage. The principal metabolite of heptachlor, heptachlor epoxide (3) is more acutely toxic than heptachlor.

Points of Attack: In animals: CNS, liver.

Medical Surveillance: Consider the points of attack in preplacement and periodic physical examinations.

First Aid: If this chemical gets into the eyes, irrigate immediately. If this chemical contacts the skin, wash with soap immediately. If a person breathes in large amounts of this chemical, move the exposed person to fresh air at once and perform artificial respiration. When this chemical has been swalllowed, get medical attention. Give large quantities of water and induce vomiting. Do not make an unconscious person vomit.

Personal Protective Methods: Wear appropriate clothing to prevent any reasonable probability of skin contact. Wear eye protection to prevent any reasonable probability of eye contact. Employees should wash immediately when skin is wet or contaminated. Work clothing should be changed daily if it is possible that clothing is contaminated. Remove nonimpervious clothing immediately if wet or contaminated. Provide emergency showers.

Respirator Selection:

5 mg/m^3: SA/SCBA

25 mg/m^3: SAF/SCBAF

500 mg/m^3: SA:PD,PP,CF

700 mg/m^3: SAF:PD,PP,CF

Escape: GMOVPPest/SCBA

Disposal Method Suggested: Incineration ($1500°F$, 0.5 sec minimum for primary combustion; $3200°F$, 1.0 sec for secondary combustion) with adequate scrubbing and ash disposal facilities.

References

(1) U.S. Environmental Protection Agency, *Heptachlor: Ambient Water Quality Criteria,* Washington, DC (1980).

(2) U.S. Environmental Protection Agency, *Heptachlor,* Health and Environmental Effects Profile No. 108, Office of Solid Waste, Washington, DC (April 30, 1980).

(3) U.S. Environmental Protection Agency, *Heptachlor Epoxide,* Health and Environmental Effects Profile No. 109, Office of Solid Waste, Washington, DC (April 30, 1980).

(4) National Cancer Institute, *Bioassay of Heptachlor for Possible Carcinogenicity,* Technical Report Series No. 9, Bethesda, Maryland (1977).

(5) Sax, N.I., Ed., *Dangerous Properties of Industrial Materials Report, 1,* No. 8, 76-78, New York, Van Nostrand Reinhold Co. (1981).

(6) See Reference (A-61).

(7) United Nations Environment Programme, *IRPTC Legal File 1983,* Vol. II, pp VII/451-- 54, Geneva, Switzerland, International Register of Potentially Toxic Chemicals (1984).

n-HEPTANE

Description: $CH_3(CH_2)_5CH_3$, n-heptane, is a clear liquid which is highly flammable and volatile.

Code Numbers: CAS 142-82-5   RTECS MI7700000   UN 1206

DOT Designation: Flammable liquid.

Potential Exposure: n-Heptane is used as a solvent and as a standard in testing knock of gasoline engines. NIOSH estimates annual worker exposure to n-heptane at 10,000.

Incompatibilities: Strong oxidizers.

Permissible Exposure Limit in Air: The Federal standard is 500 ppm ($2,000$ mg/m^3). NIOSH recommends a TWA of 85 ppm and a ceiling of 440 ppm. ACGIH as of 1983/84 has set a TWA of 400 ppm ($1,600 \text{ mg/m}^3$) and an STEL of 500 ppm ($2,000 \text{ mg/m}^3$). The IDLH level is 4,250 ppm.

Determination in Air: Charcoal adsorption, workup with CS_2, analysis by gas chromatography. See NIOSH Methods, Set G. See also reference (A-10).

Permissible Concentration in Water: No criteria set.

Route of Entry: Inhalation of the vapor, ingestion, skin and eye contact.

Harmful Effects and Symptoms: *Local* — n-Heptane can cause dermatitis and mucous membrane irritation. Aspiration of the liquid may result in chemical pneumonitis, pulmonary edema, and hemorrhage.

Systemic — Systemic effects may arise without complaints of mucous membrane irritation. Exposure to high concentrations causes narcosis producing vertigo, incoordination, intoxication characterized by hilarity, slight nausea, loss of appetite, and a persisting gasoline taste in the mouth. These effects may be first noticed on entering a contaminated area. n-Heptane may cause low order sensitization of the myocardium to epinephrine.

Points of Attack: Skin, respiratory system, lungs, peripheral nervous system.

Medical Surveillance: Preplacement examinations should evaluate the skin and general health, including respiratory, liver, and kidney function.

First Aid: If this chemical gets into the eyes, irrigate immediately. If this chemical contacts the skin, wash with soap promptly. If a person breathes in large amounts of this chemical, move the exposed person to fresh air at once and perform artificial respiration. When this chemical has been swallowed, get medical attention. Do not induce vomiting.

Personal Protective Methods: Wear appropriate clothing to prevent repeated or prolonged skin contact. Wear eye protection to prevent any reasonable probability of eye contact. Employees should wash promptly when skin is wet or contaminated. Remove clothing immediately if wet or contaminated to avoid flammability hazard.

Respirator Selection:

850 ppm:	CCROV/SA/SCBA
4,250 ppm:	GMOV/SAF/SCBAF
Escape:	GMOV/SCBA

Disposal Method Suggested: Incineration.

References

(1) National Institute for Occupational Safety and Health, *Criteria for a Recommended Standard: Occupational Exposure to Alkanes,* NIOSH Document No. 77-151, Washington, DC (1977).
(2) See Reference (A-61).
(3) See Reference (A-60).

3-HEPTENE

Description: $C_3H_7CH=CHC_2H_5$ is a colorless liquid boiling at $96°C$.

Code Numbers: CAS 592-78-9

DOT Designation: —

Synonyms: 1-ethyl-2-propyl ethylene.

Potential Exposure: Those involved in use as a plant growth retardant or in organic synthesis.

Permissible Exposure Limits in Air: No standards set.

Permissible Concentration in Water: No criteria set.

Routes of Entry: Skin contact.

Harmful Effects and Symptoms: Mild skin irritant, mild irritant upon inhalation; narcotic at higher concentrations.

Personal Protective Method: Skin protection.

Respirator Selection: Canister-type mask.

Disposal Method Suggested: Incineration.

References

(1)   Sax, N.I., Ed., *Dangerous Properties of Industrial Materials Report, 2,* No. 2, 29-30, New York, Van Nostrand Reinhold Co. (1982).

HEXABORANE

Description: B_6H_{10} is a liquid which boils at $108°C$.

Code Numbers: CAS 23777-80-2

DOT Designation: —

Synonyms: Hexaboron decahydride; borohexane.

Potential Exposure: Those involved in hexaborane production or use as a high energy fuel.

Permissible Exposure Limits in Air: No standards set.

Permissible Concentration in Water: No criteria set.

Routes of Entry: Skin contact.

Harmful Effects and Symptoms: Highly irritating to the skin. Can attack central nervous system and cause drowsiness, dizziness, visual disturbances, muscle twitches and even muscle spasms.

Points of Attack: Skin, central nervous system.

Personal Protective Methods: Should wear boronhydride-resistant (non-rubber) clothing.

Respirator Selection: Should wear self-contained breathing apparatus.

Disposal Method Suggested: Incineration.

References

(1)   Sax, N.I., Ed., *Dangerous Properties of Industrial Materials Report, 3,* No. 1, 61-62, New York, Van Nostrand Reinhold Co. (1983).

HEXACHLOROBENZENE

- Carcinogen (Animal Positive) (A-62) (A-64)
- Hazardous waste (EPA)
- Priority toxic pollutant (EPA)

Description: Hexachlorobenzene, C_6Cl_6, is a solid, crystallizing in needles, MP $231°C$ and boiling at $323°$ to $326°C$.

Code Numbers: CAS 118-74-1   RTECS DA2975000   UN 2729

DOT Designation: —

Synonyms: Perchlorobenzene, HCB.

Potential Exposure: Hexachlorobenzene (HCB) appears to be distributed worldwide, with high levels of contamination found in agricultural areas devoted to wheat and related cereal grains and in industrial areas. HCB is manufactured and formulated for application to seed wheat to prevent bunt; however, most of the HCB in the environment comes from agricultural processes. HCB is used as a starting material for the production of pentachlorophenol which is marketed as a wood preservative. HCB is one of the main substances in the tarry residue which results from the production of chlorinated hydrocarbons. HCB is formed as a by-product in the production of chlorine gas by the electrolysis of sodium chloride using a mercury electrode.

HCB residues have been found in soil, wildlife, fish, and food samples collected from all over the world. In the United States, HCB residues have been reported in birds and bird eggs collected from Maine to Florida, duck tissue collected from across the country, and fish and fish eggs from the East Coast and Oregon. Animal foods, including chicken feed, fish food, and general laboratory feeds, have been found to contain HCB residues. The frequency of detection of HCB residues in domestic meats has been steadily increasing since 1972, in part because of closer scrutiny. HCB has been detected in trace amounts in only two drinking water supplies.

EPA's monitoring of human adipose tissues collected from across the United States reveals that about 95% of the population has trace HCB residues.

Permissible Exposure Limits in Air: There is no Federal standard for HCB. The ACGIH has no suggested standards either.

Permissible Concentration in Water: To protect human health: preferably zero. An additional lifetime cancer risk of 1 in 100,000 is imposed by a concentration of 7.2 ng/ℓ (0.0072 µg/ℓ).

Determination in Water: Methylene chloride extraction followed by concentration and gas chromatography with electron capture detection (EPA Method 612) or gas chromatography plus mass spectrometry (EPA Method 625).

Routes of Entry: Inhalation, ingestion, eye and skin contact.

Harmful Effects and Symptoms: The death of breast-fed infants and an epidemic of skin sores and skin discoloration were associated with accidental consumption of HCB-contaminated seed grain in Turkey in the mid-1950s. Clinical manifestations included weight loss, enlargement of the thyroid and lymph nodes, skin photosensitization, and abnormal growth of body hair.

HCB levels of up to 23 ppb in blood are believed to have contributed to enzyme disruptions in the population of a small community in southern Louisiana in 1973.

Long-term (up to 3 years) animal ingestion studies show a detectable increase in deaths at 32 ppm, cellular alteration at 1 ppm, biochemical effects at 0.5 ppm, and behavioral alteration between 0.5 and 5 ppm.

Apparently, the effective dosage to offspring is increased by exposure to the parent. A 12% reduction in offspring survival resulted when exposure to very low levels had been continuous for three generations. Teratogenic effects appear minimal.

While HCB appears to have little effect on aquatic organisms, a bioaccumulation factor of 15,000 has been demonstrated in catfish. HCB is toxic to some birds. 80 ppm caused death, and 5 ppm caused liver enlargement and other effects in quail. The half life of HCB in cattle and sheep is almost 90 days. HCB is very stable. It readily vaporizes from soil into the air; emissions to air in turn contaminate the soil.

HCB causes liver tumors and dysfunction in hamsters.

Medical Surveillance: Preplacement and regular medical examinations.

First Aid: Wash contaminated areas of body with soap and water. If swallowed use gastric lavage followed by saline catharsis.

Personal Protective Methods: Chemical safety goggles are recommended for eye protection. Respirators are required to prevent inhalation.

Disposal Method Suggested: Present control methods consist of incineration, deep-well injection, and landfill. Incineration is most effective at 1300°C and 0.25 sec (2).

References

(1) U.S. Environmental Protection Agency, *Chlorinated Benzenes: Ambient Water Quality Criteria,* Washington, DC (1980).
(2) U.S. Environmental Protection Agency, *Status Assessment of Toxic Chemicals: Hexachlorobenzene,* Report EPA-600/2-79-210g, Cincinnati, Ohio (December 1979).
(3) U.S. Environmental Protection Agency, *Hexachlorobenzene,* Health and Environmental Effects Profile No. 110, Office of Solid Waste, Washington, DC (April 30, 1980).
(4) International Agency for Research on Cancer, *IARC Monographs on the Carcinogenic Risks of Chemicals to Humans,* Lyon, France, 20, 155 (1979).
(5) See Reference (A-60).
(6) Sax, N.I., Ed., *Dangerous Properties of Industrial Materials Report, 4,* No. 1, 88–92, New York, Van Nostrand Reinhold Co. (Jan./Feb. 1984).
(7) United Nations Environment Programme, *IRPTC Legal File 1983,* Vol. 1, pp VII/88–90, Geneva, Switzerland, International Register of Potentially Toxic Chemicals (1984).

HEXACHLOROBUTADIENE

- Carcinogen (EPA-CAG)(A-40)
- Hazardous waste (EPA)
- Priority toxic pollutant (EPA)

Description: $CCl_2=CCl-CCl=CCl_2$ is a clear colorless liquid with a faint turpentinelike odor which boils at about 215°C.

Code Numbers: 87-68-3   RTECS EJ0700000   UN 2279

DOT Designation: Label should bear St. Andrews Cross (X).

Synonyms: Perchlorobutadiene, 1,3-hexachlorobutadiene.

Potential Exposure: Hexachlorobutadiene is used as a solvent for elastomers, a heat-transfer fluid, a transformer and hydraulic fluid, and a wash liquor for removing higher hydrocarbons.

Permissible Exposure Limits in Air: There is no Federal standard and ACGIH (1983/84) has proposed a TWA value of 0.02 ppm (0.24 mg/m^3) with the notation that this is a substance suspect of carcinogenic potential for man.

Permissible Concentration in Water: To protect freshwater aquatic life—90 $\mu g/\ell$ on an acute toxicity basis and 9.3 $\mu g/\ell$ on a chronic basis. To protect saltwater aquatic life—32 $\mu g/\ell$ on an acute toxicity basis. For the protection of human health—preferably zero. An additional lifetime cancer risk of 1 in 100,000 is imposed by a concentration of 4.47 $\mu g/\ell$.

Determination in Water: Methylene chloride extraction followed by concentration, gas chromatography with electron capture detection (EPA Method 612) or gas chromatography plus mass spectrometry (EPA Method 625).

Harmful Effects and Symptoms: HCB is a suspected carcinogen. Studies undertaken by Dow Chemical, U.S.A. showed some renal tubular neoplasms, some of which metastasized to the lungs, when fed to rats at 20 mg/kg/day of hexachlorobutadiene for 2 years. Other effects included decreased body weight gain and increased urinary excretion of coproporphyrin. Results of this and another study conducted at Dow reveal a dose-effect relationship for hexachlorobutadiene in rats. Higher doses caused severe kidney injury and cancer; intermediate doses caused reversible kidney injury; and low doses (0.2 mg/kg/day) caused no discernible ill effects.

There is a definite lack of information on human exposures to hexachlorobutadiene and the effects of this compound in humans.

References

(1) U.S. Environmental Protection Agency, *Hexachlorobutadiene: Ambient Water Quality Criteria,* Washington, DC (1980).

(2) U.S. Environmental Protection Agency, *Survey of Industrial Processing Data. Task 1 - Hexachlorobenzene and Hexachlorobutadiene Pollution from Chlorocarbon Processing,* report by Midwest Res. Inst., Kansas City, Mo., Report No. EPA 560/3-75-003, Off. Toxic Subst., Washington, DC (1975).

(3) U.S. Environmental Protection Agency, *Sampling and Analysis of Selected Toxic Substances. Task 1B - Hexachlorobutadiene,* EPA Rep. No. 560/6-76-015, Off. Toxic Subst., Washington, DC (1976).

(4) Laseter, J.L., et al., *An Ecological Study of Hexachlorobutadiene (HCBD),* Report No. EPA 560/7-76-010, Off. Toxic Subst., Washington, DC (1976).

(5) U.S. Environmental Protection Agency, *Hexachlorobutadiene,* Health and Environmental Effects Profile No. 111, Office of Solid Waste, Washington DC (April 30, 1980).

(6) Sax, N.I., Ed., *Dangerous Properties of Industrial Materials Report, 2,* No. 5, 71-75, New York, Van Nostrand Reinhold Co. (1982).

(7) See Reference (A-60).

(8) Parmeggiani, L., Ed., *Encyclopedia of Occupational Health & Safety,* Third Edition, Vol. 1, pp 1041–42, Geneva, International Labour Office (1983).

(9) United Nations Environment Programme, *IRPTC Legal File 1983,* Vol. I, pp VII/136–7, Geneva, Switzerland, International Register of Potentially Toxic Chemicals (1984).

HEXACHLOROCYCLOHEXANE

- Carcinogen (EPA-CAG)(A-40)
- Hazardous waste (EPA)
- Priority toxic pollutant (EPA)

Description: $C_6H_6Cl_6$ is a brownish-to-white crystalline solid with a phosgenelike odor melting at 65°C. It consists of 8 stereoisomers of which the gamma isomer is most insecticidally active and hence most important. See "Lindane."

Code Numbers: CAS 608-73-1 RTECS GV3150000 UN 2761

DOT Designation: —

Synonyms: HCH, BHC, hexaklor (in Sweden), hexacloran (in USSR), ENT 8601.

Potential Exposure: The major commercial usage of HCH is based upon its insecticidal properties. As indicated previously, the γ-isomer has the highest acute toxicity, but the other isomers are not without activity. It is generally advantageous to purify the γ-isomer from the less active isomers. The γ-isomer acts on the nervous system of insects, principally at the level of the nerve ganglia.

As a result, lindane has been used against insects in a wide range of applications including treatment of animals, buildings, man for ectoparasites, clothes, water for mosquitoes, living plants, seeds and soils. Some applications have been abandoned due to excessive residues, e.g., stored foodstuffs.

By voluntary action, the principal domestic producer of technical grade BHC requested cancellations of its BHC registrations on September 1, 1976. As of July 21, 1978, all registrants of pesticide products containing BHC voluntarily cancelled their registrations or switched their former BHC products to lindane formulations.

Permissible Exposure Limits in Air: The maximum air concentration allowed by EPA is 0.5 mg/m^3.

Permissible Concentration in Water: There are no criteria for the protection of freshwater or saltwater aquatic life from technical BHC (mixed isomers) due to insufficient data. To protect human health—preferably zero for technical product. An additional cancer risk of 1 in 100,000 is imposed by a concentration of 0.123 $\mu g/\ell$.

Determination in Water: Methylene chloride extraction followed by gas chromatography with electron capture or halogen specific detection (EPA Method 608) or gas chromatography plus mass spectrometry (EPA Method 625).

Harmful Effects and Symptoms: HCH is a persistent stomach poison and contact insecticide with some tumigant action, the activity of which is determined by the content of the γ-isomer. It is nonphytotoxic, except to cucurbits, at insecticidal concentrations, but, at higher concentrations, may cause root deformation and polyploidy. It taints certain crops seriously, especially blackcurrent and potato.

The mammalian toxicities of the isomers differ: alpha, low acute, chronic and cumulative toxicities; beta, low acute, but high chronic and cumulative toxicities; gamma, see "Lindane;" and delta, low acute and chronic toxicities but irritant to mucous membranes.

Disposal Method Suggested: A process has been developed for the destructive pyrolysis of benzene hexachloride at 400° to 500°C with a catalyst mixture which contains 5 to 10% of either cupric chloride, ferric chloride, zinc chloride, or aluminum chloride on activated carbon (A-32).

References

(1) U.S. Environmental Protection Agency, *Hexachlorocyclohexane: Ambient Water Quality Criteria,* Washington, DC (1980).

(2) U.S. Environmental Protection Agency, *Hexachlorocyclohexane,* Health and Environmental Effects Profile No. 112, Office of Solid Waste, Washington, DC (April 30, 1980).

(3) Sax, N.I., Ed., *Dangerous Properties of Industrial Materials Report, 3,* No. 1, 66-72, New York, Van Nostrand Reinhold Co. (1983).

(4) See Reference (A-60).

HEXACHLOROCYCLOPENTADIENE

- Hazardous substance (EPA)
- Hazardous waste (EPA)
- Priority toxic pollutant (EPA)

Description: C_5Cl_6 with the structural formula:

is a dense, oily liquid.

Code Numbers: CAS 77-47-4   RTECS GY1225000   UN 2646

DOT Designation: Corrosive material.

Synonyms: C-56®, HCCPD, Hex.

Potential Exposure: Hexachlorocyclopentadiene is used to produce the flame retardant chlorendic anhydride, which has applications in polyesters, and to produce chlorendic acid which is used as a flame retardant in resins.

Hexachlorocyclopentadiene is also used as an intermediate in the production of pesticides, such as aldrin, dieldrin and endosulfan (A-32).

Occupational exposures appear to constitute the only documented source of human exposure to Hex. Oral contact does not appear to be a likely mode of occupational exposure.

Dermal and inhalation exposures are recognized hazards for the following groups: workers engaged directly in Hex manufacture; those engaged in the formulation and use of other related pesticides where Hex may be present as an impurity; flame retardant workers; and those having "quasi-occupational" exposure such as sewage treatment workers, industrial hygienists, etc.

Permissible Exposure Limits in Air: There is no Federal standard. ACGIH has set a TWA of 0.01 ppm (0.1 mg/m³) and an STEL of 0.03 ppm (0.3 mg/m³) as of 1983/84. In selecting the TLV and STEL values for Hex, the ACGIH emphasizes that these particular levels were selected on the basis of preventing irritant effects rather than chronic toxicity. The USSR has recommended a tenfold lower limit (0.001 ppm) for occupational workers.

Determination in Air: Sample collection on Poropak T absorbent, desorption with hexane and analysis by gas chromatography with electron capture detection (7).

Permissible Concentration in Water: To protect freshwater aquatic life–7.0 µg/ℓ based on acute toxicity and 5.2 µg/ℓ based on chronic toxicity. To protect saltwater aquatic life–7.0 µg/ℓ based on acute toxicity. To protect human health–1.0 µg/ℓ based on organoleptic considerations and 206 µg/ℓ based on public health considerations.

Determination in Water: Methylene chloride extraction followed by concentration, gas chromatography with electron capture detection (EPA Method 612) or gas chromatography plus mass spectrometry (EPA Method 625).

Harmful Effects and Symptoms: Hex has proven to be a potent irritant. Industrial workers have experienced irritation of the eye, irritation of the upper airway passages and headaches upon exposure to Hex vapors and burns upon contact of skin with the liquid. Inhalation of this compound causes lacrimation, salivation, and gasping respiration in mice, rabbits, rats, and guinea pigs. Long term inhalation results in degenerative changes of the brain, heart, liver, adrenals, and kidneys as well as severe lung damage.

In the laboratory, Hex has exhibited toxicity to fish and mammals. Hex has been reported to be nonmutagenic in unpublished laboratory tests, and there are no data available to evaluate the carcinogenicity of the compound.

Disposal Method Suggested: Incineration after mixing with another combustible fuel. Care must be exercised to assure complete combustion to prevent the formation of phosgene. An acid scrubber is necessary to remove the halo acids produced.

References

(1) U.S. Environmental Protection Agency, *Hexachlorocyclopentadiene: Ambient Water Quality Criteria,* Washington, DC (1980).
(2) U.S. Environmental Protection Agency, *Chemical Hazard Information Profile: Hexachloropentadiene,* Washington, DC (March 15, 1977).
(3) National Institute for Occupational Safety and Health, *Information Profiles on Potential Occupational Hazards: Hexachlorocyclopentadiene,* pp. 16–18, Report PB-276,678, Rockville, Maryland (October 1977).
(4) National Academy of Sciences, *Kepone, Mirex, Hexachlorocyclopentadiene: An Environmental Assessment,* Washington, DC (1978).
(5) U.S. Environmental Protection Agency, *Reviews of the Environmental Effects of Pollutants: XII. Hexachlorocyclopentadiene,* Report No. EPA-600/1-78-047, Cincinnatti, Ohio (1978).
(6) U.S. Environmental Protection Agency, *Hexachlorocyclopentadiene,* Health and Environmental Effects Profile No. 114, Office of Solid Waste, Washington, DC (April 30, 1980).
(7) Dillon, H.K., *Development of Air Sampling and Analytical Methods for Toxic Chlorinated Organic Compounds. Research Report on Hexachlorocyclopentadiene,* Report SORI-EAS-80-057, Southern Research Institute (February, 1980).
(8) Sax, N.I., Ed., *Dangerous Properties of Industrial Materials Report, 4,* No. 2, 76–79, New York, Van Nostrand Reinhold Co. (1984).
(9) United Nations Environment Programme, *IRPTC Legal File 1983,* Vol. I, pp VII/241–2, Geneva, Switzerland, International Register of Potentially Toxic Chemicals (1984).

HEXACHLOROETHANE

- Carcinogen (Positive, NCI)(5)
- Hazardous waste (EPA)
- Priority toxic pollutant (EPA)

Description: CCl_3CCl_3 is a colorless solid with a camphorlike odor. It sublimes at $189°C$.

Code Numbers: CAS 67-72-1   RTECS KI4025000   UN (NA9037)

DOT Designation: ORM-A.

Synonyms: Perchloroethane, carbon hexachloride, HCE.

Potential Exposure: Applications of HCE are quite extensive. As a medicinal, HCE is used as an anthelmintic to treat fascioliasis in sheep and cattle. It is also added to the feed of ruminants, preventing methanogenesis and increasing feed efficiency.

HCE is used in metal and alloy production, mainly in refining aluminum alloys. It is also used for removing impurities from molten metals, recovering metals from ores or smelting products and improving the quality of various metals and alloys.

HCE is contained in pyrotechnics. It inhibits the explosiveness of methane and the combustion of ammonium perchlorate. Smoke containing HCE is used to extinguish fires.

HCE has various applications as a polymer additive. It has flameproofing qualities, increases sensitivity to radiation crosslinking, and is used as a vulcanizing agent. Added to polymer fibers, HCE acts as a swelling agent and increases affinity for dyes.

NIOSH has estimated 1,500 workers are exposed annually to HCE.

Incompatibilities: Hot iron, zinc, aluminum, alkalies.

Permissible Exposure Limits in Air: The Federal standard is 1.0 ppm (10 mg/m^3). The ACGIH has proposed a TWA of 10 ppm (100 mg/m^3) as of 1983/84 but no STEL value. They added the notation "skin" to indicate the possibility of cutaneous absorption. The IDLH level is 300 ppm.

Determination in Air: Charcoal adsorption, workup with CS_2, analysis by gas chromatography. See NIOSH Methods, Set H. See also reference (A-10).

Permissible Concentration in Water: To protect freshwater aquatic life— 118,000 µg/ℓ based on acute toxicity and 20,000 µg/ℓ based on chronic toxicity. To protect saltwater aquatic life—113,000 µg/ℓ based on acute toxicity. To protect human health—9.4 µg/ℓ to keep lifetime cancer risk below 10^{-5}.

Determination in Water: Methylene chloride extraction followed by concentration, gas chromatography with electron capture detection (EPA Method 612) or gas chromatography plus mass spectrometry (EPA Method 625).

Routes of Entry: Inhalation, skin absorption, ingestion, eye and skin contact.

Harmful Effects and Symptoms: Hexachloroethane acts primarily as a central nervous system depressant, and in high concentrations it causes narcosis. It also may be moderately irritating to the skin, mucous membranes and liver. Irritation occurs when there is an excessive amount of hexachloroethane dust in the air or when it is heated and vapors are formed. It should be noted that the low vapor pressure of this compound as well as its solid state minimize its inhalation hazards.

Points of Attack: Eyes.

Medical Surveillance: Consider the points of attack in preplacement and periodic physical examinations.

First Aid: If this chemical gets into the eyes, irrigate immediately. If this chemical contacts the skin, wash with soap immediately. If a person breathes in large amounts of this chemical, move the exposed person to fresh air at once and perform artificial respiration. When this chemical has been swallowed, get medical attention. Give large quantities of salt water and induce vomiting. Do not make an unconscious person vomit.

Personal Protective Methods: Wear appropriate clothing to prevent repeated or prolonged skin contact. Employees should wash promptly when skin is wet or contaminated and daily at the end of each work shift. Work clothing should be changed daily if it is possible that clothing is contaminated. Remove nonimpervious clothing promptly if contaminated.

Respirator Selection
<table>
<tr><td>10 ppm:</td><td>SA/SCBA</td></tr>
<tr><td>50 ppm:</td><td>SAF/SCBAF</td></tr>
<tr><td>300 ppm:</td><td>SA:PD,PP,CF</td></tr>
<tr><td>Escape:</td><td>PestResp/SCBA</td></tr>
</table>

Disposal Method Suggested: Incineration after mixing with another combustible fuel. Care must be exercised to assure complete combustion to prevent the formation of phosgene. An acid scrubber is necessary to remove the halo acids produced.

References

(1) U.S. Environmental Protection Agency, *Chlorinated Ethanes: Ambient Water Quality Criteria,* Washington, DC (1980).

(2) U.S. Environmental Protection Agency, *Chemical Hazard Information Profile: Hexachloroethane,* Washington, DC (1979).

(3) National Istitute for Occupational Safety and Health, *Profiles on Occupational Hazards for Criteria Document Priorities,* pp. 295–298, Report PB-274,073, Cincinnati, Ohio (1977).

(4) U.S. Environmental Protection Agency, *Hexachloroethane,* Health and Environmental Effects Profile No. 115, Office of Solid Waste, Washington, DC (April 30, 1980).

(5) National Cancer Institute, *Bioassay of Hexachloroethane for Possible Carcinogenicity,* Technical Report Series No. 68, Bethesda, Maryland (1978).

(6) See Reference (A-61).

(7) See Reference (A-60).

(8) United Nations Environment Programme, *IRPTC Legal File 1983,* Vol. I, pp VII/311–12, Geneva, Switzerland, International Register of Potentially Toxic Chemicals (1984).

HEXACHLORONAPHTHALENE

See "Chlorinated Naphthalenes."

References

(1) See Reference (A-61).

(2) See Reference (A-60).

HEXACHLORONORBORNADIENE

Description: 1,2,3,4,7,7-Hexachloronorbornadiene is a liquid which boils at 118° to 120°C at 0.1 mm pressure.

Code Numbers: CAS 3389-71-7

DOT Designation: —

Synonyms: HEX-BCH; hexachlorobicycloheptadiene.

Potential Exposure: HEX-BCH is used in the production of the pesticides endrin and isodrin.

Permissible Exposure Limits in Air: No standards set.

Permissible Concentration in Water: No criteria set.

Harmful Effects and Symptoms: HEX-BCH is moderately toxic; the oral LD_{50} value for rats is 776 mg/kg. It seems to produce neuromuscular effects.

Points of Attack: Neuromuscular system.

Disposal Method Suggested: Incineration or burial in a permitted landfill.

References

(1) U.S. Environmental Protection Agency, *Chemical Hazard Information Profile: 1,2,3,-4,7,7-Hexachloronorbornadiene,* Washington, DC (August 19, 1983).

HEXACHLOROPHENE

● Hazardous waste (EPA)

Description: $C_6H(OH)Cl_3CH_2C_6H(OH)Cl_3$ is a crystalline compound melting at 165°C.

Code Numbers: CAS 70-30-4   RTECS SM0700000   UN 2761

DOT Designation: —

Synonyms: 2,2'-Methylenebis(3,4,6-trichlorophenol), G-11, HCP.

Potential Exposure: HCP has been used as an antibacterial agent in a wide variety of consumer products, including soaps and deodorants. It has also been used as an antifungal agent to treat various citrus fruits and vegetables (A-2).

Permissible Exposure Limits in Air: No standard set.

Permissible Concentration in Water: A no-adverse-effect level in drinking water has been calculated by NAS/NRC as 0.008 mg/ℓ. An ADI was calculated on the basis of the available chronic toxicity data to be 0.0012 mg/kg/day.

Harmful Effects and Symptoms: The principal effect of hexachlorophene is on the central nervous system where reversible edematous vacuolation of the myelin sheaths of the white matter occurs if the plasma concentration is maintained over a long period at or above 1.2 μg/ml. Newborn humans and laboratory animals are more susceptible to this effect than adults.

Hexachlorophene does not appear to be an active carcinogen or teratogen, although long term chronic toxicity studies integrating carcinogenicity and target organ toxicity are recommended to assemble more data. NCI has completed a carcinogenesis bioassay and concluded that hexachlorophene is not a carcinogen.

Disposal Method Suggested: Incineration, preferably after mixing with another combustible fuel. Care must be exercised to assure complete combustion to prevent the formation of phosgene. An acid scrubber is necessary to remove the halo acids produced (A-31).

References

(1) U.S. Environmental Protection Agency, *Hexachlorophene,* Health and Environmental Effects Profile No. 116, Office of Solid Waste, Washington, DC (April 30, 1980).

HEXAFLUOROACETONE

Description: CF_3COCF_3 is a colorless, nonflammable gas which boils at -27°C.

Code Numbers: CAS 684-16-2   RTECS UC2450000   UN 2420

DOT Designation: Poison gas.

Synonyms: 6FK, HFA.

Potential Exposure: Hexafluoroacetone is used as a chemical intermediate. A gas at room temperature, it forms various hydrates with water which are used as solvents for resins and polymers. Other derivatives are used to make water repellent coatings for textiles and also to produce polymers.

Permissible Exposure Limit in Air: There is no Federal standard. ACGIH as of 1983/84 set a TWA of 0.1 ppm (0.7 mg/m^3) and an STEL of 0.3 ppm (2.0 mg/m^3).

Permissible Concentration in Water: No criteria set.

Harmful Effects and Symptoms: In humans, acute exposure to hexafluoroacetone produces lung irritation which may lead to pulmonary edema. Since it reacts with moisture to form a highly acidic sesquihydrate (pH <1), HFA is irritating to the eyes, nose, throat and skin in humans.
The liver, kidneys, testes, and thymus of rats showed injury after a single inhalation exposure to 200 ppm for 4 hr. Subchronic inhalation exposures of rats and dogs to hexafluoroacetone resulted in pathological changes in the testes, spleen, thymus, lymph nodes, and bone marrow. Acute dermal exposure results in eschar formation.

References

(1) U.S. Environmental Protection Agency, *Chemical Hazard Information Profile: Hexafluoroacetone,* Washington, DC (1979).
(2)   National Institute for Occupational Safety and Health, *Information Profiles on Potential Occupational Hazards,* pp 20-22, Report PB-276,678, Rockville, Maryland (Oct. 1977).
(3)   Sax, N.I., Ed., *Dangerous Properties of Industrial Materials Report, 1,* No. 4, 75-76, New York, Van Nostrand Reinhold Co. (1981).

HEXAMETHYLENEDIAMINE

Description: C$_6$H$_{16}$N$_2$, H$_2$N(CH$_2$)$_6$NH$_2$, is a colorless solid which melts at 39° to 42°C.

Code Numbers: CAS 124-09-4   RTECS MO1180000   UN 2280

DOT Designation: Corrosive material.

Synonyms: 1,6-Hexanediamine; 1,6-diaminohexane; HMDA.

Potential Exposure (2): HMDA is used as a raw material for nylon fiber and plastics; in the manufacture of oil-modified and mosture-area types of urethane coatings; in the manufacture of polyamides for printing inks, dimer acids, and textiles; and as an oil and lubricant additive (probably as a corrosion inhibitor); also used in paints and as a curing agent for epoxy resins.

Incompatibilities: Oxidizing agents.

Permissible Exposure Limits in Air: The USSR has set a TLV of 0.2 ppm (1 mg/m^3) (A-36).

Permissible Concentration in Water: A maximum of 0.01 mg/ℓ is quoted for drinking water (A-36).

Routes of Entry: Inhalation and ingestion.

Harmful Effects and Symptoms (1): Continuous 90-day inhalation of 1 mg/m^3 of diaminohexane by albino rats caused an increase in the number of reticulocytes (only at the beginning of the exposure) and an increase in the Vi antibody concentration. The animals also exhibited a decrease in the number of eosinophils, suppressed leukocytic activity, retarded growth, and a disturbance of the chronaxy correlation of the muscle antagonists. Diaminohexane at a concentration of 0.04 mg/m^3 caused similar but less pronounced changes. Diaminohexane at 0.001 mg/m^3 had no effect.

Exposure of rats to an atmosphere containing 1.25 mg/m^3 diaminohexane for 4 hr/day for 8 days decreased the threshold of neuromuscular excitability, increased blood leucocyte and liver glycogen levels, caused disorders of renal excretory capacity, and altered the phagocytic activity of neutrophils.

Diaminohexane inhibited DNA and RNA formation *in vitro* in studies using rat embryo and human amnion cell cultures.

Intraperitoneal injection of diaminohexane into rats inhibited ovarian ornithine decarboxylase activity which had been stimulated by human chorionic gonadotropin. Diaminohexane injected into mice bearing ascites-carcinoma cells powerfully decreased ornithine decarboxylase activity in the carcinoma cells.

An *in vitro* study showed that diaminohexane inhibited collagen-induced human platelet aggregation.

Occupational exposure to epoxy resins and their hardeners (including diaminohexane) was studied in 488 workers. Prolonged contact caused skin damage, allergic rhinitis, bronchial asthma, impairment of bronchial permeability, toxicoallergic hepatitis, gastritis, colitis, hypergammaglobulinemia, increased transaminase activity, and eosinophilia of the peripheral blood.

HMDA is a material of concern to EPA (1) because of its potential for nitrosamine formation.

Disposal Method Suggested: Incineration; incinerator is equipped with a scrubber or thermal unit to reduce NO$_x$ emissions.

References

(1) U.S. Environmental Protection Agency, *Chemical Hazard Information Profile: 1,6-Diaminohexane,* Washington, DC (June 6, 1978).

(2) Sax, N.I., Ed., *Dangerous Properties of Industrial Materials Report, 2,* No. 2, 30-31, New York, Van Nostrand Reinhold Co. (1982).

(3) See Reference (A-60).

HEXAMETHYLENETETRAMINE

Description: (CH$_2$)$_6$N$_4$, hexamethylenetetramine, is an odorless, crystalline solid.

Code Numbers: CAS 100-97-0   RTECS MN4725000

DOT Designation: —

Synonyms: Methenamine, hexamine, formamine, ammonioformaldehyde.

Potential Exposures: Hexamethylenetetramine is used as an accelerator in the rubber industry, as a curing agent in thermosetting plastics, as a fuel pellet for camp stoves, and in the manufacture of resins, pharmaceuticals, and explosives.

Permissible Exposure Limits in Air: There is no Federal standard for hexamethylenetetramine. The ACGIH has not set standards either.

Determination in Air: Collection by impinger (vapor) and filter (particulates), decomposition to formaldehyde, reacted with chromatropic acid and measured by spectrophotometry (A-10).

Permissible Concentration in Water: No criteria set.

Routes of Entry: Ingestion and skin contact.

Harmful Effects and Symptoms: *Local* — Very mild skin irritant.
Systemic — None. Side effects from ingestion are urinary tract irritation, skin rash, and digestive disturbances. Large oral doses can cause severe nephritis which may be fatal.

Medical Surveillance: No specific considerations are necessary.

Personal Protective Methods: If repeated or prolonged, skin exposure is likely; gloves or protective clothing may be needed.

Disposal Method Suggested: Incineration; incinerator is equipped with a scrubber or thermal unit to reduce NO_x emissions.

HEXAMETHYLPHOSPHORIC TRIAMIDE

● Carcinogen (Animal Positive, IARC)(4)

Description: Hexamethylphosphoric triamide, $[(CH_3)_2N]_3PO$, is a colorless liquid with a density of 1.03 g/ml and a boiling point of 232°C.

Code Numbers: CAS 680-31-9   RTECS TD0875000

DOT Designation: —

Synonyms: Synonyms for hexamethylphosphoric triamide include ENT 50882, hempa, hexametapol, hexamethylphosphamide, hexamethylphosphoramide, hexamethylphosphoric acid triamide, hexamethylphosphorotriamide, hexamethylphosphotriamide, HMPA, HMPT, HPT, phosphoric tris(dimethylamide), phosphoryl hexamethyltriamide, tris(dimethylamino) phosphine oxide, and tris(dimethylamino) phosphorus oxide.

Potential Exposure: Hexamethylphosphoric triamide is a material possessing unique solvent properties and is widely used as a solvent, in small quantities, in organic and organometallic reactions in laboratories. This is the major source of occupational exposure to HMPA in the United States.

It is also used as a processing solvent in the manufacture of aramid fibers. HMPA has been evaluated for use as an ultraviolet light inhibitor in polyvinyl chloride formulations, as an additive for antistatic effects, as a flame retardant, and as a deicing additive for jet fuels.

Hexamethylphosphoric triamide has also been extensively investigated as an insect chemosterilant.

It is estimated that 5,000 people are occupationally exposed to hexamethylphosphoric triamide. More than 90% of these exposures are in research laboratories.

Permissible Exposure Limits in Air: There is no current OSHA standard for hexamethylphosphoric triamide exposure. ACGIH (1983/84) classifies HMPA as an "industrial substance suspect of carcinogenic potential for man" with no suggested threshold limit value, but with the notation "skin" indicating the possibility of cutaneous absorption.

Permissible Concentration in Water: No criteria set.

Routes of Entry: Inhalation of vapors.

Harmful Effects and Symptoms: *Animal* — HMPA is known to have a variety of toxic effects on laboratory animals. Acute toxic effects seen in rats fed HMPA include kidney disease, severe bronchiectasis and bronchopneumonia with squamose metaplasia and fibrosis in lungs. In rabbits, repeated application of HMPA to the skin caused dose related weight loss, altered gastrointestinal function and apparent nervous-system dysfunction. Testicular atrophy and aspermia have been observed in rats following oral treatment with HMPA. Oral treatment with HMPA has also been highly inhibitory to testicular development in cockerels.

HMPA is known to produce mutagenic effects in fruit flies *(Drosophia melanogaster)*. However, studies of the effects of HMPA on human and mice chromosomes showed no greater frequency of HMPA induced chromosomal aberrations when compared with controls.

The E.I. duPont de Nemours and Company (DuPont) reported to the National Institute for Occupational Safety and Health in a letter dated Sept. 24, 1975, that nasal tumors (squamous cell carcinoma) have been observed in rats exposed to hexamethylphosphoric triamide. NIOSH has also been advised that DuPont has notified its customers and employees of these findings.

Preliminary results of the inhalation toxicity study of HMPA, released by DuPont, show nasal tumors in rats exposed daily to 400 and 4,000 ppb HMPA after 8 months of exposure. In some cases, the tumors originating from the epithelial lining of the nasal turbinate bones filled the nasal cavity and penetrated into the brain. No nasal tumors were reported among rats exposed to 50 ppb HMPA and controls.

Prior to the DuPont observations, the only other known report of tumors associated with exposure to HMPA was a long term feeding study by Kimbrough. While lung tumors were observed, the results of this study were inconclusive because the tumor incidence among HMPA exposed rats was not greater than among the control rats.

Human — There are no data available on the toxic effects of hexamethylphosphoric triamide in humans.

Medical Surveillance: In light of the potential risk of human exposure to this chemical in the work environment, the National Institute for Occupational Safety and Health is advising the occupational health community of the above findings. Preplacement and regular physical examinations are indicated.

Personal Protective Methods: Conventional laboratory precautions of goggles, gloves and good hood ventilation are indicated.

References

(1) National Institute for Occupational Safety and Health, *Current Intelligence Bulletin No. 6: Hexamethylphosphoric Triamide (HMPA),* Rockville, Maryland (October 24, 1975).

(2) U.S. Environmental Protection Agency, *Chemical Hazard Information Profile: Hexamethylphosphoramide,* Washington, DC (August, 1976).

(3) National Institute for Occupational Safety and Health, *Information Profiles on Potential Occupational Hazards—Single Chemicals: Hexamethyl Phosphoramide,* pp. 106–13, Rockville, Maryland (December, 1979).

(4) International Agency for Research on Cancer, *IARC Monographs on the Carcinogenic Risks of Chemicals to Humans,* Lyon, France, 15, 211 (1977).

n-HEXANE

Description: $CH_3(CH_2)_4CH_3$, is a colorless, volatile liquid and is highly flammable. It boils at 69°C.

Code Numbers: CAS 110-54-3   RTECS MN9275000   UN 1208

DOT Designation: Flammable liquid.

Potential Exposure: n-Hexane is used as a solvent, particularly in the extraction of edible fats and oils, as a laboratory reagent, and as the liquid in low temperature thermometers. Technical and commercial grades consist of 45 to 85% n-hexane, as well as cyclopentanes, isohexane, and from 1 to 6% benzene.
NIOSH estimates that 2.5 million workers are exposed annually to hexane.

Incompatibilities: Strong oxidizers.

Permissible Exposure Limits in Air: The Federal standard is 500 ppm (1,800 mg/m^3) for workroom exposure to n-hexane. NIOSH has proposed a TWA of 100 ppm (360 mg/m^3) with a 510 ppm ceiling value for 15-minute exposure. ACGIH as of 1983/84 has proposed a TWA of 50 ppm (180 mg/m^3) but no STEL for n-hexane. For other isomers, the TWA proposed is 500 ppm (1,800 mg/m^3) and the STEL is 1,000 ppm (3,600 mg/m^3). The IDLH level is 5,000 ppm.

Determination in Air: Charcoal adsorption, workup with CS_2, analysis by gas chromatography. See NIOSH Methods, Set G. See also reference (A-10).

Permissible Concentration in Water: No criteria set.

Routes of Entry: Inhalation of vapor, ingestion, eye and skin contact.

Harmful Effects and Symptoms: *Local* — Dermatitis and irritation of mucous membranes of the upper respiratory tract.

Systemic — Asphyxia may be produced by high concentrations. Acute exposure may cause narcosis resulting in slight nausea, headache, and dizziness. Myocardial sensitization to epinephrine may occur but is of low order. Peripheral neuropathy has been reported resulting from exposure to n-hexane.
Chronic exposure to hexane at airborne concentrations of 1,800 mg/m^3 and above has been associated with the development of polyneuropathy. Professional judgment would indicate that a TWA concentration of 350 mg/m^3 offers a sufficient margin of safety to protect against the development of chronic nerve disorders in workers exposed to hexane.

Points of Attack: Skin, eyes, respiratory system, lungs.

Medical Surveillance: Consider the skin, respiratory system, central and peripheral nervous system, and general health in preplacement and periodic examinations.

First Aid: If this chemical gets into the eyes, irrigate immediately. If this chemical contacts the skin, wash with soap promptly. If a person breathes in large amounts of this chemical, move the exposed person to fresh air at once and perform artificial respiration. When this chemical has been swallowed, get medical attention. Do not induce vomiting.

Personal Protective Methods: Wear appropriate clothing to prevent repeated or prolonged skin contact. Wear eye protection to prevent any reasonable probability of eye contact. Employees should wash promptly when skin is wet or contaminated. Remove clothing immediately if wet or contaminated to avoid flammability hazard.

Respirator Selection:
>   1,000 ppm: CCROV/SA/SCBA
>   5,000 ppm: GMOV/SAF/SCBAF
>   Escape: GMOV/SCBA

Disposal Method Suggested: Incineration.

References

(1) U.S. Environmental Protection Agency, *Clinical Hazard Information Profile: n-Hexane,* Washington, DC (May 13, 1977).

(2) National Institute for Occupational Safety and Health, *Criteria for a Recommended Standard: Occupational Exposure to Alkanes,* NIOSH Document No. 77-151, Washington, DC (1977).

(3) See Reference (A-61).

(4) See Reference (A-60).

(5) Parmeggiani, L., Ed., *Encyclopedia of Occupational Health & Safety,* Third Edition, Vol. 1, pp 1042-44, Geneva, International Labour Office (1983).

(6) United Nations Environment Programme, *IRPTC Legal File 1983,* Vol. II, pp VII/386-8, Geneva, Switzerland, International Register of Potentially Toxic Chemicals (1984).

HEXANOL

Description: $CH_3(CH_2)_4CH_2OH$ is a colorless liquid boiling at $157°C$.

Code Numbers: CAS 111-27-3   RTECS MQ4025000   UN 2282

DOT Designation: —

Synonyms: Amylcarbinol; pentylcarbinol; 1-hydroxyhexane.

Potential Exposure: To those using hexanol as a solvent or in the synthesis of pharmaceuticals, plasticizers and textile chemicals.

Incompatibilities: Oxidants.

Permissible Exposure Limits in Air: The USSR has set a TLV of 2.4 ppm (10 mg/m^3) (A-36).

Permissible Concentration in Water: A value of 0.03 mg/ℓ is the maximum allowable in drinking water (A-36).

Routes of Entry: Skin contact, inhalation, ingestion.

Harmful Effects and Symptoms: Causes skin and eye irritation. Believed to be moderately toxic upon ingestion.

Points of Attack: Skin, eyes, nervous system.

Personal Protective Methods: Wear goggles.

Respirator Selection: Wear canister-type mask.

Disposal Method Suggested: Incineration.

References

(1) Sax, N.I., Ed., *Dangerous Properties of Industrial Materials Report, 2,* No. 2, 32-33, New York, Van Nostrand Reinhold Co. (1982).

(2) See Reference (A-60).

1-HEXENE

Description: $CH_3(CH_2)_3CH=CH_2$ is a colorless liquid boiling at $63.5°C$.

Code Numbers: CAS 592-41-6

DOT Designation: —

Synonyms: Hexylene; butylethylene.

Potential Exposure: Those involved in its use in organic synthesis.

Incompatibilities: Oxidizing materials.

Permissible Exposure Limits in Air: No standards set.

Permissible Concentration in Water: No criteria set.

Harmful Effects and Symptoms: Moderately irritating to skin, eyes and mucous membranes.

Personal Protective Methods: Protective clothing should be worn.

Respirator Selection: Self-contained breathing apparatus should be used.

Disposal Method Suggested: Incineration.

References

(1) Sax, N.I., Ed., *Dangerous Properties of Industrial Materials Report, 1,* No. 8, 78-79, New York, Van Nostrand Reinhold Co. (1981).
(2) Sax, N.I., Ed., *Dangerous Properties of Industrial Materials Report, 3,* No. 2, 50-51, New York, Van Nostrand Reinhold Co. (1983).

sec-HEXYL ACETATE

Description: $CH_3COOCH(CH_3)CH_2CH(CH_3)CH_3$ is a colorless liquid with a mild, pleasant fruity odor. It boils at 140° to 141°C.

Code Numbers: CAS 142-92-7   RTECS AI0875000

DOT Designation: —

Synonyms: 1,3-Dimethylbutyl acetate, methylamyl acetate, methylisoamyl acetate, methylisobutyl carbinol acetate, hexyl acetate.

Potential Exposure: This material is used as a solvent in the spray lacquer industry. It is a good solvent for cellulose esters and other resins.

Incompatibilities: Nitrates, strong oxidizers, strong alkalies, acids.

Permissible Exposure Limits in Air: The Federal standard and the 1983/84 ACGIH TWA value is 50 ppm (300 mg/m^3). There is no tentative STEL value set. The IDLH level is 4,000 ppm.

Determination in Air: Charcoal adsorption, workup with CS_2, gas chromatography. See NIOSH Methods, Set D. See also reference (A-10).

Permissible Concentration in Water: No criteria set.

Routes of Entry: Inhalation, ingestion, eye and skin contact.

Harmful Effects and Symptoms: Eye irritation, headaches, narcosis.

Points of Attack: Eyes, central nervous system.

Medical Surveillance: Consider the points of attack in preplacement and periodic examinations.

First Aid: If this chemical gets into the eyes, irrigate immediately. If this

chemical contacts the skin, flush with water promptly. If a person breathes in large amounts of this chemical, move the exposed person to fresh air at once and perform artificial respiration. When this chemical has been swallowed, get medical attention. Give large quantities of salt water and induce vomiting. Do not make an unconscious person vomit.

Personal Protective Methods: Wear appropriate clothing to prevent repeated or prolonged skin contact. Wear eye protection to prevent any reasonable probability of eye contact. Employees should wash promptly when skin is wet or contaminated. Remove nonimpervious clothing promptly if wet or contaminated.

Respirator Selection:

1,000 ppm:	CCROVF
2,500 ppm:	GMOV/SAF/SCBAF
4,000 ppm:	SAF:PD,PP,CF
Escape:	GMOV/SCBA

Disposal Method Suggested: Incineration.

References

(1) See Reference (A-61).

HEXYLENE GLYCOL

Description: $(CH_3)_2C(OH)CH_2CHOHCH_3$ is a colorless liquid boiling at 198°C.

Code Numbers: CAS 107-41-5 RTECS SA0700000

DOT Designation: —

Synonyms: 2-Methyl-2,4-pentanediol, trimethyltrimethylene glycol.

Potential Exposure: Hexylene glycol is used in the formulation of hydraulic brake fluids and printing inks. It is used as a fuel and lubricant additive, as an emulsifying agent and as a cement additive.

Permissible Exposure Limits in Air: There is no Federal standard but ACGIH (1983/84) has set a ceiling value of 25 ppm (125 mg/m^3). There is no tentative STEL.

Permissible Concentration in Water: No criteria set.

Routes of Entry: Inhalation, ingestion, skin and eye contact.

Harmful Effects and Symptoms: Skin application produces mild to moderate irritation. Eye and throat irritation and respiratory discomfort were slight upon exposure to 100 ppm but more pronounced at 1,000 ppm. Ingestion produces central nervous system depression (A-34).

Points of Attack: Skin, eyes, respiratory system, central nervous system.

Medical Surveillance: Attention to points of attack in preplacement and regular physical examinations is indicated.

Personal Protective Methods: Wear rubber gloves, face shield and protective clothing.

Respirator Selection: Provide with general purpose breathing apparatus.

Disposal Method Suggested: Incineration.

References

(1)   Sax, N.I., Ed., *Dangerous Properties of Industrial Materials Report, 2,* No. 2, 33-35,
      New York, Van Nostrand Reinhold Co. (1982).
(2)   See Reference (A-60).

HYDRAZINE

- Carcinogen (Animal Positive, IARC) (2) (A-62) (A-64)
- Hazardous waste (EPA)

Description: Hydrazine (H_2N-NH_2) is a colorless, oily liquid with an ammoniacal odor. It boils at 113°C.

Code Numbers: CAS 302-01-2   RTECS MU7175000   UN 2029

DOT Designation: Flammable liquid and poison.

Synonyms: Diamide, diamine, hydrazine base.

Potential Exposure: Because of its strong reducing capabilities, hydrazine is used as an intermediate in chemical synthesis and in photography and metallurgy. It is also used in the preparation of anticorrosives, textile agents, and pesticides, and as a scavenging agent for oxygen in boiler water. Hydrazine is widely used in pharmaceutical synthesis (A-41). It is also used as a rocket fuel. NIOSH has estimated annual worker exposure as follows: hydrazine, 9,000; hydrazine dihydrochloride, 89,000; hydrazine sulfate, 2,500; hydrazine hydrobromide, 1,500; and hydrazine hydrate, 1,700.

Incompatibilities: Oxidizers, hydrogen peroxide, nitric acid; metal oxides, strong acids; porous materials.

Permissible Exposure Limits in Air: The Federal standard is 1.0 ppm (1.3 mg/m^3). NIOSH has recommended (1) a standard in terms of free base as a ceiling for any 2 hour period of 0.03 ppm (0.04 mg/m^3). The ACGIH has as of 1983/84 recommended a TWA of 0.1 ppm (0.1 mg/m^3) with the notation that hydrazine is a substance suspect of carcinogenic potential for man. The notation "skin" also indicates the possibility of cutaneous absorption. No tentative STEL value is given. The IDLH level is 80 ppm.

Determination in Air: Collection in a bubbler with HCl, reaction with p-dimethylaminobenzaldehyde, colorimetric measurement. See NIOSH Methods, Set Q. See also reference (A-10).

Permissible Concentration in Water: No criteria set, but EPA (A-37) has suggested a permissible ambient goal of 18 µg/ℓ based on health effects.

Routes of Entry: Inhalation, skin absorption, ingestion, eye and skin contact.

Harmful Effects and Symptoms: *Local* — All hydrazines have similar toxic local effects due to their irritant properties. The vapor is highly irritating to the eyes, upper respiratory tract, and skin, and causes delayed eye irritation. Severe exposure may produce temporary blindness. The liquid is corrosive, producing penetrating burns and severe dermatitis. Permanent corneal lesions may occur if the liquid is splashed in the eyes. A sensitization dermatitis may be produced.

Systemic — Inhalation of hydrazine may cause dizziness and nausea. In animals hydrazine has caused liver and kidney damage and pulmonary edema.

It has also been reported to cause adenocarcinoma of the lung and liver in animals.

Points of Attack: Central nervous system, respiratory system, skin, eyes.

Medical Surveillance: Based partly on experimental animal data, placement should include a history of exposure to other carcinogens, smoking, alcohol, medications, and family history. The skin, eye, lungs, liver, kidney, blood, and central nervous system should be evaluated. Sputum or urine cytology may give useful information. Hydrazine may be detected in the blood.

First Aid: If this chemical gets into the eyes, irrigate immediately. If this chemical contacts the skin, flush with water immediately. If a person breathes in large amounts of this chemical, move the exposed person to fresh air at once and perform artificial respiration. When this chemical has been swallowed, get medical attention. Give large quantities of water and induce vomiting. Do not make an unconscious person vomit.

Personal Protective Methods: Wear appropriate clothing to prevent any possibility of skin contact. Wear eye protection to prevent any possibility of eye contact. Employees should wash immediately when skin is wet or contaminated. Remove clothing immediately if wet or contaminated to avoid flammability hazard. Provide emergency showers and eyewash.

Respirator Selection:

           10 ppm: SA/SCBA
           50 ppm: SAF/SCBAF
           80 ppm: SAF:PD,PP,CF
           Escape: SCBA

Disposal Method Suggested: Controlled incineration with facilities for effluent scrubbing to abate any ammonia formed in the combustion process.

References

(1) National Institute for Occupational Safety and Health, *Criteria for a Recommended Standard: Occupational Exposure to Hydrazines,* NIOSH Document No. 78-172, Washington, DC (1978).
(2) International Agency for Research on Cancer, *IARC Monographs on the Carcinogenic Risks of Chemicals to Humans,* Lyon, France, 4, 127 (1974).
(3) Sax, N.I., Ed., *Dangerous Properties of Industrial Materials Report, 1,* No. 5, 63-64, New York, Van Nostrand Reinhold Co. (1981). (Hydrazine Hydrate).
(4) Sax, N.I., Ed., *Dangerous Properties of Industrial Materials Report, 1,* No. 5, 64-65, New York, Van Nostrand Reinhold Co. (1981). (Hydrazine Sulfate).
(5) See Reference (A-60).
(6) Sax, N.I., Ed., *Dangerous Properties of Industrial Materials Report, 3,* No. 4, 65-68, New York, Van Nostrand Reinhold Co. (1983).
(7) Parmeggiani, L., Ed., *Encyclopedia of Occupational Health & Safety,* Third Edition, Vol. 1, pp 1068-70, Geneva, International Labour Office (1983).

HYDRAZOBENZENE

See "1,2-Diphenylhydrazine."

HYDROGENATED TERPHENYLS

Description: $C_6H_{11}-C_6H_4-C_6H_5$ might be a typical hydrogenated terphenyl with one cyclohexane ring in a chain with two benzene rings.

Code Numbers: CAS 37275-59-5

DOT Designation: —

Potential Exposure: These materials are used as high temperature heat transfer media and as plasticizers.

Permissible Exposure Limits in Air: There is no Federal standard but ACGIH (1983/84) has adopted a TWA value of 0.5 ppm (5 mg/m^3). There is no tentative STEL value.

Permissible Exposure Limits in Water: No criteria set.

Harmful Effects and Symptoms: Potential acute hazards consist of damage to the lungs and damage to the skin and eyes from burns from the hot coolant; potential chronic hazards comprise damage to liver, kidney and blood-forming organs with the possibility of induction of metabolic disorders and cancer (A-34).

HYDROGEN BROMIDE

Description: HBr, hydrogen bromide, is a corrosive colorless gas.

Code Numbers: CAS 10035-10-6   RTECS MW3850000   UN 1048

DOT Designation: Nonflammable gas.

Synonyms: Anhydrous hydrobromic acid.

Potential Exposure: Hydrogen bromide and its aqueous solutions are used in the manufacture of organic and inorganic bromides, as a reducing agent and catalyst in controlled oxidations, in the alkylation of aromatic compounds, and in the isomerization of conjugated diolefins. It is used in the production of many drugs (A-41).

Incompatibilities: Strong oxidizers, strong caustics, metals, moisture.

Permissible Exposure Limits in Air: The Federal standard and the ACGIH 1983/84 TWA value for hydrogen bromide is 3.0 ppm (10 mg/m^3). There is no STEL proposed. The IDLH level is 50 ppm.

Determination in Air: Absorption in a bubbler with sodium hydroxide, analysis by ion-specific electrode. See NIOSH Methods, Set L. See also reference (A-10).

Allowable Concentration in Water: No criteria set.

Routes of Entry: Inhalation, ingestion, eye and skin contact.

Harmful Effects and Symptoms: Hydrogen bromide (hydrobromic acid) is less toxic than bromine, but is an irritant to the mucous membranes of the upper respiratory tract. Long term exposures can cause chronic nasal and bronchial discharge and dyspepsia. Skin contact may cause burns. Eye irritation, eye burns and temporary blindness can result from HBr exposure.

Points of Attack: Respiratory system, eyes, skin.

Medical Surveillance: Consider the points of attack in preplacement and periodic physical examinations.

First Aid: If this chemical gets into the eyes, irrigate immediately. If this

chemical contacts the skin, flush with water immediately. If a person breathes in large amounts of this chemical, move the exposed person to fresh air at once and perform artificial respiration. When this chemical has been swallowed, get medical attention. Give large quantities of water and do not induce vomiting.

Personal Protective Methods: Wear appropriate clothing to prevent any possibility of skin contact. Wear eye protection to prevent any possibility of eye contact. Employees should wash immediately when skin is wet or contaminated. Remove nonimpervious clothing immediately if wet or contaminated. Provide emergency showers and eyewash if contaminants are involved.

Respirator Selection:
    50 ppm: CCRAGF/GMAG/SAF/SCBAF
    Escape: GMAG/SCBA

Disposal Method Suggested: Soda ash/slaked lime is added to give a neutral bromide solution which is discharged to sewers or streams with water dilution.

References

(1) U.S. Environmental Protection Agency, *Chemical Hazard Information Profile: Bromine and Bromine Compounds,* Washington, DC (November 1, 1976).
(2) See Reference (A-61).
(3) See Reference (A-60).

HYDROGEN CHLORIDE

● Hazardous substance (EPA)

Description: HCl, hydrogen chloride, is a colorless, nonflammable gas, soluble in water. The aqueous solution is known as hydrochloric acid or muriatic acid and may contain as much as 38% HCl.

Code Numbers: CAS 7647-01-0 RTECS MW9610000 UN 1050

DOT Designation: Nonflammable gas.

Synonyms: Anhydrous hydrochloric acid, chlorohydric acid.

Potential Exposure: Hydrogen chloride itself is used in the manufacture of pharmaceutical hydrochlorides (A-41), chlorine, vinyl chloride from acetylene, alkyl chlorides from olefins, arsenic trichloride from arsenic trioxide; in the chlorination of rubber; as a gaseous flux for babbitting operations; and in organic synthesis involving isomerization, polymerization, alkylation, and nitration reactions.

The acid is used in the production of fertilizers, dyes, dyestuffs, artificial silk, and paint pigments; in refining edible oils and fats; in electroplating, leather tanning, ore refining, soap refining, petroleum extraction, pickling of metals, and in the photographic, textile, and rubber industries.

Incompatibilities: Most metals, alkali or active metals.

Permissible Exposure Limits in Air: The Federal standard and the 1983/84 ACGIH TWA value for hydrogen chloride is 5 ppm (7 mg/m^3) as a ceiling value. There is no STEL value. The IDLH level is 100 ppm.

Determination in Air: Collection in a bubbler using sodium acetate, analysis by ion-specific electrode. See NIOSH Methods, Set R. See also reference (A-10).

Permissible Concentration in Water: No criteria set.

Route of Entry: Inhalation of gas or mist, ingestion, eye and skin contact.

Harmful Effects and Symptoms: *Local* — Hydrochloric acid and high concentrations of hydrogen chloride gas are highly corrosive to eyes, skin, and mucous membranes. The acid may produce burns, ulceration, and scarring on skin and mucous membranes, and it may produce dermatitis on repeated exposure. Eye contact may result in reduced vision or blindness. Dental discoloration and erosion of exposed incisors occur on prolonged exposure to low concentrations. Ingestion may produce fatal effects from esophageal or gastric necrosis.

Systemic — The irritant effect of vapors on the respiratory tract may produce laryngitis, glottal edema, bronchitis, pulmonary edema, and death.

Points of Attack: Respiratory system, lungs, skin, eyes.

Medical Surveillance: Special consideration should be given to the skin, eyes, teeth, and respiratory system. Pulmonary function studies and chest x-rays may be helpful in following recovery from acute overexposure.

First Aid: If this chemical gets into the eyes, irrigate immediately. If this chemical contacts the skin, flush with water immediately. If a person breathes in large amounts of this chemical, move the exposed person to fresh air at once and perform artificial respiration. When this chemical has been swallowed, get medical attention. Give large quantities of water and do not induce vomiting.

Personal Protective Methods: Wear appropriate clothing to prevent any possibility of skin contact with liquids of pH <3 or repeated or prolonged contact with liquids of pH >3. Wear eye protection to prevent any possibility of eye contact. Employees should wash immediately when skin is wet or contaminated with liquids of pH <3. Remove nonimpervious clothing immediately if wet or contaminated with liquids of pH <3. Provide emergency showers and eyewash if liquids of pH <3 are involved.

Respirator Selection:
>   50 ppm: CCRAG/SA/SCBA
>   100 ppm: CCRABF/GMAG/SAF/SCBAF
>   Escape: GMAG/SCBA

Disposal Method Suggested: Soda ash-slaked lime is added to form the neutral solution of chloride of sodium and calcium. This solution can be discharged after dilution with water. Alternatively, hydrogen chloride can be recovered from a variety of process waste streams (A-57).

References

(1)   National Institute for Occupational Safety and Health, *Information Profiles on Potential Occupational Hazards: Hydrogen Chloride (Gas),* pp 23–28, Report PB-276,678, Rockville, Maryland (1977).
(2)   See Reference (A-61).
(3)   Sax, N.I., Ed., *Dangerous Properties of Industrial Materials Report, 1,* No. 7, 62–65, New York, Van Nostrand Reinhold Co. (1981).
(4)   See Reference (A-60).
(5)   Parmeggiani, L., Ed., *Encyclopedia of Occupational Health & Safety,* Third Edition, Vol. 1, pp 1084–85, Geneva, International Labour Office (1983).

HYDROGEN CYANIDE

- Hazardous substance (EPA)
- Hazardous waste (EPA)

Description: Hydrogen cyanide, a colorless gas or liquid which boils at 26°C, is intensely poisonous with the odor of bitter almonds. It is highly flammable and explosive and is a very weak acid. Hydrogen cyanide, HCN (together with its soluble salts), owes its toxicity to the –CN moiety and not to its acid properties. HCN vapor is released when cyanide salts come in contact with any acid.

Code Numbers: CAS 74-90-8   RTECS MW6825000   UN 1051

DOT Designation: Poison gas and flammable gas.

Synonyms: Hydrocyanic acid, prussic acid.

Potential Exposure: Hydrogen cyanide is used as a fumigant, in electroplating, and in chemical synthesis of acrylates and nitriles, particularly acrylonitrile. It may be generated in blast furnaces, gas works, and coke ovens. Cyanide salts have a wide variety of uses, including electroplating, steel hardening, fumigating, gold and silver extraction from ores, and chemical synthesis. The number of workers with potential exposure to HCH has been estimated by NIOSH to be approximately 1,000.

Incompatibilities: Bases such as caustics, amines.

Permissible Exposure Limits in Air: The Federal standard for hydrogen cyanide is 10 ppm (11 mg/m^3). NIOSH has recommended 5 mg/m^3 expressed as cyanide and determined as a ceiling concentration based on a 10 minute sampling period. The ACGIH TWA adopted as of 1983/84 is 10 ppm (10 mg/m^3). The notation "skin" is added to indicate the possibility of cutaneous absorption. There is no tentative STEL value. The IDLH level is 50 ppm.

Determination in Air: Sample is filtered, then drawn through a KOH bubbler. Analysis is by ion-specific electrode (A-10). Long duration detector tubes may be used (A-11).

Permissible Concentration in Water: A USPHS drinking water criterion for alternate source selection is 100 μg/ℓ (A-37).

Routes of Entry: Inhalation of vapor, percutaneous absorption of liquid and concentrated vapor, ingestion and eye and skin contact.

Harmful Effects and Symptoms: *Local* – Hydrogen cyanide is a mild upper respiratory irritant and may cause slight irritation of the nose and throat. There may also be irritation from skin and eye contact with the liquid. Hydrogen cyanide liquid may cause eye irritation.

Systemic – Hydrogen cyanide is an asphyxiant. It inactivates certain enzyme systems, the most important being cytochrome oxidase, which occupies a fundamental position in the respiratory process and is involved in the ultimate electron transfer to molecular oxygen. Inhalation, ingestion, or skin absorption of hydrogen cyanide may be rapidly fatal. Larger doses may cause loss of consciousness, cessation of respiration, and death. Lower levels of exposure may cause weakness, headache, confusion, nausea, and vomiting. These symptoms may be followed by unconsciousness and death.

Points of Attack: Liver, kidneys, cardiovascular system, central nervous system.

Medical Surveillance: Preplacement and periodic examinations should include the cardiovascular and central nervous systems, liver and kidney function, blood, history of fainting or dizzy spells. Blood CN levels may be useful during acute intoxication. Urinary thiocyanate levels have been used but are nonspecific and are elevated in smokers.

First Aid: If this chemical gets into the eyes, irrigate immediately. If this chemical contacts the skin, flush with water immediately. If a person breathes in large amounts of this chemical, move the exposed person to fresh air at once and perform artificial respiration. When this chemical has been swallowed, get medical attention. Give large quantities of water and induce vomiting. Do not make an unconscious person vomit.

Personal Protective Methods: Wear appropriate clothing to prevent any possibility of skin contact. Wear eye protection to prevent any possibility of eye contact. Employees should wash immediately when skin is wet or contaminated. Remove clothing immediately if wet or contaminated to avoid flammability hazard. Provide emergency showers and eyewash.

Respirator Selection:

    50 ppm: SA/SCBA
    Escape: GMS/SCBA

Disposal Method Suggested: Chemical conversion to ammonia and carbon dioxide using chlorine or hypochlorite in a basic media. Controlled incineration is also adequate to totally destroy cyanide. Alternatively, HCN can be recovered, from ammonoxidation process waste streams for example (A-58).

References

(1) National Institute for Occupational Safety and Health, *Criteria for a Recommended Standard: Occupational Exposure to Hydrogen Cyanide,* NIOSH Doc. No. 77-108 (1977).
(2) See Reference (A-61).
(3) See Reference (A-60).
(4) United Nations Environment Programme, *IRPTC Legal File 1983,* Vol. II, pp VII/391–4, Geneva, Switzerland, International Register of Potentially Toxic Chemicals (1984).

HYDROGEN FLUORIDE

- Hazardous substance (EPA)
- Hazardous waste (EPA)

Description: HF is a colorless, fuming liquid or gas with a strong, irritating odor. It boils at $19°$ to $20°C$.

Code Numbers: CAS 7664-39-3   RTECS MW7875000   UN 1052 (Anhyd.)
                                                        UN 1790 (Solution)

DOT Designation: Corrosive material.

Synonyms: Hydrofluoric acid gas, fluorohydric acid gas, anhydrous hydrofluoric acid.

Potential Exposure: Hydrogen fluoride, its aqueous solution hydrofluoric acid, and its salts are used in production of organic and inorganic fluorine compounds such as fluorides and plastics; as a catalyst, particularly in paraffin alkylation in the petroleum industry; as an insecticide; and to arrest the fermentation in brewing. It is utilized in the fluorination processes, especially in the aluminum industry, in separating uranium isotopes, in cleaning cast iron, copper, and brass, in removing efflorescence from brick and stone, in removing sand from metallic castings, in frosting and etching glass and enamel, in polishing crystal, in decomposing cellulose, in enameling and galvanizing iron, in working silk, in dye and analytical chemistry, and to increase the porosity of ceramics.

Incompatibilities: Metals, concrete, glass, ceramics.

Permissible Exposure Limits in Air: The Federal standard and the 1983/84 ACGIH TWA value is 3 ppm (2.5 mg/m^3). NIOSH has recommended a TWA of 2.5 mg/m^3 and a 15 minute ceiling value of 5.0 mg/m^3. The STEL value is 6 ppm (5 mg/m^3) as of 1980. The IDLH level is 20 ppm.

Determination in Air: Absorption in a bubbler with NaOH, analysis by ion-specific electrode. See NIOSH Methods, Set L. See also reference (A-10).

Permissible Concentration in Water: No criteria set.

Routes of Entry: Inhalation, skin absorption, ingestion, eye and skin contact.

Harmful Effects and Symptoms: Irritation of eyes, nose and throat; skin and eye burns; pulmonary edema; nasal congestion, bronchitis. See also "Fluorides."

Points of Attack: Eyes, respiratory system, skin, lungs.

Medical Surveillance: See "Fluorides."

First Aid: If this chemical gets into the eyes, irrigate immediately. If this chemical contacts the skin, flush with water immediately. If a person breathes in large amounts of this chemical, move the exposed person to fresh air at once and perform artificial respiration. When this chemical has been swallowed, get medical attention. Give large quantities of water and do not induce vomiting.

Personal Protective Methods: Wear appropriate clothing to prevent any possibility of skin contact. Wear eye protection to prevent any possibility of eye contact. Employees should wash immediately when skin is wet or contaminated. Remove nonimpervious clothing immediately if wet or contaminated. Provide emergency showers and eyewash.

Respirator Selection:
> 20 ppm: CCRSF/GMS/SAF/SCBAF
> Escape: GMS/SCBA

Disposal Method Suggested: Reaction with excess lime followed by lagooning and either recovery or landfill disposal of the separated calcium fluoride. The supernatant liquid from this process is diluted and discharged to the sewer. Alternatively, hydrogen fluoride can be recovered and recycled (A-57) in many cases.

References

(1) National Institution for Occupational Safety and Health, *Criteria for a Recommended Standard: Occupational Exposure to Hydrogen Fluoride,* NIOSH Doc. No. 76-143 (1976).
(2) U.S. Environmental Protection Agency, *Hydrofluoric Acid,* Health and Environmental Effects Profile No. 117, Office of Solid Waste, Washington, D.C. (April 30, 1980).
(3) See Reference (A-61).
(4) See Reference (A-60).
(5) Parmeggiani, L., Ed., *Encyclopedia of Occupational Health & Safety,* Third Edition, Vol. 1, pp 1085–86, Geneva, International Labour Office (1983).

HYDROGEN PEROXIDE

Description: H$_2$O$_2$, anhydrous hydrogen peroxide, is a colorless rather unstable liquid with a bitter taste. Hydrogen peroxide is completely miscible with water and is commercially sold in concentrations of 3, 35, 50, 70, and 90% solutions.

Code Numbers: CAS 7722-84-1   RTECS MX0900000   UN 2015

DOT Designation: Oxidizer and corrosive material.

Synonyms: Peroxide, hydrogen dioxide, hydroperoxide.

Potential Exposure: Hydrogen peroxide is used in the manufacture of acetone, antichlor, antiseptics, benzoyl peroxide, buttons, disinfectants, pharmaceuticals (A-41), felt hats, plastic foam, rocket fuel, sponge rubber and pesticides (A-32). It is also used in bleaching bone, feathers, flour, fruit, fur, gelatin, glue, hair, ivory, silk, soap, straw, textiles, wax, and wood pulp, and as an oxygen source in respiratory protective equipment. Other specific occupations with potential exposure include liquor and wine agers, dyers, electroplaters, fat refiners, photographic film developers, wool printers, veterinarians, and water treaters.

Incompatibilities: Oxidizable materials; iron, copper, brass, bronze, chromium, zinc, lead, manganese, silver, catalytic metals.

Permissible Exposure Limits in Air: The Federal standard and the ACGIH 1983/84 TWA value for hydrogen peroxide (90%) is 1 ppm (1.5 mg/m^3). The STEL is 2 ppm (3 mg/m^3). The IDLH level is 75 ppm.

Permissible Concentration in Water: No criteria set.

Route of Entry: Inhalation of vapor or mist, ingestion, eye and skin contact.

Harmful Effects and Symptoms: *Local* — The skin, eyes, and mucous membranes may be irritated by concentrated vapor or mist. Bleaching and a burning sensation may occur at lower levels, while high concentrations may result in blistering and severe eye injury, which may be delayed in appearance.

Systemic — Inhalation of vapor or mist may produce pulmonary irritation ranging from mild bronchitis to pulmonary edema. No chronic systemic effects have been observed.

Points of Attack: Eyes, skin, respiratory system.

Medical Surveillance: Preplacement and periodic examinations should be directed to evaluation of the general health with particular reference to the skin, eyes, mucous membranes, and respiratory tract.

First Aid: If this chemical gets into the eyes, irrigate immediately. If this chemical contacts the skin, flush with water immediately. If a person breathes in large amounts of this chemical, move the exposed person to fresh air at once and perform artificial respiration. When this chemical has been swallowed, get medical attention. Give large quantities of water and induce vomiting. Do not make an unconscious person vomit.

Personal Protective Methods: In areas where concentrated hydrogen peroxide is being used, if there is danger of spill or splash, skin protection should be provided by protective clothing, gloves, goggles, and boots. Where fumes or vapor are excessive, workers should be provided with gas masks with fullface pieces and proper canisters or supplied air respirators. Additional health hazards may occur from the decomposition of hydrogen peroxide. Oxygen, possibly at high pressure, may form, which may create an explosion hazard. Hydrogen peroxide is generally handled in a closed system to prevent contamination.

Employees should wash promptly when skin is wet or contaminated. Remove nonimpervious clothing immediately if wet or contaminated. Provide emergency showers and eyewash.

Respirator Selection:
            10 ppm: SA/SCBA
            50 ppm: SAF/SCBAF
            75 ppm: SA:PD,PP,CF
            Escape: GMS/SCBA
            Note: Do not use oxidizable sorbents.

Disposal Method Suggested: Dilution with water to release the oxygen. After decomposition, the waste stream may be discharged safely.

References

(1)   See Reference (A-61).
(2)   See Reference (A-60).
(3)   Parmeggiani, L., Ed., *Encyclopedia of Occupational Health & Safety,* Third Edition, Vol. 1, pp 1088–90, Geneva, International Labour Office (1983).
(4)   United Nations Environment Programme, *IRPTC Legal File 1983,* Vol. II, pp VII/395–7, Geneva, Switzerland, International Register of Potentially Toxic Chemicals (1984).

HYDROGEN SELENIDE

Description: H_2Se is a colorless gas with a very offensive odor.

Code Numbers: CAS 7783-07-5   RTECS MX1050000   UN 2202

DOT Designation: Flammable gas and poison.

Synonyms: Selenium hydride.

Potential Exposure: Hydrogen selenide is used in semiconductor manufacture. Also, it may be produced by the reaction of acids or water and metal selenides or hydrogen and soluble selenium compounds.

Incompatibilities: Oxidizers, acid, water, halogenated hydrocarbons.

Permissible Exposure Limits in Air: The Federal standard and the 1983/84 ACGIH TWA value is 0.05 ppm (0.2 mg/m^3). There is no STEL value. The IDLH level is 2 ppm.

Determination in Air: Collection by impinger or fritted bubbler followed by atomic absorption spectrometry (A-21).

Permissible Concentration in Water: No criteria set, but EPA (A-37) suggests a permissible ambient goal of 10 μg/ℓ (the same as for selenium) based on health effects.

Routes of Entry: Inhalation, eye and skin contact.

Harmful Effects and Symptoms: The effects of hydrogen selenide intoxication are similar to those caused by other irritating gases in industry: irritation of the mucous membranes of the nose, eyes, and upper respiratory tract, followed by slight tightness in the chest. These symptoms clear when the worker is removed from the exposed area. In some cases, however, pulmonary edema may develop suddenly after a latent period of six to eight hours following exposure. Other symptoms include garlic breath and metallic taste; nausea, vomiting and diarrhea; dizziness, lassitude and fatigue.

Points of Attack: Respiratory system, eyes.

Medical Surveillance: Consider the points of attack in preplacement and periodic physical examinations.

First Aid: If a person breathes in large amounts of this chemical, move the exposed person to fresh air at once and perform artificial respiration.

Personal Protective Methods: No particular methods are recommended by NIOSH.

Respirator Selection:

        0.5 ppm: SA/SCBA
          2 ppm: SAF/SCBAF
        Escape: GMS/SCBA

References

(1)   See Reference (A-61).
(2)   See Reference (A-60).

HYDROGEN SULFIDE

- Hazardous substance (EPA)
- Hazardous waste (EPA)

Description: H_2S, hydrogen sulfide, is a flammable, colorless gas with a characteristic rotten-egg odor and is soluble in water.

Code Numbers: CAS 7783-06-4   RTECS MX1225000   UN 1053

DOT Designation: Flammable gas and poison.

Synonyms: Sulfuretted hydrogen, hydrosulfuric acid, stink damp.

Potential Exposure: Hydrogen sulfide is used in the synthesis of inorganic sulfides, sulfuric acid, and organic sulfur compounds, as an analytical reagent, as a disinfectant in agriculture, and in metallurgy. It is generated in many industrial processes as a by-product and also during the decomposition of sulfur-containing organic matter, so potential for exposure exists in a variety of situations. Hydrogen sulfide is found in natural gas, volcanic gas, and in certain natural spring waters.

It may also be encountered in the manufacture of barium carbonate, barium salt, cellophane, depilatories, dyes, and pigments, felt, fertilizer, adhesives, viscose rayon, lithopone, synthetic petroleum products; in the processing of sugar beets; in mining, particularly where sulfide ores are present; in sewers and sewage treatment plants; during excavation of swampy or filled ground for tunnels, wells, and caissons; during drilling of oil and gas wells; in purification of hydrochloric acid and phosphates; during the low temperature carbonization of coal; in tanneries, breweries, slaughterhouses; in fat rendering; and in lithography and photoengraving.

NIOSH estimates that 125,000 workers in the United States are exposed or potentially exposed to hydrogen sulfide.

Incompatibilities: Strong oxidizers, metals.

Permissible Exposure Limits in Air: The Federal standard is a ceiling value of 20 ppm (30 mg/m^3) with a maximum peak above this value for an 8 hr shift of 50 ppm (75 mg/m^3) for a maximum duration of 10 min once only if no other measurable exposure occurs.

NIOSH recommends that exposure to hydrogen sulfide be limited to a ceiling concentration of 15 mg/m^3 (10 ppm) for 10 min. A requirement for continuous monitoring when there is a potential for exposure to hydrogen sulfide at a concentration of 70 mg/m^3 (50 ppm) or higher is also recommended. Occupational exposure to hydrogen sulfide has been defined as exposure at or above the ceiling concentration of 15 mg/m^3.

ACGIH as of 1983/84 has set a TWA value of 10 ppm (14 mg/m^3) and an STEL of 15 ppm (21 mg/m^3). The IDLH level is 300 ppm.

Determination in Air: Collection with an impinger using cadmium hydroxide; spectrophotometric analysis. See NIOSH Methods, Sec A. See also reference (A-10). NIOSH has also recommended sampling and analytical methods using a methylene blue technique adapted from the NIOSH/OSHA Standards Completion Program. The methylene blue method for sampling and analysis was selected because of its sensitivity and because it has already demonstrated a wide applicability in industry. The use of NIOSH validated detector tubes is also recommended for this purpose. For continuous monitoring, no specific instrument is recommended. However, minimum capabilities such as response time, sensitivity, and range are specified.

Permissible Concentration in Water: According to EPA (A-3), available data indicate that water containing concentrations of 2.0 µg/ℓ undissociated H$_2$S would not be hazardous to most fish and other aquatic wildlife, but concentrations in excess of 2.0 µg/ℓ would constitute a long term hazard. However, EPA (A-37) suggests a permissible ambient goal of 207 µg/ℓ based on health effects.

Routes of Entry: Inhalation of gas, ingestion, eye and skin contact.

Harmful Effects and Symptoms: *Local* — Palpebral edema, bulbar conjunctivitis, keratoconjunctivitis, and ocular lesions may occur when hydrogen sulfide comes in contact with the eyes. Photophobia and lacrimation may also develop. Direct irritation of the respiratory tract may cause rhinitis, pharyngitis, bronchitis, and pneumonia. Hydrogen sulfide may penetrate deep into the lungs and cause hemorrhagic pulmonary edema. Hydrogen sulfide's irritative effects are due to the formation of alkali sulfide when the gas comes in contact with moist tissues.

Systemic — Acute exposure may cause immediate coma which may occur with or without convulsions. Death may result with extreme rapidity from respiratory failure. Postmortem signs include a typical greenish cyanosis of the chest and face with green casts found in viscera and blood. The toxic action of hydrogen sulfide is thought to be due to inhibition of cytochrome oxidase by binding iron which is essential for cellular respiration. Subacute exposure results in headache, dizziness, staggering gait, and excitement suggestive of neurological damage, and nausea and diarrhea suggestive of gastritis.

Recovery is usually complete although rarely polyneuritis may develop as a result of vestibular and extrapyramidal tract damage. Tremors, weakness, and numbness of extremities may also occur. Physicians may observe a "rotten-egg" breath and abnormal electrocardiograms in victims. Systemic effects from chronic exposure to hydrogen sulfide have not been established.

Points of Attack: Respiratory system, lungs, eyes.

Medical Surveillance: Preplacement medical examinations should evaluate any preexisting neurological, eye, and respiratory conditions and any history of fainting seizures.

It is recommended by NIOSH that preplacement and periodic examinations (once every 3 years) be made available to all workers occupationally exposed to hydrogen sulfide. These are to include physical examinations which give particular attention to the eyes and the nervous and respiratory systems.

First Aid: If this chemical gets into the eyes, irrigate immediately. If this chemical contacts the skin, flush with water immediately. If a person breathes in large amounts of this chemical, move the exposed person to fresh air at once and perform artificial respiration.

Personal Protective Methods: Hydrogen sulfide's strong odor, noticeable at low concentrations, is a poor warning sign as it may cause olfactory paralysis, and some persons are congenitally unable to smell H_2S.

Accidental exposure may occur when workers enter sewage tanks and other confined areas in which hydrogen sulfide is formed by decomposition. In a number of cases workers enter unsuspectingly and collapse almost immediately. Workers, therefore, should not enter enclosed spaces without proper precautions.

All Federal standard and other safety precautions must be observed when tanks or other confined spaces are to be entered. In areas where the exposure to hydrogen sulfide exceeds the standards, workers should be provided with fullface canister gas masks or preferably supplied air respirators.

When liquid H_2S is involved, wear clothing to prevent skin freezing. Wear eye protection to prevent any reasonable probability of eye contact. Remove clothing immediately if wet or contaminated to avoid flammability hazard.

Respirator Selection:
>300 ppm: SAF/SCBAF
>Escape: GMAGS/SCBA

Disposal Method Suggested: Hydrogen sulfide can be recovered as such or converted to elemental sulfur or sulfuric acid (A-57).

References

(1) National Institute for Occupational Safety and Health, *Criteria for a Recommended Standard: Occupational Exposure to Hydrogen Sulfide,* NIOSH Doc. No. 77-158 (1977).

(2) U.S. Environmental Protection Agency, *Hydrogen Sulfide,* Health and Environmental Effects Profile No. 118, Office of Solid Waste, Washington, DC (April 30, 1980).

(3) See Reference (A-61).

(4) See Reference (A-60).

(5) Sax, N.I., Ed., *Dangerous Properties of Industrial Materials Report, 3,* No. 4, 68-73, New York, Van Nostrand Reinhold Co. (1983).

(6) Parmeggiani, L., Ed., *Encyclopedia of Occupational Health & Safety,* Third Edition, Vol. 1, pp 1090-91, Geneva, International Labour Office (1983).

HYDROQUINONE

Description: $C_6H_4(OH)_2$, hydroquinone, exists as colorless, hexagonal prisms. It melts at 165°C.

Code Numbers: CAS 123-31-9   RTECS MX3500000   UN 2662

DOT Designation: —

Synonyms: Quinol, hydroquinol, p-diphenol, hydrochinone, dihydroxybenzene, p-dihydroxybenzene, p-hydroxyphenol, 1,4-benzenediol.

Potential Exposure: Hydroquinone is a reducing agent and is used as a photographic developer and as an antioxidant or stabilizer for certain materials which polymerize in the presence of oxidizing agents. Many of its derivatives are used as bacteriostatic agents, and others, particularly 2,5-bis(ethyleneimino) hydroquinone, have been reported to be good antimitotic and tumor-inhibiting agents.

Incompatibilities: Strong oxidizers.

Permissible Exposure Limits in Air: The Federal standard and the 1983/84 ACGIH TWA value is 2 mg/m^3. The STEL value is 4 mg/m^3. The IDLH level is 200 mg/m^3.

Determination in Air: Collection on a filter, workup with acid, analysis by high pressure liquid chromatography. See NIOSH Methods, Set 2. See also reference (A-10). A colorimetric method may also be used (A-23).

Permissible Concentration in Water: No criteria set.

Route of Entry: Inhalation of dust, ingestion, eye and skin contact.

Harmful Effects and Symptoms: *Local* – The dust is a mild primary irritant. Skin sensitization to the dry solid is very rare but does occur on occasion from contact with its alkaline solutions. The skin may be depigmented by repeated applications of ointments of hydroquinone, but this virtually never occurs from contact with dust or dilute water solutions. Following prolonged exposure to elevated dust levels, brownish conjunctiva stains may appear. These may be followed by corneal opacities and structural changes in the cornea which may lead to loss of visual acuity. The early pigmentary stains are reversible, while the corneal changes tend to be progressive.

Systemic – Oral ingestion of large quantities of hydroquinone may produce blurred speech, tinnitus, tremors, sense of suffocation, vomiting, muscular twitching, headache, convulsions, dyspnea and cyanosis from methemoglobinemia, and coma and collapse from respiratory failure. The urine is usually green or brownish green. No systemic symptoms have been found following inhalation of hydroquinone dust.

Points of Attack: Eyes, respiratory system, skin, central nervous system.

Medical Surveillance: Careful examination of the eyes, including visual acuity and slit lamp examinations, should be carried out in preplacement and periodic examinations. Also the skin should be examined. Hydroquinone is excreted in the urine as a sulfate ester. This has not been helpful in following worker exposure to dust.

First Aid: If this chemical gets into the eyes, irrigate immediately for 15 minutes duration. If this chemical contacts the skin, flush with water. If a person breathes in large amounts of this chemical, move the exposed person to fresh air at once. When this chemical has been swallowed, get medical attention. Give large quantities of salt water and induce vomiting. Do not make an unconscious person vomit.

Personal Protective Methods: Wear appropriate clothing to prevent repeated or prolonged skin contact. Wear eye protection to prevent any possibility of eye contact with liquids of >7% concentration. Employees should wash promptly when skin is wet or contaminated. Work clothing should be changed daily if it is possible that clothing is contaminated. Provide eyewash if liquids containing >7% contaminants are involved.

Respirator Selection:
        100 mg/m³ : HiEPF/SAF/SCBAF
        200 mg/m³ : SAF:PD,PP,CF/PAPHiEF
        Escape: HiEPF/SCBA

Disposal Method Suggested: Incineration (1800°F, 2.0 sec minimum), then scrub to remove harmful combustion products.

References

(1)   National Institute for Occupational Safety and Health, *Criteria for a Recommended Standard: Occupational Exposure to Hydroquinone,* NIOSH Doc. No. 78-155 (1978).
(2)   Sax, N.I., Ed., *Dangerous Properties of Industrial Materials Report, 2,* No. 2, 35-37, New York, Van Nostrand Reinhold Co. (1982).
(3)   See Reference (A-60).
(4)   United Nations Environment Programme, *IRPTC Legal File 1983,* Vol. II, pp VII/398, Geneva, Switzerland, International Register of Potentially Toxic Chemicals (1984).

HYDROXYLAMINE

Description: $HONH_2$ is a white crystalline substance melting at 32°C; it is very hygroscopic and unstable.

Code Numbers: CAS 7803-49-8   RTECS NC2975000

DOT Designation: —

Potential Exposure: Those involved in chemical synthesis or use of hydroxylamine as a reducing agent.

Incompatibilities: Oxidants, zinc powder (A-38).

Permissible Exposure Limits in Air: No standards set.

Permissible Concentration in Water: No criteria set.

Routes of Entry: Inhalation and ingestion of dust.

Harmful Effects and Symptoms: Corrosive to skin, eyes and mucous membranes. Can cause methemoglobinemia.

Medical Surveillance: Annual physical exams including renal and hepatic (A-38).

First Aid: Flush eyes and exposed body areas with water (A-38).

Personal Protective Methods: Wear protective clothing including clean overalls, cap, gloves and goggles.

Respirator Selection: Canister-type mask recommended unless heat necessitates use of self-contained breathing apparatus.

Disposal Method Suggested: Add sodium bisulfite solution and flush to sewer; or incinerate.

References

(1)   Sax, N.I., Ed., *Dangerous Properties of Industrial Materials Report, 2,* No. 2, 37-39, New York, Van Nostrand Reinhold Co. (1982).
(2)   See Reference (A-60).
(3)   Parmeggiani, L., Ed., *Encyclopedia of Occupational Health & Safety,* Third Edition, Vol. 1, pp 1091–92, Geneva, International Labour Office (1983).

HYDROXYPROPYL ACRYLATE

Description: $CH_2=CHCOOCH_2CHOHCH_3$ is a colorless liquid.

Code Numbers: CAS 999-61-1   RTECS AT1925000

DOT Designation: —

Synonyms: 1,2-Propanediol-1-acrylate, propylene glycol monoacrylate.

Potential Exposure: This material is used as a bifunctional monomer for acrylic resins; it is used as a binder in nonwoven fabrics and may be used in the production of detergent lube oil additives.

Permissible Exposure Limits in Air: There is no Federal standard but ACGIH as of 1983/84 has adopted a TWA value of 0.5 ppm (3 mg/m^3) with the notation "skin" indicating the possibility of cutaneous absorption. There is no STEL.

Permissible Concentrations in Water: No criteria set.

Harmful Effects and Symptoms: In animal experiments, direct contact caused severe eye burns and was corrosive to the skin (A-34).

I

INDENE

Description: CH=CHCH=CHC=CCH=CHCH$_2$ is a colorless liquid boiling at 182°C.

Code Numbers: CAS 95-13-6   RTECS NK8225000

DOT Designation: —

Synonym: Indonaphthene.

Potential Exposure: Indene is used in the preparation of synthetic resins such as coumarone-indene resins. It is also used as an organic intermediate.

Permissible Exposure Limits in Air: There are no Federal standards but ACGIH (1983/84) has adopted a TWA of 10 ppm (45 mg/m^3) and set a tentative STEL of 15 ppm (70 mg/m^3).

Permissible Concentration in Water: No criteria set, but EPA (A-37) has suggested a permissible ambient goal of 621 μg/ℓ based on health effects.

Harmful Effects and Symptoms: Animal studies have shown that liver damage and occasionally splenic and renal injury result from exposure to high (800 to 900 ppm) vapor concentrations. Liquid indene defatted the skin but did not cause dermatitis. Aspiration of liquid into the lung caused pneumonitis, pulmonary edema and hemorrhage in laboratory animals (A-34).

INDENO(1,2,3-cd)PYRENE

See "Polynuclear Aromatic Hydrocarbons."

INDIUM AND COMPOUNDS

Description: Indium metal is malleable, ductile and softer than lead. The two most important indium compounds are the oxide and sulfate. Physical properties are presented below:

Chemical	Specific Gravity	Melting Point °C	Boiling Point °C	Solubility in Water
Indium	7.30^{20}	156.61	2000±10	Insol. (hot and cold)*
Indium sesquioxide	7.179	—	Volat. 850	Insol. cold
Indium sulfate	3.438	—	—	Sol. cold, very sol. hot

*Finely divided indium forms hydroxide on contact with water.

Code Numbers:

Metal	CAS 7440-74-6	RTECS NL1050000
Sulfate	CAS 13464-82-9	RTECS NL1925000

DOT Designation: —

Synonyms: None.

Potential Exposures: The uses of indium and its compounds are shown in the following table:

Compound	Use	Purpose
Indium	Dental alloy	—
	Solder and alloy industries	Produces high quality solders and braze-bonded connectors
	Automotive bearings (Europe only)	—
	Low pressure sodium lamps (Europe)	—
	Other	Nuclear reactor control rod alloys
		Catalysts
		Indium oxide fuel cells
		Cryogenic gasket material
Indium oxide	Glass	Used for coloring. A light to dark brown can be obtained, depending on the amount used.
Indium sulfate	Electroplating	Used to prepare sulfate electrolytes
Radioisotopes of indium and indium compounds	Medical	Treatment of cancer and diagnostic organ scanning

Permissible Exposure Limits in Air: The threshold limit value set by ACGIH (1983/84) for indium and its compounds is 0.1 mg/m^3. This value is recommended in view of the character and severity of injury from indium salts, especially the pulmonary effects. The STEL value is 0.3 mg/m^3.

Determination in Air: Collection on a membrane filter, analysis by anodic stripping voltammetry (A-10).

Permissible Concentration in Water: No criteria set.

Determination in Water: Neutron activation has been used to analyze for indium in seawater (2). The detection limit is 0.006 ng/ℓ. Polarography has also been used to determine indium in water with a detection limit of 1.0 μg/ℓ.

Routes of Entry: Ingestion or inhalation.

Harmful Effects and Symptoms: Indium and its compounds are considered moderately toxic and as a local irritant on contact with the skin. However, they are rated extremely toxic on ingestion. The American Conference of Governmental Hygienists has recommended a threshold limiting value of 0.1 mg/m^3 based almost entirely on experiments with animals and the severity of injury from indium salts. The data involving humans is very limited. As a result possibly too much weight is given to a Russian report that individuals exposed to indium compounds during production complained of pains in joints and bones, tooth decay, nervous and gastrointestinal disorders, heart pains and general debility.

This has not been reported in comparable United States activities.

Indium is poorly absorbed through the intestine, and oral levels of toxicity are quite high. The cells of the reticuloendothelial system phagocytize indium compound particles, and indium toxicity is apparently due to the concentration of heavy metals by these cells. Radioisotopes of indium are used for diagnostic x-ray scanning. Carrier-free isotopes can be used with relatively few toxic effects, although several cases of anaphylactic shock have occurred following injection of indium compounds.

Medical Surveillance: Preplacement and regular physical examinations are indicated.

Personal Protective Methods: Personnel who work with or are exposed to dust, salts or mists of indium and its compounds should, in the absence of exact data, wear safety equipment such as chemical safety goggles for the eyes and a respirator to avoid inhalation. It may be necessary to wear protective clothing to avoid skin contact.

Disposal Method Suggested: Indium and its compounds are rated moderately toxic. Recovered indium has a recycled value so a minimum quantity is discarded. Closed containers are required for the disposal of indium and its compounds. The environmental hazard from indium appears to result from the use and disposal of radioactive isotopes and not chemical toxicity (1).

References

(1) U.S. Environmental Protection Agency, *Preliminary Investigation of Effects on the Environment of Boron, Indium, Nickel, Selenium, Tin, Vanadium and Their Compounds, Vol. II, Indium,* Report EPA-460/2-75-005B, Washington, DC, Office of Toxic Substances (August 1975).

(2) U.S. Environmental Protection Agency, *Toxicology of Metals, Vol II: Indium,* Report EPA-600/1-77-022, Research Triangle Park, NC, pp 234-41 (May 1977).

(3) Parmeggiani, L., Ed., *Encyclopedia of Occupational Health & Safety,* Third Edition, Vol. 1, pp 1103, Geneva, International Labour Office (1983).

IODINE

Description: I_2 is a violet solid with a sharp, characteristic odor.

Code Numbers: CAS 7553-56-2   RTECS NN1575000

DOT Designation: —

Synonyms: None.

Potential Exposure: Iodine is used in the manufacture of organic chemicals, especially pharmaceuticals. It is used in the manufacture of silver iodide for use in photography. It is used as a catalyst in organic reactions.

Incompatibilities: Gaseous or aqueous ammonia, acetylene, acetaldehyde, powdered aluminum, and active metals.

Permissible Exposure Limits in Air: The Federal standard and the 1983/84 ACGIH TWA value is a 0.1 ppm (1.0 mg/m³) ceiling value. There is no STEL value set.

Determination in Air: Iodine may be determined colorimetrically using o-tolidine reagent (A-24).

Permissible Concentration in Water: No criteria set.

Routes of Entry: Inhalation, ingestion, eye and skin contact.

Harmful Effects and Symptoms: Irritation of eye and nose, lacrimation, headaches, tight chest, skin burns, cutaneous hypersensitivity, burns of the mouth, vomiting, abdominal pain, diarrhea, skin rash.

Points of Attack: Respiratory system, lungs, eyes, skin, central nervous system, cardiovascular system.

Medical Surveillance: Consider the points of attack in preplacement and periodic physical examinations.

First Aid: If this chemical gets into the eyes, irrigate immediately. If this chemical contacts the skin, wash with soap immediately. If a person breathes in large amounts of this chemical, move the exposed person to fresh air at once and perform artificial respiration. When this chemical has been swallowed, get medical attention. Give large quantities of water and induce vomiting. Do not make an unconscious person vomit.

Personal Protective Methods: Wear appropriate clothing to prevent any possibility of skin contact with liquids of >7% content or repeated or prolonged contact with liquids of <7% content. Wear eye protection to prevent any possibility of eye contact with liquids of >7% iodine content. Employees should wash immediately with soap when skin is wet or contaminated with liquids of >7% content. Remove nonimpervious clothing immediately if wet or contaminated with liquids containing >7% and promptly remove if liquid contains <7% iodine. Provide emergency showers and eyewash if liquids containing >7% contaminants are involved.

Respirator Selection:
      1 ppm: SA/SCBA
      5 ppm: SAF/SCBAF
    10 ppm: SAF:PD,PP,CF
   Escape: GMOVAGP/SCBA

Disposal Method Suggested: React with large volumes of reducing agent (hypo- or bisulfites or ferrous salts) solution, neutralize and flush to sewer with water (A-38). Alternatively, iodine may be recovered from various process waste streams (A-57).

References

(1) Sax, N.I., Ed., *Dangerous Properties of Industrial Materials Report, 1,* No. 5, 65-68, New York, Van Nostrand Reinhold Co. (1981).
(2) See Reference (A-61).
(3) See Reference (A-60).
(4) Parmeggiani, L., Ed., *Encyclopedia of Occupational Health & Safety,* Third Edition, Vol. 1, pp 1153-55, Geneva, International Labour Office (1983).

IODOFORM

Description: CHI_3 is a yellow or greenish-yellow crystalline solid melting at 119°C with a characteristic pungent odor.

Code Numbers: CAS 75-47-8   RTECS PB7000000

DOT Designation: −

Synonym: Triiodomethane.

Potential Exposure: Iodoform has been used as a topical antiinfective and in veterinary medicine as an antiseptic and disinfectant for superficial lesions.

Permissible Exposure Limits in Air: There is no Federal standard but ACGIH (1983/84) has adopted a TWA value of 0.6 ppm (10 mg/m^3) and set an STEL of 1 ppm (20 mg/m^3).

Permissible Concentration in Water: No criteria set.

Harmful Effects and Symptoms: Toxicologic data on iodoform are very limited (A-34). The material has been evaluated (1) and found not to be carcinogenic.

References

(1) National Cancer Institute, *Bioassay of Iodoform for Possible Carcinogenicity,* Technical Report Series No. 110, Bethesda, MD (1978).

IRON AND IRON COMPOUNDS

● Carcinogen (See entries under (1) Hematite; (2) Iron-Dextran Complex)

Description: Fe, iron, is a malleable, silver-grey metal. Ferric oxide is a dense, dark red powder or lumps. Hematite is the most important iron ore and is generally found as red hematite (red iron ore, mainly Fe_2O_3) and brown hematite (brown iron ore, mainly limonite, a hydrated sesquioxide of iron). Magnetic iron oxide, Fe_3O_4, is black. Iron is insoluble in water. Iron oxide is soluble in hydrochloric acid. Iron pentacarbonyl, $Fe(CO)_5$, is a yellow to red viscous liquid which boils at 103°C.

Code Numbers:

Iron dust	None	RTECS NO6851000	UN 1383
Iron(III) oxide	CAS 1309-37-1	RTECS NO7400000	–
Iron pentacarbonyl	CAS 13463-40-6	RTECS NO4900000	UN 1994

DOT Designation: Iron dust, –; iron oxide, –; iron carbonyl, flammable liquid and poison.

Synonyms: For iron oxide—ferric oxide and rouge.

Potential Exposure: Iron is alloyed with carbon to produce steel. The addition of other elements (e.g., manganese, silicon, chromium, vanadium, tungsten, molybdenum, titanium, niobium, phosphorus, zirconium, aluminum, copper, cobalt, and nickel) imparts special characteristics to the steel.

Occupational exposures occur during mining, transporting, and preparing of ores and during the production and refining of the metal and alloys. In addition, certain workers may be exposed while using certain iron-containing materials—welders, grinders, polishers, silver finishers, metal workers and boiler scalers.

Incompatibilities: Iron oxide fume—calcium hypochlorite.

Permissible Exposure Limits in Air: The Federal standard for iron oxide fume is 10 mg/m^3. There are no standards for other iron compounds. ACGIH as of 1983/84 has set a TWA of 5 mg/m^3 for iron oxide fume and 10 mg/m^3 for rouge. They also set a TWA of 1.0 mg/m^3 for soluble iron compounds and 0.1 ppm (0.08 mg/m^3) for iron pentacarbonyl. The STEL for iron oxide fume is 10 mg/m^3, for rouge 20 mg/m^3 and for iron pentacarbonyl, 0.2 ppm (1.6 mg/m^3).

Determination in Air: Iron oxide fume—collection on a filter, workup with acid, analysis by atomic absorption. See NIOSH Method, Set O. See also reference (A-10). Iron pentacarbonyl may also be detected by atomic absorption spectrometry (A-25) as may soluble iron salts.

Permissible Concentration in Water: The EPA (A-3) has suggested the following criteria—0.3 mg/ℓ for domestic water supplies and 1.0 mg/ℓ for freshwater aquatic life.

Route of Entry: Inhalation of dust.

Harmful Effects and Symptoms: *Local* — Soluble iron salts, especially ferric chloride and ferric sulfate, are cutaneous irritants and their aerosols are irritating to the respiratory tract. Iron compounds as a class are not associated with any particular industrial risk.

Systemic — The inhalation of iron oxide fumes or dust may cause a benign pneumoconiosis (siderosis). It is probable that the inhalation of pure iron oxide does not cause fibrotic pulmonary changes, whereas the inhalation of iron oxide plus certain other substances may cause injury.

On the basis of epidemiological evidence, exposure to hematite dust increases the risk of lung cancer for workers working underground, but not for surface workers. It may be, however, that hematite dust becomes carcinogenic only in combination with radioactive material, ferric oxide, or silica. There is no evidence that hematite dust or ferric oxide causes cancer in any part of the body other than the lungs.

Iron compounds derive their dangerous properties from the radical with which the iron is associated. Iron pentacarbonyl is one of the more dangerous metal carbonyls. It is highly flammable and toxic. Symptoms of overexposure closely resemble those caused by $Ni(CO)_4$ and consist of giddiness and headache, occasionally accompanied by fever, cyanosis, and cough due to pulmonary edema. Death may occur within 4 to 11 days due to pneumonia, liver damage, vascular injury, and central nervous system degeneration.

Points of Attack: For iron oxide fumes—respiratory system and lungs.

Medical Surveillance: Special consideration should be given to respiratory disease and lung function in placement and periodic examinations. Smoking history should be known. Chest x-rays and pulmonary function should be evaluated periodically especially if symptoms are present.

First Aid: For iron oxide fumes—If a person breathes in large amounts of this chemical, move the exposed person to fresh air at once and perform artificial respiration.

Personal Protective Methods: Dust masks are recommended for all workers exposed to areas of elevated dust concentrations and especially those workers in underground mines. In areas where iron oxide fumes are excessive, vapor canister masks or supplied air masks are recommended. Generally speaking, protective clothing is not necessary, but attention to personal hygiene, showering, and clothes changing should be encouraged.

Respirator Selection:
    100 mg/m^3: FuHiEP/SA/SCBA
    500 mg/m^3: HiEPF/SAF/SCBAF
    5,000 mg/m^3: PAPHiE/SA:PD,PP,CF

Disposal Method Suggested: Landfill. Alternatively, iron may be recovered from various iron-containing wastes (A-57).

References

(1) Nat. Inst. for Occup. Safety and Health, *Information Profiles on Potential Occupational Hazards, Iron and Its Compounds,* Report No. PB 276-678, Rockville, MD, pp. 143-156 (October 1977).
(2) See Reference (A-61).
(3) See also Sax, N.I., Ed., *Dangerous Properties of Industrial Materials Report, 3,* No. 4, 42-50, New York, Van Nostrand Reinhold Co. (1983) for citation to: Ferric Chloride, Ferric Sulfate Hexahydrate and Ferrous Sulfate Heptahydrate.
(4) Parmeggiani, L., Ed., *Encyclopedia of Occupational Health & Safety,* Third Edition, Vol. 1, pp 1155-56, Geneva, International Labour Office (1983).

IRON-DEXTRAN COMPLEX

● Carcinogen (Animal Positive, Human Suspected) (IARC)
● Hazardous Waste Constituent (EPA)

Description: Iron-dextran is a complex of ferric hydroxide with dextran. Dextrans are polysaccharides produced by bacterial action on sucrose. Compounds in this group are pharmaceuticals containing some form of iron associated with carbohydrates of varying complexity.

DOT Designation: —

Synonyms: Imferon®; Fenate®.

Potential Exposure: Iron-dextran complex is a parenteral form of medication used in iron-deficiency anemia in humans and baby pigs. The product for human use is a sterile, dark-brown colloidol solution in saline. The products designed for animal use are more concentrated.

The iron-dextran complex was introduced in the United States in 1957. There is no record of current U.S. production, although it is known to be produced by two manufacturers. Iron-dextran is available as two prescription items, Chromagen-D and Imferon. In 1980, 30,000 prescriptions were dispensed by pharmacies. The therapeutic dose for humans is 1–5 ml (50–250 mg iron) daily by deep intramuscular injection. Use is advised solely for those patients in whom an iron-deficiency state is chemically present and who have not responded to oral administration of iron. Clear warning of potential injection-site sarcoma is included with the physician's package insert (3).

Permissible Exposure Limits in Air: No standards set.

Permissible Concentration in Water: No criteria set.

Harmful Effects and Symptoms: Iron dextran is carcinogenic in mice and rats after subcutaneous or intramuscular injection, producing local tumors (1).

There have been case reports of sarcomas associated with injections of iron dextran. The tumors appeared at the probable site of the injections, and the similarity of the local effect in humans and animals was noted (2).

References

(1) IARC Monographs 2:161–78 (1973).
(2) *IARC Monographs on the Evaluation of the Carcinogenic Risk of Chemicals to Humans,* Supplement I, IARC, Lyon, France, p 35 (1979).
(3) See Reference (A-62). See also Reference (A-64).

ISOAMYL ACETATE

See "Amyl Acetates."

ISOAMYL ALCOHOL

Description: $(CH_3)_2CHCH_2CH_2OH$ is a colorless liquid with an alcoholic odor which causes coughing. It boils at $132°C$.

Code Numbers: CAS 123-51-3 RTECS EL5425000 UN 1105

DOT Designation: Flammable liquid.

Synonyms: 3-Methyl-1-butanol, isobutylcarbinol, isopentyl alcohol, fermentation amyl alcohol, fusel oil.

Potential Exposure: Amyl alcohols are used in the manufacture of lacquers, paints, varnishes, paint removers, shoe cements, perfumes, pharmaceuticals, chemicals, rubber, plastics, fruit essences, explosives, hydraulic fluids, ore flotation agents, in the preparation of other amyl derivatives, in the extraction of fats, and in the textile and petroleum refining industries.

Incompatibilities: Strong oxidizers.

Permissible Exposure Limits in Air: The Federal standard for 3-methyl-1-butanol (isoamyl alcohol) is 100 ppm (360 mg/m^3). There are no Federal standards for the other amyl alcohol isomers. The ACGIH has set a tentative STEL of 125 ppm (450 mg/m^3) for isoamyl alcohol as of 1983/84. The IDLH level is 8,000 ppm.

Determination in Air: Adsorption on charcoal, workup with 2-propanol in CS_2, analysis by gas chromatography. See NIOSH Methods, Set E. See also reference (A-10).

Permissible Concentration in Water: No criteria set, but EPA (A-37) has suggested an ambient level goal for all primary pentanols of 5,000 $\mu g/\ell$, based on health effects.

Routes of Entry: Inhalation, ingestion, eye and skin contact.

Harmful Effects and Symptoms: *Local* — The liquid and vapor are mild irritants to the membranes of the eyes and upper respiratory tract and skin.

Systemic — In low concentrations, amyl alcohol may cause irritation of nose and throat, nausea, vomiting, flushing, headache, diplopia, vertigo, and muscular weakness. In higher dosage, it is a narcotic.

Points of Attack: Eyes, skin, respiratory system.

Medical Surveillance: Consider possible irritant effects on skin and respiratory tract in any preplacement or periodic examinations. Amyl alcohol can be determined in blood.

First Aid: If this chemical gets into the eyes, irrigate immediately. If this chemical contacts the skin, flush with water promptly. If a person breathes in large amounts of this chemical, move the exposed person to fresh air at once and perform artificial respiration. When this chemical has been swallowed, get medical attention. Give large quantities of saltwater and induce vomiting. Do not make an unconscious person vomit.

Personal Protective Methods: Wear appropriate clothing to prevent repeated or prolonged skin contact. Wear eye protection to prevent any reasonable probability of eye contact. Employees should wash promptly when skin is wet. Remove nonimpervious clothing promptly if wet or contaminated.

Respirator Selection:

1,000 ppm:	CCROVF
5,000 ppm:	GMOV/SAF/SCBAF
8,000 ppm:	SAF:PD,PP,CF
Escape:	GMOV/SCBA

Disposal Method Suggested: Incineration.

References

(1)   See Reference (A-61).

ISOBUTYL ACETATE

See "Butyl Acetates."

ISOBUTYL ACRYLATE

Description: $CH_2=CHCOOCH_2CH(CH_3)_2$ is a clear liquid boiling at 61° to 63°C at 51 mm Hg pressure.

Code Numbers: CAS 106-63-8   RTECS AT 2100000   UN 2527

DOT Designation: —

Synonyms: 2-Methylpropyl acrylate; isobutyl propenoate.

Potential Exposure: Those using this material as a monomer in synthetic resin manufacture.

Permissible Exposure Limits in Air: No standards set.

Permissible Concentration in Water: No criteria set.

Harmful Effects and Symptoms: Is a skin and eye irritant. Has moderate ingestive toxicity.

Personal Protective Methods: Wear skin protection.

Disposal Method Suggested: Incineration.

References

(1) Sax, N.I., Ed., *Dangerous Properties of Industrial Materials Report, 2,* No. 2, 43-44, New York, Van Nostrand Reinhold Co. (1982).

ISOBUTYL ALCOHOL

See "Butyl Alcohols."

References

(1) U.S. Environmental Protection Agency, *Isobutyl Alcohol,* Health and Environmental Effects Profile No. 120, Washington, DC, Office of Solid Waste (April 30, 1980).

(2) See Reference (A-61).
(3) Sax, N.I., Ed., *Dangerous Properties of Industrial Materials Report, 2,* No. 2, 44-46, New York, Van Nostrand Reinhold Co. (1982).
(4) U.S. Environmental Protection Agency, *Chemical Hazard Information Profile: Isobutyl Alcohol,* Washington, DC (March 31, 1983).

ISOBUTYL MERCAPTAN

Description: $(CH_3)_2CHCH_2SH$ is a mobile liquid with a skunk-like odor boiling at 98°C.

Code Numbers: CAS 513-44-0   RTECS TZ7630000

DOT Designation: —

Synonyms: Isobutanethiol; 2-methyl-1-propanethiol.

Potential Exposure: Uses include: as gas odorant for detecting leaks and in organic synthesis.

Permissible Exposure Limits in Air: No standards set.

Permissible Concentration in Water: No criteria set.

Routes of Entry: Skin and eye contact, inhalation.

Harmful Effects and Symptoms: Eye irritant. Moderately toxic by ingestion or inhalation.

Personal Protective Methods: Wear skin protection.

Respirator Selection: Self-contained breathing apparatus.

Disposal Method Suggested: Incineration.

References
(1) Sax, N.I., Ed., *Dangerous Properties of Industrial Materials Report, 2,* No. 2, 48-49, New York, Van Nostrand Reinhold Co. (1982).

ISOBUTYRALDEHYDE

Description: $(CH_3)_2CHCHO$ is a colorless liquid with a pungent odor boiling at 64°C.

Code Numbers: CAS 78-84-2   RTECS NQ4025000   UN2045

DOT Designation: Grade C Flammable Liquid (U.S. Coast Guard).

Synonyms: α-methyl-propionaldehyde; isobutyl aldehyde; isopropyl formaldehyde; 2-methyl propanal.

Potential Exposure: To those using this material in organic synthesis of gasoline additives, perfumes, flavors and plasticizers.

Permissible Exposure Limits in Air: A TLV of 100 ppm (290 mg/m^3) has been proposed (A-36).

Determination in Air: By photometry (A-36).

Permissible Concentration in Water: No criteria set.

Routes of Entry: Skin absorption, inhalation, ingestion.

Harmful Effects and Symptoms: Irritates eyes, skin and respiratory tract.

Personal Protective Methods: Full skin protection required.

Respirator Selection: Self-contained breathing apparatus.

Disposal Method Suggested: Incineration.

References

(1) Sax, N.I., Ed., *Dangerous Properties of Industrial Materials Report, 2,* No. 2, 46-48, New York, Van Nostrand Reinhold Co. (1982).

ISOCYANATES

See "Isophorone Diisocyanate," "Methylenebis(4-Cyclohexyl Isocyanate)," "Methylenebis(Phenyl Isocyanate)," and "Toluene Diisocyanate."

References

(1) Nat. Inst. for Occup. Safety and Health, *Information Profiles on Potential Occupational Hazards: Organoisocyanates,* Report PB-276,678, Rockville, MD, pp 265-275 (October 1977).

ISOOCTYL ALCOHOL

Description: $C_7H_{15}CH_2OH$ is a clear liquid which boils at 182° to 195°C. It consists in practice of a mixture of isomers made by the Oxo process.

Code Numbers: CAS 26952-21-6   RTECS NS7700000

DOT Designation: —

Synonym: Isooctanol.

Potential Exposure: It is used to form phthalate, maleate, adipate and sebacate esters with the corresponding acids for use as plasticizers. It is used as a raw material for surfactants and as an antifoaming agent, emulsifier and solvent.

Permissible Exposure Limits in Air: There is no Federal standard but ACGIH as of 1983/84 has adopted a TWA of 50 ppm (270 mg/m^3). There is no STEL proposed.

Permissible Concentration in Water: No criteria set.

Personal Protective Methods: Wear rubber gloves, a face shield and a multipurpose gas mask (A-38).

Disposal Method Suggested: Incineration, preferably in admixture with a more flammable solvent (A-38).

References

(1) Sax, N.I., Ed., *Dangerous Properties of Industrial Materials Report, 2,* No. 2, 49-50, New York, Van Nostrand Reinhold Co. (1982).

ISOPHORONE

● Priority toxic pollutant (EPA)

Description: This cyclic ketone is a colorless or pale liquid with a camphor-like odor. It boils at 215°C.

Code Numbers: CAS 78-59-1   RTECS GW7700000

DOT Designation: —

Synonyms: 3,3,5-Trimethyl-2-cyclohexene-1-one,   trimethylcyclohexenone, isoacetophorone.

Potential Exposure: Isophorone is an industrial chemical synthesized from acetone and used commercially as a solvent or cosolvent for finishes, lacquers, polyvinyl and nitro cellulose resins, pesticides, herbicides, fats, oils, and gums. It is also used as a chemical feedstock for the synthesis of 3,5-xylenol, 2,3,5-trimethyl-cyclohexanol, and 3,5-dimethylaniline. Certain occupations (particularly individuals who are exposed to isophorone as a solvent) have elevated levels of exposure relative to the general population. NIOSH estimates that more than 1.5 million workers are exposed to isophorone.

Incompatibilities: Strong oxidizers.

Permissible Exposure Limits in Air: The Federal standard is 25 ppm (140 mg/m^3). ACGIH as of 1983/84 has adopted a 5 ppm (25 mg/m^3) ceiling value but no tentative STEL. The IDLH level is 800 ppm.

Detection in Air: Charcoal adsorption, workup with CS_2, analysis by gas chromatography. See NIOSH Methods, Set V. See also reference (A-10).

Permissible Concentration in Water: To protect freshwater aquatic life— 117,000 μg/ℓ, on an acute toxicity basis. To protect saltwater aquatic life— 12,900 μg/ℓ, on an acute toxicity basis. To protect human health—5,200 μg/ℓ.

Determination in Water: Methylene chloride extraction followed by exchange to toluene, gas chromatography with flame ionization detection (EPA Method 609) or gas chromatography plus mass spectrometry (EPA Method 625).

Routes of Entry: In the industrial handling of isophorone inhalation of the vapors is the most likely mode of contact, although skin and eye contact with the liquid may also occur. Because of the odor and taste of isophorone, ingestion is not expected unless by accident.

Harmful Effects and Symptoms: Irritation of eyes, nose and throat, narcosis, dermatitis, headaches, dizziness.

Points of Attack: Respiratory system, skin.

Medical Surveillance: Consider the points of attack in preplacement and periodic physical examinations.

First Aid: If this chemical gets into the eyes, irrigate immediately. If this chemical contacts the skin, wash with soap promptly. If a person breathes in large amounts of this chemical, move the exposed person to fresh air at once and perform artificial respiration. When this chemical has been swallowed, get medical attention. Give large quantities of water and induce vomiting. Do not make an unconscious person vomit.

Personal Protective Methods: Wear appropriate clothing to prevent repeated or prolonged skin contact. Wear eye protection to prevent any possibility of eye contact. Wash promptly when skin is wet or contaminated. Provide emergency eyewash.

Respirator Selection:
200 ppm: SA/SCBA
800 ppm: SAF/SCBAF
Escape: GMOV/SCBA

Disposal Method Suggested: Incineration.

References

(1) Nat. Inst. for Occupational Safety and Health, *Criteria for a Recommended Standard: Occupational Exposure to Ketones,* DHEW (NIOSH) Publ. No. 78-173, Washington, DC (1978).

(2) U.S. Environmental Protection Agency, *Isophorone: Ambient Water Quality Criteria,* Washington, DC (1980).

(3) Sax, N.I., Ed., *Dangerous Properties of Industrial Materials Report, 2,* No. 1, 108-110, New York, Van Nostrand Reinhold Co. (1982).

(4) United Nations Environment Programme, *IRPTC Legal File 1983,* Vol. I, pp VII/239-40, Geneva, Switzerland, International Register of Potentially Toxic Chemicals (1984).

ISOPHORONE DIISOCYANATE

Description: $C_{12}H_{18}N_2O_2$, 3-isocyanatomethyl-3,5,5-trimethylcyclohexylisocyanate with the structural formula

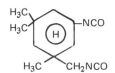

is a colorless to yellowish liquid.

Code Number: UN 2290.

DOT Designation: Poison.

Synonym: IPDI.

Potential Exposure: As a raw material for polyurethane paints, varnishes and elastomers which are exceptionally stable to atmospheric and chemical attack.

Permissible Exposure Limits in Air: There is no Federal standard but ACGIH (1983/84) has adopted a TWA of 0.01 ppm (0.09 mg/m³) with the notation "skin" indicating the possibility of cutaneous absorption. There is no STEL proposed.

Permissible Concentration in Water: No criteria set.

Harmful Effects and Symptoms: The oral toxicity of this compound is low but there are definite percutaneous and inhalation effects shown in animal studies (A-34).

2-ISOPROPOXYETHANOL

Description: $(CH_3)_2CHOCH_2CH_2OH$ is a colorless liquid boiling at 140° to 144°C.

Code Numbers: CAS 109-59-1 RTECS KL5075000

DOT Designation: —

Synonyms: Ethylene glycol isopropyl ether, β-hydroxyethyl isopropyl ether, isopropyl Cellosolve®.

Potential Exposures: This material is used as a solvent for resins in lacquers and inks.

Permissible Exposure Limits in Air: There are no Federal standards but ACGIH (1983/84) has adopted a TWA of 25 ppm (105 mg/m^3) and a STEL of 75 ppm (320 mg/m^3).

Permissible Concentration in Water: No criteria set.

Harmful Effects and Symptoms: See the entry under the related compound, n-butoxyethanol for guidance.

Points of Attack: See "n-Butoxyethanol" for guidelines on a similar compound.

Medical Surveillance: See "n-Butoxyethanol" for guidelines on a similar compound.

First Aid: See "n-Butoxyethanol" for guidelines on a similar compound.

Personal Protective Methods: See "n-Butoxyethanol" for guidelines on a similar compound.

Respirator Selection: See "n-Butoxyethanol" for guidelines on a similar compound.

Disposal Method Suggested: See "n-Butoxyethanol" for guidelines on a similar compound.

References

(1) See Reference (A-60).

ISOPROPYL ACETATE

Description: $CH_3COOCH(CH_3)_2$ is a colorless liquid with a fruity odor boiling at 90°C.

Code Numbers: CAS 108-21-4 RTECS AI4930000 UN 1220

DOT Designation: Flammable liquid.

Synonyms: Isopropyl ester of acetic acid, sec-propyl acetate.

Potential Exposure: This material is used as a solvent for nitrocellulose and other resins, fats, oils, waxes and gums. It is used in synthetic perfumes and food flavorings.

Incompatibilities: Nitrates, strong oxidizers, strong alkalies, strong acids.

Permissible Exposure Limits in Air: The Federal standard and the 1983/84 ACGIH TWA value is 250 ppm (950 mg/m^3). The STEL is 310 ppm (1,185 mg/m^3). The IDLH level is 16,000 ppm.

Determination in Air: Adsorption on charcoal, workup with CS_2, analysis by gas chromatography. See NIOSH Methods, Set E. See also reference (A-10).

Permissible Concentration in Water: No criteria set.

Routes of Entry: Inhalation, ingestion, eye and skin contact.

Harmful Effects and Symptoms: Irritation of eyes, nose and skin, dermatitis, narcosis.

Points of Attack: Eyes, skin, respiratory system.

Medical Surveillance: Consider the points of attack in preplacement and periodic physical examinations.

First Aid: If this chemical gets into the eyes, irrigate immediately. If this chemical contacts the skin, flush with water. If a person breathes in large amounts of this chemical, move the exposed person to fresh air at once and perform artificial respiration. When this chemical has been swallowed, get medical attention. Give large quantities of saltwater and induce vomiting. Do not make an unconscious person vomit.

Personal Protective Methods: Wear appropriate clothing to prevent repeated or prolonged skin contact. Wear eye protection to prevent any reasonable probability of eye contact. Employees should wash promptly when skin is wet or contaminated. Remove clothing immediately if wet or contaminated to avoid flammability hazard.

Respirator Selection:
    1,000 ppm: CCROVF
    5,000 ppm: GMOVc
  12,500 ppm: SAF/SCBAF
  16,000 ppm: SAF:PD,PP,CF
      Escape: GMOV/SCBA

Disposal Method Suggested: Incineration.

References

(1)   Sax, N.I., Ed., *Dangerous Properties of Industrial Materials Report, 1,* No. 3, 68-69, New York, Van Nostrand Reinhold Co. (1981).
(2)   See Reference (A-61).

ISOPROPYL ALCOHOL

- Carcinogen (Isopropyl oils formed as by-products in IPA manufacture by "strong-acid" process) (IARC) (A-62) (A-64)

Description: Isopropyl alcohol, $CH_3CHOHCH_3$, is a flammable liquid, boiling at 82.5°C with a slight odor resembling that of a mixture of ethanol and acetone.

Code Numbers: CAS 67-63-0   RTECS NT8050000   UN 1219

DOT Designation: Flammable liquid.

Synonyms: 2-Propanol, isopropanol, secondary propyl alcohol, dimethyl carbinol, IPA.

Potential Exposure: Isopropyl alcohol is a chemical widely used in liniments, skin lotions, cosmetics, permanent wave preparations, pharmaceuticals, and hair tonics. It is also employed as a solvent in perfumes, in extraction processes, as a preservative, in lacquer formulations, and in many dye solutions. In addition, it has been employed as an ingredient of antifreezes, soaps, and window cleaners. It may also be used as a raw material for the manufacture of acetone and various isopropyl derivatives, such as pesticides (A-32). NIOSH estimates that approximately 141,000 employees are potentially exposed to isopropyl alcohol in the United States.

Incompatibilities: Strong oxidizers.

Permissible Exposure Limits in Air: The Federal standard and the 1983/84 ACGIH TWA value is 400 ppm (980 mg/m^3). NIOSH recommends adherence to environmental workplace limits of 400 ppm as a time-weighted average concentration and 800 ppm as a ceiling concentration for up to a 10-hour workday, 40-hour workweek. The STEL is 500 ppm (1,225 mg/m^3). The IDLH level is 20,000 ppm.

Determination in Air: Charcoal adsorption, workup with 2-butanol in CS_2, analysis by gas chromatography. See NIOSH Methods, Set E. See also reference (A-10).

Permissible Concentration in Water: No criteria set, but EPA (A-37) has suggested a permissible ambient goal of 13,500 μg/ℓ based on health effects.

Routes of Entry: Inhalation of vapor, ingestion, eye and skin contact.

Harmful Effects and Symptoms: Isopropyl alcohol is a relatively innocuous chemical when compared with several other compounds currently under study by NIOSH. At low airborne concentrations of isopropyl alcohol, the only effect observed is mild irritation of eyes and skin. At much higher concentrations, it is believed, largely on the basis of animal studies, that narcosis can occur.

On the basis of several studies, NIOSH suspects that a carcinogen may have been present in the old "strong-acid" isopropyl alcohol manufacturing process. Animal studies subsequently did not support the contention that isopropanol is a carcinogen. Although the data obtained from the animal studies were frequently inconsistent, evidence points to isopropyl oil obtained from the old process as a possible carcinogen. The unequivocal identification of the carcinogen has not been made. Epidemiologic and other studies are recommended by NIOSH to determine whether a carcinogen exists in the new "weak-acid" isopropyl alcohol manufacturing process. Until such information becomes available. NIOSH is recommending that special procedures be followed during the manufacturing process.

An increased incidence of cancer of the paranasal sinuses has been found in workers in factories manufacturing isopropyl alcohol by the "strong-acid" process in which isopropyl oils were formed as by-products (5,6).

Points of Attack: Eyes, skin, respiratory system.

Medical Surveillance: NIOSH recommends that workers subject to isopropyl alcohol exposure have comprehensive preplacement medical examinations. Periodic medical examinations shall be made available on an annual basis. Particular attention will be given in these medical examinations to the skin sinuses, and respiratory system. Isopropyl alcohol and its metabolite, acetone, may be detected in blood, urine, and body tissues.

First Aid: If this chemical gets into the eyes, irrigate immediately. If this chemical contacts the skin, flush with water. If a person breathes in large amounts of this chemical, move the exposed person to fresh air at once and perform artificial respiration. When this chemical has been swallowed, get medical attention. Give large quantities of saltwater and induce vomiting. Do not make an unconscious person vomit.

Personal Protective Methods: Wear appropriate clothing to prevent repeated

or prolonged skin contact. Wear eye protection to prevent any reasonable probability of eye contact. Employees should wash promptly when skin is wet or contaminated. Remove clothing immediately if wet or contaminated to avoid flammability hazard.

Respirator Selection:

1,000 ppm:	CCROVF
5,000 ppm:	GMOVc
20,000 ppm:	GMOVfb/SAF/SCBAF
Escape:	GMOV/SCBA

Disposal Method Suggested: Incineration.

References

(1) National Institute for Occupational Safety and Health, *Criteria for a Recommended Standard: Occupational Exposure to Isopropyl Alcohol,* NIOSH Doc. No. 76-142 (1976).
(2) U.S. Environmental Protection Agency, *Chemical Hazard Information Profile: Isopropyl Alcohol,* Washington, DC (December 29, 1977), also revised edition (1979).
(3) See Reference (A-61).
(4) Sax, N.I., Ed., *Dangerous Properties of Industrial Materials Report, 2,* No. 2, 50-52, New York, Van Nostrand Reinhold Co. (1982).
(5) IARC Monographs 15:223–43 (1977).
(6) *IARC Monographs on the Evaluation of the Carcinogenic Risk of Chemicals to Humans,* Supplement I, IARC, Lyon, France, p 36 (1979).
(7) See Reference (A-62). See also Reference (A-64).

ISOPROPYL AMINE

Description: $(CH_3)_2CHNH_2$ is a colorless liquid with a pungent ammonia-like odor. It boils at $32°C$.

Code Numbers: CAS 75-31-0   RTECS NT8400000   UN 1221

DOT Designation: Flammable liquid.

Synonyms: 2-Aminopropane, monoisopropylamine.

Potential Exposures: Isopropyl amine is used in the synthesis of pharmaceuticals (A-41), pesticides (A-32), rubber accelerators, dyes and surface active agents.

Incompatibilities: Strong acids, strong oxidizers.

Permissible Exposure Limits in Air: The Federal standard and the ACGIH 1983/84 TWA value is 5 ppm ($12 mg/m^3$). The STEL is 10 ppm ($24 mg/m^3$). The IDLH level is 4,000 ppm.

Determination in Air: Collection in a bubbler using sulfuric acid, analysis by gas chromatography. See NIOSH Methods, Set K. See also reference (A-10).

Permissible Concentration in Water: No criteria set.

Routes of Entry: Inhalation, ingestion, skin absorption, eye and skin contact.

Harmful Effects and Symptoms: Irritation of eyes, nose, throat and skin, pulmonary edema, visual disturbance, skin and eye burns, dermatitis.

Points of Attack: Respiratory system, skin, eyes, lungs.

Medical Surveillance: Consider the points of attack in preplacement and periodic physical examinations.

First Aid: If this chemical gets into the eyes, irrigate immediately. If this chemical contacts the skin, flush with water immediately. If a person breathes in large amounts of this chemical, move the exposed person to fresh air at once and perform artificial respiration. When this chemical has been swallowed, get medical attention. Give large quantities of water and induce vomiting. Do not make an unconscious person vomit.

Personal Protective Methods: Wear appropriate clothing to prevent any reasonable probability of skin contact. Wear eye protection to prevent any possibility of eye contact. Employees should wash immediately when skin is wet or contaminated. Remove clothing immediately if wet or contaminated to avoid flammability hazard. Provide emergency showers and eyewash.

Respirator Selection:
- 250 ppm: CCRSF/GMS/SAF/SCBAF
- 4,000 ppm: SAF:PD,PP,CF
- Escape: GMS/SCBA

Disposal Method Suggested: Controlled incineration (incinerator is equipped with a scrubber or thermal unit to reduce NO_x emissions).

References

(1) See Reference (A-61).
(2) See Reference (A-60).

N-ISOPROPYL ANILINE

Description: $C_6H_5NHCH(CH_3)_2$ is a yellowish liquid boiling at 206°C.

Code Numbers: CAS 643-28-7   RTECS BY4200000

DOT Designation: —

Synonyms: None.

Potential Exposures: N-isopropyl aniline is used in the dyeing of acrylic fibers and as a chemical intermediate.

Permissible Exposure Limits in Air: There is no Federal standard but ACGIH (1983/84) has adopted a TWA value of 2 ppm (10 mg/m^3) and set an STEL of 5 ppm (20 mg/m^3). The notation "skin" is added indicating the possibility of cutaneous absorption.

Permissible Concentration in Water: No criteria set.

Harmful Effects and Symptoms: Acute toxicity data indicate slight skin and eye effects from direct contact. No chronic toxicity, industrial hygiene or medical data were available (A-34).

ISOPROPYL ETHER

See "Diisopropyl Ether."

ISOPROPYL GLYCIDYL ETHER

Description: $CH_2-CH-CH_2-O-CH(CH_3)_2$ is a colorless liquid which boils at 132°C.

Code Numbers: CAS 4016-14-2   RTECS TZ3500000

DOT Designation: —

Synonyms: Isopropoxymethyl-oxiran, 1,2-epoxy-3-isopropoxypropane, isopropyl epoxypropyl ether, IGE.

Potential Exposure: This material is used as a reactive diluent for epoxy resins, as a stabilizer for organic compounds, and as an intermediate for the synthesis of ethers and esters.

Incompatibilities: Strong oxidizers, strong caustics.

Permissible Exposure Limits in Air: The Federal standard and the ACGIH 1983/84 TWA value is 50 ppm (240 mg/m^3). The STEL is 75 ppm (360 mg/m^3). The IDLH level is 1,500 ppm.

Determination in Air: Charcoal adsorption, workup with CS_2, analysis by gas chromatography. See NIOSH Methods, Set F. See also reference (A-10).

Permissible Concentration in Water: No criteria set.

Routes of Entry: Inhalation, ingestion, skin and eye contact.

Harmful Effects and Symptoms: Irritation of eyes, skin and upper respiratory system, sensitization of skin.

Points of Attack: Eyes, skin, respiratory system.

Medical Surveillance: Consider the points of attack in preplacement and periodic physical examinations.

First Aid: If this chemical gets into the eyes, irrigate immediately. If this chemical contacts the skin, wash with soap immediately. If a person breathes in large amounts of this chemical, move the exposed person to fresh air at once and perform artificial respiration. When this chemical has been swallowed, get medical attention. Give large quantities of saltwater and induce vomiting. Do not make an unconscious person vomit.

Personal Protective Methods: Wear appropriate clothing to prevent repeated or prolonged skin contact. Wear eye protection to prevent any reasonable probability of eye contact. Employees should wash promptly when skin is wet or contaminated. Remove clothing immediately if wet or contaminated to avoid flammability hazard.

Respirator Selection:
  1,500 ppm: SAF/SCBAF
    Escape: GMOV/SCBA

Disposal Method Suggested: Concentrated waste containing no peroxides—discharge liquid at a controlled rate near a pilot flame. Concentrated waste containing peroxides—perforation of a container of the waste from a safe distance followed by open burning.

References

(1) Nat. Inst. for Occup. Safety and Health, *Information Profiles on Potential Occupational Hazards: Glycidyl Ethers,* Report PB-276,678, Rockville, MD, pp 116-123 (October 1977).

(2) Nat. Inst. for Occup. Safety and Health, *Criteria for a Recommended Standard: Occupational Exposure to Glycidyl Ethers,* NIOSH Doc. No. 78-166, Washington, DC (1978).

K

KEPONE®

See "Chlordecone," the approved generic name for this compound.

KEROSENE

Description: Kerosene is a pale yellow or clear, mobile liquid, composed of a mixture of petroleum distillates, having a characteristic odor. Chemically, it is composed of aliphatic hydrocarbons with 10 to 16 carbons per molecule and benzene and naphthalene derivatives.

Code Numbers: CAS 8008-20-6 RTECS OA5500000 UN 1223

DOT Designation: Combustible liquid.

Synonyms: Kerosine, coal-oil, range-oil.

Potential Exposure: Kerosene is used as a fuel for lamps, stoves, jets, and rockets. It is also used for degreasing and cleaning metals and as a vehicle for insecticides.

Permissible Exposure Limits in Air: Presently there is no Federal standard for kerosene vapor in workroom air, nor has ACGIH proposed any standards. This is due at least in part to the variable composition of kerosene.

Permissible Concentration in Water: No criteria set.

Routes of Entry: Inhalation of vapor, ingestion, skin and eye contact.

Harmful Effects and Symptoms: *Local* — The liquid may produce primary skin irritation as a result of defatting. Aspiration of liquid may cause extensive pulmonary injury. Because of its low surface tension, kerosene may spread over a large area, causing pulmonary hemorrhage and chemical pneumonitis. Kerosene mist may also cause mucous membrane irritation.

Systemic — Inhalation of high concentrations may cause headache, nausea, confusion, drowsiness, convulsions, and coma. When kerosene is ingested, it may cause nausea, vomiting, and, in severe cases, drowsiness progressing to coma, and death by hemorrhagic pulmonary edema and renal involvement.

Medical Surveillance: No specific considerations are needed.

Personal Protective Methods: Barrier creams, gloves, and protective clothing are recommended. Where workers are exposed to vapors, masks are recommended.

Disposal Method Suggested: Incineration.

References

(1) National Institute for Occupational Safety and Health, *Criteria for a Recommended Standard: Occupational Exposure to Refined Petroleum,* NIOSH Doc. No. 77-192, Washington, DC (1977).

(2) See Reference (A-60).

KETENE

Description: $H_2C=C=O$ is a colorless gas with a sharp odor.

Code Numbers: CAS 463-51-4 RTECS OA7700000

DOT Designation: −

Synonyms: Carbomethene, ethenone, keto-ethylene.

Potential Exposure: Ketene is used as an acetylating agent in cellulose acetate and aspirin (A-41) manufacture; it is used in the conversion of higher acids to their anhydrides.

Incompatibilities: Water, variety of organic compounds.

Potential Exposure Limit in Air: The Federal standard and the 1983/84 ACGIH TWA value is 0.5 ppm (0.9 mg/m^3). The STEL value is 1.5 ppm (3.0 mg/m^3). The IDLH level is 25.0 ppm.

Determination in Air: Collection in a bubbler using hydroxylammonium chloride; reaction with ferric chloride; colorimetric determination. See NIOSH Methods, Set G. See also reference (A-10).

Permissible Concentration in Water: No criteria set.

Routes of Entry: Inhalation, eye and skin contact.

Harmful Effects and Symptoms: Irritation of eyes, nose, throat and lungs; pulmonary edema.

Points of Attack: Respiratory system, lungs, eyes, skin.

Medical Surveillance: Consider the points of attack in preplacement and periodic physical examinations.

First Aid: If a person breathes in large amounts of this chemical, move the exposed person to fresh air at once and perform artificial respiration.

Personal Protective Methods: Only respirators have been specified by NIOSH (A-4). A Japanese source (A-38) recommends safety goggles, rubber gloves, face shield.

Respirator Selection:
25 ppm: SAF/SCBAF
Escape: GMOV/SCBA

Disposal Method Suggested: Incineration.

References

(1) See Reference (A-61).

L

LEAD ACETATE

- Carcinogen (Animal Positive, IARC) (A-62) (A-64)
- Hazardous Waste Constituent (EPA)

Description: $(CH_3COO)_2Pb \cdot 3H_2O$ is a colorless crystalline substance melting at 75°C.

Code Numbers:

Anhydrous	Trihydrate
CAS 301-04-2	CAS 6080-56-4
RTECS AI5250000	RTECS OF8050000
	UN 1616

DOT Designation: Poison B (IATA).

Synonyms: Sugar of lead.

Potential Exposure: Lead acetate is used as a mordant in cotton dyes, in the lead coating of metals, as a drier in paints, varnishes and pigment inks, and in medicinals such as astringents. Concentrations up to 0.6% (weight to volume), are limited for use as color additives in hair dyes. For 1978, sales of lead acetate hair dyes exceeded 1 million bottles.

Human exposure to lead acetate occurs through ingestion, inhalation, and skin absorption. Lead acetate is absorbed at about 1.5 times the rate of other lead compounds. The National Occupational Hazard Survey (NOHS) has estimated that 132,000 workers may be exposed to lead acetate. OSHA has estimated that 223,000 workers may be exposed to lead acetate. NOHS and OSHA numbers vary due to the difference in methodology used for estimating exposure.

Permissible Exposure Limits in Air: 50 $\mu g/m^3$ as an 8-hour TWA value according to OSHA. A value of 0.15 mg/m^3 has also been cited in Holland (A-60).

Permissible Concentration in Water: See "Lead-Inorganic."

Harmful Effects and Symptoms: Lead acetate is carcinogenic to rats—this compound induced benign and malignant tumors of the kidney following oral or parenteral administration. Gliomas occurred in rats given lead acetate by the oral route (2). A dose of 15 mg/kg is highly toxic to humans.

Personal Protective Methods: Avoid dust inhalation; wear U.S. Bureau of Mine approved dust mask.

Respirator Selection: See preceding.

Disposal Method Suggested: Convert to nitrate using nitric acid; evaporate, then saturate with H_2S; wash and dry the sulfide and ship to the supplier (1).

References

(1) Sax, N.I., Ed., *Dangerous Properties of Industrial Materials Report, 1,* No. 4, 79-82, New York, Van Nostrand Reinhold Co. (1981).
(2) See Reference (A-62). Also see Reference (A-64).
(3) United Nations Environment Programme, *International Register of Potentially Toxic Chemicals,* Geneva, Switzerland (1979).
(4) See Reference (A-60).
(5) United Nations Environment Programme, *IRPTC Legal File 1983,* Vol. I, pp VII/14, Geneva, Switzerland, International Register of Potentially Toxic Chemicals (1984).

LEAD ARSENATE

● Hazardous substance (EPA)

Description: A mixture of $Pb(AsO_3)_2$, $Pb_3(AsO_4)_2$, $PbHAsO_4$, $Pb(H_2AsO_4)_2$, and $Pb_2As_2O_7$, this material is a colorless, odorless powder.

Code Numbers: For $PbHAsO_4$: CAS 7784-40-9 RTECS CG0980000 UN 1617. For $Pb_3(AsO_4)_2$: CAS 10102-48-4.

DOT Designation: Poison-B.

Synonyms: For $PbHAsO_4$—arsenate of lead, dibasic lead arsenate.

Potential Exposures: Those engaged in the manufacture (A-32), formulation and application of this insecticide and veterinary tapeworm medicine.

Incompatibilities: None hazardous.

Permissible Exposure Limits in Air: The Federal standard is 0.15 mg/m^3. The STEL is 0.45 mg/m^3 [$Pb_3(AsO_4)_2$] but its abolition is under consideration by ACGIH as of 1983/84. The IDLH level is 300 mg/m^3.

Permissible Concentration in Water: See "Arsenic and Arsenic Compounds" and also "Lead—Inorganic."

Determination in Water: See "Arsenic and Arsenic Compounds" and also "Lead—Inorganic."

Routes of Entry: Inhalation, ingestion, skin and eye contact.

Harmful Effects and Symptoms: Arsenic intoxication; nausea, diarrhea; inflammation of skin and mucous membranes; lead intoxication; abdominal pain, appetite loss, constipation; tiredness, weakness, nervousness; paresthesia. A rebuttable presumption against registration for pesticide uses was issued on October 18, 1978 by EPA on the basis of oncogenicity, teratogenicity and mutagenicity.

Points of Attack: Kidneys, blood, gingival tissue, lymphatics, skin, gastrointestinal system, central nervous system.

Medical Surveillance: Consider the points of attack in preplacement and periodic physical examinations.

First Aid: If this chemical gets into the eyes, irrigate immediately. If this chemical contacts the skin, wash with soap promptly. If a person breathes in large amounts of this chemical, move the exposed person to fresh air at once and perform artificial respiration. When this chemical has been swallowed, get medical attention. Give large quantities of water and induce vomiting. Do not make an unconscious person vomit.

Personal Protective Methods: Wear appropriate clothing to prevent repeated or prolonged skin contact. Employees should wash daily at the end of each work shift. Work clothing should be changed daily if it is possible that clothing is contaminated. Remove nonimpervious clothing promptly if wet or contaminated.

Respirator Selection:
0.75 mg/m^3: DXSPest
1.5 mg/m^3: DXSQ/HiEPPest/SA/SCBA
7.5 mg/m^3: HiEPFPest/SAF/SCBAF
150 mg/m^3: PAPHiEPest/SA:PD,PP,CF
300 mg/m^3: SAF:PD,PP,CF
Escape: DXSPest/SCBA

Disposal Method Suggested: Long-term storage in large, weatherproof, and sift-proof storage bins or silos; small amounts may be disposed of in a chemical waste landfill (A-31).

References
(1) Parmeggiani, L., Ed., *Encyclopedia of Occupational Health & Safety,* Third Edition, Vol. 2, pp 1205-1206, Geneva, International Labour Office (1983).
(2) United Nations Environment Programme, *IRPTC Legal File 1983,* Vol. I, pp VII/58-9, Geneva, Switzerland, International Register of Potentially Toxic Chemicals (1984).
(3) Also see "Arsenic and Arsenic Compounds."

LEAD—INORGANIC

- Hazardous substances (Various compounds, EPA)
    Lead compounds which are classified by EPA as hazardous substances include: lead acetate, lead arsenate (see separate entry), lead chloride, lead fluoborate, lead fluoride (see also "Fluorides"), lead iodide, lead nitrate, lead stearate, lead sulfate, lead sulfide, and lead thiocyanate.
- Hazardous waste constituents (EPA)
- Priority toxic pollutant (EPA)

Description: Pb, inorganic lead, includes lead oxides, metallic lead, lead salts, and organic salts such as lead soaps, but excludes lead arsenate and organic lead compounds. Lead is a blue-grey metal which is very soft and maleable. Commercially important lead ores are galena, cerussite, anglesite, crocoisite, wulfenite, pyromorphite, matlockite, and vanadinite. Lead is slightly soluble in water in presence of nitrates, ammonium salts, and carbon dioxide.

Code Numbers: Lead metal—CAS 7439-92-1   RTECS OF7525000

DOT Designation: —

Synonyms: None.

Potential Exposures: Metallic lead is used for lining tanks, piping, and other equipment where pliability and corrosion resistance are required such as in the chemical industry in handling corrosive gases and liquids used in the manufacture of sulfuric acid; in petroleum refining; and in halogenation, sulfonation, extraction, and condensation processes; and in the building industry.

It is also used as an ingredient in solder, a filler in the automobile industry, and a shielding material for x-rays and atomic radiation; in manufacture of tetraethyllead and organic and inorganic lead compounds, pigments for paints and varnishes, storage batteries, flint glass, vitreous enameling, ceramics as a glaze, litharge rubber, plastics, and electronic devices. Lead is utilized in metallurgy and may be added to bronze, brass, steel, and other alloys to improve their characteristics. It forms alloys with antimony, tin, copper, etc. It is also used in metallizing to provide protective coatings and as a heat treatment bath in wire drawing.

Exposures to lead dust may occur during mining, smelting, and refining, and to fume, during high temperature (above 500°C) operations such as welding or spray coating of metals with molten lead. There are numerous applications for lead compounds, some of the more common being in the plates of electric batteries and accumulators, as compounding agents in rubber manufacture, as ingredients in paints, glazes, enamels, glass, pigments, and in the chemical industry.

It is estimated that approximately 783,000 industrial workers are potentially exposed to lead products.

In addition to these usual levels of exposure from environmental media, there exist miscellaneous sources which are hazardous. The level of exposure resulting from contact is highly variable. Children with pica for paint chips or for soil may experience elevation in blood lead ranging from marginal to sufficiently great to cause clinical illness. Certain adults may also be exposed to hazardous concentrations of lead in the workplace, notably in lead smelters and storage battery manufacturing plants. Again, the range of exposure is highly variable. Women in the workplace are more likely to experience adverse effects from lead exposure than men due to the fact that their hematopoietic system is more lead-sensitive than men's.

Incompatibilities: Strong oxidizers, hydrogen peroxide, active metals—sodium, potassium.

Permissible Exposure Limits in Air: The Federal (OSHA) standard for lead and its inorganic compounds was 0.2 mg/m^3 as a time-weighted average. The EPA has set a national ambient air quality standard for lead of 1.5 mg/m^3 on a 3-month average basis. The NIOSH Criteria Document recommends a time-weighted average value of 0.15 mg Pb/m^3. On November 14, 1978, OSHA set a final standard in which industries will be given 1 to 3 years to reach 0.1 mg (100 μg)/m^3 level and from 1 to 10 years to reach a final standard of 0.05 mg (50 μg)/m^3. ACGIH as of 1983/84 has set a TWA of 0.15 mg/m^3 (as Pb) and an STEL of 0.45 mg/m^3. Lead chromate is assigned a TWA of 0.05 mg/m^3 by ACGIH (1983/84) with the notation that it is a substance suspect of carcinogenic potential for man.

Determination in Air: Collection on a filter, workup with nitric acid, analysis by atomic absorption spectrometry. See NIOSH Methods, Set O. See also reference (A-10).

Permissible Concentration in Water: To protect freshwater aquatic life—

$$e^{[2.35 \ln (\text{hardness}) - 9.48]}$$

never to exceed

$$e^{[1.22 \ln (\text{hardness}) - 0.47]}$$

To protect saltwater aquatic life—668 μg/ℓ on an acute toxicity basis and 25 μg/ℓ on a chronic basis. To protect human health—50 μg/ℓ (USEPA).

Various organizations worldwide have set other standards for lead in drinking water as follows (A-65): South African Bureau of Standards, 150 μg/ℓ; World Health Organization, 100 μg/ℓ; Federal Republic of Germany (1975), 40 μg/ℓ.

Determination in Water: Digestion followed by atomic absorption or by colorimetric (dithizone) analysis or by inductively coupled plasma (ICP) optical emission spectrometry. That gives total lead; dissolved lead may be determined by 0.45 micron filtration prior to such analyses.

Routes of Entry: Ingestion of dust; inhalation of dust or fume, skin and eye contact.

Harmful Effects and Symptoms: *Local* — None.

Systemic — The early effects of lead poisoning are nonspecific and, except by laboratory testing, are difficult to distinguish from the symptoms of minor seasonal illnesses. The symptoms are decreased physical fitness, fatigue, sleep disturbance, headache, aching bones and muscles, digestive symptoms (particularly constipation), abdominal pains, and decreased appetite. These symptoms are reversible and complete recovery is possible.

Later findings include anemia, pallor, a "lead line" on the gums, and decreased hand-grip strength. Lead colic produces an intense periodic abdominal cramping associated with severe constipation and, occasionally, nausea and vomiting. Alcohol ingestion and physical exertion may precipitate these symptoms. The peripheral nerve affected most frequently is the radial nerve. This will occur only with exposure over an extended period of time and causes "wrist drop." Recovery is slow and not always complete. When the central nervous system is affected, it is usually due to the ingestion or inhalation of large amounts of lead. This results in severe headache, convulsions, coma, delirium, and possibly death. The kidneys can also be damaged after long periods of exposure to lead, with loss of kidney function and progressive azotemia.

Because of more efficient material handling methods and biological monitoring, serious cases of lead poisoning are rare in industry today.

Points of Attack: Kidneys, blood, gingival tissue, gastrointestinal system, central nervous system.

Medical Surveillance: In preemployment physical examinations, special attention is given to neurologic and renal disease and baseline blood lead levels. Periodic physical examinations should include hemoglobin determinations, tests for blood lead levels, and evaluation of any gastrointestinal or neurologic symptoms. Renal function should be evaluated.

Periodic evaluation of blood lead levels are widely used as an indicator of increased or excessive lead absorption. Other indicators are urine coproporphyrin and delta aminolevulinic acid (ALA). Erythrocytic protoporphyrin determinations may also be helpful. See also reference (10).

First Aid: If this chemical gets into the eyes, irrigate immediately. If this chemical contacts the skin, flush with soap promptly. If a person breathes in large amounts of this chemical, move the exposed person to fresh air at once and perform artificial respiration. When this chemical has been swallowed, get medical attention. Give large quantities of water and induce vomiting. Do not make an unconscious person vomit.

Personal Protective Methods: Wear appropriate clothing to prevent repeated or prolonged skin contact. Wear eye protection to prevent any reasonable probability of eye contact. Employees should wash daily at the end of each work shift. Remove nonimpervious clothing immediately if wet or contaminated.

Respirator Selection:
    0.5 mg/m^3: HiEP
    2.5 mg/m^3: HiEPF
     50 mg/m^3: PAPHiE/SA:PD,PP,CF
    100 mg/m^3: SAF:PD,PP,CF

Disposal Method Suggested: Lead oxide–chemical conversion to the sulfide or carbonate followed by collection of the precipitate and lead recovery via smelting operations. Landfilling of the oxide is also an acceptable procedure (A-31). Alternatively, it may be dissolved in HNO_3, precipitated as the sulfide and returned to a supplier for reprocessing (A-38). Processes for recovering and recycling lead from a number of industrial waste sources have been described (A-57).

References

(1) National Institute for Occupational Safety and Health, *Criteria for a Recommended Standard: Occupational Exposure to Inorganic Lead*, NIOSH Doc. No. 73-11010, Wash., DC (1973).

(2) OSHA, "Occupational exposure to lead: Final standard," *Federal Register* 43 No. 220, 52952-53014 (Nov. 14, 1978).

(3) U.S. Environmental Protection Agency, *Toxicology of Metals, Vol. II: Lead*, pp 242-300, Report EPA-600/1-77-022, Research Triangle Park, NC (May 1977).

(4) U.S. Environmental Protection Agency, *Lead: Ambient Water Quality Criteria*, Wash., DC (1980).

(5) U.S. Environmental Protection Agency, *Status Assessment of Toxic Chemicals: Lead*, Report EPA-600/2-79-210h, Cincinnati, Ohio (Dec. 1979).

(6) U.S. Environmental Protection Agency, *Reviews of the Environmental Effects of Pollutants: VII. Lead*, Report EPA-600/1-78-029, Cincinnati, Ohio (1978).

(7) U.S. Environmental Protection Agency, *Air Quality Criteria Document for Lead*, Report EPA-600/8-77-017, Research Triangle Park, NC (1977).

(8) National Academy of Sciences, *Medical and Biologic Effects of Environmental Pollutants: Lead, Airborne Lead in Perspective*, Wash., DC (1972).

(9) U.S. Environmental Protection Agency, *Lead,* Health and Environmental Effects Profile No. 121, Wash., DC, Office of Solid Waste (April 30, 1980).

(10) Nat. Inst. for Occup. Safety and Health, *A Guide to the Work Relatedness of Disease*, Revised Edition, DHEW (NIOSH) Pub. No. 79-116, pp 98-116, Cincinnati, Ohio (Jan. 1979).

(11) World Health Organization, *Lead,* Environmental Health Criteria No. 3, Geneva, Switzerland (1977).

(12) See Reference (A-60) for citations to: Lead, Lead Carbonate, Lead Chromate, Lead Naphthenate, Lead Nitrate and Lead Peroxide.

(13) Parmeggiani, L., Ed., *Encyclopedia of Occupational Health & Safety,* Third Edition, Vol. 2, pp 1200-1205, Geneva, International Labour Office (1983).

(14) United Nations Environment Programme, *IRPTC Legal File 1983,* Vol. II, pp VII/405-14, Geneva, Switzerland, International Register of Potentially Toxic Chemicals (1984).

LEAD–ORGANIC

See entries under "Tetraethyllead" and "Tetramethyllead."

LEAD PHOSPHATE

- Carcinogen (Animal Positive) (A-64)
- Hazardous Waste Constituent (EPA)

Description: Lead phosphate, $Pb_3(PO_4)_2$ is a white powder.

Code Numbers: CAS 7446-27-7   RTECS OG3675000

DOT Designation: –

Potential Exposure: Lead phosphate is used as a stabilizer in styrene and casein plastics. Lead acetate is absorbed at about 1.5 times the rate of other lead compounds.

The National Occupational Hazard Survey (NOHS) has estimated that 18,000 workers may be exposed to lead phosphate. OSHA has estimated that 28,000 and 27,000 workers may be exposed to lead phosphate dibasic and lead phosphate tribasic, respectively. NOHS and OSHA numbers vary due to the difference in methodology used for estimating exposure.

Permissible Exposure Limits in Air: 50 $\mu g/m^3$ is an 8-hour TWA according to OSHA (A-62) (A-64).

Permissible Concentration in Water: 50 $\mu g/\ell$ in drinking water (A-3) but not based on carcinogenicity.

Routes of Entry: Ingestion, inhalation, skin absorption.

Harmful Effects and Symptoms: Lead phosphate is carcinogenic to rats (A-62) (A-64).

Personal Protective Methods: Avoid direct inhalation; wear a U.S. Bureau of Mines approved dust mask.

Respirator Selection: See preceding.

References

(1)   United Nations Environment Programme, *IRPTC Legal File 1983,* Vol. II, pp VII/417, Geneva, Switzerland, International Register of Potentially Toxic Chemicals (1984).

LINDANE

- Carcinogen (Negative, NCI) (3) (Positive, EPA-CAG) (A-40)
- Hazardous substance (EPA)
- Hazardous waste (EPA)
- Priority toxic pollutant (EPA)

Description: $C_6H_6Cl_6$ is a colorless solid with a musty odor (pure material is odorless). It melts at 113°C.

Code Numbers: CAS 58-89-9   RTECS GV4900000   UN (NA 2761)

DOT Designation: ORM-A.

Synonyms: 1,2,3,4,5,6-Hexachloro-cyclohexane; γ-hexachloro-cyclohexane; benzene hexachloride, Gammexane®, γ-HCH, Gamma HCH, ENT-7796.

Potential Exposure: Lindane has been used against insects in a wide range of applications including treatment of animals, buildings, man for ectoparasites, clothes, water for mosquitoes, living plants, seeds and soils. Some applications have been abandoned due to excessive residues, e.g., stored foodstuffs.

Formulators, distributors and users of lindane represent a special risk group. The major use of lindane in recent years has been to pretreat seeds. Thus, those engaged in treatment and planting can be exposed.

Incompatibilities: None hazardous.

Permissible Exposure Limits in Air: The Federal standard and the ACGIH (1983/84) TWA value is 0.5 mg/m³. The notation "skin" is added to indicate the possibility of cutaneous absorption. The STEL is 1.5 mg/m³. The IDLH level is 1,000 mg/m³.

Determination in Air: Collection on a filter, workup with isooctane, analysis by gas chromatography. See NIOSH Methods, Set I. See also reference (A-10).

Permissible Concentration in Water: To protect freshwater aquatic life— 0.080 $\mu g/\ell$ as a 24 hour average, never to exceed 2.0 $\mu g/\ell$. To protect saltwater aquatic life—never to exceed 0.16 $\mu g/\ell$. To protect human health—preferably zero. An additional lifetime cancer risk of 1 in 100,000 is posed by a concentration of 0.186 $\mu g/\ell$. Earlier (1975) drinking water standards set by EPA limited lindane to 0.004 $\mu g/\ell$. In 1980, EPA set a maximum of 0.4 $\mu g/\ell$ in wastes.

Determination in Water: Methylene chloride extraction followed by gas chromatography with electron capture or halogen specific detection (EPA Method 608) or gas chromatography plus mass spectrometry (EPA Method 625).

Routes of Entry: Inhalation, skin absorption, ingestion, eye and skin contact.

Harmful Effects and Symptoms: Irritation of eyes, nose and throat; headaches, nausea; clonic convulsions, respiratory problems; cyanosis; aplastic anemia; skin irritations; muscular spasms

A rebuttable presumption against registration of lindane for pesticidal uses was issued on February 17, 1977 by EPA on the basis of oncogenicity, fetotoxicity, reproductive effects and acute toxicity (A-32).

Technical HCH, α- and β-HCH and lindane (γ-HCH) are carcinogenic in mice when administered orally, producing liver tumors. Studies in rats were considered inadequate (6).

Approximately 30 cases of aplastic anaemia, and 3 cases of acute myeloid leukaemia have been reported following exposure to HCH or lindane. A study of 285 workers exposed to many pesticides (including HCH and lindane) showed an apparent excess of lung tumors and one case of leukaemia; however, this cannot be attributed to exposure to HCH or lindane alone (6,7).

Points of Attack: Eyes, central nervous system, blood, liver, kidneys, skin.

Medical Surveillance: Consider the points of attack in preplacement and periodic physical examinations.

First Aid: If this chemical gets into the eyes, irrigate immediately. If this chemical contacts the skin, wash with soap promptly. If a person breathes in large amounts of this chemical, move the exposed person to fresh air at once and perform artificial respiration. When this chemical has been swallowed, get medical attention. Give large quantities of water and induce vomiting. Do not make an unconscious person vomit.

Personal Protective Methods: Wear appropriate clothing to prevent any reasonable probability of skin contact. Employees should wash immediately when skin is wet or contaminated. Work clothing should be changed daily if it is possible that clothing is contaminated. Remove nonimpervious clothing immediately if wet or contaminated. Provide emergency showers.

Respirator Selection:

5 mg/m^3:	CCROVDMPest/SA/SCBA
25 mg/m^3:	CCROVFDMPest/GMOVDMPest/SAF/SCBAF
500 mg/m^3:	PAPOVHiEP/SA:PD,PP,CF
1,000 mg/m^3:	SAF:PD,PP,CF
Escape:	GMOVPPest/SCBA

Disposal Method Suggested: For the disposal of lindane, a process has been developed involving destructive pyrolysis at 400° to 500°C with a catalyst mixture which contains 5 to 10% of either cupric chloride, ferric chloride, zinc chloride, or aluminum chloride on activated carbon (A-32).

References

(1) U.S. Environmental Protection Agency, *Hexachlorocyclohexane: Ambient Water Quality Criteria*, Wash., DC (1980).
(2) U.S. Environmental Protection Agency, *gamma-Hexachloro-cyclohexane*, Health and Environmental Effects Profile No. 113, Wash., DC, Office of Solid Waste (April 30, 1980).
(3) National Cancer Institute, *Bioassay of Lindane for Possible Carcinogenicity,* Technical Report Series No. 14, Bethesda, MD (1977).
(4) See Reference (A-61).
(5) Sax, N.I., Ed., *Dangerous Properties of Industrial Materials Report, 3,* No. 1, 62-66, New York, Van Nostrand Reinhold Co. (1983).
(6) IARC Monographs 20:1979.
(7) *IARC Monographs on the Evaluation of the Carcinogenic Risk of Chemicals to Humans,* Supplement I, IARC, Lyon, France, p 34 (1979).
(8) See Reference (A-62). Also see Reference (A-64).
(9) United Nations Environment Programme, *IRPTC Legal File 1983,* Vol. I, pp VII/230-34, Geneva, Switzerland, International Register of Potentially Toxic Chemicals (1984).

LIQUEFIED PETROLEUM GAS

Description: A mixture of propane (C_3H_8) and butane (C_4H_{10}), this is a colorless, odorless gas (foul-smelling odorant added).

Code Numbers: RTECS SE7545000   UN 1965 (or 1075)

DOT Designation: Flammable gas.

Synonyms: LPG, bottled gas.

Potential Exposure: LPG is used as a fuel and in the production of petrochemicals.

Incompatibilities: Strong oxidizers.

Permissible Exposure Limits in Air: The Federal standard and the 1983/84 ACGIH TWA value is 1,000 ppm (1,800 mg/m³). The STEL is 1,250 ppm (2,250 mg/m³). The IDLH is 19,000 ppm.

Determination in Air: Use of a combustible gas meter. See NIOSH Methods, Set G. See also reference (A-10).

Permissible Concentration in Water: No criteria set.

Routes of Entry: Inhalation, skin and eye contact.

Harmful Effects and Symptoms: Light-headedness and drowsiness.

Points of Attack: Respiratory system, central nervous system.

Medical Surveillance: Consider the points of attack in preplacement and periodic physical examinations.

First Aid: If this chemical gets into the eyes, irrigate immediately. If this chemical contacts the skin, flush with water immediately. If a person breathes in large amounts of this chemical, move the exposed person to fresh air at once and perform artificial respiration.

Personal Protective Methods: Wear appropriate clothing to prevent skin freezing. Wear eye protection to prevent any reasonable probability of eye contact. Remove clothing immediately if wet or contaminated to avoid flammability hazard.

Respirator Selection:
   10,000 ppm: SA/SCBA
   19,000 ppm: SAF/SCBAF/SA:PD,PP,CF
     Escape: SCBA

Disposal Method Suggested: Flaring using smokeless flare designs (A-31).

References

(1) See Reference (A-61).
(2) See Reference (A-60).
(3) Parmeggiani, L., Ed., *Encyclopedia of Occupational Health & Safety,* Third Edition, Vol. 2, pp 1244–48, Geneva, International Labour Office (1983).

LITHIUM HYDRIDE

Description: LiH is an off-white translucent odorless solid.

Code Numbers: CAS 7580-67-8   RTECS OJ6300000   UN 1414

DOT Designation: Flammable solid (dangerous when wet).

Synonyms: None.

Potential Exposure: Lithium hydride is used as a desiccant; it is used in hydrogen generators and in organic synthesis as a reducing agent and condensing agent with ketones and acid esters; it is reportedly used in thermonuclear weapons.

Incompatibilities: Oxidizers, halogenated hydrocarbons, acids, water.

Permissible Exposure Limits in Air: The Federal standard and the 1983/84 ACGIH TWA value is 0.025 mg/m^3. There is no STEL limit set. The IDLH level is 50 mg/m^3.

Determination in Air: Collection on a filter, analysis by atomic absorption spectrometry (A-21).

Permissible Concentration in Water: No criteria set, but EPA (A-37) has suggested a permissible ambient goal of 0.3 μg/ℓ based on health effects.

Routes of Entry: Inhalation, ingestion, eye and skin contact.

Harmful Effects and Symptoms: Burns of the eyes, skin, mouth, and esophagus; nausea; muscular twitches; mental confusion; blurred vision.

Points of Attack: Respiratory system, skin, eyes.

Medical Surveillance: Consider the points of attack in preplacement and periodic physical examinations.

First Aid: If this chemical gets into the eyes, irrigate immediately. If this chemical contacts the skin, flush with water immediately. If a person breathes in large amounts of this chemical, move the exposed person to fresh air at once and perform artificial respiration. When this chemical has been swallowed, get medical attention. Give large quantities of water and do not induce vomiting.

Personal Protective Methods: Wear appropriate clothing to prevent any possibility of contact with air of >0.1 mg/m^3 content. Wear eye protection to prevent any possibility of eye contact. Employees should wash immediately when skin is contaminated. Work clothing should be changed daily if it is possible that clothing is contaminated. Remove clothing immediately if contaminated. Provide emergency showers and eyewash if air containing >0.5 mg/m^3 is involved.

Respirator Selection:

0.1 mg/m^3:	HiEP/SA/SCBA
1.25 mg/m^3:	HiEPF/SAF/SCBAF
25 mg/m^3:	PAPHiEF
50 mg/m^3:	SAF:PD,PP,CF
Escape:	HiEPF/SCBA

Disposal Method Suggested: Lithium hydride may be mixed with sand, sprayed with butanol and then with water, neutralized and flushed to a sewer with water (A-38).

References

(1)   U.S. Environmental Protection Agency, *Chemical Hazard Information Profile: Lithium and Lithium Compounds,* Wash., DC (Sept. 1, 1976).

(2)   See Reference (A-61).

M

MAGNESIUM

Description: Magnesium is a light, silvery-white metal and is a fire hazard. It is found in dolomite, magnesite, brucite, periclase, carnallite, kieserite and as a silicate in asbestos, talc, olivine, and serpentine. It is also found in seawater, brine wells, and salt deposits. It is insoluble in water and ordinary solvents.

Code Numbers: CAS 7439-95-4 RTECS OM2100000 UN 1418 (powder) UN 1869 (pellets, turnings and ribbons)

DOT Designation: Flammable solid, dangerous when wet.

Potential Exposure: Magnesium alloyed with manganese, aluminum, thorium, zinc, cerium, and zirconium is used in aircraft, ships, automobiles, hand tools, etc., because of its lightness. Dow metal is the general name for a large group of alloys containing over 85% magnesium. Magnesium wire and ribbon are used for degassing valves in the radio industry and in various heating appliances; as a deoxidizer and desulfurizer in copper, brass, and nickel alloys; in chemical reagents; as the powder in the manufacture of flares, incendiary bombs, tracer bullets, and flashlight powders; in the nuclear energy process; and in a cement of magnesium oxide and magnesium chloride for floors.

Magnesium is an essential element in human and animal nutrition and also in plants, where it is a component of all types of chlorophyll. It is the most abundant intracellular divalent cation in both plants and animals. It is an activator of many mammalian enzymes.

Permissible Exposure Limits in Air: No standards set.

Determination in Air: Filter collection and atomic absorption analysis (A-10).

Permissible Concentration in Water: The USPHS drinking-water standards of 1925 included a maximum recommended magnesium concentration of 100 mg/ℓ. This limit was raised to 125 mg/ℓ in the 1942 and 1946 standards, but it was deleted in the 1962 standards. The USSR has not set a limit on magnesium; however, the World Health Organization (WHO) has established European and International desirable limits ranging from 30 to 125 mg/ℓ, depending on the sulfate concentration. If the sulfate exceeds 250 mg/ℓ, the magnesium is limited to 30 mg/ℓ. The WHO specifies an absolute maximum of 150 mg/ℓ for magnesium in drinking water.

In view of the fact that concentrations of magnesium in drinking water less than those that impart astringent taste pose no health problem and are more likely to be beneficial, no limitation for reasons of health is needed.

Determination in Water: Magnesium in water can be determined by atomic-absorption spectrophotometry, with a sensitivity of 15 μg/ℓ, and by photometry with a sensitivity of 100 μg/ℓ.

Harmful Effects and Symptoms: *Local* — Magnesium and magnesium compounds are mild irritants to the conjunctiva and nasal mucosa, but are not specifically toxic. Magnesium in finely divided form is readily ignited by a spark or flame, and splatters and burns at above 2300°F. On the skin, these hot particles are capable of producing second and third degree burns, but they respond to treatment as other thermal burns do. Metallic magnesium foreign bodies in the skin cause no unusual problems in man. In animal experiments, however, they have caused "gas gangrene"—massive localized gaseous tumors with extensive necrosis.

Magnesium salts at levels over 700 mg/ℓ (especially magnesium sulfate) have a laxative effect, particularly on new users, although the human body can adapt to the effects of magnesium with time.

Systemic — Magnesium in the form of nascent magnesium oxide can cause metal fume fever if inhaled in sufficient quantity. Symptoms are analogous to those caused by zinc oxide: cough, oppression in the chest, fever, and leukocytosis. There is no evidence that inhalation of magnesium dust has led to lung injury. It has been noted that magnesium workers show a rise in serum magnesium—although no significant symptoms of ill health have been identified. Some investigators have reported higher incidence of digestive disorders and have related this to magnesium absorption, but the evidence is scant. In foundry casting operations, hazards exist from the use of fluoride fluxes and sulfur-containing inhibitors which produce fumes of fluorides and sulfur dioxide.

Medical Surveillance: No specific recommendations.

Personal Protective Methods: Employees should receive training in the use of personal protective equipment, proper methods of ventilation, and fire suppression. Protective clothing should be designed to prevent burns from splatters. Masks to prevent inhalation of fumes may be necessary under certain conditions, but generally this can be controlled by proper ventilation. Dust masks may be necessary in areas of dust concentration as in transfer and storage areas, but adequate ventilation generally provides sufficient protection.

References

(1) Parmeggiani, L., Ed., *Encyclopedia of Occupational Health & Safety,* Third Edition, Vol. 2, pp 1266–68, Geneva, International Labour Office (1983) (Magnesium, Alloys & Compounds).

(2) Sax, N.I., Ed., *Dangerous Properties of Industrial Materials Report, 4,* No. 2, 79-81, New York, Van Nostrand Reinhold Co. (1984).

MAGNESIUM OXIDE FUME

Description: MgO forms a white fume.

Code Numbers: CAS 1309-48-4   RTECS OM3850000

DOT Designation: —

Synonyms: Magnesia fume.

Potential Exposure: Those involved in the manufacture of refractory crucibles, fire bricks, magnesia cements and boiler scale compounds.

Incompatibilities: Chlorine trifluoride.

Permissible Exposure Limits in Air: The Federal standard for magnesium oxide fume is 15 mg/m^3, according to NIOSH. ACGIH (1983/84) has adopted 10 mg/m^3 as a TWA. There is no STEL set, nor is the IDLH level established.

Determination in Air: Collection on a filter, workup with nitric acid, analysis by atomic absorption. See NIOSH Methods, Set O. See also reference (A-10).

Permissible Concentration in Water: No criteria set, but EPA (A-37) has suggested a permissible ambient goal for magnesium oxide of 138 μg/ℓ based on health effects.

Routes of Entry: Inhalation of fume, eye and skin contact.

Harmful Effects and Symptoms: *Systemic* — Magnesium in the form of nascent magnesium oxide can cause metal fume fever if inhaled in sufficient quantity. Symptoms are analogous to those caused by zinc oxide: cough, oppression in the chest, fever, and leukocytosis. There is no evidence that inhalation of magnesium dust has led to lung injury. It has been noted that magnesium workers show a rise in serum magnesium—although no significant symptoms of ill health have been identified. Some investigators have reported higher incidence of digestive disorders and have related this to magnesium absorption, but the evidence is scant. In foundry casting operations, hazards exist from the use of fluoride fluxes and sulfur-containng inhibitors which produce fumes of fluorides and sulfur dioxide.

Points of Attack: Respiratory system, lungs, eyes.

Medical Surveillance: No specific recommendations.

First Aid: If a person breathes in large amounts of this chemical, move the exposed person to fresh air at once and perform artificial respiration.

Personal Protective Methods: No particular measures are cited by NIOSH except respirator protection.

Respirator Selection:

> 150 mg/m^3: FuHiEP/SA/SCBA
> 750 mg/m^3: HiEPF/SAF/SCBAF
> 7500 mg/m^3: PAPHiE/SA:PD,PP,CF

Disposal Method Suggested: Sanitary landfill.

References

(1) See Reference (A-61).
(2) See Reference (A-60).

MALATHION

● Hazardous substance (EPA)

Description: Malathion, O,O-dimethyl-S-(1,2-dicarboethoxyethyl) dithiophosphate, $C_{10}H_{19}O_6PS_2$, has the following structural formula:

$$(CH_3O)_2P-S-CHCOOC_2H_5$$
$$\underset{\displaystyle S}{\overset{\displaystyle \|}{}} \quad \underset{\displaystyle CH_2COOC_2H_5}{\overset{\displaystyle |}{}}$$

It is a colorless, light to amber liquid. It melts at 3°C.

Code Numbers: CAS 121-75-5   RTECS WM8400000   UN (NA 2783)

DOT Designation: ORM-A

Synonyms: Maldison (New Zealand), Mercaptothion (South Africa), ENT 17034, Cythion®, Malathion®, Malathiazol®, Malathiazoo®.

Potential Exposure: Malathion is marketed as 99.6% technical grade liquid. Available formulations include wettable powders (25% and 50%), emulsifiable concentrates, dusts and aerosols. Malathion is used in the control of certain insect pests on fruits, vegetables, and ornamental plants. It has been used in the control of houseflies, mosquitoes, and lice, and on farm and livestock animals.

NIOSH estimates that approximately 75,000 workers in the United States are occupationally exposed to malathion in the course of manufacture, formulation and application.

Incompatibilities: Strong oxidizers.

Permissible Exposure Limits in Air: When skin exposure is prevented, exposure to malathion in the workplace shall be controlled so that employees are not exposed to malathion at a TWA concentration greater than 15 mg/m^3 of air for up to a 10-hour work shift, 40-hour workweek, according to NIOSH (1).

A TWA value of 10 mg/m^3 has been set by the American Conference of Governmental Industrial Hygienists as of 1983/84 with the notation "skin" to indicate the possibility of cutaneous absorption.

Determination in Air: Collection on a filter, workup with isooctane, analysis by gas chromatography. See NIOSH Methods, Set 1. See also reference (A-10).

Permissible Concentration in Water: The EPA has set a criterion for malathion of 0.1 μg/ℓ for freshwater and marine aquatic life.

A no-adverse-effect level in drinking water has been calculated by NAS/NRC as 0.16 mg/ℓ. An acceptable daily intake (ADI) of 0.02 mg/kg/day was also calculated.

Routes of Entry: Inhalation of vapor, skin absorption, ingestion and skin and eye contact.

Harmful Effects and Symptoms: As described and documented in detail in a NIOSH document (1), the main signs and symptoms of malathion intoxication are increased bronchial secretion and excessive salivation, nausea, vomiting, excessive sweating, miosis, and muscular weakness and fasciculations. These signs and symptoms are induced by the inhibition of functional acetylcholinesterase (AChE) in the nervous system.

NCI has done a carcinogenesis bioassay on malathion and concluded that it is not a carcinogen.

Points of Attack: Respiratory system, liver, blood, cholinesterase, central nervous system, cardiovascular system, gastrointestinal system.

Medical Surveillance:

1) Preplacement and periodic medical examinations shall include:
   (a) Comprehensive initial or interim medical and work histories.
   (b) A physical examination which shall be directed toward, but not limited to evidence of frequent headache, dizziness, nausea, tightness of the chest, dimness of vision, and difficulty in focusing the eyes.
   (c) Determination, at the time of the preplacement examination, of a baseline or working baseline erythrocyte ChE activity.

(d) A judgement of the worker's physical ability to use negative or positive pressure regulators as defined in 29 CFR 1910.134.

2) Periodic examinations shall be made available on an annual basis or at some other interval determined by the responsible physician.

3) Medical records shall be maintained for all workers engaged in the manufacture or formulation of malathion and such records shall be kept for at least one year after termination of employment.

4) Pertinent medical information shall be available to medical representatives of the U.S. Government, the employer and the employee.

Erythrocyte cholinesterase levels should be checked as noted above and as described in detail by NIOSH (1).

First Aid: If this chemical gets into the eyes, irrigate immediately. If this chemical contacts the skin, dust off solid, wash with soap promptly. If a person breathes in large amounts of this chemical, move the exposed person to fresh air at once and perform artificial respiration. When this chemical has been swallowed, get medical attention. Give large quantities of water and induce vomiting. Do not make an unconscious person vomit.

Personal Protective Methods: *Protective Clothing* — Any employee whose work involves likely exposure of the skin to malathion or malathion formulations, e.g., mixing or formulating, shall wear full-body coveralls or the equivalent, impervious gloves, and impervious footwear and, when there is danger of malathion coming in contact with the eyes, safety goggles shall be provided and worn. Any employee who applies malathion shall be provided with and required to wear the following protective clothing and equipment: goggles, whole-body coveralls, and impervious footwear.

Personal Hygiene — Employees should wash promptly when skin is wet or contaminated. Remove nonimpervious clothing promptly if wet or contaminated.

Respiratory Protection — Engineering controls shall be used wherever feasible to maintain airborne malathion concentrations below the recommended workplace environmental limit. Compliance with the workplace environmental limit by the use of respirators is allowed only when airborne malathion concentrations are in excess of the workplace environmental limit because required engineering controls are being installed or tested, when nonroutine maintenance or repair is being accomplished, or during emergencies. When a respirator is thus permitted, it shall be selected and used in accordance with NIOSH requirements (1).

Respirator Selection:

150 mg/m^3: CCROVDMFuPest/SA/SCBA
750 mg/m^3: CCROVFDMFuPest/GMOVDMFuPest/SAF/SCBAF
5,000 mg/m^3: PAPCCROVFHiEPPest/SAF:PD,PP,CF
Escape: GMOVFP/SCBA

Disposal Method Suggested: Malathion is reported to be "hydrolyzed almost instantly" at pH 12; 50% hydrolysis at pH 9 requires 12 hours. Alkaline hydrolysis under controlled conditions (0.5 N NaOH in ethanol) gives quantitative yields of $(CH_3O)_2P(S)SNa$, whereas hydrolysis in acidic media yields $(CH_3O)_2P(S)OH$. On prolonged contact with iron or iron-containing material, it is reported to break down and completely lose insecticidal activity (A-32). Incineration together with a flammable solvent in a furnace equipped with afterburner and scrubber is recommended (A-38).

References

(1) National Institute for Occupational Safety and Health, *Criteria for a Recommended Standard: Occupational Exposure to Malathion,* NIOSH Doc. No. 76-205, Wash., DC (June 1976).
(2) See Reference (A-61).
(3) United Nations Environment Programme, *IRPTC Legal File 1983,* Vol. II, pp VII/ 739–41, Geneva, Switzerland, International Register of Potentially Toxic Chemicals (1984).

MALEIC ANHYDRIDE

- Hazardous substance (EPA)
- Hazardous waste (EPA)

Description: $(CHCO)_2O$ with the structural formula

is a white crystalline solid with an acrid odor which melts at $53°C$ and boils at $202°C$.

Code Numbers: CAS 108-31-6   RTECS ON3675000   UN 2215

DOT Designation: ORM-A

Synonyms: 2,5-Furanediene; cis-Butenedioic anhydride; Toxilic anhydride.

Potential Exposure: Maleic anhydride is used in the manufacture of unsaturated polyester resins, in the manufacture of fumaric acid, in alkyd resin manufacture and in the manufacture of pesticides (malathion, maleic hydrazide and captan for example) (A-32).

Incompatibilities: Strong oxidizers, alkali metals, caustics, amines at $>150°F$.

Permissible Exposure Limits in Air: The Federal standard and the ACGIH 1983/84 TWA value is 0.25 ppm (1 mg/m³). There is no STEL value, nor is there an IDLH level set.

Permissible Concentration in Water: No criteria set.

Routes of Entry: Inhalation, ingestion, eye and skin contact.

Harmful Effects and Symptoms: Subacute inhalation of maleic anhydride can cause severe headaches, nosebleeds, nausea, and temporary impairment of vision. It can also lead to conjunctivitis and corneal erosion.

Patterson et al (2) state that repeated exposure to concentrations above 1.25 ppm has caused asthmatic responses in workers. Allergies have developed so that lower concentrations of maleic anhydride can no longer be tolerated. An increased incidence of bronchitis and dermatitis has also been noted among workers with long-term exposure to maleic anhydride. One case of pulmonary edema in a worker was reported (2).

Points of Attack: Eyes, respiratory system, skin.

Medical Surveillance: Consider the points of attack in preplacement and periodic physical examinations.

First Aid. If this chemical gets into the eyes, irrigate immediately. If this chemical contacts the skin, wash with soap immediately. If a person breathes in large amounts of this chemical, move the exposed person to fresh air at once and perform artificial respiration.

When this chemical has been swallowed, get medical attention. Give large quantities of water and induce vomiting. Do not make an unconscious person vomit.

Personal Protective Methods: Wear appropriate clothing to prevent repeated or prolonged skin contact. Wear eye protection to prevent any possibility of eye contact.

Employees should wash promptly when skin is wet or contaminated. Remove nonimpervious clothing promptly if wet or contaminated. Provide emergency eyewash.

Respirator Selection:

12.5 ppm:	CCROVDM/GMOVDM/SAF/SCBAF
250 ppm:	PAPHiEOVF
500 ppm:	SAF:PD,PP,CF
Escape :	GMOVFP/SCBA

Disposal Method Suggested: Controlled incineration: care must be taken that complete oxidation to nontoxic products occurs.

References

(1) U.S. Environmental Protection Agency, *Chemical Hazard Information Profile: Maleic Anhydride,* Wash., DC (Aug. 1, 1978).

(2) Patterson, R.M. et al, *Assessment of Maleic Anhydride as a Potential Air Pollution Problem,* U.S. NTIS, PB Report, PB-258363 (1976).

(3) U.S. Environmental Protection Agency, *Maleic Anhydride,* Health and Environmental Effects Profile No. 122, Wash., DC, Office of Solid Waste (April 30, 1980).

(4) See Reference (A-61).

(5) Sax, N.I., Ed., *Dangerous Properties of Industrial Materials Report, 2,* No. 3, 79-81, New York, Van Nostrand Reinhold Co. (1982).

(6) See Reference (A-60).

(7) United Nations Environment Programme, *IRPTC Legal File 1983,* Vol. II, pp VII/418-19, Geneva, Switzerland, International Register of Potentially Toxic Chemicals (1984).

MALONONITRILE

● Hazardous waste (EPA)

Description: $NCCH_2CN$ is a yellow, odorless crystalline substance melting at $30°$-$32°C$.

Code Numbers: 109-77-3   RTECS OO3150000   UN 2647

DOT Designation: —

Synonyms: Malonic dinitrile; Cyanoacetonitrile; Dicyanomethane

Potential Exposure: Malononitrile is used in the following applications: as a lubricating oil additive, for thiamine synthesis, for pteridine-type anticancer agent synthesis, and in the synthesis of photosensitizers, acrylic fibers, and dyestuffs. It has also been used in the treatment of various forms of mental illness (1).

Permissible Exposure Limits in Air: Because malononitrile is about three times as toxic as isobutyronitrile, NIOSH recommends that employee exposure to malononitrile not exceed 3 ppm (8 mg/m^3) as a TWA limit for up to 10-hour workshift in a 40-hour work week (2).

Determination in Air: Malononitrile may be determined in air by colorimetric or gas chromatographic methods (2).

Permissible Concentration in Water: No criteria set.

Harmful Effects and Symptoms: Malononitrile has effects on the cardiovascular, renal, hepatic and central nervous systems. This compound can take effect after inhalation, dermal contact or ingestion. No carcinogenic, mutagenic or teratogenic effects have been reported (1).

Medical Surveillance: See Acetonitrile for guidance on methods which apply to malononitrile as well (2).

First Aid: See Acetonitrile for guidance on methods which apply to malononitrile as well (2).

Personal Protective Methods: See Acetonitrile for guidance on methods which apply to malononitrile as well (2).

Respirator Selection: See Acetonitrile for guidance on methods which apply to malononitrile as well (2).

References

(1) U.S. Environmental Protection Agency, *Malononitrile,* Health and Environmental Effects Profile No. 123, Wash., DC, Office of Solid Waste (April 30, 1980).
(2) National Institute for Occupational Safety and Health, *Criteria for a Recommended Standard . . . Occupational Exposure to Nitriles,* U.S. DHEW (NIOSH) Report No. 78-212, Bethesda, MD (1978).

MANEB

● Carcinogen (Animal Suspected, IARC) (1)

Description: [S–CNHCH$_2$CH$_2$NHC–S–Mn]$_x$ is a yellow crystalline solid which decomposes before melting.

Code Numbers: CAS 12427-38-2   RTECS OP0700000   UN 2210

DOT Designation: Spontaneously combustible.

Synonyms: Manganese 1,2-ethanediylbis(carbamodithioate) complex; ENT 14875; MEB; Manzate®; Dithane-22®.

Potential Exposure: Those involved in manufacture, formulation and application of this fungicide.

Permissible Exposure Limits in Air: No standards set.

Permissible Concentration in Water: A no-adverse effect level in drinking water has been determined by NAS/NRC to be 0.035 mg/ℓ. An acceptable daily intake (ADI) of 0.005 mg/kg/day has been calculated for maneb.

Harmful Effects and Symptoms: Maneb is low in acute toxicity and does not present alarming properties during long-term administration to experimental animals, except at very high dosages. However, it is a material of concern because of evidence of mutagenic and teratogenic effects as well as the possibility of nitrosation to carcinogenic nitrosamines, according to NAS/NRC (A-2).

A rebuttable presumption against registration of maneb for pesticide uses was issued by EPA on August 10, 1977 on the basis of oncogenicity, teratogenicity and hazard to wildlife.

Disposal Method Suggested: Maneb is unstable to moisture and is hydrolyzed by acids and hot water. It decomposes at about 100°C but may spontaneously decompose vigorously when stored in bulk (A-32).

References

(1)   International Agency for Research on Cancer, *IARC Monographs on the Carcinogenic Risks of Chemicals to Humans,* Lyon, France, 12, 137 (1976).

(2)   United Nations Environment Programme, *IRPTC Legal File 1983,* Vol. II, pp VII/422–3, Geneva, Switzerland, International Register of Potentially Toxic Chemicals (1984).

MANGANESE AND COMPOUNDS

Description: Mn, manganese, is a reddish-grey or silvery, soft metal. The most important ore containing manganese is pyrolusite. Manganese may also be produced from ferrous scrap used in the production of electric and open-hearth steel. Manganese decomposes in water and is soluble in dilute acid.

Code Numbers: (Manganese metal): CAS 7439-96-5   RTECS OO9275000

DOT Designation: –

Synonyms: None.

Potential Exposure: Most of the manganese produced is used in the iron and steel industry in steel alloys, e.g., ferromanganese, silicomanganese, Manganin, spiegeleisen, and as an agent to reduce oxygen and sulfur content of molten steel. Other alloys may be formed with copper, zinc, and aluminum. Manganese and its compounds are utilized in the manufacture of dry cell batteries (MnO_2), paints, varnishes, inks, dyes, matches and fireworks, as a fertilizer, disinfectant, bleaching agent, laboratory reagent, drier for oils, an oxidizing agent in the chemical industry particularly in the synthesis of potassium permanganate, and as a decolorizer and coloring agent in the glass and ceramics industry.

Organomanganese compounds such as methylcyclopentadienyl manganese tricarbonyl (MMT) have been proposed as supplements and/or replacements for tetraethyllead (TEL) as an antiknock in gasoline.

Exposure may occur during the mining, smelting and refining of manganese, in the production of various materials, and in welding operations with manganese-coated rods.

Manganese normally is ingested as a trace nutrient in food. The average human intake is approximately 10 mg/day.

Permissible Exposure Limits in Air: The Federal standard for manganese dust and compounds is 5 mg/m³ as a ceiling value. The Illinois Environmental

Health Resource Center recommends an environmental standard for particulate manganese of 0.006 $\mu g/m^3$. ACGIH (1983/84) has set a TWA of 1.0 mg/m³ for manganese tetroxide and manganese fume as well as an STEL of 3.0 mg/m³ for manganese fume.

The ACGIH has set a TWA value of 0.2 mg/m³ for manganese methylcyclopentadienyl tricarbonyl and an STEL of 0.6 mg/m³ as of 1983/84. This compound is capable of cutaneous absorption, as is cyclopentadienyl manganese tricarbonyl for which ACGIH has set a TWA of 0.1 mg/m³ and an STEL of 0.3 mg/m³.

The IDLH level for manganese is 10,000 ppm.

Determination in Air: Collection on a filter, workup with HCl, analysis by atomic absorption. See NIOSH Methods, Set A. See also reference (A-10).

Permissible Concentration in Water: The EPA (A-3) has suggested the following criteria:

50 $\mu g/\ell$ for domestic water supplies (welfare).
100 $\mu g/\ell$ for protection of consumers of marine mollusks.

No specific criterion for manganese in agricultural waters is proposed. In select areas, and where acidophilic crops are cultivated and irrigated, a criterion of 200 $\mu g/\ell$ is suggested for consideration.

Determination in Water: The manganese detection limit by direct flame atomization is 2 $\mu g/\ell$. However, solvent extraction is used for many determinations. Analytic conditions are more critical for the extraction of manganese than for most other metals, because many manganese-chelate complexes are unstable in solution. With pH control and immediate analysis after extraction, accurate determinations are possible.

When the graphite furnace is used to increase sample atomization, the detection limit is lowered to 0.01 $\mu g/\ell$ according to NAS/NRC.

Routes of Entry: Inhalation of dust or fume; limited percutaneous absorption of liquids; ingestion.

Harmful Effects and Symptoms: *Local* — Manganese dust and fumes are only minor irritants to the eyes and mucous membranes of the respiratory tract, and apparently are completely innocuous to the intact skin.

Systemic — Chronic manganese poisoning has long been recognized as a clinical entity. The dust or fumes (manganous compounds) enter the respiratory tract and are absorbed into the blood stream. Manganese is then deposited in major body organs with a special predilection for the liver, spleen, and certain nerve cells of the brain and spinal cord. Among workers there is a very marked variation in individual susceptibility to manganese. Some workers have worked in heavy exposure for a lifetime and have shown no signs of the disease; others have developed manganese intoxication with as little as 49 days of exposure.

The early phase of chronic manganese poisoning is most difficult to recognize, but it is also important to recognize since early removal from the exposure may arrest the course of the disease. The onset is insidious, with apathy, anorexia, and asthenia. Headache, hypersomnia, spasms, weakness of the legs, arthralgias, and irritability are frequently noted. Manganese psychosis follows with certain definitive features: unaccountable laughter, euphoria, impulsive acts, absent-mindedness, mental confusion, aggressiveness, and hallucinations. These symptoms usually disappear with the onset of true neurological disturbances, or may resolve completely with removal from manganese exposure.

Progression of the disease presents a range of neurological manifestations that can vary widely among individuals affected. Speech disturbances are common: monotonous tone, inability to speak above a whisper, difficult articulation, incoherence, even complete muteness. The face may take on masklike quality, and handwriting may be affected by micrographia. Disturbances in gait and balance occur, and frequently propulsion, retropropulsion and lateropropulsion are affected, with no movement for protection when falling. Tremors are frequent, particularly of the tongue, arms, and legs. These will increase with intentional movements and are more frequent at night. Absolute detachment, broken by sporadic or spasmodic laughter, ensues, and as in extrapyramidal affections, there may be excessive salivation and excessive sweating. At this point the disease is indistinguishable from classical Parkinson's disease.

Chronic manganese poisoning is not a fatal disease although it is extremely disabling.

Manganese dust is no longer believed to be a causative factor in pneumonia. If there is any relationship at all, it appears to be as an aggravating factor to a preexisting condition. Freshly formed fumes have been reported to cause fever and chills similar to metal fume fever.

Points of Attack: Respiratory system, central nervous system, lungs, blood, kidneys.

Medical Surveillance: Preemployment physical exams should be directed toward the individual's general health with special attention to neurologic and personality abnormalities. Periodic physical examinations may be required as often as every two months. Special emphasis should be given to behavioral and neurological changes: speech defects, emotional disturbances, hypertonia, tremor, equilibrium, difficulty in walking or squatting, adiadochokinesis, and handwriting.

There are no laboratory tests which can be used to diagnose manganese poisoning.

First Aid: If a person breathes in large amounts of manganese or its compounds, move the exposed person to fresh air at once and perform artificial respiration. When this chemical has been swallowed, get medical attention. Give large quantities of water and induce vomiting. Do not make an unconscious person vomit.

Personal Protective Methods: In areas where the ceiling value standards are exceeded, dust masks or respirators are necessary. Education in the use and necessity of these devices is important.

Respirator Selection:

25 mg/m³:	DM/DMXS
50 mg/m³:	DMXSQ/DMFu
50 mg/m³:	FuHiEP/SA/SCBA
250 mg/m³:	HiEPF/SAF/SCBAF
5,000 mg/m³:	PAPHiE/SA:PD, PP, CF
10,000 mg/m³:	SAF:PD, PP, CF

Disposal Method Suggested: Manganese metal—sanitary landfill. Manganese chloride or sulfate—chemical conversion to the oxide followed by landfilling, or conversion to the sulfate for use in fertilizer.

References

(1) Illinois Institute for Environmental Quality, *Airborne Manganese Health Effects and Recommended Standard,* Doc. No. 75-18, Chicago, Ill. (Sept. 1975).

(2) National Academy of Sciences, *Manganese,* (in a series on medical and biologic effects of environmental pollutants), Wash., DC (1973).

(3) Nat. Inst. for Occup. Safety and Health, *Information Profile on Potential Occupational Hazards,* pp 157-168, Report PB-267 678, Rockville, MD (Oct. 1977).

(4) National Academy of Sciences, *Medical and Biologic Effects of Environmental Pollutants: Manganese,* Wash., DC (1973).

(5) See Reference (A-61).

(6) Sax, N.I., Ed., *Dangerous Properties of Industrial Materials Report, 1,* No. 2, 44-45, New York, Van Nostrand Reinhold Co. (1980).

(7) See Reference (A-60) under Manganese Acetate and Manganese Dioxide.

(8) U.S. Environmental Protection Agency, *Chemical Hazard Information Profile: Methylcyclopentadienyl Manganese Tricarbonyl,* Wash., DC (October 21, 1983).

(9) Parmeggiani, Ed., *Encyclopedia of Occupational Health & Safety,* Third Edition, Vol. 2, pp 1279-82, Geneva, International Labour Office (1983).

(10) United Nations Environment Programme, *IRPTC Legal File 1983,* Vol. II, pp VII/420-21, Geneva, Switzerland, International Register of Potentially Toxic Chemicals (1984).

MCPA

Description: $H_3CC_6H_3ClOCH_2CO_2H$ is a colorless crystalline solid melting at 118° to 119°C.

Code Numbers: CAS 94-74-6   RTECS AG1575000

DOT Designation: —

Synonyms: [(4-chloro-o-tolyl)oxy] acetic acid; (4-chloro-2-methylphenoxy) acetic acid; Metaxon (in USSR); MCP (in Japan); Agroxone®; Agritox®; Cornox M®.

Potential Exposure: Those involved in the manufacture (A-32) formulation and application of this herbicide.

Permissible Exposure Limits in Air: No standards set.

Permissible Concentration in Water: A no-adverse effect level in drinking water has been calculated by NAS/NRC to be 0.009 mg/ℓ.

Harmful Effects and Symptoms: In a 90-day feeding study of MCPA in rats, growth retardation and increased kidney:body-weight ratios were observed at 400 ppm or more. The 50-ppm dietary content of MCPA was considered to be the no-adverse-effect content for rats by the authors. In another 90-day feeding study in Charles River rats (1), significant growth decrease was observed with technical MCPA at 100 ppm, and histopathologic alterations of liver and kidneys were seen in both sexes at 25 ppm or higher. In a later study with the same rat strain, no abnormalities were seen after 90 days in animals fed technical MCPA at 4, 8, and 16 mg/kg/day (note that 4 mg/kg/day is approximately equivalent to a dietary content of 25 ppm).

Some dogs that received daily oral doses of technical MCPA over a 13-week period died, and all showed severe weight loss at 50 mg/kg/day, whereas more moderate weight losses but no mortalities occurred at 25 mg/kg/day (1). In another 13-week study, decreased testicular weight and histopathologic changes of the testes and prostate were seen in dogs fed technical MCPA at 640 ppm.

There is a complete lack of data on human toxicity related to MCPA, according to NAS/NRC.

References

(1) U.S. Environmental Protection Agency, *Initial Scientific and Minieconomic Review No. 21: MCPA,* Washington, DC, Office of Pesticide Programs (1975).

(2) United Nations Environment Programme, *IRPTC Legal File 1983,* Vol. I, pp VII/6, Geneva, Switzerland, International Register of Potentially Toxic Chemicals (1984).

MELAMINE

Description: $NC(NH_2)NC(NH_2)NC(NH_2)$ is a white crystalline solid melting (with decomposition) at 345° to 360°C.

Code Numbers: CAS 108-78-1 RTECS OS0700000

DOT Designation: —

Synonyms: Cyanuramide; Cyanurotriamide; Cyanurotriamine; 2,4,6-Tri-aminotriazine.

Potential Exposure: Those involved in melamine manufacture from urea; those involved in its use in melamine-formaldehyde resin manufacture for coatings, laminates, adhesives and molding compounds.

Permissible Exposure Limits in Air: No standards set.

Permissible Concentration in Water: No criteria set.

Routes of Entry: Inhalation and ingestion (A-38).

Harmful Effects and Symptoms: Irritates eyes, skin and mucous membranes. Causes dermatitis in humans. The oral LD_{50} for rats is 3.2 mg/kg. Kidney injury may occur (A-38).

Points of Attack: Eyes, skin, mucous membranes.

First Aid: Remove contaminated clothing, wash skin with soap and water; flush eyes with water; rinse mouth with water (A-60).

Personal Protective Methods: Wear protective gloves, safety goggles; use local exhaust or breathing protection; do not eat, drink or smoke while working with melamine (A-60).

References

(1) U.S. Environmental Protection Agency, *Chemical Hazard Information Profile Draft Report: Melamine,* Washington, DC (December 29, 1982).

(2) Parmeggiani, L., Ed., *Encyclopedia of Occupational Health & Safety,* Third Edition, Vol. 2, p 1319, Geneva, International Labour Office (1983).

MELPHALAN

- Carcinogen (Animal Positive, NCI) (Human Suspected, IARC) (A-64)
- Hazardous Waste Constituent (EPA)

Description: $C_{13}H_{18}Cl_2N_2O_2$ with the following structural formula

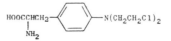

forms solvated crystals from methanol which decompose at 182° to 183°C.

Code Numbers: CAS 148-82-3   RTECS AY3675000

DOT Designation: −

Synonyms: 3-⟨p-[Bis(2-chloroethyl)amino]phenyl⟩alanine; phenylalanine mustard; alanine nitrogen mustard, Alkeran®.

Potential Exposure: This drug is used in the treatment of multiple myeloma and cancer of the ovary. It is also used in investigation on other types of cancer and as an antineoplastic in animals.

Melphalan is not produced in the United States, although one marketer in the United States has been reported. In 1979, 3,000 lb were imported.

Human exposure to melphalan occurs principally during its use in cancer treatment. FDA estimates 30,000 to 50,000 patients may be treated each year. Melphalan is administered orally or intravenously. Adult dosage is 6 mg/day, 5 days per month.

Permissible Exposure Limits in Air: No standards set.

Permissible Concentration in Water: No criteria set.

Harmful Effects and Symptoms: Melphalan is carcinogenic in mice and rats following intraperitoneal injection, producing lymphosarcomas, a dose-related increase in lung tumors in mice, and peritoneal sarcomas in rats (1).

Case reports of second primary malignancies (mainly acute leukemia) in patients treated with melphalan have been published. Epidemiological studies showed substantially increased rates of leukemia in patients treated with melphalan for multiple myeloma and ovarian cancer. Some of these patients were also treated with other alkylating agents and ionizing radiation, however sufficient numbers of patients were treated with melphalan alone to implicate it as the causal factor. Additionally the incidence of acute leukemia in patients with multiple myeloma has increased since the introduction of melphalan therapy (2) (3).

References
(1)   IARC Monographs 9:167–80, 1975.
(2)   *IARC Monographs on the Evaluation of the Carcinogenic Risk of Chemicals to Humans,* Supplement I, IARC, Lyon, France, p 37 (1979).
(3)   See Reference (A-62). Also see Reference (A-64).

MEPHOSFOLAN

Description: $C_8H_{16}NO_3PS_2$ with the formula

$$(C_2H_5O)_2P-N=\underset{\underset{S-CH_2}{\diagdown}}{C}-S-CH-CH_3$$
$$\underset{O}{\overset{\|}{}}$$

boiling at 120°C at 1.0 mm Hg pressure.

Code Numbers: CAS 950-10-7   RTECS JP1050000

DOT Designation: −

Synonyms: Diethyl-4-methyl-1,3-dithiolan-2-ylidenephosphoramidate; Dithiolane iminophosphate; Cytrolane®.

Potential Exposure: Those involved in the production, formulation and application of this insecticide.

Permissible Exposure Limits in Air: No standards set.

Permissible Concentration in Water: No criteria set.

Harmful Effects and Symptoms: This is a highly to extremely toxic material (LD_{50} for rats is 4 to 9 mg/kg).

References

(1) Sax, N.I., Ed., *Dangerous Properties of Industrial Materials Report, 3,* No. 1, 72-74, New York, Van Nostrand Reinhold Co. (1983).

2-MERCAPTOBENZOTHIAZOLE

Description: $C_7H_5NS_2$ with the structural formula

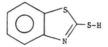

is a yellowish to tan crystalline powder with a distinct, characteristic odor melting at 177° to 178°C.

Code Numbers: CAS 149-30-4   RTECS DL6475000   UN 1228

DOT Designation: —

Synonyms: 2-Benzothiazolethiol; MBT.

Potential Exposures: 2-Mercaptobenzothiazole finds major uses in three distinct and quite different areas. It is used in the rubber industry as a vulcanization accelerator and antioxidant; in metal processing and applications (especially in the automobile industry in radiator coolants) as an anticorrosion agent; and as a fungicide and bacteriostatic agent.

Permissible Exposure Limits in Air: No standards set.

Permissible Concentration in Water: No criteria set.

Routes of Entry: Inhalation and ingestion.

Harmful Effects and Symptoms: The history of human experience with the adverse effects resulting from exposure to MBT and its derivatives is an account which details numerous cases of allergic contact dermatitis as the only consequence of exposure. Several retrospective investigations have established that MBT, particularly as a component in rubber products, is one of the most common contact allergens known today.

The potent action on the central nervous system of numerous derivatives of benzothiazole, including MBT, has only been shown to occur in animals. These effects can be manifested as either a central stimulation (convulsions) or as selective depression (flaccid paralysis, mental depression, lack of spontaneous motor activity).

Some subcutaneous tests on mice are suggestive of carcinogenicity (A-63).

First Aid: Solutions of NaMBT are alkaline and, therefore, are more dangerous than the solids with respect to skin contact and breathing of any spray from them. Splashproof goggles and protective rubber clothing are recommended for handling such solutions, in addition to the suggestions above. Also, an eyewash fountain and safety shower should be available at the handling site.

Personal Protective Methods: Respirators approved by the U.S. Bureau of Mines for nuisance dust and safety spectacles are recommended in the event of excessive dustiness in handling the compounds. Otherwise, no special handling in use is specified by the manufacturers of these compounds, other than ordinary measures of personal hygiene. See "First Aid" above also.

Respirator Selection: See "Personal Protective Methods" above.

Disposal Method Suggested: The recommended disposal method is burial in a landfill. Incineration is not recommended unless provision can be made to insure that SO_2 and nitrogen oxides will not be emitted to the atmosphere.

References

(1) U.S. Environmental Protection Agency, *Investigation of Selected Potential Environmental Contaminants: Mercaptobenzothiazoles,* Report EPA-560/2-76-006, Washington, DC, Office of Toxic Substances (June 1976).

(2) See Reference (A-60).

MERCAPTOBENZOTHIAZOLE DISULFIDE

Description: $C_{14}H_8N_2S_4$ has the structural formula

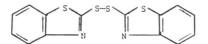

This compound forms pale yellow needle-like crystals melting at $180°C$.

Code Numbers: CAS 120-78-5   RTECS DL4550000

DOT Designation: —

Synonyms: 2,2'-Dithiobis-benzothiazole; Dibenzothiazyl disulfide, MBTS.

Potential Exposure: Those involved in MBTS manufacture or use as a rubber vulcanization accelerator. NIOSH data indicate that 72,000 workers may be exposed to MBTS.

Permissible Exposure Limits in Air: No standards set.

Permissible Concentration in Water: No criteria set.

Routes of Entry: Inhalation of dust and vapors.

Harmful Effects and Symptoms: Causes skin and mucous membrane inflammation and sensitization. Long-term inhalation may cause lung inflammation and fibrogenicity. Functional reproductive disorders have been noted in women working with MBTS. MBTS also puts a stress on the nervous system.

Points of Attack: Skin, mucous membranes, lungs.

References

(1) U.S. Environmental Protection Agency, *Chemical Hazard Information Profile: Mercaptobenzothiazole Disulfide (MBTS),* Washington, DC (May 27, 1983).

MERCURY—ALKYL AND ARYL

- Hazardous waste (phenylmercury acetate)(EPA)
- Priority toxic pollutant (methylmercury, EPA)

Description: Methylmercury compounds—methylmercury dicyandiamide, $CH_3HgNHC(NH)NHCN$, soluble in water. Ethylmercury compounds—ethylmercuric chloride, C_2H_5HgCl, insoluble in water; ethylmercuric phosphate, $(C_2H_5HgO)_3PO$, soluble in water; N-(ethylmercuric)-p-toluenesulfonanilide, $C_6H_5N(HgC_2H_5)SO_2C_6H_4CH_3$, practically insoluble in water. Phenylmercury compounds—phenylmercuric acetate, slightly soluble in water.

Code Numbers:

Methylmercury dicyandiamide	CAS 502-39-6	RTECS OW1750000	UN 2777
Ethylmercuric chloride	CAS 107-27-7	RTECS OV9800000	UN 2777
Phenylmercuric acetate	CAS 62-38-4	RTECS OV6475000	UN 2777

DOT Designation: Poison B

Synonyms: *Methylmercury Compounds* — Methylmercury Dicyandiamide: cyano(methylmercury)guanidine, Panogen®.

Ethylmercury Compounds — Ethylmercuric chloride, Ceresan®, ethylmercuric phosphate, New Ceresan®, and N-(ethylmercuric)-p-toluenesulfonanilide, Ceresan M®.

Phenylmercury Compounds — Phenylmercuric acetate, PMA.

Potential Exposure: These compounds are used in treating seeds for fungi and seedborne diseases, as timber preservatives, and disinfectants.

The aryl mercury compounds such as phenylmercury are primarily used as disinfectants, fungicides for treating seeds, antiseptics, herbicides, preservatives, mildewproofing agents, denaturants for ethyl alcohol, germicides, and bactericides.

Incompatibilities: Strong oxidizers such as chlorine.

Permissible Exposure Limits in Air: The Federal standard is 0.01 mg/m³ as an 8-hour TWA with an acceptable ceiling of 0.04 mg/m³.

The ACGIH has set a TWA of 0.01 mg/m³ and an STEL of 0.03 mg/m³ for mercury alkyls. A TWA of 0.1 mg/m³ (but no STEL) value has been proposed for aryl mercury compounds by ACGIH as of 1983/84. The IDLH level is 10 mg/m³ for mercury alkyls.

Determination in Air: Collection on solid sorbent followed by determination of flameless atomic absorption spectrophotometry (A-10).

Permissible Concentration in Water (Methylmercury): To protect freshwater aquatic life—0.016 µg/ℓ as a 24-hr average, never to exceed 8.8 µg/ℓ. To protect saltwater aquatic life—0.025 µg/ℓ as a 24-hr average, never to exceed 2.8 µg/ℓ. To protect human health—0.2 µg/ℓ.

Determination in Water: Flameless atomic absorption.

Routes of Entry: Inhalation of dust, percutaneous absorption.

Harmful Effects and Symptoms: *Local* — Alkyl mercury compounds are primary skin irritants and may cause dermatitis. When deposited on the skin, they give no warning, and if contact is maintained, can cause second-degree burns. Sensitization may occur.

Systemic — The central nervous system, including the brain, is the principal target tissue for this group of toxic compounds. Severe poisoning may produce irreversible brain damage resulting in loss of higher functions.

The effects of chronic poisoning with alkyl mercury compounds are progressive. In the early stages, there are fine tremors of the hands, and in some cases, of the face and arms. With continued exposure, tremors may become coarse and convulsive; scanning speech with moderate slurring and difficulty in pronunciation may also occur. The worker may then develop an unsteady gait of a spastic nature which can progress to severe ataxia of the arms and legs. Sensory disturbances including tunnel vision, blindness, and deafness are also common.

A later symptom, constriction of the visual fields, is rarely reversible and may be associated with loss of understanding and reason which makes the victim completely out of touch with his environment. Severe cerebral effects have been seen in infants born to mothers who had eaten large amounts of methylmercury-contaminated fish.

Points of Attack: Central nervous system, kidney, eyes, skin.

Medical Surveillance: Preplacement and periodic physical examinations should be concerned particularly with the skin, vision, central nervous system, and kidneys. Consideration should be given to the possible effects on the fetus of alkyl mercury exposure in the mother. Constriction of visual fields may be a useful diagnostic sign. (See Mercury—Inorganic.)

Blood and urine levels of mercury have been studied, especially in the case of methylmercury. A precise correlation has not been found between exposure levels and concentrations. They may be of some value in indicating that exposure has occurred, however.

First Aid: If this chemical gets into the eyes, irrigate immediately. If this chemical contacts the skin, wash with soap immediately. If a person breathes in large amounts of this chemical, move the exposed person to fresh air at once and perform artificial respiration. When this chemical has been swallowed, get medical attention. Give large quantities of water and induce vomiting. Do not make an unconscious person vomit.

Personal Protective Methods: Wear appropriate clothing to prevent any possibility of skin contact. Wear eye protection to prevent any possibility of eye contact. Employees should wash immediately when skin is wet or contaminated. Work clothing should be changed daily if it is possible that clothing is contaminated. Remove nonimpervious clothing immediately if wet or contaminated. Provide emergency showers and eyewash.

Respirator Selection:
0.5 mg/m^3: SAF/SCBAF
10 mg/m^3: SA:PD, PP, CF
Escape : GMSP/SCBA

References

(1) Nat. Inst. for Occup. Safety and Health, *Information Profiles on Potential Occupational Hazards: Organomercurials,* pp 287-296, Report PB-276 678, Rockville, MD (Oct. 1977).
(2) U.S. Environmental Protection Agency, *Mercury: Ambient Water Quality Criteria,* Wash., DC (1979).
(3) See Reference (A-61) [Organo (Alkyl) Mercury].
(4) Parmeggiani, L., Ed., *Encyclopedia of Occupational Health & Safety,* Third Edition, Vol. 2, pp 1336–38, Geneva, International Labour Office (1983).

MERCURY—ELEMENTAL

- Hazardous waste (EPA)

Description: Hg is a silvery, mobile, odorless liquid. It boils at 356°-357°C.

Code Numbers: CAS 7439-97-6 RTECS OV4550000 UN 2809

DOT Designation: ORM-B

Synonyms: Quicksilver

Potential Exposure: Mercury is used as a liquid cathode in cells for the electrolytic production of caustic and chlorine. It is used in electrical apparatus (lamps, rectifiers and batteries) and in control instruments (switches, thermometers and barometers). NIOSH estimates annual worker exposure to mercury at 150,000.

Permissible Exposure Limits in Air: The Federal standard is 0.1 mg/m³ as a ceiling value. NIOSH has recommended a value of 0.05 mg/m³. ACGIH as of 1983/84 has proposed a TWA value of 0.05 mg/m³ with the notation "skin" indicating the possibility of cutaneous absorption for mercury vapor but no STEL value. The IDLH level is 28 mg/m³.

Determination in Air: Adsorption on silvered Chromosorb P, thermal desorption, analysis by flameless atomic absorption spectrometry (A-10).

Permissible Concentration in Water: No criteria set. See Mercury—Inorganic.

Routes of Entry: Inhalation, skin absorption, eye and skin contact.

Harmful Effects and Symptoms: Coughing, chest pains, dyspnea, bronchitis, pneumonia; tremors, insomnia, irritability, indecision; headaches, fatigue, weakness; stomatitis; salivation; gastrointestinal disturbance, anorexia, weight loss, proteinuria; irritation of eyes and skin.

Points of Attack: Skin, respiratory system, central nervous system, kidneys, eyes.

Medical Surveillance: Consider the points of attack in preplacement and periodic physical examinations.

First Aid: If this chemical gets into the eyes, irrigate immediately. If this chemical contacts the skin, wash with soap promptly. If a person breathes in large amounts of this chemical, move the exposed person to fresh air at once and perform artificial respiration. When this chemical has been swallowed, get medical attention. Give large quantities of water and induce vomiting. Do not make an unconscious person vomit.

Personal Protective Methods: Wear appropriate clothing to prevent repeated or prolonged skin contact. Employees should wash promptly when skin is wet or contaminated. Work clothing should be changed daily if it is possible that clothing is contaminated. Remove nonimpervious clothing promptly if wet or contaminated.

Respirator Selection:
> 1 mg/m³: SA/SCBA
> 5 mg/m³: SAF/SCBAF
> 28 mg/m³: SA:PD,PP,CF
> Escape: GMS/SCBA

Disposal Method Suggested: Accumulate for sale or for purification and reuse (A-38). See also (A-57).

References

(1) U.S. Environmental Protection Agency, *Toxicology of Metals, Vol. II: Mercury,* pp. 301-344, Report EPA-600/1-77-022, Research Triangle Park, NC (May 1977).

(2) World Health Organization, *Mercury,* Environmental Health Criteria No. 1, Geneva, Switzerland (1976).

(3) Sax, N.I., Ed., *Dangerous Properties of Industrial Materials Report, 1,* No. 3, 70-72, New York, Van Nostrand Reinhold Co. (1981).

(4) See Reference (A-61).

(5) See Reference (A-60).

(6) Parmeggiani, L., Ed., *Encyclopedia of Occupational Health & Safety,* Third Edition, Vol. 2, pp 1332–35, Geneva, International Labour Office (1983).

(7) United Nations Environment Programme, *IRPTC Legal File 1983,* Vol. II, pp VII/424–31, Geneva, Switzerland, International Register of Potentially Toxic Chemicals (1984).

MERCURY—INORGANIC

- Hazardous substances (several salts, EPA)
  Mercury salts which are classified by EPA as hazardous substances include: mercuric cyanide, mercuric nitrate, mercuric sulfate, mercuric thiocyanate, and mercurous nitrate.
- Hazardous waste (EPA)
- Priority toxic constituents pollutant (EPA)

Description: The more commonly found mercuric salts (with their solubilities in water) are $HgCl_2$ (1 g/13.5 ml water), $Hg(NO_3)_2$ (soluble in a small amount of water), and $Hg(CH_3COO)_2$ (1 g/2.5 ml water). Mercurous salts are much less soluble in water. $HgNO_3$ will solubilize only in 13 parts water containing 1% HNO_3. Hg_2Cl_2 is practically insoluble in water. Because of this mercurous salts are much less toxic than the mercuric forms.

Code Numbers:
Mercuric chloride: CAS 7487-94-7   RTECS OV9100000   UN 1624

DOT Designation: *Mercuric Chloride* — Poison B

Synonyms: *Mercuric Chloride* — bichloride of mercury; corrosive mercury chloride; corrosive sublimate; and mercury perchloride.

Potential Exposure: Inorganic mercury is utilized in gold, silver, bronze, and tin-plating, tanning and dyeing, feltmaking, taxidermy, textile manufacture, photography and photoengraving, in extracting gold and silver from ores, in paints and pigments, in the preparation of drugs and disinfectants in the pharmaceutical industry, and as a chemical reagent.

Permissible Exposure Limits in Air: The Federal standard for mercury is 0.1 mg/m^3 as a ceiling value. The NIOSH recommended standard is 0.05 mg/m^3 as a TWA. ACGIH (1983/84) concurs in this value.

Permissible Concentration in Water: To protect freshwater aquatic life—0.00057 $\mu g/\ell$ as a 24-hr average, never to exceed 0.0017 $\mu g/\ell$. To protect saltwater aquatic life—0.025 $\mu g/\ell$ as a 24-hr average, never to exceed 3.7 $\mu g/\ell$. To protect human health—0.144 $\mu g/\ell$ (USEPA) (A-53) set in 1979-80.

An earlier EPA limit on mercury in drinking water was 2.0 $\mu g/\ell$ as set in the "Red Book" in 1976.

In contrast, the World Health Organization has set 1.0 $\mu g/\ell$ as a drinking water standard (A-65) and the Federal Republic of Germany has set 4.0 $\mu g/\ell$ as of 1975.

Determination in Water: Total mercury is determined by flameless atomic absorption. Soluble mercury may be determined by 0.45 micron filtration followed by flameless atomic absorption.

Routes of Entry: Inhalation, ingestion, eye and skin contact.

Harmful Effects and Symptoms: *Local* — Mercury is a primary irritant of skin and mucous membranes. It may occasionally be a skin sensitizer.

Systemic — Acute poisoning due to mercury vapors affects the lungs primarily, in the form of acute interstitial pneumonitis, bronchitis, and bronchiolitis.

Exposure to lower levels over prolonged periods produces symptom complexes that can vary widely from individual to individual. These may include weakness, loss of appetite, loss of weight, insomnia, indigestion, diarrhea, metallic taste in the mouth, increased salivation, soreness of mouth or throat, inflammation of gums, black line on the gums, loosening of teeth, irritability, loss of memory, and tremors of fingers, eyelids, lips, or tongue. More extensive exposures, either by daily exposures or one-time, can produce extreme irritability, excitability, anxiety, delirium with hallucinations, melancholia, or manic depressive psychosis. In general, chronic exposure produces four classical signs: gingivitis, sialorrhea, increased irritability, and muscular tremors. Rarely are all four seen together in an individual case.

Either acute or chronic exposure may produce permanent changes to affected organs and organ systems.

Medical Surveillance: Preemployment and periodic examinations should be concerned especially with the skin, respiratory tract, central nervous system, and kidneys. The urine should be examined and urinary mercury levels determined periodically. Signs of weight loss, gingivitis, tremors, personality changes, and insomnia would be suggestions of possible mercury intoxicaton.

Urine mercury determination may be helpful as an index of amount of absorption. Opinions vary as to the significance of a given level. Generally, 0.1 to 0.5 mg Hg/ℓ of urine is considered significant. See also reference (6).

First Aid: If swallowed, wash stomach with 5% formaldehyde sodium sulfoxylate solution, then 2% sodium carbonate solution. Administer dimercaprol. Watch electrolyte balance carefully (A-38).

Personal Protective Methods: In areas where the exposures are excessive, respiratory protection shall be provided either by full-face canister-type mask or supplied air respirator, depending on the concentration of mercury fumes. Above 50 mg Hg/m^3 requires supplied-air positive-pressure full-face respirators. Full-body work clothes including shoes or shoe covers and hats should be supplied, and clean work clothes should be supplied daily. Showers should be available and all employees encouraged to shower prior to change to street clothes. Work clothes should not be stored with street clothes in the same locker. Food should not be eaten in the work area.

Respirator Selection: See Reference (1).

Disposal Method Suggested: Mercuric Chloride—incineration followed by recovery/removal of mercury from the gas stream. Mercuric Nitrate—same as mercuric chloride. Mercuric Sulfate—same as mercuric chloride (A-31).

Alternatively, mercury compounds may be recovered from brines, sludges and spent catalysts (A-57).

References

(1) National Institute for Occupational Safety and Health, *Criteria for a Recommended Standard: Occupational Exposure to Inorganic Mercury,* NIOSH Doc. No. 73-11024 (1973).

(2) U.S. Environmental Protection Agency, *Toxicology of Metals, Vol. II: Mercury,* Report EPA-600/1-77-022, Research Triangle Park, NC, pp 301-344, (May 1977).

(3) U.S. Environmental Protection Agency, *Mercury: Ambient Water Quality Criteria,* Wash., DC (1980).

(4) U.S. Environmental Protection Agency, *Status Assessment of Toxic Chemicals: Mercury,* Report EPA-600/2-79-210i, Cincinnati, OH (Dec. 1979).

(5) U.S. Environmental Protection Agency, *Mercury,* Health and Environmental Effects Profile No. 124, Wash., DC, Office of Solid Waste (April 30, 1980).

(6) Nat. Inst. for Occup. Safety and Health, *A Guide to the work Relatedness of Disease,* Revised Edition, DHEW (NIOSH) Publ. No. 79-116, Cincinnati, OH, pp 117-133, (Jan. 1979).

(7) Sax, N.I., Ed., *Dangerous Properties of Industrial Materials Report, 1,* No. 3, 70, New York, Van Nostrand Reinhold Co. (1981). (Mercuric Acetate).

(8) See Reference (A-60) for citations under: Mercuric Acetate, Mercuric Bromide, Mercuric Chloride, Mercuric Nitrate, Mercuric Oxide, Mercuric Sulfate and Mercuric Thiocyanate.

(9) United Nations Environment Programme, *IRPTC Legal File 1983,* Vol. II, pp VII/434-6, Geneva, Switzerland, International Register of Potentially Toxic Chemicals (1984).

MESITYL OXIDE

Description: $CH_3COCH{=}C(CH_3)_2$ is a clear, pale yellow, or colorless liquid with a strong peppermint odor. It boils at 130°C.

Code Numbers: CAS 141-79-7   RTECS SB4200000   UN 1229

DOT Designation: Flammable liquid

Synonyms: 4-Methyl-3-pentene-2-one; Isobutenyl methyl ketone; Methyl isobutenyl ketone; Isopropylidene acetone.

Potential Exposure: Mesityl oxide is used as a solvent for cellulose esters and ethers and other resins in lacquers and inks. It is used in paint and varnish removers and as an insect repellent.

NIOSH has estimated that fewer than 500 workers are exposed annually.

Incompatibilities: Oxidizers.

Permissible Exposure Limits in Air: The Federal standard is 25 ppm (100 mg/m^3). The ACGIH as of 1983/84 has adopted a TWA of 15 ppm (60 mg/m^3) and an STEL of 25 ppm (100 mg/m^3). The IDLH level is 500 ppm.

Determination in Air: Charcoal adsorption, workup with CS_2, analysis by gas chromatography. See NIOSH Methods, Set B. See also reference (A-10).

Permissible Concentration in Water: No criteria set.

Routes of Entry: Inhalation, ingestion, eye and skin contact.

Harmful Effects and Symptoms: Irritation of eyes, skin and mucous membranes; narcosis, coma.

Points of Attack: Eyes, skin, respiratory system, central nervous system.

Medical Surveillance: Consider the points of attack in preplacement and periodic physical examinations.

First Aid: If this chemical gets into the eyes, irrigate immediately. If this chemical contacts the skin, flush with water immediately. If a person breathes in large amounts of this chemical, move the exposed person to fresh air at once and perform artificial respiration. When this chemical has been swallowed, get medical attention. Give large quantities of salt water and induce vomiting. Do not make an unconscious person vomit.

Personal Protective Methods: Wear appropriate clothing to prevent any reasonable probability of skin contact. Wear eye protection to prevent any reasonable probability of eye contact. Employees should wash immediately when skin is wet or contaminated. Remove clothing immediately if wet or contaminated to avoid flammability hazard. Provide emergency showers.

Respirator Selection:
> 1,000 ppm: CCROVF/GMOV/SAF/SCBAF
> 5,000 ppm: SAF:PD, PP, CF
> Escape: GMOV/SCBA

Disposal Method Suggested: Incineration.

References

(1) Nat. Inst. for Occup. Safety and Health, *Criteria for a Recommended Standard: Occupational Exposure to Ketones,* NIOSH Doc. No. 78-173, Wash., DC (1978).
(2) See Reference (A-60).

META-SYSTOX

See "Demeton-Methyl."

METHACRYLIC ACID

Description: $CH_2{=}C(CH_3)COOH$ is a colorless liquid, melting at $15°{-}16°C$.

Code Numbers: CAS 79-41-4   RTECS OZ2975000   UN 2531

DOT Designation: Corrosive material.

Synonyms: 2-Methylpropenoic acid.

Potential Exposure: Those involved in the production of the material or its alkyl esters as monomers or comonomers for synthetic resins for the production of plastic sheets, moldings and fibers.

Permissible Exposure Limits in Air: There is no Federal standard but ACGIH as of 1983/84 has proposed a TWA of 20 ppm ($70 mg/m^3$). There is no STEL value.

Permissible Concentration in Water: No criteria set.

Harmful Effects and Symptoms: Direct contact of liquid methacrylic acid with eyes can result in blindness; skin contact can cause skin corrosion. Medical reports of acute exposures have revealed no respiratory symptoms—only skin responses and a severe corneal burn. It is definitely less irritating than acrylic acid (A-34).

Personal Protective Methods: Wear rubber gloves, face shield and overalls (A-38).

Respirator Selection: Wear all-purpose gas mask.

Disposal Method Suggested: Incineration.

References

(1)   See Reference (A-60).

METHACRYLONITRILE

● Hazardous waste (EPA)

Description: $CH_2=CH(CH_3)CN$ is a colorless liquid boiling at 90°C.

Code Numbers: CAS 126-98-7   RTECS UD1400000

DOT Designation: —

Synonyms: 2-cyano-1-propene; isopropenylnitrile; 2-methylpropenenitrile; isopropene cyanide and methylacrylonitrile.

Potential Exposure: This material is used as a monomer in the preparation of polymeric coatings and elastomers.

Permissible Exposure Limits in Air: There are no Federal standards but ACGIH (1983/84) has set a TWA of 1 ppm (3 mg/m³) with the notation "skin" indicating the possibility of cutaneous absorption. The STEL value is 2 ppm (6 mg/m³).

Permissible Concentration in Water: No criteria set.

Harmful Effects and Symptoms: Methacrylonitrile is comparable to acrylonitrile in acute toxicity. It is only mildly irritating to the eyes and skin. In a 90-day study of beagles, there was no effect at 3.2 ppm; at 8.8 ppm there was a transitory elevation of serum transaminase levels; at 13.5 ppm, central nervous system effects were observed after 45 days manifested by convulsions and loss of motor control in the hind legs, diarrhea having been noted on the fifth day (A-34).

First Aid: Irrigate eyes with water; wash contaminated areas of body with soap and water; if inhaled, apply artificial respiration and oxygen. Use amyl nitrite capsules (A-38).

Personal Protective Methods: Wear rubber gloves and overalls.

Disposal Method Suggested: Add alcoholic NaOH, then sodium hypochlorite. After reaction, flush to sewer with water (A-38).

METHOMYL

● Hazardous Waste (EPA)

Description: $C_5H_{10}N_2O_2S$, Methyl-N-{[(methylamino)carbonyl] oxy}eth-animidothioate, with the structural formula

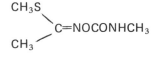

is a white crystalline solid with a slight sulfurous odor which melts at 78°-79°C.

Code Numbers: CAS 16752-77-5 RTECS AK2975000 UN (NA 2757)

DOT Designation: —

Synonyms: Lannate®

Potential Exposure: Those involved in the manufacture (A-32), formulation or application of this broad-spectrum insecticide which is used on many vegetables, field crops, certain fruit crops, and ornamentals.

Permissible Exposure Limits in Air: There is no Federal limit but ACGIH (1983/84) has set a TWA of 2.5 mg/m^3 with the notation "skin" indicating the possibility of cutaneous absorption. There is no tentative STEL value.

Permissible Concentration in Water: No criteria set.

Harmful Effects and Symptoms: Methomyl is readily absorbed through inhalation or dermal exposure and is almost completely eliminated from the body within 24 hours. Chronic toxicity studies in rats and dogs show that no effects occur below 100 ppm. Methomyl inhibits the activity of cholinesterase in the body. Studies have shown that methomyl is not carcinogenic in rats and dogs or mutagenic in the Ames bioassay. However, a different type of bioassay showed mutagenic activity at a methomyl concentration of 50 ppm. A potential product of the reaction of methomyl with certain nitrogen compounds in the environment or in mammalian systems is nitrosomethomyl, which is a potent mutagen, carcinogen, and teratogen.

References

(1) U.S. Environmental Protection Agency, *Methomyl,* Health and Environmental Effects Profile No. 125, Wash., DC, Office of Solid Waste (April 30, 1980).
(2) Sax, N.I., Ed., *Dangerous Properties of Industrial Materials Report, 2,* No. 5, 79-81, New York, Van Nostrand Reinhold Co. (1982).
(3) United Nations Environment Programme, *IRPTC Legal File 1983,* Vol. I, pp VII/19-20, Geneva, Switzerland, International Register of Potentially Toxic Chemicals (1984).

METHOXYCHLOR

- Hazardous substance (EPA)

Description: $H_3COC_6H_4CH(CCl_3)C_6H_4OCH_3$ is a colorless to tan solid with a slight fruity odor. It melts at 89°C.

Code Numbers: CAS 72-43-5 RTECS KJ3675000 UN (NA 2761)

DOT Designation: ORM-E

Synonyms: 2,2-bis(p-Methoxyphenyl)-1,1,1-trichloroethane, ENT 1716, DMTD, Marlate®.

Potential Exposure: Methoxychlor was introduced as an insecticide in 1945. It is a close relative of DDT and has been used as an insecticide of very low mammalian toxicity for home and garden, on domestic animals for fly control, for elm bark-beetle vectors of Dutch elm disease, and for blackfly larvae in streams. Methoxychlor is registered for about 87 crops—alfalfa; nearly all fruits and vegetables; corn, wheat, rice, and other grains; beef and dairy cattle; and swine, goats, and sheep—and for agricultural premises and outdoor fogging.

Thus, those engaged in manufacture, formulation and application of the material as well as people in application areas may be exposed.

Incompatibilities: Strong oxidizers.

Permissible Exposure Limits in Air: The Federal standard is 15 mg/m^3. The ACGIH as of 1983/84 has set a TWA level of 10 mg/m^3 but no STEL value. The IDLH level is 7,500 mg/m^3.

Determination in Air: Filter collection, isooctane extraction, analysis by gas chromatography using electrolytic conductivity detection (A-10).

Permissible Concentration in Water: The EPA (A-3) has recommended the following criteria:

100 µg/ℓ for domestic water supply (health).
0.03 µg/ℓ for freshwater and marine aquatic life.

A no-adverse effect level in drinking water has been calculated by NAS/NRC (A-2) to be 0.7 mg/ℓ.

Routes of Entry: Inhalation, ingestion.

Harmful Effects and Symptoms: None known in humans; in animals: trembling, convulsions, kidney, liver damage. NCI has conducted a carcinogenic bioassay and found methoxychlor to be noncarcinogenic.

Points of Attack: None known

Medical Surveillance: No special considerations indicated.

First Aid: If this chemical contacts the skin, wash with soap. If a person breathes in large amounts of this chemical, move the exposed person to fresh air at once. When this chemical has been swallowed, get medical attention. Give large quantities of water and induce vomiting. Do not make an unconscious person vomit.

Personal Protective Methods: Wear appropriate clothing to prevent repeated or prolonged skin contact. Employees should wash promptly when skin is wet or contaminated. Remove nonimpervious clothing promptly if wet or contaminated.

Respirator Selection:
150 mg/m^3: CCROVDMFuPest/SA/SCBA
750 mg/m^3: CCROVFDMFuPest/GMOVDMFuPest/SAF/SCBAF
7,500 mg/m^3: PAPOVHiEPPest/SA:PD, PP, CF

Disposal Method Suggested: $(CH_3OC_6H_4)_2CHCCl_3$ is like DDT in that it is resistant to heat and oxidation. By some accounts it is less readily dehydrochlorinated by alkali than is DDT, but other data indicate little difference. The dehydrochlorination is catalyzed by heavy metals. Methoxychlor is dechlorinated by refluxing with sodium in isopropyl alcohol. It is described as resistant to ultraviolet light, but other studies have shown that it breaks down rapidly under UV in hexane solution. Incineration is recommended by the MCA as the disposal method for methoxychlor (A-32).

References

(1) See Reference (A-61).
(2) United Nations Environment Programme, *IRPTC Legal File 1983,* Vol. I, pp VII/332–4, Geneva, Switzerland, International Register of Potentially Toxic Chemicals (1984).

2-METHOXYETHANOL

See "Ethylene Glycol Monomethyl Ether and Acetate."

4-METHOXYPHENOL

Description: $HOC_6H_4OCH_3$ is a white to tan flaky cyrstalline substance melting at $52°-53°C$.

Code Numbers: CAS 150-76-5 RTECS SL7700000

DOT Designation: –

Synonyms: Hydroquinone monomethyl ether and p-Hydroxyanisole.

Potential Exposure: This compound is used in the manufacture of anti-oxidants, pharmaceuticals, plasticizers and dyestuffs. It is used as a stabilizer and UV inhibitor in various polymers.

Permissible Exposure Limits in Air: There is no Federal standard but ACGIH as of 1983/84 has proposed a TWA of 5 mg/m^3. There is no STEL value.

Permissible Concentration in Water: No criteria set.

Harmful Effects and Symptoms: This material has an oral toxicity (LD$_{50}$) for rats of 1,600 mg/kg which puts it in the slightly toxic category.

METHYL ACETATE

Description: CH_3COOCH_3 is a colorless liquid with a fruity odor. It boils at 57°F.

Code Numbers: CAS 79-20-9 RTECS AI9100000 UN 1231

DOT Designation: Flammable liquid.

Synonyms: Acetic acid, methyl ester; methyl acetic ester; methyl ethanoate.

Potential Exposure: Methyl acetate is used as a solvent in lacquers and paint removers and as an intermediate in pharmaceutical manufacture (A-41).

Incompatibilities: Nitrates, strong oxidizers, strong alkalies, strong acids.

Permissible Exposure Limits in Air: The Federal standard and the 1983/84 ACGIH TWA value is 200 ppm (610 mg/m^3). The STEL value is 250 ppm (760 mg/m^3). The IDLH level is 10,000 ppm.

Determination in Air: Adsorption on charcoal, workup with CS$_2$, analysis by gas chromatography. See NIOSH Methods, Set D. See also reference (A-10).

Permissible Concentration in Water: No criteria set.

Routes of Entry: Inhalation, ingestion, skin and eye contact.

Harmful Effects and Symptoms: Irritation of nose and throat; headaches, drowsiness; optic atrophy.

Points of Attack: Respiratory system, skin, eyes.

Medical Surveillance: Consider the points of attack in preplacement and periodic physical examinations.

First Aid: If this chemical gets into the eyes, irrigate immediately. If this chemical contacts the skin, flush with water promptly. If a person breathes in large amounts of this chemical, move the exposed person to fresh air at once and perform artificial respiration. When this chemical has been swallowed, get medical attention. Give large quantities of salt water and induce vomiting. Do not make an unconscious person vomit.

Personal Protective Methods: Wear appropriate clothing to prevent repeated or prolonged skin contact. Wear eye protection to prevent any reasonable probability of eye contact. Employees should wash promptly when skin is wet or contaminated. Remove clothing immediately if wet or contaminated to avoid flammability hazard.

Respirator Selection:

     1,000 ppm: CCROV/SA/SCBA
     5,000 ppm: GMOVc
  10,000 ppm: GMOVfb/SAF/SCBAF/SAF:PD,PP,CF
     Escape: GMOV/SCBA

Disposal Method Suggested: Incineration.

References

(1) See Reference (A-61).
(2) See Reference (A-60).

METHYL ACETYLENE

Description: $CH_3C{\equiv}CH$ is a colorless gas with a sweet odor.

Code Numbers: CAS 74-99-7   RTECS UK4250000   UN 1004

DOT Designation: −

Synonyms: Propyne, allylene

Potential Exposure: This material may be used as a liquid rocket propellant, in admixture with propadiene as an industrial cutting fuel.

Incompatibilities: Strong oxidizers, chlorine, copper, copper alloys.

Permissible Exposure Limits in Air: The Federal standard and the 1983/84 ACGIH TWA value is 1,000 ppm (1,650 mg/m^3). The STEL value is 1,250 ppm (2,040 mg/m^3). The IDLH level is 11,000 ppm.

Determination in Air: Collection by charcoal tube, analysis by gas liquid chromatography (A-13).

Permissible Concentration in Water: No criteria set.

Routes of Entry: Inhalation.

Harmful Effects and Symptoms: Drowsiness, unconsciousness.

Points of Attack: Central nervous system.

Medical Surveillance: Consider the points of attack in preplacement and periodic physical examinations.

First Aid: If a person breathes in large amounts of this chemical, move the exposed person to fresh air at once and perform artificial respiration.

Personal Protective Methods: Wear appropriate clothing to prevent skin freezing. Wear eye protection to prevent any reasonable probability of eye contact. Remove clothing immediately if wet or conaminated to avoid flammability hazard.

Respirator Selection:
> 10,000 ppm: SA/SCBA
> 11,000 ppm: SAF/SCBAF
> Escape: GMOV/SCBA

Disposal Method Suggested: Incineration.

References

(1) See Reference (A-61).
(2) See Reference (A-60).

METHYL ACETYLENE/PROPADIENE MIXTURE

Description: This mixture of C_3H_4 isomers, $CH_3C{\equiv}CH$ and $H_2C{=}C{=}CH_2$, is a colorless gas with a strong, characteristic, foul odor.

Code Numbers: RTECS UK4920000   UN 1060

DOT Designation: Flammable gas.

Synonyms: MAPP gas; Methyl acetylene-allene mixture; Propyne-allene mixture; propyne:propadiene mixture.

Potential Exposure: This mixture is used as an industrial cutting fuel.

Incompatibilities: Strong oxidizers, copper alloys ($>$67% Cu).

Permissible Exposure Limits in Air: The Federal standard and the ACGIH 1983/84 TWA value is 1,000 ppm (1,800 mg/m^3). The STEL value is 1,250 ppm (2,250 mg/m^3). The IDLH level is 20,000 ppm.

Determination in Air: Collection by charcoal tube, analysis by gas liquid chromatography (A-13).

Permissible Concentration in Water: No criteria set.

Routes of Entry: Inhalation, skin and eye contact.

Harmful Effects and Symptoms: Excitement, disorientation; drowsiness, unconsciousness; frostbite.

Points of Attack: Central nervous system, skin, eyes.

Medical Surveillance: Consider the points of attack in preplacement and periodic physical examinations.

First Aid: If this chemical gets into the eyes, irrigate immediately. If this chemical contacts the skin, flush with water immediately. If a person breathes in large amounts of this chemical, move the exposed person to fresh air at once and perform artificial respiration.

Personal Protective Methods: Wear appropriate clothing to prevent skin freezing. Wear eye protection to prevent any reasonable probability of eye contact. Remove clothing immediately if wet or contaminated to avoid flammability hazard.

Respirator Selection:
> 5,000 ppm: GMSc
> 10,000 ppm: SA/SCBA
> 20,000 ppm: GMSfb/SAF/SA:PD, PP, CF/SCBAF
>   Escape: GMS/SCBA

Disposal Method Suggested: Incineration.

References

(1) See Reference (A-61).

METHYL ACRYLATE

Description: $CH_2=CHCOOCH_3$ is a clear, colorless liquid with a sharp, fruity odor which boils at $80°C$.

Code Numbers: CAS 96-33-3   RTECS AT2800000   UN 1919

DOT Designation: Flammable liquid.

Synonyms: Methyl propenoate

Potential Exposure: Methyl acrylate is used as a monomer in the manufacture of polymers for plastic films and textile, paper and leather coating resins. It is also used as a pesticide intermediate (A-32) and in pharmaceutical manufacture (A-41).

Incompatibilities: Nitrates, oxidizers, peroxides, polymerizers, strong alkalies, moisture.

Permissible Exposure Limits in Air: The Federal standard and the 1983/84 ACGIH TWA value is 10 ppm (35 mg/m^3). There is no STEL value set. The IDLH level is 1,000 ppm.

Determination in Air: Charcoal adsorption, workup with CS_2, analysis by gas chromatography. See NIOSH Methods, Set D. See also reference (A-10).

Permissible Concentration in Water: No criteria set.

Routes of Entry: Inhalation, eye and skin contact.

Harmful Effects and Symptoms: Irriration of skin, eyes and upper respiratory system.

Points of Attack: Respiratory system, eyes, skin.

Medical Surveillance: Consider the points of attack in preplacement and periodic physical examinations.

First Aid: If this chemical gets into the eyes, irrigate immediately. If this chemical contacts the skin, flush with water immediately. If a person breathes in large amounts of this chemical, move the exposed person to fresh air at once and perform artificial respiration. When this chemical has been swallowed, get medical attention. Give large quantities of salt water and induce vomiting. Do not make an unconscious person vomit.

Personal Protective Methods: Wear appropriate clothing to prevent any possibility of skin contact. Wear eye protection to prevent any reasonable probability of eye contact. Employees should wash immediately when skin is wet or contaminated. Remove clothing immediately if wet or contaminated to avoid flammability hazard. Provide emergency showers.

Respirator Selection:
> 75 ppm: CCROV/SA/SCBA
> 500 ppm: CCROVF/GMOV/SAF/SCBAF
> 1,000 ppm: SAF:PD, PP, CF
> Escape: GMOVF/SCBA

Disposal Method Suggested: Incineration.

References

(1) See Reference (A-61).
(2) See Reference (A-60).

METHYLAL

Description: $CH_3OCH_2OCH_3$ is a colorless liquid with a pungent odor which boils at 44°C.

Code Numbers: CAS 109-87-5 RTECS PA8750000 UN 1234

DOT Designation: Flammable liquid.

Synonyms: Dimethyoxymethane; Methyl formal; Formal; Dimethylacetal formaldehyde.

Potential Exposure: Methylal is used as a specialty fuel and as a solvent in adhesives and protective coatings.

Incompatibilities: Strong oxidizers, acids.

Permissible Exposure Limits in Air: The Federal standard and the 1983/84 ACGIH TWA value is 1,000 ppm ($3,100$ mg/m^3). The STEL value is 1,250 ppm ($3,875$ mg/m^3). The IDLH level is 10,000 ppm.

Determination in Air: Charcoal adsorption, workup in hexane, analysis by gas chromatography. See NIOSH Methods, Set F. See also reference (A-10).

Permissible Concentration in Water: No criteria set.

Routes of Entry: Inhalation, ingestion, skin and eye contact.

Harmful Effects and Symptoms: Mild irritation of the eyes and upper respiratory system; skin irritation; anaesthesia.

Points of Attack: Skin, respiratory system, central nervous system.

Medical Surveillance: Consider the points of attack in preplacement and periodic physical examinations.

First Aid: If this chemical gets into the eyes, irrigate immediately. If this chemical contacts the skin, flush with water promptly. If a person breathes in large amounts of this chemical, move the exposed person to fresh air at once and perform artificial respiration. When this chemical has been swallowed, get medical attention. Give large quantities of salt water and induce vomiting. Do not make an unconscious person vomit.

Personal Protective Methods: Wear appropriate clothing to prevent repeated or prolonged skin contact. Wear eye protection to prevent any reasonable probability of eye contact. Employees should wash promptly when skin is wet or contaminated. Remove clothing immediately if wet or contaminated to avoid flammability hazard.

Respirator Selection:
    10,000 ppm: SA/SCBA
        Escape: GMOV/SCBA

Disposal Method Suggested: Concentrated waste containing no peroxides: discharges liquid at a controlled rate near a pilot flame. Concentrated waste containing peroxides: perforation of a container of the waste from a safe distance followed by open burning.

References
(1)   See Reference (A-61).

METHYL ALCOHOL

● Hazardous waste (EPA)

Description: CH₃OH, methyl alcohol, is a colorless, volatile liquid with a mild odor. It boils at 64°-65°C.

Code Numbers: CAS 67-56-1   RTECS PC1400000   UN 1230

DOT Designation: Flammable liquid

Synonyms: Methanol, carbinol, wood alcohol, wood spirit.

Potential Exposure: Methyl alcohol is used as a starting material in organic synthesis of chemicals such as formaldehyde, methacrylates, methyl amines, methyl halides, ethylene glycol and pesticides (A-32) and as an industrial solvent for inks, resins, adhesives, and dyes for straw hats. It is an ingredient in paint and varnish removers, cleaning and dewaxing preparations, spirit duplicating fluids, embalming fluids, antifreeze mixtures, and enamels and is used in the manufacture of photographic film, plastics, celluloid, textile soaps, wood stains, coated fabrics, shatterproof glass, paper coating, waterproofing formulations, artificial leather, and synthetic indigo and other dyes.

It has also found use as an extractant in many processes, an antidetonant fuel-injection fluid for aircraft, a rubber accelerator, and a denaturant for ethyl alcohol.

NIOSH estimates that approximately 175,000 workers in the United States are potentially exposed to methyl alcohol.

Incompatibilities: Strong oxidizers.

Permissible Exposure Limits in Air: The Federal standard and the 1983/84 ACGIH TWA value is 200 ppm (260 mg/m³). NIOSH recommends adherence to the present Federal standard of 200 ppm, 262 mg of methyl alcohol per cubic meter of air as a time-weighted average for up to a 10-hour workday, 40 hour workweek. In addition NIOSH recommends a ceiling of 800 ppm, 1,048 mg of methyl alcohol per cubic meter of air as determined by a sampling time of 15 minutes. The notation "skin" indicates the possibility of cutaneous absorption. Methanol's international hygienic standards are as follows: West Germany and Sweden, 260 mg/m³ (8-hr TWA); East Germany and Czechoslovakia, 100 mg/m³ (8-hr TWA); U.S.S.R., 5 mg/m³ (acceptable ceiling concentration).

The STEL is 250 ppm (310 mg/m³). The IDLH level is 25,000 ppm.

Determination in Air: Adsorption on silica, workup with water, analysis by gas chromatography. See NIOSH Methods, Set E. See also reference (A-10).

Permissible Concentration in Water: No criteria set, but EPA (A-37) has suggested a permissible ambient goal of 3,600 μg/ℓ based on health effects.

Routes of Entry: Inhalation of vapor, percutaneous absorption of liquid, ingestion, eye and skin contact.

Harmful Effects and Symptoms: *Local* — Contact with liquid can produce defatting and a mild dermatitis. Methyl alcohol is virtually nonirritating to the eyes or upper respiratory tract below 2,000 ppm, and it is difficult to detect by odor at less than this level.

Systemic — Methyl alcohol may cause optic nerve damage and blindness. Its toxic effect is thought to be mediated through metabolic oxidation products, such as formaldehyde or formic acid, and may result in blurring of vision, pain in eyes, loss of central vision, or blindness. Other central nervous system effects result from narcosis and include headache, nausea, giddiness, and loss of consciousness. Formic acid may produce acidosis. These symptoms occur principally after oral ingestion and are very rare after inhalation.

Points of Attack: Eyes, skin, central nervous system, gastrointestinal system.

Medical Surveillance: Consider eye disease and visual acuity in any periodic or placement examinations, as well as skin and liver and kidney functions.

Special tests which may be used include: Determination of methyl alcohol in blood, and methyl alcohol and formic acid in urine; estimation of alkali reserve which may be impaired because of acidosis following accidental ingestion.

First Aid: If this chemical gets into the eyes, irrigate immediately. If this chemical contacts the skin, flush with water promptly. If a person breathes in large amounts of this chemical, move the exposed person to fresh air at once and perform artificial respiration. When this chemical has been swallowed, get medical attention. Give large quantities of salt water and induce vomiting. Do not make an unconscious person vomit.

Personal Protective Methods: Wear appropriate clothing to prevent repeated or prolonged skin contact. Wear eye protection to prevent any reasonable probability of eye contact. Employees should wash promptly when skin is wet. Remove clothing immediately if wet or contaminated to avoid flammability hazard.

Respirator Selection:

> 2,000 ppm: SA/SCBA
> 10,000 ppm: SAF/SCBAF
> 25,000 ppm: SAF:PD, PP, CF
> Escape: SCBA

Disposal Method Suggested: Incineration.

References

(1) National Institute for Occupational Safety and Health, *Criteria for a Recommended Standard: Occupational Exposure to Methyl Alcohol,* NIOSH Doc. No. 76-148, Wash., DC (1976).
(2) U.S. Environmental Protection Agency, *Chemical Hazard Information Profile: Methanol,* Wash., DC (July 11, 1977) (Revised 1979).
(3) U.S. Environmental Protection Agency, *Methyl Alcohol,* Health and Environmental Effects Profile No. 126, Wash., DC, Office of Solid Waste (April 30, 1980).
(4) See Reference (A-61).
(5) Parmeggiani, L., Ed., *Encyclopedia of Occupational Health & Safety,* Third Edition, Vol. 2, pp 1356–58, Geneva, International Labour Office (1983).
(6) United Nations Environment Programme, *IRPTC Legal File 1983,* Vol. II, pp VII/455–58, Geneva, Switzerland, International Register of Potentially Toxic Chemicals (1984).

METHYLAMINE

● Hazardous substance (EPA)

Description: CH_3NH_2 is a colorless gas with an ammonialike odor, at low concentrations fishy odor, liquid under pressure. It boils at $-6°C$.

Code Numbers: CAS 74-89-5   RTECS PF6300000   UN 1061 (anhydrous) UN 1235 (aqueous solution)

DOT Designation: Flammable gas.

Synonyms: Monomethylamine, anhydrous methylamine.

Potential Exposure: Monomethylamine is a starting material for N-oleyltaurine, a surfactant, and p-N-methylaminophenol sulfate, a photographic developer. It has possible uses in solvent extraction systems in the separation of aromatics from aliphatic hydrocarbons. It is also used in the synthesis of many different pharmaceuticals (A-41), pesticides (A-32) and rubber chemicals.

Incompatibilities: Mercury, strong oxidizers.

Permissible Exposure Limits in Air: The Federal standard and the ACGIH 1983/84 TWA value is 10 ppm (12 mg/m^3). There is no STEL value proposed. The IDLH level is 100 ppm.

Determination in Air: Collection by impinger or fritted bubbler, colorimetric analysis (A-18). Alternatively, silica gel adsorption, elution with methanolic acid, derivatization with benzaldehyde and gas chromatographic analysis (A-10).

Permissible Concentration in Water: No criteria set.

Routes of Entry: Inhalation, ingestion, skin absorption, eye and skin contact.

Harmful Effects and Symptoms: Irritation of eyes and respiratory system; coughs; burns; dermatitis; conjunctivitis.

Points of Attack: Respiratory system, eyes, skin.

Medical Surveillance: Consider the points of attack in preplacement and periodic physical examinations.

First Aid: If this chemical gets into the eyes, irrigate immediately. If this chemical contacts the skin, flush with water immediately. If a person breathes in large amounts of this chemical, move the exposed person to fresh air at once and perform artificial respiration. When this chemical has been swallowed, get medical attention. Give large quantities of water and induce vomiting. Do not make an unconscious person vomit.

Personal Protective Methods: Wear appropriate clothing to prevent any possibility of skin contact with liquid methylamine. Wear eye protection to prevent any possibility of eye contact. Employees should wash immediately when skin is wet or contaminated. Remove clothing immediately if wet or contaminated to avoid flammability hazard. Provide emergency showers and eyewash.

Respirator Selection:
>   100 ppm: SAF/SCBAF
>   Escape: GMSF/SCBAF

Disposal Method Suggested: Controlled incineration (incinerator is equipped with a scrubber or thermal unit to reduce NO_x emissions).

References

(1) U.S. Environmental Protection Agency, *Chemical Hazard Information Profile: Methylamines,* Wash., DC (May 1, 1978).
(2) See Reference (A-61).
(3) See Reference (A-60).

METHYL n-AMYL KETONE

Description: $CH_3COC_5H_{11}$ is a clear colorless liquid with a mild, banana-like odor which boils at 151°C.

Code Numbers: CAS 110-43-0   RTECS MJ507500   UN 1110

DOT Designation: Combustible liquid.

Synonyms: n-Amyl methyl ketone; 2-heptanone.

Potential Exposure: Methyl amyl ketone is used as a solvent for nitrocellulose in lacquers and as a relatively inert reaction medium.
NIOSH has estimated annual worker exposure at 67,000.

Incompatibilities: Strong acids, alkalies, oxidizers.

Permissible Exposure Limits in Air: The Federal standard is 100 ppm (465 mg/m^3). The ACGIH has adopted as of 1983/84 a TWA of 50 ppm (235 mg/m^3) and an STEL of 100 ppm (465 mg/m^3). The IDLH level is 4,000 ppm.

Determination in Air: Charcoal adsorption, workup with CS_2, analysis by gas chromatography. See NIOSH Methods, Set B. See also reference (A-10).

Permissible Concentration in Water: No criteria set.

Routes of Entry: Inhalation, ingestion, skin and eye contact.

Harmful Effects and Symptoms: Irritation of eyes and mucous membranes; headaches; narcosis, coma; dermatitis.

Points of Attack: Eyes, skin, respiratory system, central nervous system, peripheral nervous system.

Medical Surveillance: Consider the points of attack in preplacement and periodic physical examinations.

First Aid: If this chemical gets into the eyes, irrigate immediately. If this chemical contacts the skin, wash with soap. If a person breathes in large amounts of this chemical, move the exposed person to fresh air at once. When this chemical has been swallowed, get medical attention. Give large quantities of salt water and induce vomiting. Do not make an unconscious person vomit.

Personal Protective Methods: Wear appropriate clothing to prevent repeated or prolonged skin contact. Wear eye protection to prevent any reasonable probability of eye contact. Employees should wash promptly when skin is wet or contaminated. Remove clothing if wet or contaminated.

Respirator Selection:
          1,000 ppm: CCROV/SA/SCBA
          4,000 ppm: GMOV/SAF/SCBAF
            Escape: GMOV/SCBA

Disposal Method Suggested: Incineration.

References

(1) Nat. Inst. for Occup. Safety and Health, *Criteria for a Recommended Standard: Occupational Exposure to Ketones,* NIOSH DOC. No. 78-173, Wash., DC (1978).

N-METHYLANILINE

Description: $C_6H_5NH(CH_3)$ is a yellow to light brown liquid with a weak, ammonialike odor. It boils at 195°-196°C.

Code Numbers: CAS 100-61-8   RTECS BY4550000   UN 2294

DOT Designation: Label should bear St. Andrews Cross (X).

Synonyms: Monomethylaniline

Potential Exposure: The material is used as an intermediate in organic synthesis, as a solvent and as an acid acceptor.

Incompatibilities: Strong acids, strong oxidizers.

Permissible Exposure Limits in Air: The Federal standard is 2.0 ppm (9 mg/m^3). The ACGIH as of 1983/84 has adopted an STEL of 0.5 ppm (2.0 mg/m^3) and an STEL of 1.0 ppm (5.0 mg/m^3); the notation "skin" indicates the possibility of cutaneous absorption. The IDLH level is 100 ppm.

Determination in Air: Collection in a bubbler using sulfuric acid, workup with NaOH, analysis by gas chromatography. See NIOSH Methods, Set K.

Permissible Concentration in Water: No criteria set.

Routes of Entry: Inhalation, skin absorption, ingestion, skin and eye contact.

Harmful Effects and Symptoms: Weakness, dizziness, headaches, dyspnea, cyanosis; methemoglobinemia; pulmonary edema; liver and kidney damage.

Points of Attack: Respiratory system, liver, kidneys, blood.

Medical Surveillance: Consider the points of attack in preplacement and periodic physical examinations.

First Aid: If this chemical gets into the eyes, irrigate immediately. If this chemical contacts the skin, wash with soap immediately. If a person breathes in large amounts of this chemical, move the exposed person to fresh air at once and perform artificial respiration. When this chemical has been swallowed, get medical attention. Give large quantities of water and induce vomiting. Do not make an unconscious person vomit.

Personal Protective Methods: Wear appropriate clothing to prevent any reasonable probability of skin contact. Wear eye protection to prevent any reasonable probability of eye contact. Employees should wash immediately when skin is wet or contaminated. Remove nonimpervious clothing immediately if wet or contaminated.

Respirator Selection:
          20 ppm: SA/SCBA
         100 ppm: SAF/SAF:PD,PP, CF/SCBAF
          Escape: GMS/SCBA

Disposal Method Suggested: Controlled incineration whereby oxides of nitrogen are removed from the effluent gas by scrubber, catalytic or thermal device.

References

(1) See Reference (A-61).
(2) See Reference (A-60).

METHYL BROMIDE

- Carcinogen (potential, EPA)(1)
- Hazardous waste (EPA)
- Priority toxic pollutant (EPA)

Description: CH_3Br, methyl bromide, is a colorless liquid or gas with a chloroformlike odor at high concentration. It boils at $3°-4°C$.

Code Numbers: CAS 74-83-9 RTECS PA4900000 UN 1062

DOT Designation: Poison gas

Synonyms: Bromomethane, monobromomethane, Embafume®, Iscobrome®, Rotox®.

Potential Exposure: Methyl bromide: The primary use of methyl bromide is as an insect fumigant for soil, grain, warehouses, mills, ships, etc. It is also used as a chemical intermediate and a methylating agent, a refrigerant, a herbicide, a fire extinguishing agent, a low-boiling solvent in aniline dye manufacture, for degreasing wool, for extracting oils from nuts, seeds, and flowers, and in ionization chambers. It is used as an intermediate in the manufacture of many drugs (A-41).

Incompatibilities: Aluminum or strong oxidizers.

Permissible Exposure Limits in Air: The Federal standard for methyl bromide is 20 ppm ($80 mg/m^3$) as a ceiling value. The ACGIH (1983/84) has adopted a TLV for methyl bromide of 5 ppm ($20 mg/m^3$) as a TWA with an STEL value of 15 ppm ($60 mg/m^3$). The notation "skin" is added to indicate the possibility of cutaneous absorption. The IDLH level is 2,000 ppm.

Determination in Air: Charcoal adsorption, workup with CS_2, analysis by gas chromatography. See NIOSH Methods, Set J. See also reference (A-10).

Permissible Concentration in Water: To protect human health—preferably zero. An additional lifetime cancer risk of 1 in 100,000 is posed by a concentration of 1.9 $\mu g/\ell$.

Determination in Water: Inert gas purge followed by gas chromatography with halide specific detection (EPA Method 601) or gas chromatography plus mass spectrometry (EPA Method 624).

Routes of Entry: Inhalation, percutaneous absorption, ingestion, skin and eye contact.

Harmful Effects and Symptoms: *Local* — Methyl bromide is irritating to the eyes, skin, and mucous membranes of the upper respiratory tract. In cases of moderate skin exposure, there may be an itching dermatitis, and in severe cases, vesicles and second-degree burns. Methyl bromide may be absorbed by leather, resulting in prolonged skin contact.

Systemic — High concentrations of methyl bromide may cause lung irritation which may result in pulmonary edema and death. Acute exposure to methyl bromide may produce delayed effects. Onset of symptoms is usually delayed from 30 minutes to 6 hours; the first to appear are malaise, visual disturbances, headaches, nausea, vomiting, somnolence, vertigo, and tremor in the hands. The tremor may become more severe and widespread, developing into epileptiform-type convulsions followed by coma and death due to pulmonary or circulatory failure or both. A period of delirium and mania may precede convulsions, but convulsions have been reported without any other warning symptoms. Kidney damage may occur; permanent brain damage may result.

In chronic poisoning, the effects of methyl bromide are usually limited to the central nervous system: lethargy, muscular pains; visual, speech, and sensory disturbances; and mental confusion being the most prominent complaints.

Points of Attack: Central nervous system, respiratory system, lungs, skin, eyes.

Medical Surveillance: Evaluate the central nervous system, respiratory tract, and skin in preplacement and periodic examinations.

First Aid: If this chemical gets into the eyes, irrigate immediately. If this chemical contacts the skin, flush with water immediately. If a person breathes in large amounts of this chemical, move the exposed person to fresh air at once and perform artificial respiration. When this chemical has been swallowed, get medical attention. Give large quantities of salt water and induce vomiting. Do not make an unconscious person vomit.

Personal Protective Methods: Wear appropriate clothing to prevent any possibility of skin contact. Wear eye protection to prevent any reasonable probability of eye contact. Employees should wash immediately when skin is wet or contaminated. Remove nonimpervious clothing immediatly if wet or contaminated. Provide emergency showers.

Respirator Selection:

200 ppm:	SA/SCBA
1,000 ppm:	SAF/SCBAF
2,000 ppm:	SA:PD, PP, CF
Escape:	GMOV/SCBA

Disposal Method Suggested: The recommended disposal procedure is spray the gas into the fire box of an incinerator equipped with an afterburner and scrubber (alkali).

References

(1) U.S. Environmental Protection Agency, *Halomethanes: Ambient Water Quality Criteria,* Wash., DC (1980).

(2) U.S. Environmental Protection Agency, *Bromomethane,* Health and Environmental Effects Profile No. 29, Wash., DC, Office of Solid Waste (April 30, 1980).

(3) See Reference (A-61).

(4) See Reference (A-60).

(5) Parmeggiani, L., Ed., *Encyclopedia of Occupational Health & Safety,* Third Edition, Vol. 1, pp 330–31, Geneva, International Labour Office (1983) (Bromomethane).

(6) United Nations Environment Programme, *IRPTC Legal File 1983,* Vol. II, pp VII/437–40, Geneva, Switzerland, International Register of Potentially Toxic Chemicals (1984).

METHYL n-BUTYL KETONE

Description: $CH_3CO(CH_2)_3CH_3$ is a colorless liquid with a characteristic odor, boiling at 127°-128°C.

Code Numbers: CAS 591-78-6   RTECS MP1400000

DOT Designation: −

Synonyms: 2-hexanone, MBK; butyl methyl ketone.

Potential Exposure: The material is used as a solvent. NIOSH has estimated annual worker exposure at 222,000.

Incompatibilities: Strong oxidizers.

Permissible Exposure Limits in Air: The Federal standard is 100 ppm (410 mg/m^3). NIOSH (1) has recommended a TWA of 1 ppm (4 mg/m^3). The ACGIH has adopted a TWA of 5 ppm (20 mg/m^3) but no STEL value as of 1983/84. The IDLH level is 5,000 ppm.

Determination in Air: Charcoal adsorption, workup with CS_2, analysis by gas chromatography. See NIOSH Methods, Set M. See also reference (A-10).

Permissible Concentration in Water: No criteria set.

Routes of Entry: Inhalation, skin absorption, ingestion, skin and eye contact.

Harmful Effects and Symptoms: Irritation of eyes and nose; peripheral neuropathy; weakness, paresthesias; dermatitis; headaches, drowsiness.

Points of Attack: Central nervous system, skin, respiratory system.

Medical Surveillance: Preplacement examinations should evaluate skin and respiratory conditions. In the case of methyl n-butyl ketone, special attention should be given to the central and peripheral nervous systems.

First Aid: If this chemical gets into the eyes, irrigate immediately. If this chemical contacts the skin, wash with soap immediately. If a person breathes in large amounts of this chemical, move the exposed person to fresh air at once and perform artificial respiration. When this chemical has been swallowed, get medical attention. Give large quantities of water and induce vomiting. Do not make an unconscious person vomit.

Personal Protective Methods: Wear appropriate clothing to prevent any reasonable probability of skin contact. Wear eye protection to prevent any reasonable probability of eye contact. Employees should wash promptly when skin is wet or contaminated. Remove clothing immediately if wet or contaminated to avoid flammability hazard.

Respirator Selection:
　　　　1,000 ppm: CCROV/SA/SCBA
　　　　5,000 ppm: GMOV/SAF/SCBAF
　　　　　　Escape: GMOV/SCBA

Disposal Method Suggested: Incineration.

References

(1) Nat. Inst. for Occup. Safety and Health, *Criteria for a Recommended Standard: Occupational Exposure to Ketones,* NIOSH Doc. No. 78-173, Wash., DC (1978).

(2) See Reference (A-60).

(3)   See Reference (A-68) (as 2-Hexanone).
(4)   United Nations Environment Programme, *IRPTC Legal File 1983,* Vol. II, pp VII/389, Geneva, Switzerland, International Register of Potentially Toxic Chemicals (1984).

METHYL CHLORIDE

- Carcinogen (potential, EPA)(1)
- Hazardous waste (EPA)
- Priority toxic pollutant (EPA)

Description: CH_3Cl , is a colorless liquefied gas with a faint, sweet odor.

Code Numbers: CAS 74-87-3   RTECS PA6300000   UN 1063

DOT Designation: Flammable gas.

Synonyms: Monochloromethane, chloromethane.

Potential Exposure: Methyl chloride is used as a methylating and chlorinating agent in organic chemistry. In petroleum refineries it is used as an extractant for greases, oils, and resins. Methyl chloride is also used as a solvent in the synthetic rubber industry, as a refrigerant, and as a propellant in polystyrene foam production. In the past it has been used as a local anesthetic (freezing). It is an intermediate in drug manufacture (A-41).

Incompatibilities: Chemically active metals: sodium, potassium, powdered aluminum, zinc, magnesium.

Permissible Exposure Limits in Air: The Federal standard is 100 ppm (210 mg/m^3) as an 8-hour TWA, with an acceptable ceiling concentration of 200 ppm; acceptable maximum peaks above the ceiling of 300 ppm are allowed for 5 minutes' duration in a 3-hour period. The ACGIH (1983/84) TWA value is 50 ppm (105 mg/m^3) and the STEL is 100 ppm (205 mg/m^3). The IDLH is 10,000 ppm.

Determination in Air: Charcoal tube collection, CH_2Cl_2 workup, gas chromatography; NIOSH Method Set 2. See also reference (A-10).

Permissible Concentration in Water: To protect human health—preferably zero. An additional lifetime cancer risk of 1 in 100,000 is posed by a concentration of 1.9 $\mu g/\ell$.

Determination in Water: Gas chromatograph (EPA Method 601) or gas chromatography and mass spectrometry (EPA Method 624).

Routes of Entry: Inhalation and skin or eye contact.

Harmful Effects and Symptoms: *Local* — Skin contact with the discharge from the pressurized gas may cause frostbite. The liquid may damage eyes.

Systemic — Signs and symptoms of chronic exposure include staggering gait, difficulty in speech, nausea, headache, dizziness, and blurred vision. Vomiting has also occurred in some cases. These effects may be observed following a latency period of several hours.

Acute exposure is much like chronic except that the latency period is shorter and the effects more severe. Coma or convulsive seizures may occur. Acute poisoning predominantly depresses the central nervous system, but renal and hepatic damage may also occur. Recently noted in these cases is the depression of bone marrow activity. Recovery from severe exposure may take as long as 2 weeks.

Points of Attack: Liver, kidneys, skin, central nervous system.

Medical Surveillance: Preplacement and periodic examinations should give careful consideration to a previous history of the central nervous system, and to renal or hepatic disorders.

First Aid: If this chemical gets into the eyes, irrigate immediately. If this chemical contacts the skin, flush with water immediately. If a person breathes in large amounts of this chemical, move the exposed person to fresh air at once and perform artificial respiration.

Personal Protective Methods: Wear appropriate clothing to prevent wetting or freezing of skin. Wear eye protection to prevent any reasonable probability of eye contact. Remove clothing immediately if wet or contaminated to avoid flammability hazard.

Respirator Selection:

        1,000 ppm: SA/SCBA
        5,000 ppm: SAF/SCBAF
       10,000 ppm: SA:PD, PP, CF
          Escape: SCBA

Disposal Method Suggested: Controlled incineration with adequate scrubbing and ash disposal facilities.

References

(1) U.S. Environmetal Protection Agency, *Halomethanes: Ambient Water Quality Criteria,* Report PB-296,797 Wash., DC (1980).
(2) Nat. Inst. for Occup. Safety and Health, *Information Profiles on Potential Occupational Hazards: Methyl Chloride,* Report PB-276-678, Rockville, MD, pp 29-36 incl. (Oct. 1977).
(3) U.S. Environmental Protection Agency, *Chloromethane,* Health and Environmental Effects Profile No. 48, Wash., DC, Office of Solid Waste (April 30, 1980).
(4) See Reference (A-61).
(5) See Reference (A-60).
(6) Parmeggiani, L., Ed., *Encyclopedia of Occupational Health & Safety,* Third Edition, Vol. 1, pp 464–65, Geneva, International Labour Office (1983).

METHYL CHLOROFORM

See "1,1,1-Trichloroethane."

METHYL CYANOACRYLATE

Description: $CH_2 = C(CN)COOCH_3$ is a colorless liquid.

Code Numbers: CAS 137-05-3   RTECS AS7000000

DOT Designation: −

Synonyms: Eastman 910 Monomer®; Adhere®.

Potential Exposure: Those engaged in the manufacture of this material and its formulation and application into quick-setting, high-strength adhesive cements.

Permissible Exposure Limits in Air: There is no Federal standard but ACGIH

(1983/84) has adopted a TWA value of 2 ppm (8 mg/m^3) and set an STEL of 4 ppm (16 mg/m^3).

Permissible Concentration in Water: No criteria set.

Harmful Effects and Symptoms: No serious adverse effects have been encountered in simulated workbench exposure. An odor threshold exists at 1-3 ppm; nasal irritation is encountered at 3 ppm and eye irritation at 5 ppm (A-34).

METHYLCYCLOHEXANE

Description: $C_6H_{11}CH_3$ is a colorless liquid with a faint benzenelike odor boiling at 101°-102°C.

Code Numbers: CAS 108-87-2   RTECS GV6125000   UN 2296

DOT Designation: Flammable liquid

Synonyms: Cyclohexylmethane, hexahydrotoluene; and MCH.

Potential Exposure: MCH is used commercially as a solvent for cellulose derivatives particularly with other solvents, and as an organic intermediate in organic synthesis. It is one of the components found in jet fuel.

Incompatibilities: Strong oxidizers.

Permissible Exposure Limits in Air: The Federal standard is 500 ppm (2,000 mg/m^3). The ACGIH (1983/84) has set a TWA value of 400 ppm (1,600 mg/m^3) and an STEL of 500 ppm (2,000 mg/m^3). The IDLH level is 10,000 ppm.

Determination in Air: Charcoal adsorption, workup with CS_2, analysis by gas chromatography. See NIOSH Methods, Set G. See also reference (A-10).

Permissible Concentration in Water: No criteria set.

Routes of Entry: Inhalation, ingestion, eye and skin contact.

Harmful Effects and Symptoms: Lightheadedness, drowsiness; skin irritation; irritation of nose and throat.

Points of Attack: Respiratory system, skin.

Medical Surveillance: Consider the points of attack in preplacement and periodic physical examinations.

First Aid: If this chemical gets into the eyes, irrigate immediately. If this chemical contacts the skin, wash with soap promptly. If a person breathes in large amounts of this chemical, move the exposed person to fresh air at once and perform artificial respiration. When this chemical has been swallowed, get medical attention. Do not induce vomiting.

Personal Protective Methods: Wear appropriate clothing to prevent repeated or prolonged skin contact. Wear eye protection to prevent any reasonable probability of eye contact. Employees should wash promptly when skin is wet or contaminated. Remove clothing immediately if wet or contaminated to avoid flammability hazard.

Respirator Selection:

        1,000 ppm: CCROV
        5,000 ppm: GMOVc/SA/SCBA
       10,000 ppm: GMOVfb/SA:PD, PP, CF/SAF/SCBAF
          Escape: GMOV/SCBA

Disposal Method Suggested: Incineration.

References

(1) U.S. Environmental Protection Agency, *Chemical Hazard Information Profile: Methyl-cyclohexane,* Wash., DC (1979).
(2) See Reference (A-61).
(3) See Reference (A-60).

METHYLCYCLOHEXANOL

Description: $H_3CC_6H_{10}OH$ is a straw-colored liquid with a weak, coconut oil odor. The mixture of 2-, 3- and 4-methylcyclohexanols boils at $155°$-$180°C$.

Code Numbers: CAS 25639-42-3 RTECS GW0175000 UN 2617

DOT Desnigation: Flammable liquid.

Synonyms: Hexahydrocresol; Hexahydromethylphenol.

Potential Exposure: Methylcyclohexanol is used as a lacquer solvent, as a blending agent in textile soaps and as an antioxidant in lubricants (A-35).

Incompatibilities: Strong oxidizers.

Permissible Exposure Limits in Air: The Federal standard is 100 ppm (470 mg/m^3). The ACGIH as of 1983/84 has adopted a TWA of 50 ppm (235 mg/m^3) and an STEL of 75 ppm (350 mg/m^3). The IDLH level is 10,000 ppm.

Determination in Air: Adsorption on charcoal, desorption with methylene chloride, gas chromatographic analysis (A-10).

Permissible Concentration in Water: No criteria set.

Routes of Entry: Inhalation, skin absorption, ingestion, skin and eye contact.

Harmful Effects and Symptoms: Headaches, irritation of eye and respiratory system, dermatitis.

Points of Attack: Respiratory system, eyes and skin. In animals—central nervous system, liver and kidneys.

Medical Surveillance: Consider the points of attack in preplacement and periodic physical examinations.

First Aid: If this chemical gets into the eyes, irrigate immediately. If this chemical contacts the skin, wash with soap promptly. If a person breathes in large amounts of this chemical, move the exposed person to fresh air at once and perform artificial respiration. When this chemical has been swallowed, get medical attention. Give large quantities of water and induce vomiting. Do not make an unconscious person vomit.

Personal Protective Methods: Wear appropriate clothing to prevent repeated or prolonged skin contact. Wear eye protection to prevent any reasonable probability of eye contact. Employees should wash promptly when skin is wet or contaminated. Remove nonimpervious clothing promptly if wet or contaminated.

Respirator Selection:
  500 ppm: SA/SCBA
  5,000 ppm: SAF/SCBAF
  10,000 ppm: SAF:PD,PP,CF
  Escape: GMOV/SCBA

Disposal Method Suggested: Incineration.

References

(1)   See Reference (A-61).

2-METHYLCYCLOHEXANONE

Description: $H_3CC_6H_9O$ is a colorless liquid with a weak peppermint odor. It boils at 165°C.

Code Numbers: CAS 583-60-8 RTECS GW1750000 UN 2297

DOT Designation: Flammable liquid.

Synonyms: o-Methylcyclohexanone.

Potential Exposure: Methylcyclohexanone is used as a solvent and rust remover.

Incompatibilities: Strong oxidizers.

Permissible Exposure Limits in Air: The Federal standard is 100 ppm (460 mg/m³). The notation "skin" indicates the possibility of cutaneous absorption. The ACGIH as of 1983/84 has set a TWA of 50 ppm (230 mg/m³) and an STEL of 75 ppm (345 mg/m³). The IDLH level is 2,500 ppm.

Determination in Air: Collection in an adsorption tube, workup with acetone, analysis by gas chromatography. See NIOSH Methods, Set 2. See also reference (A-10).

Permissible Concentration in Water: No criteria set.

Routes of Entry: Inhalation, ingestion, skin and eye contact.

Harmful Effects and Symptoms: None known in humans. In animals—drowsiness, irritation of eyes, nose, throat and skin.

Points of Attack: In animals—pulmonary system, lungs, liver, kidneys, skin.

Medical Surveillance: Consider the points of attack in preplacement and periodic physical examinations.

First Aid: If this chemical gets into the eyes, irrigate immediately. If this chemical contacts the skin, wash with soap promptly. If a person breathes in large amounts of this chemical, move the exposed person to fresh air at once and perform artificial respiration. When this chemical has been swallowed, get medical attention. Give large quantities of water and induce vomiting. Do not make an unconscious person vomit.

Personal Protective Methods: Wear appropriate clothing to prevent repeated or prolonged skin contact. Wear eye protection to prevent any reasonable probability of eye contact. Employees should wash promptly when skin is wet or contaminated. Remove nonimpervious clothing promptly if wet or contaminated.

Respirator Selection:
      1,000 ppm: CCROVF
      2,500 ppm: GMOV/SAF/SCBAF
      Escape: GMOV/SCBA

Disposal Method Suggested: Incineration.

References
(1) See Reference (A-61).

4,4'-METHYLENEBIS(2-CHLOROANILINE)

- Carcinogen (Animal positive, IARC)(2)
- Hazardous waste (EPA)

Description: $(C_6H_3ClNH_2)CH_2(C_6H_3ClNH_2)$, 4,4'-methylenebis(2-chloroaniline or MOCA, is a yellow to light grey-tan pellet and is also available in liquid form.

Code Numbers: CAS 101-14-4   RTECS CY1050000

DOT Designation: —

Synonyms: MOCA, 4,4'-diamino-3,3'-dichlorodiphenylmethane, 4,4'-methylene-2,2'-dichloroaniline, MBOCA, Cl-MDA, DAPCM.

Potential Exposure: MOCA is primarily used in the production of solid elastomeric parts. Other uses are as a curing agent for epoxy resins and in the manufacture of crosslinked urethane foams used in automobile seats and safety padded dashboards; it is also used in the manufacture of gun mounts, jet engine turbine blades, radar systems, and components in home appliances.

Permissible Exposure Limits in Air: MOCA is included in the Federal standard for carcinogens; all contact with it should be avoided. The ACGIH (1983/84) recommends a TWA of 0.02 ppm (0.22 mg/m³) with the notation that it is a substance suspect of carcinogenic potential for man. It is also noted that cutaneous absorption is possible.

Determination in Air: Sample collection by glass fiber filter followed by silica gel sorbent, desorption with methanol, analysis by high-performance liquid chromatography (A-10).

Permissible Concentration in Water: No criteria set.

Routes of Entry: Inhalation; percutaneous absorption.

Harmful Effects and Symptoms: *Local* — None reported.
Systemic — Feeding experiments with rats produced liver and lung cancer (3). No tumors were found in experiments with dogs. No tumors or other illness has been reported from chronic exposure in man except a mild cystitis which subsided within a week.

Medical Surveillance: Preplacement and periodic examinations should include a history of exposure to other carcinogens, alcohol and smoking habits, use of medications, and family history. Special attention should be given to liver size and function and to any changes in lung symptoms or x-rays.

Personal Protective Methods: These are designed to supplement engineering controls and to prevent all contact with skin and the respiratory tract. Protective clothing and gloves should be provided, and also appropriate-type dust masks or supplied air respirators. On exit from a regulated area, employees should shower and change into street clothes, leaving the protective clothing and equipment at the point of exit, to be placed in impervious containers at the end of the work shift for decontamination or disposal.

References

(1) Linch, A.L. O'Connor, G.B., Barnes, J.R., Killian, A.S. Jr., and Neeld, W.E. Jr., "Methylene-bis-ortho-chloroaniline (MOCA): Evaluations of Hazards and Exposure Control," *Am. Ind. Hyg. Assoc. J.* 32:802 (1971).

(2) International Agency for Research on Cancer, *IARC Monographs on the Carcinogenic Risks of Chemicals to Humans,* Lyon, France 4, 65 (1974).

(3) See Reference (A-62). Also see Reference (A-64).

METHYLENEBIS(4-CYCLOHEXYL ISOCYANATE)

Description: $OCNC_6H_{10}CH_2C_6H_{10}NCO$

Code Numbers: CAS 5124-30-1

DOT Designation: −

Synonyms: None

Potential Exposure: Those involved in the manufacture of this compound or its use in the production of light-stable polyurethane resins.

Permissible Exposure Limits in Air: There is no Federal standard but ACGIH (1983/84) has adopted a ceiling value of 0.01 ppm (0.11 mg/m³).

Permissible Concentration in Water: No criteria set.

Harmful Effects and Symptoms: The only adverse industrial effects reported (A-34) have been skin irritation and sensitization; in at least one case this resulted from brief vapor exposure.

4,4'-METHYLENEBIS(N,N-DIMETHYL)ANILINE

- Carcinogen (Animal Positive) (A-64)
- Hazardous Waste Constituent (EPA)

Description: $C_{17}H_{22}N_2$ with the formula $(CH_3)_2N-C_6H_4-CH_2-C_6H_4-N(CH_3)_2$ is a crystalline compound melting at 90° to 91°C.

Code Numbers: RTECS BY5250000

DOT Designation: −

Synonyms: Michler's base; Tetramethyldiaminodiphenylmethane.

Potential Exposure: Michler's base is used as an intermediate in dye manufacture and as an analytical reagent in the determination of lead.

The Society of Dyers and Colorists reported that six dyes and one pigment can be prepared from the chemical. In 1979, only one of the dyes, Basic Yellow 2, was being produced commercially by two companies.

Michler's base has been commercially produced in the United States since at least 1921. EPA estimated five producers and importers in two regions, with domestic production reported as 370,000 lb, and no imports (Toxic Substances Control Act (TSCA), Chemical Substance Inventory, 1979, public record). U.S. production of Michler's base was reported to the U.S. International Trade Commission as greater than 10,000 lb in 1980.

According to the CPSC staff, it is possible that residual levels of Michler's base may be present in the final consumer product. Exposure even to trace amounts may be a cause for concern. This concern is based on experience with other dyes derived from aromatic amines. However, data describing the actual levels of impurities in the final product are currently not available. The potential for exposure is greatest among workers in the dye and chemical manufacturing industries. In 1974, the National Occupational Hazard Survey estimated that 1,600 workers were potentially exposed to Michler's base.

Permissible Exposure Limits in Air: No standards set.

Permissible Concentration in Water: No criteria set.

Harmful Effects and Symptoms: 4,4'-Methylenebis(N,N-dimethyl)benzenamine, administered in the feed, was carcinogenic in male and female Fischer 344 rats, inducing thryroid follicular-cell carcinomas, and in female B6C3F1 mice, causing a combination of hepatocellular carcinomas and hepatocellular adenomas. There was no conclusive evidence that 4,4'methylenebis(N,N-dimethyl)benzenamine was carcinogenic in male B6C3F1 mice (1). An IARC working group considered that there was limited evidence for the carcinogenicity of this substance in experimental animals (2). In view of another evaluation the evidence can be regarded as sufficient (3).

References

(1) National Cancer Institute, *Bioassay of 4,4'-Methylenebis(N,N-dimethyl)benzenamine for Carcinogenicity,* Technical Report Series No. 186, DHEW Publication No. (NIH) 78-1742, Bethesda, Maryland (1979).

(2) International Agency for Research on Cancer, *IARC Monographs on the Evaluation of the Carcinogenic Risk of Chemicals to Humans,* Vol. 27, Lyon, France: IARC, p 119-24 (1982).

(3) Griesemer, R.A., and C. Cueto, "Toward a Classification Scheme for Degrees of Experimental Evidence for the Carcinogenicity of Chemicals for Animals," In: R. Montesano, H. Bartsch, and L. Tomatis (eds.), *Molecular and Cellular Aspects of Carcinogen Screening Tests, IARC Scientific Publications, No. 27,* Lyon, France: International Agency for Research on Cancer, p 259-81 (1980).

METHYLENEBIS(PHENYL ISOCYANATE)

Description: $OCNC_6H_4CH_2C_6H_4NCO$ occurs in the form of white to light yellow odorless flakes.

Code Numbers: CAS 101-68-8   RTECS NQ8805000   UN 2489

DOT Designation: None.

Synonyms: MDI; 4,4-diphenylmethane diisocyanate; methylenebis(4-phenyl isocyanate); 4,4-diisocyanodiphenylmethane.

Potential Exposure: MDI is used in the production of polyurethane foams and plastics.

Incompatibilities: Strong alkalies, acids, alcohol.

Permissible Exposure Limits in Air: The Federal standard and the 1983/84

ACGIH TWA value for MDI is 0.02 ppm (0.2 mg/m^3) as a ceiling value. There is no tentative STEL value set. The IDLH level is 10 ppm.

Determination in Air: Collection by impinger or fritted bubbler, diazotization and coupling to give colored complex, analysis by spectrophotometer (A-10).

Permissible Concentration in Water: No criteria set.

Routes of Entry: Inhalation, ingestion, eye and skin contact.

Harmful Effects and Symptoms: Irritation of eyes, nose and throat; coughing, pulmonary secretions; chest pains, dyspnea, asthma.

Points of Attack: Respiratory system, eyes, lungs.

Medical Surveillance: Preplacement and periodic medical examinations should include chest roentgenograph, pulmonary function tests, and an evaluation of any respiratory disease or history of allergy. Periodic pulmonary function tests may be useful in detecting the onset of pulmonary sensitization.

First Aid: If this chemical gets into the eyes, irrigate immediately. If this chemical contacts the skin, wash with soap immediately. If a person breathes in large amounts of this chemical, move the exposed person to fresh air at once and perform artificial respiration. When this chemical has been swallowed, get medical attention. Give large quantities of water and induce vomiting. Do not make an unconscious person vomit.

Personal Protective Methods: Wear appropriate clothing to prevent any reasonable probability of skin contact. Wear eye protection to prevent any possibility of eye contact. Employees should wash promptly when skin is wet or contaminated. Work clothing should be changed daily if it is possible that clothing is contaminated. Remove nonimpervious clothing promptly if wet or contaminated.

Respirator Selection:
<pre>
    1 ppm: SAF/SCBAF
   10 ppm: SAF:PD,PP,CF
  Escape: GMOVP/SCBA
</pre>

Disposal Method Suggested: Controlled incineration (oxides of nitrogen are removed from the effluent gas by scrubbers and/or thermal devices).

References
(1)   See Reference (A-61).

METHYLENE CHLORIDE

- Carcinogen (potential, EPA)(2)
- Hazardous waste (EPA)
- Priority toxic pollutant (EPA)

Description: CH_2Cl_2, methylene chloride, is a nonflammable, colorless liquid with a pleasant aromatic odor noticeable at 300 ppm. (This, however, should not be relied upon as an adequate warning of unsafe concentrations.) It boils at 40°C.

Code Numbers: CAS 75-09-2   RTECS PA8050000   UN 1593

DOT Designation: ORM-A

Synonyms: Dichloromethane, methylene dichloride, methylene bichloride.

Potential Exposure: Methylene chloride is used mainly as a low-temperature extractant of substances which are adversely affected by high temperature. It can be used as a solvent for oil, fats, waxes, bitumen, cellulose acetate, and esters. It is also used as a paint remover and as a degreaser. NIOSH estimates that 70,000 workers are exposed to methylene chloride.

Incompatibilities: Strong oxidizers, strong caustics, chemically active metals, such as aluminum or magnesium powders; sodium, potassium.

Permissible Exposure Limits in Air: The current Federal standard for methylene chloride is 500 parts of methylene chloride per million parts of air (ppm) averaged over an eight-hour work shift, with an acceptable ceiling level of 1,000 ppm and a maximum peak concentration of 2,000 ppm for 5 minutes in any two-hour period. NIOSH has recommended that the permissible exposure limit be reduced to 75 ppm averaged over a work shift of up to 10 hours per day, 40 hours per week, with a ceiling level of 500 ppm averaged over a 15-minute period. NIOSH further recommends that permissible levels of methylene chloride be reduced where carbon monoxide is present.

A major consideration in the recommended NIOSH standard is concomitant exposure to carbon monoxide since a major metabolic product of methylene chloride is carbon monoxide. NIOSH recommends that occupational exposure to methylene chloride be controlled so that workers are not exposed in excess of 75 ppm (261 mg/m^3) determined as a time-weighted average concentration for up to a 10-hour workday, 40-hour workweek, or to peak concentrations in excess of 500 ppm ($1,740$ mg/m^3) as determined by any 15-minute sampling period in the absence of CO exposure >9 ppm.

Where the time-weighted average occupational exposure to carbon monoxide exceeds 9 ppm (the current Federal air pollution standard, 40 CFR 50.8), NIOSH recommends that occupational exposure to methylene chloride be controlled so that workers are not exposed to methylene chloride and carbon monoxide in combinations which exceed unity according to the following equation:

$$\frac{C(CO)}{L(CO)} + \frac{C(CH_2Cl_2)}{L(CH_2Cl_2)} \leqslant 1$$

C(CO) = TWA exposure concentration of carbon monoxide, ppm
L(CO) = recommended TWA exposure limit of carbon monoxide = 35 ppm
C(CH$_2$Cl$_2$) = TWA exposure concentration of methylene chloride, ppm
L(CH$_2$Cl$_2$) = recommended TWA exposure limit of methylene chloride =
         75 ppm

The ACGIH has proposed a TWA of 100 ppm (360 mg/m^3) and an STEL of 500 ppm ($1,700$ mg/m^3) as of 1983/84. The IDLH level is 5,000 ppm.

The Dutch Chemical Industry (A-60) has set a maximum allowable concentration of 200 ppm (700 mg/m^3).

Determination in Air: Adsorption on charcoal, workup with CS$_2$, analysis by gas chromatography. See NIOSH Methods, Set J. See also reference (A-10).

Permissible Concentration in Water: To protect human health—preferably zero. An additional lifetime cancer risk of 1 in 100,000 results at a level of 1.9 μg/ℓ.

Determination in Water: Inert gas purge followed by gas chromatography with halide specific detection (EPA Method 601) or gas chromatography, plus mass spectrometry (EPA Method 624).

Routes of Entry: Inhalation of vapors, percutaneous absorption of liquid, ingestion, skin and eye contact.

Harmful Effects and Symptoms: *Local* — Repeated contact with methylene chloride may cause a dry, scaly, and fissured dermatitis. The liquid and vapor are irritating to the eyes and upper respiratory tract at higher concentrations. If the liquid is held in contact with the skin, it may cause skin burns.

Systemic — Methylene chloride is a mild narcotic. Effects from intoxication include headache, giddiness, stupor, irritability, numbness, and tingling in the limbs. Irritation to the eyes and upper respiratory passages occurs at higher dosages. In severe cases, observers have noted toxic encephalopathy with hallucinations, pulmonary edema, coma, and death. Cardiac arrythmias have been produced in animals but have not been common in human experiences. Exposure to this agent may cause elevated carboxyhemoglobin levels which may be significant in smokers, or workers with anemia or heart disease, and those exposed to CO.

Points of Attack: Skin cardiovascular system, eyes, central nervous system.

Medical Surveillance: Changes in liver, respiratory tract, and central nervous system should be considered during preplacement or periodic medical examinations. Smoking history should be known; anemias or cardiovascular disease may increase the hazard.

The metabolism and excretion of methylene chloride has been thoroughly studied. Blood and expired air analyses are useful indicators of exposure. Carboxyhemoglobin levels may be useful indicators of excessive exposure, especially in nonsmokers.

First Aid: If this chemical gets into the eyes, irrigate immediately. If this chemical contacts the skin, wash with soap promptly. If a person breathes in large amounts of this chemical, move the exposed person to fresh air at once and perform artificial respiration. When this chemical has been swallowed, get medical attention. Give large quantities of salt water and induce vomiting. Do not make an unconscious person vomit.

Personal Protective Methods: Wear appropriate clothing to prevent repeated or prolonged skin contact. Wear eye protection to prevent any reasonable probability of eye contact. Employees should wash promptly when skin is wet or contaminated. Remove nonimpervious clothing promptly if wet or contaminated.

Respirator Selection:
>500 ppm: SAF/SCBAF
>Escape: GMOV/SCBA

Disposal Method Suggested: Incineration preferably after mixing with another combustible fuel; care must be exercised to assure complete combustion to prevent the formation of phosgene; an acid scrubber is necessary to remove the halo acids produced.

References

(1)   National Institute for Occupational Safety and Health, *Criteria for a Recommended Standard: Occupational Exposure to Methylene Chloride,* NIOSH Doc. No. 76-138 (1976).

(2) U.S. Environmental Protection Agency, *Halomethanes: Ambient Water Quality Criteria,* Wash., DC (1980).

(3) U.S. Environmental Protection Agency, *Dichloromethane,* Health and Environmental Effects Profile No. 74, Wash., DC, Office of Solid Waste (April 30, 1980).

(4) See Reference (A-61).

(5) Sax, N.I., Ed., *Dangerous Properties of Industrial Materials Report, 1,* No. 2, 45-47, New York, Van Nostrand Reinhold Co. (1980).

(6) Parmeggiani, L., Ed., *Encyclopedia of Occupational Health & Safety,* Third Edition, Vol. 1, pp 624-26, Geneva, International Labour Office (1983).

METHYL ETHYL KETONE

● Hazardous waste (EPA)

Description: $CH_3COCH_2CH_3$, a clear, colorless liquid with a fragrant, mint-like moderately sharp odor, boiling at 79°-80°C.

Code Numbers: CAS 78-93-3 RTECS EL6475000 UN 1193

DOT Designation: Flammable liquid.

Synonyms: MEK, 2-butanone, butanone, ethyl methyl ketone.

Potential Exposure: NIOSH has estimated worker exposure to MEK at 3,031,000, the highest for any ketone. MEK is used as a solvent in nitrocellulose coating and vinyl film manufacture, in smokeless powder manufacture, in cements and adhesives and in the dewaxing of lubricating oils (2). It is also an intermediate in drug manufacture (A-41).

Incompatibilities: Very strong oxidizers.

Permissible Exposure Limits in Air: The Federal limit and the 1983/84 ACGIH TWA value is 200 ppm (590 mg/m³). The STEL value is 300 ppm (885 mg/m³). The IDLH level is 3,000 ppm.

Determination in Air: Adsorption on charcoal, workup with CS_2, analysis by gas chromatography. See NIOSH Methods, Set A. See also reference (A-10).

Permissible Concentration in Water: No criteria set.

Routes of Entry: Inhalation, ingestion, eye and skin contact.

Harmful Effects and Symptoms: Irritation of eyes and nose; headaches, dizziness, vomiting.

Points of Attack: Central nervous system, lungs.

Medical Surveillance: Consider the points of attack in preplacement and periodic physical examinations.

First Aid: If this chemical gets into the eyes, irrigate immediately. If this chemical contacts the skin, flush with water immediately. If a person breathes in large amounts of this chemical, move the exposed person to fresh air at once. When this chemical has been swallowed, get medical attention. Give large quantities of salt water and induce vomiting. Do not make an unconscious person vomit.

Personal Protective Methods: Wear appropriate clothing to prevent repeated or prolonged skin contact. Wear eye protection to prevent any reasonable probability of eye contact. Remove nonimpervious clothing promptly if wet or contaminated. Provide emergency eyewash.

Respirator Selection:
> 1,000 ppm: CCROVF
> 30,000 ppm: GMOV/SAF/SCBAF
> Escape: GMOV/SCBA

Disposal Method Suggested: Incineration.

References

(1) National Institute for Occupational Safety and Health, *Criteria for a Recommended Standard: Occupational Exposure to Ketones,* NIOSH Doc. No. 78-173, Wash. DC (1978).
(2) U.S. Environmental Protection Agency, *Methyl Ethyl Ketone,* Health and Environmental Effects Profile No. 128, Wash., DC, Office of Solid Waste (April 30, 1980).
(3) Sax, N.I., Ed., *Dangerous Properties of Industrial Materials Report, 1,* No. 4, 85-87, New York, Van Nostrand Reinhold Co. (1981).
(4) See Reference (A-60).
(5) See Reference (A-68) (as 2-Butanone).
(6) United Nations Environment Programme, *IRPTC Legal File 1983,* Vol. I, pp VII/138–40, Geneva, Switzerland, International Register of Potentially Toxic Chemicals (1984).

METHYL ETHYL KETONE PEROXIDE

● Hazardous waste (EPA)

Description: $C_8H_{16}O_4$,

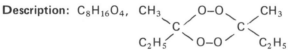

is a colorless liquid.

Code Numbers: CAS 1338-23-4   RTECS EL9450000   UN 2550

DOT Designation: Forbidden.

Synonyms: 2-Butanone peroxide, MEKP.

Potential Exposure: MEKP is used as a curing agent for thermosetting polyester resins and as a crosslinking agent and catalyst in the production of other polymers.

Permissible Exposure Limits in Air: By comparison to benzoyl peroxide and H_2O_2, a ceiling value of 0.2 ppm (1.5 mg/m³) has been set by ACGIH (1983/84).

Permissible Concentration in Water: No criteria set.

Routes of Entry: Inhalation, ingestion, skin and eye contact.

Harmful Effects and Symptoms: Given orally, by inhalation, or by intraperitoneal injection, methyl ethyl ketone peroxide causes hyperemia of the lungs with petechial or gross hemorrhage in mice and rats. Subacute exposures have been associated with mild liver damage in rats. In addition, this compound can be severely irritating to the eyes and skin. In rabbits, maximum nonirritating strengths were found to be 1.5 and 0.6% for the skin and eyes, respectively.

Mice given a total dose of ~7 mg methyl ethyl ketone peroxide developed malignant tumors, the first of which appeared after fifteen months. One subcutaneous sarcoma, three malignant lymphomas, and a pulmonary adenoma were noted in 34 of the 50 mice surviving exposure.

Chemical burns of the gastrointestinal tract, as well as residual scarring and stricture of the esophagus, were noted in an individual surviving ingestion of two ounces of a 60% methyl ethyl ketone peroxide solution.

Points of Attack: In animals—lungs, liver, eyes and skin. In humans—gastrointestinal tract.

Medical Surveillance: Consider the points of attack in preplacement and periodic physical examinations.

References

(1) U.S. Environmental Protection Agency, *Chemical Hazard Information Profile: Methyl Ethyl Ketone Peroxide,* Wash. DC (1979).
(2) Nat. Inst. for Occup. Safety and Health, *Information Profiles on Potential Occupational Hazards: Methyl Ethyl Ketone Peroxide,* pp 37-41, Report PB-276,678, Rockville, MD (Oct. 1977).
(3) Sax, N.I., Ed., *Dangerous Properties of Industrial Materials Report, 2,* No. 6, 35-37, New York, Van Nostrand Reinhold Co. (1982) (as 2-Butanone Peroxide).
(4) See Reference (A-60).

2-METHYL-5-ETHYL PYRIDINE

Description: $C_8H_{11}N$ with the following structural formula

is a liquid with an aromatic odor boiling at 74° to 75°C at 20 mm Hg pressure.

Code Numbers: CAS 104-90-5   RTECS TJ6825000   UN 2300

DOT Designation: Corrosive Material.

Synonyms: 5-ethyl-2-picoline.

Potential Exposure: In manufacture of nicotinic acid, vinylpyridine monomer, in intermediates for germicides and textile chemicals.

Permissible Exposure Limits in Air: A TLV of 0.4 ppm (2 mg/m³) has been set in the U.S.S.R. (A-36).

Permissible Concentration in Water: No criteria set.

Harmful Effects and Symptoms: Skin and eye irritant. Moderately toxic by oral and dermal routes.

Personal Protective Methods: Wear skin protection.

Respirator Selection: Canister-type mask required.

Disposal Method Suggested: Incineration.

References

(1) Sax, N.I., Ed., *Dangerous Properties of Industrial Materials Report, 2,* No. 2, 54-55, New York, Van Nostrand Reinhold Co. (1982).
(2) Sax, N.I., Ed., *Dangerous Properties of Industrial Materials Report, 3,* No. 6, 48-49, New York, Van Nostrand Reinhold Co. (Nov./Dec. 1983).

METHYL FORMATE

Description: HCOOCH$_3$ is a colorless liquid with a pleasant odor which boils at 31°-32°C.

Code Numbers: CAS 107-31-3   RTECS LQ8925000   UN 1243

DOT Designation: Flammable liquid.

Synonyms: Methyl methanoate; formic acid methyl ester.

Potential Exposure: Methyl formate is used as a solvent, as an intermediate in pharmaceutical manufacture (A-41) and as a fumigant.

Incompatibilities: Strong oxidizers.

Permissible Exposure Limits in Air: The Federal standard and the 1983/84 ACGIH TWA value is 100 ppm (250 mg/m^3). The STEL is 150 ppm (375 mg/m^3). The IDLH level is 5,000 ppm.

Determination in Air: Collection by charcoal tube, analysis by gas liquid chromatography (A-13).

Permissible Concentration in Water: No criteria set.

Routes of Entry: Inhalation, skin absorption, ingestion, skin and eye contact.

Harmful Effects and Symptoms: Irritation of eyes and nose; chest oppression, dyspnea; visual disturbance; central nervous system depression.

Points of Attack: Eyes, respiratory system, lungs, central nervous system.

Medical Surveillance: Consider the points of attack in preplacement and periodic physical examinations.

First Aid: If this chemical gets into the eyes, irrigate immediately. If this chemical contacts the skin, wash with soap immediately. If a person breathes in large amounts of this chemical, move the exposed person to fresh air at once and perform artificial respiration. When this chemical has been swallowed, get medical attention. Give large quantities of water and induce vomiting. Do not make an unconscious person vomit.

Personal Protective Methods: Wear appropriate clothing to prevent repeated or prolonged skin contact. Wear eye protection to prevent any reasonable probability of eye contact. Employees should wash promptly when skin is wet or contaminated. Remove clothing immediately if wet or contaminated to avoid flammability hazard.

Respirator Selection:
>1,000 ppm: SA/SCBA
>3,500 ppm: SA:PD, PP, CF
>5,000 ppm: SAF/SCBAF
>Escape: GMOV/SCBA

Disposal Method Suggested: Incineration.

References

(1)   See Reference (A-61).
(2)   See Reference (A-60).

5-METHYL-3-HEPTANONE

Description: $CH_3CH_2COCH_2CH(CH_3)CH_2CH_3$ is a colorless liquid with a mild fruity odor which boils at 151°C.

Code Numbers: CAS 541-85-5   RTECS MJ7350000   UN 2271

DOT Designation: Flammable liquid.

Synonyms: Ethyl sec-amyl ketone; ethyl amyl ketone; amyl ethyl ketone.

Potential Exposure: Methyl heptanone is used as a solvent for resins and as an organic intermediate.

Incompatibilities: Oxidizers.

Permissible Exposure Limits in Air: The Federal standard and the ACGIH 1983/84 TWA value is 25 ppm (130 mg/m³). There is no STEL value. The IDLH level is 3,000 ppm.

Determination in Air: Charcoal adsorption, workup with CS_2, analysis by gas chromatography. See NIOSH Methods, Set B. See also reference (A-10).

Permissible Concentration in Water: No criteria set.

Routes of Entry: Inhalation, ingestion, eye and skin contact.

Harmful Effects and Symptoms: Irritation of the eyes and mucous membranes, headaches; narcosis, coma; dermatitis.

Points of Attack: Eyes, skin, respiratory system, central nervous system.

Medical Surveillance: Consider the points of attack in preplacement and periodic physical examinations.

First Aid: If this chemical gets into the eyes, irrigate immediately. If this chemical contacts the skin, flush with water. If a person breathes in large amounts of this chemical, move the exposed person to fresh air at once and perform artificial respiration. When this chemical has been swallowed, get medical attention. Give large quantities of salt water and induce vomiting. Do not make an unconscious person vomit.

Personal Protective Methods: Wear appropriate clothing to prevent any possibility of skin contact. Wear eye protection to prevent any reasonable probability of eye contact. Employees should wash promptly when skin is wet or contaminated. Remove nonimpervious clothing promptly if wet or contaminated.

Respirator Selection:
  1,000 ppm: CCROVF/GMOV/SAF/SCBAF
  3,000 ppm: SAF:PD,PP,CF
    Escape : GMOV/SCBA

Disposal Method Suggested: Incineration.

References

(1) See Reference (A-61).
(2) See Reference (A-60).

METHYL IODIDE

● Carcinogen (animal positive, IARC)(1)

● Hazardous waste (EPA)

Description: CH_3I is a colorless liquid which turns yellow, red, or brown on exposure to light and moisture. It boils at 42°C.

Code Numbers: CAS 74-88-4   RTECS PA9450000   UN 2644

DOT Designation: —

Synonyms: Iodomethane.

Potential Exposure: Methyl iodide is used as an intermediate in the manufacture of many pharmaceuticals (A-41) and some pesticides (A-32).

Incompatibilities: Strong oxidizers.

Permissible Exposure Limits in Air: The Federal standard is 5 ppm (28 mg/m³). The ACGIH (1983/84) has adopted a TWA of 2 ppm (10 mg/m³) and an STEL of 5 ppm (30 mg/m³) with the notation that it is a substance suspect of carcinogenic potential for man and with the notation "skin" indicating the possibility of cutaneous absorption. The IDLH level is 800 ppm.

Determination in Air: Charcoal adsorption, workup with toluene, analysis by gas chromatography. See NIOSH Methods, Set H.

Permisible Concentration in Water: No criteria set.

Routes of Entry: Inhalation, ingestion, eye and skin contact.

Harmful Effects and Symptoms: Nausea, vomiting; vertigo, ataxia; slurred speech, drowsiness; dermatitis, skin blistering; eye irritation.

Points of Attack: Skin, eyes, central nervous system.

Medical Surveillance: Consider the points of attack in preplacement and periodic physical examinations.

First Aid: If this chemical gets into the eyes, irrigate immediately. If this chemical contacts the skin, dust off solid, flush with soap immediately. If a person breathes in large amounts of this chemical, move the exposed person to fresh air at once and perform artificial respiration. When this chemical has been swallowed, get medical attention. Give large quantities of salt water and induce vomiting. Do not make an unconscious person vomit.

Personal Protective Methods: Wear appropriate clothing to prevent repeated or prolonged skin contact. Wear eye protection to prevent any reasonable probability of eye contact. Employees should wash immediately when skin is wet or contaminated. Remove nonimpervious clothing immediately if wet or contaminated.

Respirator Selection:

     50 ppm: SA/SCBA
    250 ppm: SAF/SCBAF
    800 ppm: SA:PD, PP, CF
    Escape : GMOV/SCBA

References

(1)   International Agency for Research on Cancer, *IARC Monographs on the Carcinogenic Risks of Chemicals to Humans,* Lyon, France, 15, 245 (1977).

(2)   See Reference (A-61).

METHYL ISOAMYL KETONE

Description: $CH_3COCH_2CH_2CH(CH_3)_2$ is a colorless liquid with a pleasant odor boiling at 144°C.

Code Numbers: CAS 110-12-3 RTECS MP3850000 UN 2302

DOT Designation: Flammable liquid.

Synonyms: 5-Methyl-2-hexanone; MIAK.

Potential Exposure: MIAK is used as a solvent for cellulose esters, acrylics and vinyl copolymers.

Permissible Exposure Limits in Air: There is no Federal standard but ACGIH as of 1983/84 has proposed a TWA of 50 ppm (240 mg/m³). There is no STEL value.

Determination in Air: Sample collection by charcoal tube, analysis by gas liquid chromatography (A-10).

Permissible Concentration in Water: No criteria set.

Harmful Effects and Symptoms: Eye and nose irritation are encountered at low levels. Narcosis and death can result at high concentrations. The toxic behavior would be expected to resemble that of methyl isobutyl ketone closely (A-34).

METHYL ISOBUTYL CARBINOL

Description: $CH_3CHOHCH_2CH(CH_3)_2$ is a colorless liquid with a mild odor boiling at 131°-132°C.

Code Numbers: CAS 108-11-2 RTECS SA7350000 UN 2053

DOT Designation: Flammable liquid.

Synonyms: Methyl amyl alcohol; 4-methyl-2-pentanol; MIBC

Potential Exposure: MIBC is used as a solvent, in the formulation of brake fluids, and as an intermediate in organic synthesis.

Incompatibilities: Strong oxidizers.

Permissible Exposure Limits in Air: The Federal standard and the 1983/84 ACGIH TWA value is 25 ppm (100 mg/m³). The ACGIH has set an STEL of 40 ppm (165 mg/m³). The notation "skin" indicates the possibility of cutaneous absorption. The IDLH level is 2,000 ppm.

Determination in Air: Adsorption on charcoal, workup with 2-propanol in CS_2, analysis by gas chromatography. See NIOSH Methods, Set E. See also reference (A-10).

Permissible Concentration in Water: No criteria set.

Routes of Entry: Inhalation, ingestion, eye and skin contact.

Harmful Effects and Symptoms: Eye irritation, headaches, drowsiness, dermatitis.

Points of Attack: Eyes, skin.

Medical Surveillance: Consider the points of attack in preplacement and periodic physical examinations.

First Aid: If this chemical gets into the eyes, irrigate immediately. If this chemical contacts the skin, flush with water promptly. If a person breathes in large amounts of this chemical, move the exposed person to fresh air at once and perform artificial respiration. When this chemical has been swallowed, get medical attention. Give large quantities of salt water and induce vomiting. Do not make an unconscious person vomit.

Personal Protective Methods: Wear appropriate clothing to prevent repeated or prolonged skin contact. Wear eye protection to prevent any reasonable probability of eye contact. Employees should wash promptly when skin is wet or contaminated. Remove clothing immediately if wet or contaminated to avoid flammability hazard.

Respirator Selection:
> 1,000 ppm: CCROVF/GMOV/SAF/SCBAF
> 2,000 ppm: SAF:PD, PP, CF
> Escape : GMOV/SCBA

Disposal Method Suggested: Incineration.

References
(1) See Reference (A-61).
(2) See Reference (A-60).

METHYL ISOBUTYL KETONE

- Hazardous waste (EPA)

Description: $CH_3COCH_2CH(CH_3)_2$ is a colorless liquid with a pleasant odor boiling at 117°-118°C.

Code Numbers: CAS 108-10-1   RTECS SA9275000   UN 1245

DOT Designation: Flammable liquid.

Synonyms: 4-Methyl-2-pentanone; isobutyl methyl ketone; MIBK, hexone.

Potential Exposure: MIBK is used in the manufacture of methyl amyl alcohol and as a solvent in paints, varnishes and lacquers; it is also used as an alcohol denaturant and as a solvent in uranium extraction from fission products (2). NIOSH has estimated annual worker exposure at 1,850,000.

Incompatibilities: Strong oxidizers.

Permissible Exposure Limits in Air: The Federal standard is 100 ppm (410 mg/m^3). NIOSH recommends (1) a TWA of 50 ppm (200 mg/m^3). ACGIH (1983/84 has adopted a TWA of 50 ppm and an STEL of 75 ppm (300 mg/m^3). The IDLH level is 3,000 ppm, ACGIH adds the notation "skin" indicating the possibility of cutaneous absorption.

Determination in Air: Charcoal adsorption, workup with CS_2; analysis by gas chromatography. See NIOSH Methods, Set A. See also reference (A-10).

Permissible Concentration in Water: No criteria set.

Routes of Entry: Inhalation, ingestion, eye and skin contact.

Harmful Effects and Symptoms: Irritation of eye and mucous membranes; headaches; narcosis, coma; dermatitis.

Points of Attack: Respiratory system, eyes, skin, central nervous system.

Medical Surveillance: Consider the points of attack in preplacement and periodic physical examinations.

First Aid: If this chemical gets into the eyes, irrigate immediately. If this chemical contacts the skin, flush with water promptly. If a person breathes in large amounts of this chemical, move the exposed person to fresh air at once and perform artificial respiration. When this chemical has been swallowed, get medical attention. Do not induce vomiting.

Personal Protective Methods: Wear appropriate clothing to prevent repeated or prolonged skin contact. Wear eye protection to prevent any reasonable probability of eye contact. Employees should wash promptly when skin is wet or contaminated. Remove clothing immediately if wet or contaminated to avoid flammability hazard.

Respirator Selection:
        1,000 ppm: CCROVF
        3,000 ppm: GMOV/SAF/SCBAF
        Escape : GMOV/SCBA

Disposal Method Suggested: Incineration

References

(1) Nat. Inst. for Occup. Safety and Health, *Criteria for a Recommended Standard: Occupational Exposure to Ketones,* NIOSH Doc. No. 78-173, Wash., DC (1978).
(2) U.S. Environmental Protection Agency, *Methyl Isobutyl Ketone,* Health and Environmental Effects Profile No. 129, Wash. DC, Office of Solid Waste (April 30, 1980).
(3) See Reference (A-68) (as Hexone).
(4) United Nations Environment Programme, *IRPTC Legal File 1983,* Vol. II, pp VII/490–91, Geneva, Switzerland, International Register of Potentially Toxic Chemicals (1984).

METHYL ISOCYANATE

● Hazardous waste (EPA)

Description: CH_3NCO is a colorless liquid with a sharp odor which is a lachrymator (causes tears). It boils at 39°C.

Code Numbers: CAS 624-83-9   RTECS NQ9450000   UN 2480

DOT Designation: Flammable liquid and poison.

Synonyms: Isocyanatomethane

Potential Exposure: Methyl isocyanate is used in the production of polyurethane foams and plastics; it is also used as an intermediate in the manufacture of a wide variety of pesticides (A-32).

Incompatibilities: Water, rapid reaction in presence of acid, alkali, amine; iron, tin, copper, their salts, other catalysts.

Permissible Exposure Limits in Air: The Federal standard and the 1983/84 ACGIH TWA value is 0.02 ppm (0.05 mg/m³). The notation "skin" indicates the possibility of cutaneous absorption. There is no STEL value. The IDLH level is 20 ppm.

Permissible Concentration in Water: No criteria set.

Routes of Entry: Inhalation, ingestion, skin and eye contact.

Harmful Effects and Symptoms: Irritation of eyes, nose and throat; coughing, secretions, chest pain, dyspnea; asthma; skin and eye injury.

Points of Attack: Respiratory system, lungs, eyes, skin.

Medical Surveillance: Consider the points of attack in preplacement and periodic physical examinations.

First Aid: If this chemical gets into the eyes, irrigate immediately. If this chemical contacts the skin, flush with water immediately. If a person breathes in large amounts of this chemical, move the exposed person to fresh air at once and perform artificial respiration. When this chemical has been swallowed, get medical attention. Give large quantities of water and induce vomiting. Do not make an unconscious person vomit.

Personal Protective Methods: Wear appropriate clothing to prevent any possibility of skin contact. Wear eye protection to prevent any possibility of eye contact. Employees should wash immediately when skin is wet or contaminated. Remove clothing immediately if wet or contaminated to avoid flammability hazard. Provide emergency showers and eyewash.

Respirator Solution:
> 0.2 ppm: SA/SCBA
> 1 ppm: SAF/SA:PD, PP, CF/SCBAF
> 20 ppm: SAF:PD, PP, CF
> Escape: GMOV/SCBA

References
(1)   See Reference (A-61).

METHYL ISOPROPYL KETONE

Description: $CH_3COCH(CH_3)_2$ is a colorless liquid boiling at 93°C.

Code Numbers: CAS 563-80-4   RTECS EL9100000   UN 2397

DOT Designation: Flammable liquid.

Synonyms: 3-Methyl-2-butanone; isopropyl methyl ketone.

Potential Exposure: This ketone is used as a solvent.

Permissible Exposure Limits in Air: There is no Federal standard but ACGIH as of 1983/84 has adopted a TWA of 200 ppm (705 mg/m³). There is no tentative STEL value.

Permissible Concentration in Water: No criteria set.

Routes of Entry: Inhalation of vapor, skin absorption, ingestion, skin and eye contact.

Harmful Effects and Symptoms: Eye, nose, throat and skin irritation. In high concentrations, narcosis may be produced with symptoms of headache, nausea, vomiting, lightheadedness, dizziness, incoordination and unconsciousness.

Points of Attack: See "2-Pentanone," for guidance.

Medical Surveillance: See "2-Pentanone," for guidance.

First Aid: See "2-Pentanone," for guidance.

Personal Protective Methods: See "2-Pentanone," for guidance.

Respirator Selection: See "2-Pentanone," for guidance.

Disposal Method Suggested: See "2-Pentanone," for guidance.

METHYL MERCAPTAN

- Hazardous substance (EPA)
- Hazardous waste (EPA)

Description: CH_3SH is a colorless gas with a disagreeable odor like garlic. It boils at 6°C.

Code Numbers: CAS 74-93-1   RTECS PB4375000   UN 1064

DOT Designation: Flammable gas.

Synonyms: Methanethiol.

Potential Exposure: Methyl mercaptan is used in methionine synthesis, and widely as an intermediate in pesticide manufacture (A-32).

Incompatibilities: Strong oxidizers, bleaches.

Permissible Exposure Limits in Air: The Federal standard is 10 ppm (20 mg/m^3) as a 15-minute ceiling value. The ACGIH as of 1983/84 has set a TWA value of 0.5 ppm (1 mg/m^3) but no STEL value. The IDLH level is 400 ppm.

Determination in Air: Sample collection by charcoal tube, analysis by gas liquid chromatography (A-13).

Permissible Concentration in Water: No criteria set, but EPA (A-37) has suggested a permissible ambient goal of 13.8 µg/ℓ based on health effects.

Routes of Entry: Inhalation, skin and eye contact.

Harmful Effects and Symptoms: Narcosis; cyanosis; convulsions; headaches; nausea; pulmonary irritation; respiratory paralysis.

Points of Attack: Respiratory system, lungs, central nervous system.

Medical Surveillance: Consider the points of attack in preplacement and periodic physical examinations.

First Aid: If this chemical gets into the eyes, irrigate immediately. If this chemical contacts the skin, flush with water immediately. If a person breathes in large amounts of this chemical, move the exposed person to fresh air at once and perform artificial respiration.

Personal Protective Methods: Wear appropriate clothing to prevent skin freezing. Wear eye protection to prevent any reasonable probability of eye contact. Remove clothing immediately if wet or contaminated to avoid flammability hazard.

Respirator Selection:
        100 ppm: SA/SCBA
        400 ppm: SAF:PD, PP, CF/SCBAF
        Escape : GMOV/SCBA

Disposal Method Suggested: Incineration followed by effective scrubbing of the effluent gas.

References

(1)   See Reference (A-60).

METHYL METHACRYLATE MONOMER

- Hazardous substance (EPA)
- Hazardous waste (EPA)

Description: $CH_2=C(CH_3)COOCH_3$ is a colorless liquid with an acrid, fruity odor. It boils at 100°C.

Code Numbers: CAS 80-62-6   RTECS OZ5075000   UN 1247

DOT Designation: Flammable liquid.

Synonyms: Methyl 2-methyl-2-propenoate; methacrylic acid methyl ester; MME.

Potential Exposure: Virtually all of the methyl methacrylate monomer produced is used in the production of polymers such as surface coating resins, plastics (Plexiglas and Lucite), ion exchange resins and plastic dentures (1).

Incompatibilities: Nitrates, oxidizers, peroxides, polymerizers, strong alkalies, moisture.

Permissible Exposure Limits in Air: The Federal standard and the ACGIH 1983/84 TWA value is 100 ppm ($410\,mg/m^3$). The STEL is 125 ppm ($510\,mg/m^3$). The IDLH level is 4,000 ppm.

Determination in Air: Charcoal adsorption, workup with CS_2, analysis by gas chromatography. See NIOSH Methods, Set D.

Permissible Concentration in Water: A no-adverse-effect level in drinking water has been calculated by NAS/NRS to be 0.7 mg/ℓ.

Routes of Entry: Inhalation, ingestion, skin and eye contact.

Harmful Effects and Symptoms: Irritation of eyes, nose and throat; dermatitis; narcosis.

Points of Attack: Eyes, upper respiratory system, skin.

Medical Surveillance: Consider the points of attack in preplacement and periodic physical examinations.

First Aid: If this chemical gets into the eyes, irrigate immediately. If this chemical contacts the skin, flush with water promptly. If a person breathes in large amounts of this chemical, move the exposed person to fresh air at once and perform artificial respiration. When this chemical has been swallowed, get medical attention. Give large quantities of salt water and induce vomiting. Do not make an unconscious person vomit.

Personal Protective Methods: Wear appropriate clothing to prevent repeated or prolonged skin contact. Wear eye protection to prevent any reasonable probability of eye contact. Employees should wash promptly when skin is wet or contaminated. Remove clothing immediately if wet or contaminated to avoid flammability hazard.

Respirator Selection:

1,000 ppm: CCROVF
4,000 ppm: GMOV/SAF/SCBAF
Escape : GMOV/SCBA

Disposal Method Suggested: Incineration.

References

(1) U.S. Environmental Protection Agency, *Methyl Methacrylate,* Health and Environmental Effects Profile No. 130, Wash., DC, Office of Solid Waste (April 30, 1980).
(2) See Reference (A-61).
(3) See Reference (A-60).

METHYLNITROPROPYL-4-NITROSOANILINE

Description: MNNA is a grey-green powder with a rubber-like odor which decomposes on heating. It has the formula $ON-C_6H_4-NH-CH_2-C-(CH_3)_2$.
$$\begin{array}{c} | \\ NO_2 \end{array}$$

Code Numbers: CAS 24458-48-8

DOT Designation: —

Synonyms: 2-(p-Nitrosoanilinemethyl)-2-nitropropane, MNNA.

Potential Exposure: Was used as a rubber additive but use has been substantially discontinued because of instability of the compound.

Incompatibilities: Heat

Permissible Exposure Limits in Air: According to EPA (1), Monsanto tried to hold worker exposure to a TWA of 0.2 mg/m^3.

Permissible Concentration in Water: No criteria set.

Determination in Water: By liquid chromatograph equipped with variable wavelength UV detector (1).

Harmful Effects and Symptoms: Low acute toxicity is evidenced by oral LD_{50} value for rats of 2.73 g/kg. However, genotoxicity and carcinogenicity studies indicate a basis for concern (1) if there is sufficient exposure to MNNA. It does cause skin sensitization.

References

(1) U.S. Environmental Protection Agency, *Chemical Hazard Information Profile Draft Report: Methylnitropropyl-4-Nitrosoaniline,* Washington, D.C. (Dec. 27, 1983).

METHYL PARATHION

● Hazardous substance (EPA)
● Hazardous waste (EPA)

Description: Methyl parathion,

$$\begin{array}{c} S \\ \| \\ (CH_3O)_2P-O-C_6H_4-NO_2 \end{array}$$

is a crystalline solid melting at 37° to 38°C. It is used as an insecticide.

Code Numbers: CAS 298-00-0   RTECS TG0175000   UN (NA 2783)

DOT Designation: Poison B

Synonyms: Dimethyl parathion, O-p-nitrophenyl thiophosphate, Metaphos, Metacide®.

Potential Exposure: In recent years, methyl parathion has been produced in increasingly greater quantities in the United Stats and NIOSH estimates that approximately 150,000 United States workers are potentially exposed to methyl parathion in occupational settings. This applies to manufacturers (A-32), formulators and applicators.

Permissible Exposure Limits in Air: NIOSH recommends promulgation of an environmental limit of 0.2 mg of methyl parathion per cubic meter of air as a time-weighted average for up to a 10-hour workday, 40-hour workweek. ACGIH concurs as of 1983/84 and has set an STEL of 0.6 mg/m³. The notation "skin" is added to indicate the possibility of cutaneous absorption.

Determination in Air: Collection by impinger or fritted bubbler, analysis by gas liquid chromatography (1).

Permissible Concentration in Water: No criteria set. An ADI has been calculated at 0.0043 mg/kg/day by NAS/NRC based on human data on parathion. However, as NAS/NRC points out, it appears that the assumption has been made that methyl parathion is toxicologically the same as parathion and that extrapolations have been made from parathion toxicology to methyl parathion. The data on teratogenic effects of methyl parathion, however, indicate that this is not an acceptable procedure in this case.

Routes of Entry: Adherence to all provisions of the standard is required in workplaces using methyl parathion regardless of the airborne methyl parathion concentration because of serious effects produced by contact with the skin, mucous membranes, and eyes. Since methyl parathion does not irritate or burn the skin, no warning of skin exposure is likely to occur. However, methyl parathion is readily absorbed through the skin, mucous membranes, and eyes and presents a potentially great danger from these avenues of absorption. It is extremely important to emphasize that available evidence indicates that the greatest danger to employees exposed to methyl parathion is from skin contact.

Harmful Effects and Symptoms: Methyl parathion, an organophosphorus insecticide, is converted in the environment and in the body to methyl paraoxon, a potent inactivator of acetylcholinesterase, an enzyme responsible for terminating the transmitter action of acetylcholine at the junction of cholinergic nerve endings with their effector organs or postsynaptic sites.

The scientific basis for the recommended environmental limit is the prevention of medically significant inhibition of acetylcholinesterase. If functional acetylcholinesterase is greatly inhibited, a cholinergic crisis may ensue resulting in signs and symptoms of methyl parathion poisoning, including nausea, vomiting, abdominal cramps, diarrhea, involuntary defecation and urination, blurring of vision, muscular twitching, and difficulty in breathing.

Medical Surveillance: NIOSH recommends that medical surveillance, including preemployment and periodic examinations, shall be made available to workers who may be occupationally exposed to methyl parathion. Biologic monitoring is also recommended as an additional safety measure.

First Aid: If material gets on skin, wash immediately with soap and water and call a physician. (Soap with a pH above 8.0 is more effective than neutral soap.)

Personal Protective Methods: Personal protective equipment and protective clothing are recommended for those workers occupationally exposed to methyl parathion to further reduce exposure. In certain instances, such as emergency situations, NIOSH recommends that respirators be worn.

NIOSH also recommends that stringent work practices and engineering controls be adhered to in order to further reduce the likelihood of poisoning that may result from skin contact, ingestion, or from inhalation of methyl parathion.

Respirator Selection:

Concentration of Methyl Parathion		Respirator Type
2 mg/m^3 or less	(1)	Half-mask pesticide respirator
	(2)	Type C supplied-air respirator, demand type (negative pressure), with half-mask facepiece
10 mg/m^3 or less	(1)	Fullface gas mask (chin style or chest-or back-mounted type)
	(2)	Type C supplied-air respirator, demand type (negative pressure), with full facepiece
200 mg/m^3 or less	(1)	Type C supplied-air respirator, continuous-flow type, with full facepiece or suit
	(2)	Pressure-demand type respirator with full facepiece and impervious plastic shroud
Emergency (includes entry to vessels, bins, or other containers which are likely to be contaminated with methyl parathion)	(1)	Self-contained breathing apparatus, positive-pressure type, with full facepiece
	(2)	Combination supplied-air respirator, pressure-demand type, with auxiliary self-contained air supply.

Disposal Method Suggested: Incineration (1500°F, 0.5 second minimum for primary combustion; 2200°F, 1.0 second for secondary combustion) with adequate scrubbing and ash disposal facilities.

References

(1) National Institute for Occupational Safety and Health, *Criteria for a Recommended Standard: Occupational Exposure to Methyl Parathion,* NIOSH Doc. No. 77-106 (1977).

(2) United Nations Environment Programme, *IRPTC Legal File 1983,* Vol. II, pp VII/615-17, Geneva, Switzerland, International Register of Potentially Toxic Chemicals (1984).

METHYL SILICATE

Description: $(CH_3O)_4Si$ is a liquid boiling at 121°C.

Code Numbers: CAS 681-84-5   RTECS VV9800000   UN 2606

DOT Designation: Flammable liquid, poison.

Synonyms: Methyl orthosilicate; tetramethyloxysilane.

Potential Exposure: Methyl silicate is used in coating screens of television picture tubes. It may be used in mold binders and in corrosion-resistant coatings as well as in catalyst preparation and as a silicone intermediate.

Permissible Exposure Limits in Air: There are no Federal standards but ACGIH (1983/84) has proposed a TWA value of 1 ppm (6 mg/m^3) and an STEL value of 5 ppm (30 mg/m^3).

The Dutch Chemical Industry (A-60) has set a maximum allowable concentration of 5 ppm (30 mg/m^3) as a ceiling value.

Permissible Concentration in Water: No criteria set.

Harmful Effects and Symptoms: Corneal damage is the most serious hazard. Industrial experience has produced injuries ranging from eye pain to eye loss. Concentrations of 200-300 ppm for 15 minutes were required to produce minimal injury; exposure to 1,000 ppm resulted in hospitalization (A-34).

α-METHYLSTYRENE

Description: $C_6H_5C(CH_3{=}CH_2)$ is a colorless liquid with a characteristic odor. It boils at 165°C.

Code Numbers: CAS 98-83-9   RTECS WL5250000   UN 2303

DOT Designation: Flammable liquid.

Synonyms: AMS; isopropenylbenzene; 2-phenylpropene; 1-methyl-1-phenyl-ethylene.

Potential Exposure: Methylstyrene is used in the production of modified polyester and alkyd resin formulations.

Incompatibilities: Oxidizers, peroxides, halogens, catalysts for vinyl or ionic polymers: aluminum, iron chloride.

Permissible Exposure Limits in Air: The Federal standard is a 100 ppm (480 mg/m^3) ceiling. The ACGIH has adopted a TWA of 50 ppm (240 mg/m^3) and an STEL of 100 ppm (480 mg/m^3) as of 1983/84. The IDLH level is 5,000 ppm.

Determination in Air: Charcoal adsorption, workup with CS_2, analysis by gas chromatography. See NIOSH Methods, Set C. See also reference (A-10).

Permissible Concentration in Water: No criteria set.

Routes of Entry: Inhalation, ingestion, eye and skin contact.

Harmful Effects and Symptoms: Irritation of eyes, nose and throat; drowsiness; dermatitis.

Points of Attack: Eyes, respiratory system, skin.

Medical Surveillance: Consider the points of attack in preplacement and periodic physical examinations.

First Aid: If this chemical gets into the eyes, irrigate immediately. If this chemical contacts the skin, flush with water promptly. If a person breathes in large amounts of this chemical, move the exposed person to fresh air at once and perform artificial respiration. When this chemical has been swallowed, get medical attention. Do not induce vomiting.

Personal Protective Methods: Wear appropriate clothing to prevent repeated or prolonged skin contact. Wear eye protection to prevent any reasonable probability of eye contact. Employees should wash promptly when skin is wet or contaminated. Remove nonimpervious clothing promptly if wet or contaminated.

Respirator Selection:

> 1,000 ppm: CCROVF
> 5,000 ppm: GMOV/SAF/SCBAF
> Escape : GMOV/SCBA

Disposal Method Suggested: Incineration.

References

(1) See Reference (A-61).
(2) See Reference (A-60).

METRIBUZIN

Description: $C_8H_{14}N_4OS$ with the following structural formula

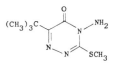

is a white crystalline solid melting at 125° to 126.5°C.

Code Numbers: CAS 21087-64-9   RTECS XZ2990000

DOT Designation: —

Synonyms: 4-amino-6-tert-butyl-3-(methylthio)-as-triazin-S(4H)-one, Lexone®, Sencor®, Sencoral®, Sengoral®, Sencorex®.

Potential Exposure: Those involved in manufacture, formulation and application of this herbicide.

Permissible Exposure Limits in Air: A TLV of 5 mg/m³ has been adopted by ACGIH as of 1983/84.

Permissible Concentration in Water: No criteria set.

Harmful Effects and Symptoms: This material is slightly toxic (LD_{50} for rats 2,200 mg/kg). No cases of poisoning have been reported in man.

References

(1) American Conference of Governmental Industrial Hygienists, *Documentation of the Threshold Limit Values: Supplemental Documentation 1982,* Cincinnati, Ohio (1982).

MEVINPHOS

● Hazardous substance (EPA)

$$\text{O}$$
$$\text{\textbardbl}$$

Description: $(CH_3O)_2P-OC(CH_3)=CHCOOCH_3$ is a pale yellow to orange high-boiling liquid.

Code Numbers: CAS 7786-34-7   RTECS GQ5250000   UN 2783

DOT Designation: Poison B.

Synonyms: 2-Carbomethoxy-1-methylvinyl dimethyl phosphate; 3-Hydroxycrotonic acid methyl ester dimethyl phosphate; ENT 22,374; Phosdrin®.

Potential Exposure: Those engaged in the manufacture, formulation and application of this contact and systemic insecticide and acaricide.

Incompatibilities: Strong oxidizers.

Permissible Exposure Limits in Air: The Federal standard is 0.1 mg/m³. The 1983/84 ACGIH TWA value is 0.01 ppm (0.1 mg/m³) and the STEL is 0.03 (0.3 mg/m³). The notation "skin" is added to indicate the possibility of cutaneous absorption. The IDLH level is 40 mg/m³.

Determination in Air: Sample collection by impinger or fritted bubbler, analysis by gas liquid chromatography (A-1).

Permissible Concentration in Water: No criteria set.

Routes of Entry: Inhalation, skin absorption, ingestion, skin and eye contact.

Harmful Effects and Symptoms: Miosis, rhinorrhea, headaches, wheezing, laryngeal spasms, salivation, cyanosis; anorexia, nausea, abdominal cramps, diarrhea; paralysis, ataxia, convulsions; low blood pressure; skin and eye irritation.

Points of Attack: Respiratory system, lungs, central nervous system, cardiovascular system, skin, blood cholinesterase.

Medical Surveillance: Consider the points of attack in preplacement and periodic physical examinations.

First Aid: If this chemical gets into the eyes, irrigate immediately. If this chemical contacts the skin, wash with soap immediately. If a person breathes in large amounts of this chemical, move the exposed person to fresh air at once and perform artificial respiration. When this chemical has been swallowed, get medical attention. Give large quantities of water and induce vomiting. Do not make an unconscious person vomit.

Personal Protective Methods: Wear appropriate clothing to prevent any possibility of skin contact. Wear eye protection to prevent any possibility of eye contact. Employees should wash immediately when skin is wet or contaminated. Remove nonimpervious clothing immediately if wet or contaminated. Provide emergency showers and eyewash.

Respirator Selection:
    1 mg/m³: SA/SCBA
    5 mg/m³: SAF/SCBAF
    40 mg/m³: SA:PD, PP, CF
        Escape: GMSOVPPest/SCBA

Disposal Method Suggested: Mevinphos is 50% hydrolyzed in aqueous solutions at an unspecified temperature in 1.4 hours at pH 11, 35 days at pH 7, and 120 days at pH 6. Decomposition is rapidly accomplished by lime sulfur (A-32). Mevinphos may also be incinerated.

References

(1) United Nations Environment Programme, *IRPTC Legal File 1983,* Vol. I, pp VII/219–21, Geneva, Switzerland, International Register of Potentially Toxic Chemicals (1984).

MICA

Description: $K_2Al_4(Al_2Si_6O_{20})(OH)_4$ (Muscovite) takes the form of colorless, odorless flakes or sheets.

Code Numbers: CAS 12001-26-2   RTECS VV8760000

DOT Designation: –

Synonyms: Muscovite; Amber mica; Roscoelite; Lepidolite; Phlogopite; Biotite; Zinnwaldite; Fluorophlogopite.

Potential Exposure: Mica is used for insulation in electrical equipment; it is used in the manufacture of roofing shingles, wallpaper and paint.

Incompatibilities: None.

Permissible Exposure Limits in Air: The Federal standard is 20 mppcf.

Permissible Concentration in Water: No criteria set.

Routes of Entry: Inhalation.

Harmful Effects and Symptoms: Pneumoconiosis; cough, dyspnea; weakness; weight loss.

Points of Attack: Lungs.

Medical Surveillance: Consider the points of attack in preplacement and periodic physical examinations.

First Aid: If this chemical gets into the eyes, irrigate immediately. If a person breathes in large amounts of this chemical, move the exposed person to fresh air at once.

Personal Protective Methods: No particular devices are suggested by NIOSH except for respirators.

Respirator Selection:
      100 mppcf: D
      200 mppcf: DXS/FuHiEP/SA/SCBA
   1,000 mppcf: HiEPF/SAF/SCBAF
 10,000 mppcf: PAPHiE/SA:PD, PP, CF

Disposal Method Suggested: Landfill.

References

(1) See Reference (A-61).

(2) Parmeggiani, L., Ed., *Encyclopedia of Occupational Health & Safety,* Third Edition, Vol. 2, pp. 1358–59, Geneva, International Labour Office (1983).

MICHLER'S KETONE

● Carcinogen (Animal Positive) (A-64)

- Hazardous Waste Constituent (EPA)

Description: $C_{17}H_{20}N_2O$ with the formula

$$(CH_3)_2N-C_6H_4-CO-C_6H_4-N(CH_3)_2$$

is a white to greenish crystalline solid melting at 172°C.

Code Numbers: CAS 90-94-8   RTECS DJ0250000

DOT Designation: —

Synonyms: Tetramethyldiaminobenzophenone.

Potential Exposure: Michler's ketone is used as a chemical intermediate in the synthesis of at least 13 dyes and pigments, especially auramine derivatives.

EPA reported 10 producers and importers in three regions, with a domestic production of 37,000 lb and imports of 37,000 lb (Toxic Substances Control Act (TSCA), Chemical Substances Inventory, 1979, public record). One domestic producer manufactured a quantity insufficient for reporting to the U.S. International Trade Commission in 1979-1980, but imports of Michler's ketone amounted to 40,000 lb in 1980.

Occupational exposure is greatest for workers in facilities that manufacture the compounds or any of the dyestuffs for which Michler's ketone is an intermediate. In 1974, the National Occupational Hazard Survey made no estimate on potential worker exposure to Michler's ketone (NIOSH). Residual levels or trace impurities of Michler's ketone may possibly be present in some dyes based on this chemical and in the final consumer product. Exposure even to trace amounts may be a cause for concern. This concern is based on the experience with other dyes derived from aromatic amines. Data are inadequate to describe the actual levels of impurities in the final product, the potential for consumer exposure, and the uptake.

Permissible Exposure Limits in Air: No standards set.

Permissible Concentration in Water: No criteria set.

Routes of Entry: Inhalation and skin absorption.

Harmful Effects and Symptoms: Michler's ketone, administered in the feed, was carcinogenic in male and female Fischer 344 rats and female B6C3F1 mice, causing hepatocellular carcinomas; in male B6C3F1 mice it produced hemangiosarcomas (1). These results can be considered as providing sufficient evidence for the carcinogenicity of Michler's ketone in experimental animals (2).

References

(1)   National Cancer Institute, *Bioassay of Michler's Ketone for Possible Carcinogenicity,* Technical Report Series No. 181, DHEW Publication No. (NIH) 78-1737, Bethesda, Maryland (1979).

(2)   Griesemer, R.A., and C. Cueto, "Toward a Classification Scheme for Degrees of Experimental Evidence for the Carcinogenicity of Chemicals for Animals," In: R. Montesano, H. Bartsch, and L. Tomatis (eds.), *Molecular and Cellular Aspects of Carcinogen Screening Tests, IARC Scientific Publications, No. 27,* Lyon, France: International Agency for Research on Cancer, pp 259-81 (1980).

MIREX

- Carcinogen (Animal Positive, IARC) (2)

Description: $C_{10}H_{12}$ with the following structural formula

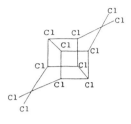

is a white crystalline solid which decomposes at $485°C$.

Code Numbers: CAS 2385-85-5   RTECS PC8225000

DOT Designation: –

Synonyms: Hexachlorocyclopentadiene dimer; dodecachloropentacyclodecane; Dechlorane®.

Potential Exposure: Those involved in the manufacture, formulation and application of the insecticide (particularly effective against fire ants). Also used as a fire retardant in plastics.

According to the National Occupational Hazard Survey, an estimated 1,000 workers were potentially exposed to Mirex. Because insecticidal use of Mirex has been discontinued, direct human exposure is small. However, residues have been detected in water, soil, food and beverages, and human tissues for as long as 12 years after exposure. These residues result in potential routes for general population exposure. Mirex is known to degrade in soil to yield Kepone (3).

Permissible Exposure Limits in Air: No standards set.

Permissible Concentration in Water: 0.001 mg/ℓ for protection of aquatic life (1).

Harmful Effects and Symptoms: This compound is moderately toxic (the LD_{50} value for rats is 300 mg/kg).

Mirex has been tested in one experiment in two strains of mice (1) and in one experiment in rats by oral administration. It has also been tested in two strains of mice by subcutaneous injection of single doses. In the studies using oral administration, it produced benign and malignant liver tumors in mice and rats of both sexes. An excess of liver tumors was also found in males of one of the two strains of mice following a single subcutaneous injection; this experiment also suggested that it produced reticulum-cell sarcomas in males of both strains (2) (3).

Disposal Method Suggested: High-temperature incineration.

References

(1) Sax, N.I., Ed., *Dangerous Properties of Industrial Materials Report, 1,* No. 2, 48, New York, Van Nostrand Reinhold Co. (1980).
(2) *IARC Monographs on the Evaluation of Carcinogenic Risk of Chemicals to Man,* Vol. 20, IARC, Lyon, France, pp 283-301 (1979).
(3) See Reference (A-62). Also see Reference (A-64).
(4) United Nations Environment Programme, *IRPTC Legal File 1983,* Vol. II, pp VII/459, Geneva, Switzerland, International Register of Potentially Toxic Chemicals (1984).

MOLYBDENUM AND COMPOUNDS

Description: Mo, molybdenum, is a silver-white metal or a greyish-black powder. Molybdenite is the only important commercial source. This ore is often associated with copper ore. Molybdenum is insoluble in water and soluble in hot concentrated nitric and sulfuric acid.

Code Numbers: Molybdenum metal: CAS 7439-98-7   RTECS QA4680000

DOT Designation: –

Synonyms: None.

Potential Exposure: Most of the molybdenum produced is used in alloys: steel, stainless steel, tool steel, cast iron, steel mill rolls, manganese, nickel, chromium, and tungsten. The metal is used in electronic parts (contacts, spark plugs, x-ray tubes, filaments, screens, and grids for radios), induction heating elements, electrodes for glass melting, and metal spraying applications. Molybdenum compounds are utilized as lubricants; as pigments for printing inks, lacquers, paints, for coloring rubber, animal fibers, and leather, and as a mordant; as catalysts for hydrogenation cracking, alkylation, and reforming in the petroleum industry, in Fischer-Tropsch synthesis, in ammonia production, and in various oxidation-reduction and organic cracking reactions; as a coating for quartz glass; in vitreous enamels to increase adherence to steel; in fertilizers, particularly for legumes; in electroplating to form protective coatings; and in the production of tungsten.

Hazardous exposures may   occur during high-temperature treatment in the fabrication and production of molybdenum products, spraying applications, or through loss of catalyst. MoO_3 sublimes above 800°C.

Incompatibilities: Soluble compounds—alkali metals, sodium, potassium, molten magnesium. Insoluble compounds—strong oxidizers.

Permissible Exposure Limits in Air: The Federal standards are—molybdenum, soluble compounds, 5 mg/m^3; molybdenum, insoluble compounds, 15 mg/m^3. The ACGIH as of 1983/84 has set a TWA for insoluble Mo compounds of 10 mg/m^3. The tentative STEL values are 10 mg/m^3 for soluble compounds, 20 mg/m^3 for insoluble compounds.

Determination in Air: Soluble compounds—collection on a filter, workup with water, analysis by atomic absorption. See NIOSH Methods, Set M. See also reference (A-10). Insoluble compounds—collection on a filter, workup with acid, analysis by atomic absorption. See NIOSH Methods, Set O. See also reference (A-10).

Permissible Concentration in Water: The USSR has established a molybdenum limit of 0.5 mg/ℓ in open water, but the WHO has not promulgated a limit. The *National Primary Drinking Water Regulations* (USEPA, 1975) do not set any limit for molybdenum but EPA (A-37) has suggested a permissible ambient goal of 70 μg/ℓ based on health effects.

Determination in Water: With atomic-absorption spectrophotometry, a detection limit of 20 μg/ℓ is attainable by direct aspiration into the flame, necessitating concentration for ordinary determinations.

When the graphite furnace is used to increase sample atomization the detection limit is lowered to 0.5 μg/ℓ.

Neutron activation may be used at even lower detection limits, according to USEPA.

Routes of Entry: Inhalation of dust or fume, ingestion (eye and skin contact for soluble compounds).

Harmful Effects and Symptoms: *Local* – Molybdenum trioxide may produce irritation of the eyes and mucous membranes of the nose and throat. Dermatitis from contact with molybdenum is unknown.

Systemic – No reports of toxic effects of molybdenum in the industrial setting have appeared. It is considered to be an essential trace element in many species, including man. Animal studies indicate that insoluble molybdenum compounds are of a low order of toxicity (e.g., disulfide, oxides, and halides). Soluble compounds (e.g., sodium molybdate) and freshly generated molybdenum fumes, however, are considerably more toxic. Inhalation of high concentrations of molybdenum trioxide dust is very irritating to animals and has caused weight loss, diarrhea, loss of muscular coordination, and a high mortality rate. Molybdenum trioxide dust is more toxic than the fumes. Large oral doses of ammonium molybdate in rabbits caused some fetal deformities. Excessive intake of molybdenum may produce signs of a copper deficiency.

Points of Attack: Soluble Compounds–respiratory system, lungs; in animals: kidneys, blood. Insoluble Compounds—none known.

Medical Surveillance: Preemployment and periodic physical examinations should evaluate any irritant effects to the eyes or respiratory tract and the general health of the worker. Although molybdenum compounds are of a low order of toxicity, animal experimentation indicates protective measures should be employed against the more soluble compounds and molybdenum trioxide dust and fumes. The normal intake of copper in the diet appears to be sufficient to prevent systemic toxic effects due to molybdenum poisoning.

First Aid: (Soluble Mo Compounds) – If this chemical gets into the eyes, irrigate immediately. If this chemical contacts the skin, flush with water. If a person breathes in large amounts of this chemical, move the exposed person to fresh air at once and perform artificial respiration. When this chemical has been swallowed, get medical attention. Give large quantities of water and induce vomiting. Do not make an unconscious person vomit.

(Insoluble Mo Compounds)–If a person breathes in large amounts of this chemical, move the exposed person to fresh air at once and perform artificial respiration. When this chemical has been swallowed, get medical attention. Give large quantities of water and induce vomiting. Do not make an unconscious person vomit.

Personal Protective Methods: (For Soluble Compounds)–Wear appropriate clothing to prevent repeated or prolonged skin contact. Wear eye protection to prevent any reasonable probability of eye contact. Employees should wash promptly when skin is wet or contaminated. Remove nonimpervious clothing promptly if wet or contaminated.

(For Insoluble Compounds) – No special protection.

Respirator Selection: (For Soluble Compounds) –

25 mg/m³:	DM
50 mg/m³:	DMXSQ/HiEP
250 mg/m³:	HiEPF/SAF/SCBAF
5,000 mg/m³:	PAPHiEF
10,000 mg/m³:	SAF:PD, PP, CF

Respirator Selection: (For Insoluble Compounds) —

    75 mg/m³: DM/DMXS
    150 mg/m³: DMXSQ/DMFu
    150 mg/m³: FuHiEP/SA/SCBA
    750 mg/m³: HiEPF/SAF/SCBAF
    7,500 mg/m³: PAPHiE/SA:PD, PP, CF

Disposal Method Suggested: Recovery is indicated whenever possible. Processes for recovery of Mo from scrap, flue dusts, spent catalysts and other industrial wastes have been described (A-57).

References

(1) U.S. Environmental Protection Agency, *Molybdenum—A Toxicological Appraisal,* Report EPA-600/1-75-004, Research Triangle Park, NC Health Effects Research Laboratory (Nov. 1975).

(2) U.S. Environmental Protection Agency, *Toxicology of Metals, Vol. II: Molybdenum,* Report EPA-600/1-77-022, Research Triangle Park, NC, pp 345-357, (May 1977).

(3) See Reference (A-61).

(4) Parmeggiani, L., Ed., *Encyclopedia of Occupational Health & Safety,* Third Edition, Vol. 2, pp 1403-4, Geneva, International Labour Office (1983).

MONOCROTOPHOS

Description: $C_7H_{14}NO_5P$, dimethyl-1-methyl-3-(methylamino-3-oxo-1-propenyl phosphate, with the formula,

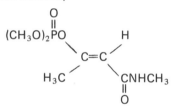

is a crystalline solid with a mild ester odor. The pure compound melts at 54°-55°C and the reddish brown technical product at 25°-30°C.

Code Numbers: CAS 6923-22-4   RTECS TC4375000   UN 2783

DOT Designation: —

Synonyms: ENT 27129; Nuvacron®; Azodrin®.

Potential Exposure: Those involved in the manufacture (A-32), formulation and application of this insecticide.

Permissible Exposure Limits in Air: There is no Federal standard but ACGIH (1983/84) has adopted a TWA value of 0.25 mg/m³. There is no tentative STEL value.

Permissible Concentration in Water: No criteria set.

Harmful Effects and Symptoms: Monocrotophos is a highly toxic, direct acting, water-soluble cholinesterase inhibitor which appears to be capable of penetration through the skin but which is excreted rapidly and does not accumulate in the body (A-34).

References

(1) United Nations Environment Programme, *IRPTC Legal File 1983,* Vol. II, pp VII/558-9, Geneva, Switzerland, International Register of Potentially Toxic Chemicals (1984).

MONOMETHYLHYDRAZINE

Description: CH_3NHNH_2 is a fuming, colorless liquid with an ammonia-like odor. It boils at $88°C$.

Code Numbers: CAS 60-34-4 RTECS MV5600000 UN 1244

DOT Designation: Flammable liquid and poison.

Synonyms: Methylhydrazine, MMH.

Potential Exposure: MMH has been used as the propellant in liquid propellant rockets; it is also used as a solvent and as an organic intermediate.

Incompatibilities: Oxides of iron, copper; manganese, lead, copper, alloys; porous materials, earth, asbestos, wood, cloth; oxidants, hydrogen peroxide, nitric acid.

Permissible Exposure Limits in Air: The Federal standard and the 1983/84 ACGIH TWA value is 0.2 ppm (0.35 mg/m^3) as a ceiling value. The notation "skin" indicates the possibility of cutaneous absorption. There is no STEL value set. The IDLH level is 5 ppm.

Determination in Air: Absorption in a bubbler with hydrochloric acid. Reaction with phosphomolybdic acid and colorimetric determination. See NIOSH Methods, Set K. See also reference (A-10).

Permissible Concentration in Water: No criteria set, but EPA (A-37) has suggested a permissible ambient goal of 5 μg/ℓ based on health effects.

Routes of Entry: Inhalation, skin absorption, ingestion, skin and eye contact.

Harmful Effects and Symptoms: Eye irritation; vomiting, diarrhea; respiratory irritation, tremors, ataxia; anoxia, cyanosis, convulsions.

Points of Attack: Central nervous system, respiratory system, liver, blood, cardiovascular system, eyes.

Medical Surveillance: Consider the points of attack in preplacement and periodic physical examinations.

First Aid: If this chemical gets into the eyes, irrigate immediately. If this chemical contacts the skin, flush with water immediately. If a person breathes in large amounts of this chemical, move the exposed person to fresh air at once and perform artificial respiration. When this chemical has been swallowed, get medical attention. Give large quantities of water and induce vomiting. Do not make an unconscious person vomit.

Personal Protective Methods: Wear appropriate clothing to prevent any possibility of skin contact. Wear eye protection to prevent any possibility of eye contact. Employees should wash immediately when skin is wet or contaminated. Remove clothing immediately if wet or contaminated to avoid flammability hazard. Provide emergency showers and eyewash.

Respirator Selection:
     5 ppm: SAF/SCBAF
     Escape : GMSF/SCBAF

Disposal Method Suggested: There are 2 alternatives (A-38): Dilute with water, neutralize with sulfuric acid, then flush to sewer with large volumes of water or incinerate with added flammable solvent in furnace equipped with afterburner and alkaline scrubber.

References

(1)   Nat. Inst. for Occup. Safety and Health, *Criteria for a Recommended Standard: Occupational Exposure to Hydrazines,* NIOSH Doc. No. 78-172, Wash., DC (1978).

(2)   Sax, N.I., Ed., *Dangerous Properties of Industrial Materials Report, 2,* No. 5, 86-90, New York, Van Nostrand Reinhold Co. (1982).

MORPHOLINE

Description: $\overline{OCH_2CH_2NHCH_2CH_2}$ is a colorless liquid with a weak ammonia-like odor. It boils at 129°C.

Code Numbers: CAS 110-91-8   RTECS QD6475000   UN 2054

DOT Designation: Flammable liquid.

Synonyms: Tetrahydro-1,4-oxazine; Diethyleneimide oxide.

Potential Exposure: Morpholine is used in the synthesis of rubber accelerators, and pharmaceuticals. It is also used as a solvent, as a boiler water additive and in the formulation of waxes, polishes and cleaners.

Incompatibilities: Strong acids, strong oxidizers.

Permissible Exposure Limits in Air: The Federal standard and the 1983/84 ACGIH TWA value is 20 ppm (70 mg/m³). The notation "skin" indicates the possiblity of cutaneous absorption. The STEL is 30 ppm (105 mg/m³). The IDLH level is 8,000 ppm.

Determination in Air: Adsorption on silica, workup with sulfuric acid, analysis by gas chromatography. See NIOSH Methods, Set K. See also reference (A-10).

Permissible Concentration in Water: No criteria set.

Routes of Entry: Inhalation, skin absorption, ingestion and skin and eye contact.

Harmful Effects and Symptoms: Visual disturbances; eye, skin and nasal irritation; coughing, respiratory irritation.

Points of Attack: Respiratory system, eyes, skin.

Medical Surveillance: Consider the points of attack in preplacement and periodic physical examinations.

First Aid: If this chemical gets into the eyes, irrigate immediately. If this chemical contacts the skin, flush with water immediately. If a person breathes in large amounts of this chemical, move the exposed person to fresh air at once and perform artificial respiration. When this chemical has been swallowed, get medical attention. Give large quantities of water and induce vomiting. Do not make an unconscious person vomit.

Personal Protective Methods: Wear appropriate clothing to prevent any possibility of skin contact with liquids of >25% content or repeated or prolonged contact with liquids of <25% content. Wear eye protection to prevent any possibility of eye contact. Employees should wash promptly when skin is wet or contaminated. Remove clothing immediately if wet or contaminated to avoid flammability hazard. Provide emergency showers and eyewash if liquids containing >15% contaminants are involved.

Respirator Selection:

1,000 ppm: SAF/SCBAF
8,000 ppm: SAF:PD,PP,CF
Escape : GMOV/SCBA

Disposal Method Suggested: Controlled incineration (incinerator is equipped with a scrubber or thermal unit to reduce NO_x emissions).

References

(1) U.S. Environmental Protection Agency, *Chemical Hazard Information Profile: Morpholine,* Wash., DC (Nov. 16, 1977).
(2) See Reference (A-61).
(3) Sax, N.I., Ed., *Dangerous Properties of Industrial Materials Report, 1,* No. 8, 82–84, New York, Van Nostrand Reinhold Co. (1981).
(4) See Reference (A-60).
(5) Parmeggiani, L., Ed., *Encyclopedia of Occupational Health & Safety,* Third Edition, Vol. 2, pp 1406–7, Geneva, International Labour Office (1983).
(6) United Nations Environment Programme, *IRPTC Legal File 1983,* Vol. II, pp VII/462, Geneva, Switzerland, International Register of Potentially Toxic Chemicals (1984).

MUSTARD GAS

- Carcinogen (Human, Suspected, IARC)
- Waste Constituent (EPA)

Description: $(ClCH_2CH_2)_2S$ is an oily liquid with a sweet odor, melting at 13° to 14°C and boiling at 215° to 217°C.

Code Numbers: CAS 505-60-2   RTECS WQ090000

DOT Designation: —

Synonyms: Bis-(2-chloroethyl)sulfide; sulfur mustard.

Potential Exposure: Mustard gas was used in chemical warfare during World War 1. Mustard gas has been tested as an antineoplastic agent, but its clinical use as a tumor inhibitor has been minimal. It is used as a model compound in biological studies on alkylating agents.

Permissible Exposure Limits in Air: No standards set.

Permissible Concentration in Water: No criteria set.

Routes of Entry: Ingestion, skin and eye contact.

Harmful Effects and Symptoms: Mustard gas is a powerful and deadly vesicant. It can cause blindness. Ulceration of skin and necrosis of the respiratory tract can occur. Ingestion can cause nausea and vomiting.

Mustard gas is carcinogenic in mice, the only species tested, after inhalation or intravenous injection producing lung tumors, and after subcutaneous injection producing local sarcomas (1).

Several studies have shown an increased mortality from respiratory tract cancer among individuals exposed to mustard gas. This mortality was greater in those with chronic occupational exposure than in those with sporadic exposure (1,2).

Points of Attack: Eyes, skin, respiratory tract.

Personal Protective Methods: Wear rubber gloves, overalls.

Respirator Selection: Wear self-contained breathing apparatus (A-38).

Disposal Method Suggested: Incineration.

References
(1)   I ARC Monographs 9:181-92, 1975.
(2)   *IARC Monographs on the Evaluation of the Carcinogenic Risk of Chemicals to Humans,* Supplement I, IARC, Lyon, France, p 38 (1979).
(3)   See Reference (A-62). Also see Reference (A-64).

N

NALED

● Hazardous substance (EPA)

Description: $(CH_3O)_2POCHBrCBrCl_2$ (with O double-bonded to P) is a colorless solid or straw-colored liquid with a slightly pungent odor. It melts at 27°C.

Code Numbers: CAS 300-76-5 RTECS TB9450000 UN (NA 2783)

DOT Designation: ORM-E.

Synonyms: 1,2-Dibromo-2,2-dichloroethyl dimethyl phosphate; bromchlophos (in South Africa); dibrom (in Denmark), BRP (in Japan); ENT-24988, Dibrom®.

Potential Exposures: Those involved in the manufacture, formulation and application of this insecticide and acaricide.

Incompatibilities: Strong oxidizers.

Permissible Exposure Limits in Air: The Federal limit is $3\,mg/m^3$. The STEL adopted by ACGIH (1983/84) is $6\,mg/m^3$. The IDLH level is $1,800\,mg/m^3$.

Determination in Air: Sample collection by impinger or fritted bubbler, anslysis by gas liquid chromatography (A-1).

Routes of Entry: Inhalation, skin absorption, ingestion, eye and skin contact.

Harmful Effects and Symptoms: Miosis, eye irritation; headaches; tightness of chest, wheezing; laryngeal spasms, salivation; cyanosis, anorexia, nausea, abdominal cramps, diarrhea; weakness, twitching, paralysis, giddiness, ataxia, convulsions; low blood pressure.

Points of Attack: Respiratory system, central nervous system, cardiovascular system, skin, eyes, blood cholinesterase.

Medical Surveillance: Consider the points of attack in preplacement and periodic physical examinations.

First Aid: If this chemical gets into the eyes, irrigate immediately. If this chemical contacts the skin, wash with soap immediately. If a person breathes in large amounts of this chemical, move the exposed person to fresh air at once and perform artificial respiration. When this chemical has been swallowed, get medical attention. Give large quantities of water and induce vomiting. Do not make an unconscious person vomit.

Personal Protective Methods: Wear appropriate clothing to prevent any reasonable probability of skin contact. Wear eye protection to prevent any possibility of eye contact. Employees should wash immediately when skin is wet or contaminated. Work clothing should be changed daily if it is possible that clothing is contaminated. Remove nonimpervious clothing immediately if wet or contaminated. Provide emergency eyewash.

Respirator Selection:

15 mg/m^3:	DMSXPest
30 mg/m^3:	DMXSQPest/FuHiEP/SA/SCBA
150 mg/m^3:	HiEPF/SAF/SCBAF
1,800 mg/m^3:	PAPHiE/SA:PD,PP
Escape:	DMXS/SCBA

Disposal Method Suggested: This pesticide is more stable to hydrolysis than dichlorvos (50% hydrolysis at pH 9 at 37.5°C in 301 minutes). It is unstable in alkaline conditions, in presence of iron, and is degraded by sunlight. About 10% hydrolysis per day is obtained in ambient water (A-32).

References

(1) U.S. Environmental Protection Agency, *Investigation of Selected Potential Environmental Contaminants: Haloalkyl Phosphates*, Report EPA 560/2-76-007, Wash., DC (Aug. 1976).

(2) United Nations Environment Programme, *IRPTC Legal File 1983,* Vol. II, pp VII/549–50, Geneva, Switzerland, International Register of Potentially Toxic Chemicals (1984).

NAPHTHALENE

- Hazardous substance (EPA)
- Hazardous waste (EPA)
- Priority toxic pollutant (EPA)

Description: $C_{10}H_8$, naphthalene, is a white crystalline solid with a characteristic "moth ball" odor. It melts at 74° to 80°C and boils at 218°C.

Code Numbers: CAS 91-20-3   RTECS QJ0525000   UN 1334 and 2304 (molten)

DOT Designation: ORM-A.

Synonyms: Naphthalin, moth flake, tar camphor, white tar, moth balls.

Potential Exposures: Naphthalene is used as a chemical intermediate or feedstock for synthesis of phthalic, anthranilic, hydroxyl (naphthols), amino (naphthylamines), and sulfonic compounds which are used in the manufacture of various dyes. Naphthalene is also used in the manufacture of hydronaphthalenes, synthetic resins, lampblack, smokeless powder, and celluloid. Naphthalene has been used as a moth repellent.

Approximately 100 million people worldwide have G6PD deficiency which would make them more susceptible to hemolytic anemia on exposure to naphthalene. At present, more than 80 variants of this enzyme deficiency have been identified. The incidence of this deficiency is 0.1% in American and European Caucasians, but can range as high as 20% in American blacks and greater than 50% in certain Jewish groups.

Newborn infants have a similar sensitivity to the hemolytic effects of naphthalene, even without G6PD deficiency.

Incompatibilities: Strong oxidizers.

Permissible Exposure Limits in Air: The Federal standard and the ACGIH (1983/84) TWA value is 10 ppm (50 mg/m^3). The STEL value is 15 ppm (75 mg/m^3). The IDLH level is 500 ppm.

Determination in Air: Adsorption on charcoal, workup with CS_2, analysis by gas chromatography. See NIOSH Methods, Set T. See also reference (A-10).

Permissible Concentration in Water: To protect freshwater aquatic life— 2,300 $\mu g/\ell$ on an acute toxicity basis and 620 $\mu g/\ell$ on a chronic basis. To protect saltwater aquatic life—2,350 $\mu g/\ell$ on an acute toxicity basis. For the protection of human health from the toxic properties of naphthalene ingested through water and through contaminated aquatic organisms, no ambient water criterion has been set due to insufficient data.

Determination in Water: Methylene chloride extraction followed by high pressure liquid chromatography with fluorescence or UV detection; or gas chromatography (EPA Method 610) or gas chromatography plus mass spectrometry (EPA Method 625).

Routes of Entry: Inhalation of vapor or dust, skin absorption, ingestion, skin and eye contact.

Harmful Effects and Symptoms: *Local* — Naphthalene is a primary irritant and causes erythema and dermatitis upon repeated contact. It is also an allergen and may produce dermatitis in hypersensitive individuals. Direct eye contact with the dust has produced irritation and cataracts.

Systemic — Inhaling high concentrations of naphthalene vapor or ingesting may cause intravascular hemolysis and its consequences. Initial symptoms include eye irritation, headache, confusion, excitement, malaise, profuse sweating, nausea, vomiting, abdominal pain, and irritation of the bladder.

There may be progressive jaundice, hematuria, hemoglobinuria, renal tubular blockage, and acute renal shutdown. Hematologic features include red cell fragmentation, icterus, severe anemia with nucleated red cells, leukocytosis, and dramatic decreases in hemoglobin, hematocrit, and red cell count. Individuals with a deficiency of glucose-6-phosphate dehydrogenase in erythrocytes are more susceptible to hemolysis by naphthalene.

Points of Attack: Eyes, blood, liver, kidneys, skin, red blood cells, central nervous system.

Medical Surveillance: Consider eyes, skin, blood, liver, and renal function in placement and follow-up examinations. Low erythrocyte glucose-6-phosphate dehydrogenase increases risk.

First Aid: If this chemical gets into the eyes, irrigate immediately. If molten naphthalene contacts the skin, flush with water immediately. Wash promptly after solution contact. If a person breathes in large amounts of this chemical, move the exposed person to fresh air at once and perform artificial respiration. When this chemical has been swallowed, get medical attention. Give large quantities of water and induce vomiting. Do not make an unconscious person vomit.

Personal Protective Methods: Wear appropriate clothing to prevent repeated or prolonged skin contact. Wear eye protection to prevent any reasonable probability of eye contact. Employees should wash promptly when skin is wet or contaminated. Work clothing should be changed daily if it is possible that clothing is contaminated. Remove nonimpervious clothing promptly if wet or contaminated.

Respirator Selection:
500 ppm: CCROVD/GMOVD/SAF/SCBAF
Escape: GMOVP/SCBA

Disposal Method Suggested: Incineration.

References

(1) U.S. Environmental Protection Agency, *Naphthalene: Ambient Water Quality Criteria*, Wash., DC (1980).
(2) Nat. Inst. for Occup. Safety and Health, *Profiles on Occupational Hazards for Criteria Document Priorities: Naphthalene*, pp 269-273, Report PB-274,073, Cincinnati, Ohio (1977).
(3) U.S. Environmental Protection Agency, *Naphthalene,* Health and Environmental Effects Profile No. 131, Wash., DC, Office of Solid Waste (April 30, 1980).
(4) See Reference (A-61).
(5) See Reference (A-60).
(6) Parmeggiani, L., Ed., *Encyclopedia of Occupational Health & Safety,* Third Edition, Vol. 2, pp 1425-26, Geneva, International Labour Office (1983).

NAPHTHAS

Description: Naphthas derived from both petroleum and coal tar are included in this group.

Petroleum naphthas are composed principally of aliphatic hydrocarbons and are termed "close-cut" fractions. "Medium-range" and "wide-range" fractions are made up of 40 to 80% aliphatic hydrocarbons, 25 to 50% naphthenic hydrocarbons, 0 to 10% benzene, and 0 to 20% other aromatic hydrocarbons.

Coal tar naphtha is a mixture of aromatic hydrocarbons, principally toluene, xylene, and cumene. Benzene, however, is present in appreciable amounts in those coal tar naphthas with low boiling points. Rubber solvent naphtha boils at 75° to 110°C, and coal tar naphtha at 160° to 220°C.

Code Numbers: CAS 8030-30-6   RTECS DE3030000   UN 1255 (petroleum) UN 2553 (coal tar) UN 1256 (solvent)

DOT Designation: Combustible liquid.

Synonyms: Coal tar naphtha—naphtha, 49° Bé-coal tar type; crude solvent coal tar naphtha; high solvent naphtha. Petroleum naphtha—aliphatic petroleum naphtha; petroleum ether (95° to 115°C); naphtha, petroleum, ligroin, benzine.

Potential Exposures: Naphthas are used as organic solvents for dissolving or softening rubber, oils, greases, bituminous paints, varnishes, and plastics. The less flammable fractions are used in dry cleaning, the heavy naphthas serving as bases for insecticides.

Incompatibilities: Strong oxidizers.

Permissible Exposure Limits in Air: The Federal standard for petroleum naphtha is 500 ppm (2,000 mg/m³); for coal tar naphtha it is 100 ppm (400 mg/m³). NIOSH (1980) has recommended a TWA for petroleum naphtha of 350 mg/m³ and for rubber solvent naphtha 400 ppm (1,600 mg/m³); ACGIH (1983/84) gives these same values for rubber solvent naphtha. ACGIH gives a TWA of 300 ppm (1,350 mg/m³) and an STEL of 400 ppm (1,800 mg/m³) for VM & P (varnish makers' and painters' naphtha). The IDLH levels for both types of naphtha is 10,000 ppm.

Determination in Air: Charcoal adsorption, workup with CS_2, analysis by gas chromatography. See NIOSH Methods, Set G. See also reference (A-10).

Permissible Concentration in Water: No criteria set.

Routes of Entry: Inhalation of vapor, ingestion, skin and eye contact. Percutaneous absorption of liquid is probably not important in development of systemic effects unless benzene is present.

Harmful Effects and Symptoms: *Local* – The naphthas are irritating to the skin, conjunctiva, and the mucous membranes of the upper respiratory tract. Skin "chapping" and photosensitivity may develop after repeated contact with the liquid. If confined against skin by clothing, the naphthas may cause skin burn.

Systemic – Petroleum naphtha has a lower order of toxicity than that derived from coal tar, where the major hazard is brought about by the aromatic hydrocarbon content. Sufficient quantities of both naphthas cause central nervous system depression. Symptoms include inebriation, followed by headache and nausea. In severe cases, dizziness, convulsions, and unconsciousness occasionally result. Symptoms of anorexia and nervousness have been reported to persist for several months following an acute overexposure, but this appears to be rare. One fraction, hexane, has been reported to have been associated with peripheral neuropathy. (See Hexane.) If benzene is present, coal tar naphthas may produce blood changes such as leukopenia, aplastic anemia, or leukemia. The kidneys and spleen have also been affected in animal experiments. (See Benzene.)

Points of Attack: Respiratory system, eyes and skin (and central nervous system in the case of petroleum naphthas).

Medical Surveillance: Preplacement and periodic medical examinations should include the central nervous system. If benzene exposure is present, workers should have a periodic complete blood count (CBC) including hematocrit, hemoglobin, white blood cell count and differential count, mean corpuscular volume and platelet count, reticulocyte count, serum bilirubin determination, and urinary phenol in the preplacement examination and at 3-month intervals. There are no specific diagnostic tests for naphtha exposure but urinary phenols may indicate exposure to aromatic hydrocarbons. It should be noted that benzene content of vapor may be higher than predicted by content in the liquid.

First Aid: If this chemical gets into the eyes, irrigate immediately. If this chemical contacts the skin, wash with soap promptly. If a person breathes in large amounts of this chemical, move the exposed person to fresh air at once and perform artificial respiration. When this chemical has been swallowed, get medical attention. Do not induce vomiting.

Personal Protective Methods: Wear appropriate clothing to prevent repeated or prolonged skin contact. Wear eye protection to prevent any reasonable probability of eye contact. Employees should wash promptly when skin is wet or contaminated. Remove clothing immediately if wet or contaminated to avoid flammability hazard.

Respirator Selection:

Coal tar naphtha:	1,000 ppm:	CCROVF
	5,000 ppm:	GMOV/SAF/SCBAF
	10,000 ppm:	SAF:PD,PP,CF
	Escape:	GMOV/SCBA

Petroleum naphtha: 1,000 ppm: CCROVF
5,000 ppm: GMOVc
10,000 ppm: GMOVfb/SAF/SCBAF
Escape: GMOV/SCBA

Disposal Method Suggested: Incineration.

References

(1)   Nat. Inst. for Occup. Safety and Health, *Criteria for a Recommended Standard: Occu-pational Exposure to Refined Petroleum,* NIOSH Doc. No. 77-192, Wash., DC (1977).
(2)   See Reference (A-61).

2-NAPHTHOL

● Hazardous Waste Constituent (EPA)

Description: $C_{10}H_8O$ with the following structural formula

is a white crystalline solid melting at 121° to 123°C.

Code Numbers: CAS 135-19-3   RTECS QL2975000   UN 2215

DOT Designation: —

Synonyms: 2-Naphthalenol; β-hydroxynaphthalene; β-naphthol.

Potential Exposure: Those involved in rubber antioxidant production and synthesis of dyes and perfumes.

Permissible Exposure Limits in Air: No standards set.

Permissible Concentration in Water: 0.4 mg/ℓ is the maximum in drinking water (A-36).

Routes of Entry: Inhalation; skin absorption.

Harmful Effects and Symptoms: Skin and eye irritant. Ingestion can cause nephritis, lens opacity, vomiting, diarrhea, abdominal pain, circulatory collapse, convulsion, hemolytic anemia and death.

Personal Protective Methods: Wear skin protection and goggles.

Respirator Selection: Self-contained breathing apparatus.

References

(1)   Sax, N.I., Ed., *Dangerous Properties of Industrial Materials Report, 2,* No. 3, 81-83, New York, Van Nostrand Reinhold Co. (1982).
(2)   Sax, N.I., Ed., *Dangerous Properties of Industrial Materials Report, 3,* No. 6, 49-52, New York, Van Nostrand Reinhold Co. (Nov./Dec. 1983).

1,4-NAPHTHOQUINONE

● Hazardous waste (EPA)

Description: $C_{10}H_6O_2$ with the structural formula

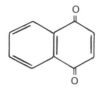

is a yellow to greenish-yellow crystalline solid melting at 123° to 126°C.

Code Numbers: CAS 130-15-4   RTECS QL7175000

DOT Designation: –

Synonyms: 1,4-Dihydro-1,4-diketonaphthalene, 1,4-naphthalenedione.

Potential Exposures: 1,4-Naphthoquinone is used as a polymerization regulator for rubber and polyester resins, in the synthesis of dyes and pharmaceuticals, and as a fungicide and algicide.

In addition to its potential entry into the environment from its manufacture, processing and uses, 1,4-naphthoquinone may also enter the environment as a degradation product of certain naphthalene derivatives. For example, the U.S. EPA in 1975 reported studies showing that the pesticide carbaryl (1-naphthyl-n-methyl-carbamate) undergoes hydrolysis to 1-naphthol.

Permissible Exposure Limits in Air: No standards set.

Permissible Concentration in Water: No criteria set. A median threshold limit value (TLM: 24 to 28 hr) of 0.3 to 0.6 mg/ℓ was listed for an unspecified species of fish (A-36).

Harmful Effects and Symptoms: The most consistent findings reported in the literature for health effects of 1,4-naphthoquinone involve hematological changes, irritant and allergenic activity, and inhibition of biochemical oxidation processes. One study found 1,4-naphthoquinone to be oncogenic. Some evidence of inhibition of *in vitro* endocrine function and of nerve activity was reported (1).

References

(1) U.S. Environmental Protection Agency, *1,4-Naphthoquinone*, Health and Environmental Effects Profile No. 132, Wash., DC, Office of Solid Waste (April 30, 1980).

(2) Sax, N.I., Ed., *Dangerous Properties of Industrial Materials Report, 4,* No. 2, 81–83, New York, Van Nostrand Reinhold Co. (1984).

α-NAPHTHYLAMINE

● Carcinogen (Human Suspected, IARC) (1)

Description: $C_{10}H_7NH_2$, alpha-naphthylamine, exists as white needle-like crystals which turn red on exposure to air, melting at 50°C.

Code Numbers: CAS 134-32-7   RTECS QM1400000   UN 2077

DOT Designation: –

Synonyms: 1-Aminonaphthalene, naphthalidine, 1-naphthylamine.

Potential Exposures: α-Naphthylamine is used in the manufacture of dyes, condensation colors, and rubber, and in the synthesis of many chemicals such as α-naphthol, sodium naphthionate, o-naphthionic acid, Neville and Winther's acid, sulfonated naphthylamines, α-naphthylthiourea (a rodenticide), and N-phenyl-α-naphthylamine.

Permissible Exposure Limits in Air: α-Naphthylamine is included in the Federal standard for carcinogens; all contact with it should be avoided.

Determination in Air: Collection on glass fiber filter and silica gel in series, elution with acetic acid in 2-propanol, analysis by gas chromatography (A-10).

Permissible Concentration in Water: No criteria set.

Routes of Entry: Inhalation and percutaneous absorption.

Harmful Effects and Symptoms: *Local* — None reported.

Systemic — It has not been established whether α-naphthylamine is a human carcinogen per se or is associated with an excess of bladder cancer due to its β-naphthylamine content. Workers exposed to α-naphthylamine developed bladder tumors. The mean latent period was 22 years compared to 16 years for β-naphthylamine. One animal experiment demonstrated papillomata, but these results have never been confirmed.

Occupational exposure to commercial 1-naphthylamine containing 4% to 10% 2-naphthylamine is strongly associated with bladder cancer in man (1). However, it is not possible at present to determine unequivocally whether 1-naphthylamine free from the 2-isomer is carcinogenic to man (1).

The carcinogenicity of 1-naphthylamine in animals is equivocal. For example, no carcinogenic effect of 1-naphthylamine was found in the hamster following oral administration; inconclusive results were obtained in mice after oral and subcutaneous administration and in dogs. 1-Naphthylamine, if carcinogenic at all, was less so to the bladder than was the 2-isomer (1). The carcinogenicity of metabolites of 1-naphthylamine [e.g., N-(1-naphthyl)-hydroxylamine] in rodents has been reported (1). N-hydroxy-1-naphthylamine is a much more potent carcinogen than N-hydroxy-2-naphthylamine.

Medical Surveillance: Placement and periodic examinations should include an evaluation of exposure to other carcinogens; use of alcohol, smoking, and medications; and family history. Special attention should be given on a regular basis to urine sediment and cytology. If red cells or positive smears are seen, cystoscopy should be done at once. The general health of exposed persons should also be evaluated in periodic examinations.

Personal Protective Methods: These are designed to supplement engineering controls and to prevent all skin or respiratory contact. Full body protective clothing and gloves should be used by those employed in handling operations. Fullface, supplied air respirators of continuous flow or pressure demand type should also be used. On exit from a regulated area, employees should shower and change into street clothes, leaving their protective clothing and equipment at the point of exit to be placed in impervious containers at the end of the work shift for decontamination or disposal. Effective methods should be used to clean and decontaminate gloves and clothing. Showers should be taken prior to dressing in street clothes.

Disposal Method Suggested: Controlled incineration whereby oxides of nitrogen are removed from the effluent gas by scrubber, catalyst or thermal device.

References

(1)   International Agency for Research on Cancer, *IARC Monographs on the Carcinogenic Risks of Chemicals to Humans,* Lyon, France, 4, 87 (1974) and 8, 349 (1975).

(2) Sax, N.I., Ed., *Dangerous Properties of Industrial Materials Report, 4,* No. 3, 79–82, New York, Van Nostrand Reinhold Co. (1984).

β-NAPHTHYLAMINE

- Carcinogen (Human positive, IARC) (1)
- Hazardous waste (EPA)

Description: $C_{10}H_7NH_2$, β-naphthylamine, is a white to reddish crystalline substance melting at 109° to 110°C.

Code Numbers: CAS 91-59-8   RTECS QM2100000   UN 1650

DOT Designation: Poison.

Synonyms: 2-Naphthylamine, 2-aminonaphthalene.

Potential Exposures: β-Naphthylamine is presently used only for research purposes. It is present as an impurity in α-naphthylamine. It was widely used in the manufacture of dyestuffs, as an antioxidant for rubber, and in rubber coated cables.

Permissible Exposure Limits in Air: β-Naphthylamine is included in the Federal standard for carcinogens; all contact with it should be avoided. ACGIH (1983/84) states that β-naphthylamine is a human carcinogen without an assigned TLV.

Determination in Air: Collection on glass fiber filter and silica gel in series, elution with acetic acid in 2-propanol, analysis by gas chromatography (A-10).

Permissible Concentration in Water: No criteria set, but EPA (A-37) has suggested an ambient level goal based on health effects of 291 µg/ℓ.

Routes of Entry: Inhalation and percutaneous absorption.

Harmful Effects and Symptoms: *Local –* β-Naphthylamine is mildly irritating to the skin and has produced contact dermatitis.

Systemic – β-Naphthylamine is a known human bladder carcinogen with a latent period of about 16 years. The symptoms are frequent urination, dysuria, and hematuria. Acute poisoning leads to methemoglobinemia or acute hemorrhagic cystitis.

2-Naphthylamine is carcinogenic, producing urinary bladder carcinomas in hamsters, dogs, and nonhuman primates, and hepatomas in mice, after oral administration (1).

Epidemiological studies have shown that occupational exposure to 2-naphthylamine, either alone or when present as an impurity in other compounds, is causally associated with bladder cancer (1) (2).

Medical Surveillance: Preplacement and periodic examinations should include an evaluation of exposure to other carcinogens; use of alcohol, smoking, and medications; and family history. Special attention should be given on a regular basis to urine sediment and cytology. If red cells or positive smears are seen, cystoscopy should be done at once. The general health of exposed persons should also be evaluated in periodic examinations.

First Aid: Wash contaminated areas of body with soap and water. If swallowed, perform gastric lavage followed by saline catharsis (A-38).

Personal Protective Methods: These are designed to supplement engineering controls and to prevent all skin or respiratory contact. Full body protective clothing and gloves should be used by those employed in handling operations. On exit from a regulated area, employees should shower and change into street clothes, leaving their clothing and equipment at the point of exit to be placed in impervious containers at the end of the work shift for decontamination or disposal. Effective methods should be used to clean and decontaminate gloves and clothing. Showers should be taken prior to dressing in street clothes.

Respirator Selection: Fullface, supplied air respirators of continuous flow or pressure demand type should be used.

Disposal Method Suggested: Controlled incineration whereby oxides of nitrogen are removed from the effluent gas by scrubber, catalyst or thermal device.

References

(1) IARC Monographs 4:97-111 (1974).
(2) *IARC Monographs on the Evaluation of the Carcinogenic Risk of Chemicals to Humans,* Supplement I, IARC, Lyon, France, p 38 (1979).
(3) Sax, N.I., Ed., *Dangerous Properties of Industrial Materials Report, 2,* No. 2, 56-58, New York, Van Nostrand Reinhold Co. (1982).
(4) See Reference (A-62). Also see Reference (A-64).
(5) Sax, N.I., Ed., *Dangerous Properties of Industrial Materials Report, 3,* No. 6, 52-55, New York, Van Nostrand Reinhold Co. (Nov./Dec. 1983).

NATURAL GAS

Description: Natural gas consists primarily of methane (85%) with lesser amounts of ethane (9%), propane (3%), nitrogen (2%), and butane (1%). Methane is a colorless, odorless, flammable gas.

Code Numbers: UN 1971/1972

DOT Designation: —

Synonyms: Marsh gas.

Potential Exposures: Natural gas is used principally as a heating fuel. It is transported as a liquid under pressure. It is also used in the manufacture of various chemicals including acetaldehyde, acetylene, ammonia, carbon black, ethyl alcohol, formaldehyde, hydrocarbon fuels, hydrogenated oils, methyl alcohol, nitric acids, synthesis gas, and vinyl chloride. Helium can be extracted from certain types of natural gas.

Permissible Exposure Limits in Air: There is no Federal standard for natural gas, methane, nitrogen, or butane. The Federal standard for propane is 1,000 ppm (1,800 mg/m^3). The ACGIH lists 800 ppm (1,900 mg/m^3) as a TWA for butane.

Permissible Concentration in Water: No criteria set.

Route of Entry: Inhalation of gas.

Harmful Effects and Symptoms: *Local* — Upon escape from pressurized tanks, natural gas may cause frostbite.
Systemic — Natural gas is a simple asphyxiant. Displacement of air by the gas may lead to shortness of breath, unconsciousness, and death from hypoxemia. Incomplete combustion may produce carbon monoxide.

Medical Surveillance: No specific considerations are needed.

Personal Protective Methods: Adequate ventilation should quite easily prevent any potential hazard.

Disposal Method Suggested: Flaring.

NICKEL AND SOLUBLE COMPOUNDS

- Carcinogen (Animal positive, IARC) (5)
- Hazardous substances (Several salts, EPA)
  Nickel compounds which are classified by EPA as hazardous substances include: nickel ammonium sulfate, nickel chloride, nickel hydroxide, nickel nitrate, and nickel sulfate.
- Hazardous waste constituents (EPA)
- Priority toxic pollutant (EPA)

Description: Ni, nickel, is a hard, ductile, magnetic metal with a silver-white color. It is insoluble in water and soluble in acids. It occurs free in meteorites and in ores combined with sulfur, antimony, or arsenic. Processing and refining of nickel is accomplished by either the Oxford (sodium sulfide and electrolysis) or the Mond (nickel carbonyl) processes. In the latter, impure nickel powder is reacted with carbon monoxide to form gaseous nickel carbonyl which is then treated to deposit high purity metallic nickel.

Code Numbers: Nickel metal—CAS 7440-02-0   RTECS QR5950000   UN 1378.

DOT Designation: Nickel powder—flammable solid.

Synonyms: For nickel metal—carbonyl nickel powder, nickel sponge, pulverized nickel, Raney nickel.

Potential Exposures: Nickel forms alloys with copper, manganese, zinc, chromium, iron, molybdenum, etc. Stainless steel is the most widely used nickel alloy. An important nickel-copper alloy is Monel metal, which contains 66% nickel and 32% copper and has excellent corrosion resistance properties. Permanent magnets are alloys chiefly of nickel, cobalt, aluminum, and iron.

Elemental nickel is used in electroplating, anodizing aluminum, casting operations for machine parts, and in coinage; in the manufacture of acid-resisting and magnetic alloys, magnetic tapes, surgical and dental instruments, nickel-cadmium batteries, nickel soaps in crankcase oils, and ground-coat enamels, colored ceramics, and glass. It is used as a catalyst in the hydrogenation of fats, oils, and other chemicals, in synthetic coal oil production, and as an intermediate in the synthesis of acrylic esters for plastics.

Exposure to nickel may also occur during mining, smelting, and refining operations.

NIOSH estimates that 250,000 U.S. workers are potentially exposed to nickel.

The route by which most people in the general population receive the largest portion of daily nickel intake is through food. Based on the available data from composite diet analysis, between 300 and 600 μg nickel per day are ingested. Fecal nickel analysis, a more accurate measure of dietary nickel intake, suggests about 300 μg per day. The highest level of nickel observed in water was 75 μg/ℓ. Average drinking water levels are about 5 μg/ℓ. A typical consumption of 2 liters daily would yield an additional 10 μg of nickel, of which up to 1 μg would be absorbed.

Occupational groups, such as nickel workers and other workers handling nickel, comprise the individuals at the highest risk. Women, particularly housewives, are at special risk to nickel-induced skin disorders. Approximately 47 million individuals, comprising the smoking population of the United States, are potentially at risk for possible cofactor effects of nickel in adverse effects on the respiratory tract.

Incompatibilities: Nickel metal and soluble nickel compounds—strong acids, sulfur, $Ni(NO_3)_2$, wood, other combustibles.

Permissible Exposure Limits in Air: The Federal standard for nickel metal and its soluble compounds is 1 mg/m^3 expressed as Ni. NIOSH recommends adherence to an exposure limit of 15 micrograms of nickel per cubic meter of air as a TWA for up to a 10-hour workday, 40-hour workweek. This limit differs greatly from the present Federal standard. ACGIH (1983/84) sets 1 mg/m^3 as a TWA for nickel metal and 0.1 mg/m^3 for soluble nickel compounds. In addition, nickel sulfide roasting fume and dust are classified under "Human Carcinogens" with a TWA value of 1.0 mg/m^3. An STEL of 0.3 mg/m^3 is set for soluble nickel compounds. There is no IDLH level set.

Determination in Air: For nickel metal and soluble Ni compounds—collection on a filter, workup with acid, analysis by atomic absorption; see NIOSH Methods, Set N. See also reference (A-10).

Permissible Concentration in Water: To protect freshwater aquatic life—

$$e^{[0.76\ ln(hardness)+1.06]}$$

as a 24 hour average, never to exceed

$$e^{[0.76\ ln(hardness)+4.02]}$$

at any time. To protect saltwater aquatic life—7.1 $\mu g/\ell$ as a 24 hour average, never to exceed 140 $\mu g/\ell$. To protect human health—13.4 $\mu g/\ell$.

Determination in Water: Digestion followed by atomic absorption or by colorimetric (heptoxime) determination or by Inductively Coupled Plasma (ICP) Optical Emission Spectrometry. This gives total nickel; dissolved nickel may be determined by the same method preceded by 0.45 micron filtration.

Routes of Entry: Inhalation of dust or fume, ingestion, eye and skin contact.

Harmful Effects and Symptoms: *Local* — Skin sensitization is the most commonly seen toxic reaction to nickel and nickel compounds and is seen frequently in the general population. This often results in chronic eczema "Nickel itch," with lichenification resembling atopic or neurodermatitis. Nickel and its compounds are also irritants to the conjunctiva of the eye and the mucous membrane of the upper respiratory tract.

Systemic — Elemental nickel (as deposited from inhalation of nickel carbonyl) and nickel salts are probably carcinogenic, producing an increased incidence of cancer of the lung and nasal passages (9). The average latency period for the induction of these cancers appears to be about 25 years (range 4 to 51 years). Effects on the heart muscle, brain, liver, and kidney have been seen in animal studies. Pulmonary eosinophilia (Loeffler's syndrome) has been reported in one study to be caused by the sensitizing property of nickel. Finely divided nickel has also shown some carcinogenic effects in rats by injection, and in guinea pigs by inhalation.

Points of Attack: Nasal cavities, lungs, skin.

Medical Surveillance: Preemployment physical examinations should evaluate any history of skin allergies or asthma, other exposures to nickel or other carcinogens, smoking history, and the respiratory tract. Lung function should be studied and chest x-rays periodically evaluated. Special attention should be given to the nasal sinuses and skin.

First Aid: If these chemicals contact the skin, flush with water immediately. If a person breathes in large amounts of this chemical, move the exposed person to fresh air at once and perform artificial respiration. When this chemical has been swallowed, get medical attention. Give large quantities of water and induce vomiting. Do not make an unconscious person vomit.

Personal Protective Methods: Full body protective clothing is advisable, as is the use of barrier creams to prevent skin sensitization and dermatitis. Employees should wash promptly when skin is wet or contaminated. Work clothing should be changed daily if it is possible that clothing is contaminated. Remove nonimpervious clothing promptly if wet or contaminated.

Respirator Selection: Respirator use conditions are recommended separately for these substances in two states: dusts or mist, and as dust, mist, and fume.

As dust or mist: 5 mg/m^3:   DM
10 mg/^3:   DMXSQ
50 mg/m^3:   HiEPF/SAF/SCBAF
$1,000 \text{ mg/m}^3$:   PAPHiE/SA:PD,PP,CF
$2,000 \text{ mg/m}^3$:   SAF:PD,PP,CF

As dust, mist, or fume: 10 mg/m^3:   FuHiEP/SA/SCBA
Other categories: same as above.

Disposal Method Suggested: Nickel antimonide, nickel arsenide, nickel selenide—encapsulation followed by disposal in a chemical waste landfill. However, nickel from various industrial wastes may also be recovered and recycled as described in the literature (A-57).

References

(1) National Institute for Occupational Safety and Health, *Criteria for a Recommended Standard: Occupational Exposure to Inorganic Nickel*, NIOSH Doc. No. 77-164 (1977).

(2) National Academy of Sciences, *Report on Medical and Biological Effects of Environmental Pollutants: Nickel*, Wash., DC (1975).

(3) U.S. Environmental Protection Agency, *Nickel: Ambient Water Quality Criteria*, Wash., DC (1980).

(4) U.S. Environmental Protection Agency, *Nickel*, Health and Environmental Effects Profile No. 133, Wash., DC, Office of Solid Waste (April 30, 1980).

(5) International Agency for Research on Cancer, *IARC Monographs on the Carcinogenic Risks of Chemicals to Humans*, Lyon, France 2, 126 (1973), 7, 319 (1974) and 11, 75 (1976).

(6) Sax, N.I., Ed., *Dangerous Properties of Industrial Materials Report, 1*, No. 1, 50-51, New York, Van Nostrand Reinhold Co. (1980).

(7) Sax, N.I., Ed., *Dangerous Properties of Industrial Materials Report, 3*, No. 3, 76-80, New York, Van Nostrand Reinhold Co. (1983).

(8) See Reference (A-61).

(9) See Reference (A-62). Also see Reference (A-64).

(10) See Reference (A-60) under Nickel Sulfate.

(11) Parmeggiani, L., Ed., *Encyclopedia of Occupational Health & Safety*, Third Edition, Vol. 2, pp 1438-40, Geneva, International Labour Office (1983).

(12) United Nations Environment Programme, *IRPTC Legal File 1983*, Vol. II, pp VII/466-69, Geneva, Switzerland, International Register of Potentially Toxic Chemicals (1984).

NICKEL CARBONYL

- Carcinogen (Animal positive, IARC) (1)
- Hazardous waste (EPA)

Description: $Ni(CO)_4$, nickel carbonyl, is a colorless, highly volatile, flammable liquid with a musty odor. It decomposes above room temperature producing carbon monoxide and finely divided nickel. It is soluble in organic solvents. It boils at 42° to 43°C.

Code Numbers: CAS 13463-39-3   RTECS QR6300000   UN 1259

DOT Designation: Flammable liquid and poison.

Synonyms: Nickel tetracarbonyl.

Potential Exposures: The primary use for nickel carbonyl is in the production of nickel by the Mond process. Impure nickel powder is reacted with carbon monoxide to form gaseous nickel carbonyl which is then treated to deposit high purity metallic nickel and release carbon monoxide. Other uses include gas plating, the production of nickel products; in chemical synthesis as a catalyst, particularly for oxo reactions (addition reaction of hydrogen and carbon monoxide with unsaturated hydrocarbons to form oxygen-function compounds), e.g., synthesis of acrylic esters, and as a reactant.

Incompatibilities: Nitric acid, chlorine, other oxidizers, combustible vapors.

Permissible Exposure Limits in Air: The Federal standard for nickel carbonyl is 0.001 ppm (0.007 mg/m^3). ACGIH (1983/84) has adopted 0.05 ppm (0.35 mg/m^3) as a TWA. There is no STEL value given. The IDLH level is 0.001 ppm.

Determination in Air: Sample collection by impinger or fritted bubbler, analysis by atomic absorption spectrometry (A-21).

Permissible Concentration in Water: No criteria set, but EPA (A-37) has suggested a permissible ambient goal of 1.4 µg/ℓ based on health effects.

Routes of Entry: Inhalation of vapor. It may be possible for appreciable amounts of the liquid to be absorbed through the skin, also ingestion and eye and skin contact.

Harmful Effects and Symptoms: *Local* — Nickel dermatitis may develop. (See "Nickel and Soluble Compounds.")

Systemic — Symptoms of exposure to the toxic vapors of nickel carbonyl are of two distinct types. Immediately after exposure, symptoms consist of frontal headache, giddiness, tightness of the chest, nausea, weakness of limbs, perspiring, cough, vomiting, cold and clammy skin, and shortness of breath. Even in exposures sufficiently severe to cause death, the initial symptoms disappear quickly upon removal of the subject to fresh air. Symptoms may be so mild during this initial phase that they go unrecognized.

Severe symptoms may then develop insidiously hours or even days after exposure. The delayed syndrome usually consists of retrosternal pain, tightness in the chest, dry cough, shortness of breath, rapid respiration, cyanosis, and extreme weakness. The weakness may be so great that respiration can be sustained only by oxygen support. Fatal cases are usually preceded by convulsion and mental confusion, with death occurring from 4 to 11 days following exposure. The syndrome represents a chemical pneumonitis with adrenal cortical suppression.

Points of Attack: Lungs, paranasal sinus, central nervous system.

Medical Surveillance: (See "Nickel and Soluble Compounds.") Note also that measurement of urinary nickel levels for several days after acute exposures may be helpful.

First Aid: If this chemical gets into the eyes, irrigate immediately. If this chemical contacts the skin, wash with soap immediately. If a person breathes in large amounts of this chemical, move the exposed person to fresh air at once and perform artificial respiration. When this chemical has been swallowed, get medical attention. Give large quantities of water and induce vomiting. Do not make an unconscious person vomit.

Personal Protective Methods: Wear appropriate clothing to prevent any possibility of skin contact. Wear eye protection to prevent any possibility of eye contact. Employees should wash immediately when skin is wet or contaminated. Remove clothing immediately if wet or contaminated to avoid flammability hazard.

Respirator Selection: Above 0.001 ppm, only SCBAF:PD,PP,CF or SAF:PD, PP,CF with aux SCBA:PD,PP should be used. Escape: GMS/SCBA.

Disposal Method Suggested: Incineration in admixture with a flammable solvent (A-38). Also, nickel carbonyl used in metallizing operations may be recovered and recycled (A-58).

References

(1) International Agency for Research on Cancer, *IARC Monographs on the Carcinogenic Risks of Chemicals to Humans*, Lyon, France 11, 75 (1976).

(2) See Reference (A-61).

NICOTINE

● Hazardous waste (EPA)

Description: $C_{10}H_{14}N_2$, (S)-3-(1-methyl-2-pyrrolidinyl)pyridine, is a pale yellow to dark brown liquid with a slight, fishy odor when warm. It boils at 266°C.

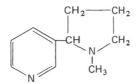

Code Numbers: CAS 54-11-5 RTECS QS5250000 UN 1654

DOT Designation: Poison B.

Synonyms: 1-Methyl-2(3-pyridyl)pyrrolidine, tetrahydronicotyrine.

Potential Exposure: Nicotine is used in some drugs and insecticides and in tanning.

Incompatibilities: Strong oxidizers, strong acids.

Permissible Exposure Limits in Air: The Federal standard and the 1983/84 ACGIH TWA value is 0.5 mg/m³ (0.07 ppm). The notation "skin" indicates the possibility of cutaneous absorption. The STEL is 1.5 mg/m³. The IDLH level is 35 mg/m³.

Determination in Air: Adsorption on resin; workup with ethyl acetate, analysis by gas chromatography. See NIOSH Methods, Set T. See also reference (A-10).

Permissible Concentration in Water: In view of the relative paucity of data on the carcinogenicity and long-term oral toxicity of nicotine, estimates of the effects of chronic oral exposure at low levels cannot be made with any confidence. It is recommended that studies to produce such information be conducted before limits in drinking water are established according to NAS/NRC (A-2).

Routes of Entry: Inhalation, skin absorption, ingestion, skin and eye contact.

Harmful Effects and Symptoms: *Observations in Man* — The symptoms of nicotine intoxication are nausea, vomiting, diarrhea, confusion, twitching, and convulsions. A dose of 40 mg of nicotine taken orally is fatal in man. Nicotine has also been shown to induce blood-sugar changes in 24 hours; 0.3 mg/kg was the lowest no-effect dose.

At high doses, nicotine is quite toxic and lethal. Nicotine is metabolized readily, principally to cotinine. Nicotine is teratogenic in mice only at high doses. Evidence on carcinogenicity is equivocal, but it is a cocarcinogen.

Symptoms of nicotine exposure include nausea, salivation, abdominal pain, vomiting, diarrhea; headaches, dizziness, disturbed hearing and vision; confusion, weakness, incoherence; paroxysmal atrial fibrillation; convulsions, dyspnea.

Points of Attack: Central nervous system, cardiovascular system, lungs, gastrointestinal system.

Medical Surveillance: Consider the points of attack in preplacement and periodic physical examinations.

First Aid: If this chemical gets into the eyes, irrigate immediately. If this chemical contacts the skin, flush with water immediately. If a person breathes in large amounts of this chemical, move the exposed person to fresh air at once and perform artificial respiration. When this chemical has been swallowed, get medical attention. Give large quantities of water and induce vomiting. Do not make an unconscious person vomit.

Personal Protective Methods: Wear appropriate clothing to prevent any possibility of skin contact. Wear eye protection to prevent any possibility of eye contact. Employees should wash immediately when skin is wet or contaminated. Remove nonimpervious clothing immediately if wet or contaminated. Provide emergency showers and eyewash.

Respirator Selection:
5 mg/m^3: SA/SCBA
25 mg/m^3: SAF/SCBAF
35 mg/m^3: SA:PD,PP,CF
Escape: GMOVPPest/SCBA

Disposal Method Suggested: Incineration.

References

(1) See Reference (A-61).
(2) Sax, N.I., Ed., *Dangerous Properties of Industrial Materials Report, 1,* No. 8, 84-85, New York, Van Nostrand Reinhold Co. (1981).
(3) See Reference (A-60).
(4) Parmeggiani, L., Ed., *Encyclopedia of Occupational Health & Safety,* Third Edition, Vol. 2, pp 1440-41, Geneva, International Labour Office (1983).

(5) United Nations Environment Programme, *IRPTC Legal File 1983,* Vol. II, pp VII/470–74, Geneva, Switzerland, International Register of Potentially Toxic Chemicals (1984).

NITRAPYRIN

Description: $C_6H_3Cl_4N$, 2-chloro-6-trichloromethylpyridine with the structural formula

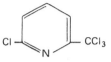

is a colorless crystalline solid melting at 62° to 63°C.

Code Numbers: CAS 1929-82-4   RTECS US7525000

DOT Designation: –

Synonyms: N-Serve®

Potential Exposures: Those involved in the manufacture (A-32), formulation and application of this nitrogen stabilizer in agricultural use.

Permissible Exposure Limits in Air: There is no Federal standard but ACGIH (1983/84) has adopted a TWA value of 10 mg/m^3 and set an STEL of 20 mg/m^3.

Determination in Air: Sample collection by charcoal tube, analysis by gas liquid chromatography (A-13).

Permissible Concentration in Water: No criteria set.

Harmful Effects and Symptoms: In animal experiments, dogs and rats were fed 15 mg/kg/day for 93 days with no adverse effects (A-34). Human exposure data have not been reported.

Disposal Method Suggested: The manufacturer of this nitrification inhibitor suggests that unwanted quantities can be disposed of by burial in a sanitary landfill (A-32).

NITRATES

Description: $M_V(NO_3)_V$ are salts with varying properties depending on the specific compounds.

Code Numbers: Sodium Nitrate—CAS 7631-99-4   RTECS WC5600000
UN 1477

DOT Designation: Sodium Nitrate—Oxidizer.

Synonyms: Vary with specific compounds.

Potential Exposure: Among the major point sources of nitrogen entry into water bodies are municipal and industrial wastewaters, septic tanks, and feedlot discharges. Diffuse sources of nitrogen include farm-site fertilizer and animal wastes, lawn fertilizer, leachate from waste disposal in dumps or sanitary landfills, atmospheric fallout, nitric oxide and nitrite discharges from automobile exhausts and other combustion processes, and losses from natural sources such as mineralization of soil organic matter.

Permissible Exposure Limits in Air: No standards set.

Permissible Concentration in Water: Nitrate in water at concentrations less than a thousand milligrams per liter is not of serious concern as a direct toxicant. It is a health hazard because of its conversion to nitrite. Nitrite is directly toxic by reaction with hemoglobin to form methemoglobin and cause methemoglobinemia. It also reacts readily under appropriate conditions with secondary amines and similar nitrogenous compounds to form N-nitroso compounds, many of which are potent carcinogens.

Epidemiological evidence on the occurrence of methemoglobinemia in infants tends to confirm a value near 10 mg/ℓ nitrate as nitrogen as a maximum concentration level for water with no observed adverse health effects, but there is littler margin of safety in this value.

Determination in Water: A variety of methods for determination of nitrate exists, but none is particularly precise, accurate or sensitive in the milligram per liter concentration range. Further development and standardization of analytical methodology will be required if standard routine determinations are to be considered reliable within the range required for proper control and assessment of health effects.

Most standard procedures for nitrate determination in the milligram per liter range are spectrophotometric. Traditionally, three types of reaction of nitrate have been used as bases: nitration of a phenolic substance to a colored derivative; oxidation of an organic substance to a colored product; and reduction of the nitrate to nitrite or ammonia, followed by reaction of the reduced nitrogenous materials to give colored substances. In addition, direct spectrophotometric determination based on ultraviolet absorption of nitrate at 273 nm is possible and becoming established. Electrochemical determination with the use of a nitrate electrode may also be feasible, but is subject to numerous interferences.

Harmful Effects and Symptoms: Two health hazards are related to the consumption of water containing large concentrations of nitrate (or nitrite): induction of methemoglobinemia, particularly in infants, and possible formation of carcinogenic nitrosamines.

Acute toxicity of nitrate occurs as a result of reduction to nitrite, a process that can occur under specific conditions in the stomach, as well as in the saliva. Nitrite acts in the blood to oxidize the hemoglobin to methemoglobin, which does not perform as an oxygen carrier to the tissues. Consequently, anoxia and death may ensue.

Healthy adults are reported to be able to consume large quantities of nitrate in drinking water with relatively little, if any, effects. Acute nitrate toxicity is almost always seen in infants rather than adults. This increased susceptibility of infants has been attributed to high intake per unit weight, to the presence of nitrate-reducing bacteria in the upper gastrointestinal tract, to the condition of the mucosa and to greater ease of oxidation of fetal hemoglobin.

Disposal Method Suggested: Sodium Nitrate—The material is diluted to the recommended provisional limit in water. [The pH is adjusted to between 6.5 and 9.1 and then the material can be discharged into sewers or natural streams (A-31)].

References

(1) National Academy of Sciences, *Medical and Biologic Effects of Environmental Pollutants: Nitrates,* Washington, DC (1978).
(2) World Health Organization, *Nitrates, Nitrites and N-Nitroso Compounds,* Environmental Health Criteria No. 5, Geneva (1978).

NITRIC ACID

● Hazardous substance (EPA)

Description: Nitric acid, HNO_3, is a colorless liquid with a characteristic choking odor which fumes in moist air. It is a solution of nitrogen dioxide, NO_2, in water and so-called fuming nitric acid contains an excess of NO_2 and is yellow to brownish-red in color.

Code Numbers: CAS 7697-37-2   RTECS QU5775000   UN 2031

DOT Designation: Oxidizer and corrosive.

Synonyms: Aqua fortis, white fuming nitric acid (WFNA), red fuming nitric acid (RFNA), hydrogen nitrate.

Potential Exposures: NIOSH estimates that 27,000 workers are potentially exposed to nitric acid. Nitric acid is the second most important industrial acid and its production represents the sixth largest chemical industry in the United States. The largest use of nitric acid is in the production of fertilizers. Almost 15% of the production goes into the manufacture of explosives, with the remaining 10% distributed among a variety of uses such as etching, bright-dipping, electroplating, photoengraving, production of rocket fuel, and pesticide manufacture (A-32).

Incompatibilities: Combustible organics; oxidizable matter—wood, turpentine, metal powder, hydrogen sulfide, etc.; strong bases.

Permissible Exposure Limits in Air: NIOSH and ACGIH (1983/84) recommend adherence to the present Federal standard of 2 ppm (5 mg/m³) as a time-weighted average for up to a 10-hour workday, 40-hour workweek. The STEL value is 4 ppm (10 mg/m³). The IDLH level is 100 ppm.

Determination in Air: Sample collection by impinger containing water, analysis by ion specific electrode (A-10).

Since many of the pulmonary disorders may result from the oxides of nitrogen, readily given off when nitric acid is dispersed in air, frequent reference is made in the criteria document (1) to simultaneous environmental monitoring for airborne nitric acid and oxides of nitrogen.

Permissible Concentration in Water: No criteria set.

Routes of Entry: Skin and eye contact, inhalation of vapors, ingestion.

Harmful Effects and Symptoms: Exposure to nitric acid represents a dual health hazard; specifically, corrosion of the skin and other tissues from topical contact and acute pulmonary edema or chronic obstructive pulmonary disease from inhalation.

Points of Attack: Eyes, respiratory system, lungs, skin, teeth.

Medical Surveillance: NIOSH recommends that workers subject to nitric acid exposure have comprehensive preplacement and annual medical examinations including a 14" x 17" posterior-anterior chest x-ray, pulmonary function tests, and a visual examination of the teeth for evidence of dental erosion.

First Aid: If this chemical gets into the eyes, irrigate immediately. If this chemical contacts the skin, flush with water immediately. If a person breathes in large amounts of this chemical, move the exposed person to fresh air at once and perform artificial respiration. When this chemical has been swallowed, get medical attention. Give large quantities of water and do not induce vomiting.

Personal Protective Methods: Wear appropriate clothing to prevent any possibility of skin contact with liquids of pH <2.5 or repeated or prolonged contact with liquids of pH >2.5. Wear eye protection to prevent any possibility of eye contact. Employees should wash immediately when skin is wet or contaminated. Remove nonimpervious clothing immediately if wet or contaminated. Provide emergency showers and eyewash if liquids of pH <2.5 are involved.

Respirator Selection:
250 mg/m^3: CCFS/GMOVS/SAF/SCBAF/SA:PD,PP,CF
    Escape: GMS/SCBA
    Note: Do not use oxidizable sorbents.

Disposal Method Suggested: Soda ash-slaked lime is added to form the neutral solution of nitrate of sodium and calcium. This solution can be discharged after dilution with water (A-31). Also, nitric acid can be recovered and reused in some cases as with acrylic fiber spin solutions (A-57).

References
(1) National Institute for Occupational Safety and Health, *Criteria for a Recommended Standard: Occupational Exposure to Nitric Acid,* NIOSH Doc. No. 76-141 (1976).
(2) Sax, N.I., Ed., *Dangerous Properties of Industrial Materials Report, 1,* No. 5, 71-72, New York, Van Nostrand Reinhold Co. (1981).
(3) See Reference (A-61).
(4) See Reference (A-60).
(5) Parmeggiani, L., Ed., *Encyclopedia of Occupational Health & Safety,* Third Edition, Vol. 2, pp 1443-45, Geneva, International Labour Office (1983).

NITRIC OXIDE

● Hazardous waste (EPA)

Description: NO is a colorless gas with a sharp, sweet odor; brown at high concentrations in air.

Code Numbers: CAS 10102-43-9   RTECS QX0525000   UN 1660

DOT Designation: Poison A, poison gas.

Synonyms: Nitrogen monoxide.

Potential Exposure: Nitric oxide is used in the manufacture of nitric acid; it is also used in the bleaching of rayon; it is a raw material for nitrosyl halide preparation.

Incompatibilities: Combustible matter, chlorinated hydrocarbons, ammonia, carbon disulfide, metals, fluorine, ozone.

Permissible Exposure Limits in Air: The Federal standard and the 1983/84 ACGIH TWA value is 25 ppm (30 mg/m^3). The STEL is 35 ppm (45 mg/m^3). The IDLH level is 100 ppm.

Determination in Air: Oxidation to NO_2, collection on triethanolamine-coated molecular sieve, desorption with triethanolamine to nitrite ion, color reaction and analysis by spectrophotometry (A-10).

Permissible Concentration in Water: No criteria set.

Routes of Entry: Inhalation, skin and eye contact.

Harmful Effects and Symptoms: Irritation of eyes, nose and throat; drowsiness; unconsciousness.

Points of Attack: Respiratory system, lungs.

Medical Surveillance: Consider the points of attack in preplacement and periodic physical examinations.

First Aid: If a person breathes in large amounts of this chemical, move the exposed person to fresh air at once and perform artificial respiration.

Personal Protective Methods: Only respirator protection is specified by NIOSH.

Respirator Selection:
> 100 ppm: CCRS/GMS/SA/SCBA
> Escape: GMS/SCBA
> Note: Do not use oxidizable sorbents.

Disposal Method Suggested: Incineration with added hydrocarbon fuel, controlled so as to produce elemental nitrogen, CO_2 and water.

References

(1) Nat. Inst. for Occup. Safety and Health, *Criteria for a Recommended Standard: Occupational Exposure to Oxides of Nitrogen*, NIOSH Doc. No. 76-149, Wash., DC (1976).
(2) World Health Organization, *Oxides of Nitrogen,* Environmental Health Criteria No. 4, Geneva, Switzerland (1977).
(3) See Reference (A-61).
(4) Sax, N.I., Ed., *Dangerous Properties of Industrial Materials Report, 1,* No. 5, 73-74, New York, Van Nostrand Reinhold Co. (1981).

NITRILOTRIACETIC ACID

● Carcinogen (Animal Positive) (A-64)

Description: $C_6H_9NO_6$ with the structure $N(CH_2COOH)_3$ is a crystalline compound melting at 230° to 235°C.

Code Numbers: CAS 139-13-9   RTECS AJ0175000

DOT Designation: −

Synonyms: NTA; triglycollamic acid; triglycine; tri(carboxymethyl)amine; aminotriacetic acid.

Potential Exposure: NTA replaced phosphates as a detergent builder in the late 1960s. In 1971, the use of NTA was discontinued. The possibility of resumed use arose in 1980. The Consumer Product Safety Commission (CPSC) stated that NTA is now used in laundry detergents in two states where phosphates are banned. FDA reported that NTA is also used as a boiler feedwater additive at a maximum use level of 5 ppm of trisodium salt. Currently, the remaining nondetergent uses of NTA are for water treatment, textile treatment, metal plating and cleaning, and pulp and paper processing.

EPA reported that the domestic production for 1979 was 70 to 75 million pounds, with 60 to 65 million pounds exported. Estimates indicate that reintroduction would increase production to 250 million pounds immediately and to 1 billion pounds in the near future, with most of this amount used domestically.

In 1974, the National Occupational Hazard Survey estimated that 14,600 workers were potentially exposed to NTA, and 46,000 workers were exposed to NTA trisodium salt. NTA levels in the U.S. drinking water prior to NTA's discontinued use in detergents was estimated by EPA to have ranged from 0.20 to 24.5 $\mu g/\ell$. NTA is rapidly degraded under aerobic conditions at temperatures above $5°C$, and biodegradation does not lead to the formation of persistent intermediates.

Permissible Exposure Limits in Air: No standards set.

Permissible Concentration in Water: No criteria set for drinking water; NTA addition to boiler feedwater is limited to 5 ppm by FDA.

Routes of Entry: One route of human exposure is ingestion of trace residues that possibly remain in processed foods.

Harmful Effects and Symptoms: Long-term feeding studies to test the monohydrate of the trisodium salt of nitrilotriacetic acid ($Na_3NTA·H_2O$) for carcinogenicity were conducted using Fischer 344 rats; and both Fischer 344 rats and B6C3F1 mice. Similar bioassays, using rats and mice, were conducted with the free nitrilotriacetic acid (NTA). NTA and $Na_3NTA·H_2O$ were shown to be carcinogenic to the urinary tract and kidneys of both rats and mice (1). Trisodium nitrilotriacetic acid was found to be carcinogenic in the kidney when administered in drinking water to male albino rats (Crl/COBS CD (SD) BR) (2).

References

(1)   National Cancer Institute, *Bioassay of Nitrilotriacetic Acid (NTA) and Nitrilotriacetic Acid, Trisodium Salt, Monohydrate (Na₃NTA·H₂O) for Possible Carcinogenicity*, Technical Report Series No. 6, DHEW Publication No. (NIH) 77-806, Bethesda, Maryland (1977).

(2)   Goyer, R.A., H.L. Falk, M. Hogan, D.D. Feldman, and W. Richter, *Renal Tumors in Rats Given Trisodium Nitrilotriacetic Acid in Drinking Water for Two Years*, J. Nat. Cancer Inst., Vol. 66, pp 869-80 (1981).

p-NITROANILINE

● Hazardous waste (EPA)

Description: $H_2NC_6H_4NO_2$ consists of yellow crystals with a pungent, faint ammonialike odor. It melts at $145°C$ and boils at $332°C$.

Code Numbers: CAS 100-01-6   RTECS BY7000000   UN 1661

DOT Designation: Poison B.

Synonyms: Azoic diazo component 37, p-aminonitrobenzene, fast red 2G base, 4-nitroaniline, PNA.

Potential Exposures: p-Nitroaniline is used as an intermediate in the manufacture of dyes, antioxidants, pharmaceuticals (A-41) and pesticides (A-32).

Incompatibilities: Strong oxidizers and reducers.

Permissible Exposure Limit in Air: The Federal standard is 1 ppm (6 mg/m^3). The ACGIH (1983/84) has set a TWA of 3 mg/m^3. The notation "skin" indicates the possibility of cutaneous absorption. There is no STEL value. The IDLH level is 300 mg/m^3.

Determination in Air: Adsorption on silica gel, elution by ethanol, analysis by gas chromatography (A-10).

Permissible Concentration in Water: No criteria set.

Routes of Entry: Inhalation, ingestion, skin and eye contact.

Harmful Effects and Symptoms: Cyanosis, ataxia, tachycardia, tachypnea, dyspnea, irritability; vomiting, diarrhea; convulsions, respiratory arrest; anemia.

Points of Attack: Blood, heart, lungs, liver.

Medical Surveillance: Consider the points of attack in preplacement and periodic physical examinations.

First Aid: If this chemical gets into the eyes, irrigate immediately. If this chemical contacts the skin, flush with water immediately. If a person breathes in large amounts of this chemical, move the exposed person to fresh air at once and perform artificial respiration. When this chemical has been swallowed, get medical attention. Give large quantities of saltwater and induce vomiting. Do not make an unconscious person vomit.

Personal Protective Methods: Wear appropriate clothing to prevent repeated or prolonged skin contact. Wear eye protection to prevent any reasonable probability of eye contact. Employees should wash immediately with soap when skin is wet or contaminated and daily at the end of each work shift. Work clothing should be changed daily if it is possible that clothing is contaminated. Remove nonimpervious clothing immediately if wet or contaminated. Provide emergency showers.

Respirator Selection:
        10 ppm: SA/SCBA
        50 ppm: SAF/SCBAF
       Escape: GMOVP/SCBA

Disposal Method Suggested: Incineration (1800°F, 2.0 seconds minimum) with scrubbing for NO$_x$ abatement.

References
(1)   See Reference (A-61).
(2)   See Reference (A-60).

5-NITRO-o-ANISIDINE

● Carcinogen (Positive, NCI) (2)

Description: $C_6H_3(OCH_3)(NH_2)(NO_2)$ is an orange-red crystalline compound melting at 118°C.

Code Numbers: CAS 99-59-2   RTECS BZ7175000

DOT Designation: —

Synonyms: 2-Amino-5-nitroanisole; 2-Methoxy-5-nitroaniline; 2-Methoxy-5-nitrobenzamine; 3-Amino-4-methoxynitrobenzene.

Potential Exposures: 5-Nitro-o-anisidine is a chemical intermediate in the production of C.I. Pigment Red 23, which is used as a colorant for commodities such as printing inks, interior latex paints, lacquers, rubber, plastics, floor coverings, paper coating, and textiles. It is also used with other C.I. coupling components to produce various hues of red, brown, yellow, and violet on cotton, silk, acetate and nylon.

Permissible Exposure Limits in Air: No standards set.

Permissible Concentration in Water: No criteria set.

Harmful Effects and Symptoms: This compound is a proven carcinogen in experimental animals (A-62) (A-64).

Points of Attack: Bladder.

Medical Surveillance: Consider the points of attack in preplacement and periodic physical examinations.

Disposal Method Suggested: Incineration (1800°F, 2.0 seconds minimum) with scrubbing for NO_x abatement.

References

(1)   U.S. Environmental Protection Agency, *Chemical Hazard Information Profile: 5-Nitro-o-Anisidine,* Washington, DC (1979).

(2)   National Cancer Institute, *Bioassay of 5-Nitro-o-Anisidine for Possible Carcinogenicity,* Technical Report Series No. 127 (1978).

NITROBENZENE

- Hazardous substance (EPA)
- Hazardous waste (EPA)
- Priority toxic pollutant (EPA)

Description: $C_6H_5NO_2$, nitrobenzene, is a pale yellow to dark brown oily liquid whose odor resembles bitter almonds (or black paste shoe polish).

Code Numbers: CAS 98-95-3   RTECS DA6475000   UN 1662

DOT Designation: Poison B.

Synonyms: Nitrobenzol, oil of mirbane, oil of bitter almonds.

Potential Exposures: Nitrobenzene is used in the manufacture of explosives and aniline dyes and as a solvent and intermediate. It is also used in shoe and floor polishes, leather dressings, and paint solvents, and to mask other unpleasant odors. Substitution reactions with nitrobenzene are used to form m-derivatives.

Pregnant women may be especially at risk with respect to nitrobenzene as with many other chemical compounds, due to transplacental passage of the agent. Individuals with glucose-6-phosphate dehydrogenase deficiency may also be special risk groups. Additionally, because alcohol ingestion or chronic alcoholism can lower the lethal or toxic dose of nitrobenzene, individuals consuming alcoholic beverages may be at risk.

Incompatibilities: Concentrated nitric acid, nitrogen tetroxide, caustic, chemically active metals like tin or zinc.

Permissible Exposure Limits in Air: The Federal standard and the 1983/84 ACGIH TWA value is 1 ppm (5 mg/m^3). The STEL is 2 ppm (10 mg/m^3). The notation "skin" indicates the possibility of cutaneous absorption. The IDLH level is 200 ppm.

Determination in Air: Adsorption on silica, workup with methanol, analysis by gas chromatography. See NIOSH Methods, Set P. See also reference (A-10).

Permissible Concentration in Water: To protect freshwater aquatic life— 27,000 μg/ℓ on an acute toxicity basis. To protect saltwater aquatic life—6,680 μg/ℓ on an acute toxicity basis. To protect humans—30 μg/ℓ based on organoleptic considerations and 19,800 μg/ℓ based on toxicity considerations.

Determination in Water: Methylene chloride extraction followed by exchange to toluene, gas chromatography with flame ionization detection (EPA Method 609) or gas chromatography plus mass spectrometry (EPA Method 625).

Routes of Entry: Inhalation, ingestion, percutaneous absorption of liquid, skin and eye contact.

Harmful Effects and Symptoms: *Local* — Nitrobenzene may cause irritation of the eyes.
Systemic — There is a latent period of 1 to 4 hours before signs and symptoms appear. Nitrobenzene affects the central nervous system producing fatigue, headache, vertigo, vomiting, general weakness, and in some cases severe depression, unconsciousness, and coma. Nitrobenzene is a powerful methemoglobin former; cyanosis appears when methemoglobin reaches 15%. Sulfhemoglobin formation may also contribute to nitrobenzene toxicity. Chronic exposure may lead to spleen and liver damage, jaundice, liver impairments, and hemolytic icterus. Anemia and Heinz bodies in the red blood cells have been observed. Alcohol ingestion may increase the toxic effects.

Points of Attack: Blood, liver, kidneys, cardiovascular system, skin.

Medical Surveillance: Preemployment and periodic examinations should be concerned particularly with a history of dyscrasias, reactions to medications, alcohol intake, eye disease, skin, and cardiovascular status. Liver and renal functions should be evaluated periodically, as well as blood and general health.

Follow methemoglobin levels until normal in all cases of suspected cyanosis. The metabolites in urine, p-nitro- and p-aminophenol, can be used as evidence of exposure.

First Aid: If this chemical gets into the eyes, irrigate immediately. If this chemical contacts the skin, wash with soap immediately. If a person breathes in large amounts of this chemical, move the exposed person to fresh air at once and perform artificial respiration. When this chemical has been swallowed, get medical attention. Give large quantities of water and induce vomiting. Do not make an unconscious person vomit.

Personal Protective Methods: Wear appropriate clothing to prevent any possibility of skin contact. Wear eye protection to prevent any reasonable probability of eye contact. Employees should wash promptly when skin is wet or contaminated and daily at the end of each work shift. Work clothing should be changed daily if it is possible that clothing is contaminated. Remove nonimpervious clothing immediately if wet or contaminated. Provide emergency showers.

Respirator Selection:
        10 ppm: CCROV/SA/SCBA
        50 ppm: CCROVF/GMOV/SAF/SCBAF
       200 ppm: SAF:PD,PP,CF
       Escape: GMOV/SCBA

Disposal Method Suggested: Incineration (1800°F, 2.0 seconds minimum) with scrubbing for NO_x abatement.

References

(1) Nat. Inst. for Occup. Safety and Health, *Information Profiles on Potential Occupational Hazards: Nitrobenzenes*, Report PB-276,678, Rockville, MD, pp 198-211 (Oct. 1977).
(2) U.S. Environmental Protection Agency, *Chemical Hazard Information Profile: Nitrobenzene*, Wash., DC (1979).
(3) U.S. Environmental Protection Agency, *Nitrobenzene: Ambient Water Quality Criteria*, Wash., DC (1980).
(4) U.S. Environmental Protection Agency, *Nitrobenzene,* Health and Environmental Effects Profile No. 134, Wash., DC, Office of Solid Waste (April 30, 1980).
(5) See Reference (A-61).
(6) See Reference (A-60).
(7) Parmeggiani, L., Ed., *Encyclopedia of Occupational Health & Safety,* Third Edition, Vol. 2, pp 1448–49, Geneva, International Labour Office (1983).
(8) United Nations Environment Programme, *IRPTC Legal File 1983,* Vol. I, pp VII/91–3, Geneva, Switzerland, International Register of Potentially Toxic Chemicals (1984).

4-NITROBIPHENYL

● Carcinogen (Animal positive, IARC) (1)

Description: $C_6H_5C_6H_4NO_2$, 4-nitrobiphenyl, exists as yellow plates or needles.

Code Numbers: CAS 92-93-3   RTECS DV5600000

DOT Designation: –

Synonyms: 4-Nitrodiphenyl, p-nitrobiphenyl, p-nitrodiphenyl, PNB.

Potential Exposures: 4-Nitrobiphenyl was formerly used in the synthesis of 4-aminodiphenyl. It is presently used only for research purposes; there are no commercial uses.

Permissible Exposure Limits in Air: 4-Nitrobiphenyl was included in the Federal standard for carcinogens; all contact with it should be avoided. ACGIH (1983/84) classifies it as a human carcinogen with no assigned TWA value.

Determination in Air: Collection on a glass fiber filter in series with silica gel, elution with 2-propanol, analysis by gas chromatography (A-10).

Permissible Concentration in Water: No criteria set, but EPA (A-37) has suggested a permissible ambient goal of 890 $\mu g/\ell$ based on health effects.

Routes of Entry: Inhalation and percutaneous absorption.

Harmful Effects and Symptoms: *Local* – None reported.

Systemic – 4-Nitrobiphenyl is considered to be a human carcinogen. This is based on the evidence that it will induce bladder tumors in dogs and that human cases of bladder cancer were reported from a mixed exposure to 4-aminodiphenyl and 4-nitrobiphenyl. These human cases were attributed to 4-aminodiphenyl because the information available at the time showed that it produced bladder tumors in dogs. 4-Aminobiphenyl may be a metabolite.

Medical Surveillance: Placement and periodic examinations should include an evaluation of exposure to other carcinogens, as well as an evaluation of smoking, of use of alcohol and medications, and of family history. Special attention should be given on a regular basis to urine sediment and cytology. If red cells or positive smears are seen, cystoscopy should be done at once. The general health of exposed persons should also be evaluated in periodic examinations.

Personal Protective Methods: These are designed to supplement engineering and to prevent all skin or respiratory contact. Full body protective clothing and gloves should be used by those employed in handling operations. Fullface, supplied air respirators of continuous flow or pressure demand type should also be used. On exit from a regulated area, employees should shower and change into street clothes, leaving their protective clothing and equipment at the point of exit to be placed in impervious containers at the end of the workshift for decontamination or disposal. Effective methods should be used to clean and decontaminate gloves and clothing.

Disposal Method Suggested: Incineration (1800°F, 2.0 seconds minimum) with scrubbing for NO_x abatement.

References

(1) International Agency for Research on Cancer, *IARC Monographs on the Carcinogenic Risks of Chemicals to Humans*, Lyon, France 4, 113 (1974).

o-NITROCHLOROBENZENE

● Carcinogen (Animal Positive) (1)

Description: o-$ClC_6H_4NO_2$ is a yellow solid melting at 32.5°C.

Code Numbers: CAS 88-73-3 RTECS CZ0875000 UN 1578

DOT Designation: Poision B.

Synonyms: 2-nitrochlorobenzene; 2-chloronitrobenzene; 2-CNB.

Potential Exposure: In production of 2-CNB or its use in dye manufacture.

Incompatibilities: Strong oxidizers.

Permissible Exposure Limits in Air: There is no U.S. standard for 2-CNB but the U.S.S.R. has a TLV of 1.0 mg/m^3.

Determination in Air: Collection on silica gel; desorption with methanol; chromatographic analysis.

Permissible Concentration in Water: A limit of 0.05 mg/ℓ has been set in the U.S.S.R.

Routes of Entry: Inhalation, skin absorption, ingestion, eye contact.

Harmful Effects and Symptoms: 2-CNB is not considered to be a very active carcinogen but is carcinogenic to male rats and both male and female mice. Cyanosis is the predominant symptom of acute toxicity.

The LD_{50} value for rats is 288 mg/kg (compared to 420 mg/kg for 4-CNB).

Points of Attack: Blood, liver, kidneys, cardiovascular system.

Medical Surveillance: Consider the points of attack in preplacement and periodic physical examinations.

First Aid: See p-Nitrochlorobenzene.

Personal Protective Methods: See p-Nitrochlorobenzene.

Respirator Selection: See p-Nitrochlorobenzene.

References

(1)   U.S. Environmental Protection Agency, *Chemical Hazard Information Profile Draft Report: 2-Chloronitrobenzene,* Washington, DC (June 13, 1983).
(2)   See Reference (A-38).
(3)   See Reference (A-39).

p-NITROCHLOROBENZENE

- Carcinogen (Animal Positive) (1)

Description: p-ClC$_6$H$_4$NO$_2$ is a yellow solid with a sweet odor. It melts at 83°C and boils at 242°C.

Code Numbers: CAS 100-00-5   RTECS CZ1050000   UN 1578

DOT Designation: Poison B.

Synonyms: PNCB, PCNB (note possible conflict with pentachloronitro-benzene), 4-chloronitrobenzene, p-chloronitrobenzene, 1-chloro-4-nitrobenzene.

Potential Exposure: PNCB is used as an intermediate in pesticide (parathion) manufacture, drug (phenacetin and acetaminophen) manufacture and in dye and antioxidant manufacture.

Incompatibilities: Strong oxidizers.

Permissible Exposure Limits in Air: The Federal standard is 1 mg/m^3. The proposed 1983/84 ACGIH TWA is 0.5 ppm (3 mg/m^3). The notation "skin" is added to indicate the possibility of cutaneous absorption. There is no STEL value set.

Determination in Air: Adsorption on silica, workup with methanol, analysis by gas chromatography. See NIOSH Methods, Set P. See also reference (A-10).

Permissible Concentration in Water: A limit of 0.02 mg/ℓ has been set in the U.S.S.R.

Routes of Entry: Inhalation, skin absorption, ingestion, eye and skin contact.

Harmful Effects and Symptoms: Anoxia, unpleasant taste; mild anemia; dizziness; weakness; nausea, vomiting, dyspnea.

Points of Attack: Blood, liver, kidneys, cardiovascular system.

Medical Surveillance: Consider the points of attack in preplacement and periodic physical examinations.

First Aid: If this chemical gets into the eyes, irrigate immediately. If this chemical contacts the skin, wash with soap immediately. If a person breathes in large amounts of this chemical, move the exposed person to fresh air at once and perform artificial respiration. When this chemical has been swallowed, get medical attention. Give large quantities of water and induce vomiting. Do not make an unconscious person vomit.

Personal Protective Methods: Wear appropriate clothing to prevent any reasonable probability of skin contact. Wear eye protection to prevent any reasonable probability of eye contact. Employees should wash immediately when skin is wet and daily at the end of each workshift. Work clothing should be changed daily if it is possible that clothing is contaminated. Remove nonimpervious clothing immediately if wet. Provide emergency showers.

Respirator Selection:
  5 mg/m^3: DMXS
  10 mg/m^3: DMXSQ/FuHiEP/SA/SCBA
  50 mg/m^3: HiEPF/SAF/SCBAF
  1,000 mg/m^3: PAPHiE/SA:PD,PP,CF
  Escape: GMXS/SCBA

Disposal Method Suggested: Incineration (1500°F, 0.5 second for primary combustion; 2200°F, 1.0 second for secondary combustion). The formation of elemental chlorine can be prevented through injection of steam or methane into the combustion process. NO_x may be abated through the use of thermal or catalytic devices (A-31).

References
(1) U.S. Environmental Protection Agency, *Chemical Hazard Information Profile Draft Report: 4-Chloronitrobenzene,* Washington, DC (June 13, 1983).
(2) See Reference (A-60).
(3) See Reference (A-39).

NITROETHANE

Description: $CH_3CH_2NO_2$ is a colorless liquid with a mild, fruity odor boiling at 114°C.

Code Numbers: CAS 79-24-3   RTECS KI5600000   UN 2842

DOT Designation: —

Synonyms: None.

Potential Exposure: Nitroethane is used as a solvent for cellulose esters, vinyl and alkyd resins, waxes, fats and dyestuffs. It has been used as a rocket propellant. It is used as an intermediate in pharmaceutical manufacture (A-41) and in pesticide manufacture (A-32).

Incompatibilities: Amines, strong acids, alkalies; strong oxidizers; hydrocarbons, other combustibles; metal oxides.

Permissible Exposure Limits in Air: The Federal standard and the 1983/84 ACGIH TWA value is 100 ppm (310 mg/m^3). The STEL is 150 ppm (465 mg/m^3). The IDLH level is 1,000 ppm.

Determination in Air: Adsorption on XAD-2 resin, desorption with ethyl acetate, analysis by gas chromatography with flame ionization detection (A-10).

Permissible Concentration in Water: No criteria set.

Routes of Entry: Inhalation, ingestion, skin and eye contact.

Harmful Effects and Symptoms: Dermatitis.

Points of Attack: Skin.

Medical Surveillance: Consider the points of attack in preplacement and periodic physical examinations.

First Aid: If this chemical gets into the eyes, irrigate immediately. If this chemical contacts the skin, wash with soap promptly. If a person breathes in large amounts of this chemical, move the exposed person to fresh air at once and perform artificial respiration. When this chemical has been swallowed, get medical attention. Give large quantities of water and induce vomiting. Do not make an unconscious person vomit.

Personal Protective Methods: Wear appropriate clothing to prevent repeated or prolonged skin contact. Wear eye protection to prevent any reasonable probability of eye contact. Employees should wash promptly when skin is wet or contaminated. Remove clothing immediately if wet or contaminated to avoid flammability hazard.

Respirator Selection:
1,000 ppm: SAF/SCBAF
Escape: SCBA

Disposal Method Suggested: Incineration, large quantities of material may require NO_x removal by catalytic or scrubbing processes.

References
(1) See Reference (A-61).
(2) See Reference (A-60).

NITROFEN

● Carcinogen (Animal Positive) (A-64)

Description: $C_{12}H_7Cl_2NO_3$ with the following structural formula

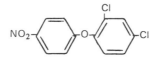

is a crystalline solid melting at 70° to 71°C.

Code Numbers: CAS 1836-75-5   RTECS KN8400000

DOT Designation: –

Synonyms: 2,4-Dichloro-1-(4-nitrophenoxy)benzene; 2,4-dichlorophenyl-p-nitrophenyl ether.

Potential Exposure: Nitrofen is a contact herbicide used for pre-and post-emergence control of annual grasses and broadleaf weeds on a variety of food and ornamental crops. Nitrofen has been applied in approximately 25 states by growers of rice, broccoli, cauliflower, cabbage, brussel sprouts, onions, garlic, and celery. Nitrofen has also been used in nurseries growing roses and chrysanthemums and on rights-of-way. The direct crop use of nitrofen for 1980 was estimated as 882,000 pounds. Nitrofen has not been used around homes and gardens.

Occupational exposure to nitrofen, primarily through inhalation and dermal contact may occur among workers at production facilities. Field handlers of the herbicide are subject to inhalation exposure during application procedures.

Permissible Exposure Limits in Air: No standards set.

Permissible Concentration in Water: No criteria set.

Routes of Entry: Inhalation, ingestion, skin absorption.

Harmful Effects and Symptoms: Nitrofen (technical grade)—a substituted diphenyl ether—administered in the feed, was carcinogenic to B6C3F1 mice, causing hepatocellular carcinomas in both sexes. There was no evidence of carcinogenicity in Fischer 344 rats (1). In another bioassay, nitrofen (technical grade) administered in the feed was found to induce hepatocellular carcinomas in B6C3F1 mice of both sexes, and hemangiosarcomas of the liver in male B6C3F1 mice. In addition, adenocarcinomas of the pancreas were induced in female Osborne-Mendel rats (2). These results can be considered as providing sufficient evidence for the carcinogenicity of nitrofen in experimental animals (3).

References

(1) National Cancer Institute, *Bioassay of Nitrofen for Possible Carcinogenicity,* Technical Report Series No. 184, DHEW Publiccation No. (NIH) 79-1740, Bethesda, Maryland (1979).

(2) National Cancer Institute, *Bioassay of Nitrofen for Possible Carcinogenicity,* Technical Report Series No. 26, DHEW Publication No. (NIH) 78-826, Bethesda, Maryland (1978).

(3) Griesemer, R.A., and C. Cueto, "Toward a Classification Scheme for Degrees of Experimental Evidence for the Carcinogenicity of Chemicals for Animals." In: R. Montesano, H. Bartsch, and L. Tomatis (eds.). *Molecular and Cellular Aspects of Carcinogen Screening Tests,* IARC Scientific Publications, No. 27, Lyon, France: International Agency for Research on Cancer, pp 259-281 (1980).

(4) United Nations Environment Programme, *IRPTC Legal File 1983,* Vol. I, pp VII/347, Geneva, Switzerland, International Register of Potentially Toxic Chemicals (1984).

NITROGEN DIOXIDE

- Hazardous substance (EPA)
- Hazardous waste (EPA)

Description: NO_2 and N_2O_4, the dimer, constitute a dark brown, fuming liquid or gas with a pungent, acrid odor. It boils at $21°C$.

Code Numbers: CAS 10102-44-0 RTECS QW9800000 UN 1067

DOT Designation: Poison A, poison gas and oxidizer.

Synonyms: Nitrogen tetroxide, NTO, dinitrogen tetroxide, nitrogen peroxide.

Potential Exposures: Nitrogen dioxide is used as an intermediate in nitric and sulfuric acid manufacture; it is used in the nitration of organic compounds; it is used as an oxidizer in liquid propellant rocket fuel combinations.

Incompatibilities: Combustible matter, chlorinated hydrocarbons, ammonia, carbon disulfide.

Permissible Exposure Limits in Air: The Federal standard is 5 ppm (9 mg/m^3). NIOSH has suggested a 1 ppm (1.8 mg/m^3) ceiling value. The Dutch chemical industry (A-60) has set a 5 ppm (9 mg/m^3) ceiling value. The IDLH level is 50 ppm. The ACGIH as of 1983/84 has suggested a TWA of 3 ppm (6 mg/m^3) and an STEL of 5 ppm (10 mg/m^3).

Determination in Air: Collection on TEA-coated molecular sieve, desorption with TEA, analysis by spectrophotometry (A-10).

Permissible Concentration in Water: No criteria set.

Routes of Entry: Inhalation, ingestion, skin and eye contact.

Harmful Effects and Symptoms: *Local* — Nitrogen oxide gases may produce irritation of the eyes and mucous membranes. Nitrogen tetroxide is an extremely corrosive liquid and may cause severe burns, ulcers, and necrosis of the skin, mucous membranes, and eye tissues.

Systemic — Exposure to high concentrations of nitrogen oxides may result in severe pulmonary irritation caused by the nitrogen dioxide.

Nitrogen dioxide at high concentrations has also been shown to cause methemoglobinemia in the dog. Typically, acute exposure may produce immediate malaise, cyanosis, cough, dyspnea, chills, fever, headache, nausea, and vomiting. Collapse and death may occur if exposure is sufficiently high. When lower concentrations are encountered, there may be only mild signs of bronchial irritation followed by a five to twelve-hour symptom-free period. Subsequently, the onset of signs and symptoms of acute pulmonary edema occur suddenly, which unfortunately may take place away from prompt medical aid.

If the acute episode is survived, bronchiolitis fibrosa obliterans may develop usually within a few days but may be latent for as long as six weeks. Victims may develop severe and increasing dyspnea which is often accompanied by fever and cyanosis. Chest roentgenogram may reveal a diffuse, reticular, and fine nodular infiltration or numerous, uniform, scattered nodular densities ranging in size from 1 to 5 mm in diameter.

Chronic exposure may result in pulmonary dysfunction with decreased vital capacity, maximum breathing capacity and lung compliance, and increased residual volume. The most common complaint is of dyspnea upon exertion. Signs include moist rales and wheezes, sporadic cough with mucopurulent expectoration, a decrease in blood pH and serum proteins, and an increase in urinary hydroxyproline and acid mucopolysaccharides. These findings are suggestive of emphysema, although they are yet inconclusive.

Points of Attack: Respiratory system, lungs, cardiovascular system.

Medical Surveillance: Preplacement and periodic examinations should be concerned particularly with the skin, eyes, and with significant pulmonary and heart diseases. Periodic chest x-rays and pulmonary function tests may be useful. Smoking history should be known. Methemoglobin studies may be of interest if exposure to nitric oxide is present. In the case of nitric acid vapor mist exposure, dental effects may be present. See also reference (2).

First Aid: If this chemical gets into the eyes, irrigate immediately. If this chemical contacts the skin, flush with water immediately. If a person breathes in large amounts of this chemical, move the exposed person to fresh air at once and perform artificial respiration. When this chemical has been swallowed, get medical attention. Give large quantities of water and do not induce vomiting.

Personal Protective Methods: Workers should not enter confined areas where nitrogen oxides may accumulate (for example, silos) without appropriate eye and respiratory protection.

Individuals should be equipped with supplied air respirators with fullface piece or chemical goggles, and enclosed areas should be properly ventilated before entering. An observer equipped with appropriate respiratory protection should be outside the area and standing by to supply any aid needed.

Wear appropriate clothing to prevent any possibility of skin contact. Wear eye protection to prevent any possibility of eye contact. Employees should wash immediately when skin is contaminated. Remove nonimpervious clothing immediately if wet or contaminated. Provide emergency showers and eyewash.

Respirator Selection:

    50 ppm: CCRSF/GMS/SAF/SCBAF
    Escape: GMS/SCBA
    Note: Do not use oxidizable sorbents.

Disposal Method Suggested: Incineration with added hydrocarbon fuel, controlled so that combustion products are elemental nitrogen, CO_2 and water.

References

(1) National Institute for Occupational Safety and Health, *Criteria for a Recommended Standard: Occupational Exposure to Oxides of Nitrogen*, NIOSH Doc. No. 76-149 (1976).

(2) Nat. Inst. for Occupational Safety and Health, *A Guide to the Work Relatedness of Disease*, Revised Edition, DHEW (NIOSH) Publ. No. 79-116, Cincinnati, Ohio pp 134-145 (Jan. 1979).

(3) World Health Organization, *Oxides of Nitrogen,* Environmental Health Criteria No. 4, Geneva, Switzerland (1977).

(4) See Reference (A-61).

(5) Sax, N.I., Ed., *Dangerous Properties of Industrial Materials Report, 1,* No. 5, 74-76, New York, Van Nostrand Reinhold Co. (1981).

NITROGEN OXIDES

See also entries under "Nitric Oxide" and "Nitrogen Dioxide."

The USEPA has set national ambient air quality standards for nitrogen oxides of 0.05 ppm (100 $\mu g/m^3$) as an annual arithmetic mean value.

References

(1) National Academy of Sciences, *Medical and Biologic Effects of Environmental Pollutants: Nitrogen Oxides,* Washington, DC (1977).

(2) U.S. Environmental Protection Agency, *Air Quality Criteria for Oxides of Nitrogen,* (Draft report in preparation; External Review Draft No. 2 being circulated - 1980) Research Triangle Park, North Carolina, Criteria and Assessment Office.

(3) Parmeggiani, L., Ed., *Encyclopedia of Occupational Health & Safety,* Third Edition, Vol. 2, pp 1457–59, Geneva, International Labour Office (1983) (Nitrogen Oxides).

NITROGEN TRIFLUORIDE

Description: NF_3 is a colorless gas with a moldy odor.

Code Numbers: CAS 7783-54-2 RTECS QX1925000 UN 2451

DOT Designation: Poison gas.

Synonyms: None.

Potential Exposure: This material has been used as an oxidizer in rocket propellant combinations.

Incompatibilities: Water, vapor oil, grease; oxidizable materials; ammonia, carbon monoxide, methane, hydrogen, hydrogen sulfide, active metals; oxides.

Permissible Exposure Limits in Air: The Federal standard and the ACGIH 1983/84 TWA value is 10 ppm (30 mg/m^3). The tentative STEL is 15 ppm (45 mg/m^3). The IDLH level is 2,000 ppm.

Permissible Concentration in Water: No criteria set.

Routes of Entry: Inhalation.

Harmful Effects and Symptoms: None known in humans; in animals—methemoglobinemia (A-4). However, a Japanese source (A-38) says NF$_3$ is corrosive to tissue and that teeth and bones are affected on long inhalation.

Points of Attack: None known in humans; in animals—blood.

Medical Surveillance: Nothing special indicated for humans.

First Aid: If a person breathes in large amounts of this chemical, move the exposed person to fresh air at once and perform artificial respiration.

Personal Protective Methods: No protective devices other than respirators are indicated by NIOSH (A-4). Rubber gloves, face shield and overalls are suggested by others (A-38), however.

Respirator Selection:

100 ppm:	SA/SCBA
500 ppm:	SAF/SCBAF
2,000 ppm:	SAF:PD,PP,CF
Escape:	GMS/SCBA

Disposal Method Suggested: Vent into large volume of concentrated reducing agent (bisulfites, ferrous salts or hypo) solution, then neutralize and flush to sewer with larger volumes of water (A-38).

References

(1) Nat. Inst. for Occup. Safety and Health, *Criteria for a Recommended Standard: Occupational Exposure to Inorganic Fluorides,* NIOSH Doc. No. 75-103, Wash., DC (1975).
(2) See Reference (A-61).

NITROGLYCERIN

See "Ethylene Glycol Dinitrate" to which similar standards and precautions apply.

References

(1) Sax, N.I., Ed., *Dangerous Properties of Industrial Materials Report, 1,* No. 4, 89-90, New York, Van Nostrand Reinhold Co. (1981).
(2) Parmeggiani, L., Ed., *Encyclopedia of Occupational Health & Safety,* Third Edition, Vol. 2, pp 1459–61, Geneva, International Labour Office (1983).

NITROMETHANE

Description: CH$_3$NO$_2$ is a colorless liquid with a mild fruity odor boiling at 101°C.

Code Numbers: CAS 75-52-5   RTECS PA9800000   UN 1261

DOT Designation: Flammable liquid.

Synonyms: Nitrocarbol.

Potential Exposure: Nitromethane is used in the production of the fumigant, chloropicrin, CCl_3NO_2 (A-32). It is also used as a racing car fuel. It is also used as an intermediate in the pharmaceutical industry (A-41).

Incompatibilities: Amines, strong acids, alkalies; strong oxidizers; hydrocarbons, other combustibles; metallic oxides.

Permissible Exposure Limits in Air: The Federal standard and the 1983/84 ACGIH TWA value is 100 ppm (250 mg/m^3). The STEL is 150 ppm (375 mg/m^3). The IDLH level is 1,000 ppm.

Determination in Air: Sample collection by charcoal tube, analysis by gas liquid chromatography (A-13).

Permissible Concentration in Water: No criteria set.

Routes of Entry: Inhalation, ingestion, skin and eye contact.

Harmful Effects and Symptoms: Dermatitis.

Points of Attack: Skin.

Medical Surveillance: Consider the points of attack in preplacement and periodic physical examinations.

First Aid: If this chemical gets into the eyes, irrigate immediately. If this chemical contacts the skin, wash with soap promptly. If a person breathes in large amounts of this chemical, move the exposed person to fresh air at once and perform artificial respiration. When this chemical has been swallowed, get medical attention. Give large quantities of water and induce vomiting. Do not make an unconscious person vomit.

Personal Protective Methods: Wear appropriate clothing to prevent repeated or prolonged skin contact. Wear eye protection to prevent any reasonable probability of eye contact. Employees should wash promptly when skin is wet or contaminated. Remove clothing immediately if wet or contaminated to avoid flammability hazard.

Respirator Selection:
    1,000 ppm: SAF/SCBAF
      Escape: SCBA

Disposal Method Suggested: Incineration, large quantities of material may require NO_x removal by catalytic or scrubbing processes.

References

(1) Nat. Inst. for Occup. Safety and Health, *Information Profiles on Potential Occupational Hazards—Classes of Chemicals: Nitroparaffins,* NIOSH Pub. No. TR-78-518, Rockville, MD, pp 199-210 (April 1978).
(2) See Reference (A-61).
(3) See Reference (A-60).

NITROPHENOLS

- Hazardous substances (EPA)
- Hazardous waste (4-Nitrophenol) (EPA)
- Priority toxic pollutants (EPA)

Description: There are three isomers of nitrophenol $NO_2C_6H_4OH$. The meta-form is produced from m-nitroaniline, and the o- and p- isomers are produced by nitration of phenol. They are colorless to slightly yellowish crystals with an aromatic to sweetish odor. They melt at 214°C (o-), 149°C (m-), and 279°C (p-).

Code Numbers:

2-Nitrophenol	CAS 88-75-5	RTECS SM2100000	UN 1663
3-Nitrophenol	CAS 554-84-7	RTECS SM1925000	UN 1663
4-Nitrophenol	CAS 100-02-7	RTECS SM2275000	UN 1663

DOT Designation: ORM-E.

Synonyms: Hydroxynitrobenzenes.

Potential Exposures: Nitrophenols are used in the synthesis of dyestuffs and other intermediates and pesticides.

Permissible Exposure Limits in Air: There is no Federal standard for nitrophenol. The ACGIH has not assigned any values either.

Permissible Concentrations in Water: To protect freshwater aquatic life— 230 $\mu g/\ell$ on an acute toxicity basis. To protect saltwater aquatic life—4,580 $\mu g/\ell$ on an acute toxicity basis. To protect human health—70 $\mu g/\ell$.

Determination in Water: Methylene chloride extraction followed by gas chromatography with flame ionization or electron capture detection (EPA Method 604) or gas chromatography pluss mass spectrometry (EPA Method 625).

Routes of Entry: Inhalation and percutaneous absorption of liquid.

Harmful Effects and Symptoms: *Local* — Unknown.
Systemic — There is very little information available on the toxicity for humans of nitrophenols. Animal experiments have shown central and peripheral vagus stimulation, CNS depression, methemoglobinemia, and dyspnea. The p-isomer is the most toxic.

Medical Surveillance: Based on animal studies, individuals with cardiovascular, renal, or pulmonary disease and those with anemia are probably more subject to poisoning by nitrophenol. Liver and renal function and blood should be evaluated in placement or periodic examinations.

First Aid: Wash contaminated areas of body with soap and water. Remove contaminated clothing immediately.

Personal Protective Methods: Nitrophenols are readily absorbed through intact skin and by inhalation; full body protective clothing is needed. Clean work clothes should be supplied daily; showers should be taken prior to changing to street clothes.

Respirator Selection: Appropriate type respirators with organic vapor canisters should be provided in areas of concentrations of dust or vapors.

Disposal Method Suggested: Controlled incineration—care must be taken to maintain complete combustion at all times. Incineration of large quantities may require scrubbers to control the emission of NO_x. Alternatively, nitrophenols may be recovered, from wastewaters, for example (A-58).

References

(1) Nat. Inst. for Occup. Safety and Health, *Information Profiles on Potential Occupational Hazards: Nitrophenols*, Report No. PB-276,678, Rockville, MD, pp 212-226 (Oct. 1977).

(2) U.S. Environmental Protection Agency, *Nitrophenols: Ambient Water Quality Criteria*, Wash., DC (1980).

(3) U.S. Environmental Protection Agency, *4-Nitrophenol*, Health and Environmental Effects Profile No. 135, Wash., DC, Office of Solid Waste (April 30, 1980).

(4) U.S. Environmental Protection Agency, *Nitrophenols,* Health and Environmental Effects Profile No. 136, Wash., DC, Office of Solid Waste (April 30, 1980).

(5) Sax, N.I., Ed., *Dangerous Properties of Industrial Materials Report, 3,* No. 3, 82-85, New York, Van Nostrand Reinhold Co. (1983). (p-Nitrophenol).

(6) See Reference (A-60) for p-Nitrophenol.

(7) United Nations Environment Programme, *IRPTC Legal File 1983,* Vol. II, pp VII/513-18, Geneva, Switzerland, International Register of Potentially Toxic Chemicals (1984).

NITROPROPANES

- Carcinogen (Suspect: 2-Nitropropane only, NIOSH) (3)
- Hazardous waste (2-Nitropropane) (EPA)

Description: $CH_3CH_2CH_2NO_2$ and $CH_3CH(NO_2)CH_3$ are colorless liquids with mild fruity odors. $CH_3CH(NO_2)CH_3$, 2-nitropropane, boils at $120°C$.

Code Numbers:

1-Nitropropane	CAS 108-03-2	RTECS TZ5075000	UN 2608
2-Nitropropane	CAS 79-46-9	RTECS TZ5250000	UN 2608

DOT Designation: Flammable liquid.

Synonyms:

1-Nitropropane: None
2-Nitropropane: Dimethylnitromethane
sec-Nitropropane
Isonitropropane
Nitroisopropane
2-NP, NiPAR S-20®

Potential Exposure: 2-Nitropropane is used in manufacturing as a solvent for organic compounds, cellulose, esters, gums, vinyl resins, waxes, epoxy resins, fats, dyes, and chlorinated rubber.

Its combustion properties have made it useful as a rocket propellant and as a gasoline and diesel fuel additive. 2-Nitropropane also has limited use as a paint and varnish remover. It serves as an intermediate in organic synthesis of some pharmaceuticals, dyes, insecticides (A-32), and textile chemicals.

It is estimated that 100,000 workers in the United States are exposed to 2-nitropropane. 2-Nitropropane is not known to occur naturally but has been detected in tobacco smoke with other nitroalkanes; the levels were found to correlate with tobacco nitrate contents. The smoke content of a filterless 85 mm United States blend cigarette was found to contain (μg): 2-nitropropane, 1.1; 1-nitropropane, 0.13; 1-nitrobutane, 0.71; nitroethane, 1.1; and nitromethane, 0.53.

Incompatibilities: Amines, strong acids, alkalies; strong oxidizers, metal oxides, combustible materials.

Permissible Exposure Limits in Air: The Federal standard for both nitropropanes is 25 ppm (90 mg/m^3). The ACGIH recommends a TWA of 15 ppm (55 mg/m^3) and a tentative STEL of 25 ppm (90 mg/m^3) for 1-nitropropane.

ACGIH recommends a TWA value as of 1983/84 of 10 ppm (35 mg/m³) [down from a previous ceiling value of 25 ppm (90 mg/m³) and an STEL of 20 ppm (70 mg/m³)] for 2-nitropropane. It declares that it is a substance suspect of carcinogenic potential for man.

The IDLH level for both compounds is 2,300 ppm.

Determination in Air: 2-Nitropropane may be determined by adsorption on Chromosorb 106, desorption with ethyl acetate and analysis by gas chromatography (A-10).

Permissible Concentration in Water: No criteria set.

Routes of Entry: Inhalation, ingestion, skin and eye contact.

Harmful Effects and Symptoms: In the case of 1-nitropropane—eye irritation, headaches, nausea, vomiting and diarrhea.

Generally observed symptoms in humans exposed to 2-nitropropane vapors include anorexia, nausea, vomiting, diarrhea, severe occipital headaches, cyanosis, and vapor irritation of the lungs. Large doses have the potential of producing methemoglobinemia and damaging the liver and kidneys. Other toxic effects on the central nervous system may arise.

Points of Attack: 1-Nitropropane—eyes, central nervous system; 2-nitropropane—respiratory system, central nervous system.

Medical Surveillance: Based on animal data, preplacement and periodic examination should consider respiratory and central nervous system effects as well as liver and kidney function. In the case of 2-nitropropane, Heinz bodies and methemoglobin levels would be of interest.

First Aid: If this chemical gets into the eyes, irrigate immediately. If this chemical contacts the skin, dust off solid, flush with water, wash with soap promptly. If a person breathes in large amounts of this chemical, move the exposed person to fresh air at once and perform artificial respiration. When this chemical has been swallowed, get medical attention. Give large quantities of water and induce vomiting. Do not make an unconscious person vomit.

Personal Protective Methods: Barrier creams or gloves to protect exposed skin and, where vapor concentrations are excessive, fullface mask with organic vapor canister or air supplied respirators are advised. Wear eye protection to prevent any reasonable probability of eye contact. Remove clothing immediately if wet or contaminated to avoid flammability hazard.

Respirator Selection:
      150 ppm: SA/SCBA
    1,250 ppm: SAF/SCBAF
    2,300 ppm: SAF:PD,PP,CF
      Escape: SCBA

Disposal Method Suggested: Incineration—large quantities of material may require NO_x removal by catalytic or scrubbing processes.

References

(1) U.S. Environmental Protection Agency, *Chemical Hazard Information Profile: 2-Nitropropane*, Wash., DC (Sept. 27, 1977).

(2) Nat. Inst. for Occup. Safety and Health, *Information Profiles on Potential Occupational Hazards—Classes of Chemicals: Nitroparaffins*, NIOSH Pub. No. TR 78-518, Rockville, MD, pp 199-210 (April 1978).

(3) National Inst. for Occup. Safety and Health, *2-Nitropropane*, Current Intelligence Bulletin No. 17, Rockville, MD (April 25, 1977).

(4) See Reference (A-61). (1-Nitropropane).
(5) Sax, N.I., Ed., *Dangerous Properties of Industrial Materials Report, 2,* No. 2, 58-59, New York, Van Nostrand Reinhold Co. (1982) . (2-Nitropane).
(6) See Reference (A-60).
(7) Sax, N.I., Ed., *Dangerous Properties of Industrial Materials Report, 4,* No. 1, 92–94, New York, Van Nostrand Reinhold Co. (Jan./Feb. 1984) (2-Nitropropane).
(8) Parmeggiani, L., Ed., *Encyclopedia of Occupational Health & Safety,* Third Edition, Vol. 2, pp 1461–63, Geneva, International Labour Office (1983) (2-Nitropropane).
(9) United Nations Environment Programme, *IRPTC Legal File 1983,* Vol. II, pp VII/678–9, Geneva, Switzerland, International Register of Potentially Toxic Chemicals (1984).

NITROSAMINES

- Carcinogens (Animal, Positive, IARC)
- Hazardous Waste Constituents (EPA)

See "N-nitrosodimethylamine" as a typical example of this class of compounds.

A large number of nitrosamines have been found by IARC to exhibit substantial evidence of carcinogenicity. They are as follows (A-40).

N-nitrosodi-n-butylamine
N-nitrosodiethanolamine
N-nitrosodiethylamine
N-nitrosodimethylamine
N-nitroso-dipentylamine
N-nitroso-diphenylamine
N-nitrosodi-n-propylamine
N-nitrosomethylethylamine
N-nitroso-N-ethylurea
N-nitroso-N-methylurea
N-nitroso-N-methylurethane
N-nitrosomethylvinylamine
N-nitrosomorpholine
N-nitrosonornicotine
N-nitrosopiperidine
N-nitrosopyrrolidine
N-nitrososarcosine

References

(1) U.S. Environmental Protection Agency, *Nitrosamines,* Health and Environmental Effects Profile No. 137, Wash., DC, Office of Solid Waste (April 30, 1980).
(2) U.S. Environmental Protection Agency, *Nitrosodiphenylamine,* Health and Environmental Effects Profile No. 138, Wash., DC, Office of Solid Waste (April 30, 1980).
(3) U.S. Environmental Protection Agency, *N-Nitrosodi-n-propylamine,* Health and Environmental Effects Profile No. 139, Wash., DC, Office of Solid Waste (April 30, 1980).
(4) World Health Organization, *Nitrates, Nitrites and N-Nitroso Compounds,* Environmental Health Criteria No. 5, Geneva, Switzerland (1978).
(5) Sax, N.I., Ed., *Dangerous Properties of Industrial Materials Report, 2,* No. 5, 90-92 New York, Van Nostrand Reinhold Co. (1982). (N-Nitrosodibutylamine).
(6) Sax, N.I., Ed., *Dangerous Properties of Industrial Materials Report, 1,* No. 2, 49-50, New York, Van Nostrand Reinhold Co. (1980). (N-Nitrosodiethylamine).
(7) Sax, N.I., Ed., *Dangerous Properties of Industrial Materials Report, 1,* No. 5, 76, New York, Van Nostrand Reinhold Co. (1981). (N-Nitrosodipentylamine).
(8) Sax, N.I., Ed., *Dangerous Properties of Industrial Materials Report, 1,* No. 5, 76-77, New York, Van Nostrand Reinhold Co. (1981). (N-Nitrosodiphenylamine).

(9)   See References (A-62) and (A-64) for: N-Nitroso-di-n-butylamine, N-Nitrosodiethanolamine, N-Nitrosodiethylamine, N-Nitrosodiphenylamine, N-Nitroso-di-N-pro-
pylamine, N-Nitroso-N-ethylurea, N-Nitroso-N-methylurea, N-Nitrosomethylvinylamine, N-Nitrosomorpholine, N-Nitrosonornicotine, N-Nitrosopiperidine, N-Nitrosopyrrolidine, and N-Nitrososarcosine.

(10)   United Nations Environment Programme, *IRPTC Legal File 1983,* Vol. I, pp VII/267,
285–6, 290–92, Geneva, Switzerland, International Register of Potentially Toxic
Chemicals (1984).

N-NITROSODIMETHYLAMINE

- Carcinogen (Human suspected, IARC) (4)
- Hazardous waste (EPA)
- Priority toxic pollutant (EPA)

Description: $(CH_3)_2NN=O$, N-nitrosodimethylamine, is a yellow liquid of
low viscosity, soluble in water, alcohol, and ether. It boils at $152^\circ C$.

Code Numbers: CAS 62-75-9   RTECS IQ0525000

DOT Designation: –

Synonyms: Dimethylnitrosamine, DMN, DMNA

Potential Exposures: DMN is used in the manufacture of dimethylhydrazine.
It has also been used as an industrial solvent and a nematocide.

A number of nitrosamines have been patented for use as gasoline and lubricant additives, antioxidants, and also pesticides. Dimethylnitrosamine (DMN)
$[(CH_3)_2NN=O]$ is used primarily in the electrolytic production of the hyper-
golic rocket fuel 1,1-dimethylhydrazine. Other areas of utility include the control of nematodes, the inhibition of nitrification in soil, use as plasticizer for
acrylonitrile polymers, use in active metal anode-electrolyte systems (high-en-
ergy batteries), in the preparation of thiocarbonyl fluoride polymers, in the
plasticization of rubber, and in rocket fuels.

Synthetic cutting fluids, semisynthetic cutting oils and soluble cutting oils
may contain nitrosamines, either as contaminants in amines, or as products from
reactions between amines and nitrite. Concentrations of nitrosamines have been
found in certain synthetic cutting oils at levels ranging from 1 ppm to 1,000
ppm. It is believed that there are 8 to 12 additives that could be responsible for
nitrosamine formation in cutting oils and that approximately 750,000 to 780,000
persons employed by more than 1,000 cutting fluid manufacturing firms are endangered, in addition to an undetermined number of machine shop workers who
use the fluids.

High levels of nitrosamines in soils (believed to arise from the use of triazine
herbicide which can combine with nitrogen fertilizer) have been reported as well
as plant uptake and leaching of dimethylnitrosamine.

N-nitrosodiethanolamine has been found in amounts ranging from 1 ng/g (1
ppb) to a high of about 48,000 ppb in about 30 toiletry products (e.g., cosmetics, hand and body lotions and shampoos.) The N-nitroso compound found
probably results from nitrosation of di- and/or triethanolamine emulsifiers by a
nitrite compound.

Permissible Exposure Limits in Air: DMN is included in the Federal standard
for carcinogens; all contact with it should be avoided. ACGIH as of 1983/84

notes that DMN is an industrial substance suspect of carcinogenic potential for man and sets no TWA or tentative STEL values. It is noted that skin absorption is possible.

Determination in Air: Adsorption on Tenax, thermal desorption with helium purges, separation by gas liquid chromatography, analysis by mass spectrometry (A-10).

Permissible Concentration in Water: To protect freshwater aquatic life—5,850 $\mu g/\ell$ on an acute toxicity basis for nitrosamines as a class. To protect saltwater aquatic life—3,300,000 $\mu g/\ell$ on an acute toxicity basis for nitrosamines as a class. For protection of human health—preferably zero. An additional lifetime cancer risk of 1 in 100,000 is posed by a concentration of 0.014 $\mu g/\ell$.

Determination in Water: Methylene chloride extraction followed by gas chromatography with nitrogen-phosphorus or reductive Hall detectors (EPA Method 607) or gas chromatography plus mass spectrometry (EPA Method 625).

Routes of Entry: Inhalation of vapor and possibly percutaneous absorption.

Harmful Effects and Symptoms: *Local* — The liquid and vapor are not especially irritating to the skin or eyes, and warning properties are poor.

Systemic — DMN is a highly toxic substance in most species, including man. Systemic effects are characterized by onset in a few hours of nausea and vomiting, abdominal cramps and diarrhea. Also headache, fever, weakness, enlargement of the liver, and jaundice may occur. Chronic exposures may lead to liver damage (central necrosis), with jaundice and ascites. There have been a number of reported cases, including severe liver injury in man and one death. Autopsy revealed an acute diffuse centrolobular necrosis. Recovery occurred in other cases.

In rats, guinea pigs, and other experimental animals, DMN is a highly potent carcinogen, producing malignant tumors, primarily of the liver and kidney, but also in the lung. Both ingestion and inhalation routes have produced tumors. These have not been reported in man, but in view of its potency in various other species, the material has been presumed to be carcinogenic in man also.

Medical Surveillance: Based on human experience and on animal studies, preplacement and periodic examinations should include a history of exposure to other carcinogens, alcohol and smoking habits, medications, and family history. Special attention should be given to liver size and function, and to any changes in lung symptoms or x-rays. Renal function should be followed. Sputum and urine cytology may be useful.

First Aid: Irrigate eyes with water. Wash contaminated areas of body with soap and water.

Personal Protective Methods: These are designed to supplement engineering controls and to prevent all contact with the skin, eyes, or respiratory tract. Full body protective clothing and gloves should be provided and also appropriate type fullface supplied air respirators of continuous flow or pressure demand type. On exit from a regulated area, employees should be required to shower before changing into street clothes, leaving their protective clothing and equipment at the point of exit, to be placed in impervious containers at the end of the work shift for decontamination or disposal.

Respirator Selection: Use all purpose canister mask (A-38).

Disposal Method Suggested: Pour over soda ash, neutralize with HCl, then flush to drain with large volumes of water (A-38).

References

(1) U.S. Environmental Protection Agency, *Chemical Hazard Information Profile: N-Nitroso Compounds*, Wash., DC (1979).

(2) U.S. Environmental Protection Agency, *Nitrosamines: Ambient Water Quality Criteria*, Wash., DC (1980).

(3) U.S. Environmental Protection Agency, *Dimethylnitrosamine*, Health and Environmental Effects Profile No. 86, Wash., DC, Office of Solid Waste (April 30, 1980).

(4) International Agency for Research on Cancer, *IARC Monographs on the Carcinogenic Risks of Chemicals to Humans,* Lyon, France 1, 95 (1972) and 17, 125 (1978).

(5) Sax, N.I., Ed., *Dangerous Properties of Industrial Materials Report, 1,* No. 2, 50-51, New York, Van Nostrand Reinhold Co. (1980). (N-Nitrosodimethylamine).

(6) Sax, N.I., Ed., *Dangerous Properties of Industrial Materials Report, 2,* No. 6, 65-69, New York, Van Nostrand Reinhold Co. (1982). (N-Nitrosodimethylamine).

(7) See Reference (A-62).

(8) United Nations Environment Programme, *IRPTC Legal File 1983,* Vol. I, pp VII/285-6, Geneva, Switzerland, International Register of Potentially Toxic Chemicals (1984).

NITROTOLUENES

- Hazardous substances (EPA)

Description: $CH_3C_6H_4NO_2$ is formed in 3 isomeric forms. The o- and m-forms are yellow liquids or solids. The p- form is a pale yellow solid. All have weak aromatic odors. They melt at $-9°C$ (o-), $15°C$ (m-) and $52°C$ (p-).

Code Numbers:

ortho:	CAS 88-72-2	RTECS XT3150000	UN 1664
meta:	CAS 99-08-1	RTECS XT2975000	UN 1664
para:	CAS 99-99-0	RTECS XT3325000	UN 1664

DOT Designation: ORM-E.

Synonyms:

ortho:	o-nitrotoluol, 2-methylnitrobenzene
meta:	m-nitrotoluol, 3-methylnitrobenzene
para:	p-nitrotoluol, 4-methylnitrobenzene

Potential Exposures: The nitrotoluenes are used in the production of toluidines and other dye intermediates. Human occupational exposures have been estimated at 8,190 by NIOSH.

Incompatibilities: Strong oxidizers, sulfuric acid.

Permissible Exposure Limits in Air: The Federal standard is 5 ppm (30 mg/m³). The ACGIH (1983/84) TWA value is 2 ppm (11 mg/m³); there is no STEL value set. The IDLH level is 200 ppm. The ACGIH adds the notation "skin" to indicate the possibility of cutaneous absorption.

Determination in Air: Adsorption on silica, workup in methanol, analysis by gas chromatography. See NIOSH Methods, Set Q. See also reference (A-10).

Permissible Concentration in Water: No criteria set, but EPA (A-37) has suggested a permissible ambient goal of 414 µg/ℓ based on health effects.

Routes of Entry: Inhalation, skin absorption, ingestion, skin and eye contact.

Harmful Effects and Symptoms: Anoxia, cyanosis, headaches; weakness, dizziness, ataxia, dyspnea, tachycardia, nausea, vomiting.

Points of Attack: Blood, skin, gastrointestinal tract, cardiovascular system, central nervous system.

Medical Surveillance: Consider the points of attack in preplacement and periodic physical examinations.

First Aid: If this chemical gets into the eyes, irrigate immediately. If this chemical contacts the skin, wash with soap immediately. If a person breathes in large amounts of this chemical, move the exposed person to fresh air at once and perform artificial respiration. When this chemical has been swallowed, get medical attention. Give large quantities of water and induce vomiting. Do not make an unconscious person vomit.

Personal Protective Methods: Wear appropriate clothing to prevent repeated or prolonged skin contact. Wear eye protection to prevent any reasonable probability of eye contact. Employees should wash promptly when skin is wet or contaminated. Remove nonimpervious clothing promptly if wet or contaminated.

Respirator Selection:
       50 ppm: SA/SCBA
      200 ppm: SAF/SCBAF/SA:PD,PP,CF
      Escape: GMOVP/SCBA

Disposal Method Suggested: Controlled incineration—care must be taken to maintain complete combustion at all times. Incineration of large quantities may require scrubbers to control the emission of NO_x.

References

(1) Nat. Inst. for Occup. Safety and Health, *Information Profiles on Potential Occupational Hazards: Nitrotoluenes,* Report PB-276-678, Rockville, MD, pp 227-240 (Oct. 1977).
(2) See Reference (A-61).
(3) Sax, N.I., Ed., *Dangerous Properties of Industrial Materials Report, 3,* No. 3, 85-88, New York, Van Nostrand Reinhold Co. (1983). (p-Nitrotoluene).
(4) See Reference (A-60) for p-Nitrotoluene.

NITROUS OXIDE

Description: N_2O is a colorless gas with a sweetish odor and taste (boiling point is $-88°C$).

Code Numbers: CAS 10024-97-2   RTECS QX 1350000   UN 1070

DOT Designation: Nonflammable gas.

Synonyms: Dinitrogen monoxide, laughing gas.

Potential Exposure: Used as an anesthetic in dentistry and surgery; used as a gas in food aerosols such as whipped cream; used in manufacture of nitrites; used in rocket fuels.

Permissible Exposure Limits in Air: 25 ppm is TLV set for waste anesthetic gases and vapors.

Permissible Concentration in Water: No criteria set.

Routes of Entry: Inhalation.

Harmful Effects and Symptoms: Is an asphyxiant. Is narcotic at high concentrations. Chronic use by dental anaesthesiologists may cause increased rates of kidney and liver disease, miscarriages and birth defects and increased incidence of cancer, but some of these effects may be due to other anesthetic gases (1).

Personal Protective Methods: Wear rubber gloves, safety glasses and protective clothing (A-38).

Respirator Selection: Wear self-contained breathing apparatus.

Disposal Method Suggested: Disperse in atmosphere (1) or spray on dry soda ash/lime with great care; then flush to sewer (A-38).

References
(1) Sax, N.I., Ed., *Dangerous Properties of Industrial Materials Report, 1,* No. 7, 66-67, New York, Van Nostrand Reinhold Co. (1981).
(2) See Reference (A-60).

NONANE

Description: $CH_3(CH_2)_7CH_3$ is a colorless liquid boiling at 151°C.

Code Numbers: CAS 111-84-2   UN 1920

DOT Designation: Flammable liquid.

Synonyms: None.

Potential Exposures: Nonane is used in the synthesis of detergents.

Permissible Exposure Limits in Air: There are no Federal standards but ACGIH (1983/84) has adopted a TWA value of 200 ppm (1,050 mg/m³) and set an STEL of 250 ppm (1,300 mg/m³).

Determination in Air: See "Octane."

Permissible Concentration in Water: No criteria set.

Routes of Entry: Inhalation, ingestion, skin and eye contact.

Harmful Effects and Symptoms: See "Octane" as a guide. ACGIH (A-34) points out acute toxicities increase sharply with the number of carbon atoms.

Points of Attack: Skin, eyes, respiratory system.

Medical Surveillance: Consider the skin, eyes and respiratory system in preplacement and periodic physical examinations.

First Aid: See "Octane" as a guide.

Personal Protective Methods: See "Octane" as a guide.

Respirator Selection: See "Octane" as a guide.

Disposal Method Suggested: Incineration.

References
(1) National Inst. for Occup. Safety and Health, *Criteria for a Recommended Standard: Occupational Exposure to Alkanes*, NIOSH Doc. No. 77-151, Wash., DC (1977).

1-NONENE

Description: $CH_3(CH_2)_6CH=CH_2$ is a colorless liquid which boils at 147°C.

Code Numbers: CAS 124-11-8

DOT Designation: —

Synonyms: 1-nonylene.

Potential Exposure: Those involved in organic synthesis of detergents and lube oil additives and plasticizers.

Permissible Exposure Limits in Air: No standards set.

Permissible Concentration in Water: No criteria set.

Harmful Effects and Symptoms: Moderately toxic when vapors are inhaled.

Respirator Selected: Wear canister-type mask.

Disposal Method Suggested: Incineration.

References

(1) Sax, N.I., Ed., *Dangerous Properties of Industrial Materials Report, 2,* No. 1, 111-112, New York, Van Nostrand Reinhold Co. (1982).
(2) Sax, N.I., Ed., *Dangerous Properties of Industrial Materials Report, 3,* No. 2, 51-52, New York, Van Nostrand Reinhold Co. (1983).

O

OCTANE

Description: C_8H_{18} is a colorless liquid with a gasolinelike odor. It boils at 125° to 126°C.

Code Numbers: CAS 111-65-9   RTECS RG8400000   UN 1262

DOT Designation: Flammable liquid.

Synonyms: Normal octane.

Potential Exposure: Octane is used as a solvent, as a fuel, and as an intermediate in organic synthesis. NIOSH estimates annual worker exposure to octane at 300,000.

Incompatibilities: Strong oxidizers.

Permissible Exposure Limits in Air: The Federal standard is 500 ppm (2,350 mg/m³). NIOSH has recommended a TWA of 75 ppm with a 385 ppm ceiling value. ACGIH has adopted as of 1983/84 a 300 ppm (1,450 mg/m³) TWA and an STEL of 375 ppm (1,800 mg/m³). The IDLH level is 3,750 ppm.

Determination in Air: Charcoal adsorption, workup with CS_2, analysis by gas chromatography. See NIOSH Methods, Set G. See also reference (A-10).

Permissible Concentration in Water: No criteria set.

Routes of Entry: Inhalation, ingestion, skin and eye contact.

Harmful Effects and Symptoms: Irritation of eyes and nose; drowsiness; dermatitis; chemical pneumonia.

Points of Attack: Skin, eyes, respiratory system.

Medical Surveillance: Consider the points of attack in preplacement and periodic physical examinations.

First Aid: If this chemical gets into the eyes, irrigate immediately. If this chemical contacts the skin, wash with soap promptly. If a person breathes in large amounts of this chemical, move the exposed person to fresh air at once and perform artificial respiration. When this chemical has been swallowed, get medical attention. Do not induce vomiting.

Personal Protective Methods: Wear appropriate clothing to prevent repeated or prolonged skin contact. Wear eye protection to prevent any reasonable probability of eye contact. Employees should wash promptly when skin is wet or contaminated. Remove clothing immediately if wet or contaminated to avoid flammability hazard.

Respirator Selection:
   1,000 ppm: CCROV
   5,000 ppm: GMOV/SA/SCBA
       Escape: GMOV/SCBA

Disposal Method Suggested: Incineration.

References

(1)   Nat. Inst. for Occup. Safety and Health, *Criteria for a Recommended Standard: Occupational Exposure to Alkanes,* NIOSH Doc. No. 77-151, Wash., DC (1977).
(2)   See Reference (A-61).
(3)   See Reference (A-60).

1-OCTANOL

Description: $CH_3(CH_2)_6CH_2OH$ is a colorless liquid (melting point 16.7°C), boiling at 194.5°C.

Code Numbers: CAS 111-87-5   RTECS RH6550000

DOT Designation: —

Synonyms: Caprylic alcohol; heptyl carbinol; primary octyl alcohol.

Potential Exposure: In the manufacture of esters and perfumes. Used in detergent and plasticizer synthesis.

Permissible Exposure Limits in Air: The U.S.S.R. has set a TLV of 1.9 ppm (10 mg/m^3) (A-36).

Permissible Concentration in Water: No criteria set.

Harmful Effects and Symptoms: May dry skin by extracting oils. Considered slightly toxic.

Personal Protective Methods: Skin protection needed. Wear rubber gloves and face shield (A-38).

Respirator Selection: Canister-type mask.

Disposal Method Suggested: Incineration.

References

(1)   Sax, N.I., Ed., *Dangerous Properties of Industrial Materials Report, 2,* No. 1, 112-113, New York, Van Nostrand Reinhold Co. (1982).
(2)   Sax, N.I., Ed., *Dangerous Properties of Industrial Materials Report, 3,* No. 2, 54-55, New York, Van Nostrand Reinhold Co. (1983).
(3)   Sax, N.I., Ed., *Dangerous Properties of Industrial Materials Report, 3,* No. 6, 55-56, New York, Van Nostrand Reinhold Co. (Nov./Dec. 1983) (2-Octanol).

1-OCTENE

Description: $CH_3(CH_2)_5CH=CH_2$ is a colorless liquid boiling at 121°C.

Code Numbers: CAS 111-66-0

DOT Designation: —

Synonyms: Capyrlene, octylene.

Potential Exposure: Those involved in organic synthesis of plasticizers and detergents.

Permissible Exposure Limits in Air: No standards set.

Permissible Concentration in Water: No criteria set.

Harmful Effects and Symptoms: May be moderately toxic when inhaled.

Respirator Selection: Wear canister-type mask.

References
(1) Sax, N.I., Ed., *Dangerous Properties of Industrial Materials Report, 2,* No. 1, 113-114, New York, Van Nostrand Reinhold Co. (1982).
(2) Sax, N.I., Ed., *Dangerous Properties of Industrial Materials Report, 3,* No. 2, 52-54, New York, Van Nostrand Reinhold Co. (1983).

OIL MIST, MINERAL

Description: This is a mist with an odor like burned lubricating oil.

Code Numbers: RTECS RI17400000

DOT Designation: −

Synonyms: Mist of white mineral oil, cutting oil; heat-treating oil; hydraulic oil; cable oil; lubricating oil.

Potential Exposure: Mineral oil is used as a lubricating oil and as a solvent for inks in the printing industry. Oil mist would be encountered in quenching of hot metal parts and in metal machining operations.

Incompatibilities: None hazardous.

Permissible Exposure Limits in Air: The Federal standard is 5 mg/m³. As of 1983/84, ACGIH has set a TWA of 5 mg/m³ (with the notation that it be sampled by a method that does not collect vapor) and a tentative STEL of 10 mg/m³.

Determination in Air: Collection on a filter, workup with chloroform, analysis by fluorescence spectrometry. See NIOSH Methods, Set S. See also reference (A-10).

Permissible Concentration in Water: No criteria set.

Routes of Entry: Inhalation.

Harmful Effects and Symptoms: None.

Points of Attack: Respiratory system, skin.

Medical Surveillance: Consider the points of attack in preplacement and periodic physical examinations.

First Aid: If this chemical contacts the skin, wash with soap. If a person breathes in large amounts of this chemical, move the exposed person to fresh air at once.

Personal Protective Methods: Wear appropriate clothing to prevent repeated or prolonged skin contact. Employees should wash promptly when skin is wet. Remove nonimpervious clothing promptly if wet or contaminated.

Respirator Selection:
$25\ mg/m^3$: MXS
$50\ mg/m^3$: MXSQ/FuHiEP/SA/SCBA
$250\ mg/m^3$: HiEPF/SAF/SA:PD,PP,CF/SCBAF
$2,500\ mg/m^3$: PAPHiE/SAF:PD,PP,CF

References

(1) United Nations Environment Programme, *International Register of Potentially Toxic Chemicals,* Geneva, Switzerland (1979).

(2) See Reference (A-61).

ORGANOMERCURY COMPOUNDS

See "Mercury—Alkyl and Aryl."

OSMIUM AND COMPOUNDS

● Hazardous waste (Osmium tetroxide) (EPA)

Description: Os, osmium, is a blue-white metal. It is found in platinum ores and in the naturally occurring alloy osmiridium. Osmium when heated in air or when the finely divided form is exposed to air at room temperature oxidizes to form the tetroxide (OsO_4), osmic acid). It has a nauseating odor.

Code Numbers:

Osmium metal:	CAS 7440-04-2	RTECS RN1100000	
Osmium tetroxide:	CAS 20816-12-0	RTECS RN1140000	UN 2471

DOT Designation: Osmium metal: —; OsO_4: Poison.

Synonyms: Osmium metal—none; Osmium tetroxide—osmic acid.

Potential Exposures: Osmium may be alloyed with platinum metals, iron, cobalt, and nickel, and it forms compounds with tin and zinc. The alloy with iridium is used in the manufacture of fountain pen points, engraving tools, record player needles, electrical contacts, compass needles, fine machine bearings, and parts for watch and lock mechanisms. The metal is a catalyst in the synthesis of ammonia, and in the dehydrogenation of organic compounds. It is also used as a stain for histological examination of tissues. Osmium tetroxide is used as an oxidizing agent and as a fixative for tissues in electron microscopy. Other osmium compounds find use in photography. Osmium no longer is used in incandescent lights and in fingerprinting.

Incompatibilities: Osmium tetroxide—hydrochloric acid, easily oxidized organic material.

Permissible Exposure Limits in Air: There is presently no Federal standard for osmium itself; the standard for osmium tetroxide is $0.002\ mg/m^3$.

The ACGIH as of 1983/84 has set a TWA for osmium tetroxide of 0.0002 ppm ($0.002\ mg/m^3$) and an STEL of 0.0006 ppm ($0.006\ mg/m^3$). The IDLH value is $1\ mg/m^3$.

Permissible Concentration in Water: No criteria set.

Routes of Entry: Inhalation of vapor or fume, ingestion, eye and skin contact.

Harmful Effects and Symptoms: *Local* — Osmium metal is innocuous, but persons engaged in the production of the metal may be exposed to acids and chlorine vapors. Osmium tetroxide vapors are poisonous and extremely irritating to the eyes; even in low concentrations they may cause weeping and persistent conjunctivitis. Longer exposure can result in damage to the cornea and blindness. Contact with skin may cause discoloration (green or black) dermatitis and ulceration.

Systemic — Inhalation of osmium tetroxide fumes is extremely irritating to the respiratory system, causing tracheitis, bronchitis, bronchial spasm, and difficulty in breathing which may last several hours. Longer exposures can cause serious inflammatory lesions of the lungs (bronchopneumonia with suppuration and gangrene). Slight kidney damage was seen in rabbits inhaling lethal concentrations of vapor for 30 minutes. Some fatty degeneration of renal tubules was seen in one fatal human case along with bronchopneumonia following an accidental overexposure.

Points of Attack: Eyes, respiratory system, lungs, skin.

Medical Surveillance: Consider the skin, eyes, respiratory tract, and renal function in preplacement or periodic examinations.

First Aid: If this chemical gets into the eyes, irrigate immediately. If this chemical contacts the skin, wash with soap immediately. If a person breathes in large amounts of this chemical, move the exposed person to fresh air at once and perform artificial respiration. When this chemical has been swallowed, get medical attention. Give large quantities of water and induce vomiting. Do not make an unconscious person vomit.

Personal Protective Methods: Wear appropriate clothing to prevent any reasonable probability of skin contact. Wear eye protection to prevent any possibility of eye contact. Employees should wash promptly when skin is wet or contaminated. Work clothing should be changed daily if it is possible that clothing is contaminated. Remove nonimpervious clothing promptly if wet or contaminated. Provide emergency eyewash.

Respirator Selection:
    0.1 mg/m^3: SAF/SCBAF
      1 mg/m^3: SAF:PD,PP,CF
        Escape: GMSHiE/SCBA

References
(1)   See Reference (A-61). (Osmium Tetroxide).
(2)   Parmeggiani, L., Ed., *Encyclopedia of Occupational Health & Safety*, Third Edition, Vol. 2, pp 1573–74, Geneva, International Labour Office (1983)

OXALIC ACID

Description: HOOCCOOH·2H$_2$O, oxalic acid in solution, is a colorless liquid. Anhydrous oxalic acid is monoclinic in form and is produced by careful drying of the crystalline dihydrate.

Code Numbers: CAS 144-62-7   RTECS RO2450000

DOT Designation: —

Synonyms: Dicarboxylic acid, ethane-di-acid, enthanedioic acid.

Potential Exposures: Oxalic acid is used as an analytic reagent and in the manufacture of dyes, inks, bleaches, paint removers, varnishes, wood and metal cleansers, dextrin, cream of tartar, celluloid, oxalates, tartaric acid, purified methyl alcohol, glycerol, and stable hydrogen cyanide. It is also used in the photographic, ceramic, metallurgic, rubber, leather, engraving, pharmaceutical, paper, and lithographic industries.

Incompatibilities: Strong oxidizers, silver.

Permissible Exposure Limits in Air: The Federal standard and the 1983/84 ACGIH TWA value is 1 mg/m^3. The ACGIH has set an STEL of 2 mg/m^3. The IDLH level is 500 mg/m^3.

Permissible Concentration in Water: No criteria set.

Routes of Entry: Inhalation of mist and, occasionally dust, skin absorption, ingestion, skin and eye contact.

Harmful Effects and Symptoms: *Local* — Liquid has a corrosive action on the skin, eyes, and mucous membranes, which may result in ulceration. Local prolonged contact with extremities may result in localized pain, cyanosis, and even gangrenous changes probably resulting from localized vascular damage.
Systemic — Chronic exposure to mist or dust has been reported to cause chronic inflammation of the upper respiratory tract. Ingestion is of lesser importance occupationally. Symptoms appear rapidly and include shock, collapse, and convulsive seizures. Such cases may also have marked kidney damage with deposition of calcium oxalate in the lumen of the renal tubules.

Points of Attack: Respiratory system, lungs, skin, kidneys, eyes.

Medical Surveillance: Evaluate skin, respiratory tract, and renal functions in placement or periodic examinations. The presence of increased urinary oxalate crystals may be helpful in evaluating oral poisoning. Determination of blood calcium and oxalate levels may also be used for this purpose.

First Aid: If this chemical gets into the eyes, irrigate immediately. If this chemical contacts the skin, flush with water promptly. If a person breathes in large amounts of this chemical, move the exposed person to fresh air at once and perform artificial respiration. When this chemical has been swallowed, get medical attention. Give large quantities of water and do not induce vomiting.

Personal Protective Methods: Wear appropriate clothing to prevent repeated or prolonged skin contact. Wear eye protection to prevent any possibility of eye contact. Employees should wash promptly when skin is wet or contaminated. Work clothing should be changed daily if it is possible that clothing is contaminated. Remove nonimpervious clothing promptly if wet or contaminated.

Respirator Selection:
50 mg/m^3: HiEPF/SAF/SCBAF
500 mg/m^3: SAF:PD,PP,CF

Disposal Method Suggested: Pretreatment involves chemical reaction with limestone or calcium oxide forming calcium oxalate. This may then be incinerated utilizing particulate collection equipment to collect calcium oxide for recycling.

References

(1) Nat. Inst. for Occup. Safety and Health, *Information Profiles on Potential Occupational Hazards: Oxalic Acid*, Report PB-276,678, Rockville, MD, pp 42-46 (Oct. 1977).
(2) See Reference (A-61).

(3) See Reference (A-60).
(4) Parmeggiani, L., Ed., *Encyclopedia of Occupational Health & Safety,* Third Edition, Vol. 2, pp 1574–75, Geneva, International Labour Office (1983).

OXYGEN DIFLUORIDE

Description: OF_2 is a colorless gas with a foul odor.

Code Numbers: CAS 7783-41-7   RTECS RS 2100000   UN 2190

DOT Designation: Poison gas.

Synonyms: Difluorine monoxide; fluorine monoxide; fluorine oxide.

Potential Exposure: Oyxgen difluoride may be used as an oxidant in missile propellant systems.

Incompatibilities: All combustible materials, chlorine, bromine, iodine, platinum, metal oxides; moist air.

Permissible Exposure Limits in Air: The Federal standard and the 1983/84 ACGIH TWA value is 0.05 ppm (0.1 mg/m^3). The STEL value is 0.15 ppm (0.3 mg/m^3). The IDLH level is 0.5 ppm.

Permissible Concentration in Water: No criteria set.

Routes of Entry: Inhalation, skin and eye contact.

Harmful Effects and Symptoms: Intractable headaches; respiratory system irritation; pulmonary edema; skin and eye burns.

Points of Attack: Lungs, eyes.

Medical Surveillance: Consider the points of attack in preplacement and periodic physical examinations.

First Aid: If this chemical gets into the eyes, irrigate immediately. If this chemical contacts the skin, flush with water immediately. If a person breathes in large amounts of this chemical, move the exposed person to fresh air at once and perform artificial respiration.

Personal Protective Methods: No protective devices other than respirators are specified by NIOSH (A-4). Long rubber gloves, goggles and protective clothing are recommended by others (A-38).

Respirator Selection:
       0.5 ppm: SA/SCBA
       Escape: GMS/SCBA
       Note: Do not use oxidizable sorbents.

Disposal Method Suggested: Spray carefully onto a soda ash/slaked lime mixture; then spray with water. Finally neutralize and flush to drain with water (A-38).

References
(1) See Reference (A-61).

OXYMETHOLONE

● Carcinogen (Human, Suspected, IARC) (A-62) (A-64)

Description: $C_{21}H_{32}O_3$ with the following structural formula

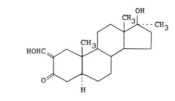

is a crystalline solid melting at 178° to 180°C.

Code Numbers: CAS 434-07-1   RTECS BV8060000

DOT Designation: —

Synonyms: 17β-Hydroxy-2-(hydroxymethylene)-17 γ-methyldihydrotestosterone; anasterone.

Potential Exposure: Oxymetholone is a synthetic steroid hormone used for adjuvant therapy for senile and postmenopausal osteoporosis. At one time, a major use was for treatment of anemias. In 1972, the FDA dropped its approval of this use pending further investigation. Oxymetholone is also used as an anabolic steroid for small animals.

Human exposure can occur principally during its use as a pharmaceutical. FDA's Bureau of Drugs estimates that 50 to 100,000 patients are treated each year with oxymetholone for postmenopausal osteoporosis. The usual adult dose of 5 to 10 mg/day is administered orally for about 3 weeks, but not exceeding 13 weeks, for a single course of therapy.

Permissible Exposure Limits in Air: No standards set.

Permissible Concentration in Water: No criteria set.

Harmful Effects and Symptoms: Ten cases of liver-cell tumors have been reported in patients with blood disorders treated for long periods with oxymetholone alone or in combination with other androgenic drugs; however, a causal relationship cannot be established. The increased risk of liver-cell tumors could be related to hepatic damage known to be caused by oxymetholone. Alternatively, patients with congenital anaemias may be at higher risk of developing these tumors, and this risk may become manifest during the extended survival resulting from oxymetholone treatment (1) (2).

References

(1)   IARC Monographs 13:131-39 (1977).
(2)   *IARC Monographs on the Evaluation of the Carcinogenic Risk of Chemicals to Humans,* Supplement I, IARC, Lyon, France, p 39 (1979).
(3)   Sax, N.I., Ed., *Dangerous Properties of Industrial Materials Report, 1,* No. 3, 73-74, New York, Van Nostrand Reinhold Co. (1981).
(4)   See Reference (A-62). Also see Reference (A-64).

OZONE

Description: O_3 is a colorless gas with a very pungent, characteristic odor, associated with electrical sparks.

Code Numbers: CAS 10028-15-6   RTECS RS8225000

DOT Designation: –

Synonyms: None.

Potential Exposure: Ozone is found naturally in the atmosphere as a result of the action of solar radiation and electrical storms. It is also formed around electrical sources such as x-ray or ultraviolet generators, electric arcs, mercury vapor lamps, linear accelerators, and electrical discharges.

Ozone is used as an oxidizing agent in the organic chemical industry (e.g., production of azelaic acid); as a disinfectant for food in cold storage rooms and for water (e.g., public water supplies, swimming pools, sewage treatment); for bleaching textiles, waxes, flour, mineral oils and their derivatives, paper pulp, starch, and sugar; for aging liquor and wood; for processing certain perfumes, vanillin, and camphor; in treating industrial wastes; in the rapid drying of varnishes and printing inks; and in the deodorizing of feathers.

Industrial exposure often occurs around ozone generating sources, particularly during inert-gas shielded arc welding.

Incompatibilities: All oxidizable materials, both organic and inorganic.

Permissible Exposure Limits in Air: The Federal (OSHA) standard and the 1983/84 ACGIH TWA value is 0.1 ppm (0.2 mg/m^3). The STEL is 0.3 ppm (0.6 mg/m^3). The USEPA has set a national ambient air quality standard for ozone of 0.12 ppm (240 μg/m^3) on a 1-hour average basis and a standard for total photochemical oxidants (expressed as ozone) of 0.08 ppm (160 μg/m^3) on a 1-hour average basis (to be exceeded not more than once a year). This was revised by EPA in 1979 (4) to 0.12 ppm.

Determination in Air: Collection by impinger containing KI/NaOH. Reaction with phosphoric and sulfamic acids and colorimetric determination. See NIOSH Methods, Set B. See also reference (A-10) and (4).

Permissible Concentration in Water: No criteria set.

Routes of Entry: Inhalation of the gas.

Harmful Effects and Symptoms: *Local* – Ozone is irritating to the eyes and all mucous membranes. In human exposures, the respiratory signs and symptoms in order of increasing ozone concentrations are: dryness of upper respiratory passages; irritation of mucous membranes of nose and throat; choking, coughing, and severe fatigue; bronchial irritation, substernal soreness, and cough. Pulmonary edema may occur, sometimes several hours after exposure has ceased. In severe cases, the pulmonary edema may be fatal.

Animal experiments demonstrate that ozone causes inflammation and congestion of respiratory tract and, in acute exposure, pulmonary edema, hemorrhage, and death.

Chronic exposure of laboratory animals resulted in chronic bronchitis, bronchiolitis, emphysematous and fibrotic changes in pulmonary parenchyma.

Systemic – Symptoms and signs of subacute exposure include headache, malaise, shortness of breath, drowsiness, reduced ability to concentrate, slowing of heart and respiration rate, visual changes, and decreased desaturation of oxyhemoglobin in capillaries. Animal experiments with chronic exposure showed aging effects and acceleration of lung tumorigenesis in lung-tumor susceptible mice.

Animal experiments further demonstrated that tolerance to acute pulmonary effects of ozone is developed and that this provided cross tolerance to other edemagenic agents. Antagonism and synergism with other chemicals also occur.

Ozone also has radiomimetic characteristics, probably related to its free-radical structure. Experimentally produced chromosomal aberrations have been observed.

Points of Attack: Eyes, respiratory system, lungs.

Medical Surveillance: Preemployment and periodic physical examinations should be concerned especially with significant respiratory diseases. Eye irritation may also be important. Chest x-rays and periodic pulmonary function tests are advisable.

First Aid: If this chemical gets into the eyes, get medical attention. If a person breathes in large amounts of this chemical, move the exposed person to fresh air at once. Administer 100% O_2.

Personal Protective Methods: In areas of excessive concentration, gas masks with proper canister and fullface-piece or goggles or the use of supplied air respirators is recommended.

Respirator Selection:

    1 ppm: CCRS/SA/SCBA
    5 ppm: CCRSF/GMS/SAF/SCBAF
    10 ppm: SAF:PD,PP,CF
    Escape: GMS/SCBA
    Note: Do not use oxidizable sorbents.

Disposal Method Suggested: Vent to atmosphere.

References

(1) Nat. Inst. for Occup. Safety and Health, *Information Profiles on Potential Occupational Hazards: Ozone*, Report PB-276,678, Rockville, MD, pp 47-50 (Oct. 1977).

(2) National Academy of Sciences, *Medical and Biologic Effects of Environmental Pollutants: Ozone and Other Photochemical Oxidants*, Wash., DC (1977).

(3) U.S. Environmental Protection Agency, *Air Quality Criteria for Ozone and Other Photochemical Oxidants*, Report EPA-600/8-78-004, Research Triangle Park, NC (1978).

(4) U.S. Environmental Protection Agency, "National Primary and Secondary Ambient Air Quality Standards," *Federal Register* 44, No. 28, 8202-8237 (Feb. 8, 1979).

(5) World Health Organization, *Photochemical Oxidants*, Environmental Health Criteria No. 7, Geneva, Switzerland (1979).

(6) See Reference (A-61).

(7) See Reference (A-60).

(8) Parmeggiani, L., Ed., *Encyclopedia of Occupational Health & Safety,* Third Edition, Vol. 2, pp 1579–80, Geneva, International Labour Office (1983).

P

PARAFFIN

● Carcinogen (Literature) (A-5)

Description: Paraffin is a white, somewhat translucent solid and consists of a mixture of solid aliphatic hydrocarbons. It may be obtained from petroleum.

Code Numbers: CAS 8002-74-2   RTECS RV0350000

DOT Designation: —

Synonyms: Paraffin wax, hard paraffin.

Potential Exposures: Paraffin is used in the manufacture of paraffin paper, candles, food package materials, varnishes, floor polishes, and cosmetics. It is also used in waterproofing and extracting of essential oils from flowers for perfume.

Permissible Exposure Limits in Air: Paraffin wax fume has no established Federal standard; however, the ACGIH (1983/84) recommends a TWA of 2.0 mg/m^3 for paraffin wax fume and an STEL of 6 mg/m^3.

Determination in Air: Sample collection by silver membrane filter, gravimetric determination (A-1).

Permissible Concentration in Water: No criteria set.

Route of Entry: Inhalation of fumes.

Harmful Effects and Symptoms: *Local* — Occasionally sensitivity reactions have been reported. Chronic exposure can produce chronic dermatitis, wax boils, folliculitis, comedones, melanoderma, papules, and hyperkeratoses.

Systemic — Carcinoma of the scrotum in wax pressmen exposed to crude petroleum wax has been documented (A-5). Other malignant lesions of an exposed area in employees working with finished paraffin are less well documented. Carcinoma of the scrotum, occurring in workmen exposed 10 years or more, began as a hyperkeratotic nevuslike lesion and developed into a squamous cell carcinoma. The lesions can metastasize to regional inguinal and pelvic lymph nodes. Paraffinoma has been reported from use of paraffin for cosmetic purposes. Perhaps paradoxically, however, no mention of carcinogenicity is made in the RTECS listing.

Medical Surveillance: Medical examinations should be concerned especially with the skin. Surveillance should be continued indefinitely.

Personal Protective Methods: Strict personal hygienic measures and protective clothing form the basis of a protective program.

Disposal Method Suggested: Incineration.

References
(1) Sax, N.I., Ed., *Dangerous Properties of Industrial Materials Report, 1,* No. 7, 69-70, New York, Van Nostrand Reinhold Co. (1981).
(2) See Reference (A-60).

PARAFORMALDEHYDE

Description: $C_3H_6O_3$ has the structural formula

This is a crystalline solid melting at 64°C.

Code Numbers: CAS 110-88-3   RTECS YM1400000   UN 2213

DOT Designation: —

Synonyms: s-trixane; s-trioxane; 1,3,5-trioxane; formaldehyde trimer; trioxymethylene; 1,3,5-trioxacyclohexane.

Potential Exposure: In polyacetal resin manufacture; as a food additive.

Permissible Exposure Limits in Air: A value of 2 ppm (3 mg/m^3) as a ceiling value is given by the Dutch (A-60). A TLV of 5 ppm is cited in Japan (A-38).

Permissible Concentration in Water: No criteria set.

Routes of Entry: Skin contact.

Harmful Effects and Symptoms: Skin irritant. Also eye irritant. Moderately toxic when inhaled. Highly toxic when ingested.

Points of Attack: Skin, eyes.

First Aid: Flush eyes with water; wash affected skin with soap and water (A-38).

Personal Protective Methods: Protect skin and eyes.

Respirator Selection: Self-contained breathing apparatus.

Disposal Method Suggested: Incineration (A-38).

References
(1) Sax, N.I., Ed., *Dangerous Properties of Industrial Materials Report, 3,* No. 3, 90-92, New York, Van Nostrand Reinhold Co. (1983).
(2) See Reference (A-60).

PARALDEHYDE

● Hazardous waste (EPA)

Description: 2,4,6-Trimethyl-1,3,5-trioxane is a colorless liquid boiling at 125°C.

Code Numbers: CAS 123-63-7   RTECS YK0525000   UN 1264

DOT Designation: Flammable liquid.

Synonyms: Paracetaldehyde, acetaldehyde trimer, s-trimethyltrioxymethylene.

Potential Exposures: Paraldehyde is used primarily in medicine. It is used frequently in delirium tremens and in treatment of psychiatric states characterized by excitement when drugs must be given over a long period of time. It also is administered for intractable pain which does not respond to opiates and for basal and obstetrical anaesthesia. It is effective against experimentally induced convulsions and has been used in emergency therapy of tetanus, eclampsia, status epilepticus, and poisoning by convulsant drugs.

Since it is used primarily in medicine, the chance of accidental human exposure or environmental contamination is low. However, paraldehyde decomposes to acetaldehyde and acetic acid; these compounds have been found to be toxic. In this sense, occupational exposure or environmental contamination is possible. Since paraldehyde is prepared from acetaldehyde by polymerization in the presence of an acid catalyst, there exists a potential for adverse effects, although none have been reported in the available literature.

Permissible Exposure Limits in Air: No standard set.

Permissible Concentration in Water: No criteria set. Data concerning the effects of paraldehyde on aquatic organisms were not found in the available literature.

Harmful Effects and Symptoms: There is no evidence in the available literature to indicate that paraldehyde, a central nervous system depressant, is carcinogenic, mutagenic, or teratogenic.

In low doses (4 to 8 ml) paraldehyde has a hypnotic effect on the central nervous system. Following chronic and acute exposures at higher concentrations, paraldehyde affects the respiratory and circulatory systems.

References

(1)   U.S. Environmental Protection Agency, *Paraldehyde,* Health and Environmental Effects Profile No. 140, Washington, DC, Office of Solid Waste (April 30, 1980).
(2)   See Reference (A-60).

PARAQUAT

Description: $C_{12}H_{14}N_2Cl_2$ with the structural formula

Paraquat is a quaternary ion which usually is used as the dichloride salt. The dichloride forms colorless crystals decomposing at 300°C.

Code Numbers: CAS 1910-42-5   RTECS DW2275000   UN 2781

DOT Designation: —

Synonyms: 1,1'-Dimethyl-4,4'-bipyridinium dichloride, Gramoxone®, Dextrone X®, Esgram®, Weedol®.

Potential Exposure: Those engaged in manufacture (A-32), formulation and application of this herbicide.

Incompatibilities: Strong oxidizers.

Permissible Exposure Limits in Air: The Federal standard is 0.5 mg/m^3. The ACGIH as of 1983/84 has set a TWA of 0.1 mg/m^3 but no STEL value for respirable sizes of particles. The IDLH level is 1.5 mg/m^3.

Determination in Air: Sample collection by mixed cellulose ester membrane filter, colorimetric analysis (A-1).

Permissible Concentration in Water: A no-adverse effect level in drinking water has been calculated by NAS/NRC to be 0.06 mg/ℓ.

Routes of Entry: Inhalation, skin absorption, ingestion, skin and eye contact.

Harmful Effects and Symptoms: Paraquat is acutely toxic to man. As a result, many accidental and suicidal deaths have been reported. It has been estimated that a lethal dose in man is about 14 ml of a 40% solution of paraquat. The symptoms of poisoning include burning of the mouth and throat, nausea and vomiting, respiratory distress, and transient effects on the kidneys, heart, and nervous system. Death is usually due to progressive fibrosis and epithelial proliferation in the lungs.

Dermal exposure to paraquat concentrates may result in severe skin irritation, while nosebleeds may result from exposure of the nasal mucosa, and several severe eye injuries have resulted from eye exposure. Absorption studies have shown that paraquat is readily absorbed through the skin of both humans and animals.

There is no effective antidote for paraquat poisoning in man, although a few patients have recovered after ingesting doses thought to be fatal. Issuance of a rebuttable presumption against registration for paraquat was being considered by EPA as of October 1980 on the basis of possible chronic effects, environmental effects, and data gaps (oncogenicity and mutagenicity).

Points of Attack: Eyes, respiratory system, heart, liver, kidneys, lungs, gastrointestinal system.

Medical Surveillance: Consider the points of attack in preplacement and periodic physical examinations.

First Aid: If this chemical gets into the eyes, irrigate immediately. If this chemical contacts the skin, flush with water immediately. If a person breathes in large amounts of this chemical, move the exposed person to fresh air at once and perform artificial respiration. When this chemical has been swallowed, get medical attention. Give large quantities of water and induce vomiting. Do not make an unconscious person vomit.

Personal Protective Methods: Wear appropriate clothing to prevent any reasonable probability of skin contact. Wear eye protection to prevent any reasonable probability of eye contact. Employees should wash immediately when skin is wet or contaminated. Remove nonimpervious clothing immediately if wet or contaminated. Provide emergency showers.

Respirator Selection:
1.5 mg/m^3: CCROVDMPest/SA/SCBA
Escape: GMOVPPest/SCBA

Disposal Method Suggested: Paraquat is rapidly inactivated in soil. It is also inactivated by anionic surfactants. Therefore an effective and environmentally safe disposal method would be to mix the product with ordinary household detergent and bury the mixture in clay soil (A-32).

References

(1) Pasi, A., *The Toxicology of Paraquat, Diquat and Morfamquat,* Bern, Switzerland, H. Huber (1978).
(2) See Reference (A-61).
(3) United Nations Environment Programme, *IRPTC Legal File 1983,* Vol. I, pp VII/131–3, Geneva, Switzerland, International Register of Potentially Toxic Chemicals (1984).

PARATHION

- Hazardous substance (EPA)
- Hazardous waste (EPA)

Description: Parathion,

$$\underset{\underset{(C_2H_5O)_2POC_6H_4NO_2}{\|}}{S}$$

is a pale yellow liquid boiling at 375°C, used as an insecticide.

Code Numbers: CAS 56-38-2   RTECS TF4550000   UN 1668, also (NA 2783).

DOT Designation: Poison-B.

Synonyms, O,O-diethyl-O-p-nitrophenyl phosphorothioate, diethyl-p-nitrophenyl monothiophosphate, DNTP, Thiophos (in U.S.S.R.), ENT-15,108, Thiopnos®, Folidol®, Bladan®, Niran®, Fosferno®.

Potential Exposure: NIOSH estimates that approximately 250,000 U.S. workers, including field workers, are potentially exposed to parathion in occupational settings. Those exposed include those engaged in manufacture (A-32), formulation and application of this insecticide.

Incompatibilities: Strong oxidizers.

Permissible Exposure Limits in Air: The Federal standard is 0.11 mg/m^3. NIOSH (1) recommends the promulgation of a new Federal standard including, in part, an environmental limit of 0.05 mg of parathion per m^3 of air as a TWA for up to a 10-hour workday, 40-hour workweek. ACGIH, as of 1983/84, set 0.1 mg/m^3 as a TWA and 0.3 mg/m^3 as an STEL value. The notation "skin" is added to indicate the possibility of cutaneous absorption. The IDLH value is 20 mg/m^3.

Determination in Air: Collection on a filter, workup with isooctane, analysis by gas chromatography. See NIOSH Methods, Set I. See also reference (A-10).

Permissible Concentration in Water: The EPA (A-3) has recommended a criterion of 0.04 µg/ℓ to protect freshwater and marine aquatic life. A no-adverse effect level in drinking water has been calculated to be 0.03 µg/ℓ by NAS/NRC (A-2).

Routes of Entry: Skin contact, ingestion, inhalation, skin absorption.

Harmful Effects and Symptoms: Parathion, an organophosphorus insecticide, is converted in the environment and in the body to paraoxon, a potent inactivator of acetylcholinesterase, an enzyme responsible for terminating the transmitter action of acetylcholine at the junction of cholinergic nerve endings with their effector organs or postsynaptic sites.

The scientific basis for the recommended environmental limit is the prevention of medically significant inhibition of acetylcholinesterase. If functional acetylcholinesterase is greatly inhibited, a cholinergic crisis may ensue. Symptoms of parathion poisoning include miosis, rhinorrhea, headaches, chest wheezing, laryngeal spasms, salivation, cyanosis, anorexia, nausea, vomiting, abdominal cramps, diarrhea, sweating, muscular fasciculation, weakness, paralysis, ataxia, convulsions, low blood pressure, dermatitis.

Points of Attack: Respiratory system, central nervous system, cardiovascular system, eyes, skin, blood cholinesterase.

Medical Surveillance: NIOSH recommends that medical surveillance, including preemployment and periodic examinations, shall be made available to workers who may be occupationally exposed to parathion. Biologic monitoring is also recommended as an additional safety measure.

First Aid: If this chemical gets into the eyes, irrigate immediately. If this chemical contacts the skin, wash with soap immediately. If a person breathes in large amounts of this chemical, move the exposed person to fresh air at once and perform artificial respiration. When this chemical has been swallowed, get medical attention. Give large quantities of water and induce vomiting. Do not make an unconscious person vomit.

Personal Protective Methods: Wear appropriate clothing to prevent repeated or prolonged skin contact. Wear eye protection to prevent any possibility of eye contact. Employees should wash immediately when skin is wet or contaminated. Work clothing should be changed daily if it is possible that clothing is contaminated. Remove nonimpervious clothing immediately if wet or contaminated. Provide emergency showers and eyewash.

Respirator Selection:
　　　　$1 \ mg/m^3$: CCROVDMFuPest/SA/SCBA
　　　　$5 \ mg/m^3$: CCROVFDMFuPest/GMOVDMFuPest/SAF/SCBAF
　　　$20 \ mg/m^3$: PAPOVHiEPPest/SA:PD,PP,CF
　　　Escape: GMOVP/SCBA

Disposal Method Suggested: One manufacturer recommends the use of a detergent in a 5% trisodium phosphate solution for parathion disposal and cleanup problems (A-32). For parathion disposal in general, however, the recommended method (A-31) is incineration ($1500°F$, 0.5 second minimum for primary combustion; $2200°F$, 1.0 second for secondary combustion) with adequate scrubbing and ash disposal facilities.

References
(1) National Institute for Occupational Safety and Health, *Criteria for a Recommended Standard: Occupational Exposure to Parathion,* NIOSH Doc. No. 76-190 (1976).
(2) See Reference (A-61).
(3) Sax, N.I., Ed., *Dangerous Properties of Industrial Materials Report, 3,* No. 3, 92-97, New York, Van Nostrand Reinhold Co. (1983).
(4) See Reference (A-60).
(5) Parmeggiani, L., Ed., *Encyclopedia of Occupational Health & Safety, Third Edition, Vol. 2,* pp 1591-94, Geneva, International Labour Office (1983).
(6) United Nations Environment Programme, *IRPTC Legal File 1983,* Vol. II, pp VII/603-6, Geneva, Switzerland, International Register of Potentially Toxic Chemicals (1984).

PARTICULATES

Particulate matter may obviously cover a very wide range of chemical compositions (1) and hence is a bit divergent from the coverage of this volume which is directed to individual hazardous and toxic chemicals.

However, suspended particulates are the subject of a national ambient air quality standard set by EPA (3). They have set a primary emission standard of 75 $\mu g/m^3$ on an annual geometric mean basis and 260 $\mu g/m^3$ on a 24-hour average basis, not to be exceeded more than once a year. They have also set secondary or background standards of 60 $\mu g/m^3$ on an annual geometric mean basis or 150 $\mu g/m^3$ on a 24-hour average basis not to be exceeded more than once a year.

A particularly hazardous and toxic particulate is particulate polycyclic organic matter (2). See the entry under "Polynuclear Aromatic Hydrocarbons" for discussion of this particular chemical class.

Since particulates as a volume pollutant are frequently associated with sulfur oxides in stack gases, the effects are often studied jointly (4).

As noted in the "Introduction" to this volume, "nuisance particulates," an ACGIH classification, are not given separate entries in this volume in the interest of conservation of space. They all are assigned the same threshold limit value (A-6) of 30 mppcf or 10 mg/m^3 of total dust containing less than 1% quartz or 5 mg/m^3 of respirable dust. These "nuisance particulates" include the following:

Aluminum oxide (Al_2O_3)	Pentaerythritol
Calcium carbonate	Plaster of Paris
Calcium silicate	Silicon
Cellulose (paper fiber)	Silicon carbide
Cement, Portland	Starch
Emery	Sucrose
Glycerin mist	Tin oxide*
Graphite (synthetic)	Titanium dioxide
Gypsum	Vegetable oil mists (except
Kaolin	castor, cashew nut, or
Limestone	similar irritant oils)
Magnesite	Zinc stearate
Marble	Zinc oxide dust
Mineral wool fiber	

*As of 1980, it was proposed to remove tin oxide from this list and and assign it a TWA of 2 ppm instead of the 10 ppm which would be a consequence of the "nuisance particulate" classification.

References

(1) Sittig, M., *Particulates and Fine Dust Removal,* Park Ridge, NJ, Noyes Data Corp. (1977).

(2) National Academy of Sciences, *Medical and Biologic Effects of Environmental Pollutants: Particulate Polycyclic Organic Matter,* Washington, DC (1972).

(3) U.S. Environmental Protection Agency, *Report to Congress on Suspended Particulates,* Report EPA-600/9-79-006, Research Triangle Park, NC (1979).

(4) World Health Organization, *Sulfur Oxides and Suspended Particulate Matter,* Environmental Health Criteria No. 8, Geneva Switzerland (1979).

PENTABORANE

Description: B_5H_9 boils at 17°C. It is a colorless, volatile liquid with an unpleasant, sweetish odor. It ignites spontaneously in air, decomposes at 150°C and hydrolyzes in water.

Code Numbers: CAS 19624-22-7   RTECS RY8925000   UN 1380

DOT Designation: Flammable liquid and poison.

Synonyms: Pentaboron nonahydride.

Potential Exposure: Pentaborane is used in rocket propellants and in gasoline additives.

Incompatibilities: Oxidizers, halogens, halogenated compounds; impure material ignites spontaneously.

Permissible Exposure Limits in Air: The Federal standard and the 1983/84 ACGIH TWA value is 0.005 ppm (0.01 mg/m^3). The STEL value is 0.015 ppm (0.03 mg/m^3). The IDLH level is 3 ppm.

Permissible Concentration in Water: No criteria set.

Routes of Entry: Inhalation, skin absorption, ingestion, skin and eye contact.

Harmful Effects and Symptoms: *Local* — Vapors are irritating to skin and mucous membranes. Pentaborane and decaborane show marked irritation of skin and mucous membranes, necrotic changes, serious kerato-conjunctivitis with ulceration.

Systemic — Pentaborane is the most toxic of boron hydrides. Intoxication is characterized predominantly by CNS signs and symptoms. Hyperexcitability, headaches, muscle twitching, convulsions, dizziness, disorientation, and unconsciousness may occur early or be delayed for 24 hours or more following excessive exposure. Slight intoxication results in nausea and drowsiness. Moderate intoxication leads to headache, dizziness, nervous excitation, and hiccups. There may be muscular pains and cramps, spasms in face and extremities, behavioral changes, loss of mental concentration, incoordination, disorientation, cramps, convulsions, semicoma, and persistent leukocytosis after 40 to 48 hours. Liver function tests and elevated nonprotein nitrogen and blood urea levels suggest liver and kidney damage.

Points of Attack: Central nervous system, eyes, skin.

Medical Surveillance: Preemployment and periodic physical examinations to determine the status of the workers' general health should be performed. These examinations should be concerned especially with any history of central nervous system disease, personality or behavioral changes, as well as liver, kidney or pulmonary disease of any significant nature. Chest x-rays and blood, liver, and renal function studies may be helpful.

First Aid: If this chemical gets into the eyes, irrigate immediately. If this chemical contacts the skin, wash with soap immediately. If a person breathes in large amounts of this chemical, move the exposed person to fresh air at once and perform artificial respiration. When this chemical has been swallowed, get medical attention. Give large quantities of water and induce vomiting. Do not make an unconscious person vomit.

Personal Protective Methods: Constant vigilance in the storage and handling of boron hydrides is required. Continuing worker education in the use of personal protective equipment is necessary even when maximum engineering safety measures are applied.

Wear appropriate clothing to prevent any reasonable probability of skin contact. Wear eye protection to prevent any possibility of eye contact. Employees should wash immediately when skin is wet or contaminated. Remove clothing immediately if wet or contaminated to avoid flammability hazard. Provide emergency showers and eyewash.

Respirator Selection:
    0.05 ppm: SA/SCBA
    0.25 ppm: SAF/SCBAF
    3 ppm: SA:PD,PP,CF
    Escape: GMS/SCBA

Disposal Method Suggested: Incineration with aqueous scrubbing of exhaust gases to remove B_2O_3 particulates.

References
(1) See Reference (A-61).

PENTACHLOROBENZENE

● Hazardous waste (EPA)
● Priority toxic pollutant (EPA)

Description: C_6HCl_5 is a colorless crystalline solid with a pleasant aroma, which melts at 86°C.

Code Numbers: CAS 608-93-5   RTECS DA6640000

DOT Designation: —

Synonym: QCB.

Potential Exposure: Pentachlorobenzene is used primarily as a precursor in the synthesis of the fungicide pentachloronitrobenzene, and as a flame retardant. Pentachlorobenzene has been identified in the effluent from a wastewater treatjent plant in southern California. Access to water can occur by industrial discharge or from the degradation of other organochlorine compounds, such as lindane (1).

Permissible Exposure Limits in Air: No standards set.

Permissible Concentration in Water: The U.S. EPA (2) has drafted a criterion of 74.0 μg/ℓ for the protection of human health.

Harmful Effects and Symptoms: Oral feeding of pentachlorobenzene to pregnant rats has produced developmental effects and decreased body weights in fetuses. No adverse reproductive or developmental effects were seen in mice following maternal administration of the compound orally. There is no information available on the mutagenic effects of pentachlorobenzene. A single study has alluded to carcinogenic effects of pentachlorobenzene in mice and lack of carcinogenic effects in dogs and rats. The details of this study were not available for evaluation.

Disposal Method Suggested: Incineration after mixing with another combustible fuel. Care must be exercised to assure complete combustion to prevent the formation of phosgene. An acid scrubber is necessary to remove the halo acids produced.

References

(1) U.S. Environmental Protection Agency, *Pentachlorobenzene,* Health and Environmental Effects Profile No. 141, Office of Solid Waste, Washington, DC (April 30, 1980).

(2) U.S. EPA, *Chlorinated Benzenes: Ambient Water Quality Criteria,* Washington, DC (1980).

PENTACHLOROETHANE

- Hazardous waste (EPA)
- Priority toxic pollutant (EPA)

Description: CCl_3CHCl_2 is a colorless, heavy, nonflammable liquid with a sweetish odor. It boils at 162°C.

Code Numbers: CAS 76-01-7   RTECS KI6300000   UN 1669

DOT Designation: Poison.

Synonyms: Ethane pentachloride, pentalin.

Potential Exposure: Pentachloroethane is used in the manufacture of tetrachloroethylene and as a solvent for cellulose acetate, certain cellulose ethers, resins, and gums. It is also used as a drying agent for timber by immersion at temperatures greater than 100°C.

Permissible Exposure Limits in Air: There is no U.S. Occupational Standard or ACGIH Threshold Limit Value for pentachloroethane. Three foreign countries, however, have set standards for this compound (1). In 1966, Germany set the maximum allowable concentration for pentachloroethane at 5 ppm or 40 mg/m^3; Rumania has a maximum allowable concentration of 30 mg/m^3, and Yugoslavia has a maximum allowable concentration of 5 ppm or 40 mg/m^3 for this compound.

Permissible Concentration in Water: To protect freshwater aquatic life— 7,240 $\mu g/\ell$ on an acute toxic basis and 1,100 $\mu g/\ell$ on a chronic basis. To protect saltwater aquatic life—390 $\mu g/\ell$ on an acute toxicity basis and 231 $\mu g/\ell$ on a chronic basis. To protect human health—no criteria derived due to insufficient data.

Harmful Effects and Symptoms: Cats and dogs have shown significant histopathlogic changes in the liver, lungs, and kidneys after inhalation of vapors from pentachloroethane. Exposure at high concentrations leads to loss of consciousness and death. Pentachloroethane has a strong narcotic effect in humans and can cause liver, lung, and kidney damage. It also acts as a local irritant to the eyes and upper respiratory tract.

Disposal Method Suggested: Incineration after mixing with another combustible fuel. Care must be exercised to assure complete combustion to prevent the formation of phosgene. An acid scrubber is necessary to remove the halo acids produced.

References

(1) Nat. Inst. for Occup. Safety and Health, *Profiles on Occupational Hazards for Criteria Document Priorities: Pentachloroethane,* Report PB-274,073, Cincinnati, OH pp 303-305 (1977).

(2) U.S. Environmental Protection Agency, *Chlorinated Ethanes: Ambient Water Quality Criteria,* Washington, DC (1980).

(3)   U.S. Environmental Protection Agency, *Chemical Hazard Information Profile Draft Report: Pentachloroethane,* Washington, DC (January 4, 1983).
(4)   United Nations Environment Programme, *IRPTC Legal File 1983,* Vol. I, pp VII/313–15, Geneva, Switzerland, International Register of Potentially Toxic Chemicals (1984).

PENTACHLORONITROBENZENE

- Carcinogen (Animal positive, IARC) (2)
- Hazardous waste (EPA)

Description: $C_6Cl_5NO_2$ forms colorless needles which melt at 146°C. Technical-grade PCNB contains an average of 97.8% PCNB, 1.8% hexachlorobenzene (HCB), 0.4% 2,3,4,5-tetrachloronitrobenzene (TCNB), and less than 0.1% pentachlorobenzene.

Code Numbers: CAS 82-68-8   RTECS DA6650000

DOT Designation: —

Synonyms: PCNB (note possible conflict with p-chloronitrobenzene), Quintozene (international pesticide generic name), Terrachlor (in Turkey), PKhNB (in U.S.S.R.), Brassicol®, Tritisan®, Folosan®, Terrachlor®.

Potential Exposure: Those engaged in manufacture, formulation and application of this soil fungicide and seed treatment chemical.

Permissible Exposure Limits in Air: No standards set.

Permissible Concentration in Water: A temporary acceptable daily intake of PCNB has been established for humans by the WHO at 0.001 mg/kg.

Determination in Water: A lower limit of detection of PCNB by gas chromatography was 0.01 ppm.

Harmful Effects and Symptoms: PCNB has relatively low acute toxicity. Although without effect in the rat and dog, PCNB appears to be carcinogenic in two strains of mice.

Most of the subchronic- and chronic-toxicity studies on PCNB have used technical-grade material, which normally contains about 1.8% HCB, but in some cases as much as 11% HCB. It is therefore not clear whether HCB and other impurities significantly contribute to the observed toxicity of PCNB. Moreover, some of the studies have involved PCNB formulations containing relatively low concentrations of the fungicide. The subchronic and chronic studies, particularly the latter, should be repeated in two species with pure PCNB. Such studies are particularly warranted because of the suspected carcinogenicity of PCNB. Additional long-term oncogenic studies should also be conducted in susceptible strains of mice and other experimental animals. In addition, the FAO/WHO has recommended further short-term studies to elucidate the difference in teratogenic activity between rats and mice; studies to explain the effects on the liver and bone marrow of dogs; and further studies on the toxicity of PCNB metabolites. A rebuttable presumption against registration of PCNB for pesticidal uses was issued on October 13, 1977 by EPA on the basis of oncogenicity (A-32).

Disposal Method Suggested: It has been observed that the product decomposes readily when burned with polyethylene. The compound is highly stable in soil in general, as would be expected on the basis of the polychlorinated aromatic structure (A-32).

References

(1) U.S. Environmental Protection Agency, *Pentachloronitrobenzene,* Health and Environmental Effects Profile No. 142, Office of Solid Waste, Washington, DC (April 30, 1980).

(2) International Agency for Research on Cancer, *IARC Monographs on the Carcinogenic Risks of Chemicals to Humans,* 5, 211 (1974).

PENTACHLOROPHENOL

- Hazardous substance (EPA)
- Hazardous waste (EPA)
- Priority toxic pollutant (EPA)

Description: C_6Cl_5OH is a light brown solid with a phenolic odor, pungent when hot. It melts at $187°$ to $189°C$.

Code Numbers: CAS 87-86-5   RTECS SM6300000   UN (NA 2020)

DOT Designation: Poison B.

Synonyms: PCP, Penchlorol, Dowicide EC-7®, Dowicide G®, Monsanto Penta®, Santobrite®.

Potential Exposure: Pentachlorophenol (PCP) is a commercially produced bactericide, fungicide, and slimicide used primarily for the preservation of wood, wood products, and other materials. As a chlorinated hydrocarbon, its biological properties have also resulted in its use as a herbicide, insecticide, and molluscicide.

Two groups can be expected to encounter the largest exposures. One involves the small number of employees involved in the manufacture of PCP. All of these are presently under industrial health surveillance programs. The second and larger group are the formulators and wood treaters. Exposure, hygiene and industrial health practices can be expected to vary from the small treaters to the larger companies.

The principal use as a wood preservative results in both point source water contamination at manufacturing and wood preservation sites and, conceivably, nonpoint source water contamination through runoff wherever there are PCP-treated lumber products exposing PCP to soil.

Incompatibilities: Strong oxidizers.

Permissible Exposure Limits in Air: The Federal standard and the 1983/84 ACGIH TWA value is 0.5 mg/m³. The STEL value is 1.5 mg/m³. The IDLH level is 150 mg/m³. The notation "skin" is added by ACGIH to indicate the possibility of cutaneous absorption.

Determination in Air: Sample collection by mixed cellulose ester membrane filter in series with ethylene glycol bubbler, analysis by high performance liquid chromatography with UV detector (A-10).

Permissible Concentration in Water: To protect freshwater aquatic life—55 µg/ℓ on an acute toxicity basis and 3.2 µg/ℓ on a chronic basis. To protect saltwater aquatic life—53 µg/ℓ on an acute basis and 34 µg/ℓ on a chronic basis. To protect human health—1,010 µg/ℓ is the criteria set by EPA based on toxicity data. A value of 30 μg/ℓ is set on an organoleptic basis. A no-adverse effect level in drinking water has been calculated by NAS/NRC to be 0.021 mg/ℓ (21 μg/ℓ).

Determination in Water: Methylene chloride extraction followed by gas chromatography with electron capture or halogen specific detection (EPA Method 608) or gas chromatography plus mass spectrometry (EPA Method 625).

Routes of Entry: Inhalation of dust, skin absorption, ingestion, eye and skin contact.

Harmful Effects and Symptoms: Symptoms of exposure include irritation of eyes, nose and throat, sneezing and coughing, weakness, anorexia, weight loss, sweating, headaches, dizziness, nausea, vomiting, dyspnea, chest pains, fever, dermatitis. Based on available and cited literature, PCP is not considered to be carcinogenic according to EPA (1). However, a rebuttable presumption against registration of PCP for pesticide uses was issued on October 18, 1978 by EPA on the basis of fetotoxicity and teratogenicity (A-32).

Points of Attack: Cardiovascular system, respiratory system, eyes, liver, kidneys, skin, central nervous system.

Medical Surveillance: Consider the points of attack in preplacement and periodic physical examinations.

First Aid: If this chemical gets into the eyes, irrigate immediately. If this chemical contacts the skin, wash with soap immediately. If a person breathes in large amounts of this chemical, move the exposed person to fresh air at once and perform artificial respiration. When this chemical has been swallowed, get medical attention. Give large quantitites of water and induce vomiting. Do not make an unconscious person vomit.

Personal Protective Methods: Wear appropriate clothing to prevent any possibility of skin contact. Wear eye protection to prevent any possiblity of eye contact. Employees should wash immediately when skin is wet or contaminated. Work clothing should be changed daily if it is possible that clothing is contaminated. Remove nonimpervious clothing immediately if wet or contaminated. Provide emergency showers and eyewash.

Respirator Selection:
    2.5 mg/m^3: CCROVDMFuPest/SA/SCBA
    25 mg/m^3: CCROVFDMFuPest/GMOVDMFuPest/SAF/SCBAF
    150 mg/m^3: PAPOVFHiEPPest/SAF:PD,PP,CF
    Escape: GMOVP/SCBA

Disposal Method Suggested: Incineration (600° to 900°C) coupled with adequate scrubbing and ash disposal facilities. Alternatively pentachlorophenol in wastewaters, for example, may be recovered and recycled (A-58).

References

(1) U.S. Environmental Protection Agency, *Pentachlorophenol: Ambient Water Quality Criteria,* Washington, DC (1980).
(2) Rao, K.R., Ed., *Pentachlorophenol: Chemistry, Pharmacology and Environmental Toxicology,* Proceedings of a Symposium, Pensacola, FL, June 1977, New York, Plenum Press (1978).
(3) U.S. Environmental Protection Agency, *Pentachlorophenol,* Health and Environmental Effects Profile No. 143, Office of Solid Waste, Washington, DC (April 30, 1980).
(4) See Reference (A-61).
(5) See Reference (A-60).

(6) Sax, N.I., Ed., *Dangerous Properties of Industrial Materials Report, 3,* No. 4, 73-77, New York, Van Nostrand Reinhold Co. (1983).

(7) United Nations Environment Programme, *IRPTC Legal File 1983,* Vol. II, pp VII/519-22, Geneva, Switzerland, International Register of Potentially Toxic Chemicals (1984).

PENTAERYTHRITOL

Description: $C(CH_2OH)_4$ is a white crystalline solid melting at 261°C.

Code Numbers: CAS 115-77-5 RTECS RZ 2490000

DOT Designation: —

Synonyms: 2,2-Bis(hydroxymethyl)-1,3-propanediol, pentaerythrite, tetrahydroxymethylmethane, tetramethylolmethane.

Potential Exposures: Pentaerythritol is used in the formation of alkyd resins and varnishes. It is used as intermediate in the manufacture of plasticizers, explosives (PETN) and pharmaceuticals (A-41).

Permissible Exposure Limits in Air: There are no Federal standards but ACGIH (1983/84) has classified pentaerythritol as a nuisance particulate with a TLV of 10 mg/m³ and a tentative STEL of 20 mg/m³.

Determination in Air: Collection on a filter and gravimetric determination (A-1).

Permissible Concentration in Water: No criteria set.

Harmful Effects and Symptoms: Feeding studies using human volunteers showed that 85% of pentaerythritol fed was eliminated unchanged in the urine within 30 hours. There are, in general, no significant effects on health by common routes of exposure even at abnormal use concentrations (A-34).

PENTANE

Description: C_5H_{12} is a colorless liquid with a gasolinelike odor which boils at 36°C.

Code Numbers: CAS 109-66-0 RTECS RZ9450000 UN 1265

DOT Designation: Flammable liquid.

Synonyms: Normal pentane.

Potential Exposure: Pentane is used in solvent extraction processes, as a blowing agent in plastics, as a fuel and as a chemical intermediate (for amyl chlorides, e.g.,) NIOSH estimates annual worker exposure to pentane at 10,000.

Incompatibilities: Strong oxidizers.

Permissible Exposure Limits in Air: The Federal standard is 1,000 ppm (2,950 mg/m³). NIOSH (1) has recommended a 120 ppm TWA on a 10-hour basis with a 610 ppm ceiling value. ACGIH has recommended as of 1983/84 a 600 ppm (1,800 mg/m³) TWA value and an STEL of 750 ppm (2,250 mg/m³). The IDLH level is 5,000 ppm.

Determination in Air: Charcoal adsorption, workup with CS_2, analysis by gas chromatography. See NIOSH Methods, Set G. See also reference (A-10).

Permissible Concentration in Water: No criteria set.

Routes of Entry: Inhalation, ingestion, skin and eye contact.

Harmful Effects and Symptoms: Drowsiness, irritation of eyes and nose, dermatitis, chemical pneumonia.

Points of Attack: Skin, eyes, respiratory system, lungs.

Medical Surveillance: Consider the points of attack in preplacement and periodic physical examinations.

First Aid: If this chemical gets into the eyes, irrigate immediately. If this chemical contacts the skin, flush with water promptly. If a person breathes in large amounts of this chemical, move the exposed person to fresh air at once and perform artificial respiration. When this chemical has been swallowed, get medical attention. Do not induce vomiting.

Personal Protective Methods: Wear appropriate clothing to prevent repeated or prolonged skin contact. Wear eye protection to prevent any reasonable probability of eye contact. Employees should wash promptly when skin is wet or contaminated. Remove clothing immediately if wet or contaminated to avoid flammability hazard.

Respirator Selection:
  5,000 ppm: GMOV/SA/SCBA
     Escape: GMOV/SCBA

Disposal Method Suggested: Incineration.

References

(1)   Nat. Inst. for Occup. Safety and Health, *Criteria for a Recommended Standard: Occupational Exposure to Alkanes,* NIOSH Doc. No. 77-151, Washington, DC (1977).
(2)   See Reference (A-61).
(3)   See Reference (A-60).

2-PENTANONE

Description: $CH_3COCH_2CH_2CH_3$ is a clear liquid with a strong odor resembling acetone and ether. It boils at $102°C$.

Code Numbers: CAS 107-87-9   RTECS SA7875000   UN 1249

DOT Designation: Flammable liquid.

Synonyms: Methyl propyl ketone, MPK, ethyl acetone.

Potential Exposure: MPK is used as a solvent in place of diethyl ketone and acetone. It has also been used in synthetic food flavoring.

Incompatibilities: Oxidizing agent.

Permissible Exposure Limits in Air: The Federal standard and the 1983/84 ACGIH TWA value is 200 ppm ($700 mg/m^3$). NIOSH (1) has recommended a TWA of 150 ppm ($530 mg/m^3$). The ACGIH has set an STEL value of 250 ppm ($875 mg/m^3$). The IDLH level is 5,000 ppm.

Determination in Air: Charcoal adsorption, workup with CS_2 analysis by gas chromatography. See NIOSH Methods, Set A. See also reference (A-10).

Permissible Concentration in Water: No criteria set.

Routes of Entry: Inhalation, ingestion, skin and eye contact.

Harmful Effects and Symptoms: Irritation of eyes and mucous membrane, headaches, dermatitis, narcosis, coma.

Points of Attack: Respiratory system, eyes, skin, central nervous system.

Medical Surveillance: Consider the points of attack in preplacement and periodic physical examinations.

First Aid: If this chemical gets into the eyes, irrigate immediately. If this chemical contacts the skin, flush with water. If a person breathes in large amounts of this chemical, move the exposed person to fresh air at once and perform artificial respiration. When this chemical has been swallowed, get medical attention. Do not induce vomiting.

Personal Protective Methods: Wear appropriate clothing to prevent repeated or prolonged skin contact. Wear eye protection to prevent any reasonable probability of eye contact. Employees should wash promptly when skin is wet or contaminated. Remove clothing immediately if wet or contaminated to avoid flammability hazard.

Respirator Selection:
```
1,000 ppm: CCROV/SA/SCBA
5,000 ppm: GMOV/SAF/SCBAF
  Escape: GMOV/SCBA
```

Disposal Method Suggested: Incineration.

References

(1) Nat. Inst. for Occup. Safety and Health, *Criteria for a Recommended Standard: Occupational Exposure to Ketones,* NIOSH Doc. No. 78-173, Washington, DC (1973).

(2) U.S. Environmental Protection Agency, *Chemical Hazard Information Profile: 2-Pentanone,* Washington, DC (December 6, 1977).

(3) See Reference (A-68).

1-PENTENE

Description: $CH_3(CH_2)_2CH{=}CH_2$ is a colorless liquid boiling at $30°C$.

Code Numbers: CAS 109-67-1 UN 1108

DOT Designation: —

Synonyms: Propylethylene; amylene.

Potential Exposure: Workers in petroleum refineries and petrochemical plants.

Incompatibilities: Strong oxidants.

Permissible Exposure Limits in Air: No standards set.

Permissible Concentration in Water: No criteria set.

Routes of Entry: Skin absorption, inhalation, ingestion.

Harmful Effects and Symptoms: Simple asphyxiant. Narcotic in high concentrations. Moderately toxic by oral and inhalation routes.

Points of Attack: Skin, eyes, respiratory tract and nervous system.

First Aid: Get to fresh air, remove contaminated clothes and flush with water (A-60).

Personal Protective Methods: Use protective gloves and safety goggles (A-60).

Respirator Selection: Wear canister mask.

Disposal Method Suggested: Incineration.

References

(1)   Sax, N.I., Ed., *Dangerous Properties of Industrial Materials Report, 2,* No. 6, 69-71, New York, Van Nostrand Reinhold Co. (1982).
(2)   Sax, N.I., Ed., *Dangerous Properties of Industrial Materials Report, 3,* No. 2, 56-57, New York, Van Nostrand Reinhold Co. (1983).
(3)   See Reference (A-60).

PERCHLOROETHYLENE

See "Tetrachloroethylene."

PERCHLOROMETHYL MERCAPTAN

Description: CCl_3SCl is a pale yellow oily liquid with a foul-smelling strong acrid odor. It boils at 149°C.

Code Numbers: CAS 594-42-3   RTECS PB0370000   UN 1670

DOT Designation: Poison B.

Synonyms: PCM, trichloromethane sulfenyl chloride, trichloromethyl sulfur chloride, perchloromethanethiol, PMM.

Potential Exposure: Perchloromethyl mercaptan is used as an intermediate in pesticide manufacture, specifically in the manufacture of Captan and Folpet (A-32).

Incompatibilities: Alkalies, amines, hot iron, hot water.

Permissible Exposure Limits in Air: The Federal standard and the 1983/84 ACGIH TWA value is 0.1 ppm (0.8 mg/m^3). There is no STEL value set. The IDLH level is 10 ppm.

Determination in Air: Sample collection by charcoal tube, analysis by gas liquid chromatography (A-13).

Permissible Concentration in Water: No criteria set.

Routes of Entry: Inhalation, skin absorption, ingestion, eye and skin contact.

Harmful Effects and Symptoms: Lacrimation, eye inflammation, irritation of nose and throat, coughing, dyspnea, deep breathing pain, coarse rales, vomiting, pallor, tachycardia, acidosis, anuria.

Points of Attack: Eyes, respiratory system, liver, kidneys, skin, lungs.

Medical Surveillance: Consider the points of attack in preplacement and periodic physical examinations.

First Aid: If this chemical gets into the eyes, irrigate immediately. If this chemical contacts the skin, wash with soap immediately. If a person breathes in large amounts of this chemical, move the exposed person to fresh air at once and perform artificial respiration. When this chemical has been swallowed, get medical attention. Give large quantities of water and induce vomiting. Do not make an unconscious person vomit.

Personal Protective Methods: Wear appropriate clothing to prevent repeated or prolonged skin contact. Wear eye protection to prevent any reasonable probability of eye contact. Employees should wash promptly when skin is wet. Remove nonimpervious clothing promptly if wet or contaminated.

Respirator Selection:

1 ppm:	SA/SCBA
5 ppm:	SAF/SCBAF
10 ppm:	SAF:PD,PP,CF
Escape:	GMOV/SCBA

Disposal Method Suggested: Incineration together with a flammable solvent in a furnace equipped with afterburner and scrubber.

References

(1)   See Reference (A-61).

PERCHLORYL FLUORIDE

Description: ClO_3F is a colorless gas with a characteristic sweet odor.

Code Numbers: CAS 7616-94-6   RTECS SD1925000   UN 1955

DOT Designation: Poison gas.

Synonyms: Chlorine oxyfluoride, chlorine fluoride oxide.

Potential Exposure: Perchloryl fluoride has been used as an oxidant in rocket propellant combinations, as an insulating gas in high voltage electrical systems, as a fluorinating agent in organic synthesis.

Incompatibilities: Combustibles, strong bases, amines, finely divided metals, oxidizable materials.

Permissible Exposure Limits in Air: The Federal standard and the 1983/84 ACGIH TWA value is 3 ppm (14 mg/m^3). The STEL is 6 ppm (28 mg/m^3). The IDLH level is 385 ppm.

Determination in Air: Sample collection by impinger or fritted bubbler, analysis by ion specific electrode (A-15).

Permissible Concentration in Water: No criteria set.

Routes of Entry: Inhalation, eye and skin contact.

Harmful Effects and Symptoms: Respiratory irritation and skin burns and in animals methemoglobinemia.

Points of Attack: Respiratory system, skin, blood.

Medical Surveillance: Consider the points of attack in preplacement and periodic physical examinations.

First Aid: If this chemical gets into the eyes, irrigate immediately. If a person breathes in large amounts of this chemical, move the exposed person to fresh air at once and perform artificial respiration.

Personal Protective Methods: Wear appropriate clothing to prevent skin freezing. Wear eye protection to prevent any reasonable probability of eye contact. Remove nonimpervious clothing immediately if wet or contaminated.

Respirator Selection:

30 ppm:	SA/SCBA
150 ppm:	SAF/SCBAF
400 ppm:	SA:PD,PP,CF
Escape:	GMS/SCBA
Note:	Do not use oxidizable sorbents.

Disposal Method Suggested: Incineration together with flammable solvent in furnace equipped with afterburner and scrubber.

References

(1)   See Reference (A-61).

PERSULFATES

See "Potassium Persulfate" as good example of this class of compounds.

PHENACETIN

See "Acetophenetidin."

PHENAZOPYRIDINE HYDROCHLORIDE

● Carcinogen (Animal Positive, NCI, IARC)

Description: $C_{11}H_{12}ClN_5$ with the following structural formula

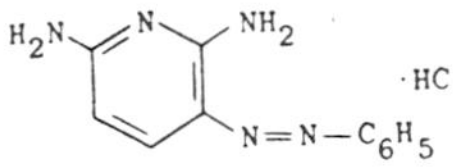

is a red crystalline compound. The fine base melts at 139°C.

Code Numbers: CAS 136-40-3   RTECS US7875000

DOT Designation: —

Synonyms: Pyridium®, Bisteril®, Pyridacil®, Uridinal®.

Potential Exposure: Phenazopyridine hydrochloride has been used for 40 years as an analgesic drug either alone or in combination with other drugs to reduce pain associated with urinary tract infections.

Exposure to phenazopyridine hydrochloride occurs during manufacture and formulation. The National Occupational Hazard Survey estimates potential occupational exposure as 2,800 persons. Patients receiving the drug are directly exposed.

Data from the National Prescription Audit reported that 4.4 million prescriptions were dispensed by retail pharmacies in 1980. The average adult dose rate is one tablet three times a day in 200 mg or 100 mg tablets.

Permissible Exposure Limits in Air: No standards set.

Permissible Concentration in Water: No criteria set.

Harmful Effects and Symptoms: Phenazopyridine hydrochloride was tested in mice and rats by oral administration (1). In female mice, it significantly increased the incidence of hepatocellular adenomas and carcinomas. In male and female rats, it induced tumors of the colon and rectum (2).

References

(1) National Cancer Institute (1978), *Bioassay of Phenazopyridine Hydrochloride for Possible Carcinogenicity,* Technical Report Series No. 99, DHEW Publication No. (NIH) 78-1349, Washington, DC, Department of Health, Education, and Welfare.
(2) *IARC Monographs on the Evaluation of Carcinogenic Risk of Chemicals to Man,* Vol. 24, IARC, Lyon, France, pp 163-173 (1980).
(3) See Reference (A-62). Also see Reference (A-64).

PHENOBARBITAL

Description: $C_{12}H_{12}N_2O_3$ with the following structural formula

is a crystalline compound melting at 174° to 178°C.

Code Numbers: CAS 50-06-6   RTECS CQ6825000

DOT Designation: —

Synonyms: 5-ethyl-5-phenylbarbituric acid; phenylethylmalonylurea; many trade names.

Potential Exposure: Used as an anticonvulsant, hypnotic and sedative.

Permissible Exposure Limits in Air: No standards set.

Permissible Concentration in Water: No criteria set.

Harmful Effects and Symptoms: This material is highly toxic to humans (1). In children, it exhibits dangerous CNS effects at 10 mg/kg. It has toxic effects on adults when 18 mg/kg is applied to the skin and causes psychological disturbances at 214 μg/kg.

Points of Attack: CNS

References

(1) Sax, N.I., Ed., *Dangerous Properties of Industrial Materials Report, 1,* No. 2, 55-57, New York, Van Nostrand Reinhold Co. (1980).

PHENOL

- Hazardous substance (EPA)
- Hazardous waste (EPA)
- Priority toxic pollutant (EPA)

Description: C_6H_5OH, phenol, is a white crystalline substance with a distinct aromatic, acrid odor. It melts at $41°C$.

Code Numbers:

CAS 108-95-2   RTECS SJ3325000   UN 1671 (Solid)
                                               UN 2312 (Molten)

DOT Designation: Poison B.

Synonyms: Carbolic acid, phenic acid, phenylic acid, phenyl hydrate, hydroxybenzene, monohydroxybenzene.

Potential Exposures: Phenol is used in the production or manufacture of explosives, fertilizer, coke, illuminating gas, lampblack, paints, paint removers, rubber, asbestos goods, wood preservatives, synthetic resins, textiles, drugs, pharmaceutical preparations, perfumes, bakelite, and other plastics (phenol-formaldehyde resins) as well as polymer intermediates (caprolactam, bisphenol-A and adipic acid). Phenol also finds wide use as a disinfectant in the petroleum, leather, paper, soap, toy, tanning, dye, and agricultural industries. NIOSH estimates that 10,000 employees are potentially exposed to phenol in the United States. This reflects the number of people that are employed in the production of phenol, formulation into products, or distribution of concentrated products. In addition, an uncertain but probably large number of people will have intermittent contact with phenol as components of medications or in the workplace as chemists, pharmacists, biomedical personnel, and other occupations.

Incompatibilities: Strong oxidizers, calcium hypochlorite.

Permissible Exposure Limits in Air: The Federal standard and the ACGIH 1983/84 TWA value is 5 ppm ($19 mg/m^3$). NIOSH recommends (1) that exposure to phenol vapor, solid, or mists be limited to no more than $20 mg/m^3$ expressed as a time-weighted average (TWA) concentration for up to a 10-hour workshift, 40-hour workweek. In addition, to protect employees from peak overexposures, NIOSH recommends that exposures be limited to no more than $60 mg/m^3$ for any 15-minute period. There is thus no substantial change in the present Federal limit of $19 mg/m^3$ (5 ppm), except for the inclusion of a ceiling value. The STEL value is 10 ppm ($38 mg/m^3$). The IDLH level is 100 ppm. ACGIH adds the notation "skin" to indicate the possibility of cutaneous absorption.

Determination in Air: Collection in a bubbler using NaOH, workup with H_2SO_4, analysis by gas chromatography. See NIOSH Methods, Set L. See also reference (A-10).

Permissible Concentration in Water: To protect freshwater aquatic life—10,200 $\mu g/\ell$, based on acute toxicity data and 2,560 $\mu g/\ell$, based on chronic toxicity data. To protect saltwater aquatic life—5,800 $\mu g/\ell$, based on acute toxicity data. For the protection of human health from phenol ingested through water and through contaminated aquatic organisms the concentration in water should not exceed 3,500 $\mu g/\ell$. For the prevention of adverse effects due to the organoleptic properties of chlorinated phenols inadvertently formed during water purification processes, the phenol concentration in water should not exceed 300 $\mu g/\ell$.

Determination in Water: Methylene chloride extraction followed by gas chromatography with flame ionization or electron capture detection (EPA Method 604) or gas chromatography plus mass spectrometry (EPA Method 625).

Routes of Entry: Inhalation of mist or vapor; percutaneous absorption of mist, vapor, or liquid, ingestion, skin and eye contact.

Harmful Effects and Symptoms: *Local* — Phenol has a marked corrosive effect on any tissue. When it comes in contact with the eyes it may cause severe damage and blindness. On contact with the skin, it does not cause pain but causes a whitening of the exposed area. If the chemical is not removed promptly, it may cause a severe burn or systemic poisoning.

Systemic — Systemic effects may occur from any route of exposure. These include paleness, weakness, sweating, headache, ringing of the ears, shock, cyanosis, excitement, frothing of the nose and mouth, dark colored urine, and death. If death does not occur, kidney damage may occur.

Repeated or prolonged exposure to phenol may cause chronic phenol poisoning. This condition is very rarely reported. The symptoms of chronic poisoning include vomiting, difficulty in swallowing, diarrhea, lack of appetite, headache, fainting, dizziness, dark urine, mental disturbances, and possibly, skin rash. Liver and kidney damage and discoloration of the skin may occur.

Points of Attack: Liver, kidney, skin.

Medical Surveillance: Consider the skin, eye, liver, and renal function as part of any preplacement or periodic examination. Phenol can be determined in blood or urine.

First Aid: If this chemical gets into the eyes, irrigate immediately. If this chemical contacts the skin, wash with soap immediately. If a person breathes in large amounts of this chemical, move the exposed person to fresh air at once and perform artificial respiration. When this chemical has been swallowed, get medical attention. Give large quantities of water and induce vomiting. Do not make an unconscious person vomit.

Personal Protective Methods: In areas where there is likelihood of a liquid spill or splash, impervious protective clothing and goggles should be worn. In areas of heavy vapor concentration, fullface mask with forced air supply should be used, as well as protective clothing, gloves rubber boots, and apron. Employees should wash immediately when skin is wet or contaminated. Work clothing should be changed daily if it is possible that clothing is contaminated. Remove nonimpervious clothing immediately if wet or contaminated. Provide emergency showers and eyewash.

Respirator Selection:
    50 ppm: CCROVDM/SA/SCBA
  100 ppm: CCROVFDM/GMOVDM/SAF/SCBAF
  Escape: GMOVP/SCBA

Disposal Method Suggested: Incineration.

References
(1) National Institute for Occupational Safety and Health, *Criteria for a Recommended Standard: Occupational Exposure to Phenol,* NIOSH Doc. No. 76-196 (1976).
(2) U.S. Environmental Protection Agency, *Phenol: Ambient Water Quality Criteria,* Washington, DC (1980).
(3) U.S. Environmental Protection Agency, *Phenol,* Health and Environmental Effects Profile No. 144, Office of Solid Waste, Washington, DC (April 30, 1980).

(4)   See Reference (A-61).
(5)   See Reference (A-60).
(6)   Sax, N.I., Ed., *Dangerous Properties of Industrial Materials Report, 3,* No. 4, 77-84, New York, Van Nostrand Reinhold Co. (1983).
(7)   Parmeggiani, L., Ed., *Encyclopedia of Occupational Health & Safety,* Third Edition, Vol. 2, pp 1671–76, Geneva, International Labour Office (1983) (Phenols & Phenolic Compounds).
(8)   United Nations Environment Programme, *IRPTC Legal File 1983,* Vol. II, pp VII/496–501, Geneva, Switzerland, International Register of Potentially Toxic Chemicals (1984).

PHENOTHIAZINE

Description: $C_{12}H_9NS$, $S(C_6H_4)_2NH$, is a greenish-yellow to greenish-gray crystalline substance which melts at 175° to 185°C.

Code Numbers: CAS 92-84-2   RTECS SN5075000

DOT Designation: —

Synonyms: Thiodiphenylamine, dibenzothiazine.

Potential Exposures: Phenothiazine is used as an anthelmintic in medicine and veterinary medicine; it is used widely as an intermediate in pharmaceutical manufacture (A-41).

Permissible Exposure Limits in Air: There is no Federal standard but ACGIH (1983/84) has adopted a TWA value of 5 mg/m³ with the notation "skin" indicating the possibility of cutaneous absorption. The STEL value is 10 mg/m³.

Permissible Concentration in Water: No criteria set.

Routes of Entry: Skin absorption, ingestion, inhalation, skin and eye contact.

Harmful Effects and Symptoms: Tonic hepatitis, hemolytic anemia, abdominal cramps and tachycardia, gastrointestinal and skin irritation, kidney damage, skin photosensitization and pruritis, hair and fingernail coloration (reddish-brown), conjunctivitis and keratitis.

Points of Attack: Skin, eyes, gastrointestinal system, liver, kidney, blood.

Medical Surveillance: Observe points of attack in preplacement and regular physical examinations.

First Aid: Flush eyes with water, wash contaminated body areas with soap and water. Flush stomach with water if swallowed (A-38).

Personal Protective Methods: Wear butyl rubber gloves, full protective clothing and protective shoes.

Respirator Selection: Wear self-contained breathing apparatus (A-38).

Disposal Method Suggested: Dissolve in combustible solvent and spray into incinerator equipped with afterburner and scrubber.

References

(1)   See Reference (A-60).
(2)   Parmeggiani, L., Ed., *Encyclopedia of Occupational Health & Safety,* Third Edition, Vol. 2, pp 1676–77, Geneva, International Labour Office (1983).

p-PHENYLENEDIAMINE

Description: $H_2NC_6H_4NH_2$ is a white to light purple or brown solid which melts at 141°C.

Code Numbers: CAS 106-50-3 RTECS SS8050000 UN 1673

DOT Designation: ORM-A.

Synonyms: p-Aminoaniline, 1,4-diaminobenzene, benzenediamine, PARA.

Potential Exposure: p-Phenylenediamine has been used in fur dyes and as a monomer in the manufacture of improved tire cords.

Incompatibilities: Strong oxidizers.

Permissible Exposure Limits in Air: The Federal standard and the 1983/84 ACGIH TWA value is 0.1 mg/m^3. The notation "skin" is added to indicate the possibility of cutaneous absorption. There is no STEL value set. The IDLH level is 25 mg/m^3.

Determination in Air: Collection by impinger or fritted bubbler, colorimetric analysis (A-26).

Permissible Concentration in Water: No criteria set.

Routes of Entry: Inhalation, ingestion, skin and eye contact.

Harmful Effects and Symptoms: Irritation of pharynx and larynx, bronchial asthma, sensitization to dermatitis.

Points of Attack: Respiratory system, skin.

Medical Surveillance: Consider the points of attack in preplacement and periodic physical examinations.

First Aid: If this chemical gets into the eyes, irrigate immediately. If this chemical contacts the skin, wash with soap promptly. If a person breathes in large amounts of this chemical, move the exposed person to fresh air at once and perform artificial respiration. When this chemical has been swallowed, get medical attention. Give large quantities of water and induce vomiting. Do not make an unconscious person vomit.

Personal Protective Methods: Wear appropriate clothing to prevent any reasonable probability of skin contact. Wear eye protection to prevent any reasonable probability of eye contact. Employees should wash promptly when skin is wet or contaminated and daily at the end of each work shift. Work clothing should be changed daily if it is possible that clothing is contaminated. Remove nonimpervious clothing promptly if wet or contaminated.

Respirator Selection:
 5 mg/m^3: SAF/SCBAF
 25 mg/m^3: SAF:PD,PP,CF
 Escape: GMS/SCBA

Disposal Method Suggested: Controlled incineration whereby oxides of nitrogen are removed from the effluent gas by scrubber, catalytic or thermal device.

References

(1) U.S. Environmental Protection Agency, *Chemical Hazard Information Profile: Phenylenediamines,* Washington, DC (June 1, 1978).

(2) See Reference (A-61).

(3)   See Reference (A-60).

(4)   United Nations Environment Programme, *IRPTC Legal File 1983,* Vol. II, pp VII/528–29, Geneva, Switzerland, International Register of Potentially Toxic Chemicals (1984).

PHENYL GLYCIDYL ETHER

Description: $C_6H_5OCH_2-CH-CH_2$ is a colorless liquid with an unpleasant sweet odor. It melts at 3° to 4°C and boils at 245°C.

Code Numbers: CAS 122-60-1   RTECS TZ3675000

DOT Designation: —

Synonyms: 1,2-Epoxy-3-phenoxypropane, PGE, glycidyl phenyl ether, phenyl epoxypropyl ether, 2,3-epoxypropylphenyl ether, phenoxymethyloxirane, phenoxypropylene oxide.

Potential Exposure: PGE is used as a reactive diluent in uncured epoxy resins to reduce the viscosity of the uncured system for ease in casting, adhesive, and laminating applications. NIOSH estimated that 8,000 workers are potentially exposed to PGE.

Incompatibilities: Strong oxidizers, amines, strong acids, strong bases.

Permissible Exposure Limits in Air: The Federal standard is 10 ppm (60 mg/m³). NIOSH has recommended a ceiling concentration of 1 ppm (6 mg/m³) as a ceiling concentration for a 15-minute sampling period. ACGIH as of 1983/84 has adopted a TWA of 1 ppm (6 mg/m³) but no STEL value. No IDLH level has been established.

Determination in Air: Adsorption on charcoal, workup with CS_2, analysis by gas chromatography. See NIOSH Methods, Set F. See also reference (A-10).

Permissible Concentration in Water: No criteria set.

Routes of Entry: Inhalation, ingestion, eye and skin contact.

Harmful Effects and Symptoms: Skin irritation and sensitivity, eye irritation, irritation of upper respiratory system, narcosis.

Points of Attack: Skin, eyes, central nervous system.

Medical Surveillance: Consider the points of attack in preplacement and periodic physical examinations.

First Aid: If this chemical gets into the eyes, irrigate immediately. If this chemical contacts the skin, dust off solid, flush with water, wash with soap promptly. If a person breathes in large amounts of this chemical, move the exposed person to fresh air at once and perform artificial respiration. When this chemical has been swallowed, get medical attention. Give large quantities of saltwater and induce vomiting. Do not make an unconscious person vomit.

Personal Protective Methods: Wear appropriate clothing to prevent any possibility of skin contact. Wear eye protection to prevent any reasonable probability of eye contact. Employees should wash promptly when skin is wet or contaminated. Remove nonimpervious clothing promptly if wet or contaminated.

Respirator Selection:
    500 ppm: SAF/SCBAF
      Escape: GMOV/SCBA

Disposal Method Suggested: Concentrated waste containing no peroxides—discharge liquid at a controlled rate near a pilot flame. Concentrated waste containing peroxides—perforation of a container of the waste from a safe distance followed by open burning.

References

(1) U.S. Environmental Protection Agency, *Chemical Hazard Information Profile: Phenyl Glycidyl Ether,* Washington, DC (1979).

(2) Nat. Inst. for Occup. Safety and Health, *Criteria for a Recommended Standard: Occupational Exposure to Glycidyl Ethers,* NIOSH Doc. No. 78-166, Washington, DC (1978).

(3) Nat. Inst. for Occup. Safety and Health, *Information Profiles on Potential Occupational Hazards: Glycidyl Ethers,* Report PB-276,678, Rockville, MD, pp 116-123, (October 1977).

(4) See Reference (A-60).

PHENYLHYDRAZINE

● Carcinogen (Suspect, Human) (A-6)

Description: $C_6H_5NHNH_2$ is a colorless to pale yellow solid or liquid with a weak aromatic odor. It melts at 20°C.

Code Numbers: CAS 100-63-0   RTECS MV8925000   UN 2572

DOT Designation: Poison.

Synonym: Hydrazinobenzene.

Potential Exposures: Phenylhydrazine is very reactive with carbonyl compounds and is a widely used reagent in conjunction with sugars, aldehydes, and ketones, in addition to its use in the synthesis of dyes, pharmaceuticals such as antipyrin, cryogenin, and pyramidone, and other organic chemicals. The hydrochloride salt is used in the treatment of polycythemia vera. NIOSH has estimated the number of workers exposed to phenylhydrazine at 5,000.

Incompatibilities: Strong oxidizers, lead dioxide.

Permissible Exposure Limits in Air: The Federal standard is 5 ppm (22 mg/m³). More recently (1) NIOSH has recommended a ceiling concentration in any 2-hour period of 0.14 ppm (0.6 mg/m³). The ACGIH has set a TWA value of 5 ppm (20 mg/m³) and an STEL of 10 ppm (45 mg/m³). The notation "skin" is added to indicate the possibility of cutaneous absorption. The IDLH level is 250 ppm. The ACGIH (1983/84) lists phenylhydrazine as "an industrial substance suspect of carcinogenic potential for man."

Determination in Air: Absorption in a bubbler containing hydrochloric acid, reaction with phosphomolybdic acid and colorimetric determination. See NIOSH Methods, Set L. See also reference (A-10).

Permissible Concentration in Water: No criteria set.

Routes of Entry: Inhalation, skin absorption, ingestion, skin and eye contact.

Harmful Effects and Symptoms: Skin sensitization, hemolytic anemia, dyspnea, cyanosis, jaundice, vascular thrombosis, kidney damage.

Points of Attack: Blood, respiratory system, liver, kidneys, skin.

Medical Surveillance: Consider the points of attack in preplacement and periodic physical examinations.

First Aid: If this chemical gets into the eyes, irrigate immediately. If this chemical contacts the skin, wash with soap immediately. If a person breathes in large amounts of this chemical, move the exposed person to fresh air at once and perform artificial respiration. When this chemical has been swallowed, get medical attention. Give large quantities of saltwater and induce vomiting. Do not make an unconscious person vomit.

Personal Protective Methods: Wear appropriate clothing to prevent any possibility of skin contact. Wear eye protection to prevent any possibility of eye contact. Employees should wash immediately when skin is wet or contaminated and daily at the end of each work shift. Work clothing should be changed daily if it is possible that clothing is contaminated. Remove clothing immediately if wet or contaminated. Provide emergency showers and eyewash.

Respirator Selection:
    250 ppm: SAF/SCBAF
    Escape: GMS/SCBA

Disposal Method Suggested: Controlled incineration whereby oxides of nitrogen are removed from the effluent gas by scrubber, catalytic or thermal device.

References

(1) Nat. Inst. for Occup. Safety and Health, *Criteria for a Recommended Standard: Occupational Exposure to Hydrazines,* NIOSH Doc. No. 78-172, Washington, DC (1978).
(2) See Reference (A-60).
(3) Parmeggiani, L., Ed., *Encyclopedia of Occupational Health & Safety,* Third Edition, Vol. 2, pp 1677–78, Geneva, International Labour Office (1983).

PHENYL MERCAPTAN

● Hazardous waste (EPA)

Description: C_6H_5SH is a water white liquid with a repulsive penetrating garliclike odor which boils at 169°C.

Code Numbers: CAS 108-98-5   RTECS DC0525000   UN 2337

DOT Designation: Flammable liquid, poison.

Synonyms: Thiophenol, benzenethiol.

Potential Exposures: Phenyl mercaptan is used as an intermediate in pesticide manufacture (A-32). It is used in solvent formulations for the removal of polysulfide sealants (A-34).

Permissible Exposure Limits in Air: There are no Federal limits but ACGIH (1983/84) has adopted a TWA value of 0.5 ppm (2.0 mg/m^3). There is no STEL value proposed.

Permissible Concentration in Water: No criteria set but 0.062 µg/ℓ imparts a faint odor to water (A-36).

Harmful Effects and Symptoms: Skin and eye irritation, dermatitis, headaches and dizziness. In animals it causes restlessness, increased respiration, incoordination, muscular weakness, skeletal muscle paralysis of the hind limbs, cyanosis, lethargy and/or sedation, respiratory depression, coma and death (A-34).

Personal Protective Methods: Wear rubber gloves and work clothes.

Respirator Selection: Use gas mask for protection (A-38).

Disposal Method Suggested: Dissolve in flammable solvent and burn in furnace equipped with afterburner and alkaline scrubber.

N-PHENYL-β-NAPHTHYLAMINE

● Carcinogen (Animal suspected, IARC) (1)

Description: $C_{10}H_7NHC_6H_5$ is a light gray powder.

Code Numbers: CAS 135-88-6   RTECS QM4550000

DOT Designation: −

Synonyms: 2-Anilinonaphthalene, Neozone D®, N-phenyl-2-naphthylamine, PBNA.

Potential Exposures: Phenyl-β-naphthylamine is used as a rubber antioxidant, as an inhibitor for butadiene, a stabilizer in lubricants and an intermediate in chemical synthesis.

Permissible Exposure Limits in Air: There are no Federal standards but ACGIH (1983/84) has designated this compound as an "industrial substance suspect of carcinogenic potential for man."

Permissible Concentration in Water: No criteria set.

Harmful Effects and Symptoms: The main problem with this compound is that β-naphthylamine, a known carcinogen, is both a contaminant in, and a metabolic product of PBNA.

Medical Surveillance (2): All employees with a potential exposure to PBNA should be placed under a medical monitoring program including history and medical examinations to detect the presence of bladder cancers and specific urine analysis for β-naphthylamine.

Personal Protective Methods (2): Protective full body clothing should be provided and its use required for employees entering the regulated area. Upon exiting from the regulated area, the protective clothing should be left at the point of exit. With the last exit of the day, the protective clothing should be placed in a suitably marked and closed container for disposal or laundering. (Laundry personnel should be made aware of the potential hazard from handling contaminated clothing.) Employees should be required to wash all exposed areas of the body upon exiting from the regulated area. Effective engineering measures include the use of walk-in hoods or specific local exhaust ventilation with suitable collectors.

Respirator Selection: Personal respiratory protective devices should only be used as an interim measure while engineering controls are being installed, for nonroutine use and during emergencies. Considering the carcinogenic potential and the lack of a standard, the appropriate personal respiratory protective measure is the use of a positive pressure supplied air respirator or a positive pressure self-contained breathing apparatus (2).

References

(1) International Agency for Research on Cancer, *IARC Monographs on the Carcinogenic Risks of Chemicals to Humans,* 16, 325, Lyon, France (1978).
(2) Nat. Inst. for Occup. Safety and Health, *Metabolic Precursors of a Known Human Carcinogen, β-Naphthylamine,* Current Intelligence Bulletin No. 16, Rockville, MD (December 17, 1976).
(3) See Reference (A-60).

PHENYLPHOSPHINE

Description: $C_6H_5PH_2$ is a crystalline solid melting at 138°C.

Code Numbers: CAS 638-21-1   RTECS SZ2100000

DOT Designation: —

Synonym: PF.

Potential Exposures: Polyphosphinate compounds used as catalysts and antioxidants disproportionate when heated to give phosphonic acid derivatives plus PF (A-34).

Permissible Exposure Limits in Air: There is no Federal standard but ACGIH (1983/84) has adopted a ceiling value of 0.05 ppm (0.25 mg/m³). There is no STEL value.

Determination in Air: Flame ionization with phosphorus detector (A-34).

Permissible Concentration in Water: No criteria set.

Harmful Effects and Symptoms: A level of 0.6 ppm is a threshold effect level for laboratory animals; hypersensitivity to sound and touch and mild hyperemia developed above this level. Above 2.2 ppm, chronic effects developed including decreases in red blood cells, dermatitis and severe testicular degeneration (which was however reversible) (A-34).

PHENYTOIN

● Carcinogen (Animal Positive, Human Suspected, IARC) (A-62) (A-64)

Description: $C_{15}H_{12}N_2O_2$ with the following structural formula

is a crystalline compound which melts at 295° to 298°C.

Code Numbers: CAS 57-41-0   RTECS MU1050000

DOT Designation: —

Synonyms: 5,5-Diphenyl hydantoin.

Potential Exposure: Phenytoin is a pharmaceutical used in the treatment of grand mal epilepsy, Parkinson's syndrome, and in veterinary medicine.

Human exposure to phenytoin occurs principally during its use as a drug. Figures on the number of patients using phenytoin are not available, but phenytoin is given to a major segment of those individuals with epilepsy. The oral dose rate is initially 100 mg 3 times per day and can gradually increase by 100 mg every 2 to 4 weeks until the desired therapeutic response is obtained. The intravenous dose is 200 to 350 mg/day.

Permissible Exposure Limits in Air: No standards set.

Permissible Concentration in Water: No criteria set.

Harmful Effects and Symptoms: Phenytoin is carcinogenic in mice after oral administration or by intraperitoneal injection, producing lymphomas and leukaemias (1).

There are case reports and epidemiological studies of lymphomas occurring in patients that received phenytoin (1); however, no excess of lymphomas was reported in a follow-up study of epilepsy patients, many of whom received phenytoin along with other antiepileptic drugs. Three recent papers report one case of malignant mesenchymoma and two cases of neuroblastoma in children with phenytoin-induced malformations. An epidemiological study looked at the frequency of use of phenytoin and of phenobarbitone in mothers of children with childhood cancers compared with mothers of normal children. While more mothers of cancer cases reported a history of epilepsy, no differences were seen in the proportion of epileptic mothers who took either phenytoin or phenobarbitone. An excess of lymphomas (6 observed, 4 expected) was seen in children of epileptic mothers, but the occurrence of brain tumors was not reported (2).

References
(1) IARC Monographs 13:201-25 (1977).
(2) *IARC Monographs on the Evaluation of the Carcinogenic Risk of Chemicals to Humans,* Supplement I, IARC, Lyon, France, p 41 (1979).
(3) See Reference (A-62). (Also see Reference (A-64).

PHORATE

• Hazardous waste (EPA)

Description: $(C_2H_5O)_2\overset{\text{S}}{\overset{\|}{P}}SCH_2SC_2H_5$ is a clear mobile liquid.

Code Numbers: CAS 298-02-2   RTECS TD9450000   UN 2783

DOT Designation: —

Synonyms: Timet (in U.S.S.R.), ENT 24,042, O,O-Diethyl-S-ethylthiomethyl phosphorodithioate, Thimet®.

Potential Exposure: Those engaged in the manufacture, formulation and application of this systemic and contact insecticide and acaricide. It is also used as a soil insecticide.

Permissible Exposure Limits in Air: There is no Federal standard but ACGIH (1983/84) has set a TWA of 0.05 mg/m^3 and an STEL of 0.2 mg/m^3. The notation "skin" is added to indicate the possibility of cutaneous absorption.

Permissible Concentration in Water: A no-adverse effect level in drinking water has been calculated by NAS/NRC to be 0.0007 mg/ℓ (A-2).

Determination in Water: Sensitive methods for analyzing residues of phorate are available. A gas-liquid chromatographic method is available that can detect 0.01 ppm in milk, and a cholinesterase-inhibition method can detect as little as 0.008 ppm in a 5-g crop sample. However, it is clear that the analytic methods available will be barely adequate for the analysis of these materials in drinking water. Particular care will have to be taken to ensure that sample sizes are great enough to allow detection of concentrations as low as 0.0007 ppm (A-2).

Harmful Effects and Symptoms: The mode of action of phorate is inhibition of acetylcholinesterase. It has high acute toxicity in laboratory animals.

References

(1) U.S. Environmental Protection Agency, *Phorate,* Health and Environmental Effects Profile No. 145, Office of Solid Waste, Washington, DC (April 30, 1980).
(2) United Nations Environment Programme, *IRPTC Legal File 1983,* Vol. II, pp VII/571–73, Geneva, Switzerland, International Register of Potentially Toxic Chemicals (1984).

PHOSGENE

- Hazardous substance (EPA)
- Hazardous waste (EPA)

Description: $COCl_2$, phosgene, is a colorless, noncombustible gas with a sweet, not pleasant odor like hay in low concentrations. In higher concentrations, it is irritating and pungent.

Code Numbers: CAS 75-44-5   RTECS SY5600000   UN 1076

DOT Designation: Poison A, poison gas.

Synonyms: Carbonyl chloride, carbon oxychloride, carbonic acid dichloride, chloroformyl chloride, combat gas.

Potential Exposures: Phosgene is used in the manufacture of dyestuffs based on triphenylmethane, coal tar, and urea. It is also used in the organic synthesis of isocyanates and their derivatives, carbonic acid esters (polycarbonates), and acid chlorides. Other applications include its utilization in metallurgy, and in the manufacture of some insecticides (A-32) and pharmaceuticals (A-41). NIOSH estimates that 10,000 workers have potential occupational exposure to phosgene during its manufacture and use.

Incompatibilities: Moisture.

Permissible Exposure Limits in Air: The Federal standard and the ACGIH 1983/84 TWA value for concentrations of phosgene in air is 0.1 ppm (0.4 mg/m³). NIOSH recommends adherence to the present Federal standard of 0.1 ppm phosgene as a time-weighted average for up to a 10-hour workday, 40-hour workweek and proposes the addition of a 0.2 ppm ceiling value for any 15-minute period. There is no STEL value set. The IDLH level is 2 ppm.

Determination in Air: Sample collection by midget impinger containing nitrobenzylpyridine which produces red color whose intensity is measured (A-10).

Permissible Concentration in Water: No criteria set.

Routes of Entry: Inhalation of gas, eye and skin contact.

Harmful Effects and Symptoms: *Local* — Conjunctivitis, lacrimation, and upper respiratory tract irritation may develop from gas. Liquid may cause severe burns.

Systemic — Acute exposure to phosgene may produce pulmonary edema frequently preceded by a latent period of 5-6 hours but seldom longer than 12 hours. The symptoms are dizziness, chills, discomfort, thirst, increasingly tormenting cough, and viscous sputum. Sputum may then become thin and foamy, and dyspnea, a feeling of suffocation, tracheal rhonchi, and grey-blue cyanosis may follow. Death may result from respiratory or cardiac failure. The hazard of phosgene is increased because at low levels (205 mg/m^3) it is lacking in warning symptoms.

Chronic exposure to phosgene may result in some tolerance to acute edemagenic doses, but may cause irreversible pulmonary changes of emphysema and fibrosis. Animal experimentation has shown an increased incidence of chronic pneumonitis and acute and fibrinous pneumonia from exposure to this agent.

Phosgene has been studied by EPA because of its mention as a possible cause for Legionnaire's Disease. Chlorofluorocarbons were reportedly combusted at burning cigarette tips with phosgene generation leading to the observed respiratory problems. However, it was found that phosgene was not the cause of Legionnaire's Disease.

Points of Attack: Respiratory system, lungs, skin, eyes.

Medical Surveillance: Preemployment medical examinations should include chest x-rays and baseline pulmonary function tests. The eyes and skin should be examined. Smoking history should be known. Periodic pulmonary function studies should be done. Workers who are known to have inhaled phosgene should remain under medical observation for at least 24 hours to insure that delayed symptoms do not occur.

First Aid: If this chemical gets into the eyes, irrigate immediately. If this chemical contacts the skin, flush with water immediately. If a person breathes in large amounts of this chemical, move the exposed person to fresh air at once and perform artificial respiration. When this chemical has been swallowed, get medical attention. Do not induce vomiting.

Personal Protective Methods: Where liquid phosgene is encountered, protective clothing should be supplied which is impervious to phosgene. Where gas is encountered above safe limits, fullface gas masks with phosgene canisters or supplied air respirators should be used. Because of the potentially serious consequences of acute overexposure and the poor warning properties of the gas to the human senses, automatic continuous monitors with alarm systems are strongly recommended.

Wear appropriate clothing to prevent any reasonable probability of skin contact. Wear eye protection to prevent any possibility of eye contact. Employees should wash immediately when skin is wet or contaminated. Remove nonimpervious clothing immediately if wet or contaminated. Provide emergency showers.

Respirator Selection:
        1 ppm: SA/SCBA
        2 ppm: SAF/SCBAF
     Escape: GMS/SCBA

References

(1)   National Institute for Occupational Safety and Health, *Criteria for a Recommended Standard: Occupational Exposure to Phosgene,* NIOSH Doc. No. 76-137 (1976).

(2)   U.S. Environmental Protection Agency, *Chemical Hazard Information Profile: Phosgene,* Washington, DC (June 13, 1977).

(3)   See Reference (A-61).

(4)   Sax, N.I., Ed., *Dangerous Properties of Industrial Materials Report, 3,* No. 3, 97-99, New York, Van Nostrand Reinhold Co. (1983).

(5)   See Reference (A-60).

(6)   Parmeggiani, L., Ed., *Encyclopedia of Occupational Health & Safety,* Third Edition, Vol. 2, pp 1678-79, Geneva, International Labour Office (1983).

(7)   United Nations Environment Programme, *IRPTC Legal File 1983,* Vol. II, pp VII/533-34, Geneva, Switzerland, International Register of Potentially Toxic Chemicals (1984).

PHOSPHATES

Description: PO_4^{3-} salts of metals such as Na_3PO_4. Properties vary with the individual compound. Trisodium phosphate, Na_3PO_4, will be chosen as an example.

Code Numbers: For Na_3PO_4: RTECS TG9490000

DOT Designation: —

Synonyms: For Na_3PO_4: Sodium phosphate; TSP.

Potential Exposure: Large sources of phosphate, which result from the widespread use of detergents from human wastes, and from surface water runoff from phosphate fertilized acreage, can cause accelerated eutrophication of freshwater aquatic systems. Detergents have been found to account for 50 to 70% of phosphate in domestic wastewater, while human wastes account for the other 30 to 50%. An estimated 1 to 5% of the phosphate applied as fertilizer reaches surface water through runoff and percolation. This amounts to about 46,000 to 230,000 metric tons (P_2O_5 content) of phosphate fertilizer per year. Municipal point sources dominate phosphate loading in the Northeast, while agricultural runoff dominates in the Southeast, Midwest, and West.

Permissible Exposure Limits in Air: No standards set.

Permissible Concentration in Water: Although a total phosphorus as phosphate criterion to control nuisance aquatic growths has not been presented, the following rationale to support such a criterion is being considered by EPA.

Total phosphate phosphorus concentrations in excess of 0.10 g/m^3 total phosphate expressed as elemental phosphorus (TTP) may interfere with coagulation in water treatment plants. When such concentrations exceed 0.025 g/m^3 at the time of the spring turnover on a volume-weighted basis in lakes or reservoirs, they may occasionally stimulate excessive or nuisance growths of algae and other aquatic plants. Algal growths impart undesirable tastes and odors to water, interfere with water treatment, become aesthetically unpleasant, and alter the chemistry of the water supply. They contribute to the phenomenon of cultural eutrophication.

To prevent the development of biological nuisances and to control accelerated or cultural eutrophication, TPP should not exceed 0.05 g/m^3 in any stream at the point where it enters any lake or reservoir or 0.025 g/m^3 within the lake or reservoir. A desired goal for the prevention of plant nuisances in streams or other flowing waters not discharging directly to lakes or impoundments is 0.1 g/m^3 TPP. Most relatively uncontaminated lake districts are known to have surface waters that contain from 0.01 to 0.3 g/m^3 TPP.

Harmful Effects and Symptoms: The harmful effects are environmental, rather than health effects.

Disposal Method Suggested: *Sodium Orthophosphates* — The material is diluted to the recommended provisional limit in water (0.05 mg/ℓ as H_3PO_4. The pH is adjusted to between 6.5 and 9.1 and then the material can be discharged into sewers or natural streams.

References

(1)   U.S. Environmental Protection Agency, *Status Assessment of Toxic Chemicals: Phosphates,* Report EPA-600/2-79-210j, Cincinnati, Ohio (December 1979).

PHOSPHINE

● Hazardous waste (EPA)

Description: PH_3, phosphine, is a colorless gas with an odor of decaying fish. Phosphine presents an additional hazard in that it ignites at very low temperatures.

Code Numbers: CAS 7803-51-2   RTECS SY7525000   UN 2199

DOT Designation: Poison A, flammable gas and poison gas.

Synonyms: Hydrogen phosphide, phosphoretted hydrogen, phosphorus trihydride.

Potential Exposures: Phosphine is only occasionally used in industry, and exposure usually results accidentally as a by-product of various processes. Exposures may occur when acid or water comes in contact with metallic phosphides (aluminum phosphide, calcium phosphide). These two phosphides are used as insecticides or rodenticides for grain, and phosphine is generated during grain fumigation. Phosphine may also evolve during the generation of acetylene from impure calcium carbide, as well as during metal shaving, sulfuric acid tank cleaning, rustproofing, and ferrosilicon, phosphoric acid and yellow phosphorus explosive handling.

Incompatibilities: Air, oxidizers, chlorine, etc., acids, moisture, halogenated hydrocarbons.

Permissible Exposure Limits in Air: 0.3 ppm (0.4 mg/m^3) is the Federal standard and the 1983/84 ACGIH value for occupational exposure to phosphine determined as TWA. The STEL value is 1 ppm (1 mg/m^3). The IDLH level is 200 ppm.

Determination in Air: Collection on silver nitrate impregnated cellulose pad to give free silver which is converted to the nitrate, then to the sulfide which is analyzed colorimetrically (A-27).

Permissible Concentration in Water: No criteria set, but EPA (A-37) has suggested a permissible ambient goal of 5.5 $\mu g/\ell$ based on health effects.

Route of Entry: Inhalation of vapor.

Harmful Effects and Symptoms: *Local* — Phosphine's strong odor may be nauseating. However, irritation to the eyes or skin is undocumented, and some authors indicate that lacrimation, if it occurs, results as a systemic effect rather than from local irritation.

Systemic — Acute effects are secondary to central nervous system depression, irritation of lungs, and damage to the liver and other organs. Most common effects include weakness, fatigue, headache, vertigo, anorexia, nausea, vomiting, abdominal pain, diarrhea, tenesmus, thirst, dryness of the throat, difficulty in swallowing, and sensation of chest pressure. In severe cases staggering gait, convulsions, and coma follow. Death may occur from cardiac arrest and, more typically, pulmonary edema, which may be latent in a manner similar to nitrogen oxide intoxication.

Chronic poisoning has been suggested by some authors and symptoms have been attributed to chronic phosphorus poisoning. However, there is evidence that phosphine may be metabolized to form nontoxic phosphates, and chronic exposure of animals has failed to produce toxic effects. Compounded with the lack of human experience and of extensive commercial usage, evidence indicates that chronic poisoning per se does not occur.

Points of Attack: Respiratory system, lungs.

Medical Surveillance: No special considerations are necessary in placement or periodic examinations, other than evaluation of the respiratory system. If poisoning is suspected, workers should be observed for 48 hours due to the delayed onset of pulmonary edema.

First Aid: If a person breathes in large amounts of this chemical, move the exposed person to fresh air at once and perform artificial respiration.

Personal Protective Methods: In areas where vapors are excessive, workers should be supplied with fullface gas masks with proper canisters or supplied air respirators.

Respirator Selection:

3 ppm:	CCRS/SA/SCBA
15 ppm:	CCRSF/GMS/SAF/SCBAF
200 ppm:	SA:PD,PP,CF
Escape:	GMS/SCBA

References

(1)   See Reference (A-61).
(2)   See Reference (A-60).
(3)   Parmeggiani, L., Ed., *Encyclopedia of Occupational Health & Safety,* Third Edition, Vol. 2, pp 1681, Geneva, International Labour Office (1983).
(4)   United Nations Environment Programme, *IRPTC Legal File 1983,* Vol. II, pp VII/535–6, Geneva, Switzerland, International Register of Potentially Toxic Chemicals (1984).

PHOSPHORIC ACID

● Hazardous substance (EPA)

Description: H_3PO_4 is a viscous colorless liquid which solidifies below about 20°C.

Code Numbers: CAS 7664-38-2   RTECS TB6300000   UN 1805

DOT Designation: Corrosive material.

Synonyms: White phosphoric acid, o-phosphoric acid, 85% phosphoric acid, m-phosphoric acid.

Potential Exposures: Phosphoric acid is used in the manufacture of fertilizers, phosphate salts, polyphosphates, detergents, activated carbon, animal feed, ceramics, dental cement, pharmaceuticals, soft drinks, gelatin, rust inhibitors, wax, and rubber latex. Exposure may also occur during electropolishing, engraving, photoengraving, lithographing, metal cleaning, sugar refining, and water-treating.

Incompatibilities: Strong caustics, most metals.

Permissible Exposure Limits in Air: The Federal standard and the 1983/84 ACGIH TWA value is 1 mg/m^3. The STEL is 3 mg/m^3. There is no IDLH level set.

Determination in Air: Collection on a cellulose membrane filter, workup with water, colorimetric determination. See NIOSH Methods, Set M. See also reference (A-10).

Permissible Concentration in Water: No criteria set.

Routes of Entry: Inhalation of mist, ingestion, eye and skin contact.

Harmful Effects and Symptoms: Irritation of upper respiratory tract, eyes and skin; burns skin and eyes; dermatitis. At 1 ppm, the Federal standard, phosphoric acid mist is irritating to unacclimated workers but is easily tolerated by acclimated workers.

Points of Attack: Respiratory system, lungs, eyes, skin.

Medical Surveillance: Consider the points of attack in preplacement and periodic physical examinations.

First Aid: If this chemical gets into the eyes, irrigate immediately. If this chemical contacts the skin, flush with water immediately. If a person breathes in large amounts of this chemical, move the exposed person to fresh air at once and perform artificial respiration. When this chemical has been swallowed, get medical attention. Give large quantities of water and induce vomiting. Do not make an unconscious person vomit.

Personal Protective Methods: Wear appropriate clothing to prevent any possibility of skin contact with liquids of >1.6% content or repeated or prolonged contact with liquid of <1.6% content. Wear eye protection to prevent any possibility of eye contact. Employees should wash immediately when skin is wet or contaminated. Work clothing should be changed daily if it is possible that clothing is contaminated. Remove nonimpervious clothing immediately if wet or contaminated. Provide emergency showers and eyewash if liquids containing >1.6% contaminants are involved.

Respirator Selection:
     50 mg/m^3: HiEPF/SAF/SCBAF
     2,000 mg/m^3: SAF:PD,PP,CF

Disposal Method Suggested: Add slowly to solution of soda ash and slaked lime with stirring then flush to sewer with large volumes of water.

References
(1) See Reference (A-61).
(2) See Reference (A-60).
(3) Sax, N.I., Ed., *Dangerous Properties of Industrial Materials Report, 3,* No. 4, 84-87, New York, Van Nostrand Reinhold Co. (1983).

PHOSPHORUS

● Hazardous material (EPA)

Description: P, phosphorus (white or yellow), melts at 44°C. Red phosphorus is excluded in that it is a nontoxic allotrope, although it is frequently contaminated with a small amount of the yellow. White or yellow phosphorus is either a yellow or colorless, volatile, crystalline solid which darkens when exposed to light and ignites in air to form white fumes and greenish light.

Code Numbers: CAS 7723-14-0   RTECS TH3500000   UN 1381

DOT Designation: Flammable solid and poison.

Synonyms: White phosphorus, yellow phosphorus.

Potential Exposures: Yellow phosphorus is handled away from air so that exposure is usually limited. Phosphorus was at one time used for the production of matches or lucifers but has long since been replaced due to its chronic toxicity. Phosphorus is used in the manufacture of munitions, pyrotechnics, explosives, smoke bombs, and other incendiaries, in artificial fertilizers, rodenticides, phosphor bronze alloy, semiconductors, electroluminescent coating, and chemicals, such as phosphoric acid and metallic phosphides.

Incompatibilities: Air, all oxidizing agents, including elemental sulfur, strong caustics.

Permissible Exposure Limits in Air: The Federal standard and the 1983/84 ACGIH TWA value is 0.1 mg/m^3. The STEL is 0.3 mg/m^3. No IDLH level has been set.

Determination in Air: Sample collection on Tenax-GC resin, extraction with xylene, analysis by gas chromatography with a phosphorus-specific flame photometric detector (A-10).

Permissible Concentration in Water: The EPA (A-3) has proposed a criterion of 0.10 µg/ℓ yellow (elemental) phosphorus for marine or estuarine waters. Further EPA (A-37) has suggested a permissible ambient goal of 1.4 µg/ℓ based on health effects.

Routes of Entry: Inhalation of vapor or fumes or mist, ingestion, skin and eye contact.

Harmful Effects and Symptoms: *Local* — Phosphorus, upon contact with skin, may result in severe burns, which are necrotic, yellowish, fluorescent under ultraviolet light, and have a garliclike odor.

Systemic — Acute phosphorus poisoning usually occurs as a result of accidental or suicidal ingestion. However, animal experiments indicate that acute systemic poisoning may follow skin burns. In acute cases, shock may ensue rapidly and the victim may succumb immediately. If acute attack is survived, an asymptomatic latency period of a few hours to a few days may follow. Death often occurs upon relapse from liver, kidney, cardiac, or vascular dysfunction or failure. Abnormal electrocardiograms, particularly of the QT, ST, or T wave phases, abnormal urinary and serum calcium and phosphate levels, proteinuria and aminoaciduria, and elevated serum SGPT are indicative signs. Vomitus, urine, and stools may be fluorescent in ultraviolet light, and a garlic odor of breath and eructations may be noted.

Chronic phosphorus poisoning is a result of continued absorption of small amounts of yellow phosphorus for periods typically of 10 years; however, exposures of as short as 10 months may cause phosphorus necrosis of the jaw, "phossy jaw". Chronic intoxication is characterized by periostitis with suppuration, ulceration, necrosis, and severe deformity of the mandible and, less often, maxilla. Sequestration of bone may occur. Polymorphic leukopenia, susceptibility to bone fracture, and failure of the alveolar bone to resorb following extractions are secondary clinical signs. Carious teeth and poor dental hygiene increase susceptibility.

Points of Attack: Respiratory system, liver, kidneys, jaw, teeth, blood, eyes, skin.

Medical Surveillance: Special consideration should be given to the skin, eyes, jaws, teeth, respiratory tract, and liver. Preplacement medical and dental examination with x-ray of teeth is highly recommended in the case of yellow phosphorus exposure. Poor dental hygiene may increase the risk in yellow phosphorus exposures, and any required dental work should be completed before workers are assigned to areas of possible exposure.

Workers experiencing any jaw injury, tooth extraction, or any abnormal dental conditions should be removed from areas of exposure and observed. Roentgenographic examinations may show necrosis; however, in order to prevent full development of sequestra, the disease should be diagnosed in earlier stages. Liver function should be evaluated periodically.

First Aid: If this chemical gets into the eyes, irrigate immediately. If this chemical contacts the skin, flush with water immediately. If a person breathes in large amounts of this chemical, move the exposed person to fresh air at once and perform artificial respiration. When this chemical has been swallowed, get medical attention. Give large quantities of water and induce vomiting. Do not make an unconscious person vomit.

Personal Protective Methods: Wear appropriate clothing to prevent any possibility of skin contact. Wear eye protection to prevent any possibility of eye contact. Employees should wash immediately when skin is wet. Remove clothing immediately if wet or contaminated. Provide emergency showers and eyewash.

Respirator Selection:
    5 mg/m^3: HiEPF/SAF/SCBAF
    200 mg/m^3: SAF:PD,PP,CF

Disposal Method Suggested: Controlled incineration followed by alkaline scrubbing and particulate removal equipment.

References

(1) See Reference (A-61).
(2) Sax, N.I., Ed., *Dangerous Properties of Industrial Materials Report, 3,* No. 4, 90-93, New York, Van Nostrand Reinhold Co. (1983).
(3) United Nations Environment Programme, *IRPTC Legal File 1983,* Vol. II, pp VII/624–26, Geneva, Switzerland, International Register of Potentially Toxic Chemicals (1984).

PHOSPHORUS OXYCHLORIDE

● Hazardous substance (EPA)

Description: POCl$_3$ is a colorless fuming liquid with a pungent odor which boils at 107°C.

Code Numbers: CAS 10025-87-3   RTECS TH4897000   UN 1810

DOT Designation: Corrosive material.

Synonym: Phosphoryl chloride.

Potential Exposures: POCl$_3$ is used in the manufacture of many pesticides (A-32), a number of pharmaceuticals (A-41) as well as plasticizers, gasoline additives and hydraulic fluid.

Permissible Exposure Limits in Air: There are no Federal standards but ACGIH as of 1983/84 has proposed a TWA value of 0.1 ppm (0.6 mg/m^3) and a STEL value of 0.5 ppm (3.0 mg/m^3).

Permissible Concentration in Water: No criteria set. (POCl$_3$ decomposes in water.)

Harmful Effects and Symptoms: Irritation of eyes and skin, skin burns, dizziness, headache, weakness, anorexia, nausea, vomiting, chest pains, cough, dyspnea, bronchitis, bronchopneumonia, pulmonary edema (A-38).

Medical Surveillance: Annual physical examinations including complete blood count, urinalysis and chest x-ray (A-38).

Personal Protective Methods: Wear long rubber gloves, goggles and protective clothing. Preclude from exposure those with diseases of skin, liver, kidneys, lungs and teeth (A-38).

Respirator Selection: Wear self-contained breathing apparatus.

Disposal Method Suggested: Pour onto sodium bicarbonate. Spray with aqueous ammonia and add crushed ice. Neutralize and pour into drain with running water (A-38).

References

(1)   Sax, N.I., Ed., *Dangerous Properties of Industrial Materials Report, 3,* No. 4, 87-88, New York, Van Nostrand Reinhold Co. (1983).

PHOSPHORUS PENTACHLORIDE

Description: PCl$_5$ is a pale yellow solid with an odor like hydrochloric acid. It sublimes at 167°C.

Code Numbers: CAS 10026-13-8   RTECS TB6125000   UN 1806

DOT Designation: Corrosive material.

Synonyms: Phosphoric chloride, phosphorus perchloride.

Potential Exposures: PCl$_5$ is used in the manufacture of agricultural chemicals, chlorinated compounds, gasoline additives, plasticizers and surfactants and in pharmaceutical manufacture (A-41).

Incompatibilities: Water, magnesium oxide, chemically active metals, sodium, potassium, alkalies.

Permissible Exposure Limits in Air: The Federal standard is 1 mg/m^3. The ACGIH (1983/84) has adopted a TWA of 0.1 ppm (1 mg/m^3) but has no STEL value. The IDLH level is 200 mg/m^3.

Determination in Air: Collection by impinger or fritted bubbler, colorimetric analysis (A-14) or analysis using ion specific electrode (A-15).

Permissible Concentration in Water: No criteria set.

Routes of Entry: Inhalation, ingestion, skin and eye contact.

Harmful Effects and Symptoms: Irritation of eyes and respiratory system, bronchitis, dermatitis.

Points of Attack: Respiratory system, lungs, eyes, skin.

Medical Surveillance: Consider the points of attack in preplacement and periodic physical examinations.

First Aid: If this chemical gets into the eyes, irrigate immediately. If this chemical contacts the skin, flush with water immediately. If a person breathes in large amounts of this chemical, move the exposed person to fresh air at once and perform artificial respiration. When this chemical has been swallowed, get medical attention. Give large quantities of water and do not induce vomiting.

Personal Protective Methods: Wear appropriate clothing to prevent any possibility of skin contact. Wear eye protection to prevent any possibility of eye contact. Employees should wash immediately when skin is wet or contaminated. Work clothing should be changed daily if it is possible that clothing is contaminated. Remove nonimpervious clothing immediately if wet or contaminated. Provide emergency showers and eyewash.

Respirator Selection:
10 mg/m^3: SA/SCBA
50 mg/m^3: SAF/SCBAF
200 mg/m^3: SAF:PD,PP,CF
Escape: GMOVP/SCBA

Disposal Method Suggested: Decompose with water forming phosphoric and hydrochloric acids. Neutralize acids and dilute if necessary for discharge into the sewer system (A-31).

References
(1) See Reference (A-61).
(2) See Reference (A-60).

PHOSPHORUS PENTASULFIDE

- Hazardous substance (EPA)
- Hazardous waste (EPA)

Description: P_2S_5, or P_4S_{10}, this is a greenish-yellow solid with an odor of rotten eggs. It melts at $275°C$.

Code Numbers: CAS 1314-80-3   RTECS TH4375000   UN 1340

DOT Designation: Flammable solid and dangerous when wet.

Synonyms: Phosphorus persulfide, regular phosphorus pentasulfide, reactive phosphorus pentasulfide, distilled phosphorus pentasulfide, undistilled phosphorus pentasulfide.

Potential Exposures: Phosphorus pentasulfide is used in the manufacture of flotation agents, insecticides (A-32) lubricating oil, additives, ignition compounds, and matches. It is also used to introduce sulfur into rubber, and organic chemicals, such as pharmaceuticals (A-41).

Incompatibilities: Water, alcohols, strong oxidizers.

Permissible Exposure Limits in Air: The Federal standard and the 1983/84 ACGIH TWA value is 1 mg/m^3. The STEL is 3 mg/m^3. The IDLH level is 750 mg/m^3.

Permissible Concentration in Water: No criteria set.

Routes of Entry: Inhalation of dust, ingestion, skin and eye contact.

Harmful Effects and Symptoms: Inhalation of fumes produced by phosphorus compounds may cause irritation of pulmonary tissues with resultant acute pulmonary edema. Chronic exposure may lead to cough, bronchitis, and pneumonia. The hazards of phosphorus pentasulfide are the same as for hydrogen sulfide to which it rapidly hydrolyzes in the presence of moisture.

Symptoms include apnea, coma, convulsions, eye irritation, lacrimation, photophobia, keratoconjunctivitis, corneal vesiculation, respiratory system irritation, dizziness, headaches, fatigue, irritability, insomnia, gastrointestinal disturbances.

Points of Attack: Respiratory system, lungs, central nervous system, eyes, skin.

Medical Surveillance: Consider the points of attack in preplacement and periodic physical examinations.

First Aid: If this chemical gets into the eyes, irrigate immediately. If this chemical contacts the skin, dust off solid, flush with water. If a person breathes in large amounts of this chemical, move the exposed person to fresh air at once and perform artificial respiration. When this chemical has been swallowed, get medical attention. Give large quantities of water and induce vomiting. Do not make an unconscious person vomit.

Personal Protective Methods: Wear appropriate clothing to prevent repeated or prolonged skin contact. Wear eye protection to prevent any reasonable probability of eye contact. Employees should wash promptly when skin is wet or contaminated. Work clothing should be changed daily if it is possible that clothing is contaminated. Remove nonimpervious clothing promptly if wet or contaminated.

Respirator Selection:
    10 mg/m^3: SA/SCBA
    50 mg/m^3: SAF/SCBAF
    750 mg/m^3: SA:PD,PP,CF
    Escape: GMSP/SCBA

Disposal Method Suggested: Decompose with water forming phosphoric acid, sulfuric acid and hydrogen sulfide. Provisions must be made for scrubbing hydrogen sulfide emissions. The acids may then be neutralized and diluted if necessary, and discharged into the sewer system (A-31).

References

(1)   See Reference (A-61).
(2)   Sax, N.I., Ed., *Dangerous Properties of Industrial Materials Report, 3,* No. 4, 89-90, New York, Van Nostrand Reinhold Co. (1983).

PHOSPHORUS TRICHLORIDE

● Hazardous substance (EPA)

Description: PCl_3 is a colorless to yellow, fuming liquid with an odor like hydrochloric acid. It boils at $75°$ to $76°C$.

Code Numbers: CAS 7719-12-2   RTECS TH3675000   UN 1809

DOT Designation: Corrosive material and poison.

Synonym: Phosphorus chloride.

Potential Exposures: Phosphorus trichloride is used in the manufacture of agricultural chemicals (A-32), pharmaceuticals (A-41), chlorinated compounds, dyes, gasoline additives, acetylcellulose, phosphorus oxychloride, plasticizers, saccharin, and surfactants.

Incompatibilities: Water, alcohol, when in contact with combustible organics; chemically active metals: sodium, potassium, aluminum, strong nitric acid.

Permissible Exposure Limits in Air: The Federal standard is 0.5 ppm (3 mg/m^3). The ACGIH as of 1983/84 has adopted a TWA of 0.2 ppm (1.5 mg/m^3 and an STEL value of 0.5 ppm (3 mg/m^3). The IDLH level is 50 ppm.

Determination in Air: Collection by impinger or fritted bubbler, colorimetric analysis (A-14) or analysis using ion specific electrode (A-15).

Permissible Concentration in Water: No criteria set.

Routes of Entry: Inhalation, ingestion, skin and eye contact.

Harmful Effects and Symptoms: Irritation of eyes, nose and throat; pulmonary edema; burns of eyes and skin.

Points of Attack: Respiratory system, lungs, eyes, skin.

Medical Surveillance: Consider the points of attack in preplacement and periodic physical examinations.

First Aid: If this chemical gets into the eyes, irrigate immediately. If this chemical contacts the skin, flush with water immediately. If a person breathes in large amounts of this chemical, move the exposed person to fresh air at once and perform artificial respiration. When this chemical has been swallowed, get medical attention. Give large quantities of water and do not induce vomiting.

Personal Protective Methods: Wear appropriate clothing to prevent any possibility of skin contact. Wear eye protection to prevent any possibility of eye contact. Employees should wash immediately when skin is wet or contaminated. Remove nonimpervious clothing immediately if wet or contaminated. Provide emergency showers and eyewash.

Respirator Selection:
          25 ppm: CCRSF/GMS/SAF/SCBAF
          50 ppm: SAF:PD,PP,CF
         Escape: GMS/SCBA
           Note: Do not use oxidizable sorbents.

Disposal Method Suggested: Decompose with water forming phosphoric and hydrochloric acids. Neutralize acids and dilute if necessary for discharge into a sewer system (A-31).

References
(1)   See Reference (A-61).
(2)   See Reference (A-60).
(3)   Sax, N.I., Ed., *Dangerous Properties of Industrial Materials Report, 3*, No. 4, 93-94, New York, Van Nostrand Reinhold Co. (1983).

PHTHALATE ESTERS

See individual entries under: "Dibutyl Phthalate," "Di(2-Ethylhexyl) Phthalate," "Dimethyl Phthalate," "Di-n-Octyl Phthalate."

References

(1) U.S. Environmental Protection Agency, *Phthalate Esters,* Health and Environmental Effects Profile No. 146, Office of Solid Waste, Washington, DC (April 30, 1980).
(2) Parmeggiani, L., Ed., *Encyclopedia of Occupational Health & Safety,* Third Edition, Vol. 2, pp 1690-93, Geneva, International Labour Office (1983).
(3) Sax, N.I., Ed., *Dangerous Properties of Industrial Materials Report, 4,* No. 2, 73-76, New York, Van Nostrand Reinhold Co. (1984) (Ethyl Phthalate).
(4) Sax, N.I., Ed., *Dangerous Properties of Industrial Materials Report, 4,* No. 3, 74-76, New York, Van Nostrand Reinhold Co. (1984) (Ethyl Phthalate).
(5) United Nations Environment Programme, *IRPTC Legal File 1983,* Vol. II, pp VII/632-37, Geneva, Switzerland, International Register of Potentially Toxic Chemicals (1984).

PHTHALIC ANHYDRIDE

● Hazardous waste (EPA)

Description: $C_8H_4O_3$, phthalic anhydride, is moderately flammable, white, lustrous, solid, with needlelike crystals. It melts at $131°C$.

Code Numbers: CAS 85-44-9   RTECS TI3150000   UN 2214

DOT Designation: None.

Synonyms: Phthalic acid anhydride, benzene-o-dicarboxylic acid anhydride, phthalandione.

Potential Exposures: Phthalic anhydride is used in the manufacture of phthaleins, benzoic acid, alkyd and polyester resins, synthetic indigo, and phthalic acid, which is used as a plasticizer for vinyl resins. To a lesser extent, it is used in the production of alizarin dye, anthranilic acid, anthraquinone, diethyl phthalate, dimethyl phthalate, erythrosin, isophthalic acid, methylaniline, phenolphthalein, phthalamide, sulfathalidine, and terephthalic acid. It has also found use as a pesticide intermediate (A-32).

Incompatibilities: Strong oxidizers.

Permissible Exposure Limits in Air: The Federal standard is 2 ppm (12 mg/m³). ACGIH as of 1983/84 has adopted 1 ppm (6 mg/m³) as a TWA and 4 ppm (24 mg/m³) as an STEL value. The IDLH level is 10,000 ppm.

Determination in Air: Collection on a filter, workup with ammonium hydroxide, analysis by high-pressure liquid chromatography. See NIOSH Methods, Set M. See also reference (A-10).

Permissible Concentration in Water: No criteria set.

Routes of Entry: Inhalation of dust, fume, or vapor, ingestion, skin and eye contact.

Harmful Effects and Symptoms: *Local* — Phthalic anhydride, in the form of a dust, fume, or vapor, is a potent irritant of the eyes, skin, and respiratory tract. The irritant effects are worse on moist surfaces. Conjunctivitis and skin erythema, burning, and contact dermatitis may occur. If the chemical is held in contact with the skin, as under clothes or shoes, skin burns may develop. Hypersensitivity may develop in some individuals. Inhalation of dust or vapors may cause coughing, sneezing, and a bloody nasal discharge. Impurities, naphthoquinone, as well as maleic anhydride, may also cause eye, skin, and pulmonary irritation.

Systemic — Repeated exposure may result in bronchitis, emphysema, allergic asthma, urticaria, and chronic eye irritation.

Points of Attack: Respiratory system, eyes, skin, liver, kidneys.

Medical Surveillance: Emphasis should be given to a history of skin or pulmonary allergy, and preplacement and periodic examinations should evaluate the skin, eye, and lungs, as well as liver and kidney functions. The hydrolysis product, phthalic acid, is rapidly excreted in the urine, although this has not been used in biological monitoring. Diagnostic patch testing may be helpful in evaluating skin allergy.

First Aid: If this chemical gets into the eyes, irrigate immediately. If this chemical contacts the skin, wash with soap promptly. If a person breathes in large amounts of this chemical, move the exposed person to fresh air at once and perform artificial respiration. When this chemical has been swallowed, get medical attention. Give large quantities of water and induce vomiting. Do not make an unconscious person vomit.

Personal Protective Methods: Wear appropriate clothing to prevent repeated or prolonged skin contact. Wear eye protection to prevent any reasonable probability of eye contact. Employees should wash promptly when skin is wet or contaminated. Work clothing should be changed daily if it is possible that it is contaminated. Remove nonimpervious clothing promptly if wet or contaminated.

Respirator Selection:
    100 ppm: HiEPF/SAF/SCBAF
    1,670 ppm: SAF:PD,PP,CF

Disposal Method Suggested: Incineration.

References

(1) U.S. Environmental Protection Agency, *Phthalic Anhydride,* Health and Environmental Effects Profile No. 147, Office of Solid Waste, Washington, DC (April 30, 1980).
(2) See Reference (A-61).
(3) See Reference (A-60).
(4) Parmeggiani, L., Ed., *Encyclopedia of Occupational Health & Safety,* Third Edition, Vol. 2, pp 1693–94, Geneva, International Labour Office (1983) (Phthalic Anhydride and some derivatives).

m-PHTHALODINITRILE

Description: $C_6H_4(CN)_2$ is a white or buff-colored odorless powder melting at 138°C.

Code Numbers: CAS 91-15-6   RTECS TI857000

DOT Designation: —

Synonyms: Phthalonitrile, o-dicyanobenzene.

Potential Exposures: This material is an intermediate for phthalocyanine pigments and dyes and for high-temperature lubricants and coatings; it may be used to produce phthalate esters.

Permissible Exposure Limits in Air: There are no Federal standards but ACGIH (1983/84) has adopted a TWA of 5 mg/m³. There is no STEL value.

Permissible Concentration in Water: No criteria set.

Harmful Effects and Symptoms: ACGIH (A-34) reports no adverse effects in a 15-year review of industrial experience and notes that the probable reason is that aromatic nitriles, unlike aliphatic nitriles, do not liberate cyanide in the body. However, a Japanese source (A-38) cites the following results of chronic exposure: headaches, amnesia, idleness, tremors of fingers, anorexia, facial pallor.

First Aid: Wash contaminated areas of body with soap and water.

Personal Protective Methods: Wear rubber gloves, an apron and overalls (A-38).

Respirator Selection: Wear a self-contained breathing apparatus (A-38).

Disposal Method Suggested: React with alcoholic NaOH; after one hour, evaporate alcohol and add calcium hypochlorite; after 24 hours flush into sewer with large volumes of water (A-38).

PICLORAM

● Carcinogen (Positive, NCI) (1)

Description: 4-Amino-3,5,6-trichloropyridine-2-carboxylic acid, with the structural formula

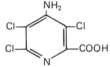

Picloram is a colorless powder with a chlorinelike odor which decomposes at 215°C without melting.

Code Numbers: CAS 1918-02-1   RTECS TJ7525000   UN 2588

DOT Designation: —

Synonyms: 4-Amino-3,5,6-trichloro-2-picolinic acid, Tordon®.

Potential Exposures: Those involved in the manufacture (A-32), formulation or application of the herbicide.

Permissible Exposure Limits in Air: There are no Federal standards but ACGIH (1983/84) has adopted a TWA value of 10 mg/m³ and an STEL value of 20 mg/m³.

Permissible Concentration in Water: No criteria set.

Harmful Effects and Symptoms: The acute oral toxicity of picloram is low but studies have shown that the material is a carcinogen (1).

Disposal Method Suggested: This chlorinated brush killer is usually formulated with 2,4-D and the disposal problems are similar. Incineration at 1000°C for 2 seconds is required for thermal decomposition. Alternatively, the free acid can be precipitated from its solutions by addition of a mineral acid. The concentrated acid can then be incinerated and the dilute residual solution disposed in an area where several years' persistence in the soil can be tolerated (A-32).

References

(1) National Cancer Institute, *Bioassay of Picloram for Possible Carcinogenicity,* Technical Report Series No. 23, Bethesda, MD (1978).

(2) United Nations Environment Programme, *IRPTC Legal File 1983,* Vol. II, pp VII/638, Geneva, Switzerland, International Register of Potentially Toxic Chemicals (1984).

2-PICOLINE

● Hazardous waste (EPA)

Description: $\overline{NC(CH_3)CHCHCHCH}$ is a colorless liquid with a strong unpleasant pyridinelike odor which boils at 129°C.

Code Numbers: CAS 109-06-8   RTECS TJ4900000   UN 2313

DOT Designation: Flammable liquid.

Synonym: 2-Methylpyridine.

Potential Exposures: This material is used as an intermediate in pharmaceutical manufacture (A-41), pesticide manufacture (A-32), and in the manufacture of dyes and rubber chemicals. It is also used as a solvent.

Permissible Exposure Limits in Air: There are no U.S. standards but the 8-hour, time-weighted average occupational exposure limit for α-picoline has been set in Russia at 5 mg/m^3 (A-36).

Permissible Concentration in Water: There are no U.S. criteria but the maximum allowable concentration in Class I waters for the production of drinking waters has been set in the Netherlands at 0.05 mg/ℓ (A-36). The EPA has suggested (A-37) a permissible ambient goal of 316 μg/ℓ based on health effects.

Harmful Effects and Symptoms: Pertinent data could not be found that defined 2-picoline as a carcinogen or a mutagen. Studies on rats indicated that the structure and composition of the liver and the structure and growth pattern of skin were disrupted in the offspring of tested rats who were given 157 mg/kg body weight daily during their pregnancy. 2-picoline has been shown to produce biochemical and physical changes in the liver, spleen, bone marrow, and lymph nodes.

References

(1) U.S. Environmental Protection Agency, *2-Picoline,* Health and Environmental Effects Profile No. 148, Office of Solid Waste, Washington, DC (April 30, 1980).

PICRIC ACID

● Priority toxic pollutant (EPA)

Description: $C_6H_2(NO_2)_3OH$, picric acid, is a pale yellow, odorless, intensely bitter crystal which is explosive upon rapid heating or mechanical shock. It melts at 122°C.

Code Numbers: CAS 88-89-1   RTECS TJ7875000   UN 0154

DOT Designation: Class A explosive.

Synonyms: Picronitric acid, trinitrophenol, nitroxanthic acid, carbazotic acid, phenol trinitrate.

Potential Exposures: Picric acid is used in the manufacture of explosives, rocket fuels, fireworks, colored glass, matches, electric batteries, and disinfectants. It is also used in the pharmaceutical and leather industries, and in dyes, copper and steel etching, forensic chemistry, histology, textile printing, and photographic emulsions.

Incompatibilities: Copper, lead, zinc, other metals, salts, plaster, concrete.

Permissible Exposure Limits in Air: The Federal standard and the ACGIH 1983/84 TWA value for picric acid is 0.1 mg/m^3. The STEL value is 0.3 mg/m^3. The notation "skin" is added to indicate the possibility of cutaneous absorption. The IDLH level is 100 mg/m^3.

Determination in Air: Collection on a mixed cellulose ester membrane filter, extraction with aqueous methanol, measurement by high performance liquid chromatography with UV detector (A-10).

Permissible Concentration in Water: To protect human health—no criteria set due to insufficient data (1).

Determination in Water: Methylene chloride extraction followed by gas chromatography with flame ionization or electron capture detection (EPA Method 604) or gas chromatography plus mass spectrometry (EPA Method 625).

Routes of Entry: Inhalation and ingestion of dust, percutaneous absorption, eye and skin contact.

Harmful Effects and Symptoms: *Local* — Picric acid dust or solutions are potent skin sensitizers. In solid form, picric acid is a skin irritant, but in aqueous solution it irritates only hypersensitive skin. The cutaneous lesions which appear usually on exposed areas of the upper extremities consist of dermatitis with erythema, papular, and vesicular eruptions. Desquamation may occur following repeated or prolonged contact. Skin usually turns yellow upon contact, and areas around nose and mouth as well as the hair are most often affected. Dust or fume may cause eye irritation which may be aggravated by sensitization. Corneal injury may occur from exposure to picric acid dust and solutions.

Systemic — Inhalation of high concentrations of dust by one worker caused temporary coma followed by weakness, myalgia, anuria, and later polyuria. Following ingestion of picric acid, there may be headache, vertigo, nausea, vomiting, diarrhea, yellow coloration of the skin, hematuria, and albuminuria. High doses may cause destruction of erythrocytes, hemorrhagic nephritis, and hepatitis. High doses which cause systemic intoxication will color all tissues yellow, including the conjunctiva and aqueous humor, and cause yellow vision.

Points of Attack: Kidneys, liver, blood, skin, eyes.

Medical Surveillance: Preplacement and periodic medical examinations should focus on skin disorders such as hypersensitivity, atopic dermatitis, and liver and kidney function.

First Aid: If this chemical gets into the eyes, irrigate immediately. If this chemical contacts the skin, wash with soap promptly. If a person breathes in

large amounts of this chemical, move the exposed person to fresh air at once and perform artificial respiration. When this chemical has been swallowed, get medical attention. Give large quantities of water and induce vomiting. Do not make an unconscious person vomit.

Personal Protective Methods: Skin protection by clothing and barrier creams can avoid the irritant and sensitizing action of picric acid. Wear appropriate clothing to prevent any reasonable probability of skin contact. Wear eye protection to prevent any reasonable probability of eye contact. Employees should wash promptly when skin is wet or contaminated and daily at the end of each work shift. Work clothing should be changed daily if it is possible that clothing is contaminated. Remove nonimpervious clothing promptly if wet or contaminated.

Respirator Selection:
0.5 mg/m^3: DMXS
1 mg/m^3: DMXSQ/FuHiEP/SA/SCBA
5 mg/m^3: HiEPF/SAF/SCBAF
100 mg/m^3: PAPHiEF/SAF:PD,PP,CF
Escape: DMXS/SCBA

Disposal Method Suggested: Controlled incineration in a rotary kiln incinerator equipped with particulate abatement and wet scrubber devices (A-31).

References

(1) U.S. Environmental Protection Agency, *Nitrophenols: Ambient Water Quality Criteria*, Washington, DC (1980).
(2) See Reference (A-61).
(3) See Reference (A-60).
(4) Parmeggiani, L., Ed., *Encyclopedia of Occupational Health & Safety* Third Edition, Vol. 2, pp 1704–5, Geneva, International Labour Office (1983).
(5) United Nations Environment Programme, *IRPTC Legal File 1983,* Vol. II, pp VII/639–40. Geneva, Switzerland, International Register of Potentially Toxic Chemicals (1984).

PINDONE

Description: $C_6H_4(CO)_2CHCOC(CH_3)_3$ is a bright yellow crystalline solid melting at $108°$ to $110°C$.

Code Numbers: CAS 83-26-1   RTECS NK6300000   UN 2588

DOT Designation: –

Synonyms: 2-Pivaloyl-1,3-indandione, Pivalyl Valone®, pivaldione (in France), pival (in Turkey), Pival®, Pivalyn®.

Potential Exposure: Those involved in manufacture (A-32), formulation and application of this insecticide and rodenticide.

Incompatibilities: None hazardous.

Permissible Exposure Limits in Air: The Federal limit and the ACGIH 1983/84 TWA value is 0.1 mg/m^3. The STEL is 0.3 mg/m^3. The IDLH level is 200 mg/m^3.

Determination in Air: Collection by impinger or fritted bubbler, analysis by gas liquid chromatography (A-1).

Permissible Concentration in Water: No criteria set.

Routes of Entry: Inhalation, ingestion.

Harmful Effects and Symptoms: Nosebleeds, excessive bleeding of minor cuts and bruises, smoky urine, black tarry stools, abdominal and back pain.

Points of Attack: Blood prothrombin.

Medical Surveillance: Consider the points of attack in preplacement and periodic physical examinations.

First Aid: If this chemical gets into the eyes, irrigate immediately. When this chemical has been swallowed, get medical attention. Give large quantities of saltwater and induce vomiting. Do not make an unconscious person vomit.

Personal Protective Methods: Work clothing should be changed daily if it is possible that clothing is contaminated.

Respirator Selection:
0.5 mg/m^3:   DMXS
1 mg/m^3:   DMXSQ/FuHiEP/SA/SCBA
5 mg/m^3:   SAF/SCBAF/HiEPF
100 mg/m^3:   PAPHiE/SA:PD,PP,CF
200 mg/m^3:   SAF:PD,PP,CF
Escape: DMXS/SCBA

References

(1)   See Reference (A-61).
(2)   United Nations Environment Programme, *IRPTC Legal File 1983,* Vol. II, pp VII/401, Geneva, Switzerland, International Register of Potentially Toxic Chemicals (1984).

PIPERAZINE DIHYDROCHLORIDE

Description: $\overline{\text{NHCH}_2\text{CH}_2\text{NHCH}_2\text{CH}_2}$ ·2HCl is a white crystalline solid.

Code Numbers: CAS 142-64-3   RTECS TL4025000   UN 2579

DOT Designation: —

Synonyms: Diethylenediamine dihydrochloride, Dowzene DHC®.

Potential Exposures: This material is used as an intermediate in the manufacture of pharmaceuticals (A-41), pesticides, rubber chemicals and fibers.

Permissible Exposure Limits in Air: There are no Federal standards but ACGIH (1983/84) has adopted a TWA value of 5 mg/m^3. There is no proposed STEL value.

Permissible Concentration in Water: No criteria set.

Harmful Effects and Symptoms: Piperazine and piperazine hydrate will cause irritation and burns of skin, eyes and respiratory system; will burn mouth if ingested (A-44).

PLATINUM AND COMPOUNDS

Description: Pt, platinum, is a soft, ductile, malleable, silver-white metal, insoluble in water and organic solvents. It is found in the metallic form and as the arsenide, sperrylite. It forms complex soluble salts such as Na_2PtCl_6.

Code Numbers:

Platinum metal	CAS 7440-06-4	RTECS TP2160000	
Chloroplatinic acid	CAS 16941-12-1	RTECS FW7040000	UN 2507

DOT Designation: Platinum metal, —; chloroplatinic acid, ORM-B.

Synonyms: None.

Potential Exposures: Platinum and its alloys are utilized because of their resistance to corrosion and oxidation, particularly at high temperatures, their high electrical conductivity, and their excellent catalytic properties. They are used in relays, contacts and tubes in electronic equipment, in spark plug electrodes for aircraft, and windings in high-temperature electrical furnaces. Platinum alloys are used for standards for weight, length, and temperature measurement. Platinum and platinum catalysts (e.g., hexachloroplatinic acid, H_2PtCl_6) are widely used in the chemical industry in persulfuric, nitric, and sulfuric acid production, in the synthesis of organic compounds and vitamins, and for producing higher octane gasoline. They are coming into use in catalyst systems for control of exhaust pollutants from automobiles. They are used in the equipment for handling molten glass and manufacturing fibrous glass; in laboratory, medical, and dental apparatus; in electroplating; in photography; in jewelry; and in x-ray fluorescent screens.

Incompatibilities: None hazardous.

Permissible Exposure Limits in Air: The Federal standard for soluble salts of platinum is 0.002 mg/m^3 expressed as Pt. There is no tentative STEL value. No IDLH level has been set. As of 1983/84, ACGIH has proposed a TWA of 1 mg/m^3 for platinum metal.

Determination in Air: For platinum soluble salts—collection on a filter, workup with acid, analysis by atomic absorption. See NIOSH Methods, Set M. See also reference (A-10).

Permissible Concentration in Water: No criteria set.

Routes of Entry: For soluble salts—ingestion, inhalation of dust or mist, skin or eye contact.

Harmful Effects and Symptoms: *Local* — Hazards arise from the dust, droplets, spray, or mist of complex salts of platinum, but not from the metal itself. These salts are sensitizers of the skin, nasal mucosa, and bronchi, and cause allergic phenomena. One case of contact dermatitis from wearing a ring made of platinum alloy is recorded.

Systemic — Characteristic symptoms of poisoning occur after 2 to 6 months' exposure and include pronounced irritation of the throat and nasal passages, which results in violent sneezing and coughing; bronchial irritation, which causes respiratory distress; and irritation of the skin, which produces cracking, bleeding, and pain. Respiratory symptoms can be so severe that exposed individuals may develop status asthmaticus. After recovery, most individuals develop allergic symptoms and experience further asthma attacks when exposed to even minimal amounts of platinum dust or mists. Mild cases of dermatitis involve only erythema and urticaria of the hands and forearms. More severe cases affect the face and neck. All pathology is limited to allergic manifestations.

EPA research efforts indicate that platinum is more active biologically and toxicologically than previously believed. It methylates in aqueous media, establishing a previously unrecognized biotransformation and distribution mechanism. Because platinum complexes are used as antitumor agents, the potential for carcinogenic activity is present; tests to clarify this aspect should be conducted. While low levels of emissions of platinum particulate have been observed from some catalyst-equipped automobiles, the major potential source of Pt is from the disposal of spent catalysts.

Points of Attack: Respiratory system, skin, eyes.

Medical Surveillance: In preemployment and periodic physical examinations, the skin, eyes, and respiratory tract are most important. Any history of skin or pulmonary allergy should be noted, as well as exposure to other irritants or allergens, and smoking history. Periodic assessment of pulmonary function may be useful.

First Aid: If this chemical gets into the eyes, irrigate immediately. If this chemical contacts the skin, flush with water immediately. If a person breathes in large amounts of this chemical, move the exposed person to fresh air at once and perform artificial respiration. When this chemical has been swallowed, get medical attention. Give large quantities of water and induce vomiting. Do not make an unconscious person vomit.

Personal Protective Methods: Wear appropriate clothing to prevent any reasonable probability of skin contact. Wear eye protection to prevent any reasonable probability of eye contact. Employees should wash promptly when skin is wet or contaminated. Work clothing should be changed daily if it is possible that clothing is contaminated. Remove nonimpervious clothing promptly if wet or contaminated.

Respirator Selection:
0.1 mg/m^3: HiEPF/SAF/SCBAF
2 mg/m^3: PAPHiEF
4 mg/m^3: SAF:PD,PP,CF

Disposal Method Suggested: Catalyst disposal is expected to be the largest contributor of Pt to the environment. The value of the metal would help to offset the cost of reclaiming the Pt from discarded catalysts. If direct vehicular emissions of Pt are found to be significant, particulate traps, which are available at reasonable cost, may provide a technological solution. In any event, recovery and recycling is the preferred technique for both health and economic reasons. Details of platinum recovery and recycling from plating wastes, platinum metal refinery effluents, spent catalysts and precious metals scrap have been published (A-57).

References

(1) U.S. Environmental Protection Agency, *A Literature Search and Analysis of Information Regarding Uses, Production, Consumption, Reported Medical Cases and Toxicology of Platinum and Palladium,* Report PB-238,546, Research Triangle Park, NC (April 1974).

(2) National Academy of Sciences, *Medical and Biologic Effects and Environmental Pollutants: Platinum Group Metals,* Washington, DC (1977).

(3) Sax, N.I., Ed., *Dangerous Properties of Industrial Materials Report, 1,* No. 3, 74-75, New York, Van Nostrand Reinhold Co. (1981).

(4) See Reference (A-61).

(5) Parmeggiani, L., Ed., *Encyclopedia of Occupational Health & Safety,* Third Edition, Vol. 2, pp 1587–88, Geneva, International Labour Office 91983) (Palladium, Alloys, & Compounds)

(6) Parmeggiani, L., Ed., *Encyclopedia of Occupational Health & Safety,* Third Edition, Vol. 2, pp 1723–24, Geneva, International Labour Office (1983) (Platinum, Alloys & Compounds).

POLYBROMINATED BIPHENYLS (PBBs)

• Carcinogen (Animal proven) (A-64).

Description: These materials are heavy, highly brominated compounds. Typical is hexabromobiphenyl, $Br_3C_6H_2-C_6H_2Br_3$ which is a solid which decomposes at 300° to 400°C. It will be used as an illustrative example of such compounds.

PBBs are produced by direct bromination of biphenyl and it could be anticipated that very complex mixtures of compounds differing from each both in number of bromine atoms per molecule and by positional isomerism are formed. The possibility also exists (analogous to the PCBs) that halogenated dibenzofurans (e.g., brominated dibenzofurans) may be trace contaminants in certain PBB formulations.

Code Numbers: CAS 36355-01-8   RTECS DV5330000

DOT Designation: —

Synonyms: PBBs, Firemaster BP-6® for hexabromobiphenyl.

Potential Exposures: The polybrominated biphenyls (PBBs) have been increasingly employed, primarily as fire retardants. For example, the PBBs are incorporated into thermoplastics at a concentration of about 15% to increase the heat stability of the plastic to which it is added. About 50% of the PBBs manufactured are used in typewriter, calculator, microfilm reader and business machine housings. One-third is used in radio and television parts, thermostats and electrical showers and hand tools and the remainder is used in a variety of other types of electrical equipment (1).

In 1973, one to two tons of PBBs, a highly toxic flame retardant, were accidentally mixed into an animal feed supplement and fed to cattle in Michigan. Contamination also resulted from traces of PBBs being discharged into the environment at the manufacturing site and at other facilities involved in handling PBBs. Approximately 250 dairy and 500 cattle farms were quarantined, tens of thousands of swine and cattle and more than one million chickens were destroyed, and lawsuits involving hundreds of millions of dollars were instituted. Before the nature of the contamination was recognized, many of the contaminated animals had been slaughtered, marketed, and eaten, and eggs and milk of the contaminated animals also consumed. Thus, large numbers of people have been exposed to PBBs. While commercial manufacture and distribution of PBBs have currently ceased, the full extent of the problem has not yet been assessed.

The main sources of PBBs today appear to be the residues remaining in and around facilities which at one time manufactured, processed, or produced products using PBBs. Preliminary indications show that the level of PBBs in the water near a Cincinnati, Ohio plant which processed PBBs is from 1 to 20 ppm. Concentrations as high as 0.15 g/m^3 have been detected in the Pine River at St. Louis even after shutdown of the PBB operation, possibly because of release of PBBs accumulated in discharge pipes or from contact with contaminated soils. Soils collected from bagging and loading areas at the Michigan plant were found to have 3,500 and 2,500 μg/kg PBBs, respectively (2).

Permissible Exposure Limits in Air: No standards set.

Permissible Concentrations in Water: No criteria set.

Harmful Effects and Symptoms: The accidental contamination in 1973 of animal feed and livestock throughout Michigan of polybrominated biphenyl flame retardants (Firemaster BP-6) has stimulated extensive studies of the potential for water contamination, transport, bioaccumulation, biological and toxicological nature of this class of environmental agent (1).

Cows from the herds in Michigan which were fed PBB-contaminated feed exhibited essentially the same symptoms as cattle tested in the laboratory. These symptoms included anorexia and weight loss, abnormal hoof growth, reduced milk production, and hepatomas; pregnant cows often delivered late by 2 to 4 weeks, with many calves stillborn or dead shortly after birth (3). The PBB composition involved was composed of 60 to 70% of 2,4,5,2',4',5'-hexabromobiphenyl, with the remainder consisting of lesser amounts of tetra-, penta-, and heptahomologs in addition to various brominated naphthalenes. It has not yet been determined which of the constituents was responsible for the toxic action that was observed.

While no immediate adverse health effects were noted in several thousand Michigan farm families which consumed milk and dairy products contaminated with PBBs, it is not possible to determine at this date any chronic or delayed effects that might be attributed to the PBBs or the potential ability of this chemical to cause birth defects.

The toxic activity of PBBs is exerted primarily on the liver and kidneys, marked by hyperplasia and necrosis. Affected to a lesser extent are the thyroid, uterus, skin, gall bladder, gastrointestinal tract, myocarcium and endocardium, and body tissues, especially fatty tissues.

Firemaster FF-1 (Firemaster BP-6 containing 2% of calcium trisilicate)—a mixture of pentabromobiphenyl, hexabromobiphenyl and heptabromobiphenyl with hexabromobiphenyl being the major component—administered by gavage produced neoplastic nodules and hepatocellular carcinomas in female Sherman strain rats (4). In another bioassay, Firemaster FF-1, also administered by gavage, was carcinogenic to Fischer 344 rats and B6C3F1 mice of each sex, inducing neoplastic nodules, hepatocellular carcinomas, and cholangiocarcinomas in rats and hepatocellular carcinomas in mice (5).

Points of Attack: Liver, kidneys, skin (3).

Medical Surveillance: Consider the points of attack in preplacement and periodic physical examinations.

Disposal Method Suggested: See "Polychlorinated Biphenyls" (entry which follows) for techniques which may be applicable.

References

(1) U.S. Environmental Protection Agency, *Summary Characterization of Selected Chemicals of Near-Term Interest,* Report EPA-560/4-76-004, Washington, DC (April 1976).

(2) U.S. Environmental Protection Agency, *Status Assessment of Toxic Chemicals: Polybrominated Biphenyls,* Report EPA-600/2-79-210k, Washington, DC (December 1979).

(3) Nat. Inst. for Occup. Safety and Health, *Information Profiles on Potential Occupational Hazards,* Report PB-276,678, Rockville, MD, pp 76-85 (October 1977).

(4) Kimbrough, R.D., D.F. Groce, M.P. Korver, and V.W. Burse, *Induction of Liver Tumors in Female Sherman Strain Rats by Polybrominated Biphenyls,* J. Nat. Cancer Inst., Vol. 66, No. 3, pp 535-542 (1981).

(5) National Toxicology Program, *NTP Technical Report on the Toxicology and Carcinogenesis Bioassay of Polybrominated Biphenyl Mixture (Firemaster FF-1),* Technical Report Series No. 244, NIH Publication No. 82-1800, Research Triangle Park, North Carolina (1982).

(6) United Nations Environment Programme, *IRPTC Legal File 1983,* Vol. II, pp VII/371, Geneva, Switzerland, International Register of Potentially Toxic Chemicals (1984).

POLYCHLORINATED BIPHENYLS (PCBs)

- Carcinogens (Animal positive, IARC) (4)
- Hazardous materials (EPA)
- Hazardous waste constituents (EPA)
- Priority toxic pollutants (EPA)

Description: $C_{12}H_{10-x}Cl_x$, diphenyl rings in which one or more hydrogen atoms are replaced by a chlorine atom. Most widely used are chlorodiphenyl (42% chlorine), containing 3 chlorine atoms in unassigned positions, and chlorodiphenyl (54% chlorine) containing 5 chlorine atoms in unassigned positions. These compounds are light, straw-colored liquids with typical chlorinated aromatic odors; 42% chlorodiphenyl is a mobile liquid and 54% chlorodiphenyl is a viscous liquid.

Polychlorinated biphenyls are prepared by the chlorination of biphenyl and hence are complex mixtures containing isomers of chlorobiphenyls with different chlorine contents. It should be noted that there are 209 possible compounds obtainable by substituting chlorine for hydrogen on from one to ten different positions on the biphenyl ring system. An estimated 40 to 70 different chlorinated biphenyl compounds can be present in each of the higher chlorinated commercial mixtures. For example, Arochlor 1254 contains 69 different molecules, which differ in the number and position of chlorine atoms.

It should also be noted that certain PCB commercial mixtures produced in the United States and elsewhere (e.g., France, Germany, and Japan) have been shown to contain other classes of chlorinated derivatives, e.g., chlorinated naphthalenes and chlorinated dibenzofurans. The possibility that naphthalene and dibenzofuran contaminate the technical biphenyl feedstock used in the preparation of the commercial PCB mixtures cannot be excluded.

Code Numbers: CAS 1336-36-3   RTECS TQ1350000   UN 2315

DOT Designation: ORM-E.

Synonyms: PCBs, chlorodiphenyls, Aroclurs®, Kanechlors®.

Potential Exposures: Chlorinated diphenyls are used alone and in combination with chlorinated naphthalenes. They are stable, thermoplastic, and nonflammable, and find chief use in insulation for electric cables and wires in the production of electric condensers, as additives for extreme pressure lubricants, and as a coating in foundry use.

Polychlorinated biphenyls (PCBs, first introduced into commercial use more than 45 years ago) are one member of a class of chlorinated aromatic organic compounds which are of increasing concern because of their apparent ubiquitous dispersal, persistence in the environment, and tendency to accumulate in food chains, with possible adverse effects on animals at the top of food webs, including man.

Incompatibilities: Strong oxidizers.

Permissible Exposure Limits in Air: The Federal standards and 1983/84 ACGIH TWA values for chlorodiphenyl (42% Cl) and chlorodiphenyl (54% Cl) are 1 mg/m³ and 0.5 mg/m³, respectively. NIOSH has recommended a level of 1.0 μg/m³ on a 10-hour TWA basis for both compounds. The STEL values adopted by ACGIH are 2 mg/m³ and 1.0 mg/m³, respectively. The IDLH levels are 10 mg/m³ and 5 mg/m³, respectively.

Determination in Air: For the 42% Cl compound, use of a filter plus bubbler followed by gas chromatography. See NIOSH Methods, Set 2. For the 54% Cl compound, use of a filter, workup with petroleum ether, analysis by gas chromatography. See NIOSH Methods, Set I. See also reference (A-10).

Permissible Concentration in Water: To protect freshwater aquatic life— 0.014 μg/ℓ as a 24-hour average. To protect saltwater aquatic life—0.030 μg/ℓ as a 24-hour average. To protect human health—preferably zero. An additional lifetime cancer risk of 1 in 100,000 results at a level of 0.00079 μg/ℓ.

Determination in Water: Gas chromatography (EPA Method 608) or gas chromatography plus mass spectrometry (EPA Method 625).

Routes of Entry: Inhalation of fume or vapor and percutaneous absorption of liquid, ingestion, eye and skin contact.

Harmful Effects and Symptoms: *Local* — Prolonged skin contact may cause the formation of comedones, sebaceous cysts, and pustules, known as chloracne. Irritation of eyes, nose, and throat may also occur. The above standards are considered low enough to prevent systemic effects, but it is not known whether or not these levels will prevent local effects.

Systemic — Generally, toxic effects are dependent upon the degree of chlorination; the higher the degree of substitution, the stronger the effects. Acute and chronic exposure can cause liver damage. Signs and symptoms include edema, jaundice, vomiting, anorexia, nausea, abdominal pains, and fatigue.

Studies of accidental oral intake indicate that chlorinated diphenyls are embryotoxic, causing stillbirth, a characteristic grey-brown skin, and increased eye discharge in infants born to women exposed during pregnancy.

Certain polychlorinated biphenyls are carcinogenic in mice and rats after oral administration, producing liver tumors (4).

A slight increase in the incidence of cancer, particularly melanoma of the skin, has been reported in a small group of men exposed occupationally to Arochlor 1254, a mixture of polychlorinated biphenyls (4,7).

Points of Attack: Skin, eyes, liver.

Medical Surveillance: Placement and periodic examinations should include an evaluation of the skin, lung, and liver function. Possible effects on the fetus should be considered.

First Aid: If this chemical gets into the eyes, irrigate immediately. If this chemical contacts the skin, wash with soap immediately. If a person breathes in large amounts of this chemical, move the exposed person to fresh air at once and perform artificial respiration. When this chemical has been swallowed, get medical attention. Give large quantities of saltwater and induce vomiting. Do not make an unconscious person vomit.

Personal Protective Methods: Wear appropriate clothing to prevent any possibility of skin contact. Wear eye protection to prevent any reasonable probability of eye contact. Employees should wash promptly when skin is wet or contaminated. Remove nonimpervious clothing promptly if wet or contaminated.

Respirator Selection:
42% Cl compound, 10 mg/m^3:	SAF/SCBAF
Escape:	GMPest/SCBA
54% Cl compound, 5 mg/m^3:	SAF/SCBAF
Escape:	GMPest/SCBA

Disposal Method Suggested: Incineration (3000°F) with scrubbing to remove any chlorine-containing products (A-31). In addition, some chemical waste landfills have been approved for PCB disposal. More recently treatment with metallic sodium has been advocated which yields a low molecular weight polyphenylene and sodium chloride.

References

(1) National Institute for Occupational Safety and Health, *Criteria for a Recommended Standard: Occupational Exposure to Polychlorinated Biphenyls,* NIOSH Doc. No. 77-225 (1977).
(2) U.S. Environmental Protection Agency, *Polychlorinated Biphenyls: Ambient Water Quality Criteria,* Washington, DC (1980).
(3) National Academy of Sciences, *Polychlorinated Biphenyls,* Washington, DC (1979).
(4) International Agency for Research on Cancer, *IARC Monographs on the Carcinogenic Risks of Chemicals to Humans,* Lyon, France, 7, 261 (1974) and 18, 43 (1978).
(5) World Health Organization, *Polychlorinated Biphenyls and Triphenyls,* Environmental Health Criteria No. 2, Geneva, Switzerland (1976).
(6) See Reference (A-62). Also see Reference (A-64).
(7) International Agency for Research on Cancer, *IARC Monographs on the Carcinogenic Risks of Chemicals to Humans,* Supplement 1, Lyon, France, p 41 (1979).
(8) Sax, N.I., Ed., *Dangerous Properties of Industrial Materials Report, 3,* No. 4, 95-100, New York, Van Nostrand Reinhold Co. (1983).
(9) Parmeggiani, L., Ed., *Encyclopedia of Occupational Health & Safety,* Third Edition, Vol. 2, pp 1753–55, Geneva, International Labour Office (1983).
(10) United Nations Environment Programme, *IRPTC Legal File 1983,* Vol. II, pp VII/644–60, Geneva, Switzerland, International Register of Potentially Toxic Chemicals (1984).

POLYNUCLEAR AROMATIC HYDROCARBONS

- Carcinogen (Benzo[a]pyrene) (Animal positive, IARC) (8)
- Hazardous wastes (EPA)
- Priority toxic pollutants (EPA)

Description: The polynuclear aromatic hydrocarbons constitute a class of materials of which benzo[a]pyrene is one of the most common and also the most hazardous.

Benzo[a]pyrene, $C_{20}H_{12}$, is a yellowish crystalline solid, melting at $179°C$. It consists of five benzene rings joined together. Other polynuclear aromatics which are discussed in separate sections in this volume are as follows: acenaphthene, fluoranthene and naphthalene. A variety of abbreviations are in common use for the polynuclear aromatics as shown below:

Abbreviation	Compound Designated
A	Anthracene
BaA	Benzo[a]anthracene (1,2-benzanthracene)
BaP (also BP)	Benzo[a]pyrene (3,4-benzopyrene)
BbFL (also BbF)	Benzo[b]fluoranthene
BeP	Benzo[e]pyrene
BjFL (also BjF)	Benzo[j]fluoranthene
BkFL (also BkF)	Benzo[k]fluoranthene (11,12-benzofluoranthene)
BPR	Benzo[ghi]perylene (1,12-benzoperylene)
CH (also CR)	Chrysene
DBA	Dibenzo[ah]anthracene (1,2,5,6-benzanthracene)
DBAc	Dibenz[a,h] and [a,j]acridine
DBC	Dibenzocarbazole

(continued)

Abbreviation	Compound Designated
DBP	Dibenzopyrene
F	Fluorene
FL (also F)	Fluoranthene
IP	Indeno[1,2,3-cd]pyrene
P	Pyrene
PA (also Phen)	Phenanthrene
PR (also Per)	Perylene

Note: These abbreviations are not endorsed by any body
such as the International Union of Chemistry; rather they
are a form of shorthand used by authors for convenience,
and they vary with the author.

Code Numbers: (For benzo[a]pyrene) CAS 50-32-8   RTECS DJ3675000

DOT Designation: —

Synonyms: PNAs, PAHs, PPAHs (Particulate Polycyclic Aromatic Hydrocarbons) and POMs (Polynuclear Organic Materials). (Benzo[a]pyrene is also known as BAP.)

Potential Exposures: PNAs can be formed in any hydrocarbon combustion process and may be released from oil spills. The less efficient the combustion process, the higher the PNA emission factor is likely to be. The major sources are stationary sources, such as heat and power generation, refuse burning, industrial activity, such as coke ovens, and coal refuse heaps. While PNAs can be formed naturally (lightning-ignited forest fires), impact of these sources appears to be minimal. It should be noted, however, that while transportation sources account for only about 1% of emitted PNAs on a national inventory basis, transportation-generated PNAs may approach 50% of the urban resident exposures.

Because of the large number of sources, most people are exposed to very low levels of PNAs. BAP has been detected in a variety of foods throughout the world. A possible source is mineral oils and petroleum waxes used in food containers and as release agents for food containers. FDA studies have indicated no health hazard from these sources.

The air pollution aspects of the carcinogenic polynuclear aromatic hydrocarbons (PAH) and of benzo[a]pyrene (BAP) in particular have been reviewed in some detail by Olsen and Haynes (1). The total emissions of benzo[a]pyrene (BAP) and some emission factors for BAP are as presented by Goldberg (2).

Permissible Exposure Limits in Air: A TLV of 0.2 mg/m³ as benzene solubles has been assigned by ACGIH. These materials are designated by ACGIH as human carcinogens.

There have been few attempts to develop exposure standards for PAHs, either individually or as a class. In the occupational setting, a Federal standard has been promulgated for coke oven emissions, based primarily on the presumed effects of the carcinogenic PAH contained in the mixture as measured by the benzene soluble fraction of total particulate matter. Similarly, the American Conference of Governmental Industrial Hygienists recommends a workplace exposure limit for coal tar pitch volatiles, based on the benzene-soluble fraction containing carcinogenic PAH.

The National Institute for Occupational Safety and Health has also recommended a workplace standard for coal tar products (coal tar, creosote, and coal tar pitch), based on measurements of the cyclohexane-extractable fraction. These standards are summarized on the following page.

Substance	Exposure Limit	Agency
Coke oven emissions	150 $\mu g/m^3$, 8 hr time-weighted average	U.S. Occupational Safety and Health Administration
Coal tar products	0.1 mg/m^3, 10 hr time-weighted average	U.S. National Institute for Occupational Safety and Health
Coal tar pitch and volatiles	0.2 mg/m^3 (benzene soluble fraction) 8 hr time-weighted average	American Conference of Governmental Industrial Hygienists

Determination in Air: Collection on a membrane filter, benzene extraction, chromatographic separation, measurement by fluorometry or using a UV detector (A-10).

Permissible Concentration in Water: A drinking water standard for PAH as a class has been developed. The 1970 World Health Organization European Standards for Drinking Water recommends a concentration of PAH not to exceed 0.2 $\mu g/\ell$. This recommended standard is based on the composite analysis of six PAHs in drinking water: fluoranthene, benzo[a]pyrene, benzo[ghi]-perylene, benzo[b]fluoranthene, benzo[k]fluoranthene, and indeno[1,2,3-cd]pyrene.

The US EPA addressed PAHs as one of the 65 priority toxic pollutants (3). They found that there was insufficient data to propose a criterion for the protection of freshwater or of saltwater aquatic life. For the protection of human health, the concentration is preferably zero. An additional lifetime cancer risk of 1 in 100,000 is posed by a concentration of 0.028 $\mu g/\ell$.

Determination in Water: Methylene chloride extraction followed by high performance liquid chromatography (HPLC) with fluorescence or UV detection or gas chromatography (EPA Method 610), or by gas chromatography plus mass spectrometry (EPA Method 625).

Routes of Entry: Inhalation of particulates, vapors.

Harmful Effects and Symptoms: Certain PNAs which have been demonstrated as carcinogenic in test animals at relatively high exposure levels are being found in urban air at very low levels. Various environmental fate tests suggest that PNAs are photo-oxidized, and react with oxidants and oxides of sulfur. Because PNAs are adsorbed on particulate matter, chemical half-lives may vary greatly, from a matter of a few hours to several days. One researcher reports that photo-oxidized PNA fractions of air extracts also appear to be carcinogenic. Environmental behavior/fate data have not been developed for the class as a whole.

It has been observed that PNAs are highly soluble in adipose tissue and lipids. Most of the PNAs taken in by mammals are oxidized and the metabolites excreted. Effects of that portion remaining in the body at low levels have not been documented.

Benzo[a]pyrene (BaP), one of the most commonly found and hazardous of the PNAs has been the subject of a variety of toxicological tests, which have been summarized by the International Agency for Research on Cancer. 50 to 100 ppm administered in the diet for 122 to 197 days produced stomach tumors in 70% of the mice studied. 250 ppm produced tumors in the forestomach of 100% of the mice after 30 days. A single oral administration of 100 mg of nine rats produced mammary tumors in eight of them. Skin cancers have been induced in a variety of animals at very low levels, and using a variety of solvents (length of application was not specified).

Lung cancer developed in 2 of 21 rats exposed to 10 mg/m^3 BaP and 3.5 ppm SO$_2$ for 1 hour per day, five days a week, for more than one year. Five of 21 rats receiving 10 ppm SO$_2$ for 6 hours per day, in addition to the foregoing dosage, developed similar carcinomas. No carcinomas were noted in rats receiving only SO$_2$. No animals were exposed only to BaP. Transplacental migration of BaP has been demonstrated in mice. Most other PNAs have not been subjected to such testing.

Medical Surveillance: Preplacement and regular physical examination are indicated for workers having contact with polynuclear aromatics in the workplace.

Personal Protective Methods: Good particulate emission controls are the indicated engineering control scheme where polynuclear aromatics are encountered in the workplace.

Disposal Method Suggested: Incineration.

References

(1)   Olsen, D.A. and Haynes, J.L., *Air Pollution Aspects of Organic Carcinogens,* Report PB-188 090, Springfield, VA, Nat. Tech. Information Service (September 1969).

(2)   Goldberg, A.J., *A Survey of Emissions and Controls for Hazardous and Other Pollutants,* Report PB-223 568, Springfield, VA, Nat. Tech. Information Service (Feb. 1973).

(3)   U.S. Environmental Protection Agency, *Polynuclear Aromatic Hydrocarbons: Ambient Water Quality Criteria,* Washington, DC (1980).

(4)   U.S. Environmental Protection Agency, *Status Assessment of Toxic Organic Chemicals: Polynuclear Aromatic Hydrocarbons,* Report EPA-600/2-79-210L, Cincinnati, OH (December 1979).

(5)   National Academy of Sciences, *Medical and Biologic Effects of Environmental Pollutants: Particulate Polycyclic Organic Matter,* Washington, DC (1972).

(6)   U.S. Environmental Protection Agency, *Health Assessment Document for Polycyclic Organic Matter,* Research Triangle Park, NC, Environmental Criteria and Assessment Office (1979).

(7)   U.S Environmental Protection Agency, *Polynuclear Aromatic Hydrocarbons,* Health and Environmental Effects Profile No. 149, Office of Solid Waste, Washington, DC (April 30, 1980).

(8)   International Agency for Research on Cancer, *IARC Monographs on the Carcinogenic Risks of Chemicals to Humans,* Lyon, France, 3, 91 (1973).

(9)   See Reference (A-62) for: Benz[a]anthracene, Benzo[b]fluoranthene and Benzo[j]-fluoranthene, Dibenz[a,h]acridine and Dibenz[j]acridine, Dibenz[a,h]anthracene, Dibenzo[c,g]carbazole, Dibenzo[a,h]pyrene, Dibenzo[a,i]pyrene], and Indeno-[1,2,3-cd]pyrene].

(10) Sax, N.I., Ed., *Dangerous Properties of Industrial Materials Report, 4,* No. 2, 35–37, New York, Van Nostrand Reinhold Co. (1984) (Acenaphthylene).

(11) Parmeggiani, L., Ed., *Encyclopedia of Occupational Health & Safety,* Third Edition, Vol. 2, pp 1755–59, Geneva, International Labour Office (1983).

POLYTETRAFLUOROETHYLENE DECOMPOSITION PRODUCTS

Description: Thermal decomposition of the fluorocarbon chain in air leads to the formation of oxidized products containing carbon, fluorine and oxygen. See the entry under "Carbonyl Fluoride" to cite one example.

DOT Designation: —

Synonyms: Teflon® decomposition products.

Potential Exposures: Those involved in exposure of fluorocarbon plastics at high temperatures. Exposure can come from smoking cigarettes contaminated with PTFE particles. So far as is known, overheating the coatings on cooking utensils has not been a problem (A-34).

Permissible Exposure Limits in Air: No TLV is recommended pending determination of the toxicity of the products, but air concentrations should be minimal.

Determination in Air: Because these products decompose in part by hydrolysis in alkaline solution, they can be quantitatively determined in air as fluoride to provide an index of exposure.

Permissible Concentration in Water: No criteria set.

Route of Entry: Inhalation.

Harmful Effects and Symptoms: Workers exposed to heated PTFE fumes have developed "polymer fume fever" which is characterized by chills, fever, tightness of the chest and other influenzalike symptoms, including pulmonary edema.

PORTLAND CEMENT

Description: Portland cement is a class of hydraulic cements whose two essential constituents are tricalcium silicate and dicalcium silicate with varying amounts of alumina, tricalcium aluminate, and iron oxide. The quartz content of most is below 1%. The average composition of regular Portland cement is as follows:

	Percent
CaO	64.0
SiO_2	21.0
Al_2O_3	5.8
Fe_2O_3	2.9
MgO	2.5
Alkali oxides	1.4
SO_3	1.7

The compounds which are contained therein are: $2CaO \cdot SiO_2$, $CaO \cdot Al_2O_3$, $3CaO \cdot SiO_2$ and $4CaO \cdot Al_2O_3 \cdot Fe_2O_3$.

Code Numbers: CAS: None   RTECS VV8770000

DOT Designation: —

Synonyms: Portland cement silicate, hydraulic cement, cement.

Potential Exposures: Cement is used as a binding agent in mortar and concrete (a mixture of cement, gravel, and sand). Potentially hazardous exposure may occur during both the manufacture and use of cement.

Incompatibilities: None hazardous.

Permissible Exposure Limits in Air: The Federal standard for Portland cement is 50 mppcf. ACGIH classifies Portland cement as a nuisance particulate with a TLV of 30 mppcf or 10 mg/m³ of total dust containing less than 1% quartz or 5 mg/m³ of respirable dust.

Permissible Concentration in Water: No criteria set.

Routes of Entry: Inhalation of dust, ingestion, skin and eye contact.

Harmful Effects and Symptoms: *Local* — Exposure may produce cement dermatitis which is usually due to primary irritation from the alkaline, hygroscopic, and abrasive properties of cement. Chronic irritation of the eyes and nose may occur. In some cases, cement workers have developed an allergic sensitivity to constituents of cement such as hexavalent chromate. It is not unusual for cement dermatitis to be prolonged and to involve covered areas of the body.

Systemic — No documented cases of pneumoconiosis or other systemic manifestations attributed to finished Portland cement exposure have been reported. Conflicting reports of pneumoconiosis from cement dust appear related to exposures that occurred in mining, quarrying, or crushing silica-containing raw materials.

Points of Attack: Respiratory system, eyes, skin.

Medical Surveillance: Preemployment and periodic medical examinations should stress significant respiratory problems, chest x-ray, pulmonary function tests, smoking history, and allergic skin sensitivities, especially to chromates. The eyes should be examined. Patch test studies may be useful in dermatitis cases.

First Aid: If this chemical gets into the eyes, irrigate immediately. If this chemical contacts the skin, wash with soap promptly. If a person breathes in large amounts of this chemical, move the exposed person to fresh air at once. When this chemical has been swallowed, get medical attention. Give large quantities of water and induce vomiting. Do not make an unconscious person vomit.

Personal Protective Methods: Wear appropriate clothing to prevent repeated or prolonged skin contact. Wear eye protection to prevent any reasonable probability of eye contact. Employees should wash promptly when skin is contaminated. Work clothing should be changed daily. Remove nonimpervious clothing promptly if contaminated.

Respirator Selection:
>   250 mppcf: D
>   500 mppcf: DXSQ/FuHiEP/SA/SCBA
>   2,500 mppcf: HiEPF/SAF/SCBAF
> 25,000 mppcf: PAPHiE/SA:PD,PP,CF

Disposal Method Suggested: Landfill. In some cases, recovery of cement from cement kiln dust or ready-mix concrete residues may be economic and technology is available (A-57).

References

(1) See Reference (A-61).
(2) Parmeggiani, L., Ed., *Encyclopedia of Occupational Health & Safety,* Third Edition, Vol. 1, pp 436–39, Geneva, International Labour Office (1983).

POTASSIUM CYANIDE

See "Cyanides."

POTASSIUM HYDROXIDE

Description: KOH is a white deliquescent solid melting at about 360°C.

Code Numbers: CAS 1310-58-3   RTECS TT2100000   UN 1813 (solid), 1814 (solution)

DOT Designation: Corrosive material.

Synonyms: Caustic potash; potassium hydrate.

Potential Exposure: In the manufacture of other potassium compounds and in the general use of KOH as an alkali.

Permissible Exposure Limits in Air: ACGIH (1983/84) has adopted a TWA value of 2 mg/m^3 as a ceiling value but has set no STEL value.

Permissible Concentration in Water: No criteria set.

Fist Aid: Remove contaminated clothing and flush affected areas with water (A-60).

Personal Protective Methods: Protective clothing, gloves and face shield should be worn (A-60).

Disposal Method Suggested: Dilute with large volume of water, neutralize and flush to sewer (A-38).

References

(1)   American Conference of Governmental Industrial Hygienists, *Documentation of the Threshold Limit Values,* Fourth Ed., p 345, Cincinnati, Ohio (1980).

(2)   See Reference (A-60).

POTASSIUM PERSULFATE

Description: K$_2$S$_2$O$_8$ is a white crystalline material which decomposes below 100°C.

Code Numbers: CAS 7727-21-1   UN 1492

DOT Designation: Oxidizer.

Synonyms: Potassium peroxydisulfate.

Potential Exposures: Potassium persulfate is used as a bleaching and oxidizing agent; it is used in redox polymerization catalysts, in the defiberizing of wet strength paper and in the desizing of textiles.

Incompatibilities: Combustible, organic or other readily oxidizable materials, sulfur, metallic dusts such as aluminum dust, chlorates and perchlorates.

Permissible Exposure Limits in Air: There is no Federal standard but ACGIH as of 1983/84 has proposed a TWA of 2 mg/m^3 (as S$_2$O$_8$) for alkali metal persulfates.

Permissible Concentration in Water: No criteria set. (Aqueous solution decomposes even at room temperature.)

Harmful Effects and Symptoms: The material reacts with moisture to give ozone and sulfuric acid causing explosions in closed containers (A-38).

Personal Protective Methods: Wear rubber gloves, a face shield and overalls. Prepare full body shields in case of severe reactions (A-38).

Disposal Method Suggested: Use large volumes of reducing agents (bisulfites, e.g.). Neutralize with soda ash and drain into sewer with abundant water (A-38).

POTASSIUM SALTS

A variety of potassium salts have been reviewed in the periodical *Dangerous Properties of Industrial Materials Report,* and in the interest of economy of space, they are simply referenced here:

	Vol	No.	Page
Potassium arsenate	3	4	101–103
Potassium arsenite	3	4	103–106
Potassium bromate	1	7	70–71
Potassium chromate	1	7	71–73
Potassium cyanate	1	7	73–74
Potassium dodecanoic acid	1	5	78

See also Reference (A-60) for listings under Potassium Bromide, Potassium Chlorate, Potassium Ferricyanide, Potassium Ferrocyanide, Potassium Fluoride, Potassium Iodate, Potassium Iodide, Potassium Nitrate, Potassium Nitrite, Potassium Periodate, Potassium Permanganate, Potassium Pyrosulfate, Potassium Sulfide, and Potassium Thiocyanate.

PROCARBAZINE AND PROCARBAZINE HYDROCHLORIDE

- Carcinogen (Animal Positive, NCI) (1)

Description: $C_{12}H_{19}N_3O$ has the following structural formula

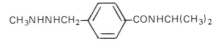

The hydrochloride melts at 223° to 226°C.

Code Numbers: CAS 671-16-9 (base)   CAS 366-70-1 (hydrochloride)
RTECS XS4550000 (base)   RTECS XS4725000 (hydrochloride)

DOT Designation: –

Synonyms: N-isopropyl-α-(2-methylhydrazino)-p-toluamide.

Potential Exposure: Procarbazine is available in capsule form. The primary use of this drug is as an antineoplastic agent in the treatment of advanced Hodgkin's disease, and oat-cell carcinoma of the lung. The hydrochloride compound is used in treatment. The FDA approved use of procarbazine hydrochloride in 1969 and indicated that the drug should be used as an adjunct to standard therapy.

Possible exposure occurs during manufacture of the drug and direct exposure during its subsequent administration to patients. The National Prescription Audit reported 1.5 million prescriptions dispensed in 1980. Some of the metabolites of procarbazine hydrochloride are both carcinostatic and carcinogenic.

Permissible Exposure Limits in Air: No standards set.

Permissible Concentration in Water: No criteria set.

Harmful Effects and Symptoms: Laboratory exposure of animals to procarbazine was studied by ip injection. In rats, malignant lymphoma, adenocarcinoma of the mammary gland, and olfactory neuroblastomas were induced in statistically significant numbers. In mice, malignant lymphoma or leukemia, olfactory neuroblastoma, alveolar/bronchiolar adenoma, and adenocarcinoma of the uterus were induced in statistically significant numbers (1).

References

(1) National Cancer Institute, *Bioassay of Procarbazine for Possible Carcinogenicity,* DHHS Publication No. (NIH) 79-819, National Technical Information Service, Springfield, VA (1979).
(2) See Reference (A-62). Also see Reference (A-64).

PROPACHLOR

Description: $C_{11}H_{14}ClNO$, $(ClCH_2CO)N(C_6H_5)CH(CH_3)_2$ is a light tan solid melting at 67° to 76°C.

Code Numbers: CAS 1918-16-7   RTECS AE1575000

DOT Designation: —

Synonyms: 2-Chloro-N-isopropyl acetanilide; Ramrod®.

Potential Exposure: Those engaged in the manufacture (A-32), formulation and application of this preemergence herbicide which is used to combat annual grasses and broad-leaved weeds in corn, soybeans, cotton, sugar cane and vegetable crops.

Permissible Exposure Limits in Air: No standards set.

Permissible Concentration in Water: NAS/NRC has calculated a no-adverse effect level in drinking water of 0.7 mg/ℓ.

Harmful Effects and Symptoms: The maximal tolerated dosage of Propachlor without adverse effect is reported as 133.3 mg/kg/day in both rats and dogs. Other workers reported slight organ pathology in rats, mice, and rabbits at 100 mg/kg/day or higher; this agrees approximately with the former data.

Apparently, no long-term toxicity studies have been completed that would contribute information on reproductive effects or carcinogenic potential of propachlor or its degradation products, which include aniline derivatives. These studies are needed.

Disposal Method Suggested: Alkaline hydrolysis would yield N-isopropyl-aniline (A-32).

PROPANE

Description: $CH_3CH_2CH_3$ is a colorless, odorless gas (foul-smelling odorant often added).

Code Numbers: CAS 74-98-6   RTECS TX2275000   UN 1978

DOT Designation: Flammable gas.

Synonyms: Dimethylmethane, propyl hydride.

Potential Exposure: Propane is used as a household, industrial and vehicle fuel; it is used as a refrigerant and aerosol propellant; it is used as an intermediate in petrochemical manufacture.

Incompatibilities: Strong oxidizers.

Permissible Exposure Limits in Air: The Federal standard is 1,000 ppm (1,800 mg/m^3). ACGIH defines propane as a simple asphyxiant and does not recommend a TLV because the limiting factor is the available oxygen. The IDLH level is 20,000 ppm.

Determination in Air: By combustible gas meter. See NIOSH Methods, Set G. See also reference (A-10).

Permissible Concentration in Water: No criteria set, but EPA (A-37) has suggested a permissible ambient goal of 120,000 µg/ℓ based on health effects.

Routes of Entry: Inhalation, skin and eye contact by liquid.

Harmful Effects and Symptoms: Dizziness, disorientation, excitation, frostbite.

Points of Attack: Central nervous system.

Medical Surveillance: Consider the points of attack in preplacement and periodic physical examinations.

First Aid: If this chemical gets into the eyes, irrigate immediately. If this chemical contacts the skin, flush with water immediately. If a person breathes in large amounts of this chemical, move the exposed person to fresh air at once and perform artificial respiration.

Personal Protective Methods: Wear appropriate clothing to prevent skin freezing. Wear eye protection to prevent any reasonable probability of eye contact. Remove clothing immediately if wet or contaminated to avoid flammability hazard.

Respirator Selection:
```
10,000 ppm: SA/SCBA
20,000 ppm: SAF/SCBAF/SA:PD,PP,CF
    Escape: SCBA
```

Disposal Method Suggested: Incineration.

References

(1)   See Reference (A-61).
(2)   See Reference (A-60).
(3)   United Nations Environment Programme, *IRPTC Legal File 1983,* Vol. II, pp VII/664–5, Geneva, Switzerland, International Register of Potentially Toxic Chemicals (1984).

PROPANE SULTONE

● Carcinogen (Animal Positive, IARC) (Human Suspect, ACGIH)

Description: $\overline{(CH_2)_3OSO_2}$ is a white crystalline compound which melts at 31°C with a foul odor.

Code Numbers: CAS 1120-71-4   RTECS RP5425000

DOT Designation: —

Synonyms: 1,2-Oxathiolane-2,2-dioxide.

Potential Exposure: Those involved in use of this chemical intermediate to introduce the sulfopropyl group ($-CH_2CH_2CH_2SO_3-$) into molecules.

Permissible Exposure Limits in Air: ACGIH (1983/84) has designated this compound "an industrial substance suspect of carcinogenic potential for man" with no numeric TWA or STEL values.

Permissible Concentration in Water: No criteria set.

Harmful Effects and Symptoms: Propane sultone is a carcinogen in rats when given by any of several routes and is locally active in both rats and mice when given subcutaneously or by skin painting in mice (1,2).

References

(1) International Agency for Research on Cancer, *Monographs on the Evaluation of Carcinogenic Risk of Chemicals to Man, 4,* 253-258, Lyon, France (1972).

(2) American Conference of Governmental Industrial Hygienists, *Documentation of the Threshold Limit Values,* 4th Ed., Cincinnati, Ohio (1980).

(3) Sax, N.I., Ed., *Dangerous Properties of Industrial Materials Report, 4,* No. 3, 82–85, New York, Van Nostrand Reinhold Co. (1984).

PROPANIL

Description: $Cl_2C_6H_3NHCOC_2H_5$ is a colorless solid melting at $92°$ to $93°C$. The technical product is a brown crystalline solid melting at $88°$ to $91°C$.

Code Numbers: CAS 709-98-8   RTECS UE4900000

DOT Designation: —

Synonyms: 3',4'-Dichloropropionanilide; Stam F-34®; Surcopur®; Rogue®.

Potential Exposure: Those involved in the manufacture, formulation and application of this contact herbicide.

Permissible Exposure Limits in Air: No standards set.

Permissible Concentration in Water: NAS/NRC (A-2) has calculated a no-adverse effect level in drinking water of 0.14 mg/ℓ.

Harmful Effects and Symptoms: Propanil is well tolerated by experimental animals on a chronic basis, and there is little or no indication of mutagenic or oncogenic properties of the compound. The highest no-adverse-effect concentration of propanil based on reproduction in the rat and acute, subchronic, and chronic studies in rats and dogs is 400 ppm in the diet. Based on these data, an ADI was calculated at 0.02 mg/kg/day (A-2).

Disposal Method Suggested: Hydrolysis in acidic or basic media yields the more toxic substance, 3,4-dichloroaniline, and is not recommended (A-32).

References

(1) United Nations Environment Programme, *IRPTC Legal File 1983,* Vol. II, pp VII/688, Geneva, Switzerland, International Register of Potentially Toxic Chemicals (1984).

PROPARGYL ALCOHOL

- Hazardous waste (EPA)

Description: $HC\equiv CCH_2OH$ is a colorless liquid with a geraniumlike odor boiling at $114°C$.

Code Numbers: CAS 107-19-7   RTECS UK5075000   UN (NA 1986)

DOT Designation: Flammable liquid and poison.

Synonyms: 2-Propyn-1-ol, ethynylmethanol.

Potential Exposures: Propargyl alcohol is used as a corrosion inhibitor, solvent, stabilizer and chemical intermediate.

Permissible Exposure Limits in Air: The Federal standard and the 1983/84 ACGIH TWA value is 1 ppm ($2 mg/m^3$) with the notation "skin" indicating the possibility of cutaneous absorption. The STEL is 3 ppm ($6 mg/m^3$).

Determination in Air: Sample collection by charcoal tube, analysis by gas liquid chromatography (A-13).

Permissible Concentration in Water: No criteria set.

Harmful Effects and Symptoms: Propargyl alcohol is an irritant to the skin and mucous membranes.

Personal Protective Methods: Wear rubber gloves, face shield and protective clothing (A-38).

Respirator Selection: Use all purpose cannister mask (A-38).

Disposal Method Suggested: Incineration.

β-PROPIOLACTONE

- Carcinogen (Animal positive, IARC) (1)

Description: $\overline{OCH_2CH_2CO}$, β-propiolactone, is a colorless liquid which slowly hydrolyzes to hydracrylic acid and must be cooled to remain stable.

Code Numbers: CAS 57-57-8   RTECS RQ7350000

DOT Designation: —

Synonyms: 2-Oxetanone, propiolactone, BPL, 3-hydroxy-β-lactone-propanoic acid.

Potential Exposures: β-Propiolactone is used as a chemical intermediate in synthesis of acrylate plastics and as a vapor sterilizing agent, disinfectant, and a viricidal agent.

Permissible Exposure Limits in Air: β-Propiolactone is included in the Federal standards for carcinogens; all contact with it should be avoided. ACGIH (1983/84) classifies the material as suspect of carcinogenic potential for man and has adopted a TWA value of 0.5 ppm ($1.5 mg/m^3$) and an STEL value of 1 ppm ($3 mg/m^3$).

Permissible Concentration in Water: No criteria set.

Routes of Entry: Inhalation of vapor and percutaneous absorption.

Harmful Effects and Symptoms: *Local* — Repeated or prolonged contact with liquid may cause erythema, vesication of the skin, and, as reported in animals, hair loss and scarring. In rodents, β-propiolactone has also produced skin papilloma and sarcoma by skin painting, subcutaneous injection, and oral administration. Tumors of the connective tissue are also suspected. Direct eye contact with concentrated liquid may result in permanent corneal opacification. Skin cancer has not been reported in man.

Systemic — The systemic effect of β-propiolactone in humans is unknown due to lack of reported cases. Acute exposure in animals has caused liver necrosis and renal tubular damage. Death has occurred following rapid development of spasms, dyspnea, convulsions, and collapse at relatively low levels (less than 5 ml/kg). β-Propiolactone has been implicated as a carcinogen by a number of animal studies which produced a variety of skin tumors, stomach tumors, and hepatoma depending on the route of administration (1)(2).

It should be noted that BPL is produced from formaldehyde and ketene which have been found to be mutagenic (1). Commercial grade BPL (97%) can contain trace quantities of the reactants.

Medical Surveillance: Based on its high toxicity and carcinogenic effects in animals, preplacement and periodic examinations should include a history of exposure to other carcinogens, alcohol and smoking habits, medication and family history. The skin, eye, lung, liver, and kidney should be evaluated. Sputum cytology may be helpful in evaluating the presence or absence of carcinogenic effects.

Personal Protective Methods: These are designed to supplement engineering controls and to prevent all contact with skin or respiratory tract. Full body protective clothing and gloves should be provided as well as fullface supplied air respirators of continuous flow or pressure demand type. Employees should remove and leave protective clothing and equipment at the point of exit, to be placed in impervious containers at the end of work shift for decontamination or disposal. Showers should be taken before dressing in street clothes.

Disposal Method Suggested: Incineration.

References

(1) International Agency for Research on Cancer, *IARC Monographs on the Carcinogenic Risks of Chemicals to Humans,* Lyon, France, 4, 259 (1974) and 15, 341 (1977).
(2) See Reference (A-62). Also see Reference (A-64).
(3) Sax, N.I., Ed., *Dangerous Properties of Industrial Materials Report, 3,* No. 2, 57-60, New York, Van Nostrand Reinhold Co. (1983).
(4) See Reference (A-60).

PROPIONIC ACID

● Hazardous substance (EPA)

Description: CH_3CH_2COOH is a colorless liquid with a pungent odor boiling at 141°C.

Code Numbers: CAS 79-09-4   RTECS UE5950000   UN 1848

DOT Designation: Corrosive material.

Synonyms: Carboxyethane, ethanecarboxylic acid, ethylformic acid, methylacetic acid.

Potential Exposures: Propionic acid is used in the manufacture of inorganic propionates and propionate esters which are used as mold inhibitors, electroplating additives, emulsifying agents, flavors and perfumes. It is an intermediate in pesticide manufacture (A-32), pharmaceutic manufacture (A-41) and in the production of cellulose propionate plastics.

Permissible Exposure Limits in Air: There is no Federal standard but ACGIH (1983/84) has adopted a TWA value of 10 ppm (30 mg/m^3) and set an STEL of 15 ppm (45 mg/m^3). The Soviet limit is reported to be 0.7 ppm (2 mg/m^3).

Permissible Concentration in Water: No criteria set.

Harmful Effects and Symptoms: Mild to moderate skin burns, mild eye redness and one report of mild cough and asthmatic response were found in medical reports of acute exposure of workers (A-34).

Points of Attack: Skin, eyes, respiratory system.

Medical Surveillance: Attention to points of attack in preplacement and regular physical examinations.

Personal Protective Methods: Wear rubber gloves, face shield and protective clothing.

Respirator Selection: Use of self-contained breathing apparatus is recommended (A-38).

Disposal Method Suggested: Incineration in admixture with flammable solvent.

References
(1)   See Reference (A-60).

PROPIONITRILE

Description: CH_3CH_2CN is a colorless liquid boiling at 97.2°C.

Code Numbers: CAS 107-12-0   RTECS UF9625000   UN 2404

DOT Designation: —

Synonyms: Cyanoethane; Ethyl cyanide; Hydrocyanic ether.

Potential Exposure: Used as a chemical intermediate.

Permissible Exposure Limits in Air: The NIOSH-recommended TLV is 6 ppm (14 mg/m^3) as an 8-hour TWA.

Permissible Concentration in Water: No criteria set.

Harmful Effects and Symptoms: Propionitrile has been shown to have acute and chronic effects in animals including formation of ulcers and teratogenicity (formation of defective embryos or fetuses). The oral LD_{50} value for rats is 50 to 100 mg/kg.

First Aid: Flush eyes with water; wash body with soap and water; use artificial respiration and oxygen if needed; administer any nitrite by inhalation (A-38).

Personal Protective Methods: Wear long rubber gloves and protective clothing. No eating and smoking in work areas (A-38).

Respirator Selection: Self-contained breathing apparatus (A-38).

Disposal Method Suggested: Alcoholic NaOH followed by calcium hypochlorite may be used as may incineration.

References

(1)   U.S. Environmental Protection Agency, *Chemical Hazard Information Profile Draft Report: Propionitrile,* Wash., DC (Sept. 2, 1983).

(2)   United Nations Environment Programme, *IRPTC Legal File 1983,* Vol. II, pp VII/696, Geneva, Switzerland, International Register of Potentially Toxic Chemicals (1984).

PROPOXUR

Description: $C_{11}H_{15}NO_3$, 2-(1-methylethoxy)phenyl methylcarbamate, with the structural formula

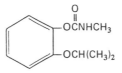

is a colorless crystalline powder with a faint characteristic odor melting at 84° to 87°C.

Code Numbers: CAS 114-26-1   RTECS FC3150000   UN 2588

DOT Designation: −

Synonyms: ENT 25671, PHC (In Japan), Baygon®, Blattanex®, Unden®.

Potential Exposures: Personnel engaged in the manufacture (A-32), formulation and application of this insecticide.

Permissible Exposure Limits in Air: There is no Federal standard but ACGIH as of 1983/84 has set a TWA value of 0.5 mg/m³ and an STEL of 2.0 mg/m³.

Permissible Concentration in Water: No criteria set.

Harmful Effects and Symptoms: Propoxur is a cholinesterase inhibitor but the effect is reversible. The compound is metabolized and excreted rapidly. In a volunteer study, a single oral dose of 1.5 mg/kg produced mild gastrointestinal symptoms which disappeared within 2 hours after ingestion (A-34).

Medical Surveillance: Blood cholinesterase levels should be checked. Exposure can be monitored by determining urinary excretion of the phenol metabolite (A-34).

References

(1)   United Nations Environment Programme, *IRPTC Legal File 1983,* Vol. I, pp VII/170–71, Geneva, Switzerland, International Register of Potentially Toxic Chemicals (1984).

n-PROPYL ACETATE

Description: $CH_3COOCH_2CH_2CH_3$ is a colorless liquid with a mild, fruity odor which boils at 101° to 102°C.

Code Numbers: CAS 109-60-4   RTECS AJ3675000   UN 1276

DOT Designation: Flammable liquid.

Synonyms: Acetic acid, n-propyl ester.

Potential Exposures: Propyl acetate is used as a perfume ingredient, component of food flavoring and solvent for cellulose ester resins. It is also used as a chemical intermediate.

Incompatibilities: Nitrates, strong oxidizers, strong alkalies, strong acids.

Permissible Exposure Limits in Air: The Federal standard and the 1983/84 ACGIH TWA value is 200 ppm (840 mg/m^3). The STEL is 250 ppm (1,050 mg/m^3). The IDLH level is 8,000 ppm.

Determination in Air: Charcoal adsorption, workup with CS_2, analysis by gas chromatography. See NIOSH Methods, Set D. See also reference (A-10).

Permissible Concentration in Water: No criteria set.

Routes of Entry: Inhalation, ingestion, skin and eye contact.

Harmful Effects and Symptoms: Irritation of eyes, nose and throat, narcosis, dermatitis.

Points of Attack: Respiratory system, eyes, skin, central nervous system.

Medical Surveillance: Consider the points of attack in preplacement and periodic physical examinations.

First Aid: If this chemical gets into the eyes, irrigate immediately. If this chemical contacts the skin, flush with water promptly. If a person breathes in large amounts of this chemical, move the exposed person to fresh air at once and perform artificial respiration.

Personal Protective Methods: Wear appropriate clothing to prevent repeated or prolonged skin contact. Wear eye protection to prevent any reasonable probability of eye contact. Employees should wash promptly when skin is wet or contaminated. Remove clothing immediately if wet or contaminated to avoid flammability hazard.

Respirator Selection:
    1,000 ppm: CCROVF
    5,000 ppm: GMOVc
    8,000 ppm: GMOVfb/SAF/SCBAF
       Escape: GMOV/SCBA

Disposal Method Suggested: Incineration.

References
(1)   See Reference (A-61).
(2)   See Reference (A-60).

n-PROPYL ALCOHOL

Description: There are two isomers of propyl alcohol—n-propyl alcohol and isopropyl alcohol. Both are colorless, volatile liquids. Isopropyl alcohol is discussed in a separate entry in this volume. n-Propanol boils at 97°C.

Code Numbers: CAS 71-23-8   RTECS UH8225000   UN 1274

DOT Designation: Flammable liquid.

Synonyms: 1-Propanol, propylic alcohol.

Potential Exposures: n-Propyl alcohol is used in lacquers, dopes, cosmetics, dental lotions, cleaners, polishes, and pharmaceuticals and as a surgical antiseptic. It is a solvent for vegetable oils, natural gums and resins, rosin, shellac, certain synthetic resins, ethylcellulose, and butyral.

Incompatibilities: Strong oxidizers.

Permissible Exposure Limits in Air: The Federal standard and the 1983/84 ACGIH TWA value for n-propyl alcohol is 200 ppm (500 mg/m^3). The STEL value is 250 ppm (625 mg/m^3). The notation "skin" is added to indicate the possibility of cutaneous absorption. The IDLH level is 4,000 ppm.

Determination in Air: Charcoal adsorption, workup with 2-propanol in CS$_2$, analysis by gas chromatography. See NIOSH Methods, Set E. See also reference (A-10).

Permissible Concentration in Water: No criteria set, but EPA (A-37) has suggested a permissible ambient goal of 6,900 μg/ℓ based on health effects.

Routes of Entry: Inhalation of vapor, percutaneous absorption, ingestion, skin and eye contact.

Harmful Effects and Symptoms: *Local* — The vapors are mildly irritating to the conjunctiva and mucous membranes of the upper respiratory tract.
Systemic — No cases of poisoning from industrial exposure have been recorded. n-Propyl alcohol can produce mild central nervous system depression.
Other reported symptoms of propyl alcohol exposure include dry cracking skin, drowsiness, headaches, ataxia, gastrointestinal pain, abdominal cramps, nausea, vomiting and diarrhea.

Points of Attack: Skin, eyes, respiratory system, gastrointestinal tract.

Medical Surveillance: No specific considerations are needed.

First Aid: If this chemical gets into the eyes, irrigate immediately. If this chemical contacts the skin, flush with water. If a person breathes in large amounts of this chemical, move the exposed person to fresh air at once and perform artificial respiration. When this chemical has been swallowed, get medical attention. Give large quantities of saltwater and induce vomiting. Do not make an unconscious person vomit.

Personal Protective Methods: Wear appropriate clothing to prevent repeated or prolonged skin contact. Wear eye protection to prevent any reasonable probability of eye contact. Employees should wash promptly when skin is wet or contaminated. Remove clothing immediately if wet or contaminated to avoid flammability hazard.

Respirator Selection:
    1,000 ppm: CCROV
    2,000 ppm: SA/SCBA
    4,000 ppm: GMOV/SAF/SCBAF
     Escape: GMOV/SCBA

Disposal Method Suggested: Incineration.

References
(1) See Reference (A-61).
(2) See Reference (A-60).

(3) U.S. Environmental Protection Agency, *Chemical Hazard Information Profile Draft Report: n-Propanol,* Wash., DC (March 31, 1983).

PROPYLENE

Description: $CH_2=CHCH_3$ is a colorless gas.

Code Numbers: CAS 115-07-1   RTECS UC6740000   UN 1077

DOT Designation: Flammable gas.

Synonym: Methylethylene.

Potential Exposures: Propylene is used in the production of polypropylene resins, isopropyl alcohol, propylene dimer and trimer as gasoline components and detergent raw materials, propylene oxide, cumene, synthetic glycerol, isoprene and Oxo alcohols.

Permissible Exposure Limits in Air: There are no Federal standards and ACGIH (1983/84) classifies propylene as a simple asphyxiant with no numerical TWA or STEL value.

Permissible Concentration in Water: No criteria set.

Harmful Effects and Symptoms: Propylene is mild in toxicity. Symptoms of propylene exposure include narcosis and irregular heartbeat (A-38).

First Aid: Administer oxygen or artificial respiration (A-38).

Personal Protective Methods: Wear rubber gloves and face protecting shield (A-38).

Respirator Selection: Wear all-purpose gas mask (A-38).

Disposal Method Suggested: Controlled incineration.

PROPYLENE DICHLORIDE

See "1,2-Dichloropropane."

PROPYLENE GLYCOL DINITRATE

Description: $CH_3CHONO_2CH_2ONO_2$ is a colorless high-boiling liquid with a disagreeable odor.

Code Numbers: CAS 6423-43-4   RTECS TY6300000

DOT Designation: —

Synonym: PGDN.

Potential Exposures: PGDN has been used as a torpedo propellant.

Permissible Exposure Limits in Air: There is no Federal standard and ACGIH as of 1983/84 has adopted a TWA value of 0.05 ppm (0.3 mg/m^3). The STEL was 0.1 ppm (0.6 mg/m^3), but consideration is being given to abolishing the STEL values.

Permissible Concentration in Water: No criteria set.

Harmful Effects and Symptoms: Human volunteers at 0.2 ppm exposure exhibited headaches and disruption in visually evoked response. At 0.5 ppm, marked impairment in balance was noted. At 1.5 ppm eye irritation occurred, in addition (A-34).

PROPYLENE GLYCOL MONOMETHYL ETHER

Description: $CH_3OCH_2CHOHCH_3$ is a colorless liquid with an ethereal odor which boils at 120°C.

Code Numbers: CAS 107-98-2   RTECS UB7700000

DOT Designation: —

Synonyms: 1-Methoxy-2-propanol, PGME.

Potential Exposures: PGME is used as a solvent for cellulose esters and acrylics. It may be used as a heat-transfer fluid.

Permissible Exposure Limits in Air: There is no Federal standard but ACGIH (1983/84) has adopted a TWA value of 100 ppm (360 mg/m^3) and set a tentative STEL of 150 ppm (540 mg/m^3).

Determination in Air: Sample collection by charcoal tube, analysis by gas liquid chromatography (A-10).

Permissible Concentration in Water: No criteria set.

Route of Entry: Inhalation.

Harmful Effects and Symptoms: PGME is an irritant and central nervous system depressant. Symptoms include irritation of eyes, nose and throat, headaches, nausea and abnormal Romberg behavior (A-39).

First Aid: Irrigate eyes with water, wash contaminated areas of body with soap and water.

Personal Protective Methods: Provide adequate ventilation; use chemical goggles.

Respirator Selection: Use chemical cartridge respirator.

Disposal Method Suggested: Incineration.

PROPYLENEIMINE

● Carcinogen (Animal positive, IARC) (2)

Description: $CH_3CH\underset{\displaystyle \diagup NH\diagdown}{\overline{}}CH_2$ is a fuming, colorless liquid with a strong ammonialike odor. It boils at 66°C.

Code Numbers: CAS 75-55-8   RTECS CM8050000   UN 1921

DOT Designation: Flammable liquid.

Synonyms: 2-Methylaziridine, 2-methylazacyclopropane, methylethyleneimine.

Potential Exposures: Propyleneimine is used in the production of polymers for use in the paper and textile industries as coatings and adhesives.

Incompatibilities: Acids, strong oxidizers.

Permissible Exposure Limits in Air: The Federal standard and the 1983/84 ACGIH TWA value is 2 ppm (5 mg/m^3). The notation "skin" is added to indicate the possibility of cutaneous absorption. There is no tentative STEL value set. The IDLH level is 500 ppm. ACGIH notes that this is "an industrial substance suspect of carcinogenic potential for man."

Permissible Concentration in Water: No criteria set.

Routes of Entry: Inhalation, skin absorption, ingestion, eye and skin contact.

Harmful Effects and Symptoms: Eye and skin burns.

Points of Attack: Eyes, skin.

Medical Surveillance: Consider the points of attack in preplacement and periodic physical examinations.

First Aid: If this chemical gets into the eyes, irrigate immediately. If this chemical contacts the skin, flush with water immediately. If a person breathes in large amounts of this chemical, move the exposed person to fresh air at once and perform artificial respiration. When this chemical has been swallowed get medical attention. Give large quantities of water and induce vomiting. Do not make an unconscious person vomit.

Personal Protective Methods: Wear appropriate clothing to prevent any reasonable probability of skin contact. Wear eye protection to prevent any possibility of eye contact. Employees should wash immediately when skin is wet or contaminated. Remove clothing immediately if wet or contaminated to avoid flammability hazard. Provide emergency showers and eyewash.

Respirator Selection:
100 ppm: SAF/SCBAF
500 ppm: SAF:PD,PP,CF
Escape: GMS/SCBA

Disposal Method Suggested: Controlled incineration (incinerator is equipped with a scrubber or thermal unit to reduce NO$_x$ emissions).

References

(1) Dermer, O.C. and Ham, G.E., *Ethyleneimine and Other Aziridines,* New York, Academic Press (1969).
(2) International Agency for Research on Cancer, *IARC Monographs on the Carcinogenic Risks of Chemicals to Humans,* Lyon, France, 9, 61 (1975).
(3) See Reference (A-61).

PROPYLENE OXIDE

- Carcinogen (Animal, Suspected, IARC) (1)
- Hazardous substance (EPA)

Description: CH$_3$CH——CH$_2$ is a colorless liquid with an etherlike odor which boils at 34° to 35°C.

Code Numbers: CAS 75-56-9 RTECS TZ2975000 UN 1280

DOT Designation: Flammable liquid.

Synonyms: 1,2-Epoxypropane, methylethylene oxide, methyl oxirane, propene oxide.

Potential Exposures: Propylene oxide is used in the production of propylene glycol and adducts as urethane foam ingredients. It is used in detergent manufacture and as a component in brake fluids.

Incompatibilities: Anhydrous metal chlorides, iron or aluminum chloride, strong acids, caustics, peroxides.

Permissible Exposure Limits in Air: The Federal standard is 100 ppm (240 mg/m^3). The ACGIH of 1983/84 has proposed a TWA of 20 ppm (50 mg/m^3) but no STEL value. The IDLH level is 2,000 ppm.

Determination in Air: Adsorption on charcoal, workup with CS_2, analysis by gas chromatography. See NIOSH Methods, Set F. See also reference (A-10).

Permissible Concentration in Water: No criteria set.

Routes of Entry: Inhalation, ingestion, skin and eye contact.

Harmful Effects and Symptoms: Irritation of eyes, upper respiratory system and lungs, irritation, blistering and burns of skin.

Points of Attack: Eyes, skin, respiratory system, lungs.

Medical Surveillance: Consider the points of attack in preplacement and periodic physical examinations.

First Aid: If this chemical gets into the eyes, irrigate immediately. If this chemical contacts the skin, flush with water immediately. If a person breathes in large amounts of this chemical, move the exposed person to fresh air at once and perform artificial respiration. When this chemical has been swallowed, get medical attention. Do not induce vomiting.

Personal Protective Methods: Wear appropriate clothing to prevent any possibility of repeated or prolonged skin contact. Wear eye protection to prevent any reasonable probability of eye contact. Employees should wash immediately when skin is wet or contaminated. Remove clothing immediately if wet or contaminated to avoid flammability hazard. Provide emergency showers.

Respirator Selection:
1,000 ppm: CCROVF/GMOV
2,000 ppm: SAF/SCBAF
Escape: GMS/SCBA

Disposal Method Suggested: Concentrated waste containing no peroxides—discharge liquid at a controlled rate near a pilot flame. Concentrated waste containing peroxides—perforation of a container of the waste from a safe distance followed by open burning (A-31).

References
(1) International Agency for Research on Cancer, *IARC Monographs on the Carcinogenic Risks of Chemicals to Humans,* Lyon, France, 11, 191 (1976).
(2) Bogyo, D.A., Lande, S.S., Meylan, W.M., Howard, P.H., and Santodonato, J., Syracuse Research Corp. Center for Chemical Hazard Assessment, *Investigation of Selected Potential Environmental Contaminants: Epoxides,* Report EPA-560/11-80-005, Washington, DC, U.S. Environmental Protection Agency (March 1980).
(3) See Reference (A-61).

(4)   See Reference (A-60).
(5)   United Nations Environment Programme, *IRPTC Legal File 1983,* Vol. II, pp VII/675–7, Geneva, Switzerland, International Register of Potentially Toxic Chemicals (1984).

n-PROPYL NITRATE

Description: $CH_3CH_2CH_2NO_2$ is a colorless to pale yellow liquid with an etherlike odor. It boils at 110° to 111°C.

Code Numbers: CAS 627-13-4   RTECS UK0350000   UN 1865

DOT Designation: Flammable liquid.

Synonyms: Nitric acid n-propyl ester.

Potential Exposures: Propyl nitrate has been used as a rocket propellant and as an ignition improver in diesel fuels.

Incompatibilities: Strong oxidizers, combustibles.

Permissible Exposure Limits in Air: The Federal standard and the 1983/84 ACGIH TWA value is 25 ppm (105 mg/m³). The STEL value is 40 ppm (170 mg/m³). The IDLH value is 2,000 ppm.

Determination in Air: Adsorption on charcoal, workup with CS_2, analysis by gas chromatography. See NIOSH Methods, Set Q. See also reference (A-10).

Permissible Concentration in Water: No criteria set.

Routes of Entry: Inhalation, ingestion, eye and skin contact.

Harmful Effects and Symptoms: None known in humans according to NIOSH (A-4) but other sources (A-38) state that vapor inhalation causes low blood pressure, hypotony and hemoglobin defect.

Points of Attack: None known.

Medical Surveillance: None indicated.

First Aid: If this chemical gets into the eyes, irrigate immediately. If this chemical contacts the skin, wash with soap promptly. If a person breathes in large amounts of this chemical, move the exposed person to fresh air at once and perform artificial respiration. When this chemical has been swallowed, get medical attention. Give large quantities of water and induce vomiting. Do not make an unconscious person vomit.

Personal Protective Methods: Wear appropriate clothing to prevent repeated or prolonged skin contact. Wear eye protection to prevent any reasonable probability of eye contact. Employees should wash promptly when skin is wet or contaminated. Remove nonimpervious clothing immediately if wet or contaminated.

Respirator Selection:
>   250 ppm: CCRS/SA/SCBA
>   1,000 ppm: CCRSF/GMS/SAF/SCBAF
>   2,000 ppm: SA:PD,PP,CF
>   Escape: GMS/SCBA
>   Note: Do not use oxidizable sorbents.

Disposal Method Suggested: Incineration; large quantities of material may require NO_x removal by catalytic or scrubbing processes (A-31). An alternative route suggested (A-38) involves pouring over soda ash, neutralizing with HCl and flushing to the drain with water.

PYRETHRINS OR PYRETHRUM

● Hazardous substance (EPA)

Description: The pyrethrins are a mixture of compounds of the structural formulas

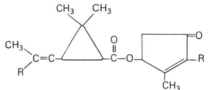

which are found in pyrethrum flowers. They have the empirical formulas $C_{20-21}H_{28-30}O_{3-5}$.

Code Numbers: CAS 80003-34-7   RTECS UR4200000   UN (NA 9184)

DOT Designation: ORM-E

Synonyms: Pyrethrin I or II, cinerin I or II, jasmolin I or II.

Potential Exposures: Those engaged in the isolation (A-32), formulation or application of these contact insecticides.

Incompatibilities: Strong oxidizers.

Permissible Exposure Limits in Air: The Federal standard and the 1983/84 ACGIH TWA value is 5 mg/m³. The STEL is 10 mg/m³. The IDLH level is 5,000 ppm.

Determination in Air: Collection by impinger or fritted bubbler, analysis by gas liquid chromatography (A-20).

Permissible Concentration in Water: No criteria set.

Routes of Entry: Inhalation, ingestion, eye and skin contact.

Harmful Effects and Symptoms: Erythema, dermatitis, sneezing, asthma. In animals convulsions, paralysis.

Points of Attack: Respiratory system, skin, central nervous system.

Medical Surveillance: Consider the points of attack in preplacement and periodic physical examinations.

First Aid: If this chemical gets into the eyes, irrigate immediately. If this chemical contacts the skin, wash with soap immediately. If a person breathes in large amounts of this chemical, move the exposed person to fresh air at once and perform artificial respiration. When this chemical has been swallowed, get medical attention. Give large quantities of water and induce vomiting. Do not make an unconscious person vomit.

Personal Protective Methods: Wear appropriate clothing to prevent repeated or prolonged skin contact. Wear eye protection to prevent any reasonable probability of eye contact. Employees should wash promptly when skin is wet or contaminated. Work clothing should be changed daily if it is possible that clothing is contaminated. Remove nonimpervious clothing promptly if wet or contaminated.

Respirator Selection:
50 mg/m³: CCROVDMFuPest/SA/SCBA
250 mg/m³: CCROVFDMFuPest/GMOVDMFuPest/SAF/SCBAF
5,000 mg/m³: PAPOVHiEPPest/SA:PD,PP,CF
Escape: GMOVPPest/SCBA

Disposal Method Suggested: Pyrethrin products could be dumped into a landfill, or buried in noncropland away from water. In each of these cases it would be better to mix the product with lime. Incineration would be an effective disposal procedure where permitted. If an efficient incinerator is not available, the product should be mixed with large amounts of combustible material and contact with the smoke should be avoided (A-32).

References
(1) See Reference (A-61).
(2) United Nations Environment Programme, *IRPTC Legal File 1983*, Vol. I, pp VII/250–51, Geneva, Switzerland, International Register of Potentially Toxic Chemicals (1984).

PYRIDINE

- Hazardous waste (EPA)

Description: C_5H_5N, pyridine, is a colorless liquid with an unpleasant odor. The odor can be detected well below 1 ppm. It is both flammable and explosive when exposed to a flame and decomposes on heating to release cyanide fumes. It boils at 115°C.

Code Numbers: CAS 110-86-1   RTECS UR8400000   UN 1282

DOT Designation: Flammable liquid.

Synonyms: Azine, azabenzene.

Potential Exposures: Pyridine is used as a solvent in the chemical industry and as a denaturant for ethyl alcohol. It is used in the manufacture of paints, explosives, dyestuffs, rubber, vitamins, sulfa drugs, and disinfectants.

Incompatibilities: Strong oxidizers, strong acids.

Permissible Exposure Limits in Air: The Federal standard and the ACGIH 1983/84 TWA value is 5 ppm (15 mg/m³). The STEL is 10 ppm (30 mg/m³). IDLH level is 3,600 ppm.

Determination in Air: Adsorption on charcoal, workup with methylene chloride, analysis by gas chromatography. See NIOSH Methods, Set L. See also reference (A-10).

Permissible Concentration in Water: No criteria set, but EPA (A-37) has suggested a permissible ambient goal of 207 µg/ℓ.

Routes of Entry: Inhalation of vapor, percutaneous absorption of liquids, ingestion, skin and eye contact.

Harmful Effects and Symptoms: *Local* — Irritation of the conjunctiva of the eye and cornea and mucous membranes of the upper respiratory tract and skin may occur. It occasionally causes skin sensitization, and photosensitization has been reported.

Systemic — Very high concentrations may cause narcosis. Repeated, intermittent, or continuous low level exposure may lead to transient effects on the central nervous system, and gastrointestinal tract. The symptoms include headache, dizziness, insomnia, nervousness, anorexia, nausea, vomiting, and diarrhea. Low back pain and urinary frequency with no changes in urine sediment or liver or renal function and complete recovery have been reported to follow exposures to about 100 ppm. Liver and kidney injury have been reported from its use as an oral medication.

Points of Attack: Central nervous system, liver, kidneys, skin, gastrointestinal system.

Medical Surveillance: Placement and periodic examinations should consider possible effects on skin, central nervous system, and liver and kidney function.

First Aid: If this chemical gets into the eyes, irrigate immediately. If this chemical contacts the skin, flush with water immediately. If a person breathes in large amounts of this chemical, move the exposed person to fresh air at once and perform artificial respiration. When this chemical has been swallowed, get medical attention. Give large quantities of water and induce vomiting. Do not make an unconscious person vomit.

Personal Protective Methods: Wear appropriate clothing to prevent any reasonable probability of skin contact. Wear eye protection to prevent any possibility of eye contact. Employees should wash immediately when skin is wet or contaminated. Remove clothing immediately if wet or contaminated to avoid flammability hazard. Provide emergency showers and eyewash.

Respirator Selection:
    250 ppm: CCROVF/GMOV/SAF/SCBAF
  3,600 ppm: SAF:PD,PP,CF
    Escape: GMOV/SCBA

Disposal Method Suggested: Controlled incineration whereby nitrogen oxides are removed from the effluent gas by scrubber, catalytic or thermal devices.

References

(1) U.S. Environmental Protection Agency, *Pyridine,* Health and Environmental Effects Profile No. 150, Office of Solid Waste, Washington, DC (April 30, 1980).

(2) See Reference (A-61).

(3) See Reference (A-60).

(4) Parmeggiani, L., Ed., *Encyclopedia of Occupational Health & Safety,* Third Edition, Vol. 2, pp 1610–12, Geneva, International Labour Office (1983).

(5) United Nations Environment Programme, *IRPTC Legal File 1983,* Vol. II, pp VII/700–702, Geneva, Switzerland, International Register of Potentially Toxic Chemicals (1984).

Q

QUINOLINE

- Carcinogen (Animal Positive) (1) (A-63)

Description: C_9H_7N has the following structural formula

Quinoline is a colorless liquid with a penetrating amine odor which boils at 238°C at atmospheric pressure.

Code Numbers: CAS 91-22-5 RTECS VA9275000 UN 2656

DOT Designation: —

Synonyms: 1-azanaphthalene; 1-benzazine; benzo(b)pyridine; chinoleine; leucoline.

Potential Exposure: In manufacture of quinoline derivatives (dyes and pesticides); in synthetic fuel manufacture. Occurs in cigarette smoke.

Permissible Exposure Limits in Air: No standards set.

Permissible Concentration in Water: No criteria set (A-36).

Routes of Entry: Inhalation, ingestion and skin absorption.

Harmful Effects and Symptoms: The oral LD_{50} for rats is 331 mg/kg which is moderately toxic. However, quinoline is a proven carcinogen in rats and mice. Irritates eyes, skin and respiratory tract (A-60).

Points of Attack: Nervous system, liver and kidneys (A-60).

First Aid: Wash affected areas with water flush, then soap solution. Remove and dispose of contaminated clothing. See a doctor.

Personal Protective Methods: Wear butyl rubber gloves, face shield and protective clothing.

Respirator Selection: Wear self-contained breathing apparatus.

Disposal Method Suggested: Incineration (A-38).

References

(1) U.S. Environmental Protection Agency, *Chemical Hazard Information Profile Draft Report: Quinoline,* Washington, D.C. (Dec. 29, 1983).

QUINONE

- Hazardous waste (EPA)

Description: $C_6H_4O_2$, quinone, exists as large yellow, monoclinic prisms; the vapors have a pungent, irritating odor. It melts at 113°C.

Code Numbers: CAS 106-51-4   RTECS DK2625000   UN 2587

DOT Designation: —

Synonyms: Benzoquinone, chinone, p-benzoquinone, 1,4-benzoquinone.

Potential Exposures: Because of its ability to react with certain nitrogen compounds to form colored substances, quinone is widely used in the dye, textile, chemical, tanning, and cosmetic industries. It is used as an intermediate in chemical synthesis for hydroquinone and other chemicals.

Incompatibilities: Strong oxidizers.

Permissible Exposure Limits in Air: The Federal standard and the 1983/84 ACGIH TWA value is 0.1 ppm (0.4 mg/m³). The STEL is 0.3 ppm (2.0 mg/m³). The IDLH level is 75 ppm.

Determination in Air: Adsorption on XAD-2 resin, desorption with ethanol/hexane, analysis by high performance liquid chromatograph with UV detector (A-10).

Permissible Concentration in Water: No criteria set.

Routes of Entry: Inhalation of vapor, ingestion, skin and eye contact.

Harmful Effects and Symptoms: *Local* — Solid quinone in contact with skin or the lining of the nose and throat may produce discoloration, severe irritation, swelling, and the formation of papules and vesicles. Prolonged contact with the skin may cause ulceration. Quinone vapor is highly irritating to the eyes. Following prolonged exposure to vapor, brownish conjunctival stains may appear. These may be followed by corneal opacities and structural changes in the cornea and loss of visual acuity. The early pigmentary stains are reversible, while the corneal dystrophy tends to be progressive.

Systemic — No systemic effects have been found in workers exposed to quinone vapor over many years.

Points of Attack: Eyes, skin.

Medical Surveillance: Careful examination of the eyes, including visual acuity and slit lamp examinations, should be done during placement and periodic examinations. Also evaluate skin.

First Aid: If this chemical gets into the eyes, irrigate immediately. If this chemical contacts the skin wash with soap immediately. If a person breathes in large amounts of this chemical, move the exposed person to fresh air at once and perform artificial respiration. When this chemical has been swallowed, get medical attention. Give large quantities of water and induce vomiting. Do not make an unconscious person vomit.

Personal Protective Methods: The skin can be damaged by contact with solid quinone, solutions, or vapor condensing on the skin, so protective clothing, gloves and boots are indicated. Wear appropriate clothing to prevent any reasonable probability of skin contact. Wear eye protection to prevent any possibility of eye contact. Employees should wash immediately when skin is wet or contaminated. Work clothing should be changed daily if it is possible that clothing is contaminated. Remove nonimpervious clothing immediately if wet or contaminated. Provide emergency showers and eyewash.

Respirator Selection:

    5 ppm: SAF/SCBAF
    75 ppm: SAF:PD,PP,CF
    Escape: GMOVP/SCBA

Disposal Method Suggested: Controlled incineration (1800°F, 2.0 seconds minimum).

References

(1) U.S. Environmental Protection Agency, *Quinone,* Health and Environmental Effects Profile No. 157, Wash., DC, Office of Solid Waste (April 30, 1980).
(2) See Reference (A-61).
(3) See Reference (A-60).
(4) United Nations Environment Programme, *IRPTC Legal File 1983,* Vol. I, pp VII/123–24, Geneva, Switzerland, International Register of Potentially Toxic Chemicals (1984).

QUINTOZENE

See "Pentachloronitrobenzene."

R

RESERPINE

- Carcinogen (Animal Positive, IARC) (1)
- Hazardous Waste Constituent (EPA)

Description: $C_{33}H_{40}N_2O_9$ has the structural formula

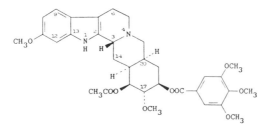

This is a crystalline substance which decomposes at 264° to 265°C. Reserpine, a pharmaceutical, is a naturally occurring substance that is isolated from the roots of the plant *Rauwolfia serpentina*.

Code Numbers: CAS 50-55-5   RTECS ZG0350000

DOT Designation: —

Synonyms: 3,4,5-Trimethoxybenzoyl methyl reserpate; many trade names.

Potential Exposure: Used as a hypertensive for humans and animals.

Permissible Exposure Limits in Air: No standards set.

Permissible Concentration in Water: No criteria set.

Harmful Effects and Symptoms: Reserpine is highly toxic to man. In humans, 0.014 mg/kg produces psychotropic effects.

Reserpine was tested in mice by oral administration; where it induced malignant mammary tumors in females and carcinomas of the seminal vesicles in males. It was tested in rats by oral administration; it increased the incidence of pheochromocytomas in males.

Thirteen case-control studies were available to the working group. Most report a relative risk of between 1 and 2 for breast cancer associated with the use of reserpine. Patients who had taken reserpine for more than 5 years had slightly higher relative risks. In 11 of the 13 studies the relative risks were not statistically significant, although pooling of the studies gave a summary relative risk of 1.2 with 95% confidence intervals 1.1 to 1.4. The possibility of confounding due to several medical care variables could not be excluded, hence none of the studies, either singly or pooled, provide conclusive evidence of a causal association.

References

(1) See Reference (A-62). Also see Reference (A-64).

RESORCINOL

● Hazardous waste (EPA)

Description: $1,3-C_6H_4(OH)_2$ is a white-to-pink crystalline compound with a characteristic odor and a sweetish taste which melts at $110°C$.

Code Numbers: CAS 108-46-3   RTECS VG9625000   UN 2876

DOT Designation: ORM-E.

Synonyms: m-Dihydroxybenzene, m-hydroxyphenol.

Potential Exposures: Resorcinol is weakly antiseptic and resorcinol compounds are used in pharmaceuticals and hair dyes for human use. Major industrial uses are as adhesives in rubber products and tires, wood adhesive resins, and as ultraviolet absorbers in polyolefin plastics. Resorcinol is also a by-product of coal conversion and is a component of cigarette smoke. Thus, substantial opportunity exists for human exposure.

Permissible Exposure Limits in Air: There are no Federal regulations but ACGIH (1983/84) has adopted a TWA value of 10 ppm (45 mg/m^3) and set an STEL of 20 ppm (90 mg/m^3).

Permissible Concentration in Water: No criteria set.

Routes of Entry: Inhalation, ingestion, skin absorption, skin and eye contact.

Harmful Effects and Symptoms: Many phenolic compounds, including resorcinol, are strong mitotic spindle poisons in plants. This evidence of mutagenic activity and the strong oncogenic activity in plants have not been adequately tested in animals to provide an understanding of the processes. In animals the only cocarcinogenic activity (in cigarette smoke condensate) demonstrated has been as a protective agent against benzo[a]pyrene carcinogenicity.

Resorcinol has been demonstrated to result in chronic toxicity: reducing growth rate in an insect species and causing chronic health complaints from workers in a tire manufacturing plant. Acute toxicity through oral, eye, skin penetration, and skin irritation has been demonstrated by all tests. Values vary in the literature and are inadequate to draw a quantitative conclusion.

Symptoms of resorcinol exposure include: conjunctivitis, dermatitis, dizziness, restlessness, tachycardia, dyspnea, drowsiness, sweating, cyanosis, hepatomegaly and jaundice, splenomegaly, unconsciousness and convulsions (A-38) (A-39).

Medical Surveillance: Perform complete blood count, urinalysis, studies of liver and kidney functions annually (A-39).

First Aid: Irrigate eyes with water, wash contaminated areas of body with soap and water. If ingested, flush stomach with water (A-38, A-39).

Personal Protective Methods: Provide adequate ventilation. Wear chemical goggles, rubber gloves and protective clothing. Preclude from exposure those individuals with liver, kidney or blood diseases.

Respirator Selection: Wear mechanical filter respirator (A-39).

Disposal Method Suggested: Dissolve in a combustible solvent and incinerate (A-38).

References

(1) U.S. Environmental Protection Agency, *Resorcinol,* Health and Environmental Effects Profile No. 152, Office of Solid Waste, Washington, DC (April 30, 1980).

(2) Sax, N.I., Ed., *Dangerous Properties of Industrial Materials Report, 1,* No. 2, 58-59, New York, Van Nostrand Reinhold Co. (1980).

(3) See Reference (A-60).

(4) United Nations Environment Programme, *IRPTC Legal File 1983,* Vol. II, pp VII/709-10, Geneva, Switzerland, International Register of Potentially Toxic Chemicals (1984).

RHODIUM METAL

Description: Rh, together with platinum, palladium, iridium, ruthenium and osmium is one of the platinum-group metals in Group VIII of the Periodic Table.

Code Numbers: CAS 7440-16-6   RTECS VI9355000

DOT Designation: —

Synonyms: None.

Potential Exposure: Rhodium has few applications by itself, as in rhodium plating of white gold jewelry or plating of electrical parts such as commutator slip rings, but, mainly, rhodium is used as a component of platinum alloys. Rhodium-containing catalysts have been proposed for use in automotive catalytic converters for exhaust gas cleanup.

Incompatibilities: None hazardous.

Permissible Exposure Limits in Air: The Federal standard and the ACGIH 1983/84 TWA value for rhodium metal fume and dusts is 0.1 mg/m^3. There is no STEL value. There is no IDLH level set.

Determination in Air: Collection on a filter, workup with acid; determination by atomic absorption. See NIOSH Methods, Set M. See also reference (A-10).

Permissible Concentration in Water: No criteria set.

Route of Entry: Inhalation.

Harmful Effects and Symptoms: None known in humans.

Points of Attack: None known.

Medical Surveillance: No special attention indicated.

First Aid: If a person breathes in large amounts of this chemical, move the exposed person to fresh air at once and perform artificial respiration. When this chemical has been swallowed, get medical attention. Give large quantities of water and induce vomiting. Do not make an unconscious person vomit.

Personal Protective Methods: No particular devices except respirators specified by NIOSH (A-4).

Respirator Selection:
   0.5 mg/m³: DM
   1 mg/m³: HiEPFu/DMXSQ/SA/SCBA
   5 mg/m³: HiEPF/SAF/SCBAF
   100 mg/m³: PAPHiE/SA:PD,PP,CF
   200 mg/m³: SAF:PD,PP,CF

Disposal Method Suggested: Recovery in view of the high economic value. Recovery techniques for recycle of rhodium in plating wastes and spent catalysts have been described (A-57).

References
(1) See Reference (A-61).
(2) Parmeggiani, L., Ed., *Encyclopedia of Occupational Health & Safety,* Third Edition, Vol. 2, pp 1939–40, Geneva, International Labour Office (1983).

RHODIUM TRICHLORIDE

Description: $RhCl_3 \cdot xH_2O$ is a red-brown odorless solid or liquid which melts at 100°C.

Code Numbers: CAS 10049-07-7   RTECS VI9275000

DOT Designation: —

Synonyms: Rhodium chloride, soluble rhodium trichloride, hydrated rhodium trichloride.

Potential Exposures: See "Rhodium Metal." In plating operations and in catalyst preparation, the metal will be used as the trichloride.

Incompatibilities: None hazardous.

Permissible Exposure Limits in Air: The Federal standard and the 1983/84 ACGIH TWA value is 0.001 mg/m³. The STEL is 0.003 mg/m³ but consideration is being given to eliminating the TWA value. There is no IDLH level set. Note: As of 1980, ACGIH has also proposed values for insoluble rhodium compounds as follows: TWA 0.1 mg/m³ but no STEL.

Determination in Air: Collection on a filter, workup with acid, analysis by atomic absorption. See NIOSH Methods, Set M. See also reference (A-10).

Permissible Concentration in Water: No criteria set.

Routes of Entry: Inhalation, ingestion, skin and eye contact.

Harmful Effects and Symptoms: Mild eye irritation.

Point of Attack: Eyes.

Medical Surveillance: Consider the point of attack in preplacement and periodic physical examinations.

First Aid: If this chemical gets into the eyes, irrigate. If this chemical contacts the skin, flush with water. If a person breathes in large amounts of this chemical, move the exposed person to fresh air at once and perform artificial respiration. When this chemical has been swallowed, get medical attention. Give large quantities of water and induce vomiting. Do not make an unconscious person vomit.

Personal Protective Methods: Wear appropriate clothing to prevent repeated

or prolonged skin contact. Wear eye protection to prevent any reasonable probability of eye contact. Employees should wash promptly when skin is wet or contaminated. Remove nonimpervious clothing promptly if contaminated or wet.

Respirator Selection:
0.01 mg/m^3: HiEP/SA/SCBA
0.05 mg/m^3: HiEPF/SAF/SCBAF
1 mg/m^3: PAPHiE/SA:PD,PP,CF
2 mg/m^3: SAF:PD,PP,CF

Disposal Method Suggested: Recovery and reclaiming wherever possible in view of high economic value. See "Rhodium Metal."

References

(1) See Reference (A-61).

RONNEL

Description: $Cl_3C_6H_2OP(S)(OCH_3)_2$ is a white-to-tan waxy solid melting at about $40°C$.

Code Numbers: CAS 299-84-3 RTECS TG0525000 UN 2588

DOT Designation: —

Synonyms: O,O-Dimethyl O-2,4,5-trichlorophenylphosporothioate, Fenchlorphos, ENT 23,284, Nankor®, Trolene®, Korlan®.

Potential Exposures: Those involved in manufacture, formulation and application of this insecticide for household and farm uses.

Incompatibilities: Strong oxidizers.

Permissible Exposure Limits in Air: The Federal standard and the 1983/84 ACGIH TWA value is 10 mg/m^3. There is no STEL value set. The IDLH level is $5,000$ mg/m^3.

Determination in Air: Collection by impinger or fritted bubbler, analysis by gas liquid chromatography (A-1).

Permissible Concentration in Water: No criteria set.

Routes of Entry: Inhalation, ingestion, skin and eye contact.

Harmful Effects and Symptoms: Cholinesterase inhibition, eye irritation, liver and kidney damage. Issuance of a rebuttable presumption against registration for ronnel for pesticidal uses was being considered by EPA as of October 1980 on the bases of possible oncogenicity, teratogenicity and fetotoxic effects (A-32).

Points of Attack: Skin, liver, kidneys, blood plasma.

Medical Surveillance: Consider the points of attack in preplacement and periodic physical examinations.

First Aid: If this chemical gets into the eyes, irrigate immediately. If this chemical contacts the skin, wash with soap promptly. If a person breathes in large amounts of this chemical, move the exposed person to fresh air at once and perform artificial respiration. When this chemical has been swallowed, get medical attention. Give large quantities of water and induce vomiting. Do not make an unconscious person vomit.

Personal Protective Methods: Wear appropriate clothing to prevent repeated or prolonged skin contact. Wear eye protection to prevent any reasonable probability of eye contact. Employees should wash promptly when skin is wet or contaminated. Work clothing should be changed daily if clothing is contaminated. Remove nonimpervious clothing promptly if contaminated or wet.

Respirator Selection:

100 mg/m³:	CCROVDMFuPest/SA/SCBA
500 mg/m³:	CCROVFDMFuPest/GMOVDMFuPest/SAF/SCBAF
5,000 mg/m³:	PAPOVHiEPPest/SA:PD,PP,CF
Escape:	GMOVPPest/SCBA

Disposal Method Suggested: Incineration with added flammable solvent in furnace equipped with afterburner and alkali scrubber.

References

(1)   See Reference (A-61).
(2)   United Nations Environment Programme, *IRPTC Legal File 1983,* Vol. II, pp VII/620, Geneva, Switzerland, International Register of Potentially Toxic Chemicals (1984).

ROSIN CORE SOLDER PYROLYSIS PRODUCTS

Description: The thermal decomposition products of wood rosin used as a soldering flux include acetone, methyl alcohol, aliphatic aldehydes, CO, CO_2, methane, ethane, abietic acid and related diterpene acids. These products vary both quantitatively and qualitatively with soldering temperature (A-34).

DOT Designation: —

Potential Exposures: Those involved in soldering operations in the production of fabricated metal products.

Permissible Exposure Limits in Air: There is no Federal standard but ACGIH (1983/84) has adopted a TWA value of 0.1 mg/m³ (as formaldehyde) and set an STEL of 0.3 mg/m³ (as formaldehyde).

Determination in Air: Collection by impinger or fritted bubbler and colorimetric analysis (A-10).

Permissible Concentration in Water: No criteria set.

Route of Entry: Inhalation.

Harmful Effects and Symptoms: Irritation of eyes, nose and throat.

Points of Attack: Eyes, nose and throat.

Medical Surveillance: Give attention to points of attack in preplacement and periodic physical examinations.

ROTENONE

Description: $C_{23}H_{22}O_6$, rotenone, with the following structural formula, is a colorless-to-red odorless solid which melts at 163°C.

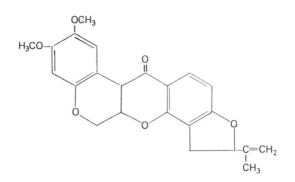

Code Numbers: CAS 83-79-4 RTECS DJ2800000 UN 2588

DOT Designation: —

Synonyms: [2R-(2a,6aα,12aα)] -1,2,12,12a-tetrahydro-8,9-dimethoxy-2-(1-methylethenyl) [1] -benzopyrano[3,4-b] furo[2,3-h] [1] -benzopyran-6(6aH)one; ENT 133; tubatoxin; derrin.

Potential Exposures: Those involved in extraction from derris root, formulation or application of this insecticide.

Incompatibilities: Strong oxidizers.

Permissible Exposure Limits in Air: The Federal standard and the 1983/84 ACGIH TWA value is 5 mg/m^3. The STEL is 10 mg/m^3. The IDLH level is 5,000 mg/m^3.

Determination in Air: Collection by impinger or fritted bubbler, analysis by ultraviolet spectrophotometry (A-28).

Permissible Concentration in Water: No criteria set.

Routes of Entry: Inhalation, ingestion, skin and eye contact.

Harmful Effects and Symptoms: Symptoms include skin, eye and pulmonary irritation; numbness of the mucous membrane; nausea, vomiting, abdominal pain; muscular tremors, incoherence, clonic convulsions; stupor. Issuance of a rebuttable presumption against registration for rotenone for pesticide uses was being considered by EPA on the basis of possible oncogenicity, mutagenicity, teratogenicity, reproductive effects, chronic toxicity and toxicity to wildlife.

Points of Attack: Central nervous system, eyes, respiratory system.

Medical Surveillance: Consider the points of attack in preplacement and periodic physical examinations.

First Aid: If this chemical gets into the eyes, irrigate immediately. If this chemical contacts the skin, wash with soap promptly. If a person breathes in large amounts of this chemical, move the exposed person to fresh air at once and perform artificial respiration. When this chemical has been swallowed, get medical attention. Give large quantities of water and induce vomiting. Do not make an unconscious person vomit.

Personal Protective Methods: Wear appropriate clothing to prevent repeated or prolonged skin contact. Wear eye protection to prevent any reasonable probability of eye contact. Employees should wash promptly when skin is wet or contaminated. Work clothing should be changed daily if it is possible that clothing is contaminated. Remove nonimpervious clothing promptly if wet or contaminated.

Respirator Selection:

$$\begin{array}{rl}
50 \text{ mg/m}^3\text{:} & \text{CCROVDMPest/SA/SCBA} \\
250 \text{ mg/m}^3\text{:} & \text{CCROVFDMPest/GMOVDMPest/SAF/SCBAF} \\
5{,}000 \text{ mg/m}^3\text{:} & \text{CCROVHiEPest/SA:PD,PP,CF} \\
\text{Escape:} & \text{GMOVPPest/SCBA}
\end{array}$$

Disposal Method Suggested: Rotenone is decomposed by light and alkali to less insecticidal products. It is readily detoxified by the action of light and air. It is also detoxified by heating; 2 hours at 100°C results in 76% decomposition. Oxidation products are probably nontoxic. Incineration has been recommended as a disposal procedure. Burial with lime would also present minimal danger to the environment (A-32).

References

(1)   See Reference (A-61).
(2)   United Nations Environment Programme, *IRPTC Legal File 1983,* Vol. I, pp VII/120, Geneva, Switzerland, International Register of Potentially Toxic Chemicals (1984).

RUBBER SOLVENT (NAPHTHA)

See "Naphthas."

S

SACCHARIN

- Carcinogen (Animal Positive, IARC) (1)
- Hazardous Waste Constituent (EPA)

Description: $C_7H_5NO_3S$ with the following structural formula

is a crystalline solid melting at 229°C.

Code Numbers: CAS 81-07-2   RTECS DE4200000

DOT Designation: —

Synonyms: 1,2-Benzisothiazolin-3-one 1,1-dioxide; o-benzoic sulfimide; o-sulfobenzoic acid imide.

Potential Exposure: Saccharin has been used as a nonnutritive sweetening agent; however, its use substantially increased after cyclamates were banned in food in 1970. The U.S. consumption pattern for all forms of saccharin has been estimated as 45% in soft drinks; 18% in tabletop sweeteners; 14% in fruits, juices, sweets, chewing gum, and jellies; 10% in cosmetics and oral hygiene products; 7% in drugs such as coating on pills; 2% in tobacco; 2% in electroplating; and 2% for miscellaneous uses.

Human exposure to saccharin occurs primarily through ingestion because of its use in many dietic foods and drinks and some personal hygiene products, including toothpastes and mouthwashes. The general public is exposed to saccharin, especially by persons required to reduce sugar intake. The National Occupational Hazard Survey and OSHA estimated that approximately 28,000 workers are occupationally exposed.

In compliance with the Delaney clause, FDA proposed to ban saccharin as a food additive in 1977 because of evidence of carcinogenicity in animals. However, final regulations are pending because of congressional action in 1977 requiring further study and labeling of saccharin. A joint FDA/NCI group, the Saccharin Working Group, was formed in 1977. A preliminary report of the epidemiologic findings has been published, but an analysis of all the information gathered is yet to be completed.

Permissible Exposure Limits in Air: No standards set.

Permissible Concentration in Water: No criteria set.

Harmful Effects and Symptoms: When given in the diet at relatively high doses, saccharin produces cancer of the urinary tract in rats (1) (2). When inserted in the bladder as an implant, saccharin causes bladder cancer in mice (1).

References

(1)   *IARC Monographs on the Evaluation of Carcinogenic Risk of Chemicals to Man,* Vol. 22, IARC, Lyon, France, pp 111-185 (1980).
(2)   See Reference (A-62). Also see Reference (A-64).
(3)   United Nations Environment Programme, *IRPTC Legal File 1983,* Vol. I, pp VII/113–16, Geneva, Switzerland, International Register of Potentially Toxic Chemicals (1984).

SAFROLE

- Carcinogen (Animal Positive, IARC) (1)
- Hazardous Waste Constituent (EPA)

Description: $C_{10}H_{10}O_2$ with the following structural formula

is a colorless to yellow liquid boiling at 232° to 234°C.

Code Numbers: CAS 94-59-7   RTECS CY2800000

DOT Designation: –

Synonyms: 4-Allyl-1,2-methylenedioxybenzene; allylcatechol methylene ether.

Potential Exposure: This compound has been used to flavor beverages and foods. It is also reported to be used in soap manufacture, perfumery, sleep aids, sedatives, and pesticides.

According to EPA (1980), nine companies reported producing and importing a total of over 2 million pounds of the compound in 1977. No production data are available in the trade literature after 1977. However, according to the United States International Trade Commission (1981), approximately 36,000 pounds of safrole were imported from Brazil during 1980.

OSHA reported that approximately 30 workers are potentially exposed. The FDA estimated exposure to safrole of the general public through food consumption was extremely low since the Agency prohibited its use in food. Minimal exposure may occur through the use of edible spices, including nutmeg and mace, which contain low levels of naturally occurring safrole.

Permissible Exposure Limits in Air: No standards set.

Permissible Concentration in Water: The compound does not pose a hazard to the general population through consumption of drinking water because safrole is insoluble in water.

Harmful Effects and Symptoms: When given in the diet, safrole produces liver cancers in male mice and in male and female rats (1).

References

(1) *IARC Monographs on the Evaluation of Carcinogenic Risk of Chemicals to Man,* Vol. 10, IARC, Lyon, France, pp 231-244 (1976).
(2) See Reference (A-62). Also see Reference (A-64).

SELENIUM AND COMPOUNDS

- Hazardous waste (Selenium Oxide) (EPA)
- Hazardous waste constituents (EPA)
- Priority toxic pollutant (EPA)

Description: Se, selenium, exists in three forms: a red amorphous powder, a grey form, and red crystals. Selenium, along with tellurium, is found in the sludges and sediments from electrolytic copper refining. It may also be recovered in flue dust from burning pyrites in sulfuric acid manufacture.

See separate listings for "Selenium Hexafluoride" and "Hydrogen Selenide."

Code Numbers:

Selenium metal	CAS 7782-49-2	RTECS VS7700000	UN 2658
Selenium dioxide	CAS 7446-08-4	RTECS VS8575000	UN 2811

DOT Designation:

Selenium metal	none
Selenium oxide	Poison B

Potential Exposure: Most of the selenium produced is used in the manufacture of selenium rectifiers. It is utilized as a pigment for ruby glass, paints, and dyes, as a vulcanizing agent for rubber, a decolorizing agent for green glass, a chemical catalyst in the Kjeldahl test, and an insecticide; in the manufacture of electrodes, selenium photocells, selenium cells, and semiconductor fusion mixtures; in photographic toning baths; and for dehydrogenation of organic compounds. It is also used in veterinary medicine and in antidandruff shampoos. Se is used in radioactive scanning of the pancreas and for photostatic and x-ray xerography. It may be alloyed with stainless steel, copper, and cast steel.

Selenium is a contaminant in most sulfide ores of copper, gold, nickel, and silver, and exposure may occur while removing selenium from these ores.

Incompatibilities: Acids, strong oxidizing agents.

Permissible Exposure Limits in Air: The Federal standard and the 1983/84 ACGIH TWA value are: selenium compounds (as Se), 0.2 mg/m^3. There is no STEL value. The IDLH level is 100 mg/m^3.

Determination in Air: Collection on a filter, workup with acid, analysis by atomic absorption. See NIOSH Methods, Set M. See also reference (A-10).

Permissible Concentration in Water: To protect freshwater aquatic life— 35 µg/ℓ as a 24 hour average, never to exceed 260 µg/ℓ for recoverable inorganic selenite. To protect saltwater aquatic life—54 µg/ℓ as a 24 hour average, never to exceed 410 µg/ℓ. To protect human health—10 µg/ℓ.

The World Health Organization (WHO) has also endorsed a 10 µg/ℓ limit for selenium in drinking water (A-65). The Federal Republic of Germany has set a value of 8 µg/ℓ.

Determination in Water: Digestion followed by atomic absorption gives total selenium. Dissolved selenium is determined by 0.45 µ filtration prior to the above analysis.

Routes of Entry: Inhalation of dust or vapor, percutaneous absorption of liquid, ingestion, eye and skin contact.

Harmful Effects and Symptoms: *Local* — Elemental selenium is considered to be relatively nonirritating and is poorly absorbed. Some selenium compounds (particularly selenium dioxide and selenium oxychloride) are strong vesicants and can cause destruction of the skin. They are strong irritants to the upper respiratory tract and eyes, and may cause irritation of the mucous membrane of the stomach. Selenium compounds also may cause dermatitis of exposed areas.

Allergy to selenium dioxide has been reported in the form of an urticarial generalized rash, and may cause a pink discoloration of the eyelids and palpebral conjunctivitis ("rose-eye"). Selenium oxide also may penetrate under the free edge of the nail, causing excruciatingly painful nail beds and painful paronychia. Selenium compounds may be absorbed through intact skin to produce systemic effects (Se sulfide in shampoo).

Selenium is considered to be an essential trace element for rats and chickens, and there is strong evidence of its essentiality in man. It is capable of antagonizing the toxic effects of certain other metals, e.g., As and Cd.

Systemic — Selenium dioxide inhaled in large quantities may produce pulmonary edema.

The first and most characteristic sign of selenium absorption is a garlic odor of the breath. This may be related to the excretion in the breath of small amounts of dimethyl selenide. This odor dissipates completely in 7 to 10 days after the worker is removed from the exposure. It cannot be relied upon as a certain guide to selenium absorption. A more subtle and earlier sign is a metallic taste in the mouth, but many workers accept this without complaint.

Other systemic effects are less specific: pallor, lassitude, irritability, vague gastrointestinal symptoms (indigestion), and giddiness. Vital organs appear to escape harm from selenium absorption but, based on the results of animal experimentation, liver and kidney damage should be regarded as possible.

Liver damage and other effects have been long recognized in livestock grazing on high selenium soils. Selenium has been mentioned for its carcinogenic, anticarcinogenic, and teratogenic effects, but, to date, these effects have not been seen in man.

Points of Attack: Upper respiratory system, eyes, skin, liver, kindeys, blood.

Medical Surveillance: Preemployment and periodic examinations should consider especially the skin and eyes as well as liver, respiratory and kidney disease and function. The fingernails should be examined.

Urinary selenium excretion has been used to indicate exposure in the environment and also occupational exposure. It varies with the Se content of the diet and geographic location. Dimethyl selenide can be determined in breath.

First Aid: If this chemical gets into the eyes, irrigate immediately. If this chemical contacts the skin, wash with soap immediately. If a person breathes in large amounts of this chemical, move the exposed person to fresh air at once and perform artificial respiration. When this chemical has been swallowed, get medical attention. Give large quantities of water and induce vomiting. Do not make an unconscious person vomit.

Personal Protective Methods: Protective clothing with special emphasis on personal hygiene (showering and care of fingernails) should help prevent skin exposure and sensitization. Masks and supplied air respirators are needed in areas where concentrations of dust and vapors exceed the allowable standards. These should be equipped with fullface plates. Work clothing should be changed daily and showering encouraged prior to change to street clothing.

Respirator Selection:
          10 mg/m³ : HiEPF/SAF/SCBAF
      100 mg/m³ : SAF:PD,PP,CF/PAPHiEF
        Escape: SCBAF/HiEPF

Disposal Method Suggested: Powdered selenium: dispose in a chemical waste landfill (A-31). When possible, recover selenium and return to suppliers (A-38).

References

(1) Rosenfeld, I., and Beath, O.A., *Selenium; Geobotany, Biochemistry, Toxicology and Nutrition,* New York, Academic Press (1964).

(2) U.S. Environmental Protection Agency, *Preliminary Investigation of Effects on the Environment of Boron, Indium, Nickel, Selenium, Tin, Vanadium and Their Compounds. Selenium,* Washington, DC (1975).

(3) National Academy of Sciences, *Selenium,* Washington, DC (1976). (Also issued by EPA Health Effects Res. Lab. as Report EPA-600/1-76-014, Research Triangle Park, NC).

(4) U.S. Environmental Protection Agency, *Selenium: Ambient Water Quality Criteria,* Washington, DC (1980).

(5) U.S. Environmental Protection Agency, *Selenium,* Health and Environmental Effects Profile No. 153, Office of Solid Waste, Washington, DC (April 30, 1980).

(6) See Reference (A-61).

(7) See Reference (A-60) for citations under Selenium Oxide, Selenium Oxychloride, Selenium Trioxide, and Selenous Acid.

(8) Parmeggiani, L., Ed., *Encyclopedia of Occupational Health & Safety,* Third Edition, Vol. 2, pp 2017–19, Geneva, International Labour Office (1983).

(9) United Nations Environment Programme, *IRPTC Legal File 1983,* Vol. II, pp VII/713–15, Geneva, Switzerland, International Register of Potentially Toxic Chemicals (1984).

SELENIUM HEXAFLUORIDE

Description: SeF_6 is a colorless gas.

Code Numbers: CAS 7783-79-1   RTECS VS9450000   UN 2194

DOT Designation: Poison gas.

Synonyms: Selenium fluoride.

Potential Exposure: Selenium hexafluoride (SeF_6) is a gas and is utilized as a gaseous electric insulator.

Incompatibilities: None hazardous.

Permissible Exposure Limits in Air: The Federal standard and the 1983/84 ACGIH TWA value is 0.05 ppm (0.2 mg/m³). An STEL value has not been set. The IDLH level is 5 ppm.

Determination in Air: Collection by impinger or fritted bubbler, analysis by atomic absorption spectrometry (A-21).

Permissible Concentration in Water: No criteria set.

Routes of Entry: Inhalation, skin and eye contact.

Harmful Effects and Symptoms: None known in humans; in animals: respiratory irritation, breathing difficulties.

Points of Attack: None known.

Medical Surveillance: Nothing special indicated.

First Aid: If a person breathes in large amounts of this chemical, move the exposed person to fresh air at once and perform artificial respiration.

Personal Protective Methods: Only respirator protection is specified by NIOSH (A-4).

Respirator Selection:

    0.5 ppm: SA/SCBA
    2.5 ppm: SAF/SCBAF
      5 ppm: SA:PD,PP,CF
    Escape: GMS/SCBA

References

(1)   See Reference (A-61).

SELENIUM SULFIDE

- Carcinogen (Animal Positive) (A-64)
- Hazardous Waste Constituent (EPA)

Description: Selenium sulfide may take the form of Se_4S_4 which is a red crystalline solid melting at $113°C$ or Se_2S_6 which is a light orange crystalline solid melting at $121.5°C$.

Code Numbers: CAS 7446-34-6   RTECS VT0525000 (for SeS)

DOT Designation: —

Potential Exposure: Selenium sulfide is used for the treatment of seborrhea especially in shampoos. The chemical is available over the counter as Selsun®, a stabilized buffered suspension. FDA reports that selenium sulfide is an active ingredient in some drug products used for the treatment of dandruff and certain types of dermatitis. A dandruff shampoo containing 1% selenium sulfide is available without a prescription and is recommended for use once or twice a week. By prescription, selenium sulfide is available in a 2.5% shampoo or lotion, with the recommended application limited to 10 minutes for 7 days to avoid the possibility of acute toxic effects. Selenium sulfide is also used topically in veterinary medicine for eczemas and dermatomycoses.

NCI estimates that substantial exposure of the population to selenium sulfide is questionable. Skin absorption has been reported only in patients with open scalp lesions. In 1974, the National Occupational Hazard Survey estimated that 8,500 workers were potentially exposed to selenium sulfide.

Selenium is widely distributed throughout the environment, occurring in groundwater, surface water, rocks, and soil. No data on the environmental occurrence of selenium sulfide are available; however, the EPA in 1976 estimated that about 700 pounds of selenium sulfide wastes are generated annually by the medicinal industry.

Permissible Exposure Limits in Air: OSHA has set a standard of 0.2 mg/m³ (as selenium).

Permissible Concentration in Water: A limit of 10 $\mu g/\ell$ of selenium has been set for domestic water supplies on a health basis (A-3).

Harmful Effects and Symptoms: Dermal application of selenium sulfide to

ICR Swiss mice was not carcinogenic, but the study was limited to 88 weeks because of test animal deaths from amyloidosis (1). An NCI skin painting study with ICR Swiss mice using Selsun®, a commercial selenium sulfide formulation, also gave negative results, and was terminated after 88 weeks because of the same amyloidosis problem (2). In another chronic skin painting study of mice and rabbits, Selsun® was also reported negative (3). However, selenium sulfide administered by gavage to F344 rats and B6C3F1 mice induced hepatocellular carcinomas in male and female rats and female mice, and alveolar/bronchiolar carcinomas and adenomas in female mice, but it was not carcinogenic to male mice (4). These results can be regarded as providing sufficient evidence of carcinogenicity in experimental animals (5).

References

(1) National Cancer Institute, *Bioassay of Selenium Sulfide for Possible Carcinogenicity (Dermal Study)*, Technical Report Series No. 197, DHHS Publication No. (NIH) 80-1753, Bethesda, Maryland (1980).

(2) National Cancer Institute, *Bioassay of Selsun® for Possible Carcinogenicity (Skin Painting Study)*, DHHS Publication No. (NIH) 80-1753, Bethesda, Maryland (1980).

(3) Stenback, F., Local and Systemic Effects of Commonly Used Cutaneous Agents: Lifetime Studies of 16 Compounds in Mice and Rabbits, *Acta Pharmacol. Toxicol.*, Vol. 41, pp 417-431 (1977).

(4) National Cancer Institute, *Bioassay of Selenium Sulfide for Possible Carcinogenicity (Gavage Study)*, Technical Report Series No. 199, DHHS Publication No. (NIH) 80-1750, Bethesda, Maryland (1980).

(5) Griesemer, R.A., and C. Cueto, Toward a Classification Scheme for Degrees of Experimental Evidence for the Carcinogenicity of Chemicals for Animals. In: R. Montesano, H. Bartsch, and L. Tomatis (eds.), *Molecular and Cellular Aspects of Carcinogen Screening Tests,* IARC Scientific Publications, No. 27, Lyon, France: International Agency for Research on Cancer, pp 259-281 (1980).

SILANES

Description: The silanes have the general formula $Si_nH_{(2n+2)}$ from which many other silane compounds are derived. Derivatives have been produced by replacing one or more of the hydrogen atoms with an inorganic or organic group. Examples of these groups are halogens, oxygen, nitrogen, metals, and various organic compounds. Some silanes of major industrial importance are organochlorosilanes, methylchlorosilanes, trichlorosilane, tetrachlorosilane, phenyl chlorosilanes, phenyl ethoxysilanes and methyl ethoxysilanes. Silane, SiH_4, is a gas with an unpleasant odor.

Code Numbers:

Silane, SiH_4	CAS 7803-62-5	RTECS VV1400000 UN 2203
Methyltrichlorosilane, CH_3SiCl_3	CAS 75-79-6	RTECS VV4550000 UN 1250

DOT Designation:

Silane	Poison gas, flammable gas
Methyltrichlorosilane	Flammable liquid

Synonyms:

Silane	Silicon tetrahydride
Methyltrichlorosilane	Trichloromethylsilane

Potential Exposure: The parent silane compounds are reported to have little application in industry; however, many of the derivates have wide use. Many are used in chemical synthesis. For example, many of the organochlorosilane compounds are used to produce silicone fluids and silicone resins. Trichlorosilane and tetrachlorosilane are used in the manufacture of transistors.

Permissible Exposure Limits in Air: There is no Federal standard. ACGIH (1983/84) has proposed a TWA for silane of 5 ppm (7 mg/m^3), but no STEL value. This is tenfold higher than previous values set by ACGIH.

Permissible Concentration in Water: No criteria set.

Routes of Entry: Inhalation, ingestion, skin contact.

Harmful Effects and Symptoms: The silanes are reported to be highly toxic by inhalation, ingestion, or skin contact, following an acute exposure. Any of the chlorosilanes can emit a highly irritating, asphyxiating vapor. HCl is liberated upon hydrolysis.

Personal Protective Methods: For silane, SiH$_4$: wear rubber gloves, fireproof clothing and face shield.

Disposal Method Suggested: Silicon tetrachloride: Pretreatment involves addition of soda ash-slaked lime solution to form the corresponding sodium and calcium salt solution. This solution can be safely discharged after dilution (A-31). Silane, SiH$_4$: Controlled burning (A-38).

References

(1) National Institute for Occupational Safety and Health, *Profiles on Occupational Hazards for Criteria Document Priorities: Silicone and its Compounds,* Report PB-274,073, Cincinnati, Ohio, pp. 55–61 (1977).

SILICA, AMORPHOUS

Description: SiO$_2$ in the form of colloidal or fused silica or silica aerogel. It is a colorless to gray, odorless powder.

Code Numbers:

Amorphous fumed silica	CAS 7631-86-9	RTECS VV7310000
Amorphous fused silica	CAS 60676-86-0	RTECS VV7320000

DOT Designation: −

Synonyms: Amorphous fumed silica: silica aerogel, silicic anhydride. Amorphous fused silica: colloidal silica, diatomaceous earth.

Potential Exposures: Those involved in the production and handling of fumed silica for paint pigments or catalysts. Those involved in mining of diatomaceous earth or fabrication of products therefrom.

Incompatibilities: Fluorine, oxygen difluoride, chlorine trifluoride.

Permissible Exposure Limits in Air: The Federal standard is 20 mppcf. The ACGIH (1983/84) has recommended a TLV of 10.0 mg/m^3 for total amorphous silica dust and 5.0 mg/m^3 for respirable dust (less than 5 μm). There is no IDLH level set.

Permissible Concentration in Water: No criteria set.

Routes of Entry: Inhalation.

Harmful Effects and Symptoms: Pneumoconiosis.

Points of Attack: Respiratory system, lungs.

Medical Surveillance: Consider the points of attack in preplacement and periodic physical examinations.

First Aid: If this chemical gets into the eyes, irrigate immediately.

Personal Protective Methods: Respirator protection is the only area specified by NIOSH.

Respirator Selection:

100 mppcf:	DM
200 mppcf:	DMXSQ/FuHiEP/SA/SCBA
1,000 mppcf:	HiEPF/SAF/SCBAF
10,000 mppcf:	PAPHiE/SA:PD,PP,CF/SAF:PD,PP,CF

Disposal Method Suggested: Sanitary landfill.

References

(1) See Reference (A-61).

SILICA, CRYSTALLINE

Description: Crystalline silica, SiO_2, is a crystalline material which melts to a glass at very high temperatures.

Code Numbers:

Silica, crystalline—Cristobalite	CAS 14464-46-1	RTECS VV7325000
Silica, crystalline—Quartz	CAS 14808-60-7	RTECS VV7330000
Silica, crystalline—Tridymite	CAS 15468-32-3	RTECS VV7335000

DOT Designation: —

Synonyms: Silica, crystalline—Quartz: Fiberglass, silica flour, silicon dioxide, silicic anhydride. It occurs in nature as agate, amethyst, chalcedony, cristobalite, flint, quartz, sand, tridymite.

Potential Exposure: NIOSH estimates that 1,200,000 workers are potentially exposed to crystalline silica in such industries as granite quarrying and cutting, foundry operations, metal, coal, and nonmetallic mining, and manufacture of clay and glass products.

Incompatibilities: Powerful oxidizers; fluorine, chlorine trifluoride, manganese trioxide, oxygen difluoride, etc.

Permissible Exposure Limit in Air: The current Federal standard for crystalline silica is (for respirable dust) 10 milligrams silica per cubic meter of air divided by the percent SiO_2 plus 2, averaged over an 8 hr work shift. The current Federal standard for total dust is 30 mg/m^3 divided by the percent SiO_2 plus 2 averaged over an 8 hr work shift. NIOSH has recommended that the permissible exposure limit be changed to 50 μg respirable free silica per cubic meter of air averaged over a work shift of up to 10 hr per day, 40 hr per week.

Uncontrolled abrasive blasting with silica sand presents such a severe silicosis hazard that NIOSH has recommended that silica sand, or other materials containing more than 1% free silica, be prohibited as an abrasive substance in abrasive blasting operations. The NIOSH Criteria Document for Crystalline Silica should be consulted for more detailed information.

The ACGIH (1983/84) has proposed intended TLV values as follows (in mg/m^3):

	Total Dust	Respirable Dust
Cristobalite	0.15	0.05
Quartz	0.3	0.1
Tridymite	0.15	0.05

Determination in Air: Collection on a filter; analysis by x-ray diffraction. See NIOSH Methods, Set S.

Permissible Concentration in Water: No criteria set.

Route of Entry: Inhalation of dust.

Harmful Effects and Symptoms: Crystalline silica can cause silicosis, a progressive and frequently incapacitating pneumoconiosis evident on x-ray and in pulmonary function testing, as well as in subjective respiratory complaints. Symptoms include coughing, wheezing, dyspnea, and impaired pulmonary function.

Points of Attack: Respiratory system, lungs.

Medical Surveillance: [See also reference (2)]. Medical examinations shall be made available to all workers subject to "exposure to free silica" prior to employee placement and at least once each 3 years thereafter. Examinations shall include as a minimum:

(a) A medical and occupational history to elicit data on worker exposure to free silica and signs and symptoms of respiratory disease;

(b) A chest roentgenogram (posteroanterior 14" x 17" or 14" x 14") classified according to the 1971 ILO International Classification of Radiographs of Pneumoconioses. [ILO U/C International Classification of Radiographs of Pneumoconioses 1971, Occupational Safety and Health Series 22 (rev). Geneva, International Labor Office, 1972.]

(c) Pulmonary function tests including forced vital capacity (FVC) and forced expiratory volume at one second (FEV$_1$) to provide a base line for evaluation of pulmonary function and to help determine the advisability of the workers using negative- or positive-pressure respirators. It should be noted that pulmonary function tests may vary significantly in various ethnic groups. For example, in black persons, the test values for the FVC should be divided by 0.85 before the percentage value is compared with normal figures;

(d) Body weight, height and age;

(e) Initial medical examinations for presently employed workers shall be offered within 6 months of the promulgation of a standard incorporating these recommendations.

The medical management of an employee with or without roentgenographic evidence of silicosis who has respiratory distress and/or pulmonary functional impairment should include full evaluation by a physician qualified to advise the employee whether he should continue working in a dusty trade.

These records shall be available to the medical representatives of the Secretary of Health, Education and Welfare, of the Secretary of Labor, of the employee or former employee, and of the employer.

Medical records shall be maintained for at least 30 years following the employee's termination of employment.

First Aid: If this chemical gets into the eyes, irrigate immediately. If a person breathes in large amounts of this chemical, move the exposed person to fresh air at once.

Personal Protective Methods: Engineering controls shall be used to maintain free silica dust exposures below the prescribed limit. "Respiratory Protection" below shall apply whenever a variance from the standard recommended is granted under provisions of the Occupational Safety and Health Act, or in the interim period during the application for a variance. When the limits of exposure to free silica prescribed cannot be met by limiting the concentration of free silica in the work environment, an employer must utilize, as provided in the Respiratory Protection section, a program to effect the required protection of every worker exposed.

Respiratory Protection — Appropriate respirators, as prescribed by NIOSH (see below) shall be provided and used when a variance has been granted to allow respirators as a means of control of exposure to routine operations and while the application is pending. Administrative controls may also be used to reduce exposure. Respirators shall also be provided and used for nonroutine operations (occasional brief exposures above the environmental standard and for emergencies); however, for these instances a variance is not required but the requirements set forth below continue to apply. Appropriate respirators as described in the table which follows shall only be used pursuant to the following requirements.

For the purpose of determining the type of respirator to be used, the employer shall measure the atmospheric concentration of free silica in the workplace when the initial application for variance is made and thereafter whenever process, work site, climate, or control changes occur which are likely to affect the free silica concentration. This requirement shall not apply when only atmosphere-supplying positive-pressure respirators are used. The employer shall ensure that no worker is exposed to free silica in excess of the standard because of improper respiratory selection, fit, use, or maintenance.

Employees experiencing breathing difficulty while using respirators shall be evaluated by a physician to determine the ability of the worker to wear a respirator.

A respiratory protective program meeting the requirements of Section 1910.134 of the Occupational Safety and Health Standards shall be established and enforced by the employer. (29 CFR 1910.134 published in the *Federal Register,* vol. 39, page 23671, dated June 27, 1974, as amended.)

The employer shall provide respirators in accordance with the table and shall ensure that the employee uses the appropriate respirator.

Respiratory protective devices in the table below shall be those approved either under 30 CFR 11, published March 25, 1972, or under the following regulations: filter-type dust, fume, and mist respirators—30 CFR 14 (Bureau of Mines Schedule 21B); and supplied air respirator—30 CFR 12 (Bureau of Mines Schedule 19B).

A respirator specified for use in higher concentrations of free silica may be used in atmospheres of lower concentrations.

Employees shall be given instruction on the use of respirators assigned to them, on cleaning respirators, and on testing for leakage.

Work Clothing — Where exposure to free silica is above the recommended environmental limit, work clothing shall be vacuumed before removal. Clothes shall not be cleaned by blowing or shaking.

Respirator Selection:

Concentrations of Free Silica in Multiples of the Standard	Respirator Type*
Less than or equal to 5X**	Single use (valveless type) dust respirator
Less than or equal to 10X	Quarter or half mask respirator with replaceable dust filter or single use (with valve) dust respirator
	Type C, demand type (negative pressure) with quarter or half mask facepiece
Less than or equal to 100X	Full facepiece respirator with replaceable dust filter
	Type C, supplied air respirator, demand type (negative pressure), with full facepiece
Less than or equal to 200X	Powered air-purifying (positive-pressure) respirator, with replaceable applicable filter***
Greater than 200X	Type C, supplied air respirator, continuous flow type (positive pressure), with full facepiece, hood, or helmet

*Where a variance has been obtained for abrasive blasting with silica sand, use only Type C continuous flow, supplied air respirator with hood or helmet.

**Where X = 50 $\mu g/m^3$.

***An alternative is to select the standard high efficiency filter which must be at least 99.97% efficient against 0.3 μm dioctyl phthalate (DOP).

Disposal Method Suggested: Sanitary landfill.

References

(1) National Institute for Occupational Safety and Health, *Criteria for a Recommended Standard: Occupational Exposure to Crystalline Silica,* NIOSH Doc. No. 75-120, Washington, DC (1974).

(2) National Institute for Occupational Safety and Health, *A Guide to the Work Relatedness of Disease,* Revised Edition, DHEW (NIOSH) Publ. No. 79-116, Cincinnati, Ohio pp. 154-167 (Jan. 1979).

(3) See Reference (A-61).

(4) Parmeggiani, L., Ed., *Encyclopedia of Occupational Health & Safety,* Third Edition, Vol. 2, pp 2033–35, Geneva, International Labour Office (1983).

SILICON

Description: Si is a nonmetallic element which is, however, often known as silicon metal. It is a crystalline or amorphous material melting at 1410°C.

Code Numbers: CAS 7440-21-3 RTECS none UN 1346

DOT Designation: Flammable solid.

Synonyms: None.

Potential Exposure: Silicon may be used in the manufacture of silanes, silicon tetrachloride, ferrosilicon, silicones. It is used in purified elemental form in transistors and photovoltaic cells.

Permissible Exposure Limits in Air: There are no Federal standards but ACGIH (1983/84) classifies silicon as a nuisance particulate with a TWA of 10 mg/m^3 and an STEL of 20 mg/m^3.

Determination in Air: Collection on a filter and gravimetric analysis (A-1).

Permissible Concentration in Water: No criteria set.

Routes of Entry: Inhalation.

Harmful Effects and Symptoms: Silicon dust does not produce significant organic disease or toxic effect when exposures are kept under reasonable control (A-34). Unpleasant deposits may be caused in eyes, ears and nasal passages and injury to the skin and mucous membranes may be caused by the dust itself or by cleansing procedures used for its removal.

SILICON CARBIDE

Description: SiC is a bluish black crystalline substance which sublimes with decomposition at 2210°C.

Code Numbers: CAS 409-21-2 RTECS none UN none

DOT Designation: —

Synonyms: Carborundum®, Crystolon®, Carbonite®, Electrolon®.

Potential Exposure: Those involved in the manufacture of silicon carbide abrasives, refractories and semiconductors. Silicon carbide fibers are also produced in fibrous form as reinforcing fibers for composite materials.

Permissible Exposure Limits in Air: There are no Federal standards but ACGIH (1983/84) has classified silicon carbide as a nuisance particulate with a TWA of 10 mg/m^3 and an STEL of 20 mg/m^3.

Determination in Air: Collection on a filter and gravimetric analysis (A-1).

Permissible Concentration in Water: No criteria set.

Routes of Entry: Inhalation.

Harmful Effects and Symptoms: Silicon carbide can alter the course of inhalation tuberculosis leading to extensive fibrosis and progressive disease (A-34).

Points of Attack: Respiratory system.

References
(1) See Reference (A-60).

SILVER AND COMPOUNDS

- Hazardous substance (Silver Nitrate, EPA)
- Hazardous waste constituents (EPA), hazardous waste (Silver Cyanide) (EPA)
- Priority toxic pollutant (EPA)

Description: Ag, silver, is a white metal and is extremely ductile and malleable, insoluble in water but soluble in hot sulfuric and nitric acids. Perhaps the most common soluble silver compound is silver nitrate, $AgNO_3$; another is silver cyanide, AgCN.

Code Numbers:

Silver metal	CAS 7440-22-4	RTECS VW3500000	UN none
Silver nitrate	CAS 7761-88-8	RTECS VW4725000	UN 1493
Silver cyanide	CAS 506-64-9	RTECS VW3850000	UN 1684

DOT Designation: Silver metal, none; silver nitrate, oxidizer; and silver cyanide, Poison B.

Synonyms: none

Potential Exposure: Silver may be alloyed with copper, aluminum, cadmium, lead, or antimony. The alloys are used in the manufacture of silverware, jewelry, coins, ornaments, plates, commutators, scientific instruments, automobile bearings, and grids in storage batteries. Silver is used in chrome-nickel steels, in solders and brazing alloys, in the application of metallic films on glass and ceramics, to increase corrosion resistance to sulfuric acid, in photographic films, plates and paper, as an electroplated undercoating for nickel and chrome, as a bactericide for sterilizing water, fruit juices, vinegar, etc., in bus bars and windings in electrical plants, in dental amalgams, and as a chemical catalyst in the synthesis of aldehydes. Because of its resistance to acetic and other food acids, it is utilized in the manufacture of pipes, valves, vats, pasteurizing coils and nozzles for the milk, vinegar, cider, brewing, and acetate rayon silk industries.

Silver compounds are used in photography, silver plating, inks, dyes, coloring glass and porcelain, etching ivory, in the manufacture of mirrors, and as analytical chemical reagents and catalysts. Some of the compounds are also of medical importance as antiseptics or astringents, and in the treatment of certain diseases, particularly in veterinary medicine.

Incompatibilities: Acetylene, ammonia, hydrogen peroxide.

Permissible Exposure Limits in Air: The Federal standard for silver metal and soluble compounds is 0.01 mg/m³. The ACGIH (1983/84) has adopted a TWA of 0.1 mg/m³ for silver metal, but no STEL value for the metal or its compounds.

Determination in Air: Collection on membrane filter, solution in nitric acid, analysis by atomic absorption spectrophotometry (A-10).

Permissible Concentration in Water: To protect freshwater aquatic life—should not exceed

$$e^{[1.72 \ln(hardness) - 6.52]} \mu g/\ell$$

at any time. To protect saltwater aquatic life—never to exceed 2.3 $\mu g/\ell$. To protect human health—50 $\mu g/\ell$ (1). (The State of Illinois has recommended that silver in drinking water be held to 0.5 $\mu g/\ell$).

Determination in Water: Digestion followed by atomic absorption or colorimetric determination (with Dithizone) or by inductively coupled plasma (ICP) optical emission spectrometry. This gives total silver. Dissolved silver may be determined by these same methods preceded by 0.45 μ filtration.

Routes of Entry: Inhalation of fumes or dust, ingestion of solutions or dust, eye and skin contact.

Harmful Effects and Symptoms: *Local* — The only local effect from metallic silver derives from the implant of small particles in the skin of the workmen (usually hands and fingers) which causes a permanent discoloration equivalent to the process of tattooing (local argyria). Silver nitrate dust and solutions are highly corrosive to the skin, eyes, and intestinal tract. The dust of silver nitrate may cause local irritation of the skin, burns of the conjunctiva, and blindness. Localized pigmentation of the skin and eyes may occur. The eye lesions are seen first in the caruncle, and then in the conjunctiva and cornea. The nasal septum and tonsillar pillars also are pigmented.

Systemic — All forms of silver are extremely cumulative once they enter body tissues, and very little is excreted. Studies on the occurrence of argyria following injection of silver arsphenamine indicate that the onset of visible argyria begins at a total dose of about 0.9 g of silver. Generalized argyria develops when silver oxide or salts are inhaled or possibly ingested by workmen who handle compounds of silver (nitrate, fulminate, or cyanide). The condition produces no constitutional symptoms, but it may lead to permanent pigmentation of the skin and eyes.

The workman's face, forehead, neck, hands, and forearms develop a dark, slate-grey color, uniform in distribution and varying in depth depending on the degree of exposure. Fingernails, buccal mucosa, toe nails, and covered parts of the body to a lesser degree, can also be affected by this discoloration process. The dust is also deposited in the lungs and may be regarded as a form of pneumoconiosis, although it carries no hazard of fibrosis. The existence of kidney lesions of consequence to renal function is improbable from occupational exposure.

Points of Attack: Nasal septum, skin, eyes.

Medical Surveillance: Special attention should be given to other sources of silver exposure, e.g., medications or previous occupational exposure. Inspection of the nasal septum, eyes, and throat will generally give incidence of pigmentation before generalized argyria occurs. This will usually be seen first in the ear lobes, face and hands.

First Aid: If this chemical gets into the eyes, irrigate immediately. If this chemical contacts the skin, flush with water. If a person breathes in large amounts of this chemical, move the exposed person to fresh air at once and perform artificial respiration. When this chemical has been swallowed, get medical attention. Give large quantities of water and induce vomiting. Do not make an unconscious person vomit.

Personal Protective Methods: Wear appropriate clothing to prevent any reasonable probability of skin contact. Wear eye protection to prevent any possibility of eye contact. Employees should wash promptly when skin is wet or contaminated. Work clothing should be changed daily if it is possible that clothing is contaminated. Remove nonimpervious clothing promptly if wet or contaminated. Provide emergency eyewash.

Respirator Selection:
0.5 mg/m^3 : HiEPF/SAF/SCBAF
10 mg/m^3 : PAPHiEF
20 mg/m^3 : SAF:PD,PP,CF

Disposal Method Suggested: Recovery, wherever possible, in view of economic value of silver. Techniques for silver recovery from photoprocessing and electroplating wastewaters have been reviewed (A-57).

References

(1) U.S. Environmental Protection Agency, *Silver: Ambient Water Quality Criteria,* Washington, DC (1980).

(2) U.S. Environmental Protection Agency, *Toxicology of Metals, Vol. II: Silver,* Report EPA-600/1-77-022, Research Triangle Park, NC, pp. 358–369 (May, 1977).

(3) U.S. Environmental Protection Agency, *Silver,* Health and Environmental Effects Profile No. 154, Washington, DC, Office of Solid Waste (April 30, 1980).

(4) See Reference (A-61).

(5) Sax, N.I., Ed., *Dangerous Properties of Industrial Materials Report, 1,* No. 1, 52-55, New York, Van Nostrand Reinhold Co. (1980).

(6) Parmeggiani, L., Ed., *Encyclopedia of Occupational Health & Safety,* Third Edition, Vol. 2, pp 2047–48, Geneva, International Labour Office (1983).

SILVEX

Description: $Cl_3C_6H_2OCH(CH_3)COOH$ is a colorless powder which melts at 179° to 181°C.

Code Numbers: CAS 73-72-1   RTECS UF8225000

DOT Designation: —

Synonyms: 2-(2,4,5-Trichlorophenoxy)propionic acid; Fenoprop (in U.K.); 2,4,5-TP; 2,4,5-TCPPA; Kuron®.

Potential Exposure: Those engaged in the manufacture (A-32), formulation and application of this herbicide.

Permissible Exposure Limits in Air: No standards set.

Permissible Concentration in Water: The EPA has recommended a criterion of 10 $\mu g/\ell$ for domestic water supplies on a health basis. A no-adverse-effect level in drinking water has been calculated by NAS/NRC to be 5.25 $\mu g/\ell$.

Harmful Effects and Symptoms: The harmful effects and symptoms of Silvex (the propionic acid ester) are very similar to those of 2,4,5-T (the acetic acid ester) (which see).

The USEPA has concluded (2) that silvex and/or its contaminant, 2,3,7,8-tetrachlorodibenzo-p-dioxin (TCDD), create a serious health risk for humans and that human exposure to silvex and/or its contaminant, TCDD, is cause for considerable concern.

The Agency has reviewed numerous studies in which industrial, academic, and government scientists have reported that TCDD and/or silvex contaminated with TCDD produce fetotoxic, teratogenic, and carcinogenic effects in test animals that have been exposed to these chemicals. The occurrence of these adverse effects in test animals following exposure to silvex and/or TCDD indicates that humans who are exposed to silvex may experience comparable effects. Concern for the health of humans who may be exposed to TCDD, and therefore silvex contaminated with TCDD is heightened because scientists have not demonstrated that there is a level of exposure that has no adverse effects in humans.

Disposal Method Suggested: (1) Mix with excess sodium carbonate, add water and let stand for 24 hours before flushing down the drain with excess water; or (2) pour onto vermiculite and incinerate with wood, paper, and waste alcohol (A-32). Recently, the EPA compared disposal procedures. They concluded that incineration was difficult and unreasonably expensive. They concluded that landspreading permitted exposure to silvex and the contaminant TCDD. The preferred method was disposal in a secure hazardous waste landfill.

References

(1) U.S. Environmental Protection Agency, *Disposal of Certain Pesticides Containing Silvex*, Federal Register 45, No. 179, 60483-60486 (Sept. 12, 1980).

(2) U.S. Environmental Protection Agency, *Decision and Emergency Order Suspending Registrations for Certain Uses and Notice of Intent to Cancel Suspended Uses: 2(2,4,5-Trichlorophenoxy Propionic Acid) (Silvex)*, Federal Register 44, No. 52, 15874-15920 (March 15, 1979).

(3) United Nations Environment Programme, *IRPTC Legal File 1983*, Vol. II, pp VII/694-5, Geneva, Switzerland, International Register of Potentially Toxic Chemicals (1984).

SIMAZINE

Description: $C_7H_{12}ClN_5$ with the structural formula

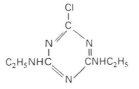

is a crystalline solid melting at 225° to 227°C.

Code Numbers: CAS 122-34-9   RTECS XY5250000

DOT Designation: —

Synonyms: 2-Chloro-4,6-bis(ethylamino)-s-triazine; CAT (in Japan); Gesatop®; Primatol®; Aquazine®.

Potential Exposure: Those involved in the manufacture, formulation and application of this preemergence herbicide.

Permissible Exposure Limits in Air: No standards set.

Permissible Concentration in Water: A no-adverse-effect level in drinking water has been calculated by NAS/NRC to be 1.505 mg/ℓ.

Harmful Effects and Symptoms: No case of poisoning in man from Simazine has been reported, although exposure to Simazine has caused acute and subacute dermatitis in the USSR, characterized by erythema, slight edema, moderate pruritus, and burning lasting 4 to 5 days (A-2).

Chronic Toxicity — Simazine fed to rats for 2 years at 1.0, 10, and 100 ppm produced no difference between treated and control animals in gross appearance or behavior. The rats fed 100 ppm had approximately twice as many thyroid and mammary tumors as the control animals, but it was stated that these were not attributable to Simazine.

A 2-year chronic-feeding study of Simazine in dogs with Simazine 80W fed at 15, 150, and 1,500 ppm showed only a slight thyroid hyperplasia at 1,500 ppm and slight increases in serum alkaline phosphatase and serum glutamic oxalacetic transaminase in several of the dogs fed 1,500 ppm.

References

(1) United Nations Environment Programme, *IRPTC Legal File 1983*, Vol. II, pp VII/773, Geneva, Switzerland, International Register of Potentially Toxic Chemicals (1984).

SOAPSTONE

Description: With the formula $3MgO \cdot 4SiO_2 \cdot H_2O$, this is an odorless crystalline solid.

Code Numbers: RTECS VV8780000

DOT Designation: —

Synonyms: Soapstone silicate; Massive talc; Steatite.

Potential Exposure: Soapstone is used as a pigment in paints, varnishes, rubber, and soap. It is used in lubricating molds and machinery. In massive form, it is used as a heat insulator.

Incompatibilities: Nonhazardous.

Permissible Exposure Limits in Air: The Federal standard is 20 mppcf.

Permissible Concentration in Water: No criteria set.

Routes of Entry: Inhalation; eye and skin contact.

Harmful Effects and Symptoms: Coughing, dyspnea; digital clubbing; cyanosis; basal crackles; acute right heart strain.

Points of Attack: Lungs, cardiovascular system.

Medical Surveillance: Consider the points of attack in preplacement and periodic physical examinations.

First Aid: If this chemical gets into the eyes, irrigate immediately. If a person breathes in large amounts of this chemical, move the exposed person to fresh air at once and perform artificial respiration.

Personal Protective Methods: Respirators are the only devices specified by NIOSH.

Respirator Selection:

      100 mppcf: D
      200 mppcf: DXSQ/FuHiEP/SA/SCBA
    1,000 mppcf: HiEPF/SAF/SCBAF
  10,000 mppcf: PAPHiEF

Disposal Method Suggested: Sanitary landfill.

References
(1)   See Reference (A-61).

SODIUM

Description: Sodium, symbol Na, is a soft silvery-white metallic element which melts at 98°C.

Code Numbers: CAS 7440-23-5   RTECS VY0686000   UN 1428

DOT Designation: Flammable solid.

Potential Exposure: Those involved in tetra-alkyl lead manufacture using lead-sodium alloy as a reactant; those using sodium as a liquid metal coolant, as a catalyst, or in the manufacture of sodium hydride, borohydride or peroxide.

Incompatibilities: Water, halogens, halogenated hydrocarbons, sulfur and sulfur compounds.

Permissible Exposure Limits in Air: No standards set.

Permissible Concentration in Water: The metal reacts with water. Sodium ion limit is 10 mg/ℓ as desirable in drinking water; 200 mg/ℓ may be injurious to humans.

Harmful Effects and Symptoms: Reacts with moisture or moist skin to give caustic burns. Will burn eyes badly also.

Personal Protective Methods: Impervious clothing should be worn to prevent caustic burns. Safety goggles and gloves should be worn.

Respirator Selection: Dust respirators may be required.

Disposal Method Suggested: Incineration with absorption of oxide fumes.

References

(1) Sax, N.I., Ed., *Dangerous Properties of Industrial Materials Report, 1,* No. 8, 85-88, New York, Van Nostrand Reinhold Co. (1981).
(2) See Reference (A-60).

SODIUM AZIDE

● Hazardous waste (EPA)

Description: Sodium azide, NaN_3, is a white crystalline solid. When heated to $275°$ to $330°C$ in air, the solid crystals decompose with the evolution of nitrogen gas, leaving a residue of sodium oxide.

Code Numbers: CAS 26628-22-8   RTECS VY8050000   UN 1687

DOT Designation: Poison B.

Potential Exposure: Sodium azide has been used for a wide variety of military, laboratory, medicine and commercial purposes. While sodium azide is not explosive, it is used extensively as an intermediate in the production of lead azide, commonly used in detonators and other explosives.

Commercial applications include use as a fungicide, nematocide, and soil sterilizing agent and as a preservative for seeds and wine. The lumber industry has used sodium azide to limit the growth of enzymes responsible for formation of brown stain on sugar pine, while the Japanese beer industry used it to prevent the growth of a fungus which darkens its product. The chemical industry has used sodium azide as a retarder in the manufacture of sponge rubber, to prevent coagulation of styrene and butadiene latexes stored in contact with metals, and to decompose nitrites in the presence of nitrates.

The largest potential exposure is that to automotive workers, repairmen and wreckers if inflatible air bags are installed on all passenger cars and if sodium azide is used as the inflation chemical (1).

Permissible Exposure Limits in Air: There is no Federal standard but ACGIH (1983/84) has set a ceiling value of 0.1 ppm ($0.3 mg/m^3$). There is no STEL value set.

Permissible Concentration in Water: No criteria set. (Sodium azide reacts with water to give hydrazoic acid.)

Harmful Effects and Symptoms: Sodium azide is a broad-spectrum, metabolic poison that interferes with oxidation enzymes and inhibits nuclear phosphorylation. Although the effects in these systems are complex, there is general agreement that azide causes a dissociation of phosphorylation and cellular respiration. For this reason parallels have been drawn to other metabolic inhibitors such as cyanide, malonitrile and fluoride.

The major effect of exposure to this chemical is a profound lowering of blood pressure. For this reason experiments have been conducted on the use of sodium azide as a hypotensive drug. Oral doses of 0.65 to 1.3 mg (approximately 0.014 mg/kg) have a rapid hypotensive effect that persists for 10 to 15 min. When this dose was administered for up to two years, it produced a substantial lowering of blood pressure to normal levels, with no noticeable side effects.

However, when a researcher accidently swallowed a 5 to 10 mg tablet of sodium azide, symptoms similar to those caused by strychnine developed—violent heart stimulation within 5 minutes, throbbing at the base of the brain and loss of consciousness for 10 minutes, followed by rapid recovery. Less severe attacks recurred during the following hour. In another instance a woman accidentally drank 1.5 ml of 10% sodium azide solution (150 mg). In 5 minutes she experienced respiratory distress and rapid pulse. After 15 minutes she experienced nausea, vomiting, violent headache, diarrhea, and other symptoms. Ten days later she continued to feel weak and dizzy.

Other effects of intoxication by sodium azide include respiratory arrest, development of convulsions, at first clonic, later tetanic, and finally heart failure.

In summary, toxic symptoms first appear in the range of 0.01 mg/kg (0.65 mg for the average person), and more serious consequences can be expected above 0.05 mg/kg (3 mg for a 70 kg person). Persons have survived one time doses of 50 to 100 mg and 150 mg (1 to 3 mg/kg). This upper figure probably represents about the maximum nonlethal dose.

Sodium azide is a potent mutagen in barley, peas, rice, and soybeans. It is also a very effective mutagen in bacteria. Its potency as a mutagen is comparable to the nitrosamines as a class. For these reasons sodium azide has been suspected of being a carcinogen. However, several studies have been performed to determine whether it is. In each instance the results were negative.

Disposal Method Suggested: Disposal may be accomplished by reaction with sulfuric acid solution and sodium nitrate in a hard rubber vessel. Nitrogen dioxide is generated by this reaction and the gas is run through a scrubber before it is released to the atmosphere. Controlled incineration is also acceptable (after mixing with other combustible wastes) with adequate scrubbing and ash disposal facilities (A-31).

References

(1) Buckheit, B., and Fan, W., *Sodium Azide in Automotive Air Bags,* National Highway Traffic Safety Administration, Washington, DC (March 30, 1978).
(2) National Institute for Occupational Safety and Health, *Profiles on Occupational Hazards for Criteria Document Priorities: Sodium Azide,* Report PB-274,073, Cincinnati, Ohio, pp. 306–308 (1977).
(3) U.S. Environmental Protection Agency, *Chemical Hazard Information Profile: Sodium Azide,* Washington, DC (August 1, 1977).
(4) Sax, N.I., Ed., *Dangerous Properties of Industrial Materials Report, 2,* No. 6, 74-78, New York, Van Nostrand Reinhold Co. (1982).
(5) See Reference (A-60).
(6) Parmeggiani, L., Ed., *Encyclopedia of Occupational Health & Safety,* Third Edition, Vol. 1, pp 1070, Geneva, International Labour Office (1983) (Hydrazoic Acid).

SODIUM BISULFITE

● Hazardous substance (EPA)

Description: NaHSO$_3$ is a white crystalline solid with a slight sulfurous odor and a disagreeable taste.

Code Numbers: CAS 7631-90-5   RTECS VZ2000000   UN (NA2693)

DOT Designation: ORM-B

Synonyms: Sodium hydrogen sulfite, sodium acid sulfite.

Potential Exposure: Sodium bisulfite is used in the digestion of wood pulp, in the tanning of leather, in the dyeing of textiles, as a photographic reducing agent, as a food preservative and as an additive in electroplating.

Permissible Exposure Limits in Air: There are no Federal standards but ACGIH as of 1983/84 has adopted a TWA value of 5 mg/m^3. There is no STEL value.

Permissible Concentration in Water: No criteria set.

Routes of Entry: Inhalation, skin and eye contact, ingestion.

Harmful Effects and Symptoms: Irritation of the eyes, skin and mucous membranes (A-34).

Points of Attack: Skin, eyes, mucous membranes.

Medical Surveillance: Attention to points of attack is indicated in regular physical examinations.

Personal Protective Methods: Wear rubber gloves, protective glasses and work clothes.

Respirator Selection: Use gas mask in areas of high dust concentration.

Disposal Method Suggested: Dump into water, add soda ash, then neutralize with HCl; flush to sewer with large volumes of water (A-38).

References

(1) See Reference (A-60).

SODIUM CYANIDE

See "Cyanides."

SODIUM FLUOROACETATE

● Hazardous waste (EPA)

Description: FCH$_2$COONa is a fluffy, colorless, odorless, hygroscopic solid (sometimes dyed black). It melts at 200°C.

Code Numbers: CAS 62-74-8   RTECS AH9100000   UN 2629

DOT Designation: —

Synonyms: Sodium monofluoroacetate, SFA, Compound 1080.

Potential Exposure: Those involved in the manufacture, formulation and application of this rodenticide or rat poison.

Incompatibilities: None hazardous according to NIOSH (A-4) but alkaline metals and carbon disulfide are cited by others (A-38).

Permissible Exposure Limits in Air: The Federal standard and the 1983/84 ACGIH TWA value is 0.05 mg/m^3. The STEL value is 0.15 mg/m^3. The notation "skin" is added to indicate the possibility of cutaneous absorption. The IDLH level is 5.0 mg/m^3.

Permissible Concentration in Water: No criteria set.

Routes of Entry: Inhalation, skin absorption, ingestion, eye and skin contact.

Harmful Effects and Symptoms: Vomiting, hallucinations, nystagmus, paresis, twitching of facial muscles, convulsions, irregular pulse, ectopic heartbeat, tachycardia, venticular fibrillation, pulmonary edema.

A rebuttable presumption against registration of sodium fluoroacetate for pesticidal uses was issued on December 1, 976 on the basis of reductions in nontarget and endangered species and because there is no human antidote (A-32).

Points of Attack: Cardiovascular system, lungs, kidneys, central nervous system.

Medical Surveillance: Consider the points of attack in preplacement and periodic physical examinations.

First Aid: If this chemical gets into the eyes, irrigate immediately. If this chemical contacts the skin, flush with water immediately. If a person breathes in large amounts of this chemical, move the exposed person to fresh air at once and perform artificial respiration. When this chemical has been swallowed, get medical attention. Give large quantities of water and induce vomiting. Do not make an unconscious person vomit.

Personal Protective Methods: Wear appropriate clothing to prevent any possibility of skin contact. Wear eye protection to prevent any reasonable probability of eye contact. Employees should wash immediately when skin is wet or contaminated. Work clothing should be changed daily if it is possible that clothing is contaminated. Remove nonimpervious clothing immediately if wet or contaminated. Provide emergency showers.

Respirator Selection:
         0.25 mg/m^3 : DMXS
         0.5 mg/m^3 : DMXSQ/HiEP/SA/SCBA
         2.5 mg/m^3 : HiEPF/SAF/SCBAF
         5 mg/m^3 : PAPHiE/SA:PD,PP,CF
         Escape: GMXS/SCBA

Disposal Method Suggested: This compound is unstable at temperatures above 110°C and decomposes at 200°C. Thus, careful incineration has been suggested as a disposal procedure by the Manufacturing Chemists Association. According to their procedure, the product should be mixed with large amounts of vermiculite, sodium bicarbonate and sand-soda ash. Slaked lime should also be added to the mixture. Two incineration procedures for this mixture are suggested. The better of these procedures is to burn the mixture in a closed incinerator equipped with an afterburner and an alkali scrubber. The other procedure suggests that the mixture be covered with scrap wood and paper in an open incinerator. (The incinerator should be lighted by means of an excelsior train) (A-32).

References
(1) See Reference (A-61).
(2) United Nations Environment Programme, *IRPTC Legal File 1983,* Vol. I, pp VII/12–13, Geneva, Switzerland, International Register of Potentially Toxic Chemicals (1984).

SODIUM HYDROXIDE

● Hazardous substance (EPA)

Description: NaOH, sodium hydroxide, is a white, deliquescent material sold as pellets, flakes, lumps, or sticks. Aqueous solutions are known as soda lye.

Code Numbers: CAS 1310-73-2 RTECS WB4900000 UN 1823

DOT Designation: Corrosive material.

Synonyms: Caustic soda, caustic alkali, caustic flake, sodium hydrate, soda lye, white caustic.

Potential Exposure: Sodium hydroxide is utilized to neutralize acids and make sodium salts in petroleum refining, viscose rayon, cellophane, and plastic production, and in the reclamation of rubber. It hydrolyzes fats to form soaps, and it precipitates alkaloids and most metals from aqueous solutions of their salts. It is used in the manufacture of mercerized cotton, paper, explosives, and dyestuffs, in metal cleaning, electrolytic extraction of zinc, tin plating, oxide coating, laundering, bleaching, and dishwashing, and it is used in the chemical industries. NIOSH estimates that 150,000 workers are potentially exposed to the alkali.

Incompatibilities: Water, acids, flammable liquids, organic halogens; metals: aluminum, tin, zinc; nitromethane and nitro compounds.

Permissible Exposure Limits in Air: The Federal standard for sodium hydroxide is 2 mg/m^3. NIOSH and ACGIH (1983/84) recommend a ceiling concentration of 2.0 mg sodium hydroxide/m^3 as determined by a sampling period of 15 min. There is no STEL value.

The recommendation for a limit on airborne workplace sodium hydroxide concentrations serves to protect against the irritation of the respiratory tract from sodium hydroxide aerosols.

The IDLH level is 200 mg/m^3.

Determination in Air: Bubbler collection and electrometric titration (A-10) or filter collection, and extraction and back titration (A-10).

Permissible Concentration in Water: There are no criteria for NaOH as such. The EPA has, however, recommended criteria for pH as follows: to protect freshwater aquatic life—pH 6.5 to 9.0; to protect saltwater aquatic life—pH 6.5 to 8.5; and to protect humans' drinking water—pH 5 to 9.

Route of Entry: Inhalation of dust or mist, ingestion, skin and eye contact.

Harmful Effects and Symptoms: *Local* – This compound is extremely alkaline in nature and is very corrosive to body tissues. Dermatitis may result from repeated exposure to dilute solutions in the form of liquids, dusts, or mists.

Extensive work practices are recommended to protect workers from local contact with sodium hydroxide. Local contact of sodium hydroxide with eyes, skin, and the alimentary tract has resulted in extensive damage to tissues, with resultant blindness, cutaneous burns, and perforations of the alimentary tract.

Systemic – Systemic effects are due entirely to local tissue injury. Extreme pulmonary irritation may result from inhalation of dust or mist. During the tissue regeneration process in the alimentary tract, some squamous cell carcinomas have developed.

Points of Attack: Eyes, respiratory system, skin, lungs.

Medical Surveillance: The skin, eyes, and respiratory tract should receive special attention in any placement or periodic examination. NIOSH recommends that workers subject to sodium hydroxide exposure have comprehensive preplacement medical examinations. Medical examinations shall be made available promptly to all workers with signs or symptoms of skin, eye, or upper respiratory tract irritation resulting from exposure to sodium hydroxide.

First Aid: If this chemical gets into the eyes, irrigate immediately. If this chemical contacts the skin, flush with water immediately. If a person breathes in large amounts of this chemical, move the exposed person to fresh air at once and perform artificial respiration. When this chemical has been swallowed, get medical attention. Give large quantities of water and do not induce vomiting.

Personal Protective Methods: Wear appropriate clothing to prevent any possibility of skin contact. Wear eye protection to prevent any possibility of eye contact. Employees should wash immediately when skin is wet or contaminated. Work clothing should be changed daily if it is possible that clothing is contaminated. Remove nonimpervious clothing immediately if wet or contaminated. Provide emergency showers and eyewash.

Respirator Selection:

    100 mg/m³ : HiEPF/SAF/SCBAF
    200 mg/m³ : PAPHiEF/SAF:PD,PP,CF
       Escape: DMXSF/SCBAF

Disposal Method Suggested: Discharge into tank containing water, neutralize, then flush to sewer with water.

References

(1)   National Institute for Occupational Safety and Health, *Criteria for a Recommended Standard: Occupational Exposure to Sodium Hydroxide,* NIOSH Doc. No. 76-105, Washington, DC (1976).
(2)   See Reference (A-61).
(3)   See Reference (A-60).
(4)   Sax, N.I., Ed., *Dangerous Properties of Industrial Materials Report, 4,* No. 3, 85–89, New York, Van Nostrand Reinhold Co. (1984).
(5)   United Nations Environment Programme, *IRPTC Legal File 1983,* Vol. II, pp VII/719–22, Geneva, Switzerland, International Register of Potentially Toxic Chemicals (1984).

SODIUM METABISULFITE

Description: $Na_2S_2O_5$ is a white crystalline powder with a sulfur dioxide odor. It may be considered the anhydride of 2 molecules of sodium disulfite.

Code Numbers: CAS 7681-57-4   RTECS UX8225000   UN (NA2693)

DOT Designation: ORM-B

Synonyms: Sodium pyrosulfite, pyrosulfurous acid disodium salt.

Potential Exposure: Sodium metabisulfite is used as an antioxidant in pharmaceutical preparations and as a preservative in foods.

Permissible Exposure Limits in Air: There is no Federal standard but ACGIH (1983/84) has adopted a TWA value of 5 mg/m^3. There is no STEL value given.

Permissible Concentration in Water: No criteria set.

Routes of Entry: Inhalation, skin and eye contact, ingestion.

Harmful Effects and Symptoms: Irritation of the eyes, skin and mucous membranes.

Points of Attack: Skin, eyes, mucous membranes.

Medical Surveillance: Consider the points of attack in physical examinations.

SODIUM SALTS

A variety of sodium salts have been reviewed in the periodical *Dangerous Properties of Industrial Materials Report* and, in the interests of economy of space in this volume, are simply referenced here.

	Vol	No.	Pages
Sodium arsenate	2	6	71–73
Sodium borate	2	6	76–78
Sodium chloride	1	5	79
Sodium chromate	1	8	88–90
Sodium dodecylbenzene sulfonate	3	1	74–81
Sodium fluoborate	1	8	90–91
Sodium fluoride	2	1	115–117
Sodium hypochlorite	3	6	69–71
Sodium lauryl sulfate	2	1	117–119
Sodium nitrite	3	6	72–75
Sodium selenite	3	6	75–77
Sodium tripolyphosphate	3	1	81–85

See also Reference (A-60) for citations on:

Sodium acetate
Sodium aluminate
Sodium amide
Sodium bicarbonate
Sodium borohydride
Sodium bromate
Sodium carbonate
Sodium chlorate
Sodium chlorite
Sodium dichromate
Sodium diethyldithiocarbamate

Sodium dithionite
Sodium formate
Sodium hexametaphosphate
Sodium hypochlorite
Sodium iodide
Sodium methoxide
Sodium nitrate
Sodium nitrite
Sodium perborate
Sodium peroxide

Sodium persulfate
Sodium pyrophosphate
Sodium selenite
Sodium silicate
Sodium sulfide
Sodium sulfite
Sodium tetraborate
Sodium thiosulfate
Sodium trichloroacetate
Sodium tripolyphosphate

SOMAN

Description: $C_7H_{16}FO_2P$ with the following structural formula is a colorless liquid boiling at 167°C.

Code Numbers: CAS 96-64-0   RTECS TA8750000

DOT Designation: —

Synonyms: Pinacolyl methyl fluorophosphonate; methylphosphonofluoridic acid-1,2,2-trimethyl propyl ester.

Potential Exposure: Used as a nerve gas.

Permissible Exposure Limits in Air: No standards set.

Permissible Concentration in Water: No criteria set.

Routes of Entry: Skin exposure, inhalation.

Harmful Effects and Symptoms: Highly toxic to humans. Estimated lethal dose for man is only 0.01 mg/kg. A very potent cholinesterase inhibitor (See Parathion).

Points of Attack: Eyes, central nervous system.

References

(1)   Sax, N.I., Ed., *Dangerous Properties of Industrial Materials Report, 1,* No. 2, 61-62, New York, Van Nostrand Reinhold Co. (1980).

SOOT

● Carcinogen (Animal Positive, IARC) (2,3)

Description: C_xH_y where x is very large compared to y; is an approximate formula for soot which probably contains a number of polycyclic aromatic hydrocarbons.

DOT Designation: —

Potential Exposure: Soots, tars, and mineral oils (including creosote, shale, and cutting oils) are largely by-products, contaminants, or wastes. These wastes result from fossil fuel processing technology such as coal carbonization and from incomplete combustion of other carbonaceous material.

Although these substances are largely by-products, contaminants, or wastes, there are commercial uses of these substances. Carbon blacks (derived from soot) are used as pigments in inks and food containers. Prior to 1976, carbon blacks were used as color additives for foods, drugs, and cosmetics.

Permissible Exposure Limits in Air: No standards set.

Permissible Concentration in Water: No criteria set.

Routes of Entry: Inhalation, ingestion.

Harmful Effects and Symptoms: Soots, coal-tars, creosote oils, shale oils and cutting oils are carcinogenic in experimental animals after skin painting or subcutaneous injection (2).

Occupational exposure to coal-soot, coal-tar and pitch, coal-tar fumes and some impure mineral oils causes cancer of several sites, including skin, lung, bladder, and gastrointestinal tract (2). Recent epidemiological data have supported these conclusions. This effect may be due to the presence of polycyclic aromatic hydrocarbons in these materials (3).

References

(1)   See Reference (A-62). Also see Reference (A-64).
(2)   IARC Monographs 3:22-42 (1973).
(3)   *IARC Monographs on the Evaluation of the Carcinogenic Risk of Chemicals to Humans,* Supplement I, IARC, Lyon, France, p 43 (1979).

STIBINE

Description: SbH_3, stibine, is a colorless gas with a characteristic disagreeable odor. It is produced by dissolving zinc-antimony or magnesium-antimony in hydrochloric acid.

Code Numbers: CAS 7803-52-3   RTECS WJ0700000   UN 2676

DOT Designation: Poison gas, flammable gas.

Synonyms: Antimony hydride.

Potential Exposure: Stibine is used as a fumigating agent. Exposure to stibine usually occurs when stibine is released from antimony-containing alloys during the charging of storage batteries, when certain antimonial drosses are treated with water or acid, or when antimony-containing metals come in contact with acid. Operations generally involved are metallurgy, welding or cutting with blow torches, soldering, filling of hydrogen balloons, etching of zinc, and chemical processes.

Incompatibilities: Acids, halogenated hydrocarbons, oxidizers, moisture.

Permissible Exposure Limits in Air: The Federal standard and the 1983/84 ACGIH TWA value is 0.1 ppm (0.5 mg/m^3). The STEL value is 0.3 ppm (1.5 mg/m^3). The IDLH level is 40 ppm.

Determination in Air: Adsorption, workup with HCl, colorimetric determination. See NIOSH Methods, Set 2. See also reference (A-10).

Permissible Concentration in Water: No criteria set. There are, however, criteria for antimony (which see).

Routes of Entry: Inhalation of gas.

Harmful Effects and Symptoms: *Local* — No local effects have been noted.

Systemic — Stibine is a powerful hemolytic and central nervous system poison. In acute poisoning, the symptoms are severe headache, nausea, weakness, abdominal and lumbar pain, slow breathing, and weak, irregular pulse. One of the earliest signs of overexposure may be hemoglobinuria. Laboratory studies may show a profound hemolytic anemia. Death is preceded by jaundice and anuria. Chronic stibine poisoning in man has not been reported.

Points of Attack: Blood, liver, kidneys, lungs.

Medical Surveillance: In preemployment and periodic examinations special attention should be given to significant blood, kidney, and liver diseases. The general health of exposed workmen should be evaluated periodically. Blood hemoglobin and urine tests for hemoglobin on persons suspected of stibine overexposure are indicated. Workers should also be advised to immediately report any red or dark urinary discoloration to the medical department. This frequently is the initial sign of stibine poisoning. (See Arsine.)

First Aid: If a person breathes in large amounts of this chemical, move the exposed person to fresh air at once and perform artificial respiration.

Personal Protective Methods: In areas where stibine gas is suspected, all persons entering or working in the area should be provided with fullface gas masks or supplied air respirators.

Respirator Selection:
>   1 ppm: SA/SCBA
>   5 ppm: SAF/SCBAF
>   40 ppm: SA:PD,PP,CF
>   Escape: GMS/SCBA

Disposal Method Suggested: Dissolve in hydrochloric acid; add water to produce precipitate; add acid to dissolve again; precipitate with H_2S; filter and dry precipitate and return to suppliers (A-38).

References

(1)   See Reference (A-61).

STODDARD SOLVENT

Description: Of the approximate formula C_9H_{20} and containing about 15% aromatic hydrocarbons, Stoddard solvent is a colorless liquid with a kerosene-like odor which boils at 150° to 200°C. See "Naphthas" for related materials.

Code Numbers: CAS 8052-41-3   RTECS WJ8925000

DOT Designation: —

Synonyms: Dry cleaning safety solvent, mineral spirits.

Potential Exposure: Stoddard solvent is used in dry cleaning, in degreasing of metal parts, and as a paint thinner.

Incompatibilities: Strong oxidizers.

Permissible Exposure Limits in Air: The Federal standard is 500 ppm (2,950 mg/m³). NIOSH (1) has recommended a TWA of 350 mg/m³ with a 15-minute ceiling value of 1,800 mg/m³. The ACGIH as of 1983/84 has proposed a TWA of 100 ppm (525 mg/m³) with a STEL of 200 ppm (1,050 mg/m³).

Determination in Air: Adsorption on charcoal, workup with CS_2, analysis by gas chromatography. See NIOSH Methods, Set G. See also reference (A-10).

Permissible Concentration in Water: No criteria set.

Routes of Entry: Inhalation, ingestion, skin and eye contact.

Harmful Effects and Symptoms: Irritation of eyes, nose and throat; dizziness; dermatitis.

Points of Attack: Skin, eyes, respiratory system, central nervous system.

Medical Surveillance: Consider the points of attack in preplacement and periodic physical examinations.

First Aid: If this chemical gets into the eyes, irrigate immediately. If this chemical contacts the skin, wash with soap promptly. If a person breathes in large amounts of this chemical, move the exposed person to fresh air at once and perform artificial respiration. When this chemical has been swallowed, get medical attention. Do not induce vomiting.

Personal Protective Methods: Wear appropriate clothing to prevent repeated or prolonged skin contact. Wear eye protection to prevent any reasonable probability of eye contact. Employees should wash promptly when skin is wet or contaminated. Remove nonimpervious clothing promptly if wet or contaminated.

Respirator Selection:

1,000 ppm:	CCROVF
5,000 ppm:	GMOV/SAF/SCBAF
Escape:	GMOV/SCBA

Disposal Method Suggested: Incineration.

References

(1) National Institute for Occupational Safety and Health, *Criteria for a Recommended Standard: Occupational Exposure to Refined Petroleum,* NIOSH Doc. No. 77-192, Washington, DC (1977).

(2) See Reference (A-61).

STREPTOZOTOCIN

- Carcinogen (Animal Positive, IARC) (A-62) (A-64)
- Hazardous Waste Constituent (EPA)

Description: $C_8H_{15}N_3O_7$ with the following structural formula

is a crystalline substance melting at 115°C.

Code Numbers: CAS 18883-66-4   RTECS LZ5775000

DOT Designation: —

Synonyms: Streptozocin.

Potential Exposure: Streptozotocin (STR), a water-soluble antibiotic, was of interest as a potential antineoplastic agent but has not been marketed in the United States.

STR is used in research for studies on diabetes because of its specific toxic action on B-cells of the pancreas, including hyperglycemia.

STR is produced by the soil microorganism *Streptomyces achromogenes*. It also has been synthesized by laboratory procedures. No production data were reported in the trade literature between 1975 and 1979.

If marketed, human exposure to STR could occur primarily through ingestion and less frequently through skin absorption and inhalation. Occupational exposure to STR is believed to be limited to pharmaceutical and research workers.

Permissible Exposure Limits in Air: No standards set.

Permissible Concenttation in Water: No criteria set.

Harmful Effects and Symptoms: Streptozotocin is carcinogenic in mice, rats and Chinese hamsters following its intravenous or intraperitoneal administration. It produces benign and malignant tumors of the liver and kidney and islet-cell tumors of the pancreas. It is carcinogenic after its administration in single doses (1).

References

(1) *IARC Monographs on the Evaluation of Carcinogenic Risk of Chemicals to Man,* Vol. 17, IARC, Lyon, France, pp 337-349 (1978).
(2) Sax, N.I., Ed., *Dangerous Properties of Industrial Materials Report, 1,* No. 5, 80, New York, Van Nostrand Reinhold Co. (1981).
(3) See Reference (A-62). Also see Reference (A-64).

STRYCHNINE

- Hazardous substance (EPA)
- Hazardous waste (EPA)

Description: $C_{21}H_{22}N_2O_2$, with the structural formula:

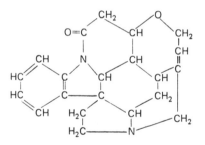

Strychnine is a colorless crystalline powder which melts at 270° to 280°C with decomposition.

Code Numbers: CAS 57-24-9   RTECS WL2275000   UN 1692

DOT Designation: Poison B.

Synonyms: Mole Death®, Mouse-Nots®.

Potential Exposure: Those involved in the extraction (from *Strychnos* seeds), formulation or application of this rodent poison.

Incompatibilities: Strong oxidizers.

Permissible Exposure Limits in Air: The Federal standard and the 1983/84 ACGIH TWA value is 0.15 mg/m³. The STEL is 0.45 mg/m³. The IDLH value is 3.0 mg/m³.

Determination in Air: Filter collection, analysis by ultraviolet spectroscopy (A-28).

Permissible Concentration in Water: No criteria set.

Routes of Entry: Inhalation of dust, ingestion, skin and eye contact.

Harmful Effects and Symptoms: Stiffness of neck and facial muscles, restlessness, apprehension, lessened acuity of perception, reflex excitability, cyanosis, tetanic convulsions, opisthotonos.

A rebuttable presumption against registration of strychnine for pesticide uses was issued by EPA on December 1, 1976 on the basis of reductions in nontarget and endangered species (A-32).

Points of Attack: Central nervous system.

Medical Surveillance: Consider the points of attack in preplacement and periodic physical examinations.

First Aid: If this chemical gets into the eyes, irrigate immediately. If this chemical contacts the skin, wash with soap promptly. If a person breathes in large amounts of this chemical, move the exposed person to fresh air at once and perform artificial respiration. When this chemical has been swallowed, get medical attention. Give large quantities of water and induce vomiting. Do not make an unconscious person vomit.

Personal Protective Methods: Wear appropriate clothing to prevent repeated or prolonged skin contact. Employees should wash promptly when skin is wet or contaminated.

Respirator Selection:
0.75 mg/m^3 : DMXS
1.5 mg/m^3 : DMXSQ/HiEP/SA/SCBA
3 mg/m^3 : HiEPF/PAPHiE/SAF/SA:PD,PP,CF/SCBAF
Escape: DMXS/SCBA

Disposal Method Suggested: Careful incineration has been recommended for disposal. Two procedures are suggested (A-32). (1) Pour or sift onto a thick layer of sand and soda ash mixture (90-10). Mix and shovel into a heavy paper box with much paper packing. Burn in incinerator. Fire may be augmented by adding excelsior and scrap wood. Stay on the upwind side. (2) Waste may be dissolved in flammable solvent (alcohols, benzene, etc.) and sprayed into fire box of an incinerator with afterburner and scrubber.

References

(1) U.S. Environmental Protection Agency, "Preliminary Notice of Determination Concluding the Rebuttable Presumption Against Registration of Pesticide Products; Notice of Availability of Position Document," *Federal Register,* 45, No. 216, 73602-8 (November 5, 1980).

(2) Sax, N.I., Ed., *Dangerous Properties of Industrial Materials Report, 2,* No. 2, 63-65, New York, Van Nostrand Reinhold Co. (1982).

(3) See Reference (A-61).

(4) United Nations Environment Programme, *IRPTC Legal File 1983,* Vol. II, pp VII/732-35, Geneva, Switzerland, International Register of Potentially Toxic Chemicals (1984).

STYRENE

● Hazardous substance (EPA)

Description: $C_6H_5CH=CH_2$, is a colorless to yellowish, very refractive, oily liquid with a penetrating odor. It boils at $145°C$.

Code Numbers: CAS 100-42-5   RTECS WL3675000   UN 2055

DOT Designation: Flammable liquid.

Synonyms: Cinnamene, cinnamol, phenethylene, phenylethylene, styrene monomer, styrol, styrolene, vinylbenzene.

Potential Exposure: Upon heating to 200°C, styrene polymerizes to form polystyrene, a plastic. It is also used in combination with 1,3-butadiene or acrylonitrile to form copolymer elastomers, butadiene-styrene rubber, and acrylonitrile-butadiene-styrene (ABS). It is also used in the manufacture of resins, polyesters, and insulators and in drug manufacture (A-41).

Incompatibilities: Oxidizers, catalysts for vinyl polymers, peroxides, strong acids, aluminum chloride.

Permissible Exposure Limits in Air: The Federal standard for styrene for an 8-hour TWA is 100 ppm (420 mg/m^3). The acceptable ceiling concentration is 200 ppm with an acceptable maximum peak of 600 ppm for a maximum duration of 5 minutes in any 3 hours. The ACGIH as of 1983/84 has recommended a TWA of 50 ppm (215 mg/m^3) and an STEL of 100 ppm (425 mg/m^3). The IDLH is 5,000 ppm.

Determination in Air: Charcoal adsorption, workup with CS_2, analysis by gas chromatography. See NIOSH Methods, Set C. See also reference (A-10).

Permissible Concentration in Water: A no-adverse effect level in drinking water has been calculated to be 0.9 mg/ℓ by NAS/NRC.

Routes of Entry: Inhalation, ingestion, skin and eye contact.

Harmful Effects and Symptoms: *Local* — Liquid and vapor are irritating to the eyes, nose, throat, and skin. The liquids are low-grade cutaneous irritants, and repeated contact may produce a dry, scaly, and fissured dermatitis.

Systemic — Acute exposure to high concentrations may produce irritation of the mucous membranes of the upper respiratory tract, nose, and mouth, followed by symptoms of narcosis, cramps, and death due to respiratory center paralysis. Effects of short-term exposure to styrene under laboratory conditions include prolonged reaction time and decreased manual dexterity.

Points of Attack: Central nervous system, respiratory system, lungs, eyes, skin.

Medical Surveillance: Consider possible irritant effects on the skin, eyes, and respiratory tract in any preplacement or periodic examinations, as well as blood, liver, and kidney function.

Mandelic acid in urine has been used as a measure of the intensity of styrene exposure.

First Aid: If this chemical gets into the eyes, irrigate immediately. If this chemical contacts the skin, flush with water. If a person breathes in large amounts of this chemical, move the exposed person to fresh air at once and perform artificial respiration. When this chemical has been swallowed, get medical attention. Do not induce vomiting.

Personal Protective Methods: Wear appropriate clothing to prevent repeated or prolonged skin contact. Wear eye protection to prevent any reasonable probability of eye contact. Employees should wash promptly when skin is wet or contaminated. Remove clothing immediately if wet or contaminated to avoid flammability hazard.

Respirator Selection:

          400 ppm: CCROV/SA/SCBA
        1,000 ppm: CCROVF
        5,000 ppm: GMOV/SAF/SCBAF
          Escape: GMOV/SCBA

Disposal Method Suggested: Incineration. In some cases, recovery and re-cycle of styrene monomer is economic and technology is available (A-58).

References

(1) Sax, N.I., Ed., *Dangerous Properties of Industrial Materials Report, 1,* No. 8, 92-95, New York, Van Nostrand Reinhold Co. (1981).
(2) Sax, N.I., Ed., *Dangerous Properties of Industrial Materials Report, 2,* No. 6, 60-64, New York, Van Nostrand Reinhold Co. (1982).
(3) See Reference (A-61).
(4) See Reference (A-60).
(5) Parmeggiani, L., Ed., *Encyclopedia of Occupational Health & Safety,* Third Edition, Vol. 2, pp 2113-15, Geneva, International Labour Office (1983).
(6) United Nations Environment Programme, *IRPTC Legal File 1983,* Vol. II, pp VII/736-37, Geneva, Switzerland, International Register of Potentially Toxic Chemicals (1984).

STYRENE OXIDE

Description: $C_6H_5CHCH_2O$ is a colorless to pale straw-colored liquid which boils at 194°C.

Code Numbers: CAS 96-09-3   RTECS CZ9625000

DOT Designation: —

Synonyms: Epoxyethylbenzene; Phenylethylene oxide; Phenyl oxirane; 1-phenyl-1,2-epoxyethane; Epoxystyrene; Styryl oxide.

Potential Exposure: Styrene oxide is used as a reactive intermediate, especially to produce styrene glycol and its derivatives. Substantial amounts are also used in the epoxy resin industry as a diluent. It may also have applications in the preparation of agricultural and biological chemicals, cosmetics, and surface coatings and in the treatment of textiles and fibers.

Styrene oxide is made in quantities in excess of a million pounds per year and, further, is a presumed metabolite of styrene which is produced in much greater quantities.

Permissible Exposure Limits in Air: No standards set.

Permissible Concentration in Water: No criteria set.

Routes of Entry: Inhalation, skin absorption, ingestion, skin and eye contact.

Harmful Effects and Symptoms: Acute human exposure to styrene oxide causes skin irritation. Some evidence suggests that styrene oxide is absorbed slowly through the skin. Styrene oxide is a presumed metabolite of styrene. Investigations have shown that the ultimate excretion product in many animals is hippuric acid. Urine of workers exposed to styrene oxide vapor contained large amounts of mandelic acid and phenylglyoxylic acid (both compounds in the proposed metabolic pathway of styrene oxide), but the hippuric acid concentrations were normal.

Of 30 male mice given thrice-weekly dermal application of 0.1 ml of a 10% solution of styrene oxide in benzene, three developed skin tumors, and one of these had a squamous cell carcinoma. Of 150 benzene-painted controls, 11 developed skin tumors, and one of these had a squamous cell carcinoma. The IARC concluded that no significant increase in skin tumors occurred.

Personal Protective Methods: Wear protective gloves and goggles (A-60).

References

(1)   U.S. Environmental Protection Agency, *Chemical Hazard Information Profile: Styrene Oxide,* Washington, DC (March 9, 1978).

(2)   See Reference (A-60).

SUBTILISINS

Description: These are proteolytic enzymes which take the form of light-colored, free-flowing powders.

Code Numbers: none

DOT Designation: —

Synonyms: Maxatase®, Alcalase®, Alk®, Protease 150®.

Potential Exposure: These materials are used as enzyme agents in laundry detergent formulations.

Permissible Exposure Limits in Air: There are no Federal standards but ACGIH (1983/84) has set a ceiling value of 0.00006 mg/m^3, based on "high volume" sampling. There is no STEL value.

Permissible Concentration in Water: No criteria set.

Routes of Entry: Inhalation of dust.

Harmful Effects and Symptoms: Irritation of skin and respiratory tract, bronchoconstriction and respiratory allergies, breathlessness, wheezing, sore throat, congested nares, headache, persistent cough (A-34).

Points of Attack: Skin and respiratory system.

Medical Surveillance: Attention to points of attack is indicated in preplacement and regular physical examinations.

Personal Protective Methods: Maximum skin protection should be provided by protective clothing.

Respirator Selection: Dust respirators are necessary.

SULFALLATE

● Carcinogen (Animal Positive) (A-64)

Description: $(C_2H_5)_2NCSCH_2C=CH_2$ is an amber liquid boiling at 128° to 130°C under 1.0 mm pressure. (with S double-bonded to C, and Cl on the C=CH$_2$ carbon)

Code Numbers: CAS 95-06-7   RTECS EZ5075000

DOT Designation: —

Synonyms: CDEC; 2-chloroallyl diethyldithiocarbamate; Vegadex®.

Potential Exposure: Sulfallate is formulated into emulsifiable concentrates, liquids, and granules.

The major use for sulfallate in the United States is as a preemergent selective herbicide to control certain annual grasses and broadleaf weeds around vegetable and fruit crops. Sulfallate has also been used for weed control among shrubbery and ornamental plants.

The 1974 National Occupational Hazard Survey provided no estimate of the number of workers potentially exposed to sulfallate. A potential for exposure exists during the manufacture and application of the herbicide. Agricultural workers have the greatest risk of sulfallate exposure, and rural residents of agricultural communities may be exposed to airborne residues of sulfallate after spraying operations. At commercially recommended rates of 4 lb/gal, the average persistence of sulfallate in soil is 3 to 6 weeks. The general population may be exposed through ingestion of residues in food crops.

Permissible Exposure Limits in Air: No standards set.

Permissible Concentration in Water: No criteria set.

Harmful Effects and Symptoms: Sulfallate, a chlorinated dithiocarbamate, administered in the feed, was carcinogenic to Osborne-Mendel rats and to B6C3F1 mice, inducing mammary gland tumors in females of both species, tumors of the forestomach in male rats, and lung tumors in male mice (1).

References

(1) National Cancer Institute, *Bioassay of Sulfallate for Possible Carcinogenicity,* Technical Report Series No. 115, DHEW Publication No. (NIH) 78-1370, Bethesda, Maryland (1978).

(2) United Nations Environment Programme, *IRPTC Legal File 1983,* Vol. I, pp VII/158, Geneva, Switzerland, International Register of Potentially Toxic Chemicals (1984).

SULFOTEP

Description: $(C_2H_5O)_2POP(OC_2H_5)_2$ is a yellow mobile liquid with a garlic-like odor.

The structure shows two $P=S$ bonds.

Code Numbers: CAS 3689-24-5   RTECS XN4375000   UN 1704

DOT Designation: Poison B.

Synonyms: Thiopyrophosphoric acid, tetraethyl ester; thiodiphosphoric acid, tetraethyl ester; sulfotepp; Dithio; Dithione; Thiotep; ENT 16273; Bladafum®; TEDP.

Potential Exposure: Those involved in the manufacture (A-32), formulation and application of this insecticide.

Incompatibilities: Strong oxidizers.

Permissible Exposure Limits in Air: The Federal standard is 0.2 mg/m³. The ACGIH (1983/84) has set a TWA of 0.2 mg/m³ and an STEL of 0.6 mg/m³. The notation "skin" is added to indicate the possibility of cutaneous absorption. The IDLH level is 35 mg/m³.

Determination in Air: Collection on impinger or fritted bubbler, analysis by gas liquid chromatography (A-1).

Permissible Concentration in Water: No criteria set.

Routes of Entry: Inhalation, skin absorption, ingestion, skin and eye contact.

Harmful Effects and Symptoms: Eye pain, impaired vision, tears; headaches, chest pains, cyanosis, anorexia, nausea, vomiting, diarrhea; local sweating; weakness; twitching; paralysis; Cheyne-Stokes respiration; convulsions; low blood pressure.

Points of Attack: Central nervous system, respiratory system, cardiovascular system.

Medical Surveillance: Consider the points of attack in preplacement and periodic physical examinations.

First Aid: If this chemical gets into the eyes, irrigate immediately. If this chemical contacts the skin, wash with soap immediately. If a person breathes in large amounts of this chemical, move the exposed person to fresh air at once and perform artificial respiration. When this chemical has been swallowed, get medical attention. Give large quantities of water and induce vomiting. Do not make an unconscious person vomit.

Personal Protective Methods: Wear appropriate clothing to prevent any possibility of skin contact. Wear eye protection to prevent any possibility of eye contact. Employees should wash immediately when skin is wet or contaminated. Remove nonimpervious clothing immediately if wet or contaminated. Provide emergency showers and eyewash.

Respirator Selection:
> 2 mg/m³ : SA/SCBA
> 10 mg/m³ : SAF/SCBAF
> 35 mg/m³ : SA:PD,PP,CF
> Escape: GMOVPPest/SCBA

Disposal Method Suggested: Incineration with added flammable solvent in furnace equipped with afterburner and alkaline scrubber.

References

(1)   See Reference (A-61).
(2)   United Nations Environment Programme, *IRPTC Legal File 1983,* Vol. II, pp VII/757–8, Geneva, Switzerland, International Register of Potentially Toxic Chemicals (1984).

SULFUR

Description: Sulfur, S, is a yellow crystalline element which melts at 119°C.

Code Numbers: CAS 7704-34-9   RTECS WS4250000   UN 1350 (solid), 2448 (molten)

DOT Designation: ORM-C

Synonyms: Brimstone.

Potential Exposure: Wide use in manufacture of wood pulp, rubber, sulfuric acid, carbon bisulfide. Used as a fungicide.

Incompatibilities: Metals, oxidants.

Permissible Exposure Limits in Air: No standards set.

Permissible Concentration in Water: No criteria set.

Harmful Effects and Symptoms: Low toxicity. Usually regarded as a nuisance dust. Irritating to the eyes. Chronic inhalation can cause irritation of the mucous membranes.

Personal Protective Methods: Safety goggles needed. Heat insulating gloves and protective clothing needed for molten sulfur (A-60).

Respirator Selection: Cannister mask or self-contained breathing apparatus needed when H_2S or SO_2 are present.

Disposal Method Suggested: Salvage for reprocessing or dump to landfill (1).

References

(1) Sax, N.I., Ed., *Dangerous Properties of Industrial Materials Report, 2,* No. 2, 65-68, New York, Van Nostrand Reinhold Co. (1982).
(2) See Reference (A-60).
(3) Parmeggiani, L., Ed., *Encyclopedia of Occupational Health & Safety,* Third Edition, Vol. 2, pp 2120–22, Geneva, International Labour Office (1983).
(4) United Nations Environment Programme, *IRPTC Legal File 1983,* Vol. II, pp VII/744–6, Geneva, Switzerland, International Register of Potentially Toxic Chemicals (1984).

SULFUR CHLORIDE

Description: S_2Cl_2, sulfur chloride, is a fuming, oily liquid with a yellowish-red to amber color and a suffocating odor. It has an added hazard since it oxidizes and hydrolyzes to sulfur dioxide and hydrogen chloride.

Code Numbers: CAS 10025-67-9 RTECS WS4300000 UN 1828

DOT Designation: Corrosive material.

Synonyms: Sulfur monochloride, sulfur subchloride, disulfur dichloride.

Potential Exposure: Sulfur chloride finds use as a chlorinating agent and an intermediate in the manufacture of organic chemicals, e.g., carbon tetrachloride, and sulfur dyes, insecticides, synthetic rubber, and pharmaceuticals. Exposure may also occur during the extraction of gold, purification of sugar juice, finishing and dyeing textiles, processing vegetable oils, hardening wood, and vulcanization of rubber.

Incompatibilities: Peroxides, oxides of phosphorus, organics; water.

Permissible Exposure Limits in Air: The Federal standard and the 1983/84 ACGIH TWA value for sulfur chloride (sulfur monochloride) is 1 ppm (6 mg/m^3). The STEL value is 3 ppm (18 mg/m^3). The IDLH level is 10 ppm.

Determination in Air: Collection by impinger or fritted bubbler, analysis by ion-specific electrode (A-15).

Permissible Concentration in Water: No criteria set.

Routes of Entry: Inhalation of vapor, ingestion, skin and eye contact.

Harmful Effects and Symptoms: *Local* – Fumes, in sufficient quantity, may cause severe irritation to eyes, skin, and mucous membranes of the upper respiratory tract.

Systemic – Although this compound is capable of producing severe pulmonary irritation, very few serious cases of industrial exposure have been reported. This is probably because the pronounced irritant effects of sulfur chloride serve as an immediate warning signal when concentration of the gas approaches a hazardous level.

Points of Attack: Respiratory system, skin, eyes, lungs.

Medical Surveillance: Preemployment and periodic examinations should give special emphasis to the skin, eyes, and respiratory system. Pulmonary function tests may be useful. Exposures may also include sulfur dioxide and hydrochloric acid. (See these compounds.)

First Aid: If this chemical gets into the eyes, irrigate immediately. If this chemical contacts the skin, flush with water immediately. If a person breathes in large amounts of this chemical, move the exposed person to fresh air at once and perform artificial respiration. When this chemical has been swallowed, get medical attention. Give large quantities of water and induce vomiting. Do not make an unconscious person vomit.

Personal Protective Methods: Wear appropriate clothing to prevent any possibility of skin contact. Wear eye protection to prevent any possibility of eye contact. Employees should wash immediately when skin is wet or contaminated. Remove nonimpervious clothing immediately if wet or contaminated. Provide emergency showers and eyewash.

Respirator Selection:
>10 ppm: CCRSF/GMS/SAF/SCBAF
>Escape: GMS/SCBA

Disposal Method Suggested: Spray carefully onto sodium ash/slaked lime mixture. Then spray with water, dilute, neutralize and flush to drain (A-38).

References
(1)   See Reference (A-61).
(2)   See Reference (A-60).

SULFUR DIOXIDE

Description: SO_2, sulfur dioxide, is a colorless gas at ambient temperatures with a characteristic strong suffocating odor. It is soluble in water and organic solvents.

Code Numbers: CAS 7446-09-5   RTECS WS4550000   UN 1079

DOT Designation: Nonflammable gas.

Synonyms: Sulfurous anhydride, sulfurous oxide.

Potential Exposure: NIOSH estimates that 500,000 workers are potentially exposed to sulfur dioxide, which is encountered in many industrial operations. Sulfur dioxide is used in the manufacture of sodium sulfite, sulfuric acid, sulfuryl chloride, thionyl chloride, organic sulfonates, disinfectants, fumigants, glass, wine, ice, industrial and edible protein, and vapor pressure thermometers. It is also used in the bleaching of beet sugar, flour, fruit, gelatin, glue, grain, oil, straw, textiles, wicker ware, wood pulp, and wool; in the tanning of leather; in brewing and preserving; and in the refrigeration industry. Exposure may also occur in various other industrial processes as it is a by-product of ore smelting, coal and fuel oil combustion, paper manufacturing and petroleum refining.

Incompatibilities: Powdered and alkali metals such as sodium, potassium.

Permissible Exposure Limits in Air: The Federal standard is 5 ppm (13 mg/m³). NIOSH has recommended lowering this standard to 0.5 ppm (1.3 mg/m³) as a TWA. NIOSH has concluded from experimental and epidemiologic studies that exposure to sulfur dioxide at the existing Federal standard of 5 ppm can cause adverse respiratory effects by increasing airway resistance in a significant number of workers. Some workers are especially sensitive to these effects.

In addition, there is evidence that the effects produced by sulfur dioxide are enhanced by airborne particulate matter and that sulfur dioxide may promote the carcinogenic action of other airborne substances. Compliance with the NIOSH recommended standard of 0.5 ppm, however, should prevent adverse effects of sulfur dioxide on the health and safety of workers. ACGIH (1983/84) has set a TWA of 2 ppm (5 mg/m³) and an STEL of 5 ppm (10 mg/m³). The IDLH level is 100 ppm.

Determination in Air: Collection by impinger or bubbler, containing H_2O_2 followed by titration with barium perchlorate (A-10) or titration of sulfuric acid formed with NaOH (A-10).

Permissible Concentration in Water: No criteria set.

Routes of Entry: Inhalation of gas; direct contact of gas or liquid phase on skin and mucous membranes.

Harmful Effects and Symptoms: *Local* – Gaseous sulfur dioxide is particularly irritating to mucous membranes of the upper respiratory tract. Chronic effects include rhinitis, dryness of the throat, and cough. Conjunctivitis, corneal burns, and corneal opacity may occur following direct contact with liquid.

Systemic – Acute overexposure may result in death from asphyxia. Survivors may later develop chemical bronchopneumonia with bronchiolitis obliterans. Bronchoconstriction with increased pulmonary resistance, high-pitched rales, and a tendency to prolongation of the expiratory phase may result from moderate exposure, though bronchoconstriction may be asymptomatic. The effects on pulmonary function are increased in the presence of respirable particles.

Chronic exposure may result in nasopharyngitis, fatigue, altered sense of smell, and chronic bronchitis symptoms such as dyspnea on exertion, cough, and increased mucous excretion. Transient stimulation of erythropoietic activity of the bone marrow has been reported. Slight tolerance, at least to the odor threshold, and general acclimatization are common. Sensitization in a few individuals, particularly young adults, may also develop following repeated exposures. There is some evidence that some individuals may be innately hypersusceptible to SO_2. Animal experimentation has also indicated that sulfur dioxide may be a possible cocarcinogenic agent.

Points of Attack: Respiratory system, skin, eyes, lungs.

Medical Surveillance: Preplacement and periodic medical examinations should be concerned especially with the skin, eye, and respiratory tract. Pulmonary function should be evaluated, as well as smoking habits, and exposure to other pulmonary irritants. See also reference (2).

First Aid: If this chemical gets into the eyes, irrigate immediately. If this chemical contacts the skin, flush with water immediately. If a person breathes in large amounts of this chemical, move the exposed person to fresh air at once and perform artificial respiration.

Personal Protective Methods: Wear appropriate clothing to prevent skin freezing. Wear eye protection to prevent any possibility of eye contact. Remove clothing immediately if wet or contaminated. Provide emergency eyewash.

Respirator Selection:
> 20 ppm: CCRS/SA/SCBA
> 100 ppm: CCRSF/GMS/SAF/SCBAF
> Escape: GMS/SCBA

Disposal Method Suggested: Pass into soda ash solution, then add calcium hypochlorite, neutralize and flush to sewer with water (A-38).

References

(1) National Institute for Occupational Safety and Health, *Criteria for a Recommended Standard: Occupational Exposure to Sulfur Dioxide,* NIOSH Doc. No. 74-111 (1974).
(2) National Institute for Occupational Safety and Health, *A Guide to the Work Relatedness of Disease,* Revised Edition, DHEW (NIOSH) Publ. No. 79-116, Cincinnati, Ohio, pp. 168–176 (January, 1979).
(3) World Health Organization, *Sulfur Oxides and Suspended Particulate Matter,* Environmental Health Criteria No. 8, Geneva, Switzerland (1979).
(4) Sax, N.I., Ed., *Dangerous Properties of Industrial Materials Report, 1,* No. 3, 78-79, New York, Van Nostrand Reinhold Co. (1981).
(5) See Reference (A-61).
(6) See Reference (A-60).

SULFUR HEXAFLUORIDE

Description: SF_6 is a colorless, odorless gas.

Code Numbers: CAS 2551-62-4   RTECS WS4900000   UN 1080

DOT Designation: Nonflammable gas.

Synonyms: Sulfur fluoride.

Potential Exposure: SF_6 is used in various electric power applications as a gaseous dielectric or insulator. The most extensive use is in high-voltage transformers. SF_6 is also used in circuit breakers, waveguides, linear particle accelerators, Van de Graaff generators, chemically pumped continuous-wave lasers, transmission lines, and power distribution substations.

Nonelectrical applications include use as a protective atmosphere for casting of magnesium alloys and use as a leak detector or in tracing moving air masses.

Several sources note that vitreous substitution of SF_6 in owl monkeys results in a greater ocular vascular permeability than that caused by saline. This implies that SF_6 could have an important use in retinal surgery (1).

Permissible Exposure Limits in Air: There is no Federal standard. ACGIH (1983/84) has adopted a TWA of 1,000 ppm (6,000 mg/m^3) and an STEL of 1,250 ppm (7,500 mg/m^3).

Determination in Air: Collection by impinger, analysis by ultraviolet spectrometry (A-1).

Permissible Concentration in Water: No criteria set.

Routes of Entry: Inhalation, skin absorption, ingestion.

Harmful Effects and Symptoms: SF_6 is considered to be physiologically inert in the pure state. Only slight effects result, regardless of the amount absorbed or the extent of exposure. In high concentrations, however, pure SF_6 can act as a simple asphyxiant by displacing the necessary oxygen.

Ordinarily, however, SF_6 does not exist in the pure state. It contains variable quantities of sulfur fluorides. In the presence of water, these sulfur fluorides can hydrolyze to yield hydrogen fluoride (HF) and oxyfluoride compounds such as sulfuryl fluoride (SO_2F_2) and thionyl fluoride (SOF_2). These compounds have much more toxic health effects. Sulfur hexafluoride may also be contaminated with more toxic sulfur compounds such as S_2F_{10}.

Personal Protective Method: Wear rubber gloves, safety goggles and coveralls (A-38).

Respirator Selection: Use of self-contained breathing apparatus is recommended.

Disposal Method Suggested: Seal unused cylinders and return to suppliers.

References
(1) U.S. Environmental Protection Agency, *Chemical Hazard Information Profile: Sulfur Hexafluoride,* Washington, DC (July 10, 1978).
(2) See Reference (A-60).

SULFURIC ACID

● Hazardous substance (EPA)

Description: H_2SO_4, concentrated sulfuric acid, is a colorless, odorless, oily liquid which is commercially sold at 93% to 98% H_2SO_4, the remainder being water. Fuming sulfuric acid (oleum) gives off free sulfur trioxide and is a colorless or slightly colored, viscous liquid. Sulfuric acid is soluble in water and alcohol.

Code Numbers: CAS 7664-93-9 RTECS WS5600000 UN 1830

DOT Designation: Corrosive material.

Synonyms: Oil of vitriol, spirit of sulfur, hydrogen sulfate.

Potential Exposure: Sulfuric acid is used as a chemical feedstock in the manufacture of acetic acid, hydrochloric acid, citric acid, phosphoric acid, aluminum sulfate, ammonium sulfate, barium sulfate, copper sulfate, phenol, superphosphates, titanium dioxide, as well as synthetic fertilizers, nitrate explosives, artificial fibers, dyes, pharmaceuticals, detergents, glue, paint, and paper. It finds use as a dehydrating agent for esters and ethers due to its high affinity for water, as an electrolyte in storage batteries, for the hydrolysis of cellulose to obtain glucose, in the refining of mineral and vegetable oil, and in the leather industry. Other uses include fur and food processing, carbonization of wool fabrics, gas drying, uranium extraction from pitchblende, and laboratory analysis. NIOSH estimates that 200,000 workers are potentially exposed to sulfuric acid.

Incompatibilities: Organics, chlorates, carbides, fulminates, picrates, metals.

Permissible Exposure Limits in Air: The Federal standard and the 1983/84 ACGIH TWA value for sulfuric acid is 1 mg/m^3. NIOSH has concluded that adherence to the present Federal standard of 1 mg/m^3 of air, in conjunction with a strong program of work practices, will protect the worker from sulfuric acid exposure for up to a 10-hour workday, 40-hour workweek over a working lifetime.
There is no tentative STEL value set. The IDLH level is 80 mg/m^3.

Determination in Air: Collection on a filter, workup with isopropyl alcohol, titration. See NIOSH Methods, Set L. See also reference (A-10).

Permissible Concentration in Water: There are no criteria for H_2SO_4 as such. The EPA has, however, recommended criteria for pH as follows: To protect freshwater aquatic life — pH 6.5 to 9.0; to protect saltwater aquatic life — pH 6.5 to 8.5; and to protect humans' drinking water — pH 5 to 9.

Routes of Entry: Inhalation of mist, ingestion, eye and skin contact.

Harmful Effects and Symptoms: *Local* — Burning and charring of the skin are a result of the great affinity for, and strong exothermic reaction with, water. Concentrated sulfuric acid will effectively remove the elements of water from many organic materials with which it comes in contact. It is even more rapidly injurious to mucous membranes and exceedingly dangerous to the eyes. Ingestion causes serious burns of the mouth or perforation of the esophagus or stomach. Dilute sulfuric acid does not possess this property, but is an irritant to skin and mucous membranes due to its acidity and may cause irreparable corneal damage and blindness as well as scarring of the eyelids and face.

Systemic — Sulfuric acid mist exposure causes irritation of the mucous membranes, including the eye, but principally the respiratory tract epithelium. The mist also causes etching of the dental enamel followed by erosion of the enamel and dentine with loss of tooth substance. Central and lateral incisors are mainly affected. Breathing high concentrations of sulfuric acid causes tickling in the nose and throat, sneezing, and coughing. At lower levels sulfuric acid causes a reflex increase in respiratory rate and diminution of depth, with reflex bronchoconstriction resulting in increased pulmonary air flow resistance. A single overexposure may lead to laryngeal, tracheobronchial, and pulmonary edema. Repeated excessive exposures over long periods have resulted in bronchitic symptoms, and rhinorrhea, lacrimation, and epitaxis. Long exposures are claimed to result in conjunctivitis, frequent respiratory infections, emphysema, and digestive disturbances.

Points of Attack: Respiratory system, eyes, skin, teeth, lungs.

Medical Surveillance: Preplacement and periodic medical examinations should give special consideration to possible effects on the skin, eyes, teeth, and respiratory tract. Pulmonary function tests should be performed.

First Aid: If this chemical gets into the eyes, irrigate immediately. If this chemical contacts the skin, flush with water immediately. If a person breathes in large amounts of this chemical, move the exposed person to fresh air at once and perform artificial respiration. When this chemical has been swallowed, get medical attention. Give large quantities of water and do not induce vomiting.

Personal Protective Methods: Wear appropriate clothing to prevent any possibility of skin contact with liquids of >1.0% content or repeated or prolonged contact with liquids of <1.0% content. Wear eye protection to prevent any possibility of eye contact. Employees should wash immediately when skin is wet or contaminated. Remove nonimpervious clothing immediately if wet or contaminated. Provide emergency showers and eyewash if liquids containing >1.0% contaminants are involved.

Respirator Selection:
      50 mg/m³: GMAGHiEP/HiEPF/SAF/SCBAF
   100 mg/m³: SAF:PD,PP,CF
     Escape: GMAGHiEP/SCBA

Disposal Method Suggested: Add slowly to solution of soda ash and slaked lime with stirring; flush to drain with large volumes of water. Recovery and re-use of spent sulfuric acid may be a viable alternative to disposal and processes are available (A-57).

References

(1) National Institute for Occupational Safety and Health, *Criteria for a Recommended Standard: Occupational Exposure to Sulfuric Acid,* NIOSH Doc. No. 74-128 (1974).

(2) Sax, N.I., Ed., *Dangerous Properties of Industrial Materials Report, 1,* No. 5, 80-83, New York, Van Nostrand Reinhold Co. (1981).

(3) See Reference (A-61).

(4) See Reference (A-60).

(5) Parmeggiani, L., Ed., *Encyclopedia of Occupational Health & Safety,* Third Edition, Vol. 2, pp 2124–26, Geneva, International Labour Office (1983).

SULFUROUS ACID 2-(p-tert-BUTYLPHENOXY)-1-METHYLETHYL-2-CHLOROETHYL ESTER

- Carcinogen (Animal Positive, IARC) (A-62) (A-64)
- Hazardous Waste Constituent (EPA)

Description: $C_{15}H_{23}ClO_4S$ with the following structural formula

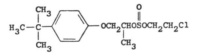

is a heavy liquid which boils at 175°C at 0.1 mm Hg pressure.

Code Numbers: CAS 140-57-8   RTECS WT2975000

DOT Designation: —

Synonyms: 2-(p-tert-Butylphenoxy)-1-methylethyl-2-chloroethyl sulfite; Aracide®; Aratron®; Niagaramite®; Ortho-Mite®; Aramite®.

Potential Exposure: Aramite is a miticide and antimicrobial agent.

Aramite was previously registered for use on 39 crops, 20 of which were fruits and nuts; but in 1970, its use was restricted to postharvest application on fruit trees (EPA).

The single producer of aramite ceased production in 1975, prior to the 1977 voluntary cancellations of registrations by all distributors. Currently, it is believed that aramite is not available in the United States. Although actual production volumes are not known, it is believed that this pesticide was not widely used. For example, only 20,000 lb were used on fruit and nuts grown in California in 1971. By 1977, only 135 lb of the formulated product were available on the U.S. market (EPA).

No data are available on the number of workers who were actually or potentially exposed to aramite during its manufacture and formulation. No estimates of direct consumer exposure to the pesticide through consumption of contaminated fruit and nut crop have been identified. The potential for exposure to the general public has been eliminated by regulatory action.

Aramite is regulated by EPA under the Federal Insecticide, Fungicide, and Rodenticide Act and the Resource Conservation and Recovery Act. The significant regulatory action was a voluntary cancellation of the active ingredient registration by the sole producer in 1975. One formulator holding remaining stockpiles was permitted to sell them through December 1977.

Permissible Exposure Limits in Air: No standards set.

Permissible Concentration in Water: No criteria set.

Harmful Effects and Symptoms: This material is slightly toxic (LD_{50} value for rats is 3,900 mg/kg) but it is carcinogenic to animals (2).

Aramite is carcinogenic in the rat and dog following its oral administration. It produced liver tumors in the rat and carcinomas of the gall bladder and biliary ducts in the dog.

Aramite was tested in two strains of mice by the oral route and produced a significant increase of hepatomas in males of one strain (A-62) (A-64).

Disposal Method Suggested: Acid or alkaline hydrolysis followed by flushing to sewer.

References

(1)   Sax, N.I., Ed., *Dangerous Properties of Industrial Materials Report, 1,* No. 3, 79-80, New York, Van Nostrand Reinhold Co. (1981).
(2)   International Agency for Research on Cancer, *IARC Monographs on the Carcinogenic Risks of Chemicals to Humans 5,* 39, Lyon, France (1974).
(3)   United Nations Environment Programme, *IRPTC Legal File 1983,* Vol. II, pp VII/749, Geneva, Switzerland, International Register of Potentially Toxic Chemicals (1984).

SULFUR OXIDES

Although sulfur dioxide is the primary oxide of concern (see entry under Sulfur Dioxide), national ambient air quality standards have been set for sulfur oxides as a class by EPA. The primary emission standards are 0.03 ppm (80 μg/m^3) as an annual arithmetic mean and 0.14 ppm (365 μg/m^3) on a 24-hour average basis, not to be exceeded more than once a year. The secondary or background standard is 0.5 ppm (1,300 μg/m^3) on a 3-hour average basis, not to be exceeded more than once a year.

References

(1)   National Academy of Science, *Medical and Biologic Effects of Atmospheric Pollutants: Sulfur Oxides,* Washington, DC (1978).
(2)   Sax, N.I., Ed., *Dangerous Properties of Industrial Materials Report, 1,* No. 5, 83-84, New York, Van Nostrand Reinhold Co. (1981). (Sulfur Trioxide).

SULFUR PENTAFLUORIDE

Description: S_2F_{10} is a colorless liquid or gas with an odor like sulfur dioxide. It boils at 29°C.

Code Numbers: CAS 5714-22-7   RTECS WS4480000

DOT Designation: —

Synonyms: Sulfur decafluoride, Disulfur decafluoride.

Potential Exposure: Sulfur pentafluoride is encountered as a by-product in the manufacture of sulfur hexafluoride which is made by the direct fluorination of sulfur or sulfur dioxide.

Incompatibilities: None hazardous.

Permissible Exposure Limits in Air: The Federal standard and the 1983/84 ACGIH TWA value is 0.025 ppm (0.25 mg/m^3). The STEL is 0.075 ppm (0.75 mg/m^3). The IDLH level is 1.0 ppm.

Permissible Concentration in Water: No criteria set.

Routes of Entry: Inhalation, ingestion, skin and eye contact.

Harmful Effects and Symptoms: None known in humans; in animals, breathing difficulties.

Points of Attack: Respiratory system, central nervous system.

Medical Surveillance: Consider the points of attack in preplacement and periodic physical examinations.

First Aid: If this chemical gets into the eyes, irrigate immediately. If this chemical contacts the skin, wash with soap immediately. If a person breathes in large amounts of this chemical, move the exposed person to fresh air at once and perform artificial respiration. When this chemical has been swallowed, get medical attention. Give large quantities of water and do not induce vomiting.

Personal Protective Methods: Wear appropriate clothing to prevent any possibility of skin contact. Wear eye protection to prevent any possibility of eye contact. Remove nonimpervious clothing immediately if wet or contaminated. Provide emergency showers and eyewash.

Respirator Selection:
    0.25 ppm: SA/SCBA
    1 ppm: SAF/SA:PD,PP,CF/SCBAF
    Escape: GMAG/SCBA

References
(1) See Reference (A-61).

SULFUR TETRAFLUORIDE

Description: SF$_4$ is a colorless gas.

Code Numbers: CAS 7783-60-0 RTECS WT4800000

DOT Designation: —

Potential Exposure: SF$_4$ is used as a fluorinating agent in making water-repellent and oil-repellent materials and lubricity improvers. It is also used as a pesticide intermediate (A-32).

Permissible Exposure Limits in Air: There are no Federal standards but ACGIH (1983/84) has adopted a TWA value of 0.1 ppm (0.4 mg/m^3) and set an STEL value of 0.3 ppm (1.0 mg/m^3).

Permissible Concentration in Water: No criteria set. (SF$_4$ reacts violently with water to give SO$_2$ and HF.)

Harmful Effects and Symptoms: Sulfur tetrafluoride is about as toxic as phosgene. It is a strong irritant. The toxic effects are attributed largely to fluorine which is released upon hydrolysis.

SULFURYL FLUORIDE

Description: SO_2F_2 is a colorless, odorless gas.

Code Numbers: CAS 2699-79-8   RTECS WT5075000   UN 2191

DOT Designation: Nonflammable gas.

Synonyms: Sulfuric oxyfluoride, Vikane®.

Potential Exposure: Sulfuryl fluoride is used as an insecticidal fumigant. It is also used in organic synthesis of drugs and dyes.

Incompatibilities: None hazardous.

Permissible Exposure Limits in Air: The Federal standard and the 1983/84 ACGIH TWA value is 5 ppm (20 mg/m^3). The STEL is 10 ppm (40 mg/m^3). The IDLH level is 1,000 ppm.

Determination in Air: Collection by impinger or fritted bubbler, analysis by gas liquid chromatography (A-1).

Permissible Concentration in Water: No criteria set.

Routes of Entry: Inhalation, eye and skin contact.

Harmful Effects and Symptoms: Conjunctivitis, rhinitis, irritation of the pharynx, paresthesias.

Points of Attack: Respiratory system, central nervous system.

Medical Surveillance: Consider the points of attack in preplacement and periodic physical examinations.

First Aid: If this chemical gets into the eyes, irrigate immediately. If a person breathes in large amounts of this chemical, move the exposed person to fresh air at once and perform artificial respiration.

Personal Protective Methods: Respirators are the only devices specified by NIOSH for protection from sulfuryl fluoride.

Respirator Selection:

50 ppm:	SA/SCBA
250 ppm:	SAF/SCBAF
1,000 ppm:	SAF:PD,PP,CF
Escape:	GMS/SCBA

Disposal Method Suggested: Addition of soda ash-slaked lime solution to form the corresponding sodium and calcium salt solution. This solution can be safely discharged after dilution (A-31). The precipitated calcium fluoride may be buried or added to a landfill. Small amounts could also be released directly to the atmosphere without serious harm (A-32).

Refereences

(1)   See Reference (A-61).

SULPROFOS

Description: $C_{12}H_{19}O_2PS_3$ with the following structural formula

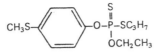

is a tan colored liquid which boils at 155°–158°C at 0.1 mm Hg.

Code Numbers: CAS 35400-43-2   RTECS TE4165000

DOT Designation: —

Synonyms: Bolstar®; Helothion®; O-Ethyl-O-(4-methylthiophenyl)-S-propyl phosphorodithioate.

Potential Exposure: Those involved in the manufacture, formulation and application of this insecticide.

Permissible Exposure Limits in Air: ACGIH (1983/84) has adopted a TLV of 1.0 mg/m³ but no STEL value.

Permissible Concentration in Water: No criteria set.

Harmful Effects and Symptoms: This is a moderately toxic compound (LD$_{50}$ for rats is 304 mg/kg). Shows typical anticholinesterase effects. No cases of poisoning in man have been recorded.

References

(1)   American Conference of Governmental Industrial Hygienists, *Documentation of the Threshold Limit Values: Supplemental Documentation 1982,* Cincinnati, Ohio (1982).

T

2,4,5-T

- Carcinogen (animal suspected, IARC) (1)
- Hazardous substance (EPA)
- Hazardous waste (EPA)

Description: $Cl_3C_6H_2OCH_2COOH$ is a colorless to tan odorless solid which melts at 154° to 155°C.

Code Numbers: CAS 93-76-5 RTECS AJ8400000 UN(NA 2765)

DOT Description: ORM-A

Synonyms: 2,4,5-Trichlorophenoxyacetic acid, Weedone®.

Potential Exposure: Those engaged in the manufacture (A-32), formulation and application of this herbicide. The EPA has issued a rebuttable presumption against registration of 2,4,5-T for pesticide uses, however.

Incompatibilities: None hazardous.

Permissible Exposure Limits in Air: The Federal standard and the 1983/84 ACGIH TWA value is 10 mg/m³. The STEL value is 20 mg/m³. The IDLH level is 5,000 mg/m³.

Determination in Air: Collection by impinger or fritted bubbler, analysis by gas liquid chromatograhy (A-1).

Permissible Concentration in Water: A no-adverse effect level in drinking water has been calculated by NAS/NRC (A-2) to be 0.7 mg/ℓ.

Routes of Entry: Inhalation, ingestion, skin and eye contact.

Harmful Effects and Symptoms: *Observations in Man* — Data compiled by Dow Chemical Company showed that 126 manufacturing personnel exposed to 2,4,5-T at an estimated 1.6 to 8.1 mg/day (0.02 to 0.12 mg/kg/day) for periods of up to 3 years developed no herbicide-related illness.

The results were entirely different in another plant, where the 2,4,5-T produced contained a high proportion of TCDD; 18% of the men suffered moderate to severe chloracne, and several cases of porphyria were found. Chromosomal analysis of 52 workers exposed for various periods up to 960 days to 2,4,5-T (containing TCDD at <1 ppm) at 1.6 to 8.1 mg/day failed to show any abnormalities.

Symptoms of 2,4,5-T exposure found in animal studies include ataxia; skin irritation; acne-like rash; blood in stools.

A rebuttable presumption against registration of 2,4,5-T for pesticidal uses was issued on April 21, 1978 by EPA on the basis of oncogenicity, teratogenicity and fetotoxicity (A-32).

Points of Attack: Skin, liver, gastrointestinal tract.

Medical Surveillance: Consider the points of attack in preplacement and periodic physical examinations.

First Aid: If this chemical gets into the eyes, irrigate immediately. If this chemical contacts the skin, wash with soap. If a person breathes in large amounts of this chemical, move the exposed person to fresh air at once and perform artificial respiration. When this chemical has been swallowed, get medical attention. Give large quantities of water and induce vomiting. Do not make an unconscious person vomit.

Personal Protective Methods: Respirators are the only protective devices specified by NIOSH (A-4).

Respirator Selection:
50 mg/m³:	DMXS
100 mg/m³:	DMXSQ/HiEP/SA/SCBA
500 mg/m³:	HiEPF/SAF/SCBAF
5,000 mg/m³:	PAPHiE/SA:PD,PP,CF
Escape:	DMXS/SCBA

Disposal Method Suggested: Two disposal procedures have been discussed (A-32) for 2,4,5-T: (1) Mix with excess sodium carbonate, add water and let stand for 24 hours before flushing down the drain with excess water; and (2) pour onto vermiculite and incinerate with wood, paper, and waste alcohol.

References

(1) International Agency for Research on Cancer, *IARC Monographs on the Carcinogenic Risks of Chemicals to Humans,* Lyon, France 15, 273 (1977).
(2) See Reference (A-61).
(3) See Reference (A-60).
(4) United Nations Environment Programme, *IRPTC Legal File 1983,* Vol. I, pp VII/16–17, Geneva, Switzerland, International Register of Potentially Toxic Chemicals (1984).

TABUN

Description: $C_5H_{11}N_2O_2P$ with the formula

$$(CH_3)_2N-\overset{\displaystyle O}{\overset{\displaystyle \|}{\underset{\displaystyle \underset{\displaystyle OC_2H_5}{|}}{P}}}-CN$$

is a fruity-smelling liquid which boils at 240°C with decomposition.

Code Numbers: CAS 77-81-6 RTECS TB 4550000

DOT Designation: —

Synonyms: Dimethylphosphoramidocyanidic acid ethyl ester; dimethylamidoethoxy phosphonyl cyanide.

Potential Exposure: In military operations as a nerve gas.

Permissible Exposure Limits in Air: No standards set.

Permissible Concentration in Water: No criteria set.

Routes of Entry: Skin absorption.

Harmful Effects and Symptoms: This compound is extremely toxic with a lethal dose for man of only 0.01 mg/kg. The primary action is on the sympathetic nervous system, causing a vasoparesis. Vapors, when inhaled, can cause nausea, vomiting and diarrhea followed by muscular twitching and convulsions. See Parathion.

Points of Attack: Sympathetic nervous system.

References

(1) Sax, N.I., Ed., *Dangerous Properties of Industrial Materials Report, 1,* No. 2, 63, New York, Van Nostrand Reinhold Co. (1980).

TALC

Description: $Mg_3SiO_{10}(OH)_2$ is an odorless solid which exists in both a non-asbestos form and a fibrous form. This entry will be concerned with the non-fibrous form.

Code Numbers: CAS 14807-96-6   RTECS VV8790000

DOT Designation: —

Synonyms: Hydrous magnesium silicate; Steatite talc; Nonfibrous talc; Non-asbestiform talc.

Potential Exposure: Talc is used in the ceramics, paint, roofing, insecticide, paper, cosmetics and rubber industires (1).

Incompatibilities: None Hazardous.

Permissible Exposure Limits in Air: The Federal standard is 20 mppcf. There is no tentative STEL value set, nor is there an IDLH level established. A maximum allowable concentration of 700 particles/cm^3 has been set by the Dutch Chemical Industry (A-60).

Permissible Concentration in Water: No criteria set.

Routes of Entry: Inhalation of dust, eye and skin contact.

Harmful Effects and Symptoms: Fibrous industrial talc causes pneumoconiosis often accompanied by chronic hypertrophic pulmonary osteoarthropathy in humans exposed for long periods of time. In experimental mammals, pure talc induces a cytogenic rather than fibrogenic effect on the lungs. Although some fibrogenic agents such as quartz and tremolite may have an additive effect on lung damage, some evidence has been found to suggest that pure talc inhibits the pathogenicity of quartz.

Many case histories are available on talc workers who developed pneumoconiosis after employment periods ranging from 1 to 37 years. The clinical signs and symptoms include dyspnea, cough, chest pain, and weakness. Most of the reports have noted the similarity of talc-induced pneumoconiosis to asbestosis. Early studies suggested that tremolite, rather than pure talc, was the primary pathogenic agent and histological evidence supporting this supposition has been advanced.

Points of Attack: Lungs, cardiovascular system.

Medical Surveillance: Consider the points of attack in preplacement and periodic physical examinations.

First Aid: If this chemical gets into the eyes, irrigate immediately.

Personal Protective Methods: Respirators are the only protective devices suggested by NIOSH (A-4).

Respirator Selection:

    100 mppcf: DM
    200 mppcf: DMXSQ/FuHiEP/SA/SCBA
  1,000 mppcf: HiEPF/SAF/SCBAF
10,000 mppcf: PAPHiEF/SAF:PD,PP,CF

Disposal Method Suggested: Landfill.

References

(1) National Institute for Occupational Safety and Health, *Information Profiles on Potential Occupational Hazards: Talc,* pp 51-53, Report PB-276, 678, Rockville, MD (October 1977).
(2) See Reference (A-61).
(3) Parmeggiani, L., Ed., *Encyclopedia of Occupational Health & Safety,* Third Edition, Vol. 2, pp 2141–42, Geneva, International Labour Office (1983).

TANNIC ACID

● Carcinogen (Animal Positive, IARC)

Description: $C_{76}H_{52}O_{46}$ is tannic acid which has a structure similar to that of corilagin, a tannin, which has the following structure:

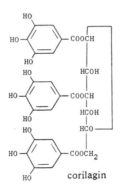

corilagin

Code Numbers: CAS 1401-55-4   RTECS WW5075000

DOT Designation: —

Synonyms: Gallotannin; gallotannic acid; Chinese tannin.

Potential Exposure: Used as a mordant in dyeing; used as a coagulant in rubber manufacture; used in manufacturing gallic acid and pyrogallol; used to clarify beer and wine; and as a boiler feedwater additive.

Permissible Exposure Limits in Air: No standards set.

Permissible Concentration in Water: No criteria set.

Harmful Effects and Symptoms: Tannic acid is moderately toxic (LD_{50} for rats is 200 mg/kg).

Personal Protective Methods: Wear skin protection.

Respirator Selection: Wear a filter mask.

Disposal Method Suggested: Incineration.

References

(1) Sax, N.I., Ed., *Dangerous Properties of Industrial Materials Report, 2,* No. 1, 119-121, New York, Van Nostrand Reinhold Co. (1982).
(2) International Agency for Research on Cancer, *IARC Monographs on the Carcinogenic Risks of Chemicals to Humans, 10,* 253, Lyon, France (1976).

TANTALUM AND COMPOUNDS

Description: Tantalum is a refractory metal in Group V-B of the periodic table. The pure metal is ductile and features a high melting point of 2996°C.

Code Numbers: Tantalum metal: CAS 7440-25-7   RTECS WW5505000

DOT Designation: —

Synonym: Tantalum 181.

Potential Exposure: Tantalum metal is used in electronic components, chemical equipment and in nuclear reactor components. Tantalum carbide is used in metal cutting tools and wear-resistant parts. Some tantalum salts are used in catalysts.

Incompatibilities: Strong oxidizers.

Permissible Exposure Limits in Air: The Federal standard and the ACGIH 1983/84 TWA value is 5 mg/m³. The STEL value is 10 mg/m³. There is no IDLH level set.

Determination in Air: Filter collection, analysis by atomic absorption spectroscopy (A-21).

Permissible Concentration in Water: No criteria set.

Routes of Entry: Inhalation of dust.

Harmful Effects and Symptoms: None known in humans.

Points of Attack: None known in humans.

Medical Surveillance: Nothing special indicated.

First Aid: If a person breathes in large amounts of this chemical, move the exposed person to fresh air at once and perform artificial respiration.

Personal Protective Methods: Respirators are the only protective devices specified by NIOSH (A-4).

Respirator Selection:

25 mg/m³:	DM
50 mg/m³:	DMXSQ/DMFu
50 mg/m³:	FuHiEP/SA/SCBA
250 mg/m³:	HiEPF/SAF/SCBAF
5,000 mg/m³:	PAPHiE/SA:PD,PP,CF
10,000 mg/m³:	SAF:PD,PP,CF

Disposal Method Suggested: Sanitary landfill if necessary, recover if possible because of economic value. Technology exists for tantalum recovery from spent catalysts, for example (A-57).

References

(1) See Reference (A-61).

(2) Parmeggiani, L., Ed., *Encyclopedia of Occupational Health & Safety,* Third Edition, Vol. 2, pp 2146–47, Geneva, International Labour Office (1983).

TDE

- Carcinogen (animal positive, IARC) (2)
- Hazardous waste (EPA)
- Priority toxic pollutant (EPA)

Description: 2,2-Bis(p-chlorophenyl)-1,1-dichloroethane is a colorless crystalline compound melting at 109° to 110°C having the structural formula shown below.

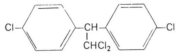

Code Numbers: CAS 72-54-8   RTECS KI0700000   UN (NA-2761)

DOT Designation: ORM-A

Synonyms: ENT 4225, DDD, Rhothane®.

Potential Exposure: Those involved in the manufacture (A-32), formulation and application of this insecticide. In an action on March 18, 1971, EPA cancelled all pesticide uses of this product which is a metabolite of DDT. Hence it is no longer manufactured commercially (A-7).

Permissible Exposure Limits in Air: No standards set.

Permissible Concentration in Water: For the protection of freshwater aquatic life, the value is 0.6 $\mu g/\ell$, based on acute toxicity. For saltwater aquatic life, the value is 3.6 $\mu g/\ell$, based on acute toxicity.

For the protection of human health, with respect to TDE, see criteria proposed for DDT.

Determination in Water: Methylene chloride extraction followed by gas chromatography with electron capture or halogen specific detection (EPA Method 608) or gas chromatography plus mass spectrometry (EPA Method 625).

Harmful Effects and Symptoms: There is some evidence that DDD is carcinogenic in mice; however, in other species, it appears to be noncarcinogenic. p,p'-DDD has been shown to be mutagenic in *Drosophila,* but not in yeast or bacteria. In cell culture, p,p'-DDD causes chromosomal breaks.

Since DDD is a metabolite of DDT, as well as a contaminant of commercial preparations of DDT, many of the effects of DDT could be mediated through DDD.

References

(1) U.S. Environmental Protection Agency, *DDD,* Health and Environmental Effects Profile No. 58, Washington, DC, Office of Solid Waste (April 30, 1980).

(2) International Agency for Research on Cancer, *IARC Monographs on the Carcinogenic Risks of Chemicals to Humans,* Lyon, France 5, 83 (1974).
(3) U.S. Environmental Protection Agency, *DDT—Ambient Water Quality Criteria,* Washington, DC (1980).
(4) United Nations Environment Programme, *IRPTC Legal File 1983,* Vol. I, pp VII/309–10, Geneva, Switzerland, International Register of Potentially Toxic Chemicals (1984).

TEDP

See "Sulfotep."

TELLURIUM AND COMPOUNDS

Description: Te, tellurium, is a semimetallic element with a bright luster which is insoluble in water and organic solvents. It may exist in a hexagonal crystalline form or an amorphous powder. It is found in sulfide ores and is produced as a by-product of copper or bismuth refining.

Code Numbers: Elemental Te—CAS 13494-80-9   RTECS WY2625000

DOT Designation: —

Synonyms: Aurum paradoxum, metallum problematum.

Potential Exposures: The primary use of tellurium is in the vulcanization of rubber. It is also used as a carbide stabilizer in cast iron, a chemical catalyst, a coloring agent in glazes and glass, a thermocoupling material in refrigerating equipment, and as an additive to selenium rectifiers; in alloys of lead, copper, steel, and tin for increased resistance to corrosion and stress, workability, machinability, and creep strength, and in certain culture media in bacteriology. Since tellurium is present in silver, copper, lead, and bismuth ores, exposure may occur during purification of these ores.

NIOSH has estimated worker exposure to tellurium and tellurium diethyl-dithiocarbamate at about 7,000.

Incompatibilities: Hazards vary depending upon specific compound.

Permissible Exposure Limits in Air: The applicable Federal standard and the 1983/84 ACGIH TWA value for tellurium is 0.1 mg/m^3. No tentative STEL value has been set nor has an IDLH level been set.

Determination in Air: Collection on a filter, workup with acid, analysis by atomic absorption. See NIOSH Methods Set N. See also reference (A-10).

Permissible Concentration in Water: No criteria set, but EPA (A-37) has suggested a permissible ambient goal of 1.4 μg/ℓ based on health effects.

Routes of Entry: Inhalation of dust or fume; percutaneous absorption from dust, ingestion, skin and eye contact.

Harmful Effects and Symptoms: *Local* — The literature contains no indication of any local effects from tellurium.

Systemic — The toxicity of tellurium and its compounds is of a low order. There is no indication that either tellurium dust or fume is damaging to the skin or lungs. Inhalation of fumes may cause symptoms, however, some of which are particularly annoying socially to the worker. The most common signs of exposure are foul (garliclike) breath and perspiration, metallic taste in the mouth, and dryness. This is probably due to the presence of dimethyl telluride. These symptoms may appear after relatively short exposures at high concentrations, or longer exposures at lower concentrations, and may persist for long periods of time after the exposure has ended. Workers also complain of afternoon somnolence and loss of appetite.

Exposure to hydrogen telluride produces symptoms of headache, malaise, weakness, dizziness, and respiratory and cardiac symptoms similar to those caused by hydrogen selenide. Pulmonary irritation and the destruction of red blood cells have been reported in studies of laboratory animals exposed to hydrogen telluride.

In other animal studies, tellurium hexafluoride was found to be a respiratory irritant which caused pulmonary edema, and metallic tellurium was shown to have a teratogenic effect on the fetus of rats.

Points of Attack: Skin, central nervous system.

Medical Surveillance: Oral hygiene and the respiratory tract should receive special attention in preplacement or periodic examinations.

First Aid: If this chemical gets into the eyes, irrigate immediately. If this chemical contacts the skin, wash with soap promptly. If a person breathes in large amounts of this chemical, move the exposed person to fresh air at once and perform artificial respiration. When this chemical has been swallowed, get medical attention. Give large quantities of water and induce vomiting. Do not make an unconscious person vomit.

Personal Protective Methods: Clean change of work clothes is necessary for hygienic purposes, and showering after each shift before change to street clothes should be encouraged. Respiratory protection is indicated in areas where exposure to hydrogen telluride and tellurium hexafluoride fumes and dust are above the allowable limits.

Respirator Selection:
    0.5 mg/m^3: DMXS
      1 mg/m^3: DMXSQ/FuHiEP/SA/SCBA
      5 mg/m^3: HiEPF/SAF/SCBAF
     50 mg/m^3: PAPHiE/SA:PD,PP,CF
        Escape: HiEP/SCBA

References

(1) U.S. Environmental Protection Agency, *Chemical Hazard Information Profile: Tellurium,* Washington, DC (1979).
(2) U.S. Environmental Protection Agency, *Toxicology of Metals, Vol. II: Tellurium,* pp 370-387, Report EPA-600/1-77-022, Research Triangle Park, NC (May 1977).
(3) See Reference (A-61).
(4) Parmeggiani, L., Ed., *Encyclopedia of Occupational Health & Safety,* Third Edition, Vol. 2, pp 2156-57, Geneva, International Labour Office (1983).

TELLURIUM HEXAFLUORIDE

Description: TeF_6 is a colorless gas with a repulsive odor.

Code Numbers: CAS 7783-80-4   RTECS WY2800000   UN 2195

DOT Designation: Poison gas

Synonyms: None.

Potential Exposures: Tellurium hexafluoride is stated (A-35) to be a by-product of ore refining.

Incompatibilities: None hazardous.

Permissible Exposure Limits in Air: The Federal standard and the 1983/84 ACGIH TWA value is 0.02 ppm (0.2 mg/m³). There is no STEL value. The IDLH level is 1.0 ppm.

Determination in Air: Adsorption on charcoal, workup with sodium hydroxide, analysis by atomic absorption. See NIOSH Methods, Set M. See also reference (A-10).

Permissible Concentration in Water: No criteria set.

Routes of Entry: Inhalation.

Harmful Effects and Symptoms: Headaches; dyspepsia; garlic odor on breath. Animal studies have shown pulmonary edema resulting from TeF₆ exposure.

Points of Attack: Respiratory system.

Medical Surveillance: Consider the points of attack in preplacement and periodic physical examinations.

First Aid: If a person breathes in large amounts of this chemical, move the exposed person to fresh air at once and perform artificial respiration.

Personal Protective Methods: Respirator protection is the only method specified by NIOSH (A-4).

Respirator Selection:
        0.2 ppm: SA/SCBA
          1 ppm: SAF/SA:PD,PP,CF/SCBAF
        Escape: GMS/SCBA

References

(1)   See Reference (A-61).

TEMEPHOS

Description: O,O'-(thiodi-4,1-phenylene) O,O,O',O'-tetramethyldi(phosphorothioate), $C_{16}H_{20}O_6P_2S_3$, is a crystalline solid melting at 30°C with the structural formula shown below.

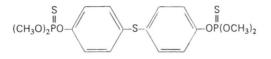

The technical product is a brown viscous liquid.

Code Numbers: CAS 3383-96-8   RTECS TF6890000   UN 2783

DOT Designation: —

Synonyms: Ent 27,165, Abate®, Abathion®, Abat®, Swebate®, Nimitex®, Biothion®.

Potential Exposure: Those involved in the manufacture (A-32), formulation and application of this insecticide.

Permissible Exposure Limits in Air: There is no Federal standard. ACGIH (1983/84) has adopted a TWA value of 10 mg/m³ and set an STEL value of 20 mg/m³.

Determination in Air: Collection on a filter and gravimetric analysis (A-1).

Permissible Concentration in Water: No criteria set. Experience in the field for a period of more than one year has shown, however, that 1 mg/ℓ in drinking water is without effect (A-34).

Determination in Water: Techniques used for residue determination include colorimetry and gas liquid chromatography (A-7) and may be applicable to water analysis.

Harmful Effects and Symptoms: Human volunteers have tolerated an oral dosage of 256 mg/man/day for 5 days or 64 mg/man/day for 4 weeks without clinical symptoms or side effects and without detectable effect on red blood cells or plasma cholinesterase (A-34).

Disposal Method Suggested: Essentially complete hydrolysis occurs upon heating in concentrated KOH for 20 minutes (A-32).

References

(1)   United Nations Environment Programme, *IRPTC Legal File 1983,* Vol. II, pp VII/609, Geneva, Switzerland, International Register of Potentially Toxic Chemicals (1984).

TEPP

● Hazardous substance (EPA)

Description: $(C_2H_5O)_2 \overset{O}{\overset{||}{P}} - O - \overset{O}{\overset{||}{P}}(OC_2H_5)_2$ is a colorless to amber liquid with a faint fruity odor.

Code Numbers: CAS 107-49-3   RTECS UX6825000   UN (NA2783)

DOT Designation: Poison-B

Synonyms: Tetraethyl pyrophosphate, ethyl pyrophosphate, Vapotone®, Nifos-T®, ENT 18 771.

Potential Exposure: Those engaged in the manufacture (A-32), formulation and application of this aphicide and acaricide.

Incompatibilities: Strong oxidizers.

Permissible Exposure Limits in Air: The Federal standard is 0.05 mg/m³. The ACGIH (1983/84) has set a TWA of 0.004 ppm (0.05 mg/m³) and an STEL of 0.01 ppm (0.2 mg/m³). The notation "skin" is added to indicate the possibility of cutaneous absorption. The IDLH level is 10 mg/m³.

Permissible Concentration in Water: No criteria set.

Routes of Entry: Inhalation, skin absorption, ingestion, skin and eye contact.

Harmful Effects and Symptoms: Eye pain, impaired vision, tears, headaches, chest pains, cyanosis, anorexia, nausea, vomiting, diarrhea, local sweating, weakness, twitching, paralysis, Cheyne-Stokes respiration, convulsions, low blood pressure.

Points of Attack: Central nervous system, respiratory system, cardiovascular system, gastrointestinal tract.

Medical Surveillance: Consider the points of attack in preplacement and periodic physical examinations.

First Aid: If this chemical gets into the eyes, irrigate immediately. If this chemical contacts the skin, flush with water immediately. If a person breathes in large amounts of this chemical, move the exposed person to fresh air at once and perform artificial respiration. When this chemical has been swallowed, get medical attention. Give large quantities of water and induce vomiting. Do not make an unconscious person vomit.

Personal Protective Methods: Wear appropriate clothing to prevent any possibility of skin contact. Wear eye protection to prevent any possibility of eye contact. Employees should wash immediately when skin is wet or contaminated. Remove nonimpervious clothing immediately if wet or contaminated. Provide emergency showers and eyewash.

Respirator Selection:
> 0.5 mg/m³: SA/SCBA
> 2.5 mg/m³: SAF/SCBAF
> 10 mg/m³: SA:PD,PP,CF
> Escape: GMOVPPest/SCBA

Disposal Method Suggested: TEPP is 50% hydrolzyed in water in 6.8 hours at 25°C, and 3.3 hours at 38°C; 99% hydrolysis requires 45.2 hours at 25°C, or 21.9 hours at 38°C. Hydrolysis of TEPP yields nontoxic products (A-32). Incineration is, however, an option for TEPP disposal (A-38).

References
(1) See Reference (A-61).
(2) United Nations Environment Programme, *IRPTC Legal File 1983,* Vol. II, pp VII/706–8, Geneva, Switzerland, International Register of Potentially Toxic Chemicals (1984).

TEREPHTHALIC ACID

Description: $HOOCC_6H_4COOH$ is a solid which sublimes at about 300°C.

Code Numbers: CAS 100-21-0   RTECS WZ0875000

DOT Designation: —

Synonyms: 1,4-benzenedicarboxylic acid; p-benzenedicarboxylic acid; TPA.

Potential Exposure: TPA is used primarily in the production of polyethylene terephthalate polymer for the fabrication of polyester fibers and films. NIOSH estimates that about 8,600 workers are exposed annually to TPA.

Permissible Exposure Limits in Air: There is no Federal standard. A value of 0.015 ppm (0.1 mg/m³) has reportedly been set by the U.S.S.R. (A-36).

Permissible Concentration in Water: No criteria set.

Harmful Effects and Symptoms: Terephthalic acid is relatively nontoxic (1). Liver and kidney function were normal in laboratory animals fed diets containing TPA. Excretion was rapid and nearly complete in the urine after oral and intratracheal administration. Dermal application did not cause irritation, and both dermal and ocular exposure resulted in negligible absorption and excretion. Cardiopulmonary effects were not found in rats, although a decrease in aortic blood pressure, stimulation of respiration, and a decreased pulmonary resistance were found.

Data on human exposure are not available (1).

Disposal Method Suggested: Incineration.

References

(1) National Institute for Occupational Safety and Health, *Information Profile on Potential Occupational Hazards—Single Chemicals: Terephthalic Acid,* pp 115-119, Report TR 79-607, Rockville, MD (December 1979).

TERPHENYLS

Description: With the formula $C_{18}H_{14}$, there are three isomeric terphenyls as follows:

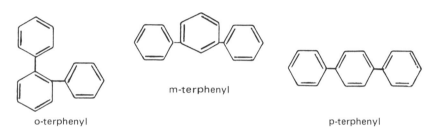

m-terphenyl

o-terphenyl   p-terphenyl

Pure terphenyl is a white, crystalline solid. The commercial grades are light yellow. All three isomers (ortho, meta, and para) are unusually stable toward heat. The properties of the terphenyls are:

	Ortho	Meta	Para
Melting point, °C	58	89	207
Boiling point, °C	332	365	385

Code Numbers: For p-terphenyl—CAS 92-94-4   RTECS WZ6475000

DOT Designation: —

Synonyms: Triphenyls, diphenylbenzenes. For p-terphenyl—4-phenylbisphenyl, p-diphenylbenzene, Santowax P®.

Potential Exposures: Terphenyl is used primarily as a heat storage and heat transfer agent. It is also used as a high-temperature lubricant, a constituent of waxes and polishes, and as a plasticizer for resin-bodied paints.

NIOSH estimates that fewer than 5,000 workers are exposed to terphenyl each year.

Incompatibilities: None hazardous.

Permissible Exposure Limits in Air: The Federal standard is a 1 ppm (9 mg/m³) ceiling value. The ACGIH (1983/84) has set a TWA ceiling value of 0.5 ppm (5 mg/m³) but no STEL value. The IDLH level is 3,500 mg/m³.

Determination in Air: Collection on a filter, workup with CS_2, analysis by gas chromatography. See NIOSH Methods, Set C. See also reference (A-10).

Permissible Concentration in Water: No criteria set. See "Polynuclear Aromatic Hydrocarbons" for related but more complex compounds.

Routes of Entry: Inhalation of dusts and mists, ingestion, skin and eye contact.

Harmful Effects and Symptoms: Irritation of eyes and skin, thermal burns.

Points of Attack: Skin, respiratory system, eyes.

Medical Surveillance: Consider the points of attack in preplacement and periodic physical examinations.

First Aid: If this chemical gets into the eyes, irrigate immediately. If this chemical contacts the skin, flush with water immediately. If a person breathes in large amounts of this chemical, move the exposed person to fresh air at once and perform artificial respiration. When this chemical has been swallowed, get medical attention. Do not induce vomiting.

Personal Protective Methods: Wear appropriate clothing to prevent any possibility of skin contact with molten terphenyls or repeated or prolonged contact with liquids. Wear eye protection to prevent any possibility of eye contact with molten material; any reasonable probability of contact with solutions. Employees should wash promptly when skin is wet or contaminated. Work clothing should be changed daily if it is possible that clothing is contaminated. Remove nonimpervious clothing promptly if wet or contaminated. Provide emergency showers and eyewash.

Respirator Selection:
       450 mg/m³: HiEPF/SAF/SCBAF
     3,500 mg/m³: PAPHiEF/SAF:PD,PP,CF
        Escape: DMXS/SCBA

Disposal Method Suggested: Incineration.

References
(1)   See Reference (A-61).

TESTOSTERONE

● Carcinogen (Animal Positive, IARC)

Description: $C_{19}H_{28}O_2$ with the following structural formula

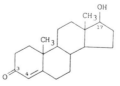

is a crystalline compound which melts at 155°C.

Code Numbers: CAS 58-22-0   RTECS XA3030000

DOT Designation: —

Synonyms: 17β-Hydroxyandrost-4-ene-3-one with a large variety of trade names.

Potential Exposure: Used as an androgenic, anabolic and estrogenic hormone for both males and females.

Permissible Exposure Limits in Air: No standards set.

Permissible Concentration in Water: No criteria set.

Harmful Effects and Symptoms: High acute toxicity. Workers engaged in manufacture and packaging have shown hormone effects; male workers have experienced breast enlargement. It is a proven animal carcinogen (2).

References
(1) Sax, N.I., Ed., *Dangerous Properties of Industrial Materials Report, 1*, No. 3, 81-82, New York, Van Nostrand Reinhold Co. (1981).

TETRABROMOETHANE

See "Acetylene Tetrabromide."

TETRACHLORODIBENZO-p-DIOXIN

- Carcinogen (EPA-CAG) (A-40) (Animal Positive, IARC) (5,6,7)
- Hazardous waste constituent (EPA)
- Priority toxic pollutant (EPA)

Description: Polychlorinated dibenzo-p-dioxins are formed in the manufacturing process of all chlorophenols. However, the amount formed is dependent on the degree to which the temperature and pressure are controlled during production.

An especially toxic dioxin, 2,3,7,8-tetrachlorodibenzo-p-dioxin (TCDD), is formed during the production of 2,4,5-TCP (trichlorophenol) by the alkaline hydrolysis of 1,2,4,5-tetrachlorobenzene. Tetrachlorodibenzo-p-dioxin has the formula $C_{12}H_4Cl_4O_2$. TCDD is a white crystalline solid with a melting point range of 302° to 305°C. Decomposition begins at 500°C and is virtually complete within 21 seconds at a temperature of 800°C. The structural formula is:

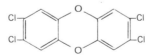

As can be anticipated, TCDD has been associated with all synthetic compounds derived from 2,4,5-TCP. This includes the widely used herbicide and defoliant 2,4,5-T (2,4,5-trichlorophenoxyacetic acid).

Code Numbers: CAS 1746-01-6   RTECS HP3500000

DOT Designation: −

Synonym: TCDD.

Potential Exposures: TCDD has no uses as such. As noted above, TCDD is an inadvertent contaminant in herbicide precursors and thus in the herbicides themselves. Thus, it is applied in herbicide formulations, but is not used per se. It has been estimated that approximately 2 million acres in the United States have been treated for weed control on one or more occasions with approximately 15 million pounds of TCDD contaminated 2,4,5-T, 2,4-D, or combinations of the two.

In the cases of human exposure to 2,4,5-TCP, the only adverse effects reported were caused by occupational exposure or accidents that occurred during the manufacture of chlorinated phenols or products derived from them.

In 1949, intermediary chemicals of the manufacturing process were released in a U.S. 2,4,5-T plant. This accident led to 117 cases of chloracne among exposed workers.

In 1953 there was an accident in a Middle Rhine factory manufacturing 2,4,5-TCP from 1,2,4,5-tetrachlorobenzene. In addition to contracting chloracne, many workers had liver cirrhosis, heat complaints, and nervous system disorders, and were depressed.

In 1958, 31 employees of a Hamburg, Germany, plant in which 2,4,5-T was made from technical 2,4,5-TCP contracted chloracne and suffered the physical and psychological symptoms associated with it. In 1961 TCDD was conclusively identified as the cause of the chloracne.

An explosion occurred in a 2,4,5-T plant in Amsterdam in 1963. Six months later, 9 of 18 men, who were attempting to decontaminate the plant, developed chloracne. All of the men had worn deep sea diving suits, and all but one wore face masks with goggles while working in the plant. Of these men, three died within 2 years. The man without the face mask or goggles was severely affected. He was unable to walk and required long-term treatment.

In 1964, workers in a 2,4,5-T plant in the United States developed chloracne from exposure to TCDD.

There was an explosion at the Coalite Co.'s 2,4,5-TCP plant in Great Britain in 1968. TCDD had accidentally been produced as the result of an exothermic reaction. Seventy-nine cases of chloracne were reported; many of them were severe.

In 1971 there was an accidental poisoning episode in the United States that affected humans, horses, and other animals. Waste oil contaminated with TCDD had been sprayed on a riding arena to control dust. Later analyses showed that the arena contained TCDD in concentrations of 31.8 to 33.0 $\mu g/g$. The most important route of entry of dioxin into the body was the skin. (This does not preclude the effects of ingesting food contaminated with dioxin from handling.)

A 6-year-old girl was the most severely affected. She had an inflammatory reaction of the kidneys and bladder bleeding that was diagnosed as acute hemorrhagic cystitis with signs of focal pyelonephritis. Nine less severely affected persons developed diarrhea, headaches, nausea, polyarthralgias, and persistent skin lesions. The girl most affected was thoroughly reexamined in 1976. Results indicated that all of her original symptoms had completely disappeared. She had grown normally and all tests, including a detailed neurological examination, were normal.

In July 1976, TCDD was accidentally released in the Seveso region of Italy. Most of the inhabitants were adversely affected. The first overt reaction was the appearance of numerous burn-like lesions on many of the inhabitants. These lesions generally receded. They were probably caused by direct contact with the sodium hydroxide and phenolic components in the fallout. However, 2½ months after the explosion, an increasing number of children and young people in the zone most affected began to develop symptoms of chloracne on their faces and bodies, a definite mark of dioxin poisoning. By November 28 people had confirmed cases of chloracne. This number rose to 38 by December and to 130 a year after the explosion. A number of the victims exposed underwent a "complete change of character": they became extremely nervous, tired, moody, and irritable, and had a marked loss of appetite.

There were a number of Seveso women who were pregnant at the time of the accident. The total number of legal and illegal abortions performed as a result of the explosion probably totaled 90. There were 51 spontaneous (as distinct from induced) abortions. A survey conducted by an epidemiological commission has shown that 183 babies were delivered in the 2 months following the accident. Eight cases of birth abnormalities have been noted among babies born to women in the Seveso area who were pregnant at the time of the explosion. However, local physicians have had difficulty relating these abnormalities directly to the explosion because the incidence of birth abnormalities was not significantly higher than the normal incidence of abnormal births.

Permissible Exposure Limits in Air: There are no numerical limits; in view of its effects, all contact should be avoided.

Permissible Concentration in Water: There are insufficient data to permit the development of criteria for the protection of freshwater or saltwater aquatic life. For the protection of human health, the concentration is preferably zero. An additional lifetime cancer risk of 1 in 100,000 is posed at a concentration of 4.55×10^{-7} $\mu g/\ell$ as of 1979. A concentration of 0.0039 ng/ℓ was estimated to limit cancer risk to one in a million by EPA in 1980.

Determination in Water: Methylene chloride extraction followed by transfer to benzene and capillary column gas chromatography/mass spectrometry with electron impact ionization (EPA Method 613) or gas chromatography plus mass spectrometry (EPA Method 625).

Routes of Entry: Skin absorption, inhalation of vapors.

Harmful Effects and Symptoms: TCDD is one of the most toxic substances known. It exhibits a delayed biological response in many species and is highly lethal at low doses to aquatic organisms, birds, and mammals, including man. It has been shown to be acnegenic, embryolethal, teratogenic, mutagenic (in certain organisms), carcinogenic, and to affect the immune responses in mammals. TCDD has also been shown to persist for 10 years after application to soils and to bioaccumulate in aquatic organisms by factors as high as 8,000-fold. These findings in conjunction with the wide distribution of contaminated products, lead to the conclusion that TCDD represents a potential hazard to both aquatic and terrestrial life.

The toxicity of a dioxin varies with the position and number of chlorines attached to the aromatic rings. Generally, the toxicity increases with increased chlorine substitution. Those dioxins that have halogens at the 2,3, and 7 positions are particularly toxic. TCDD, which has chlorine atoms at the 2,3,7, and 8 positions, is considered the most toxic of the dioxins.

TCDD is an extremely toxic compound exhibiting acute, subchronic and chronic effects in animals and humans. The liver appears to be the target organ of acute exposure. Retention of TCDD by the liver indicates that it apparently undergoes little or no metabolism.

Acute effects of exposure include chloracne, porphyria cutanea tarda, hepatotoxicity, psychological alterations, weight loss, thymic atrophy, thrombocytopenia, suppression of cellular immunity and death. TCDD is teratogenic and fetotoxic. Oral exposure of pregnant rats to 0.125 to 2.0 µg TCDD/kg/day produced fetal mortality, fetal intestinal hemorrhage and both early and late resorptions. There was found an increased incidence of cleft palate when pregnant mice were given TCDD doses of 1.0 µg/kg/day for 10 days during gestation. TCDD has been shown to be mutagenic in three bacterial systems and a potent inducer of hepatic and renal microsomal drug metabolizing enzymes.

The carcinogenic potential of tetrachlorodibenzo-p-dioxin has been established by the findings of two feeding studies. One study found that Sprague-Dawley rats fed dose levels of 5 ppt to ppb TCDD has a significant excess of tumors as compared to the controls. Another feeding study, using the same strain of rats given 0.1, 0.01, 0.001 µg TCDD/kg/day, induced a statistically significant excess of hepatocellular carcinoma in treated rats. Based on these two studies, tetrachlorodibenzo-p-dioxin is likely to be a human carcinogen.

Rats fed 2,3,7,8-tetrachlorodibenzo-p-dioxin developed a variety of malignant tumors, mainly of the respiratory tract, sebaceous glands, liver and bile ducts. (5).

Two bronchogenic carcinomas were reported (versus 0.12 expected, calculated by the Monograph Working Group) in a study of workers exposed to TCDD; however, smoking habits were not reported, only 55 of the 78 workers were traced, and they were only followed for five or six years (6). The available evidence from human studies is considered inadequate (7).

Points of Attack: Liver.

Medical Surveillance: In short, contact wth TCDD should be avoided but obviously careful preplacement and regular physical exams should be carried out in those cases where worker exposure cannot be avoided with emphasis on liver and kidney function studies (A-39).

First Aid: Irrigate eyes with water. Wash contaminated areas of body with soap and water.

Personal Protective Methods: As stated above, contact with TCDD should be avoided. When it cannot be avoided, extreme worker protection should be provided.

Respirator Selection: Use chemical cartridge respirator.

References

(1) U.S. Environmental Protection Agency, *2,3,7,8-Tetrachlorodibenzo-p-Dioxin: Ambient Water Quality Criteria,* Washington, DC (1979).
(2) U.S. Environmental Protection Agency, "Rebuttable Presumption Against Registration of Pesticide Products Containing 2,4,5-Trichlorophenol and Its Salts," *Federal Register* 43, No. 149, 34026-34054 (August 1, 1978).
(3) U.S. Environmental Protection Agency, *TCDD,* Health and Environmental Effects Profile No. 155, Washington, DC, Office of Solid Waste (April 30, 1980).
(4) Sax, N.I., Ed., *Dangerous Properties of Industrial Materials Report, 1,* No. 2, 63-64, New York, Van Nostrand Reinhold Co. (1980).
(5) IARC Monographs 15:41-102 (1977).
(6) *IARC Monographs on the Evaluation of the Carcinogenic Risk of Chemicals to Humans,* Supplement I, IARC, Lyon, France, pp 46-47 (1979).
(7) See Reference (A-62). Also see Reference (A-64).
(8) Nat. Inst for Occupational Safety & Health, *2,3,7,8-Tetrachlorodibenzo-p-dioxin,* Current Intelligence Bulletin 40, DHHS (NIOSH) Publication No. 84–104, Cincinnati, Ohio (Jan. 23, 1984).
(9) Parmeggiani, L., Ed., *Encyclopedia of Occupational Health & Safety,* Third Edition, Vol. 1, pp 638–42, Geneva, International Labour Office (1983).

TETRACHLORODIFLUOROETHANES

Description: CCl_2FCCl_2F, 1,1,2,2-tetrachloro-1,2-difluoroethane, and also CCl_3CF_2Cl, 1,1,1,2-tetrachloro-2,2-difluoroethane are both colorless liquids or solids melting at 26°C and 40° to 41°C respectively. They have a slight etherlike odor.

Code Numbers:

1,1,2,2-T-1,2-D	CAS 76-12-0	KI1420000
1,1,1,2-T-2,2-D	CAS 76-11-9	KI1425000

DOT Designation: —

Synonyms: 1,1,2,2-T-1,2-D—Refrigerant 112, Halocarbon 112, Freon 112, F-112, Ganetron 112. 1,1,1,2-T-2,2-D—Refrigerant 112a, Halocarbon 112a.

Potential Exposures: These materials are used as solvents and refrigerants.

Incompatibilities: Chemically active metals—sodium, potassium, beryllium, powdered aluminum, zinc, magnesium.

Permissible Exposure Limits in Air: Both compounds—The Federal standard and the 1983/84 ACGIH TWA value is 500 ppm (4,170 mg/m³). The STEL value is 625 ppm (5,210 mg/m³. The IDLH level is 15,000 ppm.

Determination in Air: Adsorption on charcoal, workup with CS_2, analysis by gas chromatography. See NIOSH Methods, Set I. See also reference (A-10).

Permissible Concentration in Water: No criteria set.

Routes of Entry: Inhalation, ingestion, skin and eye contact.

Harmful Effects and Symptoms: 1,1,2,2-T-1,2-D—skin irritation, conjunctivitis, pulmonary edema. 1,1,1,2-T-2,2-D—central nervous system depression; pulmonary edema; skin and eye irritation; drowsiness, dyspnea.

Points of Attack: The 1,2-difluoro compounds—lungs, skin. The 2,2-difluoro compound—respiratory system, lungs, skin.

Medical Surveillance: Consider the points of attack in preplacement and periodic physical examinations.

First Aid: If this chemical gets into the eyes, irrigate immediately. If this chemical contacts the skin, wash with soap promptly. If a person breathes in large amounts of this chemical, move the exposed person to fresh air at once and perform artificial respiration. When this chemical has been swallowed, get medical attention. Give large quantities of water for the 1,2-difluoro compound and salt water for the 2,2-difluoro compound and induce vomiting. Do not make an unconscious person vomit.

Personal Protective Methods: Both compounds—Wear appropriate clothing to prevent repeated or prolonged contact. Wear eye protection to prevent any reasonable probability of eye contact. Employees should wash promptly when skin is wet or contaminated. Remove nonimpervious clothing promptly if wet or contaminated.

Respirator Selection: Both compounds—

5,000 ppm:	SA/SCBA
15,000 ppm:	SAF/SCBAF/SAF:PD,PP,CF
Escape:	GMOV/SCBA

References

(1) See Reference (A-61).

1,1,1,2-TETRACHLOROETHANE

- Hazardous waste (EPA)
- Priority toxic pollutant (EPA)

Description: Cl_3CCH_2Cl is a colorless liquid with a boiling point of 129°C.

Code Numbers: CAS 630-20-6   RTECS KI8450000   UN 1702

DOT Designation: ORM-A

Synonyms: None.

Potential Exposures: 1,1,1,2-Tetrachloroethane is used as a solvent and in the manufacture of a number of widely used products, as are the other chloroethanes.

Incompatibilities: See 1,1,2,2-Tetrachloroethane.

Permissible Exposure Limits in Air: There are no standards for this compound unlike the 1,1,2,2-isomer which is discussed in the entry which follows.

Permissible Concentration in Water: For 1,1,1,2-tetrachloroethane the criterion to protect freshwater aquatic life is 9,320 $\mu g/\ell$ based on acute toxicity data. For saltwater aquatic life, no criterion for 1,1,1,2-tetrachloroethane can be derived using the guidelines, and there are insufficient data to estimate a criterion using other procedures. For the protection of human health, there are insufficient data to derive criteria for 1,1,1,2-tetrachloroethane.

Harmful Effects and Symptoms: Literature reporting adverse occupational exposures to this chloroethane cannot be found. Animal experiments measuring acute and subacute effects indicate, however, that chronic exposure may produce liver damage (1).

Points of Attack:

Medical Surveillance:

First Aid:

Personal Protective Methods:

Respirator Selection:

Disposal Method Suggested:

See the same categories under "1,1,2,2-Tetrachloroethane" for guidance, bearing in mind that the latter compound is a proven carcinogen and this is not so proven.

References

(1) U.S. Environmental Protection Agency, *1,1,1,2-Tetrachloroethane,* Health and Environmental Effects Profile No. 156, Washington, DC, Office of Solid Waste (April 30, 1980).

(2) U.S. Environmental Protection Agency, *Chlorinated Ethanes: Ambient Water Quality Criteria,* Washington, DC (1980).

(3) Sax, N.I., Ed., *Dangerous Properties of Industrial Materials Report, 4,* No. 3, 93–95, New York, Van Nostrand Reinhold Co. (1984).

(4) United Nations Environment Programme, *IRPTC Legal File 1983,* Vol. I, pp VII/316–17, Geneva, Switzerland, International Register of Potentially Toxic Chemicals (1984).

1,1,2,2-TETRACHLOROETHANE

- Carcinogen (positive, NCI) (5)
- Hazardous waste (EPA)

● Priority toxic pollutant (EPA)

Description: $CHCl_2CHCl_2$, tetrachloroethane, is a heavy, volatile liquid which is nonflammable and has a sweetish, chloroformlike odor. It boils at 146°C.

Code Numbers: CAS 79-34-5   RTECS KI8575000   UN 1702

DOT Designation: ORM-A

Synonyms: Sym-tetrachloroethane, acetylene tetrachloride, ethane tetrachloride.

Potential Exposures: Tetrachloroethane is used as a dry cleaning agent, as a fumigant, in cement, and in lacquers. It is used in the manufacture of tetrachloroethylene, artificial silk, artificial leather, and artificial pearls. Recently, its use as a solvent has declined due to replacement by less toxic compounds. It is also used in the estimation of water content in tobacco and many drugs, and as a solvent for chromium chloride impregnation of furs.
NIOSH estimates that 11,000 workers are exposed annually.

Incompatibilities: Chemically active metals, strong caustics, hot iron, aluminum, zinc in presence of steam. Oxidative decomposition of tetrachloroethane by ultraviolet radiation or by contact with hot metal results in the formation of small quantities of phosgene, hydrochloric acid, carbon monoxide, carbon dioxide, or dichloroacetyl chloride.

Permissible Exposure Limits in Air: The Federal standard is 5 ppm (35 mg/m^3). NIOSH has recommended a time-weighted average limit of 1 ppm as a 10 hour TWA. ACGIH has proposed a TWA value of 1 ppm (7 mg/m^3) and an STEL value of 5 ppm (35 mg/m^3) as of 1983/84. The notation "skin" is added to indicate the possibility of cutaneous absorption. The IDLH level is 150 ppm.

Determination in Air: Adsorption on charcoal, workup with CS_2, analysis by gas chromatography. See NIOSH Methods, Set I. See also reference (A-10).

Permissible Concentration in Water: To protect freshwater aquatic life—9,320 μg/ℓ on an acute toxicity basis and 2,400 μg/ℓ on a chronic basis. To protect saltwater aquatic life—9,020 μg/ℓ on an acute toxicity basis. To protect human health—preferably zero. An additional lifetime cancer risk of 1 in 100,000 is posed by a concentration of 1.7 μg/ℓ.

Determination in Water: Inert gas purge followed by gas chromatography with halide specific detection (EPA Method 601) or gas chromatography plus mass spectrometry (EPA Method 624).

Routes of Entry: Inhalation of vapor and absorption of liquid through the skin (there is some evidence that tetrachloroethane absorbed through the skin affects the central nervous system only), ingestion, and eye contact.

Harmful Effects and Symptoms: *Local* — Repeated or prolonged contact with this chemical can produce a scaly and fissured dermatitis.
Systemic — Early effects brought on by tetrachloroethane narcotic action include tremors, headache, a prickling sensation and numbness of limbs, loss of kneejerk, and excessive sweating. Paralysis of the interossei muscles of the hands and feet and disappearance of ocular and pharyngeal reflexes have also occurred due to peripheral neuritis which may develop later. Blood changes include increases in mononuclear leukocytes, progressive anemia, and a slight thrombocytosis.
Clinical symptoms following these changes are fatigue, headache, constipation, insomnia, irritability, anorexia, and nausea. Later on, liver dysfunction

may result in complaints of general malaise, drowsiness, loss of appetite, nausea, and unpleasant taste in the mouth, and abdominal discomfort. This may be followed by jaundice, mental confusion, stupor or delirium, hematemesis, convulsions, and purpuric rashes.

Pulmonary edema ascribed to capillary injury has been noted in severe cases, along with renal damage, though it is not known to what extent this contributes to the total toxic picture. Nephritis may develop and the urine may contain albumin and casts. Fatty degeneration of the myocardium has been reported only in animal experiments.

Points of Attack: Liver, kidneys, central nervous system.

Medical Surveillance: Preplacement and periodic examination should be comprehensive because of the possible involvement of many systems. Special attention should be given to liver, kidney, and bone marrow function, as well as to the central and peripheral nervous systems. Alcoholism may be a predisposing factor.

First Aid: If this chemical gets into the eyes, irrigate immediately. If this chemical contacts the skin, wash with soap promptly. If a person breathes in large amounts of this chemical, move the exposed person to fresh air at once and perform artificial respiration. When this chemical has been swallowed, get medical attention. Give large quantities of salt water and induce vomiting. Do not make an unconscious person vomit.

Personal Protective Methods: Wear appropriate clothing to prevent any possibility of skin contact. Wear eye protection to prevent any possibility of eye contact. Employees should wash immediately with soap when skin is wet or contaminated. Remove nonimpervious clothing immediately if wet or contaminated. Provide emergency showers and eyewash.

Respirator Selection:

    50 ppm: CCROV/SA/SCBA
    150 ppm: CCROVF/GMOV/SAF/SA:PD,PP,CF/SCBAF
    Escape: GMOV/SCBA

Disposal Method Suggested: Incineration, preferably after mixing with another combustible fuel. Care must be exercised to assure complete combustion to prevent the formation of phosgene. An acid scrubber is necessary to remove the halo acids produced.

References

(1)   National Institute for Occupational Safety and Health, *Criteria for a Recommended Standard: Occupational Exposure to 1,1,2,2-Tetrachloroethane,* NIOSH Doc. No. 77-121, Washington, DC (1977).

(2)   U.S. Environmental Protection Agency, *Chemical Hazard Information Profile: 1,1,2,2-Tetrachloroethane,* Washington, DC (1979).

(3)   U.S. Environmental Protection Agency, *Chlorinated Ethanes: Ambient Water Quality Criteria,* Washington, DC (1980).

(4)   U.S. Environmental Protection Agency, *1,1,2,2-Tetrachloroethane,* Health and Environmental Effects Profile No. 157, Washington, DC, Office of Solid Waste (April 30, 1980).

(5)   National Cancer Institute, *Bioassay of 1,1,2,2-Tetrachloroethane for Possible Carcinogenicity,* Technical Report Series No. 27, Bethesda, MD (1978).

(6)   See Reference (A-61).

(7)   Sax, N.I., Ed., *Dangerous Properties of Industrial Materials Report, 1,* No. 5, 84-85, New York, Van Nostrand Reinhold Co. (1981).

(8)   Sax, N.I., Ed., *Dangerous Properties of Industrial Materials Report, 2,* No. 6, 79-83, New York, Van Nostrand Reinhold Co. (1982).

(9) Sax, N.I., Ed., *Dangerous Properties of Industrial Materials Report, 3,* No. 2, 60-64, New York, Van Nostrand Reinhold Co. (1983).
(10) See Reference (A-60).
(11) Parmeggiani, L., Ed., *Encyclopedia of Occupational Health & Safety,* Third Edition, Vol. 2, pp 2161-63, Geneva, International Labour Office (1983).
(12) United Nations Environment Programme, *IRPTC Legal File 1983,* Vol. I, pp VII/318-20, Geneva, Switzerland, International Register of Potentially Toxic Chemicals (1984).

TETRACHLOROETHYLENE

- Carcinogen (positive, NCI) (3)
- Hazardous waste (EPA)
- Priority toxic pollutant (EPA)

Description: $Cl_2C=CCl_2$, tetrachloroethylene, is a clear, colorless, nonflammable liquid with a characteristic odor. The odor is noticeable at 50 ppm, though after a short period it may become inconspicuous, thereby becoming an unreliable warning signal. It boils at 121°C.

Code Numbers: CAS 127-18-4   RTECS KX3850000   UN 1897

DOT Designation: ORM-A

Synonyms: Perchloroethylene, carbon dichloride, ethylene tetrachloride, Perclene, PCE, tetrachloroethene.

Potential Exposures: Tetrachloroethylene is a widely used solvent with particular use as a dry cleaning agent, a degreaser, a chemical intermediate, a fumigant, and medically as an anthelmintic.

By far the most significant exposure to PCE occurs in industrial environments. The major uses of PCE are in textile and dry cleaning industries (69%), metal cleaning (16%), and as a chemical intermediate (12%). As with inhalation exposures, dermal exposures of significance would be primarily confined to occupational exposure.

Incompatibilities: Strong oxidizers, chemically active metals, such as barium, lithium, beryllium. Tetrachloroethylene is quite stable. However, it reacts violently with concentrated nitric acid to give carbon dioxide as a primary product.

Permissible Exposure Limits in Air: The Federal standard is 100 ppm (670 mg/m^3), as an 8-hour TWA with an acceptable ceiling concentration of 200 ppm; acceptable maximum peaks above the ceiling of 300 ppm are allowed for 5 minutes duration in a 3-hour period. NIOSH has recommended a time-weighted average limit of 50 ppm and a ceiling limit of 100 ppm determined by 15-minute samples, twice daily. Neither of these levels may provide adequate protection from potential carcinogenic effects because they were selected to prevent toxic effects other than cancer.

The ACGIH has set a TWA of 50 ppm (335 mg/m^3) as of 1983/84 but no STEL value. An intended change, however, will involve adoption of an STEL of 200 ppm (1,340 mg/m^3). The notation "skin" is added to indicate the possibility of cutaneous absorption. The IDLH level is 500 ppm. Foreign limits are lower in some cases. Thus, the German Democratic Republic has set a standard of 250 mg/m^3 and reportedly the USSR a standard of 1.0 mg/m^3. The Dutch chemical industry has set a value of 35 ppm (240 mg/m^3) (A-60).

Determination in Air: Adsorption on charcoal, workup with CS_2, analysis by gas chromatography. See NIOSH Methods, Set J. See also reference (A-10).

Permissible Concentration in Water: To protect freshwater aquatic life— 5,280 μg/ℓ on an acute toxicity basis and 840 μg/ℓ on a chronic toxicity basis. To protect saltwater aquatic life—10,200 μg/ℓ on an acute toxicity basis and 450 μg/ℓ on a chronic toxicity basis. To protect human health—preferably zero. An additional lifetime cancer risk of 1 in 100,000 is posed by a concentration of 8.0 μg/ℓ.

Determination in Water: Inert gas purge followed by gas chromatography with halide specific detection (EPA Method 601) or gas chromatography plus mass spectrometry (EPA Method 624).

Routes of Entry: Inhalation of vapor, percutaneous absorption of liquid, ingestion, skin and eye contact.

Harmful Effects and Symptoms: *Local* — Repeated contact may cause a dry, scaly, and fissured dermatitis. High concentrations may produce eye and nose irritation.

Systemic — Acute exposure to tetrachloroethylene may cause central nervous system depression, hepatic injury, and anesthetic death. Cardiac arrhythmias and renal injury have been produced in animal experiments. Signs and symptoms of overexposure include malaise, dizziness, headache, increased perspiration, fatigue, staggering gait, and slowing of mental ability. These usually subside quickly upon removal into the open air.

Perchloroethylene has been found to be carcinogenic.

Points of Attack: Liver, kidneys, eyes, upper respiratory system, central nervous system.

Medical Surveillance: Evaluate skin, and liver and kidney function, as well as central nervous system. Alcoholism may be a predisposing factor.

Breath analyses may be helpful in evaluating exposures. Workers with pre-employment histories of liver, kidney, or nervous disorders should be advised as to possible increased risk.

First Aid: If this chemical gets into the eyes, irrigate immediately. If this chemical contacts the skin, wash with soap promptly. If a person breathes in large amounts of this chemical, move the exposed person to fresh air at once and perform artificial respiration. When this chemical has been swallowed, get medical attention. Give large quantities of salt water and induce vomiting. Do not make an unconscious person vomit.

Personal Protective Methods: Wear appropriate clothing to prevent repeated or prolonged skin contact. Wear eye protection to prevent any reasonable probability of eye contact. Employees should wash promptly when skin is wet or contaminated. Work clothing should be changed daily if it is possible/probable that clothing is contaminated. Remove nonimpervious clothing promptly if wet or contaminated.

Exposure to tetrachloroethylene should not be controlled with the use of respirators except: during the time period necessary to install or implement engineering or work practice controls, in work situations in which engineering and work practice controls are technically not feasible, to supplement engineering and work practice controls when such controls fail to adequately control exposure to tetrachloroethylene, for operations which require entry into tanks or closed vessels, or in emergencies.

Respirator Selection:
    500 ppm: CCROVF/GMOV/SAF/SCBAF
     Escape: GMOV/SCBA

Disposal Method Suggested: Incineration, preferably after mixing with another combustible fuel. Care must be exercised to assure complete combustion to prevent the formation of phosgene. An acid scrubber is necessary to remove the halo acids produced. Alternatively, PCE may be recovered from waste gases (A-58) and reused.

References

(1) National Institute for Occupational Safety and Health, *Criteria for a Recommended Standard: Occupational Exposure to Tetrachloroethylene,* NIOSH Doc. 76-185 (1976).

(2) National Institute for Occupational Safety and Health, Current Intelligence Bulletin No. 20: *Tetrachloroethylene,* Washington, DC (January 20, 1978).

(3) *Bioassay of Tetrachloroethylene for Possible Carcinogenicity,* DHEW Publication No. (NH) 77-813, U.S. Department of Health, Education, and Welfare, Public Health Service, National Institutes of Health, National Cancer Institute (October 1977).

(4) U.S. Environmental Protection Agency, *Tetrachloroethylene: Ambient Water Quality Criteria,* Washington, DC (1980).

(5) U.S. Environmental Protection Agency, *Health Assessment Document for Tetrachloroethylene,* External Review Draft No. 2 in circulation, Research Triangle Park, NC, Environmental Criteria and Assessment Office (January 1980).

(6) U.S. Environmental Protection Agency, *Tetrachloroethylene,* Health and Environmental Effects Profile No. 158, Washington, DC, Office of Solid Waste (April 30, 1980).

(7) See Reference (A-61).

(8) United Nations Environment Programme, *IRPTC Legal File 1983,* Vol. II, pp VII/364–6, Geneva, Switzerland, International Register of Potentially Toxic Chemicals (1984).

TETRAETHYLLEAD

- Hazardous substance (EPA)
- Hazardous waste (EPA)

Description: $Pb(C_2H_5)_4$ is a colorless liquid with a slight musty odor. It boils at $100°C$. In commerce it is usually dyed red, orange or blue. Tetraethyllead will decompose in bright sunlight yielding needlelike crystals of tri-, di-, and monoethyllead compounds, which have a garlic odor.

Code Numbers: CAS 78-00-2 RTECS TP4550000 UN 1649

DOT Designation: Poison B

Synonyms: Tetraethylplumbane (RTECS), TEL, lead tetraethyl.

Potential Exposure: Those engaged in the manufacture, transport and blending into gasoline of this antiknock compound.

Incompatibilities: Strong oxidizers, sulfuryl chloride, potassium permanganate.

Permissible Exposure Limits in Air: The Federal standard is 0.075 mg/m^3. The ACGIH as of 1983/84 has set a TWA of 0.1 mg/m^3 with the notation that this is for the control of general room air and that biologic monitoring is essential for personnel control. The notation "skin" also indicates the possibility of cutaneous absorption. The STEL is 0.3 mg/m^3. The IDLH level is 40 mg/m^3.

Determination in Air: Collection by filter plus impinger, colorimetric analysis (A-29). Also charcoal tube collection and analysis by atomic absorption spectrometry may be used (A-1). Also adsorption on XAD-2 resin, desorption with pentane and gas chromatographic analysis with photoionization detection may be used (A-10).

Permissible Concentration in Water: No criteria set, but EPA (A-37) has suggested a permissible ambient goal of 1.4 $\mu g/\ell$ based on health effects.

Routes of Entry: Inhalation, skin absorption, ingestion, skin and eye contact.

Harmful Effects and Symptoms: *Local* — Liquid alkyllead may penetrate the skin without producing appreciable local injury. However, the decomposition products of TEL (i.e., mono-, di-, triethyllead compounds) in dust form may be inhaled and result in irritation of the upper respiratory tract and possibly paroxysmal sneezing. This dust, when in contact with moist skin or ocular membranes, may cause itching, burning, and transient redness. TEL itself may be irritating to the eyes.

Systemic — The absorption of a sufficient quantity of tetraethyllead, whether briefly at a high rate, or for prolonged periods at a lower rate, may cause acute intoxication of the central nervous system. Mild degrees of intoxication cause headache, anxiety, insomnia, nervous excitation, and minor gastrointestinal symptoms with a metallic taste in the mouth.

The most noticeable clinical sign of tetraethyllead poisoning is encephalopathy which may give rise to a variety of symptoms, which include mild anxiety, toxic delirium with hallucinations, delusions, convulsions, and acute toxic psychosis. Physical signs are not prominent; but bradycardia, hypotension, increased reflexes, tremor, and slight weight loss have been reported. No peripheral neuropathy has been observed. When the interval between the termination of (either brief or prolonged) exposure and the onset of symptoms is delayed (up to 8 days) the prognosis is guardedly hopeful, but when the interval is short (few hours), an early fatal outcome may result. Recovered patients show no residual damage to the nervous system, although recovery may be prolonged.

Diagnosis depends on developing a hisotry of exposure to organic lead compounds, followed by the onset of encephalopathy. Biochemical measurements are helpful but not diagnostic. Blood lead is usually not elevated in proportion to the degree of intoxication. Urine aminolevulinic acid and coproporphyrin excretion will show values close to normal with no correlation with the severity of intoxication. Erythrocyte protoporphyrin also remains within normal range.

Points of Attack: Central nervous sytem, cardiovascular system, kidneys, eyes.

Medical Surveillance: In both preemployment and periodic physical examinations, the worker's general health should be evaluated, and special attention should be given to neurologic and emotional disorders.

First Aid: If this chemical gets into the eyes, irrigate immediately. If this chemical contacts the skin, flush with petroleum products. If a person breathes in large amounts of this chemical, move the exposed person to fresh air at once and perform artificial respiration. When this chemical has been swallowed, get medical attention. Give large quantities of water and induce vomiting. Do not make an unconscious person vomit.

Personal Protective Methods: Wear appropriate clothing to prevent any possibility of skin contact with liquids of >0.1% content. Wear eye protection to prevent any reasonable possibility of eye contact. Work clothing should be changed daily if it is possible that clothing is contaminated with liquids of >0.1% content. Remove nonimpervious clothing immediately if wet or contaminated with liquids containing >0.1%. Provide emergency showers and eyewash if liquids containing >0.1% contaminants are involved.

Respirator Selection:
0.75 mg/m³: SA/SCBA
3.75 mg/m³: SAF/SCBAF
40 mg/m³: SA:PD,PP,CF
Escape: GMOV/SCBA

Disposal Method Suggested: Controlled incineration with scrubbing for collection of lead oxides which may be recycled or landfilled. It is also possible to recover alkyllead compounds from wastewaters (A-58) as an alternative to disposal.

References

(1)   See Reference (A-61).
(2)   See Reference (A-60).
(3)   Parmeggiani, L., Ed., *Encyclopedia of Occupational Health & Safety,* Third Edition, Vol. 2, pp 1197–99, Geneva, International Labour Office (1983).

TETRAHYDROFURAN

● Hazardous waste (EPA)

Description: C_4H_8O is a colorless liquid with an etherlike odor boiling at 66°C with the structural formula shown below.

Code Numbers: CAS 109-99-9   RTECS LU5950000   UN 2056

DOT Designation: Flammable liquid

Synonyms: Diethylene oxide, tetramethylene oxide, THF, 1,4-epoxybutane, cyclotetramethylene oxide, oxacyclopentane.

Potential Exposures: The primary use of tetrahydrofuran is as a solvent to dissolve synthetic resins, particularly polyvinyl chloride and vinylidene chloride copolymers. It is also used to cast polyvinyl chloride films, to coat substrates with vinyl and vinylidene chloride, and to solubilize adhesives based on or containing polyvinyl chloride resins.
     A second large market for THF is as an electrolytic solvent in the Grignard reaction-based production of tetraethyl and tetramethyl lead. THF is used as an intermediate in the production of polytetramethylene glycol.
     NIOSH estimates that 90,000 workers have potential exposure to THF each year.

Incompatibilities: Strong oxidizers.

Permissible Exposure Limits in Air: The Federal standard and the ACGIH 1983/84 TWA value is 200 ppm (590 mg/m³). The STEL value is 250 ppm (735 mg/m³). The IDLH level is 20,000 ppm. The USSR reportedly has a limit of 100 mg/m³.

Determination in Air: Charcoal adsorption, workup with CS_2, analysis by gas chromatography. See NIOSH Methods, Set F. See also reference (A-10).

Permissible Concentration in Water: No criteria set but EPA (A-37) has suggested a permissible ambient goal of 8,100 $\mu g/\ell$ based on health effects.

Routes of Entry: Inhalation, ingestion, skin and eye contact.

Harmful Effects and Symptoms: Irritation of eyes and upper respiratory system; nausea, dizziness, headaches. Liver and kidney damage have been reported in experimental animals.

Points of Attack: Eyes, skin, respiratory system, central nervous system.

Medical Surveillance: Consider the points of attack in preplacement and periodic physical examinations.

First Aid: If this chemical gets into the eyes, irrigate immediately. If this chemical contacts the skin, flush with water promptly. If a person breathes in large amounts of this chemical, move the exposed person to fresh air at once and perform artificial respiration. When this chemical has been swallowed, get medical attention. Give large quantities of salt water and induce vomiting. Do not make an unconscious person vomit.

Personal Protective Methods: Wear appropriate clothing to prevent repeated or prolonged skin contact. Wear eye protection to prevent any reasonable probability of eye contact. Employees should wash promptly when skin is wet or contaminated. Remove clothing immediately if wet or contaminated to avoid flammability hazard.

Respirator Selection:
```
 1,000 ppm:  CCROVF
 5,000 ppm:  CMOVc
10,000 ppm:  GMOVfb/SAF/SCBAF
20,000 ppm:  SAF:PD,PP,CF/PAPOVF
    Escape:  GMOV/SCBA
```

Disposal Method Suggested: Concentrated waste containing peroxides—perforation of a container of the waste from a safe distance followed by open burning (A-31).

References

(1) U.S. Environmental Protection Agency, *Chemical Hazard Information Profile: Tetrahydrofuran,* Washington, DC (October 21, 1977). (Revised edition issued 1979.)
(2) Nat. Inst. for Occup. Safety and Health, *Profiles on Occupational Hazards for Criteria Document Priorities,* Report PB-274,073, Cincinnati, OH, pp 314-316, (1977).
(3) Sax, N.I., Ed., *Dangerous Properties of Industrial Materials Report, 1,* No. 2, 64-65, New York, Van Nostrand Reinhold Co. (1980).
(4) See Reference (A-61).
(5) See Reference (A-60).
(6) Parmeggiani, L, Ed., *Encyclopedia of Occupational Health & Safety,* Third Edition, Vol. 2, p 2164, Geneva, International Labour Office (1983).

TETRAMETHYLLEAD

Description: $Pb(CH_3)_4$ is a colorless liquid with a slight musty odor. It boils at 110°C. In commerce it is usually dyed red, orange or blue.

Code Numbers: CAS 75-74-1   RTECS TP4725000   UN 1649

DOT Designation: —

Synonyms: Tetramethylplumbane (RTECS), lead tetramethyl, TML.

Potential Exposures: Those engaged in the manufacture, distribution and blending into gasoline of this antiknock compound.

Incompatibilities: Strong oxidizers, such as sulfuryl chloride or potassium permanganate.

Permissible Exposure Limits in Air: The Federal standard is 0.07 mg/m^3. The ACGIH as of 1983/84 has set a TWA of 0.15 mg/m^3 with the notation that this is for the control of general room air and that biologic monitoring is essential for personnel control. The notation "skin" indicates the possibility of cutaneous absorption. The STEL is 0.5 mg/m^3. The IDLH level is 40 mg/m^3.

Determination in Air: Collection by filter plus impinger, colorimetric analysis (A-29). Also charcoal tube collection and analysis by atomic absorption spectrometry may be used (A-1). Also adsorption on XAD-2 resin, desorption with pentane and gas chromatographic analysis with photoionization detection may be used (A-10).

Permissible Concentration in Water: No criteria set, but EPA (A-37) has suggested a permissible ambient goal of 2 µg/ℓ based on health effects.

Harmful Effects and Symptoms: Insomnia, bad dreams, restlessness, anxiousness; hypotension; nausea; anorexia, delirium, mania, convulsions; coma.

Points of Attack: Central nervous system, cardovascular system, kidneys.

Medical Surveillance: Consider the points of attack in preplacement and periodic physical examinations.

First Aid: If this chemical gets into the eyes, irrigate immediately. If this chemical contacts the skin, flush with petroleum product. If a person breathes in large amounts of this chemical, move the exposed person to fresh air at once and perform artificial respiration. When this chemical has been swallowed, get medical attention. Give large quantities of water and induce vomiting. Do not make an unconscious person vomit.

Personal Protective Methods: Wear appropriate clothing to prevent any possibility of skin contact with liquids of >1.06% content. Wear eye protection to prevent any reasonable probability of eye contact. Work clothing should be changed daily if it is possible that clothing is contaminated. Remove nonimpervious clothing immediately if wet or contaminated with liquids containing >1.06%. Provide emergency showers and eyewash if liquids containing >1.06% contaminants are involved.

Respirator Selection:
        0.7 mg/m^3: SA/SCBA
        3.5 mg/m^3: SAF/SCBAF
        40 mg/m^3: SA:PD,PP,CF
        Escape: GMOV/SCBA

Disposal Method Suggested: Controlled incineration with scrubbing for collection of lead oxides which may be recycled or landfilled. It is also possible to recover alkyllead compounds from wastewaters (A-58) as an alternative to disposal.

References

(1) See Reference (A-61).
(2) See Reference (A-60).

TETRAMETHYL SUCCINONITRILE

Description: $(CH_3)_2C(CN)C(CN)(CH_3)_2$ is a colorless odorless solid which melts at 170°C.

Code Numbers: CAS 3333-52-6   RTECS WN4025000

DOT Designation: —

Synonym: TMSN.

Potential Exposures: This compound is reported (A-35) to be a breakdown product of azobisisobutyronitrile which is used as a blowing agent for the production of vinyl foams.

Incompatibilities: Strong oxidizers.

Permissible Exposure Limits in Air: The Federal standard and the 1983/84 ACGIH TWA value is 0.5 ppm (3.0 mg/m^3). NIOSH is recommending a ceiling level of 6 mg/m^3 for a 15-minute exposure. The STEL is 2 ppm (9 mg/m^3). The IDLH level is 5 ppm.

Determination in Air: Adsorption on charcoal, workup with CS_2, analysis by gas chromatography. See NIOSH Methods, Set K. See also reference (A-10).

Permissible Concentration in Water: No criteria set, but EPA (A-37) has suggested a permissible ambient goal of 41 μg/ℓ based on health effects.

Routes of Entry: Inhalation, skin absorption, ingestion, skin and eye contact.

Harmful Effects and Symptoms: Headaches, nausea, convulsions, coma.

Points of Attack: Central nervous system.

Medical Surveillance: Consider the points of attack in preplacement and periodic physical examinations.

First Aid: If this chemical gets into the eyes, irrigate immediately. If this chemical contacts the skin, wash with soap promptly. If a person breathes in large amounts of this chemical, move the exposed person to fresh air at once and perform artificial respiration. When this chemical has been swallowed, get medical attention. Give large quantities of water and induce vomiting. Do not make an unconscious person vomit.

Personal Protective Methods: Wear appropriate clothing to prevent repeated or prolonged skin contact. Wear eye protection to prevent any reasonable probability of eye contact. Employees should wash promptly when skin is wet or contaminated. Work clothing should be changed daily if it is possible that clothing is contaminated. Remove nonimpervious clothing promptly if wet or contaminated.

Respirator Selection:
    30 mg/m^3: SA/SCBA
        Escape: GMOVP/SCBA

Disposal Method Suggested: Incineration—incinerator is equipped with a scrubber or thermal unit to reduce NO_x emissions.

TETRANITROMETHANE

● Hazardous waste (EPA)

Description: $C(NO_2)_4$ is a colorless to pale yellow liquid or solid with a pungent odor. It causes tears. It melts at 14°C and boils at 126°C.

Code Numbers: CAS 509-14-8   RTECS PB4025000   UN 1510

DOT Designation: Oxidizer

Synonyms: TNM, Tetan.

Potential Exposures: Tetranitromethane is used as an oxidizer in rocket propellant combinations. It is also used as an explosive in admixture with toluene (A-35).

Incompatibilities: Hydrocarbons, alkalies, metals.

Permissible Exposure Limits in Air: The Federal standard and the 1983/84 ACGIH TWA value is 1 ppm (8 mg/m³). There is no STEL value set. The IDLH level is 5 ppm.

Determination in Air: Collection by impinger using ethyl acetate, analysis by gas chromatography. See NIOSH Methods, Set Q. See also reference (A-10).

Permissible Concentration in Water: No criteria set.

Routes of Entry: Inhalation, ingestion, skin and eye contact.

Harmful Effects and Symptoms: Irritation of eyes, nose and throat; dizziness, headaches; chest pain, dyspnea; methemoglobinemia, cyanosis; skin burns.

Points of Attack: Respiratory system, eyes, skin, blood, central nervous system.

Medical Surveillance: Consider the points of attack in preplacement and periodic physical examinations.

First Aid: If this chemical gets into the eyes, irrigate immediately. If this chemical contacts the skin, wash with soap promptly. If a person breathes in large amounts of this chemical, move the exposed person to fresh air at once and perform artificial respiration. When this chemical has been swallowed, get medical attention. Give large quantities of water and induce vomiting. Do not make an unconscious person vomit.

Personal Protective Methods: Wear appropriate clothing to prevent any reasonable probability of skin contact. Wear eye protection to prevent any possibility of eye contact. Employees should wash promptly when skin is wet or contaminated. Work clothing should be changed daily if it is possible that clothing is contaminated. Remove clothing immediately if wet or contaminated to avoid flammability hazard. Provide emergency eyewash.

Respirator Selection:
> 5 ppm: CCRSF/GMS/SAF/SCBAF
> Escape: GMS/SCBA
> Note: Do not use oxidizable sorbents.

Disposal Method Suggested: Open burning at remote burning sites is not entirely satisfactory since it makes no provision for the control of the toxic effluents, NO_x and HCN. Suggested procedures are to employ modified closed pit burning, using blowers for air supply and passing the effluent combustion gases through wet scrubbers (A-31).

References
(1)   See Reference (A-61).

TETRASODIUM PYROPHOSPHATE

Description: $Na_4P_2O_7$ is a white crystalline powder melting at $880°C$.

Code Numbers: CAS 7722-88-5   RTECS UX7350000

DOT Designation: —

Synonyms: Tetrasodium diphosphate, sodium pyrophosphate, TSPP.

Potential Exposures: TSPP is used as a water softener, as a builder in synthetic detergents, as a metal cleaner, in boiler water treatment, in viscosity control of drilling muds and in textile scouring and dyeing.

Permissible Exposure Limits in Air: There is no Federal standard but ACGIH (1983/84) has adopted a TWA value of 5.0 mg/m^3. There is no STEL value.

Permissible Concentration in Water: No criteria set.

Harmful Effects and Symptoms: TSPP is basically of low toxicity but it is a somewhat alkaline material and the dust is therefore irritating to the eyes and respiratory passages (A-34).

TETRYL

Description: Tetryl, $(NO_2)_3C_6H_2N(NO_2)CH_3$, is a yellow solid. It melts at $129°C$.

Code Numbers: CAS 479-45-8   RTECS BY6300000   UN 0208

DOT Designation: Class A explosive

Synonyms: Trinitrophenylmethylnitramine, nitramine, tetranitromethylaniline, pyrenite, picrylmethylnitramine, picrylnitromethylamine, N-methyl-N,2,4,6-tetranitroaniline, tetralite.

Potential Exposures: Tetryl is used in explosives as an intermediary detonating agent and as a booster charge; it is also used as a chemical indicator.

Incompatibilities: Oxidizable materials.

Permissible Exposure Limits in Air: The Federal standard and the 1983/84 ACGIH TWA value is 1.5 mg/m^3. The STEL value is 3.0 mg/m^3. There is no IDLH level set. The notation "skin" is added to indicate the possibility of cutaneous absorption.

Determination in Air: Collection on a filter, workup with DEEA, colorimetric determination. See NIOSH Methods, Set Q. See also reference (A-10).

Permissible Concentration in Water: No criteria set.

Routes of Entry: Inhalation, skin absorption, ingestion, skin and eye contact.

Harmful Effects and Symptoms: *Local* — Tetryl is a potent sensitizer, and allergic dermatitis is common. Dermatitis first appears on exposed skin areas, but can spread to other parts of the body in fair-skinned individuals or those with poor personal hygiene. The severest forms show massive generalized edema with partial obstruction of the trachea due to swelling of the tongue, and these cases require hospitalization. Contact may stain skin and hair yellow or orange. Tetryl is acutely irritating to the mucous membranes of the respiratory tract and the eyes, causing coughing, sneezing, epistaxis, conjunctivitis, and palpebral and periorbital edema.

Systemic — Tetryl exposure may cause irritability, easy fatigability, malaise, headaches, lassitude, insomnia, nausea, and vomiting. Anemia either of the marrow depression or deficiency type has been observed among tetryl workers. Tetryl exposure has produced liver and kidney damage in animals.

Points of Attack: Respiratory system, eyes, central nervous system, skin. In animals—liver and kidneys.

Medical Surveillance: Preplacement physical examination should give special attention to those individuals with a history of allergy, blood dyscrasias, or skin, liver, or kidney disease. Periodic examinations should be directed primarily to the control of dermatitis and allergic reactions, plus any effects on the respiratory tract, eyes, central nervous system, blood, liver or kidneys.

First Aid: If this chemical gets into the eyes, irrigate immediately. If this chemical contacts the skin, wash with soap promptly. If a person breathes in large amounts of this chemical, move the exposed person to fresh air at once and perform artificial respiration. When this chemical has been swallowed, get medical attention. Give large quantities of water and induce vomiting. Do not make an unconscious person vomit.

Personal Protective Methods: Wear appropriate clothing to prevent repeated or prolonged skin contact. Wear eye protection to prevent any reasonable probability of eye contact. Employees should wash promptly when skin is wet or contaminated and daily at the end of each work shift. Work clothing should be changed daily if it is possible that clothing is contaminated. Remove nonimpervious clothing promptly if wet or contaminated.

Respirator Selection:

7.5 mg/m^3:	DMXS
15 mg/m^3:	DMXSQ/FuHiEP/SA/SCBA
75 mg/m^3:	HiEPF/SAF/SCBAF
1,500 mg/m^3:	PAPHiEPF
3,000 mg/m^3:	SAF:PD,PP,CF
Escape:	DMXS/SCBA

Disposal Method Suggested: Solution in acetone and incineration in furnace equipped with afterburner and caustic soda solution scrubber.

References

(1) See Reference (A-61).
(2) Parmeggiani, L., Ed., *Encyclopedia of Occupational Health & Safety,* Third Edition, Vol. 2, pp 2165–66, Geneva, International Labour Office (1983).

THALLIUM AND COMPOUNDS

- Hazardous substance (thallium sulfate, EPA)
- Hazardous wastes (EPA)
- Priority toxic pollutant (EPA)

Description: Tl, thallium, is a soft, heavy metal insoluble in water and organic solvents. It is usually obtained as a by-product from the flue dust generated during the roasting of pyrite ores in the smelting and refining of lead and zinc.

Code Numbers: For thallium sulfate—
CAS 7446-18-6   RTECS XG6800000   UN 1707

DOT Designation: Poison B

Synonyms: None.

Potential Exposures: Thallium and its compounds are used as rodenticides, fungicides, insecticides, catalysts in certain organic reactions, in phosphor activators, in bromoiodide crystals for lenses, plates, and prisms in infrared optical instruments, in photoelectric cells, in mineralogical analysis, alloyed with mercury in low-temperature thermometers, switches and closures, in high-density liquids, in dyes and pigments, and in the manufacture of optical lenses, fireworks, and imitation precious jewelry. It forms a stainless alloy with silver and a corrosion-resistant alloy with lead. Its medicinal use for epilation has been almost discontinued.

From the standpoint of exposure hazard, it would seem that smokers may have twice as great a level of thallium intake as nonsmokers. This suggestion is based soley on data concerning urinary thallium excretion in six people and on very limited information concerning thallium in cigars. Obviously, people occupationally exposed to thallium may constitute a special risk category, but this matter has received little attention, largely because total annual industrial production of thallium is so small, probably about 0.5 ton. In the U.S., the main source of poisoning, thallium-containing rodenticides and insecticides, has been terminated. The manufacture and distribution of these products is no longer permitted.

Incompatibilities: None hazardous.

Permissible Exposure Limits in Air: The Federal standard and the ACGIH 1983/84 TWA value for thallium (soluble compounds) is 0.1 mg Tl/m^3. There is no STEL value. The IDLH level is 20 mg/m^3. The notation "skin" is added to indicate the possibility of cutaneous absorption.

Determination in Air: Collection on a filter, workup with acid, analysis by atomic absorption. See NIOSH Methods, Set U. See also reference (A-10). See also (A-21).

Permissible Concentration in Water: To protect freshwater aquatic life— 1,400 µg/ℓ on an acute toxicity basis and 40 µg/ℓ on a chronic basis. To protect saltwater aquatic life—2,130 µg/ℓ on an acute toxicity basis. For the protection of human health from the toxic properties of thallium ingested through water and contaminated aquatic organisms, the ambient water criterion is 13.0 µg/ℓ.

Determination in Water: Digestion followed by atomic absorption measurement gives total thallium. Dissolved thallium may be determined by the same procedure preceded by 0.45 micron filtration.

Routes of Entry: Inhalation of dust and fume. Ingestion and percutaneous absorption of dust, eye and skin contact.

Harmful Effects and Symptoms: *Local* — Thallium salts may be skin irritants and sensitizers, but these effects occur rarely in industry.

Systemic — Thallium is an extremely toxic and cumulative poison. In nonfatal occupational cases of moderate or long-term exposure, early symptoms usually include fatigue, limb pain, metallic taste in the mouth and loss of hair, although loss of hair is not always present as an early symptom. Later, peripheral neuritis, proteinuria, and joint pains occur.

Occasionally, neurological signs are the presenting factor, especially in more severe poisonings. Long-term exposure may produce optic atrophy, paresthesia, and changes in pupillary and superficial tendon reflexes (slowed responses). Acute poisoning rarely occurs in industry, and is usually due to ingestion of thallium. When it occurs, gastrointestinal symptoms, abdominal colic, loss of kidney function, peripheral neuritis, strabismus, disorientation, convulsions, joint pain, and alopecia develop rapidly (within 3 days). Death is due to damage to the central nervous system.

Points of Attack: Eyes, central nervous system, lungs, liver, kidneys, gastrointestinal tract, body hair.

Medical Surveillance: Preplacement and periodic examinations should give special consideration to the central nervous system, gastrointestinal symptoms, and liver and kidney function. Hair loss may be a significant sign. Urine examinations may be helpful.

First Aid: If this chemical gets into the eyes, irrigate immediately. If this chemical contacts the skin, flush with water promptly. If a person breathes in large amounts of this chemical, move the exposed person to fresh air at once and perform artificial respiration. When this chemical has been swallowed, get medical attention. Give large quantities of water and induce vomiting. Do not make an unconscious person vomit.

Personal Protective Methods: Eating, gum chewing, and smoking should not be allowed in production areas. Wear appropriate clothing to prevent any reasonable probability of skin contact. Wear eye protection to prevent any reasonable probability of eye contact. Employees should wash promptly when skin is wet or contaminated. Work clothing should be changed daily if it is possible that clothing is contaminated. Remove clothing promptly if wet or contaminated.

Respirator Selection:
0.5 mg/m^3: DMXS (for DM only)
1 mg/m^3: DMXSQ (for DM only) HiEP/SA/SCBA
5 mg/m^3: HiEPF/SAF/SCBAF
20 mg/m^3: PAPHiEF/SAF:PD,PP,CF
Escape: GMXS/SCBA

Disposal Method Suggested: Dilute thallium solutions may be disposed of in chemical waste landfills (A-31). When possible, thallium should be recovered and returned to the suppliers (A-38).

References
(1) U.S. Environmental Protection Agency, *Thallium: Ambient Water Quality Criteria,* Washington, DC (1980).
(2) U.S. Environmental Protection Agency, *Toxicology of Metals, Vol II: Thallium,* Report EPA-600/1-77-022, Research Triangle Park, NC, pp 388–404 (May 1977).
(3) U.S. Environmental Protection Agency, *Thallium,* Health and Environmental Effects Profile No. 159, Washington, DC, Office of Solid Waste (April 30, 1980).
(4) See Reference (A-61).
(5) Sax, N.I., Ed., *Dangerous Properties of Industrial Materials Report, 4,* No. 1, 94–97, New York, Van Nostrand Reinhold Co. (Jan./Feb. 1984) (Thallous Sulfate).
(6) Parmeggiani, L., Ed., *Encyclopedia of Occupational Health & Safety,* Third Edition, Vol. 2, pp 2170–71, Geneva, International Labour Office (1983).
(7) United Nations Environment Programme, *IRPTC Legal File 1983,* Vol. II, pp VII/752–5, Geneva, Switzerland, International Register of Potentially Toxic Chemicals (1984).

THIOACETAMIDE

- Carcinogen (Animal Proven) (A-64)
- Hazardous Waste Constituent (EPA)

Description: CH_3CSNH_2 is a crystalline compound with a slight mercaptan odor melting at 113° to 114°C.

Code Numbers: CAS 62-55-5   RTECS AC8925000

DOT Designation: —

Synonyms: Acetothioamide; ethanethioamide, TAA.

Potential Exposure: TAA is used as a replacement for hydrogen sulfide in qualitative analyses. TAA has been used as an organic solvent in the leather, textile, and paper industries; as an accelerator in the vulcanization of buna rubber; and as a stabilizer of motor fuel.

Chemists and laboratory technicians are at greatest risk of exposure. In 1974, the National Occupational Hazard Survey estimated that 4,600 workers were potentially exposed to TAA.

Permissible Exposure Limits in Air: No standards set.

Permissible Concentration in Water: No criteria set.

Routes of Entry: Inhalation and skin absorption.

Harmful Effects and Symptoms: There is sufficient evidence for the carcinogenicity of thioacetamide in experimental animals (1). Thioacetamide was carcinogenic in mice and rats following administration in the diet, the only exposure route tested. It induced liver-cell tumors in Swiss mice and liver-cell and bile duct tumors in Wistar rats. No carcinogenesis was observed in hamsters given thioacetamide dissolved in distilled water by stomach tube (2).

References

(1)   International Agency for Research on Cancer, *IARC Monographs on the Evaluation of the Carcinogenic Risk of Chemicals to Humans,* Supplement 4, Lyon, France: IARC (1982).

(2)   International Agency for Research on Cancer, *IARC Monographs on the Evaluation of the Carcinogenic Risk of Chemicals to Man,* Vol. 7, pp 77-83 Lyon, France: IARC (1974).

4,4'-THIOBIS(6-tert-BUTYL-m-CRESOL)

Description: $[(CH_3)_3CC_6H_2(OH)(CH_3)_2]S$ is a gray to tan powder melting at 150°C.

Code Numbers: CAS 96-69-5   RTECS GP3150000

DOT Designation: —

Synonyms: Santowhite®, Santonox®.

Potential Exposures: This material is used as an antioxidant in neoprene and other synthetic rubbers and in polyethylene and polypropylene.

Permissible Exposure Limits in Air: There is no Federal standard but ACGIH (1983/84) has adopted a TWA value of 10 mg/m³ and set an STEL of 20 mg/m³.

Permissible Concentration in Water: No criteria set.

Harmful Effects and Symptoms: This compound is insignificantly toxic on the basis of acute oral toxicity to rats. However gastroenteritis, retarded weight gain and enlarged livers resulted in rat feeding studies (A-34).

THIOGLYCOLIC ACID

Description: $HSCH_2COOH$ is a colorless liquid with a strong unpleasant odor, boiling at 123°C.

Code Numbers: CAS 68-11-1   RTECS AI5950000   UN 1940

DOT Designation: Corrosive material.

Synonyms: Mercaptoacetic acid, Thiovanic Acid.

Potential Exposures: Thioglycolic acid is used in the formulation of permanent wave solutions and depilatories; it is used in pharmaceutical manufacture (A-41); it is used as a stabilizer in vinyl plastics.

Permissible Exposure Limits in Air: There are no Federal standards but ACGIH (1983/84) has adopted a TWA value of 1 ppm (5 mg/m³). There is no STEL value given.

Permissible Concentration in Water: No criteria set.

Harmful Effects and Symptoms: In animals—weakness, gasping respiration and convulsions; severe eye pain, conjunctival inflammation, corneal opacity and iritis; gastrointestinal irritation and liver damage; death (A-34).

First Aid: Flush eyes with water. Wash contaminated areas of body with soap and water.

Personal Protective Methods: Wear rubber gloves and coveralls. Use chemical goggles.

Respirator Selection: Use of self-contained breathing apparatus is recommended (A-38).

Disposal Method Suggested: Dissolve in flammable solvent and burn in furnace equipped with afterburner and alkaline scrubber.

References

(1) Parmeggiani, L., Ed., *Encyclopedia of Occupational Health & Safety*, Third Edition, Vol. 2, pp 2171–72, Geneva, International Labour Office (1983).

THIOTEPA

- Carcinogen (Animal Positive) (IARC) (1) (NCI) (2)
- Hazardous Waste Constituent (EPA)

Description: $C_6H_{12}N_3PS$ with the following structural formula

is a crystalline substance melting at 51.5°C.

Code Numbers: CAS 52-24-4   RTECS SZ2975000

DOT Designation: —

Synonyms: Tris(1-Aziridinyl)phosphine sulfide; triethylenethiophosphoramide.

Potential Exposure: Thiotepa has been prescribed for a wide variety of neoplastic diseases (adenocarcinomas of the breast and the ovary, superficial carcinoma of the urinary bladder, controlling intracavitary or localized neoplastic disease, lymphomas such as lymphosarcoma and Hodgkin's disease, as well as bronchogenic carcinoma). It is now largely superseded by other treatments.

There is only one known producer of Thiotepa in the United States, and production data are not reported. The National Prescription Audit reported that in 1980, 30,000 prescriptions were dispensed by retail pharmacies. An unknown quantity was ordered by hospitals.

Permissible Exposure Limits in Air: No standards set.

Permissible Concentration in Water: No criteria set.

Harmful Effects and Symptoms: Tris(1-aziridinyl)phosphine sulfide (Thiotepa) is carcinogenic in mice and rats after administration by various routes, producing a variety of malignant tumors (1) (2).

There are several reports and epidemiological studies suggesting the development of acute nonlymphocytic leukemia in patients treated with thiotepa for ovarian and other malignant tumors (3).

References

(1)   IARC Monographs 9:85-94 (1975).
(2)   National Cancer Institute, *Bioassay of Thiotepa for Possible Carcinogenicity,* Technical Report Series No. 58, DHEW Publication No. (NIH) 78-1308, Washington, DC. 168 pp.
(3)   *IARC Monographs on the Evaluation of the Carcinogenic Risk of Chemicals to Humans,* Supplement I, IARC, Lyon, France, p 44 (1979).
(4)   See Reference (A-62). Also see Reference (A-64).
(5)   Sax, N.I., Ed., *Dangerous Properties of Industrial Materials Report, 1,* No. 2, 69-70, New York, Van Nostrand Reinhold Co. (1980).

THIOUREA

● Carcinogen (Animal Positive, IARC) (2)

Description: CH_4N_2S, H_2NCSNH_2, thiourea consists of colorless, lustrous crystals with a bitter taste. It melts at $180°$ to $182°C$.

Code Numbers: CAS 62-56-6   RTECS YU2800000

DOT Designation: —

Synonyms: Thiocarbamide; 2-Thiourea; THU; Sulourea.

Potential Exposure: Thiourea is used in the manufacture of photosensitive papers, flame-retardant textile sizes and in boiler water treatment. It is also used in photography, pharmaceutical manufacture, pesticide manufacture, and in textile chemicals.

Permissible Exposure Limits in Air: No standards set.

Permissible Concentration in Water: No criteria set.

Harmful Effects and Symptoms: Thiourea has been identified as a sensitizer in people suffering from photosensitivity. Exposure occurred through handling of photocopying paper.

Thiourea caused morphological changes in human erythrocytes in vitro. Thiourea also affected the energy metabolism of erythrocytes, as evidenced by a prompt and significant fall in the ATP Level.

Thiourea is a proven carcinogen in experimental animals (A-62) (A-64).

References

(1) U.S. Environmental Protection Agency, *Chemical Hazard Information Profile: Thiourea,* Washington, DC (1979).

(2) International Agency for Research on Cancer, *IARC Monographs on the Carcinogenic Risks of Chemicals to Humans,* Lyon, France, 7, 95 (1974).

THIRAM

● Hazardous waste (EPA)

Description: $C_6H_{12}N_2S_4$ is a white or yellow crystal insoluble in water, but soluble in organic solvents. It melts at $140°C$.

Code Numbers: CAS 137-26-8 RTECS JO1400000 UN (NA-2771)

DOT Designation: ORMA-A

Synonyms: Tetramethylthiuram disulfide, bis(dimethylthiocarbamoyl)disulfide, TMTD (in USSR), Thiuram (in Japan), ENT 987, Arasan®, Tersan®, Pomarsol®, Ferna-Col®, Fernasan®.

Potential Exposures: Tetramethylthiuram disulfide is used as a rubber accelerator and vulcanizer; a seed, nut, fruit, and mushroom disinfectant; a bacteriostat for edible oils and fats; and as an ingredient in suntan and antiseptic sprays and soaps. It is also used as a fungicide, rodent repellent, wood preservative, and may be used in the blending of lubricant oils.

Incompatibilities: Strong oxidizers, strong acids, oxidizable materials.

Permissible Exposure Limits in Air: The Federal standard and the 1983/84 ACGIH TWA value for thiram (tetramethylthiuram disulfide) is 5 mg/m³. The STEL value is 10 mg/m³. The IDLH level is 1,500 mg/m³.

Determination in Air: Collection on a membrane filter, elution with chloroform, addition of cuprous iodide, spectrophotometric determination of cuprous complex (A-10).

Permissible Concentration in Water: A no-adverse-effect level in water of 0.035 mg/ℓ has been calculated by NAS/NRC (A-2).

Routes of Entry: Inhalation of dust, spray, or mist, ingestion, skin and eye contact.

Harmful Effects and Symptoms: *Local* — Irritation of mucous membranes, conjunctivitis, rhinitis, sneezing, and cough may result from excessive exposures. Skin irritation with erythema and urticaria may also occur. Allergic contact dermatitis has been reported in workers who wore rubber gloves containing tetramethylthiuram disulfide.

Systemic — Systemic effects have not been reported in the U.S. literature. Bronchitis was mentioned in one European report in workers exposed to thiram or other products during synthesis. Intolerance to alcohol has been observed in workers exposed to thiram, manifested by flushing of face, palpitation, rapid pulse, dizziness, and hypotension. These effects are thought to be due to the blocking of the oxidation of acetaldehyde. It should be noted in this connection that the diethyl homologue of this compound, tetraethylthiuram disulfide, is marketed as the drug "Antabuse" and that severe and disagreeable symptoms ensue immediately in subjects who ingest the smallest amount of ethyl alcohol after they have been "premedicated" with the drug.

Points of Attack: Respiratory system, lungs, skin.

Medical Surveillance: Preplacement and periodic medical examinations should give special attention to history of skin allergy, eye irritation, and significant respiratory, liver, or kidney disease. Workers should be aware of the potentiating action of alcoholic beverages when working with tetramethylthiuram disulfide.

First Aid: If this chemical gets into the eyes, irrigate immediately. If this chemical contacts the skin, wash with soap promptly. When this chemical has been swallowed, get medical attention. Give large quantities of water and induce vomiting. Do not make an unconscious person vomit.

Personal Protective Methods: Wear appropriate clothing to prevent any reasonable probability of skin contact. Wear eye protection to prevent any reasonable probability of eye contact. Employees should wash promptly when skin is wet or contaminated. Work clothing should be changed daily if it is possible that clothing is contaminated. Remove nonimpervious clothing promptly if wet or contaminated.

Respirator Selection:

50 mg/m³:	SA/SCBA/CCRPest
250 mg/m³:	SAF/SCBAF/CCRFPest/GMPest
1,500 mg/m³:	PAPHiE/SAF:PD,PP,CF/PAPFPest
Escape:	DMXS/SCBA

Disposal Method Suggested: Thiram can be dissolved in alcohol or other flammable solvent and burned in an incinerator with an afterburner and scrubber (A-32).

References

(1)   See Reference (A-61).
(2)   Sax, N.I., Ed., *Dangerous Properties of Industrial Materials Report, 1,* No. 5, 41-42, New York, Van Nostrand Reinhold Co. (1981).
(3)   Parmeggiani, L., Ed., *Encyclopedia of Occupational Health & Safety,* Third Edition, Vol. 2, pp 2164-65, Geneva, International Labour Office (1983).

THORIUM AND COMPOUNDS

● Carcinogen (Thorium Dioxide as Radiopaque Medium) (A-62) (A-64)

Description: Th, thorium, is a natural radioactive element insoluble in water and organic solvents. It occurs in the minerals monazite, thorite, and thorinite, usually mixed with its disintegration products.

Code Numbers: For thorium metal—
CAS 7440-29-1   RTECS XO6400000   UN (NA-9170)

DOT Designation: Radioactive and flammable solid

Synonyms: None.

Potential Exposures: Metallic thorium is used in nuclear reactors to produce nuclear fuel, in the manufacture of incandescent mantles, as an alloying material, especially with some of the lighter metals, e.g., magnesium, as a reducing agent in metallurgy, for filament coatings in incandescent lamps and vacuum tubes, as a catalyst in organic synthesis, in ceramics, and in welding electrodes.

Exposures may occur during production and use of thorium-containing materials, in the casting and machining of alloy parts, and from the fume produced during welding with thorium electrodes.

Permissible Exposure Limits in Air: Maximum permissible concentration for thorium under the Federal standard (see 20 CFR Part 20—Table 1) is 1×10^{-6} μCi/ml (air).

Permissible Concentration in Water: No criteria set.

Routes of Entry: Ingestion of liquid, inhalation of dust or gas, and percutaneous absorption.

Harmful Effects and Symptoms: *Local* — Thorium and thorium compounds are relatively inert, but some irritant effect may occur depending on the anion present. Gas and aerosols can penetrate the body by way of the respiratory system, the digestive system, and the skin.

Systemic — Thorium and its compounds are toxicologically inert on the basis of its chemical toxicity. Only 0.001% of an ingested dose is retained in the body. Thorium, once deposited in the body, remains for long periods of time. It has a predilection for bones, lungs, lymphatic glands, and parenchymatous tissues. Characteristic effects of the activity of thorium and its disintegration products are changes in blood forming, nervous and reticuloendothelial systems, and functional and morphological damage to lung and bone tissue. Only much later do illness and symptoms characteristic of chronic radiation disease appear. After a considerable time, neoplasms may occur and the immunological activity of the body may be reduced.

External radiation with gamma rays can occur from contact with material containing mesothorium, with thorium in large quantities, and with by-products that contain disintegration products of thorium. Thorium dioxide (Thorotrast) is known to cause severe radiation damage and cancer of bone, blood vessels, liver, and other organs when administered to patients for diagnostic purposes. Its use is now forbidden for introduction into body tissues. Workers in plants where thorium dioxide is produced have not experienced either chemical or radiation injury.

Medical Surveillance: Monitoring of personnel for early symptoms and changes such as abnormal leukocytes in the blood smear may be of value.

In cases of chronic or acute exposure, the determination of thorium in the urine or the use of whole body radiation counts and breath radon are useful methods of monitoring the exposure dose and excretion rates.

Personal Protective Methods: Protection of the worker is afforded by respiratory protection with either dust masks, special canister gas masks, or supplied air respirators. Protective clothing and gloves to prevent dust settling on the skin, with daily change of work clothes and showering after each shift before change to street clothes should be routine.

Disposal Method Suggested: Recovery and recycle is the preferred route.

References

(1)   See Reference (A-62).

(2)   Parmeggiani, L., Ed., *Encyclopedia of Occupational Health & Safety,* Third Edition, Vol. 2, pp 2173–75, Geneva, International Labour Office (1983).

TIN AND INORGANIC TIN COMPOUNDS

Description: Sn, tin, is a soft, silvery white metal insoluble in water. The primary commercial source of tin is cassiterite (SnO_2, tinstone). Tin tetrachloride, $SnCl_4$, is a colorless fuming liquid boiling at 114°C.

Code Numbers:

Tin metal	CAS 7440-31-5	RTECS XP7320000	
Tin tetrachloride	CAS 7646-78-8	RTECS XP8750000	UN 1827

DOT Designation: $SnCl_4$—corrosive material

Synonyms: None.

Potential Exposures: The most important use of tin is as a protective coating for other metals such as in the food and beverage canning industry, in roofing tiles, silverware, coated wire, household utensils, electronic components, and pistons. Common tin alloys are phosphor bronze, light brass, gun metal, high tensile brass, manganese bronze, die-casting alloys, bearing metals, type metal, and pewter. These are used as soft solders, fillers in automobile bodies, and as coatings for hydraulic brake parts, aircraft landing gear and engine parts. Metallic tin is used in the manufacture of collapsible tubes and foil for packaging.

Exposures to tin may occur in mining, smelting, and refining, and in the production and use of tin alloys and solders.

Inorganic tin compounds are important industrially in the production of ceramics, porcelain, enamel, glass, and inks; in the production of fungicides, anthelmintics, insecticides; as a stabilizer it is used in polyvinyl plastics and chlorinated rubber paints; and it is used in plating baths.

Incompatibilities: For inorganic tin compounds—chlorine, turpentine. For stannic chloride—water, alcohols, amines.

Permissible Exposure Limits in Air: The Federal standard for inorganic compounds excluding the oxides is 2.0 mg/m^3. The 1983/84 ACGIH TWA is 2.0 mg/m^3 including the oxide. The STEL value is 4.0 mg/m^3. The IDLH level for inorganic tin compounds is 400 mg/m^3.

Determination in Air: Collection on a filter, measurement by atomic absorption. See NIOSH Methods, Set N. See also reference (A-10).

Permissible Concentration in Water: No criteria set.

Routes of Entry: Inhalation of dust, eye and skin contact.

Harmful Effects and Symtoms: *Local* — Certain inorganic tin salts are mild irritants to the skin and mucous membranes. They may be strongly acid or basic depending on the cation or anion present.

Systemic — Exposure to dust or fumes of inorganic tin is known to cause a benign pneumoconiosis (stannosis). This form of pneumoconiosis produces distinctive progressive x-ray changes of the lungs as long as exposure persists, but there is no distinctive fibrosis, no evidence of disability, and no special complicating factors. Because tin is so radio-opaque, early diagnosis is possible.

Points of Attack: Eyes, skin, respiratory system.

Medical Surveillance: In the case of inorganic tin compounds, the skin and eyes are of particular interest. Chest x-rays may reveal that exposures have occurred.

First Aid: If this chemical gets into the eyes, irrigate immediately. If this chemical contacts the skin, wash with soap immediately. If a person breathes in large amounts of this chemical, move the exposed person to fresh air at once and perform artificial respiration. When this chemical has been swallowed, get medical attention. Give large quantities of water and induce vomiting. Do not make an unconscious person vomit.

Personal Protective Methods: Skin contact should be prevented by protective clothing. In all areas of dust concentration, dust masks should be provided, and in the case of fumes, masks with proper canisters or supplied-air respirators should be used.

Respirator Selection:

10 mg/m³:	DMXS
20 mg/m³:	DMXSQ/FuHiEP/SA/SCBA
100 mg/m³:	HiEPF/SAF/SCBAF
400 mg/m³:	PAPHiEF/SAF:PD,PP,CF
Escape:	HiEPF/SCBAF

Disposal Method Suggested: $SnCl_4$—pour onto sodium bicarbonate; spray with ammonium hydroxide while adding crushed ice; when reaction subsides, flush down drain (A-38).

References

(1) U.S. Environmental Protection Agency, *Toxicology of Metals, Vol II: Tin,* Report EPA-600/1-77-022, Research Triangle Park, NC, pp 405-426 (May 1977).
(2) See Reference (A-61).
(3) Sax, N.I., Ed., *Dangerous Properties of Industrial Materials Report, 1,* No. 3, 82-83, New York, Van Nostrand Reinhold Co. (1981).
(4) See Reference (A-60) for citations on Tin, Tin(II)Chloride, and Tin Tetrachloride.
(5) Parmeggiani, L., Ed., *Encyclopedia of Occupational Health & Safety,* Third Edition, Vol. 2, pp 2177-79, Geneva, International Labour Office (1983).
(6) United Nations Environment Programme, *IRPTC Legal File 1983,* Vol. II, pp VII/759-60, Geneva, Switzerland, International Register of Potentially Toxic Chemicals (1984).

TIN ORGANIC COMPOUNDS

Description: Tin organic compounds have the formula $Sn(R)_{4-x}$ (radical)$_x$. A typical compound is triphenyltin hydroxide which is a solid melting at 122°C. See also the entry under "Cyhexatin".

Code Numbers: For triphenyltin hydroxide—
CAS 76-87-9　RTECS WH8575000　UN 2786

DOT Designation: —

Synonyms: For triphenyltin hydroxide as one example—Fentin hydroxide, ENT 28009, hydroxytriphenylstannane, Du-Ter®.

Potential Exposure: Organotin compounds are used as additives in a variety of products and processes. Diorganotins find application as heat stabilizers in plastics, as catalysts in the production of urethane foams, in the cold curing of rubber, and as scavengers for halogen acids. Tri- and tetraorganotins are used as preservatives for wood, leather, paper, paints, and textiles and as biocides. NIOSH estimates that 30,000 employees in the United States may be exposed to organotin compounds.

Incompatibilities: Strong oxidizers.

Permissible Exposure Limits in Air: The Federal standard and the 1983/84 ACGIH TWA value is 0.1 mg/m^3. The STEL is 0.2 mg/m^3. The notation "skin" is added to indicate the possibility of cutaneous absorption. The IDLH level is 200 mg/m^3.

Determination in Air: Collection on a filter, digestion with nitric and sulfuric acids, conversion to tin tetraiodide, extraction with hexane, redissolution in acids, color development with pyrocatechol violet and colorimetric measurement (A-10).

Permissible Concentration in Water: No criteria set, but EPA (A-37) has suggested a permissible ambient goal of 1.4 μg/ℓ based on health effects.

Routes of Entry: Inhalation, skin absorption, ingestion, skin and eye contact.

Harmful Effects and Symptoms: Organic tin compounds, especially tributyl and dibutyl compounds, may cause acute burns to the skin. The burns produce little pain but may itch. They heal without scarring. Clothing contaminated by vapors or liquids may cause subacute lesions and diffuse erythematoid dermatitis on the lower abdomen, thighs, and groin of workmen who handle these compounds. The lesions heal rapidly on removal from contact. The eyes are rarely involved, but accidental splashing with tributyltin has caused lacrimation and conjunctival edema which lasted several days; there was no permanent injury.

Triphenyltin hydroxide has been subjected to a carcinogenisis bioassay by NCI and found to be not carcinogenic.

Certain organic tin compounds, especially alkyltin compounds, are highly toxic when ingested. The trialkyl and tetraalkyl compounds cause damage to the central nervous system with symptoms of headaches, dizziness, photophobia, vomiting, and urinary retention, some weakness and flaccid paralysis of the limbs in the most severe cases. Percutaneous absorption of these compounds has been postulated, but to date, deaths and serious injury have resulted only from ill-advised attempts at therapeutic use by mouth. The mechanism of action of the organotins is not clearly understood, although triethyltin is an extremely potent inhibitor of oxidative phosphorylation. Occasionally, mild organotin intoxication is seen in chemical laboratories with headache, nausea, and EEG changes.

Symptoms also include sore throat and cough, abdominal pain; skin burns and itching.

Points of Attack: Central nervous system, eyes, liver, urinary tract, skin, blood.

Medical Surveillance: For organotins, preplacement and periodic examinations should include the skin, eyes, blood, central nervous system, and liver and kidney function.

First Aid: If this chemical gets into the eyes, irrigate immediately. If this chemical contacts the skin, flush with water immediately. If a person breathes in large amounts of this chemical, move the exposed person to fresh air at once and perform artificial respiration. When this chemical has been swallowed, get medical attention. Give large quantities of water and induce vomiting. Do not make an unconscious person vomit.

Personal Protective Methods: It is important that employees be trained in the correct use of personal protective equipment. Skin contact should be prevented by protective clothing, and, especially in the case of organic tin compounds, clean work clothes should be supplied daily and the worker required to shower following the shift and prior to change to street clothes.

Respirator Selection:

1 mg/m^3:	CCROVDM/SA/SCBA
5 mg/m^3:	CCROVHiEF/SAF/SCBAF
100 mg/m^3:	SA:PD,PP,CF
200 mg/m^3:	SAF:PD,PP,CF/SCBAF:PD,PP
Escape:	GMOVP/SCBA

References

(1) National Institute for Occupational Safety and Health, *Criteria for a Recommended Standard: Occupational Exposure to Organotin Compounds,* NIOSH Doc. No. 77-115 (1977).

(2) Sax, N.I., Ed., *Dangerous Properties of Industrial Materials Report, 2,* No. 4, 92-94, New York, Van Nostrand Reinhold Co. (1982). (Fentin Hydroxide).

(3) See Reference (A-61).

(4) United Nations Environment Programme, *International Register of Potentially Toxic Chemicals,* Geneva, Switzerland (1979). (Acetoxytriphenylstannane) (Diethyldiiodostannane) (Dibutyldichlorostannane).

(5) United Nations Environment Programme, *IRPTC Legal File 1983,* Vol. II, pp VII/725-29, Geneva, Switzerland, International Register of Potentially Toxic Chemicals (1984).

TITANIUM AND COMPOUNDS

Description: Ti, titanium, is a dark-grey, lustrous metal insoluble in water. It is brittle when cold and malleable when hot. The most important minerals containing titanium are ilmenite, rutile, perovskite, and titanite or sphene.

Among the important titanium compounds are titanium dioxide (discussed in a separate, subsequent section) and titanium tetrachloride, $TiCl_4$. $TiCl_4$ is a liquid at room temperature and fumes when it contacts the moisture in the air.

Code Numbers:

Ti metal		RTECS XR1700000	UN 2546
$TiCl_4$	CAS 7550-45-0	RTECS XR1925000	UN 1838

DOT Designation: Ti metal—Flammable solid; $TiCl_4$—Corrosive material.

Potential Exposure: Titanium metal, because of its low weight, high strength, and heat resistance, is used in the aerospace and aircraft industry as tubing, fittings, fire walls, cowlings, skin sections, and jet compressors, and it is also used in surgical appliances. It is used, too, as control-wire casings in nuclear reactors, as a protective coating for mixers in the pulp-paper industry and in other situations in which protection against chlorides or acids is required, in vacuum lamp bulbs and x-ray tubes, as an addition to carbon and tungsten in electrodes and lamp filaments, and to the powder in the pyrotechnics industry. It forms alloys with iron, aluminum, tin, and vanadium of which ferrotitanium is especially important in the steel industry.

Other titanium compounds are utilized in smoke screens, as mordants in dyeing, in the manufacture of cemented metal carbides, as thermal insulators, and in heat resistant surface coatings in paints and plastics.

Permissible Exposure Limits in Air: No standards set (except for titanium dioxide, which see).

Permissible Concentration in Water: No criteria set.

Route of Entry: Inhalation of dust or fume.

Harmful Effects and Symptoms: *Local*—Titanium and titanium compounds are, for the most part, virtually inert and not highly toxic to man. Titanium tetrachloride, which is released into the air during maintenance of chlorinating and rectifying operations, is an exception. Titanium tetrachloride and its hydrolysis products are highly toxic and irritating. Skin exposure may cause irritation and burns, and even brief contact with the eyes may cause suppurating conjunctivitis and keratitis, followed by clouding of the cornea.

Systemic—During the production of titanium metal, it is possible that the air may be contaminated with chlorine, hydrogen chloride, titanium tetrachloride, and similar harmful constituents. Reports of severe lung injury caused by such exposures have been recorded; in some cases the condition resembles silicotic lungs. Reports of pulmonary fibrosis due to titanium carbide are now mostly discounted, but precautions are still recommended. Titanium tetrachloride may cause injury to the upper respiratory tract and acute bronchitis.

Medical Surveillance: Preemployment and periodic physical examinations should give special attention to lung disease, especially if irritant compounds are involved. Chest x-rays should be included in both examinations and pulmonary function evaluated periodically. Smoking history should be taken. Careful attention should be given to the eyes and the skin.

Personal Protective Methods: Employees exposed to titanium tetrachloride should wear protective clothing and respirators. In areas of dust or fumes of titanium tetrachloride, all workers should be provided with goggles and dust masks, fullface gas masks, or supplied air respirators. Clothing should be changed daily to avoid dust inhalation from clothing, and employees should be encouraged to shower before changing to street clothes.

References

(1) U.S. Environmental Protection Agency, *Toxicology of Metals, Vol II: Titanium,* pp 427-441, Report EPA-600/1-77-022, Research Triangle Park, NC (May 1977).
(2) Sax, N.I., Ed., *Dangerous Properties of Industrial Materials Report, 1,* No. 3, 83, New York, Van Nostrand Reinhold Co. (1981).
(3) See Reference (A-60) for citations on Titanium Tetrachloride and Titanium Trichloride.
(4) Parmeggiani, L., Ed., *Encyclopedia of Occupational Health & Safety,* Third Edition, Vol. 2, pp 2179-81, Geneva, International Labour Office (1983).

TITANIUM DIOXIDE

Description: TiO_2 is an odorless white powder.

Code Numbers: CAS 13463-67-7   RTECS XR2275000

DOT Designation: —

Synonyms: Titania, rutile, anatase, brookite.

Potential Exposures: Titanium dioxide is a white pigment in the rubber, plastics, ceramics, paint, and varnish industries, in dermatological preparations, and is used as a starting material for other titanium compounds, as a gem, in curing concrete, and in coatings for welding rods.

Incompatibilities: None hazardous.

Permissible Exposure Limits in Air: The Federal standard for titanium dioxide is 15 mg/m^3. ACGIH (1983/84) classifies TiO$_2$ as a nuisance particulate with a TWA of 10 mg/m^3 of total dust or 5 mg/m^3 of respirable dust; an STEL of 20 mg/m^3 has been adopted. No IDLH level is set.

Determination in Air: Collection on a filter, workup with acid, analysis by atomic absorption. See NIOSH Methods, Set O. See also reference (A-10).

Permissible Concentration in Water: No criteria set, but EPA (A-37) has suggested a permissible ambient goal for titanium compounds (as Ti) of 83 μg/ℓ, based on health effects.

Routes of Entry: Inhalation of dust.

Harmful Effects and Symptoms: Slight lung fibrosis.

Points of Attack: Lungs.

Medical Surveillance: Consider the points of attack in preplacement and periodic physical examinations.

First Aid: If a person breathes in large amounts of this chemical, move the exposed person to fresh air at once and perform artificial respiration.

Personal Protective Methods: Respirators are the only protective means specified by NIOSH (A-4).

Respirator Selection:
       75 mg/m^3:   DM(for DM only)
      150 mg/m^3:   DMXSQ(for DM only)/FuHiEP/SA/SCBA
      750 mg/m^3:   HiEPF/SAF/SCBAF
    7,500 mg/m^3:   PAPHiE/SA:PD,PP,CF

Disposal Method Suggested: Landfill.

References

(1) See Reference (A-61).
(2) Sax, N.I., Ed., *Dangerous Properties of Industrial Materials Report, 1,* No. 3, 84, New York, Van Nostrand Reinhold Co. (1981).
(3) Sax, N.I., Ed., *Dangerous Properties of Industrial Materials Report, 3,* No. 1, 85-89, New York, Van Nostrand Reinhold Co. (1983).
(4) See Reference (A-60).
(5) United Nations Environment Programme, *IRPTC Legal File 1983,* Vol. II, pp VII/761–62, Geneva, Switzerland, International Register of Potentially Toxic Chemicals (1984).

o-TOLIDINE

- Carcinogen (Animal Positive, IARC) (1) (2) (A-64)
- Hazardous Waste Constituent (EPA)

Description: $H_2NC_6H_3(CH_3)-C_6H_3(CH_3)NH_2$ is a white to reddish crystalline solid melting at 129° to 131°C.

Code Numbers: CAS 119-93-7   RTECS DD1225000

DOT Designation: —

Synonyms: Bianisidine, 4,4'-diamino-3,3'-dimethylbiphenyl, diaminoditoyl, 3,3'-dimethylbenzidine; DMB.

Potential Exposure: Over 75% of DMB is used as a dye and as an intermediate in the production of dyestuffs and pigments. According to the Society of Dyers and Colorists, more than 95 dyes are derived from DMB. Approximately 20% of DMB is used in the production of polyurethane-based high-strength elastomers, coatings, and rigid plastics. DMB has also been used in small quantities in chlorine test kits by water companies and swimming pool owners and in test tapes in clinical laboratories.

Workers potentially exposed to DMB include dye makers, repackagers of DMB and dyes, workers in toluene-diisocyanate production, and clinical and analytical chemistry laboratory personnel. Workers in a variety of occupations may be exposed to small quantities of DMB used for analytical purposes, among them water and sewage plant attendants, chemical test tape or kit makers, and swimming pool service representatives. NIOSH in 1978 estimated that fewer than 100 employees were exposed to large quantities of DMB in the United States, but as many as 200,000 may be exposed to small quantities. The National Occupational Hazard Survey in 1974 estimated that 420 workers were potentially exposed to DMB.

Permissible Exposure Limits in Air: There are no Federal standards but ACGIH (1983/84) has designated o-tolidine as an "industrial substance suspect of carcinogenic potential for man" with no numerical TLV value.

Permissible Concentration in Water: No criteria set.

Harmful Effects and Symptoms: Skin and eye irritation.

The available test results are considered to provide sufficient evidence for the carcinogenicity of 3,3'-dimethylbenzidine in experimental animals (2). Commercial 3,3'-dimethylbenzidine (o-tolidine), when given to the rat by subcutaneous injection, caused mainly Zymbal's gland carcinomas, but mammary and forestomach tumors, and miscellaneous tumors at other sites were also present. A gastric intubation experiment with rats produced mammary carcinomas, but the result is of doubtful significance because of the small number of animals involved. In feeding experiments, the commercial product did not produce tumors in hamsters. Other experiments verified 3,3'-dimethylbenzidine as a systemic carcinogen producing multiple site tumors when given subcutaneously to the rat (1).

First Aid: If this chemical gets into the eyes, irrigate immediately. If this chemical contacts the skin, flush with water and wash contaminated areas with soap. If this chemical has been swallowed, wash with large volumes of water but do not induce vomiting (A-46).

Personal Protective Methods: Wear clothing and goggles to prevent skin and eye contact.

References

(1)   International Agency for Research on Cancer, *IARC Monographs on the Evaluation of the Carcinogenic Risk of Chemicals to Humans,* Supplement 4, Lyon, France: IARC (1982).

(2)   International Agency for Research on Cancer, *IARC Monographs on the Evaluation of the Carcinogenic Risk of Chemicals to Man,* Vol 1, pp 87-91 Lyon, France: IARC (1972).

TOLUENE

- Hazardous substance (EPA)

- Hazardous waste (EPA)
- Priority toxic pollutant (EPA)

Description: $C_6H_5CH_3$, toluene, is a clear, colorless, noncorrosive liquid with a sweet, pungent, benzenelike odor. It boils at 110° to 111°C.

Code Numbers: CAS 108-88-3   RTECS XS5250000   UN 1294

DOT Designation: Flammable liquid

Synonyms: Toluol, methylbenzene, phenylmethane, methylbenzol.

Potential Exposures: Toluene may be encountered in the manufacture of benzene. It is also used as a chemical feed for toluene diisocyanate, phenol, benzyl and benzoyl derivatives, benzoic acid, toluene sulfonates, nitrotoluenes, vinyltoluene, and saccharin; as a solvent for paints and coatings; or as a component of automobile and aviation fuels.

It is estimated that 100,000 workers are potentially exposed to toluene. At present levels of exposure to toluene in the environment, available toxicological data do not suggest that any special group in the general population would be at risk. Exposure to levels of the chemical necessary to produce physiological or toxicological effects would be anticipated primarily in occupational or solvent abuse situations. Environmental contribution of toluene in such settings should be minimal.

Incompatibilities: Stong oxidizers.

Permissible Exposure Limits in Air: The Federal standard is 200 ppm as an 8-hour TWA with an acceptable ceiling concentration of 300 ppm; acceptable maximum peaks above the ceiling of 500 ppm are allowed for 10 minutes duration. NIOSH has recommended a limit of 100 ppm (TWA) with a ceiling of 200 ppm for a 10-minute sampling period. ACGIH (1983/84) cites a TWA of 100 ppm (375 mg/m^3) and an STEL of 150 ppm (560 mg/m^3). The IDLH level is 2,000 ppm.

Determination in Air: Adsorption on charcoal, workup with CS_2, analysis by gas chromatography. See NIOSH Methods, Set V. See also reference (A-10).

Permissible Concentration in Water: To protect freshwater aquatic life— 17,500 µg/ℓ on an acute toxicity basis. To protect saltwater aquatic life— 6,300 µg/ℓ on an acute toxicity basis and 5,000 µg/ℓ on a chronic basis. To protect human health—14.3 mg/ℓ.

Determination in Water: Inert gas purge followed by gas chromatography and photoionization detection (EPA Method 602) or gas chromatography plus mass spectrometry (EPA Method 624).

Routes of Entry: Inhalation of vapor, percutaneous absorption of liquid, ingestion, skin and eye contact.

Harmful Effects and Symptoms: *Local* — Toluene may cause irritation of the eyes, respiratory tract, and skin. Repeated or prolonged contact with liquid may cause removal of natural lipids from the skin, resulting in dry, fissured dermatitis. The liquid splashed in the eyes may cause irritation and reversible damage.

Systemic — Acute exposure to toluene predominantly results in central nervous system depression. Symptoms and signs include headache, dizziness, fatigue, muscular weakness, drowsiness, incoordination with staggering gait, skin paresthesia, collapse, and coma.

Points of Attack: Central nervous system, liver, kidneys, skin.

Medical Surveillance: Preplacement and periodic examinations should evaluate possible effect on skin, central nervous system, as well as liver and kidney function. Hematologic studies should also be done if there is significant contamination of the solvent with benzene.

Hippuric acid levels above 5 g/ℓ of urine may result from exposure greater than 200 ppm determined as a TWA. Blood levels can also be determined for toluene.

First Aid: If this chemical gets into the eyes, irrigate immediately. If this chemical contacts the skin, wash with soap promptly. If a person breathes in large amounts of this chemical, move the exposed person to fresh air at once and perform artificial respiration. When this chemical has been swallowed, get medical attention. Do not induce vomiting.

Personal Protective Methods: Wear appropriate clothing to prevent repeated or prolonged skin contact. Wear eye protection to prevent any reasonable probability of eye contact. Employees should wash promptly when skin is wet or contaminated. Remove clothing immediately if wet or contaminated to avoid flammability hazard.

Respirator Selection:

500 ppm:	CCROV/SA/SCBA
1,000 ppm:	CCROVF
2,000 ppm:	GMOV/SAF/SCBAF
Escape:	GMOV/SCBA

Disposal Method Suggested: Incineration.

References

(1) U.S. Environmental Protection Agency, *Chemical Hazard Information Profile: Toluene,* Washington, DC (1979).

(2) U.S. Environmental Protection Agency, *Toluene: Ambient Water Quality Criteria,* Washington, DC (1980).

(3) National Institute for Occupational Safety and Health, *Criteria for a Recommended Standard: Occupational Exposure to Toluene,* NIOSH Doc. No. 73-11023 (1973).

(4) U.S. Environmental Protection Agency, *Toluene,* Health and Environmental Effects Profile No. 160, Washington, DC, Office of Solid Waste (April 30, 1980).

(5) Sax, N.I., Ed., *Dangerous Properties of Industrial Materials Report, 2,* No. 6, 83-87, New York, Van Nostrand Reinhold Co. (1982).

(6) See Reference (A-61).

(7) See Reference (A-60).

(8) Parmeggiani, L., Ed., *Encyclopedia of Occupational Health & Safety,* Third Edition, Vol. 2, pp 2184–86, Geneva, International Labour Office (1983).

(9) United Nations Environment Programme, *IRPTC Legal File 1983,* Vol. II, pp VII/763, Geneva, Switzerland, International Register of Potentially Toxic Chemicals (1984).

TOLUENE-2,4-DIAMINE

- Carcinogen (Animal Positive, IARC) (3) (4)
- Hazardous Waste (EPA)

Description: $C_7H_{10}N_2$, $H_3CC_6H_3(NH_2)_2$ takes the form of colorless needles melting at 99°C.

Code Numbers: CAS 95-80-7   RTECS XS9675000 UN (NA 1709)

DOT Designation: ORM-A

Synonyms: 3-Amino-para-toluidine, 5-amino-ortho-toluidine, 1,3-diamino-4-

methylbenzene, 2,4-diamino-1-methylbenzene, 2,4-diaminotoluene, 2,4-diamino-toluol, 4-methyl-meta-phenylenediamine, MTD.

Potential Exposures: Toluene-2,4-diamine is a chemical intermediate for toluene diisocyanate (used in the production of flexible and rigid polyurethane foams, polyurethane coatings, cast elastomers including fabric coatings, and polyurethane and other adhesives), and for dyes used for textiles, leather, furs, and in hair-dye formulations.

Toluene-2,4-diamine can be used for the production of about 60 dyes, 28 of which are currently believed to be of commercial significance. These dyes are generally used to color silk, wool, paper and leather. Some are also used to dye cotton, fast fibers and cellulosic fibers, in spirit varnishes and wood stains, as indicators, in the manufacture of pigments, and as biological stains.

Toluene-2,4-diamine is used as a developer for direct dyes, particularly to obtain black, dark-blue and brown shades and to obtain navy-blue and black colors on leather; it is also used in dyeing furs. It was formerly used in hair-dye formulations (to produce drab-brown, drab-blond, blue and gray shades on the hair) before this use was forbidden in 1971.

Permissible Exposure Limits in Air: No standards set.

Permissible Concentration in Water: No criteria set.

Harmful Effects and Symptoms: Methemoglobinemia, central nervous system depression, and degeneration of the liver typically result from exposure to toluene-2,4-diamine. Eye irritation, skin blistering, nausea, vomiting, jaundice and anemia are reported (A-38).

2,4-Diaminotoluene is carcinogenic in rats and after its oral administration, producing hepatocellular carcinomas, and its subcutaneous injection, inducing local sarcomas (3) (4).

Points of Attack: Central nervous system, liver.

Medical Surveillance: Consider the points of attack in preplacement and periodic physical examinations.

First Aid: Irrigate eyes with water. Wash contaminated areas of body with soap and water.

Personal Protective Methods: Wear butyl rubber gloves and plastic coveralls.

Respirator Selection: Use self-contained breathing apparatus.

Disposal Method Suggested: Controlled incineration (oxides of nitrogen are removed from the effluent gas by scrubbers and/or thermal devices).

References

(1) U.S. Environmental Protection Agency, *Chemical Hazard Information Profile: Toluene-2,4-Diamine,* Washington, DC (1979).

(2) U.S. Environmental Protection Agency, *2,4-Toluenediamine,* Health and Environmental Effects Profile No. 161, Washington, DC, Office of Solid Waste (April 30, 1980).

(3) International Agency for Research on Cancer, *IARC Monographs on the Carcinogenic Risks of Chemicals to Humans,* Lyon, France 16, 83 (1978).

(4) See Reference (A-62). Also see Reference (A-64).

TOLUENE DIISOCYANATE

- Hazardous waste (EPA)

Description: 2,4-$CH_3C_6H_3(NCO)_2$ is a colorless, yellow, or dark liquid or solid with a sweet, fruity, pungent odor. It melts at 19° to 22°C and boils at 251°C.

Code Numbers: CAS 584-84-9   RTECS CZ6300000   UN 2078

DOT Designation: Poison B

Synonyms: TDI, tolylene diisocyanate, diisocyanatotoluene.

Potential Exposures: TDI is more widely used than MDI (diphenylmethane diisocyanate). Polyurethanes are formed by the reaction of isocyanates with polyhydroxy compounds. Since the reaction proceeds rapidly at room temperature, the reactants must be mixed in pots or spray guns just before use. These resins can be produced with various physical properties, e.g., hard, flexible, semirigid foams, and have found many uses, e.g., upholstery padding, thermal insulation, molds, surface coatings, shoe innersoles, and in rubbers, adhesives, paints, and textile finishes. Because of TDI's high volatility, exposure can occur in all phases of its manufacture and use. MDI has a much lower volatility, and problems generally arise only in spray applications.

It is estimated that approximately 40,000 workers are potentially exposed to TDI and that many of these exposures are in small workplaces.

Incompatibilities: Strong oxidizers, water, acids, bases, amines, etc., cause foam and spatter.

Permissible Exposure Limits in Air: The Federal standard for the 2,4 isomer of TDI is 0.02 ppm (0.14 mg/m³) as a ceiling value. However, the standard recommended in the NIOSH Criteria Document for TDI is 0.005 ppm (0.036 mg/m³) as a TWA and 0.02 ppm for any 20-minute period. ACGIH (1983/84) has adopted 0.005 ppm (0.04 mg/m³) as a TWA and an STEL value of 0.02 ppm (0.15 mg/m³). The IDLH level is 10 ppm.

Determination in Air: Collection by impinger or fritted bubbler, reduction to diamine, diazotization and coupling and colorimetric measurement (A-10).

Routes of Entry: Inhalation of vapor, ingestion and eye and skin contact.

Harmful Effects and Symptoms: *Local* — TDI and MDI may cause irritation of eyes, respiratory tract, and skin. The irritation may be severe enough to produce bronchitis and pulmonary edema. Nausea, vomiting, and abdominal pain may occur. If liquid TDI is allowed to remain in contact with the skin, it may produce redness, swelling, and blistering. Contact of liquid TDI with the eyes may cause severe irritation, which may result in permanent damage if untreated. Swallowing TDI may cause burns of the mouth and stomach.

Systemic — Sensitization to TDI and MDI may occur, which may cause an asthmatic reaction with wheezing, dyspnea, and cough. These symptoms may first occur during the night following exposure to these chemicals. Some decrease in lung function in the absence of symptoms has been observed in some workers exposed to TDI for long periods of time.

Points of Attack: Respiratory system, skin, lungs.

Medical Surveillance: Preplacement and periodic medical examinations should include chest roentgenograph, pulmonary function tests, and an evaluation of any respiratory disease or history of allergy. Periodic pulmonary function tests may be useful in detecting the onset of pulmonary sensitization. See also reference (4).

First Aid: If this chemical gets into the eyes, irrigate immediately. If this

chemical contacts the skin, wash with soap immediately. If a person breathes in large amounts of this chemical, move the exposed person to fresh air at once and perform artificial respiration. When this chemical has been swallowed, get medical attention. Give large quantities of water and induce vomiting. Do not make an unconscious person vomit.

Personal Protective Methods: Wear appropriate clothing to prevent repeated or prolonged skin contact. Wear eye protection to prevent any possibility of eye contact. Employees should wash promptly when skin is wet or contaminated. Work clothing should be changed daily if it is possible that clothing is contaminated. Remove nonimpervious clothing promptly if wet or contaminated. Provide emergency eyewash.

Respirator Selection:
- 1 ppm: SAF/SCBAF
- 10 ppm: SAF:PD,PP,CF
- Escape: GMOV/SCBA

Disposal Method Suggested: Controlled incineration (oxides of nitrogen are removed from the effluent gas by scrubbers and/or thermal devices).

References

(1) National Institute for Occupational Safety and Health, *Criteria for a Recommended Standard: Occupational Exposure to Toluene Diisocyanate,* NIOSH Doc. No. 73-11022 (1973).

(2) Nat. Inst. for Occup. Safety and Health, *Information Profiles on Potential Occupational Hazards: Organoisocyanates,* Report PB-276,678, Rockville, MD, pp 265–275, (October 1977).

(3) U.S. Environmental Protection Agency, *Toluene Diisocyanate,* Health and Environmental Effects Profile No. 162, Washington, DC, Office of Solid Waste (April 30, 1980).

(4) Nat. Inst. for Occup. Safety and Health, *A Guide to the Work Relatedness of Disease,* Revised Edition, DHEW (NIOSH) Publ. No. 79-116, Cincinnati, OH, pp 177-184, (January 1979).

(5) See Reference (A-60).

(6) United Nations Environment Programme, *IRPTC Legal File 1983,* Vol. I, pp VII/82–3, Geneva, Switzerland, International Register of Potentially Toxic Chemicals (1984).

o-TOLUIDINE

- Carcinogen (suspected human, IARC) (1) (NCI) (2) (Animal positive) (4)
- Hazardous waste (EPA)

Description: $CH_3C_6H_4NH_2$ is a colorless to pale yellow liquid with a weak aromatic odor. It boils at 199° to 200°C.

Code Numbers: CAS 95-53-4   RTECS XU2975000   UN 1708

DOT Designation: Poison

Synonyms: ortho-Aminotoluene, ortho-methylaniline, 1-methyl-1,2-aminobenzene, 2-methylaniline.

Potential Exposures: o-Toluidine is used as a dye intermediate, as an intermediate in pharmaceutical manufacture (A-41), in textile printing, in rubber accelerators.

Incompatibilities: Strong oxidizers.

Permissible Exposure Limits in Air: The Federal standard is 5 ppm (22 mg/m^3). The TWA value adopted by ACGIH is 2 ppm (9 mg/m^3). However, o-toluidine is stated to be an "industrial substance suspect of carcinogenic potential for man." The notation "skin" is added indicating the possibility of cutaneous absorption. There is no STEL value. The IDLH level is 100 ppm.

Determination in Air: Adsorption on silica, workup with n-propanol, analysis by gas chromatography. See NIOSH Methods, Set L. See also reference (A-10).

Permissible Concentration in Water: No criteria set, but EPA (A-37) has suggested an ambient water goal of 304 μg/ℓ based on health effects.

Routes of Entry: Inhalation, skin absorption, ingestion, skin and eye contact.

Harmful Effects and Symptoms: Anoxia, headaches, cyanosis, weakness, dizziness, drowsiness, microhematuria, eye burns, dermatitis.

When given in the diet, o-toluidine hydrochloride was carcinogenic in both male and female rats and mice, producing a significant increased incidence of one or more types of neoplasms (2).

Points of Attack: Blood, kidneys, liver, cardiovascular system, skin, eyes.

Medical Surveillance: Consider the points of attack in preplacement and periodic physical examinations.

First Aid: If this chemical gets into the eyes, irrigate immediately. If this chemical contacts the skin, wash with soap immediately. If a person breathes in large amounts of this chemical, move the exposed person to fresh air at once and perform artificial respiration. When this chemical has been swallowed, get medical attention. Give large quantities of water and induce vomiting. Do not make an unconscious person vomit.

Personal Protective Methods: Wear appropriate clothing to prevent any possibility of skin contact. Wear eye protection to prevent any possibility of eye contact. Employees should wash immediately when skin is wet or contaminated. Remove nonimpervious clothing immediately if wet or contaminated. Provide emergency showers and eyewash.

Respirator Selection:
      50 ppm: CCROV/SA/SCBA
   100 ppm: CCROVF/GMOV/SAF/SCBAF
    Escape: GMOV/SCBA

Disposal Method Suggested: Controlled incineration (oxides of nitrogen are removed from the effluent gas by scrubbers and/or thermal devices).

References

(1) International Agency for Research on Cancer, *IARC Monographs on the Carcinogenic Risks of Chemicals to Humans,* Lyon, France 16, 349 (1978).
(2) National Cancer Institute, *Bioassay of o-Toluidine Hydrochloride for Possible Carcinogenicity,* Technical Report Series No. 153, Bethesda, MD (1979).
(3) See Reference (A-61).
(4) See Reference (A-62). Also see Reference (A-64).
(5) See Reference (A-60).
(6) U.S. Environmental Protection Agency, *Chemical Hazard Information Profile Draft Report; o-Toluidine; o-Toluidine Hydrochloride,* Wash., D.C. (Feb. 23, 1984).
(7) United Nations Environment Programme, *IRPTC Legal File 1983,* Vol. I, pp VII/107, Geneva, Switzerland, International Register of Potentially Toxic Chemicals (1984).

TOXAPHENE

- Carcinogen (Animal positive) (IARC) (4) (NCI) (5)
- Hazardous substance (EPA)
- Hazardous waste (EPA)
- Priority toxic pollutant (EPA)

Description: $C_{10}H_{10}Cl_8$ with the structural formula shown below is an amber-colored waxy solid with a melting range of 65° to 90°C.

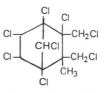

Code Numbers: CAS 8001-35-2   RTECS XW5250000   UN 2761

DOT Designation: ORM-A

Synonyms: Chlorinated camphene, camphechlor (British and international generic name except in Belgium, Canada, France, India and U.S. where toxaphene is used); polychlorcamphene (USSR generic name).

Potential Exposure: Toxaphene is a commercially produced, broad spectrum, chlorinated hydrocarbon pesticide. It was introduced in the United States in 1948 as a contact insecticide under various trade names and is currently the most heavily used insecticide in the United States, having replaced many of the agricultural applications of DDT, for which registration has been cancelled. Annual production of toxaphene exceeds 100 million pounds, with primary usage in agricultural crop application, mainly cotton. Workers in toxaphene production, formulation or application are especially at risk as are those living in areas where toxaphene is produced or used.

Incompatibilities: Strong oxidizers.

Permissible Exposure Limits in Air: The Federal limit and ACGIH 1983/84 TWA value is 0.5 mg/m³. The 1983/84 STEL value is 1.0 mg/m³. The notation "skin" is added to indicate the possibility of cutaneous absorption. The IDLH level is 200 mg/m³.

Determination in Air: Filtration from air, working up with petroleum ether and analysis by gas chromatography. See NIOSH Methods, Set F. See also reference (A-10).

Permissible Concentration in Water: To protect freshwater aquatic life— 0.013 µg/ℓ as a 24-hour average, never to exceed 1.6 µg/ℓ. To protect saltwater aquatic life—never to exceed 0.07 µg/ℓ. To protect human health—preferably zero. An additional lifetime cancer risk of 1 in 100,000 is presented by a concentration of 0.0071 µg/ℓ.

Determination in Water: Gas chromatography (EPA Method 608) or gas chromatography plus mass spectrometry (EPA Method 625).

Routes of Entry: Inhalation, skin absorption, ingestion, eye and skin contact.

Harmful Effects and Symptoms: Nausea, confusion, agitation, tremors, convulsions, unconsciousness; dry red skin. A rebuttable presumption against registration of toxaphene for pesticide uses was issued on May 25, 1977 by EPA on the basis of oncogenicity and reductions in nontarget species (A-32).

The pesticide toxaphene caused liver cancers in male and female mice given the compound in a feeding study. Test results also suggested that toxaphene caused thyroid cancers in rats (5,7).

Points of Attack: Central nervous system, skin.

Medical Surveillance: Consider the points of attack in preplacement and periodic physical examinations.

First Aid: If this chemical gets into the eyes, irrigate immediately. If this chemical contacts the skin, wash with soap promptly. If a person breathes in large amounts of this chemical, move the exposed person to fresh air at once and perform artificial respiration. When this chemical has been swallowed, get medical attention. Give large quantities of water and induce vomiting. If in a petroleum-base solution, do not induce vomiting. Do not make an unconscious person vomit.

Personal Protective Methods: Wear appropriate clothing to prevent any possibility of skin contact with liquid toxaphene or repeated or prolonged contact with solid toxaphene. Wear eye protection to prevent any reasonable probability of eye contact. Employees should wash immediately when skin is wet or contaminated with liquid toxaphene, promptly when contaminated with solid toxaphene. Work clothing should be changed daily if it is possible that clothing is contaminated with solid toxaphene. Remove clothing immediately if wet with liquid toxaphene and promptly remove if contaminated with solid toxaphene. Provide emergency showers.

Respirator Selection:
    5 mg/m^3: CCRPest/SA/SCBA
    25 mg/m^3: CCRFPest/GMPest/SAF/SCBAF
    200 mg/m^3: PAPPest/SAF:PD,PP,CF

Disposal Method Suggested: Incineration of flammable solvent mixture in furnace equipped with afterburner and alkali scrubber.

References

(1) U.S. Environmental Protection Agency, *Toxaphene: Ambient Water Quality Criteria,* Washington, DC (1980).
(2) U.S. Environmental Protection Agency, *Reviews of the Environmental Effects of Pollutants: X. Toxaphene,* Report No. EPA-600/1-79-044 (1979).
(3) U.S. Environmental Protection Agency, *Toxaphene,* Health and Environmental Effects Profile No. 163, Washington, DC, Office of Solid Waste (April 30, 1980).
(4) International Agency for Research on Cancer, *IARC Monographs on the Carcinogenic Risks of Chemicals to Humans,* Lyon, France 20, 327 (1979).
(5) National Cancer Institute, *Bioassay of Toxaphene for Possible Carcinogenicity,* Technical Report Series No. 37, Bethesda, MD (1979).
(6) Sax, N.I., Ed., *Dangerous Properties of Industrial Materials Report, 2,* No. 2, 68-70, New York, Van Nostrand Reinhold Co. (1982).
(7) See Reference (A-62). Also see Reference (A-64).
(8) United Nations Environment Programme, *IRPTC Legal File 1983,* Vol. II, pp VII/768-71, Geneva, Switzerland, International Register of Potentially Toxic Chemicals (1984).

TRIBUTYL PHOSPHATE

Description: $(C_4H_9O)_3PO$ is a colorless to pale yellow odorless liquid which boils at 293° to 294°C.

Code Numbers: CAS 126-73-8   RTECS IC7700000

DOT Designation: —

Synonyms: Phosphoric acid tributyl ester, tri-n-butyl phosphate, TBP.

Potential Exposures: Tributyl phosphate is used as an antifoaming agent and as a plasticizer. It is also used as a solvent in uranium extraction and as a solvent for cellulose esters. It may be used as a heat exchange medium and as a dielectric material.

Incompatibilities: None hazardous.

Permissible Exposure Limits in Air: The Federal standard is 5 mg/m^3. The ACGIH as of 1983/84 has adopted a TWA of 0.2 ppm (2.5 mg/m^3) and an STEL of 0.4 ppm (5 mg/m^3). The IDLH level is 1,300 mg/m^3.

Determination in Air: Collection on a filter, workup with ether, analysis by gas chromatography. See NIOSH Methods, Set P. See also reference (A-10).

Permissible Concentration in Water: No criteria set.

Routes of Entry: Inhalation of mist, eye and skin contact, ingestion.

Harmful Effects and Symptoms: Irritation of eyes, respiratory system and skin; headaches; nausea.

Points of Attack: Respiratory system, skin, eyes.

Medical Surveillance: Consider the points of attack in preplacement and periodic physical examinations.

First Aid: If this chemical gets into the eyes, irrigate immediately. If this chemical contacts the skin, wash with soap promptly. If a person breathes in large amounts of this chemical, move the exposed person to fresh air at once and perform artificial respiration. When this chemical has been swallowed, get medical attention. Give large quantities of water and induce vomiting. Do not make an unconscious person vomit.

Personal Protective Methods: Wear appropriate clothing to prevent repeated or prolonged skin contact. Wear eye protection to prevent any reasonable probability of eye contact. Employees should wash promptly when skin is wet or contaminated. Remove nonimpervious clothing promptly if wet or contaminated.

Respirator Selection:
         50 mg/m^3: SA/SCBA
        250 mg/m^3: GMOVDM/SAF/SCBAF
      1,300 mg/m^3: SA:PD,PP,CF
         Escape: GMOVP/SCBA

Disposal Method Suggested: Tributyl phosphate may be recovered from nuclear fuel processing operations (A-58).

References

(1) See Reference (A-61).
(2) See Reference (A-60).

TRICHLOROACETIC ACID

Description: Cl_3CCOOH is a colorless crystalline solid melting at 57° to 58°C.

Code Numbers: CAS 76-03-9   RTECS AJ7875000   UN 1839

DOT Designation: Corrosive material

Synonym: TCA.

Potential Exposures: TCA is used as an intermediate in pesticide manufacture (A-32) and in the production of sodium trichloroacetate which is itself a herbicide.

Permissible Exposure Limits in Air: There is no Federal standard but ACGIH (1983/84) has adopted a TWA of 1 ppm (5 mg/m^3). There is no STEL value. The Soviet standard is reported to be 0.75 ppm.

Permissible Concentration in Water: No criteria set.

Harmful Effects and Symptoms: TCA is corrosive to the skin and eyes but is not readily absorbed through the skin. The reader is recommended to consider the effects of 2,2-dichloropropionic acid (Dalapon) for analogies (A-34).

Points of Attack: Eyes, skin and respiratory tract.

First Aid: Get to fresh air, remove contaminated clothes, flush affected areas with water.

Personal Protective Methods: Wear close-fitting safety goggles, protective clothing, protective gloves.

Respirator Selection: Wear self-contained breathing apparatus.

References

(1)   See Reference (A-60).
(2)   United Nations Environment Programme, *IRPTC Legal File 1983,* Vol. I, pp VII/18, Geneva, Switzerland, International Register of Potentially Toxic Chemicals (1984).

1,2,4-TRICHLOROBENZENE

● Priority toxic pollutant (EPA)

Description: C$_6$H$_3$Cl$_3$ is a low-melting solid or liquid with a pleasant aroma. It melts at 17°C and boils at 213.5°C.

Code Numbers: CAS 120-82-1   RTECS DC2100000   UN 2321

DOT Designation: Label should bear St. Andrew's Cross (X).

Synonym: Unsym-trichlorobenzene.

Potential Exposures: 1,2,4-Trichlorobenzene is used as a dye carrier (46%), herbicide intermediate (28%) (A-32), a heat transfer medium, a dielectric fluid in transformers, a degreaser, a lubricant and a potential insecticide against termites. The other trichlorobenzene isomers are not used in any quantity.

Possible human exposure to trichlorobenzene (TCB) might occur from municipal and industrial wastewater and from surface runoff (1). Municipal and industrial discharges contained from 0.1 to 500 μg/ℓ. Surface runoff has been found to contain 0.006 to 0.007 μg/ℓ. In the National Organic Reconnaissance Survey conducted by EPA in 1975, TCB was found in drinking water at a level of 1.0 μg/ℓ.

Permissible Exposure Limits in Air: There is no Federal standard. The ACGIH has set a TWA at a ceiling of 5 ppm (40 mg/m^3) but no STEL value as of 1983/84.

Permissible Concentration in Water: To protect human health—no criterion developed due to insufficient data.

Determination in Water: Methylene chloride extraction followed by concentration, gas chromatography with electron capture detection (EPA Method 612) or gas chromatography plus mass spectrometry (EPA Method 625).

Harmful Effects and Symptoms: *Local* — Chlorinated benzenes are irritating to the skin, conjunctiva, and mucous membranes of the upper respiratory tract. Prolonged or repeated contact with liquid chlorinated benzenes may cause skin burns.

Systemic — In contrast to aliphatic halogenated hydrocarbons, the toxicity of chlorinated benzenes generally decreases as the number of substituted chlorine atoms increases. Basically, acute exposure to these compounds may cause drowsiness, incoordination, and unconsciousness. Animal exposures have produced liver damage. Chronic exposure may result in liver, kidney, and lung damage as indicated by animal experiments.

Points of Attack: Skin, eyes, liver, kidneys, lungs.

Medical Surveillance: Consider the points of attack in preplacement and periodic physical examinations.

Disposal Method Suggested: Incineration, preferably after mixing with another combustible fuel. Care must be exercised to assure complete combustion to prevent the formation of phosgene. An acid scrubber is necessary to remove the halo acids produced.

References

(1) U.S. Environmental Protection Agency, *Chlorinated Benzenes: Ambient Water Quality Criteria,* Washington, DC (1980).

(2) See Reference (A-60).

(3) Sax, N.I., Ed., *Dangerous Properties of Industrial Materials Report, 4,* No. 3, 96–99, New York, Van Nostrand Reinhold Co. (1984).

(4) United Nations Environment Programme, *IRPTC Legal File 1983,* Vol. I, pp VII/96–7, Geneva, Switzerland, International Register of Potentially Toxic Chemicals (1984).

1,1,1-TRICHLOROETHANE

● Hazardous waste (EPA)
● Priority toxic pollutant (EPA)

Description: CH_3CCl_3, 1,1,1-trichloroethane, is a colorless, nonflammable liquid with an odor similar to chloroform. It boils at 74°C.

Code Numbers: CAS 71-55-6   RTECS KJ2975000   UN 2831

DOT Designation: ORM-A

Synonym: Methyl chloroform.

Potential Exposures: In recent years, 1,1,1-trichloroethane has found wide use as a substitute for carbon tetrachloride. In liquid form it is used as a degreaser and for cold cleaning, dip-cleaning, and bucket cleaning of metals. Other industrial applications of 1,1,1-trichloroethane's solvent properties include its use as a dry-cleaning agent, a vapor degreasing agent, and a propellant.

NIOSH has estimated worker exposure to 1,1,1-trichloroethane at 2,900,000 per year.

Incompatibilities: Strong caustics, strong oxidizers, chemically active metals, such as aluminum, magnesium powders, sodium, potassium. Upon contact with hot metal or exposure to ultraviolet radiation, it will decompose to form the irritant gases hydrochloric acid, phosgene, and dichloroacetylene.

Permissible Exposure Limits in Air: The Federal standard and the 1983/84 ACGIH TWA value is 350 ppm (1,900 mg/m³). NIOSH had recommended a 350-ppm ceiling as determined by a 15-minute sampling period. NIOSH has now issued criteria for a recommended standard of 200 ppm for occupational exposures to 1,1,1-trichloroethane. This recommendation to change the standards from 350 ppm is based on central nervous system responses to acute exposures in man, cardiovascular and respiratory effects in man and animals, and the absence of reported effects in man at concentrations below the proposed limit. ACGIH has set an STEL of 450 ppm (2,450 mg/m³). The Dutch Chemical Industry has set a much lower standard (A-60) of 10 ppm (45 mg/m³). The IDLH level is 1,000 ppm.

Determination in Air: Absorption on charcoal, workup with CS_2, analysis by gas chromatography. See NIOSH Methods, Set J. See also reference (A-10).

Permissible Concentration in Water: To protect freshwater aquatic life— 18,000 µg/ℓ on an acute toxicity basis. To protect saltwater aquatic life— 31,200 µg/ℓ on an acute toxicity basis. To protect human health—1,030,000 µg/ℓ.

Determination in Water: Inert gas purge followed by gas chromatography with halide specific detection (EPA Method 601) or gas chromatography plus mass spectrometry (EPA Method 624).

Routes of Entry: Inhalation of vapor, moderate skin absorption, ingestion, skin and eye contact.

Harmful Effects and Symptoms: *Local* — Liquid and vapor are irritating to eyes on contact. This effect is usually noted first in acute exposure cases. Mild conjunctivitis may develop but recovery is usually rapid. Repeated skin contact may produce a dry, scaly, and fissured dermatitis, due to the solvent's defatting properties.

Systemic — 1,1,1-Trichloroethane acts as a narcotic and depresses the central nervous system. Acute exposure symptoms include dizziness, incoordination, drowsiness, increased reaction time, unconsciousness, and death. 1,1,1-Trichloroethane has been subjected to a carcinogenesis bioassay by NCI and found to be not carcinogenic (1).

Points of Attack: Skin, eyes, cardiovascular system, central nervous system.

Medical Surveillance: Consider the skin, liver function, cardiac status, especially arrythmias, in preplacement or periodic examinations. Expired air analyses may be useful in monitoring exposure.

First Aid: If this chemical gets into the eyes, irrigate immediately. If this chemical contacts the skin, wash with soap promptly. If a person breathes in large amounts of this chemical, move the exposed person to fresh air at once and perform artificial respiration. When this chemical has been swallowed, get medical attention. Give large quantities of salt water and induce vomiting. Do not make an unconscious person vomit.

Personal Protective Methods: 1,1,1-Trichloroethane attacks natural rubber; therefore, protective clothing of leather, polyvinyl alcohol, or neoprene is recommended. Wear appropriate clothing to prevent repeated or prolonged skin contact. Wear eye protection to prevent any reasonable probability of eye contact. Employees should wash promptly when skin is wet or contaminated. Remove nonimpervious clothing promptly if wet or contaminated.

Respirator Selection:
         500 ppm: CCROV/SA/SCBA
       1,000 ppm: CCROVF/GMOV/SAF/SCBAF
         Escape: GMOV/SCBA

Disposal Method Suggested: Incineration, preferably after mixing with another combustible fuel. Care must be exercised to assure complete combustion to prevent the formation of phosgene. An acid scrubber is necessary to remove the halo acids produced. Alternative to disposal, trichloroethane may be recovered from waste gases and liquids from various processes (A-58) and recycled.

References
(1) National Cancer Institute, *Bioassay of 1,1,1-Trichloroethane for Possible Carcinogenicity,* Carcinog. Tech. Rept. Ser. NCI-CG-TR-3, Washington, DC (1977).
(2) National Institute for Occupational Safety and Health, *Criteria for a Recommended Standard: Occupational Exposure to 1,1,1-Trichloroethane (Methyl Chloroform),* NIOSH Doc. No. 76-184, Washington, DC (1976).
(3) U.S. Environmental Protection Agency, *Chlorinated Ethanes: Ambient Water Quality Criteria,* Washington, DC (1980).
(4) U.S. Environmental Protection Agency, *1,1,1-Trichloroethane,* Health and Environmental Effects Profile No. 164, Washington, DC, Office of Solid Waste (April 30, 1980).
(5) Sax, N.I., Ed., *Dangerous Properties of Industrial Materials Report, 2,* No. 1, 124-126, New York, Van Nostrand Reinhold Co. (1982).
(6) Sax, N.I., Ed., *Dangerous Properties of Industrial Materials Report, 2,* No. 5, 81-85, New York, Van Nostrand Reinhold Co. (1982).
(7) See Reference (A-61).
(8) Parmeggiani, L., Ed., *Encyclopedia of Occupational Health & Safety,* Third Edition, Vol. 2, pp 2213-14, Geneva, International Labour Office (1983).
(9) United Nations Environment Programme, *IRPTC Legal File 1983,* Vol. I, pp VII/321-4, Geneva, Switzerland, International Register of Potentially Toxic Chemicals (1984).

1,1,2-TRICHLOROETHANE

- Carcinogen (positive, NCI) (4)
- Hazardous waste (EPA)
- Priority toxic pollutant (EPA)

Description: $CH_2ClCHCl_2$, 1,1,2-trichloroethane, is a colorless, nonflammable liquid. It boils at 113°C. It is an isomer of 1,1,1-trichloroethane but should not be confused with it toxicologically. 1,1,2-Trichloroethane is comparable to carbon tetrachloride and tetrachloroethane in toxicity.

Code Numbers: CAS 79-00-5   RTECS KJ3150000

DOT Designation: —

Synonyms: Vinyl trichloride, beta-trichloroethane.

Potential Exposures: 1,1,2-Trichloroethane is used as a chemical intermediate and as a solvent, but is not as widely used as its isomer 1,1,1-trichloroethane. NIOSH estimates worker exposures at 112,000 per year.

Incompatibilities: Strong oxidizers, strong caustics, chemically active metals, such as aluminum, magnesium powders, sodium, potassium.

Permissible Exposure Limits in Air: The Federal standard and the 1983/84 ACGIH TWA value is 10 ppm (45 mg/m^3). The STEL value is 20 ppm (90 mg/m^3). The notation "skin" is added to indicate the possibility of cutaneous absorption. The IDLH level is 500 ppm.

Determination in Air: Adsorption on charcoal, workup with CS_2, analysis by gas chromatography. See NIOSH Methods, Set J. See also reference (A-10).

Permissible Concentration in Water: To protect freshwater aquatic life— 18,000 $\mu g/\ell$ on an acute toxicity basis and 9,400 $\mu g/\ell$ on a chronic basis. To protect saltwater aquatic life—no criteria developed due to insufficient data. To protect human health—preferably zero. An additional lifetime cancer risk of 1 in 100,000 is posed by a concentration of 6.0 $\mu g/\ell$.

Determination in Water: Inert gas purge followed by gas chromatography with halide specific detection (EPA Method 601) or gas chromatography plus mass spectrometry (EPA Method 624).

Routes of Entry: Inhalation of vapor, absorption through the skin, ingestion, skin and eye contact.

Harmful Effects and Symptoms: *Local* — Irritation to eyes and nose, and infection of the conjunctiva have been shown in animals.
Systemic — Little is known of the toxicity of 1,1,2-trichloroethane since no human toxic effects have been reported. Animal experiments show 1,1,2-trichloroethane to be a potent central nervous system depressant. The injection of anesthetic doses in animals was associated with both liver and renal neurosis. 1,1,2-Trichloroethane is carcinogenic in at least one rodent species (4).

Points of Attack: Central nervous system, eyes, nose, liver, kidneys.

Medical Surveillance: Consider the skin, central nervous system, and liver and kidney function. Alcoholism may be a synergistic factor. Expired air analyses may be useful in monitoring exposure.

First Aid: If this chemical gets into the eyes, irrigate immediately. If this chemical contacts the skin, wash with soap promptly. If a person breathes in large amounts of this chemical, move the exposed person to fresh air at once and perform artificial respiration. When this chemical has been swallowed, get medical attention. Give large quantities of salt water and induce vomiting. Do not make an unconscious person vomit.

Personal Protective Methods: Wear appropriate clothing to prevent repeated or prolonged skin contact. Wear eye protection to prevent any reasonable probability of eye contact. Employees should wash promptly when skin is wet or contaminated. Remove nonimpervious clothing promptly if wet or contaminated.

Respirator Selection:
500 ppm: SAF/SCBAF
Escape: GMOV/SCBA

Disposal Method Suggested: Incineration, preferably after mixing with another combustible fuel. Care must be exercised to assure complete combustion to prevent the formation of phosgene. An acid scrubber is necessary to remove the halo acids produced.

References

(1) U.S. Environmental Protection Agency, *Chemical Hazard Information Profile: 1,1,2-Trichloroethane,* Washington, DC (August 1, 1978) (Revised issue 1979).

(2) U.S. Environmental Protection Agency, *Chlorinated Ethanes: Ambient Water Quality Criteria,* Washington, DC (1980).

(3) U.S. Environmental Protection Agency, *1,1,2-Trichloroethane,* Health and Environmental Effects Profile No. 165, Washington, DC, Office of Solid Waste (April 30, 1980).

(4) National Cancer Institute, *Bioassay of 1,1,2-Trichloroethane for Possible Carcinogenicity,* Technical Report Series No. 74, Bethesda, MD (1978).

(5) See Reference (A-61).

(6) Sax, N.I., Ed., *Dangerous Properties of Industrial Materials Report, 2,* No. 6, 88-90, New York, Van Nostrand Reinhold Co. (1982).

(7) Sax, N.I., Ed., *Dangerous Properties of Industrial Materials Report, 3,* No. 2, 66-69, New York, Van Nostrand Reinhold Co. (1983).

(8) Parmeggiani, L., Ed., *Encyclopedia of Occupational Health & Safety,* Third Edition, Vol. 2, pp 2213-14, Geneva, International Labour Office (1983).

(9) United Nations Environment Programme, *IRPTC Legal File 1983,* Vol. I, pp VII/325-27, Geneva, Switzerland, International Register of Potentially Toxic Chemicals (1984).

TRICHLOROETHYLENE

- Carcinogen (animal positive, IARC) (6), NCI (7)
- Hazardous substance (EPA)
- Hazardous waste (EPA)
- Priority toxic pollutant (EPA)

Description: $ClCH=CCl_2$, trichloroethylene, a colorless, nonflammable, noncorrosive liquid has the "sweet" odor characteristic of some chlorinated hydrocarbons. It boils at 86° to 87°C.

Code Numbers: CAS 79-01-6   RTECS KX4550000   UN 1710

DOT Designation: ORM-A

Synonyms: Ethylene trichloride, ethinyl trichloride, trichloroethene, tri, TCE.

Potential Exposures: Trichloroethylene is primarily used as a solvent in vapor degreasing. It is also used for extracting caffeine from coffee, as a dry-cleaning agent, and as a chemical intermediate in the production of pesticides, waxes, gums, resins, tars, paints, varnishes, and specific chemicals such as chloroacetic acid.

It was estimated by NIOSH in 1973 that 200,000 workers are potentially exposed to trichloroethylene and that many of these exposures are in small workplaces. In 1978, this estimate was revised to 100,000 full-time exposures to TCE with up to 3.5 million more workers subjected to continuous low levels or to brief exposures of various levels.

Incompatibilities: Strong caustics; when acidic reacts with aluminum; chemically active metals—barium, lithium, sodium, magnesium, titanium.

Decomposition of trichloroethylene, due to contact with hot metal or ultraviolet radiation, forms products including chlorine gas, hydrogen chloride, and phosgene. Dichloroacetylene may be formed from the reaction of alkali with trichloroethylene.

Permissible Exposure Limits in Air: The Federal standard is 100 ppm (535 mg/m³) as an 8-hour TWA with an acceptable ceiling concentration of 200 ppm; acceptable maximum peaks above the ceiling of 300 ppm are allowed for 5 minutes duration in a 2-hour period. The NIOSH Criteria for a Recommended Standard (1) recommends limits of 100 ppm as a TWA and a peak of 150 ppm determined by a sampling time of 10 minutes.

It was recommended by NIOSH in 1978 (2) that the permissible limit for occupational exposure to trichloroethylene be reduced and that TCE be controlled as an occupational carcinogen. Current information regarding engineering feasibility indicates that personnel exposures of 15 ppm, on a time-weighted-average, can be readily attained using existing engineering control technology. However, NIOSH does not feel that this should serve as a final goal. Rather, industry should pursue further reductions in worker exposure as advancements in technology research allow.

The ACGIH has recommended as of 1983/84 a TWA value of 50 ppm (270 mg/m³) and an STEL of 150 ppm (805 mg/m³); consideration is being given to upping the STEL to 200 ppm (1,080 mg/m³). The IDLH level is 1,000 ppm.

The Dutch chemical industry (A-60) has set a maximum allowable concentration of 35 ppm (190 mg/m³).

Determination in Air: Adsorption on charcoal, workup with CS_2, analysis by gas chromatography. See NIOSH Methods, Set J. See also reference (A-10).

Permissible Concentration in Water: To protect freshwater aquatic life— 45,000 µg/ℓ on an acute toxicity basis. To protect saltwater aquatic life— 2,000 µg/ℓ on an acute toxicity basis. To protect human health—preferably zero. An additional lifetime cancer risk of 1 in 100,000 is posed by a concentration of 27 µg/ℓ.

Determination in Water: Inert gas purge followed by gas chromatography with halide specific detection (EPA Method 601) or gas chromatography plus mass spectrometry (EPA Method 624).

Routes of Entry: Inhalation, percutaneous absorption, ingestion, skin and eye contact.

Harmful Effects and Symptoms: *Local* — Exposure to trichloroethylene vapor may cause irritation of the eyes, nose, and throat. The liquid, if splashed in the eyes, may cause burning irritation and damage. Repeated or prolonged skin contact with the liquid may cause dermatitis.

Systemic — Acute exposure to trichloroethylene depresses the central nervous system exhibiting such symptoms as headache, dizziness, vertigo, tremors, nausea and vomiting, irregular heart beat, sleepiness, fatigue, blurred vision, and intoxication similar to that of alcohol. Unconsciousness and death have been reported. Alcohol may make the symptoms of trichloroethylene overexposure worse. If alcohol has been consumed, the overexposed worker may become flushed. Trichloroethylene addiction and peripheral neuropathy have been reported.

The National Cancer Institute (NCI) in the United States has issued a "state of concern" alert, warning producers, users, and regulatory agencies that trichloroethylene administered by gastric intubation to mice induced predominantly hepatocellular carcinomas with some metastases to the lungs.

Points of Attack: Respiratory system, heart, liver, kidneys, central nervous system, skin.

Medical Surveillance: Preplacement and periodic examinations should include the skin, respiratory, cardiac, central, and peripheral nervous systems, as well as liver and kidney function. Alcohol intake should be evaluated.

Expired air analysis and urinary metabolites have been used to monitor exposure.

First Aid: If this chemical gets into the eyes, irrigate immediately. If this chemical contacts the skin, wash with soap promptly. If a person breathes in large amounts of this chemical, move the exposed person to fresh air at once and perform artificial respiration. When this chemical has been swallowed, get medical attention. Give large quantities of salt water and induce vomiting. Do not make an unconscious person vomit.

Personal Protective Methods: Wear appropriate clothing to prevent repeated or prolonged skin contact. Wear eye protection to prevent any reasonable probability of eye contact. Employees should wash promptly when skin is wet or contaminated. Remove nonimpervious clothing promptly if wet or contaminated.

Respirator Selection:

500 ppm:	CCROV/SA/SCBA
1,000 ppm:	CCROVF/GMOV/SAF/SCBAF
Escape:	GMOV/SCBA

Disposal Method Suggested: Incineration, preferably after mixing with another combustible fuel. Care must be exercised to assure complete combustion to prevent the formation of phosgene. An acid scrubber is necessary to remove the halo acids produced. An alternative to disposal for TCE is recovery and recycling (A-58).

References

(1) National Institute for Occupational Safety and Health, *Criteria for a Recommended Standard: Occupational Exposure to Trichloroethylene,* NIOSH Doc. No. 73-11025 (1973).

(2) National Institute for Occupational Safety and Health, *Special Occupational Hazard Review with Control Recommendations: Trichloroethylene,* NIOSH Doc. No. 78-130, Washington, DC (January 1978).

(3) U.S. Environmental Protection Agency, *Trichloroethylene: Ambient Water Quality Criteria,* Washington, DC (1980).

(4) U.S. Environmental Protection Agency, *Status Assessment of Toxic Chemicals: Trichloroethylene,* Report EPA-600/2-79-210m, Cincinnati, OH (December 1979).

(5) U.S. Environmental Protection Agency, *Trichloroethylene,* Health and Environmental Effects Profile No. 166, Washington, DC, Office of Solid Waste (April 30, 1980).

(6) International Agency for Research on Cancer, *IARC Monographs on the Carcinogenic Risks of Chemicals to Humans,* Lyon, France II, 263 (1976).

(7) National Cancer Institute, *Carcinogenesis Bioassay of Trichloroethylene,* Technical Report Series No. 2, Bethesda, MD (1976).

(8) Sax, N.I., Ed., *Dangerous Properties of Industrial Materials Report, 1,* No. 2, 67-69, New York, Van Nostrand Reinhold Co. (1980).

(9) Sax, N.I., Ed., *Dangerous Properties of Industrial Materials Report, 3,* No. 1, 89-94, New York, Van Nostrand Reinhold Co. (1983).

(10) Parmeggiani, L., Ed., *Encyclopedia of Occupational Health & Safety,* Third Edition, Vol. 2, pp 2214–16, Geneva, International Labour Office (1983).

(11) United Nations Environment Programme, *IRPTC Legal File 1983,* Vol. II, pp VII/367–70, Geneva, Switzerland, International Register of Potentially Toxic Chemicals (1984).

cis-N-TRICHLOROMETHYL-THIO-4-CYCLOHEXENE-1,2-DICARBOXAMIDE

See "Captan."

TRICHLOROPHENOLS

- Carcinogen (2,4,6-isomer only) (potential, EPA) (1) (A-62) (A-64)
- Hazardous substances (EPA)
- Hazardous waste (EPA) (2,4,5 and 2,4,6-trichlorophenols)
- Priority toxic pollutants (EPA)

Description: $HOC_6H_2Cl_3$ exists as 6 isomers (2,4,5-; 3,4,5-; 2,4,6-; 2,3,4-; 2,3,5- and 2,3,6-) with the most important being the 2,4,5- and 2,4,6-isomers. The 2,4,5-isomer melts at 68° to 70°C and the 2,4,6-isomer at 69.5°C.

Code Numbers:

2,4,5-TCP—CAS 95-95-4	RTECS SN1400000	UN (NA-2020)
2,4,6-TCP—CAS 88-06-2	RTECS SN1575000	UN (NA-2020)

DOT Designation: ORM-A

Synonyms: 2,4,5-TCP—Dowicide 2®, 2,4,5-TCP; 2,4-6-TCP—Dowicide 25®, 2,4,6-TCP.

Potential Exposures: 2,4,5-TCP is used to produce defoliant 2,4,5-T and related products (A-32). Also used directly as a fungicide, antimildew and preservative agent, algicide, bactericide.

2,4,6-TCP is used to produce 2,3,4,6-TCP and PCP. Used directly as germicide, bactericide, glue and wood preservative and antimildew treatment.

Incompatibilities: Perhaps the most important is the reaction of 2,4,5-trichlorophenol in alkaline medium at high temperatures to give dioxin in accordance with the equation:

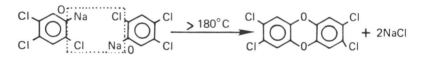

Trichlorophenol Sodium Salt   Dioxin

Permissible Exposure Limits in Air: No standards set.

Permissible Concentration in Water: For 2,4,6-trichlorophenol, to protect freshwater aquatic life—970 $\mu g/\ell$ on a chronic toxicity basis. To protect saltwater aquatic life—no criteria developed due to insufficient data. To protect human health—For 2,4,5-TCP, 2,600 $\mu g/\ell$; for 2,4,6-TCP, preferably zero. An additional lifetime cancer risk of 1 in 100,000 occurs at a level of 12 $\mu g/\ell$. These are based on organoleptic effects. A limit based on toxicological effects for 2,4,5-TCP would be 1,600 $\mu g/\ell$.

Determination in Water: Methylene chloride extraction followed by gas chromatography with flame ionization or electron capture detection (EPA Method 604) or gas chromatography plus mass spectrometry (EPA Method 625).

Harmful Effects and Symptoms: The clinical signs of acute poisoning with 2,4,5-TCP include decreased activity and motor weakness. Convulsive seizures occur but are not as severe as with the monochlorophenols.

A five percent solution of 2,4,5-TCP in sesame oil was mildly irritating to a few individuals upon prolonged contact but there was no evidence of sensitization.

The toxic effects of 2,4,5-TCP can be due to the presence of dioxin formed as described above.

Disposal Method Suggested: Incineration, preferably after mixing with another combustible fuel. Care must be exercised to assure complete combustion to prevent the formation of phosgene. An acid scrubber is necessary to remove the halo acids produced.

References

(1) U.S. Environmental Protection Agency, *Chlorinated Phenols: Ambient Water Quality Criteria,* Washington, DC (1980).

(2) U.S. Environmental Protection Agency, *2,4,6-Trichlorophenol,* Health and Environmental Effects Profile No. 168, Washington, DC, Office of Solid Waste (April 30, 1980).

(3) See Reference (A-60).

(4) Sax, N.I., Ed., *Dangerous Properties of Industrial Materials Report, 3,* No. 6, 79–81, New York, Van Nostrand Reinhold Co. (Nov./Dec. 1983).

(5) United Nations Environment Programme, *IRPTC Legal File 1983,* Vol. II, pp VII/523–25, Geneva, Switzerland, International Register of Potentially Toxic Chemicals (1984).

1,2,3-TRICHLOROPROPANE

● Hazardous waste constituent (EPA)

Description: $CH_2ClCHClCH_2Cl$ is a colorless liquid with a strong acid odor. It boils at 156°C.

Code Numbers: CAS 96-18-4   RTECS TZ9275000

DOT Designation: —

Synonyms: Allyl trichloride, glycerol trichlorohydrin, glycerin trichlorohydrin, trichlorohydrin.

Potential Exposures: Trichloropropane dissolves oils, fats, waxes, chlorinated rubber, and numerous resins. Hence it is used as a paint and varnish remover, a solvent, and a degreasing agent. NIOSH estimated that fewer than 5,000 workers are exposed per year.

Incompatibilities: Active metals, strong caustics, strong oxidizers.

Permissible Exposure Limits in Air: The Federal standard and the 1983/84 ACGIH TWA value is 50 ppm (300 mg/m³). The STEL value is 75 ppm (450 mg/m³). The IDLH level is 1,000 ppm.

Determination in Air: Adsorption on charcoal, workup with CS_2, analysis by gas chromatograhy. See NIOSH Methods, Set I. See also reference (A-10).

Permissible Concentration in Water: No criteria set.

Routes of Entry: Inhalation, skin absorption, ingestion, skin and eye contact.

Harmful Effects and Symptoms: Trichloropropane is highly toxic by inhalation and moderately toxic by skin absorption. It is a local irritant and produces a number of unpleasant sensory effects. Human subjects exposed to trichloropropane at 100 ppm found this to be an objectionable level of exposure, and all reported eye and throat irritation as well as an unpleasant odor.

Also, according to NIOSH, skin irritation, central nervous system depression and liver injury may result.

Points of Attack: Eyes, respiratory system, skin, central nervous system, liver.

Medical Surveillance: Consider the points of attack in preplacement and periodic physical examinations.

First Aid: If this chemical gets into the eyes, irrigate immediately. If this chemical contacts the skin, wash with soap promptly. If a person breathes in large amounts of this chemical, move the exposed person to fresh air at once and perform artificial respiration. When this chemical has been swallowed, get medical attention. Give large quantities of salt water and induce vomiting. Do not make an unconscious person vomit.

Personal Protective Methods: Wear appropriate clothing to prevent repeated or prolonged skin contact. Wear eye protection to prevent any possibility of eye contact. Employees should wash immediately with soap when skin is wet or contaminated. Remove nonimpervious clothing immediately if wet or contaminated. Provide emergency showers and eyewash.

Respirator Selection:
    1,000 ppm: CCROFV/GMOV/SAF/SCBAF
        Escape: GMOV/SCBA

Disposal Method Suggested: Incineration, preferably after mixing with another combustible fuel. Care must be exercised to assure complete combustion to prevent the formation of phosgene. An acid scrubber is necessary to remove the halo acids produced.

References

(1) National Institute for Occupational Safety and Health, *Profiles on Occupational Hazards for Criteria Document Priorities,* Report PB-274,073, Cincinnati, OH, pp 289–91 (1977).
(2) U.S. Environmental Protection Agency, *1,2,3-Trichloropropane,* Health and Environmental Effects Profile No. 169, Washington, DC, Office of Solid Waste (April 30, 1980).
(3) See Reference (A-61).

1,1,2-TRICHLORO-1,2,2-TRIFLUOROETHANE

Description: CCl_2FCClF_2 is a colorless liquid with an odor like carbon tetrachloride (at high concentrations). It boils at $48°C$.

Code Numbers: CAS 76-13-1   RTECS KJ4000000

DOT Designation: —

Synonyms: Fluorocarbon 113, Freon 113, F-113, Trifluorotrichloroethane, TTE.

Potential Exposures: TTE is used as a solvent and refrigerant; it is used in fire extinguishers, as a blowing agent and as an intermediate in the production of chlorotrifluoroethylene monomer by reaction with zinc.

Incompatibilities: Chemically active metals, calcium, powdered aluminum, zinc, magnesium, beryllium; contact alloys >2% Mg decomposes.

Permissible Exposure Limits in Air: The Federal standard and the 1983/84 ACGIH TWA value is 1,000 ppm (7,600 mg/m³). The STEL value is 1,250 ppm (9,500 mg/m³). The IDLH level is 4,500 ppm.

Determination in Air: Adsorption on charcoal, workup with CS_2, analysis by gas chromatography. See NIOSH Methods, Set I.

Permissible Concentration in Water: No criteria set.

Routes of Entry: Inhalation, ingestion, skin and eye contact.

Harmful Effects and Symptoms: Throat irritation, dermatitis, drowsiness.

Points of Attack: Skin, heart.

Medical Surveillance: Consider the points of attack in preplacement and periodic physical examinations.

First Aid: If this chemical gets into the eyes, irrigate immediately. If this chemical contacts the skin, wash with soap promptly. If a person breathes in large amounts of this chemical, move the exposed person to fresh air at once and perform artificial respiration. When this chemical has been swallowed, get medical attention. Give large quantities of salt water and induce vomiting. Do not make an unconscious person vomit.

Personal Protective Methods: Wear appropriate clothing to prevent repeated or prolonged skin contact. Wear eye protection to prevent any possibility of eye contact. Employees should wash promptly when skin is wet or contaminated. Remove nonimpervious clothing promptly if wet or contaminated.

Respirator Selection:
    4,500 ppm: SA/SCBA
       Escape: GMOV/SCBA

Disposal Method Suggested: Incineration, preferably after mixing with another combustible fuel. Care must be exercised to assure complete combustion to prevent the formation of phosgene. An acid scrubber is necessary to remove the halo acids produced.

References
(1) See Reference (A-61).

TRICRESYL PHOSPHATES

Description: Tricresyl phosphates, $(CH_3C_6H_4O)_3PO$, are available as the ortho-isomer (TOCP), the meta-isomer (TMCP), and the para-isomer (TPCP). The ortho-isomer is the most toxic of the three; the meta- and para-isomers are relatively inactive. The commercial product may contain the ortho-isomer as a contaminant unless special precautions are taken during manufacture. Pure tri-para-cresyl phosphate is a solid, and ortho- and meta- are colorless, oily, odorless liquids.

The tri-o-cresyl phosphate will be discussed here as the specific example of these compounds.

Code Numbers: CAS 78-30-8 RTECS TD0350000 UN 2574

DOT Designation: Poison

Synonyms: Tritolyl phosphate; TCP; phosphoric acid, tri-o-tolyl ester.

Potential Exposures: Tricresyl phosphate is used as a plasticizer for chlorinated rubber, vinyl plastics, polystyrene, polyacrylic, and polymethacrylic esters, as an adjuvant in milling of pigment pastes, as a solvent and as a binder in nitrocellulose and various natural resins, and as an additive to synthetic lubricants and gasoline. It is also used as hydraulic fluid, fire retardant and in the recovery of phenol in coke-oven wastewaters.

Incompatibilities: None hazardous.

Permissible Exposure Limits in Air: The Federal standard and the 1983/84 ACGIH TWA value for tri-o-cresyl phosphate is 0.1 mg/m^3; there is no standard for the meta- and para-isomers. The STEL for tri-o-cresyl phosphate is 0.3 mg/m^3. The IDLH level is 40 mg/m^3.

Determination in Air: Collection on a filter, workup with ether, analysis by gas chromatography. See NIOSH Methods, Set P. See also reference (A-10).

Permissible Concentration in Water: No criteria set.

Routes of Entry: Inhalation of ortho-isomer vapor or mist, especially when heated; ingestion and percutaneous absorption of liquid, as well as skin and eye contact. The widespread epidemics of poisoning that have occurred have been due to ingested ortho-isomer as a contaminant of foodstuff. There have been relatively few reports of neurological symptoms in workers handling these substances. Experimental human studies with labeled phosphorus derivatives show only 0.4% of the applied dose was absorbed.

Harmful Effects and Symptoms: *Local* — None reported.

Systemic — The major effects from inhaling, swallowing, or absorbing tricresyl phosphate through the skin are on the spinal cord and peripheral nervous system, the poison attacking the anterior horn cells and pyramidal tract as well as the peripheral nerves. Gastrointestinal symptoms on acute exposure (nausea, vomiting, diarrhea, and abdominal pain) are followed by a latent period of 3 to 30 days with the progressive development of muscle soreness and numbness of fingers, calf muscles, and toes, with foot and wrist drop. In chronic intoxication, the gastrointestinal symptoms pass unnoticed, and after a long latent period, flaccid paralysis of limb and leg muscles appear. There are minor sensory changes and no loss of sphincter control.

Points of Attack: Peripheral nervous system, central nervous system.

Medical Surveillance: Preplacement and periodic examinations should include evaluation of spinal cord and neuromuscular function, especially in the extremities, and a history of exposure to other organophosphate esters, pesticides, or neurotoxic agents. Periodic cholinesterase determination may relate to exposure, but not necessarily to neuromuscular effect.

First Aid: If this chemical gets into the eyes, irrigate immediately. If this chemical contacts the skin, wash with soap immediately. If a person breathes in large amounts of this chemical, move the exposed person to fresh air at once and perform artificial respiration. When this chemical has been swallowed, get medical attention. Give large quantities of water and induce vomiting. Do not make an unconscious person vomit.

Personal Protective Methods: Wear appropriate clothing to prevent repeated or prolonged skin contact. Employees should wash promptly when skin is wet or contaminated. Remove nonimpervious clothing promptly if wet or contaminated.

Respirator Selection:
       0.5 mg/m^3: DMXS
         1 mg/m^3: DMXSQ/FuHiEP/SA/SCBA
         5 mg/m^3: HiEPF/SAF/SCBAF
        40 mg/m^3: PAPHiE/SA:PD,PP,CF
           Escape: DMXS/SCBA

References

(1) Sax, N.I., Ed., *Dangerous Properties of Industrial Materials Report, 2,* No. 2, 73-75, New York, Van Nostrand Reinhold Co. (1982).

(2) Sax, N.I., Ed., *Dangerous Properties of Industrial Materials Report, 2,* No. 3, 83-84, New York, Van Nostrand Reinhold Co. (1982). (Tri-Para Cresyl Ester).

(3) See Reference (A-61).

(4) See Reference (A-60).

(5) Parmeggiani, L., Ed., *Encyclopedia of Occupational Health & Safety,* Third Edition, Vol. 2, pp 2216-18, Geneva, International Labour Office (1983).

TRICYCLOHEXYLTIN HYDROXIDE

See "Cyhexatin".

TRIETHYLAMINE

● Hazardous substance (EPA)

Description: $(C_2H_5)_3N$ is a colorless liquid with a fishy odor which boils at 89° to 90°C.

Code Numbers: CAS 121-44-8   RTECS YE0175000   UN 1296

DOT Designation: Flammable liquid

Synonyms: N,N-diethylethanamine, TEA.

Potential Exposures: Triethylamine (TEA) is used as a corrosion inhibitor in paint removers based on methylene chloride or other chlorinated solvents. TEA is used to solubilize 2,4,5-T in water and serves as a selective extractant in the purification of antibiotics. Octadecyloxymethyltriethylammonium chloride, an agent used in textile treatment, is manufactured from TEA.

Incompatibilities: Strong oxidizers, strong acids.

Permissible Exposure Limits in Air: The Federal standard is 25 ppm (100 mg/m^3). The ACGIH (1983/84) has adopted a TWA of 10 ppm (40 mg/m^3). The STEL is 40 ppm (160 mg/m^3). The IDLH level is 1,000 ppm.

Determination in Air: Collection in a bubbler using sulfuric acid, analysis by gas chromatography. See NIOSH Methods, Set K. See also reference (A-10).

Permissible Concentration in Water: No criteria set.

Routes of Entry: Inhalation, ingestion, skin absorption, skin and eye contact.

Harmful Effects and Symptoms: Irritation of eyes, respiratory system and skin.

Points of Attack: Respiratory system, eyes, skin.

Medical Surveillance: Consider the points of attack in preplacement and periodic physical examinations.

First Aid: If this chemical gets into the eyes, irrigate immediately. If this chemical contacts the skin, wash with soap immediately. If a person breathes in large amounts of this chemical, move the exposed person to fresh air at once and perform artificial respiration. When this chemical has been swallowed, get medical attention. Give large quantities of water and induce vomiting. Do not make an unconscious person vomit.

Personal Protective Methods: Wear appropriate clothing to prevent repeated or prolonged skin contact. Wear eye protection to prevent any possibility of eye contact. Employees should wash immediately when skin is wet or contaminated. Remove clothing immediately if wet or contaminated to avoid flammability hazard. Provide emergency showers and eyewash if liquids containing >1.0% TEA are involved.

Respirator Selection:
1,000 ppm: SAF/SCBAF
Escape: GMS/SCBA

Disposal Method Suggested: Controlled incineration (incinerator is equipped with a scrubber or thermal unit to reduce NO_x emissions).

References
(1)   U.S. Environmental Protection Agency, *Chemical Hazard Information Profile: Ethylamines,* Washington, DC (April 1, 1978).
(2)   See Reference (A-61).
(3)   See Reference (A-60).
(4)   Sax, N.I., Ed., *Dangerous Properties of Industrial Materials Report, 3,* No. 6, 81–83, New York, Van Nostrand Reinhold Co. (Nov./Dec. 1983).

O,O,O-TRIETHYL PHOSPHOROTHIOATE

● Hazardous waste constituent (EPA)

Description: $(C_2H_5O)_3PS$ is a colorless liquid with a characteristic odor boiling at 94°C (at 10 mm pressure).

Code Numbers: CAS 126-68-1   RTECS TG3675000

DOT Designation: –

Synonyms: None.

Potential Exposures: This material is used as a plasticizer, lubricant additive and antifoam agent. It is also used as a chemical intermediate.

Permissible Exposure Limits in Air: No standards set.

Permissible Concentration in Water: No criteria set.

Harmful Effects and Symptoms: There is no information available on the possible carcinogenic, mutagenic, teratogenic, or adverse reproductive effects of O,O,O-triethyl phosphorothioate. Triethyl phosphate, a possible metabolite of the compound, has shown weak mutagenic activity in *Salmonella, Pseudomonas,* and *Drosophila.* Like other organophosphates, O,O,O-triethyl phosphorothioate may be expected to produce cholinesterase inhibition in humans.

References
(1)   U.S. Environmental Protection Agency, *O,O,O-Triethyl Phosphorothioate,* Health and Environmental Effects Profile No. 170, Washington, DC, Office of Solid Waste (April 30, 1980).

TRIFLUOROBROMOMETHANE

Description: $CBrF_3$ is a colorless gas with a slight ethereal odor.

Code Numbers: CAS 75-63-8 RTECS PA5425000 UN 1009

DOT Designation: Nonflammable gas

Synonyms: Bromofluoroform, bromotrifluoromethane, Freon 13B1®, F-13B1, monobromotrifluoromethane, trifluoromonobromomethane, Halon 1301®.

Potential Exposures: This material is used as a fire extinguishing agent and as a refrigerant.

Incompatibilities: Chemically active metals—calcium, powdered aluminum, zinc, magnesium.

Permissible Exposure Limits in Air: The Federal standard and the 1983/84 ACGIH TWA value is 1,000 ppm (6,100 mg/m^3). The STEL is 1,200 ppm (7,300 mg/m^3). The IDLH level is 50,000 ppm.

Determination in Air: Adsorption on charcoal, workup with methylene chloride, analysis by gas chromatography. See NIOSH Methods, Set I. See also reference (A-10).

Permissible Concentration in Water: No criteria set.

Routes of Entry: Inhalation, eye and skin contact with the liquid.

Harmful Effects and Symptoms: Lightheadedness, cardiac arrhythmia.

Points of Attack: Heart, central nervous system.

Medical Surveillance: Consider the points of attack in preplacement and periodic physical examinations.

First Aid: If a person breathes in large amounts of this chemical, move the exposed person to fresh air at once and perform artificial respiration.

Personal Protective Methods: Respirator use is the only protective means specified by NIOSH (A-4).

Respirator Selection:
    10,000 ppm: SA/SCBA
    50,000 ppm: SAF/SA:PD,PP,CF/SCBAF
        Escape: GMOV/SCBA

Disposal Method Suggested: Incineration, preferably after mixing with another combustible fuel. Care must be exercised to assure complete combustion to prevent the formation of phosgene. An acid scrubber is necessary to remove the halo acids produced.

References
(1) See Reference (A-61).

TRIFLURALIN

● Carcinogen (Positive, NCI) (1)

Description: $C_3H_7N-C_6H_2(NO_2)_2CF_3$ is an orange crystalline solid melting at 49°C.

Code Numbers: CAS 1582-09-8 RTECS XU9275000

DOT Designation: —

Synonyms: 2,6-Dinitro-N,N-dipropyl-4-(trifluoromethyl)benzeneamine; Treflan®.

Potential Exposure: Those involved in the manufacture (A-32), formulation and application of this herbicide.

Permissible Exposure Limits in Air: No standards set.

Permissible Concentration in Water: A no-adverse effect level in drinking water has been calculated by NAS/NRC (A-2) as 0.7 mg/ℓ.

Harmful Effects and Symptoms: *Observations in Man* — No controlled studies have been conducted with dinitroaniline compounds in humans. Since 1969, 16 episodes of trifluralin poisoning have been reported. There have been no fatalities, and only one case required hospitalization. Ten of the 16 cases involved symptoms that appeared to be related to the solvent, rather than trifluralin itself. In general, adverse effects of dinitroaniline herbicides in humans have been few and minor according to NAS/NRC in 1977 (A-2). More recently, however, a rebuttable presumption against registration has been issued by U.S. EPA on the basis of oncogenicity and mutagenicity.

Disposal Method Suggested: Incineration. Trifluralin does contain fluorine, and therefore incineration presents the increased hazard of HF in the off-gases. Prior to incineration, fluorine-containing compounds should be mixed with slaked lime plus vermiculite, sodium carbonate or sand-soda ash mixture (90-10).

References

(1) National Cancer Institute, *Bioassay of Trifluralin for Possible Carcinogenicity,* Technical Report Series No. 33, Bethesda, MD (1978).
(2) Sax, N.I., Ed., *Dangerous Properties of Industrial Materials Report, 1,* No. 2, 70-71, New York, Van Nostrand Reinhold Co. (1980).
(3) United Nations Environment Programme, *IRPTC Legal File 1983,* Vol. II, pp VII/767, Geneva, Switzerland, International Register of Potentially Toxic Chemicals (1984).

TRIMELLITIC ANHYDRIDE

Description: Trimellitic anhydride, $C_9H_4O_5$, is a crystalline solid melting at 161° to 163.5°C. It is the anhydride of trimellitic acid (1,2,4-benzenetricarboxylic acid).

Code Numbers: CAS 552-30-7   RTECS DC2050000

DOT Designation: —

Synonyms: 1,3-Dihydro-1,3-dioxo-5-isobenzofurancarboxylic acid, anhydrotrimellitic acid, 4-carboxyphthalic anhydride, TMA.

Potential Exposure: NIOSH (3) estimates approximately 20,000 American workers are currently at risk of exposure to trimellitic anhydride in its various applications. TMA is used as a curing agent for epoxy and other resins, in vinyl plasticizers, paints and coatings, polymers, polyesters, agricultural chemicals, dyes and pigments, pharmaceuticals, surface active agents, modifiers, intermediates, and specialty chemicals.

Permissible Exposure Limits in Air: There is no current Occupational Safety and Health Administration (OSHA) exposure standard for trimellitic anhydride. The Amoco Chemicals Corporation, the sole domestic producer, suggests a limit of "0.05 mg/m³ or less for susceptible individuals" (1). ACGIH as of 1983/84 has recommended a TWA value of 0.005 ppm (0.04 mg/m³) but no STEL value.

Permissible Concentration in Water: No criteria set.

Routes of Entry: Skin contact or inhalation of dusts or vapors.

Harmful Effects and Symptoms: The National Institute for Occupational Safety and Health (NIOSH) recommends that trimellitic anhydride (TMA) be handled as an extremely toxic agent in the workplace. Exposure to this compound may result in noncardiac pulmonary edema (apparently without benefit of a pulmonary irritation warning), immunological sensitization, and irritation of the pulmonary tract, eyes, nose, and skin.

Respiratory symptoms were observed in fourteen workers employed in the synthesis of trimellitic anhydride. The authors suggest three distinct syndromes induced by inhalation of TMA. The first, rhinitis and/or asthma, developed over an industrial exposure period of weeks to years. After this period, the sensitized worker exhibited symptoms immediately following exposure to trimellitic anhydride dust or fume, which abated after the work exposure had stopped.

The second syndrome, termed "TMA-flu" by the workers, also required a sensitization period of exposure and was characterized by delayed onset cough, wheezing, and labored breathing starting 4 to 8 hours after a work shift and peaking at night. These respiratory symptoms were usually accompanied by malaise, chills, fever, muscle and joint aches, and appeared to be associated with relatively high exposures to trimellitic anhydride during particular work shifts.

The third syndrome, which followed initial high exposure to TMA, was primarily an irritant effect. It was characterized by a "running" nose without itching or sneezing, occasional nosebleed, cough, labored breathing, and occasional wheezing. Symptoms usually abated after 8 hours and rarely lasted into the night.

The above studies suggest harmful respiratory effects of trimellitic anhydride at relatively high concentrations, but even at lower concentrations some workers may develop an immunological sensitization over a period of time.

Data on occupational exposures to trimellitic anhydride were also obtained during a NIOSH Health Hazard Evaluation of a paint and varnish company during the manufacture of an epoxy paint (2).

Employees' symptoms and complaints were: eye irritation, nasal irritation, shortness of breath, wheezing, cough, heartburn, nausea, headache, skin irritation, and throat irritation.

Medical Surveillance: The *NIOSH Occupational Exposure Sampling Strategy Manual*, NIOSH Publication No. 77-173, may be helpful in developing efficient programs to monitor employee exposures to trimellitic anhydride. The manual discusses determination of the need for exposure measurements, selection of appropriate employees for sampling, and selection of sampling times.

Personal Protective Methods: Engineering and work practice controls should be used to minimize employee exposure to trimellitic anhydride.

To ensure that ventilation equipment is working properly, effectiveness (e.g., air velocity, static pressure or air volume) should be checked at least every three months. System effectiveness should also be checked within five days of any change in production, process, or control which might result in significant increases in airborne exposures to trimellitic anhydride.

Employers should provide appropriate protective clothing and equipment necessary to prevent repeated or prolonged skin contact with trimellitic anhydride (3).

Employers should see that employees whose clothing may have become contaminatd with trimellitic anhydride change into uncontaminated clothing before leaving the work premises.

Employers should see that nonimpervious clothing which becomes contaminated with trimellitic anhydride be promptly removed and not reworn until the trimellitic anhydride is removed from the clothing.

Employers should see that clothing contaminated with trimellitic anhydride is placed in closed containers for storage until it can be discarded or cleaned. If the clothing is to be laundered or otherwise cleaned to remove the trimellitic anhydride, the employer should tell the person performing the cleaning operation of the hazardous properties of trimellitic anhydride.

Employers should provide dust-resistant safety goggles where there is any possibility of trimellitic anhydride dust contacting the eyes.

Exposure to trimellitic anhydride should not be controlled with the use of respirators (3) except during the time period necessary to install or implement engineering or work practice controls; or in work situations in which engineering and work practice controls are technically not feasible; or to supplement engineering and work practice controls when such controls fail to adequately control exposure to trimellitic anhydride; or for operations which require entry into tanks or closed vessels; or in emergencies.

Respirator Selection: See Reference (3) for details.

Disposal Method Suggested: Incineration.

References

(1) Amoco-Industrial Hygiene Toxicology and Safety Data Sheet, Environmental Health Services, Medical and Health Services Department (July 8, 1976).

(2) National Institute for Occupational Safety and Health, Health Hazard Evaluation Determination Report No. 74-111-283.

(3) National Institute for Occupational Safety and Health, *Trimellitic Anhydride (TMA)*, Current Intelligence Bulletin No. 21, Washington, DC (February 3, 1978).

(4) U.S. Environmental Protection Agency, *Chemical Hazard Information Profile: Trimellitic Anhydride,* Washington, DC (March 3, 1978).

TRIMETHYLAMINE

Description: $(CH_3)_3N$ is a fishy, ammoniacal-smelling gas which boils at 3° to 4°C.

Code Numbers: CAS 75-50-3   RTECS YH2280000   UN 1083 (gas) UN 1297 (aqueous solution)

DOT Designation: Flammable gas.

Synonyms: Dimethylmethaneamine.

Potential Exposure: In organic synthesis of quaternary ammonium compounds; as an insect attractant; as a warning agent in natural gas.

Permissible Exposure Limits in Air: ACGIH (1983/84) has adopted a TWA value of 10 ppm (24 mg/m^3) and an STEL of 15 ppm (36 mg/m^3). The U.S.S.R. reportedly (A-36) has a TLV of 1.1 ppm (5 mg/m^3).

Permissible Concentration in Water: No criteria set.

Harmful Effects and Symptoms: This is a moderately toxic material (LD$_{50}$ for mice of 90 mg/kg) (2). Skin burns and eye irritation are produced by contact. Inhalation of vapors causes coughing and vomiting; higher concentrations cause difficult breathing and pulmonary edema.

Points of Attack: Skin, eyes, respiratory tract.

Personal Protective Methods: Wear full protective clothing including rubber gloves and boots. Wear gastight goggles.

Respirator Selection: Wear NIOSH-approved respirator or self-contained breathing apparatus.

Disposal Method Suggested: Vent to atmosphere.

References
(1) Sax, N.I., Ed., *Dangerous Properties of Industrial Materials Report, 2,* No. 2, 70-73, New York, Van Nostrand Reinhold Co. (1982).
(2) See Reference (A-60).
(3) American Conference of Governmental Industrial Hygienists, *Documentation of the Threshold Limit Values: Supplemental Documentation 1981,* Cincinnati, OH (1981).

TRIMETHYL BENZENES

Description: $C_6H_3(CH_3)_3$ exists in three isomeric forms. 1,3,5 Trimethyl—mesitylene is a colored liquid boiling at $165°C$. 1,2,4-Trimethyl—pseudocumene is a colorless liquid boiling at $169°C$. 1,2,3-Trimethyl—hemimellitene is a color-less liquid boiling at $176°C$.

Code Numbers:

Mesitylene	CAS 108-67-8	RTECS OX6825000	UN 2325
Pseudocumene	CAS 95-63-6	RTECS DC3325000	
Hemimellitene	CAS 526-73-8	RTECS DC3300000	

DOT Designations:
Mesitylene: Flammable liquid.
Pseudocumene: —
Hemimellitene: —

Synonyms:
Mesitylene: 1,3,5-Trimethylbenzene.
Pseudocumene: 1,2,4-Trimethylbenzene, psi-cumene.
Hemimellitene: 1,2,3-Trimethylbenzene.

Potential Exposures: These materials are used as solvents and in dye and perfume manufacture. Pseudocumene is used as the raw material for trimellitic anhydride manufacture. These compounds are found in diesel engine exhaust fumes.

Permissible Exposure Limits in Air: There are no Federal standards but ACGIH (1983/84) has adopted TWA values for trimethylbenzenes as a class of 25 ppm (125 mg/m³) and set STEL values of 35 ppm (170 mg/m³).

Determination in Air: Sample collection by charcoal tube, analysis by gas liquid chromatography (A-10).

Permissible Concentration in Water: No criteria set.

Routes of Entry: Inhalation, percutaneous absorption, ingestion, skin and eye contact.

Harmful Effects and Symptoms: Skin irritation; chemical pneumonitis at the site of contact when deposition of the liquid into the lungs occurs; nervousness, tension, anxiety, asthmatic bronchitis; hypochromic anemia (A-34). Conjunctivitis, headache, fatigue, nausea and narcosis are also reported (A-39).

Medical Surveillance: Perform complete blood count in connection with annual physical examinations of exposed personnel (A-39).

First Aid: Irrigate eyes with water. Wash contaminated areas of body with soap and water.

Personal Protective Methods: Wear chemical goggles, rubber gloves.

Respirator Selection: Use chemical cartridge respirator.

TRIMETHYL PHOSPHITE

Description: $(CH_3O)_3P$ is a colorless liquid with a distinctive pungent pyridine-like odor which boils at 111° to 112°C.

Code Numbers: CAS 121-45-9   RTECS TH1400000   UN 2329

DOT Designation: Flammable liquid

Synonym: TMP.

Potential Exposures: This material is used in flameproofing and as an intermediate in the manufacture of a number of pesticides (A-32).

Permissible Exposure Limits in Air: There are no Federal standards but ACGIH (1983/84) has proposed a TWA value of 2 ppm (10 mg/m^3) and an STEL value of 5 ppm (25 mg/m^3).

Permissible Concentration in Water: No criteria set.

Harmful Effects and Symptoms: This is a highly odorous material with relatively low toxicity. Skin and eye irritation appears to be the main hazards (A-34).

TRINITROBENZENE

● Hazardous waste (EPA)

Description: 1,3,5-$C_6H_3(NO_2)_3$ is a yellow crystalline solid melting at 122° to 123°C.

Code Numbers: CAS 99-35-4   RTECS DC3850000   UN 0214

DOT Designation: Explosive

Synonyms: TNB.

Potential Exposures: Trinitrobenzene is used as an explosive, and as a vulcanizing agent for natural rubber.

Incompatibilities: Initiating explosives, combustible materials, sources of fire, oxidizing materials (A-38).

Permissible Exposure Limits in Air: No standards set.

Permissible Concentration in Water: No criteria set.

Harmful Effects and Symptoms: Information on the carcinogenicity, mutagenicity, teratogenicity, or adverse reproductive effects of trinitrobenzene was not found in the available literature.

Trinitrobenzene has been reported to produce liver damage, central nervous system damage, and methemoglobin formation in animals (1). Breathing difficulties have also been reported (A-38).

Disposal Method Suggested: Dissolve in a combustible solvent and spray into an incinerator equipped with afterburner and scrubber (A-38).

References

(1) U.S. Environmental Protection Agency, *Trinitrobenzene,* Health and Environmental Effects Profile No. 171, Washington, DC, Office of Solid Waste (April 30, 1980).

TRINITROTOLUENE

Description: TNT exists in five isomers; 2,4,6-trinitrotoluene is the most commonly used. It is a colorless to pale yellow odorless solid melting at 81°C. It explodes at 240°C but burns at 295°C when not confined. TNT is a relatively stable high explosive.

Code Numbers: CAS 118-96-7   RTECS XU0175000   UN 0209

DOT Designation: Class A explosive

Synonyms: TNT, sym-trinitrotoluol, methyltrinitrobenzene.

Potential Exposures: TNT is used as an explosive, i.e., as a bursting charge in shells, bombs, and mines.

Incompatibilities: Strong oxidizers, ammonia, strong alkalies, oxidizable materials.

Permissible Exposure Limits in Air: The Federal standard is 1.5 mg/m^3, according to NIOSH. ACGIH (1983/84) has adopted 0.5 mg/m^3 as a TWA with the notation "skin" indicating the possibility of cutaneous absorption. The STEL value set is 3 mg/m^3; there is no IDLH level specified.

Determination in Air: Collection by filter, colorimetric analysis (A-1).

Permissible Concentration in Water: No criteria set.

Routes of Entry: Inhalation of dust, fume, or vapor; ingestion of dust; percutaneous absorption from dust, eye and skin contact.

Harmful Effects and Symptoms: *Local* — Exposure to trinitrotoluene may cause irritation of the eyes, nose, and throat with sneezing, cough, and sore throat. It may cause dermatitis and may stain the skin, hair, and nails a yellowish color.

Systemic — Numerous fatalities have occurred in workers exposed to TNT from toxic hepatitis or aplastic anemia. TNT exposure may also cause methemoglobinemia with cyanosis, weakness, drowsiness, dyspnea, and unconsciousness. In addition it may cause muscular pains, heart irregularities, renal irritation, cataracts, menstrual irregularities, and peripheral neuritis.

Points of Attack: Blood, liver, eyes, cardiovascular system, central nervous system, kidneys, skin.

Medical Surveillance: Placement or periodic examinations should give special considerations to history of allergic reactions, blood dyscrasias, reactions to medications, and alcohol intake. The skin, eye, blood, and liver and kidney function should be followed.

Urine may be examined for TNT by the Webster test or for the urinary metabolite 2,6-dinitro-4-aminotoluene; however, both may be negative if there is a liver injury.

First Aid: If this chemical gets into the eyes, irrigate immediately. If this chemical contacts the skin, wash with soap promptly. If a person breathes in large amounts of this chemical, move the exposed person to fresh air at once and perform artificial respiration. When this chemical has been swallowed, get medical attention. Give large quantities of water and induce vomiting. Do not make an unconscious person vomit.

Personal Protective Methods: Wear appropriate clothing to prevent repeated or prolonged skin contact. Wear eye protection to prevent any reasonable probability of eye contact. Employees should wash promptly when skin is wet or contaminated and daily at the end of each work shift.

The Webster skin test (colorimetric test with alcoholic sodium hydroxide) or indicator soap should be used to make sure workers have washed all TNT off their skins.

Work clothing should be changed daily if it is possible that clothing is contaminated. Remove nonimpervious clothing promptly if wet or contaminated.

Respirator Selection:
15 mg/m³: SA/SCBA
75 mg/m³: SAF/SCBAF
3,000 mg/m³: SAF:PD,PP,CF
Escape: GMOVP/SCBA

Disposal Method Suggested: TNT is dissolved in acetone and incinerated. The incinerator should be equipped with an afterburner and a caustic soda solution scrubber.

References

(1) U.S. Environmental Protection Agency, *Chemical Hazard Information Profile: 2,4,6-Trinitrotoluene,* Washington, DC (1979).
(2) Sax, N.I., Ed., *Dangerous Properties of Industrial Materials Report, 2,* No. 5, 93-96, New York, Van Nostrand Reinhold Co. (1982).
(3) See Reference (A-61).
(4) Parmeggiani, L., Ed., *Encyclopedia of Occupational Health & Safety,* Third Edition, Vol. 2, pp 2218–19, Geneva, International Labour Office (1983).

TRIPHENYLAMINE

Description: $(C_6H_5)_3N$ is a colorless crystalline solid melting at 127° to 129°C.

Code Numbers: CAS 603-34-9

DOT Designation: —

Synonyms: None.

Potential Exposures: Triphenylamine is a primary photoconductor and is coated on film bases (A-34).

Permissible Exposure Limits in Air: There are no Federal standards but ACGIH (1983/84) has adopted a TWA value of 5 mg/m³. There is no STEL value.

Permissible Concentration in Water: No criteria set.

Harmful Effects and Symptoms: No injuries have occurred as a result of industrial handling of this compound (A-34).

TRIPHENYL PHOSPHATE

Description: $(C_6H_5O)_3PO$ is a colorless solid with a faint, aromatic odor. It melts at 48° to 49°C.

Code Numbers: CAS 115-86-6   RTECS TC8400000

DOT Designation: −

Synonyms: Phenyl phosphate, TPP.

Potential Exposures: TPP is used to impregnate roofing paper and as a plasticizer for cellulose esters in lacquers and varnishes.

Incompatibilities: None hazardous.

Permissible Exposure Limits in Air: The Federal standard and the 1983/84 ACGIH TWA value is 3 mg/m³. The STEL is 6 mg/m³. There is no IDLH level set.

Determination in Air: Collection on a filter, workup with ether, analysis by gas chromatography. See NIOSH Methods, Set P. See also reference (A-10).

Permissible Concentration in Water: No criteria set.

Routes of Entry: Inhalation, ingestion.

Harmful Effects and Symptoms: Minor changes in blood enzymes; in animals, muscular weakness, paralysis.

Points of Attack: Blood.

Medical Surveillance: Consider the blood in preplacement and periodic physical examinations.

First Aid: If a person breathes in large amounts of this chemical, move the exposed person to fresh air at once and perform artificial respiration. When this chemical has been swallowed, get medical attention. Give large quantities of water and induce vomiting. Do not make an unconscious person vomit.

Personal Protective Methods: Respirator protection is the only technique specified by NIOSH (A-4). Note the contrast to tricresyl phosphate.

Respirator Selection:
    15 mg/m³: DMXS
    30 mg/m³: DMXSQ/FuHiEP/SA/SCBA
    150 mg/m³: HiEPF/SAF/SCBAF
    1,500 mg/m³: PAPHiE/SA:PD,PP,CF

Disposal Method Suggested: Incinerate in furnace equipped with alkaline scrubber.

References
(1) See Reference (A-61).
(2) See Reference (A-60).

TRIS-(1-AZIRIDINYL)PHOSPHINE SULFIDE

See "Thiotepa."

TRIS(2,3-DIBROMOPROPYL) PHOSPHATE

- Carcinogen (positive, NCI) (3) (IARC) (4)
- Hazardous waste (EPA)

Description: $[BrCH_2CH(Br)CH_2O]_3P=O$ has been a widely used flame-retardant additive for children's sleepwear. Commercial preparations of tris-BP can be obtained in two grades, viz, HV (high in volatiles) and LV (low in volatiles). A typical LV sample has been reported to contain the following impurities (1): 0.05% 1,2-dibromo-3-chloropropane ($BrCH_2CHBrCH_2Cl$); 0.05% 1,2,3-tribromopropane ($BrCH_2CHBrCH_2Br$); and 0.20% 2,3-dibromopropanol ($BrCH_2CHBrCH_2OH$).

Code Numbers: CAS 126-72-7   RTECS UB0350000

DOT Designation: —

Synonyms: TBPP, tris-BP, tris.

Potential Exposures: Currently, about 300 million pounds of flame-retardant chemicals are being produced mainly for use in fabrics, plastics and carpets. Approximately two-thirds of this amount are inorganic derivatives such as alumina trihydrate and antimony oxide while the remaining one-third are large numbers of brominated and chlorinated organic derivatives.

Tris-BP was applied at a 10-million pound level to fabrics used for children's clothes (sleepwear in particular) with some used as a flame retardant in other materials such as urethane foams.

A significant portion of the total used (approximately 10%) was estimated to reach the environment from textile finishing plants and launderies while most of the remainder was postulated to eventually end up on solid wastes (e.g., manufacturing waste and used clothing).

Use and exposure has greatly decreased after a ruling by the Consumer Product Safety Commission in April 1977.

Permissible Exposure Limits in Air: No standards set.

Permissible Concentration in Water: No criteria set.

Routes of Entry: TBPP was added to fabrics used for children's garments to the extent of 5 to 10% by weight. A child wearing such a garment and chewing on a sleeve or collar could easily ingest some TBPP, particularly if the garment had not been laundered before use. The effects of saliva, urine, or feces on the extractability of TBPP or on its absorption through the skin have not been measured.

Harmful Effects and Symptoms: TBPP is a mild sensitizing agent in humans. However, no allergic responses have been reported from consumer or occupational exposure.

Orally administered tris was carcinogenic to mice, causing increased incidences of tumors in livers, lungs, and stomachs of female mice and in kidneys, lungs, and stomachs of male mice. Tris was also carcinogenic in rats, causing an increased incidence of kidney tumors in both sexes (3) (6).

Although tris has been shown to be an animal carcinogen, no conclusive data regarding carcinogenicity and mutagenicity to humans are available.

The carcinogenicity of an impurity of tris-BP, e.g., dibromochloropropane should also be noted. This compound caused a high incidence of squamous carcinoma of the stomach in both rats and mice as early as 10 weeks after initiation of feeding by oral intubation.

References

(1) U.S. Environmental Protection Agency, *Summary Characterization of Selected Chemicals of Near-Term Interest,* Report EPA 560/4-76-004, Washington, DC, Office of Toxic Substances (April 1976).

(2) U.S. Environmental Protection Agency, *Status Assessment of Toxic Chemicals: Tris-(2,3-Dibromopropyl) Phosphate,* Report EPA-600/2-79-210n, Cincinnati, OH (December 1979).

(3) National Cancer Institute, *Bioassay of Tris(2,3-Dibromopropyl) Phosphate for Possible Carcinogenicity,* Technical Report Series No. 76, Bethesda, MD (1978).

(4) International Agency for Research on Cancer, *IARC Monographs on the Carcinogenic Risks of Chemicals to Humans,* Lyon, France 20, 576 (1979).

(5) U.S. Environmental Protection Agency, *Investigation of Selected Potential Environmental Contaminants: Haloalkyl Phosphates,* Report EPA-560/2-76-007, Washington, DC (August 1976).

(6) See Reference (A-62). Also see Reference (A-64).

(7) United Nations Environment Programme, *IRPTC Legal File 1983,* Vol. II, pp VII/680, Geneva, Switzerland, International Register of Potentially Toxic Chemicals (1984).

TRITOLYL PHOSPHATE

See "Tricresyl Phosphate."

TUNGSTEN AND CEMENTED TUNGSTEN CARBIDE

Description: Tungsten, symbol W, is element atomic number 74. It is a steel-gray to tin-white metal with a melting point of $3410°C$. Tungsten carbide, WC, is a gray powder melting at $2780°C$. Cemented tungsten carbide is a mixture consisting of 85 to 95% WC and 5 to 15% cobalt.

Code Numbers:

Tungsten	CAS 7440-33-7	RTECS YO7175000
Cemented WC	CAS 12718-69-3	RTECS YO7525000

DOT Designation: —

Synonyms: Tungsten—Wolfram. Cemented WC—"Hard metal."

Potential Exposure: NIOSH estimates that at least 30,000 employees in the United States are potentially exposed to tungsten and its compounds, based on actual observations in the National Occupational Hazards Survey.

It has been stated that the principal health hazards from tungsten and its compounds arise from inhalation of aerosols during mining and milling operations. The principal compounds of tungsten to which workers are exposed are ammonium paratungstate, oxides of tungsten (WO_3, W_2O_5, WO_2), metallic tungsten, and tungsten carbide. In the production and use of tungsten carbide tools for machining, exposure to the cobalt used as a binder or cementing substance may be the most important hazard to the health of the employees. Since the cemented tungsten carbide industry uses such other metals as tantalum, titanium, niobium, nickel, chromium, and vanadium in the manufacturing process, the occupational exposures are generally to mixed dust.

Potential occupational exposures to sodium tungstate are found in the textile industry, where the compound is used as a mordant and fireproofing agent, and in the production of tungsten from some of its ores, where sodium tungstate is an intermediate product. Potential exposures to tungsten and its compounds are also found in the ceramics, lubricants, plastics, printing inks, paint, and photographic industries.

Permissible Exposure Limits in Air: Occupational exposure to insoluble tungsten shall be controlled so that employees are not exposed to insoluble tungsten at a concentration greater than 5 mg/m^3 of air, measured as tungsten, determined as a TWA concentration for up to a 10-hour work shift in a 40-hour workweek. An STEL value of 10 mg/m^3 has been set by ACGIH (1983/84).

Occupational exposure to soluble tungsten shall be controlled so that employees are not exposed to soluble tungsten at a concentration greater than 1 mg/m^3 of air, measured as tungsten, determined as a TWA concentration for up to a 10-hour work shift in a 40-hour workweek. An STEL value of 3 mg/m^3 has been set by ACGIH (1983/84).

Occupational exposure to dust of cemented tungsten carbide which contains more than 2% cobalt shall be controlled so that employees are not exposed at a concentration greater than 0.1 mg/m^3 of air, measured as cobalt, determined as a TWA concentration for up to a 10-hour work shift in a 40-hour workweek.

Occupational exposure to dust of cemented tungsten carbide which contains more than 0.3% nickel shall be controlled so that employees are not exposed at a concentration greater than 15 μg nickel/m^3 air determined as a TWA concentration, for up to a 10-hour workshift in a 40-hour workweek as specified in NIOSH's *Criteria for a Recommended Standard for Occupational Exposure to Inorganic Nickel.*

Determination in Air: By neutron activation analysis (1). Also by membrane filter collection, water extraction for soluble tungsten, acid ash for insoluble tungsten and atomic absorption analysis using a nitrogen oxide-acetylene flame (A-10). See also (2).

Permissible Concentration in Water: No criteria set, but EPA (A-37) has suggested a permissible ambient goal of 14 μg/ℓ based on health effects.

Detection in Water: By neutron activation analysis (1).

Routes of Entry: Skin contact and inhalation of dusts.

Harmful Effects and Symptoms: Insoluble tungsten compounds and cemented tungsten carbide may cause transient or permanent lung damage and skin irritation, while soluble tungsten compounds have the potential to cause systemic effects involving the gastrointestinal tract and central nervous system. No carcinogenic, mutagenic, teratogenic, or reproductive effects in humans have been reported. Compliance with the appropriate recommended environmental limits should eliminate the hazards associated with tungsten compounds and cemented tungsten carbide, except for a few individuals who may become sensitized to cobalt or nickel and have adverse reactions upon exposure to extremely small amounts of cemented tungsten carbide.

Medical Surveillance: Preplacement medical screening is recommended to identify any preexisting pulmonary conditions that might make a worker more susceptible to exposures in the work environment. Periodic medical examinations will aid in early detection of any occupationally related illnesses which might otherwise go undetected because of either delayed toxic effects or subtle changes. Maintenance of medical records for a period of 30 years is recommended.

Personal Protective Methods: Exposures to tungsten and its compounds in occupational environments can best be prevented by engineering controls and good work practices. Since tungsten compounds and dusts from cemented tungsten carbide affect chiefly the respiratory system, measures are recommended that will reduce the atmospheric concentrations of tungsten in the work atmosphere. Adoption of these measures during normal operations will also minimize the possibility of skin contact or accidental ingestion.

In addition to using sound engineering controls, employers should institute a program of work practices which emphasizes good sanitation and personal hygiene. These practices are important in preventing skin and respiratory irritation caused by tungsten compounds or cemented tungsten carbide.

Respirators should not be used as a substitute for proper engineering controls in normal operations. However, during emergencies and during nonroutine repair and maintenance activities, exposures to airborne dusts or mists of tungsten compounds or cemented tungsten carbide might not be reduced either by engineering controls or by administrative measures to the levels specified. If this occurs, then respiratory protection may be used only: during the time necessary to install or test the required engineering controls; for operations such as maintenance and repair activities causing brief exposure at concentrations above the TWA concentration limits; and during emergencies when airborne concentrations may exceed the TWA concentration limits.

Eye protection shall be provided in accordance with 29 CFR 1910.133 for operations, such as grinding, which produce and scatter particulates into the air.

While most workers do not experience skin irritation as a result of exposure to tungsten compounds or cemented tungsten carbide, there are some who develop sensitivity. Fingerless gloves may be used during grinding of hard metal to protect the hands from abrasion. Protective sleeves of dustproof material may be worn to prevent impact of hard-metal dust on the skin of the arms. In the absence of such gloves and sleeves, creams protective against abrasion may be applied liberally to the hands and arms to minimize contact of the skin with hard-metal dust. When skin irritation occurs, these workers should be referred to a physician for appropriate protective and therapeutic measures. When abrasive dust of tungsten carbide is likely to contact major parts of an employee's body, the employee should wear closely woven coveralls provided by the employer. The coveralls should be laundered frequently to minimize mechanical irritation from dust in the cloth.

Respirator Selection:

Concentration Range (mg/m³, as tungsten)	Respirator Type Approved Under Provisions of 30 CFR 11
. . . . For Soluble Tungsten Compounds*	
≤10**	Half-mask dust and mist respirator Supplied-air respirator with half-mask facepiece
≤50**	Full facepiece dust and mist respirator Supplied-air respirator will full facepiece Self-contained breathing apparatus with full facepiece
>50**	Powered air-purifying respirator (positive pressure) with high-efficiency filter*** Supplied-air respirator with full facepiece, hood, or helmet, continuous-flow or other positive-pressure type Self-contained breathing apparatus with full facepiece in pressure demand or other positive pressure mode

(continued)

Concentration Range (mg/m³, as tungsten)	Respirator Type Approved Under Provisions of 30 CFR 11

. . . . For Insoluble Tungsten Compounds† . . . .

≤25**	Single-use dust respirator Dust respirator with quarter-mask facepiece
≤50**	Half-mask dust respirator Supplied-air respirator with half-mask facepiece
>50**	Full facepiece dust respirator Supplied-air respirator with full facepiece, hood, or helmet Powered air-purifying respirator (positive pressure) with high efficiency filter*** For abrasive-blasting with tungsten carbide, supplied-air respirator with hood or helmet operated in pressure-demand or other positive pressure mode or with continuous flow

*Tungsten hexachloride in contact with water decomposes to tungsten and chlorine. Therefore, the dust and mist respirator is not satisfactory if the chlorine gas concentration exceeds the permissible level. When this occurs, the minimum acceptable respirator is a full facepiece respirator with a combination dust, mist, and chlorine canister.

**When an employee informs the employer that eye irritation occurs while wearing a respirator, the employer shall provide an equivalent respirator with full facepiece, helmet, or hood and ensure that it is used.

***A high-efficiency filter is defined as having a penetration of <0.03% when tested against a 0.3 µm DOP aerosol.

†In areas where ammonium paratungstate is used, a canister or cartridge which will remove ammonia is needed in addition to a particulate filter.

Disposal Method Suggested: Recovery of tungsten from sintered metal carbides, scrap and spent catalysts has been described (A-57) as an alternative to disposal.

References

(1) U.S. Environmental Protection Agency, *Toxicology of Metals, Vol II: Tungsten,* Report EPA-600/1-77-022, Research Triangle Park, NC, pp 442–453 (May 1977).

(2) National Institute for Occupational Safety and Health, *Criteria for a Recommended Standard: Occupational Exposure to Tungsten and Cemented Tungsten Carbide,* NIOSH Doc. No. 77-127 (September 1977).

(3) Parmeggiani, L, Ed., *Encyclopedia of Occupational Health & Safety,* Third Edition, Vol. 2, pp 2225–26, Geneva, International Labour Office (1983).

TURPENTINE

Description: Turpentine is the oleoresin from species of *Pinus pinacea* trees. The crude oleoresin (gum turpentine) is a yellowish, sticky, opaque mass and the distillate (oil of turpentine) is a colorless, volatile liquid with a characteristic odor. Chemically, it contains: alpha-pinene, beta-pinene, camphene, monocyclic terpenes, and terpene alcohols. It has the approximate formula $C_{10}H_{16}$ and boils at 150° to 180°C.

Code Numbers: CAS 8006-64-2   RTECS YO8400000   UN 1299

DOT Designation: Combustible liquid

Synonyms: Gum spirits, turps, wood turpentine, spirits of turpentine, sulfate wood turpentine, steam-distilled turpentine, gum turpentine.

Potential Exposures: Turpentines have found wide use as chemical feedstock for the manufacture of floor, furniture, shoe, and automobile polishes, camphor, cleaning materials, inks, putty, mastics, cutting and grinding fluids, paint thinners, resins, and degreasing solutions. Recently, alpha- and beta-pinenes, which can be extracted, have found use as volatile bases for various compounds.

Incompatibilities: Strong oxidizers, chlorine.

Permissible Exposure Limits in Air: The Federal standard and the 1983/84 ACGIH TWA value for turpentine is 100 ppm (560 mg/m³). The STEL value is 150 ppm (840 mg/m³). The IDLH level is 1,900 ppm.

Determination in Air: Adsorption on charcoal, workup with CS_2, analysis by gas chromatography. See NIOSH Methods, Set G. See also reference (A-10).

Permissible Concentration in Water: No criteria set.

Routes of Entry: Inhalation of vapor and percutaneous absorption of liquid are the usual paths of occupational exposure. However, symptoms have been reported to develop from percutaneous absorption alone. Ingestion and eye and skin contact are also possible routes of entry.

Harmful Effects and Symptoms: *Local* — High vapor concentrations are irritating to the eyes, nose, and bronchi. Aspiration of liquid may cause direct lung irritation resulting in pulmonary edema and hemorrhage. Turpentine liquid may produce contact dermatitis. Eczema from turpentine is quite common and has been attributed to the autooxidation products of the terpenes (formic acid, formaldehyde, and phenols). This hypersensitivity usually develops in a small portion of the working population. Liquid turpentine splashed in the eyes may cause corneal burns and demands emergency treatment.

Systemic — Turpentine vapor in acute concentrations may cause central nervous system depression. Symptoms include headache, anorexia, anxiety, excitement, mental confusion, and tinnitus. Convulsions, coma, and death have been reported in animal experiments.

Turpentine vapor also produces kidney and bladder damage. Chronic nephritis with albuminuria and hematuria has been reported as a result of repeated exposures to high concentrations. Predisposition to pneumonia may also occur from such exposures. Recovery usually takes from a few days to a few weeks. Several animal experiments of chronic low-level exposure have produced no ill effects to the central nervous system, kidneys, bladder, or blood.

Points of Attack: Skin, eyes, kidneys, respiratory system.

Medical Surveillance: Consideration should be given to skin disease or skin allergies in any preplacement or periodic examinations. Liver, renal, and respiratory disease should also be considered.

First Aid: If this chemical gets into the eyes, irrigate immediately. If this chemical contacts the skin, wash with soap promptly. If a person breathes in large amounts of this chemical, move the exposed person to fresh air at once and perform artificial respiration. When this chemical has been swallowed, get medical attention. Do not induce vomiting.

Personal Protective Methods: Wear appropriate clothing to prevent repeated or prolonged skin contact. Wear eye protection to prevent any reasonable probability of eye contact. Employees should wash promptly when skin is wet or contaminated. Remove clothing immediately if wet or contaminated to avoid flammability hazard.

Respirator Selection:
<pre>
    1,000 ppm: CCROVF
    1,900 ppm: SAF/SCBAF/GMOV
      Escape: GMOV/SCBA
</pre>

Disposal Method Suggested: Incineration.

References

(1)   See Reference (A-61).
(2)   Sax, N.I., Ed., *Dangerous Properties of Industrial Materials Report, 2,* No. 2, 75-76, New York, Van Nostrand Reinhold Co. (1982).
(3)   See Reference (A-60).
(4)   Parmeggiani, L., Ed., *Encyclopedia of Occupational Health & Safety,* Third Edition, Vol. 2, pp 2229, Geneva, International Labour Office (1983).

U

URANIUM AND COMPOUNDS

● Hazardous substances (uranyl nitrate and acetate)(EPA)

Description: U, uranium, is a hard, silvery white amphoteric metal and is a radioactive element. In the natural state, it consists of three isotopes: ^{238}U (99.28%), ^{234}U (0.006%), and ^{235}U (0.714%). There are over 100 uranium minerals; those of commercial importance are the oxides and oxygenous salts. The processing of uranium ore generally involves extraction then leaching either by an acid or a carbonate method. The metal may be obtained from its halides by fused salt electrolysis.

Code Numbers:

Uranium metal	CAS 7440-61-1	RTECS YR3490000	UN (NA 9175)
Uranyl nitrate	CAS 13520-83-7	RTECS YR3850000	UN (NA 9177)

DOT Designations:

Uranium metal: Radioactive material and flammable solid.
Uranium nitrate: Radioactive and oxidizer.

Synonyms:

Uranium: None.
Uranyl nitrate Bis(nitrato)dioxouranium.

Potential Exposures: The primary use of natural uranium is in nuclear energy as a fuel for nuclear reactors, in plutonium production, and as feeds for gaseous diffusion plants. It is also a source of radium salts. Uranium compounds are used in staining glass, glazing ceramics, and enamelling, in photographic processes, for alloying steels, and as a catalyst for chemical reactions, radiation shielding, and aircraft counterweights.

Uranium presents both chemical and radiation hazards, and exposures may occur during mining, processing of the ore, and production of uranium metal.

Incompatibilities:

Uranium	CO_2, CCl_4, HNO_3
Uranium hydride	Strong oxidizers, H_2O, halogenated hydrocarbons
Uranyl nitrates	Combustibles
Uranium hexafluoride	Water

Permissible Exposure Limits in Air: The Federal standards are: uranium (soluble compounds) 0.05 mg/m^3; uranium (insoluble compounds) 0.25 mg/m^3 according to NIOSH. ACGIH, in 1983/84, cites 0.2 mg/m^3 for both soluble and insoluble natural U compounds and an STEL value of 0.6 mg/m^3 for both soluble and insoluble compounds. The IDLH levels are 20 mg/m^3 for soluble compounds, and 30 mg/m^3 for insoluble compounds.

909

Determination in Air: Collection on a filter and fluorometric analysis (A-30).

Permissible Concentration in Water: No criteria set, but EPA (A-37) has suggested a permissible ambient goal of 3 μg/ℓ based on health effects.

Routes of Entry: Inhalation of fume, dust, or gas, ingestion, skin and eye contact. The following uranium salts are reported to be capable of penetrating intact skin: uranyl nitrate, $UO_2(NO_3)_2 \cdot 6H_2O$; uranyl fluoride, UO_2F_2; uranium pentachloride, UCl_5; uranium trioxide (uranyl oxide), UO_3; sodium diuranate [sodium uranate(VI), $Na_2U_2O_7 \cdot H_2O$]; ammonium diuranate [ammonium uranate(VI), $(NH_4)_2U_2O_7$]; uranium hexafluoride, UF_6.

Harmful Effects and Symptoms: *Local* — No toxic effects have been reported, but prolonged contact with skin should be avoided to prevent radiation injury.

Systemic — Uranium and its compounds are highly toxic substances. The compounds which are soluble in body fluids possess the highest toxicity. Poisoning has generally occurred as a result of accidents. Acute chemical toxicity produces damage primarily to the kidneys. Kidney changes precede in time and degree the effects on the liver. Chronic poisoning with prolonged exposure gives chest findings of pneumoconiosis, pronounced blood changes, and generalized injury.

It is difficult to separate the toxic chemical effects of uranium and its compounds from their radiation effects. The chronic radiation effects are similar to those produced by ionizing radiation. Reports now confirm that carcinogenicity is related to dose and exposure time. Cancer of the lung, osteosarcoma, and lymphoma have all been reported.

For soluble uranium compounds: lacrimation, conjunctivitis; short breath, coughing, chest rales; nausea, vomiting; skin burns; casts in urine, albuminuria, high blood urea nitrogen; lymphatic cancer.

For insoluble uranium compounds: dermatitis; cancer of lymphatic and blood-forming tissues.

Points of Attack: For soluble uranium compounds: respiratory system, blood, liver, lymphatics, kidneys, skin, bone marrow. For insoluble uranium compounds: skin, bone marrow, lymphatics.

Medical Surveillance: Special attention should be given to the blood, lung, kidney, and liver in preemployment physical examinations. In periodic examinations, tests for blood changes, changes in chest x-rays, or for renal injury and liver damage are advisable. Uranium excretion in the urine has been used as an index of exposure. Whole body counting may also be useful.

First Aid: If this chemical gets into the eyes, irrigate immediately. If this chemical contacts the skin, flush with water immediately in the case of soluble U compounds; soap wash promptly, in the case of insoluble compounds. If a person breathes in large amounts of these compounds, move the exposed person to fresh air at once and perform artificial respiration. When these compounds have been swallowed, get medical attention. Give large quantities of water and induce vomiting. Do not make an unconscious person vomit.

Personal Protective Methods: *Soluble U Compounds, Especially UF₆* — Wear appropriate clothing to prevent any possibility of skin contact with UF_6. Wear eye protection to prevent any possibility of eye contact. Employees should wash immediately when skin is wet or contaminated with UF_6 and daily at the end of each workshift. Work clothing should be changed daily if it is possible that clothing is contaminated with UF_6. Remove nonimpervious clothing immediately if wet or contaminated with UF_6. Provide emergency showers and eyewash if UF_6 is involved.

Insoluble U Compounds — Wear appropriate clothing to prevent repeated or prolonged skin contact. Wear eye protection to prevent any possibility of eye contact. Employees should wash promptly when skin is wet or contaminated. Work clothing should be changed daily if it is possible that clothing is contaminated. Remove nonimpervious clothing promptly if wet or contaminated. Provide emergency eyewash.

Respirator Selection:

Soluble uranium compounds
    2.5 mg/m³: HiEPF/SA/SCBA
  For halides: CCRAGFHiE/GMAGHiEP
    20 mg/m³: PAPHiE (not for halides)/SAF:PD,PP,CF
    Escape: HiEPAG (halides)/GMAGP (halides only)/SCBA
Insoluble uranium compounds
    2.5 mg/m³: FuHiEPFS/SA/SCBA
  12.5 mg/m³: HiEPF/SAF/SCBAF
    30 mg/m³: PAPHiEF/SAF:PD,PP,CF
    Escape: HiEP/SCBA

Disposal Method Suggested: Recovery for reprocessing is the preferred method. Processes are available for uranium recovery from process wastewaters and process scrap (A-57).

References

(1) U.S. Environmental Protection Agency, *Toxicology of Metals, Vol. II: Uranium,* Report EPA-600/1-77-022, pp 454-472, Research Triangle Park, NC (May 1977).

(2) See Reference (A-61).

(3) Sax, N.I., Ed., *Dangerous Properties of Industrial Materials Report, 2,* No. 2, 78-79, New York, Van Nostrand Reinhold Co. (1982). (Uranyl Acetate).

(4) Sax, N.I., Ed., *Dangerous Properties of Industrial Materials Report, 4,* No. 1, 99–102, New York, Van Nostrand Reinhold Co. (Jan./Feb. 1984). (Uranyl nitrate).

(5) Parmeggiani, L., Ed., *Encyclopedia of Occupational Health & Safety,* Third Edition, Vol. 2, pp 2237–39, Geneva, International Labour Office (1983).

URETHANE

● Carcinogen (Animal Positive, IARC) (2)

Description: $H_2NCOOC_2H_5$ is a colorless, odorless crystalline powder which melts at 48° to 50°C.

Code Numbers: CAS 51-79-6   RTECS FA8400000

DOT Designation: —

Synonyms: Ethyl carbamate.

Potential Exposure: It is used as a chemical intermediate in the preparation of amino resins. It may be reacted with formaldehyde to give crosslinking agents which impart wash-and-wear properties to fabrics. It has also been used as a solubilizer and cosolvent in the manufacture of pesticides, fumigants and cosmetics. It was formerly used in the treatment of leukemia. It occurs when diethylpyrocarbonate, a preservative used in wines, fruit juices and soft drinks, is added to aqueous solutions.

Permissible Exposure Limits in Air: No standards set.

Permissible Concentration in Water: No criteria set.

Routes of Entry: Inhalation, ingestion, skin and eye contact.

Harmful Effects and Symptoms: Nausea, vomiting, drowsiness. Gastro-enteric hemorrhages have occurred after prolonged administration. Large doses (more than 3 g/day orally) have made debilitated patients more prone to hepatitis or fatal hepatic necrosis.

The primary effect in animals of an acute exposure to urethane is bone marrow depression. Focal degeneration of the brain, central nervous system depression and vomiting are sometimes produced.

Urethane is a proven carcinogen in experimental animals (A-62) (A-64).

Disposal Method Suggested: Controlled incineration (incinerator is equipped with a scrubber or thermal unit to reduce NO_x emissions).

References

(1) U.S. Environmental Protection Agency, *Chemical Hazard Information Profile: Urethane,* Washington, DC (1979).
(2) International Agency for Research on Cancer, *IARC Monographs on the Carcinogenic Risks of Chemicals to Humans,* Lyon, France, 7, 111 (1974).

V

VALERALDEHYDE

Description: $CH_3(CH_2)_3CHO$ is a colorless liquid boiling at 102° to 103°C.

Code Numbers: CAS 110-62-3 RTECS YV3600000 UN 2058

DOT Designation: Flammable liquid.

Synonyms: Amyl aldehyde, butyl formal, pentanal.

Potential Exposures: Valeraldehyde is used in food flavorings. It is also used in the acceleration of rubber vulcanization.

Permissible Exposure Limits in Air: There are no Federal standards, but ACGIH (1983/84) has adopted a TWA value of 50 ppm (175 mg/m^3). There is no tentative STEL value set.

Permissible Concentration in Water: No criteria set.

Harmful Effects and Symptom: Valeraldehyde is relatively nontoxic systemically (A-34). It exhibits the narcotism and irritation common to all aldehydes, but in mild form (A-38).

Personal Protective Methods: Wear rubber gloves and full protective clothing (A-38).

Disposal Method Suggested: Incineration.

VANADIUM AND COMPOUNDS

- Hazardous substances (vanadium pentoxide, vanadyl sulfate) (EPA)
- Hazardous wastes (ammonium vanadate, vanadium pentoxide) (EPA)

Description: V, vanadium, is a light-grey or white, lustrous powder or fused hard lump insoluble in water. It is produced by roasting the ores, thermal decomposition of the iodide, or from petroleum residues, slags from ferrovanadium production, or soot from oil burning.

Vanadium pentoxide dust (V_2O_5) is a yellow-orange powder or dark-gray flakes and is odorless. Vanadium pentoxide fume (V_2O_5) is a finely-divided particulate dispersed in air.

Code Numbers:

Vanadium metal	CAS 7440-62-2	RTECS YW1355000	
V_2O_5 dust	CAS 1314-62-1	RTECS YW2450000	UN 2862
V_2O_5 fume	CAS - None	RTECS YW2460000	UN 2862

913

DOT Designation: Vanadium pentoxide: ORM-E.

Synonyms: Vanadium metal – none; vanadium pentoxide – vanadic acid anhydride, vanadia.

Potential Exposures: Most of the vanadium produced is used in ferrovanadium and of this the majority is used in high speed and other alloy steels with only small amounts in tool or structural steels. It is usually combined with chromium, nickel, manganese, boron, and tungsten in steel alloys.

Vanadium pentoxide (V_2O_5) is an industrial catalyst in oxidation reactions; is used in glass and ceramic glazes; is a steel additive; and is used in welding electrode coatings. Ammonium metavanadate (NH_4VO_3) is used as an industrial catalyst, a chemical reagent, a photographic developer; and in dyeing and printing. Other vanadium compounds are utilized as mordants in dyeing, in insecticides; as catalysts, and in metallurgy.

Since vanadium itself is considered nontoxic, there is little hazard associated with mining; however, exposure to the more toxic compounds, especially the oxides, can occur during smelting and refining. Exposure may also occur in conjunction with oil-fired furnace flues.

Incompatibilities: None hazardous.

Permissible Exposure Limits in Air: The Federal standards are: V_2O_5 dust, 0.5 mg/m^3; V_2O_5 fume, 0.1 mg/m^3. NIOSH has recommended a value of 0.05 mg/m^3 for both vanadium pentoxide dust and fume as a 15-minute ceiling value. ACGIH has recommended a TWA of 0.05 mg/m^3 for both dust and fume as of 1983/84, but no STEL value. The IDLH level is 70 mg/m^3 for both vanadium pentoxide dust and fume.

Determination in Air: Both vanadium pentoxide dust and fume may be collected on a filter, worked up with sodium hydroxide, and analyzed by atomic absorption. See NIOSH Methods, Set 1 for vanadium pentoxide dust and Set 2 for vanadium pentoxide fume. See also reference (A-10).

Permissible Concentration in Water: A limit of 0.1 mg/ℓ has been suggested in the U.S.S.R. as a maximum permissible limit for water basins. There is no U.S. standard for vanadium in drinking water, but EPA (A-37) has suggested a permissible ambient goal of 7 µg/ℓ based on health effects.

The lack of data on acute or chronic oral toxicity is not surprising because of the extremely low absorption of vanadium from the gastrointestinal tract. Inhaled vanadium can produce adverse health effects, but the available evidence does not indicate that vanadium in drinking water is a problem.

Determination in Water: With a vanadium detection limit of 40 µg/ℓ, conventional flame atomization lacks sensitivity for direct determination in most samples. Both vanadium(IV) and vanadium(V) are extracted from an aqueous solution at a pH of 3.8 with a cupferron-methylisobutylketone system. With direct sampling, the graphite furnace can be used to increase sample atomization with a detection limit of 5 µg/ℓ.

Routes of Entry: Inhalation of dust or fume, ingestion, skin and eye contact.

Harmful Effects and Symptoms: *Local* – Vanadium compounds, especially vanadium pentoxide, are irritants to the eyes and skin. The initial eye symptoms are profuse lacrimation and a burning sensation of the conjunctiva. Skin lesions are of the eczematous type which itch intensely. In some cases, there may be generalized urticaria. Workers may also exhibit greenish discoloration of the tongue. This same discoloration may be detectable on the butts of cigarettes smoked by vanadium workers.

Systemic — Vanadium compounds are irritants to the respiratory tract. Entrance to the body is through inhalation of dusts or fumes. Serous or hemorrhagic rhinitis, sore throat, cough, tracheitis, bronchitis, expectoration, and chest pain, may result after even a brief exposure. More serious exposure may result in pulmonary edema and pneumonia which may be fatal. Individuals who recover may experience persistent bronchitis resembling asthma, and bouts of dyspnea; however, no chronic lung lesions have been described.

The results of experimental biochemical studies show that vanadium compounds inhibit cholesterol synthesis and the activity of the enzyme cholinesterase. A variety of other biochemical effects have been noted experimentally, but these have not been reported in relation to occupational exposures. Slightly lower cholesterol levels in blood were noted in one report, but this seems of doubtful significance.

Points of Attack: For vanadium pentoxide dust and fume: respiratory system, lungs, skin, eyes.

Medical Surveillance: Preemployment and periodic physical examinations should emphasize effects on the eyes, skin, and lungs. Urinary vanadium excretion may be useful as an index of exposure.

First Aid: V_2O_5 *Dust* — If this chemical gets into the eyes, irrigate immediately. If this chemical contacts the skin, wash with soap promptly. If a person breathes in large amounts of this chemical, move the exposed person to fresh air at once and perform artificial respiration. When this chemical has been swallowed, get medical attention and induce vomiting. Do not make an unconscious person vomit.

V_2O_5 *Fume* — If a person breathes in large amounts of this chemical, move the exposed person to fresh air at once and perform artificial respiration.

Personal Protective Methods: V_2O_5 *Dust* — Wear appropriate clothing to prevent repeated or prolonged skin contact. Wear eye protection to prevent any reasonable probability of eye contact. Employees should wash promptly when skin is wet or contaminated. Remove nonimpervious clothing promptly if wet or contaminated.

V_2O_5 *Fume* — No special techniques are indicated by NIOSH (A-4) except respirator protection.

Respirator Selection:

V_2O_5 Dust
   25 mg/m³: HiEPF/SAF/SCBAF
   70 mg/m³: PAPHiEF/SAF:PD,PP,CF
   Escape: HiEP/SCBA

V_2O_5 Fume
   5 mg/m³: HiEPF/SAF/SCBAF
   70 mg/m³: PAPHiEFF/SAF:PD,PP,CF
   Escape: HiEP/SCBA

Disposal Method Suggested: Vanadium pentoxide may be disposed of in a sanitary landfill (A-31).

References

(1) National Institute for Occupational Safety and Health, *Criteria for a Recommended Standard: Occupational Exposure to Vanadium,* NIOSH Document No. 77-222 (1977).

(2) National Academy of Sciences, *Medical and Biologic Effects of Environmental Pollutants: Vanadium,* Washington, DC (1974).

(3) See Reference (A-61). (Vanadium Pentoxide Dust and Fume).

(4) Sax, N.I., Ed., *Dangerous Properties of Industrial Materials Report, 2,* No. 2, 81-84, New York, Van Nostrand Reinhold Co. (1982). (Vanadium Oxychloride and Pentoxide).

(5) Sax, N.I., Ed., *Dangerous Properties of Industrial Materials Report, 2,* No. 1, 127-128, New York, Van Nostrand Reinhold Co. (1982).

(6) See Reference (A-60) for citations to Vanadium Pentoxide, Vanadium Sulfide, Vanadium Tetrachloride, and Vanadium Trichloride.

(7) Parmeggiani, L., Ed., *Encyclopedia of Occupational Health & Safety,* Third Edition, Vol. 2, pp 2240-41, Geneva, International Labour Office (1983).

VERMICULITE

Description: $3MgO \cdot Fe_2Al_2O_3 \cdot 3SiO_2$ is a platelike crystalline material which melts at about 1500°C.

Code Numbers: CAS 1318-00-9

DOT Designation: —

Synonyms: Jeffersite; Magnesium aluminum iron silicate.

Potential Exposure: When the naturally occurring vermiculite mineral is heated to 870° to 1095°C, it is exfoliated whereupon it expands to 6 to 20 times its original size. The expanded material is used as an insulating material, in lightweight concrete aggregates, as an additive in fertilizers, as a seed bed for plants, as an insecticide carrier, an absorbent in the removal of strontium—90 from milk, and in the removal of phosphates from wastewater, as a filler for paint, rubber, and plastics, as an animal feed additive, in litter for hatcheries, and as a packing material.

Permissible Exposure Limits in Air: No standards set.

Permissible Concentration in Water: No criteria set.

Harmful Effects and Symptoms: Bloody pleural effusions have occurred in workers in plants where vermiculite is exfoliated for use as a carrier for fertilizer and pesticide chemicals.

Disposal Method Suggested: Vermiculite used as an absorbent may be commercially regenerated. When not recycled, vermiculite is disposed of via landfill. Any fine dust generated during the derivation process normally is washed down and enters the wastewater stream. Being insoluble, this would be filtered or settle out in treatment.

References

(1) U.S. Environmental Protection Agency, *Chemical Hazard Information Profile: Vermiculite,* Washington, DC (1979).

VINYL ACETATE

● Hazardous substance (EPA)

Description: $CH_3COOCH = CH_2$ is a colorless, flammable liquid which boils at 73°C.

Code Numbers: CAS 108-05-4   RTECS AK0875000   UN 1301

DOT Designation: Flammable liquid.

Synonyms: Acetic acid vinyl ester, vinyl acetate monomer, ethenyl ethanoate.

Potential Exposure: Vinyl acetate is used primarily in polymerization processes to produce polyvinyl acetate, polyvinyl alcohol and vinyl acetate copolymer. The polymers, usually made as emulsions, suspensions, solutions or resins are used to prepare adhesives, paints, paper coatings and textile finishes. Low molecular weight vinyl acetate is used as a chewing gum base. NIOSH estimates that 70,000 workers annually are exposed to vinyl acetate in the U.S.

Permissible Exposure Limits in Air: There is no Federal standard for vinyl acetate. NIOSH (1) recommends a ceiling value of 4 ppm (15 mg/m^3) measured over a 15-minute period. The ACGIH (1983/84) has set a TWA value of 10 ppm (30 mg/m^3) and an STEL value of 20 ppm (60 mg/m^3). There is no IDLH level set.

Determination in Air: Adsorption on Chromosorb 107, thermal desorption, analysis by gas chromatograph with a flame ionization detector (1). See also reference (A-10).

Permissible Concentration in Water: No criteria set.

Routes of Entry: Inhalation, ingestion, skin and eye contact.

Harmful Effects and Symptoms: Vinyl acetate vapor at concentrations below 250 mg/m.^{3} is a primary irritant to the upper respiratory tract and eyes and the liquid may irritate the skin to the point of vesiculation. The irritations reported have all been reversible, however, and there are no known residual systemic effects (1).

Points of Attack: Eyes, skin, upper respiratory system.

Medical Surveillance: Particular attention should be paid to upper respiratory tract, eyes and skin in preplacement and periodic physical examinations.

First Aid: Flush skin or eyes with water after contact.

Personal Protective Measures: For eye protection: chemical safety goggles or face shields with goggles. For skin protection: gloves, aprons, suits and boots to prevent contact with liquid vinyl acetate.

Respirator Selection: Priority should be given to engineering controls to reduce exposure, but details of respirator selection are given by NIOSH (1).

Disposal Method Suggested: Incineration.

References

(1) National Institute for Occupational Safety and Health, *Criteria for a Recommended Standard: Occupational Exposure to Vinyl Acetate,* NIOSH Publication No. 78-205, Washington, DC (1978).

(2) See Reference (A-60).

(3) U.S. Environmental Protection Agency, *Chemical Hazard Information Profile Draft Report: Vinyl Acetate,* Wash., D.C. (April 23, 1984).

VINYL BROMIDE

● Carcinogen (Animal Positive) (3)

Description: CH_2=CHBr is a colorless gas which boils at about 16°C.

Code Numbers: CAS 593-60-2   RTECS KU8400000   UN 1085

DOT Designation: Flammable gas.

Synonyms: Bromoethylene, bromoethene.

Potential Exposure: It is used as an intermediate in organic synthesis and for the preparation of plastics by polymerization and copolymerization. The major use of vinyl bromide is in the production of flame-retardant synthetic fibers. An example of this is SEF, a modacrylic fiber produced by Monsanto. The formula for SEF is 79 to 81% acrylonitrile, 9% vinyl bromide, 8% vinylidene chloride, and 2 to 4% other. SEF is used primarily in children's sleepwear. It is produced in a batch polymerization operation with a suspension polymerization medium and a wet spinning process. This method of production would probably preclude residual vinyl bromide monomer in the finished product. NIOSH has estimated worker exposure at 26,000 annually.

Permissible Exposure Limits in Air: There is no Federal standard. The ACGIH as of 1983/84 has set a TWA value of 5 ppm (20 mg/m^3) with the notation that vinyl bromide is "an industrial substance suspect of carcinogenic potential in man." NIOSH (1978) has recommended an exposure standard of 1 ppm. The Dutch chemical industry (A-60) has set a much higher maximum allowable concentration of 250 ppm (1,100 mg/m^3) which is a bit surprising.

Permissible Concentration in Water: No criteria set.

Harmful Effects and Symptoms: Vinyl bromide is considered a moderately toxic substance. In high concentrations, vinyl bromide may produce dizziness, disorientation, and sleepiness in humans. As noted above, ACGIH has designated vinyl bromide as "suspect of carcinogenic potential for man."

Points of Attack: In animals: liver, lung, breast, zymbal gland, lymphatic system (2).

References

(1) U.S. Environmental Protection Agency, *Chemical Hazard Information Profile: Vinyl Bromide,* Washington, DC (January 30, 1978).
(2) National Institute for Occupational Safety and Health, *Vinyl Halides: Carcinogenicity,* Current Intelligence Bulletin No. 28, Washington, DC (September 21, 1978).
(3) Huntingdon Research Center, *HRC Project 7511-253; 18-Month Sacrifice Pathology Report (Vinyl Bromide),* New York (June 26, 1978).
(4) Sax, N.I., Ed., *Dangerous Properties of Industrial Materials Report, 2,* No. 2, 87-88, New York, Van Nostrand Reinhold Co. (1982).

VINYL CHLORIDE

- Carcinogen (Human Positive, IARC)(5)
- Hazardous waste (EPA)
- Priority toxic pollutant (EPA)

Description: $CH_2{=}CHCl$, vinyl chloride, is a flammable gas at room temperature (boils at $-14°C$), and is usually encountered as a cooled liquid. The colorless liquid forms a vapor which has a pleasant, ethereal odor.

Code Numbers: CAS 75-01-4   RTECS KU9625000   UN 1086

DOT Designation: Flammable gas.

Synonyms: Chloroethylene, chloroethene, monochloroethylene.

Potential Exposures: Vinyl chloride is used as a vinyl monomer in the manufacture of polyvinyl chloride and other resins. It is also used as a chemical intermediate and as a solvent (1).

The hazard of vinyl chloride was originally believed to primarily concern workers employed in the conversion of VCM to PVC who may receive a particularly high exposure of VCM in certain operations (e.g., cleaning of polymerization kettles) or a long-term exposure to relatively low concentrations of VCM in the air at different factory sites. Much larger populations are now believed to be potentially at risk, including: producers of VCM; people living in close proximity to VCM or PVC producing industries; users of VCM as propellant in aerosol sprays; persons in contact with resins made from VCM; consumers of food and beverage products containing leachable amounts of unreacted VCM from PVC packaged materials; and ingestion of water containing unreacted VCM leached from PVC pipes.

NIOSH (2) estimates definite worker exposure to vinyl chloride at 27,000 and probable worker exposure at 2,200,000. About 1,500 workers were employed in monomer synthesis and an additional 5,000 in polymerization operations. As many as 350,000 workers were estimated to be associated with fabrication plants. By 1976 it was estimated that nearly one million persons were associated with manufacturing goods derived from PVC.

Human exposure to VC occurs primarily through inhalation and less frequently through skin absorption. More than 3.5 million workers are potentially exposed to this chemical. In addition, 4.6 million people are potentially exposed who live within 5 miles of industrial sites at which environmental emissions could occur. Air emissions near production facilities contained 3.1 to 12.5 ppb of VC. Very low levels of exposure may occur from leaching into food and beverages (0.05 to 25 ng/kg) or medical products from unreacted VC remaining in polyvinyl chloride.

Permissible Exposure Limits in Air: The Federal standard for exposure to vinyl chloride sets a limit of 1 ppm (2.6 mg/m^3) over an 8-hour period, and a ceiling of 5 ppm averaged over a period not exceeding 15 minutes. ACGIH (1983/84) lists vinyl chloride as a human carcinogen, and gives a proposed TWA value of 5 ppm (10 mg/m^3) but no STEL value.

Determination in Air: Adsorption on activated carbon, desorption with carbon bisulfide, analysis by gas chromatograph with flame ionization detector (A-10). See also reference (7).

Permissible Concentration in Water: No criteria have been determined for the protection of freshwater or saltwater aquatic life due to insufficient data. For the protection of human health: preferably zero. An additional lifetime cancer risk of 1 in 100,000 is posed by a concentration of 20 μg/ℓ.

Determination in Water: Inert gas purge followed by gas chromatography with halide specific detection (EPA Method 601) or gas chromatography plus mass spectrometry (EPA Method 624).

Routes of Entry: Vinyl chloride gas is absorbed by inhalation, skin absorption has been suggested, but experimental evidence is presently lacking.

Harmful Effects and Symptoms: *Local* — Vinyl chloride is a skin irritant, and contact with the liquid may cause frostbite upon evaporation. The eyes may be immediately and severely irritated.

Systemic — Vinyl chloride depresses the central nervous system causing symptoms which resemble mild alcohol intoxication. Lightheadedness, some nausea, and dulling of visual and auditory responses may develop in acute exposures. Death from severe vinyl chloride exposure has been reported.

Chronic exposure of workers involved in reactor vessel entry and hand cleaning may result in the triad of acroosteolysis, Raynaud's phenomenon, and sclerodermatous skin changes. Chronic exposure may also cause hepatic damage.

Vinyl chloride is regarded as a human carcinogen (2) (9) and a cause of angiosarcoma of the liver. Excess cancer of the lung and lymphatic and nervous systems has also been reported. Experimental evidence of tumor induction in a variety of organs, including liver, lung, brain and kidneys, as well as nonmalignant alterations such as fibrosis and connective tissue deterioration indicate the multisystem oncogenic and toxicologic effects of vinyl chloride.

Points of Attack: Liver, brain and hemato-lymphopoietic system.

Medical Surveillance: Preplacement and periodic examinations should emphasize liver function and palpation. Liver scans and grey-scale ultrasonography have been useful in detecting liver tumors. Medical histories should include alcoholic intake, past hepatitis, exposure to hepatotoxic agents, drugs and chemicals, past blood transfusions, past hospitalizations. Radiographic examinations of the hands may be helpful if acroosteolysis is suspected. Longterm follow-up of exposed persons is essential as in the case of other carcinogens.

First Aid: Irrigate eyes with running water. Wash contaminated areas of body with soap and water.

Personal Protective Methods: Where vinyl chloride levels cannot meet the standard, workers should be required to wear respiratory protection, either airsupplied respirator or, if the level does not exceed 25 ppm, a chemical-cartridge or cannister-type gas mask. In hazard areas, proper protective clothing to prevent skin contact with the vinyl chloride or polyvinyl chloride residue should be worn.

Respirator Selection: Use of self-contained breathing apparatus is recommended (A-38).

Disposal Method Suggested: Incineration, preferably after mixing with another combustible fuel. Care must be exercised to assure complete combustion to prevent the formation of phosgene. An acid scrubber is necessary to remove the halo acids produced. A variety of techniques have been described for vinyl chloride recovery from PVC latexes (A-58)(6).

References

(1) U.S. Environmental Protection Agency, *Scientific and Technical Assessment Report on Vinyl Chloride and Polyvinyl Chloride,* Office of Research and Development, Washington, DC (June 1975).

(2) National Institute for Occupational Safety and Health, *Vinyl Halides—Carcinogenicity,* Current Intelligence Bulletin No. 28, Washington, DC (September 21, 1978).

(3) U.S. Environmental Protection Agency, *Vinyl Chloride: Ambient Water Quality Criteria,* Washington, DC (1980).

(4) U.S. Environmental Protection Agency, *Chloroethene,* Health and Environmental Effects Profile No. 45, Office of Solid Waste, Washington, DC (April 30, 1980).

(5) International Agency for Research on Cancer, *IARC Monographs on the Carcinogenic Risks of Chemicals to Humans, 7,* 291 (1974) and 19, 377 (1979), Lyon, France.

(6) Sittig, M., *Vinyl Chloride and PVC Manufacture: Process and Environmental Aspects,* Noyes Data Corp., Park Ridge, NJ (1978).

(7) U.S. Environmental Protection Agency, "National Emission Standards for Hazardous Air Pollutants," *Federal Register, 45,* No. 224, 76346-54 (November 18, 1980).

(8) Sax, N.I., Ed., *Dangerous Properties of Industrial Materials Report, 1,* No. 3, 85-87, New York, Van Nostrand Reinhold Co. (1981).

(9) See Reference (A-62). Also see Reference (A-64).

(10) Parmeggiani, L., Ed., *Encyclopedia of Occupational Health & Safety,* Third Edition, Vol. 2, pp 2256–60, Geneva, International Labour Office (1983).

(11) United Nations Environment Programme, *IRPTC Legal File 1983,* Vol. I, pp VII/350–54, Geneva, Switzerland, International Register of Potentially Toxic Chemicals (1984).

4-VINYL-1-CYCLOHEXENE

Description: C_8H_{12}, $C_6H_9CH=CH_2$ is a flammable liquid with a boiling point of 128°C.

Code Numbers: CAS 100-40-3   RTECS GW6650000

DOT Designation: —

Synonyms: Cyclohexenylethylene; 1-Ethenylcyclohexene; 1,2,3,4-tetrahydrostyrene.

Potential Exposure: 4-Vinyl-1-cyclohexene is used as an intermediate for the production of vinylcyclohexene dioxide, which is used as a reactive diluent in epoxy resins. Previous uses of 4-vinyl-1-cyclohexene include comonomer in the polymerization of other monomers and for halogenation to polyhalogenated derivatives which are used as flame retardants.

Permissible Exposure Limits in Air: No standards set.

Permissible Concentration in Water: No criteria set.

Harmful Effects and Symptoms: Workers exposed to 4-vinyl-1-cyclohexene experienced keratitis, rhinitis, headache, hypotonia, leucopenia, neutrophilia, lymphocytosis, and impairment of pigment and carbohydrate metabolism (2).

Disposal Method Suggested: Incineration.

References

(1) U.S. Environmental Protection Agency, *Chemical Hazard Information Profile: 4-Vinyl-1-Cyclohexene,* Washington, DC (1979).

(2) International Agency for Research on Cancer, *IARC Monographs on the Carcinogenic Risks of Chemicals to Humans,* Lyon, France, 11, 277 (1976).

VINYL CYCLOHEXENE DIOXIDE

● Carcinogen (Animal Positive, IARC)(2)

Description: $C_8H_{12}O_2$ with the structural formula:

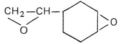

is a colorless liquid which boils at 108° to 109°C.

Code Numbers: CAS 106-87-6   RTECS RN8640000

DOT Designation: —

Synonyms: 3-(Epoxyethyl)-7-oxabicyclo[4.1.0]heptane,   vinyl cyclohexene diepoxide, 1,2-epoxy-4-(epoxyethyl)cyclohexane.

Potential Exposures: This material is used as a monomer in the production of epoxy resins for coatings and adhesives. Hence, those engaged in the production or use of this compound are at risk.

Permissible Exposure Limits in Air: There is no Federal standard. ACGIH (1983/84) has set a TWA value of 10 ppm (60 mg/m^3) but no STEL value. ACGIH goes on to designate vinyl cyclohexene dioxide as an "industrial substance suspect of carcinogenic potential for man."

Permissible Concentration in Water: No criteria set.

Harmful Effects and Symptoms: Vinyl cyclohexene dioxide is irritating to the eyes and skin of rabbits. It can cause acute respiratory tract irritation and congestion of the lungs. In addition to its irritant properties, vinyl cyclohexene dioxide has been associated with testicular atrophy, leucopenia, and necrosis of the thymus. Several studies have demonstrated carcinogenicity in rodents, but there are no studies of the effects of this compound on humans.

Disposal Method Suggested: Concentrated waste containing no peroxides: discharge liquid at a controlled rate near a pilot flame. Concentrated waste containing peroxides: perforation of a container of the waste from a safe distance followed by open burning.

References

(1) National Institute for Occupational Safety and Health, *Information Profiles on Potential Occupational Hazards: Vinyl Cyclohexene Dioxide*, pp 54-57, Report PB-276,-678, Rockville, MD (October 1977).

(2) International Agency for Research on Cancer, *IARC Monographs on the Carcinogenic Risks of Chemicals to Humans*, 11, 141, Lyon, France (1976).

(3) Parmeggiani, L., Ed., *Encyclopedia of Occupational Health & Safety*, Third Edition, Vol. 2, p 2262, Geneva, International Labour Office (1983).

VINYL ETHER

Description: $(CH_2=CH)_2O$ is a volatile liquid which boils at 28.4°C.

Code Numbers: CAS 109-93-3   RTECS XZ6700000

DOT Designation: Flammable liquid.

Synonyms: Ethenyloxy ethene; divinyl ether; divinyl oxide.

Potential Exposure: In use as an inhalation anesthetic; in formulation of copolymers with vinyl chloride.

Permissible Exposure Limits in Air: No standards set.

Permissible Concentration in Water: No criteria set.

Harmful Effects and Symptoms: Is a mild chronic irritant. Prolonged exposure is said to cause liver damage.

Personal Protective Methods: Wear skin protection.

Respirator Selection: Use self-contained breathing apparatus.

Disposal Method Suggested: Allow to evaporate or incinerate. Beware of explosive peroxides in old containers.

References

(1) Sax, N.I., Ed., *Dangerous Properties of Industrial Materials Report, 1*, No. 7, 78-79, New York, Van Nostrand Reinhold Co. (1981).

VINYL FLUORIDE

Description: $CHF{=}CH_2$ is a colorless gas which boils at $-72°C$.

Code Numbers: RTECS YZ7351000   UN 1860

DOT Designation: Flammable gas.

Synonyms: Fluoroethylene, fluoroethene.

Potential Exposures: Vinyl fluoride's primary use is as a chemical intermediate to make polyvinyl fluoride (Tedlar®) film. Polyvinyl fluoride film is
characterized by superior resistance to weather, high strength, and a high dielectric constant. It is used as a film laminate for building materials and in
packaging electrical equipment. Polyvinyl fluoride film poses a hazard, so it
is not recommended for food packaging. Polyvinyl fluoride evolves toxic fumes
upon heating.

Permissible Exposure Limits in Air: There is no Federal standard. NIOSH
has recommended a TWA value of 1 ppm with a 5 ppm ceiling for 15-minute
exposure. ACGIH, as of 1983/84, has proposed no values for vinyl fluoride.

Permissible Concentration in Water: No criteria set.

Harmful Effects and Symptoms: Vinyl fluoride is considered nontoxic;
however, when it reaches high concentrations, the oxygen content of the air
can reach a critically low level. The approximate lethal concentration of vinyl
fluoride to most animals has been estimated at 800,000 ppm.

References

(1)   U.S. Environmental Protection Agency, *Chemical Hazard Information Profile: Vinyl
Fluoride,* Washington, DC (January 30, 1978).

(2)   See Reference (A-60).

VINYLIDENE BROMIDE

Description: $CH_2{=}CBr_2$ is a liquid which boils at $92°C$.

DOT Designation: —

Synonyms: 1,1-Dibromoethylene.

Potential Exposure: There are no known commercial uses of this compound
as of 1980.

Permissible Exposure Limits in Air: No standards set.

Permissible Concentration in Water: No criteria set.

Harmful Effects and Symptoms: Vinylidene bromide has been tested in fed
and fasted rats for short-term inhalation toxicity by measuring survival 24 hours
after a 4-hour exposure period. The vapors were respiratory tract irritants; they
presumably contained HBr gas, a known irritant which the researchers suspect
may have caused the observed injury and death. The no-effect concentration was
set at 46 ppm, since neither the fed nor the fasted rats died at this exposure
level. At 98 ppm, two of five fasted rats died; none of five fed rats dies. At 471
ppm, three of five fed and five of five fasted rats were killed. The probable cause
of death was acute pulmonary edema and hemorrhage.

References

(1)   U.S. Environmental Protection Agency, *Chemical Hazard Information Profile: Vinylidene Bromide,* Washington, DC (January 30, 1978).

VINYLIDENE CHLORIDE

- Carcinogen (EPA, CAG)(A-40)
- Hazardous substance (EPA)
- Hazardous waste (EPA)
- Priority toxic pollutant (EPA)

Description: Vinylidene chloride is a volatile liquid, boiling at 31.7°C at 760 mm, with a mild, sweet odor resembling that of chloroform.

Code Numbers: CAS 75-35-4   RTECS KV9275000   UN 1303

DOT Designation: Flammable liquid.

Synonyms: 1,1-Dichloroethylene, DCE, VDC.

Potential Exposures: Vinylidene chloride is used in the manufacture of 1,1,1-trichloroethane (methyl chloroform). However, the manufacture of poly-vinylidene copolymers is the major use of VDC. The extruded films of the co-polymers are used in packaging and have excellent resistance to water vapor and most gases. The chief copolymer is Saran® (polyvinylidene chloride/vinyl chloride), a transparent film used for food packaging. The films shrink when exposed to higher than normal temperatures. This characteristic is advantageous in the heat-shrinking of overwraps on packaged goods and in the sealing of the wraps.

Applications of VDC latexes include mixing in cement to produce high-strength mortars and concretes, and as binders for paints and nonwoven fabrics providing both water resistance and nonflammability. VDC polymer lacquers are also used in coating films and paper. VDC is also used to produce fibers. Monofilaments, made by extruding the copolymer, are used in the textile indus-try as furniture and automobile upholstery, drapery fabric, outdoor furniture, venetian-blind tape, and filter cloths.

NIOSH estimates definite worker exposure to vinylidene chloride at 6,500 and probable worker exposure to vinylidene chloride at 58,000.

Permissible Exposure Limits in Air: There are no Federal standards. ACGIH has adopted a threshold limit value of 10 ppm (40 mg/m³) as of 1983/84 and an STEL of 20 ppm (80 mg/m³). Consideration is being given to lowering the TWA to 5 ppm (40 mg/m³). NIOSH has gone further and recommended an exposure standard of 1 ppm, the same as for vinyl chloride.

Determination in Air: Adsorption on charcoal, desorption with carbon bi-sulfide, analysis by gas chromatography (A-10).

Permissible Concentration in Water: To protect freshwater aquatic life: 11,600 µg/ℓ on an acute toxicity basis for dichloroethylenes as a class. To pro-tect saltwater aquatic life: 224,000 µg/ℓ on an acute basis for dichloroethylenes as a class; 1,700 µg/ℓ as a 24-hour average, never to exceed 3,900 µg/ℓ. To protect human health: preferably zero. An additional lifetime cancer risk of 1 in 100,000 is posed by a concentration of 0.33 µg/ℓ.

Determination in Water: Inert gas purge followed by gas chromatography with halide specific detection (EPA Method 601) or gas chromatography plus mass spectrometry (EPA Method 624).

Route of Entry: Inhalation of vapor.

Harmful Effects and Symptoms: Aspects of the reported carcinogenicity of vinylidene chloride appear conflicting and indicate sex, species, and strain specificity. Laboratory studies have demonstrated that exposure by inhalation to vinylidene chloride caused angiosarcoma of the liver and other cancers in animals. Angiosarcoma of the liver was induced in mice exposed to 55 ppm vinylidene chloride. At lower levels, exposure to vinylidene chloride (25 ppm) has induced adenocarcinomas of the kidney. The table below (1) presents a summary of tumors in animals exposed to these vinyl halides.

Species	Site	Tumor
Mouse	liver	angiosarcoma
	lung	bronchioalveolar adenoma
	kidney	adenocarcinoma
Rat	mesenteric lymph node	angiosarcoma
	breast	mammary tumor
	zymbal gland	carcinoma

Personal Protective Methods: The primary requirement for reduction of exposure to VDC would be to limit emissions through improved housekeeping procedures in the industry. Beyond that, the use of gloves and protective clothing is indicated and a fullface mask should be used in areas of excessive vapor concentrations.

Disposal Method Suggested: Incineration, preferably after mixing with another combustible fuel. Care must be exercised to assure complete combustion to prevent the formation of phosgene. An acid scrubber is necessary to remove the halo acids produced.

References

(1) The National Institute for Occupational Safety and Health, *Vinyl Halides—Carcinogenicity,* Current Intelligence Bulletin No. 28, Washington, DC (September 21, 1978).
(2) U.S. Environmental Protection Agency, *Dichloroethylenes: Ambient Water Quality Criteria,* Washington, DC (1980).
(3) U.S. Environmental Protection Agency, *Status Assessment of Toxic Chemicals: Vinylidene Chloride,* Report EPA-600/2-79-210a, Cincinnati, OH (December 1979).
(4) U.S. Environmental Protection Agency, *1,1-Dichloroethylene,* Health and Environmental Effects Profile No. 71, Office of Solid Waste, Washington, DC (April 30, 1980).
(5) U.S. Environmental Protection Agency, *Dichloroethylenes,* Health and Environmental Effects Profile No. 73, Office of Solid Waste, Washington, DC (April 30, 1980).
(6) Sax, N.I., Ed., *Dangerous Properties of Industrial Materials Report, 2,* No. 6, 92-94, New York, Van Nostrand Reinhold Co. (1982).
(7) United Nations Environment Programme, *IRPTC Legal File 1983,* Vol. II, pp VII/356-58, Geneva, Switzerland, International Register of Potentially Toxic Chemicals (1984).

VINYLIDENE FLUORIDE

Description: $CH_2=CF_2$ is a colorless gas with a faint ethereal odor.

Code Numbers: RTECS KW0560000 UN 1959

DOT Designation: —

Synonyms: 1,1-Difluoroethylene, 1,1-difluoroethene, Genetron® 1132A.

Potential Exposures: Vinylidene fluoride is used in the formulation of many polymers and copolymers such as chlorotrifluoroethylene-vinylidene fluoride (Kel F®), perfluoropropylene-vinylidene fluoride (Viton®, Fluorel®), polyvinylidene fluoride, and hexafluoropropylene-tetrafluoroethylene-vinylidene fluoride. It is also used as a chemical intermediate in organic synthesis. NIOSH has estimated 32,000 workers are exposed annually.

Permissible Exposure Limits in Air: There is no Federal standard. NIOSH has proposed a TWA of 1 ppm and a 5 ppm ceiling value for 15-minute exposure. ACGIH has proposed no values.

Permissible Concentration in Water: No criteria set.

Harmful Effects and Symptoms: Vinylidene fluoride is considered toxic by inhalation. The lowest lethal concentration is 128,000 ppm for a 4-hour exposure. Vinylidene fluoride has been reported to be nontoxic to rats at 800,000 ppm.

References

(1) U.S. Environmental Protection Agency, *Chemical Hazard Information Profile: Vinylidene Fluoride,* Washington, DC (January 30, 1978).

VINYL TOLUENE

Description: $CH_3C_6H_4CH=CH_2$ is a colorless liquid with a strong, disagreeable odor. It boils at $168°C$. It consists of mixed meta- and para-isomers.

Code Numbers: CAS 25013-15-4   RTECS WL5075000   UN 2618

DOT Designation: Flammable liquid.

Synonyms: Methylstyrene, tolylethylene.

Potential Exposure: Vinyl toluene is used as a solvent and an organic intermediate.

Incompatibilities: Oxidizing agents, catalysts for vinyl polymerization such as peroxides, strong acids, aluminum chloride.

Permissible Exposure Limits in Air: The Federal standard is 100 ppm (480 mg/m^3). ACGIH as of 1983/84 has proposed a TWA of 50 ppm (240 mg/m^3) and an STEL value of 100 ppm (485 mg/m^3). The IDLH level is 5,000 ppm.

Determination in Air: Adsorption on charcoal, workup with CS_2, analysis by gas chromatography. See NIOSH Methods, Set C. See also reference (A-10).

Permissible Concentration in Water: No criteria set.

Routes of Entry. Inhalation, eye and skin contact, ingestion.

Harmful Effects and Symptoms: Irritation of eyes, skin and upper respiratory system; drowsiness.

Points of Attack: Eyes, skin, respiratory system.

Medical Surveillance: Consider the points of attack in preplacement and periodic physical examinations.

First Aid: If this chemical gets into the eyes, irrigate immediately. If this chemical contacts the skin, flush with soap promptly. If a person breathes in large amounts of this chemical, move the exposed person to fresh air at once and perform artificial respiration. When this chemical has been swallowed, get medical attention. Do NOT induce vomiting.

Personal Protective Methods: Wear appropriate clothing to prevent repeated

or prolonged skin contact. Wear eye protection to prevent any reasonable probability of eye contact. Employees should wash promptly when skin is wet or contaminated. Remove nonimpervious clothing promptly if wet or contaminated.

Respirator Selection:

400 ppm:	CCROV/SA/SCBA
1,000 ppm:	CCROVF
5,000 ppm:	GMOV/SAF/SCBAF
Escape:	GMOV/SCBA

Disposal Method Suggested: Incineration.

References

(1) See Reference (A-61).
(2) See Reference (A-60).

VM & P NAPHTHA

See "Naphthas" for a notation on VM & P (Varnish Makers' & Painters') Naphtha.

W

WARFARIN

- Hazardous waste (EPA)

Description: $C_{19}H_{16}O_4$ with the structural formula

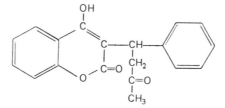

It is a colorless, odorless crystalline solid melting at 161°C.

Code Numbers: CAS 81-81-2 RTECS GN4550000 UN 2476

DOT Designation: —

Synonyms: Coumafene (in France), Zoocoumarin (in U.S.S.R.), 3-(α-acetonylbenzyl)-4-hydroxycoumarin, Warf 42.

Potential Exposures: Those engaged in the manufacture, formulation and application of this rodenticide or rat poison.

Incompatibilities: Strong oxidizers.

Permissible Exposure Limits in Air: The Federal standard and the 1983/84 ACGIH TWA value is 0.1 mg/m³. The STEL value is 0.3 mg/m³. The IDLH level is 200 mg/m³.

Permissible Concentration in Water: No criteria set.

Routes of Entry: Skin absorption, ingestion, inhalation, skin and eye contact.

Harmful Effects and Symptoms: Hematoma, back pain, epistaxis, bleeding lips, mucous membrane hemorrhage, abdominal pain, vomiting, fecal blood, petechial rash, abnormal hematology.

Points of Attack: Blood, cardiovascular system.

Medical Surveillance: Consider the points of attack in preplacement and periodic physical examinations.

First Aid: If this chemical gets into the eyes, irrigate immediately. If this chemical contacts the skin, wash with soap promptly. If a person breathes in

large amounts of this chemical, move the exposed person to fresh air at once and perform artificial respiration. When this chemical has been swallowed, get medical attention. Give large quantities of water and induce vomiting. Do not make an unconscious person vomit.

Personal Protective Methods: Wear appropriate clothing to prevent repeated or prolonged skin contact. Employees should wash promptly when skin is wet or contaminated. Work clothing should be changed daily if it is possible that clothing is contaminated. Remove nonimpervious clothing promptly if wet or contaminated.

Respirator Selection:
$0.5 \ mg/m^3$: DMXS
$1 \ mg/m^3$: DMXSQ/HiEP/SA/SCBA
$5 \ mg/m^3$: HiEPF/SAF/SCBAF
$100 \ mg/m^3$: PAPHiE/SA:PD,PP,CF
$200 \ mg/m^3$: SAF:PD,PP,CF
Escape: DXS/SCBA

Disposal Method Suggested: Incineration.

References

(1) See Reference (A-61).
(2) United Nations Environment Programme, *IRPTC Legal File 1983,* Vol. I, pp VII/201– 2, Geneva, Switzerland, International Register of Potentially Toxic Chemicals (1984).

WELDING FUMES

Description: The chief components of welding fumes are ordinarily oxides of iron, zinc, aluminum, or titanium. Other fumes are generally made up of slag particles containing iron, manganese and silicon. When stainless steels are arc welded, the fumes may contain chromium and nickel compounds. Some electrodes contain fluoride coatings or cores and the fumes can thus contain more fluorides than oxides. Fumes from welding cadmium- or lead-coated steel can contain those coating elements.

DOT Designation: –

Potential Exposures: Those involved in arc welding operations or oxy-gas welding of iron or aluminum.

Permissible Exposure Limits in Air: There are no Federal standards but ACGIH (1983/84) has set a TWA value of $5 \ mg/m^3$ for welding fumes as total particulates, not otherwise classified.

Determination in Air: Collection on a filter and gravimetric analysis (A-1).

Permissible Concentration in Water: No criteria set.

Routes of Entry: Inhalation.

Harmful Effects and Symptoms: The harmful effects and symptoms are those associated with the particular chemicals and elements involved (which see).

Personal Protective Methods: Most welding, even with primitive ventilation, does not produce exposures inside the welding helmet above $5 \ mg/m^3$. That which does, should be controlled.

WOOD DUST

Description: Wood dust (an ACGIH term) corresponds perhaps most closely to the industrial classification of wood flour. Wood flour is produced in 40 to 60 mesh and 70 to 80 mesh fractions, for example.

DOT Designation: —

Synonym: Sawdust.

Potential Exposures: Wood dust is produced incidentally in sawmill operations and in wood sawing and sanding operations in furniture manufacture. When produced as an industrial wood flour product, it is used as an absorbent for nitroglycerin in dynamite manufacture and as a filler in plastics, linoleum and paperboard.

Permissible Exposure Limits in Air: There are no Federal standards but ACGIH has proposed as of 1983/84 a TWA value of 1 mg/m^3 for hardwood (beech and oak, e.g.) dust, and a value of 5 mg/m^3 for softwood dust. No STEL value is given for hardwood dust; a value of 10 mg/m^3 is given for softwood dust.

Permissible Concentration in Water: No criteria set.

Route of Entry: Inhalation.

Harmful Effects and Symptoms: Dermatitis, respiratory disease and cancer are the three main categories. Dermatitis may be primary irritant dermatitis from mechanical or chemical irritation; sensitization of the skin may follow. Respiratory diseases include suberosis, maple bark disease and sequiosis. Cancers of the larynx, tonsils, tongue and lung have been reported as arising from inhalation of wood dust in the furniture industry (A-34).

X

XYLENES

- Hazardous substance (EPA)
- Hazardous waste (EPA)

Description: $C_6H_4(CH_3)_2$, xylene, exists in three isomeric forms, ortho-, meta- and para-xylene. Commercial xylene is a mixture of these three isomers and may also contain ethylbenzene as well as small amounts of toluene, trimethylbenzene, phenol, thiophene, pyridine, and other nonaromatic hydrocarbons. m-Xylene is predominant in commercial xylene and shares physical properties with o-xylene in that both are mobile, colorless, flammable liquids. p-Xylene, at a low temperature (13° to 14°C), forms colorless plates or prisms.

Code Numbers: CAS 1330-20-7 RTECS ZE2100000 UN 1307

DOT Designation: Flammable liquid.

Synonyms: Xylol, dimethylbenzene.

Potential Exposures: Xylene is used as a solvent; as a constituent of paint, lacquers, varnishes, inks, dyes, adhesives, cements, cleaning fluids and aviation fuels; and as a chemical feedstock for xylidines, benzoic acid, phthalic anhydride, isophthalic, and terephthalic acids, as well as their esters (which are specifically used in the manufacture of plastic materials and synthetic textile fabrics). Xylene is also used in the manufacture of quartz crystal oscillators, hydrogen peroxide, perfumes, insect repellants, epoxy resins, pharmaceuticals, and in the leather industry.

NIOSH estimates that 140,000 workers are potentially exposed to xylene.

Incompatibilities: Strong oxidizers.

Permissible Exposure Limits in Air: The Federal standard and the ACGIH 1983/84 TWA value is 100 ppm (435 mg/m³) for all isomers. NIOSH recommends adherence to the present Federal standard of 100 ppm as a time-weighted average for up to a 10-hour workday, 40-hour workweek. NIOSH also recommends a ceiling concentration of 200 ppm as determined by a sampling period of 10 minutes. The STEL value is 150 ppm (655 mg/m³). The notation "skin" is added to indicate the possibility of cutaneous absorption. The IDLH level is 10,000 ppm.

Determination in Air: Adsorption on charcoal, workup with CS_2, analysis by gas chromatography. See NIOSH Methods, Set U. See also reference (A-10).

Permissible Concentration in Water: No criteria set but EPA (A-37) has suggested a permissible ambient goal of 6,000 µg/ℓ based on health effects.

Routes of Entry: Inhalation of vapor and, to a small extent, percutaneous absorption of liquid. Also ingestion and skin and eye contact.

Harmful Effects and Symptoms: *Local* — Xylene vapor may cause irritation of the eyes, nose, and throat. Repeated or prolonged skin contact with xylene may cause drying and defatting of the skin which may lead to dermatitis. Liquid xylene is irritating to the eyes and mucous membranes, and aspiration of a few milliliters may cause chemical pneumonitis, pulmonary edema, and hemorrhage. Repeated exposure of the eyes to high concentrations of xylene vapor may cause reversible eye damage.

Systemic — Acute exposure to xylene vapor may cause central nervous system depression and minor reversible effects upon liver and kidneys. At high concentrations xylene vapor may cause dizziness, staggering, drowsiness, and unconsciousness. Also at very high concentrations, breathing xylene vapors may cause pulmonary edema, anorexia, nausea, vomiting, and abdominal pain.

Points of Attack: Central nervous system, eyes, gastrointestinal tract, blood, liver, kidneys, skin.

Medical Surveillance: Preplacement and periodic examinations should evaluate possible effects on the skin and central nervous system, as well as liver and kidney functions. Hematologic studies should be done if there is any significant contamination of the solvent with benzene.

Although metabolites are known, biologic monitoring has not been widely used. Hippuric acid or the ether glucuronide of o-toluic acid may be useful in diagnosis of m-, p- and o-xylene exposure, respectively.

First Aid: If this chemical gets into the eyes, irrigate immediately. If this chemical contacts the skin, wash with soap promptly. If a person breathes in large amounts of this chemical, move the exposed person to fresh air at once and perform artificial respiration. When this chemical has been swallowed, get medical attention. Do not induce vomiting.

Personal Protective Methods: Wear appropriate clothing to prevent repeated or prolonged skin contact. Wear eye protection to prevent any reasonable probability of eye contact. Employees should wash promptly when skin is wet or contaminated. Remove clothing immediately if wet or contaminated to avoid flammability hazard.

Respirator Selection:

1,000 ppm:	CCROVF
500 ppm:	GMOV/SAF/SCBAF
10,000 ppm:	SAF:PD,PP,CF
Escape:	GMOV/SCBA

Disposal Method Suggested: Incineration.

References

(1) National Institute for Occupational Safety and Health, *Criteria for a Recommended Standard: Occupational Exposure to Xylene,* NIOSH Doc. No. 75-168 (1975).
(2) See Reference (A-61).
(3) Sax, N.I., Ed., *Dangerous Properties of Industrial Materials Report, 1,* No. 7, 79-81, New York, Van Nostrand Reinhold Co. (1981). (m-Xylene).
(4) Sax, N.I., Ed., *Dangerous Properties of Industrial Materials Report, 3,* No. 3, 88-92, New York, Van Nostrand Reinhold Co. (1983). (p-Xylene).
(5) See Reference (A-60).
(6) Parmeggiani, L., Ed., *Encyclopedia of Occupational Health & Safety,* Third Edition, Vol. 2, pp 2335–36, Geneva, International Labour Office (1983).

(7)   United Nations Environment Programme, *IRPTC Legal File 1983*, Vol. II, pp VII/789–94, Geneva, Switzerland, International Register of Potentially Toxic Chemicals (1984).

m-XYLENE-α-DIAMINE

Description: $H_2NCH_2C_6H_4CH_2NH_2$ is a liquid, melting at 15°C and boiling at 200°C.

Code Numbers: CAS 1477-55-0   RTECS PF8970000

DOT Designation: −

Synonyms: m-Phenylene bis (methylamine)

Potential Exposures: Those involved in its manufacture or in its use as an intermediate, in the manufacture of polyamide resins, for example.

Permissible Exposure Limits in Air: There is no Federal standard but ACGIH (1983/84) has adopted a ceiling value of 0.1 mg/m^3. They have set no tentative STEL value. The notation "skin" is added indicating the possibility of cutaneous absorption.

Permissible Concentration in Water: No criteria set.

Harmful Effects and Symptoms: Acute oral and skin irritation and sensitization and gastrointestinal irritation have been observed in animals (A-34). Phenylene diamine is suggested by ACGIH as a closely related chemical worthy of comparison.

3,5-XYLENOL

Description: $C_8H_{10}O$ with the formula

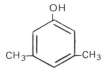

is a crystalline compound melting at 64°C.

Code Numbers: CAS 108-68-9   RTECS ZE6475000

DOT Designation: −

Synonyms: 1-Hydroxy-3,5-dimethylbenzene; 3,5-dimethylphenol.

Potential Exposure: To those involved in xylenol use in pharmaceutical, pesticide, plasticizer, fuel and lubricant additive, rubber chemical and dyestuff manufacture.

Permissible Exposure Limits in Air: No standards set.

Permissible Concentration in Water: 0.25 mg/ℓ is the maximum allowable in drinking water (A-36).

Routes of Entry: Skin absorption.

Harmful Effects and Symptoms: This is a slightly toxic material (LD$_{50}$ for rats is 608 mg/kg). It is a skin irritant and can cause dermatitis. Subacute doses can damage kidneys, liver, pancreas, spleen and may cause edema of the lungs. Chronic poisoning causes digestive disturbances, nervous disorders and skin eruptions.

Points of Attack: Eyes, skin, digestive system.

Personal Protective Methods: Wear skin protection.

Respirator Selection: Wear self-contained breathing apparatus.

Disposal Method Suggested: Incineration.

References

(1) Sax, N.I., Ed., *Dangerous Properties of Industrial Materials Report, 1,* No. 7, 81-82, New York, Van Nostrand Reinhold Co. (1981).

(2) Sax, N.I., Ed., *Dangerous Properties of Industrial Materials Report, 4,* No. 1, 102–106, New York, Van Nostrand Reinhold Co. (Jan./Feb. 1984).

XYLIDINES

Description: $(CH_3)_2C_6H_3NH_2$ is a pale yellow to brown liquid with a weak, aromatic amine odor. It boils at 213° to 226°C. There are actually 6 xylidine isomers.

Code Numbers: CAS 1300-73-8   RTECS ZE8575000   UN 1711

DOT Designation: Poison.

Synonyms: Aminodimethylbenzene, dimethylaniline, dimethylphenylamine.

Potential Exposures: Xylidines are used in dye manufacture and as intermediates in pharmaceutical manufacture (A-41).

Incompatibilities: Strong oxidizers, hypochlorite bleaches.

Permissible Exposure Limits in Air: The Federal standard is 5 ppm (25 mg/m^3). The ACGIH (1983/84) proposes a TWA of 2 ppm (10 mg/m^3) but no STEL value. The notation "skin" is added to indicate the possibility of cutaneous absorption. The IDLH level is 150 ppm.

Determination in Air: Adsorption on silica, workup with n-propanol, analysis by gas chromatography. See NIOSH Methods, Set L. See also reference (A-10).

Permissible Concentration in Water: No criteria set, but EPA (A-37) has suggested a permissible ambient concentration of 345 μg/ℓ.

Routes of Entry: Inhalation, skin absorption, ingestion, skin and eye contact.

Harmful Effects and Symptoms: Anoxia, cyanosis; lung, liver and kidney damage.

Points of Attack: Blood, lungs, liver, kidneys, cardiovascular system.

Medical Surveillance: Consider the points of attack in preplacement and periodic physical examinations.

First Aid: If this chemical gets into the eyes, irrigate immediately. If this chemical contacts the skin, wash with soap immediately. If a person breathes in large amounts of this chemical, move the exposed person to fresh air at once

and perform artificial respiration. When this chemical has been swallowed, get medical attention. Give large quantities of water and induce vomiting. Do not make an unconscious person vomit.

Personal Protective Methods: Wear appropriate clothing to prevent any possibility of skin contact. Wear eye protection to prevent any possibility of eye contact. Employees should wash immediately when skin is wet or contaminated. Remove nonimpervious clothing immediately if wet or contaminated. Provide emergency showers and eyewash.

Respirator Selection:

50 ppm:	CCROV/SA/SCBA
150 ppm:	CCROVF/GMOV/SAF/SA:PD,PP,CF/SCBAF
Escape:	GMOV/SCBA

Disposal Method Suggested: Incineration; oxides of nitrogen are removed from the effluent gas by scrubber, catalytic or thermal device.

References

(1) See Reference (A-61).
(2) See Reference (A-60).

Y

YTTRIUM AND COMPOUNDS

Description: Yttrium, Y, is an element in Group III-B of the Periodic Table. It is very similar to the rare earth metals. It has an atomic weight of 89 and a melting point of 1509°C.

Code Numbers: YCl_3: CAS 10361-92-9 RTECS ZG3150000

DOT Designation: –

Potential Exposure: Yttrium metal has a low cross section for neutron capture and is very stable at high temperatures. Further, it is very inert toward liquid uranium and many liquid uranium alloys. Thus, it may well have applications in nuclear power generation. The metal is usually prepared by reduction of the halide with an active metal such as calcium.

Incompatibilities: Yttrium nitrate: combustible materials.

Permissible Exposure Limits in Air: The Federal standard and the 1983/84 ACGIH TWA value is 1 mg/m^3. The STEL is 3 mg/m^3. No IDLH level has been set.

Determination in Air: Collection on a filter, workup with acid, analysis by atomic absorption. See NIOSH Methods, Set N. See also reference (A-10).

Permissible Concentration in Water: No criteria set.

Routes of Entry: Inhalation of dusts, ingestion, skin and eye contact.

Harmful Effects and Symptoms: Eye irritation in humans. In animals— lung and possible liver damage.

Points of Attack: Eyes, lungs.

Medical Surveillance: Consider the points of attack in preplacement and periodic physical examinations.

First Aid: If this chemical gets into the eyes, irrigate immediately. If this chemical contacts the skin, wash with soap. If a person breathes in large amounts of this chemical, move the exposed person to fresh air at once and perform artificial respiration. When this chemical has been swallowed, get medical attention. Give large quantities of water and induce vomiting. Do not make an unconscious person vomit.

Personal Protective Methods: Wear eye protection to prevent any possibility of eye contact. Provide emergency eyewash.

Respirator Selection:

5 mg/m³: DMXS
10 mg/m³: DMXSQ/FuHiEP/SA/SCBA
50 mg/m³: HiEPF/SAF/SCBAF
500 mg/m³: PAPHiE/SA:PD, PP, CF

Disposal Method Suggested: Recovery is indicated wherever possible. Specifically, processes are available for yttrium oxysulfide recovery from color television tube manufacture (A-57).

References

(1) See Reference (A-61).

Z

ZINC CHLORIDE

- Hazardous substance (EPA)
- Priority toxic pollutant (EPA)

Description: $ZnCl_2$, zinc chloride, consists of white hexagonal, deliquescent crystals.

Code Numbers: CAS 7646-85-7   RTECS ZH1400000   UN 2331

DOT Designation: ORM-E

Synonyms: Butter of zinc.

Potential Exposure: Zinc chloride is used as a wood preservative, for dry battery cells, as a soldering flux, and in textile finishing, in vulcanized fiber, reclaiming rubber, oil and gas well operations, oil refining, manufacture of parchment paper, dyes, activated carbon, chemical synthesis, dentists' cement, deodorants, disinfecting and embalming solutions, and taxidermy. It is also produced by military screening-smoke devices.

NIOSH has estimated annual worker exposure to zinc chloride at 1,600,000.

Incompatibilities: None hazardous.

Permissible Exposure Limits in Air: The Federal standard and the 1983/84 ACGIH TWA value for zinc chloride fume is 1 mg/m³. The STEL is 2 mg/m³. The IDLH level is 2,000 mg/m³.

Determination in Air: Collection by filter, analysis by atomic absorption (A-27).

Permissible Concentration in Water: *Freshwater Aquatic Life* — For zinc, the criterion to protect freshwater aquatic life as derived using the Guidelines is: 47 µg/ℓ as a 24 hour average and the concentration should not exceed

$$e^{[0.83 \ln(\text{hardness}) + 1.95]} \mu g/\ell$$

at any time.

Saltwater Aquatic Life — For saltwater aquatic life, 58 µg/ℓ as a 24 hr average, never to exceed 170 µg/ℓ.

Human Health — For the prevention of adverse effects due to the organoleptic properties of zinc, the current standard for drinking water of 5,000 µg/ℓ was adopted for ambient water criterion. The WHO (A-65) has also endorsed the 5,000 µg/ℓ limit for zinc in drinking water. However, the Federal Republic of Germany has a 2,000 µg/ℓ limit (A-65).

Determination in Water: Digestion followed by analysis by atomic absorption or by colorimetric analysis (using dithizone) or by inductively coupled plasma (ICP) optical emission spectrometry. The above gives total zinc; soluble zinc may be determined by 0.45 micron filtration followed by the above methods.

Routes of Entry: Inhalation of dust and fumes; ingestion, skin and eye contact.

Harmful Effects and Symptoms: *Local* — Solid zinc chloride is corrosive to the skin and mucous membranes. Aqueous solutions of 10% or more are also corrosive and cause primary dermatitis and chemical burns, especially at sites of minor trauma. Aqueous solutions are also extremely dangerous to the eyes, causing extreme pain, inflammation, and swelling, which may be followed by corneal ulceration.

Zinc chloride may produce true sensitization of the skin in the form of eczematoid dermatitis. Ingestion of zinc chloride may cause serious corrosive effects in the esophagus and stomach, often complicated by pyloric stenosis.

Systemic — There are no reports of inhalation of zinc chloride from industrial exposure. All reported experience with inhaled zinc chloride is based on exposures caused by military accidents. In all of those cases, there was severe irritation of the respiratory tract. In the more severe cases, acute pulmonary edema developed within two to four days following exposure. The fatalities reported were due to severe lung injury with hemorrhagic alveolitis and bronchopneumonia. In human experimentation with concentrations of 120 mg/m^3, there were complaints of irritation of the nose, throat, and chest after 2 minutes. With exposure to 80 mg/m^3 for 2 minutes, the majority of subjects experienced slight nausea, all noticed the smell, and one or two coughed.

Points of Attack: Respiratory system, lungs, skin, eyes.

Medical Surveillance: In preemployment and periodic physical examinations, special attention should be given to the skin and to the history of allergic dermatitis, as well as to exposed mucous membranes, the eyes, and the respiratory system. Chest x-rays and periodic pulmonary function studies may be helpful. Smoking history should be known. Measurement of urinary zinc excretion may be useful.

First Aid: If a person breathes in large amounts of zinc chloride fume, move the exposed person to fresh air at once and perform artificial respiration.

Personal Protective Methods: Employees exposed to zinc chloride should be given instruction in personal hygiene, and in the use of personal protective equipment. Goggles should be provided in areas where splash or spill of liquid is possible. In areas with excessive dust or fume levels, respiratory protection by use of filter type dust masks or air supplied respirators with fullface pieces should be required. In areas where danger of spills or splashes exists, skin protection should be provided with rubber gloves, face shields, rubber aprons, gantlets, suits, and rubber shoes.

For zinc chloride fume, no specific methods (aside from respirator protection) are specified by NIOSH (A-4).

Respirator Selection: For Zinc Chloride Fume —

10 mg/m^3: FuHiEP/SA/SCBA
50 mg/m^3: HiEPF/SAF/SCBAF
$1,000 \text{ mg/m}^3$: PAPHiE
$2,000 \text{ mg/m}^3$: SAF:PD, PP, CF
Escape : HiEP/SCBA

Disposal Method Suggested: Dump in water; add soda ash and stir, then neutralize and flush to sewer with water (A-38). Alternatively, zinc chloride may be recovered from spent catalysts and used acrylic fiber spinning solutions (A-57).

References

(1) U.S. Environmental Protection Agency, *Zinc: Ambient Water Quality Criteria,* Wash., DC (1980).
(2) U.S. Environmental Protection Agency, *Toxicology of Metals, Vol. II: Zinc,* Report EPA-600/1-77-022, Research Triangle Park, NC, pp 473-87, (May 1977).
(3) Sax, N.I., Ed., *Dangerous Properties of Industrial Materials Report, 1,* No. 7, 90-92, New York, Van Nostrand Reinhold Co. (1981).
(4) See Reference (A-61).
(5) See Reference (A-60).
(6) United Nations Environment Programme, *IRPTC Legal File 1983,* Vol. II, pp VII/801- 2, Geneva, Switzerland, International Register of Potentially Toxic Chemicals (1984).

ZINC CHROMATE

● Carcinogen (IARC)(2)
● Priority toxic pollutant (EPA)

Description: $ZnCrO_4$ is a yellow crystalline powder. There may be various zinc chromates with other ZnO to Cr_2O_3 ratios in addition to this 2:1 compound.

Code Numbers: CAS 13530-65-9   RTECS GB3290000

DOT Designation: –

Potential Exposure: Zinc chromate is used as a pigment in surface coatings and linoleum. It is also used to impart corrosion resistance to epoxy laminates.

Permissible Exposure Limits in Air: There are no Federal standards but ACGIH (1983/84) has designated zinc chromate (along with lead chromate) as an "industrial substance suspect of carcinogenic potential for man" with an assigned TWA value of 0.05 mg/m^3. There is no STEL value.

Determination in Air: See "Chromium and Compounds."

Permissible Concentration in Water: Since zinc chromate may consist of compounds with various ZnO/Cr_2O_3 ratios, it is best simply to refer to the EPA water quality criteria cited in the sections of the volume dealing with "Chromium" and with "Zinc Chloride."

Determination in Water: See "Zinc Chloride" and "Chromium."

Harmful Effects and Symptoms: The primary effect is the carcinogenic effect of chromium (2).

References

(1) Nat. Inst. for Occup. Safety and Health, *Criteria for a Recommended Standard: Occupational Exposure to Chromium,* NIOSH Doc. No. 76-129, Rockville, MD, (1976).
(2) International Agency for Research on Cancer, *IARC Monographs on the Carcinogenic Risks of Chemicals to Humans,* Lyon, France, 2, 100 (1973).
(3) Sax, N.I., Ed., *Dangerous Properties of Industrial Materials Report, 1,* No. 7, 92-94, New York, Van Nostrand Reinhold Co. (1981).

ZINC OXIDE

Description: ZnO, zinc oxide, is an amorphous, odorless, white or yellowish-white powder, practically insoluble in water. It is produced by oxidation of zinc or by roasting of zinc oxide ore.

Code Numbers: CAS 1314-13-2   RTECS ZH4810000

DOT Designation: —

Synonyms: Zinc white, flowers of zinc.

Potential Exposure: Zinc oxide is primarily used as a white pigment in rubber formulations and as a vulcanizing aid. It is also used in photocopying, paints, chemicals, ceramics, lacquers, and varnishes, as a filler for plastics, in cosmetics, pharmaceuticals, and calamine lotion. Exposure may occur in the manufacture and use of zinc oxide and products, or through its formation as a fume when zinc or its alloys are heated. NIOSH estimates that 50,000 workers have potential exposure to zinc oxide.

Incompatibilities: Chlorinated rubber.

Permissible Exposure Limits in Air: The Federal standard and the 1983/84 ACGIH TWA value for zinc oxide fume is 5 mg/m^3. NIOSH recommends adherence to the present Federal standard of 5 mg/m^3 as a time-weighted average for up to a 10-hour workday, 40-hour workweek. NIOSH also recommends a ceiling concentration of 15 mg/m^3 as determined by a sampling time of 15 minutes. The STEL value is 10 mg/m^3. There is no IDLH level set. In contrast to zinc oxide fume, zinc oxide dust is simply classified as a nuisance particulate by ACGIH with a TLV of 10 mg/m^3.

Determination in Air: Collection on a polymer membrane filter, analysis by x-ray diffraction (A-10). Atomic absorption may also be used.

Permissible Concentration in Water: See "Zinc Chloride" for EPA criteria on zinc.

Determination in Water: See "Zinc Chloride" for EPA methods of analysis for zinc.

Routes of Entry: Inhalation of dust or fumes.

Harmful Effects and Symptoms: *Local* — When handled under poor hygienic conditions, zinc oxide powder may produce a dermatitis called "oxide pox." This condition is due primarily to clogging of the sebaceous glands with zinc oxide and produces a red papule with a central plug. The area rapidly becomes inflamed and the central plug develops into a pustule which itches intensely. Lesions occur in areas of the skin that are exposed or subject to heavy perspiration. These usually clear, however, in a week to ten days with good hygiene and proper care of secondary infections.

Systemic — The syndrome of metal fume fever is the only important effect of exposure to freshly formed zinc oxide fumes and zinc oxide dusts of respirable particle size. The fumes are formed by subjecting either zinc or alloys containing zinc to high temperatures. Typically, the syndrome begins four to twelve hours after sufficient exposure to freshly formed fumes of zinc oxide. The worker first notices the presence of a sweet or metallic taste in the mouth, accompanied by dryness and irritation of the throat. Cough and shortness of breath may occur, along with general malaise, a feeling of weakness, fatigue, and pains in the muscles and joints. Fever and shaking chills then develop. Fever can range from 102° to 104°F. Profuse sweats develop and the fever subsides. The entire episode runs its course in 24 to 48 hours.

During the acute period, there is an elevation of the leukocyte count (rarely above 20,000/mm³), and the serum LDH may be elevated. Chest x-rays are not diagnostic.

Metal fume fever produces rapid development of tolerance or short-lived relative immunity. This may be lost, however, over a weekend or holiday, and the worker may again develop the complete syndrome when he returns to work if fume levels are sufficiently high. There are no sequelae to the attacks.

Other possible systemic effects of zinc oxide are in doubt. Cases of gastro-intestinal disturbance have been reported, but most authorities agree there is no evidence of chronic industrial zinc poisoning.

Points of Attack: Respiratory system, lungs.

Medical Surveillance: Preemployment and periodic physical examinations should be made to assess the status of the general health of the worker. Examinations are also recommended following episodes of metal fume fever or intercurrent illnesses. Zinc excretion in urine can be used as an index of exposure.

First Aid: If a person breathes in large amounts of this chemical, move the exposed person to fresh air at once and perform artificial respiration.

Personal Protective Methods: Employees should receive instruction in personal hygiene and in the causes and effects of metal fume fever. Workers exposed to zinc oxide powder should be supplied with daily clean work clothes and should be required to shower before changing to street clothes. In cases of accident or where excessive fume concentrations are present, gas masks with proper canister or supplied air respirators should be provided.

Respirator Selection:
    50 mg/m³: FuHiEP/SA/SCBA
    250 mg/m³: HiEPF/SAF/SCBAF
    2,500 mg/m³: PAPHiE/SA:PD, PP, CF

Disposal Method Suggested: Landfill.

References
(1)   National Institute for Occupational Safety and Health, *Criteria for a Recommended Standard: Occupational Exposure to Zinc Oxide,* NIOSH Doc. No. 76-104 (1976).
(2)   See Reference (A-61).

ZINC SALTS

A variety of zinc salts have been reviewed in the periodical "Dangerous Properties of Industrial Materials Report," and in the interests of economy of space in this volume are simply referenced here:

	Vol	No.	Pages
Zinc acetate	1	7	88–90
Zinc ammonium chloride	4	2	91–93
Zinc borate	4	2	93–96
Zinc bromide	4	2	96–98
Zinc carbonate	4	2	98–100
Zinc cyanide	4	2	100–102

	Vol	No.	Pages
Zinc fluoborate	1	7	88-90
Zinc fluoride	3	6	83-85
Zinc fluosilicate	3	6	85-88
Zinc formate	4	1	106-108
Zinc hydrosulfite	4	1	108-110
Zinc nitrate	2	2	89-92
Zinc phenol sulfonate	4	1	110-112
Zinc sulfate	2	2	92-94

In addition see Parmeggiani, L., Ed., *Encyclopedia of Occupational Health & Safety,* Third Edition, Vol. 2, pp 2340-42, Geneva, International Labour Office (1983).

ZIRAM

- Carcinogen (Animal Positive) (Animal Suspected) (IARC) (A-63)

Description: $\left[(CH_3)_2N\overset{\overset{\displaystyle S}{\|}}{C}-S\right]_2Zn$ is a white odorless powder which melts at 240°C.

Code Numbers: CAS 137-30-4 RTECS ZH0525000

DOT Designation: –

Synonyms: Bis(dimethyldithiocarbamate)-zinc; Dimethylcarbamodithoic acid zinc salt; Zinc dimethyldithiocarbamate; many trade names.

Potential Exposure: Those involved in the manufacture, formulation or application of this fungicide and rubber accelerator. NIOSH estimates that 28,000 workers are exposed to ziram, mainly in the rubber industry.

Permissible Exposure Limits in Air: No standards set.

Permissible Concentration in Water: No criteria set.

Harmful Effects and Symptoms: The acute LD_{50} value for rats is 1,400 mg/kg which is slightly toxic. It causes irritation of skin, eyes and respiratory tract in humans, however, and may be a cholinesterase inhibitor in humans. It is an animal carcinogen.

Points of Attack: Skin, eyes, respiratory tract.

Disposal Method Suggested: May be decomposed by strong acids or alkalis.

References

(1) U.S. Environmental Protection Agency, ***Chemical Hazard Information Profile Draft Report; Ziram,*** Washington, DC (September 30, 1983).

(2) United Nations Environment Programme, ***IRPTC Legal File 1983,*** Vol. II, pp VII/799-800, Geneva, Switzerland, International Register of Potentially Toxic Chemicals (1984).

ZIRCONIUM AND COMPOUNDS

- Hazardous substances (several compounds, EPA)

  Zirconium compounds which are classified by EPA as hazardous substances include: Zirconium nitrate; zirconium potassium fluoride; zirconium sulfate; and zirconium tetrachloride.

Description: Zr, zirconium, is a greyish-white, lustrous metal in the form of platelets, flakes, or a bluish-black, amorphous powder. It is never found in the free state; the most common sources are the ores zircon and baddeleyite. It is generally produced by reduction of the chloride or iodide. The metal is very reactive, and the process is carried out under an atmosphere of inert gas. The powdered metal is a fire and explosive hazard.

Zirconium tetrachloride, $ZrCl_4$, is a white crystalline solid which sublimes at about $330°C$.

Code Numbers:

Zr metal	CAS 7440-67-7	RTECS ZH7070000	UN 2008
$ZrCl_4$	CAS 10026-11-6	RTECS ZH7175000	UN 2503

DOT Designations: Zr metal: Flammable solid; $ZrCl_4$: Corrosive material.

Synonyms: None.

Potential Exposure: Zirconium metal is used as a "getter" in vacuum tubes, a deoxidizer in metallurgy, and a substitute for platinum; it is used in priming of explosive mixtures, flashlight powders, lamp filaments, flash bulbs, and construction of rayon spinnerets. Zirconium or its alloys (with nickel, cobalt, niobium, tantalum) are used as lining materials for pumps and pipes, for chemical processes, and for reaction vessels. Pure zirconium is a structural material for atomic reactors, and alloyed, particularly with aluminum, it is a cladding material for fuel rods in water-moderated nuclear reactors. A zirconium-columbium alloy is an excellent superconductor.

Zircon ($ZrSiO_4$) is utilized as a foundry sand, an abrasive, a refractory in combination with zirconia, a coating for casting molds, a catalyst in alkyl and alkenyl hydrocarbon manufacture, a stabilizer in silicone rubbers, and as a gem stone; in ceramics it is used as an opacifier for glazes and enamels and in fritted glass filters. Both zircon and zirconia (zirconium oxide, ZrO_2) bricks are used as linings for glass furnaces. Zirconia itself is used in die extrusion of metals and in spout linings for pouring metals, as a substitute for lime in oxyhydrogen flame, as a pigment, and an abrasive; it is used, too, in incandescent lights, as well as in the manufacture of enamels, white glass, and refractory crucibles.

Other zirconium compounds are used in metal cutting tools, thermocouple jackets, waterproofing textiles, ceramics, and in treating dermatitis and poison ivy.

Incompatibilities: $ZrCl_4$: water, moist air, alkali metals; ZrH_2: strong oxidizers.

Permissible Exposure Limits in Air: The Federal standard and the 1983/84 ACGIH TWA value for zirconium compounds is 5 mg/m^3 as Zr. The STEL value is 10 mg/m^3. The IDLH level is 500 mg/m^3.

Determination in Air: Collection on a filter, workup with acid, analysis by atomic absorption. See NIOSH Methods, Set M. See also reference (A-10).

Permissible Concentration in Water: No criteria set.

Routes of Entry: Inhalation of dust or fume, eye and skin contact.

Harmful Effects and Symptoms: *Local* — No ill effects from industrial exposure to zirconium have been proven. A recent study from the USSR, however, reports that some workers exposed to plumbous titanate zirconate developed a mild occupational dermatitis associated with hyperhydrosis of the hands. This condition was accompanied by subjective complaints of vertigo, sweet taste in the mouth, and general indisposition. These workers were also said to have elevated thermal and pain sensitivity, and electric permeability of the horny layer, along with increased sweating, and reduced capillary resistance.

Zircon granulomas were reported in the U.S. as early as 1956. This condition arose from the use of deodorant sticks in the axillae, but it was resolved when use was stopped. Zircon is no longer used as a deodorant. Because of a possible allergic sensitivity reaction, individuals who have experienced granulomas from zirconium should avoid dust and mist.

Systemic — Inhalation of zirconium dust and fumes has caused no respiratory or other pathological problems. Animal experiments, however, have produced interstitial pneumonitis, peribronchial abscesses, peribronchiolar granuloma, and lobular pneumonia.

Points of Attack: Respiratory system, skin.

Medical Surveillance: No special considerations are needed.

First Aid: If this chemical gets into the eyes, irrigate immediately. If this chemical contacts the skin, wash with soap. If a person breathes in large amounts of this chemical, move the exposed person to fresh air at once and perform artificial respiration. When this chemical has been swallowed, get medical attention. Give large quantities of water and induce vomiting. Do not make an unconscious person vomit.

Personal Protective Methods: Employees should be trained in the correct use of personal protective equipment. In areas of dust accumulation or high fume concentrations, respiratory protection is advised either by dust mask or supplied air respirators. Skin protection is not generally necessary, but where there is a history of zircon granuloma from deodorants, it is probably advisable.

Respirator Selection:
$$25 \text{ mg/m}^3: \text{DMXS}$$
$$50 \text{ mg/m}^3: \text{DMXSQ/HiEP/SA/SCBA}$$
$$250 \text{ mg/m}^3: \text{HiEPF/SAF/SCBAF}$$
$$500 \text{ mg/m}^3: \text{SA:PD, PP, CF/PAPHiE}$$
$$\text{Escape} : \text{HiEP/SCBA}$$

Disposal Method Suggested: $(ZrCl_4)$ — Mix waste with sodium bicarbonate; spray with ammonia, and add crushed ice. When reaction subsides, flush down the drain (A-38).

References

(1) See Reference (A-61).

(2) Sax, N.I., Ed., *Dangerous Properties of Industrial Materials Report, 2,* No. 2, 94-96, New York, Van Nostrand Reinhold Co. (1982). (Zirconium-95 and Zirconium sulfate).

(3) Sax, N.I., Ed., *Dangerous Properties of Industrial Materials Report, 3,* No. 4, 107-109, New York, Van Nostrand Reinhold Co. (1983). (Zirconium Potassium Fluoride).

(4) Sax, N.I., Ed., *Dangerous Properties of Industrial Materials Report, 3,* No. 4, 109-111, New York, Van Nostrand Reinhold Co. (1983). (Zirconium Tetrachloride).

(5) Sax, N.I., Ed., *Dangerous Properties of Industrial Materials Report, 3,* No. 6, 88-90, New York, Van Nostrand Reinhold Co. (Nov./Dec. 1983) (Zirconium Nitrate).

(6) Sax, N.I., Ed., *Dangerous Properties of Industrial Materials Report, 3,* No. 6, 90-92, New York, Van Nostrand Reinhold Co. (Nov./Dec. 1983) (Zirconium Sulfate).

(7) Parmeggiani, I., Ed., *Encyclopedia of Occupational Health & Safety,* Third Edition, Vol. 2, pp 2342-44, Geneva, International Labour Office (1983) (Zirconium & Hafnium).

Carcinogen Index

The following list includes suspected and potential, as well as proven carcinogens.

Acetamide
Acetophenetidin
Acrylonitrile
Actinomycin D
Adriamycin
Aflatoxins
Aldrin
2-Aminoanthraquinone
4-Aminobiphenyl
3-Amino-9-ethylcarbazole
1-Amino-2-methylanthraquinone
4-Amino-2-nitrophenol
3-Amino-1,2,4-triazole
o-Anisidine hydrochloride
Antimony trioxide
Arsenic and arsenic compounds
Asbestos
Auramine
Azobenzene

Benz[a]anthracene
Benzene
Benzidine
Benzo[a]pyrene
Benzyl chloride
Beryllium and compounds
Bis(2-chloroisopropyl) ether
Bis(chloromethyl) ether
Bromodichloromethane
Bromoform

1,2-Diphenylhydrazine
Direct Black 38
Direct Blue 6

Epichlorohydrin
Ethylene dibromide
Ethylene dichloride
Ethyleneimine
Ethylene thiourea

N-2-Fluorenyl acetamide
Formaldehyde

Hematite
Heptachlor
Hexachlorobenzene
Hexachlorobutadiene
Hexachlorocyclohexane
Hexachloroethane
Hexamethylphosphoric triamide
Hydrazine

Iron-Dextran
Isopropyl oils

Lead acetate
Lead phosphate
Lindane

Maneb
Melphalan
Methyl bromide
Methyl chloride
4,4'-Methylenebis(2-chloroaniline)
4,4'-Methylenebis(N,N-dimethyl)aniline
Methylene chloride
Methyl iodide
Michler's ketone
Mirex
Mustard gas

α-Naphthylamine
β-Naphthylamine
Nickel and soluble compounds
Nickel carbonyl
Nitrilotriacetic acid
5-Nitro-o-anisidine
4-Nitrobiphenyl
o-Nitrochlorobenzene
p-Nitrochlorobenzene
Nitrofen
2-Nitropropane

Nitrosamines
N-Nitrosodimethylamine

Oxymetholone

Paraffin
Pentachloronitrobenzene
Phenazopyridine hydrochloride
Phenylhydrazine
Phenyl-β-naphthylamine
Phenytoin
Picloram
Polybrominated biphenyls
Polychlorinated biphenyls
Polynuclear aromatic hydrocarbons
Procarbazine
Propane sultone
β-Propiolactone
Propyleneimine
Propylene oxide

Quinoline

Reserpine

Saccharin
Safrole
Selenium sulfide
Soot
Streptozotocin
Sulfallate
Sulfurous acid-2-(p-t-butylphenoxy)-1-methylethyl-2-chloroethyl
    ester

2,4,5-T
Tannic acid
TDE
Testosterone
Tetrachlorodibenzo-p-dioxin
1,1,2,2-Tetrachloroethane
Tetrachloroethylene
Thioacetamide
Thiotepa
Thiourea
Thorium dioxide
o-Tolidine
Toluene-2,4-diamine
o-Toluidine
Toxaphene
1,1,2-Trichloroethane
Trichloroethylene
2,4,6-Trichlorophenol

Trifluralin
Tris(2,3-dibromopropyl) phosphate

Urethane

Vinyl bromide
Vinyl chloride
Vinyl cyclohexene dioxide
Vinylidene chloride

Zinc chromate
Ziram